LAROUSSE

Dictionary of

SCIENTISTS

LAROUSSE Dictionary of SCIENTISTS

Editor
Hazel Muir

LARROUSSE
Larousse plc
43–45 Annandale Street, Edinburgh EH7 4AZ
Larousse Kingfisher Chambers Inc.,
95 Madison Avenue, New York, New York 10016

First published by Larousse plc 1994

10 9 8 7 6 5 4 3 2 1

British Library Cataloguing in Publication Data
for this book is available from the British Library

Library of Congress Catalog Card Number: 94-75739

ISBN 0-7523-0002-4

The publisher would like to thank Jack Weigel, Coordinator of the Science Library at the University of Michigan/Ann Arbor, for his invaluable assistance in reviewing the material for this volume.

Cover illustration: Newton by William Blake

Illustration provided by e.t. archive

Typeset by Pillans & Wilson Ltd, Edinburgh

Printed in Great Britain by Clays Ltd, St Ives plc

Contents

Contributors

Astronomy
Professor H A Brück
Former Astronomer Royal for Scotland
Professor Emeritus of Astronomy
University of Edinburgh

Dr Mary Brück
Former Senior Lecturer in Astronomy
University of Edinburgh

Dr David Hughes
Reader in Astronomy
Department of Physics
University of Sheffield

Botany
Dr Robert R Mill
Scientific Officer
Royal Botanic Garden
Edinburgh

Dr Colin Will
Chief Librarian
Royal Botanic Garden
Edinburgh

Biology
Professor Maggi Allan
Department of Genetics
and Director of Molecular Medicine Unit
Department of Medicine
Columbia University
New York

Dr M James C Crabbe
Reader
University of Reading

Mary Jennings
Cambridge Wellcome Unit for the History of Medicine
Department of History and Philosophy of Science
University of Cambridge

Dr David F Smith
Wellcome Trust Research Fellow
Wellcome Unit for the History of Medicine
University of Glasgow

Christopher D Todd
Senior Lecturer
Gatty Marine Laboratory
University of St Andrews

Dr David C Watts DSc
University Reader in Biochemistry
United Medical and Dental Schools of Guy's and St Thomas's Hospitals
London

Chemistry
Dr Matthew J Almond
Lecturer
Department of Chemistry
University of Reading

A R Butler
Reader in Chemistry
University of St Andrews

Jean Jones
Freelance Historian of Science and Exhibition Organizer

Dr John Shorter
Emeritus Reader in Chemistry
University of Hull

Computer Science
Dr Geoffrey Tweedale
Senior Research Fellow
Manchester Metropolitan University

Earth Science
Dr Graham Durant
Curator
Hunterian Museum
University of Glasgow

R A S Ratcliffe
Formerly of the Meteorological Office and the Royal Meteorological Society

Dr C A Williams
Geophysicist
Bullard Laboratories
University of Cambridge

Engineering
Ronald M Birse
Historian of Science and Technology

Mathematics
Jeremy J Gray
Senior Lecturer in Mathematics
Open University
Milton Keynes

Medicine
Professor William F Bynum
Wellcome Institute for the History of Medicine
and University College London

Dr Roy Porter
Reader in Social History of Medicine
Wellcome Institute for the History of Medicine

Dr E M Tansey
Historian of Modern Medical Science
Wellcome Institute for the History of Medicine
and Honorary Archivist of the Physiological Society

Physics
Dr John G Ashman
Research Student
University of Bradford

Dr Craig Buttar
Senior Experimental Officer
Department of Physics
University of Sheffield

Dr Simon Gage
Edinburgh International Science Festival

Dr Willem D Hackmann
Assistant Curator
Museum of the History of Science
University of Oxford

Dr Ben Marsden
Honorary Research Fellow
History of Science Unit
University of Kent at Canterbury

Richard M Sillitto
Reader Emeritus in Physics
University of Edinburgh

Zoology
Professor R J Berry
Department of Biology
University College London

Arthur W Ewing
Former Senior Lecturer in Zoology
University of Edinburgh

Dr Adrian Lister
SERC Advanced Fellow and Lecturer
Department of Biology
University College London

Preface

This book aims to provide a comprehensive and up-to-date single-volume dictionary of scientists who have made their mark in the traditional natural sciences (physics, chemistry, biology, astronomy, Earth sciences) and their applications, and includes an extensive selection of mathematicians. Some space is devoted to the most celebrated engineers and physicians whose success lay in the application of science, and a few of those whose enthusiasm for communicating science has sparked new public interest. Many of the early philosophers whose investigations were to herald the dawn of scientific thought have also been included. In general, selection favours those who made important discoveries or theoretical advances, rather than those renowned for their skills as teachers or administrators, and social scientists have not been included. All Nobel prizewinners (Physics, Chemistry and Physiology or Medicine) since the first award in 1901 up to 1993 are here, and are listed separately at the end of the book for reference.

Most entries have been expanded and updated from the database of *Chambers Biographical Dictionary*; in addition we have added around 500 new entries, many of which describe living scientists who have recently been working at the frontiers of research. It is the modern selection which has been the most difficult to make—while it is relatively easy to assess with hindsight the achievements of previous generations, it is often difficult to compare the significance of recent contributions and to know whether they will stand the test of time in our ever-accelerating scientific age. However, we believe that a contemporary selection is a vital component of this type of biographical reference source, and adds an interesting flavour of the way in which scientific methods have changed over the years.

In the dictionary, each scientist is listed alphabetically according to the name he or she is best known by. Cross-references to other scientists who have entries in the book are indicated by the use of bold type. Often where work has been carried out by a number of collaborators, details of the work have been given in one entry to which the others cross-refer. This has allowed some space to be saved, but does not imply that the longer entry has been devoted to the more important scientist. The length of each entry will not necessarily reflect the eminence of the scientist in question, as often the more modern the work, the greater the space that is required to outline the more complex issues involved.

Acknowledgements

For various help and advice we would like to thank Professor Frank E Close and Professor Malcolm Longair. Thanks also to all the contributors to *Chambers Biographical Dictionary* whose work formed the basis of this dictionary.

Abbreviations used in this dictionary

AC	alternating current
b	born
c.	circa
Caltech	California Institute of Technology
CBE	Commander of the British Empire
d	died
DBE	Dame Commander of the British Empire
DC	direct current
fl	flourished
FRS	Fellow of the Royal Society
KBE	Knight Commander of the British Empire
MIT	Massachusetts Institute of Technology
OBE	Officer of the British Empire
SI	Système International (international standard units)

A

ABBE, Ernst
(1840–1905)

German physicist, born in Eisenach. He became professor at the University of Jena in 1870, and in 1878 director of the astronomical and meteorological observatories there. Consulted by the Jena optical instrument maker Carl Zeiss in the 1860s, Abbe embarked on a comprehensive study of the fundamental factors limiting the resolution of microscopes, and of the parameters which must be controlled in the manufacturing process to achieve the theoretical performance. He developed instruments for measuring refractive indices of glass, so that the glass manufacturer's output could be monitored, and a focometer for measuring the focal lengths of lenses so that the performance of the optical workshop could be controlled. The arrangement known in microscopy as 'Abbe's homogeneous immersion' made it possible to obtain maximum benefit from **Amici**'s invention of the immersion objective. Zeiss took Abbe into partnership, and following Zeiss's death in 1888 Abbe became owner of the company. Abbe had very advanced ideas about the securing of good relationships between management and workers in industry, and exercised these ideas in the setting up of the Carl Zeiss Foundation. Arising out of his work on the microscope, Abbe founded the diffraction theory of optical imaging, from which the modern techniques of optical image processing and image enhancement have developed. His literary style was poor, and the recognition of the importance of his ideas owed much to subsequent interpreters.

ABEGG, Richard Wilhelm Heinrich
(1869–1910)

German chemist, one of the first scientists to perceive the chemical significance of the electron, helping to develop the theory of valence which is fundamental to modern chemistry. Born in Danzig, he studied at Kiel, Tübingen and Berlin universities, and moved to Göttingen in 1894 as an assistant to **Nernst**. He later taught in Breslau, where he was professor at the Technische Hochschule from 1899 until his death. Following the discovery of the electron by **J J Thomson** in 1897, Abegg realized that the 'valence' of an element—that is the manner in which it combines chemically with other elements—is determined by the number of electrons in the outer shell of its atom, not by the total number of electrons in the atom. He observed that the inert properties of the rare gases arise due to the stable configuration of eight outer electrons. Abegg is also remembered for his 'rule of eight', stating that each atom has a 'normal' valency and a 'contravalence', the sum of the two being equal to eight. He did significant work on osmotic pressures, the freezing point of dilute solutions, and the dielectric constant of ice. He died in a ballooning accident at Koszalin, Poland.

ABEL, Sir Frederick Augustus
(1827–1902)

English chemist who invented cordite in collaboration with Sir **James Dewar**. He was born in London, and studied at the Royal Polytechnical Institution and the Royal College of Chemistry, where he was appointed an assistant in 1846. In 1852 he moved to the Royal Military Academy, Woolwich, where he lectured in chemistry before being appointed ordnance chemist in 1854 and chemist to the War Department in 1858. During his long and distinguished career at Woolwich, which lasted more than 30 years, he developed high explosives and smokeless gunpowder, and studied explosions in mines and those caused by dust. The invention of cordite, which quickly became the standard propellant used by the British army, took place in 1889. Abel also devised a new method to manufacture guncotton which prevented it from exploding spontaneously, and invented a device, the 'Abel tester', for determining the flash point of petroleum. He was elected FRS in 1860, knighted in 1883 and made a baronet in 1893. He died in London.

ABEL, John Jacob
(1857–1938)

American biochemist, born in Cleveland, Ohio. He studied at Johns Hopkins University and widely in Europe before returning to Johns Hopkins as its first Professor of Pharmacology (1893–1932). He was the first to determine the molecular weight of the steroid cholesterol and its bile acid derivative, cholic acid. In 1897 he isolated from the adrenal gland the hormone adrenaline (epinephrine), the most powerful haemostatic agent then known, and showed that it was identical to 'suprarenine'. However, analysis of 30 different samples showed variable purity. Adrenaline was first crystallized as the ammonium salt by **Takamine** in 1902. In 1892 Abel discovered carbonic acid in horse urine and associated it with urea production, and in 1914 he joined a cellophane sac to an artery and demonstrated that salicylic acid could be diffused from the blood at a rate comparable with the kidney. Using this technique of dialysis he showed that blood contains amino acids, thereby pioneering dialysis for the treatment of kidney disease. These experiments also demonstrated the use of hirudonin, an extract of leaches, to prevent blood coagulation. In 1927 Abel's research team crystallized insulin, showed it to be a protein and demonstrated that its hypoglycaemic action was not caused by contaminants.

ABEL, Niels Henrik

(1802–29)

Norwegian mathematician, born in Finnøy. He showed mathematical genius by the age of 15, entered Oslo University in 1821, and in 1823 proved that there is no algebraic formula for the solution of a general polynomial equation of the 5th degree. Such a formula had been sought ever since the cubic and quartic equations had been solved in the 16th century by **Cardano** and others. He developed the concept of elliptic functions independently of **Jacobi**, and pioneered its extension to the theory of Abelian integrals and functions; this became a central theme of later 19th-century analysis, although his work was not fully understood, or even published, in his lifetime. Abel emphasized the analogy of the elliptic functions with the familiar trigonometric functions, the new functions being complex functions with two distinct periods. They were influential in the development of complex analysis. His theory of Abelian integrals was a remarkable and unexpected generalization of old results of **Leonhard Euler** on sums of integrals, and was the spur to the later work of **Riemann** and **Weierstrass**. Abel also did early work on the rigorous theory of the binomial series.

ABELSON, Philip Hauge

(1913–)

American physical chemist who played a major part in the development of the atomic bomb. Born in Tacoma, Washington, he was educated at Washington State College and the University of California at Berkeley. He worked at the Carnegie Institution, Washington DC, from 1939 to 1941. In 1940 he assisted **McMillan** to bombard uranium with neutrons, which led to the creation of a new element, a silvery metal christened neptunium. This was the first element heavier than uranium and the first to be made synthetically, though small quantities of it were subsequently found in nature. In 1941 Abelson transferred to the Naval Research Laboratory in Washington DC, becoming principal physicist in 1945. From 1941 he worked on the Manhattan atomic bomb project, devising a cheap method for making uranium hexafluoride and then developing the gas diffusion process for separating ^{235}U, the fissile isotope of uranium, from its more abundant but non-fissile isotope, ^{238}U. This process supplied the enriched uranium required for the atomic bomb. In 1946 Abelson moved back to the Carnegie Institution, where he was appointed Director of the Geophysics Laboratory in 1953 and President in 1971. He worked on the development of nuclear submarines and also promoted the peaceful uses of radioactive isotopes. He himself used the method to identify amino acids in fossils of various ages and fatty acids in rocks more than 1 000 million years old.

ABNEY, Sir William de Wiveleslie

(1844–1920)

English chemist and educationist, principally remembered for his contributions to the science of photography, including stellar photography. He was born in Derby, educated at the Royal Military Academy at Woolwich, and joined the Royal Engineers in 1861. Subsequently he became an instructor, and later head, at the School of Military Engineering. In 1877 he moved to the Department of Science and Art in South Kensington and rose to become Assistant Secretary to the Board of Education (1899), and after his retirement in 1903, its principal scientific adviser. Abney invented a new gelatine emulsion which made instantaneous photography viable and also an emulsion-coated paper from which modern photographic papers have been developed. One of the pioneers of stellar photography, he discovered how to make photographic plates that were sensitive to red and infrared light, thus initiating many major advances. In 1885 he co-authored a seminal paper on the infrared spectra of organic substances. He was elected FRS in 1876, knighted in 1900, and died in Folkestone.

ABRAHAM, Sir Edward Penley

(1913–)

English biochemist, born in Southampton. Educated at Queen's College, Oxford, he received a Rockefeller Foundation travelling fellowship (1938–9) and worked in Stockholm before joining the Sir William Dunn School of Pathology at Oxford to study wound shock. There he joined **Chain** and

Florey to purify penicillin. Chain had isolated penicillin as a powder, and while trying to prepare a sample suitable for X-ray analysis by **Dorothy Hodgkin**, Abraham found that it readily crystallized as the sodium salt, thereby providing the crucially sought confirmation of its purity. During 1950–60 he isolated the antibiotic cephalosporin C from the mould *Cephalosporium acremonium*; like penicillin it acts on the bacterial cell wall and is inhibited by penicillinase. Cephalosporin antibiotics are used against bacteria that cause throat infections, boils, abscesses, diphtheria and typhoid fever, particularly for patients allergic to penicillin, but unfortunately possess undesirable side effects and may cause allergies and kidney damage. Abraham was elected FRS in 1958, and served as Professor of Chemical Pathology at Oxford from 1964 to 1980. He was knighted in 1980.

ACHARD, Franz Karl
(1753–1821)

German agricultural chemist, born in Berlin. He was **Marggraf**'s assistant at the Academy of Sciences, Berlin, and followed up Marggraf's discovery of sugar in beet by perfecting a process for its extraction on a commercial scale. He founded the first beet sugar factory at Kunern, Silesia, in 1801. Later he was commissioned by Frederick the Great to improve the way in which tobacco was processed. Achard is also remembered for his work in physical chemistry, particularly on boiling points and the effect of heat on thermal conductivity. He died in Kunern.

ACHESON, Edward Goodrich
(1856–1931)

American chemist, born in Washington, Pennsylvania, the son of William Acheson, a merchant and later manager of a blast furnace. After schooling for only three years, he became a timekeeper at a blast furnace in 1873. From 1880 to 1881 he carried out research on electric lamps as an assistant to **Edison**. After 1884 he worked independently to develop the electric furnace for the conversion of carbon into diamonds, without success. In 1891 he developed the manufacture of silicon carbide (carborundum), an extremely useful abrasive, and he later devised a new method of making artificial graphite (1896). In 1899 he founded the Acheson Graphite Co, which led to the manufacture and sale of graphite products: solid graphite for electrodes and crucibles, and colloidal solutions for lubricating purposes and inks. He held 69 patents.

ADAMS, Sir John Bertram
(1920–84)

English nuclear physicist and founder member of CERN, born in Kingston, Surrey. After his school education he joined the Siemens Research Laboratory at Woolwich before working on wartime radar development. His work on short-wavelength systems took him to Harwell, where he engineered the world's first major postwar particle accelerator (the 180 mega-electronvolt cyclotron) in 1949. At CERN, the European centre for nuclear research in Geneva, where he became Director-General in 1960, he engineered the 25 giga-electronvolt proton synchroton (PS) in 1954. In 1961 he was recalled to the UK to establish the laboratory at Culham for research on controlled nuclear fusion. He was later appointed controller of the new Ministry of Technology in 1964, and in 1966 a researcher for the UK Atomic Energy Authority. He subsequently returned to Geneva (1969–76) to mastermind the building of the 450 giga-electronvolt super proton synchroton (SPS). He became Director-General of CERN for the second time from 1976 to 1980, during which the SPS was converted to a proton–anti-proton collider; this led to the discovery of the Z^0 particle by **Rubbia**. Both the PS and SPS are still operating at CERN, having their own experimental programmes and as part of the Large Electron–Positron (LEP) collider complex. Adams was elected FRS in 1963, and knighted in 1981.

ADAMS, John Couch
(1819–92)

English astronomer, born in Lidcot, Cornwall. He entered St John's College, Cambridge, in 1839, graduated as Senior Wrangler and first Smith's Prizeman (1843), and was elected to a fellowship at his college. He took up the problem of the unexplained irregularities in the motion of the planet Uranus, assuming that these were due to the existence of an unknown perturbing body in the space beyond Uranus, and by September 1845 he had derived elements for the orbit of such a body. Meanwhile in France **Le Verrier** had tackled the same problem, and making the same assumption as Adams, derived both the mass and the orbit of the hypothetical new body (1846). Le Verrier transmitted his results to the Berlin Observatory where **Galle** searched for and discovered (23 September 1846) an object whose motion defined it as a planet. A major controversy arose between France and England on the question of priority in this great discovery, but in the course of time it was recognized that Adams and Le Verrier had each solved the problem independently. The new planet received the name Neptune. In 1847 Adams was offered but declined a knighthood when Queen Victoria visited Cambridge, and in 1848 he was awarded the Copley Medal of the Royal Society. In 1858 he was appointed Lowndean Professor of Astronomy and Geometry at Cambridge, where he spent the rest of his life pursuing problems in mathematical astronomy: his work on the motion of the Moon earned him the Gold Medal of the Royal Astronomical Society (1866). As president of that society he had the satisfaction of presenting its Gold Medal to Le Verrier for his theories on the motions of the four great planets (1876). Three years after his death a portrait medallion of Adams was placed in Westminster Abbey near the grave of **Isaac Newton**.

ADAMS, Walter Sydney
(1876–1956)

American astronomer, born in Antioch, Syria, where his parents were missionaries. After studying at Dartmouth College, Massachusetts, the University of Chicago and the Yerkes Observatory, Adams went in 1904 with **Hale** to help establish the Mount Wilson Observatory in the mountains behind Pasadena, California. He was Deputy Director there from 1913 to 1923 and subsequently became Director. He was responsible for the design and installation of the 2.54 and 5.08 metre telescopes at Mount Wilson and Palomar. Adams discovered the spectroscopic method of distinguishing between main-sequence and giant stars. This technique enabled stellar spectra to be related to luminosity. In a long-term collaborative project, he used this to measure the distance to 6000 stars. He also used spectroscopic methods to measure the temperature, pressure and density of the light-emitting material in sunspots. In 1915 he obtained the first spectrum of the white dwarf star Sirius B, and showed that it was hotter than the Sun, contrary to expectations. He also found that it had a density 40000 times that of water. In 1925 Adams measured the gravitational redshift induced in the light leaving the surface of Sirius B, adding yet further support to **Einstein**'s general theory of relativity. Adams's 1934 observations of Mars showed that the atmosphere had less than 0.1 per cent oxygen. In 1941 he discovered spectral lines introduced by interstellar cyanogen molecules; these have an excitation temperature between 2 and 3 kelvin and gave an early clue to the existence of the microwave background radiation in the universe.

ADANSON, Michel
(1727–1806)

French botanist, born in Aix-en-Provence. He grew up in Paris, where he studied theology, classics and philosophy, and attended classes in the Jardin du Roi, taught by **Bernard Jussieu**, who influenced his approach to plant classification. He travelled to Senegal in 1748, and during six years there he collected thousands of botanical and zoological specimens. On his return he published *Histoire naturelle du Sénégal* (1757) and *Les Familles naturelles des plantes* (1763–4). In the latter work he analysed the theoretical basis for the classification of plants in natural order, ie reflecting their real similarities and differences. He was critical of **Linnaeus**'s reliance on a single character—fructification—as the basis for natural classification, and suggested instead that trial and experience should determine the most significant characters by which particular groupings of plants should be made. By giving detailed critical descriptions of each family, he anticipated the modern multi-character approach in taxonomy. The baobab genus of African savannah trees, *Adansonia*, is named after him.

ADDISON, Thomas
(1793–1860)

English physician, born near Newcastle and educated at Edinburgh University (MD 1815) and London. He held positions at several hospitals and dispensaries, including Guy's Hospital, where for many years he was the leading medical teacher and diagnostician. He collaborated with **Bright** on an unfinished textbook of medicine (1839), and left outstanding descriptions of many diseases and their pathological signs, including pneumonia, tuberculosis, 'Addison's disease' (insufficiency of the suprarenal capsules), and 'Addison's anaemia' (now known as pernicious anaemia). His *On the Constitutional and Local Effects of Disease of the Suprarenal Capsules* (1855) was a model of clinical investigation; he noted the peculiar bronzing of the skin of patients suffering from the effects of adrenal insufficiency, and through postmortem examinations, showed that symptoms were associated with destruction of the gland by tuberculosis, cancer or another cause. He hid a nervous, retiring personality behind a haughty exterior, and preferred the wards and postmortem room of the hospital to the more lucrative possibilities of private practice.

ADRIAN, Edgar Douglas, 1st Baron

(1889–1977)

English physiologist, born in London, and one of the founders of modern neurophysiology. He trained in the Physiological Laboratory, Cambridge where, after medical studies at St Bartholomew's Hospital London and war service, he became a lecturer, and also a Fellow of Trinity College, Cambridge. Much influenced by his tutor **Lucas**, he devoted his career to the study of the nervous system, and carried out important work in designing and building equipment to amplify and record the minute electrical impulses of nerves. In collaboration with **Zotterman** and then with **Bronk**, he recorded the electrical activity of single, isolated nerve fibres, from sensory and subsequently motor nerves. He showed that there is only one kind of nervous impulse, neural information being conveyed by variations in the frequency at which those impulses are transmitted, the 'frequency code' which is a fundamental characteristic of nerves. In an extensive examination of different sensory systems in many different animals, he investigated the mechanisms of peripheral functioning of receptors and sense organs in response to a stimulus, and then followed the processes of information transmission into the central nervous system, where he studied the recording and analysing of sensory information. He also developed techniques to study and understand the gross electrical activity of the brain, electroencephalography (EEG), used clinically for the study of epilepsy and other brain disorders. For his work on the function of neurons he shared the 1932 Nobel Prize for Physiology or Medicine with **Sherrington**. He was appointed Professor of Physiology at Cambridge (1937–51), Master of Trinity College, Cambridge (1951–65), and Vice-Chancellor (1957–67) and subsequently Chancellor, of the university (1968–75). He also served in many other capacities, including those of Foreign Secretary (1945–50) and President (1950–5) of the Royal Society.

AEPINUS, Franz Ulrich Theodosius

(1724–1802)

German mathematician and physicist, born in Rostock. He studied medicine and mathematics at Jena and Rostock, where he subsequently taught mathematics. He became director of the observatory at Berlin in 1755, and later Professor of Physics in St Petersburg (1757–98). Aepinus studied the thermoelectric properties of the mineral tourmaline, identifying a similarity between its attractive force when heated and the properties of magnetism. His *Tentamen theoriae electricitatis et magnetismi* (1759) was an early application of mathematics to the theory of electric and magnetic forces, and induction.

AGASSIZ, Alexander Emmanuel Rodolphe

(1835–1910)

Swiss–American oceanographer and marine zoologist, born in Neuchâtel, Switzerland, son of **Jean Agassiz**. He went to the USA in 1849 to join his father and studied engineering and zoology at Harvard. He directed his engineering skills towards ways of using steel wire rope for dredging and deep soundings, greatly speeding up these operations. He also invented mechanical strain equalizers which facilitated hauling apparatus on board, a new double-edged dredge and a closing tow net for mid-depth sampling. He was a believer in the 'azoic' theory that no life was possible beyond shallow depths in the sea, though he was denied evidence to the contrary by the continuing failure of his mid-depth nets. Between 1877 and 1880 he sailed aboard the *Blake* in the Atlantic and on the *Albatross* in the Pacific, studying the Gulf Stream, the plankton abundance and the dependence of the bottom fauna upon it. He resurrected and became the chairman of the Calumet and Hecla copper mines in Michigan which financed his scientific travels; these included visits to reefs around the world in search of evidence on reef theory, on which he disagreed with **Charles Darwin**. From 1873 to 1885 he was curator of the Harvard Museum of Comparative Zoology founded by his father, which contains both their collections.

AGASSIZ, Jean Louis Rodolphe

(1807–73)

Swiss–American naturalist and glaciologist, born in Môtier-en-Vully. He studied at the Medical School of Zürich and at the universities of Heidelberg and Munich. His main interest was zoology and while still a student he published *The Fishes of Brazil* (1829). He was an enthusiast, a pre-eminent teacher and public expositor. In 1832 he was appointed Professor of Natural History at the University of Neuchâtel where he became interested in glaciers. His careful observations of evidence for the glacial transportation of rock material brought the recognition that glaciers are mobile. He traced their previous extents in the Alps, and surmised from this the theory of ice ages, with global cooling and the past effects upon flora and fauna. He can be considered the father of glacial theory for having given credibility to the previously unacceptable ideas of **Horace Bénédict de**

Saussure and **Hutton**. After a highly successful Lowell lecture tour in Boston in 1846, he was appointed Professor of Natural History at the Lawrence Scientific School at Harvard. In 1859 Agassiz founded the Museum for Comparative Zoology at Harvard, to which he donated his collections. He became an oceanographer in 1851, taking an interest in coral reefs. His publications included *Études sur les glaciers* (1840) and *Système Glacière* (1847).

AGNESI, Maria Gaetana
(1718–99)

Italian mathematician and scholar, born in Milan, the daughter of a professor of mathematics at Bologna and the eldest of 21 children. Educated privately, she was a child prodigy, speaking six languages by the age of 11. She published books on philosophy and mathematics, and her mathematical textbook *Istituzioni analitiche* (1784) became famous throughout Italy. One of the few women mathematicians to gain a reputation before the 20th century, she assimilated the work of many different authors and developed new mathematical techniques. She is best known for her description of a versed sine curve, which following an early mistranslation of Italian, became known as the 'witch of Agnesi'.

AGRICOLA, Georgius, Latin name of Georg Bauer
(1494–1555)

German mineralogist and metallurgist, born in Glauchau, Saxony. He was rector of the school at Zwickau (1518–22). Later, as a practising physician in Chemnitz, his interest in the link between medicine and minerals led him to devote himself to the study of mining. He was among the first to recognize the erosive power of rivers in forming landscapes, but his ideas about the ability of exhausted mines to regenerate themselves did not stand the test of time. He was author of *De Natura Fossilum* (1546), in which a systematic classification of minerals was attempted, and his *De Re Metallica* (1555, translated by US President Herbert Hoover in 1912), is a valuable and detailed record of 16th-century mining, ore-smelting and metal working which remained a standard work for almost four centuries.

AHLQUIST, Raymond Perry
(1914–83)

American pharmacologist, born in Missoula, Montana. He studied at the University of Washington, Seattle, graduating with a PhD in pharmaceutical chemistry in 1940. He became an assistant professor at South Dakota State College (1940–4) and then moved to the Medical College of Georgia, becoming Professor of Pharmacology (1948–77), and Charbonnier Professor (1977–83). Ahlquist's major contribution was published in 1948, a paper that explained the diverse biological actions that were caused by the catecholamines (adrenaline and noradrenaline) as due to two different populations of cell receptors (the alpha- and beta-adrenoceptors), which displayed differential sensitivity to these two principal catecholamines. Ahlquist's proposal, the dual receptor concept, provided an explanation for the effects that had been observed by physiologists and pharmacologists for over forty years, but was originally rejected for publication. It contradicted an erroneous but widely accepted theory postulated by many others, including **Walter Cannon**, and took many years to gain acceptance. The potential clinical utility of the concept was recognized by Sir **James Black** in his development of beta-blockers.

AIKEN, Howard Hathaway
(1900–73)

American mathematician and computer engineer, born in Hoboken, New Jersey. He grew up in Indianapolis, Indiana, and was educated at Wisconsin, where he majored in electrical engineering. After working for Westinghouse for 10 years, he returned to academia, first as a student at the University of Chicago and then on the staff at Harvard (1939–61). Frustrated by the toil involved in solving complex differential equations, he succeeded in persuading IBM to sponsor the building of a calculating machine. Together they built the Automatic Sequence-Controlled Calculator (ASCC), or Harvard Mark I. Completed in 1943 and weighing 35 tons, this was the world's first program-controlled calculator, which Aiken regarded as the realization of **Babbage**'s dream. According to one physicist, it sounded like a 'roomful of ladies knitting'. A Mark II was built in 1947, but by then Aiken's design—which was essentially a giant mechanical calculator—was being overtaken by the development of the stored-program computer. Aiken later became professor at Miami (1961–73).

AIRY, Sir George Biddell
(1801–92)

English astronomer, born in Alnwick, Lincolnshire. He entered Trinity College, Cambridge, in 1819, graduated as Senior Wrangler and Smith's Prizeman (1823) and was elected a Fellow of Trinity College

(1824). He became Lucasian Professor of Mathematics in 1826, and Plumian Professor of Astronomy and Director of the Cambridge Observatory in 1828. His researches on optics, in which he applied the wave theory of light to analysis of the structure of images of point sources formed by lenses, earned him the Copley Medal of the Royal Society (1831), and his laborious investigations in planetary theory were recognized by the Royal Astronomical Society's award of its Gold Medal (1833). Two years later he was appointed Astronomer Royal and Director of the Greenwich Observatory. Finding this establishment in a poor and ineffective state, Airy saw to its complete reorganization, transforming it into the finest meridian observatory in the world. He established magnetic and meteorological departments in Greenwich and used submarine telegraphy to determine the longitudes of various observatories internationally, in this way achieving worldwide acceptance of the Greenwich zero meridian. He pioneered the transmission of telegraphic time signals for the railways. Also important was his determination of the mean density of the Earth through pendulum experiments at the top and bottom of deep mines. He was President of the Royal Society in 1871, four times President of the Royal Astronomical Society, and received honours and decorations from all over the world, including the Prussian *Pour le Mérite* and membership of the French Legion of Honour. He was knighted in 1872.

AIRY SHAW, Herbert Kenneth
(1902–85)

English botanist, born in Woodbridge, Suffolk. He entered Corpus Christi College, Cambridge, in 1921 to read classics but graduated in botany, having already developed a special interest in taxonomy. He began unpaid work at Kew in 1925 and was not appointed to a salaried post there until 1929. He first worked on European and oriental plants with William Bertram Turrill, but his interest began to shift in the 1930s towards tropical Asian botany. He began working on tropical Euphorbiaceae, on which he became a leading authority. He also became an expert on nomenclatural matters. A devout Christian, he resigned in 1951 to edit a Christian journal, but returned to Kew in 1960. Over the following 20 years he published more than 250 papers, together with regional treatments of Euphorbiaceae for Siam (Thailand) (1972), Borneo (1975), New Guinea (1980), Sumatra (1981), Central Malaysia (1982), the Philippines (1983) and Australasia (1980). He also edited the 7th and 8th editions (1966, 1973) of **John Willis**'s *Dictionary of the Flowering Plants and Ferns* and compiled a glossary of Russian botanical terms. He had a lifelong interest in entomology and wrote many papers on the subject.

AITKEN, John
(1839–1919)

Scottish physicist, born in Falkirk. He studied marine engineering at Glasgow University, but on completion of his apprenticeship, the onset of ill-health compelled him to abandon his engineering career and his interests shifted towards scientific research, in particular to aspects of the physical sciences. His early work included investigations into the human perception of colour and aspects of the formation and melting of ice, including glacier motion. His largest and most significant body of work related to the processes involved in the boiling of liquids and the condensation of vapours. Aitken showed that water vapour in the atmosphere does not condense to form clouds unless it has a solid or liquid nucleus to condense upon. This brought into prominence the importance of dust, both visible and microscopic, to the formation of droplets of rain and mist. A skilled designer and maker of experimental equipment, he developed the dust counter, a device to measure the dust content of air. Using this equipment he investigated the meteorological and industrial influences on the production of dust. His work on the formation of dews showed that the liquid that condenses as dew on cold surfaces comes mainly from the ground and not from the air. He was a Fellow of the Royal Society of both London and Edinburgh, the latter publishing a résumé of his life's work entitled *Collected Scientific Papers* (1923, edited by Cargill G Knott).

AITKEN, Robert Grant
(1864–1951)

American astronomer, born in Jackson, California. Aitken took his degree at Williams College, where he forsook his plans to enter the ministry and turned to astronomy. He joined the Lick Observatory in 1895, and was director there from 1930 to 1935. He undertook (assisted by William Hussey) a massive survey of double stars between 1899 and 1915, discovering more than 4 500 new ones. Aitken concentrated on determining their motions and orbits. His discoveries gained him the Gold Medal of the Royal Astronomical Society in 1932. He published *Binary Stars* (1918) and *New General Catalogue of Double Stars* (1932).

AKI, Keiiti

(1930–)

Japanese–American seismologist, born in Yokohama. He was educated at the University of Tokyo where he received his BS in 1952 and a PhD in geophysics in 1958. He was a Research Fellow at Caltech (1958–60), then returned to Tokyo (1960–2, 1964–6) before moving permanently to the USA in 1966, becoming a US citizen in 1976. He was Professor of Geophysics and then R R Shrock professor at MIT (1966–84). Since 1984 he has been W M Keck Professor of Geological Sciences and Science Director at the Southern California Earthquake Center. He introduced the concept of seismic moment, a measure of the extent of an earthquake (1966), then in 1967 a scaling law for seismic spectra. He pioneered strong motion seismology which has been of great importance to civil engineering (1968) and in 1969 discovered coda waves, which describe the ground vibration after primary waves have spread out from the earthquake source. He also developed seismic tomography (three-dimensional seismic modelling) techniques (1974), and produced the first quantitative analysis of the earth tremors accompanying a volcanic eruption. His Aki–Larner method allowed theoretical calculation of seismic motion in a layered medium with an irregular interface. He published the popular textbook *Quantitative Seismology: Theory and Methods* (2 vols) in 1980.

ALBERTUS MAGNUS, St, Count of Bollstädt

(c.1200–1280)

Philosopher and theologian, known as the 'Doctor Universalis', born in Lauingen. He studied in Padua, entered the newly founded Dominican order, and taught in the schools of Hildesheim, Ratisbon and Cologne. From 1245 to 1254 he lectured in Paris, and in 1254 he became provincial of the Dominicans in Germany. In 1260 he was named bishop of Ratisbon. Albertus excelled all his contemporaries in the width of his learning. In his scientific works he was a faithful follower of **Aristotle** as presented by Jewish, Arabian and western commentators; in voluminous commentaries on the writings of Aristotle he comprehensively documented 13th-century European knowledge of the natural sciences, mathematics and philosophy. He also engaged in alchemy, although his works express doubts about the possibility of transmutation of the elements, and gave a detailed description of the element arsenic.

ALCMAEON

(fl 520 BC)

Greek philosopher, born in Crotona, Italy. Though writing on medical subjects, Alcmaeon was a philosopher and almost certainly not a physician. Like other pre-Socratics, he displayed wider interests that included astrology and meteorology and standard philosophical questions like the immortality of the soul. A pupil of **Pythagoras**, he made an early statement of the Pythagorean doctrine that health was a positive state that depended on maintenance of humoral opposites (dry/wet, hot/cold, bitter/sweet) in their proper balance. Alcmaeon was concerned with the internal causes of diseases. Rather like the followers of **Hippocrates**, he divided these causes into disorders of environment (climatic conditions), of nutrition and of lifestyle (excessive or insufficient exercise, etc). It appears that Alcmaeon was an early pioneer of dissection.

ALDER, Kurt

(1902–58)

German chemist, born in Königshülte (now Chorźow, Poland). He studied chemistry at the universities of Berlin and Kiel, and his PhD supervisor was **Diels**. He remained at Kiel where he became professor in 1934, and in collaboration with Diels, he reported (1928) a facile reaction between a diene and a compound with an activated double bond to give a cyclic product (the Diels–Alder reaction). This reaction has had widespread use in synthetic organic chemistry and has been particularly useful in the preparation of natural products. For its discovery, Alder and Diels were jointly awarded the Nobel Prize for Chemistry in 1950. In 1936 Alder joined I G Farbenindustrie but returned to academic life in 1940 as professor in Cologne, where he remained until his death.

ALEKSANDROV, Pavel Sergeevich

(1896–1982)

Russian mathematician, born in Bogorodsk. He studied at Moscow and became professor there in 1929. The leader of the Soviet school of topologists, he developed many of the methods of combinatorial or algebraic topology. He developed both global and local homology and cohomology theories capable of dealing with spaces of any number of dimensions, and following hints dropped by **Noether**, was able to define the Betti numbers in terms of suitable homology groups. He was a lifelong friend of the German topologist **Hopf**, and their book *Topologie* (1935) was a landmark in the development of the subject.

ALEMBERT, Jean le Rond d'
See **D'ALEMBERT, Jean le Rond**

ALFVÉN, Hannes Olof Gösta
(1908–)

Swedish theoretical physicist, a pioneer of plasma physics, born in Norrköping. Educated at Uppsala University, he joined the Royal Institute of Technology, Stockholm (1940), becoming Professor of Electronics in 1945 and Professor of Plasma Physics in 1964. He moved to the University of California in 1967. Alfvén carried out pioneering research on ionized gases, or plasmas, and their behaviour in electric and magnetic fields. He made important contributions to the theory describing how magnetic fields can sometimes become 'frozen' into plasmas, effectively binding the ions to the magnetic field lines passing through. He also predicted the existence of 'Alfvén waves' in plasmas (1942), which were later observed. His theories have led to important advances in the analysis of the motion of particles in the Earth's magnetic field and stellar structure, and in the attempts to develop nuclear fusion reactors. He shared the Nobel Prize for Physics with **Néel** in 1970.

ALHAZEN (Ibn al-Haytham)
(c.965–c.1040)

Arab mathematician, born in Basra. He wrote a work on optics (known in Europe in Latin translation from the 13th century) giving the first account of atmospheric refraction and reflection from curved surfaces, and the construction of the eye. He constructed spherical and parabolic mirrors, and it was said that he spent a period of his life feigning madness to escape a boast he had made that he could prevent the flooding of the Nile. In later life he turned to mathematics and wrote on **Euclid**'s treatment of parallels and on **Apollonius**'s theory of conics.

AL-KHWARIZMI, Abu Ja'far Muhammad ibn Musa
(c.800–c.850)

Arab mathematician who wrote in Baghdad on astronomy, geography and mathematics. He wrote an early Arabic treatise on the solution of quadratic equations, synthesizing Babylonian solution methods with Greek-style proofs of their correctness for the first time. His writings in Latin translation were so influential in transmitting Indian and Arab mathematics to medieval Europe that the methods of arithmetic based on the Hindu (or so-called Arabic) system of numeration became known in medieval Latin, by corruption of his name, as 'algorismus', from which comes the English 'algorithm'; the word 'algebra' is derived from the word *al-jabr* in the title of his book on the subject.

ALLEN, Sir Geoffrey
(1928–)

English chemical physicist and industrial chemist. He received his PhD at University of Leeds, and following two years at the National Research Council in Ottawa, Canada, he joined University of Manchester where he eventually became Professor of Chemical Physics. He then moved to professorships at Imperial College of Science and Technology. Allen's principal scientific interests lie in the chemistry and physics of polymers. In 1981 he joined Unilever after 25 years of academic life, and since then he has held various industrial appointments including Director of Unilever and Courtaulds, and Executive Adviser to Kobe Steel Ltd. Through the Society of Chemical Industry he is known as a spokesman for the subject of chemistry to the general public. He was Chairman of the Science Research Council in 1977, and in 1980 he was responsible for the inclusion of engineering in the title to form the Science and Engineering Research Council. He was elected FRS in 1976, knighted in 1979, and is a Vice-President of the Royal Society and a Senior Vice-President of the Institute of Materials.

ALLEN, James Alfred Van
See **VAN ALLEN, James Alfred**

ALPHER, Ralph Asher
(1921–)

American physicist, born in Washington DC, known for his theoretical work concerning the origin and evolution of the universe. After studying at George Washington University, he spent World War II as a civilian physicist, and later worked at Johns Hopkins University and in industry. Together with **Bethe** and **Gamow**, he proposed in 1948 the 'alpha, beta, gamma' theory which suggests the possibility of explaining the abundances of chemical elements as the result of thermonuclear processes in the early stages of a hot, evolving universe. These ideas were developed to become part of the 'Big Bang' model of the universe. Also in 1948, he predicted that a hot 'Big Bang' must have produced intense electromagnetic radiation which would have 'cooled' (or redshifted), and this background radiation was in fact observed in 1964 by **Penzias** and **Robert Wilson**.

ALPINO, Prospero, Latin Prosper Alpinus
(1553–1616)

Italian botanist and physician, born in Marostica in the republic of Venice. He spent three years as physician to the Venetian consul in Cairo, Georgio Emo; during this period he observed the sexual fertilization of the date palm, in which male and female flowers are on different trees, and described 57 species, wild or cultivated, in Egypt. He wrote *De Plantis Aegypti Liber* (1592), which contains a number of original woodcuts. Whereas most contemporary botanical works were herbals, describing the medicinal uses of plants, this book is an early example of a Flora—a description of the plants found in a particular area. In this and his *De Medecina Egyptorum* (1591) he brought the coffee plant and the banana, both of which he had seen growing near Cairo, to European attention for the first time. In 1594 he became lecturer in botany at Padua, and director of the botanic garden there (1603), where he grew many of the plants he had collected in Egypt.

ALTER, David
(1807–81)

American physicist and inventor, born in Westmoreland, Pennsylvania. He graduated from the Reformed Medical College in New York City in 1831. A prodigious inventor, he worked on electric clocks, invented an electric telegraph (1836) and discovered methods of purifying bromine and producing coal oil from coal. One of the earliest investigators of spectra, he suggested that each element has a characteristic spectrum, and pioneered the use of the spectroscope in determining the chemical constitutions of gases and vaporized solids. His theories were later confirmed by **Bunsen** and **Kirchhoff**.

ALTMAN, Sidney
(1939–)

Canadian biologist, born in Montreal. He was educated at MIT and the University of Colorado, Boulder, and started his career as a teaching assistant at Columbia University (1960–2). After holding several fellowships, he became assistant then associate professor (1971–80) at Yale University, where since 1980 he has been Professor of Biology. In the early 1970s, Altman studied the process by which an RNA transcript of DNA 'matures' to form transfer RNA, the amino acid carrier in protein synthesis. He found that the precursor RNA was specifically cleaved at both ends by ribonuclease P, an enzyme which contains both RNA and protein components and is bound to the ribosome. Although he initially thought both components were essential for enzyme activity, parallel studies by **Cech** (1982) led him to discover in 1983 that the RNA component of ribonuclease P alone catalyses the maturation of transfer RNA. Other examples of RNA catalysis have since been described, giving further confirmation of one of the many predictions of **Crick** and suggesting a possible route for the emergence of a stable, self-replicating living system prior to evolution. For his pioneering research, Altman shared the 1989 Nobel Prize for Chemistry with Cech. He contributed to and edited *Transfer RNA* (1978).

ALTOUNYAN, Roger Edward Collingwood
(1922–87)

British physician and medical pioneer, inventor of the anti-asthma drug Intal and the 'spinhaler'. Born in Syria of Armenian–English extraction, he spent his summer holidays with his four sisters in the Lake District, where they met the author Arthur Mitchell Ransome (1884–1967) and became the real-life models of the children in his *Swallows and Amazons* series of adventure books. After qualifying as a doctor at the Middlesex Hospital in London, he practised for a while in Aleppo, Syria; he returned to England in 1956 to join a pharmaceutical company, where he worked in his own time to develop the drug Intal to combat asthma, from which he suffered. He was thus an early experimental subject. A pilot and flying instructor during World War II, he developed the spinhaler device to inhale the drug, based on the aerodynamic principles of aircraft propellors.

ALVAREZ, Luis Walter
(1911–88)

American experimental physicist and Nobel prize winner of exceptionally wide-ranging talents. Born in San Francisco, he studied physics at the University of Chicago where he built one of the first Geiger counters in the USA and together with **Compton** used it to study cosmic rays. He joined **Lawrence** at Berkeley University in 1936 where he worked on nuclear physics; here he discovered electron capture by nuclei, tritium radioactivity and the magnetic moment of the neutron. During World War II he worked at MIT, where he invented a radar guidance system for landing aircraft in conditions of poor visibility. After the war he returned to Berkeley where he became Professor of Physics in 1945. There he worked on developing the bubble chamber technique invented by **Glaser** in 1956 for studying particle interactions. The first bubble chambers were small, a few inches in diameter, but Alvarez was convinced that this was a useful technique and built a series of larger bubble chambers culminating in a 'large' bubble chamber 72 inches long. He used these to carry out a range of particle physics experiments in which a large number of 'resonances', particles that live only for a very short time, were identified. These results helped to show a pattern in the properties of the resonances which ultimately led to the quark model invented by **Gell-Mann** and **Zweig**. In 1968 Alvarez was awarded the Nobel Prize for Physics for this work. He applied physics and ingenuity to a variety of problems: he used cosmic X-rays to show that Chephren's Egyptian pyramid had no undiscovered chambers; he showed that only one killer was involved in the assassination of President Kennedy; and he founded two companies to make optical devices, including the variable-focus spectacle lens which he invented for his own use. With his geologist son Walter (b 1940) he studied the catastrophe of 65 000 000 years ago which killed the dinosaurs. Based on radiotracer analysis, they proposed that its cause was the impact on Earth of an asteroid or comet.

AMAGAT, Émile Hilaire
(1841–1915)

French experimental physicist, born in Saint-Satur. He received his doctorate in Paris in 1872. After teaching mathematics and physics at Freiburg, Cluny and Lyons (1877–92), he was examiner at the École Polytechnique. He became a full member of the Academy of Sciences in 1902 and four years later was elected President of the French Physical Society. Amagat was renowned for his accurate experimental work on fluid statics, especially the behaviour of gases and liquids under extreme conditions of temperature and pressure. He used a church tower in Lyons to house a mercury manometer which measured pressures of up to 80 atmospheres. Later a column of mercury 327 metres high, constructed inside a shaft at a coal mine near Saint-Étienne, enabled him to determine the compressibility of nitrogen up to 430 atmospheres. With newly designed apparatus he eventually reached 3 000 atmospheres and recorded the marked deviations of gases from **Boyle**'s law under such conditions. He carried out associated studies into solid elasticity, compressibility of mercury, density of gases and their saturated vapours, and variations of temperature of fusion, again under great pressures. **Bridgman** was to push the limits still higher.

AMBARTSUMIAN, Viktor Amazaspovich
(1908–)

Soviet–Armenian astrophysicist, born in Tiflis (Tbilisi), Georgia. Educated at the University of Leningrad (now St Petersburg), he was Professor of Astrophysics there (1934–44) and from 1944 at Yerevan, where he founded the Byurakan Astronomical Observatory which became one of the Soviet Union's most important observatories. He devised theories of young star clusters and methods for computing the mass ejected from nova stars. In the early days of radio astronomy, a strong radio source in the constellation Cygnus had been associated with a pair of galaxies which were thought to be closely connected. One prominent theory was that a collision between the two galaxies could be a possible cause of the radio emission, and that this mechanism might account for the radio brightness of many other extragalactic sources. However, Ambartsumian demonstrated in 1955 that such processes could not provide the energy required and proposed instead that enormous explosions occur within the cores of galaxies, of the same form as those producing supernovae, but on a galactic scale. In 1939 he wrote *Theoretical Astrophysics* (English translation 1958), which was widely influential.

AMDAHL, Gene Myron
(1922–)

American computer scientist and entrepreneur, born in Flandreau, South Dakota. After taking a degree in electrical engineering at South Dakota State University, he became project manager at IBM in Poughkeepsie in 1952, the year in which he was awarded a PhD in physics at the University of Wisconsin. Resigning in 1955, Amdahl was re-hired by IBM in 1960 to manage the company's advanced data processing systems. He became a leader in the design of several IBM computers, most notably the phenomenally successful System/360. In 1970 he resigned again and founded the Amdahl Corporation for producing computers compatible with large models of System/370, IBM's successor to System/360. Although successful, after a drop in revenues, Amdahl resigned as chairman of the company and set up Trilogy Ltd in 1980 to compete with both the Amdahl Corporation and IBM.

AMICI, Giovanni Battista
(1786–1863)

Italian optician, astronomer and natural philosopher, born in Modena. He constructed optical instruments and perfected his own alloy for telescope mirrors. Around 1812 he began making improvements in the microscope which, at that time, was a non-achromatic system. In 1827 he produced the dioptric, achromatic microscope that bears his name. He produced the first (water) immersion objective 13 years later. In 1844 he made high-power objectives utilizing meniscus lenses (concave on one face and convex on the other) to correct defects in images caused by spherical aberrations; two years later he published important work on the sexual process in plants, which these optical advances had made accessible to observation. He was appointed director of the Florence observatory in 1835.

AMONTONS, Guillaume

(1663–1705)

French instrument-maker and experimental physicist, born in Paris. Amontons was incurably deaf from adolescence. The son of a lawyer, he studied geometry, mechanics and physical science, then drawing, surveying and architecture in order to qualify himself for employment in public works. He designed many instruments including a hygrometer (1687), a 'folded' barometer (1688), a conical nautical barometer (1695) and various air thermometers. To determine longitude at sea he suggested in 1695 a new clepsydra (or water clock). In 1699, recently elected to the Academy of Sciences, he published a study of mechanical friction; and experiments relating to his 'fire-mill' (a hot-air engine) showed that a constant volume of air heated from room temperature to that of boiling water increased its pressure by one third. In 1702 Amontons found that for any constant mass of air, the increase in pressure was proportional to temperature rise, irrespective of initial pressure. **Gay-Lussac** later demonstrated explicitly the general relationship between temperature and volume for all gases. In 1703 Amontons defined the 'extreme cold' of his thermometric apparatus as that point at which the air would exert no pressure, hinting at an unattainable absolute zero of temperature. **Fahrenheit** was particularly indebted to his investigation of the thermal expansion of mercury (1704). Amontons died from gangrene in 1705.

AMPÈRE, André Marie

(1775–1836)

French mathematician, chemist and physicist, whose name is given to the basic SI unit of electric current (ampere, amp). He was born in Lyons, where his father was a wealthy merchant. Soon after his birth, the family moved to the nearby village of Poleymieux; today the house is a national museum. Following the educational theories of Rousseau, he was allowed to educate himself by means of his father's considerable library. He had a phenomenal memory, and his mathematical abilities became apparent at an early age. He was appointed to a lectureship in mathematics at the École Polytechnique in Paris in 1803, appointed Inspector-General of the Imperial University in 1808, and elected to the Chair of Experimental Physics at the Collège de France in 1824. Although he contributed to a number of fields (notably what today we would call physical chemistry), he became best known for laying the foundations of the science of electrodynamics through his theoretical and experimental work, following **Oersted**'s discovery in 1820 of the magnetic effects of electric currents, in *Observations électro-dynamiques* (1822) and *Théories des phénomènes électro-dynamiques* (1830). He derived an expression for the force between small elements of two current-carrying conductors, showing that the force is inversely proportional to the square of the distance between them. Ampère was elected FRS in 1827. Of sensitive disposition, the execution of his father in the civil disturbances of 1793 resulted in a prolonged nervous breakdown. The death of his first wife in 1804, and a catastrophic second marriage, were blows from which he never fully recovered.

ANAXAGORAS

(c.500–428 BC)

Greek philosopher, born in Clazomenae. For 30 years he taught in Athens, where he had many illustrious pupils. His scientific speculations may have provided the pretext for his prosecution for impiety (he held that the Sun, Moon and stars were huge incandescent rocks), and he was banished from Athens for life. He withdrew to Lampsacus, on the Hellespont, and died there. His most celebrated, but obscure, cosmological doctrine was that matter is infinitely divisible into particles, which contain a mixture of all qualities, and that mind (*nous*) is a pervasive formative agency in the creation of material objects.

ANAXIMANDER

(c.611–546 BC)

Greek philosopher, born in Miletus, successor and perhaps pupil of **Thales**. He posited that the first principle was not a particular substance like water or air but the *apeiron*, the infinite or indefinite. He is reputed to have used the gnomon (a sundial with a vertical rod) to measure the lengths of the seasons, by fixing the times of the equinoxes and solstices. He is credited with drawing the first map of the inhabited world, and he recognized that the Earth's surface must be curved, though he visualized it as a cylinder rather than a sphere. He was aware that the heavens appeared to revolve about the Pole Star, and therefore pictured the sky as a complete sphere. No trace of his scientific writings has been found, but he is credited with many imaginative scientific speculations, for example that the Earth is unsupported and at the centre of the universe, that living creatures first existed in the waters of the earth, and that human beings must have developed from some lower species that more quickly matured into self-sufficiency.

ANAXIMENES
(d c.500 BC)

Greek philosopher, born in Miletus, and the third of the three great Milesian thinkers, succeeding **Thales** and **Anaximander**. No biographical details are known about him. He posited that the first principle and basic form of matter was air, which could be transformed into other substances by a process of condensation and rarefaction. He also believed that the Earth and the heavenly bodies were flat and floated on the air like leaves.

ANDEL, Tjeerd Van
See **VAN ANDEL, Tjeerd**

ANDERSON, Carl David
(1905–)

American physicist, born in New York. He studied at Caltech under **Millikan**, and in 1932 discovered the positron, the positively charged electron-type particle, thus confirming the existence of antimatter predicted by **Dirac**. He did notable work on gamma and cosmic rays, and was awarded the 1936 Nobel Prize for Physics (jointly with **Victor Hess**). Later he confirmed the existence of intermediate-mass particles known as μ-mesons or muons.

ANDERSON, John Stuart
(1908–90)

English chemist, born in London. He was educated at Imperial College, London, where he obtained his BSc and PhD (1931), the latter for work on nickel carbonyl and nickel nitrosyl compounds. He then spent a year at Heidelberg where his research opened up the chemistry of the metal nitrosyl complexes. Following a move to Melbourne University in 1938, he developed an interest in lanthanide and actinide chemistry which eventually led to a method for the recovery of the element protactinium from the Springfields nuclear separation process. He also studied the lower halides of zirconium and hafnium (developing a potential hafnium separation process), and the partition of minor lanthanide elements between minerals in the course of metallurgical processes. This work led naturally to an interest in solids and in later years his major discoveries were in the areas of solid-state and high-temperature chemistry. During a spell at the Atomic Energy Research Establishment at Harwell he worked on the stoichiometry of uranium oxides. At Oxford, where he was the first Professor of Inorganic Chemistry (1963–75), he worked on lanthanide carbides at temperatures up to 2 000 kelvin. Here he also developed electron microscopy as a means to determine the local structure of crystals. With **Emeleus** he wrote *Modern Aspects of Inorganic Chemistry* (1938), a textbook which had a considerable effect on the teaching of inorganic chemistry. He was elected FRS in 1953, and received the Davy Medal of the Royal Society in 1973. He won the Royal Society of Chemistry solid-state award in 1974 and the Longstaff Medal in 1976.

ANDERSON, Philip Warren
(1923–)

American physicist, born in Indianapolis. He studied antenna engineering at the Naval Research Laboratories during World War II and then at Harvard, where he received his PhD in 1949. As a student under **van Vleck** he worked on pressure broadening in microwave and infrared spectra. He joined Bell Telephone Laboratories in 1949 carrying out theoretical work on the electrical properties of disordered systems. From considerations of wave propagation in disordered materials he demonstrated that it is possible for an electron in these materials to be trapped in a small region (1958). This process, later known as 'Anderson localization', became a corner stone to the understanding of the electronic behaviour of disordered materials and contributed to their extensive exploitation in applications that today include amorphous silicon solar cells, thin film transistors and xerography. Anderson investigated minimagnets and revealed the microscopic origin of magnets in bulk materials. His work also included theoretical studies of superfluidity found in helium-4 below 2 kelvin, and he clarified the meaning of the **Josephson** effect. He was appointed Assistant Director of Bell Telephone Laboratories in 1974, moving the following year to Princeton University to become Professor of Physics. For his theoretical work on the electronic structure of magnetic and disordered systems he received the 1977 Nobel Prize for Physics jointly with **Mott** and van Vleck.

ANDERSON, Tempest
(1846–1913)

English ophthalmic surgeon, traveller and vulcanologist, born in York. He was educated at St Peter's, York, and University College London. After developing a particular interest in the eye, he became an ophthalmic surgeon at York County Hospital. He made his own cameras and lenses, and enthusiastically promoted the use of photography in geology. He travelled extensively, including visits to Martinique, St Vincent, other islands of the West Indies, and to the Philippines, compiling an early photographic record of volcanic phenomena. With **Flett**, he was commissioned by the Royal Society to investigate the recent eruptions of La Soufrière and Mt Pelée. His vulcanological works included *Volcanic Studies in Many Lands* (1903), and he published many articles in medical journals.

ANDRADE, Edward Neville da Costa
(1887–1971)

English physicist, born in London. After studying at London, he received his PhD from Heidelberg University (1911). He continued his studies at Cambridge and Manchester, and became Professor of Physics at the Artillery College in Woolwich (1920–8) and University College London (1928–50). For part of World War II he was scientific advisor to the director of scientific research at the Ministry of Supply. Andrade was known for his work on grain growth in metals and the viscosity of liquids. From 1950 until his retirement in 1952 he was Director of the Royal Institution and the Davy–Faraday Laboratory.

ANDREWS, Ernest Clayton
(1870–1948)

Australian geologist, born in Sydney. He studied at Sydney University under Sir Edgeworth David. His specialities included the physiographical evolution of the Pacific Rim coastlines, particularly eastern Australia and the western USA, and the glaciers of New Zealand. He also published papers on the flora of these areas and studied the coral reef formations of Fiji and Tonga. His work on the geology of *Broken Hill district* (1922) was of great economic importance. He held many offices in the scientific bodies of the USA, New Zealand and Australia, and was President of the Australian Association for the Advancement of Science from 1930 to 1932.

ANDREWS, Roy Chapman
(1884–1960)

American naturalist and explorer, born in Beloit, Wisconsin. After graduating from Beloit College, he joined the staff of the American Museum of Natural History, New York, and later became Director there (1935–42). He is best known as the discover of fossil dinosaur eggs and fossil mammals in Mongolia, but also made many valuable contributions to palaeontology, archaeology, botany, zoology, geology and topography. He explored Alaska before World War I, and took part in several expeditions to central Asia sponsored by the Museum of Natural History. His published works included *Across Mongolian Plains* (1921), *The Ends of the Earth* (1929), *Meet Your Ancestors* (1945), *Heart of Asia* (1951) and *In the Days of the Dinosaur* (1959).

ANDREWS, Thomas
(1813–85)

Irish physical chemist, born in Belfast. From 1828 onwards he studied, in succession, chemistry in Glasgow and Paris, and medicine in Dublin, Belfast and Edinburgh. Andrews then practised as a physician in Belfast and taught chemistry at the Royal Belfast Academical Institution. In 1845 he was appointed Vice-President of the recently established Queen's College, Belfast, and from 1849 to 1879 he occupied the Chair of Chemistry there. He was awarded the Royal Medal of the Royal Society in 1844 and elected FRS in 1849. Throughout his career he was constantly engaged in research. His most important work (around 1860) was on the pressure, volume and temperature relationships of carbon dioxide, which established the existence of the critical temperature of a gas, above which it cannot be liquefied. He also made extensive and precise studies of heats of chemical reaction, and demonstrated that ozone (discovered by **Schönbein** in 1840) is an allotrope of oxygen.

ANFINSEN, Christian Boehmer
(1916–)

American biochemist, born in Monessen, Pennsylvania. He was educated at Harvard, and from 1939 to 1944 was American Scandinavian Foundation Visiting Investigator at the Carlsberg Laboratory, a leading institute of protein chemistry, where he was influenced by **Linderstrøm-Lang**. During 1947–8 Anfinsen worked with **Theorell** at the Medical Nobel Institute before moving to the National Institutes of Health in Bethesda, Maryland, where he served as head of laboratories from 1950 to 1982. At the time when **Sanger** was elucidating the primary structure of the protein insulin, Anfinsen developed the combined application of enzymic and chemical hydrolysis for the preliminary fragmentation of the protein under study and the chromatographic identification of peptides that were joined by disulphide (cystinyl) bonds. He concluded that the 128 amino acids of ribonuclease formed a single peptide chain with four disulphide bonds and a single N-terminus. The final primary sequence of ribonuclease reflected the combined research of several workers including **Moore** and **Stein**, with whom he shared the Nobel Prize for Chemistry in 1972. From 1955 he studied the secondary and tertiary structures of ribonuclease, as defined by Linderstrøm-Lang, by observing the effects of partial proteolytic digestion and urea denaturation. Impressed by the recent discovery that part of a protein hormone could be removed without loss of biological activity (1948), Anfinsen extensively surveyed the importance of various regions of ribonuclease in refolding and reoxidizing the fully reduced denatured enzyme to restore activity. This led him to discover an enzyme that assists reoxidation of the cysteine residues to give the correct disulphide interactions. Anfinsen related his observations to molecular biology and evolution in *The Molecular Basis of Evolution* (1959). Since 1958 he has been a co-editor of the important review series *Advances in Protein Chemistry*.

ÅNGSTRÖM, Anders Jonas
(1814–74)

Swedish physicist, born in Lödgö. He was keeper of the observatory at Uppsala (from 1843), and became Professor of Physics at the University of Uppsala in 1858, a position he held until his death. From 1867 he was secretary to the Royal Society at Uppsala. He wrote on heat, magnetism, and especially optics; the angstrom unit, for measuring wavelengths of light, is named after him. In a paper which he presented to the Swedish Academy in 1853, Ångström reported the finding that an incandescent gas emits the wavelengths which the same gas will absorb when cold. In subsequent studies of the solar spectrum, he measured the wavelengths of around 1 000 **Fraunhofer** lines, and implied that he believed hydrogen to be present in the Sun. His son, Knut J Ångström (1857–1910), was also a noted Uppsala physicist, and made important studies of solar radiation.

ANNING, Mary
(1799–1847)

English fossil collector, born in Lyme Regis. She was the daughter of a carpenter and vendor of fossil specimens who died in 1810, leaving her to make her own living. In 1811 she discovered in a local cliff the fossil skeleton of an ichthyosaur, now in the Natural History Museum, London. She also discovered the first pleisiosaur (1821) and the first pterodactyl, *Dimorphodon* (1828). She did much to advance knowledge by her diligence and aptitude in collecting specimens.

APOLLONIUS OF PERGA
(fl 250–220 BC)

Greek mathematician, known as 'the Great Geometer'. He was the author of the definitive ancient work on conic sections which laid the foundations of later teaching on the subject. The *Conica* is essentially a unified treatment of conics, with no special treatment of ellipses, parabolas and hyperbolas; some later writers claim that the work is written in a Greek version of coordinate geometry. Four of its books survive in Greek, three in Arabic translation only (translated by **Halley**), and the eighth and last is lost. Apollonius also wrote on various geometrical problems, including that of finding a circle touching three given circles, but most of these works are lost and we know about them only through later commentators. He put forward two descriptions of planetary motion, one in terms of epicycles and the other in terms of eccentric motion.

APPELMAN, Evan Hugh
(1935–)

American chemist, born in Chicago. He obtained his PhD from University of California at Berkeley in 1960, and since then has worked at the Argonne National Laboratory in Illinois. He is known for his synthesis of the long-elusive perbromate ion (BrO_4^-) and the archetypal hypofluorous acid, HOF, as well as more generally for his work on fluorine chemistry. These compounds filled long-standing 'gaps' in the systematic chemistry of the elements of the periodic table.

APPLETON, Sir Edward Victor
(1892–1965)

English physicist, born in Bradford. He showed exceptional promise from an early age, and matriculated at the University of London at the age of 16. Trained at St John's College, Cambridge, he was appointed assistant demonstrator in experimental physics at the Cavendish Laboratory in 1920. His researches on the propagation of wireless waves led to his appointment as Wheatstone Professor of Physics at London University (1924–36). In 1936 he returned to Cambridge as Jacksonian Professor of Natural Philosophy. In 1939 he became Secretary of the

Department of Scientific and Industrial Research, and in 1949 he was appointed Principal and Vice-chancellor of Edinburgh University. His work revealed the existence of a layer of electrically charged particles in the upper atmosphere (the Appleton layer) which plays an essential part in making wireless communication possible between distant stations, and was also fundamental to the development of radar. He was elected FRS in 1927, appointed President of the British Association for the Advancement of Science in 1953, knighted in 1941, and awarded the Nobel Prize for Physics in 1947 for his contributions 'in exploring the ionosphere'.

ARAGO, (Dominique) François (Jean)
(1786–1853)

French scientist and statesman, born in Estagel, near Perpignan. At 17 he entered the École Polytechnique; in 1804 he became secretary to the observatory, and in 1830 its chief director. He took a prominent part in the July Revolution (1830), and as a member of the Chamber of Deputies voted with the extreme left. In 1848 he became a member of the provisional government, but refused to take the oath of allegiance to Napoleon III after the events of 1851–2. His achievements were mainly in the fields of astronomy, magnetism and optics. Arago observed the solar eclipse of July 1842 from Perpignan. The coronal streamers were clearly visible, and he referred to them as 'hanks of thread in disorder'. He developed a polarimeter and used it to observe the polarization of cometary light. From measurements made of comet 1819 II and Halley (in 1835), he concluded that comets are not self-luminous, but simply reflect sunlight. Arago did much to encourage **Le Verrier** in his mathematical endeavours to discover Neptune. He is to be specially remembered for his great compendium of astronomy, *Astronomie Populaire*. This did much to extend the scientific *Aufklärung* (enlightenment) of the European middle classes. In studies of magnetism he gave an early demonstration of a magnetic field produced by the flow of an electric current round a conducting coil. He also speculated on the nature of light, propounding first the particle theory and later the wave theory of **Fresnel**.

ARBER, Agnes, née Robertson
(1879–1960)

English botanist, born in London and educated at University College London, and Newnham College, Cambridge. She began her career as research assistant to the plant anatomist Ethel Sargant, from whom she learned the technique of using serial sections to study plant anatomy. From 1903 to 1909 she worked at University College London, studying gymnosperms. In 1909 she married Edward Arber, demonstrator in palaeobotany at Cambridge. At **Seward**'s suggestion, she began studying early printed herbals. The outcome was her first and most widely read book, *Herbals, Their Origin and Evolution* (1912), which became the standard work. She was also interested in Johann Wolfgang von Goethe and the philosophy of biology, and published *Goethe's Botany* (1946), *The Natural History of Plant Form* (1950), *The Mind and the Eye* (1954) and *The Manifold and the One* (1957). Her main contributions, however, were to comparative plant anatomy, especially of the monocotyledons. She published 84 papers during 1902–57, and three books: *Water Plants: a Study of Aquatic Angiosperms* (1920), *Monocotyledons: A Morphological Study* (1925), where she illuminated the phyllode theory of the origin of the monocotyledonous leaf, and *The Gramineae: A Study of Cereal, Bamboo and Grass* (1934).

ARBER, Werner
(1929–)

Swiss microbiologist, born in Gränichen. In 1949 he commenced his studies at the Swiss Federal Institute of Technology. In 1953 he moved to the University of Geneva as a research assistant in the department of biophysics, where he helped develop techniques for the preparation of bacteriophages (the viruses which attack bacteria), obtaining his PhD in 1958. In 1962 he discovered the mechanism of host-cell induced variation, involving two types of enzymes, endonucleases and methylases. After a period at the University of Southern California, he returned to Geneva and then moved to Basle as Professor of Molecular Biology from 1970. Arber proposed that when bacteria defend themselves against attack by phage, they use selective enzymes, which he called 'restriction enzymes', which cut the phage DNA and do so at specific points in the DNA chain. Such enzymes clearly gave the option of securing short lengths of DNA; if these could then be joined in specific ways, then securing an 'unnatural DNA' should be possible, and with it entry to the new field of so-called genetic engineering. Through the efforts of many groups, especially in the USA, this was brought to full fruition in the 1970s, with valuable results such as the preparation of monoclonal antibodies for use in clinical diagnosis and treatment of disease. Arber shared the Nobel Prize for Physiology or Medicine in 1978 with **Hamilton Smith** and **Nathans**.

ARCHIMEDES

(c.287–212 BC)

Greek mathematician, born in Syracuse, the most celebrated of the ancient mathematicians and one of the greatest of all times. He almost certainly studied at Alexandria. In popular tradition he is remembered for the construction of siege-engines against the Romans and the Archimedean screw still used for raising water; he demonstrated the powers of levers in moving large weights, and in this context made his famous declaration 'Give me a firm place to stand, and I will move the Earth'. His importance in mathematics, however, lies in his discovery of formulae for the areas and volumes of spheres, cylinders, parabolas, and other plane and solid figures, in which the methods he used anticipated the theories of integration to be developed 1 800 years later. He also used mechanical arguments involving infinitesimals as a heuristic tool for obtaining the results prior to rigorous proof. He founded the science of hydrostatics, studying the equilibrium positions of floating bodies of various shapes. In legend he discovered 'Archimedes principle' while in the bath and ran into the street with a cry of 'Eureka!' ('I have found it'); the principle states that a wholly or partly immersed body in a fluid displaces a weight of fluid equal to its own weight. His astronomical work is lost. His work, some of which only survives in Arabic translation, combines an amazing freedom of approach with enormous technical skill in the details of his proofs. He was killed at the siege of Syracuse by a Roman soldier whose challenge he supposedly ignored while immersed in a mathematical problem.

ARGAND, Aimé

(1755–1803)

Swiss physicist and chemist, born in Geneva, who lived for a time in England. In 1784 he invented the Argand burner, the first scientifically designed oil-burner for the purpose of illumination. It consisted of a cylindrical wick supported between two concentric tubes, the circular flame being supplied with air on both sides, within a glass chimney which protected it from extraneous draughts. Argand lamps were soon brought into use for lighthouse illumination, and were later improved by **Fresnel** who introduced burners with as many as four concentric wicks.

ARGAND, Jean-Robert

(1768–1822)

Swiss mathematician, born in Geneva. He gave his name to the Argand diagram, in which complex numbers are represented by points in the plane. By profession a bookkeeper, his work was largely independent of that of the more famous mathematicians of his time.

ARGELANDER, Friedrich Wilhelm August

(1799–1875)

German astronomer, born in Memel, East Prussia. Having first studied law at the University of Königsberg, he changed to the field of astronomy under the influence of **Bessel** whom he assisted with his meridian stellar observations. In 1823 he became director of a new observatory at Abo in Finland, later moving to a chair of astronomy at the University of Helsinki. In 1836 he became Professor of Astronomy at the University of Bonn, where he remained for the rest of his life. There, with the help of his friend King Frederick William IV of Prussia, he founded an observatory which long remained a model establishment of its kind. Argelander's greatest achievement was the remarkable survey of the positions and brightnesses of all stars in the northern hemisphere brighter than 9th magnitude, and of many fainter ones, a total of 324 198 which are catalogued in the *Bonner Durchmusterung* and plotted on celestial charts. This pioneer survey of the northern sky, later extended to part of the southern hemisphere by Argelander's colleagues Eduard Schönfeld and Adalbert Kruger, has been of great importance for stellar astronomy and the basis of subsequent catalogues.

ARISTARCHOS OF SAMOS

(fl c.270 BC)

Greek astronomer who worked in Alexandria. He is famous for his theory of the motion of the Earth, maintaining not only that the Earth revolves on its axis but that it travels in a circle around the Sun, anticipating the theory of **Copernicus**. He was also a practical astronomer; he used a method for determining the relative distances of the Sun and Moon which involved observing the angular separation of the two bodies in the sky at the Moon's first quarter. His result, that the Sun is about 20 times more distant than the Moon, though greatly in error due to the crudeness of his method, was the first attempt to make this important observation. He inferred correctly that as the Sun and Moon are almost of the same apparent size, their dimensions are in proportion to their distances.

ARISTOTLE
(384–322 BC)

Greek philosopher and scientist, one of the most important and influential figures in the history of Western thought. He was born at Stagira, a Greek colony on the peninsula of Chalcidice, the son of the court physician to the King of Macedon (who was father of Philip II and grandfather of Alexander the Great). In 367 BC he went to Athens and was first a pupil then a teacher at **Plato**'s Academy, where he remained for 20 years until Plato's death in 347 BC. Speusippus succeeded Plato as head of the Academy and Aristotle left Athens for 12 years. He spent time at Atarneus in Asia Minor (where he married), at Mytilene, and in about 342 BC he was appointed by Philip of Macedon to act as tutor to his son Alexander. He finally returned to Athens in 335 BC to found his own school (called the Lyceum due to its proximity to the temple of Apollo Lyceius), where he taught for the following 12 years. His followers became known as 'peripatetics', supposedly because of his restless habit of walking up and down while lecturing. Alexander the Great died in 323 BC and there was a strong anti-Macedonian reaction in Athens; Aristotle was accused of impiety, and perhaps with the fate of Socrates in mind, he took refuge at Chalcis in Euboea, where he died the following year. Aristotle's writings represented an enormous, encyclopedic output over virtually every field of knowledge: logic, metaphysics, ethics, politics, rhetoric, poetry, biology, zoology, physics and psychology. The bulk of the work that survives actually consists of unpublished material in the form of lecture notes or students' textbooks, which were edited and published by Andronicus of Rhodes in the middle of the 1st century BC. Even this incomplete corpus is extraordinary for its range, originality, systematization and sophistication. In Greek biology, Aristotle represented the highest point; the successors of his era (such as **Pliny** the Elder) were little more than uncritical compilers. He began the history of zoological taxonomy, incorporating considerable factual detail in his writings, much of it from the Hippocratic School but a considerable amount based apparently on personal observation. His main work of descriptive zoology is *Historia animalium*, but there are many references also in *De partibus*, *De generatione* and other works. Although interested in biodiversity, he had no interest in classification *per se*, and nowhere lists the main taxonomic groups he recognized. He developed a *scala naturae*, although he was clearly opposed to the concept of evolution. His descriptions of a developing chick marked the founding of embryology. Major Aristotelian themes of natural philosophy included the theory that the Earth is the centre of the eternal universe. He taught that everything beneath the orbit of the Moon is composed of earth, air, fire and water, and is subject to generation, destruction, qualitative change and rectilinear motion; everything above the orbit of the Moon was composed of ether, and subject to no change but circular motion. Aristotle believed that all material things can be analysed in terms of their matter and their form, the latter factor constituting their essence, and that many scientific explanations are properly teleological. His work exerted an immense influence on medieval philosophy (especially through St Thomas Aquinas), Islamic philosophy (especially through **Averroës**), and indeed on the whole Western intellectual and scientific tradition. In the Middle Ages, he was referred to simply as 'the Philosopher', and the uncritical and religious acceptance of his doctrines was to hamper the progress of science until the Scientific Revolution of the 16th and 17th centuries. The works most widely read today include the *Metaphysics* (the book written 'after the *Physics*'), *Nicomachean Ethics*, *Politics*, *Poetics*, the *De Anima* and the *Organon* (treatises on logic).

ARMSTRONG, Edwin Howard
(1890–1954)

American electrical engineer and inventor, born in New York City. He graduated from Columbia University in 1913, having already discovered the principle of the feedback circuit, an important advance in the design of early radio receivers. Several other engineers had made the same discovery almost simultaneously, however, and the resulting patent litigation was not resolved until 1934, when the Supreme Court awarded priority to **De Forest**. During World War I he became interested in methods of detecting aircraft, and in the course of his research he devised the superheterodyne circuit, which became the basis for amplitude modulation radio receivers. By 1939, as Professor of Electrical Engineering at Columbia University (1935–54), he had perfected the frequency-modulation system of radio transmission which virtually eliminated the problem of interference from static. After World War II he developed the technique of multiplexing, which allowed several different signals to use the same carrier-wave frequency. These and many others of his inventions were universally adopted and could have brought him great satisfaction, but he was a contentious man forever engaged in lawsuits over real or imagined slights, and in a fit of depression he took his own life.

ARMSTRONG, Henry Edward
(1848–1937)

English chemist, born in London. He studied chemistry at the Royal College of Chemistry and began research with Sir Edward Frankland, but received his PhD from Leipzig in 1870 for work with **Kolbe**. He returned to the UK to become professor at the London institutions, which later became Imperial College, and remained there until his retirement in 1911. He did pioneering work in a number of areas of organic chemistry, particularly the structure and reactions of benzene and naphthalene compounds. He also speculated on the state of ions in aqueous solution. However, he is most frequently remembered for the inspirational nature of his teaching. He was elected FRS in 1876.

ARNOLD, Joseph
(1782–1818)

English botanist, born in Beccles, Suffolk. He studied medicine at Edinburgh, graduating in 1807 with a thesis on dropsy. In his youth he showed an interest in botany and contributed articles to the *Gentlemen's Magazine*. In 1808 he entered service with the British navy as a ship's surgeon and was surgeon on the *Hindostan* to Sydney. William Bligh (1754–1817), the captain on the return voyage, promised Arnold an introduction to **Banks**; this offer seems never to have been taken up. In January 1815 he sailed on the *Northampton* as surgeon to female convicts bound for Botany Bay. During this voyage he collected insects in South America and Australia for his friend Alexander McLeay (1767–1848). Unfortunately, his collections and most of his journals were lost on the return voyage when the *Indefatigable* caught fire at Batavia. He visited Java with Sir Thomas Stamford Raffles, who later invited Arnold to accompany him to Sumatra as naturalist. On 19 May 1818 at Pulau Lebar (Sumatra), Arnold discovered the largest flower known, measuring a yard across and weighing 15 pounds. This was named *Rafflesia arnoldii* by **Robert Brown**. He died shortly afterwards, of fever, at Padang in Sumatra.

ARP, Halton Christian
(1927–)

American astronomer, born in New York City. After receiving his first degree at Harvard and his doctorate at Caltech, he spent the majority of his scientific life at the Mount Wilson and Palomar Observatory. In 1956 he was the first to establish the relationship between the maximum luminosity of novae and the rate at which this luminosity declines. In 1966 he produced an *Atlas of Peculiar Galaxies*, and in the same year he suggested that the redshifts associated with distant galaxies are not related to velocity, with the implication that **Hubble**'s law cannot be used to measure distances; this view never became widely accepted. Arp has done much to relate radio galaxies to their optical counterparts. He is also famous for his work on the high-luminosity variable stars (RV Tauri stars) that are found in globular clusters. In addition he studies pulsating novae, particularly those in the Andromeda galaxy.

ARRHENIUS, Svante August
(1859–1927)

Swedish physical chemist, born in Wijk, near Uppsala. In 1876 he entered the University of Uppsala, where he studied both chemistry and physics, but found that conditions there for pursuing such studies were unsatisfactory and in 1881 he migrated to Stockholm. His doctoral thesis on the experimental determination and theoretical interpretation of the electrical conductivities of dilute solutions of electrolytes was barely accepted by the University of Uppsala, whose scientific establishment was apparently outraged by its novel ideas. Arrhenius was fortunately already highly regarded by **Ostwald**, **van't Hoff** and other prominent physical chemists, and he was awarded a travelling scholarship by the Academy of Sciences. He spent five years visiting the laboratories of Europe in which physico-chemical studies were pursued. In 1891 Arrhenius became a lecturer in the Stockholm Högskola; he was promoted to professor in 1895. In 1902 he received the Davy Medal of the Royal Society, and he was elected a Foreign Member of the society in 1911. He was awarded the Nobel Prize for Chemistry in 1903. He became Director of the Nobel Institute of Physical Chemistry in Stockholm in 1905, a post he held until a few months before his death in 1927. The theory of electrolytic dissociation, as expounded by Arrhenius in his thesis, was clarified, extended and consolidated in a paper in 1887. In addition to explaining electrolytic conductivity, the theory rationalized many other properties of electrolyte solutions, notably osmotic pressure (studied by van't Hoff) and catalysis by acids (studied by Ostwald). The other main contribution of Arrhenius to physical chemistry was his formulation in 1889 of the dependence of the rate coefficient of a chemical reaction upon temperature (the Arrhenius equation). In later life he applied the methods of physical chemistry to the chemistry of living matter, and he was also interested in astrophysics, particularly the origins and destinies of stars and planets.

ARSONVAL, Jacques-Arsène d'
(1851–1940)

French physicist, born in Borie into an ancient French noble family. Initially he studied classics, but then followed family tradition by deciding on a medical career. He attended the lectures of **Bernard**, and became his assistant (1873–8). He was director of the laboratory of biological physics at the Collège de France from 1882, and professor from 1894. In 1910 he moved to a new laboratory at Nogent-sur-Marne, established with funds raised from public subscription, and retired in 1931. Arsonval invented the reflecting galvanometer named after him, and also the first electrically controlled constant-temperature incubator for embryological and bacteriological research, but he became best known for his experiments with high-frequency oscillating currents for electromedical purposes. He used the heating effect on living tissues of such currents in a form of electrotherapy known for many years as 'd'Arsonvalization', changed to 'diathermy' in the 1920s. He was made a knight of the Legion of Honour in 1884 and received the Grand Cross in 1931.

ARTIN, Emil
(1898–1962)

Austrian mathematician, born in Vienna. He studied in Leipzig, and taught at Göttingen and Hamburg before emigrating to the USA in 1937, where he held posts at Indiana and Princeton before returning to Hamburg in 1958. His work was mainly in algebraic number theory and class field theory, and has had great influence on modern algebra. After **Hilbert** had written his decisive summary of 19th-century number theory in 1894, attention was focused on what is known as Abelian class field theory. Takagi, the first modern Japanese mathematician, brought this line of research to a conclusion, and on the basis of his work Artin then solved the most important problem raised by Hilbert: the existence of a general reciprocity law. This discovery by Artin completed a line of inquiry begun by **Gauss** which was central to the theory of numbers. Artin also wrote a classic description of modern **Galois** theory. Recently his work on braids (a part of knot theory) caught the attention of particle physicists.

ASELLI, Gasparo
(1582–1626)

Italian physician, born in Cremona of an old patrician family. Aselli studied medicine at the University of Pavia before practising in Milan. Famed for his surgical skills, he served from 1612 to 1620 as the head surgeon of the Spanish army in Italy. His surgical experience clearly re-established an interest in anatomical investigation that had initially been sparked by his teacher, Giambattista Carcano-Leone, himself the pupil of the pre-eminent **Falloppio**. Aselli's most notable work was the discovery of the lacteal vessels of the intestine. Vivisecting a dog in 1622, he noted milky filaments in the abdomen which he called white veins or 'lacteals'. Pondering the significance of these vascular structures, Aselli recognized that the chylous vessels possessed an absorbent function in respect of digestion. He was successful in investigating the chylous vessels in divers experimental animals, though he tended to confuse the lacteals with the lymphatics of the liver; the first to trace their course to the thoracic duct was **Pecquet** in 1651.

ASHBURNER, Michael
(1942–)

English geneticist. He was educated at Churchill College, Cambridge, where he received a BA in 1964, a PhD in 1968 and an ScD in 1978. He remained at Cambridge as a research assistant from 1966 to 1968. He then became a university demonstrator (1968–73), lecturer (1973–80), Reader in Developmental Genetics (1980–91) and Senior Research Fellow at Churchill College (1980–90). Elected FRS in 1990, he has been Professor of Biology at the University of Cambridge since 1991, and throughout his career has held many research fellowships and visiting professorships abroad. Ashburner is best known for his work on the definition and regulation of the heat shock genes of *Drosophila melanogaster*, the fruit fly. Heat shock genes are activated by a short burst of increased temperature, and are therefore a useful model system to study gene control.

ASHBY, Eric, Baron Ashby of Brandon, Suffolk
(1904–92)

British-born Australian botanist and educator, born in London. Educated at London and Chicago universities, he was appointed Professor of Botany at Sydney University in 1938. From 1947 to 1950 he held the Chair of Botany at Manchester University, where he was also director of the botanical laboratories. He became President and Vice-Chancellor of Queen's University, Belfast (1950–9), and Vice-Chancellor of Cambridge University (1967–9). He was Master of Clare College, Cambridge, from 1959 to 1975. Throughout his long career as educator and administrator, he was deeply involved in experimental biology and environmental matters, publishing numerous books and papers on these subjects. He chaired the Royal Commission on Environmental Pollution (1970–3) and was knighted in 1956. He was created a life peer in 1973. Ashby was a member of the University Grants Commission (1959–67), and President of the British Association for the Advancement of Science in 1963. His most significant publications were *Portrait of Haldane* (1974), a biography of **J B S Haldane**; *Reconciling Man with the Environment* (1978) with Mary Anderson; and *The Politics of Clean Air* (1981).

ASHWORTH, John Michael

(1938–)

English microbiologist, born in Luton. He was educated at Exeter College, Oxford, and received his PhD at the University of Leicester. He remained at Leicester as a research demonstrator, lecturer and reader until 1974, when he accepted the Chair of Biology at the University of Essex. Around this time he was seconded to the Cabinet Office, becoming Under-Secretary of the Cabinet Office in 1979, and Chief Scientist of the Central Policy Review Staff from 1976 until 1981, when he was elected as Vice-Chancellor of the University of Salford. He was instrumental in the revitalization of the university, and held the position until 1990, when he became Director of the London School of Economics. He has served on many national boards and committees, and as Director of the Granada Group. His key scientific contribution has been in the field of cell differentiation, using the cellular slime moulds (organisms similar to fungi, inhabiting similar environments) as 'model' organisms. His work concentrated on integrating genetical, biochemical and cell biological approaches and helped illuminate the complexity of the control systems that comprise the 'development programme' of these organisms. For his contributions in this field he was awarded the Colworth Medal of the Biochemical Society in 1972.

ASTBURY, William Thomas

(1889–1961)

English X-ray crystallographer, born in Longton, Stoke-on-Trent. He studied at Cambridge and with Sir **William Bragg**'s team at University College London (1920–1) before becoming a lecturer in textile physics at the University of Leeds (1928), where he held the new Chair of Biomolecular Structure from 1945. In 1926 he began taking X-ray diffraction photographs using, not crystals, but natural protein fibres such as those of hair, wool and horn. Using photographic techniques he had helped to develop, he showed that diffraction patterns could be obtained which changed when the fibre was stretched, or wet. On this basis he classified fibrous proteins into the keratin–myosin–elastin group and the collagen group comprising the proteins of connective tissue, such as tendons, cartilage, fish scales and gelatin. The latter group does not stretch, though they contract considerably upon soaking in hot water. Astbury's interpretation of their molecular structures proved incorrect, but his pioneer work laid the basis for important work which followed, most notably **Pauling** and R B Corey's discovery that the contracted structure of hair corresponded to the protein alpha helix while the beta-keratin stretched structure, found naturally in silk fibroin, corresponded to the anti-parallel pleated sheet. The alpha helix was also found to occur in haemoglobin by **Perutz**, and together with the parallel pleated sheets, constitutes the protein secondary structure in the classification of **Linderstrøm-Lang**. Astbury probably coined the phase 'molecular biology', and with Florence Bell he attempted the first hypothetical structure for the key genetic material DNA (1938).

ASTON, Francis William

(1877–1945)

English physicist, born in Birmingham. Educated at Birmingham and Cambridge universities, he was noted for his work on isotopes. Following work done by **J J Thomson**, Aston developed the mass spectrograph (1919), with which he identified 212 naturally occurring isotopes. He stated the 'whole number rule' which observes that all isotopes have very nearly whole number masses relative to the defined mass of the ^{16}O isotope. He won the Nobel Prize for Chemistry in 1922 for this work. The Aston dark space, in electronic discharges, is named after him.

ATANASOFF, John Vincent

(1903–)

American physicist and computer pioneer, born in Hamilton, New York, the son of an electrical engineer who had emigrated from Bulgaria. He was educated at the University of Florida, Iowa State College and the University of Wisconsin, where he received a PhD in physics in 1930. Aged 27 he became a theoretical physicist at Iowa State College. In 1942, with the help of Clifford Berry, a talented graduate student, he built an electronic calculating machine—the ABC (Atanasoff–Berry-Computer)—one of the first calculating devices utilizing vacuum tubes. In 1941, on a visit that was to have important repercussions, **Mauchly** visited

Atanasoff and discussed his work. Neither Atanasoff nor Iowa State applied for patents: Atanasoff failed to appreciate the importance of his work and, caught up in World War II, he let the project drop and dismantled the ABC in 1948. He took no further interest in building computers, though his ideas did enter the mainstream of computer development through Mauchly, who was influenced by Atanasoff in constructing the ENIAC. In a controversial sequel in 1972, a landmark court case involving the ENIAC patents ruled that Atanasoff—not **Eckert** and Mauchly—was the true originator of the electronic digital computer.

ATIYAH, Sir Michael Francis
(1929–)

English mathematician, born in London. Educated in Egypt and at Manchester Grammar School, he graduated from Trinity College, Cambridge, in 1952. After lecturing at Cambridge and Oxford he became Savilian Professor of Geometry at Oxford (1963–9), and in 1966 was awarded the Fields Medal (the mathematical equivalent of the Nobel Prize). After three years at Princeton, he returned to Oxford as Royal Society Research Professor in 1973. He became Master of Trinity College and President of the Royal Society in 1990. One of the most distinguished British mathematicians of his time, he has worked on algebraic geometry and algebraic topology. He worked on K-theory, a cohomology theory for vector bundles, and then on differential operators between bundles, which led him to the index theory of differential operators, a rich blend of analysis and topology. His work has been taken up by many collaborators and students in the UK and abroad. With his students, Russian collaborators, and with Raoul Bott at Harvard, he worked on the theory of 'instantons' (which give a mathematical description of objects which behave like both particles and waves) and on **Yang**–Mills theory. Most recently he has taken up the mathematics of conformal field theory, where he has been particularly concerned with bridging the gap between mathematicians and physicists. He was elected FRS in 1962 and knighted in 1983.

AUBLET, Jean Baptiste Christophe Fusée
(1723–78)

French botanist, explorer and humanist, born in Salon, near Arles, Provence. Even as a boy, Aublet had a passion for botany and ran away to Spain in search of plants. He worked for an apothecary in Granada for a year before being traced. He spent nine years establishing a garden of medicinal plants in Mauritius (1753–61), and then spent two years in Guyana (then French Guiana). There he began a pioneer study of the tropical forest and made extensive collections. These formed the basis for his *Histoire des plantes de la Guiane française* (4 vols, 1775) and laid the foundation of tropical American forest botany. On his return from Guyana he spent several months exploring Haiti. In 1760 he fell in love with a Negro slave, Armelle, whom he bought. He later became the first secular slavery abolitionist and the *Histoire* contains a supplementary chapter, *Observations sur les nègres esclaves*. His interest in racial and ethnic problems led Oliver Fuller Cook to give the name *Ethnora maripa* to the famous Maripa palm, which Aublet had discovered.

AUDUBON, John James
(1785–1851)

American ornithologist and bird artist, born in Haiti. After his mother's early death, his father took him to France. He was sent to the USA in 1804 to look after his father's property near Philadelphia, but in 1807 he sold up and migrated to Kentucky, where he opened a general store in Louisville and elsewhere (1808), but was eventually declared bankrupt (1819). He then embarked on an ambitious project to make a comprehensive catalogue of every species of bird in America, and spent some years travelling down the Ohio and the Mississippi. In 1826 he took his work to Europe in search of a publisher, and 1827 saw the publication in England of the first of the 87 portfolios of his massive *Birds of America* (1827–38). It eventually comprised coloured plates of 1 065 birds in life size; between 1840 and 1844 he produced a 'miniature' edition in seven volumes which became a bestseller. He returned to America and prepared drawings for *The Viviparous Quadrupeds of North America* (1845–9, with John Bachman), which was completed by his two sons. The National Audubon Society, dedicated to the conservation of birds in the USA, was founded in his honour in 1866.

AUENBRUGGER, Leopold
(1722–1809)

Austrian physician, born in Graz, the son of an innkeeper. As a boy he learned that by tapping on the sides of winebarrels he could determine the fluid level, at the point where the sound changed character. As a physician in Vienna, he introduced the technique of percussion, whereby the presence of dense structures in the chest or abdomen could be diagnosed. His short book describing the diagnostic technique, *Inventum novum* ('New Invention', 1761), made little impact on medical practice until its value was appreciated by French physicians in the early 19th century. As a very successful private practitioner, Auenbrugger was also a highly cultured man who wrote the libretto for one of Salieri's operas.

AUER, Karl, Baron von Welsbach
(1858–1929)

Austrian chemist, born in Vienna, who invented the gas mantle which bears his name and carried out important work on the rare metals. Auer isolated the elements neodymium and praesodymium. Continuing his researches, he found that a mixture of thorium nitrate and cerium nitrate becomes incandescent when heated. He used this mixture to impregnate gas mantles, thereby making gas lighting cheaper and more efficient. His mantle is still used in kerosene lamps. Auer also made osmium filaments for use in electric light bulbs. These lasted longer than their carbon counterparts and were the forerunners of the cheaper tungsten filaments used today. In addition he developed the cerium–iron alloy known as 'Auer metal' or 'mischmetal'. The first improvement over the flint and steel in use since medieval times, it is still used to strike sparks in cigarette lighters and gas appliances.

AUERBACH, Charlotte
(1899–)

German–British geneticist, born in Krefeld. She was educated in Berlin, before attending university courses in Berlin, Würzburg and finally Freiburg (under **Spemann**), graduating in 1925. She then took up school teaching before starting her PhD course at the Kaiser Wilhelm Institute under Otto Mangold. In 1933 all Jewish students were forbidden to enter the university and she moved to Edinburgh, completing her thesis at the Institute of Animal Genetics. In the late 1930s she studied mutation with **Hermann Müller**, and was the first to discover chemical mutagenesis, arising from her work on the effects of nitrogen mustard and mustard gas on *Drosophila*. Chemical mutagenesis thereafter became her main research, with particular emphasis on the biological side and the kinds of mutations induced. She was appointed as a lecturer in genetics (1947) and reader (1957) at Edinburgh, and served as professor from 1967 to 1969. She was one of the very first to work out how chemical compounds cause mutations and to compare the differences between the actions of chemical mutagens and X-rays, contributing many papers and books on the subject. These include *Genetics in the Atomic Age* (1956), *The Science of Genetics* (1961) and *Mutation Research* (1976). She was elected FRS in 1957 and awarded the Royal Society's Darwin Medal in 1976.

AUGER, Pierre Victor
(1899–)

French physicist, born in Paris. He was educated at the École Normale Supérieure and later became professor at the University of Paris. Auger discovered that an atom can de-excite from a state of high energy to a lower energy state non-radiatively, by losing one of its own electrons rather than emitting a photon. Electrons emitted in this way are known as Auger electrons. Sources of Auger electrons are useful for calibration of nuclear detectors as the electrons are monoenergetic, whereas beta electrons are emitted with a range of energies. He went on to investigate the properties of the neutron following its discovery by **Chadwick**, and later worked on cosmic-ray physics. He discovered extended air showers, also known as Auger showers, where the interaction of cosmic rays with Earth's upper atmosphere produces cascades of large numbers of secondary particles; he showed that the primary particles impinging on the upper atmosphere have energies in the region of 10^{15}–10^{20} giga-electronvolts. The origin of these high-energy cosmic rays is still not understood today.

AVERROËS, Ibn Rushd
(1126–98)

The most famous of the medieval Islamic philosophers, born in Cordova, son of a distinguished family of jurists. He was himself Kadi (judge) successively at Cordova, Seville and in Morocco, and wrote on jurisprudence and medicine in this period as well as beginning his huge philosophical output. In 1182 he became court physician to Caliph Abu Yusuf, but in 1185 was banished in disgrace (for reasons now unknown) by the caliph's son and successor. Many of his works were burnt, but after a brief period of exile he was restored to grace and lived in retirement at Marrakesh until his death. The most numerous and the most important of his works were the *Commentaries on Aristotle*, many of them known only through their Latin (or Hebrew) translations, which greatly influenced later Jewish and Christian writers and offered a partial synthesis of Greek and Arabic philosophical traditions.

AVERY, Oswald Theodore
(1877–1955)

Canadian-born American bacteriologist, responsible for a key step in the genesis of molecular biology. Born in Halifax, Nova Scotia, he studied medicine at Colgate University and spent his career at the Rockefeller Institute Hospital, New York (1913–48). He soon became an expert on pneumococci, and in 1928 was intrigued by a claim that a non-virulent, rough-coated strain could be transformed into the virulent smooth strain in mouse serum, by the mere presence of some of the dead (heat-killed) smooth bacteria. Avery confirmed this result, and went on in 1944 to show that the transformation is actually caused by a deoxyribonucleic acid (DNA), a chemical present in the dead bacteria. Cautiously, he did not go on to suggest that the informational molecules which carry the whole reproductive pattern of any living species (the genes) are simply DNA; this idea emerged slowly around 1950 and forms the central concept of molecular biology.

AVOGADRO, (Lorenzo Romano) Amedeo (Carlo)
(1776–1856)

Italian (Piedmontese) physicist and chemist, born in Turin. The son of a lawyer, civil servant, and one-time senator of Piedmont in the Sardinian Kingdom, Amedeo succeeded his father as Count of Quaregna in 1787. He trained in his father's profession, graduating as doctor of ecclesiastical law in 1796. Whilst busy as a lawyer and public administrator he acquired a scientific education. From 1806 he abandoned his official positions completely to concentrate on science, and soon became Professor of Mathematics and Physics at the College of Vercelli (1809). In 1819 he was elected to the Turin Academy of Sciences. When the first Italian Chair of Mathematical Physics was established at Turin one year later, Avogadro was appointed to the post, which he held until 1822 and again from 1834 until his retirement in 1850. He published widely on physics and chemistry, founding his work upon an increasingly unpopular caloric theory of heat elaborated most extensively in his *magnum opus*, *Fisica de' corpi ponderabili* (1837–41). In 1811, seeking to explain **Gay-Lussac**'s law of combining gaseous volumes (1809), he formulated the famous hypothesis that equal volumes of all gases contain equal numbers of molecules when at the same temperature and pressure (Avogadro's law). He also introduced the idea of a polyatomic molecule. The hypothesis was practically ignored for around 50 years, a neglect partially explained by Avogadro's social, intellectual and geographical isolation. Initially close to French scientific traditions, he was increasingly out of touch with the main currents of European thought. But in 1860 at the Karlsruhe Chemical Congress, **Cannizzaro** circulated a work which attempted to systematize inorganic chemistry using a restatement of Avogadro's hypothesis in contemporary terms. Universal acceptance did not come until the 1880s.

AXEL, Richard
(1946–)

American molecular biologist, born in New York City. He graduated MD from Johns Hopkins University in 1970 and AB *magna cum laude* from Columbia University in 1967. He was an intern in the Department of Pathology of Columbia University College of Physicians and Surgeons (1970–1), a Visiting Fellow at the Department of Pathology at Columbia (1971–2) and Assistant Professor of Pathology at the Institute for Cancer Research at Columbia (1972–8). He is currently Professor of Pathology and Biochemistry and a member of the Institute of Cancer Research at Columbia University. In 1975 Axel showed that when DNA is combined with cellular proteins as chromatin, it can be cleaved at specific regions within this chromatin by staphylococcal nucleases. Regions of chromatin specifically attacked by nucleases were defined as hypersensitive, and such regions were later shown to contain genes which are transcriptionally active. This was an essential step in understanding the series of events which lead to gene activation and regulation. In 1979 he was responsible, with M Wigler and Saul Silverstein, for the transformation of mammalian tissue culture cells with a cloned viral gene. This technique made it possible to mutate cloned genes in specific ways, re-introduce the mutated genes into mammalian cells and then determine the effect of specific mutations on gene activity. This technology has been the basis for our understanding of the regulation of many gene types and is the forerunner of the technique by which cloned genes are reintroduced into mammalian embryos; it may therefore be used therapeutically to cure certain genetic diseases.

AXELROD, Julius
(1912–)

American pharmacologist, born in New York City. He qualified in biology and chemistry, and worked for some years as a laboratory analyst before entering research. He obtained his PhD degree at the age of 45. In 1949 he joined the section of heart chemistry of the National Institutes of Health (Associate Chemist 1949–50; Chemist 1950–2; Senior Chemist 1953–5) and then moved as chief of the pharmacological section of the Clinical Sciences Laboratory at the National Institutes for Mental Health (1955–84). His inquiries focused on the chemistry and pharmacology of the nervous system, especially the role of the catecholamines, adrenaline and noradrenaline. Originally intrigued by a report that abnormalities of their metabolism might be implicated in schizophrenia, he examined the chemical transformations that they underwent in normal neural tissue. He determined in great detail the mechanisms of action of noradrenaline in nerve cells and discovered a new enzyme, catechol-o-methyl transferase (COMT), an essential regulator of noradrenaline in the body. His work accelerated investigations into the links between brain chemistry and psychiatric disease, and in the search for psychoactive drugs. He was joint winner of the 1970 Nobel Prize for Physiology or Medicine with **Ulf von Euler** and **Katz**.

AYRTON, Hertha, originally Phoebe Sarah Marks

(1854–1923)

English physicist, born in Portsea, near Portsmouth, and educated in mathematics at Girton College, Cambridge. She is best known for her work on the motion of waves and formation of sand ripples, and the behaviour of the electric arc. Her collected papers describing her work on the latter were published as *The Electric Arc* (1902). After extensive research on arc lamps (with her husband **William Ayrton**), including cinema projector lamps and search lights, she took out several patents; her improvements to search light technology were put into practice in aircraft detection during both world wars. During World War I she invented the Ayrton fan for dispersing poison gases—this invention was later adapted to various other applications, including improvement of ventilation in mines. Ayrton was nominated for fellowship of the Royal Society in 1902, but refused on the grounds that she was a married woman. She received the society's Hughes Medal in 1906.

AYRTON, William Edward

(1847–1908)

English engineer and inventor, born in London. He studied mathematics at University College London, and electricity at Glasgow under **Kelvin**. He joined the Indian telegraph service in 1868, and soon devised a method of detecting faults which was of great benefit in the maintenance of overland telegraphic communication. In 1873 he was appointed to the Chair of Natural Philosophy and Telegraphy at the new Imperial Engineering College in Tokyo, where he established laboratories for the teaching of applied electricity, the first of their kind. Returning to London in 1879 he became Professor of Physics and Electrical Engineering successively at the City and Guilds of London Institute, Finsbury Technical College (1881), and the Central Technical College, South Kensington (1884). With his colleague John Perry (1850–1920) he invented the first absolute block system for electric railways (1881), the first electric tricycle (1882), and many electrical measuring instruments. They published jointly some 70 scientific and technical papers between 1876 and 1891. His first wife, Matilda Chaplin (1846–83), was a pioneer woman doctor; his second, **Hertha Ayrton**, continued his work on the electric arc and other inventions.

B

BAADE, (Wilhelm Heinrich) Walter
(1893–1960)

German-born American astronomer, born in Schröttinghausen. He studied at the universities of Münster and Göttingen, and from 1919 to 1931 worked at the Hamburg Observatory at Bergedorf. In 1931 he moved to the Mount Wilson Observatory in California. Baade's major interest concerned the stellar content of various systems of stars. He discovered the existence of two discrete stellar types or 'populations', characterized by blue stars in spiral galaxies and fainter red stars in elliptical galaxies. In 1944, helped by the wartime blackout in the Pasedena area, he succeeded in resolving into stars with the 100 inch telescope the centre of the Andromeda galaxy M31 and of its two companions M32 and NGC205. In the 1950s, using the 200 inch telescope on Mount Palomar which had come into operation in 1948, he began a systematic survey of the positions of the recently discovered radio sources which led to a number of optical identifications. His work on stellar systems earned him the Gold Medal of the Royal Astronomical Society in 1954. After retirement from California he became Gauss Professor at the University of Göttingen (1959).

BABBAGE, Charles
(1792–1871)

English mathematician, born in London. Educated at Trinity and Peterhouse colleges, Cambridge, he spent most of his life attempting to build two calculating machines. The first, the 'difference engine', was intended for the calculation of tables of logarithms and similar functions by repeated addition performed by trains of gear wheels. A small prototype model was described to the Royal Astronomical Society in 1822, and earned the award of the society's first Gold Medal. Babbage was then granted money by the government to build a full-sized machine, but in 1842, after some £17 000 of public money and £6 000 of his own had been invested without any substantial result, government support was withdrawn (an unfinished portion of the machine is now in the Science Museum, London). Meanwhile Babbage had conceived the plan for a much more ambitious machine, the 'analytical engine', which was designed not just to compute a single mathematical function, but which could be programmed by punched cards, like those in the Jacquard loom, to perform many different computations. The cards were to store not only the numbers, but also the sequence of operations to be performed. The idea was too ambitious to be realized by the mechanical devices available at the time, but can now be seen to be the essential germ of the electronic computer of today, and Babbage can be regarded as the pioneer of modern computers. He held the Lucasian Chair of Mathematics at Cambridge (1828–39), though he delivered no lectures.

BABCOCK, Harold Delos
(1882–1968)

American physicist and astronomer, born in Edgerton, Wisconsin. He was educated at the University of California, and joined the staff of the Mount Wilson Observatory (now part of Hale Observatories) in 1909, and became best known for his studies on the magnetic fields of the Sun and other stars. With his son, Horace Welcome Babcock (b 1912), he invented in 1951 the solar magnetograph which made possible detailed observations of the Sun's magnetic field, and resulted in the discovery of magnetically variable stars. In 1959 he announced that the Sun reverses its magnetic polarity periodically. His investigations of the magnetic field of the star 78 Virginis provided a link between the electromagnetic and the relativity theories. His son became Director of the Hale Observatories in 1964, and as well as investigating stellar magnetism, studied the glow of the night sky and the rotation of galaxies.

BABCOCK, Stephen Moulton
(1843–1931)

American agricultural chemist, sometimes dubbed 'the father of scientific dairying', born near Bridgewater, New York State. A farmer's son, at certain periods in his life he also farmed. He studied at Tufts College, the University of Göttingen and Cornell. He then taught at Cornell, moving in 1882 to become Chief Chemist at the New York Experimental Station at Geneva, and in 1887 he was appointed Professor of Agricultural Chemistry at the University of Wisconsin and Chief

Chemist to the Wisconsin Agricultural Experiment Station. He became the station's Assistant Director in 1901. He devised the 'Babcock test' for measuring fat in milk, his method being to centrifuge equal quantities of milk and sulphuric acid and then allow the fat, which had been separated out, to rise up a specially calibrated tube. Because this test could be carried out by unskilled operatives, it was soon widely used and much improved the quality of dairy produce. From 1907 onwards Babcock studied the effect of selective diets on cattle; the importance of accessory food factors ('vitamins') emerged from this work and was developed more fully by **Frederick Gowland Hopkins**. Babcock also invented an instrument to measure viscosity in liquids, and carried out research on milk, sugar and fat solvents.

BABINET, Jacques

(1794–1872)

French physicist, born in Lusignan. He proposed the definition of the angstrom unit in terms of the wavelength of the red line in the spectrum of cadmium. In 1837 he enunciated in diffraction theory the principle, now named after him, that the diffraction pattern of a given screen is similar to that produced by the complementary screen, where the transparent and opaque parts are reversed. He also invented an elegant instrument for measuring the polarization of light, the Babinet compensator (1849).

BACKUS, John

(1924–)

American computer programmer. Educated at Columbia University, where he took a Master's degree in mathematics, Backus joined IBM in 1950 as a computer programmer. By 1953 he headed a small group of programmers who were developing an assembly language for IBM's 701 computer. Once this was complete, Backus sent a memo to his boss, **Hurd**, suggesting the development of a compiler and higher level language for the IBM 704. This was approved as the FORTRAN (FORmula TRANslation) project, which was not completed until 1957. FORTRAN was a landmark: for the first time, software was available which opened the computer up to the non-specialist and allowed computers to talk to each other. It soon led the way to other computer languages, such as COBOL, ALGOL (Backus was involved in the design of ALGOL 60) and—one of the most popular of all—BASIC. By 1963 Backus had joined IBM Research.

BACON, Francis, Baron Verulam of Verulam, Viscount St Albans

(1561–1626)

English philosopher and statesman, born in London, the younger son of Sir Nicholas Bacon. He entered Trinity College, Cambridge, and in 1576 Gray's Inn, being called to the bar in 1582. He became an MP in 1584. On the accession of James VI and I (1603) he sought royal favour by extravagant professions of loyalty; by planning schemes for the union of England and Scotland, and by making speeches in parliament to prove that the claims of the king and parliament could be reconciled. For these services he was knighted (1603) and made a commissioner for the union of Scotland and England. In 1612 he offered to manage parliament for the king and, in this capacity, was promoted in 1613 to the attorney-generalship. In 1616 he became a privy councillor, in 1617 lord keeper, and in 1618 Lord Chancellor, being raised to the peerage. In 1621, however, he was accused of taking bribes in his capacity as a judge; the evidence was incontrovertible, and it was the end of his political career. He died five years later deep in debt. Bacon's attempt to review and classify all branches of human knowledge he called 'the grand instauration of the sciences', and as a first step towards this objective he published in 1605 the *Advancement of Learning*; he never succeeded in bringing this scheme to completion. His philosophy was also described in *De Augmentis Scientiarum* (1623), a Latin expansion of the *Advancement*, and *Novum Organum* (1620). In this he repudiated the deductive logic of **Aristotle** and his followers, and stressed the importance of experiment and inductive reasoning in interpreting nature, as well as the necessity for proper regard for any possible evidence which might run counter to any held thesis. He described heat as a mode of motion, and light as requiring time for transmission, but he was generally behind the scientific knowledge of his time. His greatness consists in his insistence on the facts, that man is the servant and interpreter of nature, that truth is not derived from authority, and that knowledge is the fruit of experience; and in spite of the defects of his method, the impetus he gave to future scientific investigation is indisputable. He was the practical creator of scientific induction. As a writer of English prose and a student of human nature, he is seen to best advantage in his essays. In his fanciful *New Atlantis* he suggested the formation of scientific academies, and by the year 1645 weekly meetings were being held which led to the foundation in 1660 of the Royal Society of London for Improving Natural Knowledge.

BACON, Roger
(c.1214–92)

English philosopher and scholar, probably born near Ilchester, Somerset. He studied at Oxford and Paris, and began to gain a reputation for diverse and unconventional learning in philosophy, magic and alchemy which led to the soubriquet *Doctor Mirabilis* ('Wonderful Teacher'). Around 1247 he joined the Franciscan Order, and soon afterwards returned to Oxford to develop his interests in experimental science. He attempted the compilation of an encyclopedia of universal knowledge, but even in his day the task was beyond the capabilities of any one man. He seems to have been uncomfortably outspoken, quarrelsome and uncompromising in his views, and he suffered rejection, censorship and eventually imprisonment at the hands of the Order for the heresy of his 'suspected novelties'; he died in Oxford soon after his release from 14 years of incarceration by the Order in Paris. He has been (mistakenly) credited with scientific inventions like the magnifying glass and gunpowder, but he certainly published some remarkable speculations about lighter-than-air flying machines, mechanical transport on land and sea, the circumnavigation of the globe, and the construction of microscopes and telescopes. His views on the primacy of mathematical proof and on experimentalism have often seemed strikingly modern, and despite surveillance and censorship from the Franciscans he published many works on mathematics, philosophy and logic whose importance was only recognized in later centuries.

BAEKELAND, Leo Hendrik
(1863–1944)

Belgian-born American chemist, one of the founders of the plastics industry. He was born in Ghent and studied there and at other Belgian universities, and became Professor of Physics and Chemistry in Brugge in 1885. He emigrated to the USA in 1889 and founded a chemical company to manufacture one of his inventions, photographic printing paper which could be used with artificial light. Subsequently he made the first synthetic phenolic resin from the condensation products of phenol and formaldehyde. The condensation of organic molecules had been described by **Baeyer** in 1872, but it was Baekeland who first put this phenomenon to industrial use. The resin, known as Bakelite, replaced hard rubber and amber as an insulator and its success led to the founding of the Bakelite Corporation in 1910. Baekeland wrote on many topics in organic chemistry and electrochemistry, and also on the reform of patent law. He was elected President of the American Chemical Society in 1924, and received many academic honours at home and abroad. He died in Beacon, New York.

BAER, Karl Ernst Ritter von
(1792–1876)

Estonian-born German naturalist and pioneer in embryology. Born in Piep of a wealthy family of Prussian origins, Baer graduated in medicine at Dorpat in 1816 and later studied at Berlin, Vienna and Würzburg. He was appointed professor at Königsberg (1817–34), where he undertook his most important research, and from 1834 taught at St Petersburg. From 1826 he investigated the mammalian ovary, especially the small follicles discovered therein by **Graaf** in 1673. Through dissecting a colleague's pet dog, Baer showed that the Graafian follicle contained a microscopic yellow structure which was the egg (ovum). This piece of research finally established that reproduction necessarily involved an egg (ovum) rather than merely, as thought since the Greeks, the mingling of male and female seminal fluids. Investigating embryo development, Baer was the first to differentiate the notochord, the gelatinous cord that in vertebrates becomes the backbone and skull; he also drew attention to the neural folds which develop into the central nervous system. Investigating morphogenesis, Baer stressed that the embryos of various species initially share highly analogous and even indistinguishable forms. Differentiation does not occur till later. This was to be known as the 'biogenetic law': in embryonic development, general characters appear before special ones. In the development process, Baer showed that the embryo of a higher creature passes through stages resembling phases in the unfolding of lower animals. This idea (formulaically stated as ontogeny recapitulating phylogeny) proved important to comparative anatomy, embryology and evolutionary theory. Baer himself was no evolutionist, but his teleological notion of *Entwicklungsgeschichte* (development history) was influential in paving the way for evolutionary styles of thinking.

BAEYER, (Johann Friedrich Wilhelm) Adolf von
(1835–1917)

German chemist, born in Berlin. His early chemical education was with **Bunsen** in Heidelberg and he also studied with **Kekulé**. In 1858 he presented a thesis on cacodyl compounds. From 1960 he taught at a technical school in Berlin until 1872 when he moved to the University of Strassburg. In

1875 he succeeded **Liebig** in the chair at Munich, where he remained until his retirement. His research covered a range of topics, but he is best known for his work on the structure and synthesis of the blue dye indigo. He later turned his attention to molecules with distorted bond angles, and this resulted in the Baeyer strain theory which explained the instability of certain molecules and the non-planar nature of molecules such as cyclohexane. His work on indigo led, in 1882, to the view that a molecule can adopt two different structures in equilibrium depending on the position of a hydrogen atom. This phenomenon is known as tautomerism. Baeyer was awarded the Nobel Prize for Chemistry in 1905.

BAILEY, Liberty Hyde

(1858–1954)

American horticulturist and botanist, born in South Haven, Michigan. He became assistant to **Asa Gray** at Harvard University. He was appointed Professor of Horticulture and Landscape Gardening at Michigan State (1885) and Cornell (1988), and founded the Bailey Hortorium of New York State College in 1920. This was a remarkable private herbarium of 125 000 plants and a library of 3 000 books. Much of his work was concerned with the identification of cultivated plants. He published many papers on *Carex* (sedges), *Rubus*, *Vitis* (vines), brassicas and palms. He also experimented on crosses and varieties within the genus *Cucurbita*, the important genus which includes melons, squashes and cucumbers. He edited various works such as the *Standard Cyclopedia of Horticulture* (1914–17), later revised into the single-volume *Hortus*, and coined the term 'cultivar'. His *Manual of Cultivated Plants* was published in 1923. He was President of the American Association for the Advancement of Science, of the Botanical Society of America, the International Congress of Plant Sciences, and the American Society of Horticultural Science.

BAILY, Francis

(1774–1844)

English astronomer, born in Newbury, Berkshire. He started his working life as a banker and went on the stock exchange in 1799. Astronomy was a hobby encouraged by his acquaintance with **Priestley**. In 1820 he was a founder member and the first Vice-President of the Royal Astronomical Society. Elected FRS in 1821, he retired from business in 1825 and fitted his house in Tavistock Place, London, with an observatory. Baily's accurate catalogue of nearly 3 000 star positions led to the award of the Gold Medal of the Royal Astronomical Society in 1827. On 15 May 1836 he observed a total eclipse of the Sun from Jedburgh, Scotland. Just before the Sun disappears behind the Moon the sunlight appears as a brilliant cluster of spots. Baily's description of this phenomenon was so graphic that these appearances of residual segments of the solar disc between the mountains of the Moon have become known as Baily's beads. His description of the solar corona after the 1842 eclipse was the first realistic description of this region of the solar outer atmosphere. Baily also studied ancient eclipses, and knowing the rate at which the orbital parameters of the Moon changed, he dated the eclipse that occurred during the battle between the Medes and the Lydians as 30 September 610 BC. He improved **Cavendish**'s apparatus, and after a series of delicate measurements between 1838 and 1842, he concluded that the Earth has a mean density of 5.66 times that of water.

BAIRD, John Logie

(1888–1946)

Scottish electrical engineer and television pioneer, born in Helensburgh. He studied electrical engineering in Glasgow at the Royal Technical College (now the University of Strathclyde) and the University of Glasgow. Poor health compelled him to give up the post of engineer to the Clyde Valley electric power company, and after a brief career as a sales representative and an unsuccessful business venture in the West Indies, he settled in Hastings (1922) and began research into the possibilities of television. Hampered by continued ill-health and lack of financial support, he nevertheless built his first television apparatus almost entirely from scrap materials, and in 1926 gave the first demonstration in London of a television image. The following year he transmitted primitive television pictures over the telephone lines from London to Glasgow, and in 1928 the signals he broadcast from a station in Kent were picked up in the USA. His 30 line mechanically scanned system was adopted by the BBC in 1929, being superseded in 1936 by his 240 line system. In the following year the BBC chose a rival 405 line system with electronic scanning made by Marconi-EMI. Other lines of research initiated by Baird in the 1920s included radar and infrared television ('Noctovision'); he continued his research up to the time of his death and succeeded in producing three-dimensional and coloured images (1944) as well as projection onto a screen and stereophonic sound.

BAKER, Herbert Brereton
(1862–1935)

English physical chemist, born in Livesey, near Blackburn. He was educated at Balliol College, Oxford, where his tutor was **Dixon**, to whom he acted as assistant (1883–5). From 1886 to 1904 he was a schoolmaster, mainly at Dulwich College, where he contrived to engage in both teaching and research, and he then returned to Oxford as Dr Lee's Reader in Chemistry at Christ Church. He was Professor of Chemistry at Imperial College, London, from 1912 to 1932. Baker was elected FRS in 1902, appointed CBE in 1917, received the Davy Medal of the Royal Society in 1923, and was President of the Chemical Society (1926–8). His researches were almost entirely on the effect of intensive drying on chemical systems and he made many remarkable observations, eg that in the complete absence of moisture, phosphorus or sulphur may be distilled in oxygen without reaction. The interpretation of much of his work was controversial.

BALARD, Antoine Jérôme
(1802–76)

French chemist, born in Montpellier, where he began life as an apothecary. His investigations into sea-water and marine life led him to devise a test for iodine which he discovered turns blue in the presence of starch. This very simple and sensitive test is still used today. The same investigations led to the isolation of bromine, which he recognized as an element, around 1825. Such success by a young apothecary caused a stir in scientific circles, not least because the similarities between iodine, bromine and chlorine showed clearly that there are 'families' of elements. Balard taught for a while at the University of Montpellier, where in 1834 he identified hypochlorous acid and chlorine monoxide while working on the chemistry of bleaching. He also studied bleaching powder and methods of extracting soda and potash from sea-water. He was appointed to the Chair of Chemistry at the Sorbonne in 1842 and at the Collège de France in 1851. He was known for his generosity to colleagues and students, and was the mentor of **Pasteur**. He died in Paris.

BALDWIN, Jack Edward
(1938–)

English chemist, born in Lewes. After having obtained both a BSc and PhD from Imperial College, London, he joined the staff as assistant lecturer (1963) and lecturer (1966). He then moved to Pennsylvania State University (1967–9) and MIT (1969–72). After a brief tenure of the Daniell Chair of Chemistry at King's College, London, he returned to MIT (1972–8). He is currently Waynflete Professor of Chemistry at Oxford University. Baldwin is best known for his studies of the biosynthesis of penicillins, in which he has used modern biological techniques as well as classical organic chemistry. He has also contributed significantly to the study of the porphyrins. As well as his chemical studies he has done much to promote the importance of academic organic chemistry to the political establishment. He was elected FRS in 1978.

BALFOUR, Francis Maitland
(1851–82)

Scottish embryologist. Grandson of the Marquis of Salisbury and younger brother of Arthur Balfour, the future prime minister, Francis Balfour came from an illustrious family. Edinburgh-born, he showed precocious promise in natural history while still at Harrow School. A student at Trinity College, Cambridge, he received a first in natural science in 1871, and embarked upon physiological research under **Foster**, producing in 1878 notable work on elasmobranch development that displayed the qualities of precision and close observation for which he later became renowned. In the same year he was elected FRS. After completing a two-volume *Treatise on Comparative Embryology* (1880), he was awarded a medal by the Royal Society in 1881, and in the following year Cambridge created for him a chair in animal morphology. A keen climber, he died in a fall on an unconquered peak in the Chamonix district before the year was out. One of numerous embryologists examining morphogenesis in the light of **Baer**'s research programme and of **Charles Darwin**'s evolutionary theory, Balfour was distinguished by painstaking microscopic accounts of the development process and by his aversion to the grander phylogenetic philosophizings of contemporaries such as **Haeckel**.

BALFOUR, John Hutton
(1808–84) and
Sir Isaac Bayley
(1853–1922)

Scottish botanists, both born in Edinburgh. John Hutton Balfour studied medicine at Edinburgh and Paris. He was appointed Professor of Botany at Glasgow University in 1841. In 1845 he became Professor of Botany at the University of Edinburgh, and Regius Keeper of the Royal Botanic Garden in Edinburgh. He was one of the driving forces behind the establishment of the Botanical Society of Edinburgh (now the Botanical Society of Scotland), which aimed among other things to create a botanical library and herbarium for its members. An inspiring teacher and field worker, he wrote several textbooks. He retired in 1879, and after a brief interlude he was succeeded in both posts by his son Isaac Bayley Balfour in 1888. Isaac Bayley studied botany in Edinburgh and in Germany before becoming Professor of Botany in Glasgow, and later Sherardian Professor of Botany at Oxford, before taking up the Edinburgh posts. He was naturalist to the transit of Venus expedition to Rodriguez island in 1880. During this expedition, he explored the remote island of Socotra, bringing back and introducing to cultivation *Begonia socotrana*, one of the parents of many of today's *Begonia* hybrids. He also travelled in China. He founded the journal *Annals of Botany*. Under his leadership, the Royal Botanic Garden Edinburgh was transformed into one the world's great gardens. He was knighted in 1920.

BALL, John
(1818–89)

Irish botanist and alpinist, born in Dublin. Taken to Switzerland when seven years old, he was profoundly affected by the view of the Alps from the Jura. The following year, at Ems, he spent much time trying to measure the height of the hills with a mountain barometer. After a period at Cambridge University, Ball visited Sicily and published a valuable paper on its botany. In 1852 he became Liberal MP for Carlow and advocated many measures which later became law, including the disestablishment of the Church of Ireland. As Colonial Under-Secretary from 1855 to 1857, he influenced the government in its decision to prepare a series of Floras of the British colonies and possessions. He was the first President of the Alpine Club (1857) and author of the *Alpine Guide* (1863–8). In 1871 he accompanied Sir **Joseph Dalton Hooker** and George Frederick Maw to Morocco and investigated the flora of the Great Atlas. His *Spicilegium Florae Maroccanae* (1877–8) was the earliest work on the Moroccan flora. In 1882 he visited South America. Ball proposed a theory on the antiquity of the alpine flora, and one that most endemic South American plants originated on a hypothetical ancient mountain range in Brazil.

BALL, Sir Robert Staywell
(1840–1913)

Irish mathematician, born in Dublin, son of the secretary of Dublin Zoo. He obtained a scholarship to Trinity College, Dublin, where he studied mathematics and experimental physics. In 1865 he became tutor to Lord Rosse's son at Parsonstown (now Burr), the home of the then largest telescope in the world. 1867 saw Ball appointed to the Chair of Applied Mathematics and Mechanics at the Royal College of Science, Dublin, and in 1874 he became the Astronomer Royal for Ireland and Andrews Professor at Trinity College. He was elected FRS in 1873 and knighted in 1886. In 1890 he moved to Cambridge as Lowndean Professor and Director of the Observatory, and remained there until he died. Ball's main scientific interest was in the movement of rigid bodies about fixed points, ie the mathematical theory of screws. He was also interested in the connection between this theory and the quaterion theory of the linear vector function. Astronomically, Ball is remembered as a great public lecturer—'a lecturer whose science and wit and playfulness combined can absolutely rivet any audience from a savant to a little child' (H Montagu Butler, 1913). He was also a great literary popularizer. His book *The Story of the Heavens* was published between 1885 and 1913, and his *Popular Guide to the Heavens* was published in many editions between 1892 and 1955.

BALMER, Johann Jakob
(1825–98)

Swiss physicist, born in Lausanne. He was self-educated and spent most of his life teaching in a girls school, but late in life became interested in spectra. In 1855, Balmer produced a simple formula for the frequencies of lines in the visible part of the spectrum of hydrogen. This was deduced empirically, but a full explanation of the form of the relation later became clear through **Niels Bohr**'s theory of atomic structure.

BALTIMORE, David
(1938–)

American molecular biologist, born in New York City, joint winner of the 1975 Nobel Prize for Physiology or Medicine with **Temin** and **Dulbecco**. As a high school student, Baltimore met Temin at the Jackson Laboratory, Bar Harbor, Maine. He received a BA in chemistry from Swarthmore College, Pennsylvania, and a PhD at Rockefeller University, New York. In 1972 became professor at MIT. In 1990 he was appointed President of Rockefeller University, but he resigned in 1991 following a controversial legal battle over an allegedly fraudulent paper in the journal *Cell*; Baltimore's name had appeared on the paper, although he himself was not charged with misconduct. Over the period 1965–8 he worked on viral genetics with Dulbecco at the Salk Institute, California. Temin had suggested that certain viruses, like the Rous sarcoma virus which stores its genetic code in RNA, could insert viral genes into the DNA of the cells which the viruses infected. This suggestion was contrary to the then-accepted idea that DNA produced RNA which then formed protein. Baltimore discovered an enzyme which could make DNA from RNA, called 'reverse transcriptase', in 1970. Viruses with RNA as the genetic code, such as the virus associated with AIDS, are now known as 'retroviruses'. With reverse transcriptase, scientists can manipulate the genetic code. Concerned about the misuse of genetic engineering, Baltimore supported a moratorium on research in the mid-1970s. In the 1980s he chaired a US National Academy of Sciences committee on AIDS.

BANACH, Stefan

(1892–1945)

Polish mathematician, born in Krakow. He studied at Lvov, where he became a lecturer in 1919 and professor in 1927. During World War II he was forced to work in a German institute for research on infectious diseases; although he returned to work at Lvov University after the war, his health was ruined and he died less than a year later. Banach is regarded as one of the founders of functional analysis, and his book *Théorie des opérations linéaires* (1932) remains a classic. He founded an important school of Polish mathematicians which emphasized the importance of topology and real analysis. His name is attached to a class of infinite dimensional linear spaces, whose elements are usually functions, in which a concept of length is defined. These spaces occupy an important place in the study of analysis, and are increasingly applied to problems in physics, especially in particle physics, where they are natural generalizations of the formalism of **Hilbert** and **von Neumann**.

BANKS, Sir Joseph

(1744–1820)

English botanist, born in London, and educated at Harrow, Eton, and Christ Church, Oxford. In 1766 he made a voyage to Newfoundland collecting plants, and between 1768 and 1771 accompanied James Cook's expedition round the world in a vessel, the *Endeavour*, equipped at his own expense. He later took part with Solander in an expedition to Iceland. In 1778 he was elected President of the Royal Society, an office which he held for 41 years. His friendship with King George III led him to persuade the King to turn Kew Gardens into a botanical research centre. He founded the African Association, and the colony of New South Wales owed its origin mainly to him. His desire to cultivate the Pacific bread-fruit in the West Indies led to the *Bounty* expedition under Captain Bligh. He was also responsible for the introduction of the mango from Bengal, and many fruits of Ceylon and Persia, and he suggested the growing of the Chinese tea plant in India. His name is commemorated in the genus *Banksia*. His significance lies in his far-reaching influence, rather than through any single personal contribution to science. He facilitated the work of others, partly through his wealth, which he dispensed generously in scientific causes, and also through his official positions. He was a strong advocate of the need for close international contact between scientists, even in time of war, such as the French and American revolutions. Banks was made a baronet in 1781.

BANTING, Sir Frederick Grant

(1891–1941)

Canadian physiologist, born in Alliston, Ontario. He grew up on a small farm, and studied medicine at Toronto University. During World War I he served as a surgeon in the Canadian army Medical Corps. Soon after he returned, he established a surgical practice, but as this was not very successful, he became a demonstrator at the Western University Medical School where he conducted his first medical research. Several attempts had already been made by other workers around the world to prepare a pancreatic extract for the treatment of diabetes, but they had all met with little success. Banting came up with a possible new method and was advised to approach **Macleod**, Professor of Physiology at Toronto and an expert on carbohydrate metabolism, who allowed him to use his laboratory. With **Best**, a recent graduate in physiology and biochemistry, and later with the biochemist **Collip**, a practical method was devised and extracts of pancreas of sufficient purity were produced to allow the first successful clinical trials in January 1922. Soon afterwards a means of commercial preparation of insulin was devised and it became the principle means of treating diabetes. Banting was made Professor of Medical Research at Toronto in 1923 and was awarded the 1923 Nobel Prize for Physiology or Medicine jointly with Macleod. Banting shared his part of the prize with Best. He established the Banting Research Foundation in 1924, and the Banting Institute at Toronto in 1930. He was knighted in 1934. A pioneer in aviation medicine, he was killed in a wartime air crash.

BÁRÁNY, Robert

(1876–1936)

Austrian-born physician and otologist, born in Vienna. He graduated from the University of Vienna in 1900 and undertook further studies in internal medicine, psychiatric neurology, neurology and surgery in Frankfurt, Heidelberg, Freiburg and Vienna. In 1903 he joined the staff of the University of Vienna ear clinic. He extended earlier work on the inner ear which had been carried out on animals, and pioneered the study in humans of the inner ear's balancing apparatus. He was able to correlate dizziness with objective bodily factors such as eye movements, and the reactions of muscles. He proved the connection between the balancing apparatus and the brain, making it possible for equilibrium disturbances and vertigo to be investigated systematically. Bárány volunteered for service during World War I, hoping to be able to put his ideas about the treatment of brain wounds to the test. In 1915, while in a Russian prisoner-of-war camp in Siberia, news reached him that he had been awarded the 1914 Nobel Prize for Physiology or Medicine. He was released and in 1917 became director of the oto-rhino-laryngology clinic at the University of Uppsala in Sweden.

BARCROFT, Sir Joseph

(1872–1947)

Irish physiologist, born in Newry, County Down. He read natural sciences at King's College, Cambridge, and graduated in 1897. He stayed in the Physiological Laboratory at Cambridge for his entire career, becoming professor in succession to **John Langley** in 1926. He retired in 1937, although he continued to work in the laboratory until appointed head of the newly created Animal Physiology Unit of the Agricultural Research Council in 1942. His early academic career was disrupted during World War I, when he worked in chemical warfare at Porton Down. His research work concentrated on understanding respiratory function, studies that included whole animal physiological experiments; observations on humans at extremes of altitude on expeditions in the Andes and in simulated laboratory conditions; and important biochemical work on the respiratory pigment haemoglobin and its binding and dissociation with oxygen. Many of these contributions are summarized in *The Respiratory Function of the Blood* (1st edition 1914; 2nd edition in two volumes, 1925–8) and in 1934 he published *Features in the Architecture of Physiological Functions*, which summarized his views of integrative physiology. In the final years of his life, apart from wartime service again at Porton, he focused on the physiology of the developing foetus. Using a Caesarean section technique, Barcroft measured foetal blood volume and placental blood flow, the transfer of gases across the placental membrane, the control of respiration and movement and growth parameters of the foetus, findings that he encapsulated in his final book, *Researches on Pre-Natal Life* (1946). He was elected FRS in 1910 and knighted in 1935.

BARDEEN, John

(1908–)

American physicist, born in Madison, Wisconsin. He studied electrical engineering at Wisconsin and worked as a geophysicist at the Gulf Research Laboratories for three years, before obtaining his PhD in mathematical physics at Harvard (1936). He joined a new solid-state physics group at Bell Telephone Laboratories in 1945. Together with **Brattain** and **Shockley** he developed the point-contact transistor (1947), for which they shared the Nobel Prize for Physics in 1956. Their early device, made by placing two metal contacts only 0.005 centimetres apart on a germanium surface, paved the way for the electronics revolution that continues today. Bardeen was professor at Illinois University (1951–75), and with **Cooper** and **Schrieffer** he was awarded the Nobel Prize for Physics again in 1972 for developing the Bardeen–Cooper–Schrieffer, or BCS, theory. The observation that the temperature at which a metal becomes superconducting is inversely proportional to its atomic mass suggested to Bardeen that the effect of the vibrations of the atoms in the metal lattice on the electrons (electron–phonon interactions) must be involved. Cooper meanwhile had shown that in a metal two electrons could form a resonant state, known as a 'Cooper pair', and these ideas provided the first satisfactory theory to explain why certain metals lose their resistance to electric current at low temperatures. Bardeen was the first recipient of two Nobel physics prizes.

BARGER, George

(1878–1939)

English chemist, born in Manchester and educated in Holland and at King's College, Cambridge, graduating in chemistry and botany. He studied chemical aspects of pharmacology, chemotherapy, hormones, and vitamins, and achieved the synthesis of many biologically important compounds. His career began in 1903 at the Wellcome Physiological Research Laboratories, where he carried out fundamental work on the chemical structure and synthesis of adrenaline. He also collaborated extensively with **Dale**, and with Dale moved to the new National Institute for Medical Research in 1914, leaving in 1919 to become Professor of Medicinal Chemistry at Edinburgh University. With Dale, and starting with an analysis of the properties of extracts of the fungus ergot, he joined in the chemical and physiological characterization of many biologically active compounds, including histamine and a range of chemicals that stimulated the sympathetic nervous system, the so-called 'sympathomimetic' amines, including noradrenaline. In 1927 following the isolation by **Harington** of the active principle of the thyroid gland, thyroxine, Barger succeeded in determining its structure and achieved a chemical synthesis. He was also associated with his pupil **Todd** (Dale's son-in-law) in determining the structure and synthesis of vitamin B_1. Shortly before his death he was appointed Regius Professor of Chemistry at Glasgow, but he never took up the position.

BARKHAUSEN, Heinrich Georg

(1881–1956)

German physicist, born in Bremen, the son of a district judge. In 1911 he was appointed Professor of Low-Current Technology in the Technische Hochschule, Dresden. This was the first chair anywhere devoted to the relatively new field of electrical communications. He carried out fundamental research on electron tubes and electrical oscillations, developed with Karl Kurz the Barkhausen–Kurz oscillator (an electron tube capable of continuous-wave oscillations at ultra-high frequencies which was the forerunner of the present-day microwave tubes), and wrote comprehensive books on both subjects. In 1919 he discovered that the magnetization of iron proceeds in discrete steps and he devised a loudspeaker system to render this discontinuity audible. This phenomenon is now known as the Barkhausen effect. In 1928 he was awarded the Heinrich Hertz Medal, and in 1933 the Morris Liebmann Memorial Prize of the Institute of Radio Engineers, of which he was Vice-President in 1935. After World War II he returned to his beloved Dresden to help with the reconstruction of his Institute of High-Frequency and Electron-Tube Technology, destroyed by bombing in 1945.

BARKLA, Charles Glover

(1877–1944)

English physicist, born in Widnes, Lancashire. He studied at the universities of Liverpool and Cambridge, and at the latter the chapel of King's College was often full to hear his baritone solos. His first researches at Cambridge were measurements of the transmission velocity of electromagnetic waves in conductors of different materials and dimensions. He soon left this to work on X-rays, an area of research he pursued for 40 years. He became Professor of Physics at London (1909–13) and Professor of Natural Philosophy at Edinburgh (1913–44). Barkla found that the extent of scattering of X-rays by gases of different densities was proportional to the molecular weight, and deduced from this that more massive atoms contain more electrons. This was the first indication that there is a correlation between an element's position in the periodic table and the number of electrons it contains—these were the first moves towards the concept of the atomic number. In 1904 he demonstrated that X-rays could be polarized and in doing so established that they are transverse waves. A large part of his work concerned secondary X-rays—the X-rays emitted by substances placed in the path of an X-ray beam. He found that this radiation was of two types, one consisting of X-rays scattered but unchanged in quality from the incident beam and the other a fluorescent radiation characteristic of the scattering substance. Further research revealed the fluorescent radiation to be of two characteristic types, labelled L and K. By 1911, after barely 10 years of research on secondary X-rays, he was recognized as a world leader in the field and in 1917 was awarded the Nobel Prize for Physics.

BARLOW, Peter

(1776–1862)

English natural philosopher, born in Norwich. Largely self-educated, in 1801 he obtained a post as a teacher of mathematics at the Royal Military Academy, Woolwich. For the next 15 years he devoted himself to pure science, devised mathematical tables and made useful studies in applied physics. His *New Mathematical Tables* (1814) were reprinted as late as 1947 as *Barlow's Tables*. He then became interested in the strength of construction materials, and carried out extensive series of tests on timber, cast and wrought iron and other materials, publishing the results from 1817 onwards. As a result of this work he was frequently consulted on structural problems by architects and civil engineers, notably by Thomas Telford on the design of the Menai Strait suspension bridge, completed in 1826. He also worked on the strength of ships' timbers, on tidal engineering, and on ships' magnetism and its correction (for which he received the Copley Medal of the Royal Society in 1825). The

Barlow lens is a negative achromat used as an astronomical eyepiece and in photography.

BARNARD, Christian Neethling

(1922–)

South African surgeon, born in Beaufort West. He graduated from Cape Town Medical School. After a period of research in the USA, he returned to Cape Town in 1958 to work on open-heart surgery and organ transplantation. In December 1967 at Groote Schuur Hospital he performed the first successful human heart transplant. The recipient, Louis Washkansky, died of pneumonia 18 days later, drugs given to prevent tissue rejection having heightened the risk of infection. A second patient, Philip Blaiberg, operated on in January 1968, survived for 594 days.

BARNARD, Edward Emerson

(1857–1923)

American astronomer, born in Nashville, Tennessee. Born into a poor family, he had little education in his early years, but became experienced in photographic techniques during work in a portrait studio and developed a strong amateur interest in astronomy. After discovering a number of comets and becoming skilled in astronomical work, he became both a teacher and a student at the observatory of Vanderbilt University. He moved to Lick Observatory in 1887 and was appointed Professor at Yerkes Observatory of the University of Chicago in 1895. Following a systematic photographic survey of the sky, he correctly concluded with **Maximilian Wolf** that those areas of the Milky Way which appear to be devoid of stars, or 'black nebulae', are in fact clouds of obscuring matter. His wide-ranging research included studies of novae, binary stars and variable stars. He discovered the fifth satellite of Jupiter (1892), later named Amalthea, and identified the star with the greatest known apparent motion across the sky, now known as Barnard's star (1916).

BARROW, Isaac

(1630–77)

English mathematician and divine, born in London. He was educated at Charterhouse and Trinity College, Cambridge, where he was elected a Fellow in 1649. His royalist sympathies and leanings towards Arminianism prevented him from obtaining the professorship of Greek until 1660. He travelled abroad (1655–9), became Professor of Geometry at Gresham College, London (1662), and the first Lucasian Professor of Mathematics at Cambridge (1663), but he resigned in 1669 to become Royal Chaplain and was succeeded by **Isaac Newton**. He founded the library of Trinity College, Cambridge, when he became Master in 1673. Barrow published Latin versions of **Euclid** and **Archimedes**, and lectures on optics (which were soon superseded by Newton's), as well as extensive theological works and sermons. In his original work in mathematics he anticipated aspects of the theories of differential calculus, which began to develop at the end of the 17th century.

BARTHOLIN, Caspar, the Younger

(1655–1738)

Danish anatomist and politician, son of **Thomas Bartholin** the Elder. Born in Copenhagen, he studied medicine in Holland and France. He became an expert anatomist. He was the first to describe the greater vestibular glands in the female reproductive system ('Bartholin's glands') and the larger salivatory duct of the sublingual gland ('Bartholin's duct'). Caspar Bartholin was the last and possibly the most eminent of a line of Bartholins. His grandfather, the first Caspar Bartholin (1585–1629) studied medicine at Copenhagen, Basle and at Padua with **Fabrizio**, returning in 1611 to a chair at Copenhagen and increasingly specializing in theological studies. Caspar Bartholin the Elder was the father of two eminent scholar-scientists, Thomas Bartholin the Elder and **Erasmus Bartholin**.

BARTHOLIN, Erasmus

(1625–98)

Danish physician, physicist and mathematician, son of Caspar Bartholin the Elder (1585–1629) and brother of **Thomas Bartholin** the Elder. He studied medicine at Leiden and Padua, and was appointed Professor of Medicine and Mathematics at Copenhagen in 1656. In 1669 he discovered that when an object is viewed through Iceland feldspar (calcite), a double image is produced, but he was unable to explain this. **Huygens**, **Isaac Newton** and **Fresnel** all contributed to the explanation of the phenomenon of double refraction, which is basic to the understanding of polarization.

BARTHOLIN, Thomas, the Elder

(1616–80)

Danish physician and mathematician, son of Caspar Bartholin the Elder, and father of **Caspar Bartholin** the Younger and Thomas Bartholin the Younger. A fine mathematician and experimental physiologist, he made important discoveries concerning the thoracic ducts and lymphatic system in dogs. Use of microscopes enabled him to explore the capillary system. He was the first to describe the human lymphatic system, which he studied independently of the Swede **Rudbeck**, and he defended the theory of blood circulation introduced by **William Harvey** (1578–1657). He was Professor of Mathematics at Copenhagen University (1646–8), then Professor of Anatomy (1648–61). As personal physician to King Kristian V of Denmark, Bartholin was of great importance in the reform of the Danish medical system and medical education. He produced the first Danish pharmacopoeia, established the first Danish scientific journal and wrote numerous literary, philosophical and historical works.

BARTLETT, Neil

(1932–)

English chemist, born in Newcastle upon Tyne. He was educated at the University of Durham. Following a period as a schoolmaster he moved to Canada to take up an academic appointment at University of British Columbia in 1958. During 1966–9 he worked at the Bell Telephone Laboratories, New Jersey, and since 1969 he has been Professor of Chemistry at University of California, Berkeley. Bartlett is famous for his discovery in 1962 of the first chemical compound of the noble gas xenon, thus disproving the supposition that noble gas compounds do not exist. His discovery followed his formation of the compound $O_2^+Pt_6^-$ upon accidental exposure of PtF_6 to air and his realization that since O_2 and Xe have similar ionization potentials, a synthesis of $Xe^+PtF_6^-$ would be feasible. He was elected FRS in 1973, and won the Chemical Society Corday–Morgan Medal (1962) and the American Chemical Society Award for Distinguished Service to Inorganic Chemistry (1989).

BARTON, Sir Derek Harold Richard

(1918–)

English chemist, born in Gravesend. He received his undergraduate and postgraduate training at Imperial College, London, and obtained his PhD in 1942 for work under the direction of Sir Ewart Jones. After two years in military intelligence and a year with Albright and Wilson, he returned to Imperial College as assistant lecturer and then ICI Fellow. In 1949 he spent a year at Harvard while **Robert Woodward** was on sabbatical, and produced a seminal paper on the relationship between conformation and chemical reactivity for which he was awarded the Nobel Prize for Chemistry in 1969 (shared with **Hassel**). In 1950 he moved to Birkbeck College, and worked on a number of structural problems in triterpenoid and steroid chemistry. This continued when he moved to the Chair of Organic Chemistry in Glasgow in 1955 and back to Imperial College in 1957, but by 1960, X-ray crystallography had largely replaced degradative studies in the determination of structure, and Barton turned his attention to synthetic and biosynthetic work. Important results were obtained in a number of areas, particularly in the biosynthesis of steroids. He also pioneered the use of photochemical reactions in synthesis. In 1977 he was appointed Director of the French National Centre for Scientific Research (CNRS) Institute for the Chemistry of Natural Substances in Gif-sur-Yvette and he became active in designing new reactions, some involving free radicals and others ligand-coupling reactions based on bismuth. After 11 very productive years in France, in spite of much administrative work and time given to the appreciation of *haute cuisine*, he was appointed Distinguished Professor at Texas A & M University where his research has continued unabated. During the course of his career he has received nearly every honour possible for an organic chemist, including election to the Royal Society in 1954. He was knighted in 1972 and made an Officer of the French Legion of Honour in 1985.

BARTRAM, John

(1699–1777)

American botanist, born in Marple, near Darby, Pennsylvania, and educated locally. Considered the 'father of American botany', **Linnaeus** called Bartram 'the greatest natural botanist in the world'. Bartram inherited an uncle's farm, but sold this and purchased another on the banks of the Schuylkill river at Kingsessing, near Philadelphia. There, he became a successful small farmer and also built up an unrivalled collection of North American plants, which he began selling to European botanists and horticulturists through the English woollen draper and botanist Peter Collinson (1694–1768). This successful business allowed Bartram to visit Virginia, the Allegheny Mountains, the Carolinas and elsewhere in search of plants. In 1743 the British Crown commissioned him to visit the Indian tribes of the 'League of Six Nations'; the results were published as *Observations on the Inhabitants, Climate . . . from Pensilvania to Onondago, Oswego and the Lake Ontario, in Canada* (1751). In 1765 he was named King's Botanist, and explored

Florida where he discovered *Franklinia altamaha* (named after his friend **Benjamin Franklin**). His son William (1759–1823) was also a botanist; his bestselling *Travels* (1791) strongly influenced English Romanticism. He also published *Observations on the Creek and Cherokee Indians, 1789* (1853).

BARY, Heinrich Anton de

See **DE BARY, Heinrich Anton**

BASOLO, Fred

(1920–)

American chemist, born in Coello, Illinois. He was educated at the University of Illinois, and became Professor of Chemistry at Northwestern University, Illinois, in 1958. Famous for his work as an organometallic chemist, he has also been author or co-author of many important chemical works including *Mechanisms of Inorganic Reactions* (1958) and *Co-ordination Chemistry* (1984).

BASOV, Nikolai Gennadiyevich

(1922–)

Russian physicist who invented masers and lasers, born in Voronezh. He served in the Red Army during World War II, and studied in Moscow. He joined the Lebedev Physics Institute in Moscow as a laboratory assistant in 1948, became deputy director (1958–73), and was appointed director in 1973. His work in quantum electronics—more specifically, the interaction between matter and incident electromagnetic waves—provided the theoretical basis for the development of the maser (Microwave Amplification by Stimulated Emission of Radiation) in 1955. In 1958 he proposed the use of semiconductors for the creation of lasers (Light Amplification by Stimulated Emission of Radiation), subsequently successfully producing numerous types of these devices (1960–5) using a variety of methods to excite the materials to a state in which they produced the laser effect, including exposure to electron beams and light sources (optical pumping). In 1968 he used powerful lasers to produce thermonuclear reactions. For his work on amplifiers and oscillators used to produce laser beams he was awarded the 1964 Nobel Prize for Physics jointly with his colleague **Prokhorov** and the American physicist **Townes**.

BASSI, Agostino Maria

(1773–1856)

Italian biologist and pioneer bacteriologist, born in Lodi. Educated at Pavia, his work on animal diseases partly anticipated that of **Pasteur** and **Koch**. As early as 1835 he showed after many years' work that a disease of silkworms (muscardine) is fungal in origin, that it is contagious, and that it can be controlled. He proposed that some other diseases are transmitted by micro-organisms.

BASTIAN, Henry Charlton

(1837–1915)

English biologist. Born in Truro, Cornwall, and privately educated, he entered University College London in 1856 and studied medicine. After serving at St Mary's Hospital in London as an assistant physician, he was elected Professor of Pathological Anatomy at University College in 1867, rising to become Hospital Physician (1871), and Professor of Clinical Medicine (1887–95). He developed a large private practice while simultaneously, from 1868 to 1902, holding an appointment at the National Hospital, Queen Square; he was also Crown referee in cases of insanity (1884–98). With growing eminence, numerous honours followed, including a Civil List annual pension of £150. Blessed with a powerful analytic mind and displaying wide-ranging medical, biological and bacteriological interests (he argued for spontaneous generation), Bastian's lasting fame derives from his pioneering researches in clinical neurology, conducted at the Queen Square Hospital. From 1869, he published a series of major papers on the localization of the brain centres controlling speech, investigating the neurological basis of such pathological forms as word blindness (alexia) and word deafness. *A Treatise on Aphasia and Other Speech Defects* (1898) was the definitive statement of his work. His biomedical approach to mental disorder was expressed in *The Brain as an Organ of Mind* (1880).

BATES, Henry Walter

(1825–92)

English naturalist, born in Leicester. With his friend **Wallace** he left to explore the Amazon in 1848, and after Wallace's return to England in 1852, he continued alone until 1859 when he returned with 14 700 specimens, including almost 8 000 species of insect new to science. In 1861 he published *Contributions to an Insect Fauna of the Amazon Valley*, in which he described the phenomenon now known as Batesian mimicry. In this, harmless, edible species of animal are found resembling others which may be distantly related and are distasteful or poisonous, and thus gain protection from predators. This discovery came during the Darwinian controversy and provided strong evidence in favour of natural selection. His travels are described in *The Naturalist on the River Amazon* (2 vols, 1863). In 1864 he became assistant secretary of the Royal Geographical Society.

BATESON, William
(1861–1926)

English geneticist, born in Whitby, Yorkshire, and educated in natural sciences at Cambridge. He introduced the term 'genetics' in 1909, and became the UK's first Professor of Genetics at Cambridge (1908–10). He left to become Director of the new John Innes Horticultural Institution (1910–26). His interest in heredity was stimulated by a trip across the Asian Steppes (1886–7) to investigate the relationship between variation in environment, salinity in lakes, and variation in populations of shellfish. He went on to produce the first English translation of **Mendel**'s work. Bateson believed in discontinuous variation, and was in sharp and acrimonious disagreement with **Pearson** and the biometricians, who advocated a statistical, continuous approach to hereditary variation. Bateson showed that some genes are inherited together, a process now known as 'linkage'. He played a dominant part in establishing Mendelian ideas, but was a major opponent of chromosome theory. In Bateson's view, there was no material link between the chromosome and any physical characteristic of the organism. Although an ardent evolutionist, he was opposed to **Charles Darwin**'s theory of natural selection, as the small changes demanded by the theory seemed insufficient to account for the evolutionary process. He wrote *Mendel's Principles of Heredity: A Defence* (1902) and *Problems of Genetics* (1913).

BATTERSBY, Sir Alan Rushton
(1925–)

English chemist, born in Leigh. Having obtained his BSc from the University of Manchester (1946), he moved to the University of St Andrews where he received his PhD in 1949. From 1948 to 1953 he was on the staff at St Andrews and during this period he spent time at the Rockefeller Institute, New York, and at the University of Illinois, before moving to the University of Bristol in 1954. In 1962 he was invited to the newly created second Chair of Organic Chemistry at the University of Liverpool. From 1969 to 1992 he was professor at the University of Cambridge. During the period 1954–68 he worked on the chemistry of alkaloids and terpenes, but he is most famous for research on the haem pigments, cytochromes, chlorophylls and vitamin B_{12}. He has shown that the building block for all these naturally occurring substances, porphobilinogen, is polymerized by the enzyme deaminase to give a linear tetrapyrrole followed by an enzyme-catalysed ring closure. He has also identified several of the precorrins on the biosynthetic pathway leading to vitamin B_{12}. Battersby was elected FRS in 1966 and knighted in 1992. He retired from the chair at Cambridge in 1992, but remains Professorial Fellow of St Catharine's College.

BAUER, Ferdinand Lucas
(1760–1826) and
Franz Andreas
(1758–1840)

Two of three illustrious brothers, the other being Josef Anton Bauer (b 1756), born in Feldsberg, then in Austria, now part of Moravia. All three specialized in natural history subjects, particularly plants. In 1780 Franz and Ferdinand moved to Vienna and began working for Baron Nicolaus von Jacquin, director of the university's botanic garden. They jointly illustrated Jacquin's *Icones Plantarum Rariorum* (1753–91). Ferdinand was engaged by **Sibthorp** as a botanical illustrator to accompany him on a Mediterranean voyage, and he illustrated Sibthorp and Smith's *Flora Graeca* (1806–40). He was appointed to the Flinders expedition to Australia in 1801, working under the direction of **Robert Brown**, and returning in 1806 with a portfolio of illustrations, including many new plants and animals. Some of his paintings were published as *Illustrationes Florae Novae-Hollandiae* in 1813, the same year in which he returned to Austria, where he remained until his death. Franz was engaged, at the suggestion of **Banks**, as botanical painter to King George III at Kew, where he remained for 50 years. In a very successful career there he produced hundreds of illustrations, and taught painting and botany. His *Illustrations of Orchidaceous Plants* (1830–8) demonstrates a rare combination of scientific accuracy and aesthetic qualities.

BAUER, Georg
See **AGRICOLA, Georgius**

BAUHIN, Caspar or Gaspard
(1560–1624)

Swiss botanist and physician, born in Basle, younger brother of Jean Bauhin, who later published *Historia Universalis Plantarum*. He studied at Basle, Padua, Montpellier, Paris and Tübingen. He was appointed Professor of Anatomy and Botany at Basle and compiled a medical textbook, *Theatrum Anatomicum* (1605). He was also the author of *Prodromus Theatri Botanici* (1620), in which he named the potato *Solanum tuberosum*, the name which it retains today. His more important work *Pinax Theatri Botanici* (1623), containing descriptions of over 6 000 plants, was much used by **Linnaeus** and is still important as a compendium of many of the plants known in the 17th century. The comprehensive nature of this work, together with Bauhin's exhaustive bibliographical references, ensures that it is still of value today as a source for early plant descriptions and names. The arrangement of the plants within the book conforms to no single organizing principle, and characters such as morphology are mixed with others derived from habit or economic uses. His arrangement was criticized by **Morison**, who nevertheless adopted its nomenclature, as did **Ray**. At the lower taxonomic level, related plants are grouped together into units which resemble modern genera. Within these units, plants are systematically named.

BAUMÉ, Antoine
(1728–1804)

French chemist, born in Senlis. The owner of a chemical manufacturing business, he purified saltpetre, manufactured sal ammoniac and helped to improve porcelain manufacture. He invented the hydrometer, an instrument for determining the relative densities of liquids, and devised the relative density scale which bears his name. He was appointed professor at the School of Pharmacy in 1752 and a member of the Academy of Sciences in 1773. He died in Charenton-le-Pont.

BAYER, Johann
(1572–1625)

German astronomer and celestial mapmaker, born in Rhain, Bavaria. By profession a lawyer but with a keen interest in astronomy, he published in 1603 in Nuremberg a celestial atlas, *Uranometria*, in which the positions of nearly a thousand stars are depicted in addition to a similar number recorded in **Tycho Brahe**'s famous catalogue. Bayer added 12 new constellations to the 48 defined by Ptolemy in the 2nd century; this part of the atlas incorporated observations of the Dutch navigator Petrus Theodori (d 1596). Bayer also introduced the mode of designating stars in each constellation in order of magnitude by letters of the Greek alphabet, a system which remains in use for naked-eye The *Uranometria* distinguished some non-stellar objects, for example the star cluster Praesepe which **Galileo** later resolved with his telescope.

BAYES, Thomas
(1702–61)

English mathematician, born in London, the son of Joshua Bayes, one of the first six Non-Conformist ministers to be publicly ordained in England. In 1731 he became Presbyterian minister in Tunbridge Wells. He is principally remembered for his posthumously published *Essay towards solving a problem in the doctrine of chances* (1763), in which he was the first to study the idea of statistical inference, and to estimate the probability of an event from the frequency of its previous occurrences. Although his mathematical results are now a standard part of statistical theory, there is still controversy about the circumstances of their application.

BAYLISS, Sir William Maddock
(1860–1924)

English physiologist, born in Wednesbury. Bayliss was apprenticed to a local general practitioner before beginning his studies at University College London (UCL). He received a BSc in 1882, but failed his second MB in anatomy, and after that he began to concentrate on physiological studies. He went to Oxford in 1885 and received a degree in physiology in 1888 before returning to UCL to serve under **Sharpey-Schäfer**. Bayliss remained at UCL for the rest of his life, from 1912 as Professor of General Physiology. Much of his research was conducted in collaboration with **Starling**. They worked on electrophysiology, the vascular system and intestinal motility. In studies of pancreatic secretion they showed that the discharge is induced by a chemical substance which they called secretin, produced in the duodenum and carried to the pancreas in the blood. During World War I Bayliss worked on wound shock. His celebrated book *Principles of General Physiology* first appeared in 1914 and became known as the 'textbook for professors'. In 1903 he took out an action for libel against Stephen Coleridge, secretary of the National Antivivisection Society, who had accused Bayliss of carrying out experiments on unanaesthetized animals. He won £2 000 damages and presented the sum for the furtherance of research in physiology. He was elected FRS in 1903, and knighted in 1922.

BEACH, Frank Ambrose
(1911–)

American comparative psychologist and endocrinologist, born in Emporia, Kansas. He was educated at Kansas State University and at the University of Chicago. He was research assistant to Karl Spencer Lashley at

Harvard (1935–6) and then worked at the American Museum of Natural History in New York at its Department of Experimental Biology, which under his direction became the Department of Animal Behavior (1936–46). He subsequently became Professor of Psychology first at Yale (1946–58) and then at the University of California at Berkeley (1958–78). He was one of the first American biologists to appreciate the work of European ethologists, and he liaised between the ethologists and some American comparative psychologists. His research has been concerned with the hormonal regulation of reproductive behaviour and he emphasized the importance of understanding behaviour as a prerequisite for endocrine studies. He demonstrated that sensory and experiential factors during ontogeny affect the hormonal regulation of behaviour, and developed the concept of 'gender role' to account for aspects of human sexuality. He put forward the idea that the control of sexual behaviour has changed during the course of evolution from being under predominantly hormonal control in lower vertebrates to sensory and environmental ones in higher mammals. He wrote *Hormones and Behavior* (1948) and *Sex and Behavior* (1965).

BEADLE, George Wells
(1903–89)

American biochemical geneticist. Born on a farm in Wahoo, Nebraska, he became interested in agricultural genetics as a student. He studied the genetics of maize and the fruit fly (*Drosophila*), and popularized the use of the bread mould *Neurospora* in research. At Stanford University (1937–46) he collaborated closely with **Tatum**. They irradiated spores of *Neurospora*, isolated mutants and investigated their nutritional requirements. They developed the idea that specific genes control the production of specific enzymes; for example, spores unable to grow unless provided with vitamin B_6 were found to have a mutation in the gene which metabolized vitamin B_6. Beadle and Tatum shared the 1958 Nobel Prize for Physiology or Medicine with **Lederberg**. In the 1950s, as President of the American Association for the Advancement of Science (AAAS), Beadle worked for more openness in scientific research and the AAAS agreed not to hold meetings in cities which practised racial segregation. He was professor at Caltech (1946–61), and President of Chicago University from 1961 to 1968.

BEALE, Lionel Smith
(1828–1906)

English physiologist and microscopist, born in London. The son of a surgeon, Beale was apprenticed to a surgeon–apothecary before entering King's College, London, to study medicine. Graduating MB in 1851, he established a private laboratory and set himself up as a teacher of microscopy. Just two years later, he was elected Professor of Physiology and Morbid Anatomy at the College, becoming Professor of Pathological Anatomy in 1869, and Professor of the Principles and Practice of Medicine in 1876. His distinguished teaching and research career was rewarded with many honours. He was President of the Microscopical Society (1879–80). Beale was a leading histologist, a master of the techniques of vital staining and a fine investigator of cellular tissue. 'Beale's cells' (the pyriform nerve ganglion cells) commemorate his researches. He produced numerous popular books on practical microscopy. Beale attracted greatest notice, however, for his opposition to the reductionist biological theories associated with **T H Huxley**, and for his deep suspicions towards **Charles Darwin**'s evolutionary theory. He also opposed the new bacteriology, contending that disease germs were not autonomous micro-organisms but minute particles of degraded protoplasm.

BEAUFORT, Sir Francis
(1774–1857)

British naval officer and hydrographer, born in Navan, County Meath. He joined the Royal Navy in 1787 and was wounded in action in 1795. On shore service in 1803–4, he established a line of telegraphs from Dublin to Galway. In 1805 he was engaged in surveying Rio de la Plata and later the Greek Archipelago, and he assisted in suppressing piracy in the Levant. He was hydrographer to the navy from 1829 to 1855, and was elected a Fellow of the Royal Society. His main contributions to meteorology were twofold. In 1806, he systematized the recording of wind speed at sea by relating wind speed to sea state and the amount of sail a sailing ship could carry. His scale was from 0 to 12, force 12 representing a wind that 'no canvas can withstand'. The scale was modified to relate to wind over land, mainly with reference to the motion of branches of trees and to structural damage. Beaufort's second contribution was to designate 24 letters to represent different meteorological states, eg 's' to represent snow. This allows very concise descriptions of the weather on a given day and the notation has proved very useful for observers. He was knighted in 1848.

BEAUMONT, William
(1785–1853)

American army surgeon, born in Lebanon, Connecticut. His pioneering study on *Diges-*

tion (1833) was based on experiments with a young Canadian patient, Alexis St Martin, who was suffering from a gunshot wound which had left a permanent opening in his stomach, and which Beaumont treated. Between 1824 and 1833 Beaumont used St Martin as a willing experimental subject. He confirmed the chemical nature of digestion, established that the presence of food stimulated the secretion of gastric juice and directly observed the role of alcohol in the causation of gastric inflammation (gastritis). His *Experiments and Observations on the Gastric Juice and the Physiology of Digestion* (1833) was the first major American contribution to physiology. It was reprinted in Edinburgh and translated into German. Beaumont retired from the army in 1839 and went into general practice in St Louis, Missouri.

BECHE, Sir Henry Thomas de la

See **DE LA BECHE, Sir Henry Thomas**

BECHER, Johann Joachim

(1635–82)

German chemist and physician, born in Speyer, Palatinate. As a child during the turmoil of the Thirty Years War, he had little formal education, but he was appointed Professor of Medicine at Mainz in 1666. He was subsequently physician to the Elector of Bavaria in Munich, where he was also involved in commercial projects. He later spent some time in Vienna, but in 1678 he fell into disgrace with his patrons; he fled to Holland and in 1680 to England. He died in London shortly after. In 1667 Becher published a classification of substances, particularly minerals, under the title *Physicae subterraneae*, in which he expounded a three 'element' theory of matter. This was later elaborated into the phlogiston theory by **Georg Stahl**. Becher published many other works, including *Tripus hermeticus* (1680), which contained a catalogue of chemical apparatus, and he also practised alchemy, undertaking experiments designed to transmute sand into gold.

BECKER, Wilhelm

(1907–)

Swiss astronomer, born in Münster, and educated at the universities of Münster (1927) and Berlin (1932). Between 1932 and 1953 he was an astronomer at observatories of Munich, Potsdam, Vienna and Hamburg before being appointed Professor of Astronomy and Director of the Astronomical Institute at Basle (1953–77), where he remained until his retirement. Becker's chief field of activity has been stellar photographic photometry. He was a pioneer of the three-colour system, whereby the brightnesses of stars observed through three different colour filters are compared. The method, applied to clusters of young stars, allows their distances to be determined. Studies of the distribution in space of these clusters have contributed substantially to the mapping of the spiral arms of the Milky Way.

BECKMANN, Ernst Otto

(1853–1923)

German chemist, born in Solingen. He entered the University of Leipzig in 1875 and on graduation became an assistant at the Technische Hochschule in Brunswick. In 1883 he returned to Leipzig, and after appointments in Giessen and Erlangen, became director of the newly founded Kaiser Wilhelm Institute in 1912. He is best known for his discovery of a reaction by which ketoximes are converted into amides, and for the accurate thermometer, used to determine molecular weights by freezing point depression or boiling point elevation, which bears his name.

BECQUEREL, Alexandre-Edmond

(1820–91)

French physicist, born in Paris, son and assistant of **Antoine César Becquerel**. At 18 he was admitted to both the École Polytechnique and the École Normale Supérieure, but he decided instead to join his father at the Natural History Museum of Paris. In 1852 he was appointed to the Chair of Physics at the Conservatoire des Arts et Métiers, taught chemistry at the Société Chimique de Paris (1860–3), and eventually succeeded his father at the Museum in 1878. He did research into electricity, magnetism and optics. He measured the properties of electric currents, and demonstrated in 1843 that **Joule**'s law governing the production of heat generated by an electric current applied not only to solids, but also to liquids. He measured the electromotive force of the voltaic pile by means of an electrostatic balance designed by his father. He also investigated diamagnetism, the magnetic properties of oxygen and solar radiation. In the course of his experiments on light, he constructed the 'actinometer' (an instrument that determined the intensity of light by measuring the amount of electric current produced in photochemical reactions), and a phosphoroscope while investigating luminescence.

BECQUEREL, Antoine César
(1788–1878)

French physicist, born in Châtillon-sur-Loing where his father was the royal lieutenant. He chose a military career, joining the Corps of Engineers after graduating from the École Polytechnique. He left military service after the fall of Napoleon (1815), and devoted himself to science. Becquerel investigated the electrical properties of minerals, and was the first to use electrolysis as a means of isolating metals from their ores. In 1837 he was awarded the Copley Medal of the Royal Society, and the following year he became professor at the Natural History Museum of Paris (1838). He wrote a great number of scientific papers, many with his son, but also with **Ampère** and **Biot**. He wrote to **Faraday** concerning diamagnetism, but failed to reach any general conclusions, and invented several laboratory instruments, such as an electromagnetic balance and a differential galvanometer.

BECQUEREL, Antoine Henri
(1852–1908)

French physicist, born in Paris, he studied at the École Polytechnique and the school of bridges and highways. In 1875 he started teaching at the Polytechnique and later he succeeded his father **Alexandre-Edmond Becquerel** in the Chair of Physics at the Natural History Museum. He was an expert in fluorescence and phosphorescence, continuing the work of his father and grandfather. Following the discovery of X-rays by **Röntgen**, Becquerel investigated fluorescent materials to see if they also emitted X-rays. He exposed a fluorescent uranium salt, pitchblende, to light and then placed it on a wrapped photographic plate. He found that a faint image was left on the plate, which Becquerel believed was due to the pitchblende emitting the light it had absorbed as a more penetrating radiation. However, by chance, he left a sample that had not been exposed to light on top of a photographic plate in a drawer. He noticed that the photographic plate also had a faint image of the pitchblende. After several chemical tests he concluded that these 'Becquerel rays' were a property of atoms. He had, by chance, discovered radioactivity and prompted the beginning of the nuclear age. His work led to the discovery of radium by **Marie** and **Pierre Curie** and he subsequently shared with them the 1903 Nobel Prize for Physics.

BEDDOES, Thomas
(1760–1808)

English physician and writer, born in Shifnal. He studied medicine and became Reader in Chemistry at Oxford, but his sympathies with the French Revolution led to his resignation (1792). From 1798 to 1801 he carried on at Clifton (Bristol) a 'Pneumatic Institute' for the cure of diseases by the inhalation of gases, with **Humphry Davy** his assistant. He wrote on political, social, and medical subjects and edited the works of John Brown, founder of the Brunonian system of medicine according to which all diseases are divided into the sthenic, depending on an excess of excitement, and the asthenic; the former to be removed by debilitating medicines, and the latter by stimulants.

BEDNORZ, (Johannes) Georg
(1950–)

German physicist, born in West Germany. He graduated at Münster (1976) and was awarded his PhD by the Swiss Federal Institute of Technology in Zürich (1982). He subsequently joined the IBM Zürich research laboratory and with **Alex Müller**, who had considerable experience in the field of superconductors, investigated a new range of material types based on oxides. This inspired departure from the more conventional materials based on mixtures of different metals (intermetallic compounds) yielded results in 1986. They announced that they had observed superconductivity at a temperature 12 kelvin higher than the current record of 23 kelvin, which had been held by an alloy of germanium and niobium since 1973. At a stroke a whole new family of superconducting compounds had been revealed. Laboratories around the world rushed to reproduce and improve upon these results, and within two years observations of superconductivity at 90 kelvin in materials made of mixtures of yttrium, barium and copper oxide were confirmed. With superconductors now operating above the boiling point of liquid nitrogen, an inexpensive and plentiful coolant, the practical applications of superconducting devices were set to multiply enormously. Bednorz was awarded the 1987 Nobel Prize for Physics jointly with Müller only one year after the announcement of their discovery.

BEER, Sir Gavin Rylands de
See **DE BEER, Sir Gavin Rylands**

BEHRING, Emil von
(1854–1917)

German bacteriologist and pioneer in immunology, born in Hansdorf, West Prussia. He entered the gymnasium at Hohenstein in East Prussia in 1865. Enrolling in the Army Medical College in Berlin in 1874, he obtained his medical degree in 1878, and in 1881 moved to Poznań as an assistant surgeon. During this period he became interested in disinfectants, particularly iodoform. In 1888 he went to join **Koch**'s Institute of Hygiene in Berlin. His major contribution was the development of serum therapy against tetanus and diphtheria (1890) in the first application of the science of serology, in collaboration with **Kitasato**. Diphtheria was one of the most common fatal diseases affecting children at the time, and Behring's work was instrumental in counteracting the disease. In later years, he tried to develop a tuberculosis antitoxin, but had to admit defeat. Much of his work in this area concerned the relationship between tuberculosis occurring in humans and that in cattle. Unlike Koch, he believed the forms to be identical. Although today they are not considered the same, the cattle form may be transmitted to humans, and Behring's recommendations for reducing the occurrence of the disease in animals and for disinfecting milk were important public health measures. He later became Professor of Hygiene at Halle (1894–5) and at Marburg (from 1895), and was awarded the first Nobel Prize for Physiology or Medicine in 1901. During World War I, the tetanus vaccine developed by him helped to save so many German lives that he received the Iron Cross, very rarely awarded to a civilian.

BEILBY, Sir George Thomas
(1850–1924)

Scottish industrial chemist, born in Edinburgh and educated at Edinburgh University. He worked as a chemist with oil shale, gold and chemical companies. In 1881 he designed a continuous retort which increased the yield of paraffin and ammonia from shale. He discovered an economical way of producing potassium cyanide, which is needed to extract gold from low grade ores, and the first factory was established at Leith, near Edinburgh. Beilby also studied flow in solids and explained why metals harden with cold working. He founded the Fuel Research Station at East Greenwich and was a member of many government committees. He was knighted in 1916, and died in London.

BEILSTEIN, Friedrich Konrad
(1838–1906)

German–Russian chemist, born in St Petersburg of German parentage. He studied in Heidelberg, Göttingen and Paris, and in 1860 became assistant to **Wöhler** in Göttingen. From 1866 until his retirement he was professor at the Technical Institute in St Petersburg. His name is synonymous with his *Handbuch der organischen Chemie*, which was first published in 1880. It aims to be a complete catalogue of all organic compounds. Beilstein himself compiled the second and third editions but subsequent editions have been prepared by the Deutsche Chemische Gesellschaft. The most recent edition is in English. It is a work of great importance to all organic chemists, although the cost of each volume is now a substantial barrier to its wide use.

BÉKÉSY, Georg von
(1899–1972)

Hungarian-born American physicist and physiologist whose career was devoted to the ear and the physics and physiology of hearing. Born in Budapest, he studied chemistry in Berne, Switzerland, and after military service studied optics at the University of Budapest for his doctoral degree. Failing in the postwar depression to find a suitable academic appointment, he worked as a telephone research engineer in Hungary (1924–46). In 1932 he was given a concurrent appointment at Budapest University and in 1940 was created Professor of Experimental Physics. In 1946 **Zotterman** invited him to Stockholm, and the following year he moved to Harvard. He retired in 1966, when he became professor at the University of Hawaii, where he remained until his death. He won the 1961 Nobel Prize for Physiology or Medicine, for his discoveries about the physical mechanisms of stimulation within the cochlea of the ear, which explained how people distinguished sounds; much of this work is summarized in his *Experiments in Hearing* (1960). His work was not confined to the experimental analysis of the ear and the transmission of aural information to the brain; he was also interested in developing instrumentation, and his work contributed techniques for measuring deafness, for improving surgery of the ear and in working towards restoring hearing.

BEL, Joseph Achille Le
See **LE BEL, Joseph Achille**

BELL, Alexander Graham
(1847–1922)

Scots-born American inventor, born in Edinburgh, son of Alexander Melville Bell. Educated at Edinburgh and London, he worked as assistant to his father in teaching elocution (1868–70). In 1871 he moved to the USA and became Professor of Vocal Physiology at Boston (1873), devoting himself to the teaching of deaf-mutes and to spreading his father's system of 'visible speech'. After experimenting with various acoustic devices he produced the first intelligible telephonic transmission with the famous words to his assistant 'Mr Watson, come here—I want you', on 5 June 1875. He was granted three patents relating to his invention between 1875 and 1877, successfully defended them against Elisha Gray and others, and formed the Bell Telephone Company in 1877. In 1880, with the proceeds of the Volta Prize awarded to him by the French government, he established the Volta Laboratory for research into deafness, and his work there resulted in the photophone (1880) and the graphophone (1887). He also founded the journal *Science* (1883). After 1897 his principal interest was in aeronautics: he encouraged **Samuel Langley**, and invented the tetrahedral kite.

BELL, Sir Charles
(1774–1842)

Scottish anatomist, surgeon and neurophysiological pioneer, born in Edinburgh, the son of a Church of England clergyman. Bell developed his interest in medicine from his elder brother, John, who became a celebrated surgeon and anatomist. Charles studied at Edinburgh University, moving to London in 1804, where he became proprietor of an anatomy school and rose to prominence as a surgeon. In 1812 he was appointed surgeon to the Middlesex Hospital, in 1828 becoming one of the co-founders of the Middlesex Hospital Medical School. An interest in gunshot wounds led him to treat the wounded from the battle of Corunna; after Waterloo, he organized a hospital in Brussels. A fine draftsman, he was energetic in teaching anatomy to artists. His *Essays on the Anatomy of Expression in Painting* (1806) illustrated the science of physiognomy. Knighted in 1831, in 1836 he was appointed Professor of Surgery at Edinburgh. Today he is remembered for his pioneering neurophysiological researches, first set out in his *Idea of a New Anatomy of the Brain* (1811). Bell demonstrated that, far from being single units, nerves consist of separate fibres sheathed together. Above all he showed that fibres convey either sensory or motor stimuli but never both—fibres transmit impulses only in one direction. His explorations of the sensorimotor functions of the spinal nerves triggered a bitter and prolonged priority dispute with the French physiologist **Magendie**. Bell's experimental work led to the discovery of the long thoracic nerve (Bell's nerve). We speak of 'Bell's palsy' as a result of his demonstration that lesions of the seventh cranial nerve could create facial paralysis.

BELL, (Chester) Gordon
(1934–)

American computer scientist and designer. He studied electrical engineering at MIT, where he was awarded a master's degree in 1957. He later worked at the Engineering Speech Communication Laboratory at MIT (1959–60), before joining **Olsen**'s Digital Equipment Corporation in 1960. At DEC he was Manager of Computer Design (1960–6) and Vice-President of Engineering (1972–83), and became chiefly known for his design of the VAX (Virtual Address Extension) superminicomputers. Ranging from small desktop machines to computer clusters that could compete with mainframes, DEC's VAXes shared the same software and could transfer data over networks. They were the basis of DEC's fortunes in the early 1980s. Bell was Professor of Computer Science at Carnegie Mellon University from 1966 to 1972. He was also assistant director of the National Science Foundation from 1986, co-founder of the Stardent Computer Company, and co-founder of the Boston Computer Museum of which he became a director.

BELL, John Stewart
(1928–90)

Irish nuclear physicist, born in Belfast. He was educated at Queen's University, Belfast, where he obtained degrees in physics and mathematical physics. After working at Harwell, he went to the University of Birmingham to study for his PhD under **Peierls**. In his thesis, he proved the 'CPT theorem'. This is a powerful theorem, that can be derived from very general assumptions such as 'cause precedes effect', which states that to ensure that particles and antiparticles have the same mass and lifetime, a theory must be symmetric under the combination of three separate symmetries: C-charge conjugation symmetry which says that particles and antiparticles should behave in the same way; P-parity which is a mirror symmetry that implies 'left-handed' and 'right-handed' particles should behave in the same way; and T-time invariance that implies that an initial state can be derived from a final state by reversing the direction of motion and all the components of the system. After completing his PhD he joined the theory division of CERN (the European centre for nuclear research) in Geneva. There he worked on accelerator physics, and with **Steinberger**, the relationship between the observation of CP violation and implications for the CPT theorem. He also made important contributions to the foundations of quantum mechanics. He showed that predictions of certain hidden variable theories, the theories which say that we only appear to live in a quantum-mechanical universe because some information is not available to us, were bounded by inequalities. Work done by experimenters used these inequalities to show that the world we live in is truly quantum mechanical. Bell's work was honoured by the award of the Dirac Medal of the Institute of Physics and the Heineman Prize of the American Physical Society, and he was elected FRS in 1972.

BELL, Thomas
(1792–1880)

English naturalist, born in Poole, Dorset. A dental surgeon at Guy's Hospital (1817–61), he lectured in zoology and became Professor of Zoology at King's College, London, in 1836. He was Secretary of the Royal Society, President of the Linnaean Society and first President of the Ray Society (1844). His *British Stalk-eyed Crustacea* (1853) remains a standard work on British crabs and lobsters. He edited the *Natural History and Antiquities of Selborne* (1877), by **White**, whose house he purchased.

BELL BURNELL, (Susan) Jocelyn, née Bell
(1943–)

English radio astronomer, born in York, co-discoverer of the first pulsar. She was educated at the universities of Glasgow and Cambridge, where she received her PhD in 1968. She later joined the staff of the Royal Observatory, Edinburgh, and became the manager of their James Clerk Maxwell Telescope on Hawaii. She was awarded the Herschel Medal of the Royal Astronomical Society in 1989, and since 1991 has been Professor of Physics at the Open University. In 1967 she was a research student at Cambridge working with **Hewish** when they noticed an unusually regular radio signal on the 3.7 metre wavelength radio telescope. This turned out to be the first discovery of a pulsar (PSR 1919 + 21, with a period of 1.337 seconds). Within the first few weeks of discovery the correct conclusions were drawn; the object is stellar as opposed to a member of the solar system, and is a condensed neutron star. The papers by Franco Pacini (1967) and **Gold** (1968) laid the theoretical foundation.

BENACERRAF, Baruj
(1920–)

Venezuelan–American immunologist, born in Caracas. In 1940 his family moved to the USA and two years later he entered the Medical College of Virginia. As an asthma sufferer in childhood, Benacerraf became interested in immune hypersensitivity. In 1949 he took a research job at the Broussais Hospital, Paris, and in 1956 he returned to the USA to a post at the New York University School of Medicine. Here he became Professor of Pathology in 1960. In New York, his research concentrated upon the cells involved the body's defence against antigens—substances which are foreign to the body. He found that some animals responded to simple synthetic antigens by producing antibodies while others did not, and went on to show that this was genetically determined. He named the genes involved the immune-response genes. Later work clarified the role of the genetically determined structures on the surfaces of cells that regulate immunological responses to diseased cells and in organ transplants. For this work he shared the 1980 Nobel Prize for Physiology or Medicine with **Dausset** and **George Snell**. Benacerraf moved to the National Institutes of Health in 1968 and to Harvard Medical School in 1970.

BENEDEN, Edouard Joseph Louis-Marie van
(1846–1910)

Belgian cytologist and embryologist, born in Liège. In 1872 he travelled to Brazil and returned with a collection of a number of specimens, especially worms. He worked in his father's laboratory at Ostend on protozoa, nematodes, tunicates and later mammals. He showed that the endoderm layer of the hydroid cell comes from the female parent and the ectoderm layer from the male, and in 1877 proposed the creation of the phylum Mesozoa, to cater for the transition between single-celled and multi-celled organisms. In 1887 he demonstrated the constancy of the number of chromosomes in the cells of an organism, decreasing during maturation and restored at fertilization.

BENIOFF, Victor Hugo
(1899–1968)

American geophysicist, born in Los Angeles of a Russian father and Swedish mother. He was educated at Pomona College, California (AB 1921), in physics and astronomy. He first worked at Lick Observatory (1923–4) and in 1924 moved to the Carnegie Institution in Washington where he developed a system of seismic recording drums. These gave an unprecedented accuracy of 0.1 second and formed the basis of the now famous Caltech seismic network. His bent for instrument design continued with the variable reluctance seismograph and linear strain seismograph (in service from 1931). These made possible the precise determination of seismic travel-time, the discovery of new seismic phases, extended the recordable magnitudes down to teleseismic (distant) events and enabled first motion determinations, which give the directions of the fault breaks caused by earthquakes. Later these instruments were recommended for monitoring nuclear tests. He received his PhD at Caltech in 1935 for work on these inventions. When the Seismological Laboratory was transferred from Carnegie to Caltech in 1937, Benioff became assistant professor in seismology and professor from 1950, and collaborated with **Gutenberg** and **Charles Richter**. From about 1950 (15 years prior to other workers) he became interested in general problems in earthquake mechanisms and global tectonics, developed several new analytical techniques and amassed evidence of earthquakes around the Pacific. In 1954 he published evidence of epicentres deepening away from the trench and extending down to 700 kilometres below Japan (similar to those identified by **Wadati**). These seismically active down-going crustal slabs beneath trenches are now known as Wadati–Benioff zones.

BENNET, Abraham
(1750–99)

English physicist and an early pioneer of electricity, born in Taxal, Cheshire, the son of a schoolmaster. In 1775 he was ordained in London, and appointed to curacies first at Tideswell and then in Wirksworth, Derbyshire, where he remained for the rest of his life. He held several other posts at the same time, including that of domestic chaplain to the Duke of Devonshire, perpetual curate of Woburn and librarian to the Duke of Bedford. A meticulous experimenter, his main interest was in atmospheric electricity which he tried to relate to the weather. In the course of these investigations of extremely small electric charges, he invented in 1786 the gold-leaf electroscope which he described to **Priestley** as 'the most sensible electrometer yet constructed', and which was based on the 'portable electrometer' of Tiberius Cavallo (1749–1809), an Italian who had settled in London. He also invented a 'doubling process' which, operating on the principle of the induction machine, allowed him to measure extremely small atmospheric charges. These experiments were described in his *New Experiments on Electricity*, published in 1789, the year in which he was elected FRS. He also worked on magnetism, using spider's-web filaments to suspend magnetic needles.

BENSON, Sidney William
(1918–)

American physical chemist, born in New York City. He was educated at Columbia University, New York and Harvard. After holding junior posts at Harvard and at the City College of New York, he moved to the University of Southern California (USC), advancing from Assistant Professor to Professor of Chemistry (1943–63). From 1963 to 1976 he was Chairman of the Department of Kinetics and Thermochemistry at Stanford Research Institute. In 1976 he returned to USC as professor, and from 1977 he was Director of the Hydrocarbon Research Institute. Benson has held many visiting professorships and consultancies, and has received many international honours and awards, including the Langmuir Award of the American Chemical Society (1986) and the Polanyi Medal of the Royal Society of Chemistry (1986). From 1967 to 1983 he was Editor-in-Chief of the *International Journal of Chemical Kinetics*. His contributions to physical chemistry have been wide-ranging: kinetics, photochemistry, laser chemistry, heterogeneous catalysis, theory of liquid structure, statistical mechanics and thermochemistry. He is best known for work in the area covered by his book *Thermochemical Kinetics* (1968).

BENTHAM, George
(1800–84)

English botanist, nephew of Jeremy Bentham, born in Stoke, Plymouth. During 1805–7 he lived in St Petersburg, where he acquired a knowledge of Russian and Scandinavian languages. He spent much of his youth in France, where his interest in botany developed, and studied at Tours. He studied philosophy with John Stuart Mill at Montpellier, where Bentham's father was a farm manager. In 1826 he returned to England and published *Catalogue des plantes indigènes des Pyrenées*. From 1826 to 1832 he was his uncle's secretary and partly rewrote Jeremy's work on naval administration. His publication of *Outlines of a New System of Logic* (1827) was a financial disaster. During 1829–40 he was Secretary of the Royal Horticultural Society. He lived in London until 1842, and published *Labiatarum genera et Species* (1832–6). He then moved to Herefordshire and wrote *Plantas Hartwegianas* (1839–57) and *Botany of the Voyage of H.M.S. Sulphur* (1844–6). Returning to London, he published *Handbook of the British Flora* (1858), *Flora Australiensis* (7 vols, 1863–78) and *Genera Plantarum* (3 vols, 1862–83); the last remains a standard work.

BENTLEY, Charles Raymond

(1929–)

American geophysicist and glaciologist, born in Rochester, New York. He studied physics at Yale (BS 1950) and geophysics at Columbia (PhD 1959), where he participated in research cruises in the Atlantic with **Ewing**. He was research geophysicist at Columbia (1952–6), spent two years in Antarctica and participated in the particular scientific effort for the International Geophysical Year (1957–8). Since then he has worked at the university of Wisconsin-Madison, becoming A P Crary Professor there in 1968. In west Antarctica he discovered that the ice rests on a floor far below sea-level; thus the region would be open ocean if the ice melted. His interest in Antarctica has continued with pioneering work on applying geophysical techniques to determine the physical characteristics of the ice. He is one of the founders of radioglaciology, pioneering methods of using electromagnetic waves to measure the thickness and properties of the ice. Using seismic and gravity techniques, he has contributed to the understanding of Antarctic isostasy and crustal structure. In the 1980s his research group discovered and investigated the soft, deformable sediment beneath a west Antarctic ice stream that is now believed to be a critical factor in local glacial dynamics.

BENZER, Seymour

(1921–)

American geneticist, born in New York City. He studied physics at Purdue University, Indiana, and taught biophysics there until 1965, when he moved to Caltech. He was a member of the Phage Group, set up by **Delbrück**, **Hershey** and **Luria** to encourage the use of phage as an experimental tool, and isolated more than 1 000 mutants of the rII phage virus of the bacterium *Escherichia coli*. He showed that genes and proteins were co-linear, ie that a change in a gene led to a change in the protein it coded for. He also introduced the term 'cistron' for the smallest section of a DNA strand comprising a functional gene. In 1961 he showed that some sections of the DNA strand are more susceptible to mutations than would be expected by chance; these sites became known as 'hot spots', and were later found to be associated with modified nucleic acids in the DNA strand. Since the 1960s, Benzer has moved away from bacterial genetics to research the genetics underlying the behaviour of the fruit fly, *Drosophila*.

BÉRARD, Jacques Étienne

(1789–1869)

French chemist, born in Montpellier. The son of a chemical manufacturer, he assisted in the Arcueil laboratory of the influential chemist **Berthollet** from 1807. Bérard soon became a member of the distinguished scientific Society of Arcueil. Returning to Montpellier in 1813 (where he was visited by **Humphry Davy**), he studied medicine and gained his MD (1817). He was appointed Professor of Chemistry at Montpellier in the faculties of pharmacy and then medicine (1832). A republican political stance may have delayed his elevation to Dean which he assured in 1846. Bérard's first experimental research, on salts and solubility, was undertaken under Berthollet's direction. Later he investigated 'chemical rays' (ultraviolet), and showed that radiant heat, like light, could be polarized. His best-known work, the experimental determination of the specific heats of gases with François Delaroche, established Bérard amongst the élite Parisian chemists. He was awarded a prestigious prize in 1821 for work on the ripening of fruit taking into account the action of oxygen.

BERG, Paul

(1926–)

American molecular biologist, born in Brooklyn, New York City. He was educated at Pennsylvania State and Western Reserve universities, although his studies were interrupted by World War II, which he spent in the US navy. He later became Professor of Biochemistry at Stanford (from 1959) and at Washington (from 1970). Since 1985 he has been Director of the Beckman Center of Molecular and Genetic Medicine at Stanford. In the 1960s, Berg purified several transfer RNA molecules (tRNAs—each tRNA carries a specific amino acid through the cell to the ribosome for assembly into a protein). In the following decade, he worked with simian virus 40 and the bacterial virus lambda phage, and developed techniques to cut and splice genes from one organism into another. Concerned about the effects of mixing genes from different organisms, Berg organized a year-long moratorium on genetic engineering experiments, and in 1975 chaired an international committee to draft guidelines for such studies. These techniques are now widely used in biological research. In 1978 Berg enabled gene transfer between cells from different mammalian species for the first time. He shared the 1980 Nobel Prize for Chemistry with **Sanger** and **Walter Gilbert**.

BERGER, Hans
(1873–1941)

German psychiatrist who invented electroencephalography (EEG), born in Neuses bei Coburg. He studied medicine at Jena University, where he remained for the rest of his career, becoming Professor of Psychiatry in 1919. His research attempted (mostly unsuccessfully) to establish relationships between psychological states and various physiological parameters, such as heartbeat, respiration and the temperature of the brain itself. In the course of this work, he placed electrical recording equipment on the surface of the skull; from the mid-1920s, he recorded what became known as 'brain waves'. Although precise correlations between electrical brain activity and psychic processes has not emerged, Berger's electroencephalograph, particularly after the work of individuals such as **Adrian**, has become a useful tool of research and diagnosis into brain functions and diseases.

BERGERON, Tor Harold Percival
(1891–1977)

Swedish meteorologist, born in Godstone, near London, of Swedish parents. After graduating from Stockholm University in 1916, he joined the Bergen School under **Vilhelm Bjerknes** (1922) and obtained his doctorate at Oslo University (1928). Bergeron researched the occlusion process whereby air in the warm sector of a depression is forced upwards. He also described the 'Arctic front' and explained different types of fronts, including the formation of waves on a stationary front, by studying surface and upper air observations in detail. An inspired teacher, his visits to the USA, Russia, Yugoslavia and elsewhere did much to promote the Bergen School ideas internationally. He returned to the Swedish Meteorological and Hydrological Institute in 1936 and later became professor at Uppsala (1947–61). With a collaborator, he was responsible for the currently accepted theory that the usual process for the initiation of rain is for ice crystals and water vapour to coexist in clouds. Later he studied the effects of topography on precipitation, both on the mesoscale and on the scale of mountain ranges and synoptic systems, as well as the precipitation mechanism in hurricanes. He received the International Meteorological Organisation Prize (1966) and was awarded the Symons Gold Medal of the Royal Meteorological Society.

BERGIUS, Friedrich
(1884–1949)

German chemist, born in Goldschmieden, near Breslau (now Wrocław, Poland). After studying chemistry at Leipzig, he received his doctorate in 1907 and became, successively, assistant to **Nernst**, **Haber** and **Bodenstein**. He was appointed assistant lecturer at Hanover in 1909 and began work on the effect of high pressures and temperatures on wood, conditions necessary for its conversion into coal. He also developed a process for the conversion of coal into oil. He continued this work when he became head of the research laboratory of the Goldschmidt company in Essen in 1914. Finding himself unable to continue these studies after World War I, he turned his attention to the acidic hydrolysis of wood. After World War II he founded a company in Madrid and in 1947 he became scientific adviser to the government of Argentina; he died in Buenos Aires. He shared the 1931 Nobel Prize for Chemistry with **Bosch** for his work on coal formation.

BERGMAN, Torbern Olof
(1735–84)

Swedish chemist and physicist, born in Catherineberg, West Gothland. He studied at the University of Uppsala, first theology and law, and then natural science. After graduating in 1758 he began to teach mathematics and physics at the university, where he became Professor of Chemistry and Mineralogy in 1767. Ill-health forced him to retire in 1780 and he died of tuberculosis. His earliest publications were in physics (light and electricity) and he seems largely to have been self-taught in chemistry. Bergman was a pioneer in analytical chemistry, particularly of minerals. His most important paper, however, was his *Essay on Elective Attractions* (1775), in which he developed the concept of 'elective affinity'. During 1775–83 he compiled extensive affinity tables for acids and bases. His failing health led to the work being abandoned in an incomplete state, but it stimulated much later work on chemical affinity. He was an adherent of the phlogiston theory developed by **Georg Stahl**.

BERGSTRÖM, Sune Karl
(1916–)

Swedish biochemist, born in Stockholm. He studied medicine and chemistry at the Karolinska Institute in Stockholm, and taught chemistry at Lund (1948–58) before returning to the Karolinska Institute as professor (1958–81). In 1952, following the development of techniques of introducing radioactive atoms into cholesterol by **Konrad Bloch**, Bergström demonstrated the formation of the bile acid ester, taurocholic acid, from cholesterol and made major contributions towards solving the metabolic pathway involved by observing the consequences of injecting hypothetical intermediates into a rat. He studied the metabolism of pythocholic acid from snake bile (1960), and found it to be derived from cholesterol by the same general metabolic pathway. He also isolated and purified the prostaglandins, complex fatty acids produced by nearly all mammalian cells. Their physiological effects include lowering blood pressure and the promotion of contraction by intestinal and uterine muscle, for which purpose they are employed to promote foetal abortion. Together with his former student **Samuelsson** and **Vane**, Bergström was awarded the 1982 Nobel Prize for Physiology or Medicine for his work on prostaglandins.

BERNAL, John Desmond
(1901–71)

Irish crystallographer, born in Nenagh, County Tipperary. Educated by the Jesuits at Stonyhurst College, Lancashire, he won a scholarship to Emmanuel College, Cambridge, and from the first showed himself a polymath (with the lifelong nickname 'Sage'). He developed modern crystallography and was a founder of molecular biology, pioneering work on the structure of water. He progressed from a lectureship at Cambridge to become Professor of Physics and then Professor of Crystallography at Birkbeck College, London (1937–68). It was here that he did his major pioneering work, taking X-ray photographs of biologically important molecules, amino acids, proteins, sterols and nucleoproteins. He was convinced that from an understanding of their physical molecular structure would come a clearer insight into the way the living processes worked. His researches were dominated by one purpose, namely the search for the origin of life and he included among his major works *The Origin of Life* (1967). His wartime service, in close association with Lord Louis Mountbatten in Combined Operations, involved abortive attempts at creating artificial icebergs to act as aircraft carriers, as well as working on bomb tests and contributing substantially to the scientific underpinning of the invasion of the European continent. A communist from his student days, he was active in international peace activity during the Cold War and supported **Lysenko** in the USSR when his destruction of Soviet genetics drove **J B S Haldane** out of the British Communist party. Bernal's hopes for communism's possibilities for science were first shown in his *The Social Function of Science* (1939) and *Marx and Science* (1952).

BERNARD, Claude
(1813–78)

French physiologist, born near Villefranche, Beaujolais. The child of poor vineyard workers, Bernard was educated in church schools and at 19 apprenticed to a Lyons pharmacist. After an unsuccessful attempt to shine as a dramatist, he abandoned the theatre and opted for a professional qualification, choosing medicine. Trained in Paris, he became assistant at the Collège de France to the physiologist **Magendie**, and in this work developed a passion for experimental medicine. Marriage to a rich Paris physician's daughter who brought with her an ample dowry obviated the need for Bernard to practise medicine and allowed him to pursue his researches (the marriage was miserable). In 1855 he succeeded to Magendie's Paris post. Thereafter his career was a stream of promotions and honours, including the Legion of Honour (1867); at heart, Bernard remained a country lover, and he returned as often as possible to his beloved Beaujolais. His greatest contribution to physiological theory was the notion that life requires a constant internal environment (*milieu intérieur*); cells function best within a narrow range of osmotic pressure and temperature when bathed in a fairly constant concentration of chemical constituents. A fearless vivisector, Bernard made wide-ranging experimental discoveries in many areas of physiology. He studied the neurophysiological action of paralysing poisons like curare and demonstrated their use in experimental medicine. Exploring digestion, he researched the operation of enzymes in gastric juice and demonstrated the role of nerves in controlling gastric secretion. He showed how carbohydrates needed to be transformed into sugars before they could be absorbed. He probed the role of bile and pancreatic juice in the digestion of proteins. On the basis of such researches he concluded that nutrition is an intricate process, involving various stages and chemical rather than physical transformations. He discovered glycogen, probed sugar production by the liver and the problem of diabetes. He also discovered the vasomotor and vasoconstrictor nerves, and examined the inhibitory action of the vagus nerve upon the heart. Bernard further developed **Lavoisier**'s ideas on animal heat. He showed that the oxidation producing animal heat is indirect, occurring in all tissues and not simply in the lungs. Bernard stands as one of the founders of modern physiological research. Combining experimental skill with a partiality for theory, he was notably innovative and one of the great masters of productive research. His *Introduction to the Study of Experimental Medicine* (1865) is a classic account of the new biology, concerned to lay bare the elementary conditions of the phenomena of life, and to demonstrate the absence of qualitative differences between the normal and the pathological. Diseases, Bernard claimed, have no ontological reality; they are merely distortions and extreme forms of regular physiological processes and states.

BERNOULLI, Daniel
(1700–82)

Swiss mathematician, born in Groningen, son of **Jean Bernoulli**. He studied medicine and mathematics at the universities of Basle, Strasbourg and Heidelberg, and became Professor of Mathematics at St Petersburg (1725). There he took up the Petersburg paradox in probability, a coin-tossing game which intuitively should command a small entry price but in theory demands an infinite one. In 1732 he returned to Basle to become Professor of Anatomy, then was appointed to a professorship in botany, and finally physics. He worked on trigonometric series, mechanics and vibrating systems. In his best-known work he pioneered the modern field of hydrodynamics. His *Hydrodynamica* (1738) explored the relationships between pressure, density and velocity in flowing fluids, and anticipated the kinetic theory of gases, pointing out that pressure could result from the bombardment of a surface by particles of matter and that pressure would increase with increasing temperature. He solved a differential equation proposed by Jacopo Riccati, now known as Bernoulli's equation.

BERNOULLI, Jacques or Jakob
(1654–1705)

Swiss mathematician, born in Basle, brother of **Jean Bernoulli**. He studied theology and travelled in Europe before returning to Basle (1682), where he became professor in 1687. He investigated infinite series, the cycloid, transcendental curves and the logarithmic spiral. His analysis of the catenary, the curve formed by a non-elastic flexible string when suspended at each end, was applied to the design of bridges. In 1690 he applied **Leibniz**'s newly discovered differential calculus to a problem in geometry, and introduced the term 'integral'. His posthumously published *Ars conjectandi* (1713) was an important contribution to probability theory, and included his discovery of the 'law of large numbers', his permutation theory and the 'Bernoulli numbers', coefficients found in exponential series. A logarithmic spiral was at his request carved on his tombstone in Basle cathedral, with the motto in Latin 'Though changed I rise the same'.

BERNOULLI, Jean or Johann
(1667–1748)

Swiss mathematician, born in Basle, younger brother of **Jacques Bernoulli**. He studied medicine and graduated in Basle (1694), but turned to mathematics, and became professor at Groningen (1695) and Basle (1705). He wrote on differential equations, both in general and with respect to particular curves, such as the clamped beam and the hanging chain, finding the length and area of curves, isochronous curves and curves of quickest descent. He was a quarrelsome man and a bitter rivalry developed between him and his brother. After his brother's death he took the leading role in **Leibniz**'s side in the priority dispute with **Isaac Newton** regarding the invention of the calculus. He founded a dynasty of mathematicians which continued for two generations, and was employed by the Marquis de l'Hospital to help him write the first textbook on the differential calculus.

BERNSTEIN, Richard Barry
(1923–90)

American physical chemist, born in Long Island, New York. He was educated at Columbia University, New York, and from 1944 to 1946 worked on the Manhattan Project to separate uranium-235. From 1948 to 1953 he was Assistant Professor of Chemistry at Illinois Institute of Technology, Chicago, and from 1953 to 1963 he advanced from Assistant Professor to Professor of Chemistry at the University of Michigan. After 1963 he was professor in succession at the universities of Wisconsin (Madison) and Texas (Austin), Columbia University, and finally the University of California (Los Angeles) from 1983 until his death. Bernstein received many awards, including the Hinshelwood lectureship at Oxford (1980), the Debye Award of the American Chemical Society (1981), the Willard Gibbs Medal of the same society (1989) and a National Medal of Science (1989). His researches were in chemical kinetics and reaction dynamics by molecular beam scattering and laser techniques. By these techniques he elucidated much detail of what occurs when molecules collide and react. He co-authored (with Raphael Levine) *Molecular Reaction Dynamics* (1974), and also wrote *Chemical Dynamics via Molecular Beam and Laser Techniques* (1982). In 1955 he was the pioneer of 'femtochemistry', the study of reactions which take place on timescales of the order of 10^{-12} seconds.

BERT, Paul

(1833–86)

French physiologist and statesman, born in Auxerre. He studied in Paris from 1853, obtaining licentiates in law and in natural sciences, and later an MD (1863). Bert was the student and préparateur of **Bernard** from 1863 to 1866. After short periods as professor at the University of Bordeaux and at the Museum of Natural History, he succeeded Bernard as Professor of Physiology at the Sorbonne in 1869. He entered politics during the Franco-Prussian War, after which he joined the Liberal Republican party and was elected in 1872 to the Chamber of Deputies. He took a great deal of interest in education, and as Minister of Public Instruction fought for free, compulsory and secular elementary education. Bert's early work included studies of animal transplantation and comparative physiology of respiration. He moved on to work on the effect of changes in atmospheric pressure, altitude and the composition of the air upon blood gases. He showed the importance of the oxygen content of the air and the relative unimportance of barometric pressure. This work had important implications for deep-sea diving and understanding the problems of breathing at high altitudes. His *La Pression Barométrique* (1878) was translated in 1943 because of its importance for aviation medicine.

BERTHELOT, (Pierre Eugène) Marcellin

(1827–1907)

French chemist and politician, born in Paris. His scientific education was acquired largely in the private laboratory of Pelouze, and he became demonstrator at the Collège de France in 1851. He defended his thesis on the chemistry of glycerin in 1854, and in 1858 graduated as a pharmacist. In 1859 he was appointed to the Chair of Organic Chemistry at the École de Pharmacie but continued to work in a laboratory at the Collège de France. His research covered many topics, but within organic chemistry, he is best known for his work on the synthesis of alcohols, aromatic compounds and turpentine derivatives. In a pioneering study, Berthelot and Péan de Saint-Gilles examined the kinetics of esterification, the results of which were significant in the formulation by **Guldberg** and **Waage** of the law of mass action. His later experimental work laid the foundations for the science of thermochemistry. In the last stage of his professional life he wrote extensively on the history of early chemistry. He also did much to initiate the chemical analyses of archaeological objects. During the siege of Paris in the Franco-Prussian War (1870–1), Berthelot played a leading role in the defence of the city and was elected to the Senate in 1871. After a number of minor posts he became Minister of Education for two years (1886–7), and very briefly, Foreign Minister (1895). However, he never achieved in politics the same eminence as in his scientific work. He received the Legion of Honour in 1861 and succeeded **Pasteur** as Secretary of the French Academy of Sciences in 1889.

BERTHOLLET, Claude Louis, Comte de

(1749–1822)

French chemist, born in Talloires, Savoy, who made major contributions to the revolutionary advances in chemistry which took place in the second half of the 18th century. He studied medicine at Turin and Paris, later becoming private physician to Phillip, Duke of Orleans. Berthollet helped **Lavoisier** with his researches on gunpowder and also with the creation of a new system of chemical nomenclature, which is still in use today. He accepted Lavoisier's antiphlogistic doctrines, but disproved his theory that all acids contain oxygen by analysing hydrocyanic acid. Berthollet was the first chemist to realize that there is a connection between the manner in which a chemical reaction proceeds and the mass of the reagents; this insight led others to formulate the law of definite proportions later stated by **Proust**. He also demonstrated that chemical affinities are affected by the temperature and concentration of the reagents. In the course of his studies of chlorine, he failed to find a way to use potassium chlorate in gunpowder, but made a successful bleach by dissolving chlorine in a solution containing sulphuric acid and various salts. He made the results public so that the milk and agricultural land reserved for traditional bleachfields could be put to better use, and the use of chlorine in bleaching spread rapidly in France and abroad. Following **Priestley**'s discovery that ammonia is composed of hydrogen and nitrogen, Berthollet made the first accurate analysis of their proportions. He was elected to the Academy of Sciences in 1781. During the French Revolution, Berthollet helped to reorganize the coinage, agricultural policy and higher education. When the Academy was reconstituted as the Institut de France in 1793, Berthollet was one of the first members to be elected. In 1796 he led a delegation to Italy to choose works of art for the galleries in Paris. After accompanying Napoleon to Egypt he remained there for two years, helping to reorganize the educational system. He voted for Napoleon's deposition in 1814, and on the restoration of the Bourbons, was created a count. He died in Arcueil.

BERZELIUS, Jöns Jacob

(1779–1848)

Swedish chemist who made fundamental advances in atomic theory and electrochemistry. He was born in Väfversunda in East Götland. He studied medicine at Uppsala, and worked as an unpaid assistant in the College of Medicine at Stockholm before succeeding to the Chair of Medicine and Pharmacy in 1807. Subsequently he travelled abroad to meet **Humphry Davy**, **Oersted**, **Berthollet** and other leading chemists. In 1815 he was appointed Professor of Chemistry at the Royal Caroline Medico-Chirurgical Institute in Stockholm, retiring in 1832. Soon after **Volta**'s invention of the electric battery, Berzelius began in 1802 to experiment with the voltaic pile, which was to revolutionize physics and chemistry by providing scientists with a reliable, continuous source of current for the first time. Working with Wilhelm Hisinger he discovered, by 1803, that all salts are decomposed by electricity. Berzelius went on to suggest that all compounds, organic as well as inorganic, are made up of positive and negative components, a theory which laid the foundations for our understanding of radicals. In 1803 he and Hisinger discovered cerium. Later Berzelius also discovered selenium and thorium; he was the first person to isolate silicon, zirconium and titanium, and made detailed studies of other rare metals. His greatest achievement, however, was his contribution to atomic theory. His work with salts led him to consider whether elements were always present in compounds in fixed proportions, a question much debated at the time, eg by Berthollet and **Proust**. To this end he analysed some 2000 compounds between 1807 and 1817. Matching the idea of constant proportions with **Dalton**'s atomic theory, and persuaded of the central importance of oxygen from his studies of **Lavoisier**'s work, he drew up a table of atomic weights using oxygen as a base. In 1818 he published the weights of 45 of the 49 elements then known, revising this table in 1828. In the course of this research he devised the modern system of chemical symbols to replace the chaotic system which had grown piecemeal over a long period of time. Berzelius also made significant contributions to organic chemistry, for example by his analysis of organic acids. He studied isomerism and catalysis, phenomena which both owe their names to him, as does protein which Berzelius recognized as fundamental to all life and named from the Greek *proteios*, meaning 'primary'. A pioneer of gravimetric analysis, Berzelius has had few rivals as an experimenter. As a result of the poverty of his early years, he had to improvise much of his apparatus, and some of his innovations are still standard laboratory equipment, eg wash bottles, filter paper and rubber tubing. He also greatly improved blowpipe techniques. Despite the fact that he suffered ill-health for most of his life, he also managed to write extensively and his works were translated into many languages. He wielded enormous influence until the last two decades of his life, which were marred by controversies with younger chemists. Berzelius's achievements brought him many honours. He was elected to the Stockholm Royal Academy of Sciences in 1808 and became its secretary in 1818. He was awarded the Gold Medal of the Royal Society of London, and in 1835 he was created a baron by Charles XIV. He died in Stockholm.

BESSEL, Friedrich Wilhelm

(1784–1846)

German mathematician and astronomer, born in Minden. Starting as a ship's clerk, in 1810 he was appointed director of the observatory and professor at Königsberg. He catalogued stars, and was the first to identify the nearest stars and determine their distances. He predicted the existence of a planet beyond Uranus as well as the presence of dark stars, and investigated subtle perturbations of a planet on the motion of the Sun. In the course of this work he systematized the mathematical functions involved, which today bear his name, although **Fourier** had worked upon them earlier. He also corresponded with **Gauss** on such matters as the possibly non-Euclidean nature of space.

BESSEMER, Sir Henry

(1813–98)

English metallurgist and inventor, born in Charlton, near Hitchin in Hertfordshire. A self-taught man, he learned metallurgy in his father's type foundry, and at the age of 17 set up his own business in London to produce small castings and art work. Three years later he exhibited at the Royal Academy, having at the same time improved his education by part-time study, and patented several inventions. In 1840 he developed a method for the production of bronze powder and 'gold' paint, at a fraction of the cost of the long-established German process; the profits from this enterprise enabled him to set up his own small iron-works at St Pancras in London. In 1855, as a result of his efforts to find a method of manufacturing stronger gun-barrels for use in the Crimean War (1853–6), he patented an economical process by which molten pig-iron can be turned directly into steel by blowing air through it in a 'Bessemer converter'. Further essential improvements were initiated by **Mushet** (around 1856) and **Sidney Gilchrist Thomas** (1878). Bessemer established a steelworks at Sheffield in 1859, specializing at first in armaments, and later expanding to meet the worldwide demand for steel rails, locomotives and bridges. He was elected FRS in 1877 and knighted in 1879. Other English steelmasters were reluctant to accept Bessemer's process, but in the USA entrepreneurs like Andrew Carnegie made a fortune from it.

BEST, Charles Herbert

(1899–1978)

Canadian physiologist, born in West Pembroke, Maine, the son of a physician. Best graduated in physiology and biochemistry from the University of Toronto in 1921. As a research student at Toronto in 1921 he helped **Banting** to isolate the hormone insulin, used in the treatment of diabetes. Banting divided his share of the 1923 Nobel Prize for Physiology or Medicine with Best. In 1925 Best obtained a medical qualification and in 1929 succeeded **Macleod** as Professor of Physiology at Toronto. When Banting was killed in 1941, Best replaced him as Director of the Banting and Best department of medical research. Besides his involvement with the isolation of insulin, he also enjoyed considerable success during his later career. He discovered choline (a vitamin that prevents liver damage) and histaminase (the enzyme that breaks down histamine), and introduced the use of the anticoagulant, heparin. He continued to work on insulin, showing in 1936 that the administration of zinc with insulin can prolong its activity.

BETHE, Hans Albrecht

(1906–)

German-born American physicist, born in Strassburg (now Strasbourg, France). The son of an academic, he was educated at the universities of Frankfurt and Munich. He taught in Germany until 1933 when he moved first to England and then to the USA, where he held the Chair of Physics at Cornell (1935–75) until his retirement. During World War II he was director of theoretical physics for the atomic bomb project based at Los Alamos. In 1939 he proposed the first detailed theory for the generation of energy by stars through a series of nuclear reactions known as the carbon cycle. This is a series of six nuclear reactions involving carbon, nitrogen and oxygen by which protons are converted to helium and energy is released. Our Sun is not hot enough for the carbon cycle to play an important role in its energy production, but hotter stars generate much of their energy in this way. In 1967 Bethe was awarded the Nobel Prize for Physics for this work. He also contributed with **Alpher** and **Gamow** to the 'alpha, beta, gamma' theory of the origin of the chemical elements during the early development of the universe. He later worked on models of supernovae.

BEVAN, Edward John

(1856–1921)

English industrial chemist, born in Birkenhead. After a private education he studied chemistry at Owens College, Manchester, under **Roscoe**. Subsequently he and a fellow pupil from his Manchester days, **Cross**, set up as industrial analysts and consultants with an office in London. Cross and Bevan became known as the leading experts in the field of cellulose chemistry. Building on the work of **Swan** and **Chardonnet**, they patented a process for making viscose yarn in 1892 and film in 1894. In both cases wood pulp was dissolved in caustic soda solution and carbon bisulphide. It was then extruded into a bath of sulphuric acid, either through very fine nozzles or a thin slit. The cellulose regenerated in the acid, forming either thread or 'cellophane' film. This process laid the foundations of the synthetic textile industry. Cross and Bevan also initiated many improvements in paper-making.

BEVERTON, Raymond John Heaphy

(1922–)

English marine biologist, born in London. After graduating from Downing College, Cambridge (1947), he began his career as Research Officer at the Fisheries Laboratory, Lowestoft (1947–59), and subsequently became Deputy Director there (1959–65). He was later appointed Secretary and Chief Executive of the newly established Natural Environment Research Council (1965–80) and Professor of Fisheries Ecology at the University of Wales Institute of Science and Technology (1984–7). Beverton's early work at Lowestoft concerned the North Sea plaice population; he was impressed by its remarkable constancy, despite extremely high and variable natural mortality rates of early post-larval stages of fish. He concentrated his efforts in elucidating density-dependent control mechanisms (eg mortality rate changes) in the early life-history stages. In predicting the maximum sustainable yield from a natural stock one of the major difficulties lay in the problem of growth overfishing—how to allow smaller fish to achieve their adult growth potential before harvesting them. This was effectively modelled in Beverton's classic text *On the Dynamics of Exploited*

Fish Populations (1957, with Sidney J Holt); his 'yield per recruit' and subsequent 'stock and recruitment' models provided fisheries scientists with a convenient and elegant means of predicting the commercial fishing pressure that natural stocks could withstand. He was appointed CBE in 1968 and elected FRS in 1975.

BHABHA, Homi Jehangir
(1909–66)

Indian physicist, born in Bombay and educated there before entering Cambridge (1927) where his tutor was **Dirac**. In India during World War II, a readership was created for him at the Indian Institute of Science in Bangalore, where he became professor in 1941, the same year as he was elected FRS. In 1945 he became Director of the Tata Institute for Fundamental Research in Bombay and then Director of the Indian Atomic Energy Commission. When this was reorganized as the Department of Atomic Energy (1948), he was its first secretary. In the same year he was appointed Director of the Atomic Energy Research Centre, renamed the Bhabha Atomic Research Centre after his death in a plane crash. In his early career he derived a correct expression for the cross-section (probability) of scattering positrons by electrons, a process now known as Bhabha scattering. He co-authored a classic paper on the theory of cosmic-ray showers (1937), which described how primary cosmic rays (mainly high-energy protons from outer space) interact with the upper atmosphere to produce the particles observed at ground level. This paper also demonstrated that a very penetrating component of cosmic rays observed underground could not be electrons. These particles were identified as muons (1946). In 1938, in a letter to *Nature*, Bhabha was the first to conclude that the lifetimes of fast, unstable cosmic-ray particles, such as muons, would be increased due to the time dilation effect that follows from **Einstein**'s special theory of relativity. The experimental verification of this is often cited as one of the most straight-forward pieces of evidence supporting special relativity.

BICHAT, Marie François Xavier
(1771–1802)

French physician, born in Thoirette, Jura. Like his father, Bichat studied medicine, first in Lyons, but his education was interrupted by military service during the Revolution. Bichat settled in Paris in 1793 at the height of the Terror. From 1797 he taught medicine, from 1801 working at the Hôtel-Dieu, Paris's huge hospital for the poor. His greatest contribution to medicine and physiology was his perception that the diverse organs of the body contain particular tissues or (his favourite word) 'membranes'. He described 21 such membranes, including connective, muscle, and nerve tissue. Bichat maintained that in the case of disease in an organ, it was generally not the whole organ but only certain tisues which are affected. He did not make use of the microscope, which he distrusted; for this reason, his analysis of tissues did not include any perception of their cellular structure. Bichat established the centrality of the study of tissues (histology). Performing his researches with great fervour during the last years of his short life—he performed over 600 postmortems—Bichat may be seen as a bridge between the morbid anatomy of **Morgagni** and the later cell pathology of **Virchow**. His lasting importance lay in simplifying anatomy and physiology, by showing how the complex structures of organs could be grasped in terms of their elementary tissues.

BIELA, Wilhelm von
(1782–1856)

Austrian infantry captain, born in Rossia. He began to study astronomy at Prague's Charles University in 1815, after the conclusion of the Napoleonic Wars. In 1826 he was garrisoned in Josefovm, eastern Bohemia, and on 28 February discovered (probably by accident) the comet which bears his name. The comet had previously been discovered by Jacques Leibax Montaigne at its 1772 apparition and by **Pons** in 1825. **Gauss** computed the orbit, and the period eventually turned out to be 6.75 years.

BIERMANN, Ludwig
(1907–86)

German astronomer, born in Hamm and educated at the University of Berlin, where he graduated in 1932 and was later a lecturer (1938–45). He led the astrophysics division of the Max Planck Institute for Physics and Astrophysics, first at Göttingen (1947–58) and as the Institute's Director at Munich (1858–75). Biermann was a theoretician whose earliest researches were in the field of stellar atmospheres, leading to studies of magnetic phenomena in sunspots and the solar corona. His most striking achievement was his theoretical prediction (1951) of the existence of a continuous stream of high-speed particles flowing from the Sun, later called the solar wind, as the driving force behind the ionic tails of comets. The reality of the solar wind was confirmed *in situ* by instruments aboard the Soviet Luna rockets in 1959. Biermann was awarded the Gold Medal of the Royal Astronomical Society in 1974.

BIFFEN, Sir Rowland Harry
(1874–1949)

English botanist and geneticist, born in Cheltenham. Educated at Cambridge, he travelled in Brazil and the West Indies studying natural sources for rubber. He was appointed as a demonstrator in botany at Cambridge in 1898 and in 1908 became the first Professor of Agricultural Botany there, a post which he retained until 1931. He was also Director of the Plant Breeding Institute at Cambridge from 1912 to 1931. Using Mendelian genetic principles, he pioneered the breeding of hybrid rust-resistant strains of wheat. He wrote *The Auricula*, published posthumously in 1951, and was knighted in 1925.

BIGELOW, Jacob
(1787–1879)

American physician and botanist, born in Sudbury, Massachusetts. He was educated at Harvard and Pennsylvania, and practised medicine in Boston. He held several professorships at Harvard, and was associated with the compilation of the single-word nomenclature of the American Pharmacopoeia of 1820, subsequently adopted in England. He published *Florulae Bostoniensis*, a study of all of the plants growing within a 10 mile radius of Boston. With Francis Booth he explored New Hampshire and Vermont to extend the range of his work. The enlarged and corrected work became a standard manual of botany for many years. His three-volume *American Medical Botany* (1817–20) described the medicinal properties of many indigenous species. The genus *Bigelovia* is named after him.

BINNIG, Gerd Karl
(1947–)

German physicist, developer of the scanning tunnelling microscope, born in Frankfurt. He was educated at Goethe University in Frankfurt and on completion of his PhD joined IBM's Zürich research laboratories (1978), where he became a group leader in 1984. In his first year there he started work with **Rohrer** on a new form of high-resolution electron microscope that made use of the tunnelling electron current between a scanning needle with a tip of just one atom and the surfacc of a sample to profile the sample's surface. Within only three years they had built a microscope that could show differences in vertical position as small as 0.1 angstrom, one-thirtieth the size of an average atom, and resolve features horizontally separated by 6 angstroms. For this work Binnig and Rohrer shared the 1986 Nobel Prize for Physics with **Ruska**, who had invented the electron microscope some 55 years earlier. The scanning tunnelling electron microscope, an important piece of research equipment, allows atom by atom inspection of surfaces and has been particularly useful in the development of ever smaller electronic circuits.

BIOT, Jean-Baptiste
(1774–1862)

French physicist and astronomer, born in Paris. His father was keen for him to have a career in commerce, but Biot instead enlisted in the army in the artillery. On leaving, he entered the École Polytechnique where he was soon noticed for his outstanding mathematical abilities. After several teaching posts, he was appointed Professor of Mathematical Physics at the Collège de France in 1800, and became a senior member of the Academy of Sciences in 1808. He made a balloon ascent with **Gay-Lussac** to study magnetism at high altitudes in 1804. Two years later he was appointed assistant astronomer at the Bureau des Longitudes and he travelled to Spain with **Arago** to determine the length of a degree of longitude. He invented a polariscope and established the fundamental laws of the rotation of the plane of polarization of light by an optically active substance. For his work on optical activity he was awarded the Royal Society's Rumford Medal in 1840. Other subjects he investigated were the refractive indices of gases, the composition of meteorites, the electrochemical behaviour of the voltaic pile, and the distribution of electricity on the surface of regular and irregular spheriods, which led to a laboratory demonstration device known as 'Biot's apparatus', consisting of a polished brass ball on an insulated stand, with two close-fitting brass hemispheres. His son, Édouard Constant (1803–50), was a Chinese scholar.

BIRCH, Arthur John
(1915–)

Australian chemist, born in Sydney. Having graduated from the University of Sydney in 1936, he obtained an 1851 Exhibition scholarship to work with **Robinson** at Oxford, where he obtained his DPhil in 1941. After research fellowships at Oxford, he moved to Cambridge where he remained until his appointment as professor at the University of

Sydney in 1952. In 1955 he was appointed professor at the University of Manchester, but he returned to Australia in 1967 to the National University in Canberra. He retired as its Foundation Professor of Organic Chemistry in 1980. He used the study of natural products to explore a number of mechanistic and synthetic themes. He is most famous for the Birch reduction using metal-ammonia solutions, a technique of great value in the synthesis of a number of natural products, notably 19-nortestosterone. He studied the biosynthesis of phenolic natural products. A member of a number of important Australian and Commonwealth scientific committees, he was elected FRS in 1958.

BIRCH, Francis

(1903–88)

American physicist and geophysicist, born in Washington DC. He was educated at Harvard, gaining an SB (1924) in electrical engineering, an AM (1929) and a PhD in physics (1932). He first worked for the New York Telephone Company (1924–6), and later returned to Harvard where he became assistant professor in geophysics (1937–43) and associate professor of geology (1943–9). During World War II he was recruited as an engineer by the Radiation Laboratory at MIT, moving to Los Alamos (1942–4) to work on the atomic bomb which was dropped on Hiroshima. He rose to the rank of Commander in the US Naval Reserve. In 1949 he became Sturgis Hooper Professor of Geology at Harvard and Emeritus Professor in 1974. His major contributions to geophysics have been on the response of materials at extremely high pressures and temperatures, and relating them to the composition of the Earth's interior.

BIRCHALL, (James) Derek

(1930–)

English industrial chemist, born in Leigh, Lancashire, who joined ICI in 1957. Birchall's early discoveries included the mechanism of the effect of ferricyanide ion on the growth and nucleation of solids. His group at ICI have been responsible for numerous innovations including novel inorganic fire-extinguishing agents, inorganic fibres, fillers for elastomers and new phosphate complexes. He has also worked on the toxic effect of aluminium, the enhancement of such toxicity by silicon and the effect of the balance of these two elements. Elected FRS in 1982, he received an OBE in 1990. Other awards include the Royal Society of Chemistry awards for Solids and Materials Processing (1990) and Materials Science (1991), and the Royal Society of Chemistry Industrial Lectureship (1991–2). He is currently Professor of Inorganic Chemistry at Keele University.

BIRD, Adrian Peter

(1947–)

English molecular biologist. He was educated at the University of Sussex and the University of Edinburgh, where he obtained his PhD in 1971. After postdoctoral research at Yale and the University of Zürich, he joined the scientific staff of the Medical Research Council Mammalian Genome Unit in Edinburgh in 1975, and later became acting director there (1986–7). From 1988 to 1990 he was Senior Scientist at the Institute for Molecular Pathology in Vienna, and in 1990 he returned to Edinburgh as Buchanan Professor of Genetics. He was elected FRS in 1989. Bird discovered the phenomenon known as 'CG islands'. In DNA, CG sequences are unevenly distributed in the genome; they are present at 10 to 20 times their average density in regions 1000 to 2000 nucleotides long called CG islands. These islands surround the promoters of 'housekeeping genes', those genes which code for proteins required in all cell types as opposed to 'luxury function' genes which are only expressed in specific cell types. The mammalian genome contains an estimated 30000 CG islands, mostly marking the ends of transcription units. Since it is possible to clone specifically the DNA surrounding the CG islands, it is therefore relatively easy to identify and characterize 'housekeeping' genes.

BIRKELAND, Kristian Olaf Bernhard

(1867–1917)

Norwegian physicist, born in Christiania (now Oslo). He studied physics in Paris, Geneva, and briefly under **Heinrich Hertz** in Bonn. In 1898 he was appointed to the Chair of Physics at the University of Christiania. Later he moved to Cairo, partly for health reasons, but also to resume his astronomical observations. He died in Tokyo while returning to Norway by a roundabout route because of World War I. Birkeland demonstrated the electromagnetic nature of the aurora borealis, and in 1903 developed a method for obtaining nitrogen from the air with Samuel Eyde (1866–1940), the founder of the Norwegian electrochemical industry. This was as a consequence of a concern at this time that the natural nitrate (fertilizers) deposits were being depleted rapidly, and it was also a requirement of the growing explosives industry. The key ingredient nitric acid could be obtained by 'fixing' atmospheric nitrogen. The Birkeland–Eyde process was the first large-scale process which exploited the cheap electricity generated by the Norwegian hydro-electric plants. It was superseded shortly before World War I by the Haber–Bosch process.

BIRKHOFF, George David

(1884–1944)

American mathematician, born in Overisel, Michigan. He studied at Harvard and Chicago, and was professor at Wisconsin (1902–9), Princeton (1909–12) and Harvard (1912–39). He had many contacts with European mathematicians and was regarded as the leading American mathematician of the early part of the 20th century. In 1913 he proved '**Poincaré**'s last theorem', which Poincaré had left unproven at his death. This was a crucial step in the global geometrical analysis of motions determined by differential equations, opening the way to modern topological dynamics. Later he developed ergodic theory, which has its roots in such problems as Poincaré's theory of the stability of the solar system and the statistical analysis of motion in thermodynamics. In ergodic theory the methods of probability theory are applied to the path of a particle which it would be impossible to analyse exactly (the motion of a molecule of oxygen in a box, for example). Motion is said to be ergodic if the particle visits every region of the space in which it is confined with certainty.

BISHOP, (John) Michael

(1936–)

American molecular biologist and virologist, born in York, Pennsylvania. He was educated at Gettysburg College where he obtained a BA in 1957, and at Harvard Medical School where he was awarded his MD in 1962. He began his career as an intern then resident in general medicine at Massachusetts General Hospital (1962–4) and subsequently became a research associate in virology at the National Institutes of Health in Washington DC from 1964 to 1966. He remained in Washington as assistant then associate professor until he moved to San Francisco in 1972. He has been Professor of Microbiology and Immunology at the University of California Medical Centre since 1972, and Director of the G W Cooper Research Foundation since 1981. Bishop has received many major awards, including the Nobel Prize for Physiology or Medicine in 1989 (jointly with **Varmus**) for his discovery of oncogenes. Oncogenes are normal cellular genes whose product has some role in the normal growth and development of all mammalian cells. However, certain faults in oncogene regulation, for example, overproduction as a result of the influence of a virus, or faulty expression due to a mutation within the gene, can severely damage the growth and differentiation of the affected cell type and thus cause cancer. An understanding of the function of oncogenes is therefore a crucial step in combating all types of cancer.

BJERKNES, Jacob Aall Bonnevie

(1897–1975)

Norwegian–American meteorologist, born in Stockholm. He started his career as Carnegie assistant with his father **Vilhelm Bjerknes** at Leipzig (1916) and contributed to their joint paper *On the Structure of Moving Cyclones* (1919), produced while at the Bergen Geophysical Institute where he later became professor. He realized cyclones contained two lines of convergence, later recognized as warm and cold fronts, and produced cross-sections of a cyclone and discussed their life cycle (1921–2). He also postulated that cyclones move parallel to the isobars in the warm sector. In 1940 he formed the meteorological school for US Air Force weather officers at the University of California and was made the first Professor of Meteorology at the University of California at Los Angeles. He began research into problems of ocean–atmosphere interaction in 1959, discovering that changes in wind stress could cause changes in ocean currents and conversely, that unusual sea temperature anomalies would result in atmospheric changes. In particular, he studied tropical sea temperature anomalies in the Pacific and showed that they affected the subtropical anticyclones and hence the temperate latitude westerlies. He was active internationally, and was awarded the Symons Gold Medal of the Royal Meteorological Society (1940) and the International Meteorological Organisation Prize (1959).

BJERKNES, Vilhelm Friman Koren

(1862–1951)

Norwegian mathematician, meteorologist and geophysicist, born in Christiania (now Oslo). After graduating from the University of Christiania (MS 1888), he worked in Bonn for two years as assistant and collaborator to **Heinrich Hertz**, then returned to Norway where he received his PhD in 1892. He was appointed lecturer (1893), then Professor of Applied Mechanics and Mathematical Physics (1895) at the University of Stockholm. Bjerknes initially collaborated with his father (a mathematics professor) on theories of hydrodynamical forces. While continuing this work at Stockholm, he produced his famous circulation theorem (1898); he applied his thorough knowledge of hydrodynamics to a study of atmospheric and ocean processes, and realized that it is necessary to take into account both thermodynamic and dynamic processes for a complete understanding of weather systems. He devised equations which enabled both the thermal energy and that due to baroclinicity to be calculated for a developing cyclone. In 1907 he returned Norway, and collaborated with **Svedrup** on dynamic meteorology. In 1912 he and Svedrup moved to Leipzig where Bjerknes became Professor of Geophysics, though soon after **Nansen** persuaded him to set up a geophysical institude in Bergen, where until 1926, he spent the most productive years of his life, writing his classic work *On the Dynamics of the Circular Vortex with Application to the Atmosphere and to Atmospheric Vortex and Wave Motion* (1921).

BJERRUM, Niels Janniksen
(1879–1958)

Danish physical chemist, born in Copenhagen. He obtained his doctorate at the University of Copenhagen in 1908 and worked in various continental laboratories in Leipzig, Zürich, Paris and Berlin. In 1914 he became Professor of Chemistry in the Royal Veterinary and Agricultural College, Copenhagen, a post he held until his retirement in 1949. Bjerrum made major contributions to several areas of physical chemistry. His early work on the complex ions of chromium led him to suggest in 1909 that strong electrolytes are completely dissociated in solution and that interionic forces are important. In 1916 he formulated these ideas in greater detail. Such considerations led in 1923 to the quantitative theory of **Debye** and **Hückel**. Bjerrum later showed that various defects of this theory could be remedied by allowing for the intervention of ion-pairing. In 1925 he devised a more explicit derivation of the equation suggested by **Brønsted** for salt effects on the rates of reactions involving ions. He also worked on the quantum theory of specific heats of gases, on the thermodynamics and experimental techniques for the measurement of voltages of electrochemical cells, on the application of physicochemical methods to agricultural problems, and on acid–base equilibria, particularly those involving zwitter-ions, molecular entities which carry both negative and positive integral charges.

BJORKEN, James David
(1934–)

American theoretical physicist, born in Chicago. Educated at MIT and Stanford University, he has held posts at Stanford University, Stanford Linear Accelerator Center (1959–79 and since 1989) and Fermilab (1979–89). Bjorken's best-known work has been in deep inelastic scattering theory, which describes what is observed when high-energy leptons (eg electrons) are scattered by nucleons (protons and neutrons). On the basis of current algebra, developed by **Gell-Mann**, he predicted in the early 1960s that for large momentum transfers, ie close collisions, the deep inelastic cross-section would 'scale'. Scaling means that the structure functions used to describe the internal structure of nucleons would only depend on dimensionless variables, in direct contrast with elastic scattering where the cross-section falls dramatically at high momentum transfers. The experimental data taken as part of **Jerome Friedman**, **Henry Kendall** and **Richard Taylor**'s research, for which they were awarded the 1990 Nobel Prize for Physics, was in good agreement with Bjorken's predictions. **Feynman** interpreted the structure function data as being simply momentum distributions of point-like constituents (which he called partons, now known as quarks) moving inside the nucleons.

BLACK, Sir James Whyte
(1924–)

Scottish pharmacologist, born in Uddingston. After graduating in medicine at St Andrews University he became an assistant lecturer in physiology there (1946–7), a lecturer at the University of Malaya and then senior lecturer at Glasgow Veterinary School (1950–8). Following appointments at ICI Pharmaceuticals (1958–64) and the Wellcome Research Laboratories (1978–84) as Director of Therapeutic Research, he was appointed Professor of Pharmacology and departmental head at University College London (1973–7); since 1984 he has held similar posts at King's College Medical School, London. In 1962 while at ICI, he discovered the drug netherlide (Alderlin®) which, by binding to β adrenergic receptors, blocks the action of catecholamines in changing cardiac tension without inducing the sympathomimetic activity (which causes brachycardia) associated with other drugs. This opened the way to new treatments for certain types of heart disease (angina, tachycardia) and led to his development of the safer, more effective drug propranolol. At the Wellcome Laboratories Black synthesized burimamide and cimetidine, new drugs with which he distinguished two classes of histamine receptor, enabling specific suppression of stomach acid secretion for the treatment of ulcers. He was elected FRS in 1976 and knighted in 1981, and shared with **Elion** and **Hitchings** the 1988 Nobel Prize for Physiology or Medicine.

BLACK, Joseph
(1728–99)

Scottish chemist who discovered carbon dioxide and the phenomena of latent and specific heat, one of the most influential chemists of his generation. He was born in Bordeaux, where his father was a wine merchant, and after attending school in Belfast studied medicine at Glasgow and Edinburgh universities. In 1756 Black succeeded **Cullen** as Professor of Medicine at Glasgow, and 10 years later he moved to the Chair of Medicine and Chemistry at Edinburgh. In his MD thesis of 1754, Black showed that the causticity of lime and the alkalis is due to the loss of 'fixed air' (carbon dioxide), which is present in limestone and the carbonates of these alkalis, but which is driven off by heating. Black found that this new air supported neither life nor combustion, and with this discovery was the first person to realize that there are gases other than air, a fundamental advance which was to dramatically affect the future development of chemistry. At the same time, his experimental method, in which he weighed the amounts of carbonates and alkalis accurately at every stage, laid the foundations of quantitative analysis. Black then turned his attention to changes of state. Having observed that ice takes a long time to melt, even on a sunny day, he showed experimentally that solids need heat to turn them into a liquid at the same temperature and also that the heat required for any given mass is unique to every substance. In the same way, liquids need characteristic amounts of heat to turn them into gases. Black called the heat necessary for changes of state 'latent heat'. He also discovered that the same weights of different substances require different but constant amounts of heat to raise them through the same temperature difference, a phenomena he christened 'specific heat'. Black was a successful physician and industrial consultant and, in his Edinburgh days, was widely consulted on problems such as bleaching and dyeing, iron making, ore analysis, fertilizers and water supplies. Famed as a teacher, his chemistry classes drew students from all over the UK and from Europe and America, contributing largely to the lustre of the university in the greatest days of the Scottish Enlightenment. He died in Edinburgh.

BLACKETT, Patrick Maynard Stuart, Baron
(1897–1974)

English physicist, born in London. Educated at Dartmouth College, he served in the Royal Navy during World War I. He then entered Magdalene College, Cambridge, and studied physics at the Cavendish Laboratory. He was the first to photograph, in 1925, nuclear collisions involving transmutation. He was professor at London University (1933–7) and Manchester University (1937–53), and subsequently returned to London as professor at the Imperial College of Science and Technology (1953–65). Blackett was awarded the Nobel Prize for Physics in 1948 for developing the Wilson cloud chamber and his use of it to confirm the existence the positron (the antiparticle of the electron). He pioneered research on cosmic radiation, and in World War II, operational research. He also contributed to the theories of particle pair production, and the discovery of 'strange' particles, so called because they contain a 'strange' quark.

BLACKMAN, Frederick Frost
(1866–1947)

English botanist, born in Lambeth, London. While at Mill Hill School, he started a herbarium, so showing an early interest in botany. He studied medicine at St Bartholomew's Hospital, graduating in 1885. However he then went to Cambridge and studied natural sciences at St John's College. Thereafter he worked at the Cambridge Botany School (1891–1936) where his lucid lectures attracted many students to plant physiology. He was renowned for his fundamental research on the respiration of plants (published in two series of papers: *Experimental Researches in Vegetable Assimilation and Respiration*, begun in 1895, and *Analytical Studies in Plant Respiration*, commencing 1928), and on the limiting factors affecting their growth. His first paper described a complex apparatus for precisely determining the quantities of carbon dioxide in small samples of a gaseous mixture. Using this, he showed conclusively for the first time that the bulk of the exchange of CO_2 between leaves and air was via the stomata. In 1928 he published a classic paper on the effects of change in the partial pressure of oxygen, in which he coined the term oxidative anabolism for the process whereby the products of glycolysis are used for the reformation of carbohydrates.

BLAKESLEE, Albert Francis

(1874–1954)

American botanist, born in Genesco, New York, and educated at the East Greenwich (Rhode Island) Academy, Wesleyan University and Harvard University. At Harvard, he discovered heterothallism in *Mucor* (bread mould), showing that sexual reproduction is successful only between individuals of different physiological types. He continued his work on moulds in Germany before returning to Harvard for a year and then working as professor of botany at Connecticut Agricultural College, Storrs. He distinguished himself there as a teacher, researcher, observer and organizer, and pioneered a method of distinguishing between laying and non-laying hens by the colour of their beaks and feet. There, he first became interested in the genus *Datura*; this would ultimately become a lifelong devotion. From 1915 to 1942 he worked at Cold Spring Harbor, teaching genetics and continuing breeding experiments in *Datura*. In so doing, he found the first *Datura* haploid, the first sectorial chimaera, and the first species hybrid. In 1937 he established that treatment with colchicine can bring about polyploidy in plants. From 1936 he was Director of the Carnegie Station for Experimental Evolution, and in 1942 he moved to Smith College Genetics Experiment Station, Northampton, Massachusetts, as its first director.

BLALOCK, Alfred

(1899–1964)

American surgeon, born in Culloden, Georgia. He received his medical education at Johns Hopkins University, and his postgraduate training there and at Vanderbilt University Hospital. He joined the staff at Vanderbilt (1925–41) and Johns Hopkins (1941–64), where he pioneered the surgical treatment of various congenital defects of the heart and its associated blood vessels, many of which could be recognized by the presence of cyanosis in infants, and performed the first 'blue baby' operation, cooperating with the paediatrician **Taussig**. He also did important experimental work on the pathophysiology of surgical shock and its treatment by transfusion of whole blood or blood plasma, and was the first to treat *myasthenia gravis* by removal of the thymus gland. His collected scientific papers were published in 1966, edited by one of his pupils.

BLOCH, Felix

(1905–83)

Swiss-born American physicist, born in Zürich. He originally intended to be an engineer and entered the Swiss Federal Institute of Technology. Later he switched to physics and went to the University of Leipzig where he obtained his PhD in 1928. For his PhD, he solved Schrödinger's equation for the motion of electrons in a regular lattice of positive ions to explain the conduction of metals. This gave rise to the band model of solids which forms the basis of much of solid-state physics. He left Germany for the USA in 1933 and became Professor of Theoretical Physics at Stanford University (1934–71). During World War II he worked on radar, and after the war he developed the technique of nuclear magnetic resonance (NMR). This method has several important applications and has become a useful tool in chemistry and biology; it can provide information about nuclei, including the magnetic moment and angular momentum, and it is sensitive to the environment of the nuclei. He was awarded the 1952 Nobel Prize for Physics jointly with **Purcell** for this work. He was the first Director-General of CERN, the European centre for nuclear research in Geneva (1954–5).

BLOCH, Konrad Emil

(1912–)

German-born American biochemist, born in Neisse (now Nysa in Poland). Educated at the Technische Hochschule, Munich, and Columbia University, he emigrated to the USA in 1936. In 1954 he was appointed the first Professor of Biochemistry at Harvard University. Bloch recognized that glucose could be formed from pyruvate but not acetate, except in certain bacteria. This finding underlies our present-day understanding that, in animals, fatty acids cannot be converted into sugars, giving instead ketone bodies. In 1943 he revealed the direct metabolic relationship between cholesterol and bile acids by administering radioactively labelled cholesterol to a dog and isolating cholic acid from the urine. In 1950 he began investigating how many of the 27 carbon atoms of cholesterol originated from acetate. During this work it was found that the mould *Neurospora*, which forms the steroid ergosterol, required acetate for growth. Further studies of the nutritional requirements of acetate-dependent bacteria resulted in the recognition of mevalonic acid as the first-formed building block, and **Cornforth** and George Popják implicated an acyclic polyisoprenoid precursor (identified by others as squalene—a compound first isolated from shark liver oil) in forming the ring structure. For his work on cholesterol, Bloch shared the 1964 Nobel Prize for Physiology or Medicine with **Lynen**.

BLOEMBERGEN, Nicolaas
(1920–)

Dutch-born American physicist, born in Dordrecht. Educated at the universities of Utrecht and Leiden, he received his PhD in 1946 then joined the staff of Harvard, where he has been Gordon McKay Professor of Applied Physics since 1957. His early interest was in nuclear magnetic resonance which he studied under Nobel laureate **Purcell**. He later pioneered methods of three-level and multi-level pumping to energize masers, introducing a modification to **Townes**'s early design enabling the maser to work continuously rather than intermittently. His work on non-linear interactions of radiation with matter yielded a theoretical framework which still remains in place. He also investigated methods of using laser light to selectively excite and break a single bond in a molecule. For his many years of research that contributed to the development of the laser, an invaluable tool for probing the structure of matter, he shared the 1981 Nobel Prize for Physics with **Schawlow** and **Kai Siegbahn**. His books included *Nonlinear Optics* (1965).

BLUMBERG, Baruch Samuel
(1925–)

American biochemist, born in New York City, joint winner of the 1976 Nobel Prize for Physiology or Medicine with **Gajdusek**. He studied at Columbia University and at Oxford, and became Professor of Biochemistry at the University of Pennsylvania in 1964. Blumberg discovered the 'Australia antigen' in 1964 and reported its association with hepatitis B. The finding was very rapidly applied to screening blood donors. His study of the distribution of the HBV virus, as it is known, in the population revealed that apparently healthy people could carry and transmit the live virus, and led to ethical and employment problems associated with screening nurses, physicians and other welfare employees. HBV is widespread in Vietnam, Thailand and elsewhere in south-east Asia, and further problems emerged over the adoption of Vietnamese children in America. In 1969 Blumberg introduced a protective vaccine, now widely used. It was also discovered that persistent infection with HBV is associated with nearly all primary hepatocellular carcinomas; effective treatment has now been achieved.

BLUME, Karel Lodewijk
(1796–1862)

German botanist and physician, born in Brunswick. He was appointed as head of the vaccination programme in Java in 1818, combining this post with the assistant directorship of Buitzenborg (now Bogor) Botanic Garden until it closed in 1826. Returning to the Netherlands in 1829, he was the founder and first director of the Rijksherbarium, first in Brussels and then in Leiden. While in Java he collected widely, most of the material following him to Europe. With Philipp Franz von Siebold he founded the Royal Dutch Society for the Advancement of Horticulture (1842). He was the author of many important works on the flora of Java, including *Enumeratio Plantarum Javae* (1827–8), *Flora Javae* (1828), and the four-volume *Rumphia* (1835–48). His name is commemorated in the genus *Blumea*, and in the title of a botanical journal, also called *Blumea*. His work was an essential basis for later studies of the plant geography of this botanically very important region.

BODE, Johann Elert
(1747–1826)

German astronomer, born in Hamburg, where he founded (1774) the *Astronomisches Jahrbuch*. He became Director of the Berlin Observatory (1786), where he remained until a year before his death. He is best known for the empirical rule known as Bode's law which expresses the proportionate distances of the planets from the Sun in terms of a sequence of numbers. The rule, alternatively called the Titius–Bode law, was first discovered in 1766 by Johann Daniel Titius, an astronomer of Wittenberg, and brought into use by Bode. It proved useful in stimulating a search for a missing planet between Mars and Jupiter where the asteroids were actually found to lie. Bode is also remembered as the one who proposed the name Uranus for that planet discovered by **William Herschel** in 1781.

BODENSTEIN, Ernst August Max
(1871–1942)

German physical chemist, born in Magdeburg. He studied under **Viktor Meyer** at Heidelberg and **Nernst** at Göttingen, obtaining his doctorate in 1893. He later worked with **Ostwald** in Leipzig, becoming Titular Professor there in 1904. In 1906 he moved to Berlin, but in 1908 he was appointed Professor of Physical Chemistry at the Technische Hochschule in Hanover. He returned to the University of Berlin in 1923 as Professor and Director of the Institute for Physical Chemistry, a post he held until his retirement in 1936. Bodenstein may be regarded as a founding-father of gas kinetics, in which he developed great experimental skill. Much of his work concerned the reactions between hydrogen and the halogens. In the case of the iodine system, his results appeared to show an almost pure bimolecular mechanism (later it emerged that this was an over-simplification), while the results for bromine and particularly for chlorine were much more complicated. Their interpretation required the concept of chain reactions developed by **Christiansen** and **David Chapman**'s steady-state treatment of reaction intermediates. These and related studies laid foundations for the researches of **Hinshelwood** and **Semenov** and of many later workers. Bodenstein also made important studies of heterogeneous reactions.

BODMER, Sir Walter Fred
(1936–)

English geneticist, born in Frankfurt, Germany, and educated at Clare College, Cambridge, where he received his BA in 1956, and an MA and PhD in 1959. He remained at Cambridge as a demonstrator in the genetics department during 1960–1. He then moved to the department of genetics at Stanford University, and later became Professor of Genetics at the University of Oxford (1970–9). He was appointed Director of Research at the Imperial Cancer Research Fund in 1979 and since 1991 has held the post of Director-General there. During his career he has also served as Chairman of the BBC Science Consultative Group (1981–7), President of the Royal Statistical Society (1984–5) and Vice-President of the Royal Institution (1981–2). He was elected FRS in 1974 and knighted in 1986. Bodmer has published extensively on the genetics of the HLA histocompatibility system. This system is responsible for distinguishing self from non-self in the animal body and an understanding of its complexities has been vital in the remarkable progress in transplant surgery. He has also published extensively on somatic cell genetics, cancer genetics and human population genetics.

BOERHAAVE, Hermann
(1668–1738)

Dutch physician and botanist, born in Voorhout, near Leiden. He studied theology and oriental languages, and took his degree in philosophy in 1689, but in 1690 he began the study of medicine, and in 1709 was appointed Professor of Medicine and Botany at Leiden. This post included supervision of the botanic garden, to which he added more than 2 000 species in the first 10 years, often utilizing men in the service of the Dutch East and West Indies companies as seed and plant collectors. He distributed many of these introductions to fellow botanists in many other countries. He published catalogues of the garden, describing the floral structure of the plants. The two works on which his medical fame chiefly rests, *Institutiones Medicae* (1708) and *Aphorismi de Cognoscendis et Curandis Morbis* (1709), were translated into various European languages, and even into Arabic. He pointed out that both plants and animals show the same law of generation, and by 1718 he was teaching sex in plants, his international stature ensuring widespread acceptance of these ideas. In 1724 he also became Professor of Chemistry, and his *Elementa Chemiae* (1724) is a classic. Meanwhile patients came from all parts of Europe to consult him, earning him a fortune.

BOGILIUBOV, Nikolai Nikolaevich
(1909–)

Soviet mathematical physicist, born in Nizhny Novgorod. He was educated at the Academy of Sciences in the Ukraine and subsequently worked there and at the Soviet Academy of Sciences in Moscow. Later he became Director of the Joint Institute for Nuclear Research in Dubna (1965). He developed the mathematical technique of changing variables in quantum field theory known as the Bogiliubov transformation. This has been used in many areas of research, such as particle physics. He has also contributed to the development of the theory of superconductivity, where a material loses all resistance to electrical flow, usually when cooled to very low temperatures.

BOHM, David Joseph
(1917–92)

American theoretical physicist, born in Wilkes-Barre, Pennsylvania. He was educated at Pennsylvania State College, where he graduated in 1939, and under **Oppenheimer** at the University of California at Berkeley. After participating in the Manhattan Project to develop the first atomic bomb, he became assistant professor at Princeton University in 1947, but in 1951 his contract was discontinued following allegations of Marxist activities due to his refusal to testify as to whether his colleagues were members of the Communist Party. He moved to Brazil as professor at the University of São Paulo (1951–5), and later to the UK as a Research Fellow at Bristol University (1957–61) and Professor of Theoretical Physics at Birkbeck College, London (1961–83). During his early career on the Manhattan Project, Bohm was instrumental in developing important techniques to describe oscillations in plasmas (high-temperature ionized gases), and his later application of these techniques resulted in outstanding contributions to our understanding of the behaviour of electrons in metals. Following his move to Princeton, he developed what was to become a lifelong interest in quantum mechanics; his textbook *Quantum Theory* (1951) was one of the clearest expositions of the subject ever written. He went on to propose radical new approaches to quantum theory, questioning the conventional view that the formalism necessarily leads to indeterminism and the peculiar requirement of the presence of an observer for a quantum entity to exist. Although he received little recognition for his alternative 'pilot-wave' theory for around 40 years due to its unorthodox assumption that particles can exert an instantaneous influence on each other across huge distances, his concepts have recently been revived. In later work Bohm investigated many of the philosophical problems associated with modern physics, and the nature of thought and consciousness. He was elected FRS in 1990.

BOHR, Aage Niels
(1922–)

Danish physicist, born in Copenhagen, the son of the Nobel prize-winning physicist **Niels Bohr**, and himself a Nobel laureate. Educated at the University of Copenhagen and the University of London, he worked from 1946 at his father's Institute of Theoretical Physics in Copenhagen where he became Professor of Physics in 1956. From 1963 to 1970 he was also director there and from 1975 to 1981 he was director of Nordita (the Nordic Institute for Theoretical Atomic Physics). Together with **Mottelson** he developed the collective model of the nucleus. The collective model combined the quantum-mechanical shell model of the nucleus developed by **Goeppert-Mayer** and **Jensen** and the classical liquid drop model developed by Niels Bohr, **Bethe** and **Weizsäcker**, and assumed individual motion of nucleons in the field of the core as well as collective vibrations of all nucleons, as had been proposed by **Rainwater**. This led to the prediction of nuclei being deformed so that they were no longer spherical. This model has been developed and explains the properties of nuclei well. Aage Bohr shared the 1975 Nobel Prize for Physics with Mottelson and Rainwater for this work.

BOHR, Harald August
(1887–1951)

Danish mathematician, born in Copenhagen, younger brother of **Niels Bohr**. Appointed to a professorship at the College of Technology in Copenhagen in 1915, he later became professor at the University of Copenhagen (1930). With Edmund Landau he devised the Bohr–Landau theorem (1914), which describes the conditions under which the **Riemann** zeta function is equal to zero. Bohr was also responsible for the introduction of the notion of 'almost periodic' functions through his studies of representation of functions by **Dirichlet** series.

BOHR, Niels Henrik David
(1885–1962)

Danish physicist, born in Copenhagen. Educated at Copenhagen University, he moved to England to work with **J J Thomson** at Cambridge and **Rutherford** at Manchester, and later returned to Copenhagen as professor (1916). While at Manchester he developed a theory of the hydrogen atom that could explain the observed spectral lines. This was based on two recent discoveries; Rutherford's evidence for the nuclear atom and the idea that energy is quantized, as suggested by **Planck** and **Einstein**. Using the nuclear atom model, Bohr applied the restriction that the electrons orbiting the nucleus could follow only certain allowed orbits. This corresponded to a set of discrete allowed electron energy levels and the observed spectral lines could be interpreted in terms of transitions between these levels. Initially the theory, which seemed to discard classical physics, was not well received but in addition to explaining the observed lines of the hydrogen spectrum it also predicted some new lines. When they were subsequently observed the model was accepted. Bohr's model was later shown to be a solution of the **Schrödinger** equation. During World War II he escaped from German-occupied Denmark and assisted atom bomb research in the USA, returning to Copenhagen in 1945. He later worked on nuclear physics and developed the liquid drop model of the nucleus used by **Bethe** and **Weizsäcker** to explain the stability of some nuclei and nuclear fission. Like Rutherford, Bohr influenced a great many young physicists including **Fermi** and **Landau**. He was founder and Director of the Institute of Theoretical Physics at Copenhagen (1920–2), and was awarded the Nobel Prize for Physics in 1922. His son, **Aage Bohr**, won the 1975 Nobel Prize for Physics.

BOISBAUDRAN, Paul Émile Lecoq de
(1838–1912)

French chemist, born in Cognac. He received no formal education but studied course books of the École Polytechnique, carrying out experimental work in a home laboratory. At the age of 20 he entered the family wine firm and travelled extensively on business in Europe, but continued his studies of chemistry and physics. The firm did well and Boisbaudran was able to spend more time on scientific work. He pursued such work vigorously until around 1895, when his health began to fail. He was awarded the Cross of the Legion of Honour and (in 1879) the Davy Medal of the Royal Society. Boisbaudran is best known for his work in spectroscopic analysis. In *Spectres lumineux* (1874), he presented careful studies of the spectra of 35 elements. In 1875 he discovered a new element, gallium, in zinc-blende by spectroscopic methods, and recognized it as the eka-aluminium predicted by **Mendeleyev**. From 1879 he studied the rare earth elements spectroscopically and was involved in the discoveries of samarium, dysprosium, terbium, europium and gadolinium.

BOISSIER, Pierre-Edmond

(1810–85)

Swiss botanist and traveller of French descent, born in Geneva. From childhood he began identifying plants of the Vaud mountains. He studied at Geneva Academy before spending the winter of 1831–2 in Paris, where he met several botanists. After a six-month journey to Italy (1833), studying plants and shells, he decided to make botany his vocation. In 1834–6 he toured Spain, discovering many new species; the results were published as *Voyage botanique dans le midi de l'Espagne* (2 vols, 1839–45). From 1842 to 1846 he travelled in Greece, Turkey, Syria and Egypt, accompanied by his wife Lucile Butini, after whom he named two beautiful discoveries, *Omphalodes luciliae* (Boraginaceae) and the Glory-of-the-Snow *Chionodoxa luciliae* (Liliaceae). The many new plants discovered were described in *Diagnoses Plantarum Orientalium Novarum* (3 vols, 1843–59). He accumulated one of his period's best collections of Middle Eastern plants and compiled a complete *Flora Orientalis* (5 vols, 1867–84; posthumous supplement, 1888). This was the first Flora covering the entire Middle East and remains a standard work, although it has been superseded by regional and national Floras such as **Peter Davis**'s *Flora of Turkey* and **Rechinger**'s *Flora Iranica*.

BOK, Bart Jan

(1906–83)

American astronomer, born in Hoorn, the Netherlands, and educated at the universities of Leiden and Groningen. In 1929 he was awarded a fellowship which allowed him to go to Harvard, USA, where he spent 25 years, becoming Associate Director of the Harvard College Observatory to its Director **Shapley** and in 1947 Robert Wheeler Willson Professor of Astronomy. In 1957 he succeeded **Woolley** as Director of the Mount Stromlo Observatory in Australia. He was responsible for the choice of the Siding Springs Mountain in New South Wales as the location of the major southern hemisphere observatory for future large instruments. He returned to the USA in 1966 to the post of Director of the Stewart Observatory in Tucson, Arizona, which he held until his retirement in 1970. Bok's lifelong interest, pursued largely in collaboration with his wife Priscilla Fairfield Bok (1896–1975), was in the structure of our galaxy, in particular in the distribution of stars and interstellar matter in regions of potential star formation. He made a special study of small dark clouds, the Bok globules, which he showed to contain enough material for future condensations into star clusters, a result confirmed by later millimetre-wave observations.

BOLTWOOD, Bertram Borden

(1870–1927)

American radiochemist who greatly furthered knowledge of radioactive elements and their place in the decay series. Born in Amherst, Massachusetts, he grew up fatherless in an academic family, and was educated at Yale, Munich and Leipzig. He was an instructor at Yale from 1897 to 1900, and later established himself as a consulting chemist with his own laboratory. From 1904 onwards he concentrated on radiochemistry, becoming the leading American in this field and laying the foundations for the study of isotopes. In 1906 he returned to Yale as Associate Professor, becoming Professor of Radiochemistry in 1910. By showing that there is a constant ratio of radium to uranium in unaltered minerals, Boltwood confirmed the work of **Rutherford** and **Soddy** which suggested that radioactive elements decay and transmute into other elements. In 1905 he showed that when uranium decays it produces radium and that the end product of its decay is lead. He introduced Pb:U ratios as a method for dating rocks in 1907 and this, together with other radiometric dating methods developed from his work, eventually revolutionized geology and archaeology. In 1907 Boltwood discovered the element ionium, since renamed thorium-230. Despite his usually dynamic and exuberant character, depression brought on by overwork and a heavy burden of administrative duties at Yale led to his suicide at Hancock Point, Maine, in 1927.

BOLTZMANN, Ludwig
(1844–1906)

Austrian physicist, born in Vienna. Boltzmann studied at the University of Vienna where he obtained his doctorate in 1867. From 1869 he held professorships in mathematics and physics at Graz, Vienna, Munich and Leipzig. He was a popular lecturer, numbering **Nernst** amongst his many students. Although his interests were diverse, encompassing physics, chemistry, mathematics, philosophy and even aeronautics, he is most celebrated for the application of statistical methods to physics and the relation of kinetic theory to thermodynamics. In 1866 he had tried to find a general proof of the second law of thermodynamics using only mechanical principles. Two years later his study of thermal equilibrium incorporated the statistical methods recently introduced by **Maxwell** in deriving a velocity distribution for colliding gas molecules. Boltzmann extended Maxwell's theory, treating the case when external forces were present, to derive the 'Maxwell–Boltzmann distribution'. He assumed that all possible ways of apportioning energy amongst molecules were equally probable; this was his first paper on the equipartition of energy. The 'H-theorem' (1872) indicated that a gas would tend towards the equilibrium state. In 1877 he presented the famous 'Boltzmann equation' (later carved on his tombstone) relating thermodynamic 'entropy' (S) and the statistical distribution of molecular configurations (W): $S = k \log W$, where k is Boltzmann's constant. This equation showed how increasing entropy corresponded to increasing molecular randomness. Other work dealt with electromagnetism; in 1872 Boltzmann's experiments on dielectrics confirmed predictions of Maxwell's theory, which he went on to promote on the Continent. Boltzmann wrote on viscosity and diffusion (1880–2) and in 1884 derived the law for black-body radiation found experimentally by **Stefan**, his teacher in Vienna. Boltzmann's work was unified by a commitment to an atomic theory of matter but the atomists came under attack from positivists in Vienna led by **Mach**. Partly because of the unpopularity of his views, Boltzmann suffered severe depression from 1900 and tragically killed himself whilst on holiday in 1906.

BOLYAI, János
(1802–60)

Hungarian mathematician, born in Kolozsvár. He took up a military career, but retired due to ill-health in 1833. After attempting to prove **Euclid**'s parallel postulate, that the straight line which passes through a given point and is parallel to another given line is unique, he realized that it was possible to have a consistent system of geometry in which this postulate did not hold, and so became one of the founders of non-Euclidean geometry, together with **Lobachevski**. His work continued that of his father Farkas (or Wolfgang, 1775–1856), who, however, had hoped to vindicate Euclid. The poor reception of his work, published as an appendix to a book by his father, inclined him not to publish again, and credit for his discovery was largely posthumous. Bolyai believed that henceforth the nature of space was an empirical matter. No authentic picture of him survives, although there is a statue of him and his father, erected after their deaths.

BOLZANO, Bernard
(1781–1848)

Catholic theologian, philosopher and mathematician, born in Prague of Italian ancestry. He was ordained as a priest in 1804 and appointed Professor of the Philosophy of Religion in 1805, and became a popular liberal figure. He was deprived of his chair in 1819 for non-conformity, at a time of political repression in Bohemia, and his work was put on the Index. He continued to write extensively, despite poor health, leaving a large body of writing that is only now being published. In mathematics, he was a pioneer in giving a rigorous foundation to the theory of functions of a real variable, and investigating the concept of the infinite. He discovered, but did not publish, the first example of a function that is continuous everywhere it is defined but nowhere differentiable, although he was probably unable to establish this conclusion rigorously. The reception of this work in his lifetime was slight, and he shortly proceeded other, more mainstream mathematicians like **Cauchy**, who also saw clearly the need to place the calculus on proper foundations that made no appeal to geometry or mechanics; Bolzano's influence was less than he deserved.

BOND, William Cranch
(1789–1859)

American astronomer, born in Portland, Maine. He started work in the family shop as a watchmaker, and turned to astronomy after witnessing the solar eclipse of 1806. His home observatory became the best American observatory of the day. In 1839 he moved it to Harvard, and was awarded an honorary MA and the title of Observer. He became the observatory's first director. Together with his son George Phillips Bond (1825–65, Director of the Harvard Observatory 1858–65) and simultaneously with **Lassell**, he discovered

Hyperion, a satellite of Saturn. They also discovered Saturn's crêpe ring, but erroneously concluded that all Saturn's rings are liquid. He collaborated with his son in the field of celestial photography, exhibiting a daguerreotype photograph of the Moon at the Great Exhibition in London in 1851. In 1850 they photographed the planet Jupiter and the star Vega, and in 1857 the double star Mizar.

BONDI, Sir Hermann
(1919–)

Austrian mathematical physicist, born in Vienna. He moved to England in 1937 to become an undergraduate at Trinity College, Cambridge. In 1942 he started research into radar at the Admiralty Signals Establishment, and after World War II returned to Cambridge, becoming a Fellow of Trinity College. He became a British citizen in 1947, moved to King's College, London, in 1954 as Professor of Applied Mathematics, and was knighted in 1973. Bondi returned to Cambridge in 1983 to become Master of Churchill College. He served with great distinction as Director General of the European Space Research Organisation (1967–71), as Chief Scientific Advisor to the Ministry of Defence and as Chairman of the National Environmental Research Council (1980–4). Scientifically, Bondi is best known for his seminal book on cosmology published in 1952 and for his proposal, with **Hoyle** and **Gold**, that the universe is in a steady state, matter being continuously created to fill the gaps left by the expansion. In the 1950s this theory stimulated a great deal of research in cosmology. It fell out of favour when it was discovered that the universe was more dense in the past, and encountered serious difficulties in 1965 following the discovery of the microwave background radiation by **Penzias** and **Robert Wilson**. In 1962 Bondi wrote a keynote paper showing how the emission of gravitational waves is a necessary consequence of **Einstein**'s general theory of relativity.

BONHOEFFER, Karl Friedrich
(1899–1957)

German physical chemist, born in Breslau (now Wrocław, Poland). He was the son of Geheimrat Bonhoeffer, Professor of Psychiatry at the University of Berlin. His younger brother Dietrich was a Protestant pastor and theologian, implicated in the plot against Hitler in July 1944 and executed by the Nazis. Karl Bonhoeffer studied at the universities of Tübingen and Berlin; he obtained his doctorate under **Nernst** in 1922. From 1923 to 1930 he was **Haber**'s assistant at the Kaiser Wilhelm Institute for Physical Chemistry in the Dahlem district of Berlin. He later occupied chairs of physical chemistry in Frankfurt, Leipzig and (briefly) Berlin, and in 1949 he became Director of the newly founded Max Planck Institute for Physical Chemistry in Göttingen, a post he occupied until his death. Bonhoeffer made important contributions to various areas of physical chemistry. In Haber's institute he worked on 'active' hydrogen (atomic hydrogen) and 'active' nitrogen, absorption spectra, flames, reactions of OH radicals and (with **Harteck**) the separation and properties of para- and ortho-hydrogen, the forms of the H_2 molecule which differ in orientation of nuclear spin. Later he studied the kinetics of electrode processes, the passivity of metals and periodic (oscillating) reactions. He also worked on problems in physiological chemistry, which gradually became his dominant interest. In Göttingen, starting from scratch, he brought the institute to the forefront of electrochemistry in a few years.

BONPLAND, Aimé Jacques Alexandre
(1773–1858)

French botanist, born in Rochelle. He travelled with **Humboldt** in South America (1799–1804), and collected and described (but did not publish) 6000 new species of plants. He contracted malaria in the early part of this trip, and suffered for the remainder of his life from periods of inertia and lethargy brought on by the illness. He was appointed Superintendent of the Empress Josephine's gardens at Malmaison and Navarre soon after his return to Europe. After the Empress died in 1814, he decided to return to South America. Appointed Professor of Natural History at Buenos Aires in 1816, he undertook a journey up the Paraná, then territory disputed by Argentina and Paraguay; José Francia, dictator of Paraguay, arrested him, and kept him prisoner for nine years. After his release he remained in Uruguay, from where he corresponded regularly with Humboldt. He talked of bringing his herbarium back to France, but never did.

BOOLE, George
(1815–64)

English mathematician and logician, born in Lincoln, the son of a cobbler. He was largely self-taught, and although he did not receive a degree, he was appointed Professor of Mathematics at Cork in 1849. He was one of the first to direct attention to the theory of invariants. This concerns expressions in several variables that do not change when the coordinates change. For example, the expression b^2-4ac in the theory of curves defined by quadratic equations in variables x and y under rotations and translations of the axes is an invariant expression. Boole also did important work on finite differences and differential equations. This led him to think of differential operators algebraically, and gradually he was led to consider the operations of logic algebraically also. This led him to the work for which he is best remembered, his *Mathematical Analysis of Logic* (1847) and *Laws of Thought* (1854). In these he employed mathematical symbolism to express logical relations, thus becoming an outstanding pioneer of modern symbolic logic, greatly influencing the subsequent work of **Frege** and **Bertrand Russell** among others. Boolean algebra is a generalization of the familiar operations of arithmetic to quantities that obey rules analogous to those Boole proposed for sets; it is particularly useful in the design of circuits and computers.

BORDET, Jules Jean Baptiste Vincent

(1870–1961)

Belgian physiologist and an authority on serology, born in Soignies. He received his MD from the University of Brussels in 1892, and in 1894 went to Paris to work in **Metchnikoff**'s laboratory at the Pasteur Institute. In 1901 he became Director of the Pasteur Institute in Brussels, where he remained until he was succeeded by his son, Paul, in 1940. Bordet studied the mechanics of bacteriolysis, which he concluded was due to the action of two substances: a specific antibody, heat-resistant to 55 degrees Celsius and present only in immunized animals, and a non-specific, heat-labile substance found in both unvaccinated and vaccinated animals, which he identified as **Hans Buchner**'s 'alexin'. Bordet's work in this field made possible new techniques for the diagnosis and control of infectious diseases. With his brother-in-law, Octave Gengou, Bordet discovered the whooping cough bacillus, extracted an endotoxin, and prepared a vaccine (1906). In 1909 he isolated the germs for bovine peripneumonia and avian diphtheria. He was awarded the 1919 Nobel Prize for Physiology or Medicine.

BOREL, Émile Félix Édouard Justin

(1871–1956)

French mathematician and politician, born in Saint Affrique. He studied and then taught at the École Normale Supérieure, and became professor at the Sorbonne in 1909. In addition to his prolific mathematical work, he was active in politics, scientific popularization and journalism; he was a member of the Chamber of Deputies (1924–36) as a Radical Socialist, and served as Minister for the Navy (1925–40). His mathematical work was first in complex analysis, where he showed the power of ideas in Cantorian set theory in the classical theory of functions. Subsequently he worked on measure theory and probability, and argued for a traditional view of the subject in opposition to the new axiomatic approach of **Kolmogorov**. He also wrote on the theory of games (1921–7), independently of the much better known work of **von Neumann** on this subject. These interests came together during World War II, when he wrote a book on the game of bridge.

BORELLI, Giovanni Alfonso

(1608–79)

Italian mathematician and physiologist, born in Naples. The son of a Spanish infantryman, Borelli led an adventurous early life in the thick of Italian politics and military affairs. Having probably received a mathematical training at the University of Naples, he acquired a reputation as a fine mathematics lecturer, rising to hold chairs at Naples, Pisa and Messina, and pursuing studies in mathematics, geometry and observational astronomy. While living in Tuscany, he was a prominent member of the Accademia del Cimento, the first major Italian scientific society. Impressed by the quantitative studies of **Sanctorius** and by the thinking of **William Harvey** (1578–1657) and **Descartes**, he became one of the most articulate advocates of the iatrophysical school of medicine, which sought to explain all bodily functions by physical laws. These views were set out in his *De Motu Animalium*, posthumously published in 1680.

BORN, Max

(1882–1970)

German physicist, born in Breslau (now Wrocław, Poland). He was educated at the universities of Breslau, Heidelberg, Zürich and Göttingen, and was appointed Professor of Theoretical Physics at Göttingen (1921–33), lecturer at Cambridge (1933–6) and Professor of Natural Philosophy at Edinburgh (1936–53). In 1925, with his assistant **Pascual Jordan**, he built upon the earlier work of **Heisenberg** to produce a systematic quantum theory based upon matrix mechanics. He showed that the waves in **Schrödinger**'s wave equation were probability waves, so that the state of a particle (eg its energy or position) could only be predicted in terms of probabilities. From this he deduced the existence of quantum jumps between discrete states, such as when an excited electron in high energy level of an atom 'falls' back to a lower level with the emission of a photon. This led to a statistical approach to quantum mechanics. He shared the 1954 Nobel Prize for Physics with **Bothe** for their work in the field of quantum physics.

BORNMÜLLER, Joseph Friedrich Nicolaus
(1862–1948)

German botanist and plant collector, born in Hildburghausen in Thüringen, and educated there and at Leipzig. In 1880 he went to the Gärtnerlehranstalt (gardener's training college) in Potsdam Game Park. For two years he was Inspector at the Belgrade Botanic Garden. In 1895 he married and moved to Berka, near Weimar. There he met Heinrich Haussknecht (1838–1903), who had collected with and been influenced by **Boissier**. Haussknecht in turn influenced Bornmüller and on his mentor's death in 1903, Bornmüller became curator of the Haussknecht Herbarium at Weimar, a post he held until 1938. From his adolescence until his death, Bornmüller's life was dedicated to botanical exploration. He built up a large collection from many countries in Europe and the Middle East, and his botanical publications number almost 250, of which 65 concern Iran. They include *Ein Beiträge zur Kenntnis der Flora von Syrien und Paläestina* (1898), *Beiträge zur Flora der Elbursgebirge Nord-Persiens* (1904–8), *Novitiae Florae Orientalis* (1904–10), *Plantae Straussianae* (1905–10), *Florula Lydiae* (1908), *Zur Flora des Libanon und Antilibanon* (1914), *Zur Flora der nördlichen Syriens* (1917) and *Symbolae ad Floram Anatolicam* (1936). Bornmüller was one of the most remarkable botanists of his epoch, with a remarkable memory for the minutest details and a man of great wisdom. He was especially interested in the genera *Cousinia* and *Dionysia*.

BOSCH, Carl
(1874–1940)

German industrial chemist, born in Cologne. He studied organic chemistry at Leipzig but also showed an early talent for engineering. He worked for Badische Anilin und Soda Fabrik, becoming its general manager in 1910. His brother-in-law **Haber** had discovered that ammonia could be synthesized from hydrogen and nitrogen, its component elements, at high temperatures and pressures using a catalyst (the Haber process). In 1909 Bosch began work to adapt this laboratory process to commercial production, overcoming the great technical problems caused by high temperatures and pressures, the fact that the hydrogen affected the steel vessels, the need to find a better catalyst and the difficulties of separating the ammonia once it was synthesized. The industrial production of ammonia was of enormous importance to agriculture as it was made into nitrates for fertilizers; it also affected the manufacture of explosives. Bosch invented the process which bears his name, in which hydrogen is produced on an industrial scale by passing steam and water gas over a catalyst at high temperatures. His other work was on the synthesis of methanol and the production of synthetic rubber. Bosch became president of I G Farbenindustrie, which had absorbed Badische Anilin und Soda Fabrik. In 1931 he shared the Nobel Prize for Chemistry with **Bergius** for his part in the invention and development of chemical high-pressure methods, and in 1935 succeeded **Planck** as Director of the Kaiser Wilhelm Institute. He died in Heidelberg.

BOSCOVICH, Roger Joseph
(1711–87)

Croatian mathematician and astronomer, born in Ragusa (now Dubrovnik). A Jesuit, he was educated in physics and mathematics at the Collegium Romanum in Rome, where he was later appointed to the Chair in Mathematics (1740). He became Professor of Mathematics at Pavia in 1764. Boscovich devised new methods for determining the orbits and rotation axes of planets, and investigated the shape of the Earth. As a leading proponent of **Isaac Newton**'s theory of gravitation, he wrote prolifically on gravity, astronomy, optics and trigonometry.

BOSE, Sir Jagadis Chandra
(1858–1937)

Indian physicist and botanist, born in Mymensingh, Bengal (now in east Pakistan), the son of a deputy magistrate. After his schooling at St Xavier's, a Jesuit College in Calcutta, he went to London to study medicine, but transferred to Cambridge University when he was awarded a scholarship and graduated in natural science in 1884. He returned to Calcutta where he was appointed Professor of Physics at Presidency College. Bose became known for his study of electric waves, their polarization and reflection, and for his experiments demonstrating the sensitivity and growth of plants, for which he designed an extremely sensitive automatic recorder. In some of his ideas he foreshadowed **Wiener**'s cybernetics. He founded the Bose Research Institute in Calcutta for physical and biological sciences in 1917, was knighted in the same year and became the first Indian physicist to be elected FRS (1920).

BOSE, Satyendra Nath
(1894–1974)

Indian physicist, born in Calcutta and educated at Presidency College there. He became professor at Dacca University before being appointed to another chair at Calcutta in 1952. Later he was appointed National Professor by the Indian government (1959) and President of the National Institute of Sciences of India (1949–50). In 1924 he succeeded in deriving the **Planck** black-body radiation law, without reference to classical electrodynamics. **Einstein** generalized his method to develop a system of statistical quantum mechanics, now called Bose–Einstein statistics, for integral spin particles for which **Dirac** coined the term 'bosons'. Bose also contributed to the studies of X-ray diffraction and the interaction of electromagnetic waves with the ionosphere.

BOTHE, Walther Wilhelm Georg
(1891–1957)

German physicist, born in Oranienburg. He was educated under **Planck** at the University of Berlin, where he received his PhD in 1914. From 1934 he was head of the Max Planck Institute for Medical Research at Heidelberg. He developed an electric circuit for scintillation counting which replaced the laborious process of counting scintillations by eye used by **Geiger** and Ernest Marsden. He also developed the coincidence technique where two detectors are used to measure two events simultaneously, allowing two particles to be associated with each other. He used this to study cosmic rays and nuclear physics, and to show that the recoil electron and scattered photon appear simultaneously in **Compton** scattering. His work on the development of the coincidence technique in counting processes brought him the Nobel Prize for Physics in 1954, shared with **Born**. Bothe observed a penetrating radiation when beryllium was bombarded with alpha particles; he found that the rays were uncharged, and erroneously believed them to be high-energy gamma rays. The emission was correctly identified by **Chadwick** a few years later as the neutron.

BOUGUER, Pierre
(1698–1758)

French physicist, born in Le Croisic, Brittany. His father Jean was the hydrographer royal, and Pierre succeeded him in 1713 at the age of 15. His chief interest was in the problems associated with navigation and ship design. In 1735 he was sent on the famous **La Condamine** expedition to Peru to measure the length of a degree of the meridian near the equator. There from 1735 to 1742 he took part in experimental investigations of the length of the seconds pendulum at great elevations, the deviation of a plumbline through the attraction of a mountain, the altitude limit of perpetual snow, the obliquity of the ecliptic, and other scientific topics. His views on the intensity of light laid the foundations of photometry, and one of his most important discoveries was the law of absorption, sometimes unjustly known as **Lambert**'s law. In 1748 he invented a heliometer, to measure the light of the Sun and other luminous bodies.

BOURBAKI, Nicolas
(20th century)

'French mathematician', the pseudonym of a group of French mathematicians from the École Normale Supérieure, including Henri Cartan, Claude Chevalley, **Dieudonné** and **Weil**. In the 1930s they conceived the plan of writing a treatise on analysis which would set out the subject in a strictly logical development from its basic principles, in a style that was minimal, precise, rigorous and elegant, emphasizing the fundamental structures that underlie apparently diverse areas of mathematics. The project continually grew. Publication of *Éléments de mathématiques* by 'Nicolas Bourbaki' started in 1939, and although Bourbaki was dispersed by World War II, there followed books on set theory, algebra, general topology, topological vector spaces, integration, **Lie** groups and Lie algebras, among other subjects. By the 1980s the unfinished series had more or less come to a stop, but it has had great influence on mathematical attitudes in the last 40 years, some mathematicians being attracted by its abstraction and others repelled, condemning it for divorcing mathematics from its historical applications to the physical world. Many of the books have become the definitive treatment of their subjects, and the series also includes valuable historical essays.

BOUSSINGAULT, Jean Baptiste Joseph Dieudonné
(1802–87)

French agricultural chemist, born in Paris. He studied at the School of Mines at St Étienne and then served under Simón Bolívar in the South American War of Independence, remaining in South America to conduct researches into mineralogy and meteorology until 1832. He then became Professor of Chemistry at Lyons and finally Professor of Agriculture at the Conservatoire des Arts et Métiers, Paris (1839–87). He was elected to the Academy of Sciences in 1839. In a long series of experiments on his farm at Alsace between 1834 and 1876, he demonstrated that legumes increase the nitrogen in the soil by fixing atmospheric nitrogen, but that all other plants — contrary to what was believed at the time — have to absorb nitrogen from the soil. He went on to suggest how nitrogen is recycled. He further showed that all green plants absorb carbon from the atmosphere in the form of carbon dioxide. His work laid the basis for modern advances in microbiology. He died in Paris.

BOVERI, Theodor Heinrich
(1862–1915)

German biologist and pioneer of cytology, born in Bamberg, the son of a physician. He studied history and philosophy at Munich but soon changed to science, and graduated in medicine in 1885. From 1885 to 1893 he held a fellowship at the Zoological Institute in Munich and from 1893 he was Director of the Zoological–Zootomical Institute at the University of Würzburg. By 1884 something was known of cell chromosomes from **Beneden**'s work, and Boveri confirmed and extended this. Through morphological studies of the fertilization and development of eggs of the roundworm *Ascaris*, Boveri demonstrated the individuality of chromosomes. He also showed that at fertilization, the midpiece of the spermatozoon contributes a small structure called a centrosome to the fertilized egg which plays an important role in subsequent cell divisions. In experiments with sea-urchin eggs, Boveri showed that normal development requires an appropriate number of chromosomes for the species, and that each chromosome in a set is responsible for particular hereditary traits. His conception of chromosomes as independent and organized structures provided the basis for much of later genetics. Boveri suffered from bouts of depression and poor physical health, and died at the age of 53.

BOVET, Daniel
(1907–92)

Swiss-born Italian pharmacologist, born in Neuchâtel. He studied chemistry at Geneva, and worked in the Department of Chemical Therapeutics at the Pasteur Institute in Paris (1929–47) before being invited, with his wife and collaborator Philomena Nitti, to establish the Laboratory of Chemotherapeutics at the Superior Institute of Health in Rome. His successful study of drugs affecting the muscular and nervous systems stemmed from the earlier concept of isosterism, that the body's regulatory molecules acted by a 'key in lock' mechanism; by modifying the structures of the natural molecules, inhibition could be achieved. He discovered the first antihistamine drugs in 1939, classified their pharmacological action, and developed the drug Antergan®. His second major study involved drugs blocking the action of adrenaline and noradrenaline, thereby preventing hypertension (high blood pressure), one of the most common medical conditions, and vasoconstriction. This took Bovet into the realms of lysergic acid (LSD) and other hallucinogens. Visiting Brazil he became interested in the Indian neuromuscular poison curare, of which he later made synthetic analogues, notably succinylcholine, which have been much used as muscle relaxants in anaesthesia since 1950. He was awarded the 1957 Nobel Prize for Physiology or Medicine.

BOWDITCH, Henry Pickering
(1840–1911)

American physiologist, born in Boston into an old, genteel New England family. Bowditch entered Harvard in 1857, interrupting his studies to serve in the American Civil War (1861–5), and being wounded in battle in 1863. He obtained his MD from Harvard in 1868 and spent three years in Europe, studying experimental physiology and microscopy first in France (where he decried the physiology teaching) and then in Germany with such teachers as **Ludwig**. Returning to America, he obtained a teaching post at Harvard in physiology, and sought to introduce German laboratory and research methods. Bowditch went on to produce important experimental work on cardiac contraction, on the innervation of the heart, on the vascular system, and on the reflexes, including the knee-jerk. He later became a pioneer of anthropometry, emphasizing the importance of nutrition and environment (rather than heredity) in child growth. He helped build up the Harvard physiology department, instituting important reforms in medical education while Dean of the Harvard Medical School (1883–93). He was a founder of the American Physiological Society (1887).

BOWEN, Ira Sprague
(1898–1973)

American astrophysicist, born in Seneca Falls, New York. After graduating from Oberlin College (1919), he received a PhD from Caltech (1926) and remained on the staff there, becoming professor in 1931. He finally became Director of the Mount Wilson and Palomar Observatories between 1946 and 1964. Bowen solved the 'nebulium' problem in 1928. Previously the elements responsible for the green emission lines in the spectra of planetary nebulae (the glowing clouds of gas surrounding small stars in late stages of evolution) were unknown. Bowen concluded that they were caused by so-called forbidden transitions in singly and doubly ionized oxygen and in singly ionized nitrogen. In 1938 he constructed an image slicer for the use with slit spectrometers. This helped to maximize the wavelength resolution and the spatial information that was obtainable.

BOWEN, Norman Levi

(1887–1956)

Canadian geochemist and petrologist, born in Kingston, Ontario, the son of English immigrants. After early fieldwork experience with the Ontario Bureau of Mines, he moved on to MIT (1909–10) and then to the Geophysical Laboratory of the Carnegie Institution, Washington (1910–19), where he commenced experimental petrological studies, working on the plagioclase feldspar solid-solution series. After a brief spell as professor at Queen's University, Kingston (1919–20), he returned to the Geophysical Laboratory until 1937, when he was appointed professor at the University of Chicago (1937–46). Bowen demonstrated the reaction principles affecting the behaviour of crystals in a progressively cooling magma (1922), studied assimilation in magmas whereby rocks caught up in a magma can be progressively dissolved, investigated the progressive metamorphism of siliceous limestones and dolomites, and undertook experimental work to investigate the melting and solubility of silicate systems containing water, iron oxide and alkali-aluminosilicates. He also showed that the principles of physical chemistry could be applied successfully to petrological processes. His book *The Evolution of the Igneous Rocks* (1928) had a profound effect on later generations of petrologists and remains a standard text.

BOWER, Frederick Orpen

(1855–1948)

English botanist, born in Ripon. The son of a Yorkshire family of clothiers, he was educated at Repton, where, by private study, he laid the foundations of his botanical knowledge and decided to make it his career. He went to Trinity College, Cambridge, learned laboratory methods from **Sachs** at Würzburg, and went on to Strassburg in 1879, where for a year he studied under **de Bary**. In 1882 he became lecturer in botany at South Kensington, and from 1885 to 1925 he was Regius Professor of Botany at the University of Glasgow. An all-round botanist, a good organizer and an inspiring teacher, as well as an accomplished cellist devoted to chamber music, he devoted himself to research in plant morphology and built up a worldwide reputation. From 1890 he concentrated on the evolutionary morphology of pteridophytes and published three major works, marking phases in these studies: *The Origin of A Land Flora* (1908), *The Ferns* (3 vols, 1923–8) and *Primitive Land Plants* (1935). Other works included *Size and Form in Plants* (1930) and *Sixty Years of Botany in Britain* (1938), an autobiographical account of the introduction of the 'new botany' into British education, in which he had played a major part.

BOYD, William Clouser

(1903–83)

American immunochemist, born in Dearborn, Missouri. He trained at Boston University Medical School (1926–30) and remained there throughout his career, as assistant professor (1935–8), associate professor (1938–48), and Professor of Immunochemistry from 1948. From the early 1930s, Boyd studied the nature of antigen–antibody interactions, using various chemically modified proteins to delineate the groups important for eliciting an antibody response, and the quantitative relationship between antigen and antibody in forming a precipitate. His theory that in forming the precipitate the antigen becomes coated with antibody (1934) contrasted with that of a divalent antibody (two binding sites) proposed by **Pauling**. Pauling's interpretation turned out to be correct (1949), and Boyd recognized this in 1956 with his identification of two equivalent hapten sites (the reacting part of an antigen) on the antibody. However, Boyd's major contribution was the discovery that blood groups are inherited and not altered by environment (1939). He showed by genetic analysis of blood groups that human races are populations that differ in the incidence of their alleles; on this basis he divided all humanity into 13 geographically distinct races with different blood group gene profiles. He published *Fundamentals of Immunochemistry* (1943), *Genetics and the Races of Man* (1950), *Biochemistry and Human Metabolism* (1952), *Races and People* (1965) and *Introduction to Immunochemical Specificity* (1968). He also discovered antibody-like proteins (lectins) in plants (1948), and studied the blood groups of mummies.

BOYER, Herbert Wayne

(1936–)

American biochemist, born in Pittsburgh, Pennsylvania. He studied at the university there and from 1966 worked at the University of California at San Francisco, where he was appointed Professor of Biochemistry in 1976. A pioneer of genetic engineering, he showed how the DNA of a plasmid (a bacterial virus) could be joined to bacterial DNA or to the DNA from a toad (*Xenopus laevis*). Multiple copies of this hybridized DNA were obtained by allowing the plasmid to penetrate its bacterial host (*Escherichia coli*) whereupon it multiplied with it, a process known as cloning. Boyer and his collaborators went on to characterize the plasmid and reorganize its DNA structure to include features to facilitate and monitor DNA insertion and cloning. DNA hybridization also involves the use of a 'restriction endonuclease' to cut the DNA at sites, determined by the base sequence. Boyer characterized the restriction endonuclease EcoRI in detail and published an X-ray analysis of the enzyme–DNA complex (1986), showing how the DNA was cut by being constrained between the two identical subunits of the enzyme. He exploited his methods in the commercial production of insulin and other scarce biochemicals, and formed the company Genentech for this purpose in 1976.

BOYLE, Robert
(1627–91)

Anglo-Irish experimental philosopher and chemist, born in Lismore Castle, Munster. The 14th child of the Earl of Cork, Boyle was raised amongst the Anglo-Irish aristocracy. He studied at Eton from 1635 and three years later commenced a Grand Tour of Paris, Geneva and Florence. He arrived home in the summer of 1644 to find his father dead. Settling on his inherited estate of Stalbridge, Dorset, he combined studies in natural philosophy with the composition of moral and religious essays. From 1645 Boyle took an active part in the meetings of the anti-scholastic Invisible College, precursor of the Royal Society of London. He returned to Ireland briefly, dabbled in anatomical dissection (1652–4), and then moved to Oxford where he created an experimental laboratory. Ably assisted by **Hooke**, later Royal Society curator of experiments, Boyle enlisted a versatile new air-pump (the 'machina Boyleana', c.1659) to investigate the characteristics of air and of the vacuum: these investigations appeared in *New Experiments Physico-Mechanicall, Touching the Spring of the Air* (1660). A second edition (1662), responding to criticisms of Thomas Hobbes and others concerning the existence and nature of the vacuum, included 'Boyle's law', which states that the pressure and volume of gas are inversely proportional. The air-pump became a powerful symbol of the new 'experimental philosophy' promoted by the Royal Society from its foundation (1660). Boyle's voluminous writings exemplified the society's experimental programme but were, to some, intolerably prolix and invited caricature, notably from Jonathan Swift. *The Sceptical Chymist* (1661) attacked both alchemical principles (salt, sulphur and mercury) and Aristotelian elements (earth, water, air and fire). Boyle proposed elements which were primitive and simple, perfectly unmingled bodies: the practical limit of analysis and the constituents of all things. The *Origin of Forms and Qualities* (1666) elaborated this corpuscular 'mechanical philosophy', wherein natural phenomena were explained in terms of matter in motion; shape, number and motion of particles alone determined the diverse properties of substances. Boyle's researches in hydrostatics, the chemistry of colours, calcination of metals, properties of acids and alkalis, crystallography and refraction all served to demonstrate this approach. In 1668 he took up residence with his sister, Lady Ranelagh, in London, remaining there for the rest of his life. As a director of the East India Company and Governor of the Society for the Propagation of the Gospel in New England since 1661, he worked for the diffusion of Christianity, circulating at his own expense translations of the Scriptures. He endowed the Boyle Lectures, 'for proving the Christian Religion against notorious Infidels'.

BOYS, Sir Charles Vernon
(1855–1944)

English physicist, born in Wing in Rutland. He was educated at Marlborough College and the Royal School of Mines in London, and from 1881 to 1897 he was employed at the Royal College of Science, latterly as an assistant professor. He then became one of the Metropolitan Gas Referees, and at the same time continued to build up a lucrative practice as an expert witness, mainly in disputes involving patents. His many inventions included an improved torsion balance using fused quartz fibres which he prepared himself; the radio-micrometer, so sensitive that it could detect the heat from a candle a mile away; a calorimeter for use with coal gas; and a camera with a moving lens, with which he took some striking photographs of lightning flashes. He was elected FRS in 1888 and knighted in 1935.

BRADLEY, James
(1693–1762)

English astronomer, born in Sherbourn, Gloucestershire. Having embarked on an ecclesiastical career he was drawn into the field of astronomy by an uncle, and was appointed in 1721 to the Savilian Chair of Astronomy at the University of Oxford. He was responsible for significant improvements in the precision of observations of stellar positions. Searching for stellar parallax, he discovered in 1728 the phenomenon of aberration and explained it in terms of the combination of the finite speed of light and of the Earth's motion around the Sun. In 1747 he published his second major discovery, the effect of nutation, which he explained as due to the Moon's attraction on the bulging equator of the Earth's spheroid. The precision of all his observational work allowed Bradley to state that parallaxes of stars must be smaller than one arcsecond. He was appointed Astronomer Royal in 1742, and according to **Isaac Newton**'s dictum, was at the time the best astronomer in Europe.

BRAGG, Sir (William) Lawrence

(1890–1971)

Australian-born British physicist, born in Adelaide, son of Sir **William Bragg**. He was educated at Adelaide University from the age of 15, and Trinity College, Cambridge, where he discovered the Bragg law (1912), which describes the conditions for X-ray diffraction by crystals: $n\lambda = 2d \sin \theta$, where λ is the X-ray wavelength, d is the separation of the layers of atoms forming the crystal planes, θ is the angle of incidence of the X-rays and n is an integer. This was a significant advance in theoretical chemistry, leading to the first method to study the exact positions of atoms in crystal interiors. He later collaborated with his father in the study of crystals by X-ray diffraction and continued it as professor at Manchester and then at Cambridge (from 1938). Like his father, he became Director of the Royal Institution (1954–65) and did much to popularize science. They shared the 1915 Nobel Prize for Physics. Lawrence Bragg became Professor of Physics at Victoria University, Manchester (1919–37), and succeeded **Rutherford** as head of the Cavendish Laboratory in Cambridge (1938–53). There he supported **Crick** and **James Watson** in their work, using X-ray crystal studies to deduce the helical structure of DNA, so creating molecular biology and revolutionizing biological science. He was knighted in 1941.

BRAGG, Sir William Henry

(1862–1942)

English physicist, born in Westward, Cumberland, the eldest child of a former officer in the merchant navy, turned farmer. With his son, **Lawrence Bragg**, he founded X-ray crystallography. He was educated at Trinity College, Cambridge, and achieved Third Wrangler in the Mathematical Tripos, Part I (1884). He was appointed Professor of Mathematics at Adelaide, Australia (1886), but his extraordinary scientific career only really began when in 1904 he gave a lecture on radioactivity which inspired him to research into this area. He became professor at Leeds in 1909, and from 1912 worked in conjunction with his son on determining the atomic structure of crystals from their X-ray diffraction patterns. Eventually, this was to become one of the key techniques for deducing the structures of penicillin and DNA. Their efforts won them a joint Nobel Prize for Physics in 1915, the only father–son partnership to share this honour. Bragg moved to University College London the same year, and became Director of the Royal Institution in 1923. His works include *Studies in Radioactivity* (1912), *X-rays and Crystal Structure* (1915, with his son) and *The Universe of Light* (1933). During World War I he directed research on submarine detection for the Admiralty. He was knighted in 1920, probably both as recognition for his war work and his scientific eminence. He was elected a Fellow of the Royal Society in 1907, was awarded its Rumford Medal in 1916, and served as the society's president from 1935 to 1940.

BRAHE, Tycho

(1546–1601)

Danish astronomer, the greatest astronomical observer of the pre-telescope era. Born into a noble family at Knudstrup, South Sweden (then under the Danish crown), he studied mathematics and astronomy at the University of Copenhagen and then at famous centres of learning in Germany (1562–9). Since 1563 when he observed a conjunction of Jupiter and Saturn, he had realized that there were serious errors in the astronomical tables then in use, and that there was a grave need for reliable star positions. He returned to Denmark with a celestial globe acquired in Germany, and constructed his own improved positional instruments. In 1572 a brilliant nova or new star appeared in the constellation of Cassiopeia (now known to have been a supernova) which Tycho observed for many months, fixing its position with such accuracy that he could declare it to be more distant than the Moon. His publication of this finding in *De Nova Stella* (1573) made his name. In 1576 King Frederick II granted him generous funds to construct and maintain an observatory on the island of Hveen in the Sound. This magnificent establishment which Tycho named Uraniborg ('Castle of the Sky') was equipped with exquisite sextants, quadrants and armillary spheres, as well as accurate clocks, described in his *Astronomiae instauratae Mechanica* (1598). Tycho's observations of the comet of 1577 proved that it was more distant than Venus and moved around the Sun, contrary to the accepted notion that comets were local atmospheric phenomena. Work at Uraniborg over more than 20 years resulted in Tycho's catalogue of 777 stars and observations of the Sun and planets, with positions given to an accuracy of one or two arcminutes (posthumously published). In 1596 Tycho's royal patron died. Bereft of support, Tycho left Denmark and at the invitation of the emperor Rudolf II settled at the castle of Benatky near Prague (1599), where he set up some of his smaller instruments, though he made little use of them: he died two years later. During that last phase of his life he was joined by a young assistant **Kepler** who was to make fruitful use of his master's observations. Unlike his younger contemporary **Galileo**, Tycho Brahe did not subscribe to the Copernican doctrine of a Sun-centred planetary system, arguing with good scientific reasoning that his observations revealed no annual shift in the positions of stars, as would be expected by a moving observer. The explanation, of course, was that such a shift (parallax) was incomparably smaller than the smallest angle he was able to measure. The Tychonic system envisaged the planets—other than the Earth—revolving around the Sun, with that entire assembly revolving around a stationary Earth. Tycho Brahe's great talents were accompanied by a hot-tempered and arrogant manner; he lost most of his nose in a duel at the age of 19 following an argument over a mathematical detail, and wore a false silver nose for the rest of his life. His tomb in Prague is marked by his effigy clad in armour with his right hand resting on his celestial globe.

BRAMBELL, (Francis William) Rogers
(1901–70)

British zoologist, born in Sandycove, now in the Republic of Ireland. He was educated at Trinity College, Dublin. In 1930 he became Lloyd Roberts Professor of Zoology at University College of North Wales, Bangor, and he remained in that post until his retirement in 1968. Brambell's work on the reproductive cycles of mammals led to important new discoveries regarding the routes of transmission of maternal antibodies to the offspring. These investigations have been of great value for studies of disease resistance in newly born infants. He was also responsible for major improvements in the welfare of farm animals. Elected FRS in 1949, he received the Royal Medal of the Royal Society in 1964.

BRAND, Hennig
(17th century)

German alchemist who discovered phosphorus. He was born in Hamburg and was active in the second half of the century. He began his career as a military officer, and subsequently practised as a physician (perhaps without qualifications) in his native city. Turning to alchemy, he claimed to have found a substance in urine which turned silver into gold, a claim which aroused much controversy but which was supported by **Leibniz**. Around 1669 he discovered in urine a white waxy substance which glowed in the dark and which he named phosphorus ('light bearer'). He is the first scientist known to have discovered an element, the names of earlier discoverers being lost. He did not publicize his discovery and phosphorus was discovered independently by **Boyle** in 1680. Brand worked for a time as a mining consultant—a post he gained thanks to the good offices of Leibniz—and then returned to Hamburg. Nothing is known of his subsequent career.

BRANDT, Georg
(1694–1768)

Swedish chemist who discovered cobalt, born in Riddarhyttan. He studied medicine and chemistry at Leiden under **Boerhaave**, and received his MD at Rheims in 1726. The following year he was appointed director of the chemical laboratory at the Bureau of Mines, Stockholm, and in 1730 he became Assay Master of the Swedish Mint. Around 1730 he discovered cobalt. He investigated arsenic and its compounds systematically, publishing the results in 1733, and he discovered the difference between potash and soda. Brandt was also one of the first chemists to decry alchemy and to expose its fraudulent practices. He died in Stockholm.

BRANLY, Edouard
(1844–1940)

French physicist who discovered the principles of wireless telegraphy, born in Amiens. From his early education in St Quentin he progressed to gain degrees in mathematics and natural sciences, a doctorate from the Sorbonne and a medical degree from the Catholic University in Paris. He investigated electric waves, the effects of ultraviolet light waves and the electrical conductivity of gases. He invented the coherer for the reception of wireless telegraphic (radio) waves in 1890, thereby establishing the principles later developed by **Marconi**. He evolved the forerunner of receiving antennae.

BRATTAIN, Walter Houser
(1902–87)

American physicist, born in Amoy, China. He grew up on a cattle ranch in the Washington state, and was educated at the University of Oregon and at Minnesota. He joined Bell Telephone Laboratories (1929) where he carried out fundamental research on semiconductors, finding that the rectifying properties of diodes made by placing a metal contact onto a semiconductor surface are caused by electronic processes occurring at the semiconductor's surface rather than deep in the bulk of the material. After the interruption of World War II, during which he worked on the magnetic detection of submarines, he resumed research into semiconductors at the Bell laboratories working under **Shockley**. He attempted to control the number of charge carriers near the surface of semiconductors by applying an electric field, finally succeeding by using an electrolyte to apply the field. With **Bardeen** and Shockley, he developed the point-contact transistor, using a thin germanium crystal contacted by two pieces of metal only 0.005 centimetres apart. Soon afterwards, the junction transistor devised by Shockley which evolved into the silicon microchip took the dominant place it has held in electronics ever since. For his work on the transistor, Brattain shared the Nobel Prize for Physics with Bardeen and Shockley in 1956. He retired from Bell Telephone Laboratories in 1967.

BRAUN, (Karl) Ferdinand
(1850–1918)

German physicist, born in Fulda. He studied at Marburg and Berlin where he obtained his doctorate in 1872. After holding posts at the universities of Würzburg, Leipzig, Marburg, Karlsruhe, Tübingen and Strassburg, he returned to Tübingen as Professor of Physics and Director of the Physical Institute which he had founded during his previous short spell there. Although his main contributions were in pure science, he is best known for the first cathode-ray (the 'Braun tube') oscilloscope introduced in 1897. In 1909 he shared with **Marconi** the Nobel Prize for Physics for his practical contribution to wireless telegraphy. His solution to increasing transmitting range was to develop a new type of antenna circuit.

BRAUN, Wernher von
See **VON BRAUN, Wernher**

BRAUN-BLANQUET, originally BRAUN, Josias
(1884–1980)

Swiss phytosociologist and alpinist, born in Chur, the son of a Rhaetic banking family. While a young bank employee he became interested in the flora of the high Alps. He became *Privatdozent* at the École Polytechnique Fédérale de Zürich. While working at a Geneva bank in 1912, he met the botanists John Briquet, Gustave Beauverd and Paul Chenevard. Lacking any university qualifications was a serious impediment to pursuing botanical research in Switzerland, and he went to Montpellier, where he received his doctorate and married Gabrielle Blanquet in 1915. His doctoral thesis on the vegetation of the Cevennes was fundamental to his development of the sciences of geobotany and phytosociology, of which he is regarded as the founder. In 1927 he founded the Station Internationale de Géobotanique Méditerranéenne et Alpine at Montpellier. Braun-Blanquet's classification system of plant associations gained wide acceptance in Europe and elsewhere, though much less so in Britain. He published nearly 250 papers and books on the botany of the Alps, Mediterranean and Pyrenees, including *Die Vegetationsverhältnisse der Schneestufe den Rätisch-Lepontischen Alpen* (1913, for which he climbed 77 peaks over 3000 metres) and his seminal work, *Pflanzensoziologie* (1928).

BREIT, Gregory
(1899–1981)

American physicist, born in Nickolaev, Russia. He moved to the USA at the age of 16 and was educated at Johns Hopkins University. After working at Leiden, Harvard and the University of Minnesota, he joined the staff of the Carnegie Institute, Washington. His later years were spent at New York, Wisconsin and Yale universities. He was also a member of many American scientific academies and was a councillor in the Physical Society (1935–8). His major research was in quantum mechanics, nuclear physics and quantum electrodynamics in which he often collaborated with **Wigner**. Together they developed the Breit–Wigner formula, which describes cross-sections (probabilities) for resonant nuclear reactions, and the formula describing how neutrons interact with a stationary nucleus. With Merle Antony Tuve, he measured the height and density of the Earth's ionosphere by reflecting short bursts of radio waves from it, which was essentially the first use of radar imaging.

BRENNER, Sydney
(1927–)

South African–British molecular biologist, born in Germiston, South Africa, and educated at Witwatersrand University and Oxford. With the encouragement of **Crick**, he joined the staff of the Medical Research

Council (MRC) in 1957. He served as Director of the MRC Molecular Biology Laboratory in Cambridge (1979–86), then became Director of the MRC Molecular Genetics Unit there (1986–92). With Crick, Brenner worked to unravel the genetic code; starting with the finding that the nucleotide uracil synthesizes the amino acid phenylalanine, they worked out the nucleotide codes for the 20 amino acids in 1961. Brenner proposed the term 'codon' for the unit of three nucleotides which code for one amino acid. He went on to research the embryology of a simple animal, the nematode worm *Caenorhabditus elegans*; a database has since been established for storing its genetic map. Since 1992 Brenner has been a member of the Scripps Research Center at La Jolla, California. He was elected FRS in 1965.

BRESLOW, Ronald

(1931–)

American chemist, born in Rahway, New Jersey. After an undergraduate course in chemistry at Harvard he proceeded to the degree of PhD (1955) for work supervised by **Robert Woodward**. From 1955 to 1956 he worked with **Todd** in Cambridge, UK. On his return to the USA he joined the staff of Columbia University, becoming full professor in 1962 and University Professor in 1992. His early work concerned new synthetic methods in organic chemistry, but later he developed the idea of biomimetic chemistry, using some of the principles involved in enzyme catalysis to construct simpler chemical catalysts for effecting reactions with no biological parallels. He was elected to the US National Academy of Sciences in 1966 and awarded the US National Medal of Science in 1991.

BREWSTER, Sir David

(1781–1868)

Scottish physicist, born in Jedburgh. Educated for the church, he became editor of the *Edinburgh Magazine* (later the *Edinburgh Philosophical Journal*) in 1802, and of the *Edinburgh Encyclopaedia* in 1808. Brewster was interested in the study of optics, and in 1815 he observed that when light is incident on a plane glass surface at one particular angle, the light reflected is totally plane-polarized, and that measurement of the angle at which this happens enables the calculation of the refractive index of the glass (Brewster's law). In the same year he discovered stress birefringence, showing that stress on transparent materials can alter the way in which they transmit light. This is now the basis of the photoelastic method for making visible the stresses in models of engineering structures. In 1816 Brewster invented the kaleidoscope, and he later improved **Wheatstone**'s stereoscope by fitting refracting lenses. In 1818 he was awarded the Rumford Gold and Silver medals of the Royal Society for his discoveries on the polarization of light. He was one of the chief originators of the British Association for the Advancement of Science (1831), and was knighted in 1832. Brewster was appointed Principal of St Salvator and St Leonard's, St Andrews (1838), and was principal of Edinburgh University from 1859 until his death.

BRIDGMAN, Percy Williams

(1882–1961)

American physicist and philosopher of science, born in Cambridge, Massachusetts. The son of a journalist, Bridgman entered Harvard in 1900 and remained there, becoming Hollis Professor of Mathematics and Natural Philosophy (1926), Higgins Professor (1950) and, on his retirement, Professor Emeritus (1954–61). Soon after completing his PhD in 1908 he initiated experiments on the properties of solids and liquids under high pressure, research for which he was awarded the Nobel Prize for Physics in 1946. These difficult practical investigations eventually elevated the pressures attainable from 6 500 to approximately 400 000 atmospheres. Bridgman designed much of his own equipment: a seal which actually improved with higher pressures left the limits of experiment essentially dictated only by container strength. Studying thermal and electrical conductivity, compressibility, phase changes, and the often strange physical properties of liquids and solids under extreme conditions, he obtained a new form of phosphorus and demonstrated that at high pressures, viscosity increases with pressure for most liquids. Much of this research was relevant to geophysics. In a manner closely allied to his dominant experimentalist viewpoint, Bridgman became deeply concerned with the foundations of his subject and the nature of theorizing in physics. The 'operationalist' approach, laid out in *The Logic of Modern Physics* (1927), asserted the identity of any concept with the set of operations (physical and mental) involved in its experimental measurement. Bridgman hoped to bring this contentious but fruitful mode of understanding to bear upon relativity theory and quantum theory; and later to the social, political, psychological and religious domains. When in 1961 Bridgman found himself increasingly debilitated and incurably ill with cancer, he took his own life.

BRIGGS, Henry
(1561–1630)

English mathematician, born in Warley Wood, Halifax. In 1581 he graduated at St John's College, Cambridge, and he became a Fellow of the college in 1588. He was later appointed as the first Professor of Geometry at Gresham College, London (1596), and as the first Savilian Professor of Geometry at Oxford (1619). He visited **Napier** in 1616 and 1617, and with Napier's agreement proposed the use of the base 10 for logarithms instead of the natural logarithm base *e* used by Napier. This was an important simplification for the practical use of logarithms in calculation. Briggs calculated and published logarithmic and trigonometric tables to 14 decimal places.

BRIGHT, Richard
(1789–1858)

English physician, born in Bristol. He studied medicine in Edinburgh, London, Berlin and Vienna. His *Travels from Vienna through Lower Hungary* (1818) recorded interesting observations on gypsies and diplomacy in Vienna at the close of the Napoleonic period. From 1820 he was on the staff at Guy's Hospital, London, where he helped found the *Guy's Hospital Reports*, was a successful teacher, and made many careful clinical and pathological observations. His classic *Reports of Medical Cases* (2 vols, 1827–31) contain, *inter alia*, his description of kidney disease ('Bright's disease'), with its associated oedema and protein in the urine. He also left important observations on diseases of the nervous system, lungs and abdomen, and was much sought after as a consultant physician, becoming Physician Extraordinary to Queen Victoria. His projected textbook of medicine, in collaboration with **Addison**, was never completed.

BRINSTER, Ralph Lawrence
(1932–)

American molecular biologist, born in Montclair, New Jersey. He was educated at Rutgers University where he received his BS in 1953, and at the University of Pennsylvania where he obtained a VMD (1960) and PhD in physiology (1964). Since 1960 he has taught at the School of Veterinary Medicine at the University of Pennsylvania, where he became Rich King Mellon Professor of Reproductive Physiology. With **Palmiter**, he was the first person to successfully introduce a human gene into the germ line of a mouse. The procedure involved the microinjection of the human growth hormone gene into a mouse embryo and replacing the embryo into the mother's uterus. Mice produced in this way were found to be significantly larger than their normal counterparts, indicating that the human growth hormone gene had been active. This technique has now been used for a variety of mammalian genes and is a vital tool in the investigation of regulatory mechanisms controlling gene expression. It is hoped that the technique can be extended and modified to replace faulty genes in human genetic diseases such as cystic fibrosis.

BRITTEN, Roy John
(1919–)

American molecular biologist, born in Washington DC. He graduated with a BS from the University of Virginia in 1941 and received his PhD from the University of Princeton in 1951. Since then he has been a staff member in the Department of Terrestrial Magnetism at the Carnegie Institution of Washington. From 1973 to 1981 he was also Senior Research Associate at Caltech, and he was appointed Distinguished Carnegie Senior Research Associate in Biology in 1981. He is a Fellow of the American Academy of Arts and Sciences, and a member of the US National Academy of Sciences. Britten discovered, with **Davidson**, repeated DNA sequences in the genomes of higher organisms. Using recently invented nucleic acid hybridization technology, they showed that the genomes of higher organisms contain much more DNA than is required to code for specific genes giving rise to protein. This DNA is organized into unique, single-copy DNA sequences (coding for single genes), moderately repetitive DNA (coding for gene families), and highly repetitive sequences which are repeated hundreds of thousands of times in the genome. The role of this highly repetitive DNA remains unknown, though the positioning of these sequences flanking most genes has led to speculation that they may play a part in the coordinate regulation of genes.

BRITTON, Nathaniel Lord
(1859–1934)

American botanist, born in Staten Island, New York. Originally a geologist, having trained at the School of Mines, Columbia College, he became Professor of Botany at Columbia in 1891, where he reorganized the herbarium and library on taxonomic principles, and was the initiator and first director of the New York Botanical Garden (1896–1921). He edited the *Bulletin of the Torrey Botanical Club*, and published *Flora of Richmond County* in 1879. He travelled in the West Indies and Puerto Rico, and wrote *Illustrated Flora of the Northern United States, Canada and the British Possessions*

(1896–8), and *Flora of Bermuda* (1918). He was also co-author of the four-volume monographic work *The Cactaceae* (1919–23).

BROCA, (Pierre) Paul

(1824–80)

French surgeon and anthropologist, born in Sainte-Foy-la-Grande, Gironde. Educated at the University of Paris, where he received his MD in 1849, Broca became assistant professor at the Faculty of Medicine in 1853. He first located the motor speech centre in the brain (1861), since known as the convolution of Broca, and did research on prehistoric surgical operations. His anthropological investigations gave strong support to **Charles Darwin**'s theory of the evolutionary descent of man.

BROECKER, Wallace

(1931–)

American chemical oceanographer and climatologist, born in Chicago. He was educated at Columbia University (BA 1953, PhD 1958), and has remained there throughout his career, becoming assistant professor in 1959, associate professor in 1961, and professor in 1964. Since 1977 he has been Newberry Professor of Geology. Broecker's work has been broad-ranging, involving measurements of chemical elements in the oceans, salinities, upwelling and radiocarbon dating. He has pioneered techniques to track chemical isotopes over vast distances and back thousands of years in time. He formulated simple 'box' models of the ocean system, utilizing salinities, water temperatures and global wind patterns to quantify ocean circulation, both for the present and back to glacial times. More recently (1991) this information has been brought together in the form of an overview of an oceanic 'global conveyor' which may fluctuate with time, changing the Earth's climate.

BROGLIE, Louis-Victor Pierre Raymond

See **DE BROGLIE, Louis-Victor Pierre Raymond**

BRONGNIART, Alexandre

(1770–1847)

French naturalist, chemist and geologist, born in Paris. Early in his career he attempted to improve the art of enamelling and subsequently became director of the porcelain factory at Sèvres (1800). From 1808, he was professor at the Sorbonne and in 1822 he succeeded **Haüy** as Professor of Mineralogy at the Natural History Museum in Paris. He was a close associate of **Cuvier**, and together they undertook classical studies of the geological strata of the Paris Basin using the nature of fossils within the beds to map out the sequence. They thereby deduced the fundamental stratigraphical principle whereby the changing fossil record can be related to the relative age of rock strata, published in *Essai sur la géographie minéralogique des environs de Paris* (1808). They noted the alternation of freshwater and marine strata in the Tertiary rocks around Paris, and interpreted this as being the result of catastrophic processes. Brongniart's zoological interests led him to elucidate the zoological and geological relations of trilobites. His son Adolphe Théodore (1801–76) was a noted palaeobotanist.

BRONK, Detlev Wulf

(1887–1975)

American neurophysiologist, born in New York City. After naval service during World War I, he trained in both electrical engineering and physics, gaining a PhD in 1926 from the University of Michigan. Intent on applying his mathematical and physical skills to physiology he spent 1927–8 in England working with **Adrian** in Cambridge on the biophysical properties of the motor nerve fibre which they succeeded in isolating, and with **Hill** in London on temperature changes in muscles during activity. In 1930 he became Director of the Eldridge Reeves Johnson Foundation for Medical Physics at the University of Pennsylvania, where he remained (with a few breaks including military medical research in World War II) until 1949 when he was appointed President of the Johns Hopkins University. From 1953 to 1968 he was President of the Rockefeller Institute (later University). The increasing administrative responsibilities of these positions deflected him away from his physiological research which had been most productive during his period at Pennsylvania, when he had contributed significantly to the study of the autonomic nervous system, especially the mechanisms controlling cardiac function. His work was also successful in integrating physical concepts and instrumentation into experimental physiology and in promoting a biophysical approach to medical research.

BRØNSTED, Johannes Nicolaus

(1879–1947)

Danish physical chemist, born in Varde, Jutland. He trained initially as a chemical engineer at the Technical University of Denmark, but then studied chemistry at the University of Copenhagen. After graduating in 1902, he spent a period in industry before becoming an assistant at the University of Copenhagen (1905). In 1908 Brønsted was appointed to the new Chair of Physical Chemistry at the University of Copenhagen. In 1930 he became Director of a new Physico-Chemical Institute in Copenhagen. Most of his contributions to physical chemistry concerned the behaviour of solutions. His studies of the effect of ionic strength on the solubilities of sparingly soluble salts provided strong experimental support for the **Debye–Hückel** theory (1923) and his analogous studies of rates of reaction involving ions were also interpreted in terms of the same theory (1920–4). His redefinition of acids and bases as proton donors and acceptors respectively (1923) was similar to that developed independently by **Lowry** around the same time. This was connected with his recognition of general acid–base catalysis: the relation between catalytic power and acid or base strength is still known as the Brønsted equation. His reformulation of thermodynamics generated controversies, which had not been resolved at the time of his death.

BROTERO, Felix da Silva Avellar
(1744–1828)

Portuguese botanist, founder of Portuguese botanical science and regarded as the 'Portuguese **Linnaeus**'. Brotero was born in Santo Antonio do Tojàl, near Lisbon, the son of a doctor. Orphaned at the age of two, he received education in Latin, logic and metaphysics from his aunt and uncle, and from the Arrabidian monastery. Together with a friend, he went to Paris in 1778 to evade the Inquisition; he studied natural sciences there and medicine at Reims. In 1780, at the beginning of the decade that saw the outbreak of the French Revolution, Brotero returned to Portugal and was appointed Professor of Botany and Agriculture at the University of Coimbra. He transformed the Coimbra botanical garden into one of the best in Europe. Recognizing that Portugal had no published Flora, he set about compiling one. In 1788 he published *Compendio de Botanica* (2 vols), which treated the whole of botanical science then known, and was the first time such a work had been published in Portuguese. He created Portuguese botanical terminology, inventing many new terms. In his *Flora Lusitanica* (2 vols, 1804–5), a simplified version of the Linnaean classification was used. Simultaneously, he began work on a more detailed treatment of selected genera, published as *Phytographia Lusitaniae Selectior* (1st edition 1800, withdrawn because of many printing errors; 2nd edition, 3 vols, 1816–27). In 1820 he was elected to represent the province of Estremadura in the Cortez (Portuguese parliament). Possibly because of this, Brotero received little distinction in Portugal for his many achievements, although he was held in high honour abroad.

BROUNCKER, William, 2nd Viscount Brouncker of Castle Lyons
(1620–84)

Irish mathematician. Educated at Oxford, he was a founder member and first President of the Royal Society. He expressed π as a continued fraction, and found expressions for the logarithm as an infinite series. He was a friend of Samuel Pepys, often mentioned in his *Diary*. With **John Wallis** he solved **Fermat**'s questions about the misnamed **Pell**'s equation, posed in the form of a challenge to other mathematicians, giving a general method for their solution.

BROUWER, Luitzen Egbertus Jan
(1881–1966)

Dutch mathematician, born in Overschie. He showed precocious intellectual powers and at the age of 16 entered Amsterdam University, where he later became professor (1912–51). His doctoral thesis was on the foundations of mathematics, an area in which he continued to work throughout his life. He founded the intuitionist or constructivist school of mathematical logic, which does not accept the law of the excluded middle when infinite sets are involved, and in which the existence of a mathematical object can only be proved by giving an explicit method for its construction. This places severe restrictions on much modern mathematics, and has not been accepted by mathematicians in general, though it continues to interest logicians and philosophers. It has the incidental virtue of forcing mathematicians who accept it to give explicit constructions of every mathematical object they describe. Brouwer also made fundamental advances in topology, introducing the concept of simplicial approximation, the degree of a mapping, and proving the invariance of dimension, enabling spaces of many kinds to be given a dimension. An important theorem on fixed points of mappings is named after him. One of the first proofs of the intuitively obvious but tricky Jordan curve theorem (that a closed curve which does not cross itself divides the plane into two regions) is also due to him.

BROWN, Alexander Crum
See **CRUM BROWN, Alexander**

BROWN, Herbert Charles
(1912–)

American chemist, born in London. Born to Russian émigrés in London, his family moved when he was two years old to Chicago, where his father ran a hardware store. When his father died he had to drop out of high school to run the store, but eventually he enrolled in Crane Junior College and took his first chemistry course. When the college closed, he moved to Wright Junior College and then to the University of Chicago for graduate work, where he first encountered the chemistry of boron. His thesis dealt with the reduction of carbonyl compounds by diborane. Apart from some excursions into physical organic chemistry, the use of boron compounds in organic synthesis has been the dominating theme of his work and it was for this that he was awarded the Nobel Prize for Chemistry in 1979. His study of carbonium ions brought him into conflict with **Christopher Ingold** and, more significantly, with **Winstein**, who strongly supported the idea of delocalized or non-classical carbonium ions, an idea firmly rejected by Brown. The conflict has never been fully resolved, but non-classical ions probably do play some part in solution chemistry. Brown spent most of his professional life at Purdue University. His work has continued unabated since his retirement, his main interest being developing methods for the synthesis of specific optical isomers by the use of chiral auxiliaries. In 1987 he received the Priestley Medal of the American Chemical Society for his services to chemistry.

BROWN, Michael Stuart

(1941–)

American molecular geneticist, born in New York City. He was educated at the University of Pennsylvania, and since 1977 has been Paul J Thomas Professor of Genetics at the University of Texas and Director of the Center for Genetic Diseases. While a medical intern at Boston General Hospital early in his career (1966–8), he met **Joseph Goldstein**, and with Goldstein moved to Texas, where they began to work on cholesterol metabolism. Cholesterol is produced by mammalian cells as well as being taken up into cells from food, and is carried in the bloodstream by proteins called LDLs (low-density lipoproteins). Brown worked on the genetic disease hypocholesterolemia, which results in abnormally high levels of cholesterol in the bloodstream due to the failure of cells to regulate the production rate. He found that sufferers from the disease lack a receptor on their cell surfaces to which the LDLs bind, thereby stopping the production of cholesterol. In 1984 Brown and Goldstein elucidated the gene sequence which codes for the LDL receptor, and opened up the possibility of synthesizing drugs to control cholesterol metabolism. They were jointly awarded the 1985 Nobel Prize for Physiology or Medicine.

BROWN, Rachel Fuller

(1898–1980)

American biochemist, born in Springfield, Massachusetts. Educated at the University of Chicago, she began her career as a chemist at the New York State Department of Health in 1926. Brown made important studies of the causes of pneumonia and the bacteria involved. Shortly after the end of World War II, by which time some methods of controlling bacterial forms of disease had been introduced, Brown isolated the first antifungal antibiotic, Nystatin (1949). She was awarded the Pioneer Chemist Award of the American Institute of Chemists in 1975.

BROWN, Robert

(1773–1858)

Scottish botanist, born in Montrose, the son of an Episcopal clergyman. Educated at Aberdeen and Edinburgh, he served with a Scottish regiment in Ireland (1795). In 1798 he visited London, where his ability so impressed **Banks** that he was appointed naturalist to Matthew Flinders's coastal survey of Australia in 1801–5. He brought back nearly 4000 species of plants for classification. Appointed librarian to the Linnaean Society, he published *Prodromus Florae Novae Hollandiae et Insulae Van-Diemen* (1810), containing many of **Ferdinand Bauer**'s botanical illustrations. He adopted, without modifications, **Bernard Jussieu**'s natural system of plant classification, thus encouraging its general acceptance in place of **Linnaeus**'s artificial 'sexual system'. In 1810 he received charge of Banks's library and splendid collections, and when they were transferred to the British Museum in 1827 he became botanical keeper there. He is renowned for his investigation into the impregnation of plants. He was the first to note that, in general, living cells contain a nucleus, and to name it. In 1827 he first observed the 'Brownian movement' of fine particles in a liquid, significant in shaping physicists' later ideas on liquids and gases.

BROWN-SÉQUARD, Édouard

(1817–94)

French physiologist. Born in Port Louis, Mauritius, the son of a Philadelphia sea captain and a French mother, Brown-Séquard studied at Paris, receiving his MD in 1846. He practised medicine in the USA, and was briefly professor at Virginia Medical College in Virginia. Having for many years divided his time between the USA, London and Paris, and making his living through private practice, he was appointed Professor of Physiology at Harvard (1864), at the School of Medicine in Paris (1869–73), and the Collège de France from 1878, succeeding **Bernard**, upon whose research programme Brown-Séquard based many of his own investigations. He proved an ingenious physiological researcher, experimenting in particular on blood, muscular irritability, animal heat, the spinal cord, and the nervous system, where his work on the vasoconstrictive nerves proved especially significant. He also demonstrated the artificial production of epileptic states through lesions of the nervous system. A pioneer of endocrinology, he proved that removal of the adrenal glands would always produce death in animals. He claimed to have rejuvenated himself with extracts from freshly-killed dog testicles, thereby paving the way for the later experiments of **Voronoff**.

BRUCE, Sir David
(1855–1931)

Australian-born Scottish microbiologist and physician, born in Melbourne, after whom the cattle disease brucellosis is named. As an officer in the Royal Army Medical Corps (1883–1919), he identified in Malta the bacterium that causes undulant fever in humans, named *Brucella* (1887). He was Assistant Professor of Pathology in the army medical school (1889–94), and worked to improve studies in pathology. In 1895 in South Africa he discovered that the tsetse fly was the carrier of the protozoal parasite (*Trypanosoma brucei*) responsible for the cattle disease nagana, and later showed that sleeping sickness in humans was also a trypanosome disease, transmitted by the same insect. Elected FRS in 1899 and knighted in 1908, he was Commandant of the Royal Army Medical College during World War I.

BRUNEL, Isambard Kingdom
(1806–59)

English engineer and inventor, born in Portsmouth, son of Sir Marc Isambard Brunel (1769–1849). In 1823, after two years spent at the Collège Henri Quatre in Paris, he entered his father's office. He helped to plan the Thames Tunnel, and in 1829–31 designed the Clifton Suspension Bridge, which was completed only in 1864 with the chains from his own Hungerford Suspension Bridge (1841–5) over the Thames at Charing Cross. He designed the *Great Western* (1838), the first steamship built to cross the Atlantic, and the *Great Britain* (1845), the first ocean screw-steamer (now preserved at Bristol). The *Great Eastern*, until 1899 the largest vessel ever built, was constructed to his design in collaboration (strained at times) with **John Scott Russell** from whose yard in Millwall the 'Great Ship' was launched at the second attempt in January 1858, three months late, and 40 years ahead of the technology of the time. In 1833 he was appointed engineer to the Great Western Railway, and constructed all the track, tunnels, bridges and viaducts on that line. He chose the 7 foot 'broad gauge' for the track, when almost every other railway at the time had adopted George Stephenson's 4 feet 8½ inches or less, but eventually the Great Western had to conform, and the last broad gauge train ran in May 1892. Among docks and harbours constructed or improved by him were those of Bristol, Monkwearmouth, Cardiff and Milford Haven.

BRÜNNICH, Morton Thrane
(1737–1828)

Danish naturalist and mineralogist, born in Copenhagen. He originally studied oriental languages and theology. Under the influence of the writings of **Linnaeus**, he soon turned to the study of natural history. Initially he investigated the insects of Denmark and contributed entomological observations to E Pontopiddon's *Danske Atlas* (1763–81). He looked after private natural history collections in Copenhagen which contained many ornithological specimens from northern Europe. These formed the basis of his *Ornithologia Borealis* (1764), in which he described for the first time the Manx shearwater and great northern diver. In the same year he published *Entomologia*. He became a lecturer in natural history and economy at Copenhagen, where he established a natural history museum. He travelled widely in Europe studying mineralogy and natural history, and published a book on Mediterranean fish, *Icthyologia Massiliensis* (1768), and a student text, *Zoologiae Fundamenta* (1771). Although he worked on the museum collection, publishing the first volume of *A History of Animals and the Zoological Collection in the Natural History Museum of the University* (1782), most of his energies were latterly directed towards mineralogy.

BUCH, (Christian) Leopold von
(1774–1853)

German geologist and traveller of independent means, born in Stolpe, one of 13 children. At the age of 15 he was sent to Berlin to study mineralogy and chemistry, and subsequently he went to Freiburg to study under **Abraham Werner**. Schooled in Neptunian doctrine, he firmly believed that basalt formed by crystallization from water.

However, following visits to Italy and France, he radically changed his views to contradict Werner's teaching and accepted basalt as the product of volcanic activity. Fieldwork in the Canary Islands led to his proposal of a 'craters of elevation' hypothesis to account for some 'stratified' pyroclastic deposits, suggesting that entire volcanic cones were uplifted from below after their formation. This idea was much cited but ultimately refuted by **Lyell**. As a result of travels in Scandinavia (1806–8), Buch recognized the uplift of land relative to sea-level. He published the first coloured geological map of Germany in 42 sheets (1826), making important contributions to the knowledge of its Triassic and Jurassic stratigraphy, and also undertook important early studies of Alpine geology. He introduced the term gabbro and described other igneous rocks. In later years he turned his attention to palaeontology, formulating a classification of cephalopods (1829–30).

BUCHAN, Alexander

(1829–1907)

Scottish meteorologist and oceanographer, born in Kinnesswood, Kinrosshire. Educated at the Free Church College for teachers at Edinburgh and later at Edinburgh University, he became Secretary of the Scottish Meteorological Society in 1860 and held this position throughout his career. He was editor of the *Scottish Meteorological Society Journal* from 1861 and contributed 64 articles. From 1877 he was a member of the Meteorological Council which directed the operations of the Meteorological Office, and it was largely through his efforts that observatories were established at the summit of Ben Nevis (1883) and near its base at Fort William. His publications were in oceanography and biometeorology as well as meteorology. In 1868 he produced the first charts of storm tracks across the Atlantic. He prepared the meteorological report of the *Challenger* expedition of 1876, and also contributed to the oceanographic section of that report. In his major work *Report on Atmospheric Circulation* (1889), he presented global charts of monthly mean temperature, pressure and wind direction for the whole year. His studies of weather charts led him to conclude that the British climate is subject to warm and cold spells falling approximately between certain dates each year, the so-called Buchan spells. He was the first recipient of the Symons Gold Medal of the Royal Meteorological Society.

BUCHANAN, John Young

(1844–1925)

Scottish chemical oceanographer, born in Glasgow. He graduated from the University of Glasgow in 1863 and continued his study of chemistry widely in mainland Europe. He devised a stop-cock water bottle for intermediate water sampling which would retain dissolved gases, used during the *Challenger* expedition (1872–6), for which he was in charge of shipboard chemistry. He made routine determinations of specific gravity of sea-water, and analysed dissolved gases and carbon content. Buchanan produced the first reliable surface salinity and temperature map of the oceans, his distributions contradicting those of **Humboldt**. He demonstrated that vertical currents bring cold water to the surface, and from a study of temperate lakes, established the concept of the thermocline—the base of the warm surface water (1886). Later he worked for a year in Edinburgh with **Wyville Thomson** and **John Murray** but became involved in disagreements with the Treasury, who refused to employ him further. However, he continued his research on summer cruises on his own yacht *Mallard* west of Scotland (1878–82), sailing aboard cable ships and frequently with Prince Albert of Monaco. He discovered a strong easterly flowing current at 30 fathoms depth below the westerly southern equatorial current which he recognized as being a constant and important factor in oceanic circulation. With Wyville Thomson he discovered the large canyon at the mouth of the Congo river.

BUCHNER, Eduard

(1860–1917)

German chemist, born in Munich, brother of **Hans Buchner**. He began his study of chemistry at the Technische Hochschule in Munich, but because of financial difficulties, he was forced to leave and work in canneries for four years. In 1884 he resumed his studies, this time at the Bavarian Academy of Sciences. While working at the Institute of Plant Physiology, he became interested in alcoholic fermentation and showed that, contrary to **Pasteur**'s contention, the absence of oxygen is not necessary for fermentation. For these studies he was awarded the Nobel Prize for Chemistry in 1907. In 1893 he moved to the University of Kiel, and after appointments at Tübingen and Berlin, he became professor at the University of Breslau in 1909. Two years later he moved to Würzburg. In 1914 he volunteered for military service; he was killed at Focsani, Romania, in August 1917.

BUCHNER, Hans

(1850–1902)

German bacteriologist, born in Munich, brother of **Eduard Buchner**. He obtained his MD from the University of Leipzig in 1874, later becoming Professor at Munich (1880–1902) and Director of the Institute of Hygiene from 1894. In pioneering work on the proteins now known as gamma globulins, he discovered that blood serum contains protective substances against infection, which combine with invading micro-organisms to kill them by a number of mechanisms. He is also known for developing methods for studying anaerobic bacteria, those able to grow only in the absence of oxygen.

BUCKLAND, William
(1784–1856)

English geologist and clergyman, born in Tiverton. He was particularly well known for his attempts to reconcile his religious and geological beliefs. He was appointed reader in mineralogy at Oxford in 1813 and became an enthusiastic lecturer, inspiring many students including **Lyell**. In his inaugural lecture (1819) he asserted that geological observations could be used to support the Bible and religion. His published works included *Vindiciae Geologicae* (1820) and *Geology and Mineralogy Considered With Reference to Natural Theology* (1836), both serious attempts to link Biblical and geological events. Buckland undertook geological fieldwork to gain support for this 'diluvial' hypothesis, and to account for some geological phenomena with reference to the great Biblical flood of Noah. He believed that he had discovered evidence for the theory in Kirkdale cavern; this was published in *Reliquiae Diluvianae* (1823). He is credited with the discovery of the first dinosaur remains, a tooth of *Megalosaurus* (1824), and was greatly intrigued by the iguanas and fossil mammals discovered by **Charles Darwin** during the voyage of HMS *Beagle*. In 1845 he became Dean of Westminster. His son Frank Buckland (1826–80) was an important agricultural zoologist and early conservationist.

BUFFON, George-Louis Leclerc, Comte de
(1707–88)

French naturalist, born in Montbard, Burgundy, the son of a wealthy lawyer. After studying law at the Jesuit college in Dijon, he devoted himself to science, and while on a visit to England (1733) translated **Isaac Newton**'s *Fluxions* into French. In 1739 he was appointed Director of the King's Botanic Garden (Jardin du Roi) and the Royal Museum. After receiving various high honours, he was made Comte de Buffon by Louis XV. His most celebrated work is the monumental *Histoire Naturelle* (44 vols, 1749–67), in which he attempted to discuss all the then-known facts of natural science. Although he never entirely broke with the 'chain of being' ideas sanctioned by the church, his work includes many glimpses of an evolutionary perspective. He believed that all species were linked by intermediate gradations, and stated that God did not directly preoccupy himself with their minor details. Particular animal plans (eg cats) were laid down, but related species (eg lions, tigers) could arise from a common ancestor, in-keeping with their particular habitats. He also tackled geology, and in *Les Époques de la Nature* (1788) described the Earth's features as formed by processes currently observable, a striking foreshadowing of **Lyell**'s 'uniformitarian' principle published 40 years later. Buffon's works were influential up to the time of **Charles Darwin**, and he gave his name to the street on which the Natural History Museum and Botanic Garden in Paris now stand.

BULLARD, Sir Edward Crisp
(1907–80)

English geophysicist, born in Norwich. He was educated at Clare College, Cambridge, where he received a BA in physics (1929) and a PhD in geophysics for a pendulum gravity survey of the East African rift valley (1935). He then turned his attention to seismic refraction experiments on land, and on **Ewing**'s persuasion, extended this to offshore research and accomplished the first British seismic experiment at sea under sail (1938). He initiated the measurement of heat flux from the interior of the Earth (1940) and made the first effective measurement of the heat flow through oceanic crust; in each of these fields he was involved in the design, modification or construction of the instruments. During World War II he worked on degaussing ships to protect them from German magnetic mines. In Toronto as Professor of Physics (1946–9), he worked on the dynamo theory of the Earth's magnetic field (1949) then as Director of the National Physical Laboratory in London (1950–5) he employed their early computer in the first numerical approach to dynamo theory. On returning to Cambridge, as assistant director then as professor (1955–74), he brought the Department of Geodesy and Geophysics into world recognition. His computer-fit of the continents (1965) was instrumental in bringing the theory of continental drift back into favour. He was a frequent summer visitor to Scripps Institution of Oceanography, where he moved on his retirement in 1974. The geophysical laboratories at Cambridge and a fracture zone joining the Mid-Atlantic Ridge to the Scotia Arc are named after him. He was elected FRS in 1941 and knighted in 1953.

BULLEN, Keith Edward
(1906–76)

New Zealand mathematician and geophysicist, born in Auckland. He was educated at Auckland University where he studied mathematics, physics and chemistry, and received his PhD at St John's College, Cambridge (1934). After holding teaching posts in schools he became a lecturer at Auckland University College (1928–40) and Melbourne University (1940–5), and in 1946 he was appointed Professor of Mathematics at the University of Sydney. As a research student under **Harold Jeffreys**, he redetermined the differences in arrival times for primary and secondary waves and locations for the epicentres of global earthquakes, and (without a computer) recalculated the traveltime tables to four decimal places. This resulted in the *Jeffreys–Bullen Tables* (1940) which are still in use; they included the Earth ellipticity corrections suggested by **Gutenberg**. In the course of this work it became clear that the Earth has a layered structure with a dense core; around the same time, **Lehmann**, Gutenberg, **Byerley** and **Oldham** were arriving at similar conclusions. Bullen divided the Earth into seven density layers consistent with the distribution of mass and seismic shadow zones. He was elected FRS in 1949.

BUNSEN, Robert Wilhelm
(1811–99)

German chemist and physicist, born in Göttingen. He studied at the University of Göttingen and for his PhD produced a Latin dissertation on hygrometers. He succeeded **Wöhler** as professor at Cassel, went to Marburg in 1838, and after a short period at Breslau, followed **Leopold Gmelin** as professor at Heidelberg (1852), where he remained until his retirement. He was a talented experimentalist, although the eponymous burner, for which he is best known, is a modification of something developed (in England) by **Faraday**. He did invent the grease-spot photometer, a galvanic battery, an ice calorimeter and an actinometer. One of his most significant contributions to modern science was his use of spectroscopy to detect new elements. By this technique, and in collaboration with **Kirchhoff**, caesium and rubidium were discovered in the mineral waters of Dürkheim. His most important work was his study of organo-arsenic compounds such as cacodyl oxide. These compounds were seen at the time to give support to the radical theory of chemical combination proposed by **Lavoisier** and **Berzelius**. Following the partial loss of the sight of one eye during an experiment, he forbade the study of organic chemistry in his laboratory.

BURBIDGE, Geoffrey
(1925–)

English astrophysicist, born in Chipping Norton. He graduated from Bristol University in 1946, and received his doctorate from University College London. He then went to the USA, working at Harvard, Chicago, Mount Wilson and Palomar, and San Diego. In the mid-1950s, in collaboration with **Hoyle**, he tried to assess some of the possible astrophysical consequences of antimatter. His most famous paper was published in 1957 and was written in conjunction with Hoyle, **William Fowler** and his wife **Margaret Burbidge**. This paper applied nuclear physics to an astrophysical situation, and in it they solved the problem of the creation of the higher elements in evolved stars. This required both rapid neutron processes, taking place inside supernovae explosions, and slow neutron processes occurring in red giant stars. In 1967, again with his wife, Burbidge published an early and important book on quasars, these being objects which appear like faint stellar sources but which emit enormous amounts of radio energy. In 1970 he showed that the light-emitting stars in elliptical galaxies only account for 25 per cent of the total mass, highlighting the 'missing mass' mystery, which continues to this day—most of the matter in the universe cannot be detected by its radiation.

BURBIDGE, (Eleanor) Margaret, née Peachey
(1923–)

British astronomer, born in Davenport and educated at University College London (1941–7). Her lifelong interest in astronomical spectroscopy began in London where she served as Assistant Director of the university observatory (1948–51). In 1951 she moved to the USA and held appointments at Yerkes Observatory, Caltech and the University of California at San Diego, where in 1964 she was appointed Professor of Astronomy. In 1972 she became Director of the Royal Greenwich Observatory, a post which she relinquished the following year to return to her chair in California. Since 1979 she has been Director of the Center for Astrophysics and Space Science at San Diego. In collaboration with her husband **Geoffrey Burbidge**, **Hoyle** and **William Fowler**, she published the results of theoretical research on nucleosynthesis, ie the processes whereby the heavy chemical elements are built up in the cores of massive stars, a discovery of fundamental importance to physics. On the observational side Burbidge's main field of research is in the spectra of galaxies and quasars.

BÜRGI, Jost See **BYRGIUS, Justus**

BURKITT, Denis Parsons
(1911–93)
British surgeon and nutritionist, born in Enniskillen, Northern Ireland, and educated at Dublin University. Service with the Royal Army Medical Corps took him to Uganda, where he worked as a general surgeon after World War II, and began a series of clinical and epidemiological observations on a common childhood cancer found there (from 1957). It behaved as if it were infectious and subsequent research showed that the cancer—now known as Burkitt's lymphoma—was caused by a virus. Burkitt's other major contribution related the low African incidence of coronary heart disease, bowel cancer and other diseases to the high unrefined fibre in the native diet. He became one of the leading apostles of fibre in Western diets, which earned him the nickname 'the bran man'. He was elected FRS in 1976, a rare honour for a surgeon.

BURKS, Arthur Walter
(1915–)
American computer scientist and philosopher, born in Duluth, Minnesota. Educated at DePauw University and the University of Michigan, where he was awarded a PhD in 1941, he began a career in teaching. By the end of World War II he had joined **Eckert** and **Mauchly** in the construction of the ENIAC—America's major wartime computing project—and also was involved in the logical design of that machine's successor, the EDVAC. In 1946 he joined **von Neumann**'s computing team at the Institute for Advanced Study in Princeton. With von Neumann and **Goldstine**, he wrote the influential paper *Preliminary Discussion of the Logical Design of an Electronic Computing Instrument* (1946). After he left Princeton in 1948, he also conducted pioneering work with von Neumann on cellular automata. Burks became Professor of Philosophy at the University of Michigan in 1954. Author of several publications on computers and philosophy, he was co-author of *The First Electronic Computer: The Atanasoff Story* (1971), in which he examined priorities concerning the birth of the electronic digital computer.

BURNELL, (Susan) Jocelyn
See **BELL BURNELL, (Susan) Jocelyn**

BURNET, Sir (Frank) Macfarlane
(1899–1985)
Australian immunologist and virologist, born in Traralgon in eastern Victoria. Trained in medicine at Melbourne University, he graduated in 1922, and proceeded to the higher degree of MD in 1924. After postgraduate work in Melbourne, Burnet moved to London, earning his passage as a ship's surgeon, to work in bacteriological research in the Lister Institute, before returning to Melbourne in 1928 to the Walter and Eliza Hall Institute for Medical Research. He remained there until 1965, becoming Assistant Director in 1934 and Director in 1944. However, he returned to London for three years in 1931, to work at the National Institute for Medical Research under the direction of **Dale**. Here Burnet began working on viruses, and perfected the technique of cultivating viruses in living chick embryos. For the next 20 years he made important contributions to understanding the chemistry and biology of many animal viruses, especially the influenza virus. His work on viruses and immunization stimulated his interest in immunology, and from the end of the 1950s he turned his attention to immunological problems, especially the phenomenon of graft rejection. His work transformed the undertanding of how the entry of foreign substances (antigens) into the body results in the production of specific antibodies which bind and neutralize the invader. Since there is an enormous number of potential antigens, a satisfactory theory was required to account for the manufacture of these highly specific antibodies to deal with each of them. Convinced by the theories of **Jerne** that all possible antibodies already existed, Burnet postulated his 'clonal selection theory'. This suggested that antibodies were present on specialized white blood cells, the lymphocytes, each of which carried a unique antibody. When the white blood cell bound to an invading antigen, it would be stimulated to reproduce, thus producing large quantities of its unique antibody. He predicted that if an embryo was injected with an antigen, tolerance to it would be induced and no antibodies would subsequently be produced against it. **Medawar**'s work on immunological intolerance in relation to skin grafting and organ transplants provided the experimental evidence to support Burnet's theory, and in 1960 the two men shared the Nobel Prize for Physiology or Medicine. Burnet was elected FRS in 1942, knighted in 1951, and awarded the Order of Merit in 1958.

BURNSIDE, William
(1852–1927)
English mathematician, born in London. He graduated Second Wrangler from Cambridge in 1875, and eventually became Professor of Mathematics at the Royal Naval College, Greenwich (1885–1919), finding that the post

left him adequate time for research. He worked in mathematical physics, complex function theory, differential geometry and probability theory, but his lasting work was in group theory. His *Theory of Groups* (1897) was the first English textbook on the subject and is still of value, containing much original research and posing the famous Burnside problem: that finite groups of odd order have a chain of non-trivial normal subgroups. This was not solved until 1962 by Walter Feit and John Griggs Thompson, opening the way to the complete classification of finite groups, accomplished in the late 1980s.

BURY, Charles Rugeley
(1890–1968)

English physical chemist, born in Henley-on-Thames. He studied chemistry at Trinity College, Oxford, from 1908 to 1911. After acting as demonstrator in the Balliol–Trinity Laboratory and some study at Göttingen, he became an assistant lecturer in chemistry at the University College of Wales, Aberystwyth, in 1913. He joined the army soon after the outbreak of war in 1914, and on demobilization in 1919, he returned to Aberystwyth, where he served on the chemistry staff until 1943. For the last decade of his career he worked for ICI at Billingham. In Aberystwyth he made many experimental studies involving colligative properties, electrolytic conductivities and transport numbers, specific heats, viscosities, the phase rule and micelle aggregation. Bury's most notable contribution, however, was theoretical in nature. In 1921 he deduced on chemical grounds the electronic structures of the atoms of the elements. His contribution has rarely received proper recognition and has often been confused with the work of **Langmuir** and of **Niels Bohr**.

BUSH, Vannevar
(1890–1974)

American electrical engineer and inventor, born in Everett, Massachusetts. He graduated from Tufts College in 1913 and three years later was awarded a doctorate in engineering from MIT. He devoted most of his considerable research effort from 1925 to the development of mechanical, electromechanical and latterly electronic calculating machines or analogue computers, which led directly to the digital computers universally used today. His major achievement was the differential analyser, first pioneered by **Kelvin**, which was used for solving complex differential equations. Bush overcame the problems that had defeated Kelvin. In particular, he was able to incorporate into his machine a 'torque amplifier', which 'stepped up' the smallest forces of the numerous shafts and gears. An influential copy of Bush's analyser was built at Manchester University by **Hartree**. Bush also devised a cipher-breaking machine which was successful in breaking Japanese codes during World War II, and he was instrumental in setting up the Manhattan Project in 1942 which led to the American atomic bomb. His no-nonsense, straight-talking approach can be savoured in his autobiography, *Pieces of the Action* (1970).

BUTENANDT, Adolf Friedrich Johann
(1903–)

German biochemist, born in Wesermuende. He studied medicine in Göttingen and became a pupil of **Windaus**. G W Corner and Edgar Allen had recognized the corpus luteum as a secretory organ, and in 1929, Butenandt and **Doisy** independently crystallized and determined the structure of the female steroid hormone oestrone from the urine of pregnant women. In 1931 Butenandt isolated the male hormone androsterone. The work for which he became truly famous, however, was the isolation of a few milligrams of progesterone from the combined corpora lutea of no less than 50 000 pigs, and the determination of its structure. Butenandt employed microanalytical techniques pioneered by **Pregl** and was awarded the 1939 Nobel Prize for Chemistry jointly with **Ružička**, although he was forbidden to accept it by the Nazi regime. In 1936, at the Kaiser Wilhelm Institute for Biochemistry (later renamed the Max Planck Institute), Butenandt discovered the first insect hormone, ecdysone (1956), and soon afterwards, bombykol—the scent produced by female silkworms to attract the male. He was appointed Director of the Max Planck Institute in 1960.

BUTLEROV, Aleksandr Mikhailovich
(1828–86)

Russian chemist, born in Chistopol (now in Tatarskaya). His interest in chemistry was aroused at primary school and grew whilst he was at the gymnasium and university in Kazan. He taught chemistry at the university (1849–68) and was twice rector. He was a gifted teacher, but most of his research was in entomology and his first thesis concerned the distribution of butterflies around the Volga river. After 1857 he carried out further chemical research and published a number of papers on the oxidation of organic compounds. In 1857 he made a long trip abroad, meeting many eminent German and French chemists, and was converted to a view of chemical structure essentially the same as we have today: that of molecules composed of atoms with fixed valencies linked by single or multiple bonds. He became a passionate advocate of this view and used the existence of isomers to provide supporting evidence. From 1868 to 1885 he was professor at the University of St Petersburg where he studied a number of polymerization reactions, and he continued lecturing there even after his retirement. He fought tirelessly for the recognition of Russian science and was elected a full member of the St Petersburg Academy of Sciences in 1874. His many interests outside chemistry included bee-keeping, spiritualism and the higher education of women.

BUYS BALLOT, Christoph Hendrik Diederik

(1817–90)

Dutch meteorologist, born in Kloetinge. He studied and taught at Utrecht University, where he became professor in 1867. In 1854 he founded the Royal Netherlands Meteorological Institute, and he later became its director. The weather maps over much of Europe which he compiled from 1852 onwards were among the first ever weather charts, showing wind direction and speed and temperature anomalies in graded shadings. In 1857 he produced Buys Ballot's law which states that wind speed is proportional to the pressure gradient and blows at right angles to the isobars with low pressure on the left in the northern hemisphere (on the right in the southern). He organized the first service of weather forecasts and storm warnings (1860), and designed an instrument called the aeroclinoscope which indicated the position of the centre of a depression and the pressure gradient. Following the international Congress of Vienna (1873) he became president of a committee which drew up recommendations on meteorological instruments, the form of weather messages and format for recording them, and defined the different types of meteorological observing stations. His recommendations were confirmed at the Congress of Rome (1879) and greatly improved and standardized the meteorological system internationally.

BYERLEY, Perry

(1897–1978)

American geophysicist, born in Clarinda, Iowa, and educated at the University of California at Berkeley (MA 1922, PhD 1924). He became an instructor at the University of Nevada (1924–5), and spent the rest of his career at the University of California, where he was appointed head of the seismograph stations (1925). He later became Chairman of the Department of Geology and Geophysics (1949–54) and Emeritus Professor from 1965. During 1960–1 he spent a year in Cambridge where his paper on seismic *P*-wave travel-times from a Montana earthquake was instrumental in prompting **Harold Jeffreys** to begin the *Jeffreys–Bullen Tables*. Byerley's work was instrumental in establishing seismology in the USA. His greatest contribution was the study of earthquake first motions to determine the compressional and dilatational sectors which give the two possible planes along which fault movement may have taken place. He determined the directions of displacement along transform faults on the Mid-Atlantic Ridge; their significance was later recognized by **Ewing**. The seismological station at Berkeley is named after him.

BYRGIUS, Justus, or BÜRGI, Jost

(1552–1633)

Swiss mathematician and inventor, born in the canton of St Gall. A court watch-maker, he assisted **Kepler** in his astronomical work, and invented celestial globes. He compiled logarithms, but did not publish them before **Napier**.

C

CADOGAN, Sir John Ivan George
(1930–)

Welsh chemist, born in Pembrey. After graduating with a BSc and PhD from King's College, London, he spent two years at the Ministry of Supply (1954–6) before becoming a lecturer at King's College. He worked on a number of topics, including collaborative studies with Donald Hey on the reactions of free radicals in solution and, independently, on the chemistry of phosphorus compounds. At the early age of 33 he was appointed professor at the University of St Andrews (1963) and rapidly established a flourishing research school. In 1969 he moved to the chair at the University of Edinburgh. Cadogan conducted research into many areas of chemistry but possibly the most important were studies of phosphorus compounds and synthesis of reactive intermediates. He also established many links with the British chemical industry, and in 1979 became chief scientist at the BP Research Centre. In 1987 he was promoted to Director of Research. This move from academic to industrial life was a particularly successful one. Elected FRS in 1976 and knighted in 1991, he was President of the Royal Society of Chemistry from 1982 to 1984.

CAGNIARD DE LA TOUR, Charles
(1777–1859)

French engineer and physicist, born in Paris. Cagniard studied at the École Polytechnique and the École du Génie Géographe. In 1809 a favourable report on his novel heat engine was presented before the Academy of Sciences by Lazare Carnot. This 'buoyancy engine' relied on air expanding in a liquid to produce work, and was one of a family of such engines which may have influenced **Sadi Carnot**. Between 1809 and 1815 Cagniard designed a hydraulic engine and a pump, the 'cagniardelle', which used an Archimedean screw to generate a strong blast of air. Investigations between 1820 and 1823 revealed the 'critical point' of various liquids and their vapours. In 1819 Cagniard publicized his invention of the siren which he used from 1824 to 1827 to further his acoustical studies. Subsequently he worked on crystallization (1828–31) and fermentation (1836–8), concluding amidst great controversy that yeast was not inert but contained microscopic living organisms.

CAILLETET, Louis Paul
(1832–1913)

French physicist, born in Châtillon-sur-Seine, and educated at the École des Mines in Paris. Returning home to work with his father's metallurgy business, he investigated the permeability of iron to hydrogen and other gases, accounting for the unpredictable behaviour of some irons in terms of an excess of dissolved gases. In 1870 he began a series of careful measurements to determine whether real gases deviate from the behaviour predicted by the 'ideal' gas laws. From this an interest in the liquefaction of gases grew. By compressing a gas whilst cooling it, then allowing it to rapidly expand thus cooling it still further (the Joule–Thomson effect), he managed during 1877–8 to liquefy oxygen, nitrogen, carbon monoxide, hydrogen, nitrogen dioxide and acetylene. Similar success was achieved by **Raoul Pictet** around the same time. The liquefaction of these gases, previously thought to be permanent gases, consolidated the view of the role of heat in causing changes between the liquid to gas phases. Cailletet's other achievements included the installation of a 300 metre manometer on the Eiffel Tower and the study of a liquid-oxygen respiratory apparatus designed for high-altitude ascents.

CAIRNS, Hugh John Forster
(1922–)

English molecular biologist, educated at Oxford University. He progressed through a series of medical appointments at the Radcliffe Infirmary, Oxford, the Postgraduate Medical School in London and the Royal Victoria Infirmary in Newcastle until 1950, when he took up an appointment as a virologist at the Hall Institute in Melbourne, Australia. He later moved to the Virus Research Institute, Entebbe, Uganda (1952–4), and subsequently accepted a post at the Australian National University, Canberra (1955–63). He then moved to the USA and held posts at Caltech and Cold Spring Harbor, where he became Director of the Laboratory of Quantitative Biology (1963–8). He then became Professor of Biology at the State University of New York at Stony Brook (1968–73) and Head of the Imperial Cancer Research Fund Laboratory at Mill Hill, London (1973–80). Since 1980 he has been Professor of Microbiology at Harvard School of Public Health. He was elected FRS in 1974. Cairns has written and worked extensively on the factors causing the initiation and progression of cancer, and in this work has contributed extensively to the understanding of cell and molecular biology. He demonstrated that cancer develops from a single abnormal cell probably initiated by mutation of the DNA sequence, but the further progression of a cancer is dependent on multiple environmental factors such as smoking, diet and hormones, and does not require further alteration to the cell's DNA.

CALLENDAR, Hugh Longbourne

(1863–1930)

English physicist, born in Hatherop, Gloucestershire. The eldest son of a rector, Callendar went from Marlborough to Trinity College, Cambridge. He was Professor of Physics at the Royal Holloway College (1888), McGill University, Montreal (1893), University College London (1898), and the Royal College of Science (1902–30). In 1886 he described an accurate platinum resistance thermometer, later used in extensive measurements of the thermal properties of water and steam. By the end of the century this type of instrument was accepted as the international standard. In addition Callendar devised a constant-pressure air thermometer (1891) and an electrical continuous-flow calorimeter ideally designed to measure specific heats of liquids. *The Callendar Steam Tables* (1915) and his *Properties of Steam and Thermodynamic Theory of Turbines* (1920) were standard references for engineers and scientists.

CALMETTE, (Léon Charles) Albert

(1863–1933)

French bacteriologist, born in Nice. A pupil of **Pasteur** and founder of the Pasteur Institute at Saigon, he was the discoverer of an anti-snakebite serum there. In 1895 he founded the Pasteur Institute at Lille (Director, 1895–1919). He described a diagnostic test for tuberculosis known as Calmette's reaction, but is best known for the vaccine BCG (Bacillus Calmette-Guérin), for inoculation against tuberculosis, which he jointly discovered with Camille Guérin (1908). They recognized that virulent bovine tubercle bacilli became less virulent when cultured on a medium containing bile, but would still convey a certain amount of immunity to protect against infection with either bovine or human tubercle bacilli. This attenuated strain was used to produce BCG which was introduced in continental Europe around 15 years after the discovery, and later in the UK and the USA.

CALVIN, Melvin

(1911–)

American chemist, born in Minnesota of Russian immigrant parents. He became Professor of Chemistry at the University of California (1947–71) and head of the Lawrence Radiation Laboratory there (1963–80). In 1948 he helped elucidate the Thunberg–Wieland cycle by which some bacteria, unlike animals, synthesize four-carbon sugars, and hence glucose, from acetate as shown by **Konrad Bloch**. The idea that the cycle, operating in reverse, might fix carbon dioxide gas led him to investigate this process in photosynthesis (1950). When *Chlorella*, a green alga, was exposed to radioactive carbon dioxide in the dark, radioactivity was transferred to succinate, fumarate, malate and other compounds before being found in glucose. A brief spell of illumination caused the radioactivity to appear in the triose phosphates and sugar phosphates that are now associated with the pentose phosphate pathway. The outcome was the Calvin cycle whereby carbon dioxide interacts with ribulose diphosphate giving, via several reactions, various sugar phosphates and regenerating ribulose diphosphate ready for a repeat of the cycle. The pathway occurs in all photosynthesizing organisms. For this work he was awarded the Nobel Prize for Chemistry in 1961.

CAMERER, or CAMERARIUS, Joachim

(1534–98)

German botanist, son of Joachim Camerer (1500–74). Physician in Nuremberg, and author of *Hortus Medicus et Philosophicus* (1588) and *Symbola et Emblem* (1590), he was one of the most learned physicians and botanists of his age. Director of the botanic garden at Tübingen and Professor of Botany, he was renowned for his experimental paper on the male role of pollen in fertilization, a proof of sexuality in plants (*De Sexu Plantarum Epistola*, 1694). This paper was published in the Transactions of Tübingen University, a very limited-circulation journal, and was unknown to **Ray**, who had at the same time suggested (but without proof) that pollen was equivalent to sperm in animals. Camerer's microscopic studies revealed that fertilization is essential for the development of the plant embyro. He distinguished between monoecious (separate male and female flowers on the same plant), dioecious (separate male and female plants), and hermaphrodite (combined male and female flowers) plants. The complete honesty with which he recorded his experimental failures, as well as his successes, made the results of his experiments more convincing. His most distinguished descendant was Rudolph Jacob Camerer (Camerarius) (1665–1721), German physician and botanist, born in Tübingen.

CAMPBELL, William Wallace

(1862–1938)

American astronomer, born in Hancock County, Ohio. He joined the Lick Observatory in California in 1891, became its director (1901–30), and was also President of the University of California (1923–30). He is best known for his work on the radial velocities of stars. This project started in 1896, and in 1928 he published a catalogue of nearly 3 000 radial velocities. These data were subsequently used for the study of galactic rotation. In 1922 he produced confirming evidence for **Einstein**'s general theory of relativity, and the work done by **Eddington**, when he measured the bending of the beam of starlight that just skims the Sun's surface during a solar eclipse. He led seven expeditions to study solar eclipses, and elucidated the Sun's motion through our galaxy.

CAMPER, Pieter

(1722–89)

Dutch anatomist, born in Leiden. After studying medicine at Leiden, Camper practised as a physician and travelled widely before being appointed professor at Franeker (1749–61). He rose to become Professor of Anatomy and Surgery at Amsterdam (1761–3), and later at Groningen (1763–73). He wrote a series of works on human and comparative anatomy, for the first time recognizing the air spaces in the bones of birds. Camper also pioneered exploration of the intersection between anatomy, art and aesthetic appreciation. On the basis of extensive measurement of the facial angle, he attempted to give a scientific basis to physical anthropology and thereby paved the way for the racist anthropology of the 19th century. A talented artist himself, he provided the illustrations for William Smellie's notable textbook on midwifery.

CANDOLLE, Alphonse Louis Pierre Pyrame de

(1806–93)

Swiss botanist, born in Geneva, son of **Augustin Pyrame de Candolle**. He succeeded his father as professor at Geneva in 1842. He codified the methods for the study of plant geography, and tried to discover physical laws governing the distribution of plants on the earth. He published the great *Géographie botanique raisonnée* (2 vols, 1855) and *Origine des plantes cultivées* (1883). In these works he devoted considerable space to the role of temperature in geographic distribution, and established the optimum temperature range for plant growth, one of the major factors in determining the ranges of particular species. He discussed the sizes of plant families in relation to the average range of individual species within these families. He also tackled the problem of the origins of existing species, and concluded that there are proofs for the geological antiquity of many existing species, and that their creation was successive and evolutionary.

CANDOLLE, Augustin Pyrame de

(1778–1841)

Swiss botanist, born in Geneva. He studied chemistry, physics and botany at Geneva and Paris, and became Professor of Botany at Montpellier (1808) and Geneva (1817). His earliest work, on lichens (1797), was followed by *Astragalogia* (1802) and *Propriétés médicales des plantes* (1804). His new edition of *Flore française* appeared in 1805. The French government commissioned him to make a six-year botanical and agricultural survey of France (1806–12). He was the first to use the word 'taxonomy' for his classification of plants by their morphology, rather than physiology, as set out in his *Théorie élémentaire de la botanique* (1813). He defined the natural method of plant classification, a system based on observation of the whole plant, rather than by concentrating on a single character. He also derived the doctrine of 'symmetry' to describe a common basic ground-plan of floral structure. He continued his work in *Regni Vegetabilis Systema Naturale* (1818–21) and the multi-volume *Prodromus Systematis Naturalis Regni Vegetabilis* which began publication in 1824, and which greatly increased the number of recognized plant families.

CANNIZZARO, Stanislao

(1826–1910)

Italian organic chemist and legislator, born in Palermo. He studied in Palermo, Pisa and Turin. Condemned to death for his part in the Sicilian Revolution in 1848, he fled to Paris and later taught in Alessandria, Piedmont, before returning to Italy as Professor of Chemistry at Genoa in 1855. In 1860 he supported Garibaldi's Sicilian revolt. He was appointed the Chair of Chemistry at Palermo in 1861 and 10 years later, after the unification of Italy, moved to the chair at Rome.

Cannizzaro did much to coordinate organic and inorganic chemistry by showing that the same laws apply to both. He realized the significance of the discovery by **Avogadro** that equal volumes of different gases at the same temperatures and pressures contain equal numbers of molecules. His greatest achievement was to recognize the difference between atomic weight and molecular weight, a discovery fundamental to the future development of chemistry. He used this discovery to calculate the molecular weights of volatile compounds, calculating the weights of non-volatile compounds by measuring their specific heats. He then drew up a table of atomic and molecular weights using hydrogen as a base. Cannizzaro also discovered that benzaldehyde reacts with potassium hydroxide to form benzoic acid and benzyl alcohol (the Cannizzaro reaction), and gave the name 'hydroxyl' to the OH radical. An inspiring teacher and a man who devoted much energy to matters of public health and other civic duties, he was made a senator in 1871. He died in Rome.

CANNON, Annie Jump

(1863–1941)

American astronomer, born in Dover, Delaware, the daughter of a wealthy shipbuilder. She was educated at Wellesley College, graduating in 1884. She returned to the college 10 years later as an assistant in the physics department, proceeding to Radcliffe College where she took up the study of astronomy. In 1896 she joined the famous group of women astronomers on the staff of Harvard College Observatory under its director **Edward Pickering** in a major programme of classification of stellar spectra. Cannon classified the spectra of no fewer than 225 300 stars brighter than magnitude 8.5, published in the nine volumes of the *Henry Draper Catalogue*. She was already 60 years old when Pickering's successor **Shapley** decided to extend the catalogue to fainter stars upon which she classified another 130 000 stars. She received many honours, among them the Henry Draper Gold Medal of the US National Academy of Sciences and honorary doctorates from the universities of Groningen and Oxford. Annie Cannon's productive career ended only a few weeks before her death.

CANNON, Walter Bradford

(1871–1945)

American physiologist, born in Prairie du Chien, Wisconsin. He entered Harvard College in 1892, and despite the necessity of working to supplement his income, graduated four years later with credits from 22 courses. He enrolled in the Harvard Medical School and began research as a first year medical student in a pioneering attempt to study intestinal motility using X-rays. The findings of the research programme which continued from this were published in *The Mechanical Factors of Digestion* (1911). After graduating in medicine, Cannon became an instructor, assistant professor, and professor and head of the physiology department at Harvard (1906–42). In 1908 he became Chairman of the American Medical Association Committee which sought to counter the antivivisection movement. Leading on from the observation that emotional excitement causes gastrointestinal stasis, Cannon investigated the functions of the autonomic (sympathetic and parasympathetic) nervous system. The sympathetic nerves, he argued, prepared an animal for 'fight or flight', through stimulation of the adrenal glands and by increasing heart rate and blood pressure. He also proposed that the two branches of the autonomic system act together to maintain a large number of physiological functions, including body temperature and the composition of the body's fluids. He coined the term 'homeostasis' to describe this and developed the concept in *Wisdom of the Body* (1932). Through his friendship with **Pavlov**, Cannon became interested in Russia, and sought to maintain and improve communication with Soviet scientists and doctors. After Franco's victory in Spain, he helped to find posts in the USA for scientists and physicians who had fought for the republicans. He also assisted many victims of Nazi Germany. His death at the age of 73 was the result of a neoplasm caused by exposure to X-rays during his research.

CANTON, John

(1718–72)

English physicist, born in Stroud. He became a schoolmaster in London, but had little formal education as his father removed him from school to learn the family trade of broadcloth weaving. His keen interest in natural philosophy came to the notice of a wealthy neighbour and Fellow of the Royal Society, Dr Henry Miles, who had a great influence on his career, and was instrumental in Canton's election to the Royal Society in 1749. He was especially interested in atmospheric electricity, and was the first in England (1752) to confirm **Benjamin Franklin**'s conjecture about the electrical nature of lightning. He determined that clouds could be charged either positively or negatively, for which he designed his well-known experiments on electrostatic induction, invented a portable pith-ball electroscope (1754) and was the first to make powerful artificial magnets. In 1762 he demonstrated the compressibility of water. He received the Royal Society's Copley Medal in 1765.

CANTOR, Charles Robert
(1942–)

American molecular geneticist, born in Brooklyn, New York City. He was educated at Columbia University and the University of California at Berkeley, where he recieved his PhD in 1966. He has taught at Columbia University since 1966, and since 1981 has been Chairman of the Department of Genetics and Development at the university's College of Physicians and Surgeons. In 1984 Cantor developed pulse field gel electrophoresis for the separation of very large DNA molecules; this technique has been an essential tool in examining DNA structures at the chromosome level. He is currently Director of the Human Genome Project at the University of California at Berkeley. This worldwide project is designed to completely map the human genome, first by creating a series of manageable chunks with restriction enzymes and then by DNA sequencing. It will therefore be possible to learn the amino acid sequence of genes as yet unknown (possibly important disease-causing entities) and to identify DNA sequences which may be important in regulation of genetic processes.

CANTOR, Georg
(1845–1918)

Russian-born German mathematician, born in St Petersburg. He studied in Berlin, and in 1877 became Professor of Mathematics at Halle. He did important work on classical analysis, particularly in trigonometric series, where he took up the question of the circumstances in which a function is represented by a unique **Fourier** series. **Dirichlet** and **Riemann** had previously given results for functions which are continuous on a given interval; Cantor was led to consider sets of points at which functions can have behaviour that makes their Fourier series inappropriate. He found that he could repeat this construction, and obtain from one such set another, sometimes indefinitely, and this in turn led him to a highly original arithmetic of the infinite, extending the concept of cardinal and ordinal numbers to infinite sets. Central to his theory is the idea that infinite sets have the same size (or cardinality) if and only if there is a one-to-one correspondence between their members. He showed that the set of real numbers is uncountable (cannot be put in a one-to-one correspondence with the set of integers) and that the set of subsets of a set is always larger than the original set. He proposed but could not solve the problem of characterizing the cardinality of the continuum; the problem, in a more precise form, is now known to be unsolvable. Other aspects of his ideas on the theory of sets of points have become fundamental in topology and modern analysis. He was aware of paradoxes in his theory of sets which he reconciled to his own satisfaction with his belief in a God beyond human understanding. His friend **Dedekind** simultaneously developed a naive theory of sets as a foundation for mathematics. Their work was fused together around 1900 and became the setting for much subsequent work on the foundations of mathematics.

CARATHÉODORY, Constantin
(1873–1950)

Greek mathematician, born in Berlin. His father was a diplomat who served as Turkish ambassador in Brussels from 1875. Constantin attended Belgium's École Militaire (1891–5), then worked as engineer on the Asyut Dam in Egypt. In 1900 he began to study mathematics at the University of Berlin. Two years later he transferred to Göttingen and, under **Hermann Minkowski**, received his PhD in 1904. He taught in Germany and then at the new University of Smyrna established by the Greeks. When this was destroyed by the Turks in 1922, Carathéodory saved the library, transporting it to Athens where he remained until 1924 before accepting his final academic position at Munich. His research covered differential equations, the calculus of variations, the theory of real and complex functions, conformal mappings and the theory of point-set measure. He also wrote on **Einstein**'s special relativity and geometrical optics. In 1909 he proposed an alternative axiomatic structure for thermodynamics which banished the term 'heat', distancing this science from its engineering roots.

CARDANO, Girolamo
(1501–76)

Italian mathematician, naturalist, physician and philosopher, born in Pavia. He became famous as a physician and teacher of mathematics in Milan, and was appointed as Professor of Medicine at Pavia (1543) and Bologna (1562). In 1551 he visited Scotland to treat the archbishop of St Andrews, and in London cast the horoscope of Edward VI. In 1570 he was imprisoned by the Inquisition for heresy; he soon recanted and went to Rome in 1571 where he was given a pension by Pope

Pius V. He died a few weeks after finishing his candid autobiography *De propria vita*. A strange mixture of polymath and charlatan, he wrote over 200 treatises on, among other things, physics, mathematics, astronomy, astrology, philosophy, music and medicine. His most famous work was his treatise on algebra, the *Ars Magna*, in which methods for solving cubic and quartic equations algebraically were published for the first time. He was accused of plagiarism by **Tartaglia**, who had taught Cardano the method for cubics and claimed the solution of the cubic as his own, but credit for the first solution of a type of cubic should go to Scipione da Ferro, an Italian mathematician of the previous generation. Despite this, the solution is still known as Cardano's formula.

CARLSON, Chester Floyd
(1906–68)

American inventor, born in Seattle. He graduated in physics from Caltech in 1930, then took a law degree and worked as a patent lawyer in an electronics firm. He began to experiment with copying processes using photoconductivity and by 1938 had discovered the basic principles of the electrostatic 'xerography' (from the Greek *xeros* meaning 'dry' and *graphein*, 'to write') process. Between 1939 and 1944, 20 companies refused his patent. In 1947 an agreement was reached with a small photographic company, Haloid, which later became Xerox. Around 12 years later (1959) the first copier was brought onto the market, the Xerox 914. The copying process involved using high voltages to electrically charge an insulating sheet of material. This sensitized it to small amounts of light and by focusing an image of an original document onto the sheet a secondary electrostatic image of the page was created. By selectively picking up black pigment, the insulating sheet became the plate from which the copy was printed. Carlson's copiers were marketed worldwide by the Xerox Corporation, who for many years had little competition, making profits of $15 billion in 1987. His invention made him a multi-millionaire.

CARNOT, (Nicolas Léonard) Sadi
(1796–1832)

French engineer and physicist, born in Paris. Sadi was the eldest son of Lazare Carnot, major figure of the French revolutionary period, close associate of Napoleon, and distinguished writer on mechanics and politics. During his politically turbulent childhood years, Sadi (named after a Persian moralist) studied physical science, languages and music with Lazare. He entered the élite militaristic École Polytechnique in 1812 and received a rigorous mathematics-dominated scientific training. Finishing high in his class, he transferred to the École de l'Artillerie et du Génie at Metz in 1814 where he remained for three years before receiving his first military commission. From 1819, stationed in Paris with only occasional military duties, he visited factories, studied science and political economy, and became expert in the industrial economic organization of many nations. To his surprise he was recalled to active service in 1826 but less than two years later he resigned. Carnot believed that the apparent British economic ascendency stemmed from technological supremacy, particularly regarding the steam engine. His posthumously famous *Réflexions sur la puissance motrice du feu* (1824) sought to provide an accessible but general analysis of all heat-engines, much as Lazare had published a general discussion of the efficiency of machines. Sadi considered an 'ideal' heat engine. The maximum amount of work produced was independent of working substance (eg steam or air); work was obtained only when heat underwent a 'fall' between two temperatures (like water in a water-wheel), and the maximum amount of work depended only on this temperature fall. In 1824 Carnot imagined heat to be an indestructible substance (caloric), although later he believed it to be convertible into work. Despite a favourable reception from the French Academy of Sciences, Carnot's book made little impression. He died at the age of 36 from cholera. Through the graphical reinterpretation of **Clapeyron**, his work became known: taken up by **Clausius** and **Kelvin** from the late 1840s, Carnot's work became a vital resource for the science of thermodynamics.

CAROTHERS, Wallace Hume
(1896–1937)

American organic and industrial chemist who invented nylon. He was born in Burlington, Iowa, and studied at Tarkio College, Missouri, and the University of Illinois at Urbana. During 1926–8 he was an instructor at Harvard before joining the Du Pont Company at Wilmington, where he led a research team investigating the synthesis of polymers of high molecular weight. From the polymers of acetylene he helped **Nieuwland** to develop the synthetic rubber neoprene, first produced commercially in 1931. Other synthetic polymers took the form of strong and flexible fibres, which were first marketed as nylon two years after Carothers's suicide in Philadelphia. The patent for nylon, awarded posthumously, was given to the Du Pont Company.

CARPENTER, William Benjamin
(1813–85)

English biologist, born into a distinguished Bristol family—his sister Mary became the leader of the 'ragged school' movement. Carpenter studied medicine at University College London, and Edinburgh, graduating MD in 1839. His graduation thesis on the nervous system of the invertebrates included discoveries about the ganglia of arthropods. Between 1840 and 1844 he practised in Bristol, before moving to London to become Professor of Physiology at the Royal Institution and Professor of Forensic Medicine at University College (1849). A man of varied talents, Carpenter took part in marine exploration expeditions (1868–71), undertaking valuable research on marine zoology, especially the foraminifera, fossil and recent. His central work, however, lay in neurology. He explored the interfaces between neurological organization and consciousness, developing the idea of 'unconscious cerebration'—though this led him into a priority dispute with Thomas Laycock. Carpenter was no materialist, however, and he avoided asserting a purely physiological basis of mind. He gave a warm if guarded reception to **Charles Darwin**'s *Origin of Species*, while continuing to entertain religious doubts about the descent of man. Carpenter was one of the last all-round naturalists. A fluent writer, his chief works were *Principles of General and Comparative Physiology* (1839), *Principles of Human Physiology* (1846), *The Microscope and its Revelations* (1856), *Principles of Mental Physiology* (1874) and *Nature and Man* (1888).

CARREL, Alexis
(1873–1944)

American experimental surgeon, born in Lyons, France. Carrel entered the University of Lyons in 1990 as a medical student and was attached to hospitals in Lyons from 1893 to 1900, when he obtained his medical degree. His ability in surgery and anatomy became recognized during this period, after he had become interested in the surgery of blood vessels. The first successful attempts to suture blood vessels were made public in 1902. In 1904 he took up an assistantship in physiology at the University of Chicago, where he experimented with transplantation of organs, such as kidneys, in animals. Much of the later progress in this field relied upon his pioneering work. In 1906 Carrel was appointed to the Rockefeller Institute for Medical Research, where he developed techniques for tissue culture. He was awarded the 1912 Nobel Prize for Physiology or Medicine for these experiments. During World War I, Carrel served the French army and helped **Dakin** to develop 'Dakin's solution' for sterilizing deep wounds. In 1930 he began an experimental programme aimed at cultivation of organs which produced apparatus for keeping organs alive, now used in surgery of the heart and major blood vessels. In his widely read *Man and the Unknown* (1935) Carrel presented a technocratic vision of a world led by an intellectual élite. After the outbreak of World War II he returned to France, and established the Institute for the Study of Human Problems, hoping to introduce a programme of eugenics, nutrition and hygiene. He died of heart failure during the German occupation.

CARRINGTON, Richard Christopher
(1826–75)

English astronomer, born in Chelsea, London, and educated at Trinity College, Cambridge. From 1847 to 1852 he held the post of observer at the University of Durham where he worked on comets and minor planets. He later set up his own private observatory at Redhill, near Reigate in Surrey, where he carried out a substantial programme of observations, resulting in a catalogue of 3 735 stars close to the celestial pole, for which he was awarded the Gold Medal of the Royal Astronomical Society (1859). His most enduring legacy, however, is his work on the rotation of the Sun which he observed systematically over a period of seven years, noting the apparent motions of sunspots across its disc. Carrington was able to represent the Sun's rotation and its dependence on solar latitude by a formula which came to be universally adopted. Also associated with him is his unique observation (1859) of a solar flare seen in white light, the only recorded instance of the phenomenon being visible without special equipment.

CARROLL, James
(1854–1907)

English-born American physician, born in Woolwich. He emigrated in childhood to Canada and the USA. Serving as a surgeon in the American army, and in association with **Reed**, he did valuable research on yellow fever, deliberately infecting himself with the disease in the process (1900). In 1901 he was the bacteriologist on an expedition to Cuba with Reed, which succeeded in establishing the nature of the actual yellow fever pathogen. By showing it to be a micro-organism similar to that first discovered by Martinus Willem Beijerinck in 1898, they were the first to implicate a virus in human disease. In 1902 Carroll became Professor of

Bacteriology and Pathology at Columbia University and the Army Medical School.

CARSON, Hampton Lawrence
(1914–)

American evolutionist, born in Philadelphia, Pennsylvania. Educated at the University of Pennsylvania, he began his career at Washington University, where he was later appointed professor (1956–71). From 1971 to 1985 he was Professor of Genetics at the University of Hawaii. Carson is best known for his work on the evolution of new species in Hawaiian *Drosphilidae* (the fruit fly). He developed the idea of the 'founder effect' (originally proposed by **Mayr**), using chromosomal inversions as markers of relationships, devising probable phylogenetic lineages. These ideas were summarized in a *festschrift*, *Genetics, Speciation and the Founder Principle* (1989).

CARTAN, Élie Joseph
(1869–1951)

French mathematician, born in Dolomieu, the son of a blacksmith. As a child his mathematical talent attracted attention and he received a scholarship first to the Lycée and then to the École Normale Supérieure. He was professor in Paris from 1912 to 1940. One of the most original mathematicians of his time, he reformulated **Lie**'s work and, followed by **Weyl**, created the theory of Lie groups (now central to particle physics). He modernized differential geometry, and founded the subject of analysis on differentiable manifolds, which is also essential to modern fundamental physical theories. Among his discoveries are the theory of spinors, the method of moving frames and the exterior differential calculus. The novelty of his ideas and their somewhat obscure presentation delayed their understanding, and his importance was only fully appreciated during the later part of his life. He is now seen to be a seminal figure for much of the mathematics of this century. His son Henri (b 1906) was a founder of **Bourbaki**.

CARVER, George Washington
(c.1864–1943)

American scientist, born near Diamond Grove, Missouri. He was born into a Black slave family, and received little formal education in his early years. He finally graduated from Iowa State Agricultural College in 1894, and became renowned for his research into agricultural problems and synthetic products, especially from peanuts and sweet potatoes. For much of his life he worked to make Tuskegee Institute, Alabama, a means of education for the disadvantaged Black farmers of the South, and became famous as a teacher and humanitarian. He died in Tuskegee.

CASIMIR, Hendrik Brugt Gerhard
(1909–)

Dutch physicist, born in the Hague. He studied at Copenhagen and Leiden and held various research appointments before joining the Philips company at Eindhoven (1942), where he became coordinator of the research laboratories (1946) and a member of the board of management (1957–72). He worked in numerous fields including the mathematical formalism of quantum mechanics, the theory of paramagnetic relaxation and paramagnetism at very low temperatures. In 1934 he introduced an important phenomenological theory of superconductivity called the 'two-fluid model'. It was constructed on the assumption that in a superconductor electrons can exist in two states; as a superfluid that can move through the crystal without hindrance by the lattice, and as a normal fluid that experiences resistance. In this model both fluids could coexist in the superconducting state whilst in the normal state the superfluid was no longer present. Casimir has also carried out research into the theory of hyperfine structure and irreversible thermodynamics. His books include *Magnetism and very Low Temperature* (1940) and *On the Interaction between Atomic Nuclei and Electrons* (1936).

CASSINI, Giovanni Domenico
(1625–1712)

Italian–French astronomer, born in Perinaldo, near Nice (then in Italy), and educated at a school in Genoa. He became Professor of Astronomy at the University of Bologna (1650), where his determinations of the rotation periods of the planets and his tables of the motions of Jupiter's satellites (1668) brought him fame. In 1669 King Louis XIV nominated him a member of the French Academy and invited him to Paris. At the new Paris Observatory he made a host of observations of the planets Mars, Jupiter and Saturn, and discovered the division of Saturn's rings which still bears his name (1675). In the course of time he discovered four satellites of Saturn which he named Ludovici in honour of his patron, following the example of **Galileo** who had called Jupiter's satellites the Medicean stars. Some of Cassini's observations were made with 'aeriel' telescopes which were up to 150 feet long and supported by wooden towers. One of Cassini's great achievements was his determination of the distance of the planet Mars, and thereby of the distance of the Sun, from observations made simultaneously in Paris and in the French colony of Cayenne. Giovanni's son Jacques (1677–1756), who succeeded him in 1712, measured an arc of the meridian from Dunkirk to Perpignan, while Jacques's son, César François (1714–84), and grandson, Jacques Dominique (1748–1845), successively held the same post of Director of the Paris Observatory.

CASTNER, Hamilton Young
(1848–99)

American chemist who became a leading figure in British industry, born in Brooklyn, New York City. He was educated at Brooklyn Polytechnical Institute and the Columbia School of Mines. Having established himself successfully as a chemical consultant, he failed to find an American backer to develop a process he had invented to obtain sodium (used in the production of aluminium, soap and bleaching powder) from caustic soda (sodium hydroxide) by heating it with iron and carbon. He moved to Britain in 1886 where a new company financed by **Roscoe** and others established a plant at Oldbury, Birmingham, to exploit his patent. Two years later the company pioneered a second method which Castner developed for the production of sodium, this time by the electrolysis of molten sodium hydroxide. As molten sodium rose above the cathode it was ladelled off and more sodium hydroxide added. In theory the process should have been continuous, but in practice impurities in the sodium hydroxide caused many difficulties. Castner then devised a way of producing much purer sodium hydroxide by the electrolysis of brine. He overcame the difficulty of separating the sodium hydroxide solution from the original salt with his famous 'mercury rocking cell'. Mercury was used as the cathode, and by means of a rocking motion, was circulated below chambers containing brine and water. The sodium in the brine formed an amalgam with the mercury but was given up to the water when the mercury passed into the water chamber. The solution of sodium hydroxide was led off and evaporated to dryness. The purity of the sodium hydroxide was 99 per cent and this, in time, proved of great benefit to many industries. Simultaneously with the invention of the mercury rocking cell, Carl Kellner in Germany developed a similar process, also using mercury, and sold his patent to **Solvay**'s company. Rather than engage in expensive lawsuits over patents, the British and Belgian companies collaborated to found the Castner–Kellner Alkali Company in 1895. Its headquarters were at Weston Point, near Runcorn, where it manufactured sodium hydroxide according to Castner's rather than Kellner's method, and bleaching powder from the chlorine which was a by-product of the electrolysis. Castner died of tuberculosis at Saranac Lake, New York State, before he could see the fruits of his success. After the Castner–Kellner Alkali Company became very prosperous it was taken over by Sir John Tomlinson Brunner and **Mond**, becoming part of Imperial Chemical Industries in 1926.

CAUCHY, Augustin Louis, Baron
(1789–1857)

French mathematician, born in Paris. He studied to become an engineer, but ill-health forced him to retire and teach mathematics at the École Polytechnique. After the 1830 revolution he followed the Bourbon court to exile in Turin and Prague, but returned to Paris in 1838. He did important work on ordinary and partial differential equations, being one of the first to argue the need to establish conditions that ensure that such equations have solutions. He also advocated the wave theory of light following **Fresnel**'s work, and gave a substantial impetus to the mathematical theory of elasticity. He wrote an influential *Cours d'analyse* (1821) in which major steps were taken to give a rigorous foundation to real analysis. The first to give the now-standard definitions of continuity and differentiability, he is also remembered as the founder of the theory of functions of a complex variable, which was to play a leading role in the development of mathematics during the rest of the 19th century. In algebra Cauchy gave a definitive account of the theory of determinants, and developed the ideas of permutation groups which had appeared in the work of **Lagrange** and **Galois**. His *Oeuvres complètes* (1882–1970) run to 30 volumes, and so prolific was he that in exile he founded journals solely to carry the results of his research. The unfortunate outcome was that many of his best results did not receive the systematic treatment that they deserved, delaying their reception.

CAVALIERI, (Francesco) Bonaventura
(1598–1647)

Italian mathematician, born in Milan, and regarded as a disciple of **Galileo**, with whom he maintained a correspondence. He was appointed Professor of Mathematics at Bologna University in 1629. His method of 'indivisibles', published in 1635, was an ingenious way to work scrupulously with the idea of figures made up of lines in order to determine their areas. Rigour was bought at a price in complexity that others were often unwilling to pay, and intuitive methods also spread, but Cavalieri had found an original and general method which helped pave the way for the introduction of integral calculus. In studying Cavalierian techniques, **Leibniz** replaced the phrase 'omn. lin' for all the lines (of a figure) with the symbol , the initial letter of 'summa' for sum, thus creating the integral sign. Cavalieri also promoted the use of the logarithms for calculation in Italy following their introduction in the early 1600s.

CAVENDISH, Henry
(1731–1810)

English natural philosopher and chemist, born in Nice. Born into an aristocratic family, Cavendish was educated at Peterhouse College, Cambridge (1749–53). Leaving without a degree, he lived in Paris for a year and then, returning to London, equipped a laboratory and library for his own and others' use. He was an active and well-known Fellow of the Royal Society from 1760. Family wealth, augmented through a substantial inheritance at the age of 40, enabled him to devote an increasingly reclusive life entirely to scientific pursuits. In 1765 he investigated specific and latent heats, obtaining results which were only published much later. Soon afterwards he demonstrated chemical and physical methods for analysing the distinct 'factitious airs' of which normal atmospheric air was composed (1766). Amongst these were 'fixed air' (carbon dioxide), and 'inflammable air' (hydrogen) which Cavendish isolated. Interpreting his findings within the traditional phlogiston theory of combustion first propounded by **Georg Stahl**, he believed this 'inflammable air' to be phlogiston itself. In 1784 he ascertained that hydrogen and oxygen, when caused to explode by an electric spark, combined to produce water which could not therefore be an element. Similarly, in 1795 he showed nitric acid to be a combination of atmospheric gases. Furthermore, a small proportion of the air remained after prolonged sparking, an inert fraction much later identified as the element argon. The famous 'Cavendish experiment' (1798) employed a torsion balance apparatus devised by **Michell** to estimate with great accuracy the mean density of the Earth and the universal gravitational constant. Only a small part of Cavendish's researches were made known during his lifetime. In 1771 a theoretical study of electricity had appeared and later Cavendish ingeniously confirmed the inverse square law of attraction, but it was not until 1879 that his electrical manuscripts (covering statics and dynamics) were edited and published by **Maxwell**. The Cavendish Laboratory (established 1871) in Cambridge was named in his honour.

CAVENTOU, Joseph Bienaimé
(1795–1877)

French chemist, born in St Omer. He was educated in Paris, and became professor at the École de Pharmacie there. In 1817, in collaboration with **Pelletier**, he isolated (and introduced the term) 'chlorophyll'. Following the method for isolating morphine by precipitation with ammonia from an acid extract of the raw material, as described by Friedrich Sertürner (1783–1841), they also isolated strychnine and brucine from nux vomica (1819), and quinine and cinchonine from cinchona bark (1820). Cinchonine was particularly important for the treatment of fevers and the famous French physiologist François Megandie (1783–1855) introduced these new drugs into his 1821 pocket formulary. Caventou also isolated veratrine (1818) and was one of the first to extract caffeine from coffee beans (1822).

CAYLEY, Arthur
(1821–95)

English mathematician, born in Richmond, Surrey. His father was an English merchant in St Petersburg and he lived in Russia till the age of eight. He graduated from Trinity College, Cambridge, as Senior Wrangler in 1842. Called to the bar in 1849, he wrote nearly 300 mathematical papers during 14 years' practice in conveyancing. In 1863 he was elected first Sadleirian Professor of Pure Mathematics at Cambridge. His principal contributions to mathematics were in algebra, his theory of invariants and covariants, and his work on matrices. In geometry, he opened the way to the unification of metrical and projective geometry, and began the study of *n*-dimensional geometry; his collected mathematical papers fill 13 volumes. His example and his help behind the scenes did much to revive British mathematics, especially his study of curves and surfaces conducted jointly with his friend George Salmon, but reform of Cambridge itself had to wait for the next generation. He did, however, support the cause of women at Cambridge, giving lectures at the newly founded Girton College.

CAYLEY, Sir George
(1773–1857)

English amateur scientist and pioneer of aviation, born in Scarborough. He went to school in York and then became a pupil of George Walker, a scientist (FRS 1771) and also a skilled mechanic. He was a man of independent means, and was able to devote much of his life to theoretical and experimental studies of the mechanics of flight. He constructed and flew in 1808 a glider with a wing area of 300 square feet, probably the first practical heavier-than-air flying machine. Over the next 45 years he conducted thousands of model tests, in the course of which he tried and discarded flapping-wing ornithopters, biplanes and triplanes. He clearly understood the functions of vertical and horizontal tail surfaces in the control of aircraft, and he foresaw that the power to fly must come from a sufficiently light engine and an efficient airscrew. Such an engine was still half a century away when in 1853 he constructed the first successful man-carrying glider, which carried his coachman safely a few hundred metres across a valley. He also interested himself in railway engineering, allotment agriculture, and land reclamation methods, and invented a new type of telescope, artificial limbs, the caterpillar tractor and the tension wheel. He helped to found (1839) the Regent Street Polytechnic in London, and he was a sponsor of the first meeting of the British Association for the Advancement of Science in York in 1832.

CECH, Thomas
(1947–)

American biochemist, born in Chicago. He trained at the universities of California and Chicago, and after working at MIT for a period, he moved to the University of Colorado, Boulder, where he has been professor since 1983. He was the first to discover the ability of ribonucleic acid (RNA) to act as a biological catalyst. In 1977 he studied the repair mechanisms of damaged DNA and identified regions sensitive to enzyme cleavage. Working on the ciliate *Tetrahymena thermophila*, he reported on the maturation of ribosomal RNA (rRNA) by the excision of an intervening polynucleotide sequence followed by splicing together the cut ends (1979). Thus he discovered that protein-free precursor rRNA mediates its own cleavage and splicing, the RNA acting catalytically in the manner of an enzyme; however, unlike an enzyme, the molecule is modified in the process. Subsequently Cech and **Altman** independently identified other catalytic RNA species that act without self-modification, called 'ribozymes' by Cech. He has suggested that splicing is an ancient process associated with the emergence of living and evolving organisms. For these pioneer discoveries Cech shared with Altman the 1989 Nobel Prize for Chemistry. He has since extended this work to explore the action of telomerase enzymes which use an inbuilt RNA template to add short repeat sections of DNA to chromosomal DNA.

CELSIUS, Anders
(1701–44)

Swedish astronomer, born in Uppsala. Celsius taught mathematics and became Professor of Astronomy at the University of Uppsala (1730) where his grandfather, father and uncle had all held academic positions. Between 1732 and 1736 he travelled widely in Europe, visiting many centres of observational astronomy. Whilst in Nuremberg he published an aurora borealis compendium (1733). In 1736 he took part in an expedition organized by the Paris Academy of Sciences to measure an arc of meridian at a northern latitude, showing the flattening of the Earth's poles. He also published speculations on the distance of the Earth from the Sun, and measured the relative brightness of stars. Celsius was responsible for the construction (1740) and subsequent direction of the Uppsala observatory. The Celsius temperature scale originated with a mercury thermometer, described by him in 1742 before the Swedish Academy of Sciences. Two fixed points had been chosen: one (0 degrees) at the boiling point of water, the other (100 degrees) at the melting point of ice. A few years after his death, colleagues at Uppsala began to use the familiar inverted version of this centigrade scale.

CELSUS, Aulus Cornelius
(1st century)

Roman writer and physician. He compiled an encyclopedia on medicine, rhetoric, history, philosophy, war and agriculture. The only extant portion of the work is the *De Medicina*, rediscovered by Pope Nicholas V and one of the first medical works to be printed (1478). In it Celsus gives accounts of symptoms and treatments of diseases, surgical methods and medical history.

CESALPINO, Andrea, Latin Caesalpinus
(1519–1603)

Italian botanist, physician and physiologist, born in Arezzo. A pupil of Ghini, he became the most original and philosophical botanist since **Theophrastus**, whose work he revived. He was Professor of Medicine and director of the botanic garden in Pisa from 1553 to 1592, when he became Physician to Pope Clement VIII. After he had taught at the university for 30 years he published *De Plantis* (1583). In book XVI of this work he stated the basic principles of botany, with descriptions of about 1500 plants arranged according to his own scheme of classification; this was the first attempt at a scientific classification of plants, based on fructification, foreshadowing **Linnaeus**. He believed that plants deal with food in a similar way to that in which animal digestion operates, through a concept of internal and external heat. His dissection of plant tissues led to the identification of plant veins, and his experiments on plants laid the foundations for plant physiology. The account of seed structure and germination he provided was the fullest since Theophrastus. In medicine he was no less original, propounding a theory of the circulation of the blood.

CEULEN, Ludolph van
(1540–1610)

Dutch mathematician, born in Hildesheim. He devoted himself to finding the value of π and finally worked it out to 35 decimal places ('Ludolph's number'); it was inscribed on his tombstone at Leiden.

CHABANEAU, François
(1754–1842)

French chemist, born in Nontron. He began as a student of theology but was expelled on account of his views on metaphysics. Professor of Mathematics at Passy when only 17, and with little knowledge of the subject, he turned to physics and chemistry. Subsequently he became lecturer in physics and chemistry at the Real Seminario Patriótico at Vergara, Spain, where he and **Elhuyar y de Suvisa** founded the Real Escuela Metalúrgica in the 1780s. They worked together on ways of separating platinum from its compounds and making it malleable, announcing their success in 1783. Chabaneu died in Nontron.

CHADWICK, Sir James
(1891–1974)

English physicist, born near Macclesfield. He studied at the universities of Manchester, Berlin and Cambridge, and worked on radioactivity with **Rutherford**. In 1930 **Bothe** had observed a neutral penetrating radiation that resulted from the bombardment of beryllium by alpha particles. He had attributed this to high-energy gamma rays, as this was the only known neutral penetrating radiation at the time. In 1932 **Irène** and **Frédéric Joliot-Curie** repeated the experiment and found that the radiation could eject protons from paraffin targets. They interpreted this as a Compton scattering effect with a proton being ejected from the atom instead of an electron. Later the same year, Chadwick repeated the experiment and observed the effect of the radiation on hydrogen, helium and nitrogen. Instead of assuming that the radiation was gamma rays, he suggested that it was due to the neutral particle that Rutherford had proposed in 1920. His data confirmed the hypothesis and showed that the particle's mass was close to that of the proton. He named the particle the neutron and was awarded the 1935 Nobel Prize for Physics for this discovery. He went on to build Britain's first cyclotron in 1935 at Liverpool and during World War II worked on the Manhattan Project to develop the atomic bomb in the USA. He was elected FRS in 1927 and knighted in 1945.

CHAGAS, Carlos Ribeiro Justiniano
(1879–1934)

Brazilian physician and microbiologist, born in Oliveira, Minás Gerais. He studied at the Medical School of Rio de Janeiro, where he was introduced to the concepts and techniques of bacteriology and scientific medicine. After a few years in private practice, Chagas joined the staff of the Oswaldo Cruz Institute, where its founder and leading light, Oswald Cruz, befriended him. Much of his early work was concerned with malaria prevention and control. During one of his field missions, in Lassance, a village in the interior of Brazil, he first described a disease (Chagas' disease, or 'sleeping sickness') caused by a trypanosome (he named the organism *T Cruzi* after Cruz). Chagas elucidated its mode of spread through an insect vector, established the trypanosome's virulence in laboratory animals, and described its acute and chronic course in human beings. On circumstantial evidence, some historians have suggested that **Charles Darwin** suffered from Chagas' disease.

CHAIN, Sir Ernst Boris
(1906–79)

British biochemist, born in Berlin of Russian-Jewish extraction. After studying physiology and chemistry in Berlin, he taught in the biochemistry department of **Frederick Gowland Hopkins** in Cambridge (1933–5), where he identified an enzyme (phospholipase) in snake venom which caused paralysis of the nervous system. He then joined **Florey** at the Dunn School of Pathology in Oxford (1935–48) to characterize lysozyme, discovered earlier by **Alexander Fleming**, and determine its mode of action on bacteria. He proved that it was an enzyme acting on the polysaccharide of bacterial cell walls and identified N-acetyl glucosamine as the site of cleavage. Searching for other lysozyme-like enzymes, he encountered Fleming's paper on penicillin (1929). *Penicillium notatum* was available at the institute, and he found that penicillin was not an enzyme but a new small molecule. With **Abraham**, Chain achieved purification from 2 units per milligram

(1939–40) to eventually 1800 units per milligram. Clinical trials of this non-toxic antibiotic began in 1941, and it was produced commercially by Glaxo under **Lester Smith** during World War II. Fleming, Chain and Florey shared the 1945 Nobel Prize for Physiology or Medicine. Chain became Director of the International Research Centre for Chemical Microbiology in Rome (1948–61), and Professor of Biochemistry at Imperial College, London (1961–73). He was elected FRS in 1949 and knighted in 1969.

CHAMBERLAIN, Owen
(1920–)

American physicist, born in San Francisco. He was educated at Dartmouth College and at the University of Chicago, where he received his doctorate in 1949. He became professor at the University of California after working on the Manhattan atomic bomb project (1942–6) and at the Argonne National Laboratory. The existence of antimatter had been predicted by **Dirac**'s theory, and the first antiparticle (the anti-electron, or positron) had been discovered in 1932 by **Carl Anderson**. In 1955, Chamberlain, **Segrè** and their colleagues set up an experiment in which the anti-protons could be identified by their time of flight between two scintillators and by Cherenkov counters. They discovered a negatively charged particle with a mass very close to that of the proton. In a later experiment with collaborators from Rome, they were able to prove that these particles annihilated with protons, confirming that they were indeed anti-protons. In 1959 Chamberlain and Segrè were awarded the Nobel Prize for Physics for the discovery of the anti-proton.

CHAMBERLIN, Thomas Chrowder
(1843–1928)

American geologist, born in Mattoon, Illinois. He was educated at Beloit College, Wisconsin. From 1873 he worked as an assistant for the Geological Survey of Wisconsin, becoming chief geologist from 1876 to 1882. After a short spell as Professor of Geology at Columbia University (1885–7) he became President of the University of Wisconsin (1887–92), and later Professor of Geology and Director of the Walker Museum at the University of Chicago (1892–1919). Chamberlin was a master of research, and his studies of the glaciation of northern USA and his recognition of a complex interaction between geological, biological and chemical processes led him back to Precambrian glaciations and to the origin of the atmosphere. His horizons were considerably broadened at Chicago; he is best known for his work there in connection with the fundamental geology of the solar system, attacking the nebular hypothesis for the origin of the Earth. He proposed instead the planetesimal hypothesis, envisaging a large low-density embryonic Earth capable of significant gravitational contraction. Chamberlin argued strongly against **Kelvin**'s calculations of the age of the Earth (20–40 million years) and chided fellow geologists for their conservatism in extending this only to 100 million years. He was one of the earliest to suggest that climatic variations may have been caused by the changes in the amount of carbon dioxide in the atmosphere. Amongst some 250 publications, his books included *The Origin of the Earth* (1916), *The Two Solar Families* (1928) and *The Sun's Children* (1928).

CHAMISSO, Adelbert von (Louis Charles Adelaide de)
(1781–1838)

French-born German poet and biologist, born in Champagne. The French Revolution drove his parents to Prussia, and he served in the Prussian army (1798–1807). In Geneva he joined the literary circle of Madame de Staël, and later studied at Berlin. In 1815–18 he accompanied a Russian exploring expedition round the world as naturalist, and on his return he was appointed Keeper of the Berlin Botanical Garden. In 1819 he was the first to discover in certain animals what he called 'alternation of generations' (the recurrence in the life cycle of two or more forms). He wrote several works on natural history, but his fame rests partly on his poems, still more on his quaint and humorous *Peter Schlemihl* (1813), the story of the man who lost his shadow.

CHANCE, Britton
(1913–)

American biochemist, born in Wilkes-Barre, Pennsylvania. He was educated at Pennsylvania University, and at the University of Cambridge, in physical chemistry and physiology. He then became assistant professor (1941–9) and Professor of Biophysics (1949–83) at the University of Pennsylvania, where he has been Emeritus Professor of Biophysics and Physical Biochemistry since 1983. His career embraced many problems in biochemical energetics and biophysics, and he also had strong interests in developing the techniques and instrumentation necessary for such work. His best-known work was his demonstration in 1943, using sensitive spectrophotometric techniques, of the existence of a complex between an enzyme and its substrate; such complexes had long been theoretically presumed to exist as an essential stage in enzyme action but had not been detected. He studied the reactive mechanisms of several types of enzymes, and did important work on the problems of energy generation in biological systems by unravelling some of the steps in the electron-transport chain by which energy is released in cells. In addition to his biochemical studies, he utilized his technical expertise in developing and improving analytical equipment including photoelectric control units, radar timing and computing devices, sensitive spectrophotometers and non-invasive optical methods for studies of organ biochemistry. His recreational interest in yachting resulted in a gold medal at the 1952 Olympics and the patents on several automatic steering devices.

CHANDLER, Seth Carlo
(1846–1913)

American astronomer, born in Boston. After graduating from Harvard he became a private assistant to the US Coast Survey in 1864. In 1881 he resumed his scientific interests at Harvard College Observatory and designed a science observer code—a system for transmitting astronomical information, especially on the newly discovered comets, by telegraph. His most important contribution to science was the discovery of the periodic variations in latitude of points on the Earth's surface due to movement of the geographic poles. He devised an almucanatar, an instrument for relating the positions of the stars to a small circle at the zenith (1884–5). Using this, in 1891 he verified a cyclic variation in latitude by 0.3 arcseconds, with a period of 430 days (14 months); **Leonhard Euler** had predicted a 10 month period. This became known as the Chandler wobble. This work was subsequently extended by **Munk**.

CHANDRASEKHAR, Subrahmanyan
(1910–)

Indian-born American astrophysicist, born in Lahore (now in Pakistan), nephew of **Raman**. He was educated at the Presidency College, Madras, before going to Cambridge University, where he studied under **Dirac**. In 1936 he moved to the USA to work at the University of Chicago and Yerkes Observatory. He showed that in the final stages of stellar evolution, the fate of a star depends on its mass. Low-mass stars may collapse to form white dwarfs, small hot stars in which material is compressed to densities of millions of times that of ordinary matter. Chandrasekhar demonstrated the surprising property that the larger the mass of a white dwarf, the smaller its radius. He also concluded that stars with masses greater than about 1.4 solar masses will be unable to evolve into white dwarfs, and this limiting stellar mass, confirmed by observation, is known as the Chandrasekhar limit. He suggested that if the mass of a star is greater than this, it can become a white dwarf star only if it ejects it's excess mass in a supernova explosion before collapse. For his work on the late evolutionary stages of massive stars he was awarded the 1983 Nobel Prize for Physics, jointly with **William Fowler**.

CHAPMAN, David Leonard
(1869–1958)

English physical chemist, born in Wells, Norfolk. He studied at Christ Church, Oxford, under **Augustus Vernon Harcourt**, graduating in chemistry in 1893 and in physics in 1894. After a short period of school teaching, he joined the chemistry staff at Owen's College, Manchester under **Dixon** in 1897. In 1907 he became head of the Sir Leoline Jenkins Laboratory of Jesus College, Oxford, a post he held until his retirement in 1944. He was elected FRS in 1913. Chapman's earliest contributions to physical chemistry were theoretical treatments of explosion velocities in gases (using Dixon's measurements) and of electrocapillarity. Some of his equations for the former were later derived independently by Emile Jouguet and the region behind a detonation wave is known as the 'Chapman–Jouguet layer'. The electrical double layer considered in his theory of electrocapillarity is known as the 'Gouy–Chapman layer'. Chapman's main work, however, was in gas kinetics and in much of this he was ably assisted by his wife Muriel. They made important studies of the thermal and the photochemical reactions between hydrogen and chlorine. During the photochemical work Chapman devised the rotating sector method for measuring the lifetimes of chain carriers, a technique which has been widely used. He introduced the steady-state treatment into kinetics in 1913. This was later used extensively by **Bodenstein**, who is often credited with its invention.

CHAPMAN, Sydney
(1888–1979)

English physicist and geophysicist, born in Eccles, Lancashire. He was educated at the University of Manchester where he graduated in 1907, and at the University of Cambridge where he received a BA in mathematics in 1911. He became a senior assistant at the Royal Greenwich Observatory supervising the installation of the new magnetic observatory. Noticing that few of the existing magnetic data had been interpreted, he set about the task himself and in so doing began a lifelong study. In 1914 he returned to Cambridge as college lecturer, though during World War I, as a pacifist, he was sent back to Greenwich (1916–8). He was Professor of Mathematics at Manchester (1919–24), professor at Imperial College, London (1924–46), and Sedleian Professor of Natural Philosophy at Oxford (1946–53), seeking to improve the status of science at the latter. He solved problems of thermal conductivity and diffusion of gases and identified thermal diffusion (1917). During 1922–8 he produced the first satisfactory theory of magnetic storms, later known as the Chapman–Ferraro theory. In 1918 he identified a lunar atmospheric tide (a tide which had eluded **Airy**). After 1953 he took research posts in Alaska and at the High Altitude Observatory in Boulder, Colorado. He was elected FRS in 1919.

CHARCOT, Jean Martin

(1825–93)

French pathologist and neurologist, born and educated in Paris, and appointed to a position at Salpêtrière Hospital. Of working-class stock, Charcot advanced through the ranks to become the most eminent French physician of his day. His early investigations were devoted to chronic diseases such as gout and arthritis, and the diseases of old age. Increasingly, however, he turned the Salpêtrière into an international centre for the investigation of neurological diseases during his active medical service there, himself making important observations on multiple sclerosis, amyotrophic sclerosis and familial muscular atrophy. These were summarized in his *Leçons sur les maladies du système nerveux* (5 vols, 1872–93). His 'Tuesday lessons' attracted young doctors from all over the world. During the last two decades of his life, he became intrigued by hysteria and other functional disorders, and began using hypnosis in their diagnosis and treatment. He developed the notion that ideas can cause functional diseases and argued that normal people cannot be hypnotized. Thus, hypnosis could be useful as a diagnostic aid, as well as having therapeutic possibilities. He displayed to his classes the same young women frequently; it is now known many of them were gifted actresses who could feign signs at will. His lectures stimulated the young Sigmund Freud, who also translated some of Charcot's work into German. Charcot was a powerful medical politician who had many disciples; he was also a gifted artist whose pathological illustrations grace many of his works.

CHARDONNET, (Louis-Marie-) Hilaire Bernigaud, Comte de

(1839–1924)

French industrial chemist, born in Besançon. He studied in Paris at the École Polytechnique and the École des Ponts et Chaussées, where he was later appointed engineer. The manufacture of rayon (at first called 'artificial silk') is his best-known achievement. He adapted the process invented by **Swan** for making filaments from nitrocellulose by dissolving it in acetic acid, extruding the solution through very fine glass capilliaries into a coagulating fluid. The resulting thread was then woven into the world's first synthetic material. He patented the process in 1884 and five years later, after modifications to reduce the flammability of the rayon, he opened factories in his home town and in Satvar, Hungary. Chardonnet also designed the actinograph, which measures solar radiation and is used in aviation, and studied the effects of ultraviolet light on different organisms.

CHARGAFF, Erwin

(1905–)

Czech-born American biochemist, born in Czernowitz (now in the Ukraine). He studied in Vienna before spending two years at Yale (1928–30), where he isolated unusual branched-chain fatty acids and complex lipopolysaccharides from the tubercle bacillus. Chargaff returned to Berlin (1930–3), where he extended his study of bacterial lipids, and briefly visited Paris before settling at Columbia University, New York, in 1935. He was appointed Professor of Biochemistry in 1952. His initial work was on plant chromoproteins and he showed that these were composed of protein and lipid (lipoproteins), and when purified were coloured green, from spinach, and orange, from carrot. Chargaff's best-known work has been on the base composition of DNA, which he found to be characteristic of a species and identical in different tissues of the same animal. His most significant finding, of general application to living systems for the understanding of the structure of DNA as proposed by **Crick** and **James Watson** in 1953, was that the concentrations of the DNA bases were in pairs; thus the number of adenine bases is equal to the number of thymine bases and similarly guanine = cytosine. However, Crick claimed, somewhat controversially, to have been unaware of Chargaff's work.

CHARLES, Jacques-Alexandre-César

(1746–1823)

French experimental physicist, born in Beaugency. After losing a junior position in the bureau of finances in Paris, Charles made himself an expert in popular scientific display. From 1781 he gave ingenious public lectures using an extensive collection of apparatus. Collaborating with two Parisian artisans, the Robert brothers, he equipped a hydrogen balloon, dubbed the Charlière. In December 1783 Charles ascended in Paris with the elder Robert, landing 27 miles away, then continued alone for a further three miles. Charles was rewarded with lodgings in the Louvre. He subsequently became a member of the Academy of Sciences (1795), and Professor of Experimental Physics at the Conservatoire des Arts et Méticrs. He invented a megascope, a hydrometer and a goniometer (for measuring angles of crystals). Experiments carried out by Charles between 1786 and 1787 (but not published) showed that the expansion of a number of insoluble gases followed the fundamental law which, in the UK, now bears his name. **Gay-Lussac** published the general law, extended to soluble gases, in 1802.

CHARNEY, Jule Gregory
(1917–81)

American mathematician and meteorologist, born in San Francisco. He graduated from the University of California at Los Angeles in 1938 and received a PhD in meteorology in 1946. He then worked at Chicago University (1946–7), Oslo University (1947–8) and the Institute for Advanced Study in Princeton (1948–56) before being appointed Professor at MIT (1956–81). In his greatest work, Charney studied and solved the problem of baroclinic instability, ie development in a basic flow in which horizontal temperature gradients and vertical wind shear exist, and devised equations which enabled the development of depressions to be predicted. The equations could be used in a large computer to produce weather forecasts: this was first achieved in Princeton in 1954. Charney was one of the first to realize the important influence which different surfaces have on weather and climate. For example, he showed that the high albedo of desert surfaces leads to subsidence in the atmosphere and hence to a persistence of descrt conditions. He made an important contribution to the planning of the Global Atmospheric Research Program (1979) and was a leading personality in the establishment of the National Center for Atmospheric Research in Boulder, Colorado (1960). His many honours included the Symons Gold Medal of the Royal Meteorological Society (1961).

CHARNLEY, Sir John
(1911–82)

British orthopaedic surgeon, born in Bury, Lancashire, and educated at Manchester University. He served as an orthopaedic specialist during World War II, then returned to the Manchester Royal Infirmary, soon devoting himself to the technical problems associated with replacing badly arthritic hip joints. He worked on animals in a search for suitable material with which to make the artificial joints. An initial operative series using artificial joints made of Teflon gave unsatisfactory long-term results, but from 1962 he used polyethylene, and this proved highly functional. With a good cementing material, and scrupulous attention to aseptic technique, he perfected the operation which has given enhanced mobility to many people. He also pioneered other joint operations. He had great insight into the mechanical dimensions of skeletal movement, and combined the engineering and surgical approaches to his craft with much skill and enthusiasm. Elected FRS in 1975, he was knighted in 1977.

CHARPAK, Georges
(1924–)

French physicist, born in Dabrovica, Poland. He studied in France, obtaining his PhD from the Collège de France. After working at the National Centre for Scientific Research, he moved in 1959 to CERN (the European nuclear research centre in Geneva), where he developed gaseous particle detectors. Previously, particle detectors had taken the form of bubble chambers and spark chambers which recorded information on photographic film. The speed with which such detectors could receive data was limited and it was time consuming to analyse the many photographs produced. In a crucial advance in detector technology, Charpak devised the multi-wire proportional chamber, allowing large-area detectors capable of operating at high rates to be built relatively cheaply. This development revolutionized high-energy physics experiments and has played a vital role in the development of the subject. The use of such detectors has not been limited to high-energy physics; they have found application in the fields of biology, medicine and astronomy. For this work Charpak was awarded the 1992 Nobel Prize for Physics.

CHASLES, Michel
(1793–1880)

French geometer, born in Épernon. He entered the École Polytechnique in 1812 and became a military engineer. He resigned to devote himself to mathematics, taught at the

École Polytechnique (1841–51) and became Professor of Geometry at the Sorbonne in 1846. Chasles greatly developed projective geometry without the use of coordinates by means of a systematic study of cross-ratio and homographies, and many theorems on the projective geometry of conic sections and quadric surfaces are due to him. His command of geometry led him to a lifelong interest in the history of the subject, on which he wrote valuable books, but in 1867 he became involved in controversy with the Academy after claiming to have come into possession of autographs which proved that **Pascal** had anticipated **Isaac Newton**'s discovery of the law of gravitation. Ultimately, however, he admitted that these were forgeries.

CHÂTELET-LOMONT, Gabrielle Émilie, Marquise du

(1706–49)

French mathematician and physicist, born in Paris. She learned Latin and Italian with her father, and after her marriage in 1725 to the Comte du Châtelet-Lomont she studied mathematics and the physical sciences. In 1773 she met Voltaire, and became his mistress. Voltaire came to live with her at her husband's estate at Cirey, and there they set up a laboratory and studied the nature of fire, heat and light. She connected the causes of heat and light, and believed that both represented types of motion. She wrote *Institutions de physique* (1740) and *Dissertation sur la nature et la propagation du feu* (1744), but her chief work was her translation into French of **Isaac Newton**'s *Principia Mathematica*, posthumously published in 1759.

CHATELIER, Henri Louis Le

See **LE CHATELIER, Henri Louis**

CHATT, Joseph

(1914–)

English chemist, born in Horden, County Durham, and educated at Emmanuel College, Cambridge, where he received his BA (1937), PhD (1940) and ScD (1957). Short-term war projects occupied him until 1942, when he joined Peter Spence and Sons Ltd, Widnes, mainly to work on titanium chemistry and catalysis by activated alumina. In 1947 he joined ICI's new fundamental research laboratory at Welwyn, Hertfordshire. Chatt is distinguished for his theoretical and experimental contributions to the bonding of olefins, hydrogen and alkyl or aryl groups to transition metal atoms within a metal complex, and his ideas of the 'trans' effect of one ligand on a metal atom on an opposite ligand; also the 'a' and 'b' classification of metal atoms according to their binding characteristics. This work was of particular interest to the developing petrochemical industry. In 1962 he was invited by the Agricultural Research Council to establish a research unit of nitrogen fixation to study the nitrogenase reaction, believed to occur at metal centres. It was set up at the University of Sussex where Chatt was director of the unit and Professor of Chemistry (1964–80). Its widely acclaimed work involved a team of inorganic chemists and biochemists, as well as microbial geneticists. There Chatt developed preparations under ambient conditions of a new series of complexes containing molecular nitrogen ligated from the gas at ordinary temperatures. Some of these complexes upon acidification gave ammonia in up to 95 per cent yield, and provided in detail a plausible mechanism for the chemistry of the nitrogenase reaction. Chatt was elected FRS in 1961, and has received numerous national and international awards, including the CBE and the American Chemical Society Award for Distinguished Service to Inorganic Chemistry (1971).

CHEBYSHEV, Pafnutiy Lvovich

(1821–94)

Russian mathematician, born in Okatovo, the son of a retired army officer. A graduate of Moscow University, he became an assistant at St Petersburg in 1847 and later professor (1860–82). In number theory he made important contributions to the theory of the distribution of prime numbers. Chebyshev's theorem in number theory asserts that there is always a prime number between n and $2n$, a result first conjectured by Joseph Bertrand. In work on probability he proved fundamental limit theorems concerning the way in which a number determined probabilistically approaches its average value. Later he studied the theory of mechanisms and developed a method of approximating functions by polynomials—this technique has become important in modern computing. One such class of polynomials is named after him. The mathematical school that he founded at St Petersburg remained the dominant influence on Russian mathematics for the rest of the century.

CHERENKOV, Pavel Alekseyevich

(1904–)

Soviet physicist, born in Voronezh. He was educated at Voronezh University and the Soviet Academy of Sciences. In 1934 he observed blue light emission from water bombarded by gamma rays. This so-called

'Cherenkov effect' was explained by **Tamm** and **Frank** as being produced by particles travelling through a medium at velocities greater than the speed of light in that medium. The three shared the Nobel Prize for Physics in 1958. The principle was adapted in constructing a cosmic-ray counter mounted in the Sputnik III satellite, and has become important as a particle identification tool in high-energy particle physics. Cherenkov also contributed to the development and construction of electron accelerators, and to the study of the interactions of photons with nuclei and mesons.

CHEVREUL, Michel Eugène
(1786–1889)

French chemist and gerontologist, born in Angers. He studied chemistry at the Collège de France in Paris and Harvard University. Most of his working life was spent at the Museum of Natural History in Paris where he was appointed as an assistant in 1810, professor in 1830 and director in 1864. For some years he was also director of the dyeworks at the Gobelins Tapestry. Chevreul investigated the physics and psychology of colour. *De la loi du contraste simultane* (1839), which influenced many of the Impressionists, argued that our perception of the intensity and hue of any colour is conditioned by the degree of contrast with neighbouring colours. Chevreul was also a pioneer of organic analysis. He decomposed soaps made of animal fats, isolating and naming many members of the fatty acid series. His studies of the saponification process demonstrated that soaps are combinations of a fatty acid with an inorganic base, a discovery which opened up vast industries. He noticed the phenomenon of isomerism, reported that diabetic urine contains glucose, and at the end of his long life, studied the psychiatric effects of old age. In the course of his researches he devised experimental techniques which became standard practice: for example, the separation of fatty acids by their different solubilities in a given solvent, and the use of melting points to estimate purity. He died in Paris.

CHITTENDEN, Russell Henry
(1856–1943)

American physiological chemist, born in New Haven, Connecticut. In 1872 he entered Sheffield Scientific School of Yale University, planning a medical career. However, an undergraduate project in which he showed that scallop muscle contains glycogen and glycine led to a paper in **Liebig**'s *Annalen*, and advanced study at the University of Heidelberg. In 1880, after returning to the USA, Chittenden was awarded his PhD. He was appointed Professor of Physiological Chemistry (1882) and Director at Sheffield Scientific School (1898). In Germany, while working under **Kühne**, Chittenden became interested in proteolytic enzymes, and they continued to collaborate in this work when Chittenden returned to the USA. His later work was concerned with toxicology, alcohol and food additives, and he was involved in the provision of scientific evidence in a number of legal cases involving poisoning. His best-known work was concerned with human protein requirements. He experimented upon himself and a group of young men, and concluded that men could remain healthy on a diet of about 2 500 calories containing about 50 grams of protein. He hypothesized that various health problems might be the result of diets which are over-rich in protein.

CHLADNI, Ernst Florenz Friedrich
(1756–1827)

German physicist, born in Wittenberg. Chladni had Hungarian origins. He graduated in law at Leipzig in 1782. Combining music with physical science, he performed an extensive series of acoustical experiments, first publicized in 1787. He measured the variation of the velocity of sound by filling organ pipes with different gases and measuring the pitch of the note emitted; he also studied the vibration of strings, rods and plates. When his 'Chladni plates' (geometrically shaped pieces of metal or glass clamped at the centre) were sprinkled with sand and forced to vibrate with a violin bow, a surprising diversity of patterns appeared (now known as 'Chladni figures'). Travelling widely throughout Europe, he demonstrated these curious effects and performed on two variants of the glass harmonica he had designed: the clavicylinder and the euphonium. In 1809 he published a *Traité d'acoustique*. Soon afterwards the mathematician **Germain** and others proposed theoretical explanations for Chladni's sand patterns.

CHORLEY, Richard John
(1927–)

English geomorphologist, born in Minehead, Somerset. Educated at Exeter College, Oxford, he received a Fulbright scholarship to study geology at Columbia University. After various lecturing appointments in North America and the UK, he became a reader (1970–4) at Cambridge University, where since 1974 he has been Professor of Geography. He is a leader of the group which challenged traditional geography and led to the British phase of the so-called 'quantitative revolution'. He used general system theory in the study of landforms, advocated geography as human ecology and developed the use of models in explanation. His publications include *The History of the Study of Landforms* (2 vols, 1964, 1973), *Physical Geography* (1971), *Environmental Systems* (1978) and *Geomorphology* (1984).

CHRISTIANSEN, Jens Anton

(1888–1969)

Danish physical chemist, born in Vejle. In 1911 he graduated from the Polytechnic in Copenhagen and became assistant to **Sørensen** at the Carlsberg laboratory. From 1915 until his retirement in 1959, he was at the University of Copenhagen, holding various positions including the chairs in inorganic chemistry (1931–48) and physical chemistry (1948–59). His many honours included honorary membership of the Chemical Society. Christiansen's doctoral thesis (1921) contained fundamental contributions to chemical kinetics. As well as an interpretation of the kinetics of the hydrogen–bromine reaction in terms of a chain mechanism which had also been published separately, the thesis contained a collision theory of unimolecular reactions and a rate theory essentially equivalent to the later transition state theory. At that time the occurrence of reactions involving only one molecule seemed an anomaly, and Christiansen's approach to the problem has provided the basis for later, more sophisticated treatments. In transition state theory it is supposed that all reactions proceed via a highly energetic molecular entity, the activated complex, and this idea is fundamental to the understanding of rates and mechanisms of chemical change. Because the thesis was in Danish, the scientific world has been slow to appreciate Christiansen's role in these matters. His later researches embraced work on a wide variety of solution processes, including the decomposition of hydrogen peroxide, mutarotation of sugars, periodic reactions and enzyme reactions.

CLAIRAUT, Alexis Claude

(1713–65)

French mathematician, born in Paris and admitted to the French Academy of Sciences at the age of 18. He worked on celestial mechanics, including the figure of the Earth and the motion of the Moon. He took part in the French expedition to Lapland to determine the shape of the Earth, finding it flatter nearer the poles in accordance with **Isaac Newton**'s theory of gravity, but not the Cartesian theory of vortices. He successfully computed the date of the first return of **Halley**'s comet in 1759 to within a month. His work on the motion of the Moon led him first to doubt, but then to accept, Newton's theory of an inverse square law for gravity. His analysis, for which **Leonhard Euler** awarded him a prize of the St Petersburg Academy, helped establish Newton's theory both among experts and the public at large. He also wrote a popular elementary work on geometry which was only superseded when **Legendre** wrote one deliberately modelled on **Euclid**'s *Elements* at the end of the century, as part of a 'back to basics' movement.

CLAPEYRON, (Bénoit-Paul-) Émile

(1799–1864)

French civil engineer, born in Paris. He was educated at the École Polytechnique and the École des Mines. On graduation he was sent by the French government to Russia to assist in the organization of the Russian Corps of Engineers; he remained there for 10 years, working also as a teacher of mathematics in the School of Public Works at St Petersburg. After his return to France he was at first principally engaged in the construction of railways and bridges; he also designed locomotives, becoming the first to make deliberate use of the expansive action of steam in the cylinder. For the analysis of beams resting on more than two supports he developed the 'theorem of three moments', and in 1834 he published an exposition of **Carnot**'s classic but previously neglected paper *Réflexions sur la Puissance Motrice du Feu* (1824) on the power and efficiency of various types of heat engine. He joined the staff of the École des Ponts et Chaussées in 1844, and was elected to the French Academy of Sciences in 1858.

CLARK, William Mansfield

(1884–1964)

American chemist, distinguished for his work on physiological processes, born in Tivoli, New York. He was educated at Johns Hopkins University, where he later became Professor of Physiological Chemistry after some years working in government departments dealing with the dairy industry and public health. His studies of acidity in milk led him to develop, with Herbert Lubs, a reliable range of titration indicators and to write *The Determination of Hydrogen Ions* (1920). While working on dyes during his days with the public health service, he began the investigations of oxidation–reduction systems which continued for the rest of his working life and which contributed largely to our understanding of life processes. He died in Baltimore.

CLARKE, Bryan Campbell
(1932–)

English geneticist and evolutionist, born in Gatley, Cheshire. He was educated at the University of Oxford and taught at Edinburgh University. Since 1971 he has been Professor of Genetics at Nottingham University. Clarke is distinguished for studies of the ecological genetics of terrestrial snails (especially *Cepaea nemoralis*) and the elucidation of the evolution of new species in *Partula* on the Pacific Island of Moorea. He made considerable contributions to the understanding of genetical processes in natural populations of animals, refuting that most inherited variation is neutral to its possessors. This re-established the neo-Darwinian interpretation of evolution in the 1970s, following controversy produced by the discovery (through the application of electrophoresis introduced by **Lewontin** to the study of protein variation in population samples) of apparently excessive amounts of inherited variation. Clarke also developed the concept of frequency-dependent natural selection. He was elected FRS in 1981.

CLARKE, Charles Baron
(1832–1906)

English botanist, born in Andover, Hampshire, and educated at Trinity College and Queen's College, Cambridge. He was a member of a circle at Cambridge (including Henry Fawcett and Leslie Stephen) who advanced economic news. His interest in political economy was to continue throughout his life, being expressed in occasional essays and pamphlets. Clarke was always a traveller and mountaineer. He was appointed to the Bengal Educational Department in 1865, soon becoming an inspector of schools, with headquarters at Dhaka. So began his interest in the botany of the Indian subcontinent. His first 7000 botanical collections perished through shipwreck in East Bengal. From 1865 to 1874 he explored East Bengal, Sylhet, Chittagong, the Khasia Hills, the Sundarbums, the Nilgiri Hills and Sikkim (Yakla). In 1874 he was transferred to Calcutta which allowed him to collect in Chota Nagpur, and the Punjab Himalayas. In 1875 he was stationed at Darjeeling and collected in Sikkim, and in 1876 he traversed Kashmir and the Karakoram. Arriving back in England on furlough in 1877, he worked at Kew on his collections, by now numbering 25000. He also wrote some accounts for Sir **Joseph Dalton Hooker**'s *Flora of British India*. As a result he was placed on special duties to assist Hooker in completing the rest of the Flora; he compiled about 40 family treatments, published during 1879–84. He returned to India in 1883 and made further excursions in Sikkim, Assam and elsewhere. After retiring aged 55, he worked on his collections at Kew. He was an energetic, tireless, careful and exact worker and an ideal collector. He was particularly interested in Cyperaceae and Commelinaceae and published various papers on these families.

CLARKE, Frank Wigglesworth
(1847–1931)

American geochemist, born in Boston and educated at Harvard. He became Professor of Chemistry and Physics, first at Howard University (1873–4) and then at the University of Cincinnati (1874–83). As Chief Chemist to the US Geological Survey (1883–1925), he undertook numerous analyses of rocks and minerals, and compiled important lists of fundamental physical and chemical constants. He was simultaneously Honorary Keeper of Minerals at the US National Museum, Washington, where his active interest and painstaking efforts led to the excellence and comprehensiveness of the mineral collection. Clarke did much work on the recalculation of atomic weights, and he was the first to present a consistent theory of the chemical evolution of geological systems. His books included *Data of Geochemistry* (1908), *The Composition of the Earth's Crust* (1924, with Henry Stephens Washington, 1867–1934) and other texts. The new uranium mineral clarkeite was named in his honour.

CLARKE, William Branwhite
(1798–1878)

English geologist and clergyman, born in East Bergholt, Suffolk. Educated at Dedham and Cambridge, he took holy orders and became a practising cleric at Ramsholt (1821–4). He travelled widely on the European continent and became involved in the Belgian war of independence. After publishing important papers on the geology of Suffolk and Dorset, he emigrated to New South Wales, Australia, in 1839 to regain health following a severe illness. From the time of his arrival until 1844 he was in clerical charge of the area north of what is now Sydney, before becoming Minister of Willoughby (1847–70). There he became active in studying the geology and mineral reserves. He is widely credited as the first to discover gold in Australia in the alluvium of Macquarie (1841), and made the first reports of tin and an early report of diamonds. He studied the coal deposits and the occurrence of gold in granites. He also examined the Palaeozoic rocks of the Great Dividing Range and worked on the geology of Tasmania. Clarke was the first to identify

Silurian rocks in Australia, and demonstrated the Carboniferous age of the coal-bearing strata of New South Wales. His labours in officially reporting on the geology and economic potential of 108 000 square miles of territory gained him the title 'Father of Australian Geology'.

CLAUDE, Albert
(1899–1983)

Belgian–American biologist, born in Longlier. Before World War I, Claude worked in a steel mill, and during the war, for British Intelligence. As a war veteran he was permitted to enter the University of Liège to study medicine, although he had no high school diploma. He obtained a doctorate in medicine in 1928. Claude was awarded a scholarship for study at the Cancer Institute in Berlin, but was forced to leave after pointing out errors in the director's experiments. In 1929 he went to the Rockefeller Institute for Medical Research in New York to study the possibility that Rous sarcoma (a form of cancer) in chickens is of viral origin. Claude developed cell fractionalization using a high-powered centrifuge and isolated a tumour agent from cancerous cells which was known to be a constituent of viruses. Applying the technique to the study of normal cells, he separated various 'organelles'—the nucleus, mitochondria and microsomes (later known as ribosomes). He showed that mitochondria are the sites of respiration. In 1942 Claude began collaboration with the microscopist of the Interchemical Corporation in applying electron microscopy to biology. This led to important advances in understanding the structure of cells. In 1949 he became Director of the Jules Bordet Institute in Brussels. He shared the 1974 Nobel Prize for Physiology or Medicine with **Palade** and **de Duve**.

CLAUDE, Georges
(1870–1960)

French technologist, born in Paris. He graduated from the municipal school of physics and chemistry in 1889. After a period in the laboratory of an electricity generating works, he worked for the French Houston–Thompson Company (1896–1902) and later as an engineer with various industrial firms. Around 1896 he developed the method of handling and transporting acetylene by dissolving it in acetone, which made an acetylene industry practicable. In 1902 he devised a process for the liquefaction of air, in which the cooling effect is obtained by external work of expansion. He later worked on the separation and utilization of the noble gases. During World War I, Claude developed a synthetic ammonia process similar to the **Haber** process, but using much higher pressures. After 1927 he took part in experiments to generate electricity by utilizing the temperature difference between the bottom and the surface of the sea. These were not successful, but the experience with pipelines was useful for the trans-Channel pipeline PLUTO in World War II. Claude was elected to the French Academy of Sciences in 1924, but he was expelled after the war for 'collaboration' with the German occupying forces and sent to prison. He was released after four and a half years in 1949.

CLAUSIUS, Rudolf Julius Emmanuel
(1822–88)

German physicist, born in Köslin, Prussia. His father was pastor and principal of a small school which Clausius attended. Whilst studying at the University of Berlin (1840–4), his predominant interest changed from history to science. After receiving a PhD (Halle, 1847), he taught physics at the Royal Artillery and Engineering School, Berlin (1850), the Zürich Polytechnicum (1855), Würzburg (1867), and Bonn (1869). In 1850 his famous paper *Ueber die bewegende Kraft der Wärme* sought to reconcile the theories of **Carnot** and **Clapeyron** (assuming heat to be indestructible caloric) with the experimentally demonstrated interconvertibility of heat and work. Clausius postulated that heat cannot of itself pass from one body to another at a higher temperature (the second law of thermodynamics) in order to show that Carnot's theorem (all perfect engines operating between the same temperatures are equally efficient irrespective of working substance) remained valid with the rejection of the caloric theory. Between 1859 and 1865 he considered the dissipation of energy which **Kelvin** had suggested in 1852. Clausius introduced the term 'entropy' (1865) in such a way that dissipation was equivalent to entropy increase. Now the two laws of thermodynamics could be stated succinctly: the total energy of the universe is constant; and the entropy of the universe tends to a maximum. He studied electrolysis, suggesting in 1857 that molecules were made up of continually interchanging atoms (or ions) with electric force merely directing the interchange. Within the kinetic theory of gases he calculated the mean speed of gas molecules, ignoring collisions (1857), and introduced the analytically useful concepts of mean free path and effective radius (1858).

CLERK, Sir Dugald
(1854–1932)

Scottish mechanical engineer, born in Glasgow. He became an engineering apprentice at the age of 15, at the same time continuing his education by part-time study at the West of Scotland Technical College, then from 1871 to 1876 at Anderson's College, Glasgow, and in Leeds. With the intention of becoming a chemical engineer he studied the properties of petroleum oils under **Edward Thorpe** in Leeds, until on a visit to Glasgow he was fascinated by a Lenoir-type gas engine at work in a joiner's shop. From that moment in 1877 he resolved to devote himself to research on the theory and design of gas engines. In 1881 he patented a gas engine working on the two-stroke principle which became known as the Clerk cycle, extensively used for large gas engines and later for small petrol engines. He was employed by several of the most notable manufacturers of gas engines in Scotland and England, and in 1888 joined his friend George Marks to form the partnership of Marks and Clerk, consulting engineers and patent agents. He was elected a Fellow of the Royal Society in 1908, and received its Royal Medal in 1924. He was knighted in 1917.

CLIFFORD, William Kingdon
(1845–79)

English mathematician, born in Exeter. He entered King's College, London, at the age of 15, and then Trinity College, Cambridge (1863), where he graduated as Second Wrangler in 1867. In 1871 he became Professor of Applied Mathematics at University College London. He remained there until his early death from tuberculosis. He was the first British mathematician to appreciate the work of **Riemann**, which he translated in part and extended in some respects; a theorem on special sets of points on a Riemann surface still bears his name. The so-called Clifford–Klein problem concerns the nature of geometry and was solved by the introduction of a geometry on the surface of a torus. He wrote on projective and non-Euclidean geometry, and on the philosophy of science, and his book *The Common Sense of the Exact Sciences* was completed by **Pearson** in 1885. He had a reputation as an excellent lecturer on science to popular audiences.

COCKCROFT, Sir John Douglas
(1897–1967)

English nuclear physicist, born in Yorkshire. Educated at the universities of Manchester and Cambridge, he became Jacksonian Professor of Natural Philosophy at Cambridge (1939–46). In 1932, with **Walton**, he induced the first artificial disintegration of a nucleus by bombarding a lithium nucleus with protons accelerated across a potential of 710 kilovolts. This was the first successful use of a particle accelerator, and by studying the energies of the two alpha particles produced they were able to verify **Einstein**'s theory of mass–energy equivalence. The development of the particle accelerator was crucial for the understanding of the substructure of nuclei and later the nucleons themselves. Cockcroft and Walton were awarded the 1951 Nobel Prize for Physics for this work. Cockcroft later assisted in the design of much special experimental equipment for the Cavendish Laboratory, including the cyclotron. During World War II, he was Director of Air Defence Research (1941–4) and of the Atomic Energy Division of the Canadian National Research Council (1944–6). He became the first director of the UK's Atomic Energy Research Establishment at Harwell in 1946. He was appointed Master of Churchill College, Cambridge (1959), elected FRS in 1936 and knighted in 1948.

COCKERELL, Sir Christopher Sydney
(1910–)

English radio-engineer and inventor of the hovercraft, born in Cambridge. He graduated in engineering from Cambridge in 1931, and after a short period with an engineering firm in Bedford returned to Cambridge to study radio engineering, which he had become interested in as a hobby. He joined the Marconi company in 1935, and was engaged in the design of VHF transmitters and direction finders. He continued working in that field, and later on the development of radar, during World War II. In 1950 he left Marconi and established a boat hire business on the Norfolk Broads, but it was not long before his inventive mind was working on the problem of reducing the drag on a ship's hull by means of air lubrication. It was not a new concept, but Cockerell was the first to devise a means of making it a practical proposition. After many unsuccessful attempts he built in 1955 a balsa-wood model powered by a model aircraft engine which could reach a speed of 12 miles per hour over land or water. Two years later he realized that a flexible skirt would be the answer to the problem of keeping a stable cushion of air under a hovercraft riding the waves of the open sea. After many difficulties in generating the necessary backing, the prototype SR-N1, built by Saunders-Roe of flying-boat fame, and weighing only 7 tonnes, made the crossing of the English Channel in July 1959. Since 1974 Cockerell has also been actively interested in the commercial development of wavepower devices. He was elected FRS in 1967 and knighted in 1969.

COHEN, Seymour Stanley
(1917–)

American biochemist, born and educated in New York City. He joined the staff of the University of Pennsylvania where he was appointed Professor of Biochemistry (1954–71), and moved to the University of Denver as Professor of Microbiology in 1971. He returned to New York to accept a chair at the State University in 1976. In 1940 he reported the isolation from lung tissue and properties of thromboplastin (thrombokinase), an important enzyme of the blood clotting system which converts prothrombin to the clott-forming protein, thrombin. He also found that it was a phospholipoprotein, and that the phosphate and lipid components were combined in the form of a phospholipid which was essential for catalytic activity. Having previously studied and collaborated with **Chargaff**, he clarified the electrophoretic differences between DNA and RNA and showed in the late 1940s that T phage (a bacterial virus), inactivated by ultraviolet light, regained activity in proportion to the resynthesis of the phage DNA. This and much similar work made a valuable contribution to the growing belief at that time that DNA played a key part in heredity.

COHEN, Stanley
(1922–)

American biochemist, born in Brooklyn, New York City. Educated at Brooklyn College, Oberlin College in Ohio, and the University of Michigan, he held posts at the universities of Colorado and Washington before moving to Vanderbilt University in 1959. From 1967 to 1986 he was Professor of Biochemistry there. Following **Levi-Montalcini**'s discovery of the substance now known as nerve growth factor that promotes the development of sympathetic nerves, Cohen contributed to the isolation of the compound. As a development of this work he went on to isolate a further cell growth factor, named epidermal growth factor (EGF), which was found to accelerate some aspects of natural development in newborn mice. In further studies he demonstrated the wide range of effects of this compound on various developmental processes in the body, and elucidated the mechanisms by which it is absorbed by and interacts with individual cells. In 1986 he was awarded the Nobel Prize for Physiology or Medicine jointly with Levi-Montalcini for their work on growth factors.

COHN, Ferdinand Julius
(1828–98)

German botanist and bacteriologist, born in Breslau (now Wrocław, Poland). Barred as a Jew from taking the degree examinations at Breslau, he went to Berlin, where he obtained his doctorate in botany at the age of 19. Professor of Botany at Breslau from 1859 and founder of the Institute of Plant Physiology, the world's first institute specializing in plant physiology, he is regarded as the father of bacteriology in that he was the first to account it a separate science, to define bacteria, and to designate the group as plants. He observed sexual formation of spores in the fungal genera *Sphaeroplea* and *Pilobolus*. He did important research in plant pathology, and worked with **Koch** on anthrax. He founded the journal *Beiträge zur Biologie der Pflanzen*. Through his experiments on the effects of heat on bacteria, he identified bacterial spores. His work was a major factor in the overthrow of the theory of spontaneous generation. He published the textbook *Die Pflanze* in 1882.

COHNHEIM, Julius Friedrich
(1839–84)

German pathologist, born in Demmin, Pomerania (now in Poland). He graduated in medicine from Berlin, and served for a year as assistant to **Virchow** — he was probably his most famous pupil — before himself obtaining a chair at Kiel. He later moved to Breslau and Leipzig. Cohnheim became one of the leading pathologists of the age. His early work was in histology, soon after graduating devising new techniques for sectioning fresh tissue through freezing, and later methods of staining sections with a gold solution. From 1867 he published a brilliant string of studies on infection, showing by experiments upon frogs how the blood vessels responded in the early stages of inflammation, and proving that the white cells (leucocytes) passed through capillary walls where inflammation was occurring, later degenerating to become pus corpuscles. **Metchnikoff** was amongst many later workers who confirmed and extended these studies. Cohnheim importantly advanced understanding of tuberculosis, perhaps the most rampant cause of death in early industrial society. He was convinced that tuberculosis was infectious. He demonstrated his views by injecting rabbit eyes with tuberculous material and then tracing the development of the tuberculous process through the cornea. Cohnheim also made investigations into heart disease, examinining obstruction of the coronary artery and deducing that the resulting lack of oxygen led to myocardial infarction (heart attack). An immaculate microscopist and experimentalist, Cohnheim worked on a whole range of diseases, including cancer. His *Lectures on General Pathology* (1877) formed the finest contemporary account of the discipline.

COLBERT, Edwin Harris
(1905–)

American palaeontologist, born in Clarinda, Iowa. He was educated at NW Missouri State College, the University of Nebraska and Columbia University. He carried out research into vertebrate palaeontology at the American Museum of Natural History (1930–66), the Academy of Natural Sciences of Philadelphia (1937–48) and the Northern Arizona Society of Science and Art (1949–69), and became Professor of Vertebrate Palaeontology at Columbia University (1945–69). Colbert has been one of the foremost palaeontologists of this century and carried out fieldwork in many countries including Argentina, Israel, South Africa, India, Australia and Antarctica. He was a strong supporter of the theory of continental drift proposed by **Wegener**. During his excavations in Antarctica (1969–70), he discovered the fossil subsequently named *Lystrosaurus* which provided conclusive palaeontological support for continental drift. His *Wandering Lands and Animals* (1973) gives a popular account of the evidence for the movement of the continents. Colbert was also involved in the discovery and excavation of fossil dinosaurs in the American Mid-West. His *Evolution of the Vertebrates* (1955) is one of the standard texts.

COLDING, Ludvig August
(1815–88)

Danish engineer and physicist, born in Holbaek. Colding came from a profoundly religious family. His father (a sea captain turned farmer) was acquainted with **Oersted** who advised on Ludvig's education. After training as a carpenter, Colding entered the Copenhagen Polytechnic Institute (1837). By 1845 he had become inspector of roads and bridges, eventually rising to occupy the specially created position of Engineer of Copenhagen (1857). Municipal duties continued after he became professor at the Polytechnique Institute in 1869. He published extensively on meteorology and oceanography. More famous, however, is his independent measurement of the mechanical equivalent of heat, contemporaneously with **Joule** and others. In 1843 Colding proposed to the Danish Society of Sciences that 'force' which disappeared with friction was converted directly into heat, and much more generally, that there was a universal 'imperishability' of all natural forces. His justification explicitly allied experimental data with religion and metaphysics: the conviction of an underlying unity in nature was characteristic of the *Naturphilosophie* adhered to by his mentor Oersted.

COLLIP, James Bertram
(1892–1965)

Canadian biochemist, born in Belleville, Ontario, and educated at Toronto University. He received his PhD in 1916 for a study of acid formation in the gastric glands of the vertebrate stomach. Appointed as a lecturer in biochemistry at the University of Alberta at Edmonton (1915–17), he subsequently became Assistant Professor (1917–19) and Associate Professor (1919–28). In 1928 he became Professor of Biochemistry at McGill University, transferring in 1941 to a chair in endocrinology, and in 1947 he became Dean of Medicine at Western Ontario, where he also headed a department of medical research. His early research was on aspects of blood chemistry and the biochemical relationships between different constituents of the blood of vertebrates and invertebrates; he also examined the physiological effects of adrenaline and studied physiological mechanisms of brain stem function. Early in 1921 he was awarded a Rockefeller travelling fellowship, and went to the Toronto laboratory of **Macleod**. This coincided with the arrival there of **Banting** and **Best**, and the start of their work on the isolation and identification of insulin. Collip was drawn into the work, to try to purify their pancreatic extract sufficiently for it to be tried in clinical trials in diabetic patients. His biochemical expertise overcame some of the impurities in their procedures, and additionally his experiments alerted clinicians to the dangers of exceeding the calculated doses. The pioneer work in Toronto was rewarded with the award of the 1923 Nobel Prize to Banting and Macleod, the former sharing his half with Best, and Macleod doing likewise with Collip. In 1922 Collip returned to Edmonton, and from then onwards his career focused on biochemical aspects of endocrine function: in 1925 he made the important discovery of the active principal of the parathyroid gland, and in later years he isolated and identified hormones from the placenta. He also played an active role in establishing an administrative and academic structure to support medical research in Canada.

COLOMBO, Matteo Realdo
(1516–59)

Italian anatomist, discover of the lesser circulation of the blood. An apothecary's son, Colombo studied anatomy, medicine and surgery under **Vesalius**, succeeding him at Padua (1544) and later becoming professor at Pisa (1545). The early friendship between Vesalius and Colombo later turned to rivalry and rancour, not least when Colombo dared to point out Vesalius's errors. Colombo was later involved in vehement disputes with **Falloppio**. Colombo's volume *On Anatomy* (1559) presents anatomical descriptions based upon up-to-date autopsy and vivisection research—though without illustrations. He described the lens at the front of the eye, the pleura and peritoneum. Discussing the vascular system, he offered the most lucid description then available of the lesser circulation through the lungs. Vivisecting a dog, he made an incision into the pulmonary vein, demonstrating it contained not air, as commonly supposed, but blood. Its bright red colour led him to suggest that the lungs had rendered it 'spiritous'—or, in later parlance, oxygenated. He thus obviated the need for **Galen**'s septal pores. Colombo also undertook important researches on the nature of the heartbeat. Through his discovery of the pulmonary circuit, he was one of the diverse 16th-century anatomists whose researches paved the way for **William Harvey** (1578–1657) in his later account of the general circulation system.

COMPTON, Arthur Holly
(1892–1962)

American physicist, born in Wooster, Ohio, the son of a Presbyterian minister, whose faith he inherited. He studied at Princeton and Cambridge, England, and held posts at Washington University in St Louis, and at Chicago. Compton developed a theory to describe the interaction of X-rays with matter. He based his theory on the idea that light can be considered to consist of particles, the photons which **Einstein** had proposed to explain the photoelectric effect, and the fact that since photons had energy, they must have mass and momentum (according to mass–energy equivalence). He confirmed the theory by measuring the wavelength shift of monochromatic X-rays as a function of the angle through which they were scattered by a target. For this important test of the particle nature of light, he shared the 1927 Nobel Prize for Physics with **Charles Wilson**. A leading authority on nuclear energy, X-rays and nuclear chemistry, he was invited in 1941 to direct plutonium production for the atomic bomb. After quelling his religious doubts, he played a major part in the Manhattan Project. He was involved in building the first reactor with **Fermi** in Chicago (1942), a project which he described in *Atomic Quest* (1958).

COMRIE, Leslie John
(1893–1950)

New Zealand astronomer and pioneer in mechanical computation, born in Pukekohe. He was educated at Auckland University College, where he graduated in chemistry in 1915. After losing a leg with the New Zealand Expeditionary Forces during World War I, he studied astronomy at Cambridge University, receiving his PhD in 1923. He later accepted a teaching post in the USA, and subsequently joined HM Nautical Almanac Office (NAO) in 1926, becoming Superintendent (1930–6). At the NAO, Comrie completely revolutionized the computing methods used there by installing desk calculators and punched card machines, and more importantly, devising efficient numerical methods for use with these mechanical computing aids. Comrie was regarded as the foremost computer and table-maker of his day. A forceful man, intolerant of bureaucratic inefficiency, he eventually left the NAO and in 1936 founded his own computer service—the Scientific Computing Service Ltd. He was elected FRS in 1950, shortly before his death.

CONDAMINE, Charles-Marie de la
See **LA CONDAMINE, Charles-Marie de**

CONDON, Edward Uhler
(1902–74)

American theoretical physicist, born in Alamogordo, New Mexico, distinguished for his research in atomic spectroscopy. He was educated at the University of California and at Göttingen, and for a time worked as a news reporter; his many posts included chairs in physics at Washington and Minnesota. Measurements of the energy of alpha particles had shown that they had insufficient energy to escape from the attractive potential of the nucleus, making it difficult to explain the observed alpha decay classically. Independently of **Gamow**, Condon showed that this could be explained by the newly developed quantum mechanics, demonstrating that alpha decay is due to quantum-mechanical tunnelling of helium nuclei through the nuclear potential barrier. Using this theory he was able to derive the **Geiger**–Nuttall law that had been formulated empirically. During World War II Condon did notable work on the Manhattan Project to develop the atomic bomb, as associate director with **Oppenheimer**, although he found the necessary secrecy and security difficult to bear. He was later director of a USAAF study of unidentified flying objects from 1945 to 1951.

CONDORCET, Marie-Jean-Antoine-Nicolas de Caritat, Marquis de
(1743–94)

French mathematician, born in Ribemont, near St Quentin, the son of a cavalry officer. At 13, after distinguishing himself in the Jesuit school at Reims, he began his mathematical studies at the College of Navarre in Paris. His success was rapid and brilliant, and the high approval of **Clairaut** and **d'Alembert** determined his future. His *Essai sur le calcul intégral* (1765) won him a seat in the Academy of Sciences. He wrote five volumes of eloquent and moving obituaries of famous scientists which often amounted to intellectual biographies, and took an active part in Denis Diderot's *Encyclopédie*. At the outbreak of the Revolution he was sent by Paris to the Legislative Assembly (1791), and in 1792 became President of the Assembly. Later condemned by the extreme party, he hid for eight months—driven to change his place of concealment, he was recognized and lodged in the jail of Bourg-la-Reine, where he was found dead the next morning.

COOLIDGE, William David
(1873–1975)

American physical chemist and inventor, born in Hudson, Massachusetts. After receiving a BS from MIT, he graduated with a PhD (in physics) from University of Leipzig in 1899. After a few years teaching physical chemistry, he joined the General Electric Company in 1905. From 1908 to 1944 he advanced from Assistant Director of Research to Vice-President and Director of Research. From 1945 to 1961 he was consultant on X-rays and from 1961 to his death he was Emeritus Director of Research and Development. Among his many awards and honours was the Hughes Medal of the Royal Society (1927). In 1908 Coolidge discovered how to render tungsten ductile. This greatly increased the life of tungsten filaments in electric light bulbs. In 1916 he effected a great improvement in X-ray tubes by replacing the then usual cold aluminium cathode by a hot tungsten cathode. The 'Coolidge tube' was the prototype of modern X-ray apparatus. In association with **Langmuir** he developed the first successful submarine detection system and during World War II his research interests extended to radar, the atomic bomb, rockets and anti-submarine devices.

COOPER, Leon Neil
(1930–)

American physicist, born in New York City. Educated at Columbia University where he received his PhD in 1954, he moved to the University of Illinois to join **Bardeen** and **Schrieffer** for their successful assault on the problem of providing a theory to account for superconductivity. Cooper made a theoretical prediction that at low temperatures, pairs of electrons in a conductor could, via an interaction with the crystal lattice, feel a net attraction for each other and act in bound pairs (Cooper pairs). With his two collaborators this work was extended to show that in a superconductor, unlike in a normal conductor where electrical resistance arises from the random scattering of electrons, a cooperative state exists between all the conducting electrons such that the Cooper pairs have a common momentum which is not affected by the random scattering of individual electrons. This makes the effective electrical resistance of the material zero. This theory, which was named the BCS theory after it's three creators, won Bardeen, Cooper and Schrieffer the 1972 Nobel Prize for Physics. Cooper became assistant professor at Ohio State University in 1957, and since 1958 he has held various appointments at Brown University, Providence, where he is currently Thomas J Watson Sr Professor of Science.

COPE, Edward Drinker
(1840–97)

American palaeontologist, born to a Quaker family in Philadelphia. He studied zoology there under Joseph Leidy, the effective founder of American palaeontology, and was a curator and professor at Haverford College, Pennsylvania, by the age of 24. From 1889 he was professor at the University of Pennsylvania. Cope was an ambitious and aggressive collector of fossils. From 1868 he led a series of excavations in the American West which produced a wealth of dinosaur skeletons, especially from the badlands of South Dakota and Como Bluff, Wyoming. A famous rivalry developed between him and **Othniel Marsh**, each descending to underhand tactics in the race for important fossils. Cope dynamited fossil localities to prevent Marsh from excavating there, while Marsh's employees readdressed Cope's crates of fossils to his own laboratory. Cope wrote 1400 books and articles on his fossil discoveries, and also contributed to evolutionary theory, giving his name to two influential ideas: 'Cope's rule'—that animals have a tendency to ever increasing size during their evolution—and the Cope–Osborn theory for the origin of mammalian molars by the addition of cusps to peg-like reptilian teeth. Falling eventually into financial difficulties, Cope sold most of his vast lifelong fossil collection to the American Museum of Natural History.

COPERNICUS, Nicolas
(1473–1543)

Polish astronomer, born in Toruń, Prussia (now in Poland), and brought up after his father's death (1483) by his uncle, later bishop of Ermeland. After studying mathematics at the University of Cracow (1491–4) he went to Italy (1496) where he studied canon law and heard lectures on astronomy at the University of Bologna, while at Padua he studied medicine (1501–5). He was made a doctor of canon law by the University of Ferrara (1503), and though nominated a canon at the cathedral of Frombork (1497), he never took holy orders. Home in Poland he was his uncle's medical adviser and had various administrative duties at Frombork, where he spent the rest of his life. Beginning while in Italy he pondered deeply on what he considered the unsatisfactory **Ptolemaic** description of the world, in which the Earth was the stationary centre of the universe, and became converted to the idea of a Sun-centred universe. He set out to describe this mathematically in 1512. Copernicus hesitated to make his work public, having no wish to draw criticism from Aristotelian traditionalists or from theologians such as Martin Luther who had ridiculed him, but was eventually persuaded by his disciple **Rheticus** to publish his complete work, *De Revolutionibus Orbium Coelestium* (1543), which he dedicated to Pope Paul III. In the new system, the Earth became merely one of the planets, revolving around the Sun and rotating on its axis. The absence of any apparent movement of the stars caused by the Earth's annual motion was interpreted as due to the great size of the sphere of the stars. The transfer of the centre of the system from the Earth to the Sun in the new arrangement greatly simplified the geometry of the planetary system, though it did not get rid of all the epicycles of Ptolemy's model, a step which had to await **Kepler**. Copernicus was already old and ill by the time the book was printed, and was unaware that it carried an anonymous and unauthorized 'Preface to the Reader', presenting the work as a hypothesis rather than a true physical reality, written by Andreas Osiander, a Lutheran pastor of Nuremberg who supervised the last stages of the printing. Osiander's misguided intention was to forestall criticism of the heliocentric theory. The first printed copy of Copernicus's treatise, a work which fundamentally altered man's vision of the universe, reached its author on his death bed. Later banned by the Catholic church, *De Revolutionibus* remained on the list of forbidden books until 1835.

COREY, Elias James
(1928–)

American chemist, born in Methuen, Massachusetts. After undergraduate and graduate education at MIT, he became an instructor at the University of Illinois (1951). In 1953 he became professor there, but in 1959 he moved to a professorship at Harvard, where he was appointed Sheldon Emery Professor in 1965. He has made mechanistic and structural studies of over 100 natural products, and developed strategies for asymmetric synthesis. However, the work for which he is best known is the computer-aided analyses of synthetic problems. Retrosynthetic analysis, the way a target molecule might be broken down into simpler, readily available compounds, is a valuable technique in designing organic syntheses. Corey showed that the application of computers to retrosynthetic analysis can greatly assist the process. For this work he was awarded the Nobel Prize for Chemistry in 1990. He has received many other awards and honours, including the National Medal of Science (1988).

CORI, Carl Ferdinand
(1896–1984)

Czech-born American biochemist, born in Prague, who shared the 1947 Nobel Prize for Physiology or Medicine with his wife **Gerty Cori** (and **Houssay**)—only the third husband-and-wife team to do so after the **Curie**s in 1903 and the **Joliot-Curie**s in 1935. They married in Prague after he graduated in medicine there in 1920. In 1922 they emigrated to the USA, and both became professors at Washington University in St Louis from 1931. As with **Hans Krebs**, accurate quantitating was the key to Carl Cori's success. In 1936 he isolated glucose 1-phosphate from frog muscle and showed that this ester was formed by reaction of inorganic phosphate with the storage polysaccharide, glycogen. He found the enzyme that catalysed this reaction, glycogen phosphorylase, in muscle, heart, brain and liver. He obtained it in crystalline form in 1942, and recognized that it had both inactive and active forms and required a prosthetic group, adenylic acid (later shown by **Emil Fischer**, **Edwin Krebs** and **Earl Sutherland** to be an allosteric activator). Cori's theory that the same enzyme catalysed glycogen synthesis was proved incorrect, however, by **Leloir**. Cori also identified the isomerase that converts glucose 6-phosphate to glucose 1-phosphate, and in 1951 described an alpha-1,6-glucosidase, important in removing side chains in glycogen breakdown, and used it to determine the length of the main chain and side branches of several polysaccharides. After Gerty's death, Carl Cori worked at the Massachusetts General Hospital from 1967.

CORI, Gerty Theresa Radnitz
(1896–1957)

American biochemist, born in Prague, Czechoslovakia. She trained in medicine at the German University of Prague, and married her fellow student **Carl Cori** upon graduating. She worked at the Vienna Children's Hospital (1920–2) and then emigrated with her husband to the USA. She was employed at the State Institute for the Study of Malignant Disease, Buffalo, New York (1922–31) and then moved to the Medical School at the Washington University in St Louis (Research Associate in Pharmacology 1931–43; Research Associate Professor of Biochemistry 1943–7; Professor of Biochemistry 1947–57). With her husband she conducted research into carbohydrate metabolism, and their close collaboration makes it difficult to assess their contributions individually. They elucidated the process whereby glycogen, the stored form of carbohydrate, was enzymatically broken down to glucose, liberating energy in the process. Their detailed analysis, identifying and characterizing the enzymes responsible for each stage of the process, also enabled them to reconstruct the pathways whereby glycogen was synthesized and stored in the body. Their investigations into carbohydrate metabolism were very broad: they studied the effects of many hormones including insulin, adrenaline and pituitary extracts, and examined glycogen and glucose metabolism in biochemically abnormal circumstances, such as in tumours and in inherited metabolic diseases. The latter work led Gerty Cori to the first demonstration that glycogen storage disease could be caused by abnormalities or deficits in enzymes, a link with her very first position in paediatrics in Vienna. Gerty and Carl shared the Nobel Prize for Physiology or Medicine with **Houssay** in 1947, only the third husband-and-wife team to do so after the **Curies** in 1903 and the **Joliot-Curie**s in 1935.

CORIOLIS, Gustave Gaspard
(1792–1843)

French physicist, born in Paris. He was educated at the École Polytechnique, where he became professor in 1816. Intrigued by the problem of motion above a spinning surface, he considered the problem from around 1835, and in this work identified the 'Coriolis force'. This apparent force acting on objects moving across the Earth's surface results from the Earth's rotation. In the northern hemisphere, the path of an object appears deflected to the right, in the southern hemisphere to the left. It is responsible for wind and ocean current patterns, and is applicable to rotating systems generally.

CORMACK, Allan MacLeod
(1924–)

South African-born American physicist, born in Johannesburg. He studied physics and engineering at Cape Town University and did postgraduate work at Cambridge. He worked as a medical physicist at Groote Shuur Hospital in Johannesburg before moving to the USA, where he held various appointments at Tufts University, Medford, Massachusetts, being appointed full professor in 1964. His work pioneered the development of computerized axial X-ray tomography scanning (CAT), which enables detailed X-ray pictures of 'slices' of the human body to be produced. He shared the 1979 Nobel Prize for Physiology or Medicine with the English electrical engineer **Hounsfield**, who had independently developed a similar device.

CORNFORTH, Sir John Warcup
(1917–)

Australian chemist, born in Sydney. He entered Sydney University in 1933 to study chemistry. One year of postgraduate work in Sydney followed; he won a scholarship for further work at Oxford with **Robinson** and obtained his doctorate in 1941. He took part in the wartime effort to synthesize the new drug penicillin. He also studied the biosynthesis of cholesterol and other steroids. In 1962 he and George Popják were appointed co-directors of the Milstead Laboratory of Chemical Enzymology of Shell Research Ltd; Cornforth was also appointed professor at Warwick University. Cornforth and Popják studied in detail the stereochemistry of the interaction between an enzyme and its substrate. They also studied biological oxidation–reduction reactions. In 1968 Popják moved to the University of California and an extremely fruitful collaboration was terminated. Cornforth has also collaborated extensively with Hermann Eggerer on the stereochemistry of enzyme action, developing the chiral methyl group. In 1975 he was awarded the Nobel Prize for Chemistry, which he shared with **Prelog**, for all his work on the chemistry of enzyme action. During the period 1975–82 he was Royal Society Fellow at the University of Sussex, and he was knighted in 1977. At the age of 10 he had developed otosclerosis and, within a decade, became totally deaf. The triumph of his work, in spite of this handicap, is due in no small measure to the support and assistance of his wife.

CORRENS, Carl Franz Joseph Erich
(1864–1933)

German botanist and geneticist, born in Munich. Educated at Munich under **Nägeli** and at Tübingen, he became professor at

Münster in 1909. From 1914 he was the first director of the Kaiser Wilhelm Institute for Biology in the Dahlem district of Berlin, and with **de Vries** and Erich von Tschermak-Seysenegg was a rediscoverer of **Mendel**'s law of heredity. He proved that sex was inherited in a Mendelian fashion, through experiments with *Bryonia*. He was also able to demonstrate cytoplasmic (non-nuclear) inheritance. His unpublished manuscript was destroyed by bombing in 1945, along with the herbarium and library of the botanic garden at Berlin.

COSTER, Dirk

(1889–1950)

Dutch physicist, born in Amsterdam. After being a primary school teacher, he became a student at Leiden University, before taking a degree in electrical engineering at Delft Technological University. He then returned to Leiden for his doctoral thesis before moving on to Lund University (1922–3). He then joined **Niels Bohr** at his institute in Copenhagen. In 1923, Coster and **Hevesy** discovered the naturally occurring element hafnium (atomic number 72) in zirconium compounds as suggested by Bohr. Coster then returned to the Netherlands where he was assistant to **Lorentz** at the Teyler Laboratory in Haarlem, before accepting a chair in physics and meteorology at Groningen, which he held until 1949.

COTES, Roger

(1682–1716)

English mathematician, born in Burbage, near Leicester. He was educated at St Paul's School, London, and Trinity College, Cambridge, where he became a Fellow (1705) and Plumian Professor of Astronomy and Natural Philosophy (1706). In 1713 he took holy orders. He collaborated with **Isaac Newton** in revising the second edition of Newton's *Principia* and contributed a preface defending Newton's methodology. 'Had Cotes lived', said Newton, 'we might have known something'. His posthumously published *Harmonia mensurarum* (1722) contains work on logarithms and integration.

COTTON, Frank Albert

(1930–)

American chemist, born in Philadelphia. He graduated from Temple University in 1951 and obtained his PhD in 1955 from Harvard for a thesis under the direction of Sir **Geoffrey Wilkinson**. He then went to MIT, and has been Robert A Welch Distinguished Professor of Chemistry at Texas A & M University since 1973. Cotton is famous as a transition metal chemist and has studied most aspects of the subject. His earlier work centred upon metal carbonyls, a particular interest being a study of the chemically non-rigid behaviour of such molecules using the temperature dependence of nuclear magnetic resonance (NMR) line shapes to monitor the processes. Another line of his research has concentrated on the preparation and structural characterization of compounds with multiple metal–metal bonds. He discovered the first quadruple bond in 1963. He is author or co-author of several widely used university textbooks including *Advanced Inorganic Chemistry* (1962, with Wilkinson) and *Chemical Applications of Group Theory* (3rd edition 1990). Royal Society of Chemistry Centenary Lecturer in 1973–4, he was the first recipient of the American Chemical Society Award in Inorganic Chemistry in 1962 and won the Baekeland Medal in 1963.

COULOMB, Charles Augustin de

(1736–1806)

French physicist, born in Angoulême into a wealthy family. During his youth the family moved to Paris. An argument over career plans with his mother caused Coulomb to follow his father to Montpellier where his father had become penniless through financial speculations, but he later returned to Paris to complete his education at the École du Génie. After a long period of service in the Corps du Génie in Martinique, he returned to France in 1779, and held several public positions which he gave up at the outbreak of the Revolution. He was forced to leave Paris but returned in 1795 when he was elected a member of the new Institut de France, and was appointed Inspector-General of Public Instruction (1802–6). His experiments on mechanical resistance resulted in 'Coulomb's law' of the relationship between friction and normal pressure (1779), but he has become best known for the torsion balance for measuring the force of magnetic and electrical attraction (1784–5). With 'Coulomb's law' he observed that the force between two small charged spheres is proportional to the product of the charges divided by the square of the distance between them. The SI unit of quantity of charge is named after him.

COULSON, Charles Alfred

(1910–74)

English theoretical chemist, born in Dudley, Worcestershire. He was educated at Trinity College, Cambridge, where he took the Mathematics Tripos (Wrangler, 1931) and then the Natural Sciences Tripos (1932). He worked with **Ralph Fowler** and later **Lennard-Jones**, who introduced him to molecular orbital theory, which dominated his later researches. In 1938 he became Senior Lecturer in Mathematics at University College, Dundee. In World War II he was a conscientious objector. From 1945 to 1947 he held an ICI research fellowship at the Physical Chemistry Laboratory, Oxford, and from 1947 to 1952 he was Professor of Theoretical Physics at King's College, London. In 1952 he was appointed Rouse Ball Professor of Mathematics at Oxford. In 1972 he became Oxford's first Professor of Theoretical Chemistry, although already suffering from the illness from which he died in 1974. The Chair of Theoretical Chemistry at Oxford now bears his name. His many honours included the Davy Medal of the Royal Society (1970) and the Faraday Medal of the Chemical Society (1968). Coulson was a lifelong and prominent Methodist and was Vice-President of the Methodist Conference in 1959–60. He was chairman of OXFAM from 1965 to 1971. In spite of the variety of subjects specified in the titles of his various appointments, Coulson's research interests were almost entirely within theoretical chemistry, concerning the application of molecular orbital theory to chemical bonding and the electronic structures of molecules. He published over 400 papers. Probably his most important contribution was his definition of fractional bond order and the relation of this to bond length. His extension of quantum-mechanical methods to the treatment of giant molecules such as graphite and diamond was pioneering work, and his book *Valence* (1952) was highly influential. Much of his work predated the sophisticated calculations which are now possible by means of computers, but Coulson's work helped greatly to establish the basis upon which such calculations are performed. He also wrote books and articles on science and religion, eg *Science and Christian Belief* (1958).

COUPER, Archibald Scott
(1831–92)

Scottish chemist, born in Kirkintilloch. The son of the owner of a large cotton weaving business, he was educated at home and then in classics at Glasgow University, and logic and metaphysics at Edinburgh. He moved to Berlin in 1852 to study a number of topics, including chemistry. In 1856 he travelled to Paris to study with **Wurtz** and became interested in notions of chemical structure. He wrote a paper for presentation to the French Academy of Sciences on chemical structures, but because of unnecessary delay by Wurtz, it appeared after the publication of **Kekulé**'s paper on the same topic. Kekulé immediately attacked Couper's paper and claimed priority. This caused Couper considerable distress and he returned to Scotland in 1858, where he became assistant to **Lyon Playfair**. However, mental illness intervened and he spent the rest of his life in retirement. Although his publishing career lasted only one year, he introduced a number of important principles in the understanding and representing of chemical structures. So powerful was his insight that **Butlerov** credited Couper with establishing the central principles of our modern understanding of the structure of organic molecules.

COURANT, Richard
(1888–1972)

German-born American mathematician, born in Lublinitz. He studied in Breslau, Zürich and, as a pupil of **Hilbert**, at Göttingen, where he became professor in 1920, founding the Mathematics Institute there in 1929. In 1933 he was forced by the Nazis to resign and after a year at Cambridge, he moved to the USA where he became professor at New York University (1934), and Director of the Institute of Mathematical Sciences (later the Courant Research Institute) from 1953 to 1958. His work in applied analysis, particularly in partial differential equations and the **Dirichlet** problem, was always motivated by its physical applications. His textbook *Methoden der mathematischen Physik* (1924–7), written jointly with Hilbert, immediately became a classic. *What is mathematics?* (1941, written with H Robbins) attempts to explain mathematics to the layman.

COURNAND, André Frédéric
(1895–88)

French-born American physician, born in Paris. He was educated at the Sorbonne, emigrating to the USA in 1930, where he became a citizen in 1941. A specialist in cardiovascular physiology, he was awarded the Nobel Prize for Physiology or Medicine in 1956 jointly with **Forssman** and **Dickenson Richards** for developing cardiac catheterization. This consisted of threading a catheter through an arm or leg vein into the right atrium of the heart, so that blood samples could be taken, and recordings made. The technique made it possible to study heart functions in health and disease, and modifications of it are now important in treating heart disease. From 1934, he was on the academic staff of Columbia University.

COURTOIS, Bernard
(1777–1838)

French chemist who discovered iodine. He was born in Dijon, and studied pharmacy in Auxerre and chemistry in Paris, later working in the laboratory at the École Polytechnique and at the Thénard laboratory. While investigating opium with **Guyton de Morveau** he isolated morphine, the first alkaloid known. In 1804 he took over the management of his father's factory which made saltpetre from seaweed ash. Accidentally in 1811 he added too much sulphuric acid to the ash, and produced a violet gas which condensed into dark crystals. Friends announced his discovery of iodine at the Institute de France in 1813. In the 1820s Courtois abandoned the ailing saltpetre industry and attempted unsuccessfully to make a living by preparing and selling compounds of iodine. He died in poverty in Paris.

COUSTEAU, Jacques Yves

(1910–)

French naval officer and underwater explorer, born in Saint André, Gironde. He was educated at Stanislas, Paris, and the Navy Academy, Brest. He served in the Resistance during World War II, for which he was made a Commander of the Legion of Honour and awarded Croix de Guerre with Palm. As Lieutenant de Vaisseau (1939–43) he was partly responsible for the invention of the aqualung diving apparatus (1943). In 1946 he founded the French navy's undersea research group, and in 1947 he made a world record free dive of 91 metres. In 1950 he became commander of the oceanographic research ship *Calypso* from which he made the first underwater film, at a depth of 46 metres. Having retired from the navy in 1956, he was appointed Director of the Musée Océanographique de Monaco (1957–88). His other achievements have included the development of an underwater television, assisting **Piccard** in the development of the bathyscaphe, designing a diving saucer capable of descending to more than 180 metres, and promoting the Conshelf saturation dive programme which investigated the possibilities of undersea living (1962–5). He is best known for this popularization of marine biology with his many, much-acclaimed films, including *The Undersea World of Jacques Cousteau* (1968–76), and *Lilliput in Antarctica* (1990). His books include *The Living Sea* (1963).

COWAN, Clyde Lorrain Jr

(1919–74)

American physicist, born in Detroit. He was educated at the University of Missouri and Washington University, where he received his PhD in 1949. He became a group leader at Los Alamos Scientific Laboratory (1949–57), and served as Professor of Physics at George Washington University (1947) and the Catholic University of America (1948–74). Together with **Reines**, Cowan demonstrated the existence of nature's most elusive particle; the neutrino. The first observations were made in 1953, and more definitive confirming experimental evidence was produced in 1956. In 1930 **Pauli** had proposed the existence of the neutrino to explain the apparent violation of energy and momentum during beta decay, introducing the idea that an undetected low-mass neutral particle that interacts only very weakly with matter is involved in the reaction. To overcome the detection problems caused by the neutrino's weak interaction with matter, Cowan and Reines required a source of large numbers of neutrinos. In an experiment to study the emissions from a nuclear reactor, they confirmed the existence of the neutrino, detecting three neutrino interactions per hour. This corresponded to a total emission of over a million million per square centimetre per second.

COWLEY, Alan Herbert

(1934–)

English chemist, born in Manchester and educated at Manchester University where he received his BSc (1955), MSc (1956) and PhD (1958). During 1958–60 he was a Postdoctoral Fellow at the University of Florida and he then spent one year at ICI, Billingham. Most of his research has been carried out in the USA at the University of Texas at Austin, where he was appointed Robert A Welch Professor of Chemistry in 1989. A synthetic inorganic chemist, his current research interests centre on the synthesis, structure determination and reactivity of organometallic compounds and compounds of main group elements. One of the major themes to his research is the preparation of compounds with multiple bonding between either heavier main group elements or main group elements and transition metals within a complex. Other classes of compound which he has studied extensively are main group complexes with low coordination numbers or where a form of multiple bonding known as pi-bonding is exhibited between the main group atom and its ligands. Significant contributions have also been made in the design and synthesis of precursors for electronic materials. Cowley was elected FRS in 1988, and was Royal Society of Chemistry Centenary Lecturer in 1986–7.

COX, Allan

(1927–87)

American geophysicist, born in Santa Ana, California. He pursued his education during service in the US Merchant Marine (1945–8), he joined the US army (1951–3). A summer job with the US Geological Survey encouraged him to return to Berkeley to study geology and geophysics (BA 1956, MA 1957 and PhD 1959); during this time he was inspired by the teaching of **Byerley**. He then joined the US Geological Survey and set up a successful palaeomagnetic laboratory. He later accepted a professorship at Stanford University (1967), where he became Green Professor of Geophysics in 1974 and Dean in 1979. He studied all aspects of palaeomagnetism and collected rock samples from around the world, determining their age and magnetism. This produced evidence for many reversals of the Earth's magnetic field at random intervals back in geological time; the implied geomagnetic reversal timescale was published in a celebrated paper in *Nature* (1963). This led to further work on dating the age of the sea-floors using lineated magnetic anomalies. Cox's *Plate Tectonics and Geomagnetic Reversals* (1973) investigated the quantitative connection between plate tectonics and continental geology.

CRAM, Donald James
(1919–)

American chemist, born in Chester, Vermont. He studied chemistry at Rollins College, Florida, and at the University of Nebraska, after which he worked for the chemical company Merck. He then took a doctorate at Harvard (1945–7) and became an instructor at the University of California at Los Angeles (UCLA). He was appointed assistant professor at UCLA in 1948, full professor in 1956 and S Winstein Professor in 1985. For his early work he received the American Chemical Society Award for creative work in synthetic organic chemistry. His most highly praised work began in 1972, when he described the synthesis of chiral crown ethers containing binaphthyls. He reasoned that as cyclic polyethers (crown ethers) bind metallic cations, they should also bind protonated amines. Realizing that chiral crown ether should discriminate between the enantiomers of a chiral amine, he eventually achieved dramatic chiral separation, a matter of great importance in the synthesis of enantiomerically pure materials for biological purposes. He also introduced the informative description host–guest chemistry. Since 1972 he has synthesized a large number of novel three-dimensional host compounds. For his work on guest–host chemistry he shared the 1987 Nobel Prize for Chemistry with **Lehn** and **Pedersen**.

CRAY, Seymour R
(1925–)

American computer designer, born in Chippewa Falls, Wisconsin. Educated at Minnesota University, in 1950 he was awarded degrees in both electrical engineering and mathematics. His name is synonymous with 'supercomputers' used in military, weather-forecasting and advanced engineering design applications. He established himself at the forefront of large-scale computer design through his work at Engineering Research Associates (later Remington Rand, Sperry Rand UNIVAC Division) and Control Data Corporation (CDC). In 1972 he left CDC and organized Cray Research Inc in Chippewa Falls to develop and market the most powerful computer systems available. The Cray 1, delivered in 1976, cost $8 million and was the world's fastest computer. Cray had a special talent for putting computer circuits close together, so increasing a computer's speed. By the 1980s the Cray 1 and its later derivatives dominated the supercomputer market, with an estimated 70 per cent share. After serving first as President and then as Chairman of Cray Research, he severed his connections with the company in 1989 to head the Cray Computer Corporation (a separate company in Colorado Springs) and to work on the Cray 3 supercomputer.

CRICK, Francis Harry Compton
(1916–)

English molecular biologist, born near Northampton. He received a BSc in physics at University College London. During World War II he worked on the construction of mines for the British Admiralty, and after the war he joined the laboratory of **Perutz**, to work on the structure of proteins. In the early 1950s, in Cambridge, he met **James Watson** and together they worked on the structure of DNA. 1953 saw the publication of their model of a double-helical molecule, consisting of two chains of nucleotide bases (adenine, thymine, guanine and cytosine) wound round a common axis in opposite directions. The structure they proposed suggested a mechanism for the reproduction of the genetic material and the genetic code, on which Crick continued to work for the next decade. He worked at the Laboratory of Molecular Biology, Cambridge, from 1949 to 1977, when he became Kieckhefer Professor at the Salk Institute, California. There he carried out research into the visual systems of mammals, and the connections between brain and mind. With Watson and **Wilkins** he was awarded the Nobel Prize for Physiology or Medicine in 1962. His autobiography, *What Mad Pursuit*, was published in 1988.

CRILE, George Washington

(1864–1943)

American surgeon and physiologist, born in Chili, Ohio. He received a BA from Northwestern Ohio Normal School in 1884 and an MD from the University of Wooster in 1887. He became interested in surgical shock (abnormally low blood pressure) when a friend died following an emergency operation after an accident. While building up a busy practice he began animal experiments on surgical shock and published his first monograph on the subject in 1899. In 1900 he became Clinical Professor and in 1911 Professor of Surgery at the Western Reserve School of Medicine. He was founder and first director of the Cleveland Clinic Foundation (1921–40). Crile continued working on surgical shock, publishing *Blood Pressure in Surgery* (1903) and *Anemia and Resuscitation* (1914). He viewed the prevention of shock as the most important principle and advocated atraumatic surgery using safe anaesthetics. He emphasized the need to monitor blood pressure during surgery and popularized the apparatus used for this, and was one of the first to regularly use blood transfusions and adrenaline as means of combating shock. He devised a 'shockless' method of anaesthesia (which he called 'anoci-association') which aimed to separate the operative site from the nervous system. This gave excellent results although based on the erroneous 'kinetic theory'—the idea that surgical shock originates in the nervous system. He developed several operations for the endocrine glands, though some of his later physiological speculations were rather eccentric.

CROLL, James

(1821–90)

Scottish physicist and geologist, born in Cargill, Perthshire. He received an elementary school education during early struggles against ill-health and poverty, but his science was wholly self-taught. Successively millwright, insurance agent and keeper of the museum of Anderson's College, Glasgow, he was appointed to the Scottish Geological Survey (1867–81) and came into contact with many of the best geologists of the age. He was broadly supportive of **Kelvin**'s calculations of the age of the Earth, and rejected the mathematically unconstrained vast length of time required by uniformitarianism. Croll's main interests were in changes in climate over geological time and in glacial geology. He argued that the influence of ocean currents and the changing distribution of land and water have a great effect on climates. Among his works were *Climate and Time* (1875), *Discussions on Climate and Cosmology* (1886), *The Philosophical Basis of Evolution* (1890) and *Stellar Evolution and its Relations to Geological Time* (1890).

CROMPTON, Rookes Evelyn Bell

(1845–1940)

English engineer, born near Thirsk, a pioneer of electric lighting and road transport. While still a schoolboy he designed and built first a model and then a full-size steam road locomotive, continuing with this work during and after army service in India. His road steamers were technically successful, but could not compete with the rapid development of the railways in the second half of the 19th century. On his return to Britain he became interested in the generation and distribution of electricity for lighting, and set up the firm of Crompton and Company to manufacture generating equipment. In 1886 he founded the Kensington and Knightsbridge Electric Supply Company with the aim of supplying direct current for electric lighting, on which he became an international authority. His firm had meanwhile expanded into the manufacture of many kinds of electrical machinery and instruments, and in 1927 it became Crompton Parkinson Limited. He strongly supported standardization in industry and was involved in the establishment of the National Physical Laboratory and what is now the British Standards Institution. He was elected FRS in 1933.

CRONIN, James Watson

(1931–)

American physicist, born in Chicago. He was educated at the University of Chicago, where he received his PhD in 1955. After working at the Brookhaven National Laboratory he moved to Princeton where he became Professor of Physics in 1965, and he held a similar post at Chicago from 1971. In 1956 **Tsung-Dao Lee** and **Yang** had proposed that parity (P) may not be conserved in weak interactions between subatomic particles; this was proved experimentally by **Wu** and her collaborators. Another symmetry, known as charge conjugation C, was also found to be broken by the weak interaction. However there is a very general theorem, known as the CPT theorem, that states that the interactions between particles must be symmetric under the combined symmetry of C, P and T (time reversal). Since C and P were violated by the weak interactions it was assumed that the interactions were symmetric under combined symmetry of CP. To test this assumption, together with **Fitch** and others, Cronin made a study of the decay of the strange neutral meson known as the kaon (1964). They made the startling discovery that CP is not conserved in the weak decay of the neutral kaons. This important result implies that T must also be violated to preserve the combined symmetry of CPT. This is still not understood today, but the idea of CP violation has been used to explain the domination of matter over antimatter in the universe. Cronin and Fitch shared the 1980 Nobel Prize for Physics for this work.

CRONQUIST, Arthur
(1919–92)

American botanist, born in San José, California, and educated at Utah State University and the University of Minnesota. Based at the New York Botanical Garden, he was an authority on the family Compositae (Asteraceae) and wrote treatments of it for several American Floras and for the *Flora of the Galapagos Islands*. One of his major interests was the phylogeny of flowering plants, particularly at and above the level of family. In this field he wrote *The Evolution and Classification of Flowering Plants* (1968, 1988) but his *magnum opus* was *An Integrated Classification of Flowering Plants* (1981). This monumental work, one of the most important 20th-century publications in plant taxonomy, effectively supersedes the classification of **Engler**. The 'Cronquist system' of plant classification has been adopted in many later works. Either alone or as co-author, he also prepared several standard Floras, including *Vascular Plants of the Pacific North-West* (5 parts, 1955–69), *Manual of Vascular Plants of Northeastern United States and Adjacent Canada* (1963, 1991) and *Intermountain Flora* (6 vols, 1972), as well as several textbooks on botany.

CRONSTEDT, Axel Fredrik, Baron
(1722–65)

Swedish metallurgist and mineralogist, born in Turinge. He studied mathematics at Uppsala and served in the army (1741–3) before embarking upon a career in mining and metallurgy. He first isolated nickel (1751) and noted its magnetic properties. Cronstedt is renowned for his *Essay towards a System of Mineralogy* (1758), in which minerals and stones were distinguished for the first time and chemical composition was advocated as the primary method of classification of minerals. He proposed a new classification based on the action of fire, water and oil on specimens, and introduced mineral analysis by the blowpipe.

CROOKES, Sir William
(1832–1919)

English chemist and physicist, noted for his study of cathode rays. Crookes was born in London and became a pupil and then assistant of **Hofmann** at the Royal College of Chemistry. He then superintended the meteorological department of the Radcliffe Observatory, Oxford, and in 1855 lectured on chemistry at the College of Science in Chester. The following year he set up a private laboratory in London and made his living as a chemical consultant and editor of scientific and photographic journals. In 1861, following up the spectroscopic discoveries of **Bunsen** and **Kirchhoff**, Crookes discovered thallium by the bright green line in its spectrum. While weighing very small quantities of it in a highly evacuated vessel in order to negate changes in barometric pressure, he noticed that a pivoted rod rotated in a strong light. The principle of the 'radiometer' was soon employed in many scientific instruments. In 1878 Crookes began investigating cathode rays in high-vacuum tubes of his own design (now known as Crookes tubes). He observed that the radiation made the walls of the tube fluoresce, but that if an object was placed in the tube, it cast a shadow, proving that the radiation must travel in straight lines. He also showed that it must be charged, since it was deflected by a magnet. This work led directly to the discovery of the electron by **J J Thomson** in 1897. Crookes's other achievements were many and various. He was an authority on sanitation; he discovered the sodium amalgamation process; invented the spinthariscope, which measures scintillation from alpha radiation; promoted electric lighting; and was also the author of works on dyeing, calico-printing, sugar, beetroot, wheat and diamonds. One of his greatest achievements was to popularize science and to make contemporary discoveries comprehensible to the general public, especially through *Chemical News* which he founded in 1859 and thereafter edited. Following the death of his brother he also devoted much time to investigations of spiritualism and psychic phenomena. Knighted in 1897, Crookes was a prominent, and often controversial, public figure. He was elected FRS in 1863 and served as President of the Royal Society from 1913 to 1915. He died in London.

CROSS, Charles Frederick
(1855–1935)

English industrial chemist, born in Brentford, Essex. He studied in London, Zürich and finally in Manchester, where **Roscoe** was his teacher and **Bevan** a fellow student. In 1885 he joined with Bevan to form Cross and Bevan Research and Consulting Chemists, specializing in the chemistry of cellulose. They worked with Swedish companies to find better ways of producing wood pulp for paper-making. They also began to investigate **Swan**'s process for making filaments for light bulbs from nitrocellulose and **Chardonnet**'s process for making thread from the same material. Swan, and Chardonnet following him, had dissolved nitrocellulose in acetic acid. In 1892 Cross and Bevan patented a method in which cellulose, usually in the form of wood pulp, was dissolved in caustic soda and carbon bisulphide. The solution was then extruded through very fine jets into a bath of sulphuric acid where the cellulose was regenerated in the form of yarn. This was woven into rayon fabrics and 'silk' stockings. Two years later Cross and Bevan patented a similar process for making viscose film ('cellophane') in which the dissolved cellulose was extruded through a thin slit rather than jets. Cross joined C H Stearn in 1898 to form the Viscose Spinning Syndicate which amalgamated with Courtaulds, the silk manufacturers, in 1902 and began large-scale production in 1905. He was elected FRS in 1917, and died in Hove, Sussex.

CRUM BROWN, Alexander
(1838–1922)

Scottish chemist, born in Edinburgh. He was educated in London and at Edinburgh University where he studied medicine. In 1862 he went to Germany to study with **Kolbe**. On his return he joined the chemistry department in Edinburgh, becoming professor in 1869, a position he held until 1908. His main interest was chemical structure, and he invented the 'ball and stick' graphical presentation of organic molecules. Double bonds were shown as double parallel lines. The system was similar to that of **Couper**, also a Scot. Crum Brown also systematized the nomenclature of disubstituted benzene compounds. He believed that ultimately, chemistry would become a branch of applied mathematics, but he made no substantial contribution to mathematical chemistry. He was also interested in physiology, particularly the sensations of vertigo. He was elected FRS in 1879 and served as President of the Chemical Society from 1891 to 1893.

CULLEN, William
(1710–90)

Scottish chemist and physician, born in Hamilton. He studied mathematics before obtaining a post as ship's surgeon on a vessel bound for the West Indies. He then practised medicine in Hamilton, and attended medical classes in both Edinburgh and Glasgow. In 1747 he began teaching chemistry in Glasgow with great success; he was appointed to the Chair in Medicine in 1755, and taught chemistry, materia medica and medicine in Edinburgh. He was the leading light of the Edinburgh Medical School during its golden age. His teachings were full of shrewd observations, neatly presented in the form of a coherent system. He had hundreds of grateful and professionally successful pupils. He expounded his clinical ideas primarily through the nosologically arranged *First Lines of the Practice of Physic* (1778–9), frequently reprinted and translated during the next half-century. He emphasized the importance of the nervous system in the causation of disease, coining the word 'neurosis' to describe a group of nervous diseases. He was a close friend of David Hume and other major figures of the Scottish Enlightenment.

CUNNINGHAM, Allan
(1791–1839)

English botanist and explorer, born in Wimbledon, Surrey. He became clerk to the Curator of Kew Gardens, and then plant collector for **Banks**, first in Brazil and then, in 1816, in New South Wales. While searching for new specimens, Cunningham made many valuable explorations of the hinterland of New South Wales and Moreton Bay, Queensland. He also visited New Zealand and Norfolk Island, returning to Kew in 1831 to classify his specimens. He was offered the post of Colonial Botanist for New South Wales, but turned it down in favour of his younger brother Richard Cunningham (1793–1835). When his brother was killed by aborigines, he accepted the renewed invitation and returned to Sydney in 1837. He found that his duties included managing what he termed the 'Government Cabbage Garden' and growing vegetables for Governor Gipps's table, so he resigned and left for New Zealand, but returned to Sydney six months later in bad health and died there. His writings and most of his collections are preserved at Kew, and many indigenous Australian trees now bear his name.

CURIE, Marie
(originally **Marya**), née **Skłodowska**
(1867–1934)

Polish-born French radiochemist, born in Warsaw. She was brought up in poor surroundings after her father, who had studied mathematics at the University of St Petersburg, was denied work for political reasons. After brilliant high school studies, she worked as a governess for eight years, during which time she saved enough money to send her sister to Paris to study. In 1891 she also went to Paris where she graduated in physics from the Sorbonne (1893) taking first place; she then received an Alexandrovitch Scholarship from Poland which allowed her to study mathematics. Marie met **Pierre Curie** in 1894 and they married the following year. In 1896, **Antoine Henri Becquerel** had discovered the radioactive properties of uranium; Marie Curie decided to study this phenomenon for her doctoral thesis topic. She used an apparatus for measuring very small electrical currents, built by her husband, to search for elements that emitted ionizing radiations. In this way she discovered that thorium is also radioactive, and she showed that the radioactivity of uranium was an atomic property, rather than the result of interactions between the element and another substance. In subsequent research she discovered that the radioactivity of the minerals pitchblende and chalcolite was more intense than could be explained by the uranium and thorium content alone, and from this deduced that these minerals must contain new radioactive elements. Pierre left his work on piezoelectricity to help in the laborious process of isolating the new elements by fractional crystallization. No precautions against radioactivity were taken, as the harmful effects were not known at that time. In 1898 Pierre and Marie announced the discovery of a new element, which they named polonium in honour of Marie's native country, and later the same year they announced the discovery of radium. In 1903 Marie presented her doctoral thesis (the first advanced scientific research degree to be awarded to a woman in France), and in the same year she was awarded the Nobel Prize for Physics with Pierre and Becquerel for their work on radioactivity. It was around this time that the Curies began to suffer from symptoms later ascribed to radiation sickness. In 1904 Pierre was awarded a new chair in physics at the Sorbonne, but in 1906 he was killed in a street accident; Marie succeeded him as Professor of Physics. She continued her work with radioactivity and in 1911 was awarded the Nobel Prize for Chemistry for her discovery of polonium and radium. During World War I she developed X-radiography and then became director of the research department at the newly established Radium Institute in Paris (1918–34). She died of leukaemia, probably due to her long exposure to radioactivity. Her daughter **Irène Joliot-Curie** and son-in-law **Frédéric Joliot-Curie** followed in her footsteps in radiochemistry and also received the Nobel Prize for Chemistry.

CURIE, Pierre
(1859–1906)

French physicist, born in Paris and educated there at the Sorbonne, where he became an assistant teacher in 1878. He was appointed laboratory chief at the School of Industrial Physics and Chemistry in 1882, and remained there until in 1904 he was appointed to a new chair in physics at the Sorbonne. In studies of crystals with his brother Jacques, he discovered piezoelectricity in 1880; they observed a small electric current being produced when certain crystals were mechanically deformed and vice versa. Such crystals have found many uses in modern technology. The Curies used a piezoelectric crystal to construct an electrometer, capable of measuring very small electric currents, and this was later used by Pierre's wife **Marie Curie** in her investigations of radioactive minerals. In studies of magnetism, Pierre showed that a ferromagnetic material loses this property at a certain temperature (the Curie point) specific to the substance involved; for this work on magnetism he was awarded his doctorate in 1895. Another of his important results in magnetism was 'Curie's law', that the magnetic susceptibility of a paramagnetic material is inversely proportional to the absolute temperature. From 1898 he worked with his wife on radioactivity, and showed that the rays emitted by radium contained electrically positive, negative and neutral particles. With his wife and **Antoine Henri Becquerel** he was awarded the Nobel Prize for Physics (1903). He was killed in a street accident shortly after he accepted a new chair in physics at the Sorbonne.

CURTIS, Heber Doust
(1872–1942)

American astronomer, born in Muskegon, Michigan. He studied classics at Michigan University and then went on to teach classics, becoming Professor of Latin at Napa College, California. In 1897 he changed academic direction and became Professor of Mathematics and Astronomy at the University of the Pacific. He then worked at Lick and at their observatory in Chile. In 1920 he became the Director of the Allegheny Observatory and in 1930 Director of Michigan University Observatory. From 1902 to 1909 he worked on **Campbell**'s radial velocity programme, and thereafter he concentrated on the photography and investigation of spiral nebulae. He was convinced that spiral nebulae were isolated independent star systems, and that they lay beyond our own galaxy. Curtis noticed the dark lines on the rim of a nebula viewed edge-on. He equated this to a disc of dust and gas, and recognized that it was similar to the zone of avoidance seen in our own Milky Way system. Curtis opposed **Shapley** in the 'great debate' of April 1920; this debate discussed the scale of the universe, ie the size of our galaxy and the nature of the spiral nebulae. Curtis was proved correct when, in 1924, **Hubble** demonstrated that the spiral Andromeda nebula lay well beyond our galaxy.

CURTIS, William
(1747–99)

English horticulturist and botanist, born in Alton, Hampshire. He grew a collection of British plants on a plot of ground at Bermondsey, and established a botanic garden at Lambeth in 1777. He also established nurseries at Brompton and Chelsea, as well as becoming demonstrator and later Director of the Chelsea Physic Garden (1772–7). His *Flora Londinensis* (1777) was intended to be a list of all the plants growing within 10 miles of London. Unfortunately, due to the high costs of publication, this project was never fully realized. His study of grasses led to the publication of *Enumeration of British Grasses* (1787) and *Hortus Siccus Gramineus* (1802). He began publication of the *Botanical Magazine* in 1787. This was the first periodical devoted exclusively to plants, and while more horticultural than botanical, it was extremely influential in bringing a worldwide variety of plants to the attention of those who could afford to cultivate them. The work consisted of a detailed description of a plant, together with a hand-coloured plate for each species. Under its new title of *Kew Magazine* it continues publication to this day, in a format not much changed from Curtis's original, but alas no longer hand-coloured. His *Lectures in Botany* were published in 1802.

CUSHING, Harvey Williams
(1869–1939)

American neurosurgeon, born in Cleveland, Ohio. Educated at Yale and Harvard, he graduated in 1895 and undertook postgraduate training at the Massachusetts General Hospital and at Johns Hopkins Hospital, also making a European trip to study with **Kocher** in Switzerland and with **Sherrington** in England. He became Professor of Surgery at Harvard (1912–32), served with the Army Medical Corps during World War I, and in 1933 was appointed Sterling Professor of Neurology at Yale until his retirement in 1937. A talented and innovative neurosurgeon, much of his success was dependent on the important new techniques and procedures he developed to control blood pressure and bleeding during surgery, and he discovered a novel operative approach to the pituitary gland. The method was used successfully to remove tumours growing at the base of the pituitary, which pressed on the optic nerves and caused blindness, and Cushing wrote detailed descriptions and a classification of many types of intracranial tumours. His interest in the pituitary extended to a detailed study of its activity, especially in disease states. He characterized the effects of underactivity, which caused dwarfism in a growing child, and of overactivity, which caused a form of gigantism in adults. Cushing was also interested in the history of medicine and won a Pulitzer Prize in 1926 for his biography of the Canadian physician, Sir William Osler. He bequeathed his extensive collection of books to the Yale medical library.

CUVIER, Georges Léopold Chrétien Frédéric Dagobert, Baron
(1769–1832)

French anatomist, known as the father of comparative anatomy and palaeontology. Born in Montbéliard, he studied for the ministry at Stuttgart, but his love for zoology was confirmed during residence as a tutor on the Normandy coast (1788–94). In 1795, through the influence of **Étienne Geoffroy Saint-Hilaire**, he was appointed Assistant Professor of Comparative Anatomy at the Jardin des Plantes of the Museum of Natural History in Paris, and in 1789 Professor of Natural History at the Collège de France. After the Restoration he was made Chancellor of the University of Paris, and admitted to the cabinet by Louis XVIII. His opposition to royal measures restricting the freedom of the press lost him the favour of Charles X, but under Louis-Philippe he was made a peer of France in 1831, and in the following year Minister of the Interior. In his plans for national education, in his labours for the French Protestant Church and in his scientific work, he was indefatigable. He formulated the principle of correlation of parts, proposing that different structures and organs are functionally interrelated, and created in perfect adaptation to particular environments. In this he came into conflict with Geoffroy Saint-Hilaire, who believed that the anatomy and structure preceded the way of life. He rejected the 18th-century scheme where animals were arranged in a linear sequence with man at the pinnacle, and put forward a classification based on four major distinct animal types. Extending **Linnaeus**'s system of classification, he divided the animal kingdom into Mollusca (molluscs), Radiata (starfish), Articulata (arthropods) and Vertebrata (vertebrates). Although subsequently modified, this scheme anticipates the modern division of the animal kingdom into phyla. Cuvier's knowledge of animal structure was unsurpassed, and his classifications based on diagnostic features developed a legendary though ultimately stifling authority which persisted for decades after his death. His knowledge of comparative anatomy allowed him to reconstruct the form of fossil creatures, and he was the first to classify them together with living species. A militant anti-evolutionist, he accounted for the fossil record by positing 'catastrophism': a series of extinctions due to periodic global floods after which new forms of life appeared. Among Cuvier's more important works are *Leçons d'anatomie comparée* (1801–5), *L'Anatomie des mollusques* (1816), *Les Ossements fossiles des quadrupèdes* (1812), *Histoire naturelle des poissons* (1828–49), written with the French zoologist Achille Valenciennes, and *Le Règne animal distribué d'après son organisation* (1817).

D

DAGLEY, Stanley
(1916–)

English microbiologist, born in Burton-on-Trent. As a student in the late 1930s at Trinity College, Oxford, he worked with **Hinshelwood** on the growth and metabolism of *Aerobacter aerogenes*, and showed that carbon dioxide is a necessary metabolite for this bacterium. His training with Hinshelwood gave him expertise in both kinetic and metabilic areas of biochemistry. He took up a lectureship in biochemistry at the University of Leeds in 1947, becoming reader in 1952 and professor in 1962. He later moved to the USA, where he was appointed to the Chair of Biochemistry at the University of Minnesota in St Paul in 1966, and worked on bacterial citrate lyase, the enzyme that converts citrate into oxaloacetate and acetic acid. He later became interested in chemical pollutants, and the metabolism of fission products from substituted catechols. In 1969 he won the Horace T Morse award for outstanding contributions to undergraduate education.

DAGUERRE, Louis Jacques Mandé
(1789–1851)

French photographic pioneer, inventor of the 'daguerreotype'. Born in Cormeilles, he was a scene painter for the opera in Paris. From 1826 onwards, and partly in conjunction with **Niepce**, he perfected his daguerreotype process in which a photographic image is obtained on a copper plate coated with a layer of metallic silver sensitized to light by iodine vapour. This greatly reduced the exposure time required to produce an image from around eight hours for Niepce's original method to around 25 minutes.

DAINTON, Frederick Sydney, Baron Dainton of Hallam Moors
(1914–)

English physical chemist and administrator, born in Sheffield. He was educated at St John's College, Oxford, and Sidney Sussex College, Cambridge. He was appointed as a university demonstrator at Cambridge (1944) and H O Jones Lecturer in Physical Chemistry (1946). From 1950 to 1965 he was Professor of Physical Chemistry at the University of Leeds, from 1965 to 1970 Vice-Chancellor of the University of Nottingham, and from 1970 to 1973 Dr Lee's Professor of Chemistry at Oxford. He was Chairman of the University Grants Committee (1973–8) and of the British Library Board (1978–85). Dainton has been chairman/president of many bodies including the Faraday Society (1965–7), the Chemical Society (1972–3), the Association for Science Education (1967) and the British Association for the Advancement of Science (1980). His numerous honours include the Tilden (1950) and Faraday (1973) medals of the Chemical Society and the Davy Medal of the Royal Society (1969). Knighted in 1971, he was created a life peer in 1986. His contributions to physical chemistry have been in the areas of chemical kinetics, photochemistry and radiation chemistry, on which he has published many papers. His books include *Chain Reactions* (1956).

DAKIN, Henry Drysdale
(1880–1952)

English chemist, born in London. He was trained in Marburg under **Albrecht Kossel** with whom he isolated arginase, a key enzyme of **Hans Krebs**'s urea cycle, before returning to work at the Lister Institute where he independently synthesized adrenaline (1906), patented separately by a German firm at this time. He carried out extensive research on the oxidation processes of the body (1908–12), and made the influential observation that *in vitro*, many substances upon oxidation by hydrogen peroxide yield products similar to those found by oxidation in living cells. After World War I, Dakin emigrated to the USA and joined the staff of the Rockefeller Institute, New York, where he made early contributions to the understanding of protein structure, showing that 80 per cent of the gelatin molecule consists of amino acids and that differences in amino acid sequence could explain differences in structure (1919–20). In 1931 he isolated a pyrrole intermediate in heme biosynthesis; the metabolic path was not established until 1946 after radioisotopically labelled precursor molecules became available. His most enduring contribution came from his study of antiseptics; 'Dakin's' or the 'Carrel–Dakin' solution (a 0.5 per cent solution of sodium hypochlorite) was widely used for treating wounds during the two world wars and is still used extensively as a safe, cheap sterilizing agent today. He was elected FRS in 1917.

DALE, Sir Henry Hallett

(1875–1968)

English physiologist and pharmacologist, born in London. Dale studied natural sciences at Trinity College, Cambridge, and after a brief period in physiological research with **John Langley**, qualified in medicine from St Bartholomew's Hospital in 1902. He worked for **Starling** and **Bayliss** at University College London, before accepting a position in 1904 at the Wellcome Physiological Research Laboratories, private laboratories associated with the pharmaceutical firm Burroughs, Wellcome and Co. Around 10 years later Dale joined the newly created Medical Research Committee (later Council) as Head of the Department of Biochemistry and Pharmacology, at the National Institute for Medical Research (NIMR). Retaining his research commitments he was also appointed first Director of the NIMR in 1928, and retired in 1942. He was President of the Royal Society during 1940–5, when he also served as Secretary to the Scientific Advisory Committee to the War Cabinet. His work focused on the physiology and pharmacology of naturally occurring chemicals, work which had started with the chance observation in 1904 that an extract of ergot of rye, a fungus, reversed the effects of adrenaline. Adrenaline had recently been shown to mimic the effects of the sympathetic nervous system, and in collaboration with his colleague **Barger**, Dale analysed several compounds for such 'sympathomimetic' activities. Further analyses of ergot, with **Ewins**, showed that it contained many pharmacologically active compounds, including histamine, tyramine and acetylcholine. Dale's later experiments provided evidence that acetylcholine occurred naturally in animals, and played an important role in the transmission of nerve impulses across the synapse. For this work he shared the 1936 Nobel Prize for Physiology or Medicine with **Loewi**. Dale also did pioneering work in endocrinology, and closely associated with his research were his wider concerns with the standardization of drugs. He served on many advisory and regulatory committees at both national and international levels, and was knighted in 1932.

D'ALEMBERT, Jean le Rond

(1717–83)

French philosopher and mathematician, born in Paris, the illegitimate son of Mme de Tencin and the Chevalier Destouches. Brought up as a foundling, he was given an annuity from his father and studied law, medicine and mathematics at the Collège Mazarin. In 1743 he published *Traité de Dynamique*, developing the mathematical theory of Newtonian dynamics around the concept of energy rather than force. It includes the principle named after him, which enables Newtonian mechanics to apply when the motion is constrained to lie on a surface. Later he worked on fluid motion, partial differential equations, the motion of vibrating strings and celestial mechanics. He also published many short memoirs on analysis. Until 1758 he was Denis Diderot's principal collaborator on his *Encyclopédie*, of which he was scientific editor, and he wrote the *Discours Préliminaire* (1751), proclaiming the philosophy of the French enlightenment. His fame spread through Europe, but he refused invitations from Frederick the Great of Prussia and Catherine II of Russia and remained in Paris.

DALÉN, Nils Gustav

(1869–1937)

Swedish physicist and engineer, born in Stenstorp. He graduated as a mechanical engineer from the Chalmers Institute in Göteborg (1896), and after a year of further training in Zürich returned to Sweden to experiment on hot-air turbines, compressors and air-pumps. He was a member of the engineering firm Dalén and Alsing (1900–5) founded to exploit inventions, progressing to a position as works manager for the Swedish Carbide and Acetylene Company, becoming its managing director in 1909 under a new company name, the Swedish Gas Accumulator Company. He successfully used acetylene to light unattended navigational aids. By dissolving acetylene in acetone then forcing this under a pressure of 10 atmospheres into a porous material in a steel vessel, he managed to make the transport of acetylene sufficiently safe for practical use on lighthouses and buoys. He also invented a valve that permitted the light to be flashed with pulses whose duration could be used to identify the light source's location. To conserve acetylene, he produced a sun-valve that extinguished the flame during daylight hours. Dalén was awarded the 1912 Nobel Prize for Physics for his numerous inventions. He was prevented from attending the award ceremony by a serious accident in the same year in which he lost his sight.

DALTON, John

(1766–1844)

English chemist and natural philosopher, born in Eaglesfield, near Cockermouth, Cumberland. The son of a Quaker weaver, he received his early education from his father and at the Quaker school at Eaglesfield. Dalton started teaching in the village at the age of 12, but in 1781 he joined his elder brother as an assistant at a school in Kendal run by their cousin George Bewley. In 1785 Bewley retired and the Dalton brothers carried on the school by themselves. In 1793 Dalton moved to Manchester, where he spent the rest of his life. He was appointed teacher of mathematics at New College in Moseley Street (a predecessor of Manchester College, Oxford) and continued in this post for around six years. Thereafter he supported himself as a private teacher, while carrying out his scientific researches. Through these he acquired an international reputation and was associated with several of the early meetings of the British Association for the Advancement of Science. Manchester grew proud of him and when he died he was given a substantial public funeral. The entrance to Manchester Town Hall is flanked by statues of Dalton and **Joule**. Although Dalton's fame today rests almost entirely on his atomic theory, he carried out a wide range of research. In 1787 he began keeping a meteorological journal, which he continued all his life, recording over 200 000 observations. In 1794 he described colour blindness (Daltonism), exemplified partly by his own case and that of his brother. In his chemical and physical researches Dalton was a somewhat crude experimentalist, but the results he obtained were adequate to direct him to his atomic theory. The ideas of the theory arose as a result of his meteorological observations, and of his studies of the physical properties of the atmosphere and other gases. Of particular importance were his studies of partial pressures in mixtures of gases, showing that in a mixture of gases, each gas exerts the same pressure as it would if it were the only gas present in the given volume (Dalton's Law), and of the solubilities of gases in water. This led to the interpretation of chemical analyses in terms of the relative weights of the atoms of the elements involved and to the laws of chemical combination. These developments occurred during 1803–5, but the first full account was given by Dalton in 1808 in his *New System of Chemical Philosophy*. His atomic theory recognized that all matter is made up of combinations of atoms, the atoms of each element being identical. He concluded that atoms could be neither created nor destroyed, and that chemical reactions take place through the rearrangement of atoms.

DAM, Carl Peter Henrik

(1895–1976)

Danish biochemist, born and educated in Copenhagen. He worked under **Pregl** in Austria, **Schoenheimer** in Germany and **Karrer** in Zürich before moving to the USA (1940), where he taught at the University of Rochester (1942–5) and became a member of the Rockefeller Institute for Medical Research (1945). He was also appointed Professor of Biochemistry *in absentia* at the Polytechnic Institute, Copenhagen (1941–65). His discovery of the anti-haemorrhagic agent, vitamin K, first reported in 1934, arose from experiments to determine if the biosynthesis of cholesterol occurred in newborn chicks. This proved correct, but the chicks developed delayed blood coagulation if kept on a sterol-free diet for more than two or three weeks. Protection was afforded by green leaves and hog liver, and from the former, a fat-soluble vitamin was isolated. Vitamin K is required for the conversion of prothrombin to thrombin, a key step in the blood clotting pathway. Dam shared the 1943 Nobel Prize for Physiology or Medicine with the US biochemist **Doisy** who determined the composition of vitamin K. Dam later investigated a complex oxidative phenomenon involving vitamin E, selenium and polyunsaturated fatty acids that caused plasma exudation from blood capillaries.

DANA, James Dwight

(1813–95)

American mineralogist, crystallographer and geologist, born in Utica, New York. After graduating from Yale, he joined the US navy for a short time as a teacher of mathematics to midshipmen to avail himself of the opportunity of a cruise in the Mediterranean (1833–4). This voyage resulted in his first scientific paper in 1835, an observation *On the condition of Vesuvius*. In 1836 he became assistant in chemistry to his future father-in-law Benjamin Silliman, with whom he became editor of the *American Journal of Science* from 1846 until his death. He was appointed Professor of Natural History (1849–64) and Professor of Geology and Mineralogy (1864–90) at Yale. Dana was a scientific observer on a US exploring expedition visiting the Antarctic and Pacific (1838–42) during which his ship was wrecked, and following an overland trek to safety involving travelling down the Sacremento river to San Francisco, he reported the 'probable occurrence of gold in California' six years before its discovery. His travels presented an unrivalled opportunity for observation of geological and zoological phenomena which served as a stimulus for his subsequent research. In 1837 he published the *System of Mineralogy*, a large scholarly work which appeared when he was only 24 years of age and which was updated throughout his life; the resultant 5th edition (1868) is a monumental treatise which remains useful today.

His 400 publications also included *Manual of Mineralogy* (1848), two treatises on corals, *Manual of Geology* (1863), *Textbook of Geology* (1864), *Corals and Coral Islands* (1872) and *Hawaiian Volcanoes* (1890).

DANDY, James Edgar

(1903–76)

English botanist, born in Preston and educated at Preston Grammar School and Downing College, Cambridge. During 1925–7 he worked at Kew and from 1927 to 1966 at the Department of Botany, British Museum (Natural History), where for 10 years he was Keeper of Botany. He developed a lifelong interest in Magnoliaceae and became the world authority on them. A keen angler, he also became an expert on water plants, especially the pondweeds (Potamogetonaceae) and their monocotyledonous allies known as the 'Helobieae'. He was so knowledgeable in the art of correct botanical nomenclature that many botanists consulted him on the subject; 'Ask Dandy' became a byword. Other professional interests included the families Loasaceae and Saxifragaceae, and the plants of the West Indies and Sudan. He also had a deep interest in British botany and compiled two important works, *List of British Vascular Plants* (1958) and *The Sloane Herbarium* (1958), as well as *Watsonian Vice-Counties of Great Britain* (1969), an essential tool for recording the British flora. A great perfectionist, much of his work, including a monograph of *Magnolia*, remains unpublished, although he spent much time editing the work of other botanists. He is commemorated in the genus *Dandya* (Liliaceae).

DANIELL, John Frederick

(1790–1845)

English chemist and meteorologist, born in London. Little is known about his education, but he showed an early interest in science. Accordingly he was employed in a sugar refinery run by a relative and he effected improvements in the processes. However, he soon left the business and held no definite appointment until he became Professor of Chemistry at the newly founded King's College, London, in 1831. Around 1815 he began his publications on meteorology and in 1823 published his *Meteorological Essays*. In 1820 he invented a hygrometer and in 1830 a pyrometer, for which he was awarded the Royal Society's Rumford Medal (1832). In 1835 he began the investigation of voltaic cells and in particular the reasons for their rapid loss of voltage. This led to a constant voltage (Daniell) cell, for which he was awarded the Royal Society's Copley Medal (1837). He collapsed and died at a council meeting of the Royal Society, of which he was Foreign Secretary.

DANIELLI, James Frederic

(1911–84)

British cell biologist, born in Wembley, London, and educated in chemistry at University College London. After postdoctoral studies at Princeton University (1933–5), he returned to University College and from 1938 onwards he worked at Cambridge, at the Marine Biological Association's laboratory at Plymouth. He was appointed to a readership in cell physiology at the Royal Cancer Hospital in London, before accepting the Chair of Zoology at King's College, London in 1948. In 1953 he became both Dean of Biological Sciences at Sussex University and Professor of Medicinal Biochemistry at Buffalo State University in New York, although he soon moved permanently to New York, where he enjoyed a varied research and administrative career until retirement in 1980. Danielli's major contributions were in understanding the structure and function of cell membranes, and he worked with **Davson** on theoretical and experimental aspects of membrane structure. They proposed that membranes were composed of molecules of lipids and proteins arranged somewhat like a sandwich, and in 1943 they published *The Permeability of Natural Membranes*, a classic account of transport across biological membranes. Their model incorporated physical, mathematical, chemical and biological measurements, and it laid the foundations of understanding for a generation of physiologists. Much of Danielli's subsequent work developed from his interests in cell physiology, including the mechanisms of oedema, the effect of vitamin C on the rate of wound healing, and processes of cell metabolism and aging.

DARLINGTON, Cyril Dean

(1903–81)

English cytologist and geneticist, born in Chorley, Lancashire, and educated at South Eastern Agricultural College, Wye. He began work at the John Innes Horticultural Institution, London (1923), was founding head of its cytology department in 1937, and director of the institute from 1939 to 1953. In 1953 he became Sherardian Professor of Botany at Oxford. In 1929–30 he travelled to Iran and Caucasia to investigate the origins of various species of tulip and cherry. He proposed 60 genetic terms and seven major theories, including the idea that chromosomes themselves were the objects of evolution and selection. In later life he studied the genetics of man and society, concluding that genius, or innovation, was the result of outbreeding and the environment. He was the author of *Chromosomes and Plant Breeding* (1932), *Recent Advances in Cytology* (1932), *The Evolution of Genetic Systems* (1939), *The Facts of Life* (1953), *Darwin's Place in History* (1960), *The Evolution of Man and Society* (1969) and *The Little Universe of Man* (1981). He exposed Stalin's murder of Russian geneticists and was a strong critic of committees and the establishment.

DART, Raymond Arthur
(1893–1988)

Australian-born South African anatomist, born in Toowong. He was educated at the universities of Queensland and Sydney, and graduated in medicine from Sydney in 1917. He spent 1919–22 at University College London, and became Professor of Anatomy at Witwatersrand University in Johannesburg (1923–58). In 1925 Dart described the Taung skull, an ape-like infant part-skull found in a mine in Taung, Bechuanaland, which was brought to him by a student during the summer of 1924. Dart considered the skull to belong to an intermediate between anthropoids and man, which he named *Australopithecus africanus* (southern African ape). Further work by Dart indicated that, during evolution, bipedalism preceded brain expansion, a view that has been supported by **Johanson**.

DARWIN, Charles Robert
(1809–82)

English naturalist, born in Shrewsbury, the originator (with **Wallace**) of the theory of evolution by natural selection. The grandson of **Erasmus Darwin** and of Josiah Wedgwood, he was educated at Shrewsbury grammar school, studied medicine at Edinburgh University (1825–7), and then, with a view to the church, entered Christ's College, Cambridge, in 1828. Already at Edinburgh he was a member of the local Plinian Society; he took part in its natural history excursions, and read before it his first scientific paper — on Flustra or sea-mats. His biological studies seriously began at Cambridge, where the botanist John Stevens Henslow encouraged his interest in zoology and geology. He was recommended by Henslow as naturalist to HMS *Beagle*, then about to start for a scientific survey of South American waters (1831–6) under its captain, **Fitzroy**. He visited Tenerife, the Cape Verde Islands, Brazil, Montevideo, Tierra del Fuego, Buenos Aires, Valparaiso, Chile, the Galapagos, Tahiti, New Zealand, Tasmania and the Keeling Islands; it was there that he started his seminal studies of coral reefs. During this long expedition he obtained the intimate knowledge of the fauna, flora and geology of many lands which equipped him for his later many-sided investigations. By 1846 he had published several works on his geological and zoological discoveries on coral reefs and volcanic islands — works that placed him at once in the front rank of scientists. He formed a friendship with **Lyell**, was Secretary of the Geological Society from 1838 to 1841, and in 1839 married his cousin, Emma Wedgwood (1808–96). From 1842 he lived at Downe, Kent, among his garden, conservatories, pigeons and fowls. The practical knowledge thus gained (especially as regards variation and interbreeding) proved invaluable; private means enabled him to devote himself unremittingly, in spite of continuous ill-health, to science. At Downe he addressed himself to the great work of his life — the problem of the origin of species. After five years collecting the evidence, he 'allowed himself to speculate' on the subject, and drew up in 1842 some short notes, enlarged in 1844 into a sketch of conclusions for his own use. These embodied in embryo the principle of natural selection, the germ of the Darwinian theory; but with constitutional caution, Darwin delayed publication of his hypothesis, which was only precipitated by accident. In 1858 Wallace sent him a memoir on the Malay Archipelago, which, to Darwin's alarm, contained in essence the main idea of his own theory of natural selection. Lyell and **Joseph Dalton Hooker** persuaded him to submit a paper of his own, based on his 1844 sketch, which was read simultaneously with Wallace's before the Linnaean Society on 1 July 1858, neither Darwin nor Wallace being present at that historic occasion. Darwin now set to work to condense his vast mass of notes, and put into shape his great work on *The Origin of Species by Means of Natural Selection*, published in November 1859. That epoch-making work, received throughout Europe with the deepest interest, was violently attacked and energetically defended, but in the end succeeded in obtaining recognition (with or without certain reservations) from almost all competent biologists. From the day of its publication, Darwin continued to work at a great series of supplemental treatises: *The Fertilisation of Orchids* (1862), *The Variation of Plants and Animals under Domestication* (1867), and *The Descent of Man and Selection in Relation to Sex* (1871), which derived the human race from a hairy quadrumanous animal belonging to the great anthropoid group, and related to the progenitors of the orang-utan, chimpanzee and gorilla. In it Darwin also developed his important supplementary theory of sexual selection. Later works were *The Expression of the Emotions in Man and Animals* (1873), *Insectivorous Plants* (1875), *Climbing Plants* (1875), *The Effects of Cross and Self Fertilisation in the Vegetable Kingdom* (1876), *Different Forms of Flowers in Plants of the same Species* (1877), *The Power of Movement in Plants* (1880) and *The Formation of Vegetable Mould through the action of Worms* (1881). Though not the sole originator of the evolution hypothesis, nor even the first to apply the conception of descent to plants and animals, Darwin was the first thinker to gain for that concept a wide acceptance among biological experts. By adding to the crude evolutionism of Erasmus Darwin, **Lamarck** and others his own specific idea of natural selection, he supplied to the idea a sufficient cause, which raised it at once from a hypothesis to a verifiable theory. He also wrote a biography of Erasmus Darwin (1879). He was buried in Westminster Abbey. His son, Sir Francis (1848–1925), also a botanist, became a reader in botany at Oxford (1888) and produced Darwin's *Life and Letters* (1887–1903). Another son, Sir George Howard (1845–1913) was Professor of Astronomy at Cambridge (1883–1912), and was distinguished for his work on tides, tidal friction and the equilibrium of rotating masses.

DARWIN, Erasmus
(1731–1802)

English physician and poet, born near Newark in Nottinghamshire. He studied medicine at Cambridge and Edinburgh, and at Lichfield (1756–81) became a popular physician and prominent figure on account of his ability, his radical and freethinking opinions, his poetry, his eight-acre botanical garden, and his imperious advocacy of temperance in drinking. After his second marriage in 1781, he settled in Derby, where he founded a Philosophical Society. By his first wife he was grandfather of **Charles Darwin**; by his second, of **Galton**. He anticipated **Lamarck**'s views on evolution, and also those of his own grandson. He edited translations of **Linnaeus**'s *Systema Vegetabilum* and *Genera Plantarum*, published respectively as *A System of Vegetables* (1783) and *Families of Plants* (1821). He wrote a long verse work, *The Botanic Garden* (1789). His chief prose works were *Zoonomia, or the Laws of Organic Life* (1794–6) and *Phytologia, or Philosophy of Agriculture and Gardening* (1800).

DAUBRÉE, Gabriel Auguste
(1814–96)

French economic geologist, mineralogist and mining engineer, born in Metz. After studying at the École Polytechnique in Paris, he was admitted to the Corps des Mines in 1834. He was appointed Professor of Geology and Mineralogy (1838) and later Dean (1852) at the University of Strasbourg. He became Professor of Mineralogy and Geology (1862–72) and subsequently Director at the École des Mines in Paris, and was appointed Inspector-General of French Mines in 1867. Daubrée published more than 300 memoirs chiefly on geological and mineralogical subjects with notable studies of meteorites, minerals and experimental petrology. He studied the permeability of rocks with particular reference to thermal waters and the rocks through which they flow, and was noted for the long-term and sometimes dangerous experiments which he conducted in order to ascertain to what extent it was possible to imitate the natural production of rocks.

DAUSSET, Jean Baptiste Gabriel Joachim
(1916–)

French immunologist, born in Toulouse, the son of a physician. Dausset entered the University of Paris to study medicine in the late 1930s, and on the outbreak of World War II joined the French medical corps. He obtained his MD in 1945. Dausset's wartime experience led to an interest in transfusion responses and the way in which they can lead to antibody production. He became Director of the French National Transfusion Centre laboratories, and joined the Faculty of Medicine in Paris in 1958. In 1978 he became Professor of Experimental Medicine at the Collège de France. Dausset discovered that abnormal responses occur among patients who have had many transfusions or who have had certain drug treatments, and are caused by reactions against white blood cells. He began to turn his mind towards the general problem of tissue-rejection. In the mid-1960s he suggested that tissue rejection in humans may be under the control of a group of genes which became known as the HLA, or human-lymphocyte-antigen group, similar to the major histocompatability complex (MHC) which **George Snell** had identified in mice. In 1967 Dausset began experiments on skin grafting among members of the same family. This led to the 'tissue typing' which greatly reduces rejection risks in transplant surgery. He shared the 1980 Nobel Prize for Physiology or Medicine with Snell and **Benacerraf**.

DAVAINE, Casimir Joseph
(1812–82)

French physician and microbiologist, born in St-Amand-les-Eaux. After studying at Tournai and Lille, he went to Paris in 1830 to embark on a medical course. He practised medicine in Paris, and while he never held an official university position, he contributed a steady stream of important experimental papers, mostly concerned with the role of micro-organisms in the causation of human and animal diseases. He developed procedures to identify parasitical worms, and first identified the anthrax bacillus in the blood of animals dying from anthrax. He was an advocate of the germ theory of disease at the Academy of Medicine before it was taken up by **Pasteur**, who always appreciated Davaine's work.

DAVID, (Père) Armand
(1826–1900)

French naturalist and Lazarist missionary, born in Espelette in the Pyrenees. An avid naturalist from early childhood, in his teens he spent many hours daily in the mountains, collecting specimens. He was educated locally and at the Grand Séminaire de Bayonne. His missionary calling also developed early and he entered the Lazarist Order of St Vincent de Paul (1848), hoping to go to China. Instead, he went to Savona, Italy (1851–61) where he taught science, continued his education and made collections which formed the foundation of the Savona Natural History Museum. After 10 years he had become expert in geography, geology and mineralogy as well as a brilliant all-round naturalist. He was then ordained and sent to China as a missionary. Between 1866 and 1874 he explored the Beijing plain, Mongolia, Tibet and central China, sending specimens to the Natural History Museum, Paris. Some of the most interesting of his many botanical discoveries, catalogued in *Plantae Davidianae* (1884–6), were in the protectorate of Moupin, where he discovered the unique handkerchief tree named after him (*Davidia involucrata*). Among other introductions to western horticulture were *Pinus armandii* and *Astilbe davidii*. However, he is best known for one of his zoological discoveries, Père David's deer (*Elaphurus davidianus*).

DAVIDSON, Eric Harris
(1937–)

American molecular biologist, born in New York City. He graduated BA from the University of Pennsylvania in 1958, and received a PhD from Rockefeller University in New York in 1963. He was a research associate (1963–5) and assistant professor at Rockefeller (1965–71) before moving to Caltech as Associate Professor of Developmental Molecular Biology (1971–4) and Professor (1974–81); since 1981 he has been Norman Chandler Professor of Cell Biology there. He is a member of the US National Academy of Sciences, and is the author of numerous publications on DNA sequence organization, gene expression during embryonic development and gene regulation. Together with **Britten**, he elucidated the nature of genome organization in higher animals and showed that in addition to specific DNA sequences which code for protein-producing genes, the genome has enormous stretches of 'junk' DNA. This extra DNA is organized into moderately repetitive sequences which are a few thousand base pairs in length and are repeated a few times, and into short (few hundred base pairs) sequences which are repeated hundreds of thousands of times in the genome. The function of such highly repetitive DNA sequences remains unknown, but it has been suggested that they may be involved in the coordinate regulation of the single-copy genes with which they are associated.

DAVIS, Peter Hadland
(1918–92)

English botanist, born in Weston-super-Mare and educated at Nash House, Burnham-on-Sea, Bradfield College and Maiden Erleigh, Reading. His first botanical publication, on the Cheddar Pink, appeared in *The Times* when he was 15. His apprenticeship at Will Ingwersen's Alpine Plant Nursery, East Grinstead, in 1937 nurtured an all-consuming passion for plants. He began botanizing in the Middle East in 1938 and eventually visited every Mediterranean country (except Albania), as well as Brazil and Malaysia, collecting some 70000 specimens. After war service he studied botany at Edinburgh University and began a distinguished career (1950–85) in its botany department. In 1961 he began preparing *The Flora of Turkey and the East Aegean Islands* (10 vols, 1965–88), the first Flora covering Turkey since **Boissier**'s *Flora Orientalis* and widely regarded as a model Flora. He also published more than 100 papers on Mediterranean and south-west Asian botany and plant taxonomy, as well as vivid accounts of his earlier expeditions. With **Heywood**, he was joint author of *Principles of Angiosperm Taxonomy* (1963), another landmark of 20th-century botanical literature. A voracious collector of Wemyss ware, with Robert Rankine he wrote the definitive monograph, *Wemyss Ware: a Decorative Scottish Pottery* (1986).

DAVIS, Raymond Jr

(1914–)

American chemist and astrophysicist, born in Washington. He was educated at the University of Maryland and at Yale, where he received his PhD in 1942. He worked as a senior chemist at Brookhaven National Laboratory (1948–84) and since 1984 has been professor in the department of astronomy of the University of Pennsylvania. Davis devised the first experiment to detect neutrinos emitted from the core of the Sun. It consists of essentially a large tank of dry cleaning fluid placed in a deep mine, in which the neutrinos interact with chlorine atoms to produce minute quantities of radioactive argon atoms. The number of reactions which have taken place can be calculated from the observed rate of radioactive decay of argon. This experiment has been running since 1969. All their results have shown that there are about a third of neutrinos coming from the Sun that are expected from precise theoretical solar models. Recently other experiments have started to observe solar neutrinos to study this discrepancy.

DAVIS, William Morris

(1850–1934)

American geomorphologist, born in Philadelphia and educated at Harvard. After a short spell as an astronomer in Cordoba, Argentina (1871–2), he became Professor of Physical Geography at Harvard (1875–1912). Early in his career, Davis developed an interest in Triassic geology, and extensive travels took him to Europe, the West Indies, South Africa, Australia, New Zealand and other countries. He participated in an expedition to Turkestan (1903) with Raphael Pumpelly (1837–1923), and undertook wide-ranging studies of the role of rain in erosion, development of rivers, glacial erosion, the formation of coral reefs, arid landscapes, and the elevation and subsidence of land masses. He reinterpreted some geomorphological features in Europe and stimulated much debate with his ideas of the development of British, French and German rivers. He also introduced the concept of cycles of erosion, and made significant contributions to meteorology.

DAVISSON, Clinton Joseph

(1881–1958)

American physicist, born in Bloomington, Illinois. He was educated at Chicago and Princeton, where he was instructor in physics before taking up industrial research at the Bell Telephone Laboratories. In 1927, with **Germer**, Davisson was observing electron scattering from a block of nickel. When their vacuum system accidentally broke down, the target was oxidized. To correct this they slowly heated up the target and when they continued the experiment, they found that the results were completely different. Instead of the decrease of intensity with scattering angle that they had previously observed, they found the familiar peaks and troughs of a diffraction pattern. They had observed the diffraction of electrons, described by **de Broglie**'s theory of the wave nature of particles, and they found that the observed wavelengths agreed with those predicted. This accidental discovery was of crucial importance in the development of the quantum theory of matter. In 1937 he shared the Nobel Prize for Physics with **George Paget Thomson**.

DAVSON, Hugh

(1909–)

English physiologist, born and educated in London. He graduated in chemistry and biology from University College London (UCL) in 1931, and with a fellow student **Danielli** began studying the cell permeability of erythrocytes, the red blood cells. Poor employment opportunities in pure science directed him towards a project, funded by the Medical research Council, in which he studied the nature of the fluids of the eye in relation to glaucoma. He spent 1936 in the USA on a Rockefeller research fellowship, working on cell permeability and returned to UCL to gain an MSc, and then a DSc shortly after being appointed Professor of Physiology at Dalhousie University in 1939. His master's thesis became the basis of *Permeability of Natural Membranes*, co-written with Danielli (1942). Davson returned to the UK during World War II to contribute to research at Porton Down and in 1945 he began working once more on the eye fluids, and then on the cerebro-spinal fluid, at University College and the Institute of Ophthalmology in London. He retired in 1975, but since then has held academic appointments at the National Institutes of Health, Bethesda, and King's College and St Thomas's Hospital, London, where he continues to work on transport mechanisms in the brain. Davson has contributed significantly to physiology as the author of important textbooks including *The Physiology of the Eye* (1949), *Textbook of General Physiology* (1951) and several volumes on the cerebro-spinal fluid. He has edited or co-edited the multi-volume series *The Eye*, and new editions of **Starling**'s *Principles of Human Physiology*.

DAVY, Edward
(1806–85)

English physician and scientist, born in Ottery St Mary, Devon, the eldest son of a surgeon. After a medical education, he established in 1829 the firm of Davy & Co, supplying scientific (in particular chemical) apparatus, including some of his own inventions, such as 'Davy's blow-pipe' for chemical analysis, 'Davy's improved mercurial trough' for gas chemistry, and 'Davy's diamond cement' for repairing broken china. He invented the electric relay and deserves to stand alongside **Wheatstone** and Sir William Fothergill Cooke (1806–79) as one of the inventors of wireless telegraphy. He lectured and wrote many papers on the subject, and demonstrated his system over a mile-long wire in Regents Park, London. He later emigrated to Adelaide, South Australia (1838), where he involved himself in civic affairs and continued his experiments, on subjects including starch production and the smelting of copper. In 1853 he moved to Victoria where he made an unsuccessful attempt at farming, after which he returned to medicine which he practised for the rest of his life, and became involved in local affairs. In recognition of his earlier achievements he was made an honorary member of the Society of Telegraph Engineers in 1885.

DAVY, Sir Humphry
(1778–1829)

English chemist, one of the leading scientific figures of his day, noted for his pioneering work in electrochemistry and many other fields. The son of a woodcarver, he was born in Penzance, Cornwall, and was apprenticed to a surgeon and apothecary. In 1797 he began to teach himself physics and chemistry. His talent for both experiment and speculative thought soon became obvious and in 1798 **Beddoes** employed him as an assistant at the Pneumatic Institute in Bristol. Here Davy experimented with several newly discovered gases and discovered the anaesthetic effect of laughing gas (nitrous oxide). He also showed that heat can be transmitted through a vacuum and suggested that it is a form of motion. In 1799 he published *Researches, Chemical and Physical*, which led in 1801 to his appointment as assistant lecturer in chemistry at the Royal Institution. At once his lectures became hugely popular, his eloquence and the novelty of his experiments attracting large audiences. As a result he gained sufficient funding for research into electrochemistry, a new branch of chemistry made possible by the invention of the electric battery by **Volta**. Davy decomposed chemical compounds by passing an electrical current through their solutions. This research excited much attention and was later continued by Davy's successor **Faraday**. Davy also confirmed that water is composed of hydrogen and oxgen by electrolyzing distilled water in a special apparatus which prevented by-products from forming. He suggested that all chemical activity is electrical, but it was **Berzelius** who synthesized Davy's work and his own work into a coherent system. In 1807 Davy discovered that the alkalis and alkaline earths are compound substances formed by oxygen united with metallic bases. He isolated the metals sodium and potassium by electrolysis of fused salts, and barium, strontium, calcium and magnesium by distilling off the mercury from their amalgams. Following up the work of **Courtois**, he showed that fluorine and chlorine are elements related to iodine. His work on the compounds of chlorine helped to refute **Lavoisier**'s theory that all acids contain oxygen. He also proved that diamond is a form of carbon. Another important area of his work was agriculture. His *Elements of Agricultural Chemistry* (1813) was the first book to apply chemical principles systematically to farming. Davy resigned from the Royal Institution in 1812; in the same year he was knighted and married a wealthy Scottish widow. From 1813 to 1815 the Davys travelled on the Continent taking the young Michael Faraday with them as chemical assistant and valet. In 1815, after investigating the causes of explosions in mines and finding that firedamp (methane) only ignites at high temperatures, Davy invented the celebrated safety lamp which bears his name. A wire gauze surrounding the flame conducts the heat away, thus preventing the danger point from being reached. Coal production thereafter increased markedly as deeper, more gaseous, seams could be mined. Davy's reputation in his lifetime exceeded even his considerable achievements, not the least of which were to popularize science and to interest industrialists in scientific research. He was one of the founders of the Athenaeum Club and of the Zoological Society, which in its turn founded London Zoo. He was made a baronet in 1812. His last two years were spent on the Continent in failing health and he died in Geneva.

DAWKINS, (Clinton) Richard
(1941–)

British ethologist, born in Nairobi, Kenya. Educated at Oxford, he taught at the University of California at Berkeley before returning to Oxford (1970) where he is a Fellow of New College. His main research interests have been in theoretical modelling of ethology. His major contribution to date is his ability to expound complex evolutionary ideas so as to make them explicable to fellow biologists and laypersons alike. In *The Selfish Gene* (1976), he shows how natural selection can act on individual genes rather than at the individual or species level. He also describes how apparently altruistic behaviour in animals can be of selective advantage by increasing the probability of survival of genes controlling this behaviour. The ways in which randomly occurring small genetic changes or mutations accumulate to form the basis for evolution are discussed in *The Blind Watchmaker* (1986). *The Extended Phenotype* (1982), a more advanced book, argues that genes can have effects outside the bodies that contain them, and explores the implications of this idea for parasitology.

DAWKINS, Sir William Boyd
(1837–1929)

Welsh geologist, palaeontologist and anthropologist, born at Buttington vicarage, near Welshpool. After graduating from Jesus College, Oxford, he joined the Geological Survey in 1861, and for eight years surveyed the Wealden and other formations in Kent and the Thames Valley. He became curator at the Manchester Museum in 1870 on the recommendation of **T H Huxley**, and first Professor of Geology at Manchester in 1874. He commenced his exploration of caves and their deposits in 1859 with the exploration of Wookey Hole, Somerset, a cave in which he found evidence of its occupation at different times by hyenas and by man. During 1875–8 he explored Cresswell Crags in Derbyshire, and was again able to establish a succession of animal and human inhabitants. He subsequently undertook important studies of Pleistocene vertebrate remains from around Britain, and discovered Pliocene vertebrate remains, including mastodon and elephant remains, in a cave near Buxton (1903). Dawkins wrote *Cave Hunting; or, Caves and the Early Inhabitants of Europe* (1874) and *Early Man in Britain* (1880). He was knighted in 1919.

DAWSON, Sir John William
(1820–99)

Canadian geologist, born in Pictou, Nova Scotia. He studied at Edinburgh, and subsequently devoted himself to the study of the natural history and geology of New Brunswick and Nova Scotia, where he was superintendent of education (1850–5). From 1855 to 1893 he was principal of McGill University, Montreal. Dawson was an authority on fossil plants, the principal proponent of the organic nature of eozoon, and was systematically opposed to **Charles Darwin**'s theories. In 1851 he discovered some of the earliest known terrestrial vertebrate fossils inside Carboniferous fossil tree stumps at Joggins, Nova Scotia. His publications included *Acadian Geology* (1855), *The Story of Earth and Man* (1873), *Origin of the World* (1877), *Fossil Men* (1878) and *Relics of Primeval Life* (1897). He was knighted in 1884. His son George Mercer Dawson (1849–1901) was also a distinguished geologist who undertook much pioneering work in British Columbia and Yukon, where Dawson City was named after him, and became Director of the Geological Survey of Canada in 1875.

DEACON, Sir George Edward Raven
(1906–84)

English physical oceanographer, born in Leicester. He was educated at King's College, London, where he received a BSc in chemistry in 1926 and a teaching diploma in 1927. He was appointed in 1927 as hydrologist to the Discovery Committee, a government agency studying the sustainability of whaling in the Falkland Island Dependencies, and sailed to South Georgia aboard RRS *William Scoresby*. Deacon made further studies of the chemistry of the Southern Ocean from RRS *Discovery II* between 1930 and 1937. His measurements revealed the Antarctic convergence, where cold water dips beneath warmer sub-Antarctic water, and that Antarctic bottom water extends northwards into all the major oceans. His *Discovery* report, *The Hydrology of the Southern Ocean* (1937), earned him a DSc (1939) and fellowship of the Royal Society (1944). During World War II he was seconded to the Admiralty, and from 1944 led a group at the Admiralty Research Laboratory in Teddington which discovered a method of analysing ocean waves. This group became part of the new National Institute of Oceanography at Wormley, Surrey (1949), with Deacon as director. He was knighted in 1971. Deacon had a major influence on the national and international development of marine science. The Institute of Oceanographic Sciences (formerly NIO) Laboratory is named after him.

DE BARY, Heinrich Anton
(1831–88)

German botanist, born in Frankfurt, sometimes described as the founder of modern mycology. Successively Professor of Botany at Freiburg, Halle and Strassburg, he was the first rector of its recognized university. He studied the morphology and physiology of the fungi, discovering many of the complexities of their life cycles, and the Myxomycetae (slime moulds). He described the plasmodium of slime moulds (mobile masses of living matter resulting from the fusion of individual cells) as a mass of protoplasm resembling the circulating protoplasm of plant cells, yet devoid of cell walls. He noted that the plasmodium possesses nuclei and a differentiated external layer, the plasmalemma. He published *Comparative Anatomy of Ferns and Phanerogams* in 1877. This comprehensive treatment established the principal features of plant anatomy, and the systematic terminology used is, in most respects, still in use.

DE BEER, Sir Gavin Rylands
(1899–1972)

English zoologist, born in London. He served in both world wars and between them graduated from Oxford and then taught there (1923–38). After World War II he became Professor of Embryology in London, and from 1950 to 1960, Director of the British Museum (Natural History). In *Introduction to Experimental Embryology* (1926) and *Development of the Vertebrate Skull* (1935), he discussed the processes whereby tissues are derived during the course of development. He was particularly interested in the origins and development of odontoblasts (putative bone tissue) and cartilage. In his *Embryos and Ancestors* (1940), he put forward the idea of paedomorphosis as a factor in evolution. This is the process where an animal becomes sexually mature at an earlier stage of development. De Beer had an interest in historical problems and brought his biological knowledge to bear on them. Thus he reconstructed Hannibal's route across the Alps via pollen analysis and glaciology. He was awarded the Darwin Medal of the Royal Society and the Linnaean Society Gold Medal, both in 1958, and knighted in 1954.

DE BROGLIE, Louis-Victor Pierre Raymond, 7th Duke
(1892–1987)

French physicist, born in Dieppe. He studied history, but service at the Eiffel Tower radio station during World War I initiated his interest in science, and he took a doctorate in physics at the Sorbonne (1924). He was Professor of Physics there from 1925 to 1962. Influenced by **Einstein**'s work on the photoelectric effect which he interpreted as showing that waves can behave as particles, de Broglie put forward the converse idea—that particles can behave as waves. He suggested that an electron, for example, could exhibit wave-like properties and that the associated wavelength would be given by Planck's constant divided by the electron's momentum. The waves were detected experimentally by **Davisson** and **Germer** in 1927, and separately by **George Thomson**, thus confirming the ideas developed by **Heisenberg** and **Born**. De Broglie was awarded the Nobel Prize for Physics in 1929 for this work.

DEBYE, Peter Joseph Wilhelm, originally Petrus Josephus Wilhelmus Debije
(1884–1966)

Dutch–American physicist and physical chemist, born in Maastricht. He studied electrotechnology in the Technische Hochschule in Aachen, but was more attracted by physics. When **Sommerfeld**, who had been his teacher in theoretical physics, moved from Aachen to Munich in 1906, Debye accompanied him as assistant. He completed his doctorate at Munich in 1908 with a thesis on the diffraction of light. Debye was Professor of Theoretical Physics at Zürich (1911–12) and Utrecht (1912–14) universities, and Professor of Theoretical and Experimental Physics at Göttingen (1914–20). He was Director of the Physical Institutes at the Federal Institute of Technology (ETH) in Zürich (1920–7), at Leipzig University (1927–34) and of the Kaiser-Wilhelm-Gesellschaft in the Dahlem district of Berlin (1934–40). In Berlin he suffered increasing political interference in the institute, and in 1940 he took advantage of delivering the Baker Lectures at Cornell University to remain permanently in the USA. He was Chairman of the Cornell Chemistry Department from 1940 until his retirement in 1950, thereafter continuing active in science until his death. Debye's earliest important work was his modification of the theory of the specific heats of crystalline solids (pioneered by **Einstein**) to give a temperature dependence closer to that observed experimentally (1911). There followed his work on dielectric constants and molecular dipole moments (1912); the conventional unit of dipole moment is known as the Debye. In the 1920s and 1930s the measurement of dipole moments became much used as a means of investigating the details of chemical bonding. In the Debye–**Hückel** theory of strong electrolytes (1923), their behaviour was related quantitatively to electrostatic forces between ions. On several occasions during his career Debye pioneered extensions of X-ray diffraction from its restricted application to the structures of large single crystals. Thus the Debye–Scherrer X-ray diffraction powder method was developed in 1916–20 and the theory of X-ray scattering by gaseous molecules was conceived in 1925. Experimental studies of X-ray diffraction by gases and liquids were made in 1929–33, and for this work Debye was awarded the 1936 Nobel Prize for Chemistry. Later he made important studies by means of electron diffraction (1938). Debye also provided the theoretical treatments for the electro-optical Kerr effect (1925), adiabatic demagnetization (1926, much applied in the attaining of temperatures close to absolute zero) and thermal diffusion (1939, with applications to isotope separation). After his move to the USA he worked on light scattering related to molecular and media structures (1944–66) and many aspects of polymer behaviour (1945–66). Debye may properly be regarded as one of the really great scientists of the first half of the 20th century. His many honours and awards included the Rumford Medal of the Royal Society (of which he became a Foreign Member in 1933), the Faraday Medal of the Chemical Society (1933), and several medals of the American Chemical Society.

DEDEKIND, Julius Wilhelm Richard

(1831–1916)

German mathematician, born in Brunswick. He wrote his doctoral thesis at the University of Göttingen under **Gauss** in 1852, but the real influence on him was **Dirichlet**, who led Dedekind into number theory. From 1854 to 1858 he taught at Göttingen, then in Zürich, and he returned to Brunswick in 1862 as professor at the Polytechnic. He gave one of the first precise definitions of the real number system, and did important work in number theory which led him to introduce many concepts which have become fundamental in all modern algebra, in particular that of an 'ideal', building on the work of **Kummer**. Decisive acceptance of his work came after his death with the work of **Noether**. With his friend **Georg Cantor** he did much to found mathematics on the naive concept of a set. For example, real numbers are defined as sets of rational numbers, and ideal numbers as sets of algebraic numbers. He also made important contributions in the early history of lattice theory.

DE DUVE, Christian René

(1917–)

English-born Belgian biochemist, born in Thames Ditton, Surrey. He studied medicine at Louvain, worked with **Theorell** to determine haemoglobin and myoglobin concentrations in muscle, and worked with **Carl Cori** on glucose 6-phosphatase and blood sugar regulation. Returning to Louvain (1947) he became Professor of Biochemistry in 1951, and he also held a Chair of Biochemistry at Rockefeller University, New York, from 1962. From 1947 he explored the new technique of differential centrifugation, separating a tissue homogenate into its separate organelles by centrifugation at different speeds. Fortuitously using the enzyme acid phosphatase as a control marker while attempting to locate the site of carbohydrate metabolism in the cell, he found that its activity, initially low, increased on storage by being released from a degrading cell organelle. Thus he discovered lysosomes, small organelles which contain acid phosphatase and other degradative enzymes. Lysosome malfunction may result in a metabolic disease, such as cystinosis. De Duve discovered peroxisomes (oxidative organelles) in a similar way. For discoveries relating to the structure and biochemistry of cells he shared the 1974 Nobel Prize for Physiology or Medicine with **Albert Claude** and **Palade**. He published *A Guided Tour of the Living Cell* in 1984.

DEE, John

(1527–1608)

English alchemist, geographer and mathematician, born in London. Educated in London, Chelmsford, and at St John's College, Cambridge, he became one of the original fellows of Trinity College, Cambridge (1546). He earned the reputation of a sorcerer by his mechanical beetle in a representation of Aristophanes's *Peace*, and the following year he fetched from the Low Countries sundry astronomical instruments. As astrologer to Queen Mary I, he was imprisoned but acquitted on charges of compassing her death by magic (1555). Queen Elizabeth showed him considerable favour, making him warden of Manchester College in 1595. For most of his life he worked for the Muscovy Company and was concerned with the search for the North-West Passage to the Far East, aiding the exploration by his navigational and geographical knowledge. He wrote numerous works on logic, mathematics, astrology, alchemy, navigation, geography and the calendar (1583), and translated mathematical works from Arabic. His most important book was the first English edition of **Euclid**'s *Elements*, translated by Henry Billingsley and with a preface by Dee setting out the nature and purpose of mathematics. He died in poverty and was buried in Mortlake Church, London. His eldest son Arthur (1579–1651) was likewise an alchemist.

DE FOREST, Lee

(1873–1961)

American physicist and inventor, born in Council Bluffs, Iowa. His father was a Congregational minister who became President of the Negro Talladega College in Alabama. He was educated at Yale and Chicago. A pioneer of radio and wireless telegraphy, he patented more than 300 inventions and is known as the 'father of radio' in the USA. He introduced the grid into the thermionic valve (1906), and invented the 'audion' and the four-electrode valve. Initially he did not fully appreciate the possibilities of the triode, which can act both as a detector and amplifier as well as a rectifier, and was crucial in the development of radio. The radio circuits made possible with the triode and its successors made the simple crystal sets obsolete in the 1920s. De Forest also did much early work on sound reproduction and on television, and received his last patent at the age of 84 for an automatic dialling device. He was a founder member of the Institute of Radio Engineers (1912), and was awarded its Medal of Honour (1915). Among his other honours was the Cross of the Legion of Honour from France.

DE GENNES, Pierre-Gilles
(1932–)

French theoretical physicist. He was educated at the École Normale Supérieure in Paris, and became professor at the University of Orsay (1961–71). Since 1971 he has been Professor of Solid State Physics at the Collège de France, and since 1976 Director of the College of Industrial Physics and Chemistry in Paris. De Gennes was awarded the 1991 Nobel Prize for Physics for outstanding work on the behaviour of molecules in substances undergoing phase transitions, and for increasing our understanding of polymers and liquid crystals. He elucidated the mechanism by which polymers have the ability to flow, describing mathematically through the 'spaghetti model' how the tangled long-chain molecules can move along their own lengths; this led to a totally new theory of polymer elasticity and viscosity. In studies of liquid crystals, he showed how their optical properties can be altered through the application of a weak alternating current.

DEHMELT, Hans Georg
(1922–)

German-born American physicist, born in Görlitz. He joined the German army and was captured by US forces in 1945. Later he studied at Göttingen University, where he received his PhD in 1950. He moved to the USA in 1952, becoming professor at the University of Washington in 1961. Dehmelt developed a device known as the Penning trap, which uses electromagnetic fields to allow ions and electrons to be isolated in a restricted space for study over long periods of time. Using this trap, he measured the magnetic moment of an electron to an unprecedented accuracy of four parts in a trillion. The device has also allowed very accurate measurement of the energy levels in atoms, which in turn will lead to increases in the accuracy of time measurement by atomic clocks. For this work Dehmelt shared one-half of the 1989 Nobel Prize for Physics with **Paul**, who had developed a similar technique (the other half of the prize was awarded to **Norman Ramsey**).

DEISENHOFER, Johann
(1943–)

German–American molecular biologist, born in Zusamaltheim, Bavaria. He graduated in physics at Munich (1971) and then worked at the Max Planck Institute for Biochemistry in Martinsried until 1988, when he became Regental Professor and Professor of Biochemistry at the University of Texas Southwestern Medical Center in Dallas. He initially worked under **Huber**, and collaborated with him in studies of the crystallographic X-ray analysis of the structure of biological macromolecules. From 1974 he studied bovine trypsin and its interaction with pancreatic trypsin inhibitor. In 1976 he began a series of studies on immunoglobulin structure, particularly the stem (Fc) region of this Y-shaped molecule, identifying receptor and effector sites for other biomolecules and the nature of the interaction with the polysaccharide component of the molecule. He later collaborated with **Michel** and Huber to determine the structure of the membrane-bound photosynthetic reaction centre of the purple bacterium *Rhodopseudomonas viridis*, work for which they shared the 1988 Nobel Prize for Chemistry. Deisenhofer has also studied (by low-angle X-ray analysis) another membrane-bound enzyme, cytochrome-c oxidase, indicating the presence in the molecule of alpha helical structures at right angles to the membrane surface; in addition he has contributed, with Huber, to studies on the energetics of protein–DNA interactions.

DE LA BECHE, Sir Henry Thomas
(1796–1855)

English stratigrapher and geologist, born near London. He entered the Military School at Great Marlowe in 1810 and resided for a time in Switzerland and France, studying the natural phenomena of the Alps. During the years 1822–6 he described in detail the secondary strata in the neighbourhood of Lyme Regis. In 1832 he commenced mapping the geology of Dorset and Devon, and in 1835 he was appointed to extend the survey into Cornwall, marking the beginning of the first national Geological Survey of which he became the first director. His endeavours led to the establishment of the Mining Record Office (1839), the Museum of Practical Geology (1841), and the School of Mines and Science (1853). He became President of the Geological Society of London in 1847. De la Beche published many works including *Manual of Geology* (1831), *Researches in Theoretical Geology* (1834), regional memoirs of parts of southern England and the first account of the geology of Jamaica (1834). He was knighted in 1842.

DE LA HIRE, Philippe

(1640–1718)

French engineer, born in Paris. A mathematician as well as a keen experimenter, he was employed for some years on geodesic survey work. In 1682 he joined the Collège Royal where he taught mathematics, and five years later he became professor at the Royal Academy of Architecture. His most notable work was the *Traité de Méchanique* (1695) in which he correctly analysed the forces acting at various points in an arch, making use of geometrical techniques now generally known as graphic statics. He showed that the cycloid is the most efficient form for gear teeth, and applied it in the design of a train of gears transforming the rotary motion of a windmill into the reciprocating action of a water-pump. Among his many inventions were improvements to the astrolabe, the sundial and the surveying level, to which he added a telescopic sight. He also devised a machine capable of showing the configurations of past and future eclipses. In 1678 he was admitted to the French Royal Academy of Sciences.

DE LA RUE, Warren

(1815–89)

British astronomer and physicist, born in Guernsey. He was educated in Paris, and entered his father's business—the manufacture of paperwares—for which his inventive genius devised many new processes, including an envelope-making machine. He was one of the first printers to adopt electrotyping. His particular strength was to perfect scientific instruments to enable accurate observations to be made. He improved the Daniell constant silver chloride cell, and did research on the discharge of electricity in gases. A pioneer of celestial photography, he invented the photoheliograph which permitted mapping the Sun's surface photographically, and perfected the technique of taking stereoscopic plates first of the Moon, then of the Sun. By means of this method, he showed that sunspots are depressions in the Sun's atmosphere. His reflecting telescope and camera were donated to Oxford University. De la Rue was elected FRS (1850), and to the Chemical Society and the Royal Astronomical Society, serving as the latter's president from 1864 to 1866. He fully joined into the scientific life of the capital, with membership to the Royal Microscopical Society, the Royal Institution and the London Institution.

DELBRÜCK, Max

(1906–81)

German biophysicist, born in Berlin. He studied atomic physics at Göttingen, where he received his PhD in 1930. Working with **Niels Bohr** at Copenhagen in 1932, he was influenced by Bohr's ideas about life and biology, and came to believe that the study of biology might lead to new laws in physics. These ideas are expressed in *What is Life*, written by the émigré physicist **Schrödinger** in Dublin in 1945, and widely read by physicists. Delbrück moved to Berlin in 1935 to work with **Meitner**, and in 1937 emigrated to the USA, where he held appointments at Caltech; he remained in the USA throughout World War II. In 1947 he was invited by **Beadle** to become Professor of Biology at Caltech, a post which he held until 1977. In the 1940s Delbrück began working on the genetics of the phage virus, a simple organism with a protein coat surrounding a coil of DNA which infects the bacterium *Escherichia coli*. Independently of **Hershey**, he discovered in 1946 that viruses can exchange genetic material to create new types of virus; such recombinations were previously believed to be impossible in such primitive organisms. Together with **Luria** they set up the Phage Group, to encourage the use of phage as an experimental tool. The three were awarded the 1969 Nobel Prize for Physiology or Medicine for their work in viral genetics. Delbrück continued to develop his interest in philosophical questions, and published *Mind Over Matter* in 1985.

D'ELHUYAR Y DE SUVISA, Don Fausto

See **ELHUYAR Y DE SUVISA, Don Fausto d'**

DELISLE, Joseph Nicholas

(1688–1768)

French astronomer, born in Paris and educated at the University of Paris. He first attracted notice by an interesting though erroneous theory that the Sun's corona is produced by diffraction of light around the Moon (1715). In 1717, a meeting with Peter I of Russia in Paris led to an invitation from the Empress Catherine to St Petersburg (1745), where he founded an observatory and a school of navigation and cartography. He returned to Paris in 1747 to become astronomer to the navy. Delisle's main interest was in problems associated with the Sun, in particular with the apparent movements of Mercury and Venus when in transit across its disc. He worked out (1743) an alternative method of observing transits of Venus to that first used by **Halley** for finding the distance to the Sun. Delisle's method was widely used in the transits of 1874 and 1882 but in spite of the most elaborate preparations the results achieved were disappointing.

DEL RIO, Andrés Manuel

(1764–1849)

Spanish geologist and mineralogist, born in Madrid. He was educated at the San Isidoro College and Alcala de Henares University, where he studied experimental physics. After graduating in 1781, he continued his work at the Real Academia de Minas de Almaden with a subsidy from Charles III. He then spent four years in Paris and attended **Abraham Werner**'s lectures on mineralogy at Freiburg, where he became a friend of **Humboldt**. He also studied mining at Schemnitz and at mines in Saxony and England before returning to Paris to study chemistry with **Lavoisier**. When Lavoisier was arrested, he fled to England. In 1794 he travelled to Mexico to take up a post as Professor of Mineralogy at the newly founded Colegio de Mineria. He discovered a new metallic element, panchromium, subsequently known as vanadium (1801). In 1829 he went into exile in Philadelphia following the Mexican War of Independence. Del Rio worked on the origin of mineral veins, paragenesis of sulphide minerals and the effects of trace elements. He was author of the first textbook of mineralogy published in the Americas, *Elementos de Orictognosia* (1795).

DE LUC, Jean André

(1727–1817)

Swiss chemist, meteorologist, geologist and traveller, born in Geneva. He received an excellent education in mathematics and natural science, and was the first person to correctly measure the heights of mountains using the effects of heat and pressure on a thermometer. Following a breakdown of commerce with France and the resultant collapse of his business, he settled in England in 1773 and became reader to Queen Charlotte, and an associate of **William Herschel** and **Hunter**, who sponsored him for the fellowship of the Royal Society which he received in 1773. De Luc travelled widely in Europe writing on stratigraphy and geomorphology, and introduced the word *géologie* in 1778. As a close associate of George III, he was sent on a secret mission to Brunswick (1797), ostensibly to take up the post of Professor of Natural History at the University of Göttingen, but in reality to attempt to enlist the support of the Duke of Brunswick in influencing revolutionary events in Germany. De Luc established fundamental points in geology, but was also a highly successful experimenter in various other branches of natural philosophy. He made significant meteorological observations and demonstrated that heat was lost when ice melted, subsequently becoming embroiled in a controversy with **Joseph Black** and involving **Watt** concerning priority in the discovery of the concept of latent heat. He was a critic of theories of **Hutton**. His publications included *Cosmology and Geology* (1803) and an elementary treatise on geology (1809).

DEMARCAY, Eugène Anatole

(1852–1903)

French chemist, born in Paris. He studied at the École Polytechnique where he later became professor. In 1896 he discovered the element europium by spectrum analysis and also gave spectroscopic proof of the existence of radium. In addition he studied the volatility of metals at low temperatures and pressures, designed a machine which achieved low temperatures by compressing gases and then allowing them to expand (the principle behind the modern refrigerator), and developed a method to separate the rare earths by fractional crystallization in aqueous solution. He died in Paris.

DEMOCRITUS

(c.460–c.370 BC)

Greek philosopher, born in Abdera in Thrace and supposedly known as 'the laughing philosopher' in the ancient world because of his wry amusement at human foibles. He was one of the most prolific of ancient authors, publishing many works on ethics, physics, mathematics, cosmology and music, but only fragments of his actual writings (on ethics) survive, and he must be judged by the (often critical) references to his works in the writings of **Aristotle**, **Theophrastus**, Epicurus and others. He is best known for his physical speculations, and in particular for the atomistic theory he developed from Leucippus, to the effect that the world consists of an infinite number of minute particles whose different characteristics and combinations account for the different properties and qualities of everything in the world, animate as well as inanimate. He was an important influence on Epicurus and Lucretius.

DE MOIVRE, Abraham
(1667–1754)

French mathematician, born in Vitry, Champagne. A Protestant, he came to England around 1686, after the revocation of the Edict of Nantes, and supported himself by teaching. **Isaac Newton**'s *Principia* whetted his devotion to mathematics and he became known to the leading mathematicians of his time. In 1697 he was elected FRS, and he helped the Royal Society to decide the famous contest between Newton and **Leibniz** on the origins of the calculus in Newton's favour. His principal work was *The Doctrine of Chances* (1718) on probability theory, but he is best remembered for the fundamental formula on complex numbers known as De Moivre's theorem, later simplified and reproved by **Leonhard Euler**, that relates the exponential and trigonometric functions.

DE MORGAN, Augustus
(1806–71)

English mathematician, born in Madura, Madras Presidency, the son of an Indian army colonel. Educated at several English private schools, he 'read algebra like a novel'. At Trinity College, Cambridge, he graduated Fourth Wrangler (1827), and in 1828 he became the first Professor of Mathematics at University College London. He resigned this office in 1831, but resumed it from 1836 to 1866. He was one of the founders of the London Mathematical Society and became its first president in 1865. He wrote a number of mathematical textbooks, but his most important work was in symbolic logic. He argued for an algebraic interpretation of the rules of logic, thus making logic part of mathematics. He also had a deep knowledge of the history of mathematics and contributed 850 articles to the *Penny Cyclopaedia* (1833–44).

DENTON, Sir Eric James
(1923–)

English marine biologist, born in Bridport. He was educated at St John's College, Cambridge, and the University of Aberdeen. Denton undertook research on radar between 1943 and 1946, before moving to the Biophysics Research Unit at University College London. He lectured in physiology at Aberdeen University (1948–56), and worked at the Plymouth Laboratory of the Marine Biological Association from 1955 until his retirement in 1987. He was also Secretary of the Marine Biological Association of the United Kingdom and Director of the Plymouth Laboratory (1975–87). Elected a Fellow of the Royal Society (1964), he served as a Member of the Council (1984–5), receiving the society's Royal Medal in 1987. His notable research has focused on varying aspects of the physiology of marine animals, including the visual and acoustic physiology of fish, and luminescence and camouflage of oceanic species. In the 1950s and 1960s he undertook pioneering studies of buoyancy regulation and control amongst cephalopod molluscs (squids, cuttlefish and nautiloids) and determined the roles of gas and ionic regulation in buoyancy. Buoyancy amongst cephalopods was found to be variously attributable to the reduction of heavy substances (eg $CaCO_3$) through the replacement of heavy ions by lighter ions such as Na^+ and Cl^-, whilst maintaining body fluids isotonic with sea-water; increasing light substances (such as fats/oils); and the use of gas floats. Conversely, Denton revealed that gas/fluid balances were the adaptations displayed in cuttlefish and nautiloids. His work on sound has concerned the swim bladder-acousticolateralis systems of clupeoid fish (the herring family), especially in relation to their schooling behaviour. He was knighted in 1987 and awarded the International Biology Prize in 1989; this prize has been presented annually in Japan and Denton's was the first award in the field of marine biology.

DESAGULIERS, John Theophilus
(1683–1744)

French-born British experimental philosopher, born in La Rochelle. Seeking refuge after the revocation of the Edict of Nantes (1685), Desaguliers's Huguenot parents settled in England. By 1694 Desaguliers was assisting his father teaching in Islington. He matriculated at Christ Church, Oxford, where he received his BA in 1709. The following year, now a deacon, he was appointed lecturer in experimental philosophy at Hart Hall. In 1713 he moved to London, where he assumed the duties of curator and demonstrator of experiments at the Royal Society. Desaguliers's experimental lectures in mechanical philosophy and electricity (advocating, substantiating and popularizing the work of **Isaac Newton**) attracted a wide audience, secured the favour and financial support of George I, and earned him the Royal Society's prestigious Copley Medal three times. A prolific author and translator, in 1734 he brought out *A Course of Experimental Philosophy*, defining the subject and illustrating its salutary utilitarian nature with descriptions of mechanical devices. Desaguliers himself invented scientific instruments (eg a planetarium) and made improvements to machines, including Thomas Savery's steam engine and an air-pump used as a ventilator at the House of Commons. He was a prominent Freemason.

DESARGUES, Girard

(1591–1661)

French mathematician, born in Lyons. By 1626 he was in Paris, and he took part as an engineer in the siege of La Rochelle in 1628. Desargues founded the use of projective methods in geometry, inspired by the theory of perspective in art and a thorough grasp of **Apollonius**'s *Conics*, and introduced the idea that parallel lines 'meet at a point at infinity'. He based his approach on the use of certain projective invariants, configurations of points or lines that are not altered by a projection. These appear throughout the study of conics. An opaque style of writing and a habit of circulating his ideas only obscurely greatly hindered the reception of his ideas (although they were taken up by **Pascal**, then only 16), and mostly they were independently rediscovered by others, such as **de la Hire** and **Isaac Newton**. From 1645 he began a new career as an architect in Paris and Lyons. Some of his ideas were published in a work by his friend Abraham Bosse, including the theorem on two triangles in perspective. One of his works may be completely lost, and his most famous was circulated in an edition of 50 copies of which only one, and one copy, survive.

DESCARTES, René

(1596–1650)

French philosopher and mathematician, undoubtedly one of the great figures in the history of Western thought and usually regarded as the father of modern philosophy. He was born near Tours in a small town now called la-Haye-Descartes, and was educated from 1604 to 1614 at the Jesuit College at La Flèche. He did in fact remain a Catholic all his life, and he was careful to modify or even suppress some of his later scientific views, for example his sympathy with **Copernicus**, following **Galileo**'s condemnation by the Inquisition in 1634. He studied law at Poitiers, graduating in 1616; then from 1618 he enlisted at his own expense for private military service, mainly in order to travel and to have the leisure to think. He later wrote that when in Germany with the army of the Duke of Bavaria one winter's day in 1619 he had an intellectual vision in a 'stove-heated room': he conceived a reconstruction of the whole of knowledge, into a unified system of certain truth modelled on mathematics, based on physics and reaching via medicine to morality, all supported by a rigorous rationalism. From 1618 to 1628 he travelled widely in Holland, Germany, France and Italy; then in 1628 returned to Holland where he remained, living quietly and writing until 1649. Few details are known of his personal life, but he did have an illegitimate daughter called Francine, whose death in 1640 at the age of five was apparently a terrible blow for him. He published most of his major works in this period, the more popular ones in French, the more scholarly ones first in Latin. The *Discourse de la Méthode* (1637), the *Meditationes de prima Philosophia* (1641) and the *Principia Philosophiae* (1644) set out the fundamental Cartesian doctrines: the method of systematic doubt; the first indubitably true proposition, *cogito ergo sum* ('I think therefore I am); the idea of God as the absolutely perfect being; and the dualism of mind and matter. His theory of astronomy, which explained planetary motion by means of vortices surrounding the Sun, was eventually refuted by **Isaac Newton**. In mathematics he made his most lasting contribution: he reformed algebraic notation and helped found coordinate geometry. This enables geometrical problems to be reformulated and even solved algebraically. In 1649 he left Holland for Stockholm on the invitation of Queen Christina, who wanted him to give her tuition in philosophy. These lessons took place three times a week at 5 am and were especially taxing for Descartes whose habit of a lifetime was to stay in bed meditating and reading until about 11 am. He contracted pneumonia and died. He was buried in Stockholm, but his body was later removed to Paris and eventually transferred to Saint-Germain-des-Prés.

DESFONTAINES, René Louiche

(1750–1833)

French botanist, born in Tremblay, Brittany. Locally educated at first, after stealing some apples he was sent to college at Rennes. He then studied medicine at Paris, where as well as gaining distinction in that field he showed a strong leaning towards botany. He became one of the favourite students of Louis Guillaume Le Monnier, Professor of Botany at the Jardin des Plantes. He explored the Barbary Coast and the Atlas Mountains in 1783; the botanical results of this journey were published as *Flora Atlantica* (1798–9). He returned to France in 1785 and the following year was named a member of the Legion of Honour. In 1796, he published *Mémoire sur l'Organisation des Monocotylédones ou Plantes à une Feuille Séminaire*, in which he demonstrated the tremendous difference between monocotyledons and dicotyledons. Filled with a desire to link botany and agriculture, he edited and published *Histoire des Arbres et Arbrisseaux* (1809). He described numerous new plant genera and also wrote *Choix de Plantes du Corollaire des Instituts de Tournefort* (1803) and *Tableau de l'Étude de Botanique* (1804) as well as a vocabulary of the language spoken by the inhabitants of the Atlas Mountains (1830).

DE SITTER, Willem

(1872–1934)

Dutch astronomer and cosmologist, born in Sneek, Friesland. He studied mathematics at the University of Groningen, but later became an astronomer at Cape Town Observatory. After returning to Groningen and working as assistant to **Kapteyn**, he was appointed Director and Professor of Astronomy at the University of Leiden (1908), and from 1919 was also director of the observatory there. He studied the distributions and motions of stars, and the dynamics of the Galilean satellites of Jupiter. His interest in **Einstein**'s theory of general relativity, published in 1916, led to its publicity in Britain and other English-speaking countries, with important consequences for cosmology. Einstein had solved his equations of general relativity to produce a description of a static universe with curved space, the curvature being constant in time. De Sitter demonstrated that an expanding universe of constantly decreasing curvature emerged as another possible solution. As opposed to Einstein's static concept ('matter with no motion'), he characterized the universe as an expanding curved space–time continuum of 'motion with no matter'. **Hubble**'s discovery of the recession of distant galaxies added great weight to this theory. Later, modification of the de Sitter solution produced a much simpler description known as the Einstein–de Sitter universe.

DESLANDRES, Henri Alexandre

(1853–1948)

French spectroscopist and astronomer, born in Paris. Educated at the École Polytechnique, he graduated in 1874 and entered the army but resigned in 1881 to pursue his interest in science. His first field of research was laboratory spectroscopy at the École Polytechnique with Marie-Alfred Cornu. He worked on the band spectra of diatomic molecules, and from a sytematic study of the pattern of spectrum lines found a formula for their frequencies, known as Deslandres' law (1885). In 1889 he took up astronomical spectroscopy on joining the Paris Observatory where new spectroscopic facilities were specially created for him. His greatest successes were in the field of solar spectroscopy. He invented his velocity spectrograph (1891), an instrument which allowed photographs of successive strips of the Sun to be taken in the light of a particular spectrum line, on the same principle as the spectroheliograph being independently developed by **Hale**. In 1897 Deslandres moved to the astrophysical observatory at Meudon near Paris of which he became director on the death of **Janssen** in 1908. In 1926 the Paris and Meudon observatories were merged under his directorship, two years before his retirement. Deslandres was elected to the French Academy of Sciences in 1901 and was awarded the Gold Medal of the Royal Astronomical Society in 1913.

DESMAREST, Nicolas

(1725–1815)

French vulcanologist and geomorphologist, born in Souleines, and educated at the college of the Oratorians at Troyes and Paris. His first scientific paper was an essay on the supposed former land connection between France and England (1751), and in the wake of the 1755 Lisbon earthquake, he became interested in earthquakes and seismic wave propagation. He is best known for his studies and geological map of the volcanic region of the Auvergne (1774), where he clearly demonstrated that columnar basalts, the origin of which was in dispute, formed by cooling of molten rock. In this work he introduced the term basalt-lava, drawing together the previously held distinction between basalt and lava. He recognized the erosive action of streams to produce valleys, and that landforms can develop through the systematic degradation of geological formations. His early career promoting knowledge of manufacturing processes involved him in cloth, cheese and paper production, and ultimately lead to his appointment by the king as Inspector-General and Director of Manufactures of France (1788). He retained this post until the Revolution, when his political friends were executed and he was jailed.

DE VRIES, Hugo Marie

(1848–1935)

Dutch botanist and geneticist, born in Haarlem, the son of a Dutch prime minister. The first instructor in plant physiology in the Netherlands, he studied at Leiden, Heidelberg and Würzburg, and became Professor of Botany at Amsterdam (1878–1918). From 1890 he devoted himself to the study of heredity and variation in plants, significantly developing Mendelian genetics and evolutionary theory. His major work was *Die Mutationstheorie* ('The Mutation Theory', 1901–3). He described and correctly interpreted the phenomenon of plasmolysis, the process whereby the cytoplasm in plant cells shrinks away from the walls as a result of water loss from the cells. He introduced methods for studying osmotic and turgor properties of plant cells which have become standard techniques. In 1885 he showed that the plasmalemma bounding the vacuole in plant cells is semi-permeable, allowing the passage only of small molecules.

DEWAR, Sir James
(1842–1923)

Scottish chemist and physicist, born in Kincardine-on-Forth. From 1859 he studied at the University of Edinburgh, and later he was assistant to **Lyon Playfair** and to Playfair's successor **Crum Brown**. Dewar spent part of 1867 in the laboratory of **Kekulé** in Ghent. He became a lecturer in chemistry at the Royal (Dick) Veterinary College in Edinburgh in 1869. In 1875 he was appointed Jacksonian Professor of Natural Philosophy at Cambridge, a post he held until his death in 1923. From 1877 onwards, however, he resided at the Royal Institution in London and commuted to Cambridge as necessary. At the Royal Institution Dewar was Fullerian Professor of Chemistry and the first director of the new Davy–Faraday Laboratory from 1896. For 46 years his main interests and activities were at the Royal Institution rather than Cambridge. He gave many of the Friday evening discourses and series of lectures for young people, always illustrated by striking demonstrations. His many awards and honours included the Rumford, Davy and Copley medals of the Royal Society, of which he became a Fellow in 1877. He was President of the Chemical Society (1897–9), and was knighted in 1904. Dewar's Edinburgh researches were mainly in organic chemistry and he devised the structure for benzene known as the Dewar formula. He also worked on photochemistry, thermal dissociation, the temperature of the Sun and specific heats. The rift with Cambridge was probably caused by Dewar's dissatisfaction with the research facilities. After the move to London, Dewar's interests turned to the area with which he is generally associated: the properties of matter at very low temperatures and the liquefaction of gases. In the 1890s Dewar's target was the liquefaction of hydrogen, which he accomplished in 1898 by cooling the gas at 200 atmospheres to −200 degrees Celsius in a bath of liquid air and allowing it to expand through a fine nozzle. During this work he invented the vacuum flask, later widely known as the Dewar flask. Among the many other topics studied by Dewar were the properties of fluorine and of metal carbonyls and the behaviour of soap films and bubbles. Towards the end of his life, and to within a few days of his death, Dewar made measurements of the radiation from different parts of the sky in an improvised observatory at the Royal Institution.

DEWAR, Michael James Steuart
(1918–)

British chemist, born in Ahmednagar, India, of Scottish parents. He was educated in classics and science at Winchester College, then specialized in chemistry at Balliol College, Oxford (1936–40). For his DPhil he worked on explosives as part of the war effort, and during a postdoctoral fellowship with **Robinson**, he studied the chemistry of penicillin. At the end of World War II he obtained an ICI fellowship at Oxford, and worked on tropolone synthesis and the benzidine rearrangement. He also started work on his celebrated book *The Electronic Theory of Organic Chemistry* (1949). From 1945 to 1951 he worked as a physical chemist for the textile manufacturer Courtaulds; during this period he became increasingly interested in the application of molecular orbital theory to organic chemistry. In 1951 he became Professor of Chemistry at Queen Mary College, London, and made important progress in using these calculations to explain a number of phenomena in organic chemistry. In 1959 he moved to the University of Chicago and started his development of programmes for semi-empirical molecular orbital calculations which have proved so simple but so powerful in providing insight into mechanistic organic chemistry. This work was continued at the University of Texas at Austin (1963–90) and latterly at the University of Florida. Dewar was elected FRS in 1960, and became a member of the US National Academy of Sciences in 1983.

D'HÉRELLE, Felix
(1873–1949)

French–Canadian bacteriologist, born in Montreal. He studied there and worked in Central America, Europe and Egypt before holding a chair at Yale from 1926 to 1933. A competitor of **Twort**, he was independently the discoverer in 1915 of bacteriophage, a type of virus which infects bacteria. Thereafter he tried to use 'phage' therapeutically, but without any significant successes. However, phage later proved of great value in research, and along with Twort he can be regarded as one of the founders of molecular biology.

DIAMOND, Jared Mason
(1937–)

American physiologist and ecologist, born in Boston. He was educated at Harvard and the University of Cambridge, where he received his PhD in 1961. Since 1968 he has been professor at the University of California Medical Center. Diamond has contributed significantly to the study of ecological diversity through his studies on islands, particularly on their bird faunas, following up the theory of island biogeography proposed by **MacArthur** and **Edward Wilson**. He distinguished organisms which spread readily and easily ('super-tramps') from those which are less mobile, and calculated the turnover of species (ie extinction versus colonization) on a number of Pacific islands. His *Rise and Fall of the Third Chimpanzee* (1991) is an influential work of popular biology.

DICKE, Robert Henry
(1916–)

American physicist, born in St Louis, Missouri. He studied physics at Princeton and Rochester, and spent his career at Princeton as Professor of Physics from 1957 and Albert Einstein Professor of Physics from 1975. Independently of **Alpher** and **Gamow**, he deduced in 1964 that a 'Big Bang' origin of the universe should have left an observable remnant of microwave radiation. He had just started searching for this radiation when he heard that it had been detected by **Penzias** and **Robert Wilson**. In the 1960s he carried out important work on gravitation, proposing that the gravitational constant G slowly decreases with time (the Brans–Dicke theory, 1961). After a critical review of **Eötvös**'s work on showing that inertial mass is equal to gravitational mass—Einstein's equivalence principle—he verified this to one part in 10^{11}.

DICKSON, Leonard Eugene
(1874–1954)

American mathematician, born in Independence, Iowa. He studied at the University of Texas, and taught at Chicago for most of his life, doing much to make that university a leading centre for research in mathematics in the USA. He did important work in group theory, finite fields and linear associative algebras, and discovered all the families of finite simple groups. His *History of the Theory of Numbers* (1919–23) is encyclopedic on the subject.

DIELS, Otto
(1876–1954)

German chemist, born in Hamburg. The son of a distinguished classical scholar, he studied chemistry at Berlin (1895–9) under **Emil Fischer** and was on the staff there from 1899 to 1916. In 1916 he became professor at Kiel. He investigated a number of reactions, including the dehydrogenation of cholesterol and the bile acids, but he is most famous for the discovery of the reaction of an activated olefin with a diene to give a cyclic structure with a predictable stereochemistry, a reaction of enormous synthetic value. It was discovered in collaboration with **Alder**, and they shared the 1950 Nobel Prize for Chemistry for this work.

DIESEL, Rudolf Christian Karl
(1858–1913)

German engineer, born in Paris. He studied at the Munich Polytechnic under Carl von Linde, and after graduating in 1880 he joined Linde's firm in Paris as a refrigeration engineer. In 1890 he moved with the same firm to a new post in Berlin, and it was around this time that he began to develop the idea of a new type of internal combustion engine, obtaining the German patent for it in 1892. Supported by Krupp of Essen and Maschinenfabrik of Augsburg, he set about constructing a 'rational heat motor', demonstrating the first practical compression–ignition engine in 1897. It was displayed at the Munich Exhibition of 1898 and attracted worldwide interest. One of its main advantages was that the diesel engine achieved an efficiency about twice that of comparable steam engines. He spent most of his time after 1899 at the factory which was specially built at Augsburg to manufacture his diesel engines, but he suffered persistent ill-health and was not a business man at heart. Demand for his engines was such, however, that he could not help becoming rich and apparently successful, but in 1913, at the height of his fame, he vanished from the Antwerp–Harwich mail steamer, and was presumed drowned.

DIEUDONNÉ, Jean Alexandre
(1906–92)

French mathematician, born in Lille. He studied at the École Normale Supérieure, and held chairs in Rennes, Nancy, Chicago, the Institut des Hautes Études Scientifiques, and finally Nice (1964–70). One of the leading French mathematicians of his generation, he worked in many areas of abstract analysis, **Lie** groups, and algebraic geometry. His *Éléments d'analyse* (1960–82) in nine volumes carries on the French tradition of the definitive treatise on analysis. As a founder of the **Bourbaki** group, his ideas on the presentation of mathematics, laying great stress on precise abstract formulation and elegance, have marked out a distinctively French school of mathematical writing whose influence has lasted for some 50 years. He did much to make the **Grothendieck**'s remarkable work available in a readable form. With **Weil** he stimulated Bourbaki to write on the history of mathematics, and in later life wrote the only histories of algebraic geometry and algebraic topology to come up to the present day.

DIGGES, Leonard

(1520–?59)

English applied mathematician, known for his valuable work in surveying, navigation and ballistics. He was probably self-educated, but his books on surveying and navigation went through many editions in the 16th century. His work in ballistics, based on his own experiments, appeared as *Stratioticos* (1579), published by his son Thomas (d 1595). He took part in Thomas Wyatt's rebellion in 1554, and was condemned to death, but later he was pardoned and fined.

DILLENIUS, or DILLEN, Johann Jacob

(1687–1747)

German botanist and botanical artist, born in Darmstadt, and educated at Giessen. He moved to England in 1721, at the behest of William Sherard, a diplomat and botanist, to assist in a revision of **Bauhin**'s *Pinax*. He produced the magnificent *Hortus Elthamensis* in 1732, an account of the plants growing in Sherard's physician brother James's garden at Eltham. Sherard endowed the Sherardian Chair of Botany at Oxford, one of the conditions being that Dillenius was to be its first occupant, which came to pass in 1734. Dillenius wrote and illustrated the first monographic study of mosses, *Historia Muscorum* (1741). This work contains some of the first highly detailed illustrations and descriptions of some 600 different moss species, and such was the accuracy of these illustrations that many authors, including his friend **Linnaeus**, were able to use them in later works on moss taxonomy. He named for the first time many of the major moss genera, including *Bryum*, *Hypnum*, *Mnium* and *Sphagnum*. He provided accurate observations of morphology, together with a critical delimitation of species, but his ideas were not innovatory, and were sometimes inaccurate. For example, he erroneously formed the view that the capsule (sporogonia) in mosses was the equivalent of the male flower in higher plants.

DINES, William Henry

(1855–1927)

English meteorologist, born in Oxshott, Surrey. He graduated from Christ's College, Cambridge, in 1881. He never held a professional post, but was director of experiments of the upper air for the Meteorological Office from 1905 to 1922. As a result of the Tay bridge disaster of 1879, he became a member of the Wind Force Committee of the Royal Meteorological Society and designed the pressure tube anemometer which bears his name (1901). In that year he began upper air investigations from his home using kites and later extended this work by taking aerological observations at sea near west Scotland using a meteorograph he had constructed. In 1907 he started regular balloon ascents to further his upper air work, and using the results of about 200 ascents he worked out correlations between pressure and temperature at the surface and at various heights. From an extensive statistical analysis of his observations, Dines concluded that a cyclonic circulation resulted from dynamical processes in the upper layers of the troposphere or the lower stratosphere, rather than from thermal processes nearer the surface. After World War I he collaborated with **Lewis Fry Richardson** on studies of solar and terrestrial radiation, and designed and improved a number of meteorological instruments, including the Dines radiometer (1920).

DIOPHANTUS

(fl 3rd century)

Greek mathematician who lived at Alexandria. Little of his work has survived; the largest work is the *Arithmetica* which deals with the solution of problems about numbers and, in contrast to earlier Greek work, uses a rudimentary algebraic notation instead of a purely geometric one. For example, one problem asks for two numbers so that in each case the square of the first added to the second gives a square. In a modern version of his notation, one number would be x and the other $2x + 1$. In many problems the solution is not uniquely determined, and these have become known as Diophantine problems. Typically, a Diophantine problem yields an algebraic equation which has only finitely many integer solutions, if any. Diophantus's work was rediscovered in the 16th century, and later editions of it inspired **Vieta** to offer an algebraic account of mathematics which, if historically implausible, was nonetheless significant in early 17th-century mathematics. The study of Diophantus's work inspired **Fermat**, who was well versed in Vieta's ideas, to take up number theory in the 1600s with remarkable results.

DIRAC, Paul Adrien Maurice

(1902–84)

English mathematical physicist, born in Bristol, and educated at the universities of Bristol and Cambridge. He completed his doctoral thesis in 1926, by which time he had solved many problems in quantum mechanics. In the same year **Schrödinger** completed his wave equation, but Dirac preferred the matrix approach of **Born**, **Pascual Jordan** and **Heisenberg**. He started work on his own interpretation of quantum mechanics based on a transformation theory, following the basis for the theory of relativity. In 1928 he produced his relativistic wave equation, which corrected the failure of Schrödinger's equation to explain the electron spin (intrinsic angular momentum) discovered by **Uhlenbeck** and **Goudsmit** in 1925. This equation had negative energy solutions which he later interpreted as antimatter states (1930). He also predicted that a photon of sufficient energy could produce an electron–positron pair, and vice versa, which was confirmed experimentally by **Carl Anderson** in 1932. In 1930 Dirac published the classic work *The Principles of Quantum Mechanics* and in the same year he was elected FRS. Lucasian Professor of Mathematics at Cambridge (1932–69), he was awarded the Nobel Prize for Physics in 1933 with Schrödinger for their work in quantum theory. He then went on to work on quantum electrodynamics (QED) in which he predicted the existence of the magnetic monopole (not yet discovered); he used this to explain why the charge on all particles is a multiple of the electron's charge. In further work on QED he produced many interesting ideas, such as his proposal in 1950 that the basic representation of particles should not be point-like, but string-like, an idea which is now gaining support following the work of **Michael Green**, John H Schwarz and Edward Witten. He became Professor of Physics at Florida State University in 1971, where he worked until his death.

DIRICHLET, (Peter Gustav) Lejeune
(1805–59)

German mathematician, born in Düren, the son of the local postmaster. He showed a precocious interest in mathematics and entered the Collège de France in Paris in 1822. After a private education, he became Extraordinary Professor at Berlin in 1828, and succeeded **Gauss** as professor at Göttingen in 1855. His main work was in number theory, **Fourier** series, and boundary value problems in mathematical physics. In number theory he was the first to show that in any arithmetic progression where the first term and the difference between successive terms have no common factor, there are infinitely many primes. He also helped to make the work of Gauss accessible, and to make number theory a central branch of mathematics for the first time. The principle named after him asserts that a harmonic function defined on a simply connected region is determined by its values on the boundary. This result, despite its strong physical plausibility, was not proved until well after his death. He was the leading figure in the drive to bring rigorous methods into mathematics, his works laying great stress on conceptual clarity, and by his teaching and example he was the dominant influence on many leading German mathematicians of the next generation.

DIXON, Harold Baily
(1852–1930)

English physical chemist, born in London. He was educated at Christ Church, Oxford, initially pursuing classical studies with little success, but through the influence of **Augustus Vernon Harcourt** he changed to natural science and graduated in 1875. He was elected to a fellowship at Trinity College and through this and later appointments he taught and researched for some years in the Balliol–Trinity Laboratory. In 1887 Dixon was appointed Professor of Chemistry at Owen's College, Manchester (later the Victoria University of Manchester). He held this chair until his retirement in 1922, and played a major role in the development of the chemistry department and the university. Dixon was elected FRS in 1886; he was Bakerian Lecturer of the Royal Society in 1893 and received its Royal Medal in 1913. He was President of the Chemical Society from 1909 to 1911, and was appointed CBE in 1918. Dixon's researches were almost entirely on gaseous explosions and flames. He did much work on rates of propagation of explosion waves and devised an ingenious method of photographing explosion flames. He found that the velocity of explosion waves depended on various factors, but was generally in the region of 2 000 to 3 000 metres per second, much higher than had previously been believed.

DJERASSI, Carl
(1923–)

American chemist, born in Vienna, Austria. He attended a number of colleges in the USA during the early 1940s before receiving a PhD from the University of Wisconsin (1945). Initially he worked for Ciba Pharmaceuticals (1945–9) and he then joined Syntex research laboratories in Mexico City. From 1952 to 1959 he was on the staff of Wayne State University but returned to Syntex during a period of academic leave (1957–9). He then moved to Stanford University but maintained his close contacts with Syntex where he was appointed Vice-President and President of Research (1960–72). Djerassi has worked throughout his career on the chemistry of natural products. While with Syntex he pioneered the development of oral contraceptives, but he made equally important contributions to the application of a number of spectroscopic techniques (mass spectrometry and circular dichroism) to many classes of compounds. He has received a great number of honours from bodies all over the world, including the Priestley Medal of the American Chemical Society. Latterly he has turned his attention to literature, writing novels and a book of poems. He founded an artists' colony in California in memory of a daughter who committed suicide.

DÖBEREINER, Johann Wolfgang

(1780–1849)

German chemist, born in Hof an der Saale, Bavaria. A pharmacist by training and a friend of Johann Wolfgang von Goethe, he taught himself physics and chemistry. He was appointed Assistant Professor at Jena in 1810 and Professor in 1819, a post which he held until 1849. In Döbereiner's lamp, which he invented in 1823, he made use of his discovery that hydrogen (produced in the lamp by the action of sulphuric acid on zinc) burns on contact with a platinum sponge. This discovery was taken up by **Berzelius** who went on to formulate a general theory of catalysis. Döbereiner's greatest contribution to chemistry, however, was his observation that there is a relationship between the atomic weights of calcium, barium and strontium. His subsequent attempt to draw up a table of elements, grouping them in triads, was the forerunner of other attempts to find an underlying pattern in matter, culminating in **Mendeleyev**'s periodic table. He died in Jena.

DOBZHANSKY, Theodosius

(1900–75)

American geneticist, born in Nemirov, the Ukraine. He was a student of the population geneticist Sergei Chetverikov, and studied zoology at Kiev and taught genetics in Leningrad. In 1927 Dobzhansky emigrated to the USA, bringing the Russian populationist approach to **Thomas Hunt Morgan**'s laboratory at Columbia University, New York. He taught at Caltech (1929–40), Columbia (1940–62), Rockefeller University, New York (1962–70), and at the University of California, Davis (1971–5). He showed that the genetic variability within a population is large, and includes many potentially lethal genes (recessives) which nevertheless confer versatility when the population is exposed to environmental change. He applied his ideas to the concept of race in man, defining races as Mendelian populations differing in gene frequencies, but with more diversity within populations than between populations. He advocated human diversity and equality. His most significant texts were *Genetics and the Origin of Species* (1937), a synthesis of evolutionary theory, Mendelian genetics and Darwinian natural selection, and *Genetics and the Evolutionary Process* (1970). His popular text *Heredity, Race and Society* (1946) sold more than one million copies.

DOHRN, Anton

(1840–1909)

German zoologist, born in Szczecin, Poland. He studied zoology at Königsberg, Bonn, Jena and Berlin, and received his doctorate at Jena under **Haeckel** who, along with **Charles Darwin**, proved to exert lasting influences on Dohrn's scientific philosophy. In 1865, whilst in Helgoland with Haeckel, Dohrn developed plans for the founding of a marine research institute; in 1870 the notion of financing such an institute by means of a public aquarium took him to Naples. There the city authorities granted him a free plot of land in the Royal Park (presently Villa Comunale) on condition that he fund construction of the building himself. The Stazione Zoologica, the world's first international marine research institute, was completed in 1873; three quarters of the building costs were funded personally by Dohrn with the remainder loaned by friends. Dohrn directed the Stazione up until his death, and was succeeded in the post by his son Reinhart, and grandson Peter. As a German citizen, Reinhart Dohrn was prohibited from owning property as a result of World War I, and the Stazione became a semi-private institution. In 1983 it came under the full state supervision of the Italian Ministry of Public Instruction. The worldwide reputation of the Stazione has focused on the embryology and physiology of marine animals and some 52 Nobel laureates have been associated with the laboratory.

DOISY, Edward Adelbert

(1893–)

American biochemist, born in Hume, Illinois. Educated at Harvard, he became Director of the Department of Biochemistry at St Mary's Hospital, St Louis (1924–65). In 1924, with the famous American embryologist Edgar Allen, Doisy demonstrated that the liquor folliculi of the ovaries and follicle cells produced sexual maturation in oestrus by means of a hormone and he devised a way of quantifying its potency. The discovery that the female sex hormone oestrone was elevated in urine in pregnancy enabled Doisy to isolate crystalline oestrone (1929, at the same time as **Butenandt**), oestriol (1930) and oestradiol (1935). In his Porter Lectures at the University of Kansas (1936), he delineated the four stages of endocrinology as recognition of gland, detection of the hormone, its extraction and purification, and finally structure and synthesis. He published *Sex Hormones* (1936) and *Sex and Internal Secretions* (1939). In 1939 he isolated and characterized the antihaemmorrhagic agents vitamin K_1 (from alfalfa) and vitamin K_2 (from fish meal). For this work he shared the 1943 Nobel Prize for Physiology or Medicine with **Dam**. Doisy's other publications embraced insulin and blood buffers.

DOLLFUS, Audouin Charles
(1924–)

French planetary astronomer, born in Paris. He studied at the Lyceé Janson–de Sailly and then received his doctorate in mathematical sciences at the University of Paris. Since 1946 he has worked in the Astrophysical Section of the Meudon Observatory. Dolfus pioneered the measurement of the polarization of planetary light; to this end he made the first French stratospheric balloon ascent. Comparing the polarization of the light reflected from Mars with the polarization signature of several hundred terrestrial minerals, he found that the best fit was to pulverized limonite, Fe_2O_3. The polarization was found to vary as a function of the Martian season. It was erroneously concluded that this was due to seasonal changes in a form of microscopic plant life; it is now known to be caused by seasonal winds blowing sand over rock. At an observatory 12000 feet high in the Swiss Alps, Dolfus compared the spectral infrared water bands in the light reflected from Mars and the Moon. He concluded that the total water content of the Martian atmosphere was only sufficient to form a layer 0.02 centimetres deep on the surface. In 1966, when the rings of Saturn were viewed edge-on from Earth, Dolfus discovered Janus, a satellite which is 151472 kilometres from the planet.

DOLLO, Louis Antoine Marie Joseph
(1857–1931)

Belgian palaeontologist, born in Lille. After studying there, he became an assistant (1882) and keeper of mammals (1891) at the Royal Museum of Natural History in Brussels. In 1893 he enunciated Dollo's law of irreversibility in evolution which states that complex structures, once lost, are not regained in their original form. While this is generally true, exceptions are found. For example, the eye of snakes appears to have re-evolved from secondarily blind burrowing forms.

DOLLOND, John
(1706–61)

English optician, born in London of Huguenot parentage. A silk weaver to trade, in 1752 he turned to optics, and devoted himself with the help of his son Peter (1738–1820) to the development of an achromatic telescope. Achromatic lenses were first made by Chester More Hall in the period 1729–33, apparently for his own use, but the first achromatic telescope was made by the Dollonds in 1758, following up a suggestion made to them by **Klingenstierna** of Uppsala. Hall and some associates brought an action against the Dollonds on the grounds of the priority of their earlier but unexploited work; the action was thrown out by the courts. Early in 1961 John Dollond was appointed Optician to King George III, only a few months before his death.

DOLOMIEU, Déodat Guy Gratet de
(1750–1801)

French geologist, soldier and traveller, born in Dolomieu, Dauphiné. He travelled extensively in Italy, Sicily, Portugal, the Alps and the Pyrennes, writing on earth tremors in Calabria and Italian volcanoes. He accompanied Napoleon's expedition to Egypt (1799) but was imprisoned when returning to France. In 1800 he was freed from imprisonment in Sicily to become professor at the Natural History Museum in Paris on the recommendation of **Haüy**, but his health had been damaged by his suffering in prison and he died shortly afterwards. The mineral dolomite is named after him and by extension the Dolomite mountain range of Italy.

DOMAGK, Gerhard Johannes Paul
(1895–1964)

German biochemist, born in Lagow (now in Poland). He graduated in medicine at Kiel (1921) after service in World War I, and taught at Greifswald and Münster before becoming Director of the I G Farbenindustrie Laboratory for Experimental Pathology and Bacteriology at Wuppertal-Elberfeld in 1927. In a carefully programmed search for new dyes and new drugs, Domagk particularly sought a treatment against streptococci which caused widespread and generally lethal infections such as erisipelas and meningitis. One azo dye, itself ineffective, became of potent benefit upon adding a simple sulphonamide group. This substance, prontosil, only worked in the living animal and was found by **Bovet** to be converted to sulfanilamide in the body, a discovery which ushered in a new age in chemotherapy. Domagk's daughter, in 1932, was among the first to be cured by the new wonder drug. In 1939 his original acceptance of the Nobel Prize for Physiology or Medicine was cancelled upon instruction from the German government; Domagk finally received the award, but not the remuneration, in 1947.

DON, George
(1764–1814)

Scottish botanist, born in Muirhead. He established a botanic garden and tree nursery in Forfar, and collected many of Scotland's rare arctic–alpine plants, including many from Glen Clova. His *Account of Plants of Forfarshire* was published in 1813. After being appointed Superintendent of the Royal Botanic Garden in Edinburgh in 1792, he attended medical classes, and returned to Forfar in 1795 to practise medicine. He began publishing *Herbarium Britannicum* in 1804, intending to publish 100 plates per year. His practice and his nursery business suffered through his botanical work, and he became bankrupt. He died in extreme poverty, but his two sons David and George also became botanists. David Don (1800–41) later became Librarian to the Linnaean Society in succession to **Robert Brown**, and Professor of Botany at King's College, London. He published *Prodromus Florae Nepalensis* in 1825, while his elder brother George (1798–1856) worked in the Chelsea Physic Garden, and collected in Brazil, Sierra Leone and the West Indies for the Royal Horticultural Society. George junior's major work was *A General System of Gardening and Botany* (1832–8), but he also wrote on his Sierra Leone discoveries, and on *Allium* and *Combretum*.

DONATI, Giambattista
(1826–73)

Italian astronomer, born in Pisa. He was educated at the University of Pisa, and became an assistant at the observatory in Florence in 1852. In 1864 he succeeded **Amici** as director there. He discovered the brilliant comet ('Donati's comet') of 1858. Noted for his researches on stellar spectra, he was the first to observe the spectrum of a comet. In August 1864, using a spectroscope with a 41 centimetre objective lens feeding a 60 degree prism, he drew the spectrum of comet Temple (1864 II). He erroneously identified what are now known to be the three molecular Swan bands of C_2 as being due to metals. He was the first to notice that distant comets had spectra similar to the Sun's.

DONDERS, Franciscus Cornelis
(1818–89)

Dutch oculist and ophthalmologist. Born into a talented family from Tilburg, Donders studied at the military medical school at Utrecht and later at Utrecht University; in 1840 he obtained his MD from Leiden, returning to Utrecht where, appointed Extraordinary Professor, he pursued researches in physiological chemistry, notably on plant physiology and the problem of animal and vegetable heat. Aided in part by **Helmholtz**'s invention of the ophthalmoscope, Donders established himself as a specialist in diseases of the eye, setting up a polyclinic for eye diseases at the university. He improved the efficiency of spectacles by the introduction of prismatic and cylindrical lenses, and wrote extensively on eye physiology. In 1862 he was appointed Professor of Physiology at Utrecht.

DONNAN, Frederick George
(1870–1956)

Irish physical chemist, born in Colombo, Ceylon (now Sri Lanka). He studied science at Queen's College, Belfast, and then worked at laboratories in Leipzig (under **Wislicenus** and **Ostwald**), Berlin (under **van't Hoff**) and London (under **Ramsay**). After junior teaching positions at University College London (UCL) and in Dublin, Donnan became the first Brunner Professor of Physical Chemistry at the University of Liverpool from 1904 to 1913, and then Professor of Chemistry at UCL (succeeding Ramsay) until his retirement in 1937. His many honours included the Davy Medal of the Royal Society (1928), of which he was a Fellow from 1911, and the Longstaff Medal of the Chemical Society (1924). He was President of the Faraday Society from 1924 to 1926 and of the Chemical Society from 1937 to 1939. He was appointed CBE in 1920. Donnan's researches were concerned almost entirely with the physical chemistry of solutions. He had an international reputation as a colloid chemist and is remembered particularly for his thermodynamic analysis of equilibria across membranes (around 1911), which has found extensive application in connection with living cells.

DOPPLER, Christian Johann
(1803–53)

Austrian physicist, born in Salzburg. The son of a stonemason, he would have followed his father's craft if he had not suffered from ill-health. He showed early mathematical ability. After studying at the Polytechnic Institute in Vienna, he was appointed Professor of Mathematics and Accounting at the

State Secondary School in Prague. In 1851 he was appointed Professor of Physics at the Royal Imperial University of Vienna, the first such position to be created in Austria. 'Doppler's principle', which he enunciated in a paper in 1842 when he was Professor of Elementary Mathematics and Practical Geometry at the State Technical Academy in Prague, explains the variation of frequency observed, higher or lower than that actually emitted, when a vibrating source of waves and the observer respectively approach or recede from one another. The first experimental verification was performed in Holland in 1845, using a locomotive drawing an open car with several trumpeters. The Doppler effect applies not only to sound but to all forms of electromagnetic radiation. In the case of astronomy, the changes of the spectral wavelengths of approaching or receding celestial bodies provide important evidence for the concept of an expanding universe.

DORN, Friedrich Ernst
(1848–1916)

German chemist, born in Guttstadt, East Prussia. Educated at Königsberg, he taught physics at Darmstadt and Halle. In 1900 he noticed that radium apparently becomes less radioactive if swept with a current of gas. This led him to the discovery of a radioactive gas which is emitted by radium as part of its decay processes. He called it 'niton' but it is now known as radon. Its properties were investigated by **Ramsay**. Dorn died in Halle.

DOTY, Paul Mead
(1920–)

American biochemist and specialist in arms control. He was educated at Pennsylvania State College, Columbia University and the University of Cambridge, where he graduated BS, MA and PhD respectively. He held short-term posts at the Brooklyn Polytechnic Institute and Notre Dame University, then became assistant professor (1948–50) and associate professor (1950–6) at Harvard. He has been full professor there since 1956, and has been Consultant to the Arms Control and Disarmament Agency of the National Security Council since 1973. In 1961 Doty discovered DNA renaturation, establishing the specificity and feasibility of nucleic acid hybridization. A very special feature of DNA is that two of the constituent bases (adenine and guanine) form chemical bonds (hybridize) to the other two bases, thymidine and cytosine. This is the basis for the formation of the double helix and for the transcription of DNA to its mirror image RNA, and allows the diagnostic use of radioactively labelled specific nucleic acids to 'find' their counterpart sequence of DNA or RNA *in vitro*. This powerful technology is the basis for all experimental manipulation of nucleic acids.

DOUGLAS, David
(1798–1834)

Scottish botanist and plant collector, born in Scone, Perthshire. He shunned schooling, preferring to play truant in the countryside. When he reached the age of 10, his father removed him from school and apprenticed him as a gardener at Scone Palace. In 1820 he began working at Glasgow Botanic Garden. Sir **William Jackson Hooker** had just become professor of botany; he and Douglas went on many collecting trips and Douglas became expert in preparing herbarium specimens. In 1823 he became a plant collector for the Horticultural Society of London, and left on an expedition to North America where he began collecting in Ontario and the north-east USA. In 1825–6 he explored the Columbia river, falling in love with a Chinook princess, and discovered many new species such as *Aster douglasii*, *Garrya elliptica* and *Paeonia brownii*. The most famous of the many plants Douglas introduced into cultivation was the Douglas fir (*Pseudotsuga taxifolia*); others included the flowering currant (*Ribes sanguineum*), the quamash (*Camassia quamash*) and the California poppy (*Eschscholzia californica*). After some months in London writing his journal (unpublished until 1914), Douglas sailed again for western North America and explored southern California (1830–2), where he discovered *Nemophila menziesii* (baby blue-eyes) and *Calochortus venustus* (Mariposa tulip), and then travelled up the Fraser and Simpson Rivers in western Canada. He had ambitions to traverse Siberia, but instead sailed to Hawaii. There, he climbed Mauna Kea and died tragically in a cattle pit; he was reputedly gored to death by a bull in the trap, but murder cannot be discounted.

DOUGLASS, Andrew Ellicott
(1867–1962)

American astronomer, and 'father of dendrochronology', born in Windsor, Vermont. After research work at the Lowell Observatory at Flagstaff, Arizona, he became Professor of Physics and Astronomy at Arizona University (1906) and later Director of the Stewart Observatory (1918–38). He investigated the relationship between sunspots and climate by examining and measuring the annual growth-rings of long-lived Arizona pines and sequoias; he noted that variations in their width corresponded to specific climatic cycles, creating patterns which can be discerned in timbers from prehistoric archaeological sites and providing a time-sequence for dating purposes. He coined the term 'dendrochronology' ('tree-dating') in his *Climatic Cycles and Tree Growth* (3 vols, 1919–36).

DOVE, Heinrich Wilhelm
(1803–79)

German climatologist, born in Liegnitz, Silesia (now in Poland). Professor of Natural Philosophy at Berlin University (1845–79) and Director of the Royal Prussian Meteorological Institute (1848–79), he was a great collector and organizer of meteorological observations. He established the first comprehensive network of observations over Europe, enabling specific weather situations to be studied in detail and providing an early climatology. From these and other observations, he produced monthly mean maps of isotherms over much of the globe forming the basis of maps of temperature in many early atlases. In his famous book entitled *The Law of Storms* (1857), he postulated that the general circulation consisted basically of an equatorial wind current and a polar wind current. He regarded temperate latitude depressions as results of the conflict between the two currents. He used the fact that winds change from an equatorial direction to a polar one as a storm passed as evidence for his theory; this was probably the initial idea behind the polar front ideas of **Vilhelm Bjerknes**. Dove also investigated in detail some remarkable storms, notably that of 23–29 January 1850. His method of analysis using graphs of pressure and temperature along lines of latitude and longitude was later used by **Fitzroy** to investigate the Royal Charter storm of 1859.

DRAPER, Henry
(1837–82)

American astronomer and pioneer of astronomical photography, born in Prince Edward County, Virginia. His father was the chemical physicist **John William Draper** who, using **Daguerre**'s process, took what is probably the oldest surviving photographic portrait (of his sister Dorothy) in 1840, and in the same year photographed the Moon. Henry graduated in medicine at City University, New York, and taught natural science and later physiology there from 1860 to 82. He retired in 1882 in order to devote himself full-time to astronomical research. With a 71 centimetre (28 inch) reflecting telescope, he applied the new technique of photography to astronomy and in 1872 obtained a photograph of the spectrum of the star Vega. In 1874 he directed the photographic section of the US commission that observed the transit of Venus in that year. He was the first to notice that long photographic exposures (2–3 hours) at the focus of a large telescope revealed fainter stars than could be seen with the naked eye. After his death his widow established a fund to support spectroscopic work at Harvard, the result being the *Henry Draper Catalogue* (9 vols, 1918–24) of stellar spectra.

DRAPER, John William
(1811–82)

British-born American author and scientist, born in St Helens, near Liverpool. Son of an itinerant Methodist preacher, Draper began medical studies in London in 1829. In 1831 he emigrated to Virginia. Financially aided by his sister Dorothy, he qualified in medicine in 1836, subsequently teaching chemistry at the University of New York. Between 1850 and 1873 he was head of its medical department. Draper was an early pioneer of photography. In 1840 he made what is conceivably the oldest surviving photographic portrait, a picture of his sister, which required an exposure of 65 seconds. In the same year, he photographed the Moon and thereby initiated astronomical photography. In 1850 he made the first microphotographs, which served as illustrations of a volume called *Physiology*. He also made early ultraviolet and infrared photographs of the Sun, describing Fraunhofer lines in the solar spectrum. In 1841 he formulated Draper's law: the principle that only absorbed radiation can produce chemical change. He also established that all solids become incandescent at an identical temperature, and if heated enough, afford a continuous spectrum. Intellectually, he fell under the influence of Auguste Comte and his positivism, as is evident from his *A History of the Intellectual Development of Europe* (1863). A prolific author, today he is best remembered for his impassioned polemical studies of the history of science, especially *History of the Conflict Between Religion and Science* (1874) which claimed to show how scientific progress had been impeded by religious, especially Roman Catholic, bigotry.

DREYER, John Louis Emil
(1852–1926)

Danish astronomer, born in Copenhagen, who spent his life working in Ireland. In 1874 he was appointed as an assistant at the observatory of William Parson (Third Earl of Rosse) at Birr Castle, Ireland. Four years later he moved to Dunsink, and in 1882 he became Director of the Armagh Observatory, retiring in 1916. Dreyer is remembered for his catalogue of nebulae and star clusters. The *New General Catalogue* (a revised, corrected and enlarged version of a catalogue first published by Sir **John Herschel**) was published in 1888, with supplements in 1895 and 1908. Many nebulae are still referred to by their NGC number. Dreyer also made two great literary contributions to astronomy, these being his biography of **Tycho Brahe** (1890) and his *History of Planetary Systems from Thales to Kepler* (1906). The latter was the first authoritative analysis of the history of astronomy. Dreyer was awarded the Gold Medal of the Royal Astronomical Society in 1916.

DRIESCH, Hans Adolf Eduard

(1867–1941)

German physiologist and philosopher, born in Bad Kreuznach, the son of a wealthy merchant. From 1886 to 1889 he studied zoology at the universities of Freiburg and Jena. He spent the following 10 years travelling and working as an amateur scientist, but during this time conducted his most important experiments, mainly at the Zoological Station in Naples. He is best known for an experiment which was carried out in 1891 involving the separation of the two cells produced by the first division of a fertilized sea-urchin egg. He showed that each cell could form a whole larva. This provided evidence opposed to the theory of preformation — the idea that in the fertilized egg a whole individual already exists in miniature — and helped to open new ways forward for experimental embryology. At this time Driesch favoured mechanistic explanations of development, but by 1895 he was a convinced vitalist. He began to concentrate more on philosophy than science and performed his last experiment in 1909, the year that he joined the faculty of natural science at Heidelberg. He was later appointed Professor of Philosophy at Heidelberg (1911), Cologne (1920) and Leipzig (1921).

DRUCE, George Claridge

(1850–1932)

British botanist, born in Potterspury, Northamptonshire, and educated by his guardian and two local ministers. In 1866 he was apprenticed to a Northampton chemist firm. By daily rising early and walking to his employer's house four miles away, he collected many field observations which laid the foundation of his botanical work. Awakening with a feverish cold one day in 1872, he decided to form a herbarium; by his death, this had grown to 250000 specimens from every British county. In 1879 he purchased a chemist's shop in Oxford. There, he helped to found the Ashmolean Museum and began his best work, a series of county Floras covering the Upper Thames Valley (*Flora of Oxfordshire*, 1886; *Flora of Berkshire*, 1897; *Flora of Buckinghamshire*, 1926; *Flora of Northamptonshire*, 1930). He also compiled *British Plant List* (1908) and wrote many papers on British botany; his final work was *The Comital Flora of the British Isles* (1932). He was extremely active in the Botanical Exchange Club and a prominent Freemason. He also served on Oxford City Council (1892–1932; mayor, 1900). A well-liked man, on his 80th birthday extra postal staff were needed at Oxford to deliver over 800 items.

DUBOIS, Marie Eugène François Thomas

(1858–1940)

Dutch palaeontologist, born in Eijsden. He studied medicine in Amsterdam and taught there from 1899. His interest in the 'missing link' between apes and man took him to Java (1887), where in the 1890's he found the fossil hominids named *Pithecanthropus erectus* (Java Man) and which he claimed to be the missing link. His view was contested and even ridiculed; when in the 1920s it eventually became widely accepted, Dubois began to insist that the fossil bones were those of a giant gibbon, a view that he maintained until death.

DU BOIS-REYMOND, Emil Heinrich

(1818–96)

German physiologist, and discoverer of neuro-electricity. His father was a Swiss teacher who had settled in Berlin — the family was French-speaking. Talent ran in the family: his brother Paul (1831–89) was a mathematician, who made contributions to the theory of functions. Du Bois-Reymond studied a wide range of subjects in Berlin for two years before he finally chose a medical training. Working under **Johannes Müller**, he graduated in 1843, and plunged into research on animal electricity and especially on electric fishes. All through his career he was closely associated with the leading German investigators of human physiology: **Schwann**, **Schleiden**, **Ludwig** and also the physicist **Helmholtz**. He succeeded Müller as Professor of Physiology in 1858, and was appointed the head of the new Physiological Institute which first opened in Berlin in 1877. Du Bois-Reymond's importance lay in his investigations of the physiology of muscles and nerves, and in his demonstrations of electricity in animals. He was successful in introducing improved techniques for measuring such effects, first investigated by **Galvani**. By 1849 he had evolved a delicate multiplier for measuring nerve currents. Thanks to his highly sensitive apparatus, he was able to detect an electric current in ordinary localized muscle tissues, notably contracting muscles. He observantly traced it to individual fibres, finding that their interior electrical potential was negative with regard to the surface. He demonstrated the existence of electrical currents in nerves, correctly arguing that it would be possible to transmit nerve impulses chemically. Du Bois-Reymond's experimental methods proved the basis for almost all future work in electrophysiology. He held trenchant views about scientific metaphysics. He denounced the vitalistic doctrines that were especially prominent amongst German scientists, and denied that nature contained mystical life-forces independent of matter.

DUBOS, René Jules
(1901–82)

French-born American bacteriologist, born in Saint-Brice. He became an American citizen in 1938 and worked at Rockefeller University in New York City from 1927. His doctoral thesis dealt with soil micro-organisms, and this remained his field of interest throughout his career. In 1939 he isolated an antibacterial substance from *Bacillus brevis* and named it tyrothricin. This was the first commercially produced antibiotic. Later, it was found to be a mixture of several polypeptides. Although Dubos's compounds were not particularly effective in themselves, they raised interest in penicillin, leading **Waksman** to isolate streptomycin and others to produce the broad-spectrum tetracyclin antibiotics of the late 1950s. In 1969 he won the Pulitzer Prize for his book *So Human An Animal*.

DU CHÂTELET, Gabrielle Émilie
See **CHÂTELET-LOMONT, Gabrielle Émilie, Marquise du**

DUCHENNE, Guillaume Benjamin Amand
(1806–75)

French physician, born in Boulogne-sur-Mer and educated at Douai and Paris. After 11 years as a general practitioner in Boulogne, he returned to Paris and devoted himself to the physiology and diseases of muscles. A pioneer of electrophysiology and electro-therapeutics, he did important work on polio-myelitis, locomotor ataxia and a common form of muscular (Duchenne's) dystrophy. He also developed a method of taking small pieces of muscle (biopsy) from patients for microscopical examination. Although he never held a formal hospital appointment, he worked at the Salpêtrière Hospital, where **Charcot** was his patron.

DUFAY, Charles François de Cisternay
(1698–1739)

French chemist, born in Paris. Descended of a family with a long military tradition, he joined the army at the age of 14. An extended trip to Rome with his father in 1721 marked the end of his military service, and after a protracted selection process he was appointed adjunct chemist at the Academy of Sciences in 1723, rising to the position of director in 1733. His early research on the properties of mercurial phosphorus showed him to be a thorough and methodical investigator of natural phenomena. He published three papers on magnetism reporting his observations of natural magnetism, the conditions that gave rise to magnetism and the effect of distance on the strength of the force between magnets. Dufay's most important work concerned electrostatics, identifying for the first time the difference between positive and negative electricity and recognizing that they exhibit repulsive as well as attractive forces. He worked in many areas including double refraction in crystals, the colouring of artificial gems, the heat of slaked lime and plane geometry, and invented a fire-pump. He was the author *Six mémoires sur l'électricité* (1733) and *Mémoires sur le baromètre luminieux* (1723).

DUHAMEL DU MONCEAU, Henri-Louis
(1700–82)

French chemist, botanist and technologist, born in Paris. A student of **Antoine Laurent de Jussieu**, he devoted himself to a wide range of problems in agriculture and industry, and was elected to the French Academy of Sciences in 1728. In 1732 he was appointed Inspector-General of the Navy, with special responsibility for the timber used in ships. He proved the distinction between what we now call potassium and sodium salts; he also showed that soda can be made from rock salt, and improved the manufacture of starch, soap and brass. He studied plant and animal diseases, and effectively promoted many innovations in agriculture, most importantly the introduction of Jethro Tull's methods into France. He died in Paris.

DUHEM, Pierre-Maurice-Marie
(1861–1916)

French physicist, historian and philosopher of science, born in Paris. His father, a commercial traveller, and his mother were devout Catholics. Duhem's profound religious faith, republican politics and uncompromising character isolated him from the liberal-dominated Parisian scientific mainstream. He achieved distinction at the École Normale Supérieure, but his doctoral thesis (introducing thermodynamic potentials in physics and chemistry) questioned the work of establishment figures and was not approved. Ultimately respected for work in hydrodynamics, elasticity, physical chemistry and chemical thermodynamics, and regarded as the creator of 'energetics' or general thermodynamics in France, Duhem spent his academic career in the provinces, at Lille (1887), Rennes (1893–4) and then Bordeaux. From 1900 he became increasingly interested in the history of science: the incomplete *magnum opus The World System* (1913–17) charted the mechanics of the medieval scholastics, whilst serving to substantiate his forthright philosophical views. *The Aim and Structure of Physical Theory* (1906) proposed a formalistic philosophy, where science should neither aspire to perfect knowledge of underlying reality, nor reconstruct nature in mechanical terms: Duhem scathed the school of British physics, reliant upon elaborate mechanical models, created by **Kelvin** and **Maxwell**, though he applauded the 'abstractive' theory of **Rankine**. For Duhem, any physical theory was a system of mathematical propositions deduced from a small number of principles, representing experimental laws and observed phenomena as simply and completely as possible. His prime example was thermodynamics. Duhem was associated with the energeticist school of **Ostwald**, and he coincided on many points with the positivists, particularly in their opposition to atomism.

DULBECCO, Renato
(1914–)

Italian virologist, born in Catanzaro. He studied medicine at Turin, and worked for the Resistance movement during World War II. In 1947 he emigrated to the USA, securing appointments at Indiana University and then at Caltech. From 1972 to 1977 he was Assistant Director of Research at the Imperial Cancer Research Fund, London. He then returned to the USA as Professor of Pathology and Medicine at the University of California (1977–81), and since 1977 he has been Research Professor at the Salk Institute, La Jolla. Dulbecco demonstrated how certain viruses can transform some cells into a cancerous state, such that those cells grow continuously, unlike normal cells. For this discovery, he was awarded the 1975 Nobel Prize for Physiology or Medicine, jointly with his former students, **Baltimore** and **Temin**. With H Ginsberg, he published *Virology* in 1980.

DULONG, Pierre Louis
(1785–1838)

French chemist and physicist, born in Rouen. In 1801 he entered the École Polytechnique, but he soon withdrew and turned to medicine. Apparently without ever obtaining formal qualifications, he practised in a poor area of Paris. Dulong then moved into chemistry and was assistant first to **Thénard** and then to **Berthollet**. Thereafter he held numerous teaching, administrative and examining posts. He was Professor of Chemistry at the Faculty of Sciences in Paris from 1820 and Professor of Physics at the École Polytechnique from 1820 to 1830. He was President of the Academy of Sciences in 1828. His earliest researches were on reversible reactions and oxalates. In 1811 he discovered the highly explosive compound nitrogen trichloride. He later worked on the oxides and oxyacids of nitrogen and phosphorus, and was an early advocate of the hydrogen theory of acids. Much of his later work concerned heat; this included his collaboration with the mathematical physicist **Petit**, which led to their rule of the constancy of the product of atomic weight and specific heat for solid elements (1819). This proved of value in the controversy surrounding the establishment of atomic weights.

DUMAS, Jean Baptiste André
(1800–84)

French chemist, born in Alais. He started his professional life as an apothecary in Alais, but in order to improve his prospects, he moved in 1816 to Geneva where he studied pharmacy, chemistry and botany. In 1823 he returned to France, where he was engaged to teach chemistry at the École Polytechnique and at the Athenaeum to adults in the evening. During the following years he acquired a number of academic appointments which he held concurrently. He also lectured occasionally at the Collège de France and gave instruction in a private chemical laboratory. In 1840 he became the editor of *Annales de chimie et de physique*. His chemical work embraced a wide range of topics, many of them of practical value. Although primarily a chemist, he made important contributions to the understanding of animal and plant physiology. However, his most important contribution to science was his attempts to classify organic compounds. He developed a simple method for the determination of vapour density, and thus relative molecular mass, but interpretation of the data was confused by the lack of distinction between an atom and a molecule. In 1834 he proposed that many organic compounds are formed by substitution of hydrogen by another element; eg chloromethane is produced when chlorine replaces one of the hydrogen atoms in methane. This led him to develop a theory of 'types' in which he proposed that the methane type is retained although chlorine replaces hydrogen and the physical properties are very different. The type theory was developed further by **Laurent** and **Gerhardt**. Dumas also had a political career. After the French revolution of 1848 he was elected to the legislative assembly. He was Minister of Agriculture (1850–1) and later became a senator. He was also Vice-President and President of the Paris Municipal Council.

DUNHAM, Sir Kingsley Charles
(1910–)

English geologist and mineralogist, born in Sturminster Newton, Dorset. He was educated at Hatfield College, Durham University. After a short spell working with the New Mexico Bureau of Mines he became a geologist with HM Geological Survey of Great Britain (1935) and chief petrographer there in 1948. From 1950 he was Professor of Geology at Durham University and subsequently he was appointed Director of the Institute of Geological Sciences (1967–75). During his career he received many awards and honours, and published extensively in mineralogy. His books and memoirs include *Geology of the Organ Mountains* (1935), *Geology of the North Pennine Orefield* (2 vols, 1948, 1985) and *Geology of Northern Skye* (1966, with F W Anderson). He was knighted in 1972.

DUNNING, John Ray
(1907–75)

American physicist, born in Shelby, Nebraska. He studied at Wesleyan University, Nebraska, and at Columbia University, New York, where he worked from 1933, becoming professor in 1950. In 1940 Dunning confirmed **Niels Bohr**'s hypothesis that the slow neutrons required to maintain nuclear chain reactions would only result in the fission of ^{235}U, the less abundant isotope of uranium. This meant that ^{235}U had to be separated from ^{238}U to obtain fissionable material for the atomic bomb project. Dunning led one the groups working on the isotope separation problem. His group developed the technique of gaseous diffusion where uranium is turned to highly volatile uranium hexafluoride (UF_6), which is then passed through a diffusion filter. The lighter ^{235}U diffuses through the filter faster than the heavier ^{238}U, leading to an increase in the ^{235}U concentration. Although a technically difficult process, it provided most of the ^{235}U used in the atomic bombs dropped on Japan.

DUTROCHET, (René Joachim) Henri
(1776–1847)

French physiologist, born in Néon. Born in prosperous circumstances, his childhood was afflicted by a club foot; after medical men failed to cure it, it was finally completely corrected by a local healer. He qualified in medicine at Paris, becoming an army medical officer physician and serving as personal physician to Joseph Bonaparte of Spain. After suffering typhoid in the Peninsular War, he retired from active service, spending the rest of his life conducting researches in animal and particularly plant physiology. Dutrochet believed that life processes were to be explained exclusively in physicochemical terms. He postulated that cellular respiration was essentially identical in animals and plants. In 1832 he isolated stomata, the small openings on the surface of leaves, which were later to be understood as the entry points for atmospheric exchange in plants. Building upon **Ingen-Housz**'s recognition that plants assimilate carbon dioxide and give off oxygen, Dutrochet grasped that the agent for this was the green pigment, chlorophyll. He pioneered studies of the production of heat during plant growth and was also the first to make wide-ranging studies of osmosis. This, he recognized, was the cause of sap movement in plants.

DUVE, Christian René de
See **DE DUVE, Christian René**

DYSON, Freeman John
(1923–)

English–American theoretical physicist, born in Crowthorne. He was educated in mathematics at the University of Cambridge, and worked in bomber command during World War II. In 1947 he went to Cornell where he worked with **Feynman** and **Bethe** on quantum electrodynamics, the application of quantum theory to interactions between electromagnetic radiation and particles. He was appointed professor there in 1951 and then moved to the Institute for Advanced Studies at Princeton in 1953. He became a US citizen in 1951. While at Cornell, Dyson showed that the techniques used by Feynman, **Schwinger** and **Tomonaga** in quantum electrodynamics were mathematically equivalent. His other work in physics has included studies of ferromagnetism, statistical mechanics and phase transitions. He was also involved in the design of the Orion spacecraft and the Triga nuclear reactor. He was elected FRS in 1952, and has been awarded the Royal Society's Hughes Medal and the Max Planck Medal of the German Physical Society.

E

ECCLES, Sir John Carew
(1903–)

Australian neurophysiologist, born in Melbourne. He studied medicine there, before undertaking postgraduate studies, on a Rhodes Scholarship, in **Sherrington**'s department of physiology in Oxford in 1925. He stayed in Oxford until 1937, achieving a permanent position as a university demonstrator and also as Tutorial Fellow at Magdalen College in 1934. While in Britain, he collaborated with Sherrington in papers on neural inhibition, and proposed that the process of neurotransmission at synaptic junctions in the nervous system was an electrical phenomenon, rather than chemical mechanism then postulated by **Loewi** and **Dale**. He returned to Australia as Director of the Kanematsu Institute of Pathology at Sydney (1937–44), moved to New Zealand as Professor of Physiology at Otago University (1944–51) and then to the Australian National University at Canberra (1952–68). On reaching compulsory retirement age in 1968 he moved to the State University of New York at Buffalo, from which he retired in 1975. Amongst many contributions, he established the relationship between inhibition of nerve cells and repolarization of a cell's membrane, much of which was related to, and dependent upon, the findings of Sir **Alan Hodgkin** and Sir **Andrew Huxley**, with whom he shared the 1963 Nobel Prize for Physiology or Medicine for discoveries concerning the functioning of nervous impulses. Eccles has made several additional significant contributions to the neurosciences. He recorded the depolarization of a post-synaptic muscle fibre in response to a neural stimulus, which he termed the EPSP (excitatory post-synaptic potential); and most notably he identified inhibitory neurons and demonstrated the role of inhibitory synapses in controlling and regulating the flow of information within the nervous system. His experiments in electrophysiology were providing increasing evidence that the work of Loewi and Dale and their associates on chemical neurotransmission was substantially correct, and that his own hypothesis of electrical transmission was flawed. These coincided with him meeting in 1944 the philosopher Karl Popper who emphasized that science was deductive, not inductive, and that a failed scientific hypothesis was successful scientifically in that it indicated that the truth lay elsewhere. Eccles has emulated the example of Sherrington with widespread interests in the arts and philosophy. He has written several books on neurophysiology, including *The Neurophysiological Basis of Mind* (1953), and has collaborated with Popper on *The Self and its Brain*, a particularly powerful assessment of neurobiology, consciousness, and the philosophy of self. He was elected FRS in 1941 and knighted in 1958.

ECKERT, John Presper
(1919–)

American engineer and inventor, born in Philadelphia. He graduated in electronic engineering at the University of Pennsylvania where he remained for a further five years as a research associate. From 1942 to 1946, with **Mauchly**, he worked on the Electronic Numerical Integrator and Computer (ENIAC), one of the first modern computers. It weighed 30 tons and contained 18000 vacuum tubes, but it led directly to Eckert's and Mauchly's EDVAC (Electronic Discrete Variable Computer), a more advanced, path-breaking stored-progam machine. Convinced that such computers had commercial potential, they founded their own company in 1946 to market their next machine, UNIVAC (Universal Automatic Computer). Though their commercial venture was not successful and was sold in 1950 to Remington Rand, UNIVAC marked the beginning of the modern US data-processing industry: in 1951 it became one of the first computers to be sold commercially. Eckert himself has been granted more than 85 patents for his electronic inventions.

EDDINGTON, Sir Arthur Stanley
(1882–1944)

English astronomer and founder of modern astrophysics, born in Kendal in Westmoreland. In 1898 he entered Owens College, Manchester, where he graduated in physics. From Manchester he went to Trinity College, Cambridge, where he graduated as Senior Wrangler in the Mathematical Tripos (1904). In 1907 he was Smith's Prizeman and was elected a Fellow of Trinity College; however, he left Cambridge on being appointed Chief Assistant at the Royal Observatory Greenwich in 1906. In 1913 he was appointed Plumian Professor of Astronomy and Experimental Philosophy at Cambridge in succession to Sir George Darwin, and in the following year became also Director of the university observatory which was to be his home for the rest of his life. In the same year appeared his first book, *Stellar Movements and the Structure of the Universe*, which dealt with the kinematics and dynamics of stars in the Milky Way. In 1916 Eddington's interest shifted to the problem of the physical constitution of the stars. He showed that stars are gaseous throughout and that the state of their equilibrium is conditioned by the effects of radiation pressure as well as gas pressure in their interiors. Assuming the perfect gas law he deduced a theoretical relationship between the mass of a star and its total output of radiation. He justified that assumption by pointing out that high temperatures in the interiors of stars would have the effect of stripping away part of the electron shells from the atoms, reducing them in size and allowing them to be closer together without upsetting the assumptions of the gas laws. Eddington suggested at the same time that extreme values of density could well exist in stars like white dwarfs and that this effect had in fact been revealed in the spectrum of the companion of Sirius. This and many other investigations such as the pulsation theory of Cepheid variable stars were published in his masterpiece *Internal Constitution of the Stars* (1926). Concurrently with these researches, Eddington had become deeply interested in **Einstein**'s theory of relativity. He has been described as 'the apostle of relativity' in Britain, being the first British scientist to appreciate the importance of Einstein's work. He led a British expedition to the island of Principe, West Africa, on the occasion of the total solar eclipse of 29 May 1919, when he verified one of the predictions of Einstein's theory, the deflection of starlight at the edge of the Sun. In 1920 Eddington published a non-mathematical account of the theory of relativity, *Space, Time and Gravitation* which he extended to his *Mathematical Theory of Relativity* (1923). He wrote a series of scientific books for the layman; *The Nature of the Physical World* (1928) indicates his conviction that the true foundation of physics lies in the theory of knowledge, and that fundamental constants of nature can be derived from pure theory without recourse to observation or experiment. Eddington's *Fundamental Theory* expounding this topic was posthumously published in 1946. In 1947 the Royal Astronomical Society instituted the Eddington Medal to be awarded for outstanding work on theoretical astronomy; the first recipient was the cosmologist **Lemaître** in 1953. Eddington was awarded the Gold Medal of the Royal Astronomical Society in 1924, was knighted in 1930 and received the Order of Merit in 1938. He was President of the Royal Astronomical Society (1921–3) and at the time of his death was President of the International Astronomical Union.

EDELMAN, Gerald Maurice
(1929–   )

American biochemist, born in New York City. He was educated at the University of Pennsylvania and at Rockefeller University, where he has spent his entire research career, which embraces a considerable diversity of interest. He became Professor of Biochemistry in 1966. Before monoclonal antibodies became available he exploited techniques pioneered by **Anfinsen** to separate the light and heavy peptide subunits of an antibody, purified the light chain from a human multiple myeloma, and reported the complete amino acid sequence (1969). He also analysed the repeat nature of the antibody structure, postulating its three-dimensional relationships, and the nature of the subunit interactions, subsequently investigating the number of antibody forms (chain classes) in different vertebrates. This work, together with **Rodney Porter**'s studies in England, enabled a picture of a typical Y-shaped human immunoglobulin (IgG) antibody molecule to be established. For these discoveries they shared the 1972 Nobel Prize for Physiology or Medicine. Around this time Edelman published on the structure of beta microglobulin and isolated the first DNA ligase (1968). His DNA interest continued in the 1970s with the development of a system for studying yeast plasmid DNA replication *in vitro*. He has also worked on nerve growth factor and the molecular embryology of cell adhesion.

EDISON, Thomas Alva
(1847–1931)

American inventor and physicist, born in Milan, Ohio, the most prolific inventor the world has ever seen. Expelled from school for being retarded, he became a railroad newsboy on the Grand Trunk Railway, and soon printed and published his own newspaper on the train, the *Grand Trunk Herald*. During the Civil War (1861–5) he worked as a telegraph operator in various cities, and invented an electric vote-recording machine. In 1871 he invented the paper ticker-tape automatic repeater for stock exchange prices, which he then sold in order to establish an industrial research laboratory at Newark, New Jersey, which moved in 1876 to Menlo

Park and finally to West Orange, New Jersey, in 1887. He was now able to give full scope to his astonishing inventive genius. He took out more than 1 000 patents in all, including the gramophone (1877), the incandescent light bulb (1879), and the carbon granule microphone as an improvement for **Alexander Graham Bell**'s telephone. Amongst his other inventions were a megaphone, the electric valve (1883), the kinetoscope (1891), a storage battery, and benzol plants. In 1912 he produced the first talking motion pictures. He also discovered thermionic emission, formerly called the 'Edison effect'. During World War I he worked on a variety of military devices such as periscopes, flame throwers and torpedoes.

EDWARDS, Robert Geoffrey

(1925–)

British physiologist who pioneered the basic experiments in reproductive physiology that enabled *in vitro* fertilization ('test-tube babies') to be developed, in collaboration with the obstetrician **Steptoe**. After military service in World War II he was educated at the universities of Wales (1948–51) and Edinburgh (1951–7) and following a one-year research fellowship at Caltech he became a member of staff at the National Institute of Medical Research, Mill Hill (1958–62). He moved to the University of Glasgow (1962–3) and to Cambridge University (1963–89), where he became Ford Foundation Reader in Physiology (1969–85) and Professor of Human Reproduction (1985–9). His experimental researches focused on the mechanisms of human fertility and infertility, and the process of conception. In collaboration with Steptoe, whom he met in 1968, his specific scientific expertise contributed substantially to the successful development of the *in vitro* fertilization programme. Edwards was able to analyse and then recreate the conditions necessary for the egg and sperm to survive outside the womb, by achieving the rapid transfer of the oocyte to an optimally developed culture medium. He discovered the factors that would facilitate the ripening of immature eggs, and he provided the appropriate artificial conditions to facilitate successful fertilization and subsequent maturation of the embryo to the 8–16 cell stage, before its reimplantation into the uterus. In 1971 Edwards and Steptoe first attempted to reimplant a fertilized egg into a volunteer patient, but it was not until July 1978 that the first healthy baby was born as a result of their research. With Steptoe he established the Bourne Hallam Clinics, of which he became Scientific Director (1988–91).

EGAS MONIZ, António Caetano de Abreu Freire

(1874–1955)

Portuguese neurosurgeon and diplomat, born in Avanca. Professor of Neurology at Coimbra (from 1902) and Lisbon (1911–44), he did important work on the use of dyes in the X-ray localization of brain tumours and developed prefrontal lobotomy for the control of schizophrenia and other mental disorders. In 1949 he shared the Nobel Prize for Physiology or Medicine with **Walter Hess**. The early promise which lobotomy seemed to some to show was not substantiated and the operation, which often produces serious long-term side-effects, has fallen into disrepute. Egas Moniz also had a successful political career; he was a deputy in the Portuguese parliament (1903–17), Foreign Minister (1918), and led the Portuguese delegation to the Paris Peace Conference.

EHRENBERG, Christian Gottfried

(1795–1876)

German naturalist, born in Delitzsch, Prussian Saxony. Educated at the University of Berlin where he became professor in 1839, he travelled in Egypt, Syria, Arabia and central Asia. His works on microscopic organisms founded a new branch of science, and he discovered that phosphorescence in the sea is caused by living organisms.

EHRENFEST, Paul

(1880–1933)

Austrian physicist, born in Vienna. He studied at Göttingen and Vienna, where he completed his doctorate under **Boltzmann** in 1904. With his wife, the mathematician Tatyana Alexeyevna Afanassjewa, he collaborated on papers recasting the statistical mechanics of Boltzmann and **Gibbs**. Their classic exposition of the foundations of this subject appeared in 1911. The following year **Lorentz** resigned the Chair of Theoretical Physics at Leiden, and on his recommendation, Ehrenfest was appointed. Nicknamed 'Uncle Socrates', he was an inspired teacher. **Fermi** was one of his many students; **Einstein** and **Niels Bohr** were amongst his close friends. Ehrenfest quickly appreciated the significance of **Planck**'s energy quanta, and demonstrated the comprehensibility of the early quantum-mechanical theory and its relation to classical physics. His 'adiabatic principle' assumed a foundational status within quantum theory, showing that a logical interpretation was possible which maintained connections with statistical mechanics, a vital achievement during quantum theory's uncertain infancy. Plagued by feelings of personal and intellectual inadequacy, he committed suicide.

EHRLICH, Paul
(1854–1915)

German bacteriologist, born into a Jewish family in Strehlen, Silesia (now Strzelin, Poland). He entered the University of Breslau in 1872, and a year later transferred to the University of Strassburg, completing his medical degree at the University of Leipzig in 1878. He developed new dyes with specific affinities for different cell types, including the first staining procedure for the tubercle bacillus which made use of its acid-fast characteristics. He was appointed Head Physician in Friedrich von Frerichs's clinic at the Charité Hospital in Berlin. As a result of his experiments, he contracted tuberculosis, and spent two years with his family in Egypt to recover. Following his recovery, he went to work under **Koch**. He studied tuberculosis and cholera, and played a prominent role in the introduction of a diphtheria antitoxin. A pioneer in haematology and chemotherapy, Ehrlich's unique contribution was to conceptualize the interactions between cells, antibodies and antigens as essentially chemical responses. He recognized the need to look systematically for chemicals which attack and destroy disease-causing micro-organisms, without harming human cells. In an intensive search for such 'magic bullets', he tested hundreds of compounds to try to find one which would kill the trypanosomes which cause diseases such as sleeping sickness, discovering one partially effective compound, trypan red. He was joint winner, with **Metchnikoff**, of the 1908 Nobel Prize for Physiology or Medicine for his contributions to immunity and serum therapy. However, his greatest discovery was yet to come—one of his arsenic compounds (number 606) was later found to be extremely effective against spirochetes, the micro-organism which causes syphilis, at that time a much-feared disease; in 1910 he announced the discovery of 'salvarsan', the complete cure.

EIGEN, Manfred
(1927–)

German physical chemist, born in Bochum. He was educated at the University of Göttingen, gaining a doctorate in 1951. After a brief period as a research assistant in the university, he became an assistant (1953) and then Research Fellow (1958) at the Max Planck Institute for Physical Chemistry in Göttingen. In 1962 he was appointed head of a separate department of biochemical kinetics in the institute, and since 1964 he has been Director of the Max Planck Institute for Biophysical Chemistry, also in Göttingen. Eigen's many awards and honours include the Nobel Prize for Chemistry (1967) and foreign membership of the Royal Society (1973). Much of his early research work was in the study of extremely rapid reactions in solution, reactions which are so fast that they had been regarded as effectively instantaneous. To this end, Eigen developed two techniques: one is the so-called temperature-jump method and the other involves the use of ultrasonic waves. It was for their respective researches on very rapid reactions that Eigen, **Norrish** and **George Porter** shared the Nobel Prize. Eigen's interest in rapid reactions became increasingly concerned with enzymic and other chemical processes of living organisms. In recent years his work has extended to physico-chemical aspects of evolution and genetics.

EIJKMAN, Christiaan
(1858–1930)

Dutch physician and pathologist, born in Nijkerk. He investigated beri-beri in the Dutch East Indies (now Indonesia), and was the first to produce a dietary deficiency disease experimentally (in chickens) and to propose the concept of 'essential food factors', later called vitamins. He showed that the substance (now known as vitamin A) which protects against beri-beri is contained in the husks of grains of rice, which are removed when the rice is polished. He carried out clinical studies on prisoners in Java and showed that unpolished rice could cure the disease. After his return from the Dutch East Indies, he became Professor of Public Health and Forensic Medicine at the University of Utrecht in 1898. He shared the 1929 Nobel Prize for Physiology or Medicine with Sir **Frederick Gowland Hopkins**.

EINSTEIN, Albert
(1879–1955)

German–Swiss–American theoretical physicist, who ranks with **Galileo** and **Isaac Newton** as one of the great conceptual revisors of our understanding of the universe. Born in Ulm, Bavaria, of Jewish parents, he was educated at Munich, Aarau and Zürich. He took Swiss nationality in 1901, was appointed examiner at the Swiss Patent Office (1902–5), and began to publish original papers on the theoretical aspects of problems in physics. His first important paper was on Brownian motion, and the second concerned the photoelectric effect. In 1902 **Lenard** had observed that electrons could be ejected from the surfaces of certain metals when illuminated by ultraviolet light, that the number of electrons emitted would be proportional to the intensity of the light, and that the effect only occurred using light above a certain critical frequency. This was difficult to explain within the bounds of classical physics, but Einstein was able to account for it by introducing **Planck**'s idea that light is composed of photons, quanta or packets of energy, each of energy $h\nu$, where h is Planck's constant and ν is the frequency of light. For this work, one of the results that heralded the development of quantum theory, he was awarded the 1921 Nobel Prize for Physics. Einstein went on to achieve world fame through his special and general theories of relativity (1905 and 1916). The special theory, describing relative motion at constant velocities, provided a new system of mechanics which accommodated **Maxwell**'s electromagnetic field theory, as well as the hitherto inexplicable results of the **Michelson–Morley** experiments on the speed of light. Einstein showed that in the case of relative motion involving velocities approaching the speed of light, puzzling phenomena such as decreased size and mass, and changes in the pace of time can be observed. One further conclusion of the theory was that for a given mass (m) there is an equivalent energy (E) given by $E = mc^2$, where c is the speed of light. His general theory incorporated the effects of gravity and acceleration, and introduced the result that gravitational fields can bend the path of light itself and change its frequency. The theory accounted for the slow rotation of the elliptical path of the planet Mercury, which Newtonian gravitational theory had failed to do. In 1909 a special professorship was created for Einstein at Zürich, in 1911 he became professor at Prague, in 1912 he returned to Zürich and from 1914 to 1933 he was Director of the Kaiser Wilhelm Physical Institute in Berlin. By 1930 his best work was complete. After Hitler's rise to power he left Germany and from 1934 lectured at Princeton, USA, becoming an American citizen and professor at Princeton in 1940. In September 1939 he wrote to President Roosevelt warning him of the possibility that Germany would try to make an atomic bomb, thus helping to initiate the Allied attempt (the Manhattan Project). After World War II he urged international control of atomic weapons and protested against the proceedings of the un-American Activities Senate Subcommittee which had arraigned many scientists. He spent the rest of his life trying, by means of his unified field theory (1950), to establish a merger between electromagnetic and gravitational forces under one set of determinate laws; his attempt was not successful. In 1952 he was invited to become the second President of Israel, but declined. His works included *About Zionism* (1930) and *Why War* (1933, with Sigmund Freud).

EINTHOVEN, Willem
(1860–1927)

Dutch physiologist, born in Semarang, Dutch East Indies (now Indonesia), the son of a physician. Einthoven became a medical student at Utrecht in 1879, received a PhD in medicine in 1885 and the following year became Professor of Physiology at Leiden. He developed the string (or Einthoven) galvanometer, a sensitive current-measuring device and other apparatus for the recording of electrocardiograms. He made electrocardiograms of many patients with heart disease which he compared with records of heart sounds and murmurs. This was done via a cable 1.5 kilometres long connecting his laboratory to the university hospital. He was awarded the 1924 Nobel Prize for Physiology or Medicine.

EKEBERG, Anders Gustaf
(1767–1813)

Swedish chemist and mineralogist, born in Stockholm. He was educated at Uppsala and became an assistant professor there in 1794 and full professor in 1799, the same year that he was elected a member of the Royal Swedish Academy of Sciences. In the 1790s he braved the hostility of his superiors to introduce the theories of **Lavoisier** into Sweden. Around 1795 he began investigating yttria, a newly discovered heavy metal from the quarry at Ytterby, Sweden, and in 1802 he found that it contained another hitherto unknown heavy metal. This he called 'tantalum', a reference to the tantalizing work of coaxing its oxide to react with an acid. One of his greatest distinctions was his role as the teacher of **Berzelius**. Partially deaf from childhood, Ekeberg was blinded in one eye in 1801 by an exploding flask in the laboratory and increasingly suffered from infirmity. He died in Uppsala.

EKMAN, Vagn Walfrid
(1874–1954)

Swedish oceanographer, born in Stockholm, the son of an oceanographer. Educated at Uppsala University, he worked at the International Laboratory for Oceanographic Research in Oslo (1902–8) before returning to Sweden, where he was appointed Professor of Mathematical Physics at Lund (1910–39). Observations made by **Nansen** during the *Fram* expedition had revealed that the path of drifting Arctic sea ice did not follow the prevailing wind direction, but deviated 45 degrees to the right (in the northern hemisphere), while at the centre of a current gyre there is 'dead' water. Ekman interpreted these observations by applying the mathematical theory of **Vilhelm Bjerknes**, and explained them as due to the **Coriolis** force, an effect of the Earth's rotation. He also showed that the general motion of near-surface water is the result of interaction between the surface wind force, the Coriolis force and the frictional effects between the different water layers. The resulting variation of water velocity with depth is known as the Ekman spiral.

ELHUYAR Y DE SUVISA, Don Fausto d'

(1755–1833)

Spanish chemist and metallurgist, born in Logrono. He studied medicine in Paris and mining in Freiberg, returning to Spain to teach mineralogy and geology at the Real Seminario Patriótico in Vergara. In the 1780s he and **Chabaneau** founded the Real Esuela Metalúrgica and worked together to extract platinum and make it malleable. He collaborated with his brother Juan José in experiments to isolate tungsten (then known as wolfram) from wolframite. Subsequently the government sent him to be Director of Mines in Mexico, where he succeeded in revitalizing the mining industry until interrupted by the War of Independence in 1810. In 1821 he returned to Spain as Director-General of Mines and did much to reform mining law. He died in Madrid.

ELION, Gertrude Belle

(1918–)

American biochemist, born in New York City. She graduated from Hunter College in 1937 and completed a master's degree at New York University before briefly teaching in a high school. World War II opened up laboratory jobs to women, and she joined Burroughs Wellcome in 1944 as a research associate of **Hitchings**. She progressed through the company to become Head of Experimental Therapy (1967–83), and since 1983 has been Emeritus Scientist. With Hitchings she worked extensively on drug development, and with their 'anti-metabolite' philosophy, they initially synthesized compounds that inhibited DNA synthesis, hoping that these could prevent the rapid growth of cancer cells. Their investigations of the chemistry of purines and pyrimidines, components of DNA, resulted in them jointly holding 18 pharmaceutical patents related to these two compounds, Elion concentrating primarily on pyrimidine chemistry. From their work came drugs active against leukaemia and malaria, drugs used in the treatment of gout and kidney stones, and also drugs that suppressed the normal immune reactions of the body, vital tools in transplant surgery. In the 1970s they produced an anti-viral compound, acyclovir, active against the herpes virus, which preceeded the successful development by Burroughs Wellcome of AZT, the anti-AIDS compound. In 1988 Elion and Hitchings shared, with Sir **James Black**, the Nobel Prize for Physiology or Medicine.

ELSASSER, Walter Maurice

(1904–)

German-born American physicist, born in Mannheim. He obtained his PhD from Göttingen University, spent some time at Frankfurt University then left Germany in 1933 for the Sorbonne, Paris. In 1936 he moved to the USA working first at Caltech (1936–41) and then for the US Signal Corps in war research on radar (1941–6). He subsequently held professorships at the universities of Pennsylvania, Utah, Princeton and Maryland until his retirement in 1974. He suggested in 1939 a theory to account for the Earth's magnetic field. Considering the Earth as having a core of molten iron above the Curie temperature and therefore no longer able to retain any permanent magnetism, he suggested that the Earth's rotation sets up eddy currents in the liquid core causing it to behave as an electromagnet. His theory provided an explanation of the terrestrial permanent magnetic field and the presence of secular variation. He was also interested in the relationship between the biological sciences and quantum mechanics. He is the author of *The Physical Foundation of Biology* (1958).

ELSTER, Johann Phillipp Ludwig Julius

(1854–1920)

German physicist, born in Bad Blankenburg, a collaborator of **Geitel**. They studied together first in Heidelberg and then in Berlin, and both were appointed teachers of mathematics and physics at the Herzoglich Gymnasium in Wolfenbüttel, near Brunswick. They produced the first photoelectric cell and photometer and a Tesla transformer, but refused to take out patents, believing that such inventions should benefit all. Among other achievements, they determined in 1899 the charge on raindrops from thunderclouds, showed that lead in itself is not radioactive, and that radioactive substances producing ionization cause the conductivity of the atmosphere. They were called 'the Castor and Pollux of physics'.

ELTON, Charles Sutherland
(1900–91)

English ecologist, born in Liverpool. He studied at Liverpool College and New College, Oxford, where he spent most of his career (1936–67). His four Arctic expeditions in the 1920s and his use of trappers' records for fur-bearing animals led to his classic books on animal ecology, and his talents were turned to reduction of food loss in World War II through his studies of rodent ecology. His work on animal communities led to recognition of the ability of many animals to counter environmental disadvantage by change of habitats, and to use of the concepts of 'food chain' and 'niche'. His books included *Animal Ecology* (1927), *Animal Ecology and Evolution* (1930) and *The Pattern of Animal Communities* (1966). The first two of these effectively established animal ecology as a distinct discipline, separate from earlier studies in ecology which concentrated mainly on plants. *Animal Ecology* was written at the behest of **Julian Huxley**, his Oxford mentor and organizer of the first Spitzbergen expedition; it was completed in 85 days. Evelyn Hutchinson described it as one of the great biological books of the century, providing much of the foundation for modern ecology. Elton was pre-eminently an excellent naturalist, who regarded ecology as 'scientific natural history' rather than a branch of physiology; he repeatedly emphasized the need to study the whole animal communities living in particular habitats. He exercised a formative influence on animal ecology through his founding and long editorship of the *Journal of Animal Ecology*, and was elected FRS in 1953.

ELVEHJEM, Conrad Arnold
(1901–62)

American biochemist, born in McFarland, Wisconsin. He studied and spent most of his career at Wisconsin University, ultimately becoming President from 1958 to 1962. Pellagra, a human dietary disease particularly affecting the skin and associated with a high-corn low-protein diet, was of widespread concern in the USA at that time. In 1935 Elvehjem used chicks rather than rats as experimental animals, and showed that liver extracts cured pellaga-like symptoms caused by a steroid-limited diet, but flavin, strongly proposed by other workers, did not. In 1937 he confirmed and extended his original finding by curing the related disease, black tongue in dogs. A year later, with collaborators, he identified nicotinamide (vitamin B_6) and showed that both it and nicotinic acid cured the disease, and they correctly anticipated its efficacy for curing pellagra in man. Elvehjem also showed that certain elements are essential in animal nutrition in trace levels, including copper (necessary in the formation of haemoglobin), cobalt (in vitamin B_{12}) and zinc (enzyme cofactor).

EMELEUS, Harry Julius
(1903–)

English chemist, born in Poplar, London, and educated at Imperial College, London. He then worked at the Technische Hochschule in Karlsruhe, Germany (1926–9), and at Princeton University (1929–31). He was appointed to the staff at Imperial College in 1931, and later became Professor of Inorganic Chemistry at Cambridge (1945–70). Emeleus is distinguished for his work in main group inorganic chemistry, especially the chemistry of fluorine compounds. He published with **John Anderson** the textbook *Modern Aspects of Inorganic Chemistry* (1938). Elected FRS in 1946 and awarded a CBE in 1958, he won the Lavoisier Medal of the French Chemical Society, the Stock Medal of the German Chemical Society and the Davy Medal of the Royal Society (1962).

EMERSON, Gladys Anderson
(1903–)

American biochemist, born in Caldwell, Kansas. She was educated at Oklahoma College for Women, and later received a fellowship to study nutrition and biochemistry at Berkeley. Following the identification of vitamin E by Herbert Evans, Emerson first succeeded in isolating it in a pure form. At the Merck Pharmaceutical Company in New Jersey she studied the role of vitamin B complex deficiencies in diseases such as arteriosclerosis. She also investigated the possible dietary causes of cancer. In 1956 she became Professor of Nutrition at the University of California at Los Angeles.

EMILIANI, Cesare
(1922–)

Italian marine geologist and palaeoclimatologist, born in Bologna and educated at the university there. After working in oil- and gas-related research in Italy (1946–8), he moved to the USA to become a research associate in geochemistry (1950–6) and associate professor (1957–63) at the University of Chicago. Since 1963 he has been Professor of Marine Geology at the University of Miami, where he has undertaken important studies of marine geochemistry and climatic variation.

EMPEDOCLES

(fl c.450 BC)

Greek philosopher and poet, from Acragas (now Agrigento) in Sicily, who by tradition was also a doctor, statesman and soothsayer. He attracted various colourful but apocryphal anecdotes—such as the story that he jumped into Mount Etna's crater to support his own prediction that he would one day be taken up to heaven by the gods. His philosophy reflects both the Ionian and the Eleatic traditions, but we have only fragments of his writings from two long poems: *On Nature* describes a cosmic cycle in which the basic elements of earth, air, fire and water periodically combine and separate under the influence of dynamic forces akin to what humans might call 'love' and 'hate'. This notion was taken up and developed by **Aristotle** and continued to influence chemical theories for more than 2000 years. *Purifications* has a Pythagorean strain and describes the fall of man, and the transmigration and redemption of souls. Empedocles was noted for his keen observation and was the first to demonstrate that air has weight. He was also aware of the possibility of an evolutionary process, and believed that some creatures less well adapted to life on Earth had perished in the past.

ENCKE, Johann Franz

(1791–1865)

German astronomer, born in Hamburg and educated at the University of Göttingen where he studied mathematics under **Gauss**. As a young man he fought in the German war of liberation against Napoleon (1813–15). In 1816 he became assistant to Bernhard von Lindenau at the Seeberg Observatory near Gotha; he succeeded him as director in 1822. Later he moved to Berlin (1825), where he superintended the building of a new observatory which had been promoted by **Humboldt**, and became Professor of Astronomy at the university. He remained in Berlin for the rest of his life. His principal work was concerned with facilitating computations of the movements of comets and asteroids, and included a method of calculating the gravitational influences of the planets on the motion of comets. On investigating the orbit of a comet discovered by **Pons** in Marseilles in 1818, he demonstrated that the same comet had been observed on previous returns by, among others, **Caroline Herschel**, and deduced that it moved around the Sun in an elliptic orbit with a period of only 3¼ years. Encke's comet, as it is called, has the shortest known period of any comet. Encke was twice awarded the Gold Medal of the Royal Astronomical Society (1823, 1830).

ENDERS, John Franklin

(1897–1985)

American bacteriologist, born in West Hartford, Connecticut. He studied literature at Harvard, but his interests switched to science, and he received a PhD in bacteriology. He researched antibodies for the mumps virus, and in 1946 founded a laboratory for poliomyelitis research in Boston. He shared with **Robbins** and **Weller** the 1954 Nobel Prize for Physiology or Medicine for the cultivation of polio viruses in human tissue cells, thus greatly advancing virology and making possible the development of a polio vaccine by **Salk**. In 1962 Enders developed an effective vaccine against measles.

ENDLICHER, Stephan Ladislaus

(1804–49)

Austrian botanist, born in Pressburg (now Bratislava, Czechoslovakia). In 1823 he entered the archiepiscopal seminary in Vienna, but he left in 1826 though he continued to live in Vienna. After working at the Hofbibliothek, he became keeper of the botany department at the Naturaliencabinet in 1836, and Professor of Botany at the University of Vienna and director of the botanic garden in 1840. Endlicher formulated a system of plant classification, published as *Genera Plantarum* (1836–41, 5 supplements 1840–50). His other botanical works include *Flora Posoniensis*, a Flora of the environs of Poznań, Poland (1830); *Prodromus Florae Norfolkiae* (1833, a catalogue of the plants collected by **Ferdinand Bauer** on Norfolk Island); *Sertum Cabulicum* (1836, with Eduard Fenzl, dealing with plants collected in Afghanistan and Pakistan by Johann Martin Honigberger), *Enumeratio Plantarum* (1837, with Fenzl, **Bentham** and Heinrich Wilhelm Schott), *Iconographia Generum Plantarum* (1839–41) and *Synopsis Coniferarum* (1847). Endlicher was also a noted sinologist; he published a book on the elements of Chinese grammar and presented a fount of Chinese types to the Austrian national printers. He lived in straitened circumstances, as he lectured only in summer and so received half pay.

ENGLER, (Heinrich Gustav) Adolf

(1844–1930)

German systematic botanist, leader of the Berlin school of plant taxonomy and phytogeography. Born in Sagan, Lower Silesia (now Zagan, Poland), he was educated at Breslau where his first post was schoolmaster at St Maria Magdalene Gymnasium (1866–71). Spells as keeper of the Royal Herbarium, Munich (1871–8) and Professor of Botany and Director of the Botanic Garden, Kiel (1878–84) followed. At Kiel, he wrote *Entwicklungsgeschichte der Pflanzenwelt* (1879–82) and founded the journal *Botanische Jahrbücher für Systematik, Pflanzengeschichte und Pflanzengeographie* in 1881; this journal is still published today. During 1884–9 he was Professor of Botany at Breslau; with **Prandtl**, he began *Die natürliche Pflanzenfamilien* (1887–1915). In 1889 he became Professor of Botany at Berlin. Here at last he was director of a garden worthy of his genius. He developed the present botanic garden in the Dahlem district of Berlin, and began *Das Pflanzenreich* (107 vols, 1900–53, including his monograph of *Saxifraga*, 1916–19). In 1892 he published *Syllabus der Pflanzenfamilien*, his system of classification; for many years this was the most widely used. In later life he travelled extensively, circumnavigating the globe when nearly 70. He also edited (with Carl Georg Oscar Drude) the series on phytogeography *Die Vegetation der Erde* (15 vols, 1896–1923).

EÖTVÖS, Roland von, Baron

(1848–1919)

Hungarian physicist, born in Pest (now Budapest), son of the writer and statesman Josef von Eötvös. Educated at the universities of Königsberg and Heidelberg, where he received his PhD in 1870, he was appointed to a professorship at Budapest University in 1872. In his early research on liquids he devised the 'Eötvös law', approximately describing the relationship between surface tension, molar volume and temperature for liquids. Subsequently he concentrated on research into gravitation; he devised the sensitive Eötvös torsion balance in 1888 for the measurement of minute gravitational differences in land masses, and demonstrated conclusively **Galileo**'s assertion that all bodies have the same acceleration in a gravitational field. His work revealed the equivalence of gravitational and inertial mass, a vital stepping-stone in **Einstein**'s later development of the general theory of relativity. Eötvös served briefly as Minister of Public Instruction, resigning in 1895 to devote his time to physics teaching at Budapest University.

EPSTEIN, Sir (Michael) Anthony

(1921–)

English microbiologist, born in London. He was educated at Trinity College, Cambridge, and after working at Middlesex Hospital Medical School at the close of World War II, where he became a reader in experimental pathology in 1965, he moved to the University of Bristol as professor and head of the pathology department (1968). On relinquishing this position in 1985, he moved to Oxford, taking up a fellowship at Wolfson College. In 1964 he discovered a new human herpes virus, known as the Epstein–Barr virus, which causes infectious mononucleosis and has been implicated in some forms of human cancer, notably **Burkitt**'s lymphoma and nasopharyngeal carcinoma. This was the first virus to be shown to have an association with cancer in man, and its discovery was a major factor in stimulating the current vast research efforts on the viral origins of human tumors. Epstein has received many national and international awards, including the Paul Ehrlich and Ludwig Darmstaedter Prize and Medal, the Bristol–Myers Award, the Prix Griffuel and the Gairdner International Award. He was elected FRS in 1979, knighted in 1991, and from 1986 to 1991 served as Foreign Secretary and Vice-President of the Royal Society.

ERASISTRATUS OF CEOS

(b 3 BC)

Greek physician. Born into a medical family on the island on Ceos (Chios), he studied medicine in Athens and then on Cos, later moving to Alexandria where he founded a school of anatomy. Erasistratus was a prolific author whose writings are now known only indirectly, largely through **Galen**'s accounts. He had an inquiring mind, advocating the conduct of experiments upon birds and small mammals. On the basis of extensive vivisection and dissection of animals, accompanied by postmortems on humans, he built up an extensive grasp of human and comparative anatomy. Physiologically, he combined a corpuscular theory with belief in the pneuma (spirit), rejecting notions of occult forces. Erasistratus held strong opinions on the digestive process. He denied **Aristotle**'s view that digestion was a process similar to cooking, and also rejected the theory likening it to fermentation. Rather he contended that food, once in the stomach, was torn to pieces and pulped by the peristaltic motion of the gastric muscles. Erasistratus devoted attention to the structure of the brain and the cardiovascular system, tracing arteries and veins to the heart. Interested in nervous organization, like **Herophilus**, he clearly recognized the difference between sensory and motor nerves. On the basis of such investigations, he is considered one of the founders of modern medicine.

ERATOSTHENES

(c.276–194 BC)

Greek mathematician, astronomer and geographer, born in Cyrene. He became the head of the great library at Alexandria, and was the most versatile scholar of his time, known as 'pentathlos' or 'all-rounder'. He measured the obliquity of the ecliptic and the circumference of the Earth with considerable accuracy. In mathematics he invented a method, the 'sieve of Eratosthenes', for listing the prime numbers less than any given number, and a mechanical method of duplicating the cube. He also wrote on geography, chronology and literary criticism, but only fragments of all this work remain.

ERCKER, Lazarus

(c.1530–1593)

Bohemian metallurgist, born in Annaberg, Saxony (now in Germany). He held a variety of posts in mines and mints, and latterly served under Emperor Rudolf II as chief superintendent of the mines of the Holy Roman Empire and Bohemia. He was knighted in 1586. Reputed to be both well informed and a careful observer, his extensive influence was due to his book *Beschreibung allerfürnemisten mineralischen Ertzt und Berckwerksarten* ('Description of Leading Ore Processing and Mining Methods', 1574). This was the first systematic account of analytical and metallurgical chemistry, describing how to test alloys and metallic compounds, and how to separate and refine the metals in them. It was translated into many languages and remained the leading manual until around the middle of the18th century. Ercker probably died in Prague.

ERDTMAN, Otto Gunnar Elias

(1897–1973)

Swedish palynologist, born in Hjorted, Småland. He received a PhD from Stockholm University. For many years he could not gain academic employment and instead worked as a high school teacher, pursuing pollen research during his leisure time. With his brother, he developed and published in 1933 the acetolysis method, still used as a standard technique in the preparation of pollen for microscopy. Towards the end of this period he published *An Introduction to Pollen Analysis* (1943). In 1944 he was appointed to a Stockholm school, at which he was soon relieved of teaching duties and allowed to set up a small palynological laboratory. Later, the Swedish Research Council financed the palynological laboratory of which he became director; now at Stockholm University, it rapidly became the foremost of its kind in the world. In 1952 he published *Pollen Morphology and Plant Taxonomy: Angiosperms*, the first of four volumes covering all spores (1952, 1957, 1965, 1971). He coined many standard palynological terms. In 1954 he founded the palynological journal *Grana Palynologica* and in 1970 initiated the *World Pollen and Spore Flora* project. Another major work, *Handbook of Palynology*, was published in 1969. He was also a good artist, a flautist and a lover of poetry.

ERLANGER, Joseph

(1874–1965)

American physiologist, born in San Francisco. Educated at the Johns Hopkins Medical School, he began to do physiological experiments whilst a student, and published his first paper before graduating. This brought him to the attention of William Howell, Professor of Physiology, who offered him an assistant professorship, and he was subsequently appointed Professor of Physiology at Wisconsin (1906–10) and then at Washington University, St Louis (1910–46). His early career was devoted to studying the heart and the circulation, in particular in analysing the conducting systems of the heart. During World War I he concentrated his attentions on different problems, including the treatment of wound shock, a problem that concerned several contemporary physiologists, including **Bayliss** and **Walter Cannon**. In 1921 he began collaborating with **Gasser** in analysing fundamental properties of the neural conduction of impulses. Their major discovery was that the velocity of the impulse was proportional to the diameter of the nerve fibre, and in 1944 they shared the Nobel Prize for Physiology or Medicine. Their development of the necessary equipment, especially the cathode-ray tube, for the prosecution of their researches, heralded a new electronic age in neurophysiology.

ERLENMEYER, Richard August Carl Emil

(1825–1909)

German chemist, born in Wehen. He entered the University of Giessen in 1845 as a medical student, but on hearing **Liebig** lecture, he decided to study chemistry, first in Giessen and later in Heidelberg. At Heidelberg he came under the influence of **Kekulé** and accepted a cyclic structure for benzene. He became professor at Munich Polytechnic School in 1868 and remained there until his retirement in 1883. His researches concerned mainly organic syntheses, and he prepared a number of important compounds, including guanidine, isobutyric acid and tyrosine. For his synthetic work he invented the conical flask which bears his name. Erlenmeyer came to a view of chemical structure similar to that of **Butlerov**. He edited *Zeitschrift für Chemie und Pharmazie* and *Annalen der Chemie*.

ERNST, Richard Robert
(1933–)

Swiss physical chemist, born in Winterthur. He was educated at the Swiss Federal Institute of Technology (ETH), Zürich, obtaining his doctorate in 1962. From 1962 to 1968 he worked as a research scientist with Varian Associates in Palo Alto, California, and then joined the staff of the ETH, becoming full professor in 1976. Since 1978 he has been a consultant to Spectrospin AG and vice-president of the board of directors. His many awards and honours include the Nobel Prize for Chemistry (1991). This was awarded for innovations in nuclear magnetic resonance (NMR) spectroscopy. NMR has been an important structural tool for determination of molecular structure in organic chemistry for some 40 years, but during the first 20 years its applications were based on a restricted range of magnetic nuclei, mainly those of hydrogen and fluorine. Since 1966 Ernst has increased enormously the sensitivity of the instrumentation, and his development of 'Fourier transform NMR' has extended the range of nuclei that may be used. These now include carbon-13, which constitutes only 1 per cent of natural carbon. A further extension which he pioneered was 'two-dimensional Fourier transform NMR', which is of particular value in probing very large molecules, such as are often involved in biology and medicine.

ESAKI, Leo
(1925–)

Japanese physicist, born in Osaka. He was educated at Tokyo University, working for his doctorate on semiconductors. In 1957, working at the Sony Corporation in Tokyo, he investigated conduction by quantum-mechanical 'tunnelling' of electrons through the potential energy barrier of a germanium p–n diode. He used the effect to construct a device with diode-like properties, the tunnel (or Esaki) diode. It was characterized by its use of heavily doped semiconductors and very narrow junction widths of only 100 angstroms. The diode demonstrated 'negative resistance'; under certain regimes of current and voltage the current decreased with increasing voltage. The properties of very fast speeds of operation, small size, low noise and low power consumption, give Esaki diodes widespread application in computers and microwave devices. He shared the Nobel Prize for Physics in 1973 for work on tunnelling effects with **Josephson** and **Giaever**. He joined IBM in the USA in 1960 and now works at their Thomas J Watson Research Center in New York, where his research interests include man-made semiconductor superlattices.

ESCHENMOSER, Albert
(1925–)

Swiss chemist, born in Erstfeld, Uri. He enrolled at the Swiss Federal Institute of Technology in Zürich in 1949 and has remained there ever since, obtaining a doctorate in 1951 for work with **Ružička**. He became *Privatdozent* in 1956. He first came to prominence when his group joined with that of **Robert Woodward** in the laboratory synthesis of vitamin B_{12}. The two halves of the molecule (one made in Zürich and the other in Boston) were successfully joined to give a few micrograms of synthetic vitamin B_{12} in 1976. Since then he has worked on prebiotic chemistry (the type of chemistry which might have occurred to give rise to molecules now characteristic of living things). He has also synthesized molecules analogous to DNA but containing non-naturally occurring sugars. He has received prizes and awards from all over the world, including the Davy Medal of the Royal Society (1978).

ESKOLA, Pentti Elias
(1883–1964)

Finnish petrologist and geochemist, born in Lellainen. He was educated at the University of Helsinki with which he subsequently had a long association as professor (1924–53), and undertook major studies on Finnish Precambrian rocks culminating in his book *The Precambrian of Finland* (1963). His principal papers were related to his studies of the mineral facies of rocks, of granites and of mantle gneiss domes. He worked on eclogites in Norway (1920), and his time spent there in association with **Victor Goldschmidt** laid the foundation for his mineral facies theory. His subsequent work with **Norman Bowen** at the Geophysical Laboratory in Washington DC (1921) led him to a number of studies in mineral phase equilibria, and convinced him that primary granite magmas exist and that they are not metasomatic in origin.

ESPY, James Pollard
(1785–1860)

American meteorologist, born in Westmoreland County, Pennsylvania. He obtained a degree at Transylvania University, Lexington, and later studied law. He taught in the Franklin Institute and became so impressed with the meteorological writings of **Dalton** and **Daniell** that he began to observe and study the weather. After a trip to Europe in 1840 when he addressed the British Association on his theory of storms, he returned to America and was appointed Chief of the Meteorological Bureau of the War Department (1842–57). He organized a service of daily synchronous weather observations and from these compiled charts for 1100 days which he analysed to deduce the main characteristics of cyclones and tornadoes. He found that there was low pressure in the centre, that air moved towards the centre and that the whole system moved with the upper wind. He also proved that in a column of warm rising air, the latent heat of condensation of water vapour might exceed the cooling due to expansion and hence the warm air might remain warmer than its environment and continue to rise. He correctly believed that this process occurred in a cyclone.

EUCLID
(fl 300 BC)

Greek mathematician. He taught in Alexandria, where he appears to have founded a mathematical school. His *Elements* of geometry, in 13 books, is the earliest substantial Greek mathematical treatise to have survived, and is probably better known than any other mathematical book, having been printed in countless editions; with modifications and simplifications it was still being used as a school textbook in the earlier part of the 20th century. It was the first mathematical book to be printed and has stood as a model of rigorous mathematical exposition for centuries, though this aspect of it has been severely criticized by **Bertrand Russell** among others. The *Elements* begins with the geometry of lines in the plane, including **Pythagoras**'s theorem, goes on to discuss circles, then ratio and incommensurable magnitudes (conjecturally the work of **Eudoxus**). Older material on numbers is then discussed, including the proof that there are infinitely many primes. Book X, the longest, deals with an analysis of certain kinds of lengths, and the geometry of three dimensions is explored, culminating in a proof that there are only five regular solids. He wrote other works on geometry, including the theory of conics, and on astronomy, optics and music, some of which are lost.

EUDOXUS OF CNIDUS
(408–353 BC)

Greek mathematician, astronomer and geographer. Thought to have been a member of **Plato**'s Academy, he spent over a year studying in Egypt with the priests at Heliopolis, and formed his own school in Cyzicus. He made many advances in geometry, and it is possible that most of Books V, VI, and XII of **Euclid**'s *Elements* is largely his work. These books show, in part, how to extend the Greek theory of ratio to incommensurable magnitudes. Eudoxus drew up a map of the stars and compiled a map of the known areas of the world. He correctly recalculated the length of the solar year, and his philosophical theories are thought to have had a great influence on **Aristotle**.

EULER, Leonhard
(1707–83)

Swiss mathematician, born in Basle, where, though destined by his father for theology, he studied mathematics under **Jean Bernoulli**. In 1727 he went to St Petersburg to join Bernoulli's sons at the Academy of Sciences newly founded by Catherine II, where he became Professor of Physics (1731) and then Professor of Mathematics (1733). In 1738 he lost the sight of one eye. In 1741 he moved to Berlin at the invitation of Frederick the Great to be Director of Mathematics and Physics in the Berlin Academy, but he returned to St Petersburg in 1766 after a disagreement with the king, and remained in Russia until his death. He was a giant figure in 18th-century mathematics, publishing over 800 different books and papers, mostly in Latin, on every aspect of pure and applied mathematics, physics and astronomy. In analysis he studied infinite series and differential equations, introduced or established many new functions, including the gamma function and elliptic integrals, and created the calculus of variations. His *Introductio in analysin infinitorum* (1748) and later treatises on differential and integral calculus and algebra remained standard textbooks for a century, and his notations such as e and $i = \sqrt{-1}$ have been used ever since. In mechanics Euler studied the motion of rigid bodies in three dimensions, the construction and control of ships, and celestial mechanics. For the princess of Anhalt-Dessau he wrote *Lettres à une princesse d'Allemagne* (1768–72) giving a non-technical outline of the main physical theories of the time. He had an amazing technical skill with complicated formulae and an almost unerring instinct for the right answer, though he was less concerned with the questions of rigour which would occupy later generations. He had a prodigious memory, which enabled him to continue mathematical work though nearly blind, and he is said to have been able to recite the whole of Virgil's *Aeneid* by heart.

EULER, Ulf Svante von
(1905–83)

Swedish pharmacologist, born in Stockholm, son of Nobel laureate **Euler-Chelpin**. After qualifying in medicine from the Karolinska Institute, he joined the Department of Pharmacology there as Research Fellow (1926–30), before travelling on a Rockefeller Fellowship to work with, amongst others,

Dale and Heymans (1930–1). Especially in Dale's laboratory, he began studying the effects of neuroactive substances in the autonomic nervous system, and with John Gaddum he discovered a new active factor, a polypeptide, which they named 'substance P'. Returning to the Karolinska, he continued research on biologically active substances and in 1935 isolated a group of lipids he called prostaglandins. After further visits to London to work with George Lindor Brown and **Hill**, he reverted to the study of neurally active chemicals and isolated and characterized the principal transmitter of the sympathetic nervous system, noradrenaline, in the early 1940s. For the next 30 years he continued to study noradrenaline and its chemical relatives in many situations, including renal function, and physiological shock and stress, which stimulated an interest in aviation medicine. Appointed Professor of Physiology at the Karolinska (1939–71), he shared the 1970 Nobel Prize for Physiology or Medicine with **Axelrod** and **Katz**.

EULER-CHELPIN, Hans Karl August Simon von
(1873–1964)

Swedish biochemist, born in Augsburg, Germany. After studying in Berlin, Göttingen and Paris, he became a lecturer in physical chemistry at Stockholm in 1900, and was appointed Professor of Chemistry and Director of the Institute for the Biochemistry of Vitamins in 1929. Almost the forgotten man of fermentation, his interests were ubiquitous and collaborators and publications prolific, oriented towards elucidating the chemistry and kinetics of eg catalase, urease and peptidases as well as zymase (yeast extract causing fermentation), and in particular, saccharase. He showed that zymase was markedly activated by vitamin A (carrot extract) and vitamin B (in milk and lemon juice), carried out partial purification of the system by alcohol fractionation and aluminium oxide adsorption, and investigated the role of cozymase I and II (NAD and NADP). He also compared the pH and temperature optima as well as effects of inhibitors, such as metal ions, iodine and analine, on the saccharases from yeast and jejunum, recognizing that they are two different enzymes. With **Harden**, Euler-Chelpin was awarded the Nobel Prize for Chemistry in 1929 for researches on enzymes and fermentation.

EUSTACHIO, Bartolommeo
(1520–74)

Italian anatomist, born in San Severino (Ancona). A physician's son, Eustachio studied medicine in Rome. After serving as personal physican to the Duke of Urbino and other notables, Eustachio taught anatomy at the Collegia della Sapienza in Rome. While sympathetic to tradition and a supporter of **Galen**, he nevertheless made considerable studies of the thoracic duct, larynx, adrenal glands, the teeth, and above all the kidneys. From 1552 he was involved in the production of a remarkable series of anatomical illustrations, but these were never published. His name is remembered for the precise account he left of the Eustachian canal (auditory tube) of the ear, and also of the Eustachian valve in the foetus. His most important work was the *Opuscula Anatomica* (1564).

EVANS, Meredith Gwynne
(1904–52)

English physical chemist, born in Atherton. After undergraduate and postgraduate work at the University of Manchester, he became an assistant lecturer in chemistry. During 1934–5 he worked with **Hugh Taylor** at Princeton, returning to a full lectureship at Manchester, to which **Michael Polanyi** had recently moved as Professor of Physical Chemistry. In 1939 Evans went to the University of Leeds as Professor of Inorganic and Physical Chemistry, but his first six years there were disrupted by wartime activities. He was elected FRS in 1947. In 1949 he succeeded Polanyi at Manchester. Evans's considerable scientific reputation, acquired during a relatively short career, was in the physical chemistry of reaction mechanisms, an interest much stimulated by his association with Polanyi. His contributions were to kinetics and thermodynamics of reactions, transition state theory, applications of quantum mechanics and the study of polymerization.

EVANS, Oliver
(1755–1819)

American inventor, born in Newport, Delaware. He left school at the age of 14 and was apprenticed to a wagon-maker. By 1777 he had invented a high-speed machine for assembling the wire toothed combs used in carding textile fibres, and three years later he joined his brothers at Wilmington, where they operated a flour mill. He made such effective improvements to the machinery and its control gear that in the end only one person was needed to oversee the whole of the continuous production line. Meanwhile he had engaged in a number of profitable business ventures, so that when he moved to Philadelphia he was able to devote his time to improving the very primitive steam engines then coming into use. Most engineers at that time followed the lead of **Watt** in rejecting the use of high-pressure steam because of the practical difficulties and the danger of explosion. Evans in the USA and Richard Trevithick in England, independently of each other, were convinced that the difficulties, if not the dangers, could be overcome, and by 1802 he had built an engine using steam at a pressure of 50 psi. For several years he tried to harness the power of steam for road vehicles without much success, although his amphibious steam dredging machine of 1804 is considered to have been the first American steam-powered road vehicle. Most of the 50 or so engines he built up to the time of his death were stationary engines, and he succeeded in raising working steam pressures to more than 200 psi. Several of his engines and boilers were destroyed by explosions, however, and shortly before his death fire destroyed his Mars Works in Philadelphia. In spite of these disasters, however, he was largely responsible for the more rapid adoption of high-pressure steam in America than in Britain during the early years of the 19th century.

EWING, (William) Maurice
(1906–74)

American marine geologist, born in Lockney, Texas, and raised on a farm. After an education at the Rice Institute, Houston, he secured a teaching post at Lehigh University, Pennsylvania. During World War II he took leave of absence from Lehigh to become a research associate at the Woods Hole Oceanographic Institution, where he produced a manual for the US navy on sound transmission in sea-water. In 1944 he joined the geology department at Columbia University and established the Lamont Geological Observatory, becoming director there in 1949. He resigned on a matter of principle in 1972, but was able to continue his work at the University of Texas. Ewing pioneered marine seismic techniques which he used to show that the ocean crust is much thinner (5–8 kilometres thick) than the continental crust (around 40 kilometres thick), and discovered the global extent of mid-ocean ridges. He also discovered the deep central rift in the Mid-Atlantic Ridge (1957). His work showed that the ocean sediment thickness increases with distance from the mid-ocean ridges, supporting the sea-floor spreading hypothesis proposed by **Harry Hess**, a key component of plate tectonic theory. He also published papers on seismology, the effects of nuclear explosions, heat flow, petrology and palaeontology. A highly honoured, prolific author for nearly 50 years, he produced 368 publications not including classified wartime reports.

EWINS, Arthur James
(1882–1957)

English chemist, born in Norwood, London, and educated at Chelsea Polytechnic. In 1914, working with **Dale**, he isolated the neurotransmitter acetylcholine. Moving to the pharmaceutical firm of May and Baker he conducted the researches which resulted in the preparation of sulphapyridine (M&B 693) which opened a new era of sulphonamide therapy, effective against a series of previously intractable diseases (puerperal sepsis, erysipelas, pneumonia, mastitis and meningitis) as well as providing the first successful treatment of gonorrhea—a perennial social problem. Related drugs were produced in the USA (sulfadiazine, sulfasoxazole), and in 1941, 1700 tons of sulphonamides were administered to 10 to 15 million people with a drug-related death rate of only one in 1600. The finding that sulphonamides lowered the blood sugar level led to the discovery of tolbutamide and carbutamide for the control of blood sugar levels in diabetes. Other sulphonamides inhibit the enzyme carbonic anhydrase, and are used in kidney disease to increase urine secretion and lower high blood pressure. Ewins was elected FRS in 1943.

EYRING, Henry
(1901–81)

Mexican-born American physical chemist, born in Colonia Juarez, Chihuahua. He trained as a mining engineer at the University of Arizona, but for his PhD he changed to chemistry at the University of California, Berkeley, where he graduated in 1927. From 1927 to 1929 he was an instructor at the University of Wisconsin and in 1929–30 he held a research fellowship at the Kaiser Wilhelm Institute for Physical Chemistry in Berlin. He returned to Berkeley for a year, but from 1931 to 1946 he was on the chemistry faculty at Princeton, advancing to full professor in 1938. In 1946 Eyring moved to the University of Utah as Professor of Chemistry and Graduate Dean. His many honours and awards included the Debye Award of the American Chemical Society (1964), and he was president of the society in 1965. Eyring became interested in chemical kinetics during his association with Farrington Daniels at Wisconsin and he was further stimulated in Berlin by **Michael Polanyi**, with whom he pioneered quantum-mechanical calculation of potential energy surfaces for simple reactions. This was a first step towards calculating the rate of a chemical reaction from the forces between and motions of the atomic nuclei and electrons constituting the molecules of reactants and products. He thus began work on transition state theory, to which he made major contributions for many years. After **Urey** isolated deuterium in 1932, Eyring worked extensively on isotope effects in kinetics. Later he was interested in bioluminescence and other biological topics, and in the theory of the structure of liquids. His several books were influential, particularly *The Theory of Rate Processes* (1941), co-authored with Samuel Glasstone and Keith Laidler.

F

FABRE, Jean Henri

(1823–1915)

French entomologist, born in St Léon, Aveyron. He taught in schools at Carpentras, Ajaccio and Avignon before retiring to Sérignan in Valcluse. He is remembered for his detailed and carefully observed accounts of insect behaviour and natural history which resulted in the *Souvenirs Entomologiques* (10 vols, 1879–1907). These dealt with the activities of insects such as scarab beetles, ant lions and parasitic wasps. From his observations of the latter he realized that many of the behaviours, such as the wasp's method of capturing and immobilizing its prey, are inherited and not learned. He also isolated alizarin, a red dye used as a biological stain, from the madder plant.

FABRICIUS, David

(1564–1617)

German astronomer and clergyman, born in Esens. He was a skilled observer and his 1602–4 observations of the positions of Mars (together with those made by **Tycho Brahe**) were used by his friend **Kepler** in his analysis of planetary orbits. He discovered the first known variable star (Mira, in the constellation Cetus) in 1596. It was of third magnitude in that year, but had faded from view by 1597; it reappeared in the early 17th century. His son Johannes (1587–?1615) was famous for his pioneering observations of sunspots, their discovery being announced publicly in June 1611 (**Galileo** observed sunspots towards the end of 1610 but made no formal announcement until May 1612). David Fabricius was pastor at Resterhaave and Osteel in East Friesland, where he was murdered by one of his parishioners.

FABRICIUS, Johann Christian

(1745–1808)

Danish entomologist, born in Tondern, Schleswig. In 1775 he became Professor of Natural History at Kiel and he simultaneously held the Chair of Economics and Finance. He was a student of **Linnaeus** and was himself one of the founders of entomological taxonomy, using the mouthparts of insects as the basis of his classification. He developed some advanced ideas concerning evolution. For example, he suggested that new species might arise through hybridization. Although pre-dating **Lamarck**, he believed that inherited change could result from environmental effects. He wrote many works including *Systema Entomologicae* (1775) and *Entomologica Systematica* (1792–8).

FABRIZIO, Girolamo, Latin **Fabricius ab Aquapendente**

(c.1533–1619)

Italian physician, born in Aquapendente, near Orvieto. Fabrizio studied medicine at Padua under the foremost anatomist, **Falloppio**, whom he succeeded in 1562. He spent his academic career at Padua, being centrally involved in the erection of the university's superb anatomical theatre (still preserved today). He rose to prominence as a physician and surgeon. His energetic skills in dissection and experimentation won him lasting renown in many spheres of physiology. He published extensively on surgery, reviewing treatments for different kinds of wounds. He also conducted a major series of embryological studies, illustrated by high-quality engravings. In 1600 he wrote a comparative study of the foetus in various animals, describing in later publications the formation of the chick in the hen's egg from the sixth day. Fabrizio drew special attention to the mechanisms for the sustenance of the foetus during its intrauterine life. His well-illustrated descriptions mark the birth of embryology as a novel department of biology. Fabrizio is mainly remembered today for his thorough studies of the valves of the veins. He believed the purpose of these valves was to decelerate the flow of blood from the heart, ensuring thereby an even distribution of blood through the body. Essentially **Galenic** in his theoretical commitments, he failed to grasp the true action of the valves; that was left to his pupil, **William Harvey** (1578–1657), who nevertheless drew heavily on his studies.

FABRY, Marie Paul Auguste Charles

(1867–1945)

French physicist, born in Paris, who graduated from the École Polytechnique in Paris, as did his brothers, Eugène, a mathematician, and Louis, an astronomer. He became professor at Marseilles (1904) and the Sorbonne (1920), and was appointed the first director of the latter's Institute of Optics. Inventor of the Fabry–Perot interferometer, he is also known for his researches into light in connection with astronomical phenomena, and with Henri Buisson he confirmed experimentally the **Doppler** effect for light in the laboratory. Previously, such measurements had been made using stellar sources. He was also interested in the popularization of science and increasing the public's scientific understanding.

FAHRENHEIT, (Gabriel) Daniel
(1686–1736)

German instrument-maker, born in Danzig (now Gdańsk, Poland). Fahrenheit was born into a merchant family. After the death of his parents in 1701 he was sent to Amsterdam, where he learned the trade of instrument-maker. From 1707 he travelled widely in Europe, but by 1717 he had settled in Amsterdam. There he produced high-quality meteorological instruments, supplying eminent Dutch scholars. He devised an accurate alcohol thermometer (1709) and a commercially successful mercury thermometer (1714). In 1708 he had visited the astronomer **Roemer** in Copenhagen and adopted what he believed to be Roemer's practice of taking thermometric fixed points as the temperatures of melting ice and of the human body. Fahrenheit eventually chose a scale with these points calibrated at 32 and 96 degrees. The zero was at the freezing point of ice and salt. He did not use the boiling point of water as a fixed point: experiments revealed a variation of its temperature with pressure, and he suggested this as a principle for the construction of barometers. In 1724 he noticed (by chance) the supercooling of water.

FAIRBAIRN, Sir William
(1789–1874)

Scottish engineer, born in Kelso. In 1804 his family moved to Newcastle upon Tyne and he was apprenticed to an engine-wright at North Shields, where he also studied mathematics and made the acquaintance of George Stephenson. Moving to London in 1811 he joined the Society of Arts, and invented a steam excavator and a sausage-making machine, neither of them much of a commercial success. By 1817, however, he had established an engineering works in Manchester making machinery for water-wheels and cotton-mills, and within a few years he had gained a reputation as one of the most capable engineers in the country. From about 1840 until the year of his death the Manchester works also built over 400 locomotives for service in Britain and overseas. In 1830 he took a lead in the building of iron boats; his works at Millwall in London (1835–49) had more than 1000 employees and turned out hundreds of vessels, but in the end had to be sold off at a considerable loss. For Stephenson's railway bridge over the Menai Strait (1850) he developed, after extensive model and full-scale testing, the rectangular wrought-iron tubes ultimately adopted; the Britannia Bridge's two main spans of 140 metres were not surpassed for the next 25 years. Fairbairn aided **Joule** and **Kelvin** from 1851 in geological investigations, and guided the experiments of a government committee (1861–5) on the use of iron for defensive purposes. He was elected FRS in 1850 and made a baronet in 1869.

FAIRCLOUGH, John Whitaker
(1930–)

English computer scientist. He graduated from Manchester University with a BSc in technology in 1954, when the institution's pioneering computer work—led by Sir **Frederic Calland Williams** and **Kilburn**—was being turned to commercial purposes by Ferranti Ltd. In 1954 he joined Ferranti and moved to the USA to develop and sell computer components. Finding this unsatisfying, in 1957 he joined IBM's Poughkeepsie laboratory to work on the Stretch project. In 1958 he returned to England to join IBM's Hursley laboratory, where he managed what was known as the SCAMP project. Though this project was later scrapped, Fairclough's work led to an important technological feature in IBM's highly successful System/360 machines: the control store, the theory of which had been outlined by **Wilkes**. Fairclough later managed the development of IBM Model 40, which was the first System/360 model, tested and first shipped to a customer in 1965. He became chairman of IBM's UK laboratories in 1983, and was Chief Scientific Adviser to the Cabinet from 1986 to 1990.

FAJANS, Kasimir
(1887–1975)

Polish-American physical chemist, born in Warsaw. He was educated at the universities of Leipzig and Heidelberg obtaining his doctorate in 1909, and later did research at Zürich and with **Rutherford** at Manchester (1910–11). From 1911 to 1917 he was on the staff of the Technische Hochschule, Karlsruhe, and from 1917 to 1935 he worked at the University of Munich, as Director of the Institute for Physical Chemistry from 1932 to 1935. The political situation in Germany led to his departure to Cambridge and then to the USA. From 1936 until his retirement in 1957 he was Professor of Chemistry at the University of Michigan, Ann Arbor. Among his numerous awards and honours was honorary membership of the Royal Institution. In 1912 he discovered the radioactive displacement law at around the same time as **Soddy** and Alexander Smith Russell. One statement of this law is that the emission of an alpha particle decreases the mass of the nucleus by four units and its charge by two units, whereas the emission of a beta particle leaves the mass of the nucleus unchanged. Later he pioneered the use of adsorption indicators in precipitation titrations. Fajans also formulated the factors that govern whether an element tends to form ionic or covalent compounds in terms of ideas which became known as Fajans's rules (1924). He also contributed extensively to thermochemistry and photochemistry. Fajans wrote several influential books, including *Radioactivity and Latest Developments in the Study of the Chemical Elements* (1919) and *Radioelements and Isotopes, Chemical Forces, and Optical Properties of Substances* (1931).

FALCONER, Hugh

(1808–65)

Scottish botanist and palaeontologist, born in Forres. After studying medicine at Aberdeen and Edinburgh, he went to India as assistant surgeon at the Bengal Medical Establishment (1830). He became superintendent at the botanic gardens at Saharanpur in 1832 and discovered many extinct Lower Pliocene fossil vertebrates in the Siwalik Hills, including the giraffe-like *Sivatherium* and the largest known tortoise, the eight-foot-long *Colossochelys*. He also conducted the first experiments on growing tea in India. Falconer later returned to Britain to recover from illness (1842) and wrote on Indian botany and palaeontology, arranged Indian fossils in the British Museum and East India House, and prepared his great work *Fauna Antiqua Sivalensis* (1846–9), although only the first nine parts of a total of 13 were issued. He returned to India in 1848 as superintendent of the botanic garden and Professor of Botany at the medical college in Calcutta. He retired from the Indian Service in 1855, returning once again to England where he undertook studies of the Pleistocene faunas of Britain, Sicily, France and Gibraltar. His *Palaeontological Memoirs and Notes* were published in 1868.

FALLOPPIO, Gabrielle

(1523–62)

Italian anatomist, born in Modena. Falloppio first intended to become a priest, but later developed interests in medicine and was taught anatomy by **Vesalius** in Padua, becoming Professor of Anatomy at Pisa (1548) and Padua (1551). He extended while correcting Vesalius's work. His discoveries included structures in the human ear and skull, explorations in the field of urology, and researches into the female genitalia. He coined the term vagina and described the clitoris. He was the first to describe the tubes leading from the ovary to the uterus. He failed to grasp the function of the Fallopian tubes; it was over two centuries later before it was recognized that ova are formed in the ovary, passing down these tubes to the uterus. Falloppio also carried out investigations on the larynx, the eye, muscular action and respiration. He was the teacher of **Fabrizio**.

FARADAY, Michael

(1791–1867)

English chemist and physicist, creator of classical field theory. Born in Newington near London, the son of a blacksmith, he was apprenticed to a bookbinder whose books sparked his interest in science. In 1813, after applying to **Humphry Davy** for a job, he was taken on as his temporary assistant, accompanying him soon after on an 18 month European tour during which he met many eminent scientists and gained an irregular but invaluable scientific education. In 1827 he succeeded Davy in the Chair of Chemistry at the Royal Institution, in the same year publishing his *Chemical Manipulation*. His early publications on physical science include papers on the condensation of gases, limits of vaporization and optical deceptions. He was the first to isolate benzene, and he synthesized the first chlorocarbons. His great life work, however, was the series of *Experimental Researches on Electricity* published over 40 years in *Philosophical Transactions* of the Royal Society, in which he described his many discoveries, including electromagnetic induction (1831), the laws of electrolysis (1833) and the rotation of polarized light by magnetism (1845). He received a pension in 1835 and in 1858 was given a house in Hampton Court by Queen Victoria. As adviser to the Trinity House in 1862 he advocated the use of electric lights in lighthouses. Greatly influential on later physics, he nevertheless had no pupils and worked with only one long-suffering assistant. He is generally considered the greatest of all experimental physicists.

FARRER, Reginald John

(1880–1920)

English botanist, plant-collector and writer, born in Clapham, Yorkshire. On account of a hare lip and speech peculiarity, he was educated at home, before going to Balliol College, Oxford. By the age of 8, he could dissect a flower. He travelled extensively in Europe, Japan, China (1914–16, with William Purdom) and Upper Burma (1919–20, with Euan Cox; he died of diphtheria at Nyitadi) in search of plants. Farrer introduced many species into cultivation, and his herbarium collection is distinguished by the expressiveness and detail of his field notes. A prolific author, he wrote five horticultural works: *My Rock-Garden* (1907), *Alpines and Bog-Plants* (1908), *In a Yorkshire Garden* (1909), *The Rock-Garden* (1912) and *The English Rock-Garden* (1919); six travel books: *The Garden of Asia* (1904), *In Old Ceylon* (1907), *Among the Hills* (1911), *The Dolomites, King Laurin's Garden* (1913), *On the Eaves of the World* (2 vols, 1917) and *Rainbow Bridge* (posthumous, 1921); two plays, of which *Vasanta the Beautiful* is the better known; and six novels which, unlike his travel writing, are largely forgotten. From his student days he had an interest in Liberal politics; in 1911 he stood unsuccessfully as Liberal candidate for Ashford, Kent.

FECHNER, Gustav Theodor
(1801–87)

German physicist, psychologist and philosopher, born in Gross-Särchen, near Halle. His father was a Lutheran preacher. He studied medicine at the University of Leipzig but never practised, instead taking up physics as a career and becoming Professor of Physics at Leipzig in 1834. He worked mainly on galvanism (the original term for current electricity), electromagnetism and colour. He subsequently became interested in the connections between physiology and psychology, explored in his *Elemente der Psychophysik* (1860), and helped to formulate the Weber–Fechner law relating stimuli to sensations. He also published *Das Büchlein vom Leben nach dem Tode* (1836) and *Vorschule der Aesthetik* (1876). His psychophysics was to later evolve into experimental psychology at the hands of Wilhelm Max Wundt and others.

FELL, Dame Honor Bridget
(1900–86)

British cell biologist. Educated at the University of Edinburgh where she received her PhD in 1924, she became Director of Strangeways Research Laboratory in Cambridge (1929–70). She was Foulerton Research Fellow of the Royal Society from 1941 to 1967, and Royal Society Research Professor from 1963 to 1967. Fell greatly advanced biochemical study through her investigations using the organ culture method; she demonstrated that excess vitamin A would destroy intercellular material in the explanted cartilage and bones of foetal mice, with the implication that such organ cultures could be widely used in studies of the physiological effects of vitamins and hormones. In later life she investigated the pathogenesis of arthritis. Elected FRS in 1952, she was made a DBE in 1963.

FELLER, William
(1906–70)

Yugoslav-born American mathematician, born in Zagreb. He studied at Zürich and Göttingen, and after teaching at Kiel, left Germany in 1933 for Stockholm. In 1939 he emigrated to the USA, holding chairs at Brown University (1939–45), Cornell (1945–50) and Princeton (1950–70). His work in probability theory introduced new rigour without losing sight of practical applicability, and his textbook *Introduction to probability theory and its applications* (1950) is an enormously influential classic. This work was unique in starting from first principles, yet it contains original research often leading to surprising results, and is packed with practical examples.

FELSENFELD, Gary
(1929–)

American molecular biologist, born in New York City. He graduated AB from Harvard University in 1951 and obtained his PhD from Caltech in 1955. During 1954–5 he was a postgraduate National Science Foundation Fellow at Oxford University, and since 1961 he has been Chief of the Physical Chemistry Section of the Molecular Biology Laboratory at the National Institute for Digestive and Kidney Diseases in Washington DC. He is a Fellow of the American Association for the Advancement of Science, and a member of the US National Academy of Sciences and the American Academy of Arts and Sciences. Felsenfeld's best-known work has been on the association of regulatory protein molecules with chromatin, involving extensive use of the technique of DNA footprinting. In this technique, regulatory protein molecules are bound to a specific gene plus flanking sequence; the DNA sequence is then sequenced and the region to which protein has bound shows up on the sequencing gel as a blank space, or footprint. Using this method he has investigated the precise binding location and the interaction between protein molecules which regulate the activity of the globin genes.

FERGUSON, James

(1710–76)

Scottish astronomer, born in Keith, Banffshire, the son of a labourer. The starry sky which he observed on clear winter's nights as a shepherd boy inspired him to study the movements of the Moon and planets, and to demonstrate these movements by means of ingenious mechanical models. He eventually was able to make a living from writing and lecturing. His books and his 'machines', many of which survive, aroused in the general public a widespread interest in astronomy. According to **Caroline Herschel**, Ferguson's books were among those which influenced the great **William Herschel** to take up the study of astronomy as his life's work.

FERMAT, Pierre de

(1601–65)

French mathematician, born in Beaumont. He studied law at Toulouse, where he became a councillor of parliament. His passion was mathematics, most of his work being communicated in letters to friends containing results without proof. His correspondence with **Pascal** marks the foundation of probability theory. He studied maximum and minimum values of functions in advance of the differential calculus, and wrote an unpublished account of the conic sections, extending **Vieta**'s notation to two variables. He is best known for his work in number theory, proofs of many of his discoveries being first published by **Leonhard Euler** a hundred years later. His 'last theorem' is the most famous unsolved problem in mathematics; it states that there are no positive integers x, y and z with $x^n + y^n = z^n$ if n is greater than 2. It is not known what proof, if any, Fermat had of this result, but he did discover a valid proof for the case $n = 4$. In optics Fermat's principle was the first description of a variational principle in physics; it states that the path taken by a ray of light between two given points is the one in which the light takes the least time compared with any other possible path.

FERMI, Enrico

(1901–54)

Italian–American physicist, born in Rome. He studied at the Scuola Normale Superiore in Pisa and later obtained a scholarship to study at Göttingen and Leiden. In Göttingen he studied under **Born** and met **Heisenberg** and **Pauli**, and in Leiden he worked with **Ehrenfest**. At Rome in 1927 he was appointed to the first Chair of Theoretical Physics in Italy. He worked on modifying the classical theory of statistical mechanics developed by **Einstein** and **Satyendra Nath Bose** to take into account the Pauli exclusion principle. This was later further modified by **Dirac** to take into account ideas of quantum mechanics, and the resulting Fermi–Dirac distribution is a fundamental part of statistical physics and can explain a wide variety of phenomena from semiconductors to neutron stars. In 1934 Fermi presented a theory to describe the beta decay of nuclei using the neutrino proposed by Pauli. This was able to explain the energy spectrum of the emitted beta particles and the lifetimes of the nuclei. With **Segrè** and a research group he also studied induced radioactivity by bombarding elements with neutrons. This led to the discovery that slow neutrons, neutrons that have been passed through an absorber such as paraffin to reduce their energy, are much more efficient than high-energy neutrons in initiating nuclear reactions. Their discovery was an important step in the development of nuclear power and weapons. Fermi was awarded the 1938 Nobel Prize for Physics. Fearing for the safety of his Jewish wife in the light of Italy's anti-Semitic legislation, he went straight from the Nobel prize presentation in Stockholm to the USA, where he became professor at Columbia University (1939). In 1942 Fermi built the world's first nuclear pile in a disused squash court and produced the first controlled chain reaction. After World War II he continued in nuclear research. In 1952 he discovered 'resonances' in the invariant mass peak when he scattered pions off protons, indicating that very short-lived fundamental particles had been produced. The element fermium was named after him.

FERNEL, Jean François

(?1497–1558)

French physician. The son of an innkeeper at Montdidier, he turned to medicine only after some years studying philosophy, mathematics and astrology. He soon became a popular medical teacher in Paris and his reputation as a physician soared when he saved the life of the Dauphin's (later Henri II) mistress. He preferred his teaching and scholarship to court life, however, and despite his essential adherence to Galenism, he was an astute observer whose many writings synthesized 16th-century medical orthodoxy. Although he had in his early life been an advocate of astrological influences on health and disease, he later renounced these views in favour of a more naturalistic causative framework. He coined the Latin words which became 'physiology' and 'pathology'. His *magnum opus*, the *Universa medicina* (1567), was edited by a disciple after his death.

FERREL, William

(1817–91)

American mathematician, born in Bedford County, Pennsylvania. He attended Marshall College, Pennsylvania, and graduated from Bethany College, West Virginia, in 1844. He taught in Missouri and Kentucky before obtaining a scientific post at the American Nautical Almanac in 1858. In 1867 he joined the US Coast Survey and in 1882 became professor in the US army's signal service. He was elected a member of the US National Academy of Sciences in 1868. Ferrel gave the first mathematical formulation of atmospheric motions on a rotating earth and applied his theory to the general circulation of both the atmosphere and the oceans (1859–60). He accepted **Espy**'s theory that the energy of cyclones is largely due to the latent heat of condensation when air ascends, and went on to show that differential heating is the initial cause of both cyclones and the general circulation. He produced suggested models of three-dimensional motion in a cyclone and also indicated that the general circulation in each hemisphere could be considered on similar lines. He attempted to analyse quantitatively the effect of horizontal temperature differences on the horizontal pressure field at different levels in the atmosphere. From this work he derived the concept of the thermal wind which gives the relationship between horizontal temperature gradient and the change of wind with height. He also dealt with tidal theory, and in 1880 derived a mechanical tide predictor.

FESSENDEN, Reginald Aubrey

(1866–1932)

Canadian-born American radio engineer and inventor, born in East Bolton, Quebec. Educated in Canada, he moved first to Bermuda, where he developed an interest in science while acting as Principal of the Whitney Institute, and then to New York (1886) where he met **Edison** and became the chief chemist in his research laboratories in New Jersey. By 1892 he had returned to academic life, first at Purdue University and then as Professor of Electrical Engineering at the University of Pittsburgh (1893–1900), where he began to pursue his major research interest in radio communication. Of his many patents (over 500), the one of most fundamental importance was his invention of amplitude modulation; on Christmas Eve, 1906, he used this to broadcast what was probably the first American radio programme from the transmitter he had built at Brant Rock, Massachusetts. Another of his discoveries was the heterodyne effect, soon developed into the superheterodyne circuit that rapidly became an integral part of the design of radio receivers. Among his other patents were the sonic depth finder, the loop-antenna radio compass and submarine signalling devices.

FEYNMAN, Richard Phillips

(1918–88)

American physicist, born New York City. He studied at MIT and Princeton, where he received his PhD in 1942. After overcoming moral doubts he went to work on the atomic bomb project at Los Alamos; there he was made head of the computing division at a time when computers were people. After World War II he went to Cornell where he was appointed professor and worked with **Bethe** on quantum electrodynamics, the application of quantum theory to interactions between electromagnetic radiation and particles. Disliking the abstract way in which quantum mechanics had been developed, he devised his own pictorial way of describing quantum processes. This was the 'path integral approach' in which a process is described as the sum over all its possible histories, and it has proved to be a very powerful theoretical tool. Using this he further developed quantum electrodynamics and introduced 'Feynman diagrams'; these provide a pictorial representation of particle interactions and are a powerful calculational tool. For his work on quantum electrodynamics he was awarded the Nobel Prize for Physics in 1965 together with **Schwinger** and **Tomonaga**. Quantum electrodynamics is the most successful physical theory ever developed, having been found to be correct to one part in a billion, and is the model on which other quantum field theories are based. After a year in Brazil, Feynman moved to Caltech. Working with **Gell-Mann**, he created the V–A model of weak interactions which can describe parity conservation violation. He also did important work on understanding how deep inelastic scattering can reveal the structure of protons and on the properties of liquid helium. In addition to being a great theoretical physicist, he was also a great communicator at all levels and one of science's most colourful characters.

FIBIGER, Johannes Andreas Grib

(1867–1928)

Danish pathologist, born in Silkeborg, the son of a physician. Fibiger graduated in medicine from the University of Copenhagen in 1890 after which he conducted bacteriological research, for which he was awarded a doctorate in 1895. In 1900 he was appointed Professor and Head of the Institute of Pathological Anatomy at Copenhagen.

Cancer research at this time was inhibited by the lack of an animal model, but Fibiger showed that it was possible to induce stomach cancer in rats by feeding them with cockroaches carrying the parasite *Spiroptera neoplastica*. After World War I he became the first European scientist to employ the technique developed by Japanese scientists of inducing skin cancer by painting rabbit's ears with coal tar. He made many studies of coal tar cancer and was awarded the 1926 Nobel Prize for Physiology or Medicine.

FIBONACCI, Leonardo
(c.1170–c.1250)

Italian mathematician, also known as 'Leonardo of Pisa'. The first outstanding mathematician of the Middle Ages, he was responsible for popularizing the modern decimal system of numerals, which originated in India. His main work *Liber abaci* ('The Book of Calculations', 1202) illustrates the virtues of the new numeric system, showing how it can be used to simplify highly complex calculations; the book also includes work on geometry, the theory of proportion and techniques for determining the roots of equations. His greatest work, the *Liber quadratorum* ('The Book of Square Numbers', 1225), contains contributions to number theory and is dedicated to his patron, the Holy Roman Emperor Frederick II. He discovered the 'Fibonacci sequence' of integers in which each number is equal to the sum of the preceding two (1, 1, 2, 3, 5, 8, ...) introducing it in terms of a breeding population of rabbits.

FICK, Adolph Eugen
(1829–1901)

German physiologist, born in Kassel. Fick studied medicine at Marburg, obtaining his doctorate in 1851 with a thesis on astigmatism. He was appointed to a chair at Zürich and later at Würzburg. He displayed talents in many departments of physiology. With **Du Bois-Reymond**, **Schwann**, **Ludwig** and the physicist **Helmholtz**, he was one of the proponents of the new physical and materialistic orientation of German biomedicine, dismissing *Naturphilosophie* as nonsense. He was especially concerned that medicine should develop quantitative methods, and was energetic in developing new measuring technology. Fick pursued extensive investigations into the physics of vision, analysing the blind spot in the eye. He researched into haemodynamics, calculating cardiac output, and attempted to measure the heat generated by muscles. A law of diffusion in liquids was named after him, when he discovered that the mass of solute diffusing through unit area per second is proportional to the concentration gradient.

FINSEN, Niels Ryberg
(1860–1904)

Danish physician and scientist, born in the Faroe Islands, the son of the Islands' governor. He graduated in medicine from the University of Copenhagen in 1891 and was appointed demonstrator in anatomy, but soon he abandoned academic medicine to pursue an interest in the therapeutic uses of light. His early investigations concerned the light-induced inflammation of the skin occurring in patients with smallpox. He showed that these problems were caused by the blue and ultraviolet parts of the spectrum, but that red and infrared rays promoted healing. He later developed a method of treating lupus vulgaris, a form of tuberculosis of the skin, with intense ultraviolet light. An institute for the study of phototherapy was formed in Copenhagen in 1896 and placed under Finsen's direction. For many years he suffered restrictive pericarditis and he died aged 44, soon after he was awarded the 1903 Nobel Prize for Physiology or Medicine.

FISCHER, Edmond Henri
(1920–)

American biochemist, born in Shanghai, China. He was educated at the universities of Geneva, Montpellier and Basle, and moved to the USA in 1953. In 1961 he became professor at the University of Washington, Seattle, where he has been Professor Emeritus since 1990. His earliest work involved studies of the enzyme phosphorylase, and with **Edwin Krebs**, he showed in 1955 that phosphorylation–dephosphorylation processes are involved in the activation of glycogen phosphorylase by adenylic acid. This fundamental mechanism regulates a wide variety of processes from muscle contraction to the expression of genes, and for this work they were jointly awarded the 1992 Nobel Prize for Physiology or Medicine.

FISCHER, Emil Hermann
(1852–1919)

German chemist, born in Euskirchen, Prussia. He was the son of a Protestant merchant and studied chemistry in Bonn. He worked with **Baeyer** in Strassburg and Munich. In 1882 he became professor in Erlangen, then in Würzburg (1882) and finally he succeeded **Hofmann** in Berlin (1892), where he remained until his death. He made important studies of the chemistry of sugars. After his discovery of phenylhydrazine in 1875, he found that it reacted with aldehydes to give phenylhydrazones. By a series of related reactions, phenylhydrazine reacts with a simple sugar to give an osazone and this permits interconversion of simple sugars. This, together with studies of optical activity, led to the elucidation of the structures of the 16 possible aldohexoses (which include glucose). Textbook diagrams of the 16 isomers are known as Fischer projections. The frequent use of phenylhydrazine impaired his health and probably shortened his life, but it was for this work that he was awarded the Nobel Prize for Chemistry in 1902. With his cousin Otto Fischer he elucidated the structure of rosaniline dyes. He also made significant discoveries concerning the structures of caffeine and related compounds. From 1899 he turned his attention to proteins and later to tanins. At the height of his powers he was considered the greatest living organic chemist, but he led a very simple and uneventful life.

FISCHER, Ernst Otto
(1918–)

German organic chemist, known for his work on organometallic compounds. He was born in Munich, the son of a professor of physics, and educated at the Munich Institute of Technology where he subsequently spent most of his career, interrupted only by his years in the German army. In 1951, and independently of Sir **Geoffrey Wilkinson** with whom he shared the Nobel Prize for Chemistry in 1973, he deduced the structure of the remarkable synthetic compound ferrocene. Concluding that its molecule consists of a sandwich of two carbon rings with an iron atom centrally placed between them, he confirmed his theory by X-ray crystal analysis. This novel and peculiar class of organometallic sandwich compounds now numbers thousands. Fischer himself synthesized compounds from arenes, olefins, carbenes and carbonyls.

FISCHER, Hans
(1881–1945)

German chemist, born in Frankfurt. He received his PhD in chemistry from Marburg University in 1904 and qualified in medicine from Munich University in 1908. Medical work in Munich was followed by chemical research in Berlin at **Emil Fischer**'s institute. From Berlin he moved to Innsbruck (1916) to become Professor of Medical Chemistry and then back to Munich as Professor of Organic Chemistry in 1921. He remained there actively pursuing research until his death. His most important researches concerned the structure of the naturally occurring pigments haemin and chlorophyll. The crowning glory of his work was the synthesis of haemin, and for this he was awarded the Nobel Prize for Chemistry in 1930.

FISHER, Sir Ronald Aylmer
(1890–1962)

English statistician and geneticist, the leading figure in biological and agricultural statistics in the first half of the 20th century, born in East Finchley, London. Educated at Harrow, he graduated in mathematics at Cambridge. In 1919 he became a statistician at the Rothamsted Agricultural Research Institute. There he developed his techniques for the design and analysis of experiments which he expounded in his classic work *Statistical Methods for Research Workers* (1925), and which have become standard in medical and biological research. He also worked on genetics and evolution, and studied the genetics of human blood groups, elucidating the Rhesus factor. He became Professor of Eugenics at University College London (1933–43) and Professor of Genetics at Cambridge (1943–57). Fisher was knighted in 1952. After his retirement, he moved to Australia.

FITCH, Val Logsdon
(1923–)

American physicist, born in Merriman, Nebraska. He was originally interested in chemistry but after being sent to Los Alamos to work on the atomic bomb his interests switched to physics. He was educated at McGill and Columbia universities, and in 1954 he moved to Princeton where he became professor in 1960. Using the Nevis cyclotron, Fitch and **Rainwater** studied muonic atoms. These are atoms where an orbital electron is replaced by its heavier relative the muon. They observed that one of the spectral lines (the K-line) had a different energy from that predicted, and interpreted this as as due to the nuclear radii being smaller than had previously been believed. This was later verified in the experiments of **Hofstadter**. In 1964 together with **Cronin** and others, Fitch observed the non-conservation of the combined symmetry of parity and charge conjugation in the weak decays of neutral kaons. For this work Fitch and Cronin shared the 1980 Nobel Prize for Physics.

FITZGERALD, George Francis
(1851–1901)

Irish physicist, born in Dublin and educated at home by the sister of the mathematician **Boole**, before going to Trinity College, Dublin, to study mathematics and experimental science. In 1881 he became Professor of Natural and Experimental Philosophy, a post which he held until his death. He was one of the first physicists to take **Maxwell**'s electromagnetic theory seriously. He made important discoveries in this field and also in electrolysis and cathode rays, but his name is associated with the 'Fitzgerald–**Lorentz** contraction' suggested to account for the negative result of the **Michelson–Morley** experiment. It was one of the steps that eventually led to **Einstein**'s theory of relativity, which provided a new physical description of the contraction of bodies in the direction of motion when moving at high speed relative to an observer. Elected FRS in 1883, he was awarded the Royal Society's Royal Medal in 1889.

FITZROY, Robert

(1805–65)

English naval officer and meteorologist, born at Ampton Hall, Suffolk. He was educated at Royal Naval College (1819–28) and as Commander of the *Beagle*, he surveyed the coasts of South America (1828–30). In 1831 he circumnavigated the globe in the *Beagle* accompanied by **Charles Darwin**, with whom he collaborated in publishing *Narrative of the Surveying Voyages of HMS* Adventure *and* Beagle (1839). He was made Governor of New Zealand (1843–5) and was elected FRS in 1851. In 1854 he was attached to the meteorological department of the Board of Trade, and he became the first Director of the Meteorological Office in 1855. He set up a network of telegraph stations for rapid collection of meteorological observations and was a pioneer in making weather charts. For these he introduced a set of symbols for wind speed and direction, pressure and temperature and he invented the term 'synoptic chart'. Using these charts he began a system of gale warnings for shipping. He went on to produce weather forecasts for the press, but these attracted considerable opposition. He wrote *The Weather Book* (1863) which contains pictures of storms remarkably like present-day satellite pictures. However, he regarded storms as merely the mechanical interference of two air currents and did not consider energy sources. He analysed the famous Royal Charter storm of 1859. He also invented the 'Fitzroy barometer', and was awarded the Gold Medal of the Royal Geographical Society.

FIZEAU, Armand Hippolyte Louis

(1819–96)

French physicist, born in Paris into a wealthy family. His father was Professor of Internal Pathology at the Paris Faculty of Medicine. After interrupting his medical studies because of poor health, he moved into the physical sciences, and was greatly influenced by the lectures of **Arago**. During the early period of his research he collaborated with **Foucault**, another medical student who decided on a career in physics. They met while Fizeau was working on improving the Daguerreotype process, and were the first to obtain a detailed photographic image of the Sun (1845). In 1849 Fizeau was the first to measure the velocity of light by a laboratory experiment in which a ray of light was cut by a toothed wheel, producing intermittent flashes. The velocity of light could then be calculated from the speed of the rotation and the distance. His figure (about 315 000 kilometres per second) was not as accurate as that produced by astronomical calculations, but he showed the practicability of this approach, which was then improved by others, notably by Alfred Cornu in the 1870s. Fizeau's adaptation of this technique to measure the velocity of electricity in a wire was less successful. He also demonstrated the use of the shift in light frequency (the 'redshift') in determining a star's velocity along the line of sight. Unknown to him, **Doppler** had already published this effect, but had not fully understood the implications. The British astronomer **William Huggins** was the first to determine the velocity of a star relative to the Earth by this method (1868).

FJORTOFT, Ragner

(1913–)

Norwegian mathematician, born in Oslo. He was educated at the University of Oslo where he studied mathematics and theoretical meteorology. After weather forecasting at Bergen under Sverre Petterssen in 1939, he became interested in the possibility of numerical weather prediction. With **Charney** at Princeton, in 1949 he organized and carried out computer integrations of the non-linear barotropic equation. He became professor at the University of Copenhagen in 1951 and Director of the Norwegian Meteorological Institute (1955–78). In important original work on dynamic meteorology, he devised stability criteria for a baroclinic circular vortex with respect to axially symmetric perturbations and considered the stability of a barotropic vortex with respect to arbitrary perturbations: he established that instability requires the absolute vorticity to be a maximum within the flow. Finally he considered more general wave disturbances in a baroclinic atmosphere and obtained stability criteria. A paper which he published on boundary conditions for limited area models became a classic. Later he became interested in statistical meteorology and devised probability models for forecasting extremes of precipitation and ocean waves.

FLAMMARION, Nicolas Camille
(1842–1925)

French astronomer, born in Montigny-le-Roi. Under the directorship of **Le Verrier** he became an apprentice astronomer at the age of 16 at the Paris Observatory. He left in 1882 to establish *L'Astronomie*, a monthly magazine. He founded the observatory of Juvisy-sur-Orge in 1883 and the French Astronomical Society in 1887. Flammarion is also remembered as a popularizer of astronomy; among his many books, *L'Astronomie Populaire* (1879) is the best known and was translated into many languages. He was convinced that there were many lifeforms in the universe, a proposition that he developed in his first book *La Pluralité des Mondes Habités* (1862).

FLAMSTEED, John
(1646–1719)

English astronomer, born in Denby near Derby, the only son of a maltster. Following early astronomical studies privately pursued and a spell at Cambridge (1871–4), he was appointed to a commission concerned with the finding of longitude at sea. The commission's report induced King Charles II to found a national observatory at Greenwich which was built in 1675–6 with Flamsteed as Director and first Astronomer Royal. With a salary of only £100, no assistant and imperfect instruments assembled at his own expense, Flamsteed was in a difficult position until in the 1680s his private financial circumstances improved. He then acquired from an outstanding instrument-maker Abraham Sharp a mural arc with which he started an immense programme of stellar positional observations. Aiming at the highest possible accuracy, Flamsteed was slow with the reductions of his observations much to the annoyance of **Isaac Newton** who claimed that he needed them for the perfection of his lunar theory. After much commotion, the *Historia Coelestis* embodying the first Greenwich star-catalogue was printed in 1712 under the editorship of **Halley**, Savilian Professor of Astronomy at Oxford. Flamsteed, denouncing the production as surreptitious, burnt 300 copies of it. He pressed for an adequate publication of his work but died before its completion in 1725 as the *Historia Coelestis Britannica*. Its three volumes were supplemented by the *Atlas Coelestis*, published in 1729 by Abraham Sharp and Flamsteed's assistant Joseph Crosthwait.

FLAVELL, Richard Anthony
(1945–)

British molecular biologist, educated at the University of Hull where he obtained his PhD in 1970. He then became Royal Society European Fellow at the University of Amsterdam, and Postdoctoral Fellow at the University of Zürich. During 1973–9 he was Wetenschappelijk Medewerker at the University of Amsterdam, and he subsequently became Head of the Laboratory of Gene Structure and Expression at the National Institutes of Medical Research, Mill Hill, London (1979–82). He was President of Biogen Corporation (1982–8), and since 1988 has been professor at Yale University School of Medicine. He was elected FRS in 1984. Flavell is best known for his contribution to our understanding of the structure and expression of the human globin genes. An especially important facet of this work has been the elucidation of the molecular defects implicated in the thalassaemias, a group of inherited anaemias. He has shown that for certain thalassaemias, the genetic defect involves the total deletion of a specific globin gene while in other types of this group of diseases, the defect resides at large distances from the affected gene. Work of this kind on naturally occurring mutations has been invaluable in identifying DNA sequences which regulate globin gene control. Such work has been extended to use gene transfection or transgenic animals to formally prove the role of precise DNA elements and thus to allow the introduction of gene therapy as a treatment for these anaemias.

FLECK, Sir Alexander, Baron Fleck
(1889–1968)

Scottish physical and industrial chemist, born in Glasgow. He attended Glasgow University and lectured there for two years before working on radium with **Soddy**. In 1917 he was appointed Chief Chemist to the Castner–Kellner Co at Wallsend-on-Tyne. In 1937 he became Chairman of the Billingham division of Imperial Chemical Industries and in 1953 chairman of the whole company; he retired from ICI in 1960. He was also Chairman of Scottish Agricultural Industries (1947–51) and Deputy Chairman of African Explosives and Chemical Industries Ltd (1953–60). He was also a prominent member of many government committees and chaired the committee which investigated the nationalized coal industry in 1953–5. Elected FRS in 1955, he was knighted in 1955 and created a baron in 1961. He died in London.

FLEMING, Sir Alexander
(1881–1955)

Scottish bacteriologist, and the discoverer in 1928 of penicillin, born on a farm in Loudoun, Ayrshire. He was educated in Kilmarnock, and became a shipping clerk in

London for five years before matriculating (1902) and embarking on a brilliant medical studentship, qualifying as a surgeon at St Mary's Hospital, Paddington, where he spent the rest of his career. It was only by his expert marksmanship in the college rifle team, however, that he managed to find a place in Sir **Almroth Wright**'s bacteriological laboratory there. As a researcher he became the first to use anti-typhoid vaccines on human beings, and pioneered the use of salvarsan against syphilis, a treatment introduced by **Ehrlich** in 1910. In 1922, while trying unsuccessfully to isolate the organism responsible for the common cold, he discovered lysozyme, an enzyme present in tears and mucus that kills some bacteria without harming normal tissues. While this was not an important antibiotic in itself, as most of the bacteria killed were non-pathogenic, it inspired his search for other antibacterial substances. In 1928 by chance exposure of a culture of staphylococci he noticed a curious mould, penicillin, which he found to have unsurpassed antibiotic powers. Unheeded by colleagues and without sufficient chemical knowledge, he had to wait 11 years before two brilliant experimentalists at the William Dunn School of Pathology at Oxford, **Florey** and **Chain**, with whom he shared the 1945 Nobel Prize for Physiology or Medicine, perfected a method of producing the volatile drug. Fleming was appointed Professor of Bacteriology at London in 1938. He was elected FRS in 1943, and knighted in 1944.

FLEMING, Sir John Ambrose
(1849–1945)

English physicist and electrical engineer, born in Lancaster. He studied in London at University College and the Royal College of Chemistry, at the same time working for two years as a stockbroker's clerk. He became a science master at Cheltenham College (1874–7) and then won an entrance exhibition to St John's College, Cambridge, where he studied under **Maxwell**. After three years at University College, Nottingham, he was appointed as a consultant to the Edison Electric Light Company, and he also served for 26 years as consultant to the Marconi Wireless Telegraph Company. During his tenure of the Chair of Electrical Engineering at University College London (1885–1926), he invented in 1904 the thermionic rectifier or Fleming valve, which for half a century was a vital part of radio, television and early computer circuitry, until superseded by the transistor diode in the early 1950s. Fleming was also a pioneer in the application of electricity to lighting and heating on a large scale. He was elected FRS in 1892, and knighted in 1929.

FLEMMING, Walther
(1843–1905)

German biologist, born in Sachsenberg. Flemming studied medicine in five German universities, going on to become Professor of Anatomy at Kiel. Flemming is renowned for his investigations of cell division (mitosis), though he also made significant advances in microscope techniques. Developing the new aniline dyes as microscopic stains, and deploying improved microscopes, he found that dispersed fragments in an animal cell nucleus became strongly coloured; he called this substance chromatin. He noted that, in cell division, the chromatin granules combined to constitute larger threads; in 1888 **Waldeyer-Hartz** was to name these chromosomes. Flemming further demonstrated that elementary nuclear division as described by **Remak** was not universal; the more typical type of cell division Flemming named mitosis. In this process, the chromosomes divided lengthwise, and the indistinguishable halves moved to opposite sides of the cell. The cell then divided, giving two daughter cells with as much chromatin as the original. Flemming provided a superb account of the process in 1882. He was unaware of **Mendel**'s work. The application of Flemming's ideas to genetics was not to come for another 20 years.

FLETT, Sir John Smith
(1869–1947)

Scottish petrologist, born in Kirkwall, Orkney. He had a sparkling career as a student at the University of Edinburgh obtaining an MA at the age of 19, a BSc in natural science in 1892 and a medical degree in 1894. He practised medicine for a short time before turning to geology and becoming assistant to James Geikie (1839–1915) at the University of Edinburgh, where he subsequently became a lecturer in petrology. He was associated with the Geological Survey from 1901, initially as petrographer, then as assistant to the director in Scotland (from 1911) and finally as director (1920–35). Flett wrote on Scottish petrology (*The Old Red Sandstone of the Orkneys*, 1898) and the vulcanology of the West Indies (*Report on the eruptions of Soufrière, in St Vincent and on a Visit to Mt Pelée in Martinique*, 1903). He also made important contributions to the study of the geology of Cornwall. His *History of the Geological Survey of Great Britain* was published in 1937, and he was knighted in 1925.

FLEXNER, Simon
(1863–1946)

American microbiologist and medical administrator, born in Louisville, Kentucky.

He studied medicine in the local medical school, but his passion for medical research was awakened by **Welch** at Johns Hopkins University. After another year's study in Europe he joined Welch's department of pathology before moving to Pennsylvania University (1899–1903), and then to the newly established Rockefeller Institute for Medical Research as director of laboratories (1903–35). Among important contributions to bacteriology, virology and immunology he isolated the dysentery bacillus (1900), developed a serum for cerebrospinal meningitis (1907), and led the team that determined the cause of poliomyelitis. Equally importantly, he shaped the Rockefeller Institute into a powerful and productive centre of medical research. He encouraged both John Rockefeller Sr and Jr to establish research fellowships in the natural sciences, and edited for many years (1905–46) the outstanding American periodical of medical research, *Journal of Experimental Medicine*.

FLOREY, Howard Walter, Baron Florey of Adelaide and Marston

(1898–1968)

Australian pathologist, born and educated in Adelaide. He studied physiology as a Rhodes scholar under **Sherrington** at Oxford, and pathology in Cambridge, becoming a lecturer there in 1927. His early researches included an analysis of the effectiveness of lysozyme, a naturally occurring antibacterial component of tears, saliva and nasal secretions. He was appointed Professor of Pathology at Sheffield (1931–5) and then at Oxford (1935–62), where he headed the Sir William Dunn school of pathology. Here his work on lysozyme had developed to such an extent that he required biochemical assistance, and he appointed a refugee from Nazi Germany, **Chain**. With **Heatley** and **Abraham**, they examined and synthesized a wide range of antibacterial compounds. In 1938 they began work on penicillin, an antibacterial substance produced by a mould, and first reported in 1929 by **Alexander Fleming**, but which was then regarded as unsuitable and too unstable for routine clinical use. Together the Oxford scientists succeeded in isolating sufficient penicillin to enable them to report on its biological properties, especially its marked effect on staphylococcus bacteria, and its low toxicity. These factors indicated its practical therapeutic potential and with the help of a small grant from the Medical Research Council they started, during the early months of World War II, to assay and purify penicillin for clinical use. By 1941 they had carried out successful tests on nine patients, and hampered by the paucity of resources in wartorn Britain, Florey and Heatley travelled to the USA to persuade pharmaceutical companies to assist in the development of large-scale production methods for penicillin. Enough penicillin was ready in time to treat casualties in the D-Day battles in Normandy, where it proved highly successful in combating previously fatal bacterial infections. For this vital work, Florey shared the 1945 Nobel Prize for Physiology or Medicine with Fleming and Chain. He was provost of Queen's College, Oxford, from 1962, and was made a life peer in 1965. He never forgot his Australian heritage, and did much to found the Australian National University at Canberra.

FLORIANI, Carlo

(1940–)

Italian chemist, born in Cremona, Northern Italy. He received his PhD in organic chemistry from University of Milan in 1965 and has since worked at various institutions: Cyanamid European Research, Geneva, the University of Pisa, Columbia University (New York) and the University of Lausanne, where he is now professor. His research is concerned with the development of new synthetic methodologies in inorganic chemistry, and in particular with the organometallic and coordination chemistry of early transition metals. A particular interest is the reactivity of small molecules such as formaldehyde, dioxygen and carbon monoxide with transition metal complexes; within this area he has discovered routes to carbon dioxide activation. Such discoveries are immensely important in opening up the chemistry of this widely available but unreactive molecule. He has also promoted a renaissance in copper (I)–carbon monoxide chemistry, and developed planned syntheses of molecular aggregates which have the ability to store and release electrons. He also discovered some novel reaction pathways of the porphyrin skeleton. He was Centenary Lecturer of the Royal Society of Chemistry in 1989–90.

FLORY, Paul John

(1910–85)

American physical chemist, born in Sterling, Illinois. He was educated at Ohio State University, and in 1934 obtained a research post under **Carothers** at Du Pont, Wilmington. In 1938 he moved to a basic science research laboratory at the University of Cincinnati, but in 1940 returned to industry at the Linden Laboratory of Standard Oil. In 1943 he became leader of the Goodyear fundamental research group, in 1948 professor at Cornell University (after giving the Baker Lectures there), and in 1957 Executive

Director of the Mellon Institute. From 1961 until his retirement in 1976 he held chairs at Stanford University. He was awarded the Nobel Prize for Chemistry in 1974. His many other honours and awards included the Debye Award (1969) and Priestley Medal (1974) of the American Chemical Society. Flory's main research area was the physical chemistry of polymers. By the 1930s many polymers were known, but their nature as macromolecules had only just been recognized. Flory's work brought them within the scope of kinetics, thermodynamics and statistical mechanics. This led to detailed understanding of their properties and those of their solutions, and of the chemical reactions involved in their formation and breakdown. From the 1950s Flory also worked on liquid crystal behaviour.

FLOURENS, Pierre Jean Marie

(1794–1867)

French physiologist, born in Maureilhan, near Béziers. Flourens studied medicine at Montpellier, qualifying in 1813. He proceeded to Paris where he gained the patronage of **Cuvier**. The two became close associates and, on Cuvier's death in 1832, many of his appointments fell to Flourens. He rose to become Secretary of the Academy of Sciences (1833) and professor at the Collège de France (1835). In 1838 he was elected to the Chamber of Deputies, and in 1846 nominated a peer of France. Flourens achieved great eminence in his own lifetime for his neurophysiological investigations; he was amongst the first to demonstrate the functions of the different sections of the brain. From 1820 he began working on the central nervous system, performing vivisection experiments on pigeons and dogs. He discovered that vision depended on the cerebral cortex. The ablation of parts of it produced blindness on the opposite side. Removal of the cerebellum produced loss of coordination of movement. Flourens also found that respiration was controlled by a centre in the medulla oblongata; damage to the semicirculate canals of the ear caused balance loss. His researches paved the way for the later work on cerebral localization of **Hitzig**, **John Jackson** and David Ferrier. Flourens was a vitriolic foe to the pseudoscience of phrenology as developed by Franz Joseph Gall and Johann Christoph Spurzheim, and in later life he became a trenchant opponent of **Charles Darwin**'s theory of evolution. In 1847 he demonstrated that trichloromethane was an effective anaesthetic for small animals; a short time later the Scottish obstetician **James Young Simpson** first used it for human patients in childbirth.

FLOWER, Sir William Henry

(1831–99)

English anatomist and zoologist, born in Stratford-upon-Avon. He studied medicine at University College London and the Middlesex Hospital. After serving as a surgeon in the Crimean War, he was appointed as a demonstrator in anatomy at the Middlesex Hospital and then in 1858, as curator of the hospital's museum. In 1861, on the recommendation of **T H Huxley**, he became conservator of the Hunterian Museum in London where he reorganized the collections with zeal and originality, and in 1870, following Huxley's retirement, he was appointed Hunterian Professor of Comparative Anatomy and Physiology. From 1884 until 1898 he was the first Director of Natural History at the British Museum where he revolutionized museum displays, making them accessible to both scholars and the public alike, and introduced evolutionary theory for the first time. He was knighted in 1892.

FOCK, Vladimir Alexandrovich

(1898–1974)

Soviet theoretical physicist, born in St Petersburg. In 1919, having survived military service at the front in World War I, he went to the State Optical Institute of the Petrograd (later Leningrad, now St Petersburg) University, where he graduated in 1922. He worked simultaneously in various scientific and educational institutes, becoming professor at Leningrad University in 1932. His most important work was in quantum mechanics. In the 1920s he generalized the **Schrödinger** wave equation to the relativistic case (simultaneously achieved independently by Oskar Klein), with the resulting equation known as the Klein–Fock equation. Later he developed **Hartree**'s approach to the quantum mechanics of multi-particle systems (1930), which allowed the wave equation to be solved for atoms with more than one electron (the Hartree–Fock technique). Fock also studied general relativity and solved the problem of motion of many-body systems in this theory (1939). He also demonstrated that **Isaac Newton**'s equations and the law of universal gravity follow directly from the general theory of gravity for finite masses. For his work, Fock received many awards including the Lenin Prize (1960), and he was awarded the title 'Hero of Socialist Labour' (1968).

FOLKERS, Karl August

(1906–)

American biochemist, born in Decatur, Illinois. He studied at Illinois, Wisconsin and Yale and later directed research at Merck and

Co, at the Stanford Research Institute, and at Texas University. The isolation of the anti-pernicious anaemia factor, cyanocobalamin (vitamin B_{12}), first revealed in 1926 by treatment with raw liver, was achieved simultaneously by Folkers's team at Merck (1948) and by **Lester Smith** in England. Cyanocobalamin was also found to be identical to 'animal protein factor', a growth factor found only in animal products and thought for some time to be a vitamin. In 1956 Folkers's team isolated mevalonic acid (a key intermediate in the biosynthesis of terpenes and steroids) from a commercial yeast by-product, and in 1970 they reported the structure of porcine thyrotrophin-releasing hormone (the hypothalamic hormone that stimulates release of thyroxine from the thyroid gland) as Glu His Pro (NH_2). Folkers also worked on the synthesis and structure of a range of antibiotics.

FORBES, Edward

(1815–54)

British naturalist, born in Douglas, Isle of Man. He studied medicine at the University of Edinburgh but spent each vacation in natural history pursuits, often dredging at sea. In 1836 he gave up medicine altogether to devote himself completely to the natural sciences. In 1841 he joined the crew of the *Beacon* as naturalist during the survey around parts of Asia Minor. He later became Professor of Botany at King's College, London (1843), and shortly afterwards was appointed Curator of the Museum of the Geological Society of London. In 1844 he became the first palaeontologist to HM Geological Survey, and when the School of Mines was established in London (1851), he became Professor of Natural History. In 1854 he succeeded Robert Jameson (1774–1854) as Regius Professor of Natural History at Edinburgh University, but his tenure was short-lived as he died unexpectedly later that year at the youthful age of 39. Forbes was a versatile natural historian and produced important monographs on coelenterates, echinoderms and molluscs. He undertook important studies of Quaternary changes in molluscan faunas along British coasts, of relict floras of the British highlands, and of molluscan faunas of the Aegean. He also made formative observations in oceanography; his observations of depth-related communities in the sea effectively laid the foundations for the sciences of biogeography and palaeoecology.

FORBES, James David

(1809–68)

Scottish physicist and glaciologist, born in Edinburgh, the youngest son of Sir William Forbes of Pitsligo. He was privately educated until the age of 16 when he entered Edinburgh University. Here, despite an early interest in science, he pursued legal studies following his father's wishes. On receiving a modest inheritance on his father's death (1828), he abandoned law and returned to science. He became professor at Edinburgh University (1833–60) and Principal of St Andrews College (1860–8). In 1834, using **Melloni**'s instruments, he discovered the polarization of radiant heat transmitted through tourmaline and thin mica plates, and that heat could be circularly polarized by two reflections in a Fresnel rhomb of salt. This contributed to the concept of a continuous radiation spectrum. After 1840 his interests turned to geology and to glaciers. He visited the Alps with **Jean Agassiz** in 1841, later causing a lifelong controversy with him by claiming priority in noting that the surface of a glacier moves faster than the ice beneath it, and that glacier velocity is directly related to the steepness of the slope. He also postulated that a glacier is a viscous body whose movement is due to the mutual pressure of its parts. Forbes fought successfully for reforms in Scottish higher education, including the instigation of degree-level examinations.

FORD, Edmund Brisco

(1901–88)

English geneticist, born in Papcastle, Cumberland. He was educated at Waldham College, Oxford, and taught at Oxford University throughout his career, retiring as Professor of Ecological Genetics and Director of the Genetics Laboratory in 1969. He was stimulated by **Julian Huxley**, his tutor at Oxford, and with whom he carried out pioneering work on gene action rates. He wrote *Mendelism and Evolution* (1931), which can be considered to be the first major published work leading to the neo-Darwinian synthesis. He worked with **Fisher** on a number of field studies of microevolution in Lepidoptera, and was responsible for the first experimental evidence of modifying dominance of an interested trait, confirming Fisher's theory of the evolution of dominance, breeding from heterozygous individuals which manifested respectively less or greater effect of an inherited variation. He showed that the expression of the trait was under genetic control, and that selection could change its inheritance in the direction of either dominance or recessivity. He also used genetic polymorphism to show from field studies that the maintenance of different inherited forms of a character in the same population often resulted from natural selection acting on different phenotypes. This overturned earlier notions about the weakness of natural selection by repeatedly demonstrating strong selective pressures. Ford's New Naturalist books on *Butterflies* (1945) and *Moths* (1955) were influential; his own scientific achievements are described in *Ecological Genetics* (1964). He was elected FRS in 1946.

FOREST, Lee De See **DE FOREST, Lee**

FORREST, George
(1873–1932)

Scottish plant collector and botanist, born in Falkirk. Employed in the Royal Botanic Garden of Edinburgh by Sir **Isaac Bayley Balfour**, he was a tough and resourceful individual. His first expedition to Yunnan and Tibet in 1904–6 was underwritten by Sir Arthur Bulley, of Bees Nurseries in Liverpool. Despite narrowly escaping death at the hands of Tibetan revolutionaries, he returned to the area several times in succeeding years between 1910 and 1930, sponsored by groups of backers, including members of the newly formed Rhododendron Society. He trained and employed local labour to collect seed, which enabled him to send home several hundred pounds (weight) of seed from each trip. His explorations in Upper Burma and China, particularly the Lichiang range on the Mekong–Salwin divide, resulted in the discovery and introduction of many plants, including several rhododendrons and primulas, which he described in articles in *Gardeners' Chronicle* and other publications. Two of his most significant introductions are *Pieris formosa* variety *forrestii* and the widely grown *Gentiana sino-ornata*. He died at Teng-Yueh in the Yunnan province of China. Plants named after him include *Rhododendron forrestii* and *Gentiana forrestii*.

FORRESTER, Jay Wright
(1918–)

American computer engineer, born in Anselmo, Nebraska. Educated at Nebraska University, where he took a degree in engineering, he became a pioneer in the development of computer storage devices. In 1944 at MIT he began work on a US navy analogue flight trainer. By 1947 this had turned into a high-speed digital electronic computer — the Whirlwind — the largest computer project of the late 1940s to early 1950s. During the construction of the Whirlwind, Forrester in 1949 devised the first magnetic core store (memory) for an electronic digital computer. By 1953 a core memory was installed in the Whirlwind. From 1951 to 1956 he was a founder and director of the Digital Computer Laboratory. He has written several books, including *Industrial Dynamics* (1961), *Principles of Systems* (1968) and *World Dynamics* (1971).

FORSCHAMMER, Johann Georg
(1794–1865)

Danish chemical oceanographer, born in Husum. In 1815 he entered Kiel University to study physics and chemistry. He moved to Copenhagen in 1818 and investigated the coal and iron deposits in Bornholm, gaining a PhD in 1820. He attended lectures by **Oersted**, whom he succeeded as Director of the Polytechnic Institute in Copenhagen. In 1831 he was appointed Professor of Mineralogy and Geology at the University of Copenhagen. Forschammer determined the major components of sea-water and established that the relative concentrations of the major dissolved constituents are almost constant, thus the salinity of any sample could be determined by measurement of any single major component, a breakthrough for future analysts. He detected the presence of many elements and postulated the idea of a geochemical balance in *On the Components of Sea-water* (1859). He related this geochemical balance to a sedimentary cycle of material washed into the sea by rivers and realized that the quantity of the elements is not proportional to the quantity introduced, but inversely proportional to the time the organo/chemical particles spend in the soluble state; this led to the concept of residence times. His *Danmarks Geognostiske Forhold* (1835) was the first work on the structural geology of Denmark, for which he was termed the 'father of Danish geology'.

FORSSMAN, Werner
(1904–79)

German physician and surgeon, born in Berlin. He graduated at Berlin University and was an army doctor until 1945; subsequently he practised urological surgery at various places including Bad Kreuznach and Düsseldorf. He became known for his pioneering work in the late 1920s on cardiac catheterization, in which he carried out dangerous experiments on himself. He abandoned them in the face of criticism. He was awarded the 1956 Nobel Prize for Physiology or Medicine jointly with **Cournand** and **Dickenson Richards**, who had extended Forssman's original techniques and demonstrated their clinical and experimental usefulness. By then, it was too late for Forssman to catch up with subsequent advances in cardiology and to contribute further to the procedure he had pioneered. His autobiography (1974) offers an attractive portrait of the man and his work.

FORTUNE, Robert

(1813–80)

Scottish horticulturist and plant collector, born in Kelloe in Edrom parish, Berwickshire. After early horticultural training in Berwickshire, at Moredun near Edinburgh and at Edinburgh's Royal Botanic Garden, in 1840 he was employed by the Horticultural Society at Chiswick. They sent him to China in 1842 to collect plants and he visited Hong Kong, Chusan and Shanghai. In 1848 the East India Company asked him to help in expanding tea production in India's North West Provinces. This entailed a second three-year expedition to China, collecting tea seeds and plants, quickly followed by a third in search of black tea and yet another (1858–9) for the US Patent Office. His last journey (1861–2) was to Japan and northern China. From his expeditions, Fortune introduced many well-known plants to British gardens, including the Japanese anemone, the Chinese fan palm (*Trachycarpus fortunei*), the winter jasmine, the Japanese golden-rayed lily (*Lilium auratum*) and the Chusan daisy, progenitor of today's pompom chrysanthemums. His travel notes were published as *Three Years Wanderings in the Northern Provinces of China* (1847), *Two Visits to the Tea Mountains of China* (2 vols, 1853), *A Residence Among the Chinese* (1857) and *Yedo and Peking* (1863).

FOSTER, Sir Michael

(1836–1907)

English physiologist, born in Huntingdon. Educated at University College London, he graduated in classics in 1854, and then studied medicine, coming under the particular influence of William Sharpey, Professor of Anatomy and Physiology. In 1858 Foster qualified and after taking an MD the following year, he visited European medical centres and began physiological research on the mechanisms of the snail's heart beat, before ill-health obliged him to take a voyage as ship's surgeon in 1860. The following year he joined his father in general practice in Huntingdon, remaining until 1867 when he returned to University College as lecturer in practical physiology. In 1870 Trinity College in Cambridge created a Praelectorship in Physiology, to which Foster was appointed, and from which he built the influential Cambridge School of Physiology. In 1883 the university established its first Chair in Physiology for Foster, from which he retired in 1903. As a research physiologist he studied the mechanisms of the heartbeat in some detail, concentrating on cold-blooded animals which had a slower heart rate, and analysing the component roles of intrinsic rythmicity of the heart cells and external neural regulation. More significant however are Foster's roles in establishing and promoting the study of physiology. In Cambridge he gathered around him colleagues such as **John Langley**, **Gaskell**, **Francis Balfour** and **Walter Morley Fletcher**, and attracted students who included **Sherrington** and **Dale**. He was a founder member of the Physiological Society, the founder-editor and owner for some years of the *Journal of Physiology*, and took a large part in founding the International Congress of Physiology, of which he was elected honorary perpetual president in 1901. He published in 1876 the first edition of *Text-Book of Physiology* which was widely read and translated. He served as Biological Secretary of the Royal Society (1881–1903), and was Member of Parliament for the University of London (1900–5).

FOUCAULT, Jean Bernard Léon

(1819–68)

French physicist, born in Paris. He first entered on a career as a physician, but found this to be impossible as he came to detest the sight of blood. He turned to experimental physics and determined the velocity of light by the revolving mirror method originally proposed by **Arago**, and also proved that light travels more slowly in water than in air (1850); subsequently he showed that the ratio of the speeds in the two media is the inverse of the ratio of their respective refractive indices. This was convincing evidence of the wave nature of light as opposed to the corpuscular theory, and earned Foucault his doctorate. In 1851, by means of a freely suspended pendulum more than 200 feet long, he convincingly demonstrated the rotation of the Earth to a large crowd in a Paris church. In 1852 he constructed the first gyroscope, in 1857 the Foucault prism and in 1858 he improved the mirrors of reflecting telescopes.

FOURCROY, Antoine François, Comte de

(1755–1809)

French chemist, born in Paris where he studied medicine, qualifying in 1780. He became professor at the Jardins des Plantes in 1784, and from 1786 onwards promulgated the revolutionary chemical theories of **Lavoisier**, both in the classroom and in print. He improved methods of analysing mineral waters, discovered the double salts of ammonia and magnesia, and studied the physiology of muscles. With **Vauquelin** he discovered iridium and made extensive chemical investigtions of animal organs and fluids, isolating urea in 1808. Before the

Revolution he was one of the leaders of the scientific community and one of the founders of the influential journal *Annales de Chimie*. He also helped Lavoisier, **Berthollet** and **Guyton de Morveau** to develop a new system of chemical nomenclature. During the revolution he was a member of the Committee of Public Instruction and the Committee of Public Safety, working to reorganize higher education and munitions manufacture, and helping to establish the Institute National des Sciences et des Arts. In 1802 Napoleon appointed him Director-General of Public Instruction and in 1808 made him a Count of the Empire. He died in Paris.

FOURIER, Jean Baptiste Joseph, Baron de

(1768–1830)

French mathematician, born in Auxerre. The son of a tailor, Fourier was orphaned at the age of eight. He revealed his talent for mathematics during his education at a military school and an abbey, and later took an active part in promoting the Revolution. In 1795 he joined the staff of the École Normale in Paris, newly formed to train senior teachers, where his success led to the offer of the Chair of Analysis at the École Polytechnique. He accompanied Napoleon during the invasion of Egypt in 1798, and on his return in 1802 was made Prefect of Isère in Grenoble and created baron in 1808. After 14 years at Grenoble, he resigned to rejoin Napoleon during the Hundred Days. He was later made a member of the Academy of Sciences of Paris (1817), becoming joint secretary with **Cuvier** in 1822. He died in 1830 of a disease contracted in Egypt. Fourier introduced the expansion of functions in trigonometric series, now known as Fourier series. This ended a long period of controversy on the subject and it became generally accepted that almost any function of a real variable can be expressed as a series containing the sines and cosines of integral multiples of the variable: for example, any complex musical sound can be represented as the sum of many individual pure frequencies. This method has become an essential tool in mathematical physics and a major theme of analysis. His *Théorie analytique de la chaleur* (1822) applied the technique to the solution of partial differential equations to describe heat conduction in a solid body. His work did receive some criticism, however, as he failed to produce a general proof that the Fourier series actually converges to the value of the function involved; this difficulty was not satisfactorily resolved until almost a century later.

FOWLER, Sir Ralph Howard

(1889–1944)

English mathematician and physicist, born in Roydon, Essex. He studied at Trinity College, Cambridge, where he graduated in 1911. After being wounded during World War I, he worked on military research under **Hill**, and later returned to Cambridge where he became Plummer Professor of Theoretical Physics in 1932. His most important contributions were to statistical mechanics and quantum theory. His applications of statistical mechanics to high-temperature gases led to important advances in the theory of stellar structure, and he was first to suggest that **Fermi** and **Dirac**'s work implied that the gas in white dwarf stars exists in a 'degenerate' state. Fowler's lively interest in the development of quantum theory had a strong influence on many students, including Dirac. He married **Rutherford**'s daughter Eileen in 1921, and was knighted in 1942.

FOWLER, William Alfred

(1911–)

American physicist, born in Pittsburgh, Pennsylvania. He studied at Ohio State University and obtained his PhD for work on radioactive nuclides produced by proton bombardment from Caltech (1936). He became professor there in 1946. Fowler experimentally measured the probabilities or 'cross-sections' of reactions in the carbon cycle of nuclear reactions proposed by **Bethe** and the proton–proton cycle. As these cross-sections are very small at stellar energies, he made detailed measurements at low energies and extrapolated to higher energies. **Hoyle** had pointed out that for helium to be converted into heavier elements in the stellar core, an excited state of helium must exist. Fowler established the existence of this state. This was a crucial link in the theory of stellar evolution which he developed with Hoyle, and **Geoffrey** and **Margaret Burbidge**, which explained the synthesis of heavy elements in the stellar cores. Previously it had been believed that all elements had been created in the Big Bang. Fowler continued to work on the details of stellar nucleosynthesis, including solar neutrino flux calculations. For his work on stellar evolution and nucleosynthesis he shared the 1983 Nobel Prize for Physics with **Chandrasekhar**.

FRACASTORO, Girolamo

(1483–1553)

Italian scholar and physician, born in Verona. In 1502 he became Professor of Philosophy at Padua, but also practised successfully as a physician at Verona. He excelled as geographer, astronomer and mathematician. He wrote on the theory of music and left a Latin poem on the 'new' venereal disease, *Syphilis sive morbus Gallicus* (1530), from which the word 'syphilis' is derived. His works on contagion developed the older notions of the 'seeds of disease' and pointed to the importance of *fomites* (clothes, beddings, etc) in the spread of certain diseases. He also left treatises on botany and on the role of sympathy in the natural order.

FRAENKEL-CONRAT, Heinz
(1910–)

American biochemist, born in Breslau, Germany (now Wrocław, Poland). He studied medicine there and biochemistry in Edinburgh before moving to the USA in 1936, and returned to Europe to work with **Linderstrøm-Lang** in Copenhagen and **Sanger** and **Rodney Porter** before joining the staff of the University of California at Berkeley in 1952, becoming Professor of Virology in 1958 and later Professor of Molecular Biology. Working with viruses, he showed that active tobacco mosaic virus, able to infect plants, could be reconstituted from its inactive protein and nucleic acid components, and that it was a 'living chemical' and a basic unit in the new science of molecular biology (1955). His later work involved the development of methods for the sequence analysis of RNA and chemical modification in relation to protein–RNA interactions, as well as the conformational structure and modification of enzymes, particularly trypsin and its inhibition by ovomucoid.

FRANCIS, Edward Howel
(1924–)

Welsh geologist, born in south Wales, and educated at the University of Wales in Swansea. He became a field geologist with the Geological Survey, initially in Scotland (1949–62), then moved on to become District Geologist for North East England (1962–77), North Wales (1967–70) and then Assistant Director for Northern England and Wales (1971–7). From 1977 until his retirement in 1989 he was Professor of Geology at the University of Leeds. Francis has published many memoirs and papers on coalfields, palaeovolcanic rocks and general stratigraphy, mainly of Britain. His studies of the interactions between magma and sediment, particularly with reference to the Carboniferous rocks of the Scottish Midland Valley, were published as his presidential addresses of the Geological Society of London (1982, 1983).

FRANCK, James
(1882–1964)

German-born American physicist, born in Hamburg. He was educated in Heidelberg and Berlin, and became Professor of Physics at Göttingen University (1920). He left Germany in 1933 in protest against Nazi policies and eventually settled in the USA where he became Professor of Physical Chemistry at the University of Chicago (1938–49). He is most famous for his research with **Gustav Hertz** into the laws governing the transfer of energy between molecules, for which they were jointly awarded the Nobel Prize for Physics in 1925. They showed that mercury atoms would only absorb a fixed amount of energy from bombarding electrons, demonstrating the quantized nature of the electron energy levels of the atom. Franck was also one of the formulators of the Franck–Condon principle which permitted the prediction of the most favoured vibrational transitions in a bond system. He later worked on the development of the nuclear bomb in World War II at Los Alamos, but headed the Franck Committee of scientists who urged that the bomb should not be used.

FRANK, Ilya Mikhailovich
(1908–90)

Soviet physicist, born in St Petersburg. He was educated at Moscow State University, and in 1944 he was appointed Professor of Physics there, after working for four years at the State Optical Institute. By 1937, working with **Cherenkov** and **Tamm**, they were able to explain the emission of radiation known as the 'Cherenkov effect'. They showed that the effect arises when a charged particle traverses a medium when moving at a speed greater than the speed of light in that medium. The effect is dramatically visible in the blue glow in a uranium reactor core containing heavy water. For this work Cherenkov, Frank and Tamm shared the 1958 Nobel Prize for Physics.

FRANKLAND, Sir Edward
(1825–99)

English chemist, born in Churchtown, Lancashire. He was attracted to chemistry while apprenticed to a pharmacist and later studied under **Lyon Playfair** in London, **Bunsen** in Marburg, and **Liebig** in Giessen. From 1847 to 1851 he was Science Master at Queenwood School, Hampshire, and he then became the first Professor of Chemistry at the newly founded Owen's College, Manchester. Frankland returned to London in 1857 to lecture at St Bartholomew's Hospital, and from 1863 to 1865 he was Professor of Chemistry at the Royal Institution in succession to **Faraday**. In 1865 he followed **Hofmann** at the Royal School of Mines, where he remained for 20 years. Frankland's pioneering work in organometallic chemistry around 1850 led to his development of the theory of valency, which underlies all structural chemistry. With **Lockyer** he studied the solar spectrum, and in 1868 they jointly discovered helium in the Sun's atmosphere. In applied chemistry Frankland did important work on water supply and sanitation. He was elected FRS in 1853, received the Copley Medal of the Royal Society in 1894, and was knighted in 1897.

FRANKLIN, Benjamin
(1706–90)

American statesman and scientist, youngest son and 15th child of a family of 17, born in Boston. He was apprenticed at 12 to his brother James, a printer, who started a newspaper, the *New England Courant*, and later Benjamin assumed the paper's management. He later established his own successful printing house in Philadelphia, and in 1729 he purchased the *Pennsylvania Gazette*. In 1732 he commenced the publication of *Poor Richard's Almanac*, which attained an unprecedented circulation. In 1736 Franklin was appointed Clerk of the Assembly, in 1737 Postmaster of Philadelphia, and in 1754 Deputy Postmaster-General for the colonies, a post which took him to London on diplomatic service for a number of years. He participated actively in the deliberations which resulted in the Declaration of Independence on 4 July 1776. Franklin was US minister in Paris till 1785, when he returned to Philadelphia, and was elected President of the state of Pennsylvania. In 1788 he retired from public life. Despite this highly eventful political career, Franklin made many important contributions to science. In 1746 he commenced his famous researches in electricity which earned his election to the Royal Society. He brought out fully the distinction between positive and negative electricity in his theory of a single electric fluid; he suggested a method of proving that lightning and electricity are identical (first performed by Thomas François Dalibard at Marley, France, in 1752); and he suggested the protecting of buildings by lightning-conductors. Further, he discovered the course of storms over the North American continent; the course of the Gulf Stream, its high temperature, and the use of the thermometer in navigating it; and the various powers of different colours to absorb solar heat.

FRANKLIN, Rosalind Elsie
(1920–58)

English X-ray crystallographer, born in London. She studied physical chemistry at Cambridge and held a research post at the British Coal Utilization Research Association (1942–6), where her work was important in establishing carbon fibre technology. At the Central Government Laboratory for Chemistry in Paris (1947–50), she became experienced in X-ray diffraction techniques. She returned to London in 1951 to work on DNA at King's College. She produced excellent X-ray diffraction pictures of DNA which were published in the same issue of *Nature* (1953) in which **James Watson** and **Crick** proposed their double-helical model of DNA. Finding it difficult to cooperate with **Wilkins**, who was also working on DNA at King's College, Franklin left to join **Bernal**'s laboratory at Birkbeck College, London, to work on tobacco mosaic virus. She contracted cancer and died in 1958, four years before she could be awarded the 1962 Nobel Prize for Physiology or Medicine jointly with Watson, Crick and Wilkins for the determination of the structure of DNA.

FRASCH, Hermann
(1851–1914)

German-born American industrial chemist, born in Gailsdorf, Württemberg. He emigrated to the USA in 1868 and worked in Philadelphia, Cleveland and London, Ontario, as a chemist and oil worker. In the 1880s he developed a process for removing sulphur from petroleum, since known as the Frasch process. In 1891 he patented a method (also known as the Frasch process) for extracting sulphur from deep deposits using superheated steam. The sulphur, which melts at 116 degrees Celsius, is pumped to the surface 99 per cent pure. Frasch founded the Union Sulphur Company, which became the largest sulphur-mining company in the world, and was subsequently also Director of the International Sulphur Refineries of Marseilles. He died in Paris.

FRASER DARLING, Sir Frank
(1903–79)

English ecologist and conservationist, born in Chesterfield. After attending agricultural college at Sutton Bonnington and working as a clean milk advisor in Buckingham, he obtained a doctorate from Edinburgh University's Department of Animal Genetics. A Leverhume Research Fellowship (1930–4) allowed him to study the ecology of the red deer (*A Herd of Red Deer*, 1937) and he subsequently carried out research on the behaviour of sea birds. He showed that the breeding success of colonial birds is enhanced through stimulation by other members of the species, a phenomenon now known as the 'Fraser–Darling effect'. Residence on the remote Scottish island of Rona during the years of World War II convinced him of the importance of living in ecological balance with the environment, and he became one of the early protagonists for conservation. He was Director of the West Highland Survey (1944–50), a senior lecturer in Ecology and Conservation at Edinburgh (1953–8) and Vice-President of the Conservation Foundation, Washington DC (1959–72). He carried out official ecological surveys in East Africa and Alaska, and advised government bodies on conservation and the setting up of national parks. He also served as a member on the Royal Commission on Environmental Pollution (1970–2). His many books included *Island Years* (1940), *Natural History in the Highlands and Islands* (1947) and *Wilderness and Plenty* (1970), based on the 1969 Reith Lectures. He was knighted in 1970.

FRAUNHOFER, Joseph von
(1787–1826)

German physicist, born in Straubing, Bavaria. He started work in his father's decorative glass workshop in 1797, but after his father's death in 1798 his guardians apprenticed him to a Munich mirror-maker and glass-cutter. In 1806 he entered the optical workshop of the Munich Philosophical Instrument Company, where he worked under Pierre Guinard, a Swiss-born master glass-maker. He soon outstripped his tutor, and his skill in glass-making, allied with his scientific knowledge and insight enabled him to transform the fortunes of the firm; by 1811 he had become a director of the company. In 1823 he was appointed Director of the Physics Museum of the Bavarian Academy of Sciences. Fraunhofer made considerable advances in the design of achromatic doublet lenses, and showed how to minimize the spherical aberration of such doublets. Using his new lens types and improved mechanical designs, he developed the prism spectrometer into a precision instrument with which he discovered the dark lines in the Sun's spectrum which now bear his name (1814–17). He instituted the practice of using a set of these dark lines to specify fixed points in the spectrum, for measurements of the refractive indices and dispersive powers of glasses. In 1821, while studying optical diffraction, he invented the transmission diffraction grating, and subsequently the reflection grating. Diffraction phenomena observed at very large distances from the diffracting aperture are known as Fraunhofer diffraction. Fraunhofer also invented the technique of testing lenses by examining the 'Newton's rings' interference fringes produced between the lens surface under test and a 'test plate' whose curvature is accurately known. His work laid the foundation for Germany's subsequent supremacy in the design and manufacture of optical instruments.

FREGE, (Friedrich Ludwig) Gottlob
(1848–1925)

German logician, mathematician and philosopher, born in Wismar. He was educated at Jena, where he spent his whole professional career, becoming professor in 1879. He worked in comparative obscurity in his lifetime, though both **Bertrand Russell** and Ludwig Wittgenstein noticed his originality and he is now regarded as the founding father of modern mathematical logic and the philosophy of language. A particular technical contribution to logic was his theory of quantification. His main works are *Begriffschrift* (1879), *Die Grundlagen der Arithmetik* (1884), and *Die Grundgesetze der Arithmetik* (2 vols, 1893, 1903). These have all been translated into English, as has a collection of his still influential philosophical essays analysing such basic logical concepts as meaning, sense and reference. In 1902 he abandoned his ambitious attempt to derive the whole of arithmetic from logic after Russell produced a devastating paradox which undermined it. He became generally depressed by the poor reception of his ideas and wrote little in his last 20 years.

FRÉMY, Edmond
(1814–94)

French chemist, born in Versailles. A student of **Gay-Lussac**, he became Professor of Chemistry at the École Polytechnique and later at the Museum of Natural History. He is chiefly remembered for his work on fluorine, which he attempted to isolate, also preparing anhydrous hydrogen fluoride and many of its salts. He wrote on the synthesis of rubies and worked on the ferrates, the colouring of flowers and the saponification of fats. He died in Paris.

FRESENIUS, Karl Remigius
(1818–97)

German analytical chemist, born in Frankfurt. He was educated in Bonn and studied under **Liebig** at Giessen, becoming Liebig's assistant after he received his doctorate. Before moving to Giessen, he devised a system for separating and identifying unknown substances. This was a major advance for at the time there was still no coherent approach to qualitative analysis. Fresenius first converted the sample into a sulphide, using hydrogen sulphide as the reagent. Then, in the knowledge that different sulphides behave differently but characteristically in certain conditions, he used various agents in a systematic manner to precipitate them out. Fresenius's system was translated into many languages, appearing in English as *Elementary Instruction in Qualitative Analysis* in 1841. It soon became standard laboratory practice all over the world, and although it has been largely superseded by more sophisticated techniques, it is still sometimes taught in schools and used in modern laboratories in a modified form. In 1845 Fresenius published a book on quantitative analysis, and in the same year he was appointed Professor of Chemistry at the Agricultural Institute at Wiesbaden. There he founded a school of analytical chemistry which became world famous and trained chemists from many countries. He died in Wiesbaden.

FRESNEL, Augustin Jean

(1788–1827)

French physicist, born in Broglie. Head of the department of public works in Paris, his intensive study of the problem of projecting well-defined beams of light led to the celebrated multi-facetted lighthouse lens (the Fresnel lens). His experimental and theoretical investigations into the interference, diffraction and polarization of light, coupled with his facility with new mathematical ideas, contributed massively to the establishment of the undulatory theory of light, in a form obtained by combining **Huygens**'s wave hypothesis with **Thomas Young**'s principle of interference. The earlier part of Fresnel's career was subject to interruptions resulting from the political upheavals associated with the changing fortunes of Napoleon. His most brilliant papers were a series relating polarization phenomena to Young's hypothesis of transverse waves; these were published in 1818–21, by which time the political scene had greatly stabilized. He also invented a special prism (Fresnel's rhomb) to produce circularly polarized light. Fresnel's *Oeuvres Complètes*, published in three volumes in the 1860s, contain practically everything that was known in optics up to the time of his death.

FRIEDMAN, Herbert

(1916–)

American astrophysicist, born in New York City. Educated at Brooklyn College and Johns Hopkins University, he spent his career at the US Naval Research Laboratory in Washington. He carried out pioneering work in the use of rockets in astronomy, and in the study of astronomical X-ray sources. From the 1940s he initiated the use of rockets carrying detectors to study X-rays from space, which cannot penetrate the atmosphere to be analysed at ground level. In 1949 he began investigations of the recently discovered X-ray activity of the Sun, producing the first X-ray and ultraviolet photographs of the Sun in 1960. Following **Rossi**'s discovery in 1962 of the first non-solar X-ray source, Friedman showed that one such source in the constellation Taurus coincided with the remnant of a luminous supernova in the Crab nebula (1964). After this early work, X-ray astronomy developed as an important area of astrophysics: so also has the use of rockets, to carry astronomical instruments above the absorbing layer of the Earth's atmosphere.

FRIEDMAN, Jerome Isaac

(1928–)

American physicist, born in Chicago and educated at the university there. He worked at the University of Chicago, Stanford University and MIT, where he became professor in 1967. In 1980 he became Director of the Laboratory of Nuclear Science. In the 1960s, with **Henry Kendall** and **Richard Taylor**, he led a group at the Stanford linear accelerator investigating electron scattering from nucleons (protons and neutrons). They confirmed **Bjorken**'s prediction for scaling of the structure functions and established that the quarks have spin ½. In conjunction with analogous neutrino experiments with the Gargamelle bubble chamber at CERN (the European nuclear research centre in Geneva), they also established that the quarks have fractional charges of $+\frac{2}{3}$ and $-\frac{1}{3}$ times the charge on the electron as had been postulated by **Gell-Mann** and others. The supreme achievement of Friedman, Kendall and Taylor's group was in providing the first incontrovertible evidence for quarks as real, dynamic entities rather than abstract mathematical concepts. For this work they won the 1989 W K H Panofsky prize and the 1990 Nobel Prize for Physics.

FRIEDMANN, Alexsandr Alexandrovich

(1888–1925)

Russian mathematician and cosmologist, born in St Petersburg. He studied at the university there and later became a lecturer. After World War I he was professor at Perm for two years but returned to St Petersburg in 1920 to carry out research at the Academy of Sciences. In 1922 Friedmann solved the equations in **Einstein**'s general theory of relativity in several ways. His solutions indicated that not only is the universe the same everywhere at a given time, but that its size and density are varying as a function of time. Friedmann's fame was posthumous and was due to the renewed interest in cosmology engendered by **Lemaître** and **de Sitter**.

FRIES, Elias Magnus
(1794–1878)

Swedish botanist, born in Femsjö in southwest Småland, and educated at Växjö Gymnasium and the University of Lund. After graduating, he had various unpaid or low-salaried posts at the University of Lund; nevertheless, by the end of his first 10 years there, he had become an internationally renowned mycologist and in 1835 was made Professor of Botany at Uppsala. From the beginning, he worked intensively on both fungi and lichens, interests which may have stemmed from the extremely rich fungal flora of his local countryside. His first mycological publication was *Observationes Mycologicae* (2 vols, 1815–18). His greatest work, *Systema Mycologicum* (3 vols, 1821–32), is now officially decreed the starting point of most fungal nomenclature. His works on lichens include *Licheneum Dianome Nova* (1821), *Lichenes Sueciae Exsiccati* (1824–7) and *Lichenographia Europaeae Reformata* (1831), an authoritative account of all known European lichens. He also studied the phanerogam genera *Hieracium*, *Carex* and *Salix*. Early in his career he formulated a new classification system (*Systema Orbis Vegetabilis*, 1825) in which each taxon was divided into four subdivisions. Later, he stated that the true relationships of species can only be learned by empirical observations of living specimens in their natural surroundings.

FRISCH, Karl von
(1886–1982)

Austrian ethologist and zoologist, born in Vienna. He studied medicine there before abandoning it for zoology which he studied at Munich, where he obtained his doctorate (1910), and at Trieste. After teaching at several universities he settled at Munich, where in 1932 he established the Zoological Institute. His early work was concerned with vision in fish, and in 1910 he showed that they perceive colours and that their visual acuity is superior to that of humans. However, he is mainly remembered for his work on honey bees. By training honey bees to forage from an artificial food source he demonstrated that they are able to distinguish odours, tastes and colours, and that the honeybee's visual spectrum allows it to see ultraviolet light. He also described how hive bees communicate the location of a source of food by means of dances on the comb. If the food is within 150 yards of the hive, the returning worker bee performs a 'round dance' which stimulates others to forage in the vicinity of the hive; for more distant sources the bee does a 'waggle dance' whose tempo and orientation with reference to the Sun's position provides information about the location of the food. In 1949 he further showed that bees can still use the Sun compass when the Sun is obscured by clouds by making use of the pattern of polarized light in the sky. In 1973 he shared the Nobel Prize for Physiology or Medicine with other pioneers of ethology, **Konrad Lorenz** and **Tinbergen**. His books include *The Dancing Bees* (1927, translated 1954) and *Animal Architecture* (1974).

FRISCH, Otto Robert
(1904–79)

Austrian–British physicist, born in Vienna and educated at Vienna University. In 1930 he became assistant to **Stern** and worked with him and **Segrè** on diffraction experiments. In 1933, under the Nazi racial laws, Stern and Frisch left Germany; Frisch began work with **Blackett** at Imperial College, London, before moving to **Niels Bohr**'s institute in Copenhagen. In 1939 with **Meitner** (his aunt) he correctly interpreted **Hahn**'s observation of the splitting of the uranium nucleus under neutron bombardment as due to what later became known as nuclear fission. He then confirmed this by repeating Hahn's experiment and observing the fission fragments. During World War II he worked at Birmingham University, where he and **Peierls** studied the possibility of nuclear chain reactions; he calculated the amount of uranium-255 required for a chain reaction and predicted the amount of energy which would be released. This work led to his involvement in the British and American atom bomb projects, and he worked for a time at Los Alamos as a trouble shooter for the Manhattan Project. After the war he returned to England to become head of the nuclear physics division of the Atomic Energy Research Establishment at Harwell, and later accepted a chair at Cambridge University. He received an OBE (1948) and was elected FRS in 1948. He wrote *Meet the Atoms* (1947).

FROBENIUS, Ferdinand Georg
(1849–1917)

German mathematician, born in Berlin, the son of a parson. He studied at Göttingen and Berlin, where he took his doctorate in 1870, and from 1875 to 1892 taught at Zürich, before returning to Berlin as professor. After early work on the theory of differential equations, he founded the theory of group representations, using it both to clarify the notion of an abstract group and also to derive properties of groups inaccessible by more direct methods. Representation theory later became essential in quantum mechanics, and a major theme of 20th-century mathematics.

FUCHS, Klaus Emil Julius
(1911–88)

German-born British physicist and spy, born in Russelsheim near Frankfurt. He was educated at Kiel and Leipzig universities. Escaping from Nazi persecution to Britain in 1933, he was interned on the outbreak of World War II, but released and naturalized in 1942. From 1943 he was one of the most brilliant of a group of British scientists sent to the USA to work on the atom bomb. In 1946 he became head of the theoretical physics division at Harwell. In March 1950 he was sentenced to 14 years' imprisonment for disclosing nuclear secrets to the Russians over a six-year period; in 1951 he was formally deprived of British citizenship. On his release in June 1959 he became an East German citizen and worked at East Germany's nuclear research centre until his retirement in 1979.

FUKUI, Kenichi
(1918–)

Japanese chemist, born in Nara prefecture. After graduating in industrial chemistry at Kyoto University in 1941, he worked for three years in the Army Fuel Laboratory. He then joined the staff of Kyoto University and obtained a PhD in engineering in 1948. He was made Professor of Hydrocarbon Physical Chemistry in 1951 and has remained at Kyoto ever since. Initially he studied a diverse set of topics, but gradually his interest in the way atomic structure affects the course of a chemical reaction eclipsed all his other interests. He found that he could describe a chemical reaction as the interaction of just two of the electronic orbitals of the reacting molecules. These he called frontier orbitals. He then determined that the course of a chemical reaction was fixed, in part, by the symmetry of the frontier orbitals. He published his conclusions couched in highly mathematical terms, and it was not until **Robert Woodward** and **Hoffmann** produced their rules for the conservation of orbital symmetry that the value of Fukui's approach became appreciated. Frontier orbital theory has been widely used in rationalizing organic reactivity and preceded sophisticated computer calculations. Fukui shared the Nobel Prize for Chemistry with Hoffmann in 1981. Since then he has continued to study the significance of frontier orbitals in more complex reactions. He has received many honours and was President of the Japanese Chemical Society from 1983 to 1984.

FUNK, Casimir
(1884–1967)

Polish-born American biochemist, born in Warsaw. He studied in Berlin and Bern, and worked as a research assistant at the Lister Institute, London (1910–13), where he attempted to isolate vitamin B_1, a cure for the dietary disease beri-beri. Incorrectly believing it was an amine, he suggested the general name 'vitamine' which was altered to 'vitamin' in 1920 by Sir Jack Drummond. Funk became head of the biochemical department at the Cancer Hospital Research Institute (1913–15), emigrated to the USA in 1915, and later headed research institutes in Warsaw (1923–7) and Paris (1928–39). In 1929 he achieved a crude extract of the male sex hormone androsterone from human urine. Androsterone was finally isolated in a pure form by **Butenandt** in 1931.

G

GABOR, Dennis
(1900–79)

Hungarian-born British physicist, born in Budapest. After receiving a doctorate in engineering in Berlin (1927) he worked there as a research engineer, but left Germany in 1933. After spending some years with the British Thompson Houston Company in Rugby—during which time he wrote the first book on the electron microscope—he was appointed to a readership at Imperial College, London (1948), and later became Professor of Applied Electron Physics (1958–67). An inventor rather than a discoverer, he did important work on communications theory and information theory throughout his career, but he is best remembered for conceiving in 1947 the technique of (and the name) holography; this is a method of photographically recording and reproducing three-dimensional images, and for this he was awarded the Nobel Prize for Physics in 1971. The invention came about during work on another of his long-term preoccupations; the improvement of the resolution of the electron microscope. Gabor also worked on television technology, optical processing methods for improving image quality, and an acoustical technique analogous to holography for detecting dense objects in water or under ground. In the late 1960s he developed an acute interest in the socio-political and environmental questions raised by the Club of Rome, an international group which first raised the alarm about the clash between economic expansionism and limitations imposed by the Earth's finite resources. He visited many countries to lecture on these topics.

GADOLIN, Johan
(1760–1852)

Finnish chemist, born in Turku. He studied at Turko, Finland, and Uppsala, Sweden, becoming Professor of Chemistry at Uppsala from 1797 to 1822. Travels on the Continent put him in touch with some of the leading scientists of the day, and he soon accepted **Lavoisier**'s discoveries about combustion and his system of chemical nomenclature. He is remembered for his investigations of the rare earths, analysing a new black mineral from Ytterby, Sweden, and isolating from it a rare earth mineral, yttria, in 1794. This proved to be an important step towards identifying the remaining undiscovered elements: over the next century yttria was found to contain the oxides of nine new rare earth elements, leading eventually to the establishment of the whole series. Around 30 years after Gadolin's death at Stockholm, one of these was discovered by Jean Charles Galissard de Marignac and **Boisbaudran**, who named it gadolinium in his honour.

GAFFKY, Georg Theodor August
(1850–1918)

German bacteriologist, born in Hanover. He was educated at the University of Berlin, where after service in the Franco-Prussian War, he obtained his MD in 1873. He served as **Koch**'s assistant from 1880 to 1885. After holding the Chair of Hygiene at the University of Giessen (1888–1904), he succeeded Koch in Berlin as director of the institute which was renamed the Koch Institute in 1912. He isolated and obtained a pure culture of the typhoid bacillus for the first time (1884), and accompanied Koch on his trips to Egypt and India (1883–4), during which the vibrio responsible for cholera was discovered. He made a further visit to Egypt in 1897 to work on the bubonic plague.

GAHN, Johan Gottlieb
(1745–1818)

Swedish chemist and mineralogist, born in Voxna, Gävleborg. He studied at Uppsala, and after some years as a laboratory assistant there, he was sent by the College of Mining to the famous copper mine at Falun. Here he so improved smelting methods and the use of by-products that industrialists and scholars flocked to learn from him. During the American Revolution, Falun supplied the colonists with copper to sheathe their ships. Gahn isolated metallic manganese, developing a way to prepare it on a larger scale. He also discovered selenium, and in conjunction with **Scheele**, found phosphoric acid in bones. Gahn died in Falun. The mineral gahnite (zinc aluminium oxide) is named after him.

GAJDUSEK, Daniel Carleton
(1923–)

American virologist, born in Yonkers, New York. He studied physics at Rochester and medicine at Harvard. He spent much time in Papua New Guinea, studying the origin and dissemination of infectious diseases amongst the Fore people, especially a slowly developing lethal viral disease called *kuru*. He showed that it was spread through cannibalism, as some women and children ritually ate the brains of dead *kuru* victims. The causative agent of *kuru* was the first of the 'slow viruses' to be identified; these viruses have since been implicated in other diseases, such as sheep scrapie, 'mad cow' disease (bovine spongiform encephalopathy), and Creutzfeldt–Jacob dementia of human beings. Gajdusek shared the 1976 Nobel Prize for Physiology or Medicine with **Blumberg**.

GALEN (Claudius Galenus)
(c.130–c.201 AD)

Greek physician, born in Pergamum in Mysia, Asia Minor, where his father was an architect. He studied medicine there and at Smyrna, Corinth and Alexandria. He was chief physician to the gladiators in Pergamum from AD 157, then moved to Rome and became friend and physician to the emperor Marcus Aurelius. He was also physician to emperors Commodus and Severus. Galen was a voluminous writer on medical and philosophical subjects. The work extant under his name consists of more than 80 genuine treatises and some 15 commentaries on **Hippocrates**. Galen saw himself as completing and perfecting the medical ideas of Hippocrates, whom he greatly admired. He had, however, a low opinion of most other doctors past and contemporary, and much of his writing is polemical in character. Despite the veneration he reserved for Hippocrates, Galen was an active experimentalist, dissecting animals and using the anatomical and physiological information thus obtained in constructing his theories of human bodily structures and functions. This occasionally led him to describe structures, such as the rete mirabile (a vascular network in the brain of some animals), which later doctors assumed to be present in human beings. In addition to the humoral theory of Hippocrates (which Galen accepted), Galen elaborated a physiological system whereby the body's three principal organs—heart, liver and brain—were central to living processes. The liver took ingested food and converted it to blood. The blood then went via the vena cava to the heart, where it was impressed with vital spirits. Some of it also went to the brain, having passed through the septum of the heart; in the brain further refinement into animal spirits took place. Galen was a shrewd and successful practitioner, who, if he is to be believed, cured many patients. He put great store on the pulse, which he used as a diagnostic aid. Galen also admired **Plato**'s philosophy. Although not a Christian, he was a monotheist and thus his work was easily assimilated into Christian orthodoxy in the centuries after his death. His *De usu partium* ('The uses of the parts') was in essence a hymn to the creator, whereby the organs of the body were seen as perfectly adapted to the functions which they served. For many centuries he was venerated as the standard authority on medical matters, the man who had perfected the medical systems of antiquity.

GALILEO (Galileo Galilei)
(1564–1642)

Italian astronomer, mathematician and natural philosopher. Born in Pisa, the son of a musician, he matriculated at Pisa University (1581) where he accepted the Chair of Mathematics in 1589. His discovery (1582) of the isochronism of the pendulum, which indicated the value of the pendulum as a timekeeper, was made while watching the swinging of a lamp in the cathedral of Pisa. In the study of falling bodies, Galileo showed that contrary to the Aristotelian belief that the rate at which a body falls is proportional to its weight, all bodies would fall at the same rate if air resistance were not present. He also showed that a body moving along an inclined plane has a constant acceleration, and demonstrated the parabolic trajectories of projectiles. In 1592 he moved to the University of Padua, where his lectures attracted pupils from all over Europe. He made his first contribution to astronomy in 1604 when a bright new star appeared in the constellation Ophiuchus and was shown by him to be more distant than the planets, thus confirming **Tycho Brahe**'s conclusion that changes take place in the celestial regions beyond the planets. In 1609 he reinvented the telescope, having heard an account of that device recently constructed in Holland, and perfected a refracting telescope which led to many astounding astronomical revelations published in his *Sidereus Nuncius* ('Sidereal Messenger', 1610). These included the mountains of the Moon, the multitude of stars in the Milky Way, and quite particularly, the existence of Jupiter's four satellites, the 'Medicean stars', named in honour of his future patron. The book was received with great acclaim, and Galileo was appointed 'Chief Mathematician and Philosopher' by the Grand Duke of Tuscany. On a visit to Rome in 1611 he was elected a member of the Accademia dei Lincei and feted by the Jesuit mathematicians of the Roman College. Further discoveries included the phases of Venus, spots on the Sun's disc, the Sun's rotation, and Saturn's appendages (though not then recognized as a ring system). These brilliant researches made Galileo confident that he might express his conviction in the truth of the heliocentric **Copernican** system which had been proscribed as heretical by the ecclesiastical authorities and which Cardinal Bellarmine had formally asked him in 1616 to abstain from advocating. In his second book, *The Assayer* (1623), Galileo apparently followed that advice, but in his later controversial book *A Dialogue on the Two Principal Systems of the World* (1632), he defended the Copernican system, in disregard of Bellarmine's admonition. The sale of the book was prohibited and Galileo was cited to Rome by the Inquisition. He was detained, finally examined by the Inquisition and under threat of torture recanted. After several days in the custody of the Inquisition he was relegated first to Villa Medici, then to the house of his friend the archbishop of Siena, and finally allowed to live under house arrest in his own home at Arcetri, near Florence; there he continued his researches and completed his *Discourses on the Two New Sciences* (1638), in many respects his most valuable work, in which he discussed at length the principles of mechanics. His last telescopic discovery, that of the Moon's librations, was made in 1637 only a few months before he went blind. He continued working until his death on 8 January 1642. The sentence passed on him by the Inquisition was formally retracted by Pope John Paul II on 31 October 1992.

GALLE, Johann Gottfried
(1812–1910)

German astronomer, born in Pabsthaus, near Wittenberg, and educated at Berlin University, graduating in mathematics and physics in 1833. He taught for two years in a school before being invited by **Encke**, his former teacher, to become his assistant at the newly established Berlin Observatory where Encke was director. After 16 years in this post he became director of the observatory at Breslau where he spent the next 40 years (1851–91). He took a special interest in comets, discovered three new ones and for many years computed ephemerides of comets and minor planets for the *Astronomisches Jahrbuch*. His most dramatic discovery, made in Berlin, was of the planet Neptune whose existence had been theoretically predicted and whose expected position had been calculated by **Le Verrier**. Following a request from Le Verrier, Galle searched the specified area of sky and discovered on 23 September 1846 an uncharted object which was confirmed from its motion as the new planet. In 1872 he proposed the use of asteroids rather than regular planets for determinations of the solar parallax on account of their point-like images, a suggestion which bore fruit in a successful international campaign in 1888–9. The method was last used at the closest approach of the minor planet Eros in 1930–1.

GALOIS, Évariste
(1811–32)

French mathematician, born in Bourg-la-Reine. He entered the École Normale Supérieure in 1829, but was expelled in 1830 due to his extreme republican sympathies. Politically active, he was imprisoned twice, and was killed in a duel. His mathematical reputation rests on fewer than 100 pages of work of original genius, much published posthumously: a memoir on the solubility of equations by radicals, and a mathematical testament written the night before his death giving the essentials of his discoveries on the theory of algebraic equations and Abelian integrals. Some of his results had been independently obtained by **Niels Henrik Abel**, but Galois gave a theoretical setting for them that proved exceptionally fertile. The brevity and obscurity of his writing delayed the understanding of his work, but gradually after its publication by **Liouville** in 1846 it came to be seen as a cornerstone of modern algebra in which the concept of a group first became of central importance.

GALTON, Sir Francis
(1822–1911)

English scientist, cousin of **Charles Darwin**. Born in Birmingham to wealthy parents, Galton studied medicine at the Birmingham Hospital and King's College, London, and graduated from Trinity College, Cambridge, in 1844, when his father's death left him with an ample fortune. In 1846 he travelled in North Africa, undertaking a major expedition in 1850 in South Africa, and publishing his *Narrative of an Explorer in Tropical South Africa* and the *Art of Travel* (1855). In 1856 he was elected FRS, and for the rest of his long life he adopted the style of a London-based gentleman scientist. Galton's investigations covered many domains. He was, for instance, a pioneer in meteorology, discovering and naming the anticyclone and publishing, in *The Times* in 1875, the first newspaper weather map. He was also an early enthusiast for the use of fingerprinting in crime detection; colour blindness intrigued him. What unified his researches was an accent upon quantification. He trailblazed the use of statistical techniques, in 1888 presenting to the Royal Society a method for calculating correlation coefficients. An early convert to the evolutionary thinking of his cousin Darwin, Galton considered the respective roles of environment and heredity in shaping human and animal populations. To resolve the nature/nurture conundrum, he selectively bred plants and animals, and made special studies of the medical histories of identical twins. Growing convinced of the key role of heredity, Galton coined the term eugenics to designate the science of creating superior offspring and of preventing inferior populations. He expounded his hereditarian and eugenic ideas in *Hereditary Genius* (1869), *English Men of Science: their Nature and Nurture* (1874) and *Natural Inheritance* (1889), and endowed a eugenics chair at London University. He was knighted in 1909.

GALVANI, Luigi
(1737–98)

Italian physiologist, born in Bologna. Galvani studied in his home town, in 1768 becoming a lecturer in anatomy, and from 1782 Professor of Obstetrics. He is famous for the discovery of animal electricity. This he did by chance. He noted that dead frogs undergoing drying by being fixed to an iron fence by brass skewers suffered convulsions. He then showed that paroxysms followed if a frog was part of a circuit involving metals. Galvani believed that electricity of a hitherto unknown sort (animal electricity, or galvanism) was generated in the material of the muscle and nerve. This was later proved erroneous by **Volta** who in 1800 devised the voltaic pile and resolved the problem by showing that the current arose from the metals, not the frog. Galvani's name lives on in the word 'galvanized' meaning stimulated as if by electricity, and in the galvanometer, used from 1820 to detect electric current.

GAMOW, George
(1904–68)

Russian-born American physicist, born in Odessa, the son of a teacher. He made important advances in both cosmology and molecular biology. He was educated at Leningrad University, where he was later appointed Professor of Physics (1931–4). After research at Göttingen, Copenhagen and Cambridge, he moved to the USA as Professor of Physics at George Washington University (1934–55) and at Colorado (1956–68). In 1928 he showed, independently of **Condon** and Gurney, that alpha decay could be explained by quantum-mechanical tunnelling of alpha particles through the nuclear potential barrier. This was one of the first applications of quantum mechanics to nuclear processes. In 1946 he proposed the Big Bang theory of the creation of the universe and coined the term 'ylem' to describe the primeval hot dense sea of neutrons. In 1948, together with **Alpher** and **Bethe**, he developed a theory to explain the creation of all the elements during the Big Bang. Although the Alpher–Bethe–Gamow theory has been superseded by the stellar nucleosynthesis model of **William Fowler**, **Hoyle**, and **Geoffrey** and **Margaret Burbidge**, and the hot dense sea of neutrons is now believed to have been a hot dense sea of elementary particles such as quarks, the Big Bang model is still the accepted model for the creation of the universe. From this theory, Gamow predicted that background radiation exists throughout the universe as a remnant of the Big Bang. This was later discovered by **Penzias** and **Robert Wilson**. In molecular biology he made a major contribution to the problem of how the order of the four nucleic acid bases in DNA chains governs the synthesis of proteins from amino acids. He realized that short sequences of the bases could form a 'code' capable of carrying information directing the synthesis of proteins, a proposal shown by the mid-1950s to be correct. In addition to his research work, Gamow created the character of Mr Tompkins which he used to communicate the ideas of relativity, quantum mechanics and particle physics to the general public.

GARROD, Sir Archibald Edward
(1857–1936)

English physician, born in London into a prominent medical and scientific family. He was educated at Oxford and St Bartholomew's Hospital, London. He held appointments at several London hospitals, including, eventually, St Bartholomew's, where he did important work on arthritis, co-authored a textbook on children's diseases and, above all, investigated what he first called *Inborn Errors of Metabolism* (1909). The model for this concept was alkaptonuria, an inherited metabolic disorder in which an acid is excreted in quantities in the urine, causing it to blacken on standing. The other rare hereditary conditions which Garrod discussed were albinism, cystinuria and pentosuria. Garrod's work was an early application of the new Mendelian genetics in the study of human disease. In his *The Inborn Factors in Disease* (1931), he developed his ideas of biochemical individuality. During World War I, he served in Malta as a colonel in the Royal Army Medical Corps, where he developed an interest in classical archaeology. A pioneer of clinical investigation, Garrod succeeded **Osler** as Regius Professor of Medicine at Oxford in 1920.

GARSTANG, Walter
(1868–1949)

English zoologist and marine biologist, born in Blackburn. Educated at Oxford, he joined the staff of the Plymouth marine station when it was founded in 1888 and became Naturalist in Charge of Fisheries Investigations for the Marine Biological Association (1897–1907). From 1902 to 1907 he was convenor of the Over-Fishing Committee of the International Council for Exploration of the Sea and directed research into the problem of depleted plaice stocks in the North Sea. He became Professor of Zoology at Leeds University in 1908. There he carried out embryological and life history studies on marine organisms. He was an outspoken opponent of **Haeckel**'s biogenetic law at a time when it was generally accepted, and showed that larval forms had evolved due to their having different selective forces acting upon them than on the adults. His criticisms were propounded in a series of papers cumulating in *The Theory of Recapitulation: a Critical Re-statement of the Biogenetic Law*, delivered to the Linnaean Society in 1921. He developed the concept of 'paedomorphosis' to describe the phenomenon whereby larval adaptations affect the form of the adult. His best-known example occurs in gastropod snails, where torsion of the body relative to the shell and viscera is of advantage to the larva but apparently disadvantageous to the adult. His studies of the development of sea-squirts led him to propose the idea that the chordates have evolved from motile larval forms. His appreciation of the difficulty of communicating such concepts led him to write a series of humorous verses in which he expounded his theories.

GASKELL, Walter Holbrook
(1847–1914)

English physiologist, born in Naples and educated in Cambridge at Trinity College, initially in mathematics and then in the Physiological Laboratory, where he came under the influence of **Foster**. He qualified in medicine after clinical training at University College London, but never practised, returning instead to physiology. He travelled to Leipzig where he studied with **Ludwig**, and returned to Cambridge in 1875 where he was to remain for the rest of his life. Supported by private means, he also held a university lectureship from 1883 until 1914, and was a Fellow of Trinity Hall from 1889. His initial research topic, suggested by Foster, was a study of vasomotor action, and he correctly identified by physiological experimentation several nerves which controlled dilation of the blood vessels. In the 1880s he turned his attention to problems of the heart itself, particularly the question of the heartbeat. Gaskell's work, undertaken on cold-blooded vertebrates which had a slower heartbeat than mammals, provided convincing evidence that the muscles of the heart exhibited inherent rhythmicity, independent of external influences. This was significant testimony to support the 'myogenic' theory of the heart's action, in contrast to the opposing 'neurogenic' view. Gaskell's experiments led him to a more extensive examination of the effects of the autonomic nervous system, such as the nerves that control the heart, lungs and viscera, and his detailed anatomical and physiological work revealed the presence of two major complementary branches of the autonomic system, the sympathetic and the parasympathetic systems. These studies raised important questions about the evolution and origin of the autonomic nervous system, and for the rest of his life Gaskell devoted himself almost completely to the problem of vertebrate evolution.

GASSENDI, Pierre
(1592–1655)

French philosopher and scientist, born in Champtercier, Provence. Ordained as a priest in 1616, he became Professor of Philosophy at Aix in 1617 and Professor of Mathematics at the Collège Royal in Paris in 1645. **Kepler**, **Galileo** and **Mersenne** were among his friends. He was a strong advocate of the experimental approach to science and tried to reconcile an atomic theory of matter (based on the Epicurean model) with Christian doctrine. He may be best known as an early critic of **Descartes** in the Fifth of the *Objections* (1642) to the *Meditations*, but he also wrote on Epicurus, **Tycho Brahe** and **Copernicus**. His other works include *Exercitationes Paradoxicae adversus Aristoteleos* (1624), *Institutio Astronomica* (1647) and *Syntagma Philosophicum* (published posthumously in 1658).

GASSER, Herbert Spencer
(1888–1963)

American physiologist, born in Plattville, Wisconsin. He graduated in medicine from Johns Hopkins University in 1915 and moved to the department of physiology at Washington University in St Louis in 1916, to rejoin **Erlanger**, an instructor from his student years. Thus began a fruitful collaboration in neurophysiology that was to last for many years and which was to result in their shared Nobel Prize for Physiology or Medicine in 1944. Gasser's earliest Washington collaboration, however, was with H S Newcomer, a former colleague from Wisconsin; together they built an electronic amplifier that greatly magnified the weak electrical signals that could be recorded from nerve fibres. This advance, closely followed by similar developments by **Adrian**, was enhanced by the adaptation of the cathode-ray oscillograph to visualize, record and study, the magnified form of the action potential of the nerve. These powerful tools allowed new ways of dissecting and analysing the nature and function of the nerve fibres, and Gasser and Erlanger devoted the major parts of their lives to these problems. In 1931 Gasser moved to Cornell University as Professor of Pharmacology, and in 1935 he was appointed as Director of the Rockefeller Institute for Medical Research in New York where he shouldered a heavy administrative responsibility. His personal research continued into the properties of isolated nerve trunks, and the functional differentiation of nerve fibres. In particular his work had important bearing on understanding the physiology of sensation.

GATES, William Henry
(1955–)

American computer software scientist and entrepreneur, born in Seattle, Washington state, the son of a lawyer. In 1973 he graduated from Seattle High School and then registered at Harvard University in 1975. In that year, however, Gates was approached by a colleague, Paul Allen, who was a young computer programmer who worked outside Boston. Allen showed Gates a copy of the journal *Popular Electronics*, which described the world's first microcomputer kit for the Altair 8800, designed by Micro Instrumentation & Telemetry Systems (MITS). Allen and Gates offered to write the first microcomputer BASIC (Beginner's All-purpose Symbolic Instruction Code) for the Altair and delivered it six weeks later. Allen promptly became the software director for MITS; Gates dropped out of Harvard and became a freelance software writer. In 1976 they founded Microsoft Corporation in Redmond, Washington state, with Gates becoming chairman. Microsoft's growth was phenomenal, with its operating systems (MS-DOS, 1981) and applications programs (such as Windows, 1984) becoming computer industry standards by the end of the 1980s. By 1991 the revenues of Microsoft Corporation were $1.8 billion and Gates was the wealthiest man in the USA.

GAUSS, Carl Friedrich
(1777–1855)

German mathematician, astronomer and physicist, born in Brunswick to poor parents. His great mathematical precocity came to the notice of the Duke of Brunswick who paid for his education at the Collegium Carolinum at Brunswick and the University of Göttingen. In 1807 he became Director of Göttingen Observatory, where he remained for the rest of his life. He is generally regarded as one of the greatest mathematicians of all time. A notebook kept in Latin by him as a youth was discovered in 1898 showing that, from the age of 15, he had conjectured and often proved many remarkable results, including the prime number theorem. In 1796 he announced that he had found a ruler and compass construction for the 17 sided polygon, and in 1801 published his *Disquisitiones arithmeticae*, containing wholly new advances in number theory. The same year he was the first to rediscover the asteroid Ceres found by **Piazzi** in 1800 and since lost behind the Sun. This led him to the study of celestial mechanics, on which he published a treatise in 1809, and to statistics where he was the first to use the method of least squares. From 1818 to 1825 he directed the geodetic survey of Hanover, and this and his astronomical work involved him in much heavy routine calculation, leading to his study of the theory of errors of observation. Nevertheless he found time for much work on pure mathematics, including differential equations, the hypergeometric function, the curvature of surfaces, four different proofs of the fundamental theorem of algebra, six of quadratic reciprocity and much else in number theory. In physics he studied the Earth's magnetism and developed the magnetometer in conjunction with **Wilhelm Eduard Weber**, and gave a mathematical theory of optical systems of lenses. Manuscripts unpublished until long after his death show that he had made many other discoverics including the theory of elliptic functions that had been published independently by **Niels Henrik Abel** and **Jacobi**, and had come to accept the possibility of a non-Euclidean geometry of space, first published by **Bolyai** and **Lobachevski**.

GAUTIER, Hubert
(1660–1737)

French civil engineer, born in Nîmes. He was one of the first to recognize the importance of applying scientific principles to the design and execution of engineering works. After nearly 30 years as chief government engineer of the Province of Languedoc, in 1716 he was appointed inspector of the newly created Corps des Ponts et Chaussées, responsible for virtually all public works throughout France. Around the same time he published two classic textbooks, the *Traité des Chemins*

(1715) and the *Traité des Ponts* (1716), in which he summarized ancient and contemporary engineering practice, emphasizing the importance of strict supervision, careful testing of materials, and observance of the principles of structural mechanics. One of the most important of his recommendations was to reduce the width of bridge piers, in order to avoid excessive obstruction of waterways and the creation of dangerously high water levels in times of flood.

GAY-LUSSAC, Joseph Louis
(1778–1850)

French chemist and physicist, born in Saint-Léonard, Haute Vienne. He was educated at the École Polytechnique and the École des Ponts et Chaussées. He became assistant to **Berthollet** in 1800 and subsequently held various posts including Professor of Chemistry at the École Polytechnique (from 1810), Professor of Physics at the Sorbonne (1808–32), Professor of Chemistry at the National Museum of Natural History (from 1832), superintendent of the government gunpowder factory (from 1818) and Chief Assayer to the Mint (from 1829). He became an Academician (1806), member of the chamber of deputies (1831) and member of the upper house (1839). Gay-Lussac was elected an Honorary Fellow of the Chemical Society in 1849. His earliest research work was on the expansion of gases with temperature increases, and he discovered independently the law which in Britain is commonly known as **Charles**'s law. In 1804 he made balloon ascents in association with **Biot** to make magnetic and atmospheric observations, and in 1805–6 he travelled with **Humboldt**, making measurements of terrestrial magnetism. In 1808 he published his important law which states that when chemical combination occurs between gases, the volumes of those consumed and of those produced, measured under standard conditions of temperature and pressure, are in simple numerical ratio. This was based on work which he had begun with Humboldt in 1805. From around 1808 Gay-Lussac's work became more purely chemical, and much of it was done in collaboration with **Thénard**. Their work included the isolation and investigation of sodium, potassium, boron and silicon, extensive studies of the halogens (involving controversy with Sir **Humphry Davy**), and the improvement of methods of organic analysis. His last great pure research was on prussic acid and cyanogen, and their derivatives. During the later part of his career, Gay-Lussac did much work as a technical adviser to industry.

GEBER
(14th century)

Spanish alchemist. Nothing is known about him except his name as author of books on chemical and alchemical theory and practice. He took the name 'Geber' (Latin for 'Jabir') from Jabir ibn Haiyan, a famous Islamic alchemist and physician of the 8th century. It was through Geber's works, for example *Summa Perfectionis Magisterii*, that the discoveries of the early Arab chemists, along with many basic laboratory techniques, were relayed to Europe. For example, Geber described how to purify chemical compounds, how to prepare nitric and sulphuric acids, and how to construct a laboratory furnace. He also dealt with the supposed transmutation of base metals into gold. His works, written in Latin, were translated into several languages and continued to be influential until the 16th century. Because of their clarity, and in many respects their accuracy, they helped to make alchemy more respectable.

GEER, Baron Gerhard Jacob de
(1858–1943)

Swedish Quaternary geologist, born in Stockholm, the son of a prime minister of Sweden. Educated at Uppsala, he was Professor of Geology there (1897–1924) and founded the Geochronological Institute of the University of Stockholm, serving as its first director from 1924. He was himself a member of parliament from 1900 to 1905. He first took an interest in Quaternary geology as a student, and later, following fieldwork in Spitzbergen, he turned his attention to local Quaternary deposits and devised a novel and valuable method for dating by comparing sequences of varves (the annual deposits of sediment under glacial meltwater). He was able to decipher an annual chronology reaching back some 15 000 years from the present day. He successfully correlated the Swedish varve sequence with others from the Himalayas, Iceland, Newfoundland, Canada, Argentina, New Zealand and elsewhere, demonstrating global climatic events and greatly advancing knowledge of the later geological history of the last Ice Age. The method was eventually enhanced by the use of increasingly sophisticated radioisotope methods.

GEGENBAUR, Karl
(1826–1903)

German comparative anatomist and apostle of evolution theory. Born in Würzburg, he was educated in his home university, graduating in medicine in 1851. He became professor at Jena in 1856 and at Heidelberg in 1873.

There he attracted distinguished students, including Richard and **Oscar Hertwig**. He won a towering contemporary reputation for his phenomenal knowledge of vertebrate musculature and his expertise in comparative anatomy and morphology. On this basis Gegenbaur became one of the early and leading German advocates of evolutionary theory—notwithstanding a heart-felt commitment to Catholicism. On evolutionary questions, he was a close friend and supporter of **Haeckel**. He regarded morphological data and thinking the surest foundation for the demonstration of the truth of the evolutionary theory. His chief work, *Comparative Anatomy* (translated 1878), threw much light on the evolution of the skull from his study of cartilaginous fishes.

GEHRING, Walter Jacob
(1939–)

Swiss geneticist, born in Zürich. He was educated at the Realgymnasium, Zürich, the University of Zürich and Yale University. He was Associate Professor of Anatomy and Molecular Biophysics at Yale University from 1969 to 1972, and has been Professor of Genetics and Developmental Biology at the University of Basle since 1972. His career for many years was concerned with the genetics of *Drosophila melanogaster*, the fruit fly, and in the mid-1980s he discovered a short regulatory DNA sequence which he called the homeobox. Certain mutations in the fruit fly affect whole developmental pathways; for example, antennae can be converted to limbs by a mutation called antennapoedia. Such mutations are called homeotic mutants and are controlled by homeobox sequences. It was later found that similar control sequences occur in mammals, thus encouraging new lines of research to understand the biology and molecular biology of such major developmental pathways.

GEIGER, Hans Wilhelm
(1882–1945)

German physicist, born in Neustadt-an-der-Haardt and educated at the University of Erlangen where he received his PhD in 1906. He then worked under **Rutherford** at Manchester (1906–12). With Rutherford, he devised a means of detecting alpha particles in 1908. The instrument consisted of a gas-filled tube with a wire at high electric potential down the centre. Passage of an α-particle produced ionization in the gas, and this would result in a short burst of current which could be measured. They used the device to show that α-particles carry two units of charge. Soon after, Geiger and Ernest Marsden demonstrated that when α-particles are incident on a thin metal foil (gold, silver and copper were used), they are occasionally deflected through large angles. This led to Rutherford to propose that the atom has a compact nuclear core surrounded by electrons. With Rutherford, Geiger also showed that two α-particles are emitted in the radioactive decay of uranium, and with J M Nuttall he demonstrated the linear relationship between the logarithm of the range of α-particles and the radioactive time constant of the emitting nucleus, now called the Geiger–Nuttall rule. In 1912 he became head of the Physikalisch Technische Reichsanstalt in Berlin, where he obtained confirmatory evidence of the **Compton** effect. He later became professor at Kiel University (1925), where he and Walther Müller made improvements to the particle counter, resulting in the modern form of the Geiger–Müller counter, which also detects electrons and ionizing radiation.

GEIKIE, Sir Archibald
(1835–1924)

Scottish geologist, born in Edinburgh and educated at the university there. In 1855 he was appointed to the Geological Survey in Scotland on the recommendation of **Hugh Miller**, and in 1867 he became its director. From 1871 to 1881 he was Professor of Geology at Edinburgh University. He subsequently became Director-General of HM Geological Survey (from 1882) and also head of the Geological Museum of London. Geikie undertook notable work on the Old Red Sandstone, volcanic geology and the scenery of Scotland, and did much to encourage microscopic petrography. His many books included *Textbook of Geology* (1882), *The Ancient Volcanoes of Great Britain* (1897), *The Founders of Geology* (1905), and other works on geology and the history of science. He was elected FRS at the young age of 29 and knighted in 1907. His brother James Geikie (1839–1915) was also an accomplished geologist and prehistorian who became Professor of Geology at Edinburgh University (1882–1914).

GEITEL, Hans Friedrich
(1855–1923)

German physicist, born in Brunswick, who published almost all his work with **Elster**. They invented the first practical photoelectric cell, a photometer and a Tesla transformer. Geitel received honorary doctorates from the universities of Göttingen and Brunswick, and after his retirement he was made an honorary professor at Brunswick.

GELFAND, Izrail Moiseyevich
(1913–)

Russian mathematician, born in Krasnye Okny. He studied in Moscow, and became professor at Moscow State University in 1943. The leader of an important school of Soviet mathematicians, he has worked mainly in **Banach** algebras, the representation theory of **Lie** groups, important in quantum mechanics, and in generalized functions, used in solving the differential equations that arise in mathematical physics. In particular, his work has led to essential mathematical methods used in symmetry theories for fundamental particles. Gelfand has also contributed to the life sciences through his mathematical studies of neurophysiology and cell biology.

GELL-MANN, Murray
(1929–)

American theoretical physicist, born in New York City. He entered Yale University when he was only 15 years old, and graduated in 1948. He received his doctorate from MIT and spent a year at the Institute of Advanced Studies in Princeton before joining the Institute for Nuclear Studies at Chicago University, where he worked with **Fermi**. He became Professor of Theoretical Physics at Caltech in 1956. In 1953 Gell-Mann and Japanese physicist Kazuhiko Nishijima independently explained the properties of some of the proliferation of new subatomic particles by assigning them a property, or quantum number, conserved by the strong nuclear force and lacked by earlier known particles. Gell-Mann named this property 'strangeness'. Using this concept, with two other quantum properties called 'up' and 'down', he and Yuval Ne'eman (independently) used 'strangeness' to group mesons, nucleons (neutrons and protons) and hyperons into multiplets of 1, 8, 10 and 27. The members of the multiplets were then related by symmetry operations, specifically unitary symmetry of dimension 3, or SU(3). They were then able to form predictions in the same way that **Mendeleyev** had about chemical elements, and concluded that the hyperon (Ω^-) with strangeness number 3 was missing; by using the properties of the rest of the multiplet, they predicted its mass and decay modes (1962). It was discovered in 1964 within 0.2 per cent of the predicted mass. Their book on this work was entitled *The Eightfold Way* (1964), a pun on the Buddhist eightfold route to nirvana. Gell-Mann and **Zweig** introduced the concept of quarks, of which there would be three types, one for each of the new quantum numbers. This meant that quarks would have fractional electric charges compared to the electron: $+\frac{2}{3}e$ for the up quark and $-\frac{1}{3}e$ for the down and strange quarks (later confirmed experimentally by **Jerome Friedman** and others). These make up the other nuclear particles (hadrons). For this work he was awarded the Nobel Prize for Physics in 1969.

GENNES, Pierre-Gilles de
See **DE GENNES, Pierre-Gilles**

GENTH, Frederick Augustus
(1820–93)

German-born American mineralogist, chemist and collector, born in Wächtersbach, near Hanau. He was educated at Heidelberg, Giessen and the University of Marburg, where he received his PhD in 1846. He spent three years as a chemical assistant to **Bunsen** before emigrating to the USA, where he established an analytical laboratory in Philadelphia. There he devoted himself to commercial analysis, research and chemistry teaching. During 1849–50 he was for a while superintendent of the Washington Mine, North Carolina, but returned thereafter to his laboratory. He was appointed Professor of Chemistry at the University of Pennsylvania (1872–88) and was chief chemist and mineralogist to the 2nd Pennsylvania Geological Survey. Genth worked in mining geology for much of his life, publishing reports on *Corundum, its Alterations and Associated Minerals* (1873), *The Mineralogy of Pennsylvania* (1875) and on the *Minerals of North Carolina* (1871, 1881). He also undertook important research on meteorites, chemistry and rare minerals. He produced an important monograph on the ammonium–cobalt compounds (1856), and discovered 23 new mineral species; the mineral genthite is named in his honour.

GEOFFROY SAINT-HILAIRE, Étienne
(1772–1844)

French zoologist, born in Étampes. In 1793 he became Professor of Zoology at the Museum of Natural History in Paris and began its great zoological collection. In 1798 he was a member of the scientific commission accompanying Bonaparte to Egypt; in 1809 he was appointed Professor of Zoology at the Faculty of Sciences in Paris. Through a comparative study of embryonic forms he proposed a number of general laws. His law of development stated that organs do not arise or disappear suddenly during evolution but appear gradually, and that this accounts for vestigial structures. He further said that one organ grows at the expense of another, and that the equivalent organs and structures are to be found in all animals. His disagreement with **Cuvier** (1769–1832) on this point was a celebrated controversy at the time. Although it is impossible to embrace the invertebrates into such a holistic scheme, the idea perhaps created a sympathetic climate for Darwinism. In *Philosophie anatomique* (2 vols, 1818–20) he dealt with one of his other major interests, teratology, the study of animal malformations. He also wrote *L'histoire naturelle des mammifères* (1820–42), *Philosophie zoologique* (1830) and *Études progressives d'un naturaliste* (1835).

GEOFFROY SAINT-HILAIRE, Isidore
(1805–61)

French zoologist, born in Paris, son of **Étienne Geoffroy Saint-Hilaire**. In 1824 he became assistant naturalist at the National Museum of Natural History in Paris and he later succeeded his father as Professor of Comparative Anatomy (1837) and Professor of Zoology (1850) at the Faculty of Sciences of Paris. Continuing the work of his father, he studied teratology, the analysis of anatomical abnormalities, and in 1832–7 published *Histoire générale et particulière des anomalies de l'organisation chez l'homme et les animaux* (4 vols), a work on monstrous forms. He was interested in the domestication of animals and studied the way in which domestic animals adapt to different climates. His investigation into the domestication of alien animals in France appeared in *Domestication et naturalisation des animaux utiles* (1854). In 1852 he published the first volume of his *Histoire naturelle générale des règnes organiques*, but he died before completing the third volume.

GERHARDT, Charles Frédéric
(1816–56)

French chemist, born in Strasbourg. To prepare him to run a factory producing white lead, Gerhardt was sent, by his father, to the Polytechnicum in Karlsruhe to study chemistry and later to a commercial college in Leipzig. However, the owner of his lodgings in Leipzig encouraged his scientific leanings and he refused to join his father in running the factory. After a violent quarrel he joined a regiment of lancers, but left after a short time to study chemistry with **Liebig** (1836–7). After a final but unsuccessful attempt at reconciliation with his father he left for Paris to study with **Dumas**. He then secured a staff appointment (1841) in Montpellier, but the remoteness from Paris led to difficulties and in 1851 he resigned and returned to the capital. In 1854 he was offered two chairs in Strasbourg. Only two years later he was suddenly seized by illness and died within a week. His outstanding contribution to chemistry was the 'theory of types' in which he proposed that all organic molecules belong to one of a small number of families or types. Thus ethanol is of the water type with one hydrogen atom replaced by the ethyl 'radical'. The type theory of Gerhardt played a significant part in bringing some semblance of order to organic chemistry, but it was unable to rationalize the reactions of organic compounds.

GERMAIN, Sophie
(1776–1831)

French mathematician, born in Paris. Self-educated until the age of 18, she studied lecture notes procured from the newly established École Polytechnique, to which women were not admitted. In the guise of a male student named Le Blanc, she submitted a paper on analysis which so impressed **Lagrange** that he became her personal tutor. During a career in which she corresponded with **Legendre** and **Gauss**, she gave a more generalized proof of **Fermat**'s 'last theorem' than had previously been available, and developed a mathematical explanation of the **Chladni** figures in response to a challenge from the French Academy of Sciences. She went on to derive a general mathematical description of the vibrations of curved as well as plane elastic surfaces. Her *Recherches sur la théorie des surfaces élastiques* was published in 1821; she also wrote philosophical works such as *Pensées diverses*, published posthumously.

GERMER, Lester Halbert
(1896–1971)

American physicist, born in Chicago. While on the research staff of the Western Electric Co (1917–53) he observed with **Davisson** the diffraction of electrons by a crystal (1927) confirming **de Broglie**'s wave theory of matter.

GESNER, Conrad,
Latin **Conradus Gesnerus**
(1516–65)

Swiss naturalist and physician, born in Zürich. In 1537 he became Professor of Greek at Lausanne and published a Greek–Latin dictionary, and in 1541 he was appointed Professor of Philosophy and Natural History at Zürich. He published 72 works, and left 18 others in progress. His *Bibliotheca Universalis* (1545–9) contained the titles of all the books then known in Hebrew, Greek and Latin, with criticisms and summaries of each. This work has led to him being described as 'the father of bibliography'. His *Historia Animalium* (1551–8) attempted to describe all animals then known. He collected over 500 plants not recorded by the ancients, and was preparing a third major work at the time of his early death; the beautiful and accurate engraved illustrations for this work were lost until the 18th century, but were then reprinted, most recently in eight volumes published between 1973 and 1980. The drawings include details of flowers, fruit and seed, this last sometimes being observed with a magnifying glass. He also wrote on medicine, mineralogy and philology. His correspondence with many other botanists was published by Jean Bauhin, elder brother of **Caspar Bauhin**, in 1591. In it, he first alludes to the concepts of genus and species, and stresses the significance of flowers, fruit and seed in identification and discrimination.

GIACONNI, Ricardo
(1931–)

American astrophysicist, born in Genoa, Italy. After obtaining a doctorate at the University of Milan, he emigrated to the USA in 1956, becoming a research associate at the University of Bloomington, Indiana. In 1958 he moved to Princeton. With **Rossi** in 1960, he was the first to propose the use of a paraboid of revolution as a space X-ray telescope mirror. In June 1962 a sounding rocket that contained Geiger counters and which spent only a few minutes above the Earth's absorbing atmosphere was launched by a team of X-ray astronomers headed by Giaconni to search for X-rays emitted by flourescence from the Moon's surface. Accidentally they discovered the first extrasolar source of X-rays, the star Scorpious X-1. They also discovered the extragalactic background radiation. Giacconi's group built the first orbiting X-ray detector, a small astronomy satellite that was named Uhuru. This led to the discovery of a host of X-ray stars. In 1966 Giaconni showed that absorption of X-rays by interstellar gas and dust limited studies at long X-ray wavelengths. He has recently taken up the post of Director of the European Southern Observatory.

GIAEVER, Ivar
(1929–)

Norwegian-born American physicist, born in Bergen. He studied electrical engineering in Trondheim, served in the Norwegian army (1952–3), then emigrated to Canada in 1954 to work for the General Electric Company. After moving in 1956 to the General Electric Research and Development Center in Schenectady, New York, he examined tunnelling effects in superconductors. Using structures consisting of a sandwich of aluminium, aluminium oxide (an insulator) and lead, he measured their current–voltage characteristics at 4 kelvin, a temperature at which lead superconducts but aluminium does not. He found that the tunnel current–voltage relationship was highly non-linear, and this observation resulted in a simple technique of measuring superconductor energy gaps. His subsequent research on fine detail in the current–voltage characteristics yielded information that led to a great advance in the understanding of the nature of superconductors. His field of work, of great value in microelectronics, had previously been the subject of related work by **Esaki** and was later further developed by **Josephson**, and all three men shared the Nobel Prize for Physics in 1973 for their contributions. Giaever's research interests later shifted to immunology.

GIAUQUE, William Francis
(1895–1982)

Canadian–American physical chemist, born in Niagara Falls, Ontario. He received his chemical education at the University of California, Berkeley, where he later advanced from an instructor to full professor (1922–62); he was Emeritus Professor from 1962. He was awarded the Nobel Prize for Chemistry in 1949. A member of the US National Academy of Sciences, he received the Gibbs Medal of the American Chemical Society in 1951. His researches were devoted to studying the properties of matter at very low temperatures, with the particular purpose of testing the third law of thermodynamics, that the entropy of a perfect crystal approaches zero as its temperature approaches absolute zero. Giauque developed low-temperature calorimetry and cryogenic apparatus; he was the first to use the magnetic method for attaining temperatures to within one degree of absolute zero, and he invented the carbon thermometer for measurements at temperatures in the liquid helium range. In connection with testing the third law he also developed spectroscopic methods for determining entropy values. In 1929 Giauque and Herrick Johnston discovered the isotopes oxygen-17 and -18, present in oxygen at only a few parts per thousand.

GIBBS, Josiah Willard
(1839–1903)

American theoretical physicist, born in New Haven, Connecticut. His father was a professor of sacred literature. Gibbs graduated from Yale College in 1858. A thesis treating the design of gears exhibited his geometrical expertise and earned Gibbs the first Yale engineering doctorate (1863). In 1866 he commenced three years of foreign study, attending lectures in mathematics and physics in Paris, Berlin and Heidelberg. His academic career continued with an (initially) unsalaried appointment to the Yale Chair of Mathematical Physics (1871); this he retained until his death, in spite of more lucrative offers from the new Johns Hopkins University. Thermodynamics was his main topic of scientific inquiry: two papers published in 1873 emphasized the fundamental nature of the entropy of a system, and provided powerful graphical and geometrical methods (eg the 'thermodynamic surface') for analysing the thermodynamic properties of substances. What became his most famous work appeared in the vast memoir *On the Equilibrium of Heterogeneous Substances* (1876–8). First published in the local Connecticut journal (but circulated more widely by Gibbs and later translated into French and German), this paper introduced a concept of 'chemical potential' which has been foundational for physical chemistry. Studies of the electromagnetic theory of light, advocacy of the use of vectors, and the publication of a book of *Elementary Principles in Statistical Mechanics* (1902) show the diversity of Gibbs's activities.

GILBERT, Grove Karl
(1843–1918)

American geomorphologist, stratigrapher, structural geologist and cartographer, born and educated in Rochester, New York. After early work at Cosmos Hall, New York (1863–9), he joined the Ohio Geological Survey (1869–71). He worked on the 'Wheeler' survey of territories west of the 100th meridian (1874–6) and surveyed large areas of Nevada, Utah, New Mexico and Arizona. In subsequent survey work in the Henry Mountains (1875–6), he recognized the nature of the intrusions named laccoliths. On joining the US Geological Survey he worked in Utah and studied the ancient lakes of the Great Basin. His *Monograph on Lake Bonneville* (1890) describes a fascinating history of the Pleistocene climate and hydrography of the Great Basin, and further discusses the subsequent deformation of the old shore levels as throwing light on the problem of isostatic readjustments of the Earth's crust. He also made notable studies in glacial geology, the history of the Niagara river and recession of the falls.

GILBERT, Walter
(1932–)

American molecular biologist, born in Boston. He studied physics and mathematics at Harvard and Cambridge, UK. From 1959 he taught physics at Harvard, where he remained as Professor of Biochemistry (from 1968) and Professor of Molecular Biology (from 1972). In 1978 he founded the genetic engineering company Biogen NV, and served as its chairman for three years. During the 1960s he isolated the repressor molecule, which **Monod** and **Jacob** had suggested formed part of the operon system, controlling the switching on and off of genes through its ability to bind to a segment of DNA on the chromosome. Using methods developed by **Sanger**, he went on to describe the nucleotide sequence of DNA to which the repressor molecule binds. For this work, Gilbert shared the 1980 Nobel Prize for Chemistry with Sanger and **Berg**. Since the late 1980s he has been a vigorous supporter of the Human Genome Initiative, a project to map and sequence all genes in the human body.

GILBERT, or GYLBERDE, William
(1544–1603)

English physician and geophysicist, born in Colchester. He graduated from St John's College, Cambridge (BA 1561), where he was elected to a fellowship in 1561, and received an MA in 1564 and MD in 1969. In 1573 he moved to a practice in London and became a Fellow of the Royal College of Physicians, for whom he was censor (1581–8), treasurer for nine years and president from 1600. He was appointed physician to Queen Elizabeth (1601), and briefly before his death to King James VI and I (1603). He found time for amateur scientific experiments and took a particular interest in magnetism. He was the first to consider the Earth as one vast spherical magnet, and studied the attraction of magnets, the direction relative to the Earth's magnetic field, declination and the use of declination to determine latitude at sea. He published *De Magnete, Magneticisque Corporibus, et de Magno Magnete Tellure, Physiologia Nova* in 1600, the first great physical book to be published in England. In it he conjectured that terrestrial magnetism and electricity (produced by rubbing amber) were two allied emanations of a single force. Having distinguished the magnetic and amber effects, he demonstrated that many other substances when rubbed exhibited the same phenomenon, and these he called 'electrics'; all others were 'non-electrics'. From this Sir Thomas Browne derived the term 'electricity' in 1646. Gilbert devised his versorium (a light metallic needle turning on a vertical axis) to test for the amber effect. He also demonstrated how the magnetic strength of lodestones (natural magnets) could be increased by 'arming' them with soft-iron pole pieces. The gilbert unit of magnetomotive power is named after him. He bequeathed all his books, apparatus, instruments and mineral collections to the Royal College of Physicians, though all perished during the Great Fire of London in 1666. Some of his papers were published posthumously including *De Mundo Nostri Sublunari Philosophia Nova* (1651).

GILCHRIST, Percy Carlyle
(1851–1935)

English metallurgist, born in Lyme Regis, Dorset. He studied at Felsted School and the Royal School of Mines, where he became a Murchison Medallist. He took a post as chemist first at Cwm Avon ironworks in south Wales, and then at Blaenavon ironworks. There he reluctantly agreed to the requests of his cousin, **Sidney Gilchrist Thomas**, that he should assist in carrying out the experimental work necessary to test and develop Thomas's ideas on dephosphorization. This was done by smelting phosphoric iron ores using a furnace lined with a basic material, such as magnesium oxide, which would combine with and remove the phosphate impurities in the iron ore. The Gilchrist–Thomas process greatly increased the potential steel production of the world by making possible the use of the large European phosphoric iron ore fields. After Thomas's early death, Gilchrist became active in the steel industry, working hard to promote basic steelmaking both by **Bessemer** converter and open-hearth furnace.

GILL, Sir David
(1843–1914)

Scottish astronomer, born in Aberdeen and educated at Marischal College, Aberdeen (1858–60), where he was inspired by the teaching of **Maxwell**. Obliged for family reasons to leave university without completing his degree, he started life as a watchmaker. His serious scientific career did not begin until 1872, when he was put in charge of Lord Lindsay's private observatory at Dunecht, Aberdeenshire. He achieved recognition for his high-precision determinations of the solar distance made on expeditions to Mauritius (1874) and Ascension Island (1877), the latter privately organized by himself with his wife as companion. Soon afterwards he was appointed HM Astronomer at the Royal Observatory, Cape of Good Hope (1879–1907). On the Mauritius expedition Gill put into practice the suggestion by **Galle** of using minor planets for solar distance determination, and extended the method in a cooperative effort with astronomers at other stations, obtaining a distance to the Sun (1889) which remained the standard for astronomical ephemerides until 1968. He initiated a photographic survey of the southern sky and, as president of an international astrographic congress convened in Paris in 1887, became responsible for the immense projects of the *Carte du Ciel* and the *Astrographic Catalogue*; some 22 000 photographs were taken in the Cape's allotted part in this undertaking. Amongst Gill's other contributions was the geodetic survey of South Africa. He was twice awarded the Gold Medal of the Royal Astronomical Society (1882 and 1908), of which he became President after his retirement (1909–11). He was knighted in 1900.

GILLESPIE, Ronald James
(1924–)

English–Canadian chemist, born in London. Gillespie received a BSc (1945), PhD (1949) and DSc (1957) from University of London. He is famous, along with **Nyholm**, for modifying the valence shell electron pair repulsion (vsepr) theory of **Sidgwick** which may be used for predicting the structures of simple molecules. This modification states that lone pairs of electrons have a greater repulsive effect than bonding pairs, thus explaining the distortion of the bond angles of molecules such as water and ammonia from the regular tetrahedral angle. Gillespie has received several awards including the Ramsay Medal (1949), the Chemical Society Harrison Memorial Medal (1954), the Canadian Centennial Medal (1967), the Chemical Institute of Canada Medal (1977) and the Silver Jubilee Medal (1978). He was elected FRS in 1977.

GILMOUR, John Scott Lennox
(1906–86)

English botanist of Scottish parentage, born in London and educated at Uppingham and Clare College, Cambridge. He was curator of the herbarium at Cambridge (1930–1) but then went to Kew (1931–46), becoming Assistant Director in 1946. From 1946 to 1951 he was Director of the Royal Horticultural Society's Gardens at Wisley. He then returned to Cambridge, where he was director of the botanic gardens from 1951 to 1973; he instituted a period of expansion, including a gardener training course. Together with James Wylie Gregor, he was one of the original proponents of the 'deme' terminology, a flexible system using informal nomenclature to classify experimental taxonomic categories. He was one of the founder members of the Classification Society (later Systematics Association) in 1937. Gilmour was a philosopher, poet, musician, bibliophile, athlete and humanist, as well as a botanist. He published many papers on the theory and philosophy of systematics, as well as *Wild Flowers of the Chalk* (1947) and *Wild Flowers* (1954, with Stuart Max Walters). He chaired many committees on horticultural nomenclature and was influential in the establishment of the International Code of Nomenclature of Cultivated Plants, in whose first edition (1953) the term 'cultivar' was first used.

GINZBURG, Vitaly Lazerevich
(1916–)

Soviet astrophysicist, born in Moscow. After graduating from the University of Moscow in 1938, he joined the Physics Institute of the Academy of Sciences. Since 1942 he has been head of the sub-department of theoretical physics there. Ginzburg's first success was in 1940, when he applied quantum theory to the study of Cherenkov radiation. He subsequently suggested that the non-thermal radio emission produced in our galaxy is due to synchrotron radiation produced by relativistic electrons gyrating in the galactic magnetic field. Ginzburg (with S I Syrovatskii) also proposed in 1964 (in a classic book entitled *The Origin of Cosmic Rays*) that all cosmic rays come from within our own galaxy, originating from regions near the galactic nucleus and then diffusing into the halo. The original cosmic rays were thought to consist of only heavy nuclei.

GLAISHER, James
(1809–1903)

English meteorologist, born in Rotherhithe, London. Glaisher was largely self-educated but very single-minded in the pursuit of knowledge. After taking part in the trigonometrical survey of Ireland he became assistant to the Director of Cambridge University Observatory. In 1835 he went to Greenwich Observatory and in 1840 became Superintendent of the Magnetic and Meteorological Department. In 1845 he produced tables for the calculation of dewpoints of the air which became standard. He was later invited to produce reports on the meteorology of England for the Registrar-General's quarterly returns (1849), which he did for 54 years. For this he established about 40 observing stations, regularly tested their instruments and inspected them at quarterly intervals. The observations were converted to mean values by the use of Glaisher's own barometrical, hygrometrical and diurnal range tables. He was a founder member in 1850 of the British Meteorological Society (which later became the Royal Meteorological Society). In 1851 he produced a report on the Great Exhibition and in this he was believed to be the first to suggest that books might be copied on microfilm. He is best known for 29 balloon ascents between 1862 and 1866, which were carried out in every month of the year and with different weather conditions. The main aim was to measure temperature, dewpoint and wind at different heights, and the highest ascent is believed to have reached 30 000 feet. He was a member of many learned societies and was elected FRS in 1849.

GLASER, Donald Arthur
(1926–)

American physicist, born in Cleveland, Ohio. He was educated at the Case Institute of Technology in Cleveland and at Caltech. He was appointed professor at the University of Michigan (1949–59) before becoming professor at the University of California at Berkeley. Glaser was awarded the 1960 Nobel Prize for Physics for inventing the 'bubble chamber' for observing the paths of elementary particles. In this chamber, ionizing particles pass through a superheated liquid and leave tracks of vapour bubbles along the particle trajectory; the tracks can then be photographed allowing analysis. Bubble chambers were used to discover many subatomic particles and reached their pinnacle in 1971 with the construction of 'Gargamelle', a thousand-tonne detector. This was used in neutrino experiments to discover the 'neutral currents' predicted by **Steven Weinberg** and **Salam**. Bubble chambers have now largely been superseded by electronic or gas detectors capable of providing data immediately. From 1964 Glaser's research interests have been in the application of physics to molecular biology.

GLASHOW, Sheldon Lee
(1932–)

American physicist, born in New York City. He studied at Cornell, Harvard, Copenhagen and Geneva, before becoming Professor of Physics at Harvard in 1967. Glashow developed one of the first models to describe simultaneously two of the four forces of nature; the electromagnetic and weak forces. A unified 'electroweak' force theory was later successfully applied to leptons (electrons and neutrinos) by **Steven Weinberg** and **Salam**, and Glashow subsequently developed their theory to apply to other particles, such as baryons and mesons, by introducing a new particle property known as 'charm'. He was a major contributor to the theory (now known as quantum chromodynamics) of strong forces between elementary particles, including the interaction that binds protons and neutrons to form a nucleus. It assumes that strongly interacting particles are made of quarks and that 'gluons' bind the quarks together. He shared the 1979 Nobel Prize for Physics with Salam and Weinberg for their contributions to 'the standard model' of all particle interactions.

GLAUBER, Roy Jay
(1925–)

American theoretical physicist, born in New York City. He was a member of the theoretical physics division at Los Alamos (1944–6), and obtained his BS (1946) and PhD (1949) at Harvard. He was appointed as a lecturer at Caltech (1951–2), and in 1962 became Professor of Physics at Harvard. Since 1976 he has been Mallinckrodt Professor of Physics there. Glauber established the theoretical foundations of quantum optics in two epic papers in 1963, and also made pioneering contributions to nuclear physics (the optical model for the description of nuclear reactions) and to statistical mechanics (the kinetic Ising model, which provides a simplified treatment of the generation of a bulk magnetic moment in matter by the alignment of the atomic magnetic moments). He has been a member of the Advisory Board of the Program for Science and Technology for International Security at MIT.

GLAZEBROOK, Sir Richard Tetley
(1854–1935)

English physicist, born in Liverpool, and educated at Dulwich College, Liverpool College and Trinity College, Cambridge. He was appointed Assistant Director of the Cavendish Laboratory (1891), Principal of University College, Liverpool (1895), and Director of the National Physical Laboratory (1900–19). During his career he also held the presidencies of the Physical Society (1903–5), the Optical Society (1904–5, 1911–12), the Institution of Electrical Engineers (1906), the Faraday Society (1911–13) and the Institute of Physics (1919–21). Glazebrook wrote several physics textooks and is known for his work on electrical standards. He was elected FRS in 1882, and knighted in 1917.

GMELIN, Johann Georg
(1709–55)

German botanist, natural historian and geographer, born in Tübingen. An extremely gifted child, in 1727 he graduated in medicine aged 18. Only four years later he had become Professor of Chemistry and Natural History at St Petersburg. In 1733 he took part in a lengthy scientific expedition to eastern Siberia, exploring the Irtysh, Ob and Tom rivers, the Transbaikal region and the Lena river. After a disastrous fire near Yakutsk in 1736 which destroyed most of his collections and books, he continued northwards along the Yenisei river then south to the Caspian Sea and Ural Mountains, returning to St Petersburg in 1743, nearly 9½ years after setting out. Gmelin perceived that the Yenisei river marked a natural boundary between Europe and Asia, and became the first person to determine that the Caspian Sea lay at a lower level than the Black and Mediterranean seas. He published the expedition's botanical results as *Flora Sibirica* (4 vols, 1747–69). His account of the journey, *Riesen durch Sibirien* (4 vols, 1751–2), was banned in Russia because in it Gmelin had severely criticized the establishment. In 1747 he married and returned to Tübingen, where from 1749 he was Professor of Botany, Chemistry and Medicine.

GMELIN, Leopold
(1788–1853)

German chemist and physiologist, son of Johann Friedrich Gmelin, born in Göttingen. He studied in Germany and Italy and taught at the University of Heidelberg, where he became Professor and Director of the Chemical Institute in 1817, posts that he held until 1851. Gmelin's interests ranged from physiology—he wrote his thesis on the black pigmentation in the eyes of cattle and later studied the chemistry of digestion—to chemistry and mineralogy. He worked towards a definition of organic chemistry, suggesting that an organic compound must contain carbon and hydrogen, and he introduced the terms 'ester' and 'ketone'. One of the pioneers of physiological chemistry, he prepared uric acid, formic acid and potassium ferricyanide (Gmelin's salt). He also developed a test which shows the presence of bile pigments (Gmelin's test). Renowned as a teacher, he is even more famous as author of the mighty *Handbuchen der theoretischen Chemie* (1817–19) which had grown to 10 volumes by 1870. The organic section was subsequently dropped, but the inorganic section continued as *Gmelin's Handbuch der anorganische Chemie*. Gmelin died in Heidelberg.

GMELIN, Samuel Theophilus (Samuel Gottlieb)
(1745–74)

German botanist, born in Tübingen, nephew of **Johann Georg Gmelin**. He became Professor of Botany at St Petersburg (1767) and explored the Don, the Volga and the Caspian Sea. His two most important works were *Historia Fucorum* (1768), dealing with the brown alga *Fucus* and its allies, and the account of his travels, *Reise durch Russland*, published in four volumes in the 1770s and 1780s. The fourth volume of the *Reise* was published posthumously by his friend **Pallas**. Samuel Gmelin also edited volumes 3 and 4 of Johann Georg Gmelin's *Flora Sibirica* after his uncle's death.

GODDARD, Robert Hutchings

(1882–1945)

American physicist, rocket engineer and inventor, born in Worcester, Massachusetts. He graduated BSc at Worcester Polytechnic Institute in 1908, and three years later received the degree of PhD in physics from Clark University, Worcester, where he taught physics for almost 30 years. From the age of 17 he was fascinated by the idea of space travel, and began experiments with rockets designed to explore the Earth's upper atmosphere, and to compare the power of different propellants. In 1919 he published *A Method of Reaching Extreme Altitudes*, but in the theory of rocketry he had been anticipated by **Tsiolkovsky**; Goddard was, however, the first to carry out experiments which proved that the theory could be translated into practice. In 1923 he tested the first liquid-fuel rocket motor, using petrol and liquid oxygen; three years later his first liquid-fuel rocket was launched, and by 1929, working largely in isolation with a small grant from the Smithsonian Institution, he had developed the first instrument-carrying rocket able to make observations in flight. Receiving $50 000 from the Guggenheim Foundation, in 1930 he launched rockets at 500 miles per hour to 2 000 feet, then in 1935 exceeded the speed of sound. He developed jet vanes and gyroscopic control 10 years before the Germans, and described the prospects for electric propulsion, a solar-powered generator and electrostatically accelerated jets of ionized gas; in all, he was granted 214 patents. In 1937 one of his rockets attained an altitude of 3 kilometres (1.8 miles), but even during World War II the US government showed little interest in his continuing work. When German rocket experts were brought to the USA after the war they were astonished that Goddard, from whose published work they had learnt so much, was virtually ignored in his own country. Only after his death was he given due recognition for his pioneering work, which led directly to NASA and the US space exploration programme.

GÖDEL, Kurt

(1906–78)

Austrian-born American logician and mathematician, born in Brno, Czechoslovakia. He studied and taught in Vienna, then emigrated to the USA in 1940 and joined the Institute for Advanced Study at Princeton. He stimulated a great deal of significant work in mathematical logic and propounded one of the most important proofs in modern mathematics: Gödel's theorem, published in 1931, demonstrated the existence of formally undecidable elements within any formal system of arithmetic (like **Bertrand Russell**'s in the *Principia Mathematica*). This result put an end to Hilbert's influential hopes of giving a truly rigorous foundation to all mathematics on essentially finite terms. His own philosophy of mathematics became strongly Platonist. His contact with **Einstein** led him to discover novel solutions to Einstein's field equations in which there are closed geodesics, which seem to flout causality and invite the physical possibility of time travel.

GODWIN, Sir Harry

(1901–85)

English botanist, born in Holmes, Rotherham, Yorkshire and educated at Long Eaton in Derbyshire, University College, Nottingham, and Clare College, Cambridge. There he was greatly stimulated by **Tansley**'s ecological work. From 1923 to 1968 he was on the staff of Cambridge, as Professor of Botany from 1960 to 1968. He was also the first director of the sub-department of quaternary research at the Cambridge Botany School (1948–66). A pioneer of British pollen analysis, he established the relationship between peat stratigraphy and the pollen zones based on relative abundance of pollen of different tree species. He is also known for his contributions to the science of radiocarbon dating, and concerning changes in land/sea levels in recent geological time. In 1930 he wrote *Plant Biology*, whose emphasis on physiology rather than morphology and evolution secured a deeper understanding of the basic role of the green plant, and with a general recognition of the scientific and social value of ecological studies. His major published work was *The History of the British Flora* (1956); he also published many papers in scientific journals. He was editor of *Journal of Ecology* (1948–56) and joint editor of *New Phytologist* (1931–61). Elected FRS in 1945, he was knighted in 1970.

GOEPPERT-MAYER, Maria, née Goeppert

(1906–72)

German-born American physicist, born in Kattowitz (now Katowice in Poland). She graduated at Göttingen in 1930, emigrated to the USA and taught at Johns Hopkins University, where her husband, Joseph

Mayer, was Professor of Chemical Physics. From 1960 she held a chair at the University of California. She developed the shell model of the nucleus. This is based on the fact that certain nuclei are very stable, having 'magic numbers' of protons and neutrons. Drawing an analogy with atomic physics in which a closed shell of electrons leads to stable atoms, eg the noble gases, a model of the nucleus was developed. Some initial problems with the theory were resolved by discussions with **Fermi** who pointed out that spin-orbit coupling should be taken into account. A similar model was developed in Germany by **Jensen**. Goeppert-Mayer shared the 1963 Nobel Prize for Physics with **Wigner** and Jensen.

GOLD, Thomas

(1920–)

Austrian-born American astronomer, born in Vienna. After leaving Austria during Hitler's rise to power in central Europe, he studied at Cambridge and worked in the UK before moving to the USA in 1956. While in England he worked with **Bondi** and **Hoyle** on the steady-state theory of the origin of the universe (1948), which proposes that the universe is uniform in space and unchanging in time, its expansion being fuelled by constant spontaneous creation of matter. The theory was later to be displaced through new evidence, but it was a valuable contribution to cosmology. In 1959 Gold became Director of the Center for Radiophysics and Space Research at Cornell University, and in 1968 he suggested the currently accepted theory that pulsars (discovered by **Bell Burnell** and **Hewish** in that year) are rapidly rotating neutron stars, dense collapsed stars which produce beams of radio waves from their poles which appear as radio pulses on Earth. His unorthodox theory of the origin of petroleum and natural gas on Earth proposed that some deposits arise from gas trapped in the Earth's interior from the time of the planet's formation.

GOLDBERGER, Joseph

(1874–1929)

Hungarian-born American physician and epidemiologist, born in Girált. He went to the USA as a child and qualified in medicine (1895) at the Bellevue Hospital Medical College in New York City. After private medical practice he joined the US Public Health Service (1899), where he investigated the mechanisms of the spread of a number of infectious diseases, including measles, typhus and yellow fever. His brilliant epidemiological studies of pellagra within institutions in the southern USA demonstrated that pellagra is a nutritional disorder caused by an unbalanced diet and cured by the addition of fresh milk, meat or yeast. He used himself in a series of well-designed experiments which demonstrated that, contrary to contemporary opinion, pellagra was not an infectious disease. The deficiency was later shown to be niacin, one of the vitamins of the B complex.

GOLDSCHMIDT, Hans

(1861–1923)

German chemist, born in Berlin. He studied under **Bunsen** at Heidelberg. In 1905 he invented the highly inflammable mixture of finely divided aluminium powder and magnesium ribbon known as thermite. The high temperatures obtained from this mixture make it useful for welding and for reducing metals such as chromium, manganese and cobalt from their oxide ores. It has also been used in incendiary bombs. Around 1910, Goldschmidt and **Stock** developed a commercial process for extracting beryllium, a rare metal which is used in the manufacture of hard alloys and some scientific instruments. Goldschmidt died in Baden-Baden.

GOLDSCHMIDT, Richard Benedikt

(1878–1958)

German biologist, born in Frankfurt. He was appointed biological director of the Kaiser Wilhelm Institute, Berlin (1921), and in 1935 went to the USA where he became Professor of Zoology at California University (1936–58). He experimented on the role of the X chromosomes (the sex chromosomes) using butterflies, and promulgated the now-discredited idea that the chromosomes themselves and not the individual genes are the units of heredity. One of his more important contributions, using the Gipsy Moth, was to show that much geographical variation is genetic and not environmental in origin. Also, by means of environmental shocks such as high temperature, he created phenocopies in fruit flies. These demonstrated that effects of environmental factors could be made to mimic some of the effects of genetic mutations, except that they would not be inherited. His books included *Physiological Genetics* (1938), *The Material Basis of Evolution* (1940) and *Theoretical Genetics* (1955).

GOLDSCHMIDT, Victor Moritz

(1888–1947)

Swiss-born Norwegian geologist and crystallographer, born in Zürich. He moved to Christiania (now Oslo) with his family in 1905 when his father Heinrich Jacob Goldschmidt was appointed Professor of Chemistry at the university there. After studying chemistry, geology and mineralogy at Oslo, and various travels, he became Professor and Director of the Mineralogical Institute there in 1914. From 1916 to 1922 he was Chairman of the Materials Supply Committee of the Norwegian government. In 1927 he took up the Chair of Mineralogy at Göttingen, initially commuting by air from Oslo, but he resigned the Norwegian position and settled in Göttingen in 1929. In 1935, being Jewish, he was forced to leave Germany by the Nazis, and returned to Oslo, where he was immediately granted a chair. When Germany invaded Norway in 1940, Goldschmidt worked secretly for the Allies for some time. In the end he fled to Sweden and then to Britain, where he worked at the Macaulay Institute and at Rothamsted. In 1946 he returned to Norway, his health broken, and he died in 1947. Goldschmidt's earliest work was on the petrology of southern Norway. Following the development of X-ray crystallography by **William** and **Lawrence Bragg**, Goldschmidt undertook an extensive X-ray study of the binary compounds of the elements. This led to his rationalization of crystal structures in terms of ionic radii and polarizabilities. His interests in crystallography and petrology came together in researches on the distribution of the elements in the Earth; his massive book on *Geochemistry* was published posthumously (1954).

GOLDSTEIN, Eugen
(1850–1930)

German physicist, born in Gleiwitz, Upper Silesia (now Gliwice, Poland). He was educated at Ratobor Gymnasium and the University of Breslau before going to the University of Berlin (1878) where he worked with **Helmholtz**. He then spent his career as a physicist at the Potsdam observatory. In 1876 he showed that cathode rays could cast sharp shadows, and that the cathode rays were emitted perpendicular to the cathode surface. This led to the production of concave cathodes, giving focused cathode rays which were useful in many experiments. In 1880 he discovered that cathode rays could be bent by magnetic fields. In 1886 Goldstein published his discovery of 'Kanalstrahlen', literally canal rays, known as positive rays in English. These emerged from channels or holes in anodes in low-pressure discharge tubes. His student **Wien** showed that these rays could be deflected by electric and magnetic fields, had a charge-to-mass ratio 1000 times that of cathode rays, and hence were positively charged particles of atomic mass. This apparatus was developed into the mass spectrograph by **J J Thomson**, **Aston** and others. Another student of Goldstein's, **Stark**, was able to demonstrate that light from positive rays showed a **Doppler** shift in the first clear-cut demonstration of an optical Doppler shift from a terrestrial source.

GOLDSTEIN, Joseph Leonard
(1940–)

American molecular geneticist, born in Sumter, South Carolina. He graduated with a medical degree from the University of Texas, Dallas, in 1966. As an intern at Massachusetts General Hospital, Boston, he met **Michael Brown** at the start of a highly productive collaboration. In 1972 he returned to the University of Texas to become Head of the Division of Medical Genetics, Professor of Internal Medicine (from 1976), and Professor of Medical Genetics (from 1977). With Brown, he has worked on cholesterol metabolism in the human body, studying individuals with the inherited genetic disease familial hypercholesterolemia, which causes high levels of cholesterol in the blood. Cholesterol is carried in the blood forming a complex molecule with low-density lipoproteins (LDLs). Normally, cells in the liver pick up the cholesterol from this complex thereby decreasing the level of cholesterol; Goldstein found that in patients with hypocholesterolemia, the liver cells cannot bind to the complex because they are missing a receptor site for the LDLs. In 1984 Goldstein and Brown described several mutations in the gene that codes for the LDL receptor and opened up possibilities for new drugs to combat this disease. For this work, they were jointly awarded the 1985 Nobel Prize for Physiology or Medicine.

GOLDSTINE, Herman Heine
(1913–)

American mathematician and computer scientist, born in Chicago. Educated at the University of Chicago, where he was awarded a PhD in mathematics in 1936, Goldstine became a captain in Army Ordnance during World War II. He was the Army's liaison officer at the Moore School of Electrical Engineering of the University of Pennsylvania, and was instrumental in helping the school obtain a contract for the ENIAC computer designed by **Eckert** and **Mauchly**. He was also influential in securing government backing for that machines's successor, the EDVAC, which became America's first stored-program computer. Goldstine personally introduced **von Neumann** to the EDVAC team in 1944, which had far-reaching and controversial consequences when von Neumann later wrote his famous *Draft Report* on the EDVAC, summarizing the theory of the stored-program computer. In 1946 Goldstine became an associate project director at the Institute for

Advanced Study (IAS), helping von Neumann build such a machine. After von Neumann's death, he became acting director at the IAS between 1954 and 1957, before ending his career at IBM. He has written a history of computing, *The Computer from Pascal to Von Neumann* (1972).

GOLGI, Camillo

(1843–1926)

Italian histologist, born in Corteno, Lombardy. He studied medicine at the University of Pavia and after graduating in 1865, studied psychiatry for a while. In 1872 he became Medical Director of a clinic for the incurable at Abbiategrasso, where he tried to continue the research he had already started on the microscopic anatomy of the nervous system, and to develop new staining methods for his experimental tissues. He returned to Pavia as a lecturer (1875) and then as professor (1876–1918), with a brief intermission at the University of Sienna as Professor of Anatomy (1879). Many of his most important studies were on the anatomy of the brain and other nervous tissues, and he developed a technique, still known by his name, of impregnating particular nerve cells with silver salts. Until that time the structural complexity of the central nervous system had deterred investigators, but Golgi's method, which selectively stained only a few of the nerve cells meant that they stood out from the unstained background of other cells, and could be closely examined under the microscope. He believed that the processes of the nerve cells he described formed a continuous network, and that it was through this continuum that communication within the nervous system was effected. It was the work of **Ramón y Cajal**, using an adaptation of Golgi's method, who demonstrated that nerve cells were not continuous but were discrete entities separated one from the other. The two men shared the 1906 Nobel Prize for Physiology or Medicine, although Golgi used the award ceremony as an occasion to revile Ramón y Cajal's interpretations. Golgi's work opened up a new field of research into the fine structure of the central nervous system, sense organs, muscles and glands.

GOMBERG, Moses

(1866–1947)

American chemist, born in Kirovograd, Russia. Because of anti-tsarist activity, in 1884 the Gomberg family fled from Russia to Chicago, where father and son worked at menial jobs. Moses Gomberg entered the University of Michigan in 1886. He received his PhD there and joined the faculty in 1893, remaining there for the rest of his professional life. He studied in Munich and Heidelberg, where he was successful in preparing molecules which had proved difficult to assemble because they contain bulky groups. When he returned to Michigan he attempted the synthesis of the sterically very hindered compound hexaphenylethane. The procedure he used (the reaction of triphenylmethyl chloride with silver in an inert atmosphere) gave rise to a mysterious product, the nature of which has been a matter of discussion ever since. Gomberg became convinced that he had prepared a radical (a molecular species with a carbon atom bearing one unpaired electron), but such a species was thought to be too reactive to persist in solution. Although Gomberg's synthesis is a particularly complex case, his work did alert chemists to the possibility of long-lived radicals in solution and this is now accepted as an important type of reactive intermediate. Gomberg was elected to the US National Academy of Sciences and served as President of the American Chemical Society in 1931.

GOODALL, Jane

(1934–)

English primatologist and conservationist, born in London. She worked in Kenya with the anthropologist **Louis Leakey** who in 1960 raised funds for her to study chimpanzee behaviour at Gombe in Tanzania. She obtained her PhD from Cambridge in 1965 and subsequently set up the Gombe Stream Research Centre. She has been a visiting professor at the Department of Psychiatry and Program of Human Biology at Stanford University (1971–5) and visiting professor of zoology at Dar es Salaam since 1973. Since 1967 she has been Scientific Director of the Gombe Wildlife Research Institute. With her co-workers at Gombe, she has carried out a study of the behaviour and ecology of chimpanzees which at over 30 years is the longest unbroken field study of a group of animals in their natural habitat. This research has transformed the understanding of primate behaviour by demonstrating its complexity and the sophistication of inter-individual relationships. Among her major discoveries was the ability of chimpanzees to modify a variety of natural objects such as the stems of plants to use as tools to collect termites, and sticks and rocks as missiles for defence against possible predators. She also showed that they hunt animals for meat and that the adults share the proceeds of such kills. She has been active in chimpanzee conservation in Africa and their welfare in those countries where they are extensively used in medical research. Her books include

In the Shadow of Man (1971) and *The Chimpanzees of Gombe: Patterns of Behavior* (1986). She has received many awards for conservation and for her scientific research including the Albert Schweitzer Award (1987), the Encyclopaedia Britannica Award (1989) and the Kyoto Prize for Science (1990).

GOODRICKE, John
(1764–86)

English astronomer, born in Groningen, the Netherlands. Despite being deaf and mute from an early age, he was the first, in 1782, to recognize that the star Algol (Beta Persei) is an eclipsing binary and that its periodic diminutions in brightness are due to one of the binary stars passing in front of the other. Two years later he discovered the regular variability of the stars Delta Cephei and Beta Lyrae. He was awarded the Copley Medal of the Royal Society in 1783.

GOODSIR, John
(1814–67)

Scottish anatomist, born in Anstruther, the son of a surgeon. Goodsir attended the University of St Andrews and then Edinburgh University, where he also studied under **Knox**. Specializing in surgery, he joined his father's practice and published early in dentistry. After developing, through his friend **Edward Forbes**, a passionate interest in marine biology, he was appointed in 1840 Conservator in Comparative Anatomy at the Edinburgh University Museum, which he ambitiously aimed to turn into a first-rank teaching resource. In 1846 he became Professor of Anatomy at the university. His researches attempted to apply the mathematics of form to problems of living structures; he became especially fascinated with cell biology. Growing more metaphysically inclined, he cultivated the theory of the triangle as the basic natural living form. His later life was blighted by nervous crisis and incipient paralysis.

GORGAS, William Crawford
(1854–1920)

American military doctor, born near Mobile, Alabama. He trained in medicine at Bellevue Hospital Medical College in New York, and joined the US Army Medical Corps in 1880. Following the discovery of the role of the mosquito (*Aedes aegypti*) in the transmission of yellow fever, Gorgas directed the mosquito eradication programme in Havana and the Panama Canal Zone, the latter permitting the successful construction of the canal (1904–14). Through draining marsh land, oiling ponds, and limiting the sources of stagnant water in urban areas, the number of mosquitoes was reduced with significant decrease in the incidence of mosquito-borne diseases. He was Surgeon-General of the US Army during World War I and established the systematic medical examination of recruits, greatly improving the quality of medical care among troops.

GOSSE, Philip Henry
(1810–88)

English naturalist, born in Worcester. He went to North America in 1827 and became a professional naturalist in Jamaica, with a particular interest in coastal marine biology. His *Manual of Marine Zoology* (1855–6) and *History of British Sea-anemones and Corals* (1860), written on his return to England, greatly expanded interest in marine organisms. He published *Omphalos* (1857) in opposition to evolutionary theory. His best-known work was the *Romance of Natural History* (1860–2).

GOUDSMIT, Samuel Abraham
(1902–78)

Dutch-born American physicist, born in the Hague. He studied in Amsterdam and Leiden and emigrated in 1927 to the USA, where he was professor at Michigan (1932–46) and later worked at the Brookhaven National Laboratory, Long Island (1948–70). Aged 23, he and his fellow student **Uhlenbeck** proposed the idea that electrons in atoms can have intrinsic spin angular momentum as well as orbital angular momentum to explain close doublets of spectral lines and the results of the **Stern**–Gerlach experiment. Initially this was not accepted as physicists found it difficult to ascribe rotation to electrons, but **Dirac**'s 1928 theory of relativistic quantum mechanics showed that spin is an intrinsic property of the electron. During World War II, Goudsmit headed the secret Alsos mission charged with following German progress in atomic bomb research (1944); this led to the award of the US Medal of Freedom, and to his book *Alsos* (1947).

GOULD, Benjamin Apthorp
(1824–96)

American astronomer, born in Boston. Educated at Harvard and Göttingen, he founded the *Astronomical Journal* (1849–61), was Director of the Dudley Observatory at Albany (1856–9) and in 1866 determined, by aid of the submarine cable, the difference in longitude between Europe and America. He helped to found and was director from 1868 of the National Observatory at Cordoba,

Argentina. His *Uranometria Argentina* complemented **Argelander**'s *Atlas* of the northern heavens. The southern sky atlas gave the magnitudes of 8 198 out of the 10 649 stars visible to the naked eye, and the extreme clarity of the sky enabled him to include stars as faint as seventh magnitude. Eduard Hies had included a drawing of the Milky Way in his northern sky atlas and Gould's assistants did the same for the southern Milky Way. This lead to much more attention being paid to the galaxy that we live in. In 1866 Gould carefully compared a photographic plate of the Pleiades with hand-drawn maps made 25 years previously; as there was no detectable stellar movement, he concluded that photography could be used to produce excellent star maps.

GOULD, Stephen Jay

(1941–)

American palaeontologist, born in New York City. Educated at Antioch College, Ohio, and Columbia University, he has been Professor of Geology since 1973 and Alexander Agassiz Professor of Zoology since 1982 at Harvard. His research has extended from studies of the Irish elk (a giant fossil deer from the Pleistocene epoch) to the evolution of Caribbean snails, but he is primarily known for his theoretical contributions. In an influential paper published in 1972, Gould, together with the palaeontologist Niles Eldgredge, posited the theory of 'punctuated equilibrium'. This proposed that most evolutionary change occurs rapidly during the process of allopatric speciation in small populations, when genetic change leads to the evolution of new species, and that species then persist for long periods with little or no change. Gould has also championed the idea of 'hierarchical evolution', suggesting that natural selection operates at many levels, including genes and species as well as at the traditional level of individuals. He has been critical of the 'adaptationist program', emphasizing that many characters of organisms are not 'adaptive' in the strict sense. He has popularized his ideas in a monthly column in *Natural History* magazine, and a series of collected essays including *Ever Since Darwin* (1977), *The Panda's Thumb* (1980), *Hens' Teeth and Horses' Toes* (1983), *The Flamingo's Smile* (1985), *Bully for Brontosaurus* (1991) and *Eight Little Piggies* (1993). His books have won many awards, including the 1990 Science Book Prize for *Wonderful Life* (1989), a reinterpretation of the Cambrian Burgess Shale fauna. Gould has admitted to Marxist influence in his scientific work, and has been a forceful speaker against pseudoscientific racism and biological determinism; *The Mismeasure of Man* (1981) is a critique of intelligence testing. He was also a witness in a courtroom trial concerning the teaching of evolution in American public schools.

GRAAF, Regnier de

(1641–73)

Dutch physician and anatomist, born in Schoonhoven. He studied medicine at Utrecht and Leiden, where he was a student of Franciscus Sylvius and a contemporary of **Swammerdam** (with whom he was later involved in violent priority disputes). He went into practice at Delft, but pursued researches and as early as 1664 published his notable treatise on the pancreatic juice. In 1672, on the basis of human and animal dissection, he discovered the Graafian vesicles of the female gonad, coining the term 'ovary' for the organ. Not noticing the rupture of the ovarian follicles, Graaf was not able to develop a satisfactory theory for explaining the role of the ovary; the mammalian egg was not discovered until the work of **Baer** in 1827. Graaf is rightly credited with having been one of the founders of experimental physiology.

GRAAFF, Robert Jemison Van de

See **VAN DE GRAAFF, Robert Jemison**

GRAHAM, Thomas

(1805–69)

Scottish chemist, born in Glasgow. He was educated at Glasgow and Edinburgh universities, where he was greatly influenced by the professors of chemistry, Thomas Thomson and **Hope**, respectively. Declining to enter the ministry, as his father wished, Graham began to teach chemistry in Glasgow, at first privately and then at the Mechanics Institute. In 1830 be became Professor of Chemistry at Anderson's College and in 1837 he moved to London as Professor of Chemistry at University College. He held this chair until 1854, when he was appointed Master of the Mint, a post he held until his death. Graham was elected FRS in 1836 and received the Copley Medal of the Royal Society in 1862. He was a founder member and first President of the Chemical Society (1841). His earliest work (1825) was on the solubility of gases in liquids, but his most famous research was on the diffusion of gases and related phenomena; Graham's law states that the velocity of effusion of a gas is inversely proportional to the square root of its density. His studies of diffusion in liquids led him to the distinction between crystalloids and colloids and to devise the process of dialysis for their separation. In 1833 he distinguished between the three forms of phosphoric acid. As late in his life as 1866, he studied the occlusion of hydrogen by metals and discovered the enormous absorption of hydrogen by palladium. His textbook *Elements of Chemistry* (1842) was highly influential.

GRAHAM-SMITH, Sir Francis
(1923–)

English radio astronomer. He spent most of his early research life at Cambridge, this however being disrupted between 1943 and 1946 when he worked as a radar researcher at the Telecommunications Research Establishment, Malvern. In 1964 he moved to Jodrell Bank, and in 1976 he became Director of the Royal Greenwich Observatory, where he supervised the setting up of the Northern Hemisphere Observatory at Las Palmas on the Canary Islands. He moved back to Jodrell Bank in 1981, as Director. Graham-Smith was the Astronomer Royal of England between 1982 and 1990. In 1948, with **Ryle**, he set up a two-aerial radio interferometer to study Cassiopeia A and Cygnus A. In 1950 he showed that fluctuations in radio source signals were caused by diffraction in the F region of the ionosphere, and in 1951 (with K E Machin) he suggested that solar occultations of the Crab nebula could be used as a method of measuring electron densities in the solar corona. During the 1950s he participated in a systematic survey of radio sources, this resulting in the Cambridge 3C catalogue. He made the first investigation of radio noise from above the atmosphere using a radio receiver in the satellite Aeriel II (1962). From 1964 to 1974 Graham-Smith concentrated on pulsars and in 1968 (with Andrew Lyne) he discovered that they emitted polarized radiation. This enabled him to measure the interstellar magnetic field. He was elected FRS in 1970, and knighted in 1986.

GRAM, Hans Christian Joachim
(1853–1938)

Danish bacteriologist, born in Copenhagen. In 1884 he developed the most important staining technique in microbiology, which is still in use today. The stain divides bacteria into two groups, based upon the structure of their cell walls. In *Gram-positive* organisms, the cell wall is made up almost entirely of peptidoglycans (compounds containing amino sugars and amino acids), while *Gram-negative* cell walls are more complex and flexible, containing little peptidoglycan. Gram took up a clinical career, and was appointed Professor of Medicine at Copenhagen in 1900.

GRANIT, Ragnar Arthur
(1900–)

Finnish-born Swedish physiologist, born in Helsinki to a family of Swedish origin. He was a volunteer in the Finnish army fighting for independence from Russia after the Russian Revolution, and in 1919 enrolled at Helsinki University. He studied psychology and medicine, and becoming interested in vision and visual perception, decided to specialize in neurophysiology. After graduating he was appointed instructor in physiology in 1927, and the following year visited England to work for a short period in **Sherrington**'s laboratory at Oxford, during which he also met **Adrian**, and learned of the latter's success in amplifying the small electrical signals generated by neural impulses. Granit perceived that similar techniques could be used to study the physiology of the retina, and in 1929 won a fellowship in medical physics to work at the Johnson Foundation at the University of Pennsylvania. There he worked under the supervision of **Bronk**, and met both **Wald** and **Hartline** with whom he was to share the 1967 Nobel Prize for Physiology or Medicine. After two years he returned to Helsinki, although he continued to spend periods in Oxford, and became Professor of Physiology in 1937. He entered the military medical service during the Soviet invasion of Finland in 1939 and the following year he escaped to Sweden, where he became Professor of Neurophysiology at the Karolinska Institute in Stockholm (1940–67). His analyses of retinal processing revealed that visual mechanisms were complex responses to light and dark, and he pioneered the recording of the mass response of the retina, the electroretinogram (ERG). He studied the ERGs of animals that lived in a variety of different light conditions, and the technique was widely adopted for the physiological investigation of clinical conditions. Granit's microelectrode studies enabled him to study the detailed responses of isolated retinal cells to light of different wavelengths and intensities, from which he was able to explain the mechanisms of colour discrimination. He summarized his work in visual physiology in *Sensory Mechanisms of the Brain* (1963). He has also contributed to research on the spinal cord, on pain mechanisms and on the philosophy of science, and has written a biography of Sherrington.

GRANT, Verne Edwin
(1917–)

American botanist, geneticist, experimental taxonomist and evolutionary biologist, born in San Francisco and educated at the University of California. He has taught at the

University of California, Claremont Graduate School, Texas A & M University, the University of Arizona and the University of Texas at Austin, where he is currently Emeritus Professor. From 1950 he was for many years Geneticist at Rancho Santa Ana Botanic Garden in Claremont, California, and he later became Director of Boyce Thompson Arboretum, Arizona. He is widely respected for his pioneering studies of the fertility relationships between species of the genus *Gilia* (Polemoniaceae), and of flower pollination. These studies have been reported in many scientific journals, and two books: *Natural History of the Phlox Family* (1959) and *Flower Pollination in the Phlox Family* (1965), the latter written jointly with his wife Karen Grant (b 1936). With his wife he also published *Hummingbirds and Their Flowers* (1968), the first modern study of bird-pollinated flowers. He has also published many papers and landmark texts in the fields of genetics and evolution, including *The Origin of Adaptations* (1963), *Plant Speciation* (1971), *The Genetics of Flowering Plants* (1975), *Organismic Evolution* (1977) and *The Evolutionary Process* (1985).

GRASSMANN, Hermann Günther
(1809–77)

German mathematician and philologist, born in Stettin. He came from a scholarly family and studied theology and classics at Berlin (1827–30), then became a schoolmaster in Berlin and Stettin. His book *Ausdehnungslehre* (1844) set out a new theory of *n*-dimensional geometry expressed in a novel language and notation. Its obscurity led to its almost complete neglect by mathematicians of the time, and it is only since his death that its importance has gradually been recognized; it anticipated much later work in quaternions, vectors, tensors, matrices and differential forms. From 1849 he studied Sanskrit and other ancient Indo-European languages and, unlike his mathematics, his work in Indo-European and Germanic philology met with immediate acceptance.

GRAY, Asa
(1810–88)

American botanist, born in Sauquoit, New York, and educated there and in Clinton, Iowa, before going to Fairfield Academy, to whose medical school he transferred after a year. He practised medicine briefly at Bridgewater, New York (1831–2), and began collecting plants. He had become friendly with **Torrey** and abandoned medicine to collaborate on the latter's *A Flora of North America* (1838–43). In 1836 he planned to join the Wilkes expedition but, frustrated by delays, he instead became Professor of Natural History at the University of Michigan, Ann Arbor (1838–42). He embarked on a tour of European botanical institutes, but on his return found the university almost bankrupt, unable to pay his salary. He became Professor of Natural History at Harvard (1842–73) on the understanding that he specialized in botany, and so became the USA's first paid professional botanist and eventually its leading 19th-century plant taxonomist. He became a strong Darwinian, and used **Charles Darwin**'s theory to explain the distribution of plants occurring in both eastern Asia and eastern North America as descendants of a Tertiary circumboreal flora which had retreated south during the Pleistocene glaciation. His total output numbered some 780 works, including *Manual of Botany of the Northern United States* (1848, known as 'Gray's Manual'), *Genera Florae Americae Boreali-Orientalis Illustrata* (1845–50) and *A Free Examination of Darwin's Treatise* (1861).

GRAY, Edward Whitaker
(1748–1806)

English botanist and physician, uncle of the botanist **Samuel Frederick Gray**. Edward Whitaker Gray was born, very probably in London, into a family of seedsmen whose firm had been established at Pall Mall in 1680. He was appointed librarian to the Royal College of Physicians in 1773, and from 1787 to 1806 was Keeper of the Department of Natural History at the British Museum. He arranged the collections according to the Linnaean system, and was much criticized for this. In 1788 he was made one of the first associates of the newly formed Linnaean Society of London. Elected FRS in 1779, he became Secretary of the Royal Society in 1797. He sent a collection of plants from Oporto, Portugal, to **Banks** in 1777, and was the author of *On the Class of Animals Called by Linnaeus Amphibia* (1789).

GRAY, Harry Barkus
(1935–)

American chemist, born in Woodburn, Kentucky. His early research was carried out with **Basolo** at Northwestern University. He received his PhD in 1960 and spent one year at the University of Copenhagen, where he studied the electronic structures of metal complexes. From Copenhagen he went on to Columbia University, New York, where he became professor in 1965. He is now Arnold O Beckman Professor of Chemistry at Caltech. Gray's main research interests have been in the fields of bioinorganic chemistry, inorganic photochemistry and mechanistic inorganic chemistry. His work has contributed to the understanding of the mechanism of metalloprotein electron transfer reactions in biological systems, while his photochemical studies have elucidated photo-induced electron transfer reactions of metal cluster complexes. He was awarded the American Chemical Society Award for Inorganic Chemistry (1978) and for Distinguished Service in the Advancement of Inorganic Chemistry (1984). He was Royal Society of Chemistry Centenary Lecturer in 1984–5.

GRAY, Sir James
(1891–1975)

English zoologist, born in London and educated at the University of Cambridge. Prior to World War I he studied the fertilization and embryology of echinoderms. He became a demonstrator in the zoology department at Cambridge in 1924, Reader in Experimental Zoology in 1931, and later Professor of Zoology (1937–61). Initially he investigated the cellular mechanisms underlying fertilization, cell division and growth, and he published *Text-book of Experimental Cytology* in 1931. His subsequent research interests were in the mechanics of locomotion; he believed that it was important, before investigating the physiology, to understand the engineering principles underlying movement. Much of his research dealt with swimming and the problems of size and scale. He was one of the first workers to make use of cinematography to analyse movement and used this technique in his examination of ciliary movement. His *Ciliary Movement* appeared in 1928 and *Animal Locomotion* was published in 1968. He was a member of the Advisory Committee on Fisheries Research (1932–65, chairman from 1949), President of the Marine Biological Association (1945–55) and a trustee of the British Museum (1948–60). He was knighted in 1954.

GRAY, Samuel Frederick
(1766–1828)

English botanist and pharmacologist, born in Pall Mall, London, nephew of **Edward Whitaker Gray**. He moved to Walsall, Staffordshire, in 1797 and practised as a pharmaceutical chemist until 1800. In that year he returned to London and lectured on botany and *materia medica*. He published *Supplement to the Pharmacopoeia* (1818), and with his son John Edward Gray (1800–75) wrote *Natural Arrangement of British Plants*. This ambitiously covered all British plants, both phanerogams and cryptogams. Gray's generic concept, like that of his friend **Richard Anthony Salisbury**, was narrow and controversial; the species of many well-known phanerogam genera, eg *Allium* and *Saxifraga*, were divided among numerous smaller units — most of these have been later reduced in rank or even ignored. The book is much more important to cryptogamic botanists, especially mycologists. Gray was the first person to establish fungal genera such as *Amanita*, *Lepiota* and *Corticium*. His work on algae was continued by his son John, who wrote *Handbook of British Waterweeds* (1864) but who was principally a zoologist.

GRAY, Stephen
(1666–1736)

English physicist, born in Canterbury, who followed his father's trade as a dyer, but with a thirst for education. His first scientific paper (1696) described a microscope made of a water droplet, similar to the simple glass bead microscopes made so famous by **Leeuwenhoek** in the following decade. He was one of the first experimenters in static electricity, using frictional methods to prove conduction; this work had a great influence on the electrical theory of **Dufay**.

GREEN, George
(1793–1841)

English mathematician and physicist, born in Sneinton, near Nottingham. The son of a baker and miller, he was largely self-taught. In 1828 he published a pamphlet entitled *An essay on the application of mathematical analysis to the theories of electricity and magnetism*, containing what are now known as Green's theorem and Green's functions. Green's theorem relates the integral of one quantity taken over a volume with another taken over the surface enclosing that volume. It can be used to relate the force exerted by an object at all points on a surface enclosing it, and therefore has valuable implications for potential theory. Green's functions are a valuable technical tool for solving partial differential equations, which often resist solution by any other method. Green entered Caius College, Cambridge, in 1833, published several papers on wave motion and optics, and was elected a Fellow of the college in 1839. After his death, the young William Thomson (later Lord **Kelvin**) came to hear of Green's work, and informed the French mathematician **Liouville** of it. Through Liouville, Green's theorem was made known to continental mathematicians. Thomson later edited Green's *Mathematical Works* (1850–4).

GREEN, Malcolm Leslie Hodder
(1936–)

English chemist, born in Eastleigh, Hampshire. He worked for his PhD (1960) under

Sir **Geoffrey Wilkinson** at Imperial College, London. Following a spell at Cambridge he moved to Oxford University (1963), where he is now Professor of Inorganic Chemistry. A transition metal chemist, his studies have included the 'beta' effect in the elimination of organic groups, the ways in which certain reagents preferentially attack certain sites in nucleophilic addition to organometallic cations, and the so-called 'agostic' interaction of hydrogen atoms to transition metals, in which the H atom of an alkyl group attached to a transition metal centre itself interacts with the metal. He has also studied the activation of C–H bonds by transition metals (of immense potential importance for developing the chemistry of widely available but unreactive reagents such as methane), and developed methods for reacting transition metal atoms (of even the most refractory elements) with various reagents --this has opened up studies of group 4 and 5 metals in their rare or previously unknown zero- or low-oxidation states. Awarded the Royal Society of Chemistry Corday–Morgan Medal (1974), he was elected FRS in 1985.

GREEN, Michael Boris
(1946–)

English theoretical physicist, born in London. He was educated at Cambridge University before obtaining a postdoctoral fellowship at the Institute for Advanced Study at Princeton (1970–2). From 1972 to 1977 he worked at Cambridge before becoming a Science and Engineering Council Advanced Fellow at Oxford University from 1977 to 1979, when he was appointed lecturer at Queen Mary and Westfield College, London. He became professor there in 1985, and returned to Cambridge in 1993 as John Humphrey Plummer Professor of Theoretical Physics. With John H Schwarz and Edward Witten, he was the founder of superstring theory. This is based on the idea that the ultimate constituents of nature, when inspected at very small scales, do not exist as point-like particles but as 'strings' in more than three dimensions. It was first introduced as a possible way to avoid the difficulties encountered by early unification schemes involving gravity, but now string theories are considered very good candidates for the actual laws of physics at the ultimate small scale. For this work he was awarded the Maxwell Medal by the Institute of Physics (1987), the William Hopkins Prize by the Cambridge Philosophical Society (1987) and the Dirac Medal of the International Centre for Theoretical Physics (1989). He was elected FRS in 1989.

GREGG, Sir Norman McAlister
(1892–1966)

Australian ophthalmologist, born in Burwood, Sydney. Educated at Sydney University, he served with the Royal Australian Medical Corps during World War I and won the Military Cross in France. He later studied ophthalmology in London before returning to Sydney. After an epidemic of German measles there in 1939, his research pointed to the link between the incidence of that illness in pregnancy, and cataracts and congenital heart defects in the children of affected mothers. This has led to vaccination programmes for women at risk. Gregg was knighted in 1953.

GREGOR, William
(1761–1817)

English chemist and clergyman, born in Trewarthenick, Cornwall, and educated at Cambridge. He spent his working life as a rector in Devonshire and Cornwall. He analysed local minerals, particularly the sand known as ilmenite, in which he discovered titanium. He died in Creed, Cornwall.

GREGORY, David
(1659–1708)

Scottish mathematician, nephew of **James Gregory**, born in Aberdeen. He became Professor of Mathematics at Edinburgh in 1683 and Savilian Professor of Astronomy at Oxford in 1692. He lacked the originality of his uncle, but published textbooks on geometry, astronomy (promoting **Isaac Newton**'s gravitational theories) and optics, in which he suggested the possibility of an achromatic lens.

GREGORY, James
(1638–75)

Scottish mathematician, born in Drumoak, Aberdeenshire. He graduated from Aberdeen University and went to London in 1662, and the following year published *Optica promota*, containing a description of the Gregorian reflecting telescope that he had invented in 1661. From 1664 to 1667 he was in Padua, where he published a book on the quadrature of the circle and hyperbola, giving convergent infinite sequences for the areas of these curves. In 1668 he became Professor of Mathematics at St Andrews University. However, he considered himself badly treated there and moved to the chair at Edinburgh in 1674 at double the salary; he died a year later. Much of his later work was concerned with infinite series, a term which he introduced into the language.

GREGORY, John Walter

(1864–1932)

British geologist and explorer, born in London. Early in his career at the British Museum (Natural History) (1887–1900), he undertook expeditions to North America, the West Indies, Spitzbergen and East Africa. He became the first Professor of Geology at the University of Melbourne (1900–4) and at the University of Glasgow (1904–29). In Australia he led an expedition into the 'Dead Heart' of the continent. He accompanied the first British East African Expedition, describing the Great Rift Valley of Kenya and Tanzania (1893), and subsequently visited Libya (1908), Angola (1912), India and East Africa (1916), Chinese Tibet (1922) and Peru, where he drowned in 1932. Gregory was a prolific author and wrote more than 300 geological papers on diverse topics such as corals and bryozoa, tectonics, glacial phenomena, the origin of ore deposits and the origin of oceans. His opposition to **Wegener**'s theories of continental drift was with hindsight seen to be in error. His publications included *The Dead Heart of Australia* (1906), *The Rift Valleys and Geology of East Africa* (1921), *To the Alps of Chinese Tibet* (1923), *The Structure of Asia* (1929), *Elements of Economic Geology* (1927) and *Dalradian Geology* (1931).

GREW, Nehemiah

(1641–1712)

English botanist and physician, born in Atherstone. Educated at Cambridge and Leiden, he practised at Coventry and London, and was author of *Comparative Anatomy of the Stomach and Guts* (1681) and of the pioneering *Anatomy of Plants* (1682). In this work he used the microscope to elucidate plant structure. This was the first complete account of plant anatomy, and it remained the most significant and authoritative work in this field for more than 150 years. Along with **Malpighi** he discovered some of the crucial differences between plant root and stem tissue, which arise from opposite sides of the developing embryo. He described in detail the complex folding (vernation) of unexpanded leaves in buds, an important character in plant classification. His book includes many figures illustrating plant structures, many of these illustrations being in themselves significant developments in the field of botanical illustration. Grew accepted the idea that the stamen is the male organ of the plant. The term 'parenchyma', still used today for the mass of cells making up the ground tissue of plants was coined by Grew, as was 'cambium', although he did not fully appreciate its function. As well as making enormous contributions to anatomy and physiology, his analytical studies presaged the development of phytochemistry.

GRIFFIN, Donald Redfield

(1915–)

American zoologist, born in Southampton, New York, and educated at Harvard. He taught at Cornell University (1946–53), Harvard (1953–65) and Rockefeller University (1965–86). Before World War II, while research assistant to the pioneer of insect song studies G W Pierce, he demonstrated for the first time that the ultrasound produced by bats is used in echolocation. He also studied the homing abilities of birds, particularly marine species such as Leach's petrel and the gannet. He was the first to use an aeroplane to follow individual birds to observe their actual behaviour. During World War II he studied the effects of background noise on radio communication, and investigated night vision using infrared light. After the war he continued his studies on echolocation and homing. In 1953 he demonstrated that bats also used ultrasound for locating and capturing prey during flight, and subsequently studied the behavioural aspects of this phenomenon. By the use of radar he showed in 1968 that migrating birds are able to maintain their orientation while flying blind in clouds. In 1981 he launched the study of 'cognitive ethology', an investigation of the way in which non-humans think and feel. He wrote *Listening in the Dark* (1958), *Bird Migration* (1964) and *Animal Thinking* (1984).

GRIFFITH, Sir Richard John

(1784–1878)

Irish soldier, geologist and engineer, born in Dublin. In 1800 he joined the Royal Irish Regiment, but he resigned shortly afterwards and studied chemistry, mining and geology in London and Edinburgh. He returned to Ireland in 1808, surveyed the coalfields of Leinster, and examined the Irish bogs for a government commission. His first geological map, based mostly on personal research, was published in 1815. As commissioner of valuations after the Irish Valuation Act of 1827 he created 'Griffith's valuations' for country rate assessments. He published the first complete geological map of Ireland in 1838 with a major revision in 1855, and made major contributions to the knowledge of its strata. He was knighted in 1858.

GRIGNARD, (François Auguste) Victor

(1871–1935)

French chemist, born in Cherbourg. Although he initially studied mathematics and had a low opinion of chemistry, he joined the chemistry department in Lyons, where he became 'Chef de Travaux Practiques' (1898). It was at this time that he began his work on the use of organomagnesium compounds in organic synthesis. Since then such compounds (Grignard reagents) have proved to be among the most useful and versatile reagents available for the synthesis of complex molecules, giving many different types of molecules under mild conditions. Grignard reagents have been much used in the laboratory syntheses of naturally occurring substances. For this work he received the Nobel Prize for Chemistry in 1912. He also worked at Besançon and Nancy, where his researches were particularly fruitful. During World War I he worked on war gases, and in 1919 he was appointed to the chair in Lyons, where he remained until his retirement.

GRIMALDI, Francesco Maria
(1618–63)

Italian physicist, born in Bologna, son of a wealthy silk merchant. At the age of 14 he joined the Society of Jesus, and was educated at his Order's houses at Parma, Ferrar and Bologna, where he became Professor of Mathematics (1648). At this time the Jesuits were notable teachers of the new science. Among many other contributions, Grimaldi verified **Galileo**'s laws of falling bodies, produced a detailed lunar map, and more notably, discovered diffraction of light, and researched into interference and prismatic dispersion. He was one of the first to postulate a wave theory of light.

GRISEBACH, August Heinrich Rudolph
(1814–79)

German botanist and plant geographer, born in Hanover and educated there and at Ifeld. An early interest in natural history was fostered by his mother's brother, Georg Friedrich Wilhelm Meyer (1782–1856), Professor of Botany at Göttingen. Grisebach himself went to Göttingen in 1832, but completed his studies at Berlin. His doctoral dissertation (1836) was a short work on the family Gentianaceae, which was to prove a precursor to his massive monograph, *Genera et Species Gentianearum* (1838). Inspired by **Humboldt**'s travels, in 1839 he set off on an important expedition to the Balkan Peninsula and Bithynia, then largely unexplored botanically. He was one of the first botanists to climb Ulu Dağ and analyse its vegetational zonation. Two publications, *Reise durch Rumelien und nach Brussa im Jahre 1839* (2 vols, 1843–6) and *Spicilegium Florae Rumelicae et Bithyniae* (2 vols, 1843–5), resulted. This journey laid the foundation for his phytogeographical work which culminated in *Vegetation der Erde* (1872). Later, as a result of the death of Wilhelm Gerhard Walpers, he studied Caribbean and South American botany and published many papers including another of his most important works, *Flora of the British West Indian Islands* (7 parts, 1859–64).

GROTHENDIECK, Alexandre
(1928–)

French mathematician, born in Berlin. He became a French citizen after fleeing Germany in 1941 and being for a while interned during World War II. After early important work on infinite-dimensional vector spaces, he switched to algebraic geometry, where he revolutionized the subject. His work led to a unification of geometry, number theory, topology and complex analysis, based on the complex but profound concept of the scheme. His work led to a resolution of the important **Weil** conjectures, of which the last and most troublesome was finally solved by his pupil Pierre Deligne in 1972. His work has also had profound implications for the theory of logic. He was awarded the Fields Medal (the mathematical equivalent of the Nobel Prize) in 1966. Later he became deeply involved in a pacifist, anti-militarist movement, and then in the teaching of mathematics. He has engaged in lengthy polemics about the way his work has been taken up by others, but his remarkably powerful introduction of the ideas of category theory at the very basis of algebraic geometry have had the effect of extending the language of geometry from fields to rings. In particular, it has enabled questions about the integers to be treated using the geometrical techniques hitherto available only when dealing with the rational or real numbers, thus opening up many important but previously intractable problems.

GROVE, Sir William Robert
(1811–96)

Welsh physicist and jurist, born in Swansea. He was educated privately and at Oxford, but abandoned his law career because of poor health. He concentrated instead on science, in particular on electrochemistry. He invented a new type of voltaic cell named after him (1839), and also a 'gas battery', the first fuel cell. He also invented the earliest form of filament lamp intended for use in mines. As Professor of Physics at the London Institution (1841–64), he studied electrolytic decomposition and demonstrated the dissociation of water. Thereafter he turned to the law, was raised to the bench (1871), and became a judge in the High Court of Justice (1875–87).

He was one of the original members of the Chemical Society. Elected FRS in 1840, he was one of the leaders of the Royal Society's reform movement. He was knighted in 1872.

GRUBB, Sir Howard

(1844–1931)

Irish engineer and builder of astronomical instruments, born in Dublin, son of Thomas Grubb FRS, engineer to the Bank of Ireland and also a maker of optical instruments. He studied civil engineering at Trinity College, Dublin. His father was commissioned in 1865 to build a 48 inch reflecting telescope for Melbourne, Australia, and he was given charge of the work under his father's supervision. This instrument, one of the largest in the world at the time, was declared by a committee of the Royal Society to be 'a masterpiece of engineering', and established his reputation on an even higher level than his father's. On his father's retirement in 1868, the firm moved to larger premises in Dublin, and around 1880 completed a 27 inch refracting telescope for Vienna, for some years the most notable telescope of its kind in the world. In 1900 he patented a new type of optical gun sight, and he developed and perfected the submarine periscope. In 1925, at the age of 81, he retired from active participation in the business, and it was acquired by **Parsons** and moved to Newcastle upon Tyne as the Sir Howard Grubb Parsons Company. Grubb was elected FRS in 1883 and knighted in 1887.

GUERICKE, Otto von

(1602–86)

German engineer and physicist, inventor of the vacuum-pump, born in Magdeburg. He worked in Leipzig, Helmstadt, Jena and Leiden, studying law and mathematics, mechanics and the art of fortification. An engineer in the Swedish army, he became one of the four burgomasters of Magdeburg from 1646 to 1681, elected for his service to the town as an engineer and diplomat during its siege in the Thirty Years' War. His interest in the possibility of a vacuum led him to modify a water-pump so that it would remove most of the air from a container. Such primitive vacuum pumps enabled the natural philosophers of the day to study new areas of physics. He arranged a dramatic demonstration of the effect of atmospheric pressure on a near vacuum in 1654 at Regensburg before the emperor Ferdinand III. Two large metal hemispheres were placed together and the air within pumped out; they could not then be separated by two teams of eight horses, but fell apart when the air was allowed to re-enter. He showed that in a vacuum candles cannot remain alight and small animals die, and he devised several experiments that demonstrated the elasticity of air. He also carried out some experiments in electricity and magnetism.

GUETTARD, Jean Étienne

(1715–86)

French botanist, invertebrate palaeontologist, stratigrapher and geomorphologist, born in Étampes. Grandson of an apothecary and destined to continue the family business, he studied medicine in Paris. He subsequently became a physician and accompanied the Duke of Orléans on his travels as keeper of his natural history collections. In 1746 he published a geological map of part of northern France with stratal boundaries, and he later produced another in which he attempted to link the geology of France with that of England; these were amongst some of the earliest geological maps to be prepared. Guettard demonstrated that the distribution of plants could be used as a means of mapping outcrops. He independently recognized that fossils are the remains of once-living organisms, and named some invertebrate genera. He also perceived the erosion and depositional roles of running water, and was the first to conclusively demonstrate the volcanic character of the Auvergne (1752). His mineralogical map of western Europe was published in 1780.

GUILLAUME, Charles Édouard

(1861–1938)

Swiss physicist, born in Fleurier. Educated at Neuchâtel and at the Zürich Polytechnic, in 1883 he joined the staff and eventually became Director of the Bureau of International Weights and Measures at Sèvres. In the course of the Bureau's continual efforts to improve the precision of its standards, he redetermined the volume of the litre, and investigated the effect of thermal movement on standards of length. As a result of this he began a search for a suitable material with little or no thermal expansion or contraction, and after many trials he discovered a nickel–steel alloy, which he christened 'Invar'. The use of this new material led to significant improvements in the accuracy and stability of timekeeping devices, precision instruments and standards of measurement, and for this discovery Guillaume was awarded the Nobel Prize for Physics in 1920.

GUILLEMIN, Roger (Charles Louis)

(1924–)

French–American physiologist, born in Dijon. In 1942 Guillemin graduated BSc from Dijon University and began a medical course. During the Nazi occupation he served in the

French Resistance. In 1949 he received his MD from Lyons University and went to Montreal to conduct work on the role of the hypothalamus (a structure at the base of the brain) in regulating the activity of the pituitary gland (the major endocrine gland of the body which is below the hypothalamus). In 1953 he accepted an appointment at Baylor University School of Medicine in Houston. He was promoted to the Chair of Physiology at Baylor in 1963, and in 1970 moved to the Salk Institute in San Diego. Guillemin and his colleagues were responsible for isolating and identifying the chemical structures of three hypothalamic hormones — thyrotropin-releasing hormone, which stimulates the pituituary to release thyrotrophin (which in turn stimulates the thyroid gland), growth-hormone-releasing hormone, and the hypothalamic hormone which inhibits the release of growth hormone. These discoveries have important applications, and potential applications for the treatment of various endocrinological diseases. Guillemin and **Schally** shared half of the 1977 Nobel Prize for Physiology or Medicine; the other half was awarded to **Yalow**.

GULDBERG, Cato Maximilian
(1836–1902)

Norwegian mathematician and chemist, born in Christiania (now Oslo). He entered the University of Christiania in 1854 to study mathematics and science. He became a teacher at the Royal Military Academy (1860), Professor of Applied Mathematics in the Royal Military College (1862), a lecturer at the University of Christiania (1867), and finally Professor of Applied Mathematics there (1869). He is best known for his work in collaboration with his brother-in-law **Waage**, which established the law of mass action, which states that the rate of a homogeneous chemical reaction is proportional to the concentrations of the reacting substances (1864). Their work was published originally in Norwegian and its importance was not widely appreciated for many years (the law was discovered independently by **Augustus Vernon Harcourt** and William Esson around 1864–6). Later Guldberg published several papers of a physico-chemical nature, including a discussion of the relationship between lowering of vapour pressure and depression of freezing point of solutions.

GULLSTRAND, Allvar
(1862–1930)

Swedish ophthalmologist, born in Landskrona, the son of Landskrona's chief medical officer. He studied medicine at the universities of Uppsala and Vienna, graduating from Uppsala in 1888. Gullstrand studied astigmatism for a PhD, which he received in 1890. He became chief physician, and in 1892, director, at the Stockholm Eye Clinic. Two years later he became Professor of Ophthalmology at Uppsala. Gullstrand studied physiological optics at a time when the optics of glass lenses had become well understood. The optics of the eye lens proved much more complicated, for unlike the glass lens, it changes its shape and is made up of layers of transparent fibres. Gullstrand developed mathematical formulae describing the optics of the eye which facilitated the treatment of conditions such as astigmatism and coma. He also developed new techniques for the examination of the eye including the Gullstrand ophthalascope, used to examine the eye for arteriosclerosis and diabetes mellitus. He was awarded the 1911 Nobel Prize for Physiology or Medicine.

GUNSALUS, Irwin Clyde
(1912–)

American microbiologist, born in Sully County, South Dakota. He was educated in South Dakota and at Cornell University, where he graduated with a PhD in 1940. He was appointed Professor at Indiana University in 1947, and in 1950 moved to the Chair of Microbiology at the University of Illinois, Urbana. His major work was on the metabolism of the bacteria *Enterococci*, and the discovery of the cofactors of the amino and ketoacid enzymes — pyridoxal phosphate and lipoate. He showed, using radioactively labelled glucose, that specific atoms of the glucose molecule followed different metabolic routes in a species of *Leuconostoc* bacteria (1950), the first carbon atom being metabolized to carbon dioxide, the third to ethanol methyl, and carbon atoms four to six to lactic acid. This was an early demonstration of radiolabelling in following metabolic pathways. His observation of activation by the 4-aldehyde form of vitamin B_6 showed the metabolic importance of pyridoxal-5-phosphate. Later, he identified the pyruvate oxidation factor, now known as lipoic acid, and its role in a unique reaction sequence. He was a Guggenheim Fellow in 1949, 1959 and 1968, and worked with **Ochoa** in 1950 on citrate metabolism, and later on the protein cytochrome P450. He has been assiduous in the application of physical spectroscopic techniques to aid the understanding of the structure and mechanism of the cytochrome P450 system.

GUNTER, Edmund
(1581–1626)

English mathematician and astronomer, born in Hertfordshire and educated at Westminster and Christ Church, Oxford. He received a Southwark living in 1615, but in 1619 became Professor of Astronomy at Gresham College, London, and invented many measuring instruments that bear his name; 'Gunter's chain', the 22 yard long 100 link chain used by surveyors; 'Gunter's line', the forerunner of the modern slide-rule; 'Gunter's scale', a 2 foot rule with scales of chords, tangents and logarithmic lines for solving navigational problems; and the portable Gunter's quadrant. He made the first observation of the variation of the magnetic compass, and introduced the words 'cosine' and 'cotangent' into the language of trigonometry. On some accounts it was his reliability that gave rise to the familiar American expression 'according to Gunter'; others connect the expression to the chain.

GURDON, John Bertrand

(1933–)

English geneticist, born in Dippenhall, Hampshire. He was educated at Christ Church, Oxford, where he graduated BA in 1956 and DPhil in 1960. He became a lecturer in the zoology department of the University of Oxford (1965–72) and a staff member at the Medical Research Council Laboratory of Molecular Biology, Cambridge (1972–83). Since 1983 he has been John Humphrey Plummer Professor of Cell Biology at the University of Cambridge, and since 1991, Chairman of the Wellcome Cancer Research Campaign Institute. He was elected FRS in 1971, and became a Foreign Member of the US National Academy of Sciences in 1980. A central question in biology this century has concerned the mechanism by which one cell type (the fertilized egg) gives rise to all the different cell types in the adult animal. For example, does a fully differentiated cell type such as intestine lose the information to become any other cell type, or does this information remain in a quiescent state? In 1968 Gurdon showed that transplantation of a nucleus derived from frog gut epithelium into an enucleated fertilized egg gave rise to a normal tadpole. This demonstrated that fully differentiated animal cells retain the genetic information to become any cell type under the correct environmental stimuli.

GUTENBERG, Beno

(1889–1960)

German-born American geophysicist, born in Darmstadt. He studied geophysics and mathematics at Darmstadt Technische Hochschule, and being attracted to the relatively new science of seismology, gained his doctorate from Göttingen University (1911) on the origin of microseisms. From 1913 to 1916 he was an assistant at the International Seismological Association in Strassburg (now Strasbourg, France) where he made two important contributions; he deduced from earthquake shockwaves the existence of a zone in the mantle where seismic waves travel with low velocities (1913), and in 1914 made the first correct determination of the depth to the Earth's core, which he concluded is liquid (the existence of some sort of core had already been deduced by **Oldham** in 1906). At the University of Frankfurt (1924–30) he taught seismology and worked on the structure of the atmosphere. In 1929 he moved to the USA where he accepted the Chair in Geophysics at Caltech (1930), and he became Director of the Seismological Laboratory when it moved from the Carnegie Institution to Caltech in 1937.

GUTHNICK, Paul

(1879–1947)

German astronomer, born in Hitdork on the Rhine, the son of a wine merchant. After study at the University of Bonn (1897–1901) under Friedrich Küstner, he became assistant at the old observatory in Berlin. Following three years at the Bothkamp Observatory near Kiel (1903–6), he returned to Berlin where he held observatory and university posts. In 1921 he succeeded Hermann Struve as Professor of Astronomy in the University of Berlin and Director of a new observatory at Berlin-Babelsberg. His most important work, begun in 1912, was in the field of precision photoelectric photometry, in which he was a pioneer, applying the methods developed by the physicists **Elster** and **Geitel** to astronomical problems. He was one of the first astronomers to make use of colour filters to determine both the brightnesses and colours of stars.

GUTHRIE, Samuel

(1782–1848)

American chemist and physician, born in Brimfield, Massachusetts. He studied at the College of Physicians and Surgeons, New York City, and at the University of Pennsylvania. During the War of 1812 with Britain, he served in the Army Medical Corps and later practised medicine in Sacketts Harbor, New York. He is reputed to have invented percussion priming powder. In 1830 he devised a process which rapidly converted potato starch into molasses. The following year he made chloroform by distilling chloride of lime with alcohol in a copper vessel. He died in Sacketts Harbor.

GUYTON DE MORVEAU, Baron Louis Bernard

(1737–1816)

French lawyer and chemist, born in Dijon. During much of his life he practised law, notably as a provincial prosecutor, but devoted his leisure to studying and teaching chemistry. In 1791 he was elected a member of the legislative assembly. From 1795 to 1805 he was Director of the École Polytechnique, and he was Master of the Mint from 1800 to 1814. He was made a baron of the French empire in 1810. Guyton de Morveau adopted the views of **Lavoisier** on combustion, and in the 1780s he was involved with Lavoisier, **Berthollet** and **Fourcroy** in revising chemical nomenclature and publishing *Méthode d'une nomenclature chimique* (1787). He wrote many scientific articles and several books; his interests spanned applied chemistry, metallurgy, mineralogy and balloon flight.

GYLBERDE, William

See **GILBERT, William**

H

HABER, Fritz

(1868–1934)

German physical chemist, born in Breslau (now Wrocław, Poland). The son of a dyestuffs merchant, he took up chemistry initially with a view to entering the family business. After study at the universities of Berlin and Heidelberg, he obtained his doctorate at the Technische Hochschule, Charlottenberg. There followed a period of uncertainty in which he attempted to find a satisfying career in the organic chemical industry and in the family business. In 1894, however, he became an assistant at the Technische Hochschule in Karlsruhe, and began the study of physical chemistry and its technical applications. Haber became *Privatdozent* in 1896, Extraordinary Professor in 1898, and in 1906 Professor of Physical Chemistry and Electrochemistry. In 1911 he moved to Berlin to direct the Kaiser Wilhelm Institute for Physical Chemistry and Electrochemistry, from which he resigned in 1933 in protest at the anti-Jewish policies of the Nazi regime. He accepted an invitation to work in Cambridge, but decided to winter first in Italy, and while travelling south he died at Basle in January 1934. At Karlsruhe much of his research was in electrochemistry, for example his study of the course of the electrolytic reduction of nitrobenzene, but he also worked in several other areas of physical chemistry. In 1904 he began to study the direct synthesis of ammonia from nitrogen and hydrogen gases, work which continued after his move to Berlin and which, in association with **Bosch**, led to the large-scale production of ammonia. This was important in maintaining an explosives supply for the German war effort in 1914–18. It also led to Haber receiving the Nobel Prize for Chemistry in 1918. This occasioned some criticism because Haber had been involved in the organization of gas warfare. In the 1920s he made abortive attempts to extract gold from sea-water, with a view to financing Germany's war reparations.

HADAMARD, Jacques Salomon

(1865–1963)

French mathematician, born in Versailles. Educated in Paris, he became a lecturer in Bordeaux (1893–7) and at the Sorbonne (1897–1909), and then professor at the Collège de France and the École Polytechnique until his retirement in 1937. For a long time his series of seminars, in which he displayed a mastery of many subjects, was the main event of its kind in France and a driving force for research. He was a leading figure in French mathematics throughout his career, working in complex function theory, differential geometry and partial differential equations. In 1896 he and Charles de la Vallée Poussin independently proved the definitive form of the prime number theorem, previously conjectured in cruder forms by **Legendre** and **Gauss**, and proved in a weaker form by **Chebyshev**. His discussion of the distinction between well-posed and ill-posed problems in differential equations illuminated the way in which small variations in the initial data for a problem (inherent in any measurement) should not significantly affect the answer. He lived to an exceptional age, and was still publishing mathematical work in his eighties.

HADFIELD, Sir Robert Abbott

(1858–1940)

English metallurgist and steel manufacturer, born in Sheffield. He was educated locally and trained as a chemist before joining his father's steelmaking firm in 1879, becoming chairman of Hadfields in 1888 until his death. His discovery of manganese steel in 1882, when he was only 24, established his reputation. **Mushet** had experimented with manganese, but it was generally believed that over 3 per cent manganese made steel very brittle. Hadfield, however, found a 12–13 per cent manganese alloy (the first commercial austenitic steel) which in the water-quenched condition had the happy ability to work-harden under abrasion and wear. The alloy proved ideal for tramway and railway trackwork, excavating equipment and mining machinery. Hadfield immediately followed this with research on silicon steel (which in collaboration with William Barrett he showed had remarkable electrical properties for use in transformers) and on armour-piercing projectile steels. Vain, autocratic and immensely hard working, his entrepreneurial ability placed Hadfields amongst the world's leading steel firms by World War I. A prolific technical writer and publicist, he wrote several important books on metallurgy and the development of special steels, such as *Metallurgy and Its Influence on Modern*

Progress (1925). Hadfield was knighted in 1908, elected FRS in 1909, and made a baronet in 1917.

HADLEY, George

(1685–1768)

English philosopher, born in London. Educated at Pembroke College, Oxford, he was called to the bar in 1709. He was one of a group of philosophers who gathered at the Royal Society in its early days, and his greatest contribution to meteorology was to improve on **Halley**'s theory of the general circulation. He correctly postulated that the reason air did not flow directly from the north or south into the equatorial zone was because of the rotation of the Earth. Air moving towards the equator will have a slower rotatory velocity than that of the air into which it moves. Hence a northerly wind flowing towards the equator in the northern hemisphere will become a north-easterly wind (the north-east trades), because the Earth rotates from west to east, and similarly a southerly wind in the southern hemisphere will become south-easterly (the south-east trades). This idea was published in the *Philosophical Transactions* of the Royal Society in 1735. Hadley also collected meteorological diaries for 1729 and 1730 and published a paper on the meteorology of 1731–5.

HADLEY, John

(1682–1744)

English mathematician, born in Hertfordshire. He invented a reflecting telescope (1720) and the reflecting (Hadley's) quadrant (1730). A prominent member of the Royal Society (FRS 1717), he became its Vice-President in 1728.

HAECKEL, Ernst Heinrich Philipp August

(1834–1919)

German naturalist, born in Potsdam, the son of a government lawyer. He studied medicine at Würzburg, Berlin and Vienna, but was profoundly influenced by **Charles Darwin**'s *Origin of Species* (1859) and quit medicine to study anatomy at Jena, where he later became Professor of Zoology (1862–1909). He made expeditions to the Mediterranean, Madeira, Canaries, Arabia, India and elsewhere. He wrote on the radiolarians (1862), calcareous sponges (1872) and jellyfishes (1879), and contributed *Challenger* reports on deep-sea medusae (1882), Radiolaria (1887) and Siphonophora (1888). He was the first to attempt a genealogical tree of all animals, and postulated the celebrated idea that in its embryological development, each species illustrates its phylogenetic history. Embodied in the phrase 'ontogeny recapitulates phylogeny', this became a pervasive and influential theory, though later discredited. The terms 'ontogeny' and 'phylogeny', as well as 'ecology', were of Haeckel's coining. Known as the 'German Darwin', he was a charismatic and enthusiastic ambassador for evolution, his books becoming bestsellers. These included *Generalle Morphologie* (1866), *The Natural History of Creation* (1868), *The Evolution of Man* (1874) and *Welträtsel* ('The Riddle of the Universe', 1899). Passing beyond the bounds of biology, Haeckel founded the 'monist' movement, in which evolution was used as the cornerstone of a unified philosophy of ethics, religion, politics and science. He was a champion of German nationalism, invoking natural selection to suggest that his countrymen formed a 'master race' which would outcompete 'inferior' peoples in the struggle for existence.

HAFFKINE, Waldemar Mordecai Wolff

(1860–1930)

Russian-born British bacteriologist, born in Odessa. He worked as an assistant to **Pasteur** (1889–93), and as bacteriologist to the government of India (1893–1915) he introduced his method of protective inoculation against *Vibrio cholerae*, the bacteria which cause cholera, using a heat-killed culture prepared from a highly virulent strain. In 1902 he was wrongly accused of having sent contaminated vaccine to the Punjab, but he was exonerated in 1907. He became a British subject in 1899.

HAHN, Otto

(1879–1968)

German radiochemist, born in Frankfurt. He studied at the universities of Marburg and Munich, receiving his doctorate at the former in 1901 for a thesis in organic chemistry. He was appointed as an assistant at Marburg, but soon a growing interest in radiochemistry took him to the laboratories of **Ramsay** in London (1904–5) and **Rutherford** in Montreal (1905–6). From 1906 to 1912 he was at the University of Berlin under **Emil Fischer**, and from 1912 to 1944 (with an interruption for service in World War I) he worked in the Dahlem district of Berlin at the Kaiser Wilhelm Institute for Chemistry, of which he was director from 1928. When the Kaiser Wilhelm institutes were reorganized as Max Planck institutes after World War II, Hahn became President of the Max Planck Gesellschaft in Göttingen. His researches from 1904 onwards were devoted entirely to the chemistry of the radioactive elements and their decay products. From 1907 to 1938 much of his work was done in collaboration with the Austrian physicist **Meitner**. Hahn was involved in the discovery of several new radioelements, including radiothorium, radioactinium and mesothorium, but his best-known research was on the irradiation of uranium and thorium with neutrons. This work, initially in association with Meitner and later with Fritz Strassmann, led to the discovery of nuclear fission (1938). For this Hahn received the Nobel Prize for Chemistry in 1944. Greatly upset that his discovery led to the horror of Hiroshima and Nagasaki, he became a staunch opponent of nuclear weapons.

HAHNEMANN, (Christian Friedrich) Samuel

(1755–1843)

German physician and founder of homeopathy, born in Meissen. He studied at Leipzig, and for 10 years practised medicine. After six years of experiments on the curative power of bark (the source of quinine), he came to the conclusion that drugs produce a very similar condition in healthy persons to that which they relieve in the sick. This was the origin of his famous principle, *similia similibus curantur* ('like cures like'), which he contrasted to the ordinary belief of allopathic (ie ordinary) practitioners. His own infinitesimal doses of medicine provoked the apothecaries, who refused to dispense them; accordingly he illegally gave his medicines to his patients, free of charge, and was prosecuted in every town in which he tried to settle from 1798 until 1810. He then returned to Leipzig, where he taught his system until 1821, when he was again driven out. He retired first to Köthen, and then in 1835 to Paris. He spent much time undertaking 'proving' of a number of drugs, which then entered the homeopathic pharmacopoeia. Many of these were herbal in origin, and subsequent homeopathists have continued to emphasize natural remedies. In his later years, he developed his idea that most chronic diseases are caused by the 'psora', a material he believed to be present on the surface of the skin. By the time of his death, his system had been taken up by practitioners throughout Europe and North America, although their relations with ordinary doctors were often bitter.

HALDANE, J(ohn) B(urdon) S(anderson)

(1892–1964)

British–Indian biologist, born in Oxford, the son of physiologist **John Scott Haldane**. Educated at Eton, he graduated in classics and philosophy from Oxford, switching later to genetics. He became a reader in biochemistry at Cambridge (1922–32), researching on enzymes, and then moved on to study population genetics and the mathematics of natural selection. He became Professor of Genetics at London University (1933–7), and then held the Chair of Biometry at University College London (1937–57). During World War II he collaborated in work on underwater respiratory physiology and submarine safety. Eccentric and wilful, he became an atheist during World War I, and as a committed Marxist, he was later chairman of the editorial board of the *Daily Worker* (1940–9), but left the Communist party in 1956 over the **Lysenko** controversy. In 1957 he emigrated to India (apparently in protest over Suez), adopted Indian nationality, and became Professor of the Indian Statistical Institute in Calcutta, resigning in 1961 after quarrelling with colleagues. He became head of the Orissa State Genetics and Biometry Laboratory in 1962. He wrote extensively on science and the social responsibilities of scientists, including *Animal Biology* (with **Julian Huxley**, 1927), *Possible Worlds* (1927), *Science and Ethics* (1928), *The Inequality of Man* (1932), *Fact and Faith* (1934), *Heredity and Politics* (1938) and *Science in Everyday Life* (1939).

HALDANE, John Scott

(1860–1936)

Scottish physiologist, younger brother of Richard Burdon, 1st Viscount Haldane, and father of **J B S Haldane** and Naomi Mitchison. Born in Edinburgh, he graduated in medicine at Edinburgh University in 1884. His earliest research, on the composition of the air in houses and schools, was conducted in Dundee. He became a demonstrator and reader in medicine at Oxford (1887–1913), and was elected a Fellow of New College, Oxford. He developed the famous Haldane gas analysis apparatus in 1898 and also a method for estimating the concentrations of gases in small quantities of blood. His best-known research, published with John Gillies Priestley in 1905, was concerned with the chemical control of ventilation. He emphasized the importance of the partial pressure of carbon dioxide. Haldane became an authority on the effects of industrial occupations upon respiration and served as a director of a mining research laboratory at Birmingham from 1912. He produced an important report upon causes of death in mining accidents, focusing particularly on the role of carbon monoxide. In laboratory experiments he showed that carbon monoxide binds haemoglobin and that the recovery of animals exposed to carbon monoxide can be hastened by placing them in air at a pressure greater than normal. He also conducted research into breathing in deep-sea diving and at high altitudes.

HALE, George Ellery
(1868–1938)

American astronomer, born in Chicago and educated at MIT, where he graduated in 1890. After studies at Harvard College Observatory (1889–90) he established in 1891 the Kenwood Observatory in Chicago, a private institution which became well known through its work in solar spectroscopy. In that year, simultaneously with **Deslandres** in France but independently of him, he invented the spectroheliograph, an instrument for examining the structure of the Sun's chromosphere and prominences which became of primary importance in solar physics. In 1892 he was appointed Professor of Astrophysics in the University of Chicago and in 1897 became the first director of the newly founded Yerkes Observatory near Chicago. When in 1905, following his initiative, the Carnegie Institution had established the Mount Wilson Observatory in the favourable climate of California, Hale was appointed its director. In 1906 he set up there the first tower telescope for solar research and in 1908 the 60 inch followed 10 years later by the famous 100 inch reflectors. By the time of his retirement because of illness in 1923, the Mount Wilson Observatory had become the greatest astronomical research establishment in the world. Hale's own scientific work at Mount Wilson included his discovery and measurement of magnetic fields in sunspots. Near the end of his life he provided the impetus for and was involved in the planning of the great 200 inch mirror Hale telescope on Mount Palomar, completed in 1948.

HALES, Stephen
(1677–1761)

English botanist and chemist, the 'father of plant physiology', born in Beaksbourn, Canterbury. He entered Corpus Christi College, Cambridge, in 1696. Well-grounded in all branches of contemporary science and inspired by **Isaac Newton**'s experimental philosophy, he was elected a Fellow of Corpus Christi College in 1702, and became in 1709 perpetual curate of Teddington. He was elected FRS in 1718 and a Foreign Member of the French Academy in 1753. He was one of the founder members of the Society for the Encouragement of the Arts and Manufactures and Commerce, now the Society of Arts. He was also Chaplain to Prince George, later George III. Hales's *Vegetable Staticks* (1727) was the foundation of plant physiology. He emphasized the need for precise measurement of quantities, and his experiments consistently included controls and checks on techniques, factors which set standards in the methodology of biological experimentation. His most important botanical work was on the water balance of plants, measuring for the first time root and leaf suction, and root pressure. In *Haemastaticks* (1733) he discussed the circulation of the blood and blood pressure. Besides a work on dissolving stones in the bladder, he wrote on a variety of subjects including ventilation, electricity, and the analysis of air. He also invented machines for ventilating, distilling sea-water, preserving meat and other practical applications of science. He is honoured by a monument in Westminster Abbey.

HALL, Asaph
(1829–1907)

American astronomer, born in Goshen, Connecticut. Self-educated, he had a strong interest in astronomy and was employed as an assistant at Harvard in 1857. From 1862 to 1891 he was on the staff of the Naval Observatory at Washington, and he later returned to Harvard as Professor of Astronomy. In 1877 Hall discovered the two satellites of Mars which he named Deimos and Phobos. On 7 December 1876, he discovered a white spot on Saturn and used this as a marker in order to obtain the rotation period of the planet. His value of 10 hours 14 minutes and 24 seconds is to be compared with **William Herschel**'s value of 10 hours 16 minutes obtained in 1794.

HALL, Charles Martin
(1863–1914)

American chemist, born in Thompson, Ohio, the descendant of English immigrants. Educated at Oberlin College, in 1886 he discovered (independently of **Héroult**) the first economic method of obtaining aluminium. He found that aluminium oxide, dissolved in melted cryolite, could be electrolyzed. Eventually he secured the financial support of Andrew Mellon and others and began aluminium production in 1888 in Washington, Pennsylvania. Two years later Hall became Vice-President of the Aluminum Company of America.

HALL, James
(1811–98)

American palaeontologist and stratigrapher, born in Hingham, Massachusetts. He was educated at Rensselaer Polytechnic Institute in Troy. After a short spell as a librarian, he worked as an assistant to Amos Eaton (1776–1842), and then in 1836 was appointed to the Geological Survey of New York State at the start of a long association. From 1855 to 1858 he was State Geologist of Iowa. Hall was author of 13 volumes on the palaeontology of New York State (1847–94) and many other works on its Palaeozoic fossils and stratigraphy. He undertook important studies of crinoids and other echinoderms, and he named many new fossil genera and species. He was the first President of the Geological Society of America.

HALL, Sir James
(1761–1832)

Scottish geologist, born in Dunglass, a Haddingtonshire baronet. He studied at the universities of Cambridge and Edinburgh, but did not complete a degree; the fortune inherited from his father with the baronetcy in 1776 allowed him to become financially secure and he travelled widely in Europe furthering his scientific studies independently. He initially disagreed strongly with the views of his friend **Hutton**, but subsequently became a strong supporter, publishing *Illustrations of the Huttonian Theory* (1802, with **Lyon Playfair**). In pioneering studies in experimental petrology, he melted and recrystallized local basalts and dolerites to demonstrate their igneous origin. He visited Italy in 1785, and noted the similarity between the volcanic rocks of Mt Etna and those of Scotland. Hall conducted more than 500 ingeniously devised melting experiments, and was able to demonstrate that molten magma could metamorphose existing limestones. He also produced a machine to demonstrate the folding of geological strata.

HALL, Marshall
(1790–1857)

English physician and physiologist, born in Basford, Nottinghamshire. The son of a wealthy Methodist cotton manufacturer, Hall studied medicine at Edinburgh University, obtaining his MD in 1812 and becoming medical officer at the Edinburgh Infirmary. Medico-scientific studies in Paris, Göttingen and Berlin were followed by a return to Nottingham in 1817, where he practised medicine, being elected honorary physician to the General Hospital in 1825. The next year he removed to London, where he prospered as an élite physician and was elected a Fellow of the Royal College of Physicians. He wrote copiously on many aspects of medicine, including the circulation of the blood and respiration, and developed a successful technique for resuscitating the drowned. Hall canvassed changes in clinical practice, notably opposing the immoderate blood-lettings then *de rigueur*. But the enduring reputation of this temperamental man hinges on his meticulous and innovative explorations of nervous response, notably his researches on the physiology of reflex function, which, building on the work of **Magendie** and **Charles Bell**, provided the foundation for the influential concept of the neural arc. Hall's fearless vivisection activities became the focus of public controversy.

HALLER, (Viktor) Albrecht von
(1708–77)

Swiss anatomist, botanist, physiologist and poet, born in Bern. He studied medicine, and anatomy in Tübingen, and botany at Leiden under **Boerhaave**, and started practice in Bern in 1729. There he began collecting plants to form the basis for a massive Swiss Flora. In 1736 he became Professor of Anatomy, Surgery and Medicine at the new University of Göttingen. Here he organized a botanical garden, an anatomical museum and theatre, and an obstetrical school; he also helped to found the Academy of Sciences, wrote anatomical and physiological works, and took an active part in the literary movement. He experimented with injection techniques to investigate human blood vessels, and recognized the mechanical automatism of heart muscle function. In 1753 he resigned and returned to Bern, where he became a magistrate and director of a saltworks. After this he wrote three political romances, and prepared bibliographies of botany, anatomy, surgery and medicine. His major publications were *Enumeratio Methodica Stirpium Helvetiae Indigenarum* (1742) on cryptogams, his Swiss Flora *Historia Stirpium Indigenarum Helvetiae Inchoata* (1768), *Bibliotheca Botanica* (1771–2), *Primae Lineae Physiologiae*, and *Elementa Physiologiae Corporis Humanae* (8 vols, 1757). Botanically, he was a strong supporter of the **Linnaean** view that class characters should only be taken from the fructification.

HALLEY, Edmond
(1656–1742)

English astronomer and mathematician, born in London. Educated at St Paul's School and Queen's College, Oxford, he published three papers on the orbits of the planets, on a sunspot, and on the occultation of Mars while still an undergraduate at Queen's. In 1676 he left for St Helena to make the first catalogue of the stars in the southern hemisphere (*Catalogus Stellarum Australium*, 1679). In 1680 he was in Paris with **Cassini**, observing comets. It was in cometary astronomy that he made his greatest mark. His calculation of the orbital parameters of 24 comets enabled him to predict correctly the return (in 1758, 1835 and 1910) of a comet that had been observed in 1583, and is now named after him. He was the first to make a complete observation of the transit of Mercury, and the first to recommend the observation of the transits of Venus with a view to determining the Sun's parallax. He established the mathematical law connecting barometric pressure with heights above sea-level (on the basis of **Boyle**'s law). He published studies on magnetic variations (1683), trade winds and monsoons (1686), investigated diving and underwater activities, and voyaged in the Atlantic to test his theory of the magnetic variation of the compass, which he embodied in a magnetic sea-chart (1701). Halley predicted with considerable accuracy the path of totality of the solar eclipse that was observed over England in 1715. He was the first to realize that the Moon's mean motion had a secular acceleration. He also noticed that stars such as Aldebaran, Acturus and Sirius had a proper motion, and that they had gradually changed their positions over the previous two millennia. In map-making Halley was the first to use an isometrical representation. He was also the first to predict the extraterrestrial nature of the progenitors of meteors. He encouraged **Isaac Newton** to write his celebrated *Principia* (1687), and paid for the publication out of his own pocket. With his *Breslau Table of Mortality* (1693), he laid the actuarial foundations for life insurance and annuities. In 1703 he was appointed Savilian Professor of Geometry at Oxford, where he built an observatory on the roof of his house which is still to be seen, and in 1720 he succeeded **Flamsteed** as Astronomer Royal of England.

HAMILTON, William Donald
(1936–)

English zoologist, born in Cairo, Egypt. Educated at the universities of Cambridge and London, he taught at London (1964–77), in Brazil and at the University of Michigan (1977–84) before becoming Royal Society Research Professor at Oxford in 1984. His main interests have been in evolutionary biology and his researches paved the way for the development of sociobiology. Like **Edward Wilson** his initial studies concerned insects. In 1964 he proposed his theory of 'kin selection' which accounted for the altruistic behaviour observed in animal societies. His 'inclusive fitness' concept recognized that an individual may influence the survival and successful breeding of a relative, increasing the probability of survival of shared genes, even though the individual may be sterile. A classic example of this is in the social Hymenoptera, such as honey bees. Hamilton was thus able to demonstrate the probable genetic basis of social behaviour in insects. He subsequently developed the concept of 'reciprocal altruism', arguing that natural selection favours such behaviour in social animals. His most recent research has been concerned with the role of parasites in sexual selection, demonstrating that choice of a sexual partner is affected by parasite load. His major contribution has been to provide a theoretical framework for some of the problems in modern evolutionary biology. He was elected a Fellow of the Royal Society in 1980, and received its Darwin Medal (1988) and the Scientific Medal of the Linnaean Society (1989).

HAMILTON, Sir William Rowan
(1805–65)

Irish mathematician, the inventor of quaternions, born in Dublin. At the age of nine he had a knowledge of 13 languages, and at 15 he read **Isaac Newton**'s *Principia* and began original investigations. In 1827, while still an undergraduate, he was appointed Professor of Astronomy at Dublin and Irish Astronomer Royal; he was knighted in 1835. His first published work was on optics, and led his colleague Bartholomew Lloyd to discover the unexpected phenomenon of conical refraction of light in certain crystals. He then developed a new approach to dynamics, later and independently proposed by **Jacobi**, which found favour only with the work of **Lie** and **Poincaré**, and which became of considerable importance in the 20th-century development of quantum mechanics. In 1843 he introduced quaternions after realizing that a consistent algebra of four dimensions was possible if the requirement of commutativity was dropped, and interpreted them in terms of rotations of three-dimensional space. The discovery led to much work on other abstract algebras and so proved to be the seed of much modern algebra. Because quaternions split naturally into a one- and a three-dimensional part, their discovery allowed the successful introduction of vectors into physical problems.

HAMMETT, Louis Plack
(1894–1987)

American physical chemist, born in Wilmington, Delaware. On graduating from Harvard in 1916, he worked for a year with **Staudinger** at the Swiss Federal Institute of Technology (ETH), Zürich, before returning to the USA to assist in wartime research. After World War I he worked in industry before joining the chemistry faculty of Columbia University, New York, as instructor in 1920. He received his doctorate in 1923 and thereafter climbed the academic ladder to full professor (1935). Hammett retired from Columbia in 1961, but remained active scientifically for many years. His honours included the Priestley Medal (1961) and other awards of the American Chemical Society, and honorary fellowship of the Royal Society of Chemistry. From the 1920s he specialized in applying the methods of physical chemistry to the problems of organic chemistry, and he is regarded as a founder of physical organic chemistry. This term was little used until Hammett wrote *Physical Organic Chemistry* (1940), a book which was highly influential. His main original contributions were in two areas. In structure–reactivity relationships he devised a simple relation to summarize the reactivities of meta- or para-substituted benzene derivatives. This became known as the Hammett equation and has subsequently been much modified and extended in application. Hammett was also a pioneer in studying the effects of concentrated solutions of acids on organic reactions by means of an empirical quantity, the acidity function.

HANBURY-BROWN, Robert
(1916–)

British radio astronomer, born in Aruvankadu, India. After studying engineering at Brighton Polytechnic and receiving an external degree from London University, he joined a radar research programme during World War II. In 1949 he moved to Jodrell Bank (University of Manchester). He was elected FRS in 1960, and in 1962 he took up the Chair of Astronomy at the University of Sydney, Australia. In 1951, with Cyril Hazard, he obtained the first radio map of an external galaxy — the Andromeda nebula. The resolution was very poor, and he went on to propose and construct a radio interferometer which had much greater resolution. This instrument correlated signals from separate antennae after they had passed through independent receivers and detectors. With it he measured the size of the two radio sources Cassiopeia A and Cygnus A. An optical experiment carried out by Brown and his colleague Richard Twiss in 1956 demonstrated the 'intensity interference' phenomenon at optical wavelengths — this was one of the key discoveries which sparked off the development of quantum optics in the early 1960s. In 1962, at the Narrabri Observatory, Australia, they used as interferometers two visual telescopes with 6.5 metre mirrors, each big mirror constructed of a mosaic of small hexagonal plane mirrors. The telescopes were mounted on a circular track of diameter 188 metres. With them they measured the diameters of around 30 giant stars which had not previously been accessible to direct measurement.

HANSTEEN, Kristoph
(1784–1873)

Norwegian astronomer, born in Christiania (now Oslo), where he became Professor of Mathematics in 1814. He investigated terrestrial magnetism, discovered the 'law of magnetic force' (1821) and made a scientific journey to Eastern Siberia (1828–30).

HARCOURT, Augustus George Vernon
See **VERNON HARCOURT, Augustus George**

HARCOURT, William Venables Vernon
See **VERNON HARCOURT, William Venables**

HARDEN, Sir Arthur
(1865–1940)

English chemist, born in Manchester. He worked in the Jenner (later Lister) Institute from 1897, becoming head of the biochemistry section from 1907 until his retirement in 1930. He was appointed professor by London University in 1912. Investigating the fermentation of sugars by bacteria, he confirmed **Eduard Buchner**'s finding that living yeast juice was not necessary for fermentation. He made the crucial discovery that the first step in fermentation was the phosphorylation of the sugar to form an ester (1905), and isolated fructose 1,6-bisphosphate. Later he isolated glucose 6-phosphate and fructose 6-phosphate, two other intermediates. Harden also showed that dialysis destroyed activity, thereby implicating a dialysable cofactor in fermentation, and he recognized the presence of more than one enzyme. For this work he shared the 1929 Nobel Prize for Chemistry with **Euler-Chelpin**. In addition to his classic monograph, *Alcoholic Fermentation* (1911), Harden wrote two student chemistry textbooks and analysed **Dalton**'s work. He studied vitamins and nutritional problems of the army during World War I. Harden was elected FRS in 1909, awarded the Royal Society's Davy Medal in 1935, and knighted in 1936.

HARDY, Sir Alister Clavering
(1896–1985)

English marine biologist, born in Nottingham, and educated at Oxford University. Following a period at the Stazione Zoologica in Naples, Hardy returned to England in 1921 to the appointment of Assistant Naturalist at the Fisheries Laboratory in Lowestoft. There he initiated his classic studies on feeding of all life stages of herring, and their dependence on zooplankton, and he began also to address the sampling problems conferred by 'patchiness' of planktonic populations, and latterly experiments on vertical migration of zooplankters. He was appointed Chief Zoologist (1924–7) on the *Discovery* expedition to study the plankton communities which sustain the Antarctic whaling grounds around South Georgia. Hardy's contributions to the *Discovery Reports* were based on the analysis of thousands of zooplankton and phytoplankton samples, and records of physical and chemical parameters; the results provided an unparalleled compendium of the interrelationships between the baleen whales and their zooplankton prey. He became Professor of Zoology and Oceanography at University College, Hull (1928–42), where he directed research on his plankton indicator and continuous plankton recorder; the subsequent development of these devices allowed transoceanic sampling and recording by means of a continuously moving belt of plankton silk, and led to the development of the Continuous Plankton Recorder Survey (based variously in Hull, Edinburgh and Plymouth). In 1942 he was appointed to the Regius Chair of Natural History at Aberdeen University, and he subsequently became Linacre Professor of Zoology at Oxford (1945–61). During his retirement he was founding director of the Religious Experience Research Unit at Manchester College, and was awarded the Templeton Foundation Prize for Progress in Religion ($185 000) just 11 days before his death. He was elected FRS in 1940 and knighted in 1957.

HARDY, Godfrey Harold
(1877–1947)

English mathematician, born in Cranleigh. Educated at Winchester and Cambridge, he became a Fellow of Trinity College in 1900. In 1919 he became Savilian Professor of Geometry at Oxford, and he later returned to Cambridge as Sadleirian Professor of Pure Mathematics (1931–42). Hardy was an internationally important figure in mathematical analysis, and was chiefly responsible for introducing English mathematicians to the great advances in function theory that had been made abroad. In much of his work in analytic number theory, the **Riemann** zeta function, **Fourier** series and divergent series, he collaborated with **Littlewood**. He brought the self-taught Indian genius **Ramanujan** to Cambridge and introduced his work to the mathematical world, rating it far above his own. Their greatest joint achievement was an exact formula for the partition function, which expresses the number of ways a number can be written as a sum of smaller numbers. His mathematical philosophy was described for the layman in his book *A Mathematician's Apology* (1940), in which he claimed that one of the attractions of pure mathematics was its lack of practical use. Cricket was the other great passion of his life.

HARINGTON, Sir Charles Robert
(1897–1972)

British chemist, born in north Wales. He studied at Cambridge and then in Edinburgh (1919–20), inspired by **Barger** with whom he later reported the constitution and synthesis of thyroxine (1927). Moving to the Royal Infirmary in Edinburgh, he studied protein metabolism and then spent a year in the USA with D D Van Slyke and **Dakin** before becoming a lecturer (1922–31) and Professor of Pathological Chemistry (1931–42) at University College London. From 1942 to 1946 he was Director of the National Institute for Medical Research in London. In 1926 Harington published a provisional structure for thyroxine as a tetra-iodo derivative of tyrosine and, with Barger, confirmed this the following year. In 1927 the hormone was also tested on two myxoedema patients with modest, but not lasting, success. He reported an improved method of synthesis with better yield in 1940. His numerous publications included *The Thyroid Gland; its Chemistry and Physiology* (1933). Harington's later years were mainly involved in administration. He was elected FRS in 1931, and knighted in 1948.

HARIOT, Thomas See **HARRIOT, Thomas**

HARKER, Alfred
(1859–1939)

English petrologist and structural geologist, born in Kingston-upon-Hull. He entered St John's College in 1878 at the beginning of a lifelong association with the University of Cambridge. Although physics was his principal subject, he was appointed as a university demonstrator in geology at the Sedgwick Museum (1884) and university lecturer in 1904. Following early research in north Wales and the Lake District, he undertook fieldwork in Scotland for the Geological Survey (1895–1905) whilst retaining his Cambridge post. It was in Scotland that his most important work was done, with his detailed study of the igneous rocks of Skye and Rhum published as the authoritative memoirs *The Tertiary Igneous Rocks of Skye* (1904) and *The Geology of the Small Isles of Inverness-shire* (1908). With these works he established the general succession of rocks in the igneous complexes, identified the hybrid rocks produced by the mixing of magmas, particularly those described from Marsco, Skye, and made important observations on the form of the intrusive bodies. He made great advances in the field of petrology, particularly with his advocation of the idea of petrographic provinces, his studies of metamorphism, slaty cleavage and the physics of glacial erosion. He published *On Slaty Cleavage and Allied Rock Structures* (1886), *Natural History of Igneous Rocks* (1909), *Metamorphism* (1932) and many papers on the rocks of the Lake District, north Wales, the Isle of Skye and other regions.

HARRIOT, or HARIOT, Thomas

(c.1560–1621)

English mathematician and scientist, born in Oxford, where he graduated in 1580. In 1584–5 he was mathematical tutor to Sir Walter Raleigh and was sent to survey Virginia, on which he published *A Briefe and True Report of the New Found Land of Virginia* (1588). He corresponded with **Kepler** on astronomical matters, observed **Halley**'s comet in 1607 and made observations with the newly invented telescope from 1609, as early as **Galileo**. His map of the Moon and drawings of sunspots and the satellites of Jupiter survive. He studied optics, refraction by prisms and the formation of rainbows, and gave a simple rectification of the logarithmic spiral which was immediately applicable in long-distance navigation. Most of his work was never published and remains in manuscript, although his *Artis analyticae praxis*, a treatise on algebra, was published posthumously in 1631, showing that he had developed an effective algebraic notation for the solution of equations.

HARRIS, Henry

(1925–)

Australian–British geneticist. He was educated at the University of Sydney and Lincoln College, Oxford, where he graduated MA and DPhil in 1954. He then became Director of Research of the British Empire Cancer Campaign at Sir William Dunn School of Pathology, Oxford (1954–9), and was later appointed Head of the Department of Cell Biology at the John Innes Institute (1960–3). Thereafter he became Head of the Sir William Dunn School (1963–92), and Professor of Pathology (1963–79) and Regius Professor of Medicine (1979–92) at Oxford University. He was elected FRS in 1968. In 1965 Harris successfully fused somatic mammalian cells in culture to produce the first heterokaryons (cells in which the cytoplasm but not the nuclei have fused), and later true cell hybrids in which the nuclei of the parent cells coalesce. This technology allowed conclusions to be drawn relating to the control mechanisms by which differentiation occurs in mammalian cells. For example, it has been shown that if two differentiated cell types are fused together, the hybrid cell retains the characteristics of both cell types. This implies that genes are maintained in a quiescent state in inappropriate cell types by the absence of activator molecules rather than by the presence of inhibitors. Such activator molecules have now been identified and purified.

HARRISON, Ross Granville

(1870–1959)

American biologist, born in Germantown, Pennsylvania. He entered Johns Hopkins University in 1889 and received his PhD in zoology in 1894. He joined the staff at Johns Hopkins as an instructor in anatomy in 1896, and from 1899 to 1907 was Professor of Biology. In 1907 he moved to Yale University where he was successively appointed Professor of Comparative Anatomy (1907), Professor of Biology (1927) and Emeritus Professor (1938). Harrison introduced the hanging-drop method of tissue culture (1907). Using this technique he settled a controversy about the embryological origins of nerve fibres by demonstrating that they are formed as outgrowths of nerve cells. The hanging-drop technique has proved of great value not only in embryology, but also in oncology, genetics, virology and other fields. Harrison's work using tissue grafting techniques was also very important for embryology. In one of his early experiments in this area he showed it was possible to join together parts of embryos from differently coloured frogs. This enabled him to observe the movement of the cells during subsequent development of embryos produced in this way.

HARTECK, Paul

(1902–85)

Austrian–American physical chemist, born in Vienna. After first studying in Vienna, he took his doctorate in Berlin (1926) under **Bodenstein**. In 1926 he became assistant to Arnold T Eucken in Breslau and in 1928 to **Haber** in Berlin. He spent 1933–4 at the Cavendish Laboratory in Cambridge with **Rutherford** and **Oliphant**. In 1934 Harteck became professor and director at the Institute for Physical Chemistry in Hamburg. From 1951 to 1982 he was Distinguished Research Professor of Physical Chemistry at the Rensselaer Polytechnic Institute in Troy, New York. In Haber's institute he collaborated with **Bonhoeffer** in work on atomic hydrogen, and on the separation and properties of para- and ortho-hydrogen, the forms of molecular hydrogen which differ in the orientation of nuclear spin. At the Cavendish Laboratory, Harteck demonstrated the existence of para and ortho forms of deuterium (the isotope of hydrogen of mass number 2), and with Oliphant and Rutherford obtained tritium (the isotope of hydrogen of mass number 3) for the first time by the bombardment of deuterium with deuterons. In later life Harteck made distinguished contributions to photochemistry and to the chemistry of the Earth's upper atmosphere, planetary atmospheres and interstellar space.

HARTLINE, Haldan Keffer
(1903–83)

American physiologist, born in Bloomsburg, Pennsylvania. He graduated in medicine from Johns Hopkins Medical School in 1927. After a brief period of postgraduate study in mathematics and physics, he joined **Bronk** at the Johnson Foundation of the University of Pennsylvania in 1931, where he remained until 1949, with a brief intermission at Cornell (1940–1). In 1949 Hartline moved with Bronk to Johns Hopkins, where he was appointed Professor of Biophysics, and in 1954 he again moved, with Bronk, to the Rockefeller Institute for Medical Research as head of the biophysics laboratory. When the institute became the Rockefeller University, Hartline became Professor of Biophysics, and in 1972 was named Detlev Bronk Professor at the Rockefeller until his retirement two years later. Inspired by the work of Bronk and **Adrian** in recording the electrical activity of single nerve fibres, Hartline and the psychologist Clarence Graham attempted similar experiments in the optic nerve of the horseshoe crab. They confirmed the 'frequency coding' reported by Adrian, that information about light falling on the eye was transmitted back to the brain coded by changes in the rate of impulses. From here, Hartline extended his studies to the more complex responses in the visual system of the frog, showing that it was the integrated action of all components of the visual system that produced visual perception in the brain. Returning to the horseshoe crab, Hartline analysed the several physiological stages by which an eye distinguishes shapes in work that led directly to that of **Hubel** and **Wiesel**. Some of his most important contributions were discoveries of the physiological interactions, especially those of inhibition, amongst the different types of cells in the retina. In 1967 Hartline shared with **Wald** and **Granit** the Nobel Prize for Physiology or Medicine for his work on the neurophysiology of vision.

HARTREE, Douglas Rayner
(1897–1958)

English mathematician and physicist, born in Cambridge, where he graduated after working on the science of anti-aircraft gunnery during World War I. From 1929 to 1945 he was Professor of Applied Mathematics and Theoretical Physics at Manchester, returning to Cambridge as Professor of Mathematical Physics in 1946. His work was mainly on computational methods applied to a wide variety of problems ranging from atomic physics, where he invented the method of the self-consistent field in quantum mechanics, to the automated control of chemical plants. At Manchester he developed the differential analyser, an analogue computer, and was deeply involved in the early days of the electronic digital computer.

HARVEY, William
(1578–1657)

English physician, and discoverer of the circulation of the blood. Born in Folkestone the eldest of seven sons in the family of a yeoman farmer, Harvey went to school in Canterbury, proceeding to study medicine at Caius College, Cambridge. After graduating in 1597, he moved to Padua, working under **Fabrizio**. In 1602 he set up in practice in London as a physician. Elected a Fellow of the Royal College of Physicians in 1607, two years later he was appointed physician to St Bartholomew's Hospital, and in 1615 he was Lumleian Lecturer at the College of Physicians. In 1628 he published his celebrated treatise, *Exercitatio Anatomica de Motu Cordis et Sanguinis* ('An Anatomical Exercise on the Motion of the Heart and the Blood in Animals'), in which he expounded his views on the circulation of the blood. Successively physician to James I (from 1618) and to Charles I (from 1640), he accompanied the Earl of Arundel in his embassy to the emperor in 1636, publicly demonstrating his theory at Nuremberg. A convinced royalist, he was present at the Battle of Edgehill in 1642, attending on the King; he then accompanied Charles to Oxford, becoming warden of Merton College. In July 1646, on the surrender of Oxford to the parliamentary forces, he returned to London, retired from professional life, and devoted himself entirely to his researches. His book on animal reproduction, *Exercitationes de Generatione Animalium*, appeared in 1651. He was buried at Hempstead near Saffron Walden. The key claim of Harvey's distinguished work on the cardiovascular system was that 'the blood performs a kind of circular motion' through the bodies of men and animals. Previously, the movement of the blood had been seen as a kind of bodily irrigation. After actually viewing it experimentally in animals, Harvey concluded that the heart was a muscle functioning as a pump, and that it effected the movement of the blood through the body via the lungs by means of the arteries, the blood then returning through the veins to the heart. Harvey upheld the difference between venous and arterial blood. Through experiment and dissection he demonstrated the one-way nature of the valves in the arteries and veins. He showed that in systole the heart contracted, expelling blood; the right ventricle supplied the lungs and the left ventricle provided blood for the arterial system. Blood, he insisted, flowed through the veins towards the heart. Because Harvey's views contradicted ideas central to medicine since **Galen**, he was widely ridiculed by traditionalists, notably in France. Harvey was not able to show how blood passed from the arterial to the venous system, there being no connections visible to the naked eye. He rightly supposed that the links must be too minute to see. **Malpighi** observed them with a microscope, shortly after Harvey's death. Harvey's notable *Essays on Generation in Animals* (1651) made public his embryological researches, in which he espoused epigenetic rather than preformationist views. Harvey affirmed the doctrine that every living being has its origin in an egg. A gifted experimenter and a master of patient reasoning, modern animal physiology may be said to have begun with Harvey's labours.

HARVEY, William Henry
(1811–66)

Irish botanist, born in Summerville, Limerick, and educated at Ballintore. After leaving school, Harvey worked in his father's business, and made botanical forays. In 1831, at Killarney, he discovered the moss *Hookeria laetevirens*, previously not known to exist in Ireland. This led to his becoming acquainted with Sir **William Jackson Hooker**. He sailed in 1835 for Cape Town, where he was colonial treasurer (1836–42). He worked hard on South African botany, publishing *The Genera of South African Plants* in 1838. Because the climate disagreed with his health, he returned to England in 1842. He quickly became the foremost authority of his day on algae. In 1844 he became curator of the herbarium at Trinity College, Dublin, where he was appointed Professor of Botany in 1856. Between 1853 and 1856 he travelled extensively to Ceylon, India, Australia, Fiji and elsewhere. His works on algae were *Manual of British Algae* (1841), *Phycologia Britannica* (4 vols, 1846–51), *Nereis Australis* (1847), *Nereis Boreali-Americana* (3 vols, 1851–8), *Phycologia Australica* (5 vols, 1858–63) and *Index Generum Algarum* (1860). He also published *Thesaurus Capensis* (2 vols, 1859–63) and was co-author of volumes 1–3 of *Flora Capensis* (1859–65). He opposed the views expressed in **Charles Darwin**'s *Origin of Species*.

HASSEL, Odd
(1897–1981)

Norwegian physical chemist, born in Oslo. His scientific education was at the universities of Oslo, Munich (in the laboratory of **Fajans**) and Berlin (DPhil, 1924). From 1925 until his retirement in 1964 he was on the staff of the Department of Physical Chemistry of the University of Oslo, as professor and director from 1934. He received the Nobel Prize for Chemistry jointly with **Barton** in 1969 and was an Honorary Fellow of the Royal Society of Chemistry. Hassel's most distinguished researches were carried out in the 1930s and involved the application of X-ray and electron diffraction, and the measurement of dipole moments. He elucidated the details of the molecular structure of cyclohexane and related compounds, and thereby helped to establish the concepts and procedures of 'conformational analysis'. Due largely to World War II (he was imprisoned during the German occupation of Norway) much of his work was not well known until the 1950s. Conformational analysis was extensively developed by Barton in the 1950s and 1960s, and has become a very important feature of organic chemistry. Hassel's later work was on charge-transfer complexes.

HATCH, Marshall Davidson
(1932–)

Australian biochemist, born in Perth. He studied at the universities of Sydney and California, became a research scientist for the Commonwealth Scientific Industrial Research Organisation (CSIRO) in Australia (1955–9), and then spent a further period in California before joining the Colonial Sugar Refining Co Ltd, Sydney (1961–70). Since 1970 he has been chief research scientist at the Division of Plant Industry of CSIRO in Canberra. In 1958 he published details of the glycolysis pathway by which starch and sugars form ethanol and carbon dioxide in plants. Extending this analysis, in 1961 he purified from wheatgerm acetyl-CoA carboxylase — an enzyme that 'activates' acetate with carbon dioxide in the biosynthesis of fatty acids. Later he recognized a new alternative to the **Calvin** photosynthetic pathway (1968), known as the C_4 or Hatch–Slack–Kortschak pathway, for incorporating carbon dioxide into carbohydrates. In this sequence the glycolytic intermediate, phospho*enol*pyruvate, is caboxylated to form oxaloacetate, which may then react with 2,3-phosphoglycerate to form a sugar phosphate and pyruvate. Hatch discovered the enzyme pyruvate, phosphate dikinase (1969), which resynthesizes phospho*enol*pyruvate (not possible in glycolysis) to complete the cycle. The cycle is particularly favoured at low ambient carbon dioxide concentrations and is important in tropical grasses and other plants that tolerate climatic extremes. Hatch has since explored the details of this pathway.

HAUKSBEE, or HAWKSBEE, Francis

(d 1713)

English physicist. Little is known about his origins and private life. He is called 'the elder' to distinguish him from his nephew of the same name whose dates are known (1688–1763). They have often been confused in the literature as he had similar scientific interests, and assisted his uncle until his death. Francis, the elder, is chiefly noted for his experiments on electroluminescence, static electricity and capillarity. He carried further the observations by the physician **William Gilbert** on electricity and **Boyle** on air, inventing the first glass friction electrical machine, and improved the air-pump. He was elected FRS in 1705 and was appointed as the Royal Society's Curator of Experiments.

HAUPTMAN, Herbert Aaron

(1917–)

American mathematical physicist, born in New York City. He was educated at the City College of New York, Columbia University and the University of Maryland. Working at the US Naval Research Laboratories in Washington with **Karle** during the 1950s and 1960s, he helped develop a statistical technique that radically increased the speed of methods by which X-ray crystallography mapped structures of molecules. Using this 'direct method' the time taken to establish a molecular structure from the pattern of visible dots obtained by exposing a crystal to an X-ray beam was reduced from years to days. Hauptman and Karle published their key monograph in 1953, but its importance remained unacknowledged for years. By the late 1960s, however, the repeated success of their method had ensured it's establishment as a standard crystallographic technique. It has so far been applied to molecules of up to 200 atoms, including many important biological molecules such as steroids and other drugs. In 1985, 22 years after publishing the work, Hauptman and Karle were jointly awarded the Nobel Prize for Chemistry. Hauptman became professor at the University of Buffalo in 1970 and since 1972 has continued his research on X-ray crystallography at the Medical Foundation of Buffalo.

HAUSDORFF, Felix

(1868–1942)

German mathematician, born in Breslau (now Wrocław, Poland). He studied at Leipzig and Berlin, and taught at Leipzig (1896–1910). In 1910 he moved to Bonn, where he remained until, as a Jew, he was forced by the Nazis to resign his chair in 1935; ultimately he committed suicide with his family to avoid the concentration camps. He is regarded as the founder of point set topology, and his book *Grundzüge der Mengenlehre* (1914) introduced the basic concepts of topological spaces and metric spaces which have since become part of the standard equipment of analysis and topology; the fractal dimension of a set is often called the 'Hausdorff dimension'. His work on set theory continued that of **Georg Cantor** and **Zermelo**.

HAÜY, René Just

(1743–1822)

French crystallographer and mineralogist, born in St Just, the son of a weaver. He initially studied botany and embryology before developing his interests in mineralogy and crystallography. Following the turmoil of the French Revolution, during which he was temporarily imprisoned and conscripted for military service, Haüy became Professor of Physics at the École Normale (1794) and Curator of the École des Mines, Paris (1795). It was here that he wrote his *Traité de Minéralogie*, published in 1801, the same year that he succeeded **Dolomieu** as Professor of Mineralogy at the Museum of Natural History in Paris. Haüy is widely regarded as the father of crystallography. His initial observation that calcite crystals of different form always break to produce rhomboidal fragments led him to the discovery of basic laws of crystallography, allowing crystal form to be used as an important aid in the identification of minerals. Recognizing the importance of higher education in building up the nation's prosperity, Napoleon commissioned Haüy to write a treatise on physics to be used in the lycées of France (*Traité Élémentaire de Physique*, 1803). Applying his scientific knowledge to useful ends he published *Traité des Caractères Physiques des Pierres Précieuses* (1817), a work describing the physical characteristics observable in cut gemstones arranged in such a way as to allow identification of gemstones by cutters, dealers and their customers.

HAWKING, Stephen William
(1942–)

English theoretical physicist, born in Oxford. He graduated from the University of Oxford and received his PhD from Cambridge. He was elected FRS in 1974, and became Lucasian Professor of Mathematics at Cambridge in 1980. His early research on relativity led him to study gravitational singularities such as the 'Big Bang' when the universe originated, and the 'black holes' where space–time is curved due to enormous gravitational fields. The theory of black holes, which result when stars collapse at the end of their lives, owes much to his mathematical work. In the early 1970s, Hawking and colleagues proved mathematically that the only properties conserved when an object becomes a black hole are its mass, angular momentum and electric charge. Since 1974 he has shown that black holes could actually emit thermal radiation and could evaporate. By a process predicted by theories of quantum mechanics, mass can be lost from the black hole and escape entirely from its gravitational pull. This is known as the Hawking process, and he showed that the rate of mass loss would be inversely proportional to the mass of the hole. A galactic mass would take around 10^{90} years to evaporate. His book *A Brief History of Time* (1988) is a bestselling popular account of modern cosmology. His achievements are especially remarkable because from the 1960s he has suffered from a highly disabling and progressive neuromotor disease.

HAWKSBEE, Francis See **HAUKSBEE, Francis**

HAWORTH, Adrian Harvey
(1766–1833)

English botanist and entomologist, born in Hull. He was at first articled to a solicitor but, on completion of his articles, left the legal profession. He moved to Cottingham, near Hull, and began studying botany, entomology and ornithology. Some time between 1793 and 1797 he moved to Little Chelsea, where he stayed until 1812. Shortly after moving to Chelsea he joined the Linnaean Society, and also founded the Aurelian Society in 1802. The latter was short-lived, being dissolved in 1806. He then helped to found the Entomological Society of London. In 1812 he returned to Cottingham and helped to form the Hull botanical garden; he returned to Chelsea in 1817. Haworth was an authority on succulent plants, his chief botanical work being *Synopsis Plantarum Succulentarum* (1812, with a supplement in 1819 accompanied by a revision of *Narcissus*). His entomological publications include *Prodromus Lepidopterorum Britannicorum* (1802, listing 793 British species of butterflies and moths) and *Lepidoptera Britannica* (3 parts, 1803–12, with an 'Appendix' in 1829 which was entirely botanical in content). He also wrote a poem, 'Cottingham', comprising 24 cantos of insignificant merit.

HAWORTH, Sir (Walter) Norman
(1883–1950)

English chemist, born in Chorley. He learnt most of his early chemistry from working in his father's linoleum factory and it wasn't until 1903 that he enrolled at the University of Manchester, where he studied under **William Henry Perkin Jr**. A scholarship enabled him to study with **Wallach** at Göttingen, and he then returned to Manchester to investigate terpenes. In 1911 he moved to Imperial College, London, but was there for only one year before moving to a post at the University of St Andrews. He soon discontinued his work on terpenes and joined the group at St Andrews founded by Irvine to study the chemistry of sugars. In 1920 he moved to King's College, Newcastle, and in 1925 took up the Chair of Organic Chemistry at the University of Birmingham, where he was joined by **Hirst** for his most productive period. Scurvy had been shown to be a vitamin deficiency disease, and **Szent-Györgyi** had isolated from the adrenal glands and from fruit juice a substance named hexuronic acid which was later shown to be the antiscorbutic factor, vitamin C. A sample was sent to Haworth and, with a group including Hirst, he elucidated its chemical structure and confirmed this by chemical synthesis. He called it ascorbic acid. For this work he shared the Nobel Prize for Chemistry with **Karrer** in 1937. He was elected FRS in 1928, and knighted in 1947.

HAYASHI, Chusiro
(1920–)

Japanese astrophysicist, born in Kyoto. He graduated from the University of Kyoto in 1942, and after spending time in Tokyo and Osaka, he returned to Kyoto in 1954, becoming Professor of Physics there in 1957. In 1950 he showed that in the first two seconds after the Big Bang, temperatures would exceed 10^{10} kelvin and that after that time, the ratio between the number of neutrons and protons existing in the universe would remain constant resulting in a fixed hydrogen-to-helium ratio and negligible amounts of heavier elements. In 1961 he showed that there is a minimum surface temperature for pre-main-sequence stars of a specific mass. This led to the important conclusion that, due to their opacity, these stars must be in convective equilibrium, as opposed to radiative equilibrium (as was previously thought) and must thus have large luminosities at this stage. There is thus a zone on the Hertzprung–Russell diagram, a logarithmic graph showing the variation of stellar luminosity as a function of surface temperature, through which these stars cannot pass. This is known as the Hayashi forbidden zone.

HAYDEN, Ferdinand Vandeveer
(1829–87)

American geologist, born in Westfield, Massachusetts. His father died when he was 10 years old and he lived with an uncle until, ambitious for an education, he entered Oberlin College and worked to pay his own expenses. After graduating in 1850 he went on to Albany Medical College and was awarded an MD in 1853. Following work on geological surveys in Dakota, the Badlands, Missouri and Yellowstone (1853–62), he was appointed as a surgeon in the Union Army (1862–5) and subsequently became Professor of Geology at Pennsylvania University (1865–72). He resigned from this post because of the increasingly onerous duties of his simultaneous position as head of the US Geological Survey (1867–79). Hayden was one of the great geological pioneers of western North America; on one occasion he was captured by hostile Indians but released as insane when found to be armed only with a geological hammer and pack of fossils. It is reported that the Sioux Indians gave him the name 'the man who picks up rocks running'. He was influential in securing the establishment of Yellowstone National Park.

HAYEM, Georges
(1841–1920)

French physician and pioneer of the study of diseases of the blood, born in Paris. He studied medicine in Paris, and received his MD in 1868. He became Professor of Therapy and Materia Medica in 1879, working for much of his long career at the Hôpital Tenon. He first described the platelets in the blood, and did classic work on the formation and diseases of the red and white blood cells. He also published important accounts of diseases of several organs, including the stomach, liver, heart and brain. His work was notable for the way he attempted to apply the results of experimental physiology or pathology to the clinical setting.

HEAD, Sir Henry
(1861–1940)

English neurologist, born in Stamford Hill, London. He studied natural sciences at Trinity College, Cambridge, and clinical medicine at University College Hospital, London. His interest in the functions and diseases of the nervous system dated from his student days, when he was taught by **Foster**, **Gaskell** and **John Langley**, and he studied for a short period with Ewald Hering in Prague. He became a consulting physician at the London Hospital, and is best known for his neurological research, some in collaboration with William Rivers and Gordon Holmes. His famous observations on the sensory changes in his own arm, after cutting some nerve fibres, provided important information about the physiology of sensation, and reinforced his reputation as a leading scientifically inclined neurologist. He wrote widely on aphasia (disorders of speech), in which he doubted the then prevailing views about the cortical localization of speech, and other neurological disorders. Many of his ideas were summarized in *Aphasia and Kindred Disorders of Speech* (1926) which reported on the clinical disturbances of speech that he observed in a large number of men suffering from gunshot wounds. He edited the influential neurological journal *Brain* for many years (1905–21) and also published poetry. He was knighted in 1927.

HEATLEY, Norman
(1911–)

English chemist, born in Woodbridge, Suffolk, and educated in natural sciences at St John's College, Cambridge. He graduated in biochemistry in 1933, and undertook doctoral research at the Sir William Dunn Biochemical Institute on the application of microchemical methods to biological problems. He stayed at the Dunn Institute for most of his career, apart from secondments and sabbatical leave, initially as assistant to **Chain** working on tumour development (1936–9). In 1939 he was intending to move to Copenhagen for a year's research fellowship, when the outbreak of World War II prevented him from travelling. He returned to Oxford, as assistant to **Florey**, working mainly on the production of penicillin and other antibiotics. His first major contribution was a method of assessing the strength of penicillin, followed by crucial developments and improvements in techniques for producing, extracting and preparing penicillin. In 1942, with Florey, he visited the USA seeking help from the pharmaceutical industry in developing mass production methods for penicillin, and he remained there for a year assisting in the commercial production of penicillin. After the war he was appointed a university lecturer and a college lecturer at Lincoln College. In 1990 his considerable achievements were acknowledged by the award of an honorary doctorate in medicine, the only person to receive the award in the university's 800 year history.

HEAVISIDE, Oliver
(1850–1925)

English physicist, born in London, the son of an artist and a nephew (by marriage) of **Wheatstone**. A telegrapher by training, he spent much of his life living reclusively in Devon. There he made various important advances in the study of electrical communications, and in 1902, independently of Arthur Edwin Kennelly (1861–1939), he predicted the existence of an ionized gaseous layer capable of reflecting radio waves, the 'Heaviside layer' (now known as the ionosphere), which was verified 20 years later. He was elected FRS in 1891, and became an honorary member of the Institution of Electrical Engineers in 1908, after having been struck from their roll due to his inability to pay the subscription fee.

HEDWIG, Johannes
(1730–99)

Transylvanian-born German botanist and physician born in Kronstadt (now Braşov, Romania). He studied medicine at Leipzig, and after practising as a physician became Professor of Botany there (1789). He gave special attention to cryptogams; his posthumous *Species Muscorum Frondosorum* (1801), based on natural groupings, is the internationally accepted starting point for the scientific naming of mosses. He demonstrated the close relationship between mosses and liverworts, and defined for the first time the characters which separate these groups of plants. His theoretical results were published in *Fundamentum Historiae Naturalis Muscorum Frondosorum* (1782), and he described for the first time, and essentially accurately, the life cycles of bryophytes in *Theoria Generationis et Fructificationis Plantarum Cryptogamicarum* (1784). These works were illustrated by accurate and beautiful figures, many of which were drawn from highly magnified images. He described the development of the spore-capsule (sporogonium) in bryophytes, and was one of the first to observe, and the first to illustrate, conjugation in the algae *Spirogyra* and *Chara*.

HEEZEN, Bruce Charles
(1924–77)

American oceanographer, born in Vinton, Iowa. He graduated with a BA in palaeontology from the University of Iowa in 1948. In 1947 he attended a lecture by **Ewing**, and an invitation to take part in an expedition to the Mid-Atlantic Ridge started his fruitful career in oceanography. He worked with Ewing at what was then the Lamont Geological Observatory of Columbia University (1948–77), gaining a PhD in 1957, and later moved to the university's department of geology as assistant professor (1960–4) and associate professor from 1964. Heezen excelled as an expedition leader and teacher. In his MS thesis (1952), he demonstrated the existence of turbidity currents. Such currents had been proposed by Reginald Aldworth Daly (1936) and supported experimentally by **Kuenen**, but Heezen provided the first conclusive evidence. Little was then known of the sea-floor; in pioneering work, Heezen used the new continuously recording echo-sounder to map sea-floor topography. He was the first to map Atlantic fracture zones and to note how they offset the mid-ocean ridge. With cartographer Marie Tharp, he used a new mapform to publish the soundings; their *Physiographic Diagrams* (1957–71) and *The World Ocean Floor* (1977) showed the globe-encircling continuity of the mid-ocean ridges and median valley. They discovered the co-location of earthquake epicentres with the median valley; this led Heezen to realize that ridges were tensional features and propose that new sea-floor formed there (1960).

HEISENBERG, Werner Karl
(1901–76)

German theoretical physicist, born in Würzburg. He was educated at the universities of Munich and Göttingen, before becoming Professor of Physics at Leipzig University (1927–41). He then became professor at Berlin University and Director of the Kaiser Wilhelm Institute (1941–5). From 1945 to 1958 he was director of the Max Planck Institute in Göttingen, and he later moved with the institute to Munich. In 1925 he published a paper reinterpreting classical mechanics with a systematic quantum mechanics in terms of matrices, by using a formal dualistic theory where phenomena must be describable both in terms of wave theory and quanta. This correctly predicted the observed frequencies and intensities of atomic spectral lines. In further studies of the spectrum of molecular hydrogen, Heisenberg showed that the patterns of weak and strong lines observed were due to the presence of two different forms; ortho- and para-hydrogen. For his theory of matrix mechanics and these applications, he was awarded the Nobel Prize for Physics in 1932. In his revolutionary principle of indeterminacy or uncertainty principle (1927), he showed that there is a fundamental limit to the accuracy to which certain pairs of variables (such as position and momentum) can be determined. The product of uncertainty in position with uncertainty in momentum exceeds $h/2\pi$, where h is Planck's constant. Hence, the precise measurement of a subatomic particle's position means that the uncertainty in its momentum will be large, and vice versa. A consequence of the wave description of matter, the principle may be interpreted as a result of disturbance to a system due to the act of measuring it.

HELMHOLTZ, Hermann von
(1821–94)

German physiologist and physicist, born in Potsdam, one of the last of the scientific polymaths. As his father could not afford an expensive university education, Helmholtz studied medicine for which state aid was available but only if the beneficiary was committed to eight years service as an army surgeon. During his medical studies he followed courses on physics and chemistry, and studied mathematics privately. After qualifying he was appointed surgeon to the regiment at Potsdam. His scientific career began when he was released from military duty in 1848. He was successively appointed as Professor of Physiology at Königsberg (1849), Bonn (1855) and Heidelberg (1858). In 1871 he became Professor of Physics in Berlin, and in 1887 president of the newly founded Physikalisch Technische Reichsanstalt for research in the exact sciences and precision technology, to which **Werner von Siemens** had donated a large sum of money. Helmholtz was equally distinguished in physiology, mathematics, and experimental and mathematical physics. His physiological works are principally connected with the eye, the ear and the nervous system. His work on vision (eg on the perception of colour) is regarded as fundamental to modern visual science. He invented an ophthalmoscope (1850) independently of **Babbage**. He was also important for his analysis of the spectrum, his explanation of vowel sounds, his papers on the conservation of energy with reference to muscular action, his paper on *Conservation of Energy* (1847), his two memoirs in Crelle's *Journal* on vortex motion in fluids and on the vibrations of air in open pipes, and for researches into the development of electric current within a galvanic battery. He was elected a Foreign Member of the Royal Society and in 1873 was awarded the society's Copley Medal.

HELMONT, Johannes Baptista van
(1579–1644)

Flemish chemist, physiologist and physician, born in Brussels. He studied philosophy and theology at the University of Louvain, subsequently turning to science and medicine. He travelled and studied widely, and received an MD degree in 1609. Thereafter he lived a secluded life in his estate at Vilvorde, near Brussels, and during this period did most of his scientific work. Little of this was published in his lifetime, probably due to opposition from the Catholic Church. His collected works were published by his son in 1648 under the title *Ortus Medicinae vel Opera et Opuscula Omnia*. Van Helmont occupies a position on the border of the old and the new learning. He accepted traditional beliefs in alchemy and in the intervention of supernatural agencies, and he developed a two 'element' theory of matter (water and air). In a more modern approach, he also obtained much empirical knowledge of chemistry, medicine and both human and plant physiology. Notably he distinguished the existence of different gases (a word he derived from the Greek for 'chaos') and he emphasized the importance of the balance in chemical work.

HENCH, Philip Showalter
(1896–1965)

American physician, born in Pittsburgh, where he took his medical degree. Head of the Department of Rheumatics at the Mayo Clinic in Rochester from 1926, and Professor of Medicine at Minnesota University from 1947, he discovered cortisone, widely hailed as a 'miracle drug', and shared with **Edward Kendall** and **Reichstein** the 1950 Nobel Prize for Physiology or Medicine for their work on the biology and therapeutic uses of the suprarenal hormones. Several patients severely crippled by arthritis were demonstrated with much greater freedom of movement and suffering much less pain. Unfortunately, the early improvement often did not last and a variety of side effects from high doses of steroids began to manifest themselves. The early reports of dramatic 'cures' of severe rheumatoid arthritis were thus premature, but 'steroids' such as cortisone have played an important part in modern treatments. Hench found his life dramatically changed by the award of the Nobel Prize, and some of his later claims on behalf of steroids were controversial.

HENDERSON, Thomas
(1798–1844)

Scottish astronomer, born in Dundee. Although intended for a law career, he devoted his leisure hours to astronomical calculations, and in 1831 was appointed Director of the Royal Observatory at the Cape of Good Hope. In 1832 he succeeded in determining the parallax of the star Alpha

Centauri, only three months after the very first stellar parallax, that of the star 61 Cygni, had been announced by **Bessel** of Königsberg. In 1834 he became Professor of Practical Astronomy at the University of Edinburgh and the first Astronomer Royal for Scotland. He was an indefatigable observer who in his 10 years in Edinburgh measured, with the help of only one assistant, no fewer than 60 000 positions of planets and stars. His strenuous labours caused his early death through physical exhaustion.

HENLE, Friedrich Gustav Jakob
(1809–85)

German anatomist and histologist. Born in Fürth, near Nuremberg, to Jewish parents, Henle studied medicine at the University of Bonn, moving briefly in 1830 to Heidelberg before graduating in Bonn in 1831. Developing friendship with **Johannes Müller** resulted in his appointment as Professor of Anatomy at Berlin in 1833. In 1840 he moved to Zürich, and four years later to Heidelberg. His final post was at Göttingen. Henle was an early advocate of cell theory, largely accepting the theoretical structure as laid down by **Schwann**. A fine microscopist and inspired by Müller, he made extensive studies of cartilage and fatty and fibrous tissue. In the 1860s he discovered the loop-shaped tubules in the kidney which are named after him, and wrote treatises on systematic anatomy and general pathology.

HENRY, Joseph
(1797–1878)

American physicist, born into a poor family in Albany, New York. He studied at Albany Academy, and was appointed Professor of Mathematics there in 1826. In 1832 he became Professor of Natural Philosophy at Princeton, and in 1846 first Secretary of the Smithsonian Institution. Henry was a fine and intuitive experimenter rather than a theoretician. He discovered electrical induction independently of **Faraday** and constructed the first electromagnetic motor (1829), appreciated the effects of resistance on current (formulated precisely by **Ohm** in 1827 but not generally accepted for some time), demonstrated the oscillatory nature of electric discharges (1842), and introduced a system of weather forecasting based on meteorological observations received at the Smithsonian Institution by electric telegraph. The SI unit of inductance (the henry) is named after him.

HENRY, William
(1774–1836)

English physician and chemist, born in Manchester. He studied medicine at Edinburgh University. However, his own ill-health forced him to give up practising medicine, and under the influence of his friend **Dalton**, he turned to teaching and research in chemistry. In a fit of melancholia he committed suicide in 1836. His best-known work was the study of the influence of pressure and temperature on the solubility of gases in water, which resulted in the generalization that has become known as Henry's law, that the solubility of a gas at a given temperature is proportional to its pressure (1803). Henry also wrote the very successful *Elements of Experimental Chemistry* (1801), which went through 11 editions. He was elected FRS in 1808 and was awarded the Royal Society's Copley Medal in the same year.

HENSEN, Christian Andreas Viktor
(1835–1924)

German physiologist, born in Kiel. Hensen studied medicine at Würzburg, moving on to Berlin and then back to Kiel where he graduated. He remained in Kiel, teaching anatomy and histology, becoming full Professor of Physiology in 1868. Presumably because his father had been director of the local school for the deaf and dumb, Hensen pursued physiological researches into the organs of hearing, studying the morphology of the human cochlea, and describing what are now known as Hensen's duct and Hensen's supporting cells. He published lengthy accounts in 1880 and 1902 of the physiology of hearing. His other great passion lay in marine biology, investigating the life history of the marine fauna that he named 'plankton'. He calculated the quantity of plankton at different depths in the various oceans. He also developed pioneering methods for quantifying the amount of commercially exploitable fish in the sea.

HERACLEIDES OF PONTUS AND EKPHANTUS
(c.388–c.315 BC)

Greek philosopher and astronomer, born in Heraklea, near the Black Sea. He migrated to Athens to become a pupil of Speusippus and **Plato**. Although all his writings are lost, Aetius reports that Heracleides was the first to propose that the Earth is spinning like a wheel from west to east, thinking that it was highly improbable that the immense spheres of the stars and planets could rotate once every 24 hours. It is also suggested that he took the first steps towards a heliocentric solar system, concluding that Mercury and Venus orbited the Sun rather than the Earth. He also considered the cosmos to be infinite and that the other planets were Earth-like with atmospheres.

HERAPATH, John
(1790–1868)

British natural philosopher and journalist, born in Bristol. He entered his father's profession as a maltster, but left business in 1815 to open a mathematical academy in Bristol. From 1820 he taught mathematics in Middlesex. Herapath gave up teaching in 1832, moved to Kensington, and began to write about the growing British railway network, achieving success as proprietor of *The Railway Magazine and Annals of Science* (1836–9), a journal which served as a vehicle for his scientific papers. In 1811 researches into lunar motion led him to consider heat, which he concluded was not a substance but a form of motion. This kinetic theory was presented in an ambitious *Mathematical Inquiry into the Causes, Laws, and Principal Phænomena of Heat, Gases, Gravitation, &c.*, submitted to the Royal Society in 1820 but rejected as too speculative. The controversy surrounding its publication elsewhere (1821) was fuelled by Herapath himself. His *magnum opus*, *Mathematical Physics*, appeared in 1847; it was studied by **Joule** who, in his own work on heat, almost certainly followed Herapath.

HERBIG, George Howard
(1920–)

American astronomer, born in Wheeling, West Virginia. After graduating in astronomy from the University of California at Los Angeles in 1943, he started work at the Berkeley Radiation Laboratory but was released due to ill-health. He became a graduate student at the Lick Observatory, writing his thesis on T Tauri stars; these are very young pre-main-sequence stars which have not yet started to produce energy through nuclear fusion reactions of hydrogen. T Tauri stars are in close proximity to the dust and gas clouds from which they were formed. Not only did Herbig discover a great number of these stars, but he also discovered one of the first 'Herbig–Haro' objects, small convoluted luminous nebulae that are intimately connected with newly formed stars. A total of 43 Herbig–Haro objects had been found by 1974, but the number now is over 100. Herbig discovered the unusually high abundance of lithium in T Tauri stars and concluded that this value represents the original abundance of the element in the Milky Way. Herbig also discovered the hotter, more massive stars of spectral class A and B that are associated with T Tauri stars. These hot A and B stars have emission lines in their spectra and are known as Herbig emission stars. In 1962 he suggested that the low- and intermediate-mass stars in clusters formed first, the massive O stars forming later, the onset of their high luminosity disrupting and dispersing the surrounding gas and dust. Hebig became Professor of Astronomy at the University of California at Santa Cruz in 1967.

HÉRELLE, Felix d'
See **D'HÉRELLE, Felix**

HERMITE, Charles
(1822–1901)

French mathematician, born in Dieuze. After struggling to pass examinations while an undergraduate, he finally received his degree in 1848, and was appointed to teaching posts at the École Polytechnique and the Collège de France. He later became professor at the École Normale (1869) and Professor of Higher Algebra at the Sorbonne (1870). He proved that the base of natural logarithms (e) is transcendental, ie cannot be a solution of a polynomial equation with rational coefficients. On the basis of this work, the German mathematician Ferdinand Lindemann was able to prove 10 years later that π is also transcendental. Hermite published works on the theory of numbers, then on elliptic and Abelian functions. In later life he extended the theory of elliptic functions and turned their theory towards applications. He also worked on invariant theory. A leader in French mathematical life, he was very influential behind the scenes, guiding the early careers of **Poincaré** and his son-in-law **Picard**, and setting research priorities for a generation. He was converted by **Cauchy** to Catholicism after a nearly fatal attack of smallpox, and through his friendship with German mathematicians helped diminish antagonisms caused by the Franco-Prussian War.

HERO OF ALEXANDRIA
(1st century AD)

Greek mathematician. He invented many pneumatic machines and toys, among them Hero's fountain, the aeolipile, and a double forcing-pump suitable for a fire engine. He wrote on pneumatics, mechanics and mensuration, including the formula for expressing the area of a triangle in terms of its sides.

HEROPHILUS
(fl 300 BC)

Greek anatomist, born in Chalcedon in Asia Minor, founder of the school of anatomy in Alexandria. Little is known of his life. From his writings it is clear that he was an expert in dissection. He was probably the first to anatomize human corpses, in order to compare them with those of other animals. He described the liver, spleen, sexual organs, brain, and vascular and nervous systems. He argued that it was the brain that was the centre of reason and command in the body. He comprehended that it was through the nerves that the brain exercised its control over the limbs. In the brain he made the distinction between the cerebellum from the cerebrum, and understood the functional difference between sensory and motor nerves. He emphasized the dissimilarity between arteries and veins, and in defiance of received opinion, maintained that the arteries contain blood rather than spirit (pneuma). His work on the heart stressed the involuntary nature of the pulse, showing it to be produced by the contraction and dilation of the arteries. Clinically, Herophilus was specially concerned with gynaecology. In his treatise on midwifery, he presented discerning descriptions of the ovaries, the uterus and the cervix. He was probably also interested in menstruation and its effects on health in general. Herophilus is widely thought to have laid the foundations of anatomy and physiology.

HÉROULT, Paul Louis Toussaint
(1863–1914)

French metallurgist, born in Thury-Harcourt, Normandy, the son of a tanner. He studied at the École des Mines in Paris under **Le Chatelier**, who communicated to him his interest in aluminium. In April 1886, when he was 23, Héroult registered a patent for the extraction of aluminium by the electrolysis of cryolite in a carbon-lined crucible which served as a cathode, the method used today. The American **Charles Hall** discovered the same process independently in the same year (both Hall and Héroult were born in 1863; both died in 1914). In 1907 Héroult patented an arc furnace for melting iron and steel, in which heat was generated by an electric arc which was struck between scrap iron and carbon electrodes. The first arc furnaces for melting steel to be installed both in the USA and in Britain in the first decade of the 20th century were of Héroult's design.

HERSCHBACH, Dudley Robert
(1932–)

American physical chemist, born in San José, California. He graduated in mathematics at Stanford University and then took an MS in chemistry, also at Stanford, and finally a PhD in chemical physics at Harvard. From 1959 to 1963 he was on the chemistry faculty of the University of California, Berkeley. Since 1963 he has been Professor of Chemistry at Harvard. He shared the Nobel Prize for Chemistry in 1986 with **Yuan Tseh Lee** and **John Polanyi**, for their respective contributions to chemical reaction dynamics. Among Herschbach's many honours are the Polanyi Medal of the Royal Society of Chemistry (1982) and the Langmuir Prize of the American Physical Society (1983). Reaction dynamics is concerned with the details of the atomic and molecular motions that occur in each elementary act of chemical change, and with the energy states of reactant and product molecules. In 1959 Herschbach adapted the physicists' technique of molecular beams to the study of such details. The method involved the use of intersecting beams of atoms or molecules of reactants at very low pressure and the use of special devices to detect the products. It was first applied to elucidate the details of the reaction between potassium atoms and methyl iodide molecules, and has subsequently been applied to many other systems. Herschbach's fellow laureate Lee was his collaborator at Berkeley from 1961 to 1968, and the successful development of the technique owed much to Lee's experimental skill.

HERSCHEL, Caroline Lucretia
(1750–1848)

German-born British astronomer, sister of **William Herschel**, born in Hanover. In 1772 her brother brought her to England as assistant with his musical activities, and she became his devoted collaborator when he abandoned his first career for astronomy (1782). Between 1786 and 1797 she discovered eight comets. Among her other discoveries was the companion of the Andromeda nebula (1783). In 1787 she was granted a salary of £50 a year from the king as her brother's assistant at Slough. Her part in William's observational work made her thoroughly familiar with the heavens: her *Index to Flamsteed's Observations of the Fixed Stars* and a list of errata were published by the Royal Society (1798). Following her brother's death she returned at the age of 72 to Hanover where she worked on the reorganization of his catalogue of nebulae. For this *Reduction and Arrangement in the Form of a Catalogue in Zones of all the Star Clusters and Nebulae Observed by Sir William Herschel*, though unpublished, she was awarded the Gold Medal of the Royal Astronomical Society (1828). She was elected (with **Somerville**) an honorary member of the Royal Astronomical Society (1835) and a member of the Royal Irish Academy (1838). On her 96th birthday she received a gold medal from the King of Prussia.

HERSCHEL, Sir John Frederick William
(1792–1871)

English astronomer, only child of Sir **William Herschel**, born in Slough and educated briefly at Eton, afterwards at home, and at St John's College, Cambridge, where he was Senior Wrangler and Smith's Prizeman (1813), and was made a Fellow of his college. His first award was the Copley Prize of the Royal Society for his mathematical researches in 1821. In collaboration with Sir James South, he re-examined his father's double stars (1821–3) and produced a catalogue which earned him the Lalande Prize (1825) and the Gold Medal of the Royal Astronomical Society (1826). He reviewed his father's great catalogue of nebulae in Slough (1825–33), adding 525 new ones, for which he received the Gold medals of the Royal Astronomical Society (1826) and the Royal Society (1836). To extend the survey to the entire sky he went to South Africa and set up the 7 foot Slough reflector at Feldhausen near Capetown. In four years (1834–8) there he completed a survey of nebulae and clusters in the southern skies, observing 1708 of them, the majority previously unseen. He also discovered over 1200 pairs of double stars, catalogued over a thousand objects in the Magellanic Clouds, and extended his father's star gauging exercise to southern fields. The preparation for publication of his massive southern observations occupied him for many years; these *Cape Observations* (1847) earned him the Copley Medal of the Royal Society (1847). John Herschel dominated British science for 50 years and excelled in many branches. His preferred interest was chemistry; he was a pioneer photographer, the inventor of the fixing process using hyposulphite of soda (1819) and independently of sensitized paper (1839), as well as the originator of the terms positive and negative in photography. He never occupied an academic post, supporting his researches from his private means, his one official appointment being Master of the Mint (1850–5). He was made a baronet at Queen Victoria's coronation. Among his numerous honours were the Prussian *Pour le Mérite* and membership of the French Institute. He is buried in Westminster Abbey close to the grave of **Isaac Newton**.

HERSCHEL, Sir (Frederick) William
(1738–1822)

German-born British astronomer, born in Hanover, the son of a musician who instructed his sons in the same profession. William joined the Hanoverian Guards band as an oboist and moved in 1755 to England where he built up a successful career in music, eventually settling in Bath in 1766. It was here that his interest in astronomy began. He built his own telescopes, learning to cast his own metal discs for his mirrors. In 1781 he discovered the planet Uranus, the first to be found telescopically, which he named *Georgium Sidus* in honour of King George III who a year later appointed him his private astronomer. At Slough, near Windsor, assisted by his sister **Caroline Herschel**, he continued his researches and built ever larger telescopes, up to his 40 foot long reflector (completed in 1789). He lived and worked in Slough for the rest of his life. Herschel's discoveries included two satellites of Uranus (1787) and two of Saturn (1789), but his epoch-making work lay in his studies of the stellar universe. He drew up his first catalogue of double stars (1782), later demonstrating that such objects constitute bodies in orbit around each other (1802), and observed the Sun's motion through space (1783). His famous paper *On the Construction of the Heavens* (1784), based on star counts in thousands of sample portions of the sky, produced a model of the Milky Way as a non-uniform aggregation of stars; such studies occupied him for the rest of his life. Following the publication of **Messier**'s catalogue of nebulae and star clusters (1781), he began a systematic search for such non-stellar objects which revealed a total of 2500, published in three catalogues (1786, 1789 and 1802). He distinguished different types of nebulae, realizing that some were distant clusters of stars while others were nebulosities. Herschel was knighted in 1816. The epitaph on his tomb sums up his immense influence on the course of astronomy: *Coelorum perupit claustra*—he broke the barriers of the heavens.

HERSHEY, Alfred Day
(1908–)

American biologist, born in Owosso, Michigan. He studied at Michigan State College, and from 1950 to 1974 worked at the Carnegie Institution in Washington, where from 1962 he was Director of the Genetics Research Unit. He became an expert on the viruses which infect bacteria (bacteriophage or 'phage'), and set up the Phage Group with **Luria** and **Delbrück** in the late 1940s, to encourage the use of phage as an experimental tool. At that time, it was not known whether DNA or protein was the genetic material; **Avery** had suggested in 1944 that DNA was the genetic material but experimental evidence was lacking. Working with Martha Chase in the early 1950s, Hershey radioactively labelled the protein and the DNA of the phage particles with different markers. Using a kitchen appliance, a Waring blender, to shake off any phage attached to the bacteria after infection, they then examined the bacteria and found only labelled DNA, suggesting that DNA was the material that changed the bacteria, and hence that DNA is the material which holds the genetic code. In 1969 Hershey shared the Nobel Prize for Physiology or Medicine with Luria and Delbrück.

HERTWIG, Oscar Wilhelm August

(1849–1922)

German zoologist, born in Freidberg, Hessen. He was professor at Jena and Berlin, and investigated early stages of development. He showed that it is necessary for the nuclei of egg cells and spermatozoa to fuse for fertilization to occur, and that only a single sperm cell is required. He also studied the nuclear transmission of hereditary characters and the effects of radioactivity on the germ cells. He was a critic of the germ layer theory of development which states that three cell layers in the embryo give rise to all the adult tissues.

HERTZ, Gustav Ludwig

(1887–1975)

German physicist, nephew of **Heinrich Hertz**, born in Hamburg and educated at the University of Berlin. In 1928 he was appointed Professor of Physics at the Technical University in Berlin, and he later became Director of the Siemens Research Laboratory. With **Franck**, he showed that mercury atoms would only absorb a fixed amount of energy from bombarding electrons, demonstrating the quantized nature of the electron energy levels of the atom. For this work they shared the 1925 Nobel Prize for Physics. The results provided data for **Niels Bohr** to develop his theory of atomic structure, and for **Planck** to develop his ideas on quantum theory. After World War II Hertz went to the USSR to become head of a research laboratory (1945–54), then returned to East Germany to become Director of the Physics Institute in Leipzig (1954–61).

HERTZ, Heinrich Rudolf

(1857–94)

German physicist, born in Hamburg. His father was a barrister and later a senator. He was educated privately and at the Johanneum Gymnasium, and studied at several higher institutions, unsure whether to opt for an engineering or an academic career, but in the end he decided on the latter. He moved to Berlin where he studied under **Kirchhoff** and **Helmholtz**, before becoming the latter's assistant. In 1885 he was appointed Professor of Physics at Karlsruhe, and in 1889 he became professor at Bonn. In 1887 Hertz confirmed **Maxwell**'s predictions by his fundamental discovery of 'Hertzian waves', now known as radio waves, which excepting wavelength, behave like light waves. He did not concern himself with the practical implications which were left to be developed by **Marconi** and others. In his final years he explored the theoretical implications of Maxwell's electrodynamics for physics generally. He was widely honoured for his work on electric waves, and in 1890 he was awarded the Rumford Medal of the Royal Society. He died at the young age of 36.

HERTZSPRUNG, Ejnar

(1873–1967)

Danish astronomer, born in Fredriksberg, a suburb of Copenhagen. He graduated in chemical engineering from the Technical High School in Copenhagen in 1898 and was employed for some years in this profession before entering astronomy (1902) and joining the staff of the university observatory in Copenhagen. He published in 1905 the principle of what was later to be called the Hertzsprung–Russell diagram, which shows that the luminosities of yellow stars disperse over more than four magnitudes. The diagram became the key for the theory of stellar evolution. In 1909 he obtained a position under **Schwarzschild** at the University of Göttingen, from where he moved with Schwarzschild to the Astrophysical Observatory in Potsdam. He was appointed Director of the Leiden Observatory, where his enthusiasm and example were of great benefit to astronomy in the Netherlands. He was awarded the Gold Medal of the Royal Astronomical Society in 1929 and the Bruce Medal of the Astronomical Society of the Pacific in 1937.

HERZBERG, Gerhard

(1904–)

German–Canadian physicist, born in Hamburg. He was educated at the Technische Hochschule, Darmstadt, and at the universities of Göttingen and Bristol. From 1930 he taught at Darmstadt, but he emigrated to Canada in 1935. He was Research Professor of Physics at the University of Saskatchewan from 1935 to 1945, and after a brief appointment at Yerkes Observatory, he became Director of the Division of Pure Physics of the National Research Council, Ottawa, from 1949 to 1969. Herzberg is distinguished for his applications of spectroscopic methods in astrophysics, atomic and molecular physics, and physical chemistry. Of particular importance have been his extremely precise determinations of the energy levels of atomic and molecular hydrogen and the isotopes thereof, and of radicals such as CH, CH_2, CH_3 and NH_2. His several books have been very influential, particularly *Molecular Spectra and Molecular Structure*, which appeared in four volumes over a period of 40 years (1939–79). Herzberg was awarded the Nobel Prize for Chemistry in 1971. He was elected a Fellow of the Royal Society in 1968 and received its Royal Medal in 1971. He is also an Honorary Fellow of the Royal Society of Chemistry and was awarded its Faraday Medal in 1970.

HESS, Germain Henri
(1802–50)

Swiss-born Russian chemist, born in Geneva, and taken to Russia in childhood. After studying medicine, chemistry and geology at the University of Dorpat (now Tartu, Estonia), he worked briefly with **Berzelius**. He took part in a geological expedition to the Urals and practised medicine in Irkutsk, but from around 1830 he devoted himself to chemistry. He was appointed Professor of Chemistry at the Technological Institute of St Petersburg. His earliest researches were on mineral analysis and on the natural gas of Baku, but he then turned to thermochemistry, which was at that time in a primitive state. After extensive measurements of heats of chemical reaction, he established the 'law of constant heat summation' (1838–40), which states that the heat developed in a given chemical change is constant, independent of whether the change is carried out in one stage or in several stages. This law, commonly called Hess's law, is a special case of the law of conservation of energy, but at the time this had not been clearly formulated.

HESS, Harry Hammond
(1906–69)

American marine geophysicist and geologist, born in New York City and educated at Yale (BS 1931) and Princeton (PhD 1932). After a number of short-term appointments, he returned to Princeton as instructor (1934), later becoming Professor of Geology (1948) and Blair Professor (1964) there. During the 1930s he participated in submarine gravity studies of the West Indies island arc and became a lieutenant in the US Naval Reserve. Called to active service in 1941, he rose to the rank of Rear Admiral. With the navy echo sounders switched on during manoeuvres, he discovered flat-topped seamounts which he named guyots after the shape of the Princeton campus building, built in honour of the geographer Arnold Guyot. Aware that the uniform ocean layering could only be formed by a uniform process, and accepting that convection currents exist in the mantle as proposed by **Holmes** in 1929, he described the oceans as young, ephemeral and with constant renewal by magma flowing into the mid-ocean ridges (1960). Thus he had described the sea-floor spreading process, except for the magnetic aspects. He was a strong proponent in 1957 of the 'Mohole' project to drill through the **Mohorovičić** discontinuity. Hess Guyot and Hess Deep (a deep ocean basin) in the Pacific are named after him.

HESS, Victor Francis
(1883–1964)

Austrian-born American physicist, born in Waldstein. During 1911–12, while on the staff of Vienna University, he made a number of manned balloon flights carrying ionization chambers. He demonstrated that the radiation intensity in the atmosphere increased with height, and concluded that the high-energy cosmic radiation that was responsible must originate from outer space. He also helped to determine the number of alpha particles given off by a gram of radium (1918). For his work on 'cosmic radiation' he was awarded the 1936 Nobel Prize for Physics, jointly with **Carl Anderson**. In 1938 he emigrated to the USA to become Professor of Physics at Fordham University, New York (1938–56).

HESS, Walter Rudolf
(1881–1973)

Swiss physiologist, born in Frauenfeld. He studied medicine at Lausanne, Bern, Zürich, Berlin and Kiel universities, receiving his degree from Zürich in 1906. Hess established a successful ophthalmology practice, but in 1917 gave it up in order to work on haemodynamics, a field in which he had become interested as an undergraduate. He became Professor of Physiology and Director of the Physiology Institute of the University of Zürich. He studied the regulation of blood pressure and heart rate, and their relationship to respiration, and from 1925 worked on the function of structures at the base of the brain. Hess developed methods of stimulating localized areas of the brain by means of fine needle electrodes, which permitted major advances in the study of brain function. He was able to show that stimulating different parts of the hypothalamus causes changes in body temperature, blood pressure, respiration, and also anger, sexual arousal, and sleep. In 1948 he published his classic book *The Functional Organisation of the Diencephalon*, which presented the results of all this work, and in 1949 he was awarded the Nobel Prize for Physiology or Medicine with **Egas Moniz**.

HEVELIUS, Johannes

(1611–87)

Polish–German astronomer, born in Gdańsk, the son of a prosperous brewer. After studies in law at the University of Leiden and travels in France and Germany, he constructed in the 1640s at his father's house in Gdańsk an observatory with a terrace for large quadrants and sextants, and high masts for the attachment of long telescopes. He published in 1647 his *Selenographica*, a description of the Moon with 133 copperplates of lunar features made by his own hand. Many details on the Moon were named after Earth features; some of their names such as the Appennines and Alps have survived. In the 1660s his interest turned to comets, resulting in his *Cometographia* (1668), a list of all comets observed up to that year. In 1673 appeared the first part of his major publication, *Machina Coelestis*, which contained a detailed description of his observatory. In 1678 Hevelius was visited by the King of Poland John III Sobieski and in the following year by **Halley**. In 1879 he saw his observatory with all its instruments and manuscripts go up in flames. His last work, a catalogue of 1564 stars with maps of the constellations, was published posthumously by his wife and collaborator Elizabeth in 1690.

HEVESY, George Charles von

(1885–1966)

Hungarian chemist, famous for his work on isotopic tracer techniques. He was born in Budapest, studied at Freiburg and held posts at the Technical High School in Zürich and at Manchester University, where he worked with **Rutherford** on an unsuccessful attempt to isolate the so-called radium D (later identified as an isotope of lead). The failure of these experiments later became part of the evidence that chemically identical elements exist with different atomic weights. He moved to Vienna in 1912, and after service in the Austro-Hungarian army during World War I, he moved to the Institute of Theoretical Physics in Copenhagen to join **Niels Bohr**. Here he and **Coster** searched for an element to fill the gap at atomic number 72 in the periodic table. This they discovered in 1923 and named hafnium, the Latin name for Copenhagen. In 1935 he returned to Freiburg where he began to calculate the relative abundances of the elements in the universe. In World War II he fled to Sweden, where he became professor at Stockholm. Hevesy's work on isotopes has been very influential in physics, chemistry and medicine. He worked to find ways of separating isotopes by physical means, and from 1934 onwards he pioneered the use of radioactive tracers to study chemical processes, particularly in living organisms. Using lead and phosphorus as tracers he showed that chemical changes are continually taking place in all living tissue. This work brought him the Nobel Prize for Chemistry in 1943. He died in Freiburg.

HEWISH, Antony

(1924–)

English radio astronomer, born in Fowey. He studied at Cambridge and spent his career there, becoming Professor of Radio Astronomy in 1971. In 1967 he began studying the scintillation ('twinkling') of astronomical radio sources. The analysis of the amplitude of these scintillations provided a potent means of estimating the sizes of the small radio sources that subtend less than a second of arc at the Earth's surface. He used a radio telescope consisting of 2048 dipoles at a wavelength of 3.7 metres, covering more than four acres of ground. This led him and his student **Bell Burnell** to discover the first radio sources emitting radio signals in regular pulses now known as pulsars; many others have since been discovered. They are believed to be very small and dense rotating neutron stars. Hewish shared the Nobel Prize for Physics in 1974 with his former teacher **Ryle**. Together (in 1960) they had produced a keynote paper on aperture synthesis, where radio dish telescopes a distance D apart can be used to synthesize a 'dish' of diameter D, and had used the Cambridge one-mile radio telescope to carry this out.

HEY, James Stanley

(1909–)

British physicist, born in the Lake District. He studied physics at Manchester University and from 1940 to 1952 was on the staff of the Army Operational Research Group. He then moved to the Royal Radar Establishment. He was elected FRS in 1978. In 1942, while investigating the problem of radar jamming, he discovered that the Sun is a strong radio source in the 4–8 metre wavelength region and that solar flares and sunspots are especially effective in this respect. These results were not generally circulated until after World War II. In 1946 Hey and his colleagues announced that the radio waves coming from the constellation Cygnus originated in the small discrete source that was then named Cygnus A. He was also a pioneer in the use of 5 metre wavelength radar for studying meteors. The observations of the Giacobinid shower of 1946 was one of his great successes; he was able to detect the head echo and measure the velocity of the incident meteoroids. In 1961 he constructed a radio interferometer at Malvern in which two 25 metre steerable parabolic reflectors could be moved along rails to vary the spacing.

HEYMANS, Corneille Jean François
(1892–1968)

French–Belgian physiologist, born in Ghent, the son of a professor of pharmacology. Heymans's medical studies at the University of Ghent were interrupted by World War I, during which he served, with distinction, in the Belgium army. In 1921 he completed his medical degree and a year later he was appointed as a lecturer in pharmacology at the University of Ghent. After postgraduate work in Paris, Lausanne, Vienna, London and Cleveland, Ohio, he succeeded his father as Director of the Institute of Pharmacology and Therapeutics in Ghent (1925). Heymans developed the technique of 'cross circulation'; using two anaesthetized dogs, he demonstrated that the rate of respiration is controlled by nerves. He also showed that structures in the aorta and carotid arteries contain special cells sensitive to blood pressure and blood chemicals, and that these play a monitoring role in the nervous mechanism by which respiration is controlled. He was awarded the 1938 Nobel Prize for Physiology or Medicine.

HEYROVSKÝ, Jaroslav
(1890–1967)

Czech physical and analytical chemist, born in Prague. He began to study chemistry, physics and mathematics at the Charles University, but in 1910 he moved to University College London (UCL), to study under **Ramsay** and subsequently under **Donnan**, who aroused Heyrovský's interest in electrochemistry. In 1913 he became a demonstrator at UCL and might well have spent the rest of his career there, but the outbreak of World War I in August 1914 caught him in Prague. He did war service as a dispensing chemist and radiologist in a military hospital and was simultaneously able to continue his studies, obtaining his PhD in Prague in 1918. After the war he became an assistant at the Charles University and climbed the academic ladder to Professor of Physical Chemistry by 1926. In 1950 he became director of the newly established Polarographic Institute, which was incorporated into the Czechoslovak Academy of Sciences in 1952. His first research on the electrode potential of aluminium was carried out at UCL, and his doctoral work in Prague was on the capillarity of mercury. This led him to study the behaviour of a dropping mercury electrode in electrochemical cells and ultimately to the invention of the polarograph (1922–5). His scientific effort for the rest of his life was devoted to the improvement of the polarographic technique and extending its applications. It has been widely used as an analytical technique and as a probe for the behaviour of ionic and molecular species (both inorganic and organic) with respect to oxidation and reduction in solution. Heyrovský received the Nobel Prize for Chemistry in 1959. Among his numerous other honours were foreign membership of the Royal Society (1965) and honorary fellowship of the Chemical Society (1963).

HEYWOOD, Vernon Hilton
(1927–)

British botanist, born in Edinburgh and educated at Edinburgh University and Pembroke College, Cambridge. After a distinguished early career at the botany department of the University of Liverpool (1955–68, professor 1964–8) he was Professor of Botany at the University of Reading from 1964 to 1988. Since 1987 he has been Director of the Botanical Gardens Conservation Secretariat, and from 1988 to 1992 he was Chief Scientist (Plant Conservation) and Director of Plant Science for the International Union for the Conservation of Nature and Natural Resources (IUCN), based at Kew. Heywood is a world authority on the families Compositae and Umbelliferae, and has had long-standing interests in the flora of Spain and in plant conservation. He has published nearly 200 papers and is also the author or editor of over 36 books, including *Principles of Angiosperm Taxonomy* (1963, with **Peter Davis**), one of the most important 20th-century works on the theory and practice of plant taxonomy; *Plant Taxonomy* (1967); *The Biology and Chemistry of the Umbelliferae* (1971, editor); *Flowering Plants of the World* (1978, editor); *Our Green and Living World* (1984, joint author); *The Botanic Gardens Conservation Strategy* (1989); and *International Directory of Botanic Gardens* (1990).

HIGGS, Peter Ware
(1929–)

British theoretical physicist, born in Newcastle upon Tyne. He was educated at King's College, London, and appointed to a lectureship in mathematical physics at Edinburgh (1960), where he became Professor of Theoretical Physics in 1980. Higgs developed the first field theory of particle interactions, building on ideas developed by **Schwinger** and Philip Anderson, in which the interactions which constitute the fundamental forces of nature take place via the exchange of massive gauge bosons. **Yang**'s simple gauge theories describe the interactions between elementary particles by the exchange of massless gauge bosons, eg the photon in the case of quantum electrodynamics. However, it was pointed out by Anderson that in superconductors the photon behaves as if it does have mass. Higgs developed a field theory using these ideas in which the gauge bosons could have mass. This was later used by **Steven Weinberg** and **Salam** to develop the electroweak theory of particle interactions. The so-called Higgs mechanism which allows massive gauge bosons does so by introducing another family of particles, the Higgs bosons. The detection of these particles is the goal of the next generation of particle accelerators, in particular the Large Hadron Collider at CERN in Switzerland. Higgs was awarded the Hughes Medal of the Royal Society in 1981 for this work and was elected FRS in 1983.

HILBERT, David

(1862–1943)

German mathematician, born in Königsberg (now Kaliningrad, Russia). He studied and taught at the university there until he became professor at Göttingen (1895–1930). His definitive work on invariant theory, published in 1890, was so novel that it brought the work for a generation on a subject that had occupied so many 19th-century mathematicians to a halt, and laid the foundations for modern algebraic geometry. In 1897 he published a lengthy report on algebraic number theory which was the basis of much later work. In this field he established a synthesis of the ideas of **Dedekind**, **Kummer** and **Kronecker** that opened the way to a structural analysis of the subject while playing down the explicit, if formal side. In 1899 he was the first to give abstract axiomatic foundations of geometry which made no attempt to define the 'meaning' of the basic terms but only to prescribe how they could be used. This was later to inspire him to look for a foundation of all mathematics in similar terms. With his students, he then worked on the theory of integral equations, the calculus of variations, and theoretical physics (which he claimed was too complex to be left to physicists). The concept of the 'Hilbert space' (an infinite-dimensional space) was implicit in this work, and was made explicit by Erhard Schmidt in 1906. Unease with the contemporary foundations of mathematics, and the specific proposals of **Brouwer** and **Weyl**, led him to take up mathematical logic. He proposed that all mathematics should be based on a finite set of axioms, and that a fully rigorous theory of proofs in mathematics should be given along the lines of elementary logic (strictly speaking, using a first-order language, for which a rigorous theory was available). This aim was later shown to be unattainable by **Gödel**. At the International Congress of Mathematicians in 1900 he listed 23 problems which he regarded as important for contemporary mathematics; the solutions of many of these have led to major advances, while others remain unsolved.

HILDEBRAND, Joel Henry

(1881–1983)

American chemist, born in Camden, New Jersey. He was educated at the University of Pennsylvania and did postdoctoral work in Berlin (1905–6) under **Nernst** and **van't Hoff**. From 1907 to 1913 he was an instructor in physical chemistry at the University of Pennsylvania, and he then moved to the University of California, Berkeley. By 1918 he had become a full Professor of Chemistry and he retained this position until his formal retirement in 1952. As Emeritus Professor he was active scientifically for another 30 years. He was President of the American Chemical Society in 1955 and was awarded its Priestley Medal in 1962. He was also an Honorary Fellow of the Royal Society of Edinburgh and of the Royal Society of Chemistry. His greatest scientific contribution was to the understanding of the phenomena of solubility. Certain concepts which Hildebrand devised proved very fruitful, notably 'regular solutions', 'internal pressure' of liquids and the 'solubility parameter' (commonly prefaced with his name). One of the practical outcomes of his work was the use of a mixture of helium and oxygen as an atmosphere to be breathed by divers. He wrote the very influential book *Solubility* (1924), which in later editions became *Solubility of Non-Electrolytes* and was co-authored with Robert L Scott. He was an enthusiastic teacher and among his other books was an undergraduate text *Principles of Chemistry* (1918), which continued to its 7th edition in 1964.

HILL, Archibald Vivian

(1886–1977)

English physiologist whose detailed quantitative approach to the functional problems of muscle contraction and neuronal activity was an important contribution in the development of the speciality of biophysics. Born in Bristol, he read mathematics at Trinity College, Cambridge, becoming Third Wrangler in 1907. He was then persuaded by his college tutor, the physiologist Walter Morley Fletcher, to apply his mathematical skills to physiology. He remained in Cambridge until World War I, principally working with Fletcher on the generation of heat during muscle contraction, although also developing research projects in drug kinetics and on the physics of haemoglobin. He began designing and developing sensitive equipment to make the very precise measurements necessary in his work, especially delicate thermocouples to register small changes in temperature. During World War I he served in anti-aircraft research and after demobilization was appointed Professor of Physiology at Manchester. He shared, with **Meyerhof**, the 1922 Nobel Prize for Physiology or Medicine, shortly afterwards becoming Professor of Physiology at University College London (1923), and from 1926 to 1951 Foulerton Research Professor of the Royal Society at University College. His research showed that muscles developed heat during both contraction and recovery periods, and this was related to the production and use of lactic acid. With Meyerhof's complementary biochemical examinations, Hill provided detailed evidence of the energy-generating processes associated with muscle activity. Similarly he examined heat production by nerves and provided an explanation of the transmission of nerve impulses based on physico-chemical equations. His *Trails and Trials in Physiology* (1965) summarizes several of his most significant scientific contributions. He accepted many other responsibilities outside the laboratory, serving for example as Biological (1935–45) and Foreign (1945–6) Secretary of the Royal Society, as a member of parliament for Cambridge University (1940–5) and as a forceful member of the Academic Assistance Council (later the Society for the Protection of Science and Learning), an organization that did much to help refugee scientists from central Europe. He wrote extensively on a wide range of subjects, particularly the social obligations of scientists and published a collection of such papers, *The Ethical Dilemma of Science, and Other Writings*, in 1960.

HILLIER, James
(1915–)

Canadian-born American physicist, born in Brantford, Ontario. He was educated at the University of Toronto where he obtained a BA in mathematics and physics in 1937 and his PhD in 1941. He was a major contributor to the development of the electron microscope. In 1924, **de Broglie** had postulated that particles can behave like waves. This was later proven by **Davisson** and **Germer** and led to the proposal of the electron microscope which would, at least in theory, have much greater resolution than that achievable by optical microscopes because electron waves can be produced with very short wavelength. While at Toronto, Hillier developed one of the first high-resolution electron microscopes produced, with a resolution of 100 nm. Later he moved to the USA (1940) and made his career with RCA (the Radio Corporation of America). This gave him the resources to further develop the electron microscope. In addition to the microscope's development, Hillier also maintained an interest in its applications in areas such as biology, medicine, chemistry and metallurgy, and he held visiting posts in the biology department of Princeton University and at the Sloan-Kettering Institute for Cancer Research.

HINDE, Robert Aubrey
(1923–)

English ethologist and zoologist, born in Norwich. After serving as an RAF pilot with Coastal Command during World War II, he attended St John's College at the University of Cambridge (1944–8) and obtained a doctorate from Oxford University (1950). He was curator of the Ornithological Field Station of the Cambridge University zoology department (1950–64) and has been a Fellow of St John's College since 1858. He held a Royal Society Research Professorship from 1963 to 1989. His many scientific works have been in the area of ethology and human behaviour. While the early ethologists worked mainly with birds and lower vertebrates, Hinde, after using birds to investigate problems of drive and motivation, has been concerned more with the application of ethological methods of behaviour analysis to the primates including man. In particular he has used the comparative approach in the belief that the analysis of social behaviour of primates can provide insights into that of humans. In much of his writing he has attempted to bridge the historical gap between ethology and comparative psychology. His most recent interests have been concerned with the nature of human interpersonal relationships and the relations between biology, psychology and culture. He is the author of many research papers and books including *Animal Behaviour: A Synthesis of Ethology and Comparative Psychology* (1966), *Biological Bases of Human Social Behaviour* (1974) and *Individuals, Relationships and Culture: Links between Ethology and the Social Sciences* (1987). He was elected FRS in 1974 and made a CBE in 1988.

HINSHELWOOD, Sir Cyril Norman
(1897–1967)

English physical chemist, born in London. His education was interrupted by World War

I, during which he served as a chemist at an explosives factory. In January 1919 he entered Balliol College, Oxford, and took his degree after only five terms of study. After a year as Research Fellow at Balliol, he became a Fellow and Tutor of Trinity College, an appointment he held until 1937, when he succeeded **Soddy** as Dr Lee's Professor of Chemistry. He retired from Oxford in 1964, becoming a Senior Research Fellow of Imperial College, London, where he was active until his death. His research work was largely in chemical kinetics. In the 1920s he carried out pioneering work in gas reactions and their interpretation in terms of the kinetic theory. This led to his best-known book *The Kinetics of Chemical Change in Gaseous Systems* (1926). Hinshelwood's studies of 'unimolecular' gas reactions laid the bases for much later work, and in his studies of chain reactions he built upon the work of **Bodenstein**. His work on the hydrogen–oxygen reaction was particularly notable. The chemical equation for this reaction, $2H_2 + O_2 = 2H_2O$, is one of the simplest known, but from the standpoint of kinetics and mechanism the reaction is of great complexity. Hinshelwood's work on this reaction laid foundations on which later workers have built. He also studied heterogeneous catalysis and solution kinetics, but from 1936 onwards he was increasingly interested in the kinetics of bacterial growth. This led to *The Chemical Kinetics of the Bacterial Cell* (1946) and *Growth, Function, and Regulation in Bacterial Cells* (1966, with Alistair C R Dean). Some of his ideas generated much controversy with biologists. In 1956 he was awarded the Nobel Prize for Chemistry jointly with **Semenov** for his contributions to chemical kinetics. Hinshelwood was elected FRS in 1929 and was President of the Royal Society from 1955 to 1960. President of the Chemical Society from 1946 to 1948, he was knighted in 1948 and admitted to the Order of Merit in 1960. Hinshelwood was a fine linguist, with a good knowledge of the classics; in 1959 he served as President of the Classical Association.

HIPPARCHOS

(c.180–125 BC)

Greek astronomer, born in Nicaea in Bithynia, the most outstanding astronomer of the ancient world. He made his observations from Rhodes where he spent a long time, and may also have lived in Alexandria. He compiled a catalogue of 850 stars (completed in 129 BC) giving their positions in celestial latitude and longitude, the first such catalogue ever to exist, which remained of primary importance up to the time of **Halley**. In comparing his observed star positions with earlier records he made his great discovery of the precession of the equinoxes, ie the shifting of the point of intersection of the Sun's annual path with the celestial equator. He observed the annual motion of the Sun through the sky, developed a theory of its eccentric motion, and measured the unequal durations of the four seasons. He made similar observations of the Moon's more complex motion. Following the method of eclipse observations used by **Aristarchos**, he estimated the relative distances of the Sun and Moon, and improved calculations for the prediction of eclipses. He developed the mathematical science of plane and spherical trigonometry required for his astronomical work. In the field of geography Hipparchos was the first to fix places on the Earth by latitude and longitude.

HIPPOCRATES

(?c.460–377 or 359 BC)

Greek physician, known as the 'father of medicine', and associated with the medical profession's 'Hippocratic Oath'. The most celebrated physician of antiquity, he was born and practised on the island of Cos, but little is known of him except that he taught for money. The so-called 'Hippocratic Corpus' is a collection of more than 70 medical and surgical treatises written over two centuries by his followers, and only one or two can be fairly ascribed to him. He seems to have tried to distinguish medicine proper from the traditional wisdom and magic of early societies, and laid the early foundations of scientific medicine; he was said to be good at diagnosis and prognosis, and his followers developed the theories that the four fluids or humours of the body (blood, phlegm, yellow bile and black bile) are the primary seats of disease. They conceived that excesses or deficiencies of the humours caused diseases, which were to be treated by measures (such as drugs, diet, change of life, blood-letting) which countered them. They thought that the doctor must be the servant of nature, believing that the body has natural tendencies to correct the humoral imbalance (the 'healing power of nature'). The humoral doctrine also included a notion of temperaments, whereby each person had his or her own ideal balance, which the doctor needed to take into account when planning therapy. The balance also naturally changed during the course of the life cycle, with blood the dominant humour during early life, and phlegm the one of old age. The Hippocratic writings contain many treatises which long exerted great influence. The Oath has been seen as the foundation document of Western medical ethics, and is still occasionally used in a Christianized version. However, it was as much a guild agreement as an ethical statement, though some of its prohibitions, such as the injunction against procuring abortions, appealed to later doctors. Of his works, *Airs, Waters, Places* contained shrewd observations about the geography of disease and the role of the environment in shaping the health of a community. *Epidemics III* examined epidemics in a population and offered case histories of patients with acute diseases. *The Sacred Disease* elaborated a rigorous defence of the naturalistic causes of diseases, in the context of a monograph on epilepsy, attributed to supernaturalistic influences by many in early Greek society. *Aphorisms* consisted of a series of short pithy statements, mostly about clinical situations, but beginning with the most famous, 'Life is short, the art is long'.

HIRE, Philippe de la

See **DE LA HIRE, Philippe**

HIRN, Gustave Adolphe

(1815–90)

French physicist and engineer, born in Logelbach, Alsace. His Calvinist father owned a calico factory in the industrial Mulhouse region. Hirn was educated privately until at 19 he entered the family firm as a colour chemist. With his brother he became director of a mill and there investigated steam-engine performance, aiming to increase efficiency by steam-jacketing cylinders and introducing mineral oil lubricants to reduce friction. Convinced that heat and work were interconvertible, he sought the 'exchange rate' and by 1847 had measured the mechanical equivalent of heat, independently of **Joule** and **Mayer**. His redetermination of the mechanical equivalent in 1855 gained a prize from the Berlin Physical Society. Accurate practical experiments showed that the difference between the heat leaving a steam engine in the condenser and the heat entering it through the boiler was proportional to the work done. Hirn's *Théorie Mécanique de la Chaleur* (1862) was one of the earliest textbooks on thermodynamics.

HIRST, Sir Edmund Langley

(1898–1975)

English chemist, born in Preston. He attended Madras College while his father was a minister of the Baptist Church in St Andrews, and then proceeded to St Andrews University, where he took a degree in chemistry and a PhD on the chemistry of sugars. He joined the staff at St Andrews, but left in 1923 and worked successively at the universities of Manchester, Durham and Birmingham. His first chair was in Bristol (1936) and he became professor at the University of Manchester in 1944. He ended his distinguished academic career as Forbes Professor of Organic Chemistry at the University of Edinburgh. Hirst was a gracious and charming scholar who made important contributions to the chemistry of carbohydrates. His most famous work was his collaboration with **Norman Haworth** in the laboratory synthesis of vitamin C. He was elected FRS in 1934 and knighted in 1964.

HIS, Wilhelm

(1831–1904)

Swiss biologist, born in Basle. He studied medicine at Basle, Bern, Berlin and Würzburg, qualifying in 1855, and coming under the influence of **Virchow** and **Remak**. He later studied at Paris under **Bernard** and **Brown-Séquard**. He became Professor of Anatomy in Basle and later, from 1872, at Leipzig, where he supervised the building of one of the best medical laboratories in the world. He pursued valuable studies on the lymphatic system, and made investigations into developmental processes and embryonic growth. He was involved in a lively polemic in countering **Haeckel**'s championship of the 'biogenetic law' (ontogeny recapitulates phylogeny), and stressing the distinction between the phylogenetic and the physiological. One of His's key contributions lay in the development of the microtome (1866) for cutting very thin serial sections for microscopical purposes. He used it primarily in his examination of embryos. He also used photography for anatomical purposes, and furnished the first accurate description of the human embryo. His son, also named Wilhelm (1863–1934), gave the first description of the specialized bundles of fibres in the heart, the 'bundles of His', which are integral to the electrical conducting mechanism.

HITCHCOCK, Edward

(1793–1864)

American geologist, born in Deerfield, Massachusetts. His observation of a comet in 1811 marked the beginning of a lifelong interest in science. Although he began his career as a teacher and clergyman, he became Professor of Chemistry at Amherst College (1825–45), and directed the first state-funded geological survey of Massachusetts (1830–3), which resulted in his publication of *The Geology of Massachusetts* (1833). His paper on diluvium (1843) was useful as a review of European literature, and for assuring fellow geologists that even a minister of the church did not expect Biblical literalism. He investigated the dinosaur tracks in the Connecticut Valley, which he believed to be bird tracks. He published *The Religion of Geology* (1851), *Illustrations of Surface Geology* (1857) and his major work, *Ichnology of New England*, in 1858. He co-founded the American Association of Geologists in 1840.

HITCHINGS, George Herbert
(1905–)

American biochemist, born in Hoquiam, Washington state. He studied and worked at the University of Washington (1926–8), Harvard (1928–39) and Western Reserve University Medical School (1939–42) before moving to Burroughs Wellcome, North Carolina, where he became Chief Biochemist (1945–55), Associate Research Director (1955–68) and Director (1968–77). He was also Professor of Pharmacology at Brown University from 1968 to 1980. His research in the early 1940s in collaboration with **Elion**, facilitated by a particularly sensitive microbial assay, involved the preparation and testing of purine and pyrimidine analogues as growth factors, to replace thymine or inhibitors of folic acid, and similar work involving amino acids. From these investigations emerged the folic acid antagonist, 2-aminopurine (1948), paving the way for the discovery of allopurinol (used to alleviate gout by blocking uric acid synthesis) and anti-cancer drugs, the first of which was 2,6-diaminopurine, followed in 1952 by the anti-malarial drug pyrimethamine (Duraprim®), and later, to counter drug resistance, sulfadoxine (Fansidar®). In 1954 his team synthesized the spectacularly successful anti-leukaemia drug 6-mercaptopurine, followed by azathioprine (Imuran®), an immunosuppressant enabling organ transplantation from an unrelated donor. His laboratory also produced the important anti-bacterial trimethoprim (Septra®), the anti-viral acyclovir (Zovirax®), active against herpes, and the anti-AIDS drug zidovudine (Retrovir®). Hitchings shared with Elion and **James Black** the 1988 Nobel Prize for Physiology or Medicine for these outstanding achievements.

HITTORF, Johann Wilhelm
(1824–1914)

German physicist and chemist, born in Bonn. After study in Bonn and Berlin he became *Privatdozent* at the Münster Academy. When this became a university, Hittorf was appointed Professor of Physics and Chemistry, and on the creation of separate departments of these subjects in 1879, he became director of the physics laboratories. He retired from this post due to ill-health in 1889, but after a time was able to continue research. His earliest researches were on the allotropy of phosphorus and selenium, but he is best remembered for his work on ion migration during electrolysis (1853–9). By measuring changes in electrolyte concentration in the vicinity of electrodes, he determined 'transport numbers' and ratios of 'mobilities' of ions. The later work of **Kohlrausch** complemented Hittorf's work. In association with **Plücker**, he studied the conduction of electricity through gases and the discharge-tube spectra of gases.

HITZIG, Julius Eduard
(1838–1907)

German neurologist and psychiatrist, born in Berlin and trained in medicine at the universities of Berlin and Würzburg, after originally intending to study law. He qualified in 1862 and stayed in Berlin until 1875 when he moved to Zürich as Professor of Psychiatry and director of an asylum. Numerous conflicts with the asylum authorities, due to his uncompromising and difficult personality, resulted in an early move to Halle as Professor of Psychiatry (1881–5) and as director of a psychiatric clinic (1885–1903). Hitzig's principal scientific contributions were in collaboration with Gustav Theodor Fritsch, with whom he demonstrated in 1870 that electrical stimulation of the frontal cortex in animals caused movements of the extremities on the opposite side of the body. These early studies in cerebral electrophysiology, proving the existence of specific motor control areas, stimulated research in cortical localization and brain anatomy by many investigators. Hitzig himself continued with ablation and stimulation experiments especially in the visual cortex, and he also postulated the localization of centres for mental processes. His favouring of regional localization brought him into contemporary conflict with other neurologists who believed in a more diffuse, holistic concept of brain function. His major influence has been on the development of neurological and psychiatric studies, and their integration with morphological and physiological knowledge.

HOAGLAND, Mahlon Bush
(1921–)

American biochemist, born in Boston. He studied medicine at Harvard, and was a visiting researcher at Copenhagen and Cambridge (UK) before returning to the bacteriology and immunology department of Harvard Medical School (1952–67). He later became Professor of Biochemistry at Dartmouth Medical School (1967–85), and from 1970 to 1985 was Scientific Director of the Worcester

Foundation for Experimental Biology. Hoagland worked on the cause of cancer by beryllium, liver regeneration and growth control, but his major scientific contribution was the practical confirmation of **Crick**'s 'adaptor hypothesis' proposing that in protein synthesis there is an oligonucleotide carrier for each amino acid identified by an RNA transcript of the DNA codon. In 1955–6 Hoagland showed that the first step in protein synthesis was the activation of an amino acid via an enzyme-bound aminoacyl adenylate, and in 1957 with Paul Zamecnik, he demonstrated that the amino acid is then transferred to form an ester link with a soluble form of RNA (tRNA), Crick's adaptor molecule.

HODGKIN, Sir Alan Lloyd
(1914–)

English physiologist, born in Banbury. He read natural sciences at Trinity College, Cambridge, and apart from short research breaks elsewhere and war service, has spent his entire career in Cambridge. Master of Trinity College from 1978 to 1984, he was a lecturer and Assistant Director of Research at the Physiological Laboratory (1945–52), Royal Society Research Professor (1952–69) and Professor of Biophysics (1970–81). During World War II he worked on airborne radar research, and released from military service because of **Adrian**'s intercession, Hodgkin returned to experiments initiated in the 1930s on the conduction of nerve impulses. Much of his research was in collaboration with his former student Sir **Andrew Huxley** and the two men shared the 1963 Nobel Prize for Physiology or Medicine with **Eccles**. A great deal of their work was done at the Marine Biological Association's laboratory in Plymouth, using the squid giant axon preparation described by **John Young**. Hodgkin and Huxley described in physico-chemical and mathematical terms, the mechanisms by which nerves conduct electrical impulses by the movement of electrically charged particles across the nerve membrane. These detailed experiments continued for several years, and during the 1960s Hodgkin also analysed the electrical characteristics of single muscle fibres of the frog. The techniques developed by Hodgkin and Huxley have enabled scientists to study and understand many different kinds of excitable membranes. In 1970 Hodgkin became President of the Royal Society for five years, and started to analyse the ionic mechanisms that occur in the light-sensitive receptor cells of the retina, when illuminated. He has received many honours and awards, including the Order of Merit (1973). In 1992 he published his autobiography *Chance and Design: Reminiscences of Science in Peace and War*.

HODGKIN, Dorothy Mary, née Crowfoot
(1910–)

British crystallographer, born in Cairo, Egypt. She studied chemistry at Somerville College, Oxford, moved to Cambridge to study for her PhD, and became a Fellow and Tutor at Somerville in 1934. After various appointments within the university, she became the first Royal Society Wolfson Research Professor at Oxford in 1960. She received the Nobel Prize for Chemistry in 1964, only the third woman to do so. In 1965 she was admitted to the Order of Merit, the first woman to be so honoured since Florence Nightingale. Hodgkin was elected FRS in 1947 and gave the Bakerian Lecture in 1972. She is an Honorary Fellow of the Royal Society of Chemistry and was awarded the Longstaff Medal in 1978. She was President of the International Union of Crystallography from 1972 to 1978. In X-ray crystallography studies at Cambridge, **Bernal** introduced her to the study of biologically interesting molecules, which was extremely difficult and tedious in the 1930s. With Bernal she began work on sterols and continued this after her return to Oxford. Her detailed X-ray analysis of cholesterol was a milestone in crystallography, but an even greater achievement was the determination of the structure of penicillin (1942–5). After World War II computational facilities increased; even so, the determination of the structure of vitamin B_{12}, which was her real triumph, occupied eight years (1948–56). Her later work on insulin, an even more complicated molecule, was able to use sophisticated computers.

HODGKINSON, Eaton
(1789–1861)

English engineer, born in Anderton, Cheshire. He had little formal higher education, but became one of the foremost authorities on the strength of materials. As a result of tests carried out in the engineering works of **Fairbairn**, he proposed in 1830 the famous 'Hodgkinson's beam' as the most efficient form of cast-iron beam, detailing the results of his researches in the seminal paper *Theoretical and Practical Researches to Ascertain the Strength and Best Forms of Iron Beams*. After a further series of experiments he published in 1840 a paper *On the Strength of Pillars of Cast Iron and other Materials*. Hodgkinson also collaborated with Fairbairn and Robert Stephenson on the design of the rectangular wrought-iron tubes within which trains crossed the Menai Strait in four continuous spans, two of 460 feet and two of 230 feet. The Britannia Bridge, opened in 1850, constituted a significant advance in the theory and practice of structural engineering at the time. Hodgkinson was elected FRS in 1840, and in 1847 was appointed Professor of the Mechanical Principles of Engineering at University College London.

HOFF, Jacobus Henricus van't

See **VAN'T HOFF, Jacobus Henricus**

HOFFMANN, Roald

(1937–)

American chemist, born in Złoczow, Poland. Having survived the Nazi occupation by hiding in an attic for almost a year, he and his mother (his father was killed in a breakout from a labour camp) became refugees when Poland was liberated by the Russians. After moving around Europe for some years, they and Roald's step-father (Paul Hoffmann) migrated to the USA, arriving in New York City in 1949. In 1955 he entered Columbia University and took a number of arts courses as well as chemistry. He obtained a PhD from Harvard in chemical physics in 1962, having spent nine months at the University of Moscow. He was elected a Junior Fellow at Harvard and in 1964 began a collaboration with **Robert Woodward**, in which factors controlling the way in which cyclization reactions occur when bond breaking and making occurs simultaneously were established. The results of these considerations became known as the Woodward–Hoffmann rules for the conservation of orbital symmetry and have been of remarkable predictive value, stimulating much productive experimental work. It was for this work that, along with **Fukui**, he received the 1981 Nobel Prize for Chemistry. In 1965 Hoffmann moved to Cornell University, and in 1974 he was appointed John A Newman Professor of Physical Science. His most recent work concerns the synergism of molecular orbital calculation and experiment in a number of areas of inorganic chemistry. These studies have led to an increased understanding of bonding, and to the prediction of chemical species subsequently synthesized by others. Another area of interest is the bonding of species absorbed on surfaces. He also writes popular articles on science and has hosted a TV programme on chemistry. His second book of poems was published in 1990. In addition to the Nobel Prize he has received many awards and honours, including the Priestley Medal of the American Chemical Society.

HOFMANN, August Wilhelm von

(1818–92)

German chemist, born in Giessen. After unsuccessfully studying law and philosophy at Giessen, he turned in 1843 to chemistry and became **Liebig**'s assistant. He moved briefly to Bonn in 1845, but his most successful post was as professor at the College of Chemistry in London. It was generally agreed that chemical research in England in the 1840s was at a low ebb, and Prince Albert, consort of Queen Victoria, sought to improve matters by inviting a German to fill this important post. His choice of Hofmann was inspired. His pupils and assistants included some of the most significant chemists of the century, notably **William Henry Perkin Sr**. His attractive and benevolent personality caused him to be idolized by a generation of chemists. He and his students had considerable success in extracting from coal tar some of its most valuable constituents (eg aniline) in pure form, and in exploring the chemistry of these compounds, thus preparing the way for the development of the dyestuffs industry. He supported **Gerhardt**'s theory of chemical types, later to be discredited, and described the ammonia type. He developed the process of exhaustive methylation for the conversion of amines into the corresponding olefin. Elected FRS in 1851, in 1865 he succeeded **Mitscherlich** as professor in Berlin, where he continued to exercise a profound influence over European chemistry. He was raised to the rank of nobleman of Prussia (von Hofmann) on his 70th birthday.

HOFMEISTER, Wilhelm Friedrich Benedikt

(1824–77)

German botanist, born and educated in Leipzig. His father, Friedrich Hofmeister, owned a music shop and music publishing house, and was interested in plant systematics. His friend Heinrich Gottlieb Ludwig Reichenbach encouraged Friedrich Hofmeister to start a herbarium and establish a botanical garden in extensive grounds in a Leipzig suburb. Wilhelm, after a spell in Hamburg as a music shop apprentice, returned to Leipzig in 1841. For over a decade, he combined botanical research with a full-time career in music publishing. Very myopic, he paid attention to minute detail and was exceedingly skilful in making microscopical preparations, often directly manipulating specimens where a dissecting microscope would normally be used. He became an authority on embryology and was one of the first to observe chromosomes, although he did not appreciate their significance. His most epoch-making discovery was of the 'alternation of generations' (between gametophyte and sporophyte) in plants, elaborated in his most famous work, *Vergleichende Untersuchungen der Keimung, Entfaltung und Fruchtbildung höherer Kryptogamen* . . .

und der Samenbildung der Coniferen (1851). This publication marked the transition of botanical science from the medieval to the modern period. Though pre-Darwinian, it has been considered 'the greatest broad evolutionary treatise in botany'.

HOFSTADTER, Robert
(1915–)

American physicist, born in New York City and educated at City College, New York, and Princeton University. He then worked at the Norden Laboratory Corporation (1943–6) and Princeton University, before moving to Stanford University where he became professor in 1954. He was also Director of the Stanford High Energy Physics Laboratory (1967–74). In 1948 he developed a sodium iodide–thallium scintillation counter for X-ray detection. Later at Stanford, he used the linear accelerator to probe nuclear structure with electrons. He showed that nuclear charge is constant within the core but decreases sharply at the nuclear surface, and revealed that protons and neutrons also contain inner structure (now known to be due to quarks). For this work, he shared the 1961 Nobel Prize for Physics with **Mössbauer**.

HOLLERITH, Herman
(1860–1929)

American inventor and computer scientist, born in Buffalo, New York, the son of a German immigrant. He graduated in 1879 from the School of Mines at Columbia University, and worked as a statistician on the processing of data relating to the manufacturing industries for the US census of 1880. Realizing the need for automation in the recording and processing of such a mass of data, he devised a system based initially on punched cards. Hollerith used electrical contacts made through the holes in his cards to actuate electromechanical counters, and he worked on the development of his system while employed first at MIT, and then in the US Patent Office (1884–90). He won the competition for the most efficient data-processing equipment to be used in the 1890 US census. In 1896 he established his own company, later merging it in 1911 with three others to form the Computing Tabulating Recording Company (eventually part of IBM).

HOLLEY, Robert William
(1922–)

American biochemist, born in Urbana, Illinois. Working mainly at Cornell Medical School and at the Salk Institute in California (from 1968), he was a member of the team which first synthesized penicillin in the 1940s. In 1962, with **Benzer**, using poly UG and poly UC, he identified two distinct leucyl transfer RNAs (leu tRNS) characterized by different codons, and suggested that this provided the physical basis for the degeneracy of the amino acid code suggested by **Crick**. In the bacterium *Escherichia coli* he found five differently coding leu tRNAs (1965). He secured the first pure sample of a tRNA (1 gram of leu tRNA from 90 kilograms of yeast) and in 1965 published the full molecular structure of this nucleic acid—Crick's 'adaptor molecule', which plays a central role in the cellular synthesis of proteins. He shared the 1968 Nobel Prize for Physiology or Medicine with **Khorana** and **Nirenberg**.

HOLMES, Arthur
(1890–1965)

English geologist, born in Hebburn-on-Tyne. He was awarded a scholarship to attend Imperial College, London, where he studied physics and geology under Lord **Rayleigh** (1842–1919). Rayleigh had recently demonstrated the abundant presence of radioactive elements in the Earth's crust, and encouraged Holmes to develop the uranium–lead dating method. Holmes's first book in 1913 was a review of the history of attempts to ascertain the age of the Earth. He was appointed as a demonstrator in geology at Imperial College (1912–20) and then in the postwar ferment became chief geologist for an oil company operating in Burma. He returned to Britain to become Professor of Geology at the University of Durham (1924–43) and then Regius Professor of Geology at Edinburgh (1943–56). A pioneer of geochronology, Holmes determined the ages of rocks by measuring their radioactive constituents and played a large part in gathering age data for the Precambrian, allowing a picture to emerge of how ancient orogenies gradually built up cratonic shields and continental nuclei. He was an early scientific supporter of **Wegener**'s continental drift theory, and his predictions of the amount of heat generated by radioactive decay in the Earth revealed a mechanism for continental plate movement. He used lead isotopes to demonstrate that lead ores were not generated from granitic magmas as generally supposed, undertook important petrological studies, particularly with reference to basalt and alkali volcanics, and with his wife, Doris Reynolds, he worked on the metasomatic origin of certain igneous rocks. He wrote *The Age of the Earth* (1913) and *Principles of Physical Geology* (1944), which has been one of the most successful textbooks ever written.

HOOFT, Gerard 't See **'T HOOFT, Gerard**

HOOKE, Robert

(1635–1703)

English natural experimental philosopher and architect, born in Freshwater, Isle of Wight. Educated at Christ Church, Oxford, he worked as an assistant to John Wilkins on flying machines, **John Willis** on chemical research, and **Boyle** on the construction of his air-pump. In 1662 he was appointed the first curator of experiments at the newly founded Royal Society, of which he was secretary from 1677 to 1683, and in 1665 he became Professor of Physics at Gresham College, London. In that year he published his *Micrographia*, one of the scientific masterpieces of the age, an account of his microscopic investigations in botany, chemistry and other branches of science. One of the most brilliant and versatile scientists of his day, he was also an argumentative individual who became involved in a number of controversies, including several priority disputes with **Isaac Newton**. He anticipated the development of the steam engine, discovered the relationship between the stress and strain in elastic bodies known as Hooke's law, and formulated the simplest theory of the arch, the balance-spring of watches, and the anchor-escapement of clocks. He anticipated Newton's law of the inverse square in gravitation (1678). Hooke constructed the first Gregorian or reflecting telescope, with which he discovered the fifth star in Orion and inferred the rotation of Jupiter. He also materially improved or invented the compound microscope, the quadrant, a marine barometer, and the universal joint. After the Great Fire of London (1666) he was appointed city surveyor, and designed the new Bethlehem Hospital (Moorfields) and Montague House.

HOOKER, Sir Joseph Dalton

(1817–1911)

English botanist and traveller, second child of Sir **William Jackson Hooker**, born in Halesworth, Suffolk, and educated at Glasgow High School and Glasgow University, where he studied medicine. His first post was as assistant surgeon and naturalist on HMS *Erebus* in the Southern Ocean. Hooker published the botanical results as the six-volume *The Botany of the Antarctic Voyage*, comprising *Flora Antarctica* (1844–7), *Flora Novae-Zelandiae* (1853–5) and *Flora Tasmaniae* (1855–60). In 1846 Hooker was appointed to the Geological Survey and published much valuable work on palaeobotany until 1855, when he became assistant director (to his father) at Kew. Between 1848 and 1851 he explored Sikkim, Darjeeling, eastern Nepal and Assam; he introduced many species to cultivation, especially Rhododendrons, and wrote *Rhododendrons of Sikkim-Himalaya* (1849). With Thomas Thomson, he began a too-ambitious *Flora Indica* project which he did not complete; however, the masterly phytogeographical survey of India in the single volume which appeared (1855) remains unsurpassed. He completed a more concise *Flora of British India* (7 vols, 1872–97); this remains the standard Flora for the whole Indian subcontinent. He also wrote, with **Bentham**, *Genera Plantarum* (3 vols, 1862–83), the basis of a new classification system still used, with modifications, at Kew and elsewhere. On becoming Director of Kew in 1865, Hooker continued the expansion begun by his father, including the building of the Jodrell Laboratory. At **Charles Darwin**'s suggestion, he instigated the compilation (by **Benjamin Daydon Jackson**) of a list of all scientific names of flowering plants, *Index Kewensis* (1892), which continues to be compiled today. President of the Royal Society from 1872 to 1877, he was knighted in 1877 and received the Order of Merit in 1907.

HOOKER, Sir William Jackson

(1785–1865)

English botanist, born in Norwich. He was educated at Norwich Grammar School, then learned estate management at Starston Hall. After discovering a moss new to Britain, Hooker was introduced to Dawson Turner of Great Yarmouth, a cryptogamist and banker, who became his patron and later his father-in-law. After botanizing together in Scotland in 1806, Turner introduced Hooker to **Banks**, who arranged for him to make a collection in Iceland (lost on the return voyage). After Hooker married Maria Turner, he lived at Halesworth, Suffolk. There he started his herbarium, which was to become the largest in private hands; he also produced his first five botanical works. These dealt mostly with mosses but his *British Jungermanniae* (22 parts, 1812–16) established hepaticology (the study of liverworts) as an independent discipline. In 1820 he became Regius Professor of Botany at Glasgow; during his tenure, he built up the small botanic garden, wrote *Flora Scotica* (1821) and was an extremely popular lecturer. After Banks's death (1820) he campaigned to save the Kew collections for the nation and in 1841 became Kew's first director. He expanded it from 11 acres to about 300 acres, employing William Andrews Nesfield and Decimus Burton to landscape the grounds and build new glasshouses, including the famous Palm House (1848). While at Kew he published several still standard works on ferns, including *Genera*

Filicum (12 parts, 1838–42), *Species Filicum* (5 vols, 1846–64) and *Synopsis Filicum* (1865). He was knighted in 1836.

HOPE, Thomas Charles
(1766–1844)

Scottish chemist, born in Edinburgh. He studied medicine there, and later taught chemistry in Glasgow, returning to Edinburgh in 1795 where he assisted and then succeeded **Joseph Black**. Hope confirmed an earlier but neglected observation that water reaches its maximum density close to 4 degrees Celsius, an important factor in biology, climatology and physics. In 1793 he investigated a new mineral from Strontian, in Argyll, recognizing its similarity to lime and baryta and demonstrating the characteristic red flame colour of the new element it contained. At the same time investigations were carried out independently in Germany by **Klaproth**. The new metal was later isolated by **Humphry Davy** in 1808 and named strontium. Hope was probably the first person in Britain to revolutionize his teaching to take account of the discoveries of **Lavoisier**. His lectures were enormously popular—in 1825, for example, his class numbered 575—but his effectiveness as a teacher was diminished by his failure to provide any facilities for practical work. He died in Edinburgh.

HOPF, Heinz
(1894–1971)

German mathematician, born in Breslau (now Wrocław, Poland). After military service during World War I, he studied at Berlin and Göttingen, where he met the Russian topologist **Aleksandrov** with whom he wrote the influential *Topologie* (1935). In 1931 he became professor at Zürich. One of Europe's leading topologists, widely regarded as the successor to **Weyl**, he worked on many aspects of combinatorial topology and differential geometry, including homotopy theory and vector fields.

HOPKINS, Sir Frederick Gowland
(1861–1947)

English biochemist, born in Eastbourne. After being expelled from school for truancy, Hopkins learned discipline and chemistry at the pharmacy firm of Allen and Hanbury and under the famous forensic pathologist Sir Thomas Stevenson, before commencing a brilliant career at Guy's Hospital, where he received the University of London Gold Medal in chemistry and qualified in medicine. At Guy's he worked with **Garrod** on haem pigments, and following his publication on estimating uric acid in urine, he was invited to become the first lecturer in chemical biology in Cambridge (1897). He was appointed professor there in 1914 and served as Sir William Dunn Professor from 1921 until his retirement in 1943. Hopkins, with S W Cole, isolated tryptophan in 1901, and his pioneer study of its dietary importance (1906) led to his discovery of accessory food factors, now called vitamins. He also associated lactate production in muscle with muscle contraction (1907), and discovered glutathione (1921), which linked with his general interest in biological oxidation and his observation that some biological oxidations need not be enzyme-mediated. Hopkins was elected a Fellow of the Royal Society in 1905 and awarded its Royal (1918) and Copley (1926) medals; he is the only biochemist to have become the Royal Society's President (1931). Knighted in 1925, he shared with **Eijkman** the 1929 Nobel Prize for Physiology or Medicine.

HOPKINS, Harold Horace
(1918–)

British optical physicist, born in Leicester. He was educated at the universities of Leicester and London, and worked on optical design, first in industry, then at Imperial College in London. He held the basic patents on zoom lenses, and was one of the two originators of fibre-optic technology (the other being A C S Van Heel in Holland) which now allows optical signals to be transmitted over distances of many hundreds of kilometres in fine-drawn glass fibres. He invented the rod–lens system which is now used universally in medical endoscopes. His research interests include aberration theory, optical coherence and the diffraction theory of imaging. He was appointed Professor of Applied Optics in the University of Reading in 1967, elected FRS in 1973 and was Ives medallist of the Optical Society of America in 1978.

HOPPER, Grace Murray
(1906–92)

American computer programmer, born in New York City. She was educated at Vassar College, where she was awarded a PhD in 1934, and taught in the mathematics department between 1931 and 1944. She joined the WAVES (Women Accepted for Voluntary Emergency Service) during World War II, and stayed in the Naval Reserve for the rest of her career. Hopper was drafted by the navy to join **Aiken**'s team at Harvard as a coder for the Mark I with no guidance except the coding book. She gradually developed a set of built-in routines and was eventually able to use the machine to solve complex partial differential equations using only the 72 words of storage at her disposal. One of the few women to make a major impact on the history of computing, in 1949 she joined the Eckert & Mauchly Corporation as a senior mathematician. In 1951 she conceived of a new type of internal computer program that could perform floating-point operations and other tasks automatically. The program was called a compiler, and it was designed to scan a programmer's instructions and produce (compile) a roster of binary instructions that carried out the programmer's commands. Though the compiler and high-level language were not immediately successful, Hopper's ideas spread and were influential in setting standards for software developments, such as for COBOL. She retired from the navy as a Rear Admiral at the age of 80 when she joined the Digital Equipment Corporation as a senior consultant.

HOPPE-SEYLER, (Ernst) Felix (Immanuel) (1825–95)

German physiological chemist, born in Freiburg. He studied medicine in Berlin, where he graduated MD in 1850, and became assistant under **Virchow** at the Pathological Institute, Berlin (1856–64). He was appointed Professor of Applied Chemistry at Tübingen (1861–72) and Professor of Physical Chemistry at Strassburg (1872–95). A pioneer in the application of chemical methods to understand physiological processes, Hoppe-Seyler discovered and named haemoglobin (1862), and showed that the haemoglobin in red blood cells binds oxygen which is subsequently delivered to the tissues. He also determined its absorption spectrum in blood and in combination with carbon monoxide. Like **van't Hoff**, he extended the pioneering studies by **Schönbein** (1840) on oxygen activation as the first step of oxidation processes and proposed the catalytic formation of hydrogen peroxide from water as the first active product. Hoppe-Seyler also investigated the chemical composition and functions of chlorophyll as well as the chemistry of putrefaction. In 1877 he founded the first biochemical journal, *Zeitschrift für physiologische chemie*.

HOPWOOD, David Alan
(1933–)

British geneticist, born in Kinver, Staffordshire. He was educated in Cheshire and at St John's College, Cambridge, where he received his PhD. After some years as a university demonstrator and Fellow of Magdalene and St John's colleges in Cambridge, he moved to a lectureship at the University of Glasgow in 1961. In 1968 he became the first John Innes Professor of Genetics at the University of East Anglia and head of the genetics department at the John Innes Institute. Elected FRS in 1979, he is a member of scientific academies of India, Hungary and China, and has honorary fellowships from the Federal Institute of Technology (ETH), Zürich, and the University of Manchester Institute of Science and Technology. He received a DSc from the University of Glasgow. Hopwood's major contribution has been in the understanding of the genetics of bacteria belonging to the genus *Streptomyces*; these organisms produce the great majority of the antibiotics used in human and veterinary medicine and in agriculture. His work has led to the ability to manipulate genes for antibiotic production—this has stimulated the development of novel methods to improve antibiotic production and to generate new antibiotics by genetic engineering.

HORROCKS, Jeremiah
(1619–41)

English astronomer, born in Toxteth, Liverpool. In 1632 he entered Emmanuel College, Cambridge, and in 1639 became Curate of Hoole, Lancashire, where he made the first observation of the transit of Venus (24 November 1639 according to the Julian calendar), deduced the solar parallax, corrected the solar diameter, and made tidal observations. Horrocks was an enthusiastic admirer of **Kepler** and made considerable improvements to the equation of motion of the Moon, noticing that the line of apsides was librating and that the eccentricity was changing. Both these were attributed to the effect of the solar gravitational field. He also noticed irregularities in the motion of Jupiter and Saturn. These have now been shown to be due to their mutual gravitational attraction. Erroneously Horrocks believed that comets were blown out of the Sun, their velocities decreasing as they receded, but increasing again when they started to fall back. He was the first person to undertake a continuous series of tidal observations, hoping eventually to understand the underlying causes of these variations. His 19 March 1637 observations of the lunar occultation of the stars in the Pleiades indicated that they disappeared instantaneously. He concluded that the stars had apparent diameters that were negligible and thus not capable of measurement.

HORSLEY, Sir Victor Alexander Haden
(1857–1916)

English physiologist and surgeon. Born in Kensington of a talented family—his father was a genre painter and his grandfather an organist and composer—Horsley studied medicine at University College London, becoming Professor of Physiology there in 1893. Elected FRS in 1886, he also served as Fullerian Professor at the Royal Institution (1891–3). He developed a large private practice in surgery, and was appointed to the surgical staff at the National Hospital for the Paralysed and Epileptic in Queen's Square. Knighted in 1902, he volunteered for military service at the outbreak of World War I in 1914 and died of heat exhaustion in Mesopotamia. Horsley accomplished distinguished work on the localization of brain function. Studies of cretinism enabled him to advance thyroid research, and he was one of the first to attempt surgery on a pituitary tumour. He was an early pioneer of brain surgey. Drawing on the work of Gustav Theodor Fritsch and **Hitzig** on brain localization and cerebral cortical function, he pursued research into neurological conditions through making experimental lesions in the deep parts of the brain.

HOUGHTON, Sir John Theordore
(1931–)

British physicist and meteorologist, born in Dyserth Clwyd, Wales. After graduating from Oxford University (1951) and obtaining his DPhil in 1955, he became Research Fellow at the Royal Aircraft Establishment, Farnborough (1955–8), Reader (1962) then Professor (1976) at the Department of Meteorology at the University of Oxford, and Director of the Appleton Laboratory (1979). He became Director-General of the Meteorological Office in 1983, and later its Chief Executive (1990). His important contributions to meteorology include the design (with Desmond Smith) of the selective chopper radiometer which can assess the temperature structure of the atmosphere up to 50 kilometres when flown on meteorological satellites. Further research led to the pressure modulator radiometer (1975) and stratospheric and mesospheric sounder (1978), which measure the temperature structure and distribution of some minor constituents of the atmosphere up to 90 kilometres. These instruments obtained, for the first time, global information on the structure of the stratosphere and mesosphere, and have led to greatly increased knowledge of the chemistry, radiation properties and dynamics of the whole atmosphere. Houghton was also involved in the instrumentation of NASA's first Venus orbiter, Pioneer 12 (1978). In 1990 he acted as Chairman of the Scientific Assessment Working Group of the Intergovernmental Panel on Climate Change which had a substantial influence on government policies worldwide. He was elected FRS (1972), President of the Royal Meteorological Society (1976–8), Chairman of the Joint Scientific Committee for the World Climate Research Programme (1981–4) and Symons Gold Medallist (1991). He was awarded the CBE (1983) and knighted in 1991.

HOUNSFIELD, Sir Godfrey Newbold
(1919–)

English electrical engineer, born in Newark. He studied in London at the City and Guilds College and Faraday House College, worked as an instructor in electronic and radar communications in the RAF during World War II, joined Thorn/EMI in 1951 and became head of medical systems research there in 1972. He led the team which (independently of **Cormack**) developed the technique of computer-assisted tomography (CAT scanning), which produces detailed X-ray pictures of the human body, including the soft tissues which are normally almost transparent to X-rays. The images are built up by computer from large numbers of measurements of the absorption of X-rays transmitted in different directions through the body. The EMI scanner system which resulted from this research represented a major breakthrough in the non-invasive diagnosis of disease, and won the MacRobert Award in 1972. In 1978 Hounsfield was appointed Professorial Fellow in Imaging Sciences at the University of Manchester, and he shared the 1979 Nobel Prize for Physiology or Medicine with Cormack. He was knighted in 1981.

HOUSSAY, Bernardo Alberto
(1887–1971)

Argentine physiologist, born in Buenos Aires. A precocious child, Houssay graduated from the School of Pharmacy of the University of Buenos Aires in 1901 aged 17. He went on to study medicine while working as a hospital pharmacist and holding teaching appointments. From 1909 he was Professor of Physiology at the veterinary school, and after graduating in medicine in 1911 he took on additional duties, including a private practice and the directorship of a municipal hospital service. From 1919 he was Professor of Physiology at the medical school, until dismissed by the military dictatorship of 1943 for allegedly being too pro-American. He was reinstated under a general amnesty in 1945, but was dismissed again after Juan Perón became president in 1946. He continued his research privately until reinstated in 1955. From his student days, Houssay was interested in the pituitary gland and later studied interactions between the pituitary gland and insulin. He showed that its anterior lobe produces a hormone with the opposite effect to insulin, and that removing the pituitary from a diabetic animal reduces the severity of the diabetes. This work produced fundamental insights into the working of the endocrine system and led Houssay and his disciples on to studies of a variety of feedback mechanisms between the endocrine glands. He shared the 1947 Nobel Prize for Physiology or Medicine with **Carl** and **Gerty Cori**.

HOYLE, Sir Fred
(1915–)

English astronomer and mathematician, born in Bingley, Yorkshire. Educated at Bingley Grammar School and Emmanuel College, Cambridge, he taught mathematics at Cambridge (1945–58), was Plumian Professor of Astronomy and Experimental Philosophy there (1958–72) and Professor-at-Large at Cornell University (1972–8). In 1948, with **Bondi** and **Gold**, he propounded the 'steady state' theory of the universe, which proposes that the universe is uniform in space and unchanging in time, its expansion being fuelled by constant spontaneous creation of matter. The theory was later to be displaced through new evidence, but it was a valuable contribution to cosmology. He also suggested the currently accepted scenario of the build-up to supernova explosion in stars, in which the nuclear reactions fuelled by hydrogen and helium in stars are followed by the production and nuclear 'burning' of heavier elements. A chain of reactions then leads to supernova explosions, in which the matter is ejected into space, and recycled in second-generation stars which form from the remnants. He has been a successful writer on popular science and science fiction; his books include *Nature of the Universe* (1952), *Frontiers of Astronomy* (1955) and *The Black Cloud* (1957). He was elected FRS in 1957 and knighted in 1972.

HUBBLE, Edwin Powell
(1889–1953)

American astronomer, born in Marshfield, Missouri. After studying law at the University of Chicago and as a Rhodes scholar at Oxford (1910–13), he returned to the USA but soon abandoned law to take up the study of astronomy. He held a research position at the Yerkes Observatory (1914–17), and following military service during World War I, moved in 1919 to the Carnegie Institution's Mount Wilson Observatory where he began his fundamental investigations of the realm of the nebulae. Using Cepheid variable stars as distance indicators, he succeeded in determining the distance to the Andromeda nebula (1923). He found that spiral nebulae are independent stellar systems, and that the Andromeda nebula in particular is very similar to our own Milky Way galaxy. In 1929 he announced his discovery that galaxies recede from us with speeds which increase with their distance. This was the phenomenon of the expansion of the universe, the observational basis of modern cosmology. The linear relation between speed of recession and distance is known as Hubble's law. Hubble remained on the staff of Mount Wilson until his death. The 2.4 metre aperture Hubble Space Telescope launched in 1990 was named in his honour.

HUBEL, David Hunter
(1926–)

Canadian-born American neurophysiologist, born in Windsor, Ontario. He studied medicine at McGill University in Montreal, and after a neurology residency at the Montreal Neurological Institute, was drafted into the US army. He worked from 1955 to 1958 at the Walter Reed Army Research Institute, examining the electrical activity of the brain, using cats, over a long period of time. In 1958 he joined **Kuffler**'s laboratory at Johns Hopkins University, and followed him to Harvard the following year (Assistant Professor 1960–2; Associate Professor 1962–5; Professor of Neurophysiology 1965–77; Professor of Physiology and Chemistry 1967–8; Professor of Neurobiology 1968–82). Since 1982 he has been John Franklin Enders University Professor. With **Wiesel** at Harvard Medical School, he investigated the mechanisms of visual perception at the cortical level, and they shared the 1981 Nobel Prize for Physiology or Medicine with **Sperry**. Hubel and Wiesel's work followed that of **Hartline**, **Granit** and Kuffler, who had all contributed to understanding component parts of the visual process, especially those occurring at the level of the retina. Working on anaesthetized animals Hubel and Wiesel implanted electrodes into the brain, into the region known to be involved in visual processing, then analysed individual cell responses to different types of visual stimulation. Their laborious correlation of the anatomy and physiology of the visual cortex provided a complex picture of increasing levels of sophisticated analysis of visual information by brain cells. Subsequent studies of the development of these capabilities in young animals altered routine practice in ophthalmology, as they suggested that an eye defect in a child should be corrected immediately and not left in the hope that it will correct automatically.

HUBER, Robert
(1937–)

German biophysicist. He received his early training in Munich, and remained there as a lecturer (1968–76) and associate professor from 1976. Since 1972 he has been Director of the Max Planck Institute for Biochemistry in Martinsried. A specialist in the high-resolution X-ray crystallography of biological macromolecules, Huber compared the crystal structures of the active and inactive forms of the enzyme phosphorylase, discovered by **Carl Cori**, and found them to be very similar, indicating that activation is not associated with a marked structural change (1972). In 1974 he helped elucidate the antibody structure proposed by **Rodney Porter** by determining the structure of the 'light chain' (Bence Jones) component, including the molecular arrangement of the antigen binding site. In 1978 he followed his earlier investigations (from 1972) of the digestive protease trypsin and the naturally occurring pancreatic trypsin inhibitor with a high-resolution analysis of the interaction between the two. Part of this work was carried out with his colleague at the Max Planck Institute, **Deisenhofer**. Since 1979 he has studied the unusual selenium-containing enzyme glutathione peroxidase, and contributed to understanding the energetics of protein–DNA interactions. He has also collaborated with **Michel** and Deisenhofer (from 1982) to determine the structure of the membrane-bound photosynthetic reaction centre of the purple bacterium *Rhodopseudomonas viridis*, work for which they shared the 1988 Nobel Prize for Chemistry.

HÜCKEL, Erich
(1896–1980)

German physicist and theoretical chemist, born in Berlin-Charlottenburg. He studied physics and mathematics at the University of Göttingen from 1914 to 1921. After gaining his doctorate he was briefly an assistant at Göttingen before becoming assistant to **Debye** in Zürich. In 1928–9 he spent some time in England and Denmark, and then became *Privatdozent* and instructor in physics at the Technische Hochschule, Stuttgart. From 1935 to 1937 he was Extraordinary Professor at Phillips University, Marburg, and from 1937 to 1961 he held the Ordinary Chair of Theoretical Physics there. Hückel's research interests were almost entirely in chemical physics. He was associated with Debye in treating the behaviour of strong electrolytes in terms of interionic forces (resulting in the Debye–Hückel theory, 1923). Through spending a period with **Niels Bohr** in Copenhagen, his interests moved to the quantum-mechanical treatment of organic molecules, especially those containing double bonds. In 1931 he formulated criteria for 'aromaticity' of cyclic unsaturated compounds (Hückel rule), and in 1937 he developed a procedure for calculating electron distribution and other characteristics of unsaturated compounds, Hückel molecular orbital (HMO) theory, which still finds application today.

HUDSON, William
(1734–93)

English botanist and apothecary, born and educated in Kendal. He was apprenticed to a London apothecary and received as the prize for botany a copy of **Ray**'s *Synopsis*. From 1757 to 1758 he was sub-librarian at the British Museum and his studies of the Sloane Herbarium formed the basis of his adaptation of Linnaean nomenclature to the plants that Ray had described—work which was much more accurate than that of Sir John Hill's *Flora Britannica* of 1760. In the early 1760s, Hudson practised as an apothecary in Haymarket, London. During this period, the first edition of his *Flora Anglica* was published in 1762. This was the first British botanical work to adopt the Linnaean classification system and its binomial nomenclature, and contained much original work. Hudson was later *praefector horti* to the Apothecaries' Company at Chelsea Physic Garden (1765–7). An enlarged second edition of *Flora Anglica* was published in 1778. He also studied insects and molluscs. His insect collection, and most of his herbarium, were destroyed in a fire at his Panton Street home in 1783. Thereafter he moved to Jermyn Street, and died there of paralysis.

HUGGINS, Charles Brenton
(1901–)

Canadian-born American surgeon, born in Halifax, Nova Scotia. He worked at Chicago University from 1927, where he became Professor of Surgery in 1936 and was head of the Ben May Laboratory for Cancer Research from 1951 to 1969. He has been a pioneer in the investigation of the physiology and biochemistry of the male urogenital tract, including the prostate gland. Research on dogs (a species which also can develop benign and malignant tumours of the prostate) led him to the possibility of using hormones in treating such tumours in human beings. He also worked on the use of hormones in treating breast cancer in women. He shared the 1966 Nobel Prize for Physiology or Medicine with **Rous** for their cancer research, in Huggins's case, his discovery of hormonal treatment for cancer of the prostate gland.

HUGGINS, Sir William
(1824–1910)

English astronomer, born in London, the only child of a mercer, and educated by private tutors. He went into the mercery business, while in his spare time studying the sciences which were his main interest. On finding he had sufficient means to devote himself entirely to astronomy he erected an observatory (1856)—later enhanced by instruments on loan from the Royal Society (1871)—in the garden of his house at Tulse Hill in the suburbs of London where he lived and worked for the rest of his life. In collaboration with William Allen Miller (1817–70), Professor of Chemistry at King's College, London, Huggins commenced observations of spectra of stars (1864). Their major discovery (published in 1866) that certain nebulae exhibit spectra similar to those of gases in the laboratory revealed that these objects were composed of luminous gases. Huggins was the first to observe the spectrum of a nova or new star (1868) which also showed the characteristics of a gas. His pioneering work included spectrosopy of comets, and observations of the **Doppler** shift in the spectra of stars as a means of measuring their radial motion. He and his wife and co-worker Margaret Lindsay Murray (1848–1915) were the first to make serious use of dry plate photography in astronomy (1876). Huggins twice received the Gold Medal of the Royal Astronomical Society (1867, 1885), and was awarded the Royal (1866), Rumford (1880) and Copley (1898) medals of the Royal Society. He was knighted in 1897 on the occasion of the diamond jubilee of Queen Victoria, and was among the 12 first recipients of the Order of Merit when it was instituted in 1902.

HUGHES, John
(1942–)

British pharmacologist, who took undergraduate and postgraduate degrees at the University of London, before postdoctoral research at Yale (1967–9). He was a lecturer at Aberdeen University (1969–77), and Deputy Director of its Drug Research Unit (1973–7), before becoming Reader (1977–9) and Professor (1979–82) in Pharmacological Biochemistry, Imperial College, London. In 1983 he became Director of the Parke–Davis Research Unit, Cambridge and Senior Research Fellow at Wolfson College, and from 1989 has been Honorary Professor of Neuropharmacology at Cambridge. Whilst at Aberdeen, he discovered with **Kosterlitz** the naturally existing opiate-like chemicals, enkephalins. His work since has concentrated on understanding the pharmacology and biochemistry of a wide range of naturally occuring neuroactive compounds.

HUISGEN, Rolf
(1920–)

German chemist, born in Gerolstein. After studying chemistry at the universities of Bonn and Munich he worked for a PhD (1943) with **Wieland** at Munich, where he was appointed *Dozent* in 1947. From 1949 to 1952 he was associate professor at the University of Tübingen before becoming full professor at the University of Munich. He was appointed Professor Emeritus in 1988. Huisgen has worked on a wide range of topics, but is best known for his work on cycloaddition reactions, which have proved of great value in the design of organic syntheses, and on the synthesis of heterocyclic compounds. He has received many honours including the Liebig Medal of the Gesellschaft Deutscher Chemiker (1961), the Bavarian Order of Merit (1982), and a foreign associateship of the US National Academy of Sciences (1989).

HULSE, Russell

American physicist. He received his PhD in astronomy at the University of Massachusetts in Amherst, and is now principal research physicist at Princeton University's Plasma Physics Laboratory. During a systematic search for pulsars, the rapidly rotating dense stars which appear on Earth to emit regular pulses of radio waves, he discovered with **Joseph Taylor** one interesting candidate whose pulse frequency changed periodically. The characteristics of these changes revealed that this exotic object was a pulsar in orbit of another dense neutron star. For the discovery of the first 'binary pulsar', Hulse was awarded the 1993 Nobel Prize for Physics jointly with Taylor.

HULST, Hendrik Christoffell van de
See **VAN DE HULST, Hendrik Christoffell**

HUMASON, Milton Lasell
(1891–1972)

American astronomer, born in Dodge Center, Minnesota. He is best known for his association with **Hubble** in work on the recession of the galaxies. Employed at Mount Wilson Observatory first as a janitor in 1917, he advanced to the rank of Assistant Astronomer and in 1954 to that of Astronomer at the Mount Wilson and Palomar observatories. His skilful observations of the radial velocities of hundreds of faint galaxies, begun in 1930, made with the 100 inch telescope on Mount Wilson and later with the 200 inch

Hale telescope on Mount Palomar, provided important material for the extension of Hubble's data. He was awarded the honorary degree of Doctor of Philosophy by the University of Lund in Sweden in 1950. He retired in 1957.

HUMBOLDT, (Friedrich Heinrich) Alexander, Baron von
(1769–1859)

German naturalist and traveller, born in Berlin, brother of statesman Karl Wilhelm von Humboldt (1767–1835). He studied at Frankfurt-an-der-Oder, Berlin, Göttingen, and under **Abraham Werner** in the Mining Academy at Freiberg, where he published *Flora Subterranea Fribergensis* (1793). He then held a post in the mining department in Upper Franconia, and produced a work on muscular irritability (1799). For five years (1799–1804) he and **Bonpland** explored unknown territory in South America, which led to his monumental *Voyage de Humboldt et Bonpland aux Régions Équinoxiales* (23 vols, 1805–34). In Paris he carried out experiments on the chemical constituents of the atmosphere with **Gay-Lussac** and in 1807, after a visit to Italy, he returned to Paris with Prince Wilhelm of Prussia on a political mission, and remained in France until 1827. In 1829 he explored central Asia with **Ehrenberg** and the mineralogist Gustav Rose (1798–1873), and their examination of the strata which produce gold and platinum, magnetic observations, and the geological and botanical collections are described in a work by Rose (1837–42) and in Humboldt's *Asie Centrale* (1843). The political changes of the year 1830 led to his employment in political services and during the ensuing 12 years he was frequently in Paris, where he published his *Géographie du nouveau continent* (1835–8). His work of popular science, *Kosmos* (1845–62), endeavoured to provide a comprehensive physical picture of the universe.

HUNTER, John
(1728–93)

Scottish physiologist and surgeon, born in Long Calderwood, East Kilbride. After working as a Glasgow cabinet-maker, he came to London, assisting between 1748 and 1759 at the anatomy school run by his elder brother, William Hunter, where he learnt his dissecting skills. He then studied surgery at St George's and St Bartholomew's hospitals. In 1760 he entered the army as Staff Surgeon; service on the Belleisle and Portugal expedition provided the basis for his later expertise in gunshot wounds. In 1768 he became surgeon at St George's, and in 1776 was appointed Surgeon-Extraordinary to George III. In 1790 he was made Surgeon-General to the army. An indefatigable biological and physiological researcher, he built up huge collections of specimens to illustrate the processes of plant and animal life and elucidate comparative anatomy. His museum grew to contain an astonishing 13 600 preparations; on his death, it was bought by the government and subsequently administered by the Royal College of Surgeons in London. In the field of human pathology, Hunter investigated a wide range of subjects, from venereal disease and embryology to blood and inflammation. He developed new methods of treating aneurysm, and was the first to apply pressure methods to the main trunk blood arteries. He also succeeded in grafting animal tissues. His *Natural History of Human Teeth* (1771–8) gave dentistry a scientific foundation. His biological studies included work on the habits of bees and silkworms, on hibernation, egg incubation and the electrical discharges of fish. Hunter trained many of the leading doctors and natural historians of the next generation, including **Jenner**. He was buried in the church of St Martin-in-the-Fields. He has widely been dubbed the founder of scientific surgery.

HURD, Cuthbert Corwin
(1911–)

American computer company executive and scientist, born in Estherville, Ohio. Educated at Drake University and Iowa State College, he was awarded a PhD at the University of Illinois in 1936. During World War II he worked for the Atomic Energy Commission at Oak Ridge, where he was involved in large-scale computations that led him to consider ways to optimize the use of existing calculating machines. In 1949 he joined IBM and by 1951 he was closely involved with development work on the IBM 701 (the Defense Calculator). Hurd was amongst those IBM executives who saw this computer—partly built with funding from a US government in the throes of the Korean War—as IBM's chance to enter the field of electronic digital computers. When it was delivered to Los Alamos in 1953 it was therefore a major event and a harbinger of IBM's future dominance in this field. By the following year, Hurd's department had delivered the IBM 650: inexpensive, practical, reliable and mass produced, it was a runaway success. Hurd was later appointed Director of Control Systems at IBM (1961–2).

HUSKEY, Harry Douglas
(1916–)

American computer scientist, born in Whittier, North Carolina. Educated at the University of Idaho and Ohio State University, where he received his PhD in 1943, Huskey became involved with **Eckert** and **Mauchly** on the ENIAC computer in Philadelphia around half way through the building project. He worked on input and output devices for the machine. When the ENIAC team started to break up, he accepted an invitation from the UK to join the ACE project at the National Physical Laboratory under **Turing**. Huskey was instrumental in initiating the construction of a test model, much to the annoyance of Turing, who felt that his leadership of the ACE project was thereby undermined. On returning to the USA in 1948, he joined the National Bureau of Standards, which was pursuing two different lines of development: the serial SEAC (Standards' Eastern Automatic Computer) built in Washington DC, and the parallel SWAC (Standards' Western Automatic Computer) built in Los Angeles. After working on the SWAC, in 1953 Huskey designed a 'minicomputer' which was built by the Bendix Corporation as their G15. The speed of this computer, which owed something to Turing's ACE design, made it a favourite for certain classes of engineering problem.

HUTTON, James
(1726–97)

Scottish geologist, born in Edinburgh. After a short period as lawyer's apprentice he turned to medicine which he studied in Edinburgh before going to the Continent to complete his professional training in Paris and Leiden. He received his doctorate at Leiden in 1749. He never practised medicine, and instead went to Norfolk in 1752 to devote his time to agriculture; it was there that he became interested in geology during his many walks. In 1754 he moved back to his estate in Berwickshire, and continued in rural pusuits for 14 years. From 1765 he was also a partner in a commercially successful sal ammoniac manufacture. He returned to Edinburgh in 1768 and joined an active intellectual group which included **Joseph Black**, Adam Ferguson, Sir **James Hall** and **John Playfair**. Hutton developed his theories about the Earth over a number of years as a result of many journeys into Scotland, England and Wales and finally presented his ideas before the Royal Society of Edinburgh in *A Theory of the Earth* (1785; expanded, vols i and ii 1795, vol iii 1799). In this he demonstrated that the internal heat of the Earth caused intrusions of molten rock into the crust, and that granite was the product of the cooling of molten rock and not the earliest chemical precipitate of the primeval ocean as advocated by **Abraham Werner** and others. This 'Plutonist' versus 'Neptunist' debate raged on for a considerable time after Hutton's death in spite of the admirable supportive studies of Hutton's friends Hall and Playfair (described in Playfair's *Illustrations of the Huttonian Theory of the Earth* of 1802). Hutton's system of the Earth recognized that most rocks were detrital in origin having been produced by erosion from the continents, deposited on the seafloor, lithified by heat from below and then uplifted to form new continents. The cyclicity of such processes led him to envisage an Earth with 'no vestige of a beginning and no prospect of an end'. These uniformitarian ideas attracted strong opposition from **De Luc**, **Buckland** and associates from the 'English School' of geology, and from **Kirwan**. Nevertheless, Hutton's ideas held firm to form the basis of modern geology.

HUXLEY, Sir Andrew Fielding
(1917–)

English physiologist, grandson of **T H Huxley** and half-brother of Aldous and **Julian Huxley**, born in London. He studied natural sciences at Trinity College and at the Physiological Laboratory, Cambridge. He graduated in 1938 and began neurophysiological research with his tutor, **Alan Hodgkin**, but during World War II (1940–5) he worked on operational anti-aircraft research. He became a Fellow (1941–60) and Assistant Director of Research in Physiology (1952–60) at Trinity College, Cambridge. In the Physiological Laboratory he was demonstrator (1946–50), Assistant Director of Research (1951–9) and Reader in Experimental Biophysics (1959–60). He moved to University College London, as Joddrell Professor of Physiology (1960–9) and as Royal Society Research Professor (1969–83), before returning to Cambridge as Master of Trinity College (1984–9). With Hodgkin he provided a physico-chemical explanation for the conduction of impulses in nerve fibres, using the giant axon preparation of the squid, described by **John Young**. In 1950 he changed direction, to muscle physiology, and devised a special interference microscope with which to study the contraction and relaxation of muscle fibres. From his observations and measurements of different components of the cell fibre, he postulated a 'sliding filament' theory to account for the functional and morphological changes that occur during muscular contraction. The theory suggested that different sections of the muscle fibre overlapped, thus shortening the overall length of the fibre and generating force in the direction of the shortening. Huxley served as President of the Royal Society (1980–5) and was awarded the Order of Merit in 1983. In 1963 he shared the Nobel Prize for Physiology or Medicine with Hodgkin and **Eccles**.

HUXLEY, Hugh Esmor
(1924–)

English biophysicist, born in Birkenhead. He studied natural sciences at Christ's College, Cambridge, graduating in 1943. During the remainder of World War II he worked on radar research and returned to Cambridge in 1948, where he joined the Molecular Biology Unit of the Medical Research Council (MRC) and obtained his PhD in 1952. He was awarded a Commonwealth Fund Scholarship to study biophysics at MIT (1952–4), was then appointed to the Biophysics Department of King's College, London, and returned to Cambridge in 1961, to the MRC Laboratory of Molecular Biology, of which he became Deputy Director in 1977. From the 1950s he was a central figure in developing the sliding filament model of muscle contraction, in which interdigitating muscle filaments slide past each other to produce contraction. Huxley's electron microscopic studies defined the detailed structure of the muscle cell, and with X-ray diffraction techniques he analysed the cellular location of the major proteins, actin and myosin, involved in the contractile process. Simultaneously **Andrew Huxley** (no relation) came to similar conclusions. Hugh Huxley has been Professor of Biology at Brandeis University since 1987.

HUXLEY, Sir Julian Sorell
(1887–1975)

English biologist and humanist, grandson of **T H Huxley**, brother of novelist Aldous Huxley and half-brother of **Andrew Huxley**. Educated at Eton and Balliol College, Oxford, where he won the Newdigate Prize (1908), he was professor at the Rice Institute, Texas (1913–16), and after World War I became Professor of Zoology at King's College, London (1925–7), Fullerian Professor at the Royal Institution (1926–9), and Secretary to the Zoological Society of London (1935–42). He extended the application of his scientific knowledge to political and social problems, formulating a pragmatic ethical theory of 'evolutionary humanism', based on the principle of natural selection. This lead to his enthusiastic, although later regretted, welcome for the ideas of the Jesuit mystic and palaeontologist, **Teilhard de Chardin**. He was the first Director-General of UNESCO (1946–8). His influence was based on his capacity for synthesis stimulation, inspiring the work of biologists such as **Elton** and **Ford**, rather than his own fairly meagre scientific accomplishments in animal behaviour and experimental embryology. His *Evolution: the Modern Synthesis* (1942) was undoubtedly the key work in the acceptance of the neo-Darwinian synthesis of evolution, although it was chiefly a presentation of the ideas of **Fisher**, **Dobzhansky**, **Mayr**, **George Gaylord Simpson**, **J B S Haldane** and others. His writings also included *Essays of a Biologist* (1923), *Religion without Revelation* (1927), *Animal Biology* (with Haldane, 1927), *The Science of Life* (with H G Wells, 1931), *Problems of Relative Growth* (1932), *Evolutionary Ethics* (1943), *Biological Aspects of Cancer* (1957) and *Towards a New Humanism* (1957). Huxley's life is described in his two-volume autobiography *Memories* (1970, 1973), and in *Evolutionary Studies* (1989).

HUXLEY, T(homas) H(enry)
(1825–95)

English biologist, born in Ealing, Middlesex, the son of a schoolmaster. He studied medicine at Charing Cross Hospital and entered the Royal Navy medical service. As assistant surgeon on HMS *Rattlesnake* on a surveying expedition to the South Seas (1846–50), he collected and studied specimens of marine animals, particularly plankton. From 1854 to 1885 he was Professor of Natural History at the Royal School of Mines (later the Normal School of Science) in London, and made significant contributions to palaeontology and comparative anatomy, including studies of dinosaurs, coelenterates, and the relationship between birds and reptiles. He was best known as the foremost scientific supporter of **Charles Darwin**'s theory of evolution by natural selection, writing *Zoological Evidence as to Man's Place in Nature* (1863). Known as 'Darwin's bulldog', he had a reputation as a forceful and witty writer and debater. In a celebrated debate with Bishop Samuel Wilberforce at the British Association meeting in Oxford (1860), he declared that he would rather be descended from an ape than a man who used his great gifts to obscure the truth (implying the bishop). As an educator he was a member of the London Schools Board, establishing college laboratory courses and influencing the teaching of biology and science. Later he turned to theology and philosophy, and coined the term 'agnostic' for his views. He believed strongly in the critical attitude in science, and rejected the claims of Social Darwinists that human affairs should be guided by the 'survival of the fittest'. He wrote *Lay Sermons* (1870), *Science and Culture* (1881), *Evolution and Ethics* (1893) and *Science and Education* (published posthumously in 1899). His *Collected Essays* were published in nine volumes (1893–4), and his letters were collated by his writer son Leonard Huxley in *The Life and Letters of Thomas Henry Huxley* (1900). He founded an intellectual lineage which also included his grandsons Aldous, the novelist, and **Julian Huxley**, the biologist.

HUYGENS, Christiaan

(1629–93)

Dutch physicist, born in the Hague, the second son of the poet and statesman Constantyn Huygens (1596–1687), who was secretary to the Prince of Orange, and at the centre of Dutch cultural life. Huygens studied at Leiden and Breda. He was equally at home in mathematics and in the practical aspects of instrument-making. His mathematical *Theoremata* was published in 1651. Subsequently he joined forces with a Dutch clock-maker to construct the pendulum clock based on **Galileo**'s suggestion (1657), and developed the latter's doctrine of accelerated motion under gravity. His *Horologium Oscillatorium*, published in 1673, deals with his researches on clocks, mechanics and geometry. In 1655 Huygens discovered the rings and fourth satellite of Saturn by means of a refracting telescope he constructed with his brother. In 1663 he visited England, where he was elected FRS. He discovered the laws of collision of elastic bodies at the same time as **John Wallis** and Sir Christopher Wren, and improved the air-pump. In optics he first propounded the undulatory theory of light, and discovered polarization. The 'principle of Huygens' forms a part of the wave theory. Huygens later lived in Paris (1666–81), a member of the Royal Academy of Sciences, but as a Protestant he encountered religious hostility there, and felt it prudent to return to the Hague. He was, after **Isaac Newton**, the greatest scientist of the second half of the 17th century.

HYMAN, Libbie Henrietta

(1888–1969)

American zoologist, born in Des Moines, Iowa. She studied zoology at the University of Chicago (1906–10) and remained there as research assistant until 1931. Dissatisfied with the practical texts then available, she wrote *A Laboratory Manual for Elementary Zoology* (1919) and *A Laboratory Manual for Comparative Vertebrate Anatomy* (1929). Both these texts, in the form of later editions, are still in use and their great success made Hyman financially independent. She resigned her position at Chicago, and after a period of travel in Europe, took up residence near the American Museum of Natural History in New York. She carried out research on many aspects of the biology of the lower invertebrates and became a research associate at the museum in 1937. Until her death she worked on her *magnum opus*, a series of comprehensive volumes on the invertebrates (*The Invertebrates*, 6 vols, 1940–68). These dealt with the protozoa, coelenterates, flatworms, nematodes, echinoderma and molluscs. Such single-author works, while common in the 18th and 19th centuries, are now almost impossible due to specialization and the growth of knowledge and Hyman's books may well represent one of the last examples of this type of scholarship.

HYPATIA

(c.370–415 AD)

Greek philosopher, the first notable female astronomer and mathematician, who taught in Alexandria and became head of the neoplatonist school there. She was the daughter of Theon, a writer and commentator on mathematics, with whom she collaborated, and was herself the author of commentaries on mathematics and astronomy though none of these survives. Hypatia was renowned for her beauty, eloquence and learning, and drew pupils from all parts of the Greek world, Christian as well as pagan. Cyril, archbishop of Alexandria, came to resent her influence, and she was brutally murdered by a Christian mob he may have incited to riot.

HYRTL, Joseph

(1810–94)

Austrian anatomist, son of a musician. Hyrtl studied medicine at Vienna and became Professor of Anatomy at Prague in 1837, before returning to Vienna as professor in 1845. An irascible man, he was involved in many disputes, notably with Ernst Wilhem von Brücke, and he grew embittered in later years. Nevertheless he was an outstanding teacher of anatomy, and author of a highly influential *Handbook of Topographical Anatomy* (1845). His techniques of staining and making preparations were models for the time. His chief researches were on the anatomy of the mammalian ear (from mice and elephants) and the comparative anatomy of fish.

I

INGEN-HOUSZ, Jan

(1730–99)

Dutch chemist and biologist who settled in Britain and conducted most of his important scientific work there. He was born in Breda, and qualified as a doctor after attending the universities of Louvain, Belgium and Leiden. Around 1765 he moved to England where he studied the methods of smallpox inoculation pioneered by **William Watson** and Thomas Dimsdale. Subsequently he inoculated several members of the Austrian imperial family and in 1772 was appointed personal physician to the Empress Maria Theresa. After further travel on the Continent he settled permanently in England, and died in Bowood, near Calne, Wiltshire. Ingen-Housz's early scientific work was on electricity. He was the first person to generate static electricity by pressing a leather pad against a revolving disc, rather than a cylinder or sphere. He also devised apparatus for comparing the different heat conductivities of different metals. He is most famous, however, for his work on photosynthesis. In 1771 **Priestley** noticed that green plants give off oxygen, but Ingen-Housz showed in 1779 that this only takes place in the presence of light and that the amount of oxygen generated is proportional to the intensity of the light. By 1796 he had also discovered that all parts of plants give off carbon dioxide both in darkness and light.

INGOLD, Sir Christopher Kelk

(1893–1970)

English chemist, born in London. After taking a London University degree in physics and chemistry at Hartley University College, Southampton (now University of Southampton), he proceeded to Imperial College, London, for a PhD in chemistry under Jocelyn Field Thorpe. From 1918 to 1920 he worked at Cassel Cyanamide Company, Glasgow, but returned to Imperial College as a lecturer. The brilliance of his research on organic reactions led to his appointment in 1924 to the chair at Leeds University, where he remained for six years. In 1930 he moved to University College London, and turned it into one of the finest departments in the world. He stayed there until his retirement in 1961. During those years he developed models for the ways in which organic reactions occur. He was able to describe the ground state properties of molecules in terms of the way in which groups of atoms attached to a benzene molecule attract or repel electrons of substituents. He then went on to classify organic reactions according to the composition of the transition state (the energy maximum the molecules have to overcome for reaction to occur). This provided a convincing model for what actually happens when organic reactions occur, and explained many experimental observations within one theory. The terms S_N1 (monomolecular nucleophilic substitution) and S_N2 (bimolecular nucleophilic substitution), to be found in every textbook of organic chemistry, are due to Ingold's work on nucleophilic aliphatic substitution. During the 1930s and 1940s he undertook a massive study of the mechanism of aromatic nitration, a reaction which proved to be an exemplar of so many electrophilic aromatic substitution reactions. The resolution of some of the difficulties in Ingold's work involved the use of physical techniques not then normally used by organic chemists (isotope effects, molecular spectroscopy). Another reaction which received considerable attention was electrophilic addition to the olefinic double bond. In much of his work he was ably assisted by E D Hughes and his wife, Lady Ingold. While Baker Lecturer at Cornell he composed his monumental work *Structure and Mechanism in Organic Chemistry* (1953), a book which influenced chemists all over the world. His standing was such that he was said to head the 'English school' of organic chemists. His papers are models of clarity and precision, and can be read with as much interest today as when they were first written. Elected FRS in 1924 at the early age of 30, he received countless honours, including a knighthood in 1958. It is a matter of great sadness to Sir Christopher's many admirers that he never received the Nobel Prize for Chemistry. It was richly deserved.

INGOLD, Keith Usherwood

(1929–)

Canadian chemist, born in Leeds, England. After studying chemistry at University College London, where his father **Christopher Ingold** was professor, he proceeded to Oxford University for a DPhil (1951). He then joined the National Research Council of Canada, and apart from a short time at the University of British Columbia, has remained there ever since. In 1977 he was appointed Associate Director of the Division of Chemistry, and since 1990 he has been Distinguished Research Scientist there. His interests have involved the mechanisms and kinetics of radical reactions in the condensed phase, particularly in autoxidation. The techniques and experience gained have been applied to the study of vitamin E as a radical trapping agent. He was elected FRS in 1979.

IPATIEFF, Vladimir Nikolayevich
(1867–1952)

Russian organic and industrial chemist, born in Moscow. An officer in the Russian army, he later studied at the Mikhail Artillery Academy, St Petersburg, and was professor there from 1898 to 1906. He synthesized isoprene, the basic unit of natural rubber, in 1897. Around 1900 he began to study the decomposition of alcohols in the presence of a catalyst, obtaining in this way aldehydes, esters and olefins. He particularly investigated the catalytic properties of alumina, now widely used as an industrial catalyst. From 1904 he studied the effects of high pressures on catalytic reactions and showed that they greatly increased the speed and output of the reaction. He demonstrated that it was possible to catalyse more than one reaction at a time by using two-component catalysts, now a standard practice in the petrochemical industry. During World War I he directed Russia's chemical warfare programme. He emigrated to the USA in 1930 and worked for the Universal Oil Products Company in Chicago, where he developed a process for making high-octane petrol. He died in Chicago.

ISAACS, Alick
(1921–67)

Scottish virologist, born in Glasgow, the discoverer of interferon. He studied medicine at Glasgow, did research work at Sheffield and Melbourne, and in 1950 joined the virology division of the National Institute for Research, where he became chief in 1961. His research into the way influenza viruses interacted and impeded each other's growth led him and Jean Lindemann in 1957 to isolate a substance they called interferon. A protein, interferon is produced as part of the body's response to a viral infection. It has been shown to be of some therapeutic use in certain viral diseases and several forms of cancer.

J

JACKSON, Benjamin Daydon
(1846–1927)

English botanist and bibliographer, born in Stockwell, London and educated at private schools. From an early age Jackson had strong botanical and especially bibliographical instincts. A born indexer, his greatest work was the compilation of *Index Kewensis*, an index of all names of flowering plants hitherto described; the first volume was published in 1892. This work began in 1882, as the result of a suggestion of **Charles Darwin** to Sir **Joseph Dalton Hooker**; the manuscript, when completed in 1895, weighed over one ton. An absolutely essential tool for the plant taxonomist, *Index Kewensis* continues to be updated regularly. Jackson also compiled an indispensable *Glossary of Botanical Terms* (1900), the *Catalogue of the Library of the Linnaean Society* (1925) and wrote two botanical bibliographies, *Guide to the Literature of Botany* (1881) and *Vegetable Technology* (1882). He edited and annotated facsimiles of early botanical works including Gerard's *Catalogue of Plants Cultivated in His Garden in the Years 1596–1599* (1876) and **William Turner**'s *Libellus de Re Herbario Novus* of 1538 (1877). He was an authority on Linnaeus, and at celebrations in Sweden to mark the bicentenary of Linnaeus's birth in 1907, he received the Order of Knighthood of the Polar Star.

JACKSON, John Hughlings
(1835–1911)

English neurologist, born in Providence Green, Yorkshire. He qualified in medicine from St Bartholomew's Hospital in 1856, and after working in York he returned to London in 1859, and became a lecturer in pathology at the London Hospital. He received an MD in 1860 and was appointed assistant physician to the London Hospital in 1863, later becoming full physician (1874–94), and consulting physician (1894–1911). Simultaneously he was assistant (1862–7) and full physician (1867–1906) at the National Hospital for the Paralysed and Epileptic, Queen Square. Jackson contributed extensively to the development of scientific study within neurology, and he combined considerable clinical acumen with thoughtful interpretation. He suggested that function could be localized in specific regions of the cerebral cortex, which was an important stimulus for many researchers investigating the function of the central nervous system. He investigated unilateral epileptiform seizures and the physiology of speech, and wrote extensively on optic neuritis; he actively promoted the use of the ophthalmoscope by neurologists. He also postulated that the evolution of the nervous system proceeds from the simplest, most automatic, centres, to the highest, most complex and voluntary centres. Disease he considered to be a process of dissolution, the opposite of the evolutionary process.

JACOB, François
(1920–)

French biochemist, born in Nancy. Educated at the University of Paris and the Sorbonne, he worked at the Pasteur Institute in Paris from 1950. In 1960 he became Head of the Cellular Genetics Unit at the Pasteur, and in 1964 he was appointed Professor of Cellular Genetics at the Collège de France. During the 1950s, he worked on the nutritional requirements of the bacterium *Escherichia coli*. He found that the enzyme B-galactosidase is only produced when this bacterium is grown in the presence of the sugar lactose, and suggested that genes are turned on and off by other genes which regulate the genes that code for the enzyme. He formulated the 'operon system', which consists of the structural genes that code for a protein, and a regulator gene which controls the structural genes by producing a 'repressor' molecule which binds to a specific section of DNA known as the 'operator'. The theoretical 'repressor' molecule was identified by **Walter Gilbert** some years later. With **Lwoff** and **Monod**, Jacob was awarded the 1965 Nobel Prize for Physiology or Medicine for research into cell physiology and the structure of genes. His autobiography, *The Statue Within*, was published in 1988.

JACOBI, Carl Gustav Jacob
(1804–51)

German mathematician, born in Potsdam. He was educated at Berlin, and became a lecturer at the University of Königsberg, where he was appointed Extraordinary Professor in 1827 and Ordinary Professor of Mathematics in 1829. His *Fundamenta nova* (1829) was the first definitive book on elliptic functions, which he and **Niels Henrik Abel** had independently discovered. He was a virtuoso at manipulating lengthy expressions and discovered many remarkable infinite series connected to elliptic functions which solved important problems in number theory. At Königsberg he introduced the research seminar to German mathematics. He also made important advances in the study of differential equations, the theory of numbers, and determinants. In the theory of dynamics he wrote influentially on the motion of the top and advocated a formalism, today called the Hamilton–Jacobi formalism, that was taken up by **Lie** and **Poincaré** and later became central in treatments of quantum mechanics.

JANSKY, Karl Guthe
(1905–50)

American radio engineer, born in Norman, Oklahoma. He studied at the University of Wisconsin, and joined the Bell Telephone Laboratories in 1928. While investigating the sources of interference on short-wave radio telephone transmissions, specifically in the 15 metre wavelength region used for ship-to-shore communications, he built a high-quality receiver and aerial system mounted on wheels such that it could be rotated in various directions. With this he detected a weak source of static and noticed that the background hiss maximized every 23 hours 56 minutes; aware that this is the period of rotation of the Earth, he concluded that the radiation originated from a stellar source. By 1932 he had pinpointed the source as being in the constellation of Sagitarius — the direction towards the centre of the Milky Way. He published his findings in December 1932. This discovery not only made the front page of the *New York Times* but also heralded the birth of radio astronomy. He did not pursue further work in this field himself, and it was left to **Reber** to construct the world's first radio telescope in 1937. After the end of World War II, advances in radar and other microwave techniques led to the widespread adoption of radio astronomy as a means of celestial exploration. In 1973 the unit of radio emission strength, the jansky, was named after him.

JANSSEN, (Pierre) Jules César
(1824–1907)

French astronomer, born in Paris. After a time working in a bank he gave up in order to study mathematics and physics at the Sorbonne. He had a highly adventurous life in which he undertook many scientific missions on behalf of his country: to Peru in 1857 to determine the magnetic equator, to the Azores in 1867 to make magnetic observations, and to Japan and Algeria in 1874 and 1882 respectively to observe the century's two transits of Venus. He took part in eclipse expeditions including that of 1870 when he escaped from the beleaguered Paris in a balloon. His spectroscopic observations during the total eclipse of the Sun in India (1868) revealed the gaseous nature of solar prominences; the following day came his discovery of a method of observing prominences outside total eclipse. The same method, independently arrived at by **Lockyer**, brought both men to fame: the French Academy of Sciences commemorated their discovery by striking a medallion bearing their names and profiles. In 1875 the French government set up a new astrophysical observatory in Meudon near Paris with Janssen in charge from 1876. There he concentrated on solar physics, producing the first photographs of a fine grain-like structure (granulation) on the Sun's surface. In 1890 he set up an observing station on Mont Blanc where he was able to study the solar spectrum through a minimum of the Earth's interfering atmosphere. Janssen remained Director of Meudon Observatory until his death.

JEANS, Sir James Hopwood
(1877–1946)

English physicist, astronomer and writer, born in Ormskirk, near Southport. He taught at Princeton University and Cambridge before becoming a research associate at Mt Wilson Observatory in Pasadena. One of his first important results was the development of a formula to describe the distribution of energy of enclosed radiation at long wavelength, now known as the Rayleigh–Jeans law. He also carried out important work on the kinetic theory of gases, giving mathematical proofs of the law of equipartition of energy and **Maxwell**'s law of the velocity distribution of the molecules of a gas. He made significant advances in the theory of stellar dynamics by applying mathematics to problems such as the stability of rotating bodies and the gravitational disruption of a star due to a close encounter with another. This process was invoked in his theory of the origin of the solar system, in which the mass which formed the planets was drawn off the Sun in filaments due to the gravity of a star passing nearby. The theory was replaced by a modified version of **Laplace**'s nebular theory in the 1940s. Jeans's wide-ranging research included studies of the formation of binary stars, stellar evolution, the nature of spiral nebulae and the origin of stellar energy, which he believed to be associated with radioactivity. He was best known for his popular exposition of physical and astronomical theories and their philosophical bearings, such as *The Universe around us* (1929) and *The New Background of Science* (1933). He was knighted in 1928, and awarded the Order of Merit in 1939.

JEFFREYS, Alec John

(1950–)

English molecular biologist, born in Oxford. He was educated at Merton College, Oxford, where he graduated BA in 1972, and MA and DPhil in 1975. He was European Molecular Biology Organisation Research Fellow at the University of Amsterdam (1975–7) where he worked with **Flavell** on the regulation and structure of mammalian globin genes. He subsequently moved to the department of genetics of the University of Leicester, where he has remained throughout his career as lecturer (1977–84), reader (1984–7) and Professor of Genetics since 1987. He became Wolfson Research Professor of the Royal Society in 1991, and was elected FRS in 1986. Jeffreys developed the technique of 'DNA fingerprinting', in which DNA from an individual is digested with specific enzymes known as restriction endonucleases and the resultant DNA fragments are separated by an electric current on an agarose gel. Each individual has a completely unique pattern of DNA fragments, and thus samples of blood or semen can conclusively identify an individual in much the same way as a fingerprint. This technology is now used extensively in forensic work.

JEFFREYS, Sir Harold

(1891–1989)

English mathematician, geophysicist and astronomer, born in Fatfield, Durham. He was educated at Armstrong College, Newcastle upon Tyne, and St John's College, Cambridge, where he graduated in 1913 and was elected a college Fellow in 1914. After working in dynamical astronomy he joined the Meteorological Office, as an assistant to the director, working on dynamics of the winds and oceans (1915–7). Returning to Cambridge in 1922, he held various teaching appointments and later became Plumian Professor of Astronomy and Experimental Philosophy (1946–58). In his application of classical mechanics to geophysical problems, Jeffreys discovered the discontinuity between the Earth's upper and lower mantle, found evidence for the fluid nature of the core and did much pioneering theoretical work on the shape and strength of the Earth. His analysis of seismic traveltimes (made with the use of a mechanical calculator) was published as the *Jeffreys–Bullen Tables* (1940), which remains a standard reference. This work led him to a Bayesian theory of probability applicable to a wide range of sciences. He was elected FRS in 1925 and knighted in 1953.

JENNER, Edward

(1749–1823)

English physician, the discoverer of vaccination, born in Berkeley vicarage, Gloucestershire. He was apprenticed to a surgeon at Sodbury, near Bristol, in 1770 went to London to study under **Hunter**, and in 1773 settled in Berkeley, where he acquired a large practice. He began to examine the truth of the traditions respecting cowpox (1775), and became convinced that it was efficacious as a protection against smallpox. Many distractions delayed the actual discovery of the prophylactic power of vaccination, and the crowning experiment was made in 1796, when he vaccinated James Phipps, an eight-year-old boy, with cowpox matter from the hands of Sarah Nelmes, a milkmaid, and soon afterwards inoculated him with smallpox, and showed that the boy was protected. Jenner described his early series of vaccination experiments in *An Inquiry into the Causes and Effects of the Variolae Vaccinae* (1798), a short monograph which was privately published after being rejected by the leading scientific periodical of his day. Although he mistakenly identified the cowpox with a disease of horses called grease, he demonstrated that inoculating a person with material taken from a cow with cowpox did produce immunity from smallpox. The practice of vaccination met with brief opposition, until over 70 principal physicians and surgeons in London signed a declaration of their entire confidence in it. Jenner devoted the remainder of his life to advocating vaccination; Parliament rewarded him with two large grants and Napoleon had a medal struck in his honour. Jenner also wrote pamphlets defending vaccination and published some observations on the behaviour of the cuckoo.

JENSEN, (Johannes) Hans (Daniel)

(1907–73)

German physicist, born in Hamburg. He studied physics, mathematics and philosophy at Hamburg University and was awarded his PhD in 1933. He was appointed professor at Hamburg in 1936 and at Hanover in 1941. In 1949 he became Professor of Theoretical Physics at Heidelberg. From the large amounts of data available on nuclei, certain patterns had begun to emerge. One was that if a nuclide had a 'magic number' of neutrons and or protons then it was very stable. A similar pattern is observed in atoms where a filled shell of electrons corresponds to a stable element, eg a noble gas. Jensen used this similarity to apply the ideas of atomic physics to nuclei, leading to the nuclear shell model. This was done independently by **Goeppert-Mayer** in Chicago, and for this work Jensen, Goeppert-Mayer and **Wigner** shared the 1963 Nobel Prize for Physics.

JERNE, Niels Kai
(1911–)

English–Danish immunologist, born in London of Danish parents. He studied physics at the University of Leicester for two years, before transferring to the University of Copenhagen, where he received his doctorate in medicine in 1951. He conducted research at the Danish State Serum Institute until 1956. He was Chief Medical Officer of the World Health Organization (1956–62) and an administrator at the University of Pittsburgh (1962–6). He worked for a period at the Paul Ehrlich Institute, Frankfurt (1966–9), and was founding director of the Basle Institute of Immunology (1969–80). Jerne's research addressed the problem of antibody specificity, following on from the work of **Landsteiner**. It was well known that an animal exposed to an antigen produces antibodies, but Jerne discovered that when an animal is continually exposed to an antigen over a long period, it begins to produce a new form of antibody which binds more tightly to the antigen. He later explained the development of T-lymphocytes in the thymus and formulated the network theory which views the immune system as a network of interacting lymphocytes and antibodies. He shared the 1984 Nobel Prize for Physiology or Medicine with **Milstein** and **Köhler**.

JOBS, Steven
(1955–)

American computer inventor–entrepreneur, born in San Francisco. He was educated at Reed College, Portland, but dropped out and became a computer hobbyist. With the help of Stephen Wozniak, Jobs founded the Apple Computer Company in a garage in 1976. Their brainchild, the Apple II computer (1977), helped launch the personal computer revolution. The user-friendly approach of an Apple computer, with its 'mouse' and graphics interface, soon became part of the computer landscape. Jobs's company became the fastest growing in US history; sales soared from zero to $1.5 billion in seven years. A better inventor than businessman, in 1985 Jobs was forced to step down from Apple and he founded another company, NeXT Inc, to market computers for higher education.

JOHANSON, Donald Carl
(1943–)

American palaeoanthropologist, born in Chicago of Swedish immigrant parents. Johanson obtained a BA at the university of Illinois in 1966 and conducted graduate studies at Chicago. In 1972 he became Assistant Professor of Anthropology at the Case Western Reserve University in Cleveland, and Associate Curator of Anthropology at the Cleveland Museum of Natural History. Johanson made spectacular finds of fossil hominids 3–4 million years old at Hadar in the Afar triangle of Ethiopia (1972–7, 1990–2). They included 'Lucy', a female specimen, and the 'first family', a scattered group containing the remains of 13 individuals. Lucy was named after the song 'Lucy in the sky with diamonds', played at the camp celebration following her discovery. Johanson suggested that these remains belong to a previously undiscovered species, which he named *Australopithecus afarensis* (Afar ape-man), more primitive than *Australopithecus africanus* which had been named by **Dart**. Johanson put forward the theory of 'mosaic evolution'—the idea that some parts of the body became human before others—which challenged the idea that bipedalism, large brains and tool-making developed together. He suggested that bipedalism came first by about two million years. In 1981 Johanson was founding director of the Institute of Human Origins, Berkeley, California.

JOHANSSEN, Wilhelm Ludwig
(1857–1927)

Danish botanist and geneticist, born in Copenhagen. His family was not wealthy and he could not attend university, but through his apprenticeship to a pharmacist he developed a knowledge of chemistry. Later in Germany he learnt botany. He began teaching in the late 1880s, and in 1905 became Professor of Plant Physiology at the University of Copenhagen. His work with beans, bred over many generations, led him to the 'pure line theory', that 'pure lines' are genetically identical, and that variation within them is due entirely to environmental forces. He introduced the term 'gene' for the unit of heredity, as symbols or units of calculation. He also differentiated between the genetic material of an organism, its 'genotype' and how the heredity material gets expressed in the organism, its 'phenotype'. His *Elements of Heredity*, incorporating **Mendel**'s ideas, was published in 1909; it became an influential genetics text for around 20 years.

JOLIOT-CURIE, Frédéric, originally Jean-Frédéric Joliot

(1900–58)

French physicist, born in Paris and educated at the École Primaire Supérieure Lavoiser in Paris. After military service and a job at a steel mill, he joined the Radium Institute under **Marie Curie** (1925); there he studied the electrochemical properties of polonium. He married Marie's daughter Irène (see **Irène Joliot-Curie**) in 1926, and in 1931 they began collaborating in research. In 1933–4 they produced the first artificial isotope and for this work they were awarded the Nobel Prize for Chemistry in 1935. He became professor at the Collège de France in 1937. Aware of the consequences of nuclear fission, in 1939 he persuaded the French government to purchase the world's major stocks of heavy water (used as a moderator in chain reactions) from Norway. When France was invaded by the Germans in 1940, he arranged for it to be secretly shipped to England. During World War II he became a strong supporter of the Resistance movement and a member of the Communist party. After the liberation he became High Commissioner for Atomic Energy (1946–50), a position from which he was dismissed for his political activities. He succeeded his wife as head of the Radium Institute. He was awarded the Stalin Peace Prize (1951) and was given a state funeral by the Gaullist government when he died from cancer, caused by lifelong exposure to radioactivity.

JOLIOT-CURIE, Irène, née Curie

(1897–1956)

French physicist, born in Paris, daughter of **Pierre** and **Marie Curie**. She was educated at home by her mother, and during World War I served as a radiographer in military hospitals. In 1918 she joined her mother at the Radium Institute in Paris and began her scientific research in 1921. In 1926 she married **Frédéric Joliot**, and they collaborated in studies of radioactivity from 1931. In work on the emissions of polonium, they studied the highly penetrating radiation observed by **Bothe** and demonstrated its ability to eject protons from parafin wax; the radiation emitted was in fact neutrons, but they misinterpreted their results and attributed it to a consequence of the Compton effect. **Chadwick** read their paper and built on this work in his discovery of the neutron. In 1933–4 the Joliot-Curies made the first artificial radioisotope by bombarding aluminium with alpha particles to produce a radioactive isotope of phosphorus. It was for this work that they were jointly awarded the Nobel Prize for Chemistry in 1935. Similar methods led them to make a range of radioisotopes, some of which have proved indispensable in medicine, scientific research and industry. During World War II Irène Joliot-Curie escaped to Switzerland. Back in Paris after the war, she became director of the Radium Institute in 1946 and a director of the French Atomic Energy Commission. She died from leukaemia due to long periods of exposure to radioactivity.

JOLY, John

(1857–1933)

Irish geologist and physicist, born in Offaly, King's County. In 1876 he entered Trinity College, Dublin, where he studied engineering, physics, chemistry, geology and mineralogy. In 1882 he was appointed assistant to the Professor of Engineering there, and he became Professor of Geology and Mineralogy in 1897. His early interest in mineralogy and his technical ability led him to invent a number of pieces of scientific apparatus, including the meldometer with which he produced artificial crystals, a hydrostatic balance for density determination, a steam calorimeter and in 1888 a photometer. From 1899 he became involved in the debate about the age of the Earth. His novel calculation of its age (as 80–90 million years) by measuring the sodium content of the sea (1899) was influential. The discovery of radioactivity radically altered the direction of his research, and although others subsequently argued for a much older Earth, he remained convinced of the broad validity of his earlier calculations and made only minor concessions to an increased age. He was the first geologist to recognize the significance of radioactive atoms in maintaining the heat of the Sun and in directly moulding the history of the Earth (1903), and realized that pleochroic haloes in some minerals were the product of radioactivity and could be used for dating (1907–14). In 1914 at his suggestion, the Radium Institute was founded in Dublin; there Joly became involved in developing pioneering methods in radiotherapy, including the radium treatment of cancer using emanation-filled needles. Joly was also a pioneer of colour photography. His publications included *An estimate of the Age of the Earth* (1899), *Radium and the Geological Age of the Earth* (1903), *Radioactivity and Geology* (1909) and *The Surface History of the Earth* (1924).

JONES, Sir Harold Spencer

(1890–1960)

English astronomer, born in Kensington, London. He won a scholarship from Latymer Upper School to Jesus College, Cambridge

(1908), where he obtained first-class honours degrees in mathematics (1911) and physics (1912), and was awarded the Isaac Newton studentship (1912) and second Smith Prize (1913). He was Chief Assistant at the Royal Observatory, Greenwich (1913–23), and HM Astronomer at the Royal Observatory at the Cape of Good Hope (1923–33), returning to Greenwich as tenth Astronomer Royal (1933–55). At the Cape he organized an international project to improve the value of the Earth–Sun distance utilizing the close approach of the asteroid Eros, and coordinated the reductions which took 10 years. He was awarded the Gold Medal of the Royal Astronomical Society for this work (1943). At Greenwich, following his analysis of the motions of the Sun, Moon and planets, he discovered long-term and irregular variations in the rate of the Earth's rotation (1939). This led to the concept of ephemeris time (1950), an independent system of measuring time which was adopted in the universal system of units in 1956. In 1948 he initiated the removal of the Royal Observatory from Greenwich to a country site at Herstmonceux, Sussex. The move was completed in 1958, but the Royal Greenwich Observatory was again moved in 1989 and is now in Cambridge. Jones was elected FRS in 1930, and knighted in 1943.

JORDAN, (Marie-Ennemond) Camille

(1838–1922)

French mathematician, born in Lyons. Professor at the École Polytechnique and at the Collège de France, he was the leading group theorist of his day, doing much to establish the central ideas in the then-new subject. He was the first to penetrate the profound ideas of **Galois** and to create a structure theory of finite permutation groups with applications to the theory of equations. His *Traité de substitutions* (1870) remained a standard work for many years. He applied group theory to geometry and linear differential equations, and in the 1890s his *Cours d'analyse* was an influential textbook for the French school of analysts, setting new standards of rigour. In it he attempted to derive the important, intuitively obvious, but elusive proof (today called the Jordan curve theorem) that every closed curve that does not cross itself has an inside and an outside. He also gave a fine account of the theory of complex and elliptic functions.

JORDAN, (Ernst) Pascual

(1902–)

German theoretical physicist, born in Hanover. He was educated in Hanover and Göttingen before obtaining a post at the University of Rostock where he became professor in 1935. He was subsequently appointed to chairs in physics at the universities of Berlin (1944–52) and Hamburg (1951–70). While he was assistant to **Born** and soon after in collaboration with **Heisenberg**, he helped to formulate the theory of quantum mechanics in the matrix representation, showing how light could be interpreted as being composed of discrete quanta of energy. During the 1920s and 1930s, along with Heisenberg, **Dirac** and **Pauli**, he contributed to the theories which laid the foundations for the relativistic quantum field theory of electromagnetism now known as quantum electrodynamics.

JORDANUS DE NEMORE

(fl c.1220)

Medieval French or possibly German mathematician and mechanical philosopher. He did valuable work in mechanics, but almost nothing is known of his life, except that he lived and wrote in the first half of the 13th century; even the meaning of 'de Nemore' is unknown. Some 12 books in Latin allegedly written by him were recorded by 1260, dealing with 'the science of weights', ie statics. Here he invented the idea of component forces, studied inclined planes, made the principle of mechanical work less vague, and moved towards (but did not quite reach) the concept of static moment, in contrast to the generally accepted Aristotelian explanation of statics in terms of dynamics. His approach linked **Aristotle**'s ideas in physics with the more exact mathematical approach of **Archimedes**. His ideas in mechanics must have influenced **Galileo**; he also wrote (or at least has ascribed to him) treatises on geometry, algebra and arithmetic, but the significance of these on later thought is unclear.

JOSEPHSON, Brian David

(1940–)

Welsh physicist, born in Cardiff, the discoverer of 'tunnelling' between superconductors. He studied at Cambridge, where he received his PhD in 1964, and has spent his career there, as Professor of Physics from 1974. In 1962, while a research student, he deduced theoretically the possibility of the 'Josephson effect' on electric currents in superconductors separated by a very thin insulator. He demonstrated that a current can flow between the superconductors with no applied voltage, and that when a DC voltage is applied an AC current of frequency proportional to the voltage is produced. Both effects were soon observed experimentally confirming his predictions. Josephson junctions have since been much used in research, in fast switches for computers and in SQUIDs (superconducting quantum interference devices) which are used as magnetometers in ultrasensitive geophysical measurements. The AC Josephson effect has been used to determine the constant *e/h* (the electron's charge divided by Planck's constant) and led to a quantum standard of voltage now used in many national standards laboratories. Josephson shared the 1973 Nobel Prize for Physics with **Esaki** and **Giaever**. He has an interest in the physics and theory of intelligence, and co-edited *Consciousness and the Physical World* (1980).

JOULE, James Prescott
(1818–89)

British natural philosopher, born in Salford. Shy and often ill as a child, Joule was educated by private tutors, notably the chemist **Dalton**. A laboratory in the parental home was the site of Joule's first electrical experiments. In 1838 he described his new electromagnetic engine in **Sturgeon**'s *Annals of Electricity*. Joule's belief that such machines would supersede the steam engine was revised in the light of systematic investigations into engine 'duty' or efficiency. The 'Joule effect' (1840) asserted that the heat produced in a wire by an electric current was proportional to the resistance and to the square of the current. He came to believe that heat was derived from work, whether chemical, mechanical or electrical in form: contradicting the caloric theory, heat and work were interconvertible. Painstaking experimental researches provided successive determinations of the mechanical equivalent of heat, the first announced to the peripatetic British Association for the Advancement of Science (BAAS) in 1843 and the last in 1878 carried out for the BAAS Committee on Standards. In 1845 Joule demonstrated the conversion of work into heat by agitating water with a rotating paddle wheel apparatus. This famous experiment was repeated at the 1847 BAAS meeting in Oxford where Joule met and greatly impressed **Kelvin**. Between 1853 and 1862 they collaborated on the 'porous plug' experiments showing that when a gas expands without doing external work its temperature falls (the Joule–Thomson effect). Recognition came with the award of the Royal Society's Royal (1852) and Copley (1870) medals. During the 1850s with the promotion of the kinetic theory of gases (Joule calculated the average velocity of a gas molecule in 1848), the new thermodynamics, and energy physics, his ideas were recast in terms of the principle of the conservation of energy.

JUSSIEU, Antoine Laurent de
(1748–1836)

French botanist, born in Lyons, nephew of **Bernard Jussieu**. He studied at Paris under his uncle and became professor at the Jardin des Plantes (1793–1826), which he reorganized as the National Museum of Natural History. He elaborated in his *Genera Plantarum* (1778–89) his uncle's system of 'natural' classification. At a time of many voyages of discovery, enormous numbers of plants, often representing entirely new families, were being sent back to Paris. His classification, based on many more specimens than were available previously, together with his adoption of the Linnaean nomenclature, ensured that the arrangement of families in *Genera Plantarum* was widely accepted. He wrote the section on natural classification for the *Dictionnaire des Sciences Naturelles* in 1842. Of the 100 family names used by Jussieu, 76 are still in current use. He took 11 family names from **Linnaeus**, 46 from Bernard Jussieu, six from **Adanson**, and 34 were his own. He accepted **Ray**'s division of plants into monocots and dicots, and attempted to form higher taxonomic groupings, but in this he was less successful.

JUSSIEU, Bernard
(c.1699–1777)

French botanist, born in Lyons. He was awarded doctorates at Montpellier and Paris. A demonstrator at the Jardin des Plantes (1722), he created a botanical garden at Trianon for Louis XV in 1759, with beds planted out in natural order, reflecting a system which has become the basis of modern natural botanical classification. He first suggested that polyps were animals. A teacher of genius, he inspired his pupils by a combination of the depth of his botanical knowledge and the warmth of his personality. His support for, and use of, the principle of natural classification had a decisive influence on botanical history. His studies of ferns revealed astute insights into the relationships between the flowering structures of ferns and higher plants. Studying the way in which floral parts are inserted, he was able to associate similar genera into families, although it was left to his nephew **Antoine Laurent de Jussieu** to elaborate and make explicit these relationships. His brother Antoine (1686–1758), a physician and professor of the Jardin des Plantes, edited **Tournefort**'s *Institutiones Rei Herbariae* (1719).

K

KAMB, (Walter) Barclay

(1931–)

American glaciologist, born in San José, California. He received a BS in physics (1952) and PhD in geology (1956) from Caltech, where he has remained throughout his career, becoming Professor of Geology in 1962 and Professor of Geology and Geophysics in 1963. Since 1990 he has been Rawn Professor there. In addition he was Vice-President and Provost of Caltech from 1987 to 1989. He led many glacial expeditions to Alaska, Antarctica, the Alps, Scandinavia and the Olympics. Kamb's contributions to glacial research include work on ice-streaming phenomena in the polar ice sheets and their role in climatic change. He made the first successful determination from laboratory measurements of the crystal structure in high-pressure forms of ice from 1960, and made the first observations of the basal sliding of glaciers (1964), relating it to existing theoretical concepts. He also made the first detailed observations of a glacier in full surge (1985). In addition, Kamb maintains interests in tectonics and non-hydrostatic thermodynamics.

KAMEN, Martin David

(1913–)

Canadian-born American biochemist, born in Toronto. He studied in Chicago and, suffering under McCarthyism, held posts in several American universities. In 1960 he was appointed Professor of Biochemistry at the University of California, San Diego. In its Radiation Laboratory he was the first to isolate in significant amounts the carbon isotope ^{14}C, later widely used as a biochemical tracer, and he subsequently pioneered the application of several radioisotopes in a diversity of biochemical, particularly bacterial, systems. He confirmed **van Niel**'s hypothesis that all the oxygen released in photosynthesis comes from water and not from carbon dioxide. He determined the initial fate of 'fixed' carbon dioxide in photosynthesis (1945), that bacteria, which carry out the photosynthetic conversion of carbon dioxide to carbohydrate without oxygen release, require the presence of reducing substances such as hydrogen sulphide, and that illumination increases phosphorus turnover in photosynthesis (1948). Kamen also studied nitrogen fixation and the role of molybdenum (1972), iron in porphyrin metabolism (1948), calcium exchange in squamous cell carcinoma (1948), a diversity of cytochromes, introducing dichlorophenolindophenol as an artificial electron donor (1953) and contributing to cytochrome structural studies (1980), and the iron–sulphur clusters of bacterial ferridoxins (1973). He published *Radioactive Tracers in Biology* (1947) and *Primary Processes in Photosynthesis* (1963).

KAMERLINGH ONNES, Heike

(1853–1926)

Dutch physicist, born in Groningen into a wealthy family. His father was a local manufacturer. He studied physics and mathematics at the University of Groningen, and continued his studies for some time under **Bunsen** and **Kirchhoff** at Heidelberg. He later became Professor of Physics at Leiden (1882–1923), where he established his famous low-temperature laboratory. Much of his early work attempted to test the ideas of **van der Waals** concerning the behaviour of liquids and gases over a wide range of temperatures and pressures. From high pressures he turned to low temperatures. His most noteworthy achievements were the first liquefaction (1908) and later solidification of helium, and his discovery (1911) that the electrical resistance of metals cooled to near absolute zero all but disappears, a phenomenon which he called 'supraconductivity', later changed to 'superconductivity'. In 1913 he was awarded the Nobel Prize for Physics. He was elected a Foreign Member of the Royal Society in 1916.

KAMMERER, Paul

(1880–1926)

Austrian zoologist, born in Vienna. He was educated at the University of Vienna and later joined the Institute of Experimental Biology there. His experimental work, carried out with salamanders and toads, appeared to support the view of **Lamarck** that characteristics acquired during life can be transmitted through subsequent generations. This was in direct conflict with the Darwinian view of inheritance. The best known of these results concerned the apparent acquisition of nuptial pads on the forefeet of midwife toads. Kammerer's experiments were criticized by geneticists and he visited Britain in 1923 in order to defend his experimental results. However, in 1926 G K Noble and H Przibram of the American Museum of Natural History examined material preserved from Kammerer's work in Vienna and showed that the dark swellings, which Kammerer claimed to be nuptial pads which were inherited through three generations, were due to injections of ink. Although Kammerer disclaimed any responsibility for the deception, he shot himself a few months later. Arthur Koestler gives an account of these events in his *The Case of the Midwife Toad* (1971).

KANT, Immanuel

(1724–1804)

German philosopher, one of the great figures in the history of western thought. The son of a saddler, he was born in Königsberg, Prussia (now Kaliningrad, Russia), and stayed there all his life. He studied and then taught at the university, becoming Professor of Logic and Metaphysics in 1770. He lived a quiet, orderly life and local people were said to set their watches by the time of his daily walk. His early publications were in the natural sciences, particularly geophysics and astronomy, and in an essay on Newtonian cosmology (*Allgemeine Naturgeschichte und Theorie des Himmels*, 1755) he anticipated the nebular theory of **Laplace** and predicted the existence of the planet Uranus before its actual discovery by **William Herschel** in 1781. As well as his extremely influential philosophical work, Kant also addressed some of the questions raised by Newton's mechanics. Considering the problem of forces acting over vast distances in space, he postulated that there are two fundamental forces, repulsive or elastic, and attractive or gravitational. The first required a medium in which to act, while the second could operate in free space. These ideas, as well as his rejection of traditional beliefs regarding the indivisibility of matter, had significant influence on later scientists. His most important philosophical works, the *Kritik der reinen Vernunft* ('Critique of Pure Reason', 1781), *Kritik der praktischen Vernunft* ('Critique of Practical Reason', 1788) and *Kritik der Urteilskraft* ('Critique of Judgement', 1790), were produced relatively late in his life.

KAPITZA, Peter (Pyotr Leonidovich)

(1894–1984)

Russian physicist, born in Kronstadt. He studied at Petrograd (now St Petersburg) and under **Rutherford** at Cambridge, where he became assistant director of magnetic research at the Cavendish Laboratory (1924–32). He achieved the production of strong magnetic fields, not surpassed for 30 years, to study alpha particle deflection and magnetostriction. In 1934, during a visit to Moscow, authorities barred him from returning to Cambridge and he was appointed director of the Moscow Institute for Physical Problems. Using his equipment from Cambridge which the Russian government had purchased, he engineered a helium liquefier. This was produced commercially some years later, facilitating research in low-temperature physics worldwide. He investigated the extraordinary properties of helium-2, introducing the term 'superfluid' to describe its high thermal conductivity and ability to defy gravity by flowing up and out of its container. He was dismissed from his post in 1946 for refusing to work on the atomic bomb, but reinstated in 1955. For his inventions and discoveries in the area of low-temperature physics, he was awarded the 1978 Nobel Prize for Physics jointly with **Penzias** and **Robert Wilson**. A strong believer in the free exchange of scientific ideas, in the 1970s he defended dissident physicist **Sakharov** from expulsion from the Soviet Academy of Sciences.

KAPTEYN, Jacobus Cornelius

(1851–1922)

Dutch astronomer, born in Barnefeld, Netherlands, one of 14 children of a schoolmaster. After studies of mathematics at the University of Leiden (1869–75) and a time on the staff of Leiden Observatory, he was appointed (1877) to a new Chair of Astronomy at the University of Groningen. Not having an observatory there, he founded an astronomical laboratory, the aim of which was to reduce by the best possible methods observations made elsewhere. Kapteyn's best-known work made over a period of 12 years was the reduction of photographic plates obtained at the Cape Observatory in South Africa, the *Cape Durchmusterung*, which exhibited 455 000 stars. His investigations of the motions as well as the distribution of stars in space led to the discovery of apparently preferred directions of motion or 'star-streams' (1904), later understood in terms of the differential rotation of our galaxy around a distant centre. Kapteyn was one of the founders of modern stellar statistics. He pioneered a plan to collect astronomical data of all kinds in sample fields distributed over the sky, the so-called selected areas (1906), as a means of investigating the distribution of stars in general.

KARLE, Jerome

(1918–)

American physicist, born in New York City. He studied at the City College of New York

(CCNY), Harvard and Michigan University. After working briefly on the Manhattan Project to develop the atomic bomb in the 1940s, he spent his career at the US Naval Research Laboratories in Washington, specializing in diffraction methods for studying the fine structure of crystalline matter. He shared the 1985 Nobel Prize for Chemistry with his fellow CCNY student **Hauptman** for the development of the 'direct method' for interpreting raw data from X-ray crystallography measurements. He later investigated the use of high-speed computers to produce real-time images of crystals and complex biomolecules.

KÁRMÁN, Theodore von

(1881–1963)

Hungarian-born American physicist and aeronautical engineer, born in Budapest, sometimes called the father of modern aerodynamics. He graduated as an engineer from Budapest Technical University (1902), received a PhD under **Prandtl** at Göttingen (1908), and in 1912 became Professor of Aeronautics and Mechanics at the University of Aachen and head of the Aeronautical Institute there. After visits to the USA in 1926, he became Director of the Guggenheim Aeronautical Laboratories (1930–49) and the Jet Propulsion Laboratory (1942–5) at Caltech. He founded the Aerojet Engineering Corporation in the early 1940s. In 1951 he founded the major international aerospace research organization AGARD (Advisory Group for Aeronautical Research and Development) as part of NATO. He was the first recipient of the National Medal of Science in 1963, and published major works in many fields. Several theories bear his name, such as the Kármán 'vortex street', a double line of vortices formed when air flows over a cylindrical surface. After the dramatic collapse of the Tacoma Narrows suspension bridge in the USA in 1940, he proved that oscillations due to vortex shedding in a moderate 42 miles per hour wind were the cause of its destruction.

KARRER, Paul

(1889–1971)

Swiss chemist, born in Moscow. He was the son of a Swiss dentist who practised in Russia, but he was educated in Switzerland and studied chemistry under **Alfred Werner** at the University of Zürich. After graduating he worked with organo-arsenic compounds, and this interest led him in 1912 to move to Frankfurt to work with **Ehrlich**. In 1915, after Ehrlich's death, he became director of the chemical division of Georg Speyer Haus. He remained in this position for only three years before returning to Zürich where, in 1919, he succeeded Werner as professor. He remained there for the rest of his life. All Karrer's chemical studies involved natural products. He began with amino acids and proteins, and continued with polysaccharides such as starch and cellulose. During the 1920s he developed an interest in plant pigments, a topic which was to occupy him for the rest of his career. An early triumph was his elucidation of the structure of carotene, which led to important discoveries concerning vitamin A. He also elucidated the structures of vitamins E, K and B_2 (riboflavin). For these achievements and for important studies of the chemistry of vitamin C and biotin, he shared the 1937 Nobel Prize for Chemistry with **Norman Haworth**. Karrer's chemical studies continued during the 1940s and 1950s, and included important work on the coenzyme nicotinamide-adenine dinucleotide, carotenoids and the curare-like alkaloids. He was rector at the University of Zürich from 1950 to 1952, and published the world-renowned textbook *Lehrbuch der organischen Chemie* in 1928.

KASTLER, Alfred

(1902–)

French scientist, born in Guebwiller (then in Germany). Before beginning graduate studies he taught physics at lycées in Mulhouse, Colmar and Bordeaux, leaving to join the University of Bordeaux in 1931. After two years as a lecturer at Clermont-Ferrand University he became Professor of Physics at the University of Bordeaux. He moved to the École Normale Supérieure in Paris in 1941 where he lead a research group in Hertzian spectroscopy, becoming professor in 1952. Much of his work concerned the probing of energy levels within atoms. He used a 'double resonance' method, in which both visible light and radio waves were used to excite electrons in atoms to higher energy levels. As the atoms returned to lower energy states they emitted radiation. Measurements of the degree of polarization and angular distribution of this fluorescent radiation yielded precise information about the structures of excited atomic levels. Kastler also used optical techniques to investigate magnetic sub-levels, developing a process that became known as 'optical pumping' which laid the foundations for the subsequent development of masers and lasers; for this work he was awarded the 1966 Nobel Prize for Physics. His research resulted in numerous practical applications including frequency standards, atomic clocks and highly sensitive magnetometers.

KATZ, Sir Bernard

(1911–)

British biophysicist, born and educated in Leipzig, Germany. After qualifying in medicine he left Nazi Germany in 1935, and began physiological research at University College London, initially in **Hill**'s laboratory. He started working on problems associated with the electrical stimulation of nerves and the processes of neuromuscular transmission, and received his PhD in 1938. Just before the outbreak of World War II he joined **Eccles** at the Kanematsu Institute in Sydney, Australia, where he collaborated in further neurophysiological experiments with both Eccles and **Kuffler**, before enlisting as a radar officer in the Australian Royal Air Force. In 1946 he returned to University College, where he remained for most of his career (Lecturer 1946–50; Reader in Physiology 1950–2; Professor of Biophysics 1952–78) although he spent a substantial part of the late 1940s in Cambridge and Plymouth, working with Sir **Alan Hodgkin** and Sir **Andrew Huxley** on the mechanisms by which the nerve impulse is transmitted, using the giant axon preparation of the squid, first reported by **John Young**. For the next three decades Katz's work focused on the mechanisms of neural transmission, in particular how the chemical neurotransmitter acetylcholine is stored in nerve terminals and released by neural impulses. In collaboration with Paul Fatt and Jose del Castillo, he showed that the chemical was stored in nerve terminals, in small packets called vesicles that could be observed using an electron microscope, and was released in specific portions called quanta when stimulated by the arrival of the neural impulse. For this work he shared the 1970 Nobel Prize for Physiology or Medicine with **Axelrod** and **Ulf von Euler**. He was elected FRS in 1952, and knighted in 1969.

KAUFMANN, Nicolaus

See **MERCATOR, Nicolaus**

KEELER, James Edward

(1857–1900)

American astronomer, born in La Salle, Illinois. He was educated at Johns Hopkins University (1877–81), graduating in physics. He became assistant to **Samuel Langley** at Allegheny Observatory (1881–3) doing solar research which included observations at high altitudes on Mount Whitney. He spent a year (1883–4) at universities in Germany, and a further year at Allegheny. He was then appointed to a post at the newly founded Lick Observatory in California (1885–91), where he established his reputation as an outstanding spectroscopist from his work on nebulae with bright emission lines. He was Director of the Allegheny Observatory (1891–8) and in 1898 returned to Lick Observatory as its director. His observation of the spectrum of the rings of Saturn (1895) revealed motions around the parent planet which varied with distance from it, a proof that the rings are composed of myriads of tiny objects as had been theoretically demonstrated by **Maxwell**. His most important work at Lick Observatory was on nebulae of the kind now known as galaxies. Through long-exposure photography he was able to record very faint ones, and estimated that there were 120000 nebulae in the sky capable of being photographed with the instrument. He died unexpectedly after only two years in office.

KEILIN, David

(1887–1963)

British biochemist, born in Moscow, the son of a Polish businessman. Educated in Warsaw, Liège and Paris, his career thereafter was spent in Cambridge, where he was Director of the Molteno Institute from 1931 and Quick Professor of Biology (1931–52). In 1925 he reinvestigated earlier spectrophotometric studies and recognized absorption bands attributable to three haem compounds which he called cytochrome a, b and c, and initially believed to be identical to **Warburg**'s *Atmungsferment*. In 1929, with Edward Hartree, he found in heart muscle a haemochromogen containing cytochromes a and a_3, and equated the latter with cytochrome oxidase. He later discovered a 'haemoglobin' in the root nodules of leguminous plants (1945) resulting from the symbiosis between a nitrogen-fixing bacterium (*Rhizobium*) and the plant root. With Hartree he suggested similarities between the enzymes catalase and peroxidase, based on a study of acetaldehyde production from ethanol. A keen entomologist, Keilin used insects in many of his experiments in animal biochemistry.

KEIR, James

(1735–1820)

Scottish chemist and industrialist, born in Edinburgh, the youngest of 18 children in a prosperous family. He was educated at Edinburgh High School and at Edinburgh University, where he studied medicine and met **Erasmus Darwin**, a fellow student who became a lifelong friend. He joined the army and served in the West Indies, resigning with the rank of captain in 1768. In 1770 he settled at West Bromwich, outside Birmingham, joining Darwin's circle of friends and becoming a member of the Lunar Society, an informal but influential club whose members met to exchange ideas in science and technology. Keir helped **Priestley** with some of his experiments, and unlike Priestley, was quick to accept **Lavoisier**'s ideas on combustion. In 1771 he set up a glass-making business at Stourbridge, and having observed what happens when glass cools very slowly, suggested that basalt cools in a similar manner and therefore must be igneous. In 1776 he translated the new chemical dictionary written by **Macquer**. Two years later he and Alexander Blair founded the Tipton Chemical Works, which manufactured alkali from the sulphates of potash and soda, preceding **Leblanc**'s more famous process by 40 years. For some years manager of the Boulton and Watt engineering works, Keir helped to develop a process for electroplating. He also studied gases. He was elected FRS in 1785, and died in West Bromwich.

KEKULÉ VON STRADONITZ, Friedrich August

(1829–96)

German chemist, born in Darmstadt. Having attended the gymnasium in Darmstadt, he entered the University of Giessen with the intention of becoming an architect, but after hearing **Liebig**'s lectures, he decided that he wanted to become a chemist. After some delay he continued his studies in 1849 in the chemistry department at Giessen. During 1850 he worked in Liebig's laboratory but found that he did not care for practical work and left for Paris, where he familiarized himself with **Gerhardt**'s type theory of organic chemistry. After a short period at Reichenau in Switzerland, he became assistant to John Stenhouse at St Bartholomew's Hospital in London. In 1856 he began his teaching career in Heidelberg and also built a private laboratory in his own house, and in 1858 he became professor at the University of Ghent in Belgium. During the period in London he speculated about the structure of organic molecules and proposed that molecules consist of atoms linked together by bonds according to the valency or combining power of each atom. He suggested that carbon has a valency of four, and that in many organic compounds, carbon atoms are linked together in chains, a characteristic that sets carbon apart from most other elements. It is said that some of these ideas came to him during a vision on the top of a London omnibus. His views on the tetravalency of carbon were difficult to reconcile with the known formula of benzene (C_6H_6) and while he was in Ghent, the solution—the cyclic nature of the benzene molecule—came to him in another vision while he was dozing in front of the fire. Most of his experimental work at Ghent was aimed at satisfactorily confirming his proposed structure for benzene. In 1867 he left Ghent for Bonn, continued his work on benzene and proposed delocalized rather than fixed double bonds. Around 1875 his health began to fail, but he continued to work and his paper in 1890 on the structure of pyridine (a compound analogous to benzene) is an important landmark in the development of structural organic chemistry. Even during his final years he made significant advances. Although some of his new ideas on structural organic chemistry were also put forward by others, Kekulé, by the breadth of his vision (and, indeed, his visions) made a very special contribution to the development of organic chemistry.

KELVIN, William Thomson, 1st Baron of Largs

(1824–1907)

British natural philosopher, born in Belfast. He came to Glasgow when his father was made Professor of Mathematics there in 1832. At the age of 10, William Thomson matriculated at the university and in 1841 he entered Peterhouse College, Cambridge. He completed the exacting Mathematics Tripos in 1845 as Second Wrangler, first Smith's Prizeman, and winner of the Silver Sculls rowing trophy. After acquiring experimental skills in **Regnault**'s Paris laboratory, Thomson was appointed Professor of Natural Philosophy at Glasgow (1846), a post he held until 1899 when he retired, only to re-enrol as a research student. In a career of prodigious versatility and international distinction he harmonized physical theory and engineering practice, emphasizing accurate measurement, economy and the minimization of waste. The application of continental mathematics, combined with the use of mechanical models in physical theory was a hallmark of his scientific style. Whilst a student he read **Fourier**'s *Analytic Theory of Heat*, praised it as a 'mathematical poem', and in his first articles defended its adventurous mathematical techniques. From 1842 he addressed problems in electromagnetism, developing an analogy with Fourier's macroscopic heat flow analysis and stressing contiguous action rather than forces acting at a distance. Through the British Association for the Advancement of Science, Thomson worked to establish national and international electrical standards. When the first transatlantic telegraph cable became operational in 1866, he was knighted for the major consultative role he had played. His mirror galvanometer and siphon recorder were designed as receivers for rapid transmission on such long-distance cables. Numerous electrical instruments were manufactured under patent, bringing substantial returns. In 1848 Thomson proposed his absolute (Kelvin) scale of temperature, independent of any physical substance. The following year he gave a rigorous account of **Carnot**'s theory of heat-engines, and in 1851 he reconciled Carnot's theory and the mechanical theory of heat given credence by **Joule**. Exhaustive memoirs on the 'dynamical theory of heat', contemporaneously with **Clausius** and **Rankine**, established the second law of thermodynamics. Thomson promoted the new unifying concept of 'energy' in physics, and considered the consequences of a general dissipation of the energy of the universe. Cosmological convictions fused with geophysics to generate estimates of the Earth's age now regarded as far too low, but which at the time provided temporal arguments against **Charles Darwin**'s theory of evolution. Thomson also worked on hydrodynamics in the context of the burgeoning Clyde shipbuilding industry. His yacht, the *Lalla Rookh*, served both as symbol of affluence and floating laboratory. Created Britain's first scientific peer in 1892, he is buried in Westminster Abbey, beside **Isaac Newton**.

KENDALL, Edward Calvin
(1886–1972)

American chemist, born in South Norwalk, Connecticut. After training in Canada he joined the firm of Park, Davis (1910) and was asked to isolate the active principle of the thyroid gland. He left in 1911, going first to St Luke's Hospital in New York and then to the Mayo Foundation, Rochester (1914), where he was appointed professor and head of biochemistry. In December 1914 he isolated thyroxine (33 grams of thyroxine from three tons of pig thyroid gland), and in 1917 he described its properties and physiological activity in detail. Its structure was elucidated partly by Kendall and partly by **Harington**. In collaboration, Kendall isolated cortisone (around 1936) and 29 related steroids from the adrenal cortex. In particular he studied and determined the metabolic effects of cortisone, corticosterone, 17-hydroxycorticosterone, dihydrocorticosterone and cortisol, and observed the loss of resistance to toxic chemicals caused by adrenalectomy. He prepared synthetic corticosterone in 1944 and cortisone (active against Addison's disease) in 1947. With **Hench**, he found that cortisone was effective against rheumatic fever and that cortisone plus ACTH (adrenocortical trophic hormone) was effective against rheumatoid arthritis. Kendall, Hench and **Reichstein** shared the 1950 Nobel Prize for Physiology or Medicine.

KENDALL, Henry Way
(1926–)

American physicist, born in Boston and educated at Amherst College and MIT. He later worked at Stanford and MIT, where he was appointed professor in 1967. Around 1970, with **Jerome Friedman** and **Richard Taylor**, he led a research team working at the Stanford Linear Accelerator. In a series of experiments involving accelerating electrons towards a liquid hydrogen target and measuring the energies and angles of scattered electrons, they obtained convincing experimental evidence of the existence of quarks (first proposed by **Gell-Mann** and independently by **Zweig** in 1964). In conjunction with analogous neutrino experiments at the European nuclear research centre CERN, they also confirmed predictions regarding the charges on quarks. For this work he was awarded the W K H Panofsky Prize (1989) and the Nobel Prize for Physics (1990) jointly with Friedman and Taylor.

KENDREW, Sir John Cowdery
(1917–)

English molecular biologist, born in Oxford. He was educated at Clifton College, Bristol, and Trinity College, Cambridge. Elected a Fellow of Peterhouse College, Cambridge (1947–75), he was a co-founder (with **Perutz**) and Deputy Chairman of the Medical Research Council Unit for Molecular Biology at Cambridge (1946–75). During World War II he was seconded to the Ministry of Aircraft Production and Scientific Advisor to the Allied Air Commander in Chief. He was also Scientific Advisor to the Ministry of Defence (1960–4). Kendrew carried out researches in the chemistry of the blood and determined by X-ray crystallography the structure of the muscle protein myoglobin, giving its outline in 1957 and detailed three-dimensional structure in 1959. By observing the alpha helical nature of the polypeptide chain, he was the first to confirm the structure proposed by **Pauling**. He was awarded the 1962 Nobel Prize for Chemistry jointly with Perutz. He wrote *The Thread of Life* (1966), was elected FRS in 1960 and knighted in 1974.

KEPLER, Johannes
(1571–1630)

German astronomer, born in Weilderstadt, Württemberg, the son of a mercenary soldier, and educated at the University of Tübingen where he obtained a master's degree in theology in 1591. Being a deeply religious Lutheran he originally intended to enter the church ministry, but when the Protestant Estates of Styria were looking for a Professor of Mathematics at Graz, Kepler accepted the post (1594), having been strongly influenced by the teaching of Michael Maestlin, his Professor of Mathematics at Tübingen and an enthusiastic supporter of the ideas of **Copernicus**. Among Kepler's duties at Graz was the publication of almanacs to forecast the weather and to predict favourable days for various undertakings with reference to the rules of astrology: in fact Kepler was for a time astrologer to the Duke Albrecht of Wallenstein, the great soldier of the Thirty Years' War. By temperament inclined to mysticism, Kepler was overjoyed when he thought he had found, as recorded in his first major publication, the *Mysterium Cosmographium* (1596), that the distances from the Sun of the six planets including the Earth could be related to the five regular solids of geometry, of which the cube is the simplest. He sent copies of his book to **Galileo** and **Tycho Brahe**, the greatest astronomers of the day, who responded in a friendly manner. When Kepler was later in difficulties in Graz, the latter invited him to join him at Prague. Kepler arrived in Prague in 1600, and when Tycho Brahe died in 1601 he was appointed to succeed him as imperial mathematician by the emperor Rudolf II. Among Kepler's early work at Prague was a treatise on optics (1604) and a paper on the nova of the same year which appeared in the constellation Ophiuchus ('Kepler's supernova'). His chief interest was the study of the planet Mars, for which Tycho's numerous observations provided very valuable material. He found that the movement of Mars could not be explained in terms of the customary cycles and epicycles, and published his discovery in his *Astronomia Nova* (1609), which he dedicated to the emperor. In this fundamental book Kepler broke with the tradition of more than 2000 years by demonstrating that the planets do not move uniformly in circles but in ellipses with the Sun at one focus and with the radius vector of each planet describing equal areas of the ellipse in equal times (Kepler's first and second laws). He completed his researches in dynamical astronomy 10 years later by formulating, in his *De Harmonica Mundi* (1619), his third law which connects the periods of revolution of the planets with their mean distances from the Sun. In 1627 he published the *Tabulae Rudolphinae*, named after the emperor, which contained the ephemerides of the planets according to the new laws, and also an extended catalogue of 1005 stars based on Tycho's observations. Kepler died of fever at Regensburg where he had journeyed to the Diet in quest of funds.

KERNER, Ritter von Marilaun, Anton Joseph

(1831–98)

Austrian botanist, born in Mautern. From an early age he was interested in the natural history of his local countryside but he went on to study medicine at Vienna. He began his career as a high school teacher at Budapest (1855–8) and teacher at Budapest Polytechnic (1858–60). From 1860 to 1878 he was Professor of Botany and director of the botanic garden at the University of Innsbruck. During this period he published *Das Pflanzenleben der Donaulaender* (1863) which Henry S Conard (1874–1971), translator of the English version (1950), described as 'the immediate and direct parent of all later works on plant ecology'. He also wrote *Die hybriden Orchideen der österreischischen Flora* (1865), *Novae Plantarum Species* (1870–1), *Die Vegetationsverhältnisse des mittleren und östlichen Ungarns* (1867–79) and *Monographia Pulmonariarum* (1878). From 1878 to 1898 he was director of the Vienna botanical garden, where he established a new herbarium and began his *Schedae ad Floram Exsiccatam Austro-hungaricam* (10 fascicles, 1881–1913). His *Pflanzenleben* (1886–98) was very influential and popular, being translated into English as *The Natural History of Plants* (1894–5) by Francis Wall Oliver (1864–1951). Kerner is regarded as an outstanding floristic botanist and plant geographer, and the founder of geobotanical science in Hungary.

KERR, John

(1824–1907)

Scottish physicist, born in Ardrossan, the son of a fish merchant. Educated at Glasgow University in theology, he became a lecturer in mathematics. He was one of the first research students of **Kelvin**. In 1876 he discovered the magneto-optic effect named after him; a beam of plane polarized light when reflected from the polished pole of an electromagnet will become elliptically polarized, with the major axis rotated from the original plane. The theoretical implications were later elucidated by **Fitzgerald**. Kerr was the author of *An Elementary Treatise on Rational Mechanics* (1867). He was elected FRS in 1890.

KERR, Roy Patrick

(1934–)

New Zealand mathematician, born in Kurow. After receiving an MSc in New Zealand and a PhD from Cambridge, Kerr returned to New Zealand to become Professor of Mathematics at the University of Canterbury, Christchurch. His main contribution has been in the field of astrophysics. In

1916, **Schwarzschild** had introduced the idea that when a star contracts under gravity, there will come a point at which the gravitational field is so intense that nothing — not even light — can escape, and a black hole is formed. Schwarzschild had derived a mathematical description of the properties of such an object and its gravitational distortion of the surrounding space and time, with the assumption that the black hole is not rotating. However, the condition of non-rotation is unrealistic as almost all stars are found to rotate, and it is thought that this rotation would be preserved during gravitational collapse. Kerr found a new solution to Einstein's equations taking account of the resulting angular momentum to give the 'Kerr metric', an expression which completely describes the properties of any black holes which physicists expect to exist. In later work, he formulated the Kerr–Schild solutions, which were very useful in exact solution of the equations of general relativity.

KETTLEWELL, Henry Bernard David
(1907–79)

English geneticist and entomologist, born in Howden, Yorkshire. Educated at Charterhouse School and in Paris, he studied medicine at Gonville and Caius College, Cambridge, and St Bartholomew's Hospital, London. After World War II, he forsook general medicine and from 1952 he held various posts in the genetics unit of the zoology department at Oxford University. He was primarily a first-rate naturalist, although his best-known research was concerned with industrial melanism, particularly of the peppered moth. This common moth developed a dark-coloured morph in areas where industry and dense populations caused atmospheric carbon pollution. He pointed out the survival value of the dark coloration in industrial regions and the original light coloration in rural areas, thus demonstrating the effectiveness of natural selection as an evolutionary process. These results changed the prevalent understanding because Kettlewell (in collaboration with **Tinbergen**) filmed differential predation in action, the first example of the value of film in that way. His work is summarized in the *Evolution of Mechanism* (1973).

KHARASCH, Morris Selig
(1895–1957)

American chemist, born in Kremenets, the Ukraine. After migrating with his family to the USA when he was 13 years old, he studied chemistry at the University of Chicago and graduated with a PhD in 1919. He was first a Research Fellow in Chicago and then professor at the University of Maryland, but in 1928 he returned to Chicago as professor. He remained there until his death. He is best known for his studies of addition of hydrogen halides to olefinic double bonds. He showed that, in the absence of peroxides, addition obeys the **Markovinkov** rule (hydrogen adds to the carbon atom which already has the larger number of hydrogen atoms). When peroxides are present, even in the quantities present in an old reagent, addition occurs in the opposite sense due to the formation of radicals. This work made sense of many conflicting reports in the chemical literature. Kharasch's work was important in the development of radical chemistry. During World War II he worked on polymerization mechanisms as part of the American synthetic rubber programme.

KHINCHIN, Aleksandr Yakovlevich
(1894–1959)

Russian mathematician, born in Kondrovo. He studied at Moscow University, and became professor there in 1927. With **Kolmogorov** he founded the Soviet school of probability theory; he also worked in analysis, number theory, statistical mechanics and information theory.

KHORANA, Har Gobind
(1922–)

Indian-born American molecular chemist, born in Raipur (now in Pakistan). He studied at Punjab University, receiving his BSc in 1943 and an MSc in 1945. He was awarded a PhD in organic chemistry by Liverpool University, and was a Research Fellow at Cambridge before moving to Vancouver as Head of the Department of Organic Chemistry (1952–60). During 1960–70, he was professor and co-director of the Institute of Enzyme Research at the University of Wisconsin, and since 1970, he has been Professor of Biology and Chemistry at MIT. His early work was on the biochemistry of enzymes, but in the 1960s he turned to the nucleic acids and the genetic code. He determined the sequence of the nucleic acids, also known as 'bases', for each of the 20 amino acids in the human body. Each amino acid was usually found to have a pattern of three base codes, but Khorana discovered some with more than one triplet sequence of bases, and showed that some triplets code for start/stop sequences. His work on nucleotide synthesis at Wisconsin was a major contribution to the elucidation of the genetic code. In the early 1970s, he was one of the first to artificially synthesize a gene, initially from yeast, and then later from the bacterium *Escherichia coli*. He shared the 1968 Nobel Prize for Physiology or Medicine with **Nirenberg** and **Holley**.

KIBBLE, Thomas Walter Bannerman
(1932–)

British physicist, born in Madras, India. He was educated at the University of Edinburgh, and since 1959 has worked at Imperial College, London, where he was appointed Professor of Theoretical Physics in 1970. He has worked on theoretical high-energy particle physics and quantum field theory, especially on symmetry breaking. In recent years much of his research has been on the interface between particle physics and cosmology, in particular the implications of phase transitions occurring very early in the history of the universe and of defects, such as 'cosmic strings', generated at these transitions. He was elected FRS in 1980, and received the Royal Society's Hughes Medal in 1981.

KILBURN, Tom
(1921–)

English computer scientist, born in Dewsbury. Educated at Sidney Sussex College, Cambridge, where he read mathematics, his wartime work at the Telecommunications Research Establishment at Malvern brought him into contact with Sir **Frederic Calland Williams**. When the latter moved to Manchester University in 1946, Kilburn was seconded to follow him and together they built the world's first operational stored-program computer in 1948. Kilburn helped perfect the storage device (the world's first electronic random access memory) and published the results. He was awarded his PhD in 1948. In the early 1950s, Williams handed over to Kilburn the direction of the university's computer projects, which by now included important collaborative ventures with the Manchester-based electronics firm of Ferranti. With Kilburn's help, Ferranti introduced a string of technically successful computers, such as the Mark I (1951) and the Atlas (1962). Kilburn's design for the Atlas was a high water mark in British computing, pioneering many modern concepts in paging, virtual memory and multi-programming. Kilburn became Professor of Computer Science at Manchester (1964–81) and was elected FRS in 1965.

KILBY, Jack St Clair
(1923–)

American electrical engineer, born in Jefferson City, Montana, the son of an electrical engineer who was president of the local power company. After being turned down by MIT, he went to the University of Illinois to study electrical engineering and then the University of Wisconsin, where he received a master's degree in 1950. In 1948 he joined Centrelab, a large radio and television parts manufacturer in Milwaukee, where he gained valuable experience with miniaturization and automation. In 1958 he joined Texas Instruments, an innovative Dallas firm, and almost immediately created the first monolithic integrated circuit—a circuit known as a phase-shift oscillator (a device that oscillates signals at a given rate). It was patented in 1959 and publicized by Texas Instruments as 'a semiconductor solid circuit no larger than a match head'. Further development work by **Noyce** at Fairchild resulted in the introduction of the first commercial integrated circuits in the early 1960s. Kilby became a freelance inventor after 1970 and has registered some 50 patents for his work on integrated circuits. He became Distinguished Professor of Electrical Engineering at Texas A & M University in 1978.

KINGDON WARD, Francis (Frank Kingdon-Ward)
(1885–1958)

English plant collector, geographer and author, born in Manchester and educated at Christ's College, Cambridge. In 1907 Kingdon-Ward (as he is universally known) became a teacher at Shanghai Public School and in 1909 made his first journey into the Chinese interior. This expedition changed the course of his life; in 1911 he became a professional plant collector for Bees of Liverpool. He made a total of 24 expeditions; all except the last (Ceylon, 1956–7) were to the remote borderlands of India, Myanma (Burma) and China. He introduced many beautiful plants into cultivation, including the Himalayan blue poppy (*Meconopsis betonicifolia*) and the tea-rose primula (*Primula agleniana* variety *thearosa*). He was the first to explore many of these regions, and in parts of northern Myanma has never been followed. Nearly all our knowledge of the botany of northern Myanma is owed to Kingdon-Ward's explorations and papers. Also an excellent geographer, through his intimate knowledge of the region he developed some novel theories on the geography and palaeogeography of the Himalayas, the parallel river systems of Myanma and western China and the region's phytogeographical history. He was a prolific author; his 25 books include *On the Road to Tibet* (1910), *The Land of the Blue Poppy* (1913), *In Farthest Burma* (1921), *Plant Hunter's Paradise* (1937), *Assam Adventure* (1941) and *Pilgrimage for Plants* (1960).

KINGSLAKE, Rudolf

(1903–)

English-born American optical designer, born and educated in London. He moved to the USA in 1929 to help set up the Institute of Optics at the University of Rochester. In 1937 he left the university to become director of the optical design department at the Eastman Kodak Company. There he directed a large team of optical designers who devised a wide range of optical systems, including zoom telescopes for tanks and a famous range of aerial camera lenses for survey work and mapping. On his retirement in 1967 he returned to Rochester University, from which he finally retired as Emeritus Professor in 1983. Kingslake holds many patents on the design of specific optical systems, and continues to work and to publish books and papers; his *Lens Design Fundamentals (1978) is a classic. He was Ives medallist of the Optical Society of America in 1973.*

KINSEY, Alfred Charles

(1894–1956)

American sexologist and zoologist, born in Hoboken, New Jersey. He was Professor of Zoology at Indiana from 1920 and in 1942 was the founder director of the Institute for Sex Research there for the scientific study of human sexual behaviour. Along with W B Pomeroy and C E Martin in 1948 he published *Sexual Behavior in the Human Male*, the so-called 'Kinsey Report'. This was based upon 18 500 interviews and attracted much attention from the general public as well as from fellow scientists. It appeared to show a greater variety of sexual behaviour than had previously been suspected. However, although careful statistical procedures had been followed, the report was much criticized for the interviewing techniques used. In 1953 there followed *Sexual Behavior in the Human Female*.

KIPP, Petrus Jacobus

(1808–64)

Dutch chemist, born in Utrecht. He started a business manufacturing laboratory apparatus and invented the apparatus named after him, which soon became a standard item of laboratory equipment. It consists of two glass vessels arranged in such a way that any gas produced by the action of liquid on a solid without heating—for example, carbon dioxide, hydrogen and hydrogen suphide—can be continually evolved. A representation of Kipp's apparatus appears on the arms of the Dutch Chemical Society. Kipp also invented a method of treating charcoal and pastel drawings so that they did not smudge.

KIPPING, Frederick Stanley

(1863–1949)

English chemist, born in Manchester. Having become interested in chemistry while at school, he enrolled as a student of chemistry at Owens College, Manchester, in 1879. After completing his degree he worked first for the Manchester Gas Department, but prospects for advancement were poor and so he moved to Munich (1886) to work for a PhD in **Baeyer**'s laboratory. His supervisor was **William Henry Perkin Jr** and they became close friends. After graduation he worked with Perkin at Heriot-Watt College in Edinburgh and moved to what is now Imperial College, London, in 1890. In 1897 he became professor at University College in Nottingham (now Nottingham University) and it was there that he published a series of pioneering papers on the organic compounds of silicon. His aim was to prepare silicon analogues of carbon compounds, particularly those with double bonds. In the course of this work he produced many polymeric materials which he called silicones, a name now given to all oxygen-containing organosilicon polymers. When he retired in 1936 he expressed the view that his work on silicon compounds could have no practical value. Within a few years the value of silicones as inert, water-repellent polymers had been recognized and silicones are now important commercial materials. Kipping was elected FRS in 1897 and became well known to generations of students through his book, written with Perkin, *Organic Chemistry* (1894).

KIRCHHOFF, Gustav Robert

(1824–87)

German physicist, born in Königsberg (now Kaliningrad, Russia) into a prosperous middle-class family. While still a student, he devised 'Kirchhoff's laws' for electrical circuits, used in network analysis. Professor at Heidelberg (1854–75) and Berlin (1875–86), he distinguished himself in electricity, heat, optics and especially (with **Bunsen**) spectrum analysis, which led to the discovery of caesium and rubidium (1859). More importantly it resulted in his fundamental paper in which he explained the production of the Fraunhofer lines in the solar spectrum by the absorption of the corresponding spectral wavelengths in the atmosphere of the Sun. He later formulated Kirchoff's law of radiation, according to which the ratios of the emissive to the absorptive powers were the same for all bodies at a given temperature for radiation of a given wavelength. This was the key to the whole thermodynamics of radiation which, in the hands of his successor **Planck**, would be developed into the concept of quanta. Kirchhoff's electromagnetic theory of diffraction, despite certain limitations, is still the form of diffraction theory most commonly used in optics. He was an excellent teacher, whose published lecture notes influenced several generations of German students.

KIRKWOOD, Daniel
(1814–95)

American astronomer, born in Harford County, Maryland. He became a teacher in 1833, and later Principal of Lancaster High School (1843–9). He was appointed Professor of Mathematics at Delaware (1851) and at Indiana (1856). In 1891 he became a lecturer at the University of Stanford. Kirkwood is famous for explaining, in 1866, the unequal distribution in the semi-major axes of asteroid orbits. This parameter represents the average distance between an asteroid and the Sun. Asteroids with semi-major axes of 2.5, 2.95 and 3.3 astronomical units are missing, these orbits being in resonance with Jupiter. Jovian gravitational perturbation has moved the asteroid orbits away from these 'Kirkwood gaps'. Kirkwood also used his theory to explain the gaps in the rings of Saturn, the perturbers this time being the Saturnian satellites.

KIRWAN, Richard
(1733–1812)

Irish chemist, born in Galway. He was educated at the University of Poitiers with the idea of becoming a Jesuit, but gave up the idea when he became heir to the family estates on the death of his brother. After practising briefly as a lawyer, Kirwan spent 10 years in London and was elected FRS in 1780. On his return to Ireland he helped to found the Royal Irish Academy, presiding over it from 1799 to his death. Kirwan did valuable work on chemical affinity and the composition of salts, publishing the first systematic work on mineralogy in English in 1784. He is best known, however, for his opposition to the discoveries of **Lavoisier**, instead maintaining the traditional view that when a substance burns it loses a vital essence referred to as 'phlogiston'. He also challenged the revolutionary views of the Scottish geologist **Hutton**, who in 1785 argued that the Earth was hot and that crystalline rocks were igneous in origin. Towards the end of his life, Kirwan turned to metaphysics. He died in Dublin.

KITASATO, Baron Shibasaburo
(1852–1931)

Japanese bacteriologist, born in Oguni. After graduating from the Imperial University of Tokyo (1883), he moved to Berlin to study under **Koch** and later founded in Japan an institute for infectious diseases. Arthur Nicolaier had discovered the tetanus bacillus in 1884, and by using anaerobic methods, Kitasato succeeded in isolating the first pure culture in 1889. In collaboration with **Behring**, he made the invaluable discovery of antitoxic immunity (1890). They found that after receiving injections of non-lethal doses of tetanus toxin, animals became resistant to the disease. This led to the development of treatments and immunization for tetanus and diphtheria. He later discovered *Pasteurella pestis*, the bacillus of bubonic plague (1894, independently of **Yersin** who discovered it in the same year in Hong Kong), and isolated the bacilli of symptomatic anthrax (1889) and dysentery (1898).

KJELDAHL, Johan Gustav Christoffer Thorsager
(1849–1900)

Danish chemist, born in Jagerpris. He studied chemistry at the Technical Institute in Copenhagen and was appointed as an instructor at the Agricultural College. In 1875 he set up a laboratory for the Carlsberg brewery, and when the laboratory was transformed into a research institute in 1876, he became its director, remaining in this post until his death. Kjeldahl's work was mostly in agricultural chemistry, and his great contribution was to discover a method of estimating the nitrogen content of organic substances which was quicker, cheaper and more reliable than former methods. Now known by his name, it is based on the fact that in the presence of concentrated sulphuric acid, potassium permanganate converts organic nitrogen into ammonia, which can subsequently be measured by titration. Kjeldahl died in Tisvildeleje.

KLAPROTH, Martin Heinrich
(1743–1817)

German analytical chemist who discovered many elements. He was born in Wernigerode and grew up in poverty, training as an apothecary. He moved to Berlin in 1768 and married the niece of **Marggraf**, who brought him some scientific connections and enough money to set up his own shop. From 1792 onwards he held various lectureships, becoming Germany's leading chemist and on the foundation of the University of Berlin in 1810 the Professor of Chemistry there. Between 1789 and 1803 Klaproth discovered six new elements. Working independently of **Hope** and **Humphry Davy** he investigated strontium; he also confirmed the new element discovered by **Franz Joseph Müller** and named it tellurium. Klaproth's analytical techniques were as important as his discoveries: he found ways of treating particularly insoluble compounds, made adjustments to overcome contamination from his apparatus, and insisted on reporting discrepant as well as consistent results. He was also one of the first scientists outside France to propagate the revolutionary ideas of **Lavoisier**. He died in Berlin.

KLEIN, (Christian) Felix
(1849–1925)

German mathematician, born in Düsseldorf. He studied at Bonn (1865–8), and became Professor of Mathematics at Erlangen University (1872–5). He later accepted professorships at Leipzig (1880–6) and Göttingen (1886–1913), where he spent the rest of his life and did much to make Göttingen University the world centre of mathematics. His 'Erlanger Programm', which was published in 1872, showed how different geometries could be classified in terms of group theory. His subsequent work on geometry included studies of non-Euclidean geometry, function theory (in which he developed **Riemann**'s ideas), and elliptic modular and automorphic functions, where he was in productive competition with **Poincaré** until his health collapsed from the strain. Aware that he could never attain his previous level of work, he became an influential teacher and organizer, while continuing to conduct research and supervise many graduate students. He encouraged links between pure and applied mathematics and engineering, promoted general mathematical education, and organized the *Encyklopädie der Mathematischen Wissenschaften*, published in 23 volumes between 1890 and 1930. He also wrote trenchantly on the history of mathematics.

KLINGENSTIERNA, Samuel
(1698–1765)

Swedish mathematician and scientist, born in Linköping. He studied law at Uppsala, but subsequently turned to mathematics and physics. He was appointed secretary to the Swedish Treasury, and given a scholarship to travel and study. He studied under Christian von Wolff at Marburg and **Jean Bernoulli** at Basle. He was appointed Professor of Mathematics at Uppsala, and in 1750 Professor of Physics there; he became tutor to the crown prince (later Gustav III) in 1756. Klingenstierna showed that some of **Isaac Newton**'s views on the refraction of light were incorrect, and designed telescope lenses which substantially reduced image defects arising from chromatic and spherical aberrations. By communicating his findings to **Dollond**, he contributed also to Dollond's success in constructing achromatic telescopes.

KLITZING, Klaus von
See **VON KLITZING, Klaus**

KLUG, Sir Aaron
(1926–)

Lithuanian-born English biophysicist, born in Zelvas. He moved to South Africa as a young child and studied physics at the universities of Witwatersrand and Cape Town. He became a research student in the Cavendish Laboratory of Cambridge (1949–52) before moving to London to **Bernal**'s department at Birkbeck College (Nuffield Research Fellow, 1954–7) where he worked with **Rosalind Franklin** and became particularly interested in viruses and their structure. Following Franklin's death in 1958, he became Head of the Virus Structure Research Group at Birkbeck (1958–61) and returned to Cambridge in 1962 as a Fellow of Peterhouse and member of staff at the Medical Research Council's Laboratory of Molecular Biology, becoming its director in 1986. His studies employed a wide variety of techniques, including X-ray diffraction methods, electron microscopy, and structural modelling, to elucidate the structure of viruses such as the tomato bushy stunt virus and the polio virus. From the 1970s he also applied these successful methods to the study of chromosomes, and other biological macromolecules such as muscle filaments. He was elected FRS in 1969, awarded the Nobel Prize for Chemistry in 1982 and knighted in 1988.

KLUYVER, Albert Jan
(1888–1956)

Dutch microbiologist, born in Breda. He studied at Delft University, where he received a chemical engineering degree in 1910. After graduate work on biochemical sugar determinations, he was appointed to the Chair of General and Applied Microbiology at the Technical University of Delft in 1922, a position which he held until his death. He greatly influenced studies on the chemistry, biochemistry and intermediary metabolism of micro-organisms. His most important contribution was the statement that hydrogen transfer (the process of oxidation) is a fundamental feature of all metabolic processes. During World War II he studied microbial morphology using an electron microscope. He had considerable interest in commercial applications, and collaborated with the Netherlands Yeast and Alcohol Manufacturing Company.

KNOPOFF, Leon

(1925–)

American geophysicist, born in Los Angeles. He was educated at Caltech, where he graduated with a BS in electrical engineering (1944), an MS (1946), and a PhD in physics (1949). His first appointment was at the University of Miami in Oxford, Ohio (1948–50); since 1950 he has worked at the University of California at Los Angeles (UCLA) where he became Professor of Geophysics in 1959, Professor of Physics in 1961 and research musicologist in 1963. From 1972 to 1986 he was also Associate Director of their Institute of Geophysics and Planetary Physics. Knopoff devised the first representation theorem for the full seismic wave equation (1956) and made important advances in work on the diffraction of seismic waves, eg by the core of the Earth (1959). He was the first to apply long-period seismic array data to the interpretation of geological structures (1966), and deduced that North America comprises a number of different tectonic provinces (1974). In 1967 he pioneered numerical models to simulate seismicity and geological faulting, having in 1956 demonstrated the relationship between seismic energy release and velocity of slip on the fault; later he described the universal power law for the spatial distribution of earthquakes (1980), which has implications for the geometry of faults. He also showed that the value in the Gutenberg–Richter magnitude–frequency law is a characteristic of seismicity on a complex fault system (1992).

KNOX, Robert

(1791–1862)

Scottish anatomist, born in Edinburgh to a family claiming descent from John Knox the Protestant reformer. Knox studied medicine at Edinburgh University and in London before joining the army medical service just in time to tend the wounded at Waterloo. After army service in South Africa, he studied further in Paris prior to setting up an extramural anatomy school in his home town. His huge success with students proved his undoing; his need for a substantial supply of cadavers for dissection was met through the services of the disreputable 'ressurectionists' William Burke and William Hare, who, unknown to Knox, obtained their corpses not by grave-robbing but by murder. Knox was thereafter cold-shouldered by the medical establishment and never obtained university employment, continuing to lecture privately in Edinburgh, Glasgow and London. He published extensively on anatomy and physical anthropology, expounding extreme racist views in his successful *The Races of Men* (1850). His interest in the art/medicine interface found expression in *A Manual of Artistic Anatomy for the Use of Sculptors, Painters and Amateurs* (1852). Late in life, Knox succeeded in gaining employment as pathologist to the London Cancer Hospital, settling in Hackney.

KOCH, (Heinrich Hermann) Robert

(1843–1910)

German physician and pioneer bacteriologist, born in Klausthal in the Harz. In 1862 he entered the University of Göttingen, and he received his medical degree in 1866. He settled as a physician in Rakwitz, and later practised medicine at Hanover and elsewhere. His work on wounds, septicaemia and splenic fever gained him a seat on the Imperial Board of Health in 1880. Koch proved that the anthrax bacillus was the sole cause of the disease, and demonstrated that its epidemiology was a result of the natural history of the bacterium. This work was published in 1876 and 1877; later he quarrelled with **Pasteur** over anthrax. Despite his severe myopia, further researches in microscopy and bacteriology led to his discovery on 24 March 1882 of the tubercle bacillus that causes tuberculosis, responsible for one in seven of all deaths in Europe during this period. In 1883 he was leader of the German expedition sent to Egypt and India in quest of the cholera germ; for his discovery of the cholera bacillus he received a gift of £5 000 from the government. In 1890 he produced a drug named tuberculin which he claimed could prevent the development of tuberculosis. It was found to be ineffective as a cure, but later proved useful in diagnosis. He became Professor at Berlin and Director of the Institute of Hygiene in 1885, and first Director of the Berlin Institute for Infectious Diseases in 1891. In 1896 and 1903 he was summoned to South Africa to study rinderpest and other cattle plagues; in disclosing the causes of disease and expounding the means of prevention, Koch was unsurpassed. He won the Nobel Prize for Physiology or Medicine in 1905 for his work on tuberculosis. His formulation of essential scientific principles known as 'Koch's postulates' for investigating the causes of infectious diseases established clinical bacteriology as a medical science in the 1890s.

KOCHER, Emil Theodor

(1841–1917)

Swiss surgeon, born and educated in Bern. He became professor at the University of Bern in 1871. In the first generation of surgeons able to exploit the new possibilities of anti- and aseptic techniques, Kocher developed general surgical treatment of disorders of the thyroid gland, including goitre and thyroid tumours. His observations of patients suffering the long-term consequences of removing the thyroid gland helped elucidate some of its normal functions; by the 1890s, the isolation of one of the active thyroid hormones made replacement therapy possible. Kocher's Bern clinic attracted many young surgeons from all over the world. His textbook *Operative Surgery* (1894) went through many editions and translations. He kept meticulous case records, and followed up his former patients over long periods. He advocated the practice of 'physiological surgery', whereby the surgeon's aim was to keep operative trauma to a minimum, thereby conserving as much of the patient's tissues as possible. He pioneered operations of the brain and spinal cord, and during World War I did experimental work on the trauma caused by gunshot wounds. Kocher was the first surgeon to be awarded the Nobel Prize for Physiology or Medicine (1909), for his work on the physiology, pathology, and surgery of the thyroid gland.

KÖHLER, Georges Jean Franz
(1946–)

German immunochemist, born in Munich. After receiving his doctorate at the University of Freiberg in 1974, he joined **Milstein** at the Medical Research Council Laboratory in Cambridge, UK, where they discovered how to produce hybridomas—cell lines grown in culture that were derived (created) by the fusion of an antibody-generating cell from mouse spleen with a cancer cell. Hybridomas possess the property of infinite life and generate a colony of cells (clone) that produce a single type of antibody (monoclonal antibody) against a specific antigen (foreign body). Mixed antibody-producing cells generate a confusing mixture of antibodies, and the use of hybridomas opened the way to a precise examination of antibody structure. Köhler moved to the Basle Institute of Immunology in 1976, and in 1984 he became one of three directors of the Max Planck Institute of Immune Biology in Freiberg. He continued his research by using markers for drug resistance to study the pattern of inheritance of hybridoma cells, and demonstrated that structural mutants of immunoglobulins could be formed by hybridomas (1980). More recently he has studied the carbohydrate component of immunoglobulins by using the antibiotic tunicamycin, which specifically inhibits the attachment of carbohydrate to certain lipids. Commercially produced monoclonal antibodies now provide an unambiguous and sensititive way of identifying and quantifying a wide range of substances, from the animal species of the protein used in processed food, pregnancy testing and the concentrations of drugs, hormones or other metabolites in body fluids to diagnosing and treating cancer and other diseases. For this outstanding contribution to human wellbeing, Köhler shared with Milstein and **Jerne** the 1984 Nobel Prize for Physiology or Medicine.

KOHLRAUSCH, Friedrich Wilhelm Georg
(1840–1910)

German physicist, born in Rinteln, the son of a well-known physicist, Rudolph Kohlrausch (1809–58). He held professorships of physics successively at Göttingen (1866–70), Zürich (1870–1), Darmstadt (1871–5), Würzburg (1875–88) and Strassburg (1888–95), where he succeeded August Adolph Kundt (1839–94), and in 1895 he succeeded **Helmholtz** as President of the Physikalisch Technische Reichsanstalt in Charlottenburg, Berlin. He was noted for his researches on magnetism and electricity characterized by great precision. His most important contribution was his study of the conductivity of electrolytic solutions, which led to 'Kohlrausch's law' of the independent migration of ions. He wrote *Leitfaden der praktischen Physik* (1870), one of the first textbooks on physical laboratory methods, which was translated into English.

KO HUNG (Ge Hong in *pinyin* romanization)
(c.280–340 AD)

Chinese alchemist, born near Nanjing. He came from a relatively poor family but automatically received an education in the literary and philosophical traditions of his country. Although by temperament he would have preferred a quiet life of reading and contemplation, he became a successful and stern military commander. He was a Confucianist by confession but his most famous book *Pao-p'u tze* (or *Bao-pu zi*) is more in the Taoist (Daoist) tradition. The title has been translated in a number of ways; **Needham** gives *Book of the Preservation-of-Solidarity Master*. The book is the result of Ko Hung's travels and study with several teachers. It contains accounts of methods of producing solutions of minerals (including cinnabar and gold) in order to make immortality elixirs. He also describes the apparent production of gold from other metals. Some of his claims can be understood in terms of modern chemistry.

KOLBE, (Adolph Wilhelm) Hermann
(1818–84)

German chemist, born in Elliehausen, near Göttingen. He was the son of a Lutheran pastor and studied chemistry under **Wöhler** (1838). He was assistant to **Bunsen** in Marburg (1842) and to **Lyon Playfair** (1845) at the Museum of Economic Geology in London. In 1847 he returned to Marburg, and he succeeded Bunsen in the chair in 1851. In 1865 he went to Leipzig where he remained until his retirement. He was a successful teacher and a brilliant experimentalist, but his influence was weakened by the excessive sharpness and lack of balance of his language, particularly in his opposition to the development of structural formulae over which he was completely wrong. He accomplished a number of important syntheses, and is best known for his electrolytic procedure for the preparation of alkanes. He formulated his own version of the type theory of organic chemistry which he later abandoned with great reluctance.

KOLFF, Willem Johan
(1911–)

Dutch-born American physician, developer of the artificial kidney, born in Leiden. A medical student in Leiden, he received his MD in 1946 from Groningen University. Kolff constructed his first rotating drum artificial kidney in wartime Holland and treated his first patient with it in 1943. This dialysis machine used a series of membranes to remove impurities from the blood which would ordinarily be filtered out by the healthy kidney. From 1950, when he moved to the USA, he worked primarily at the Cleveland Clinic and Utah University, developing the artificial kidney further; he was also involved in research on the heart–lung machine used during open-heart surgery.

KÖLLIKER, Rudolph Albert von
(1817–1905)

Swiss anatomist and embryologist, born in Zürich. Kölliker attended his home-town university, before proceeding to Bonn and Berlin where he was powerfully influenced by **Johannes Müller** and **Remak**, and by **Henle**, with whom he developed a lasting and fruitful comradeship. He became Professor of Anatomy in Zürich in 1845, and later moved to Würzburg (1847). His early researches were on the spermatozoa of invertebrates; specializing in Mediterranean fauna, he proceeded to publish widely on cephalopods. Later he developed expertise in microscopic work on cell structure, employing cell theory for interpreting embryonic development. He opposed **Schwann**'s doctrine of free-cell formation in the cytoblast, regarding the segmentation of the egg as a continuous production of daughter cells. His *Manual of Human Histology* (1852) set new standards of exactitude for the subject. Kölliker's emphasis upon the significance of the nucleus in cell physiology helped to establish cytology as a specialization.

KOLMOGOROV, Andrei Nikolaevich
(1903–87)

Russian mathematician, born in Tambov. He studied at Moscow State University where he graduated in 1925, and remained there throughout his career, as professor from 1931 and Director of the Institute of Mathematics from 1933. He worked on a wide range of topics in mathematics, including the theory of functions of a real variable, functional analysis, mathematical logic, and topology. He is particularly remembered for his creation of the axiomatic theory of probability in his book *Grundbegriffe der Wahrscheinlichkeitsrechnung* (1933), which he interpreted in the language of measure theory introduced by **Lebesgue**. His work with **Khinchin** on **Markov** processes was also of lasting significance; in this he formulated the partial differential equations which bear his name and which have found wide application in physics and chemistry. Kolmogorov also worked in applied mathematics, on the theory of turbulence, on celestial mechanics where he refined ideas of **Poincaré**, on information theory and in cybernetics.

KOPP, Hermann Franz Moritz
(1817–92)

German chemist and historian of chemistry, born in Hanau. He studied chemistry at Heidelberg, Marburg and Giessen. In 1841 he became *Privatdozent* at Giessen, where he was promoted to Extraordinary Professor in 1843. When **Liebig** moved to Munich in 1852, Kopp became Ordinary Professor jointly with Heinrich Will, but resigned after one year. However, he continued to work in Giessen until he was appointed professor at Heidelberg (1863), where he remained until his death. He became an Honorary Fellow of the Chemical Society in 1849. His research work was largely concerned with the relationship between the physical properties of elements or compounds and their chemical composition. Kopp made many accurate measurements of properties such as boiling points, specific heats, densities and expansion coefficients, and developed the concepts of atomic volume and molecular volume. He published a great work on the history of chemistry, *Geschichte der Chemie*, in four volumes between 1843 and 1847.

KORNBERG, Arthur
(1918–)

American biochemist, born in Brooklyn, New York City. A graduate in medicine from Rochester University, he became director of enzyme research at the National Institutes of Health (1947–52) and head of the microbiology department of Washington University (1953–9). In 1959 he was appointed professor at Stanford University. When **Crick** and **James Watson** developed their model of the DNA molecule, the details of how DNA synthesis occurred and how DNA gives rise to protein remained unknown. In studies of *Escherichia coli*, Kornberg discovered DNA polymerase, the enzyme that synthesizes new DNA from a mixture of the four deoxynucleoside triphosphates, and showed that synthesis required a DNA template and base-pairing, giving two helical strands of opposite polarity as predicted by Crick and Watson. For this work he was awarded the 1959 Nobel Prize for Physiology or Medicine jointly with **Ochoa**. Around 1972 Kornberg showed that this bacterial enzyme (polymerase I) proof-read the new DNA for accuracy and could replace by DNA the RNA 'primer' to which the new DNA was initially joined in the **Okazaki** fragment. Kornberg became the first to synthesize viral DNA (1967) and wrote *DNA Replication* (1980). He also elucidated the synthesis of the coenzymes nicotinamideadeninedinucleotide (NAD) and flavineadeninedinucleotide (FAD), showed that acylCoA was the activated form of a fatty acid, and studied the *Escherichia coli* enzymes anthranilate synthetase and phosphopholipase A.

KORNBERG, Sir Hans Leo
(1928–)

German-born British biochemist, born in Herford. After studying at the University of Sheffield, he spent two years in postdoctoral work at Yale and in New York before joining Sir **Hans Krebs**'s Medical Research Council Cell Metabolism Research Unit in Oxford (1955). In 1961 he took up the new Chair of Biochemistry at the University of Leicester, and he later moved on to the Chair of Biochemistry at Cambridge (1975); he was elected Master of Christ's College, Cambridge, in 1982. Kornberg has worked mainly in microbial metabolism and is particularly noted for the discovery of the glyoxylate cycle, which explains how bacteria and fungi grow on fatty acids, and how many types of seed convert fats to carbohydrates during germination. He has also made important contributions to knowledge of carbohydrate transport in micro-organisms. Elected FRS in 1965, he has received many honorary degrees and awards, including the Colworth Medal of the Biochemical Society and the Warburg Medal of its German counterpart. He has been a member of several research councils, Chairman of the Royal Commission on Environmental Pollution, and a trustee of the Nuffield Foundation and the Wellcome Trust. He was knighted in 1978.

KOSSEL, Albrecht
(1853–1927)

Swiss-born German physiological chemist, born in Rostock. He studied medicine at the University of Strassburg and worked under **Hoppe-Seyler**. In 1895 he became Professor of Physiology and Director of the Physiological Institute, Marburg, and he was later professor at Heidelberg (1901–23). Kossel carefully isolated nuclei, freed from cytoplasm by treatment with proteases, from the heads of spermatozoa and from avian red cells, and showed them to contain nuclein, the nucleoprotein previously found and identified as rich in phosphorus by Friedrich Miescher (1844–95), another of Hoppe-Seyler's students. Kossel separated nuclein into its two components, protein and nucleic acid; he showed that the latter was the component rich in phosphorus and contained the four DNA bases, adenine, guanine (already known), thymine and cytosine, and carried out a partial identification of their structures. He was able to explain that in a blood leukaemia, the 'guanide' found in the blood in large amounts derived from decomposed young nucleated erythrocytes. He also discovered histidine in spermatozoa (1896). In 1910 Kossel was awarded the Nobel Prize for Physiology or Medicine.

KOSSEL, Walther
(1888–1956)

German physicist, son of **Albrecht Kossel**. Appointed Professor of Physics at Kiel (1921) and Danzig (1932), he did much research on atomic physics, especially on **Röntgen** spectra, and was known for his physical theory of chemical valency.

KOSTERLITZ, Hans Walter
(1903–)

German-born Scottish pharmacologist, educated at Heidelberg, Freiburg and Berlin universities. He became an assistant in the medical department at Berlin from 1928 until 1933, working as a radiologist under **His**. In the wake of the Nazi rise to power he moved to the department of physiology, University of Aberdeen, Scotland, as he was impressed by the work of **Macleod** on carbohydrate metabolism. Kosterlitz gained British medical qualifications and a PhD and stayed in the department for many years (research worker 1934–6; teaching fellow 1936–9; lecturer 1939–45; senior lecturer 1945–55; reader 1955–68). His researches during this period concentrated on carbohydrate biochemistry although in the early 1950s he turned to the physiology and pharmacology of the autonomic nervous system, and began examining the effects of morphine and other opiate-like drugs on gastrointestinal motility. In 1968 he was appointed Professor of Pharmacology (1968–73) subsequently becoming, upon retirement, director of the university's drug addiction research unit. Here he was joined by **Hughes** and together they discovered the existence of naturally occurring opiates already suggested by the investigations of **Snyder** and **Pert**. This they did in experiments comparing the physiological effects of known opiates on strips of guinea-pig intestine, with extracts made from brain tissue tested on identical strips of intestine. They discovered that the two materials had the same effects, and that the brain extract was inhibited by naloxone, a known inhibitor of morphine. Further analysis revealed two almost identical chemicals which they named enkephalins, and quickly proved that they were powerful analgesics. Their work, in conjunction with that of Snyder and Pert accelerated hopes of producing powerful but non-addictive pain killers, and has been an important stimulus in recent studies in brain chemistry and pharmacology.

KOVALEVSKAYA, Sofya Vasilyevna
(1850–91)

Russian mathematician and novelist, born in Moscow, daughter of an artillery officer. Married to a brother of **Kovalevsky**, she made a name for herself throughout Europe as a distinguished mathematician. She received a private education in mathematics, and studied at the University of Heidelberg under **Helmholtz** (1869) and at Berlin under **Weierstrass** (1871–4), with whom she maintained a lifelong friendship. As a woman she could not obtain an academic post in Europe, but in 1884 she was appointed to a lectureship at the University of Stockholm, where she became professor in 1889. She first worked on the theory of partial differential equations, where a central result on the existence of solutions still bears her name. Subsequently she studied Abelian integrals, and their use in analysis of the motion of a top (or any other rotating body) and the structure of Saturn's rings. She was also a talented novelist and playwright whose works are still performed. Her works included *Vera Brantzova* (1895).

KOVALEVSKY, Alexandr Onufrievich
(1840–1901)

Russian embryologist, born in Dünaburg. Professor at St Petersburg, Kassan, Kiev and Odessa, he studied the embryological development of primitive animals with chordate affinities such as *Balanoglossus*, *Sagitta*, the Brachiopoda and *Amphioxus*, the latter work published in *Development of Amphioxus lanceolatus* (1887). He established for the first time that there are common elements in the patterns of development in many animals and that it is possible to use these patterns to elucidate evolutionary relationships. His work provided **Haeckel** with evidence for his influential but now discredited theory of recapitulation which states that each organism retraces its evolutionary history during its embryonic development. He was elected to the Russian Academy of Sciences in 1890.

KREBS, Edwin Gerhard
(1918–)

American biochemist, born in Lansing, Iowa. Krebs trained at the University of Illinois and with **Carl Cori** before joining the Howard Hughes Medical Institute and Department of Pharmacology at the University of Washington School of Medicine. Stimulated by Cori's discovery of the activation of glycogen phosphorylase by adenylic acid, Krebs and **Edmond Fischer** showed that phosphorylation–dephosphorylation processes are involved, catalysed by two enzymes, phosphorylase kinase and phosphorylase phosphatase, respectively (1955). In 1959 they went on to show that phosphorylase kinase is also regulated by phosphorylation–dephosphorylation and that cyclic AMP (an adenylic acid derivative discovered by **Earl Sutherland**) was involved. These initial findings led to the discovery of the phosphorylation amplification cascade of enzymes that initiates the rapid switching on of glycogen phosphorylase and other enzymes, notably pyruvate dehydrogenase, under the influence of hormones such as glucagon and adrenaline (epinephrine). Similar systems controlled by other activators, such as protein kinase C by diacylglycerol, were also subsequently discovered. His more recent work relates to the homology of the kinases (of evolutionary significance) and the properties of the phosphatases. With Fischer, Krebs was awarded the 1992 Nobel Prize for Physiology or Medicine. He was elected FRS in 1947 and knighted in 1958.

KREBS, Sir Hans Adolf
(1900–81)

German-born British biochemist, born in Hildesheim. After training in medicine and working as assistant to **Warburg** at the Kaiser Wilhelm Institute for Cell Physiology, Berlin, he emigrated to the UK (1934) and worked with **Frederick Gowland Hopkins** on redox reactions. He became a lecturer in pharmacology (1935–45) and Professor of Biochemistry at Sheffield (1945–54), then Whitley Professor of Biochemistry at Oxford (1954–67). In 1932, with K Henseleit, he described the urea cycle whereby carbon dioxide and ammonia form urea in the presence of liver slices and catalytic amounts of ornithine and citrulline (previously found only in citrus fruits). Leading on from his earlier work, showing that all the hydrogens of glucose passed through fumarate and that pyruvate had a central role in this process, he discovered that citrate also acted as a catalyst while citrate could be formed from oxaloacetate. In this way he elucidated the citric acid cycle (Krebs' cycle) of energy production (around 1943). He also discovered D-amino acid oxidase, L-glutamine synthetase, purine synthesis in birds, that ketone bodies were formed in starvation and that rat heart muscle preferentially utilized ketone bodies (1961). In 1953 he shared with **Lipmann** the Nobel Prize for Physiology or Medicine for his discovery of the citric acid cycle. He was elected a Fellow of the Royal Society in 1947 and was awarded the society's Royal (1954) and Copley (1961) medals.

KREBS, John Richard
(1945–)

English zoologist, born in Sheffield. He was educated at Oxford (1963–9) and taught at the universities of British Columbia, Vancouver (1970–3), Bangor, north Wales (1973–4) and Oxford, where he is now Royal Society Research Professor. He is also Director of the Agricultural and Food Research Council Unit of Ecology and Behaviour and the Natural Environment Research Council Unit of Behavioural Ecology in Oxford. His research has been concerned with demonstrating the interrelationship between ecology and animal behaviour. In particular he has been interested in strategies adopted by animals which maximize their fitness. Working mainly with birds, he has studied aspects of foraging for food, territoriality and sexual behaviour. With co-workers he has developed predictive mathematical models to account for the patterns of behaviour observed. His *An Introduction to Behavioural Ecology* (1981, with N B Davies) is the standard text in its field. He was elected FRS in 1984.

KREMER, Gerhard
See **MERCATOR, Gerardus**

KROGH, (Schack) August (Steenberg)
(1874–1949)

Danish physiologist, born in Grenaa, the son of a brewer. Krogh graduated with a PhD in 1903 from Copenhagen University and worked there for the rest of his career, serving as Professor of Animal Physiology from 1916 to 1945. Initially he worked on problems of respiration, arguing for the currently accepted view that the absorption of oxygen and elimination of carbon dioxide in the lung take place by diffusion. His later work was on the capillary system. He showed that the idea that blood flow through capillaries is controlled simply by blood pressure is incorrect. In a group of capillaries fed by the same arteriole some are dilated, others constricted. The greater the activity of the tissue, the greater is the proportion of its capillaries that are dilated. He won the Nobel Prize for Physiology or Medicine in 1920 for this discovery. He later showed that the capillaries are under nervous and hormonal control, and published *The Anatomy and Physiology of Capillaries* in 1922.

KRONECKER, Leopold
(1823–91)

German mathematician, born in Liegnitz, the son of a wealthy businessman. He obtained his doctorate at Berlin (1845), where he was taught by **Dirichlet** and **Kummer**, spent a period at home managing the family estate, and then returned to Berlin in 1855 where, as an active member of the Berlin Academy of Sciences, he lived as a private scholar. He worked in algebraic number theory, elliptic functions and the foundations of analysis, and lectured widely. Although he did not attract the large audiences of **Weierstrass** and Kummer at Berlin, he was widely regarded as personally accessible and enthusiastic. He was involved in a controversy with Weierstrass and **Georg Cantor** over the use of the infinite in mathematics, as he believed that mathematics should be essentially based on the arithmetic of the whole numbers. This long, simmering disagreement eventually weakened the department at Berlin. At one point he declared that 'God made the integers; all the rest is the work of man'. His work is marked by a preference for explicit constructions and algorithms over general, but more abstract methods, and is often found to be technical and complex by even the most accomplished of his successors.

KROTO, Harold Walter
(1939–)

English chemist, born in Wisbech and educated at the University of Sheffield (BSc

1961, PhD 1964). He then moved to the National Research Council, Ottawa, where he developed his interest in the electronic spectroscopy of free radicals (which he had begun in his PhD days) and in microwave spectroscopy. In 1966 he moved to Bell Telephone Laboratories, New Jersey, where he carried out Raman spectroscopic studies of liquids and quantum chemistry calculations. Kroto is distinguished for his work in detecting unstable molecules, especially those containing reactive multiple bonds, using methods such as microwave and photoelectron spectroscopy. His studies extend to molecules which exist in interstellar space. He discovered, along with astronomers at the Herzberg Research Institute of Astrophysics at Ottawa, poly-yne molecules in interstellar space — the most complex and heaviest interstellar molecules known. In 1985, together with co-workers at Rice University, Texas, he discovered the third allotrope of carbon C_{60}, known as 'buckminsterfullerene' (familiarly 'buckyballs') after the American architect Buckminster Fuller. The 'football' shape of C_{60}, built up from an array of pentagonal and hexagonal faces, has the same topology as the buildings designed by this architect. Kroto was elected FRS in 1990, and is now Royal Society Research Professor at University of Sussex.

KUENEN, Philip Henry
(1902–76)

Dutch geological oceanographer, born in Dundee, Scotland, of a Dutch father, then a physics professor, and an English mother. He was educated at Leiden, receiving his degree in 1922 and a doctorate (1925) for studies of the porphyry district of Lugano. An experimental laboratory for the study of geological problems was established; this had a lasting influence on Kuenen who remained at Leiden until 1934. In 1929–30 he participated in the Snellius expedition to the Netherlands Indies (now Indonesia), conducting surveys complimentary to **Vening Meinesz**'s work on gravity. In 1934 he was lecturer and curator at Groningen, where he became reader (1939) and professor (1946–72). He was wounded and imprisoned briefly during the German invasion (1940). Kuenen's influence was mainly in the study of the deposition of coarse sediments from turbidity currents on which he wrote 50 papers, some in collaboration with **Heezen**, and the downbuckling of the crust with compressive stresses. He postulated submarine slumps and mudflows. In this rapidly evolving area of science, his *Textbook of Geology* (1950) was soon supplanted by **Shepard**'s *Submarine Geology*.

KUFFLER, Stephen William
(1913–80)

American neurobiologist, born in Tap, Hungary. He was educated in Vienna, and graduated from its medical school in 1937. Shortly afterwards he moved to the Kanematsu Institute in Sydney, Australia, where he worked with **Katz** and **Eccles**. In 1945 he moved to the Johns Hopkins School of Medicine until 1959, and then to Harvard as Professor of Neurobiology (1959–66) and John Franklin Enders University Professor (1966–80). He contributed extensively and significantly to many areas of neurobiology. For example, in Australia he began studying the mechanisms of synaptic transmission, firstly using an isolated frog-muscle preparation, and then studying a similar preparation in Crustacea. The latter work, done in conjunction with Katz, provided important guidelines for the mechanisms of excitatory and inhibitory transmission which later biophysical analyses with intracellular electrodes confirmed and substantially extended. In 1953 Kuffler began working on retinal physiology, initially using the light sensitive cells of the retina as tools in his examination of the higher functions of the brain. However he discovered that the cells of the retina had 'receptive fields', such that some cells were maximally excited if a light spot fell in the centre of their receptive field, while other cells were maximally inhibited under the same conditions. This pioneering work on the functional organization of the retina influenced the research that **Hubel** and **Wiesel** later undertook in Kuffler's department. He also initiated an innovative electrophysiological study of glial cells, the most numerous cells in the nervous system, from which he and his co-authors concluded that the role of these cells raised many tantalizing questions. One of his collaborators on the glial work was the British physiologist John Nicholls, with whom Kuffler wrote the stimulating text *From Neuron to Brain* (1976).

KUHN, Richard
(1900–67)

German chemist, born in Vienna-Döbling. After study at the University of Vienna he worked for his doctorate with **Willstätter** in Munich. In 1926 he moved to Zürich, and in 1929 to the Kaiser Wilhelm Institute for Medical Research in Heidelberg, where he remained for the rest of his life. His early work on enzymes led to an interest in problems of stereochemistry that preoccupied him subsequently. Work on conjugated polyenes led to important studies on carotenoids and vitamin A. Later work on vitamins B_2 and B_6 and on 4-aminobenzoic acid earned him the award of the Nobel Prize for Chemistry (1938). He was forbidden by the Nazi government to accept the award, but it was presented to him after World War II. His research continued actively after the war when he worked on resistance factors effective in preventing infections in both plants and animals. He and his collaborators published over 700 scientific papers.

KÜHNE, Wilhelm

(1837–1900)

German physiologist, born in Hamburg. Kühne trained in medicine, being taught by **Virchow** and **Bernard**. In 1871 he became Professor of Physiology at Heidelberg. He achieved fame through his study of the chemistry of digestion. Having worked on trypsin from pancreatic juice, he proposed the expression 'enzyme' (Greek for 'in yeast') to describe ferments and other organic substances that actuate chemical changes. He devoted much attention to proteins, studying postmortem change in muscle and finding rigor mortis to result from the action of the protein myosin. In the 1860s he succeeded in separating various types of egg albumen. When in 1876 Franz Christian Boll located a photosensitive protein pigment in the retina of a frog's eye, Kühne assumed the analysis of this visual purple (today called rhodopsin), and showed it was bleached by light and restored in the dark. The retina works, he revealed, like a renewable photographic plate, and he obtained a pattern of crossbars of a window on the retina of a rabbit, that had been kept in the dark, exposed to a window and subsequently killed. Further understanding of rhodopsin's mode of action was not achieved until new techniques were devised in the 1930s by **Wald**.

KUIPER, Gerard Peter

(1905–73)

Dutch-born American astronomer, born in Harenkarspel. Educated in Leiden, he moved to the USA in 1933. He took an appointment at the Lick Observatory in California, then taught at Harvard (1935–6), and joined the Yerkes Observatory before moving to the McDonald Observatory in Texas in 1939. From 1960 he worked at the Lunar and Planetary Laboratory of the University of Arizona. In 1941 Kuiper pioneered the study of contact binary stars, and he also suggested a system of spectroscopic classification of white dwarf stars. He discovered two new satellites: Miranda, the fifth satellite of Uranus; and Nereid, the second satellite of Neptune (1948–9). In 1951 he proposed that there is a flattened belt of some thousand million comets (now known as the Kuiper belt) just beyond the orbit of Pluto. This is thought to be the source of the short-period comets. Kuiper was the first to realize that the planets probably formed from a nebulous cloud that initially had a mass of around 10 per cent that of the Sun. In 1944 he was the first to confirm that a planetary satellite had an atmosphere, detecting methane on Titan. A Lockheed C-141 jet aircraft fitted with a 90 centimetre infrared telescope has been named the Kuiper Airborne Observatory, and his name has also been given to the 7 500 angstrom bands in the spectrum of Neptune and Uranus. He was involved with the early American space flights, including the Ranger and Mariner missions.

KUMMER, Ernst Eduard

(1810–93)

German mathematician, born in Sorau. He studied theology, mathematics and philosophy at the University of Halle, and then taught at the gymnasium in Liegnitz (1832–42), where **Kronecker** was among his students. He became known to **Jacobi** and **Dirichlet** through his work on the hypergeometric series, and was elected a member of the Berlin Academy of Sciences in 1839. He was later appointed Professor of Mathematics at Breslau (1842–55) and subsequently at Berlin (from 1855). Kummer worked in number theory, where, in generalizing **Gauss**'s law of quadratic reciprocity, he gained a significant insight on **Fermat**'s last theorem and proved it rigorously for many new cases. The 'ideal numbers' he introduced here were later developed by **Dedekind** and Kronecker into one of the fundamental tools of modern algebra. He also worked on differential equations and in geometry, where he discovered the quartic surface named after him.

KURCHATOV, Igor Vasilevich

(1903–60)

Soviet physicist, born in Sim, Russia, and educated at Simferopol Gymnasium and the University of Crimea. He was appointed director of nuclear physics at the Leningrad Physical-Technical Institute (1938) and, before the end of World War II, of the Soviet Atomic Energy Institute. His early research was into dielectrics, ie electrically non-conducting material. However, he later began to transfer his interest to study of the atomic nucleus around 1932 in work at the Leningrad Physical-Technical Institute. Here he supervised the construction of what was the world's largest cyclotron particle accelerator. In 1934, with L I Russinov, he discovered nuclear isomers whilst irradiating bromine.

During the war he developed methods of protecting ships from mines, and carried out tests under fire, for which he was awarded the State Prize (1942). He carried out important studies of neutron reactions and was the leading figure in the building of Russia's first nuclear fission (1949) and hydrogen bombs (1953), and the world's first industrial nuclear power plant (1954). He became a member of the Supreme Soviet in 1949.

KURTÉN, Björn

(1924–88)

Finnish palaeontologist, born to a Swedish-speaking Finnish family in Vaasa on the west coast of Finland. He studied zoology and geology at the universities of Helsinki and Uppsala, and from 1955 was a lecturer at the University of Helsinki, where he became Professor of Palaeontology in 1972. His 1953 doctoral thesis *On the variation and population dynamics of fossil and recent mammal populations*, subsequently published, laid the foundations of his pioneering work on evolution in Pleistocene (Ice Age) mammal fossils. His thoroughly biological approach to palaeontology made him one of the first to explore genetical and developmental aspects of fossil populations, such as growth patterns, polymorphisms and evidence for natural selection. His *Pleistocene Mammals of Europe* (1968) became a standard work. Specializing in Carnivora, he produced numerous publications on their evolution, including *The Cave Bear Story* (1976). He travelled widely, and his visits to the University of Florida (1963–4) and Harvard (1970–1) culminated in the publication of *Pleistocene Mammals of North America* (1980, with Elaine Anderson). He loved writing and was a committed popularizer, producing countless newspaper and magazine articles and a series of semi-popular books. Beginning in the early 1940s he wrote several novels, some set in prehistory, such as *Dance of the Tiger* (1978). In 1988, shortly before his death, he was awarded UNESCO's Kalinga Prize for the popularization of science.

KUSCH, Polykarp

(1911–)

German-born American physicist, born in Blankenburg. He became a naturalized US citizen in 1922. He obtained his BS from the Case Institute of Technology, Cleveland, in 1931 and his PhD from Illinois University in 1936. Later he became Professor of Physics at Columbia University (1937–72) and at Texas (from 1972). In 1937 **Rabi** had developed the technique of molecular beam magnetic resonance. This allowed a precise test of **Goudsmit** and **Uhlenbeck**'s theory that the electron has intrinsic angular momentum and hence an associated magnetic moment. Kusch worked with Rabi on this technique, and with colleagues they measured the hyperfine splitting in atomic hydrogen and found a large discrepancy between the observed and predicted values. **Breit** suggested that this might be due to the electron's magnetic moment being different from the generally accepted value of one Bohr magneton, and Kusch confirmed this experimentally. With the experimental results of **Willis Lamb**, this led to the reformulation of quantum electrodynamics by **Feynman**, **Schwinger** and **Tomonaga**. Kusch shared with Lamb the 1955 Nobel Prize for Physics for his precise determination of the electron's magnetic moment.

L

LACAILLE, Nicolas Louis de
(1713–62)

French astronomer, born in Rumigny. After studying theology, he became a deacon before taking up astronomy. At the age of 26 he became Professor of Mathematics at the Collège Mazarin (now the Institut de France, Paris). As a geodesist he worked on the problem of the Earth's shape, and discovered that it was pear-shaped as opposed to lemon-shaped (the latter having been suggested by Jacques Cassini). From 1750 to 1754 he visited the Cape of Good Hope, where he was the first to measure a South African arc of the meridian. Lacaille charted 14 new constellations, naming them after contemporary scientific and astronomical instruments. He also compiled the first list of 42 'nebulous stars'. Lacaille's extensive catalogue of southern stars and positional data was published as *Coelum Australe Stelliferum* in 1763.

LACK, David
(1910–73)

English ornithologist, born in London and educated at Magdalene College, Cambridge. He taught at Dartington Hall from 1933, and then in 1945 moved to Oxford as Director of the Edward Grey Institute. Whilst teaching at Dartington, he published a popular book, *The Life of the Robin* (1943), which formed the basis of his reputation. He was very influential in studying the *Geospiza* finches of the Galapagos Islands (*Darwin's Finches*, 1947), and the effect of the diet and food supply on their adaptation and differentiation. While in Oxford, Lack initiated a series of long-term studies, notably of great tits in Wytham Woods, but also of a number of other species, including swifts; the results were published in a popular book, *Swifts in a Tower* (1956). He was a dominant influence in the transformation of ornithology from an observational to a scientific discipline. His most important books were *National Regulation of Numbers* (1954), *Population Studies of Birds* (1966), *Ecology Adaptations for Breeding in Birds* (1968) and *Ecology Isolation in Birds* (1971). He was elected FRS in 1951.

LA CONDAMINE, Charles-Marie de
(1701–74)

French mathematician and scientist, born in Paris. He was educated at the Collège Louis-le-Grande in Paris, and started a military career. Having developed a strong interest in science, he entered the Academy of Sciences in 1730, and was selected to take part in an expedition to Peru (1735–43) to measure the length of a degree of the meridian, with **Bouguer**. In conjuction with meridian arc measurements at the Arctic circle, the expedition's results confirmed **Isaac Newton**'s hypothesis that the Earth is flattened at the poles. On his return journey La Condamine explored the Amazon, brought back the poison curare and definite information on india-rubber and platinum. He wrote in favour of inoculation.

LACROIX, (François-Antoine-) Alfred
(1863–1948)

French mineralogist, petrologist and structural geologist, born in Mâcon where his education began. He entered the Sorbonne and the Collège de France in Paris to train in pharmacy, and concurrently attended courses in mineralogy at the Museum of Natural History. After graduating in pharmacy he turned to mineralogy, receiving a doctorate in 1889. In 1893 he became Professor of Mineralogy at the museum, a post which he held until 1936. Lacroix recognized the importance to research of an excellent mineral collection, and devoted considerable time to building up a systematic collection from around the world. His *Minéralogie de France* was published between 1893 and 1902. He carried out wide-ranging research on eruptive rocks, including studies of Mont Pelée, Martinique, conducted immediately after the devastating volcanic eruption of 1902. He studied the subsequent eruptions and recognized the 'nuée ardente', a glowing cloud type of eruption that he was first to witness. The phenomenon was described in his *La Montagne Pelée et ses éruptions* (1904) and *La Montagne Pelée après ses éruptions* (1908). His studies during a visit to Madagascar in 1911 resulted in the three-volume *Minéralogie de Madagascar* (1922–3), one of his best works. He also worked on the igneous and volcanic rocks of the Massif Centrale, Etna, Vesuvius, Antrim, Réunion, Comoro Islands, Guinea, Mali and elsewhere, and on meteorites.

LAËNNEC, René Théophile Hyacinthe
(1781–1826)

French physician, born in Quimper, Brittany. An army doctor from 1799, in 1814 he became editor of the *Journal de Médecine* and physician to the Salpêtrière, and in 1816 chief physician to the Hôpital Necker, where he invented the stethoscope in the same year. His stethoscope consisted of a simply hollowed tube of wood, with adaptations at the end to help transmit sound more easily. The familiar binaural stethoscope, with rubber tubing going to both ears, was not developed until after Laënnec's death. He demonstrated the importance of the stethoscope in diagnosing diseases of the lungs, heart and vascular systems, and introduced the basic vocabulary to describe heart and lung sounds. In 1819 he published his *Traité de l'auscultation médiate*, the fruit of three years' intense labour and including outstanding clinical and pathological descriptions of many chest diseases, such as tuberculosis, pneumonia, pleuritis and bronchitis. He followed his patients throughout their illnesses and performed postmortem examinations in fatal cases, correlating the signs and symptoms he had noticed while the patient was alive with the pathological lesions in their dead bodies. He was a pious Catholic and royalist, and his academic career after the end of the Napoleonic Wars benefited from the political conservatism which then prevailed. He died of tuberculosis, a disease he had described so brilliantly.

LAGRANGE, Joseph Louis de, Comte
(1736–1813)

Italian–French mathematician, born in Turin. From around the age of 19, he taught mathematics at the Artillery School in Turin, and gained a Europe-wide reputation for his work on the calculus of variations, celestial mechanics and the nature of sound. In 1772 he described 'Lagrangian points', the points in the plane of two objects in orbit around their common centre of gravity at which the combined gravitational forces are zero, and hence where a third particle of negligible mass can remain at rest. In 1766 he succeeded **Leonhard Euler** as director of the mathematical section of the Berlin Academy. While in Prussia he read before the Berlin Academy some 60 dissertations on celestial mechanics, number theory, and algebraic and differential equations. He returned to Paris in 1787 at the invitation of Louis XVI. Under Napoleon he became a senator and a count, and taught at the École Normale and the École Polytechnique. In 1788 he published *Traité de mécanique analytique*, one of his most important works, in which mechanics is based entirely on variational principles, giving it a high degree of elegance. His work on the theory of algebraic equations was one of the important steps in the early development of group theory, considering permutations of the roots of an equation in a systematic attempt to explain why equations of low degree have algebraic solutions. It was a crucial influence on **Galois**. Lagrange was buried in the Panthéon.

LAITHWAITE, Eric Roberts
(1921–)

English electrical engineer and inventor, born in Atherton, Yorkshire. He studied at Regent Street Polytechnic and Manchester University, where after war service in the RAF he remained until 1964, when he was appointed Professor of Heavy Electrical Engineering at the Imperial College of Science and Technology of the University of London. His principal research interest is in the linear motor, a means of propulsion utilizing electromagnetic forces acting along linear tracks; by incorporating magnetic levitation or air cushion suspension, high-speed experimental vehicles have been constructed without either wheels or conventional rotating electric motors. He was professor of the Royal Institution (1967–76), and received the Nikola Tesla Award from the Institution of Electrical and Electronic Engineers in 1986.

LAMARCK, Jean-Baptiste Pierre Antoine de Monet, Chevalier de
(1744–1829)

French naturalist and evolutionist, born in Bazentin from a long line of military horsemen. At the age of 19 he escaped from Jesuit school to join the army, where he rapidly distinguished himself in battle against the Germans. As an officer at Toulon and Monaco he became interested in the Mediterranean flora. Resigning after an injury, he held a post in a Paris bank, and meanwhile began to study medicine and botany. In 1773 he published *Flore française*, the first key to French flowers, and the following year became keeper of the royal garden (later the nucleus of the Jardin des Plantes). From 1794 he was keeper of invertebrates at the newly formed Natural History Museum. He lectured on zoology, originating the taxonomic distinction between vertebrates and invertebrates. His *Histoire des animaux sans vertèbres* appeared in 1815–22. By about 1801 he had begun to think about the relations and origin of species, expressing his conclusions in his famous *Philosophie zoologique* (1809). Lamarck broke with the notion of immutable species, recognizing that species needed to adapt to survive environmental changes, and postulating a gradual process of development from the simple to the complex. In this he foreshadowed **Charles Darwin**'s ideas, but he became best known for espousing the idea that the development or atrophy of organs by 'use or disuse' can be inherited by later generations. This 'use or disuse' was mediated by habitual behaviour relating to the animal's needs, not (except in the highest vertebrates) by conscious volition, a suggestion for which he has been unfairly chastized by evolutionists from Darwin onwards. Eventually, hard work and illness enfeebled his sight and left him blind and poor, a pathetic figure suffering in old age the taunts of the anti-evolutionist **Cuvier**.

LAMB, Hubert Horace

(1913–)

English climatologist, born in Bedford. He was educated at Trinity College, Cambridge, where he obtained a degree in natural science and geography. He joined the Meteorological Office soon after his initial training and was seconded to the Irish Meteorological Service. Throughout World War II he produced weather forecasts for transatlantic flights, and in 1946 he had a spell of duty on whaling ships in the southern ocean. These two experiences enabled him to assess the accuracy of weather charts produced from sparse data by a subsequent comparison with more complete charts. Using this expertise and archival data, he was later able to draw a series of monthly mean surface weather charts from 1750 onwards for much of the north Atlantic and Europe. His two greatest achievements were the production of a daily weather classification for Britain for each day from 1861, which has formed the basis of much climatological research, and a list of major volcanic eruptions since 1500 along with an estimate of the dust ejected into the atmosphere by each. This work has been invaluable in climate change studies. Lamb's personal efforts resulted in the establishment of the Climatic Research Unit at the University of East Anglia in 1973. Under his direction this department uncovered much detail about past climates and the way people lived in those times. Lamb believed that climate had a great bearing on history and published *Climate, History and the Modern World* in 1982. His major publication, however, has been *Climate, Present, Past and Future* (2 vols, 1972, 1977). He has also produced a book *Historic Storms of the North Sea, British Isles and Northwest Europe* (1991) giving synoptic details of storms since 1500.

LAMB, Willis Eugene

(1913–)

American physicist, born in Los Angeles and educated at the University of California. He later became professor at Columbia University, New York (1938–51), before being appointed to similar posts at Stanford (1951–6), Oxford (1956–62) and Yale (1962–74). He moved to the University of Arizona as Professor of Physics in 1974. His accurate studies of the hyperfine structure of the hydrogen spectrum showed that the two possible energy states of hydrogen, rather than being equal as predicted by **Dirac**, differed in energy by a very small amount. This became known as the 'Lamb shift', and prompted a revision of the theory of interaction of the electron with electromagnetic radiation. This revision led to the theory known as quantum electrodynamics. Lamb shared with **Kusch** the 1955 Nobel Prize for Physics for this research.

LAMBERT, Johann Heinrich

(1728–77)

Swiss mathematician, scientist and philosopher, born in Mulhouse, Alsace, and largely self-taught. He first showed how to measure scientifically the intensity of light in his *Photometria* (1760), and his philosophical work *Neues Organon* (1764) was greatly valued by **Kant**, with whom he shares early honours for discovering that the Milky Way is a disc-like region of stars, mostly at a considerable distance from us. He wrote a successful popular book on cosmology, and in mathematics he proved that the numbers π and e are irrational (ie cannot be expressed as the ratio of two whole numbers). Lambert also studied the mathematics of map projections and proved several theorems in what was later accepted as non-Euclidean geometry. Elected a member of the Berlin Academy of Sciences, he was the only scientist to present papers in all of its sections.

LAMONT, Johann von

(1805–79)

Scottish-born German astronomer and geophysicist, born in Braemar, Aberdeenshire, and educated from the age of 12 at the Scottish monastery in Regensburg, Germany. In 1827 he went to work at the then new observatory at Bogenhausen near Munich under Johannes von Soldner, whom he succeeded as director in 1835. In 1852 he was appointed to the Chair of Astronomy at the University of Munich which he held, jointly with the directorship of the observatory, until his death. He published a number of star catalogues and determined the mass of the planet Uranus from observations of the motions of its satellites. In 1840 he equipped a magnetic observatory at Bogenhausen for comprehensive terrestrial magnetism surveys. In 1850, as a result of reviewing

German magnetic records since 1835 he discovered a 10½ year magnetic cycle subsequently shown to correlate with the sunspot cycle.

LANCHESTER, Frederick William
(1868–1946)

English engineer, inventor and designer, born in Lewisham, London. His family moved to Hove when he was two years old, and he attended the Hartley Institution (later University College, Southampton). From there he won a scholarship to what is now Imperial College in London. In 1889 he joined the Forward Gas Engine Company in Birmingham, and in 1893 set up his own workshop next to theirs. Convinced that the newly invented motor car was the road vehicle of the future, he built the first experimental motor car in Britain (1895) and founded the Lanchester Engine Company in 1899, which produced the first Lanchester car in 1901. Over the next four years almost 400 Lanchester cars were sold. Turning his attention to aeronautics, he laid the theoretical foundations of aircraft design in *Aerial Flight* (2 vols, 1907–8), which was ahead of its time in describing boundary layers, induced drag and the dynamics of flight. Another of his original contributions was *Aircraft in Warfare* (1914); quantifying the numerical strength of contending military forces, this was an early essay in operational analysis. Lanchester was elected President of the Institute of Automobile Engineers in 1910, Fellow of the Royal Aeronautical Society in 1917 and FRS in 1922.

LAND, Edwin Herbert
(1909–91)

American inventor and physicist, born in Bridgeport, Connecticut. Before graduating from Harvard he left to set up his own laboratory (1932), eager to pursue his research on polarized light. At that time it was known that some organic crystals polarized light, but the crystals were often too small to be of any practical use. Land developed a method of aligning many microscopic polarizing crystals and embedding them in a clear plastic sheet. He named the polarizing film produced Polaroid and used it to manufacture camera filters and sunglasses. During World War II he conducted research leading to the development of new weapons and war materials, including plastic optical lenses for night vision equipment, new types of lightweight stereoscopic rangefinders and an infinity optical ring sight used on anti-aircraft guns and bazookas. His invention of the one-step Polaroid Land camera (1947), which developed pictures inside the camera within one minute, earned more than $5 million in its first year. Land also contributed to the theoretical understanding of the nature of colour vision including the ‘retinex’ theory (1977). He founded the Polaroid Corporation (1937), and served as chairman of the board of directors and director of research.

LANDAU, Lev Davidovich
(1908–68)

Soviet physicist, born in Baku. At the age of 14 he entered Baku University and he received his PhD from Leningrad University in 1927. He studied with **Niels Bohr** in Copenhagen, and became Professor of Physics at Moscow in 1937. In 1932, following the discovery of the neutron, he proposed the existence of neutron stars. Landau developed a theory to describe the properties of liquid helium. Below a critical temperature, helium has zero viscosity in one direction, a property known as superfluidity. Rather than trying to describe the properties of the liquid in terms of the individual atoms, he explained the behaviour in terms of the collective behaviour of the atoms in the liquid. This was later further developed by **Feynman**. Landau received the 1962 Nobel Prize for Physics for work on theories of condensed matter, particularly helium.

LANDSBERG, Helmut Erich
(1906–85)

American climatologist, born in Frankfurt. After graduating from Frankfurt, he carried out research at Pennsylvania State University from 1934 to 1941, and during the period 1941–3 provided climatological knowledge for military planning. He wrote a monograph on condensation nuclei, produced climatological charts for the National Atlas of the USA and undertook a study of the results of urbanization of the natural environment. He wrote many books and was editor of the 14 volume series *The World Survey of Climatology*. He did much to establish climatology as a physical science. He was instrumental in establishing the National Climate Centre at Asheville and devised a data processing system which greatly facilitated the supply of information for research. In 1946 he was on the Joint Research and Development Board of the USA which coordinated research activities nationally. At the Air Force Research Center in Cambridge, Massachusetts (1951–4), he organized both the constant level balloon programme and the first numerical weather prediction efforts. He was Director of the Climatology Weather Bureau (1954–65), and President of the World Meteorological Organisation Commission for Climatology from 1969 to 1978.

Landsberg was a recipient of the International Meteorological Organisation Prize and many other major awards.

LANDSTEINER, Karl

(1868–1943)

Austrian-born American serologist, immunologist and pathologist, born in Vienna. He studied chemistry in Vienna, Würzburg, Munich and Zürich before turning to medicine, and received his MD in 1891. During a series of assistantships in Vienna, he became interested in serology. Blood transfusions had often been attempted as a means of treating blood loss following injuries, but rather than aiding recovery they frequently hastened death. Landsteiner showed that people could be divided into groups according to the ability of their blood serum to make the red blood cells of other people cluster together. In 1901 he published a paper describing a technique for dividing human blood into three groups: A, B and C (later O). His colleagues found a fourth group, later termed AB. These discoveries led to the development of safe blood transfusions. In 1909 Landsteiner was appointed to the Chair of Pathology at the University of Vienna. Life for scientists was very difficult in postwar Vienna; Landsteiner left to become a prosector at a hospital in the Hague. In 1922 he joined the Rockefeller Institute in New York where he remained for the rest of his life. In 1927 Landsteiner and his colleagues discovered the M, N and MN blood factors. He was awarded the 1930 Nobel Prize for Physiology or Medicine. In 1940 he was involved in the discovery of the Rhesus (Rh) groups, and also conducted important work on poliomyelitis, being the first to isolate the poliomyelitis virus (1908).

LANGEVIN, Paul

(1872–1946)

French physicist, born in Paris. A clever student, he studied at the École Normale Supérieure, spent a year in Cambridge when the Cavendish Laboratory admitted foreign students for the first time, and came to the notice of **J J Thomson**. He returned to Paris to take his doctorate and study with **Pierre Curie**. In 1909 he was appointed Professor of Physics at the Sorbonne. He was noted for his application of electron theory to magnetic phenomena which produced an important formula relating the paramagnetic movement of molecules to their absolute temperature (1905), confirming Curie's observation that paramagnetic susceptibility varies inversely with temperature, and predicting the phenomenon of paramagnetic saturation, discovered by **Kamerlingh Onnes** in 1914. He worked on the molecular structure of gases, and during World War I he pioneered the application of sonar techniques (based on the piezoelectric transducer) to the detection of submarines. He was elected a Foreign Member of the Royal Society in 1928, and was awarded its Hughes Medal. Imprisoned by the Nazis after the occupation of France, he was later released and, though kept under surveillance at Troyes, managed to escape to Switzerland. After the liberation he returned to Paris.

LANGLEY, John Newport

(1852–1925)

English physiologist, born in Newbury. He was educated at St John's College, Cambridge, initially in mathematics, and later in natural sciences due to the influence of **Foster**. He remained in the Physiological Laboratory, Cambridge, for his entire career, as Foster's demonstrator (1875–84), Trinity College Lecturer in Natural Science (1884–1903), University Lecturer in Histology (1884–1903) and Professor of Physiology (1903–25). Broadly, his contributions were in two main areas—in the physiology of secretion, and in elucidating the anatomy and physiology of the autonomic (a word he introduced) nervous system. Especially in collaboration with **Gaskell**, but also with others and on his own, he determined the anatomical pathways and the principal physiological functions of the sympathetic and parasympathetic branches of the autonomic system, much of his research being summarized in his book *The Autonomic Nervous System, Part One* (1921), of which part two was never published. He played an influential role in the shaping of British physiology as owner and editor of the *Journal of Physiology* from 1894 until his death. He was a keen amateur sportsman, a member of many Cambridge sporting clubs, and a prominent member of the National Skating Association, on whose behalf he was often asked to adjudicate international skating competitions.

LANGLEY, Samuel Pierpont

(1834–1906)

American astronomer and aeronautical pioneer, born in Roxbury, Massachusetts. He first trained and practiced as an engineer and architect. At 30 years of age he began his astronomical career as an assistant at the Harvard College Observatory (1865–6) followed by a year teaching mathematics at the US Naval Observatory at Annapolis. In 1867 he was appointed Professor of Astronomy at Western University of Pennsylvania and Director of the Allegheny Observatory where he stayed for over 20 years. His chosen field was solar physics: he was the inventor of the bolometer (1880), an instrument which recorded the infrared radiation of the Sun quantitatively in terms of an electric current. In 1881 he mounted an expedition to Mount Whitney in the Sierra Nevada, California, to examine the absorbing effects of the Earth's atmosphere on the Sun's radiation. He analysed the infrared spectrum of the Sun, measured the radiation of the Moon and estimated its temperature. His last appointment was as Secretary of the Smithsonian Institution in Washington (1887). A celebrated pioneer of heavier-than-air mechanically propelled flying machines, he built in 1896 a steam-driven pilotless airplane which flew a distance of 42000 feet over the Potomac river.

LANGMUIR, Irving
(1881–1957)

American physical chemist, born in Brooklyn, New York City. After studying metallurgical engineering at Columbia University, New York, he worked on chemical research with **Nernst** at Göttingen. From 1906 to 1909 he taught chemistry at Stevens Institute of Technology in Hoboken, New Jersey, and then joined the General Electric Company (GEC) laboratories at Schenectady, New York, from which he retired as Associate Director in 1950. His first work at GEC was on extending the life of the tungsten filament in an electric light bulb. This led to his studies of adsorption of gases on metals, and of the kinetics and mechanism of heterogeneous reactions. The mathematical formulation of adsorption which he devised, now known as the Langmuir isotherm, is still of importance in the study of catalysis by surfaces. A spin-off from his work in this area was his invention of atomic hydrogen welding. He also contributed to the further development of the electronic theory of the atom and of chemical bonding, thus extending the pioneering work of **Niels Bohr** and **Gilbert Lewis**. Langmuir also investigated films on liquid surfaces and devised a useful piece of apparatus which became known as the Langmuir trough. In connection with his studies of electrical discharges in gases and thermionic emission, which required high vacua, he invented an improved pump, the Langmuir pump. From 1940 he was much concerned with applications of surface chemistry to wartime problems, including gas masks, aircraft icing and the generation of artificial fogs. After World War II his work led to techniques for cloud seeding to produce rain. He received the Nobel Prize for Chemistry in 1932. Langmuir was elected an Honorary Fellow of the Chemical Society in 1929 and received its Faraday Medal in 1939. He became a Foreign Member of the Royal Society in 1935.

LANKESTER, Sir Edwin Ray
(1847–1929)

English zoologist, born in London, son of the scientific writer Edwin Lankester (1814–74). He was Fellow and Tutor of Exeter College, Oxford, professor at London University and at Oxford, and from 1898 to 1907 Director of the British Museum (Natural History). His research embraced a wide range of interests including comparative anatomy, protozoology, embryology and anthropology. He was one of the first to describe protozoan parasites from the blood of vertebrates. The coccidian parasite, Lankesterella, which is related to the causative agent of malaria, is named after him and his work in this field led to an understanding of this disease. His studies of the comparative anatomy and embryology of the invertebrates led him to support **Charles Darwin**'s theory of evolution through natural selection. He also developed the idea that the ability to learn is inherited, that the trend of increasing brain size in recent fossil forms is evidence for increased intelligence and that in humans, intelligence and ability to communicate across generations leads to cultural evolution. His anthropological studies included the discovery of flint implements, and thus the presence of early man, in the Pliocene sediments from Suffolk. He was largely responsible for the founding of the Marine Biological Association in 1884 and edited the *Quarterly Journal of the Microscopical Society* from 1869 to 1920. His many books include *Comparative Longevity* (1871), *Advancement of Science* (1890) and *Science from an Easy Chair* (1910–12), and from 1900 to 1909 he edited the *Treatise on Zoology*. He was knighted in 1907.

LAPLACE, Pierre Simon, Marquis de
(1749–1827)

French mathematician and astronomer, born in Beaumont-en-Auge, Normandy, the son of a farmer. He studied at Caen, went to Paris and became Professor of Mathematics at the École Militaire where he gained fame by his researches on the inequalities in the motion of Jupiter and Saturn, and the theory of the satellites of Jupiter. In 1785 he became a member of the French Academy of Sciences, and prospered at a time when mathematics flourished under Napoleon, who had a strong interest in the science. In 1799 he entered the senate, becoming its Vice-President in 1803, and Minister of the Interior for six weeks, after which he was replaced because of an incapacity for administration. He was created marquis by Louis XVIII in 1817. His astronomical work culminated in the publication of the five monumental volumes of *Mécanique céleste* (1799–1825), the greatest work on celestial mechanics since Newton's *Principia*. His *Système du monde* (1796) was a non-mathematical exposition, of masterly clarity, of all his astronomical theories, and his famous nebular hypothesis of planetary origin occurs as a note in later editions. This proposed that the solar system originated as a massive cloud of gas, and that the centre collapsed to form the Sun, leaving outer remnants which condensed to form the planets. In his study of the gravitational attraction of spheroids he formulated the fundamental differential equation in physics which bears his name. He also founded the modern form of probability theory.

LAPWORTH, Arthur

(1872–1941)

Scottish chemist, born in Galashiels. He was educated at Birmingham University and the City and Guilds College, London, where he worked with **Henry Armstrong**. After work at the School of Pharmacy and Goldsmiths' Institute he moved to Manchester, where he became Professor of Organic Chemistry in 1913 and Professor of Inorganic and Physical Chemistry in 1922. He was one of the founders of physical organic chemistry and his ideas influenced the views of **Robinson** and **Christopher Ingold** on reaction mechanisms in organic chemistry. He also developed a number of important ideas on acid–base catalysis, emphasizing the role of protons.

LAPWORTH, Charles

(1842–1920)

English geologist, born in Faringdon. A school teacher in Galashiels from 1864, he undertook important work in elucidating the geology of the south of Scotland and the north-west Highlands. He was Professor of Geology at Birmingham (1881–1913), and wrote especially on graptolites which he used to unravel the geological structure of the Southern Uplands of Scotland. He introduced the term Ordovician into the geological literature (1879), thereby ending a dispute involving the upper parts of **Sedgwick**'s Cambrian system and the lower parts of **Murchison**'s Silurian system which Lapworth showed to be one and the same. His investigations in the north-western Highlands of Scotland helped resolve the controversy about the nature of the junction between the Cambrian and Ordovician sedimentary rocks and the overthrust Moinian. He published a memoir on southern Scotland in 1899 and his major *Monograph of British Graptolites* during 1901–18.

LARMOR, Sir Joseph

(1857–1942)

Irish theoretical physicist, born in Magheragall, County Antrim, and educated at Queen's University, Belfast, and Cambridge University. In 1880 he was Senior Wrangler in the Mathematical Tripos in the year when **J J Thomson** came second. He was awarded a Smith's Prize and elected a Fellow of St John's College. For the next five years he was Professor of Natural Philosophy at Queen's College, Galway, before returning to St John's (1895) first as a lecturer, then succeeding **Stokes** as Lucasian Professor of Mathematics in 1903. He retired in 1932, returning to Ireland. His scientific work centred on electromagnetic theory, optics, analytical mechanics and geodynamics. He was active during the final phase of classical physics and his major work concerned the interaction of charged particles with an electromagnetic field. In 1897 he demonstrated the precessive motion of the orbit of a charged particle when subjected to a magnetic field (Larmor precession), with the motion of the spin axis describing a circle. He also derived an expression for the power radiated by an accelerated electron. Larmor was elected a Fellow of the Royal Society (1892), served as secretary there (1901–12) and received its Royal (1915) and Copley (1921) medals. He was knighted in 1909.

LARTET, Edouard Arman Isidore Hippolyte

(1801–71)

French vertebrate palaeontologist, stratigrapher and prehistorian, born in St Guiraud, Gers, and educated at the college in Auch where he received a prize from Napoleon. Lartet went to Toulouse to study law, but his life changed direction when he inherited the family estate. He undertook important studies of French Tertiary and Quaternary vertebrates and discovered the fossil jawbone of an ape, *Pliopithecus*, in the Tertiary formations of Sansan, Gers (1836), which refuted the assertion of **Cuvier** that neither men nor apes could be found in a fossil state. His studies in 1860 at the prehistoric sites of Massat and Aurignac yielded conclusive proof of the contemporaneity of man and extinct animal species. His son Louis (1840–99) was a well-known stratigrapher and palaeontologist.

LASSELL, William
(1799–1880)

English astronomer, born in Bolton. He built an observatory at Starfield, near Liverpool, where he constructed and mounted a 2 foot equatorial reflecting telescope. Lassell made a mirror of specula, an alloy of copper and tin with a minute proportion of arsenic, having a highly reflective and easily polished surface. His equatorial telescope mount completely relegated the clumsy mechanism used by Sir **John Herschel** and at a stroke made reflecting telescopes easy to use. He discovered several planetary satellites, including Triton (1846) and Hyperion (1848, simultaneously with **Bond** at Harvard). He also discovered Ariel and Umbriel, satellites of Uranus (1851). Lassell realized the importance of good observing sites and moved his telescope to Malta in 1852. In 1860 he built a telescope of four times the capacity and with it discovered 600 new nebulae.

LAUE, Max Theodor Felix von
See **VON LAUE, Max Theodor Felix**

LAURENT, Auguste
(1807–53)

French chemist, born in St Maurice. After a classical education at a local college he entered the École des Mines in Paris in 1826. From 1832 to 1834 he worked as a chemist at a porcelain factory in Sèvres and in 1835 opened a private school in Paris, but it closed after a year. He then worked for a perfumery after having published a doctoral thesis on the topic of ‘radicals’ in organic chemistry. The term in this context means a group, like ethyl or benzoyl, and differs from modern usage. His ideas were quickly taken up by many chemists in both France and elsewhere but Laurent, a highly nervous man, saw imagined hostility. In 1838 he was appointed to the newly created chair in Bordeaux and in 1843 he met **Gerhardt** with whom he quickly established a close friendship. Together they incorporated Laurent's idea of a radical into the ‘type’ theory of organic chemistry. In 1850 he applied for the chair at the Collége de France but was rejected by a vote in the Academy of Sciences, probably because of his strong republican views. He died of consumption in Paris in 1853.

LAVERAN, Charles Louis Alphonse
(1845–1922)

French physician and parasitologist, born and educated in Paris. He became Professor of Military Medicine and Epidemic Diseases at the military college of Val de Grâce (1874–8, 1884–94). He studied malaria in Algeria (1878–83), discovering in 1880 the blood parasite which causes the disease. He used fresh blood slides with no stains in his microscopical work, and his announcement that he had discovered the parasite of malaria was initially greeted with scepticism, as most scientists thought the causative agent would turn out to be a bacterium. Gradually, however, other investigators began to see the same parasite in the microscopically examined blood of malarious patients not being treated with quinine. Laveran suggested that the parasite was spread through mosquito bites, but the experimental demonstration of this was not provided until the late 1890s, by **Ronald Ross** and other investigators. Laveran also did important work on other tropical diseases including sleeping-sickness, leishmaniasis and kala-azar, which he summarized in important monographs. From 1896 until his death he worked at the Pasteur Institute at Paris. In 1907 he was awarded the Nobel Prize for Physiology or Medicine for his discovery of the malaria parasite; he donated half his money to equip a laboratory for tropical medicine at the Pasteur Institute.

LAVOISIER, Antoine Laurent
(1743–94)

French chemist, one of the greatest names in science. He was born in Paris of a prosperous family and was educated at the Collège Mazarin. While training as a lawyer, he attended chemistry lectures by G F Rouelle and soon showed a genius for many branches of science. He was elected to the Academy of Sciences at an unprecedented young age in 1768 and later became one of its most influential members, playing a large part in the establishment of the metric system, its greatest achievement. From the mid-1760s onward, his scientific research was carried out in parallel with his involvement in finance, civic duties and politics—activities which shaped his career and which in the end cost him his life. In 1768 Lavoisier gained a lucrative position as a member of the Ferme Générale, a private consortium which collected taxes for the government. In 1775 he was made one of the commissioners in charge of the production of gunpowder at the Arsenal in Paris, and in 1787 he was elected to the provincial assembly of Orléans. During his career he campaigned for social reforms and when the Revolution broke out in 1789, he supported it. However, as the Revolution grew more extreme, the former members of the Ferme Générale became objects of public hatred. Lavoisier lost his position at the Arsenal in 1791, and even his record as a reformer and his distinction as a scientist could not save him from the Reign of Terror (1792–4). He was guillotined in Paris on

8 May 1794. Lavoisier's greatest achievement was to discover what happens during combustion. In 1772, dissatisfied with the long-held belief that when a substance burned it lost a vital essence, termed 'phlogiston', Lavoisier began an extensive series of experiments on burning and in 1774 followed up work by **Priestley** and **Scheele**. He showed that when phosphorus, sulphur or mercury is heated, only some of the available air is used up, and that the remainder does not support life or combustion; he therefore concluded that air is a mixture of gases. He identified the portion of the air used during combustion with the 'air' driven off when Priestley's 'red precipitate of mercury' was heated. He showed that this gas combined with carbon to form the 'fixed air' (carbon dioxide) discovered by **Joseph Black**, and that it was a constituent of acids. Synthesizing these results he realized that combustion is a process in which the burning substance combines with a constituent of the air, a gas which he named 'oxygine' ('acid maker'). His discovery, publicized in a series of papers and demonstrations from 1779 onwards, threw the scientific world into a ferment and was not generally accepted until the 1790s. His best advocate was his book *Traité élémentaire de chimie* (1789), one of the most celebrated works in chemistry. In 1783 Lavoisier showed that water, regarded from earliest times as an element, was composed only of hydrogen and oxygen, a discovery in which he was preceded by **Cavendish**. He burned alcohol and other organic compounds in oxygen, then weighed the water and carbon dioxide formed, thus laying the basis of quantitative organic analysis. Lavoisier's genius for meticulous experiment distinguished all his work and his influence on quantitative chemical method was as far reaching as his actual discoveries. He was also a pioneer of physiological chemistry, making extensive studies of fermentation, respiration (which he recognized as a form of combustion) and animal heat. He was the first person to make a clear distinction between an element and a compound. Together with **Berthollet**, **Guyton de Morveau** and **Fourcroy**, he introduced a new system of chemical nomenclature in *Méthode de nomenclature chimique* (1787) to replace the chaotic traditional system. It is still used today.

LAWES, Sir John Bennet
(1814–1900)

English agriculturist, born in Rothamstead near St Albans and educated at Oxford. In 1834 he inherited the estate of Rothamstead. Noticing that rock phosphate was only effective as a fertilizer on acid soils, he found that it was insoluble in alkaline soils and that by treating it with acid it became useful on all types of land. In 1843 he began to manufacture this 'superphosphate' at Deptford Creek. The same year he founded an agricultural research station at Rothamstead, the first in the world, and asked Henry Gilbert to take charge of his laboratory. Lawes and Gilbert worked together for 50 years and earned Rothamstead an international reputation. They studied animal feeding, grasslands and manures, disproving **Liebig**'s assertion that humus and organic matter are no benefit to the soil. During the course of their experiments they introduced a system of randomly selected trial plots: this has been almost universally adopted and gives agricultural trials their characteristic chequerboard appearance. Lawes was elected FRS in 1854 and created a baronet in 1882. He died in Rothamstead. Rothamstead Experimental Station remains at the forefront of agricultural research.

LAWRENCE, Ernest Orlando
(1901–58)

American physicist, born in Canton, South Dakota. He was educated at the universities of South Dakota, Minnesota and Yale. He was professor at Berkeley, California, from 1930, and in 1936 was appointed as the first Director of the Berkeley Radiation Laboratory (later renamed the Lawrence Berkeley Radiation Laboratory). Lawrence built the first cyclotron particle accelerator. Maintained in circular orbits by a magnetic field, particles were accelerated by repeatedly passing through an electric field. This allowed higher energies to be achieved than by the simple linear acceleration techniques developed by **Cockcroft** and **Walton**. The first cyclotron built was only a few inches in diameter but Lawrence went on to develop larger machines that could achieve higher energies. For this work he was awarded the 1939 Nobel Prize for Physics. During the early days of atomic bomb research he developed a method for separating the rare fissionable uranium-235 isotope from the common ^{238}U using a mass spectrograph. The bomb dropped on Hiroshima contained ^{235}U obtained by this technique.

LAWS, Richard Maitland
(1926–)

English mammalogist and Antarctic scientist, born in Whitley Bay, Northumberland. He was educated at Cambridge University, where he received his PhD in 1953. From 1947 to 1953 he was employed by the Falkland Islands Dependencies Survey, and he later moved to the National Institute of

Oceanography (1954–61). After a period as Director of the Nuffield Unit of Tropical Animal Ecology in Uganda (1961–8), he was appointed as Head of the Life Sciences Division of the British Antarctic Survey (1969–87), for which he served as Director from 1973 to 1987. Laws made a precise population study of large mammals (notably seals, whales and elephants), aging individuals on the basis of growth rings in the teeth and tusks. This led to studies of ecology and energy flow, particularly of the Antarctic Ocean. Through his leadership of the British Antarctic Survey and his influence in international organizations, he was a dominant figure in changing the early phase of Antarctic research into a coordinated scientific endeavour. He was elected FRS in 1980.

LAWTON, John Hartley
(1943–)

English ecologist, born in Preston, Lancashire. He was educated at Durham University, and taught at the universities of Oxford and York. Since 1989 he has been Director of the Natural Environmental Research Council's Centre for Population Biology at Imperial College, London. His early work concentrated on ecology energetics, leading to the recognition of patterns in energy use by animal populations and thus the determinants of the structure of food-webs and the ratios of predator and prey species. One of his most illuminating contributions was the recognition of the role of plant structural or architectural complexity in the control of the species richness of insect herbivore communities. This work, and his collaborative studies on the dynamics of single species and predator–prey populations, naturally led to questions on the structuring of the insect communities on plants. His showed that there is little evidence for overt horizontal, interspecific interactions and suggested that more subtle, plant-mediated competitive effects, as well as vertical interactions such as enemy-free space, play important roles. Much of his fieldwork, which is both extensive and experimental, has been conducted on bracken and its herbivores. Besides contributions to fundamental science, Lawton has directed his findings towards the biological control of this worldwide weed. He was elected FRS in 1989.

LEAKEY, Louis Seymour Bazett
(1903–72)

Kenyan archaeologist and physical anthropologist, born of missionary parents in Kabete, where he grew up with the Kikuyu tribe. Educated at St John's College, Cambridge, he graduated in archaeology and anthropology in 1926 and studied African prehistory for a PhD (1930). He was then elected a Fellow of St John's. Leakey took part in archaeological expeditions in East Africa, made a study of the Kikuyu and wrote on African anthropology. He became Curator of the Coryndon Memorial Museum at Nairobi (1945–61). His great discoveries of early hominid fossils took place at Olduvai Gorge in East Africa. In 1959 his wife **Mary Leakey** unearthed the skull of *Zinjanthropus*, subsequently reclassified as *Australopithecus* and now thought to be about 1.75 million years old. Following the discovery in 1960 of the first remains of *Homo habilis*, a smaller species of around the same age, Leakey postulated the simultaneous evolution of two different species, of which *Homo habilis* was the true ancestor of man, while *Australopithecus* became extinct; in 1967 he discovered *Kenyapithecus africanus*, fossilized remains of a Miocene ape, around 14 million years old. He also unearthed evidence of human habitation in California more than 50000 years ago.

LEAKEY, Mary Douglas, née Nicol
(1913–)

English archaeologist and anthropologist, born in London. Wife of **Louis Leakey**, her excavations and fossil finds in East Africa revolutionized ideas about human origins. Her interest in prehistory was roused during childhood trips to south-west France, where she collected stone tools and visited the painted caves around Les Eyzies. She met Louis Leakey while preparing drawings for his book *Adam's Ancestors* (1934), and moved shortly afterwards to Kenya where she undertook pioneer archaeological research (1937–42) at sites such as Olorgesailie and Rusinga Island. In 1948, at Rusinga, in Lake Victoria, she discovered *Proconsul africanus*, a 1.7 million year old dryopithecine (primitive ape) that brought the Leakeys international attention and financial sponsorship for the first time. From 1951 she worked at Olduvai Gorge in Tanzania, initially on a modest scale, but more extensively from 1959 when her discovery of the 1.75 million year old hominid *Zinjanthropus* (subsequently reclassified as *Australopithecus*), filmed as it happened, captured the public imagination and drew vastly increased funding. *Homo habilis*—a new species contemporary with, but more advanced than *Zinjanthropus*—was found in 1960 and published amidst much controversy in 1964. Perhaps most remarkable of all was her excavation in 1976 at Laetoli, 30 miles south of Olduvai, of three trails of fossilized hominid footprints which demonstrated unequivocally that our ancestors already walked upright 3.6 million years ago. Her books include *Olduvai Gorge: My Search for Early Man* (1979) and an autobiography, *Disclosing the Past* (1984). Her son Richard Leakey (b 1944) also became a distinguished palaeoanthropologist.

LEAVITT, Henrietta Swan
(1868–1921)

American astronomer, born in Lancaster, Massachusetts. The daughter of a Congregational minister, she attended Radcliffe College where she developed an interest in astronomy. She became a volunteer research assistant at Harvard College Observatory and joined the staff there in 1902, quickly becoming head of the department of photographic photometry. Like Annie Cannon, her colleague, she was very deaf. She is best known for her discovery of the period–luminosity relationship of Cepheid variable stars; whilst studying Cepheids she noticed that the brighter they were the longer their period of light variation. By 1912 she had succeeded in showing that the apparent magnitude decreased linearly with the logarithm of the period. This simple relationship proved invaluable as the basis for a method of measuring the distance of stars.

LEBEDEV, Pyotr Nikolayevich
(1866–1912)

Russian physicist, born in Moscow. From 1887 he studied at Strassburg under August Kundt, returning to Moscow University in 1891 to teach. He completed his PhD there and subsequently became Professor of Physics. His interests included the effects of light waves on molecules, hydrodynamics and acoustic waves. While working on electromagnetic waves he succeeded in constructing an extremly small vibrator source which allowed him to generate waves 4–6 millimetres long. This was 100 times shorter than those studied by **Heinrich Hertz** and 10 times shorter than those investigated by **Righi** a short time before. He made use of these short waves to make the first observations of double refraction of electromagnetic waves in crystals of rhombic sulphur. In 1898 he began experimental work on radiation pressure which was predicted by **Maxwell**'s theory of electromagnetism but was as yet undetected. A few years earlier Lebedev had suggested that solar radiation pressure causes the tails of comets to always point away from the Sun. His theory was accepted until the greater influence of the solar wind was discovered. However, overcoming many difficult practical problems he demonstrated conclusively the presence of radiation pressure, and further was able to show that Maxwell's theory was in quantitative agreement with experimental results. He progressed to work on the origins of the terrestrial magnetic field, attempting to link it to the Earth's rotation. He left the university in 1911 in protest at actions taken by the minister for education and died suddenly the following year.

LE BEL, Joseph Achille
(1847–1930)

French chemist, born in Pechelbronn, Bas-Rhin. Born of wealthy parents, he was educated at the École Polytechnique, and subsequently worked with **Balard** and with **Wurtz**. In 1874 both he and **van't Hoff** (who was also working with Wurtz) independently published papers establishing the relationship between the three-dimensional arrangement of atoms in a molecule (stereochemistry) and optical activity. After this important contribution to the progress of organic chemistry, Le Bel made few significant discoveries. He divided his time between the family factory and a private laboratory in Paris where he worked mainly on optical activity. He held no academic appointments but was appointed President of the French Chemistry Society in 1892. He became a member of the Academy of Sciences in 1929.

LEBESGUE, Henri Léon
(1875–1941)

French mathematician, born in Beauvais. He studied at the École Normale Supérieure, and taught at Rennes, Poitiers, the Sorbonne and the Collège de France. Following the work of **Borel** and René Baire (1874–1932), he developed the theory of measure and integration which bears his name, and applied it to many problems of analysis, in particular to the theory of **Fourier** series. Overcoming the defects of the **Riemann** integral, this theory has proved indispensable in all subsequent modern analysis and allowed *Kolmogorov* to produce the first rigorous theory of probability. Lebesgue's work found important applications in complex analysis, and in many areas where continuous but not differentiable functions dominate. He also wrote widely on the history of mathematics.

LEBLANC, Nicholas
(1724–1806)

French industrial chemist who developed a process for making soda from common salt which was of great importance to the chemical industries of the 19th century. He was born in Issoudun and trained as a physician, becoming private surgeon to the future Duc d'Orléans (Philippe Égalité). At that time soda (sodium carbonate), which was essential in the manufacture of glass, porcelain, paper and soap, was made from the ashes of wood or seaweed, both in short supply. In 1755 the French Academy of Sciences had offered a prize for a method of converting common salt (sodium chloride), which was plentiful, into soda. Leblanc devised a process in which the salt was treated with sulphuric acid, the resulting sulphate being roasted with limestone and coal to give a mixture of sodium carbonate and calcium sulphide. The carbonate was extracted from the insoluble sulphide with water and then crystallized out. This process, subsequently known by the name of its inventor, was perfected by 1790, but by then the French Revolution had begun and Leblanc never received the prize. In 1791 he was granted a patent, and with finance from the Duc d'Orléans, built a factory at St-Denis for its production. After the duke was guillotined in 1793, the factory was confiscated. Napoleon returned it to Leblanc in 1802 but he had no capital to revive it, and shot himself in Paris four years later. His process was widely used in the 19th century, despite its environmental offensiveness. Soda is now manufactured by the method due to **Solvay**.

LE CHATELIER, Henri Louis
(1850–1936)

French chemist and metallurgist, born in Paris. He was educated at the École Polytechnique and the École des Mines. After brief service in the Corps des Mines, he was invited in 1877 to become Professor of General Chemistry in the École des Mines. In 1887 he exchanged this chair for that of industrial chemistry and retained this post until his retirement in 1919. Parallel with this post he also held from 1898 to 1908 the Chair of Mineral Chemistry at the Collège de France and, from 1907 onwards, the chair at the Sorbonne previously occupied by **Moissan**. His awards and honours included the Bessemer Medal of the Iron and Steel Institute (1910) and the Davy Medal of the Royal Society (1916), of which he became a Foreign Member in 1913. Le Chatelier's earliest researches were on the nature and setting of cements. This work led him to consider the fundamental laws of chemical equilibrium and in 1884 to formulate the principle named after him, which states that if a change is made in pressure, temperature or concentration of a system in chemical equilibrium, the equilibrium will be displaced in such a direction as to oppose the effect of this change. He devised the platinum/platinum–rhodium thermocouple for measuring high temperatures. Le Chatelier later studied gaseous explosions in connection with safety in mines, and made many studies of the metallurgy of steel and other alloys. He invented the inverted stage metallurgical microscope.

L'ÉCLUSE, Charles de
(1525–1609)

French botanist, known as 'Carolus Clusius', born in Arras, then in French Netherlands. He studied at Louvain and Montpellier universities. He travelled in Spain, England, Hungary and the New World, and from 1593 was professor at Leiden. He published *Rariorum Aliquot Stirpium per Hispanias Observatarum Historia* in 1576. This book was an account of the plants L'Écluse had seen in Spain and Portugal on an earlier expedition, and it contained specially prepared wood-blocks. In 1583 he published a second work devoted to the plants of Austria and Hungary. The two books were collected together and republished with some additional material as *Rariorum Plantarium Historia* (1601). There is little attempt in this work to arrange the plants in systematic order, although species within the same genus are generally treated together. His botanical interests were not confined to flowering plants, and he also illustrated fungi. He corresponded widely on botany and horticulture, and he introduced the potato into cultivation in Germany and Austria. Not being a physician, he had no special interest in medicinal botany, but preferred to study plants for their own sake.

LEDER, Philip
(1934–)

American geneticist, born in Washington DC. He was educated at Harvard University, and became a research associate at the National Heart Institute and the National Cancer Institute, and Laboratory Chief of Molecular Genetics at the National Institute for Child Health and Human Development in Bethesda, Maryland (1972–80). In 1980 he became Professor of Genetics at Harvard University Medical School, and he is currently John Emory Andrus Professor of Genetics at Harvard and a senior investigator at the Howard Hughes Medical Institute. In the late 1970s, Leder worked extensively on the structure and function of the globin genes. He discovered how the multiple globin genes are arranged on the chromosome, and how the coding and non-coding regions of the genes are organized. More recently, he has worked on the function of oncogenes in transgenic mice. Oncogenes are normal cellular genes which can be activated in a variety of ways to become carcinogenic. To examine the relative influences of oncogene activation and environmental effects, activated oncogenes are injected into mouse eggs to create transgenic mice. Leder has the worldwide patent on the 'oncomouse' and has shown that the introduction of cancer to transgenic mice requires the cooperative action of more than one oncogene. He received the National Medal of Science in 1989.

LEDERBERG, Joshua

(1925–)

American biologist and geneticist, born in Montclair, New Jersey. He studied biology at Columbia, and became professor at Wisconsin (1947–59) and Stanford (1959–78), and President of Rockefeller University (1978–90). He has been director and consultant to several biotechnology companies: Procter and Gamble, Ohio, the Celanese Corp, and Alfymax, Palo Alto. With **Tatum**, he showed that bacteria can reproduce by a sexual process known as conjunction. By mating two different strains, he found that the new strain had characteristics of both parents. He also introduced the technique of 'replica plating', whereby mutants of a bacterial strain can be retrieved from the original bacterial colony. Working with **Zinder**, Lederberg made a further fundamental contribution through his description of 'transduction' in bacteria, whereby the bacterial virus transfers part of its DNA into the host bacterium; this led to the development of techniques for manipulation of genes. He served as a consultant to the US space programme in the early 1960s, and was also a consultant to the World Health Organisation on biological warfare. In 1958 he was awarded the Nobel Prize for Physiology or Medicine, jointly with Tatum and **Beadle**.

LEDERMAN, Leon Max

(1922–)

American physicist, born in New York City. He was educated at City College of New York and Columbia University, where he received his PhD in 1951 and became professor in 1958. During 1979–89 he was also Director of the Fermi National Accelerator Laboratory, and since 1989 he has been Frank L Sulzberger Professor of Physics at the University of Chicago. Studies of muon decay had indicated that not one but two types of neutrino are involved. This was believed to be due to the muon not simply being a heavy electron, but having its own identity, the 'electron charge' being carried by the electron neutrino and the 'muon charge' being carried by the muon neutrino. In 1960 **Schwartz** proposed an experiment that could establish the existence of the two distinct neutrinos, and this was performed at Brookhaven by Lederman, Schwartz, **Steinberger** and their collaborators. In 1962 they announced that they had observed 20 muon events confirming the existence of the two distinct neutrino types. This was the basis for the idea that fundamental particles come in generations: the electron and its neutrino, the muon and its neutrino and the tau-lepton and its neutrino. Lederman went on to discover the long-lived neutral kaon and the 'bottom' quark. In 1988 he was awarded the Nobel Prize for Physics together with Schwartz and Steinberger.

LEE, Tsung-Dao

(1926–)

Chinese-born American physicist, born in Shanghai. Educated at Kiangsi and at Chekiang University, he won a scholarship to Chicago in 1946 where he worked under **Teller**. He became a lecturer at the University of California, and from 1956 was professor at Columbia University, as well as a member of the Institute for Advanced Study (1960–3). The electromagnetic and strong nuclear forces are known to conserve a quantum property known as parity, but in 1956 **Feynman** and Martin Block suggested that this may not be true for the weak nuclear force. With **Yang**, Lee made a thorough analysis of all the known data in particle physics and concluded that parity was unlikely to be conserved in weak interactions, and they suggested a simple experiment which would prove it. In the same year (1956), a similar experiment by a group of physicists from Columbia University and the American National Bureau of Standards, headed by **Wu**, confirmed that the 'law' of parity is indeed violated in the case of weak interactions. For this prediction, Lee and Yang were awarded the 1957 Nobel Prize for Physics, and the Einstein Commemorative Award from Yeshiva University in the same year. Lee's other research interests have been in the areas of statistical mechanics, field theory and turbulence.

LEE, Yuan Tseh

(1936–)

American physical chemist, born in Hsinchu, Taiwan. His early scientific training was in Taiwan and he later moved to the USA where he received a PhD in chemistry (1965) under **Herschbach** at the University of California, Berkeley. He remained at Berkeley as a postdoctoral fellow until 1968. Between 1968 and 1974 he rose from assistant professor to full professor at the University of Chicago, and then returned to Berkeley as Professor of Chemistry. He shared the Nobel Prize for Chemistry in 1986 with Herschbach and **John Polanyi**. In 1986 he also received the Debye Award of the American Chemical Society. Lee's main research area is chemical reaction dynamics; in his association with Herschbach, the successful development of the molecular beam technique owed much to Lee's experimental skill.

LEEUWENHOEK, Antoni van
(1632–1723)

Dutch cloth merchant and amateur scientist, born in Delft. Apprenticed in Amsterdam and educated as a businessman, he became skilled in grinding and polishing lenses to inspect cloth fibres. He returned to Delft around 1652 and became a draper and haberdasher. Through his second wife, Cornelia Swalmius, he joined Delft's more intellectual circles. His civic posts included chamberlain to the sheriffs of Delft, alderman and wine-gauger. With his microscopes, each made for a specific investigation, he discovered microscopic animalicules (protozoa) in water everywhere (1674), bacteria in the tartar of teeth (1676), that all males produce spermatozoa—even fleas, lice and mites, and he described their copulation and life cycles. Independently, he discovered blood corpuscles (1674), blood capillaries (1683), striations in skeletal muscle (1682), the structure of nerves (1717) and plant microstructures among endless other observations. After his death, his daughter Maria auctioned his collection of 248 microscopes and 178 separate lenses. Extant lenses magnify from 30× to 266×. His discoveries, in more than 110 letters (published 1684), were described with the assistance of Delft's anatomist, Cornelis's-Gravesande, and introduced to the Royal Society by **Graaf**. Leeuwenhoek was elected FRS in 1680.

LEFSCHETZ, Solomon
(1884–1972)

Russian-born American mathematician, born in Moscow. He studied engineering in Paris before emigrating to the USA, where he worked as an engineer. After losing both his hands in an industrial accident (1910), he was forced to abandon engineering, and turned to mathematics. He took his doctorate in 1911, and taught at Kansas (1913–25), where he soon made a reputation by his work in algebraic geometry, applying the insights of **Poincaré** to develop a theory of complex algebraic varieties in *n* dimensions. This led him into the study of topology, and in 1925 he moved to Princeton where he remained until his retirement in 1953, when he became visiting professor at Brown University. Lefschetz became the leading topologist of his generation in the USA and an important theorem on the existence of fixed points of mappings bears his name. His work during World War II roused his interest in differential equations, and he continued to work on their qualitative theory, directing American mathematicians to the extensive Russian literature on the subject.

LEGENDRE, Adrien-Marie
(1752–1833)

French mathematician, born in Paris. He studied at the Collège Mazarin, and became Professor of Mathematics at the École Militaire and a member of the French Academy of Sciences (1783). In 1787 he was appointed as one of the commissioners to relate the Paris and Greenwich meridians by triangulation and in 1813 he succeeded **Lagrange** at the Bureau des Longitudes. He proposed the method of least squares in 1806 (independently of **Gauss**). His classic work *Essai sur la théorie des nombres* (1798) includes his discovery of the law of quadratic reciprocity, although its proof is flawed, and his *Traité des fonctions elliptiques* (1825) became the definitive account of elliptic integrals prior to **Niels Henrik Abel**'s and **Jacobi**'s work, which Legendre generously acknowledged. His *Éléments de géométrie* (1794) was translated into English by Thomas Carlyle. It reintroduced rigour to the teaching of elementary geometry in France, although its attempts to prove the parallel postulate (that there is only one straight line which passes through a given point and is parallel to another given line) were all misconceived.

LEHMANN, Inge
(1888–93)

Danish geophysicist, born in Copenhagen. She was educated at a school founded by **Niels Bohr**'s aunt, where girls were encouraged to study the same subjects as boys, and entered Copenhagen University in 1907 to read mathematics. After a year at Newnham College, Cambridge (1910–11), she pursued a career in insurance until 1918, though she maintained an interest and contacts in science. Her studies were resumed in 1918, and she received her degree in 1920. She was awarded her PhD at the University of Copenhagen in 1928. From 1928 until her retirement in 1953, she was chief of the seismological department of the newly founded Danish Geodetic Institute, taking responsibility for the seismological stations in Greenland. Lehmann's research involved the interpretation of seismic events detected by

European stations; in this work she discovered that the presence of a distinct inner core of the Earth was required to explain the data received from large epicentral distances. In collaboration with **Gutenberg** she also endeavoured to resolve a velocity structure compatible with **Harold Jeffreys**'s revised traveltime tables, and found a low-velocity layer at 200 kilometres depth. This fitted well with European seismic data and became accepted generally.

LEHN, Jean-Marie

(1939–)

French chemist, born in Rosheim, Bas-Rhin. After undergraduate work at the University of Strasbourg he worked with Guy Ourisson on terpene chemistry for his doctorate (1963) and this was followed by postdoctoral work with **Robert Woodward** at Harvard on vitamin B_{12} synthesis. He then joined the staff of the University of Strasbourg, becoming professor in 1970 and simultaneously professor at the Collège de France from 1979. The mechanism of transport of metal ions across cell membranes is one of the challenges of physical biochemistry, and following **Pedersen**'s work on crown ethers, Lehn has shown that metal ions can exist in a non-polar environment if contained within the cavity of a large organic molecule. Such structures are known as cryptates, and similar compounds play an important role in the transport of metal ions across biological membranes. He has synthesized a large number of cryptands initiating a new branch of organic chemistry—supramolecular chemistry—and it is for this work that he was awarded, with Pedersen and **Cram**, the Nobel Prize for Chemistry in 1987. He was created Chevalier (1983) and later Officer (1988) of the Legion of Honour.

LEIBNIZ, Gottfried Wilhelm

(1646–1716)

German mathematician and philosopher, remarkable also for his encyclopedic knowledge and diverse accomplishments outside these fields. He was born in Leipzig, the son of a professor of moral philosophy, studied there and at Altdorf, showing great precocity of learning, and in 1667 obtained a position at the court of the Elector of Mainz on the strength of an essay on legal education. There he codified laws, drafted schemes for the unification of the churches, and was variously required to act as courtier, civil servant and international lawyer, while at the same time he absorbed the philosophy, science and mathematics of the day, especially the work of **Descartes**, **Isaac Newton**, **Pascal** and **Boyle**. In 1672 he was sent on a diplomatic mission to Paris, where he met Nicolas Malebranche and **Huygens**, and he went on in 1676 to London where his discussions with mathematicians of Newton's circle led later to an unseemly controversy as to whether he or Newton was the inventor of the infinitesimal calculus. Leibniz had published his system of differential calculus in *Nova Methodus pro Maximis et Minimis* ('New Method for the Greatest and the Least') in 1684; Newton published his in 1687, though he could relate this to earlier work. The Royal Society formally declared for Newton in 1711, but the controversy was never really settled; it is Leibniz's notation that appears in modern calculus. In 1676 he visited Baruch Spinoza in the Hague on his way to take up a new, and his last, post as librarian to the Duke of Brunswick at Hanover. Here he continued to elaborate his mathematical and philosophical theories (although he did not publish them), and maintained a huge learned correspondence. He also travelled in Austria and Italy in the years 1687–90 to gather materials for a large-scale history of the House of Brunswick, and went in 1700 to persuade Frederick I of Prussia to found the Prussian Academy of Sciences in Berlin, of which he became the first president. He died in Hanover without real recognition and with almost all his work unpublished. Leibniz was perhaps the last universal genius, spanning the whole of contemporary knowledge. He made original contributions to optics, mechanics, statistics, logic and probability theory; he built calculating machines, and contemplated a universal language; he wrote on history, law and political theory; and his philosophy was the foundation of 18th-century rationalism. His best-known philosophical doctrine was that the world is composed of an infinity of simple, indivisible, immaterial, mutually isolated 'monads' which form a hierarchy, the highest of which is God; the monads do not interact causally but constitute a synchronized harmony with material phenomena.

LEISHMAN, Sir William Boog

(1865–1926)

Scottish bacteriologist, born in Glasgow. He obtained his MD from the University of Glasgow in 1886, and later became Professor of Pathology at the Army Medical College and Director-General of the Army Medical Service (1923). In 1900 he discovered the protozoan parasite (*Leishmania*) responsible for the disease known variously as kala-azar and dumdum fever. He went on to develop the widely used 'Leishman's stain' for the detection of parasites in the blood. Leishman also made major contributions to the development of various vaccines, particularly those used against typhoid, and it was as a result of his work that mass vaccination was introduced in 1914 for the British army. He was knighted in 1909, and elected FRS in 1910.

LEITH, Emmett Norman
(1927–)

American optical scientist and electrical engineer, born in Detroit and educated at Wayne State University. He held an associate professorship in electrical engineering at Wayne State from 1952, and since 1968 has been Professor of Electrical Engineering at the University of Michigan. In 1961 he had suggested that the major difficulty in developing applications of holography at that time—the fact that in **Gabor**'s arrangement two images were produced which were in line and apparently could not be separated from one another—could be overcome by using an inclined reference beam which led to separation of the twin images. This, coupled with Leith's production of the first laser hologram, revived interest in holography and quickly led to important applications. In 1985 he was awarded the Ives Medal of the Optical Society of America.

LELOIR, Luis Frederico
(1906–)

French-born Argentinian biochemist, born in Paris. Educated in Buenos Aires and at Cambridge, he worked mainly in Argentina where he set up his own Research Institute in 1947. Following unproductive work on the adrenals and diabetes, he recognized the proteolytic action of renin from the kidneys on a precursor yielding the hormone angiotensin, an important hormone which increases blood pressure (1940). He then turned to glucose metabolism and discovered the enzyme glucose 1-phosphate kinase (1949), and that the product of the reaction, glucose 1,6-bisphosphate, is a coenzyme of the glycolysis pathway enzyme, phosphoglucomutase. Extending these studies he identified galactokinase, and showed that the product, galactose 1-phosphate, is converted into glucose 1-phosphate. In the 1950s Leloir linked these reactions to the formation of the energy storage polysaccharide glycogen, and showed that the glucose is added by a stepwise transfer process in which the reactive intermediate is a nucleotide derivative, uridine diphosphateglucose (UDPG), and further, that galactose is converted to glucose by a similar mechanism. For this work, of medical significance and complimenting that of **Carl Cori**, he was awarded the Nobel Prize for Chemistry in 1970, becoming the first Argentinian to be so honoured.

LEMAÎTRE, Georges Henri
(1894–1966)

Belgian astrophysicist and cosmologist, born in Charleroi. After studies in engineering at the University of Louvain and voluntary service in World War I in which he was decorated with the Belgian *Croix de Guerre*, he turned to mathematical and physical sciences and obtained his doctorate in 1920. Three years later he was ordained as a Catholic priest and in the same year he obtained a travelling scholarship from the Belgian government. This took him to Cambridge, Harvard and MIT. In Cambridge he came under the strong influence of **Eddington**. In 1927 he published his first major paper on the model of an expanding universe and its relation to the observed redshifts in the spectra of galaxies. In the 1930s he developed his ideas on cosmology, and from 1945 onwards he put forward the notion of the 'primeval atom' which is unstable and explodes, starting what is now called the Big Bang, the beginning of the expanding universe. Lemaître received many honours from all over the world. He was the first recipient of the Eddington Medal of the Royal Astronomical Society (1953) and in 1960 became the President of the Pontifical Academy of Sciences.

LÉMERY, Nicolas
(1645–1715)

French chemist and pharmacist, born in Rouen. He studied as an apothecary there, in Paris, and in a number of other cities including Geneva and Montpellier. He returned to Paris but because he was a Protestant could not join the Guild of Apothecaries. In 1764 he bought the office of Apothecary to the King, thereafter establishing a lucrative pharmaceutical business specializing in patent medicines. He also gave immensely popular lectures on chemistry which were attended by fashionable society as well as by pharmaceutical apprentices. Religious persecution drove him out of Paris in 1683. Having qualified as a doctor at Caen and converted to Roman Catholicism in 1685, after Protestants lost all their legal rights with the Edict of Nantes, he reopened his shop. He published two books on pharmacy and a monograph on antimony, and was admitted to the Academy of Sciences in 1699. Although he discovered little that was new, his influence as a teacher was considerable. His book *Cours de chymie* (1675) had run to 31 editions by 1756 and was translated into most European languages. It gave a lucid account of chemical methods and of the pharmaceutical compounds known at the time. Lémery classified compounds as animal, vegetable or mineral in origin and explained chemical activity in the mechanistic terms common at the time: for example, he supposed that 'spiky' atoms were associated with acidity and porous atoms with alkalinity, the two fitting together when a chemical reaction took place. He died in Paris.

LENARD, Philipp Eduard Anton
(1862–1947)

German physicist, born in Pozsony, Hungary (now Bratislava, Czechoslovakia), and educated at the University of Heidelberg. He was a researcher at the universities of Bohn and Breslau before returning to Heidelberg as professor (1896–8). He then moved on to Kiel University (1898–1907) before returning to University of Heidelberg until his retirement in 1931. From 1899 he studied the emission of electrons from metals when illuminated by ultraviolet light; his observations of the photoelectric effect were explained by **Einstein** in 1905. He also studied the magnetic deflection of cathode rays and their electrostatic properties, showing that they could travel for a short distance in air and penetrate thin metal sheets. On the basis of this work he suggested that atoms contain units of both positive and negative charge. For these studies he was awarded the Nobel Prize for Physics in 1905. He was also awarded the Royal Society Rumford Medal (1904) and the Franklin Medal from the Franklin Institute (1905). His ability to contribute was sadly marred by his lifelong pathological anti-Semitism which did not allow him to accept the work of the Jewish giants in physics, and later made him a willing supporter of Hitler and 'Aryan physics'.

LENNARD-JONES, originally JONES, Sir John Edward
(1894–1954)

English physicist, theoretical chemist and administrator, born in Leigh, Lancashire. Originally John Edward Jones, he took the name Lennard-Jones on marrying Kathleen Lennard in 1925. He graduated in mathematics at Manchester in 1915 and then became a pilot in the Royal Flying Corps. After World War I he held posts at Manchester and at Cambridge before becoming Reader in Theoretical Physics at Bristol in 1925 and Professor in 1927. In 1932 he was elected Plummer Professor of Theoretical Chemistry at Cambridge. He held this position until 1953, but from 1939 to 1946 he was engaged in war work, notably as chief superintendent of armament research at Fort Halstead in Kent and (for a year after the end of the war) as Director-General of Scientific Research (Defence) in the Ministry of Supply. He was appointed Principal of the University College of North Staffordshire in 1953, but died after only one year in that position. Lennard-Jones was appointed KBE in 1946. He was elected FRS in 1933 and received the Royal Society's Davy Medal in 1953. In 1948–50 he was President of the Faraday Society. His early research at Cambridge was on intermolecular forces and was much influenced by Sir **Ralph Fowler**. His mathematical expression for such forces which bears his name is still greatly used in statistical mechanics. During a year in Göttingen in 1929, he studied quantum mechanics and began his part in the development of the molecular orbital theory introduced by **Mulliken**, of which he became a leading exponent.

LENZ, Heinrich Friedrich Emil
(1804–65)

Russian-born German physicist and geophysicist, born in Dorpat (now Tartu). He studied chemistry and physics at Dorpat University. His first post was as a geophysical observer during a scientific voyage around the world (1823–6) on the sloop *Predpriatie*. He was appointed Professor of Physics at St Petersburg (1836) and a member of the Russian Academy of Sciences. In 1834 he formulated 'Lenz's law' governing induced current, according to which the emf induced in a circuit is such as to oppose the flux change giving rise to it, and is credited with discovering the dependence of electrical resistance on temperature (**Joule**'s law). It soon became clear that these physical laws were special cases of the law of conservation of energy.

LESLIE, Sir John
(1766–1832)

Scottish natural philosopher, born in Largo, Fife. At 13 he entered St Andrews University, later transferring to the University of Edinburgh as a divinity student. On his patron's death (1787) Leslie abandoned any clerical aspirations to concentrate on science. He travelled on the Continent and in 1788–9 taught in Virginia. Following his return to Britain (1790), he was tutor to Thomas Wedgwood, son of the potter. Subsequently he devoted himself to practical scientific investigations. The resulting *Experimental Inquiry Into the Nature and Propagation of Heat* (1804), dealing with the fundamental laws of heat radiation, earned him the Royal Society's Rumford Medal (1805). Leslie invented many instruments, mainly for the study of heat and meteorology, most notably an accurate differential air thermometer. In 1810 he devised a method of obtaining very low temperatures, by evaporating water in a receiver evacuated with an air-pump but containing a drying agent. Leslie held the Edinburgh Chairs of Mathematics (1805) and Natural Philosophy (1819). His books on geometry and natural philosophy (promoting Boscovichian atomism) were designed for use by his Edinburgh students. He was knighted in 1832.

LEUCKART, Karl Georg Friedrich Rudolf
(1822–98)

German zoologist, born in Helmstedt. He studied at Göttingen, and became Professor of Zoology at Giessen (1850) and Leipzig (1869). A pioneer of parasitology, he described the complex life cycles of many parasites such as tapeworms and liver flukes. He was able to demonstrate that the disease trichiniasis in man was due to infection by a roundworm. He also showed that the radiata did not comprise a natural group and that the radial symmetry found both in the coelentrates and echinoderms (starfish) did not imply close phylogenetic relationship. Between 1863 and 1876 he wrote his great parasitological treatise *Parasites of Man* (translated 1886).

LEVENE, Phoebus Aaron Theodor, originally Fishel Aaronovich Lenin
(1869–1940)

Russian-born American biochemist, born in Sasar. He qualified in medicine in St Petersburg in 1891 and emigrated to New York in 1892. He became interested in applying chemistry to biological problems and was a founder member of the Rockefeller Institute in New York (1905), where he remained for the rest of his career. Levene's pioneer research on the chemical composition of nucleic acids established (before 1930) the nature of the sugar which defines the two types of nucleic acid (ribose in RNA and deoxyribose in DNA), and that hydrolysis yields only mononucleotides, from which he inferred the sugar phosphate diester linkage of the nucleic acid backbone. From 1909 he vigorously postulated that the structure of yeast nucleic acid is a regular repeating tetranucleotide of the four bases. Unfortunately, this was incompatible with the variability expected of a genetic material and impeded understanding of the role of nucleic acids until resolved by **Chargaff**, **James Watson** and **Crick**. Levene also published extensively on the chemistry of the sugar phosphates, opitical isomerism of organic substances, phospholipid structure, and the carbohydrate components of egg protein and the mucopolysaccharide chondroitin sulphate.

LE VERRIER, Urbain Jean Joseph
(1811–77)

French astronomer, born in St Lô, Normandy, the son of a government official. He entered the École Polytechnique in Paris in 1831 where he studied engineering and worked in the public service. In 1837 he accepted a teaching appointment in the École Polytechnique where he turned his mind to celestial mechanics. Supported by **Arago**, he became world famous when he succeeded in 1846 in calculating from the perturbations of the orbit of the planet Uranus the likely position of an unknown planet which, following his alerting of the Berlin Observatory, was found by **Galle** in September 1846, and later given the name Neptune. The discovery gave precedence to Le Verrier over **John Couch Adams** who had independently predicted the existence of the unknown planet, though both men were eventually given equal recognition. As a result of this achievement, Le Verrier was appointed to a chair in the University of Paris and in 1854 to the directorship of the Paris Observatory in succession to Arago. He was made an officer of the Legion of Honour, was awarded the Danish honour of the Dannebrog, the Copley Medal of the Royal Society and, later in life, the Gold Medal of the Royal Astronomical Society twice (1868, 1876). He was less successful in the field of politics as a senator and inspector-general of higher education. His despotic methods in directing the observatory led to his suspension in 1870, but it was lifted three years later following the death of his successor Charles-Eugene Delaunay (1816–72). Amid these difficulties he continued his researches in celestial mechanics as well as organizing the French meteorological service and setting up an international weather-warning scheme.

LEVI-CIVITA, Tullio
(1873–1941)

Italian mathematician, born in Padua. He studied in Padua and became professor there in 1897. From about 1900 he worked on celestial mechanics and subsequently on the absolute differential calculus (or tensor calculus) which became the essential mathematical tool in **Einstein**'s general theory of relativity. From 1918 to 1938 he was professor in Rome, but he was forced to retire due to Fascist laws against Jews.

LEVI-MONTALCINI, Rita

(1909–)

Italian neuroscientist, born and educated in Turin, where she graduated in medicine in 1936. She began studying the mechanisms of how nerves grow, but from 1939 onwards was prevented, as a Jew, from holding an academic position, and worked from a home laboratory. During the latter part of World War II she served as a volunteer physician. In 1947 she was invited by Viktor Hamburger to Washington University in St Louis, where she remained until 1981 (Research Associate 1947–51; Associate Professor 1951–8; Professor 1958–81), when she moved to Rome. Her work has primarily been on chemical factors that control the growth and development of cells, and she isolated, originally from mouse salivary glands, a substance now called nerve growth factor that promoted the development of sympathetic nerves. Her work continued on locating further sources of the factor, on determining its chemical nature, and examining its biological activity in isolated tissues and whole neonatal and adult animals. She revealed that there were many diverse sources of the factor, such as mouse cancer cells and snake venom glands, that it was chemically a protein, and that cells are most responsive to its effects during the early stages of differentiation. This work has provided powerful new insights into processes of some neurological diseases and possible repair therapies, into tissue regeneration, and into cancer mechanisms. In 1986 she shared the Nobel Prize for Physiology or Medicine with **Stanley Cohen**.

LEWIS, Gilbert Newton

(1875–1946)

American physical chemist, born in Weymouth, Massachusetts. He was educated at the University of Nebraska and at Harvard, taking a PhD under **Theodore Richards** in 1899. Later he worked at Leipzig (under **Ostwald**) and Göttingen (under **Nernst**). From 1905 to 1912 he was on the chemistry faculty of MIT; he subsequently moved to the University of California, Berkeley, where he remained until his death. His influence and leadership were paramount in building up the international reputation of the chemistry department at Berkeley. Lewis's earliest research work was in chemical thermodynamics and culminated in his book (co-authored with Merle Randall) *Thermodynamics and the Free Energy of Chemical Substances* (1923). More widely known, however, is his pioneering work on the electronic theory of valency, in which he developed the concept of the electron-pair bond. This work began in 1916, the same year in which **Walther Kossel** independently developed the complementary part of the theory dealing with ionic compounds. Lewis's work in this area was published in his book *Valence and the Structure of Atoms and Molecules* (1923), and the ideas were developed further by **Langmuir** and by **Sidgwick**. Lewis is also remembered for his broadening of the definition of acids and bases—an acid is an electron acceptor, a base is an electron donor—and the term 'Lewis acid' is in common use (related ideas were developed around the same time by **Brønsted** and **Lowry**). His later work was on 'heavy hydrogen' (deuterium), photochemistry, and the behaviour of substances in respect of colour, fluorescence, and phosphorescence. Lewis was elected an Honorary Fellow of the Chemical Society in 1923. In 1929 he was awarded the Davy Medal of the Royal Society, of which he became a Foreign Member in 1940.

LEWIS, Jack, Baron Lewis of Newnham

(1928–)

English chemist, born in Barrow-in-Furness, Lancashire. He was educated at the University of Nottingham where he received his PhD in 1952. His first academic appointment was at the University of Sheffield; he has since held chairs at University of Manchester, University College London and Cambridge University. He is famous as an organometallic chemist, for his work on the reactivity and magnetic and spectroscopic properties of transition metal complexes. In particular his recent work has centred upon the synthesis and characterization of a large number of 'cluster' compounds, ie those which contain several metal atoms at the corners of a polyhedron. Such compounds have at least the potential to be developed as industrially important catalysts for reactions of simple alkanes, alkenes and alkynes. Lewis was elected FRS in 1973, knighted in 1982 and created a life peer in 1989.

LEWIS, John Robert

(1924–)

English marine ecologist, born in Wallasey. He was educated at University College of Wales, Aberystwyth, where he graduated in 1949 and was awarded a PhD in 1952. He was appointed to the staff of the zoology department of Leeds University in 1954, and remained there until his retirement in 1982. In 1965 the University of Leeds Wellcome Marine Laboratory was opened on the Yorkshire coast, at Robin Hood's Bay; Lewis was in charge of the laboratory throughout its existence, as Senior Lecturer-in-Charge from 1965 and Director from 1972. Due to government cutbacks in funding, Leeds University chose to close the laboratory in 1982. Lewis's major contributions to marine biology included the publication in 1964 of *Ecology of rocky shores*; this largely descriptive and analytical text was based explicitly upon extensive data and observations from sites throughout the British Isles and is still considered an essential core text in marine ecology worldwide. The book stresses the importance of adaptations to physical factors in determining species' abundance, and distribution and community structure; the final chapter focuses specifically on the importance of biological interactions (eg competition, predation) as fundamental structuring influences. Lewis also successfully developed and chaired a European Community-funded long-term international monitoring project, COST 47, which involved collaborations amongst 11 European countries.

LEWIS, Sir Thomas
(1881–1945)

Welsh cardiologist and clinical scientist, born in Cardiff. He received his preclinical training at University College, Cardiff. In 1902 he went to University College Hospital, in London, where he remained as student, teacher and consultant until his death. **Starling** stimulated his interest in cardiac physiology and the physician **Mackenzie** awakened his curiosity about diseases of the heart. He was the first to master completely the use of the electrocardiogram, and he and his students established the basic parameters which still govern the interpretation of electrocardiograms in health and disease. Through animal experiments he was able to correlate the various electrical waves recorded by an electrocardiograph with the sequence of events during a contraction of the heart. This enabled him to use the instrument as a diagnostic aid when the heart had disturbances of its rhythm, damage to its valves or changes due to high blood pressure, arteriosclerosis and other conditions. During his later years he turned his attention to the physiology of cutaneous blood vessels and the mechanisms of pain. He conducted experiments on himself in an attempt to elucidate the distribution of pain fibres in the nervous system and to understand patterns of referred pain. He fought for full-time clinical research posts to investigate what he called 'clinical science'. This broadening of his interests was signalled when he changed the name of the journal he had founded in 1909, from *Heart* to *Clinical Science* (1933). His textbooks of cardiology went through multiple editions and translations. A compulsive worker, he spent a month's holiday each year fishing and photographing birds. He was knighted in 1921.

LEWONTIN, Richard Charles
(1929–)

American geneticist, born in New York City, one of the major figures in the development of population genetics in the second half of the 20th century. He received his postgraduate training (1951–4) under **Dobzhansky** at Columbia University, and subsequently held positions at North Carolina State College, the University of Rochester, the University of Chicago and Harvard, where he was appointed Alexander Agassiz Professor of Zoology in 1973. Lewontin is equally at home in the theoretical and experimental dimensions of his field, and has contributed substantially to both areas. The central theme to his research is the problem articulated by Dobzhansky: how to explain the distribution of variation within and among natural populations. His theoretical contributions have primarily been in the area of multi-locus theory, while his empirical work, which involves detailed studies of the genetics of natural populations of *Drosophila* (fruit fly) species, has been dominated by what he has termed 'the struggle to measure variation'. He synthesized much of the work in these fields by himself and others in his influential 1974 book, *The Genetic Basis of Evolutionary Change*. In 1966, in two seminal papers with University of Chicago biochemist J L Hubby, Lewontin introduced gel electrophoresis as a means of assaying variation in protein sequences. The widespread application of this technique resulted in the discovery of large amounts of variation, and indirectly led to Motoo Kimura's development of the neutral theory of molecular evolution—that at the molecular level, most evolutionary changes result from mutations with no selective advantages. He has continued to apply technological developments in biochemistry to population genetic problems; the first study of nucleotide sequence polymorphism was carried out in his laboratory in 1983. Lewontin has also published widely on human genetics, especially with reference to the problems of a sociobiological analysis of human behaviour (eg *Not in Our Genes: Biology, Ideology & Human Nature*, 1984), and is a politically active Marxist.

LI, Choh Hao
(1913–)

Chinese-born American biochemist, born in Canton. He studied at Nanking in 1933, and from 1935 at Berkeley under the brilliant chemical endocrinologist Herbert McLean Evans, with whom he collaborated in much of his research on pituitary hormones. He became Director of the Hormone Research Laboratory, San Francisco, and Professor of

Biochemistry there from 1950. Li isolated interstitial cell stimulating (luteinizing) hormone (1940), and pure adrenocorticotrophic hormone from sheep pituitaries (1943), and determined the amino acid sequence of bovine ACTH (adrenocortical trophic hormone) in 1961. In 1945 he isolated growth hormone (somatotrophin), and later determined its amino acid sequence (1966) and achieved its synthesis (1971); he also succeeded in isolating follicle stimulating hormone (1949) and beta-lipotrophin (1965). A crucial step in the success of this work was the development of a sensitive assay using hyphophysectomized rats, later replaced by radioimmunoassay.

LIBAU, or LIBAVIUS, Andreas
(c.1560–1616)

German chemist, alchemist and writer, born in Halle, Saxony. He studied at Wittenberg and Jena universities and graduated as a doctor in Basle. He taught poetry and history at Jena, worked as town physician and inspector of schools at Rothenburg, and in 1607 was appointed rector of the newly founded Gymnasium Casmirianum at Coberg. Of his voluminous writings the most important was the *Alchemia*, in which he pioneered an analytical approach to chemistry and gave instructions how to prepare hydrochloric acid, ammonium sulphate and other important compounds. Although he believed in the transmutation of metals into gold, he opposed the mystical aspects of alchemy such as were manifested by the followers of **Paracelsus**. The *Alchemia* was not translated out of Latin and was quoted more than it was used. It is more appreciated now as a step from alchemy to chemistry than it was in its own time. Libau died in Coberg.

LIBBY, Willard Frank
(1908–80)

American chemist, born in Grand Valley, Colorado. He studied chemistry at the University of California, Berkeley, and on completing his PhD (1933) was appointed as an instructor there. By 1941 he was associate professor. A period at Princeton on a Guggenheim Fellowship was interrupted by America's entry into World War II, during which he served in the Manhattan Project on the development of the atomic bomb. After the war he became professor in the chemistry department and Institute of Nuclear Studies of the University of Chicago. From 1954 to 1959 he served on the US Atomic Energy Commission; he later became Professor of Chemistry at the University of California, Los Angeles, and in 1962 Director of the Institute of Geophysics and Planetary Physics. Libby was awarded the Nobel Prize for Chemistry in 1960 for his development of the radiocarbon dating technique. This technique has found extensive applications in archaeology, geology and geophysics. For example, it has been used to demonstrate that the last ice age occurred between 10 000 and 11 000 years ago, rather than 25 000 years ago, as was formerly thought. Libby also demonstrated that tritium, the radioactive isotope of hydrogen, is produced in nature by cosmic-ray bombardment and that this species can also be used in a dating procedure.

LIE, (Marius) Sophus
(1842–99)

Norwegian mathematician, born in Nordfjordeide. He studied at Christiania (now Oslo) University, then supported himself by giving private lessons. After visiting **Klein** in Berlin, and studying with Klein under **Camille Jordan** in Paris, a Chair of Mathematics was created for him in Christiania. In 1886 he succeeded Klein as Professor of Mathematics at Leipzig, but he returned once more to Christiania in 1898. His study of contact transformations arising from partial differential equations led him to develop an extensive theory of continuous families of transformations, now known as Lie groups. A natural geometer, he thought it preferable to write up his work in the language of analysis; his ideas were not always clear, and it was not until the later efforts of **Cartan** and **Weyl** that modern theories of Lie groups and algebras began to develop. These theories have become a central part of 20th-century mathematics, and have important applications in quantum mechanics and the theory of elementary particles.

LIEBIG, Justus von
(1803–73)

German chemist, born in Darmstadt. He acquired an interest in chemistry as a child and in 1820 went to study in Bonn, receiving his doctorate for work at Erlangen (1822). This was followed by two years' study in Paris where he learnt techniques of analysis from **Gay-Lussac**. In 1824 he was appointed Extraordinary Professor at the University of Giessen where he set up an institute for training chemists based on his experiences in France. During this period he studied the phenomenon of isomerism, and he and **Wöhler** became close friends. He developed improved procedures for the elemental analyses of organic compounds (carbon, hydrogen, nitrogen) and these proved to be of great value to all chemists attempting to elucidate the structure of natural products. For many years he carried on a dialogue (which at times became rather acrimonious) with **Dumas** concerning type theory. In collaboration with Wöhler, he recognized the enzyme emulsin which effects the breakdown of amygdalin. Wöhler also provided (in 1837) the initial impetus for Liebig's work on the use of lead peroxide as an oxidizing agent, particularly in the oxidation of uric acid. Around this time he was at the height of his experimental prowess and his laboratory was generally well funded. Rather suddenly, he wearied of the controversy with Dumas and turned his attention to other areas of research, including agricultural chemistry and physiology. In his book *Die organische Chemie in Ihre Anwendung auf Agricultur und Physiologie* (1840), he described the process we now know as photosynthesis and considered the value of fertilizers. He suggested that the behaviour of compounds under physiological conditions ought to parallel their reactions *in vitro*. His book attracted great interest and lead to a number of improvements in agricultural practice, including the use of ammonium salts as fertilizers. Liebig's efforts to distance himself from Dumas were unsuccessful and in 1841 the latter published a lecture on nutrition containing ideas similar to those expressed by Liebig. Dumas was accused of plagiarism and old hostilities were revived. In 1852 Liebig left Giessen for Munich, where he remained until his retirement. He continued to attract students from all over the world but, in spite of improved facilities, he did not maintain in Munich the world eminence he had so readily acquired in Giessen.

LILLEHEI, Clarence Walton

(1918–)

American thoracic and cardiovascular surgeon, born in Minneapolis, Minnesota. He received his training in medicine, physiology and surgery at Minnesota University. Most of his professional career was spent in the department of surgery there, although he spent seven years (1967–74) at Cornell University Medical Center in New York City. His pioneering work on open-heart surgery was begun in the early 1950s, before the development of the pump oxygenator made such procedures more reliable. He continued to be an international figure in cardiac surgery during the heroic period of the 1960s and 1970s, and trained many disciples.

LIND, James

(1716–94)

Scottish physician, born in Edinburgh. He first served in the navy as a surgeon's mate, then after qualifying in medicine at Edinburgh, became physician at the Royal Naval Hospital at Haslar. In 1747 he conducted a classic therapeutic trial, dividing 12 patients suffering from scurvy into six groups of two, treating each group with a different remedy. The two sailors given two oranges and a lemon each day responded most dramatically. His work on the cure and prevention of scurvy helped induce the Admiralty in 1795 at last to issue the order that the navy should be supplied with lemon juice, and during the Napoleonic Wars the British navy suffered far less scurvy than the French. Lind also stressed cleanliness in the prevention of fevers, and wrote major treatises on scurvy, fevers and the diseases encountered by Europeans in tropical climates. His writings are full of sensible practical advice, couched in a broad environmental framework.

LINDBLAD, Bertil

(1895–1965)

Swedish astronomer, born in Örebro. While still a student at the University of Uppsala, he started his work on the problem of distinguishing giant and dwarf stars of the same spectral type by means of their colours (1923). A stay in California (1920–2) provided him with the opportunity of a general attack on the problem of two-dimensional stellar classification through photometric studies of the spectra of faint stars, which led to valuable results on the space density of common stars. In 1925 he proposed the fundamental idea of the rotation of our galaxy around a distant centre. Following his work on the structure of our galaxy he embarked on the study of the dynamics of external spiral galaxies. He was appointed Director of the Observatory of the Swedish Academy in Stockholm in 1927 and established a magnificent new observatory in Saltsjöbaden (1931), where he remained throughout his life. His many honours included three honorary doctorates and Gold medals from the Royal Astronomical Society (1948) and the Astronomical Society of the Pacific (1954). He was a highly esteemed President of the International Astronomical Union between 1848 and 1952.

LINDERSTRØM-LANG, Kaj Ulrik

(1896–1959)

Danish biochemist, born in Copenhagen and trained at the city technical college. He worked for a few months at the National Research Institute for Animal Husbandry in Copenhagen before joining the staff of the Carlsberg Laboratory as assistant to **Sørensen**. In 1938 Sørensen retired, and Linderstrøm-Lang became professor and head of the chemistry department. In 1926, with

Sørensen, he first titrated the dissociable protons from proteins using the hydrogen electrode to measure pH, and in 1927 he reported the use of organic solvents to lower the dissociation constant (pK) of an ionizing group whose value was too high to measure by normal methods. His work on proteolytic (digestive) enzymes led him into a major study of protein structure. By using deuterium labelling, he found that the rate at which the protons of a polypeptide exchanged with the medium was either slow (associated with hydrogen-bonded structures within the molecule) or fast, if more superficially located, and variable, reflecting protein inherent structural instability. From his own and other observations he classified globular protein structure into three divisions (1951): primary (amino acid sequence plus disulphide bridges), secondary (helices and pleated sheets based on hydrogen bonding) and tertiary (folding to form the active protein) which, together with quaternary structure (subunit association), has become the basis of all modern teaching on this subject. In 1935 Linderstrøm-Lang was elected to the Danish Royal Society as its youngest member. He became a Foreign Member of the Royal Society in 1956.

LINDLEY, John
(1799–1865)

English botanist and horticulturist, born in Catton, near Norwich, son of the nurseryman George Lindley (c.1769–1835), author of *Guide to Orchard and Kitchen Garden* (1831). Like Sir **William Jackson Hooker**, John Lindley was educated at Norwich Grammar School. When 18, he was invited by Hooker to Halesworth. Lindley was harbouring ambitions of going to Sumatra as a botanical collector and caused consternation to Hooker's housekeeper by sleeping on the floorboards instead of in his bed, to accustom himself to the hardships of a sea voyage. Instead, he became assistant to **Banks** and published *Rosarium Monographia* (1820). William Cattley, a London merchant, became his patron; Lindley published *Collectanea Botanica* (1821–5) for him and named the orchid genus *Cattleya* after him. He was for long the foremost authority on Orchidaceae, publishing *Genera and Species of Orchidaceous Plants* (1830–40), *Sertum Orchidaceum* (1838) and *Folia Orchadicea* (1852–9). From 1822 to 1862 he was employed by the Horticultural Society of London and in 1829 became the first Professor of Botany at the University of London. With Hooker and Joseph Paxton (1803–65), Lindley was instrumental in saving Kew for the nation; his part was the preparation of a report on the royal gardens at Kew (1838). Possibly his greatest work was *The Vegetable Kingdom* (1846); he also published numerous textbooks, and founded *The Gardeners' Chronicle* in 1841.

LINNAEUS, Carolus (Carl von Linné)
(1707–78)

Swedish naturalist and physician, the founder of modern scientific nomenclature for plants and animals. Born in Råshult, the son of the parish pastor, he studied medicine briefly at Lund and then botany at Uppsala, where he was appointed lecturer in 1730. He explored Swedish Lapland (1732) and published the results in *Flora Lapponica* (1737), then travelled in Dalecarlia in Sweden and went to Holland for his MD (1735). In Holland he published his system of botanical nomenclature in *Systema Naturae* (1735), followed by *Fundamenta Botanica* (1736), *Genera Plantarum* (1737) and *Critica Botanica* (1737), in which he used his so-called 'sexual system' of classification based on the number of flower parts, for long the dominant system. His major contribution was the introduction of binomial nomenclature of generic and specific names for animals and plants, which permitted the hierarchical organization later known as systematics. He returned to Sweden in 1738 and practised as a physician in Stockholm, and in 1741 became Professor of Medicine and Botany at Uppsala. In 1749 he introduced binomial nomenclature, giving each plant a Latin generic name with a specific adjective. His other important publications included *Flora Suecica* and *Fauna Suecica* (1745), *Philosophia Botanica* (1750), and *Species Plantarum* (1753). His manuscripts and collections are kept at the Linnaean Society in London, founded in his honour in 1788. In his time he had a uniquely influential position in natural history.

LIOUVILLE, Joseph
(1809–82)

French mathematician, born in St Omer, the son of an army captain. He was educated at the École Polytechnique and the École des Ponts et Chaussées, where he trained as an engineer. He subsequently taught at the École Polytechnique (1831–51), and then at the Collège de France and the University of Paris. In 1836 he founded the *Journal de Mathématiques*, which he edited for nearly 40 years and which remains one of the leading French mathematical journals. His work in analysis continued the study of algebraic function theory originated by **Niels Henrik Abel** and **Jacobi**, and he studied the theory of differential equations, mathematical physics and celestial mechanics. In algebra he helped to publicize the work of **Galois**, and in number theory he introduced new methods of investigating transcendental numbers (numbers which cannot be solutions of polynomial equations with rational coefficients), showing for the first time that there are infinitely many of them. With **Sturm** he developed a theory of linear ordinary differential equations that significantly generalized **Fourier**'s ideas. In other papers, many unpublished, he developed other important techniques in real and complex analysis; the important theorem that a complex function on the plane which is not a constant must become infinite is named after him.

LIPMANN, Fritz Albert
(1899–1986)

German-born American biochemist, born in Königsberg (now Kaliningrad, Russia). He studied medicine at Berlin, and worked at the Carlsberg Institute, Copenhagen (1932–9), before emigrating to the USA. He joined the research staff at the Massachusetts General Hospital (1941–57) and became professor at Harvard Medical School (1949–57), and at Rockefeller University, New York, from 1957. He discovered that acetyl phosphate is the phosphorus-dependent intermediate in the oxidation of pyruvate by the bacterium *Lactobacillus delbruecki*, associated phosphorylation with respiration and identified the relationship with electron transfer potential. He introduced the controversial '~' symbol to indicate a 'high-energy' phosphate bond, representing adenosine triphosphate (ATP) as ADP~P. With Novelli, Lipmann demonstrated in 1950 the formation of citric acid from oxaloacetate and acetate (the first step in the **Hans Krebs** cycle) and found that a previously unidentified thiol cofactor is required—coenzyme A, derived from the B vitamin pantothenic acid. He also showed that acetate is converted to acetyl-S-CoA via the 'activation' of CoA by ATP (1952), and opened a new chapter in the understanding of these metabolic processes. He isolated and partially elucidated the molecular structure of coenzyme A, for which he shared the 1953 Nobel Prize for Physiology or Medicine with Hans Krebs.

LIPPERSHEY, Hans
(c.1570–c.1619)

Dutch optician, born in Wesel (now in Germany). He was one of several spectacle-makers credited with the discovery that the combination of two separated long- and short-focus convex lenses can make distant objects appear nearer, and applied for a patent on this type of telescope in 1608. He also showed that if this combination is reversed it becomes a microscope. The useful development of the microscope required much more theoretical knowledge than was available until the following century, but useful telescopes were developed straight away by, among others, **Galileo**.

LIPPMANN, Gabriel Jonas
(1845–1921)

French physicist, born in Hollerich, Luxembourg, to French parents who settled in Paris. He was appointed Professor of Mathematical Physics at the Faculty of Sciences in Paris (1883), Professor of Experimental Physics at the Sorbonne (1886), and the Director of the Laboratory of Physical Research. While on a scientific mission in Germany, he began research in electrocapillarity in the laboratory of **Kirchhoff**, which led on his return to Paris to his invention of a very sensitive mercury capillary electrometer. He made many contributions to instrument design, including developing an astatic galvanometer, the coelostat which made it possible to photograph (or observe) a region of the sky for an extended period without apparent movement, and a new form of seismograph. For his technique of colour photography based on the interference phenomenon, subsequently also used by Lord **Rayleigh** (1842–1919), he was awarded the 1908 Nobel Prize for Physics. He was elected FRS the same year.

LIPSCOMB, William Nunn
(1919–)

American inorganic chemist, born in Cleveland, Ohio. He studied at Kentucky and Caltech, and was appointed Professor of Chemistry at Harvard in 1959. He deduced the molecular structures of a group of boron hydride compounds (boranes) by X-ray crystal diffraction analysis in the 1950s, and went on to develop novel theories to explain their chemical bonds. His ingenious experimental and theoretical methods were later applied by him and others to a variety of related chemical problems. He was awarded the Nobel Prize for Chemistry in 1976.

LI SHIH-CHEN
(1518–93)

Chinese pharmaceutical naturalist and biologist, regarded as the prince of pharmacists and the father of Chinese herbal medicine. A successful physician, he was appointed to the Imperial Medical Academy. He decided to produce, single-handed and without imperial authority, an encyclopedia of pharmaceutical natural history. The work took 30 years, and involved much travelling to collect and study specimens. He compiled the *Pen Tshao Kang*

Mu ('Great Pharmacopoeia'), completed in 1578 and published in 1596. It gives an exhaustive description of 1000 plants and 1000 animals, and includes more than 11000 prescriptions. It is much more than a pharmacopoeia, however, as it treats mineralogy, metallurgy, physiology, botany and zoology as sciences in their own right. By categorizing diseases, it also forms a system of medicine. He recorded many instances of the sophistication of Chinese medicine, for example the use of mercury–silver amalgam for tooth fillings, not introduced to Europe until the 19th century. He adopted a system of priority in naming plants and animals, assigning the first name as the standard term, and treating later names as synonyms.

LISSAJOUS, Jules Antoine
(1822–80)

French physicist, born in Versailles. He was educated at the École Normale Supérieure and became professor at the Collège St Louis, Paris. In 1857 he invented the vibration microscope which showed visually the 'Lissajous figures' obtained as the resultant of two simple harmonic motions at right angles to one another. This made possible very precise comparisons of frequencies, for example of tuning forks and organ pipes, and enabled Lissajous to make important contributions to the history of changes in musical standard pitch. His researches extended also to optics, and his system of optical telegraphy was used during the siege of Paris (1871).

LISTER, Joseph, Lord
(1827–1912)

English surgeon, the 'father of antiseptic surgery', born in Upton, Essex. He was the son of the microscopist Joseph Jackson Lister (1786–1869). After graduating from London University in arts (1847) and medicine (1852), he became house surgeon at Edinburgh Royal Infirmary to the surgeon James Syme, whose daughter he married in 1856. He was successively a lecturer on surgery, Edinburgh; Regius Professor of Surgery, Glasgow (from 1859); Professor of Clinical Surgery, Edinburgh (from 1869) and King's College Hospital, London (1877–93); and President of the Royal Society (1895–1900). In addition to important observations on the coagulation of the blood and the microscopical investigation of inflammation, his great work was the introduction of his antiseptic system (1867), which revolutionized modern surgery. His system was inspired by **Pasteur**'s work on the role of micro-organisms in fermentation, putrefaction and other biological phenomena. Lister began soaking his instruments and surgical gauzes in carbolic acid, a well-known disinfectant. His early antiseptic work was primarily concerned with the operative reduction of compound fractures and the excision of tuberculous joints. Both conditions would previously have been treated with amputation, since the operation mortality for conservative surgical treatment would have been very high. Antiseptic surgery led to aseptic techniques, whereby the organisms causing wound infections were, whenever possible, excluded from the surgical field altogether. Listerian procedures made it possible for surgeons to open the abdominal, thoracic and cranial cavities without fatal infections supervening. Lister himself rarely ventured into the bodily cavities, confining himself mostly to the limbs and superficial parts of the body. He was, however, a technically gifted surgeon who popularized the use of sutures made of catgut. He worked later in his life on the aetiology of wound infection and was an ardent advocate of the value of experimental science for medical and surgical practice. He was much revered in life and was the first medical man to be elevated to the peerage.

LITTLE, Clarence Cook
(1888–1971)

American pioneer of mammalian and cancer genetics, born in Brookline, Massachusetts. He was educated at Bussey Institute, an agricultural college of Harvard University, where he studied with **Sewall Wright**. In 1909, whilst still an undergraduate, he established the first inbred strain of laboratory mice, which he called DBA after the mutant alleles it carried. After military service he worked at Cold Spring Harbor (1918–22), where he began inbreeding several other strains of mice. He was appointed President of the University of Maine (1922–5) and of the University of Michigan (1925–9), then established Roscoe B Jackson Memorial Laboratory on Mount Desert Island, Maine, where he served as Director from 1929 to 1956. The establishment and use of inbred mouse strains was an essential preliminary to much biomedical discovery. Without the vision of Little and the work of the Jackson Laboratory, our understanding (and treatment) of cancer, transplantation, drug action and multigenic conditions would have been much delayed.

LITTLEWOOD, John Edensor
(1885–1977)

English mathematician, born in Rochester. He was educated at Trinity College, Cambridge, and after lecturing in Manchester (1907–10) he returned to Cambridge as a

Fellow of Trinity and remained there for the rest of his life. At this time he started to collaborate with **Godfrey Hardy**, and a stream of joint papers on summability theory, Tauberian theorems, **Fourier** series, analytic number theory and the **Riemann** zeta function followed over the next 35 years. Littlewood was elected to the Rouse Ball Chair of Mathematics at Cambridge in 1928. He retired in 1950, but was still publishing mathematical papers at the age of 85. His reminiscences, *A Mathematician's Miscellany*, were published in 1953.

LLIBOUTRY, Louis Antonin François
(1922–)

French Earth scientist and glaciologist, born in Madrid of a French Catalonian family. He lived in Madrid until the outbreak of civil war in 1936. He graduated from the École Normale Supérieure, Paris, with a BA in mathematics (1940) and MS in physics. After receiving a doctorate from Grenoble (1950) on piezomagnetism under **Néel**, he became associate professor at the University of Chile, Santiago (1951–6). Lliboutry mapped all the unknown glaciers of the Chilean Andes and participated in an expedition to conquer Monte Fitz Roy, Patagonia (1952). Back in Grenoble where he was appointed assistant professor in 1956, he renewed glaciological studies in France and studied field glacier dynamics and mass balances, with particular interest in the critical physical conditions necessary for the formation of glaciers and ice sheets. In plate tectonics he made an important study of plate driving mechanisms, and derived a predictive theorem for plate velocities relative to the mantle. He fought for the establishment of the Laboratory of Glaciology and Environmental Geophysics, of which he became head (1964–83) and which has subsequently gained world recognition.

LOBACHEVSKI, Nikolai Ivanovich
(1792–1856)

Russian mathematician, born in Nizhni Novgorod. In 1814 he became professor at the University of Kazan, where he spent the rest of his life. As Rector of the university he oversaw the erection of many new university buildings, and successfully defended the university against cholera (1830) and fire (1841). From the 1820s he developed a theory of non-Euclidean geometry in which **Euclid**'s parallel postulate (that there is only one straight line which passes through a given point and is parallel to another given line) did not hold. A similar theory was discovered almost simultaneously and independently by **Bolyai**. Lobachevski recognized that his discovery made the nature of physical space an empirical question, and he analysed observations on the parallax of stars inconclusively with a view to resolving the matter. Despite publication in various languages, his theory was too novel and its presentation too obscure to find acceptance in his lifetime, although its worth was acknowledged by **Gauss**. He also wrote on algebra and the theory of functions.

L'OBEL, or LOBEL, Matthias de
(1538–1616)

Flemish physician and naturalist, born in Lille. He studied at Montpellier under Guillaume Rondelet, who also taught **L'Écluse** and Jean Bauhin. He became physician to William the Silent, but left to come to England under the patronage of Queen Elizabeth. He superintended the medicinal garden of Lord Zouche at Hackney. Later he became botanist and physician to King James VI and I of Scotland and England. He published *Stirpium Adversaria Nova* in 1570; in this work, he devised a system of classification based on the characters of plant leaves. He made a rough classification of plants into the classes now known as Monocotyledons and Dicotyledons. The work was enlarged and republished as *Plantarum seu Stirpium Historia* (1576), which was translated into Flemish as *Kruydtboek* (1581). The blocks used to illustrate this book were taken from previous works, mainly those of L'Écluse. L'Obel's name is commemorated in the plant genus *Lobelia*.

LOCKE, John
(1632–1704)

English philosopher, born in Wrington, Somerset, and educated at Christ Church, Oxford. He reacted against the prevailing scholasticism at Oxford and involved himself instead in experimental studies of medicine and science, making the acquaintance of **Boyle** and others. In 1667 he joined the household of Anthony Ashley Cooper as his personal physician and became his adviser in scientific and political matters generally. Through Ashley he made contact with the leading intellectual figures in London and was elected FRS in 1668. When Ashley became Earl of Shaftesbury and Chancellor in 1672, Locke became Secretary to the Council of Trade and Plantations, but retired to France from 1675 to 1679. In Paris he became acquainted with the circle of **Gassendi**. After Shaftesbury's fall and death in 1683 he felt threatened and fled to Holland, where he remained until after the Glorious Revolution of 1688. Locke returned to

England in 1689, declined an ambassadorship and became Commissioner of Appeals until 1704. His major philosophical work was the *Essay Concerning Human Understanding*, published in 1690 though developed over some 20 years. This influential work was a systematic inquiry into the nature and scope of human reason, and argued that the foundations of knowledge should be based on experiment.

LOCKYER, Sir (Joseph) Norman

(1836–1920)

English astronomer, born in Rugby. On leaving school he became a clerk at the British war office to which he remained technically attached (1857–75), devoting as much time as he could spare to science. In 1868 he designed a spectroscope for observing solar prominences outside of a total eclipse and succeeded in doing this independently of **Janssen** who had used the same principle a few months earlier. Lockyer shared with Janssen the credit for this discovery, commemorated by a medallion issued by the French government. In the same year he postulated the existence of an unknown element which he named helium (the 'Sun element'), an element not found on Earth until 1895 by **Ramsay**. He also discovered and named the solar chromosphere. In 1875 he became a member of the staff of the Science Museum in South Kensington, London. His researches gave rise to unconventional ideas such as his theory of dissociation, whereby atoms were believed to be capable of further subdivision, and his meteoritic hypothesis which postulated the formation of stars out of meteoric material. Among other activities, Lockyer took part in eclipse expeditions and made surveys of ancient temples for the purpose of dating them by astronomical methods. He was the founder (1869) and first editor of the scientific periodical *Nature*, and was knighted in 1897. His solar physics observatory at South Kensington was transferred to Cambridge University in 1911, but Lockyer remained active in a private observatory which he set up in Sidmouth, Devon, until his death.

LODGE, Sir Oliver Joseph

(1851–1940)

English physicist, born in Penkhullin to a very large middle-class family, the eldest of nine. He studied at the Royal College of Science and at University College London, and in 1881 became Professor of Physics at Liverpool. In 1900 he was appointed the first Principal of the new university at Birmingham. Lodge demonstrated in 1893 that the hypothetical ether could not be carried along by moving matter, thus eliminating one possible interpretation of the Michelson–Morley experiment, discrediting the ether theory, and preparing the way to the theory of relativity. Specially distinguished in electricity, he was a pioneer of wireless telegraphy, almost anticipating **Heinrich Hertz**, and invented a coherer for the detection of radio waves (later superseded by the crystal). His scientific writings include *Signalling across Space without Wires* (1897), *Talks about Wireless* (1925) and *Advancing Science* (1931). He gave much time to psychical research (intensified after the tragic death of his son Raymond), and on this subject wrote *Raymond* (1916) and *My Philosophy* (1933). *Past Years: An Autobiography* appeared in 1931. He was elected FRS in 1887, awarded the Society's Rumford Medal (1898), and knighted in 1902.

LOEB, Jacques

(1859–1924)

German-born American biologist, born in Mayen. Educated in philosophy at Berlin, and in medicine at Strassburg, Loeb obtained his MD in 1884, and in 1886 was appointed to an assistantship at Würzburg. An interest in the philosophy of the will led to research which attempted to show, in the animal world, phenomena analogous to plant trophisms. He began to publish in this area in 1888, showing that certain caterpillars move towards light even when their food is in the opposite direction. He emigrated to the USA in 1891, and held various university appointments before becoming head of the general physiology division at the Rockefeller Institute for Medical Research (1910–24). He conducted pioneering work on artificial parthenogenesis. He demonstrated that unfertilized sea-urchin eggs, made to start segmentation by osmotic changes, could subsequently grow into larvae (1899). A champion of materialism in philosophy, of mechanistic explanations in science, and a socialist in politics, Loeb became well known to the American public. His writings included *Dynamics of Living Matter* (1906) and *Artificial Parthenogenesis and Fertilisation* (1913).

LOEWI, Otto

(1873–1961)

German pharmacologist, born in Frankfurt. Educated at Strassburg and Munich, he was appointed Professor of Pharmacology at Graz (1909–38). Forced to leave Nazi Germany in 1938, he became research professor at New York University College of

Medicine from 1940. From 1901 he worked for a time alongside **Dale** in the laboratories of **Starling** at University College London, and confirmed current theories that a chemical substance released by the nerve terminal transmitted the stimulus invoking contraction of an isolated frog heart. In 1921 he showed that the substance released by stimulating one isolated heart (Vagusstoff) caused contraction in another. He subsequently identified several possible transmitter substances and distinguished 'Vagusstoff ' as acetylcholine, rather than adrenaline or a similar substance, because (as predicted by Dale) its action was prolonged by inhibition of the enzyme that caused its breakdown (acetylcholinesterase) with eserine. Anticholinesterase drugs are now used to increase muscle strength in patients with myasthenia gravis, an inherited muscular dystrophy. Loewi shared with Dale the 1936 Nobel Prize for Physiology or Medicine for investigations on the chemical transmission of nerve impulses. He was elected a Foreign Member of the Royal Society in 1954.

LÖFFLER, Friedrich August Johann
(1852–1915)

German bacteriologist, born in Frankfurt an der Oder. He began his career as a military surgeon, became Professor at Greifswald (1888) and from 1913 was Director of the Koch Institute for Infectious Diseases in Berlin. He first cultured the diphtheria bacillus (1884), discovered by Edwin Klebs and called the 'Klebs–Löffler bacillus', discovered the causal organism of glanders and swine erysipelas (1886), isolated an organism causing food poisoning, and prepared a vaccine against foot-and-mouth disease (1899). He also presented the first evidence for the occurrence of the pathogens which we now call filterable viruses. In 1887 he wrote an unfinished history of bacteriology.

LOGAN, Sir William Edmund
(1798–1875)

Canadian geologist, born in Montreal. Educated at Edinburgh University, he spent 10 years in a London counting house before becoming book-keeper in Swansea to a copper smelting company (1828). There he made a map of the coal basin which was incorporated into the geological survey. He discovered fossil rootlets in the clay beneath coal seams and correctly asserted that the deposit represented a fossil soil (1840). In 1842 he was appointed first Director of the Geological Survey of Canada, a post which he retained until 1869. He undertook studies of the coalfields of Nova Scotia and New Brunswick, and carried out important work on the copper-bearing rocks of the Lake Superior region. His discovery in 1841 of the animal tracks at Horton Bluff, Nova Scotia, provided the first demonstration of the existence of land animals in the Upper Palaeozoic. He was knighted in 1856.

LOMONOSOV, Mikhail Vasilievich
(1711–65)

Russian scientist and writer, born in Kholmogory near Archangel. The son of a fisherman, he ran away to Moscow in search of an education and later studied at St Petersburg, at Marburg in Germany under the philosopher Christian von Wolff, and finally at Freiberg, where he turned to metallurgy and glass-making. He became Professor of Chemistry at the St Petersburg Academy of Sciences in 1745 and set up Russia's first chemical laboratory there. His experiments led to the establishment of a glassworks making coloured glass for mosaics. He was opposed to the phlogiston theory, and it has been recorded that he suggested the law of mass conservation, and anticipated **Rumford** on the kinetic theory of heat and **Thomas Young** on the wave theory of light. He also made important contributions to Russian literature. Because he advocated popular education and freedom for the serfs, Lomonosov always had to fight against prejudice, sometimes amounting to persecution. Although he came to be revered by many people in his own day, his papers were confiscated by Catherine the Great after his death in St Petersburg.

LONDON, Fritz Wolfgang
(1900–54)

German-born American physicist, born in Breslau (now Wrocław, Poland), brother of **Heinz London** and son of a mathematics professor in Bonn. He studied classics at the universities of Frankfurt and Munich and did research in philosophy leading to a doctorate at Bonn. Later he was attracted to theoretical physics and worked with **Sommerfeld** at Munich and **Schrödinger** at Zürich in 1927. He worked on the quantum theory of the chemical bond, publishing in 1927 a classic quantum-mechanical treatment of the hydrogen molecule. In 1930 he calculated the non-polar component of forces between molecules, now known as **van der Waals** or London forces. With his brother he fled from Germany to Oxford (1933) where they joined Sir **Francis Simon**'s group at the Clarendon Laboratory. Together they published important papers on superconductivity giving the London equations (1935), which gave a description of the electromagnetic behaviour of superconductors and made use of a 'two-fluid model' (similar to **Casimir**'s) for the electrons. Fritz London moved to Duke University in the USA (1939–54), continuing to work on superconductivity, and on superfluidity.

LONDON, Heinz
(1907–70)

German-born British physicist, born in Bonn, younger brother of **Fritz London**. He was educated at the universities of Bonn, Berlin, Munich and Breslau, where he worked for his PhD with Sir **Francis Simon**. The brothers fled from Germany in 1933 and joined Simon's group at the Clarendon Laboratory, Oxford, working together on conductivity. Heinz London advanced the two-fluid model of superconductors and introduced a theory for the confinment of currents in a superconductor to a surface layer characterized by a penetration depth. With his brother he published equations describing the electromagnetic behaviour of superconductors (1935), which became known as the London equations. He moved to Bristol, was briefly interned as an enemy alien in 1940, and then released to work with Simon and others on the development of the British atomic bomb. After spending two years at Birmingham he transferred to the Atomic Energy Research Establishment at Harwell (1946). There he continued work on isotope separation, which he had started during the atomic bomb project. He did much theoretical work on helium-2, and he invented a method of producing temperatures below 0.1 kelvin by mixing isotopes of helium-3 and helium-4.

LONGUET-HIGGINS, (Hugh) Christopher
(1923–)

English theoretical chemist, born in Lenham, Kent. He studied chemistry at Balliol College, Oxford, where he was a research fellow from 1946 to 1948. From 1948 to 1952 he was lecturer and then reader in theoretical chemistry at the University of Manchester, before briefly holding the Chair of Theoretical Physics at King's College, London (succeeding **Coulson**), and then in 1954 becoming Plummer Professor of Theoretical Chemistry at Cambridge (succeeding **Lennard-Jones**). From 1968 to 1974 he was Royal Society Research Professor in the Department of Artificial Intelligence at the University of Edinburgh and from 1974 to his retirement in 1988 he held a similar position at the University of Sussex. For some 20 years he made fundamental contributions to the molecular orbital theory of organic and inorganic chemistry, successfully predicting the failure of resonance to explain convincingly the reactivity of biphenylene compounds and the formation of complex molecules containing a metal and cyclobutadiene. He used symmetry arguments to predict the course of various electrocyclic reactions, thus parallelling the work of **Robert Woodward** and **Hoffmann**. Following his move to Edinburgh in 1968, he embarked on a second phase of research in which he has worked on problems of the mind, including language acquisition, music perception and speech analysis. He was elected FRS in 1958 and a foreign associate of the US National Academy of Sciences in 1968, and gave the Gifford Lectures in 1972.

LONSDALE, Dame Kathleen, née Yardley
(1903–71)

Irish crystallographer, born in Newbridge, County Kildare. She entered Bedford College, London (1919), to study mathematics, but changed to physics at the end of her first year. On graduation in 1922 she was invited by **William Bragg** to join his crystallography research team, first at University College London (UCL), and then at the Royal Institution. She remained at the Royal Institution until 1946, apart from a short period when she worked at Leeds (1929–31). In 1946 she became Reader in Crystallography in the chemistry department of UCL and in 1949 she was promoted to Professor of Chemistry. She retired in 1968. Lonsdale was appointed DBE in 1956. In 1945, when the Royal Society agreed to admit women fellows, she was one of the first two women to be elected FRS; she was awarded the society's Davy Medal in 1957. Of her many contributions to crystallography, the most celebrated was her X-ray analysis in 1929 of hexamethylbenzene and hexachlorobenzene, which showed that the carbon atoms in the benzene ring are coplanar and hexagonally arranged. She also made important contributions to space-group theory and to the study of anisotropy and disorder in crystals. She became a Quaker in 1935 and later worked tirelessly for various causes including peace, penal reform and the social responsibility of science.

LONSDALE, William
(1794–1871)

English geologist, born in Bath. He served in the army, being present at the Battle of Waterloo, but left it in 1815 and took up geology. He became assistant secretary and curator of the Geological Society of London

(1829–42), and spent a good deal of time studying the fossil corals in the vicinity of Bath. He also made a study of the fossils in north and south Devon in 1837, placing them between the Silurian and Carboniferous. This subsequently led to the establishment of the Devonian system by **Murchison** and **Sedgwick** (1839). Lonsdale employed the evolutionary concepts that **Charles Darwin** was to champion subsequently, that species undergo modification with time and that the existing forms of life are descendants, by true generation, of pre-existing lifeforms.

LORENTE DE NO, Rafael

(1902–)

Spanish-born American neurophysiologist, born in Zaragoza, and educated in medicine at the University of Madrid. After a period as a research assistant in the Cajal Institute in Madrid (1921–9), he was appointed head of the department of otolaryngology in Santander (1929–31) before travelling to the USA. Working initially as a neuroanatomist at the Central Institute for the Deaf in St Louis (1931–6), he moved to the Rockefeller Institute in New York (associate, 1936–8; associate member, 1938–41; member and professor, 1941–74). His research covered a wide range of neurophysiological and neuroanatomical problems, including the coordination of eye movements, the functional anatomy of neuron networks, and the neurophysiology of synaptic transmission.

LORENTZ, Hendrik Antoon

(1853–1928)

Dutch physicist, born in Arnhem. He studied at Leiden and initially was uncertain whether he should follow an academic career in mathematics or in physics. At the young age of 25 he was offered at Leiden (1878) the first Chair of Theoretical Physics which had been created for **van der Waals**. He remained there until 1912 when he was appointed the Director of the prestigious Teyler's Institute in Haarlem, a private research foundation dating back to the mid-18th century. He also became involved in Dutch science policy. His major contribution to theoretical physics was his electron theory with which he formulated a new interpretation of **Maxwell**'s theory. In 1904 he derived a mathematical transformation, the '**Fitzgerald**–Lorentz contraction', which explained the apparent absence of relative motion between the Earth and the (supposed) ether, and prepared the way for **Einstein**'s theories of relativity. In 1902 he was awarded, with **Zeeman**, the Nobel Prize for Physics. He was eminent in international scientific affairs, and also served as President of the League of Nations' International Committee of Intellectual Cooperation. He was made a Foreign Member of the Royal Society in 1905, and was awarded its Copley Medal in 1912.

LORENZ, Edward Norton

(1917–)

American mathematician and meteorologist, born in West Hartford, Connecticut. He was educated at Dartmouth College and Harvard University, and has worked at MIT since 1946 (as professor from 1962) apart from a short period as visiting professor at the University of California at Los Angeles (1954–5). Following studies of the dynamics of fluids in a rotating basin, he introduced a simplified model of the general circulation of the atmosphere. With different intensities of the symmetrical zonal flow and with different intensities of an asymmetric disturbance introduced into these flows, he demonstrated that the equations possessed one or two steady-state solutions, one or two stable but periodic solutions, and also aperiodic (irregular) solutions. Applying these ideas, he showed that the real atmosphere could probably vacillate from one major mode to another. It also contains periodic motions (eg the annual cycle), but there are also irregular motions introduced by the complexity of the atmospheric conditions. He went on to demonstrate an aspect of chaos theory, that forecasts produced from slightly different initial conditions (but within the limits of the observing system) diverge with time and that however perfect the numerical model, the limit of predictability of useful forecasts is about 10–14 days. With certain favourable initial conditions, however, longer predictions can be possible. He was awarded the Symons Gold Medal of the Royal Meteorological Society (1973).

LORENZ, Konrad Zacharias

(1903–89)

Austrian zoologist and ethologist, born in Vienna, the son of a surgeon. He studied in Vienna, and along with **Tinbergen**, advocated the observation of animal behaviour under natural conditions rather than in the laboratory, thus founding in the late 1930s the school of animal behaviour called ethology. Lorenz and colleagues mainly studied the behaviour of birds, fish and some insects whose behaviours contain a relatively high proportion of stereotyped elements or 'fixed action patterns'. These could be used in the same way as morphological characters as the basis for comparisons between species. Thus Lorenz was able to compare the courtship of different species of duck and suggest how the different patterns had arisen and subsequently evolved. He was the first to describe imprinting and releaser mechanisms. In the former, newly hatched birds, such as goslings and jackdaws, become imprinted on the first moving object they see and treat the object thereafter as a member of their own species and even as a potential mate. Releaser mechanisms are innate behaviours which are elicited by very specific stimuli, for example the red breast of the robin releasing aggression. Although Lorenz was criticized for the rigidity of these and other ethological concepts, the ethological approach had a profound influence on subsequent animal behaviour studies. In *On Aggression* (1963), he argued that aggressive behaviour in man is an inherited drive and can be channelled into non-destructive activities such as sport. The validity of this view of human aggression is still disputed. In addition to his scientific writings he wrote popular accounts of his work with animals such as *King Solomon's Ring* (1949) and *Man Meets Dog* (1950). He shared the 1973 Nobel Prize for Physiology or Medicine with Tinbergen and **Karl von Frisch**.

LORENZ, Ludwig Valentin
(1829–91)

Danish physicist, born in Elsinore. He trained as a civil engineer at the Technical University of Denmark in Copenhagen, going on to teach at the Danish Military Academy where he became professor in 1866. His researches concerned many areas of optics, heat and the electrical conductivity of metals. In his early work on the nature of light itself, he adopted a model based on elasticity to describe wave propagation. He eventually abandoned this and concentrated instead on finding a phenomenological description of the way light passes through matter. Pursuing this he advanced a mathematical description for light waves (1863), showing that under certain conditions double refraction will occur. The publication of his work on relating refraction and specific densities of media (1869) preceded that of **Lorentz** by one year; the result became known as the Lorentz–Lorenz formula. From observations of the scattering of sunlight in the atmosphere, Lorenz made the first fairly accurate estimate of **Avogadro**'s number (1890). Two years after the publication of **Maxwell**'s famous paper on electromagnetic theory, without knowledge of Maxwell's results, Lorenz published his own theory of electromagnetism.

LOSCHMIDT, Johann Josef
(1821–95)

Austrian chemist, born near Carlsbad, Bohemia. From a poor peasant family, he nevertheless attended high school and ultimately went to universities in Prague and Vienna. At the latter his studies included physics and chemistry. He was involved in various industrial chemical ventures, but these were financial failures and Loschmidt became bankrupt in 1854. He qualified as a school teacher and obtained a post at a Vienna Realschule. In 1866 he became *Privatdozent* at the University of Vienna and in 1868, Assistant Professor of Physical Chemistry. Thereafter Loschmidt became prominent in the scientific life of Vienna, being appointed Chairman of the Physical Chemistry Institute in 1875. His best-known work was his calculation in 1865 of the number of molecules in one millilitre of an ideal gas under standard conditions (the Loschmidt number). It has come to be recognized that he also anticipated much of the structural theory of organic chemistry usually attributed to **Kekulé** (1865). This was published privately in a booklet in 1861.

LÖVE, Áskell
(1916–) and
Doris Benta Maria, née Wahlen
(1918–)

Icelandic and Swedish botanists, born in Reykjavík and Kristianstad respectively. Both studied at the University of Lund, where they graduated in 1941 and went on to gain doctoral degrees. Áskell Löve has published numerous books and papers on the Icelandic and arctic flora including *Islenzkar Jurtir* (1945) and *Íslensk Fardflorá* (1970), on the genus *Rumex* (Polygonaceae) and on biogeography, biosystematics and plant speciation. His wife Doris Löve has interests in Pleistocene and arctic biology, plant dispersal and plant collections from Quebec and Yukon; she has translated **Vavilov**'s *The Origin and Geography of Cultivated Plants* (1992). The two are best known for their many collaborative works on cytotaxonomy of circumpolar plants, and for their theories on continental drift and the origin of the arctic–alpine and north Atlantic floras. These include *North Atlantic Biota and Their History* (1963, editors), *Plant Chromosomes* (1975), *Cytotaxonomical Atlas of the Slovenian Flora* (1974), *Cytotaxonomical Atlas of the Arctic Flora* (1975), and *Cytotaxonomical Atlas of the Pteridophyta* (1977, co-authors).

LOVE, Augustus Edward Hough
(1863–1940)

English geophysicist, born in Weston-super-Mare, the son of a surgeon. In 1882 he entered St John's College, Cambridge, later becoming a Fellow (1886–9). In 1899 he became Sedleian Professor of Natural Philosophy at Oxford. His research interests covered the theory of deformable media, fluids and solids, theoretical geophysics, the theory of electric waves and ballistics. His *Treatise on the Mathematical Theory of Elasticity* (2 vols, 1892–3) was a standard textbook for nearly 50 years. In *Some Problems in Geodynamics* (1911), Love introduced the concept of 'Love waves'—Rayleigh waves transmitted over the surface of an elastic solid—and he was first to observe seismic Love waves. He also related the periods and group velocities of surface waves, which became a powerful tool in estimating crustal thickness and led to the first evidence for the different crustal thicknesses of continents and oceans. 'Love numbers' are important parameters in tidal theory.

LOVELACE, (Augusta) Ada, Countess of, née Byron
(1815–52)

English mathematician and writer, daughter of the poet Lord Byron. Encouraged by her mother, she taught herself geometry, and was trained in astronomy and mathematics. Her tutors included **de Morgan**. She translated and annotated an article on the analytical engine of the computer pioneer **Babbage**, written by Italian mathematician L F Menabrea, adding many explanatory notes of her own. This *Sketch of the Analytical Engine* (1843) is an important source on Babbage's work. The high-level universal computer programming language ADA was named in her honour, and is said to realize several of her insights into the working of a computer system.

LOVELL, Sir (Alfred Charles) Bernard
(1913–)

English astronomer, born in Oldham Common, Gloucestershire. After graduating from Bristol University he became a physics lecturer at Manchester University. During World War II he worked for the Air Ministry Research Establishment, where he developed airborne radar for blind bombing and submarine defence. In 1951 he became Professor of Radio Astronomy at Manchester University and Director of Jodrell Bank Experimental Station (now the Nuffield Radio Astronomy Laboratories). He gave the radio Reith Lectures in 1958, taking for his subject 'The Individual and the Universe'. Elected FRS in 1955, he was knighted in 1961. Scientifically, Lovell was a pioneer in the use of radar to detect meteors and day-time meteor showers. In 1950 he discovered that the rapid oscillations in the detected intensity (or 'scintillations') of signals from galactic radio sources were produced by the Earth's ionosphere, and were not intrinsic to the sources. Lovell was the energetic instigator of the funding, construction and use of the 250 foot steerable radio telescope at Jodrell Bank, Cheshire. Plans to use the telescope to track Sputnik 1 and 2 attracted the funding which solved financial problems associated with its construction, and it was completed in 1957. From 1958 Lovell has collaborated with **Fred Whipple** in the study of flare stars. He has written several books on radio astronomy and on its relevance to life and civilization today. His works include *Science and Civilisation* (1939), *World Power Resources and Social Development* (1945), *Radio Astronomy* (1951), *Discovering the Universe* (1963), *The Story of Jodrell Bank* (1968), *Emerging Cosmology* (1980) and *Voice of the Universe* (1987).

LOVELOCK, James Ephraim
(1919–)

English chemist, born in Letchworth, Hertfordshire. He was educated at the universities of Manchester and London before joining the staff of the National Institute for Medical Research in London (1941–61). He then held various posts in the USA, including a professorship in chemistry at Baylor College of Medicine, Texas (1961–4), and since 1964 has pursued a career as an independent scientist. He invented the 'electron capture detector' (1958), a high-sensitivity device which enables measurement of pesticide residues in the environment; it was used in the first, and most subsequent, measurements of the accumulation of CFCs (chlorofluorocarbons) in the atmosphere. In 1972 Lovelock put forward his controversial 'Gaia' hypothesis; this proposes that the climate of the Earth is constantly regulated by plants and animals, to maintain a life-sustaining balance of carbon dioxide and other organic substances in the atmosphere. Gaia is now regarded as an evolving system comprising the atmosphere, oceans, surface rocks and the biota, which has the emergent property of self-regulation. It behaves as a superorganism and has kept the climate and chemistry comfortable for life. Lovelock's theories are expounded in the popular books *Gaia* (1979), *The Ages of Gaia* (1988) and *The Practical Science of Planetary Medicine* (1991).

LOWELL, Percival
(1855–1916)

American astronomer, born into a prominent Boston family, brother of political scientist Abbott Lowell and poet Amy Lowell. He was educated at Harvard. Between 1883 and 1893

Lowell travelled around the Far East. He established the Flagstaff (now Lowell) Observatory in Arizona (1894) on a site 2 000 metres above sea-level. He is best known for his observations of Mars, these resulting in a series of maps showing linear features crossing the surface. In 1907 he led an expedition to the Chilean Andes which produced the first high-quality photographs of Mars. Lowell observed seasonal changes on the surface and dark waves that seemed to flow from the poles towards the equator. The latter have now been attributed to dust storms. He popularized his ideas in a series of books which include *Mars and its Canals* (1906) and *Mars as the Abode of Life* (1910). Lowell is known for his prediction of the brightness and position of a planet that was supposedly responsible for the orbital perturbation of Neptune and Uranus. This he called planet X. In 1930, 14 years after his death, Pluto was found (discovered by **Tombaugh**).

LOWER, Richard

(1631–91)

English physician and physiologist. Born in Tremeer, near Bodmin, and springing from an old Cornish family, Lower studied at Oxford, where he came under the influence of **Thomas Willis** and several others soon to achieve prominence in the Royal Society of London. Gaining his MD in 1665, Lower followed Willis to London, setting up in medical practice and joining the Royal Society. In 1665 he demonstrated transfusion of blood from the artery of one dog to the vein of another, though attempts to use blood transfusion for practical therapeutic purposes long proved largely unavailing. Lower's *Tractatus de Corde* ('Treatise on the Heart', 1669) provides a fine account of contemporary knowledge in pulmonary and cardiovascular anatomy and physiology. Following **William Harvey** (1578–1657), Lower recognized that the heart was not dilated by spirits, acting rather as a muscular pump, with systole as the active phase and diastole the return movement. He paid attention to the colour change between dark venous blood and red arterial blood. With **Hooke**, he experimented with dogs, deducing that the red colour resulted from the mixing of dark blood with inspired air in the lungs. Lower fathomed that the function of respiration lay in adding something to the blood. After the 1670s, he concentrated on his medical practice.

LOWRY, (Thomas) Martin

(1874–1936)

English chemist, born in Bradford, West Yorkshire. He studied chemistry under **Henry Armstrong** at the City and Guilds Institute, South Kensington, and from 1896 was for some years Armstrong's assistant. In 1906 he became lecturer in chemistry at Westminster Training College, and in 1912 he moved to Guy's Hospital Medical School. From 1913 he was head of the chemistry department at Guy's, and he became the first Professor of Chemistry in any London medical school. During World War I he was engaged in the development of explosives. In 1920 Lowry became the first to hold a chair of physical chemistry at Cambridge, where he remained until his death. His earliest researches were on camphor derivatives and in particular the changes in optical rotation (mutarotation) which occur when these are treated with acids or bases as catalysts in solution. This ultimately led to his redefinition of the terms acid and base (1923) in the same way as that advocated independently by **Brønsted** in the same year. In his later days he worked extensively on optical rotatory dispersion, foreshadowing its importance many years later as a structural tool in organic chemistry. He was elected FRS in 1914, appointed CBE in 1920, and gave the Bakerian Lecture in 1921. From 1928 to 1930 he was President of the Faraday Society.

LUBBOCK, Sir John, 1st Baron Avebury

(1834–1913)

English politician and biologist, born in London, the son of the astronomer Sir J W Lubbock (1803–65). From Eton he went into his father's banking house at the age of 14, becoming a partner in 1856. He served on several educational and currency commissions, and in 1870 was returned for Maidstone as a Liberal MP, and then in 1880 for London University—from 1886 to 1900 as a Liberal-Unionist. He succeeded in passing more than a dozen important measures, including the Bank Holidays Act (1871), the Bills of Exchange Act, the Ancient Monuments Act (1882) and the Shop Hours Act (1889). He was Vice-Chancellor of London University (1872–80), President of the British Association (1881), Vice-President of the Royal Society, President of the London Chamber of Commerce, Chairman of the London County Council (1890–2), and much else. Scientifically, he is best known for his researches on primitive man and on the habits of bees and ants; he published *Prehistoric Times* (1865, revised 1913), *Origin of Civilisation* (1870) and many books on natural history. He was a neighbour of **Charles Darwin**, and a friend and counsellor over many years. He was a member of the 'X' Club (together with **T H Huxley**, **Joseph Dalton Hooker** and others) which conspired to replace the ecclesiastical establishment with a scientific one.

LUC, Jean André De

See **DE LUC, Jean André**

LUCAS, Keith

(1879–1916)

English neurophysiologist, born in Greenwich. He matriculated in classics at Trinity College, Cambridge, but soon began to study natural sciences, under the direction of his college tutor the physiologist Walter Morley Fletcher. In 1903 he began research in the physiological laboratory and was appointed university demonstrator in 1907; at Trinity College he was elected a Fellow in 1904 and lecturer in natural science in 1908. His research was focused almost entirely on the properties of nerves and muscles; he showed that the 'all-or-none' law, that a given stimulus evokes the maximum contraction or no contraction at all, applied to skeletal muscle. Lucas's pupil **Adrian** later showed this was also true for motor nerve fibres. Lucas also examined the physico-chemical nature of the nervous impulse and its propagation, and his posthumous *The Conduction of the Nervous Impulse* (1917), prepared by Adrian, summarizes much of this work. Simultaneously and synergistically with his physiological research was Lucas's talent in designing and building sensitive recording and measuring equipment, and he was a director of the Cambridge Scientific Instrument Company. In 1914 he enlisted and contributed to the experimental research department of the Royal Aircraft Factory. After qualifying as a pilot, he was killed in a mid-air collision.

LUDWIG, Karl Friedrich Wilhelm

(1816–95)

German physiologist. Born in Witzenhausen, Ludwig became a medical student in Marburg in 1834. His was a stormy student career: duelling left him with a heavily scarred lip, and friction with the university authorities drove him to study elsewhere. In 1840 he returned to Marburg and was teaching there by 1846. Later he taught in Zürich, Vienna and Leipzig. In 1865 he helped establish the famous Institute of Physiology in Leipzig. Ludwig's work proved fundamental to modern physiology. He denied there was any role for vital force, seeking explanations of living processes in the paradigms of physics and chemistry. In formulating such views, he was much influenced by his friend, the chemist **Bunsen**. Ludwig proved fruitful in devising medical instruments, especially diagnostic technology. In 1846 he developed the kymograph in 1846 and used it to garner much information about the circulation and respiration. In 1859 he designed the mercurial blood-pump, in 1867 the stream gauge, and in 1865 perfusion, a method of maintaining circulation in an isolated organ. His blood-pump allowed examination of blood gases and respiratory exchange. His research focused on the operation of the heart and kidneys, on the lymphatic system, and on salivary secretion. The problem of the operation of secretion was an enduring preoccupation. The circulation of the blood also attracted his attention. He investigated the relations of blood pressure to heart activity, and probed the role of muscles in the fluidity of the blood. He proved an immensely energetic and influential teacher.

LUMIÈRE, Auguste Marie Louis Nicolas

(1862–1954) and

Louis Jean

(1865–1948)

French industrial and physiological chemists, pioneers of motion photography, born in Besançon. Auguste was educated at the University of Bern, Switzerland, and Louis at the École Technique, La Martinière. They founded Lumière and Jougla, a firm which manufactured photographic materials, and together invented the motion picture camera in 1893 and the cinematograph, the first machine to project images on a screen, in 1895. The same year they built the first cinema, in Lyons, and produced the first film newsreels and the first 'movie' in history, *La Sortie des ouvriers de l'usine Lumière*. They worked to improve colour photography and studied colloidal substances in living organisms. Auguste Lumière also carried out research into cancer, vitamins, and oral vaccination. Both brothers were elected to the French Academy of Sciences. Louis died in Bandol and Auguste in Lyons.

LUMMER, Otto Richard

(1860–1925)

German physicist, born in Breslau (now Wrocław, Poland). He became an assistant to **Helmholtz**, and followed him to the newly established Physikalisch Technische Reichsanstalt in Berlin (1887). In 1904 he was appointed Professor of Physics at Breslau. His main field of research was optics. He was one of the discoverers of the interference effect which became known as the 'Lummer fringes' (1884), and which he used in the design of a high-resolution spectroscope. He designed the Lummer–Brodnum cube photometer (1889), and with Ferdinand Kurlbaum (1857–1927) a bolometer (1892). His most important theoretical contribution was his investigation of certain properties of black-body radiation, an essential step on the road to **Planck**'s quantum theory.

LUNEBURG, Rudolf Karl
(1903–49)

American theoretical physicist, born in Volkersheim, Germany. He received his PhD from the University of Göttingen in 1933, and after two years as a Research Fellow at the University of Leiden he moved to the USA. He held appointments at the University of New York (1935–8, 1946–8) and with the Spencer Lens Company (1938–45) before becoming Associate Professor of Mathematics at the University of Southern California in 1949. Luneburg worked on multiple scattering of neutrons, on binocular vision, and on the electromagnetic foundations of the geometrical and wave theories of optical systems. He showed very clearly that geometrical optics is not merely an approximation to the wave theory, valid in the short wave limit, but is directly derivable from **Maxwell**'s equations, on an equal footing with the wave theory of light. He also invented the Luneburg lens, which has found applications in microwave antennae.

LURIA, Salvador Edward
(1912–91)

American biologist, born in Turin, Italy. He graduated in medicine at Turin University in 1935, and went on to the Radium Institute in Paris to study medical physics, radiation and techniques of working with phage, the bacterial virus. When Italy entered World War II, Luria emigrated to the USA, where he taught at Indiana University; **James Watson** was among his students there. In 1943, with **Delbrück**, Luria showed that bacteria undergo mutations. He went on to demonstrate in 1951 that phage genes can also mutate, and that different strains of phage can exchange and recombine genes. With **Hershey** and Delbrück, Luria founded the Phage Group, committed to using phage to investigate genetics. During the repressive McCarthy era, he was refused a visa to travel to a scientific conference in Oxford (1952). In 1969 he was awarded the Nobel Prize for Physiology or Medicine, jointly with Delbrück and Hershey, for discoveries related to the role of DNA in bacterial viruses. Critical of the cost of the US defence and space budgets, he gave part of his prize money to pacifist groups. He wrote *General Virology* (1953), which became a standard textbook and published his autobiography *A Slot Machine, A Broken Test-Tube, An Autobiography* in 1985.

LWOFF, André Michel
(1902–)

French microbiologist of Russian–Polish extraction, born in Ainy-le-Château. He worked at the Pasteur Institute in Paris from 1921, becoming departmental head there in 1938, and was a member of the Resistance movement during World War II. From 1959 to 1968 he was Professor of Microbiology at the Sorbonne. Initially he studied protozoa and proposed that evolution proceeded by enzyme loss. Following the earlier (from 1925) ideas of Eugène Wollman (1883–1943), also a researcher at the Pasteur Institute, Lwoff researched the genetics of bacterial viruses (phage) and showed that when phage enters a bacterial cell, it becomes part of the bacterial chromosome and divides with it (lysogeny), in which form (prophage) its capacity to reproduce is suspended and it protects the bacterium against further invasion by the same type of virus. The spontaneous production of new phage was rare, but could be initiated in any bacterium by treatment with X-rays or other mutagens. These findings have had important implications for the development of drug resistance and for cancer research. In 1965 Lwoff was awarded the Nobel Prize for Physiology or Medicine jointly with **Jacob** and **Monod**, for their discoveries concerning genetic control of enzyme and virus synthesis. He was elected a Foreign Member of the Royal Society in 1958, and published *Problems of Morphogenesis in Ciliates; the Kinetosomes in Development, Reproduction and Evolution* (1958) and *Biological Order* (1962).

LYAPUNOV, Aleksandr Mikhailovich
(1857–1918)

Russian mathematician, born in Yaroslavl, the son of an astronomer. He studied at St Petersburg where he came under the influence of **Chebyshev**, and then taught at Kharkov University. He returned to St Petersburg as professor in 1901. Lyapunov is principally associated with important mathematical methods in the theory of the stability of dynamical systems, related to **Poincaré**'s work although largely discovered independently of it. He committed suicide after his wife's death from tuberculosis.

LYELL, Sir Charles
(1797–1875)

Scottish geologist, born in Kinnordy, Forfarshire, the eldest son of the mycologist Charles Lyell (1767–1849). Educated at Ringwood, Salisbury and Midhurst, he studied law at Exeter College, Oxford, and was called to the bar. However, at Oxford he had discovered a taste for geology under the influence of **Buckland**, and whilst still a member of the legal profession but also Secretary to the Geological Society of London, he produced his first geological memoir, concerning freshwater marls in his native county of Forfarshire (1824). It was during an excursion to France in 1828 with **Murchison** that he decided to give up law, and instead of returning to London, he travelled on to Italy and Sicily where he became immersed in geological research. On his return he completed the first volume of his *Principles of Geology* (1830), and before the second volume appeared in 1832, he had been appointed Professor of Geology at King's College, London. His summers were devoted to geological excursions throughout Britain and abroad, traversing Europe from Scandinavia to Sicily and visiting the Canary Islands. He made two long geological tours of the USA and published his observations as *Travels in North America* (1845) and *A Second Visit to the United States* (1849). He also produced an important memoir *Consolidation of lava upon steep slopes of Etna* (1858), in which he was able to finally refute the elevation crater theory of **Buch**. Lyell's authoritative *Principles of Geology* exercised a powerful influence on contemporary scientific thought. His 'uniformitarian' principle denied the necessity of stupendous convulsions, and taught that the greatest geological changes might have been produced by the forces in operation now, provided that there was sufficient time for their task. He used his own observations to great effect, cogently arguing his case. Probably no scientific work, except perhaps **Charles Darwin**'s *Origin of Species*, has exerted such a powerful influence during the lifetime of its author. *The Elements of Geology* (1838) was a supplementary publication. Lyell devoted himself with ardour to the biological side of geology and was concerned with the geographical distribution of plants and animals. When discussions arose about the nature and significance of worked flints found in the valley of the Somme, he put himself at the forefront of the debate, collecting all available information and publishing it as *The Geological Evidence of the Antiquity of Man* (1863), a publication which startled the public by its unbiased attitude towards Darwin. He was knighted in 1848.

LYMAN, Theodore
(1874–1954)

American physicist, born in Boston. After receiving his PhD from Harvard (1900) he spent a year at the Cavendish Laboratory, Cambridge, then returned to Harvard where he remained for the rest of his career, being appointed Director of the Jefferson Physical Laboratory (1910–47). His doctoral dissertation addressed the problems of using diffraction gratings to measure wavelengths in the extreme ultraviolet region of the spectrum. Overcoming considerable practical difficulties, he published his first accurate measurements for wavelengths below 2000 angstroms in 1906 and extended the known extreme ultraviolet region significantly. By 1917 he had developed equipment to detect ultraviolet light at wavelengths as short as 500 angstroms. He analysed the spectra and optical properties of many materials and in 1914 announced the discovery of the fundamental series of spectral lines for hydrogen which bears his name. This result made an important contribution to **Niels Bohr**'s development of the quantum theory of the atom.

LYNDEN-BELL, Donald
(1935–)

English astrophysicist, born in Dover. In 1960, after completing his PhD at Cambridge, he spent two years with **Sandage** at Caltech working on the dynamics of galaxies. He then returned to Cambridge, becoming a Fellow of Clare College, and since 1972 has been Professor of Astrophysics and Director of the Institute of Astronomy (1972–7, 1982–7 and since 1992). In 1962 Lynden-Bell (with Sandage and Olin J Eggen) identified a second population of stars. These Population II stars were the first to be formed, contain few metals and have little net motion around the centre of our galaxy. This group of astronomers also suggested that globular clusters were formed very early on during the early collapse of the galaxy. With Roger Wood, Lynden-Bell pioneered the study of the runaway dynamic evolution of the centres of globular clusters. This work was expanded to cover the stability criteria that apply to compressible rotating stars. In 1969 he was the first to propose that the cores of galaxies might contain supermassive black holes surrounded by gaseous accretion discs. His important review of galactic structure (1985, with Frank J Kerr) convinced the International Astronomical Union that the best value for the distance to the centre of the galaxy was 85000 parsecs and not 100000 parsecs as had been thought previously. The Royal Astronomical Society awarded Lynden-Bell the Eddington Medal in 1984. He was elected FRS in 1978.

LYNEN, Feodor Felix Konrad
(1911–79)

German biochemist, born and educated in Munich. He studied under **Wieland** (1935–7) before joining the department of chemistry at the University of Munich (1942); he later became professor there (from 1947) and Director of the Max Planck Institute for Cell

Chemistry and Biochemistry (1954–79). In 1951 he isolated coenzyme A from baker's yeast and showed that it formed acetyl-S-CoA, an important intermediate in lipid metabolism, via a thioester bond, thereby adding a new dimension to the metabolic energy concepts of **Lipmann**. He also showed that the conversion of acetyl-S-CoA to form malonyl-S-CoA, a precursor of fatty acid biosynthesis, involved the uptake of carbon dioxide by the vitamin biotin (1958). With **Ochoa**, he showed in 1953 that acyl thioesters of ethanolamine would substitute for the full coenzyme A molecule in enzyme reactions and he used these to study the pathway of fatty acid degradation and the formation of acetoacetate (ketone bodies). At the same time as **Konrad Bloch**, Lynen independently isolated isopentanyl pyrophosphate and contributed evidence towards elucidating the biosynthesis of cholesterol. He also worked on the biosynthesis of cysteine, terpenes, rubber and fatty acids. He gave the first of the newly instituted Otto Warburg Medal Lectures of the German Chemical Society in 1963, and was awarded the Nobel Prize for Physiology or Medicine in 1964 jointly with Bloch.

LYON, Mary Frances
(1925–)

English biologist, born in Norwich. She obtained a scholarship to Girton College, Cambridge University, where she graduated in 1946, received a PhD in 1950 and an ScD in 1968. In 1950 she joined the UK Medical Research Council's staff, and has worked since 1955 at their Radiobiology Unit, Harwell/Chilton. From 1962 she headed its Genetics Division, and she was Deputy Director from 1986 until 1990, when she officially retired. She is a Foreign Associate of the US National Academy of Sciences and Foreign Honoraria Member of the Genetics Society of Japan. She chaired the Committee on Standardized Genetic Nomenclature for Mice (1975–90), as well as the Mouse Genome Committee of the Human Genome Organisation (HUGO). Lyon has published on many aspects of mammalian genetics and metagenesis. Her name is particularly associated with 'Lyon hypothesis' of random inactivation of the mammalian X chromosome, which she propounded in 1961. She suggested that one of the two X chromosomes in female mammals is inactivated in early development (becoming the 'sex chromatin' found in the nucleus of normal female, but not male cells), so that females are in effect mosaics of different genetic cell lines (characterized by which of the X chromosomes is switched off). This idea has been widely confirmed and has proved to be of great value in studies on clinical genetics and imprinting. She has extended knowledge of the mammalian X, especially with respect to human–mouse homologies. Her long-term research on the t-complex region of mouse chromosome 17 has elucidated many puzzling features and made it the most thoroughly studied part of the mouse genome, and her studies of the genetic effects of low radiation doses and female germ-cell exposures have strengthened genetic risk assessment. Through her leadership of mouse committees and her work on mouse genetic compilations, she demonstrated the immense value of mouse genetics in helping us to understand the mammalian genome and to tackle the problems of hereditary disease. She was elected a Fellow of the Royal Society in 1973, and was awarded its Royal Medal in 1984.

LYOT, Bernard Ferdinand
(1897–1952)

French astronomer and inventor of the coronagraph and the monochromatic filter named after him. Born in Paris and trained as an engineer at the École Supérieure d'Électricité, he joined the staff of the Paris Observatory at Meudon (1920), where his early researches were concerned with the measurement of polarization of light from the Moon and planets. He suggested that the surface of the Moon was covered by a layer of dust as was indeed later confirmed by the exploration of astronauts. In 1931 he succeeded in observing the solar corona outside an eclipse in broad daylight from the summit of the Pic du Midi using his new coronagraph, an instrument in which scattered light was reduced to an absolute minimum. Two years later he started photography of the Sun through monochromatic polarizing filters: his pioneer cinematographic films of the movements of solar prominences taken in the light of the red hydrogen Hα line created a sensation. For his work, all of which was of the highest quality and ingenuity, he was awarded the Gold Medal of the Royal Astronomical Society (1939) and the Bruce Medal of the Astronomical Society of the Pacific (1946). He died in Egypt on his return from an expedition to Khartoum where he had observed the total solar eclipse of February 1952.

LYSENKO, Trofim Denisovich
(1898–1976)

Soviet geneticist and agronomist, born in Karlovka, the Ukraine. During the famines of the early 1930s, he promoted 'vernalization', suggesting that plant growth could be accelerated by short exposures to low temperatures. His techniques seemed to offer a rapid way of overcoming the massive food shortages then prevailing, and gained him enthusiastic political support. In 1935, with the help of a lawyer, he formulated a neo-Lamarckian theory of genetics, suggesting that environment can alter the hereditary material; he proclaimed himself the successor to the soviet plant breeder Ivan Vladimirovich Michurin. As Director of the Institute of Genetics of the Soviet Academy of Sciences (1940–65), he declared the Mendelian theory of heredity to be erroneous. Dissenting geneticists were fired from their jobs, laboratories closed, and degree curricula and certification came under Lysenko's control. **Vavilov**, his predecessor at the institute, was sent to Siberia where he died in the early 1940s. During the 1950s, Soviet physicists and mathematicians had gained status and strength with the growth of the Soviet space programme; as support grew for **Crick** and **James Watson**'s model of DNA, criticism started to be directed towards Lysenko. After Stalin's death in 1956, Lysenko increasingly lost support and was forced to resign in 1965. He was awarded the Stalin Prize in 1949 for his book *Agrobiology* (1948).

M

MacARTHUR, Robert Helmer
(1930–72)

Canadian-born American ecologist, born in Toronto. He moved to the USA at the age of 17 and studied mathematics. While working for his PhD at Yale he changed to zoology and became Professor of Biology at Princeton in 1965. His mathematical training led him to develop quantitative models in ecology, and he was particularly concerned with the factors influencing the relative abundances of species in different habitats. He investigated the relationship between habitat size and number of species present by examining the bird fauna of islands, and showed how this was maintained by the balance between immigration and extinction. These studies were published in *The Theory of Island Biogeography* (1967, with **Edward Wilson**). With Wilson he also developed the idea of life history strategies. Thus he contrasted those species exhibiting exponential growth followed by catastrophic decline which have evolved traits which maximize rate of increase (r-selected species, eg lemmings) with K-selected species having stable populations close to the carrying capacity of the habitat and with slow intrinsic growth rates (eg tigers).

MACDIARMID, Alan Graham
(1928–)

New Zealand chemist. He received two PhDs—the first at the University of Wisconsin (1953) and the second at the University of Cambridge (1955). Following a brief period at St Andrews University, he moved to the University of Pennsylvania where he is now Professor of Chemistry. Most of his chemical research has been carried out in the USA. Macdiarmid is well known for his work on many diverse compounds of silicon, such as silicon hydrides, silicon fluorides, organosilicon compounds and adducts of silicon compounds. He has also worked on other main group elements, studying sulphur–nitrogen compounds, sulphur and phosphorus fluorides, and polyacetylenes. He was Royal Society of Chemistry Centenary Lecturer in 1982–3.

MacEWEN, Sir William
(1848–1924)

Scottish neurosurgeon, born and educated in Glasgow, where he worked throughout his life. His interest in surgery was stimulated by **Lister**, then Regius Professor of Surgery at Glasgow University. He adopted and then extended Lister's antiseptic surgical techniques and pioneered operations on the brain for tumours, abscesses and trauma. In 1879 he successfully removed a tumour involving the meninges of the brain, the first time this had been performed with the survival of the patient. He developed a new procedure for repairing hernias and published on the treatment of aneurysms by acupuncture. In addition, he operated on bones, introducing methods of implanting small grafts to replace missing portions of bones in the limbs. In 1892 he was appointed to the chair which Lister had held when MacEwen was a student, and he was knighted in 1902.

MACH, Ernst
(1838–1916)

Austrian physicist and philosopher, born in Chirlitz-Turas, Moravia. Until his teens he was educated almost entirely by his father. After a few years at the local gymnasium he entered the University of Vienna. By 1860 his doctorate was complete, and he taught mathematics and physics before being appointed Professor of Mathematics at Graz in 1864. In 1867 he moved to Prague, where he was Professor of Experimental Physics for 28 years. Finally, in 1895, he was elected to the Chair of History and Theory of the Inductive Sciences at Vienna. A crippling stroke in 1897 forced his resignation from active research in 1901, but he continued to write and served as a member of the Austrian parliament. Between 1873 and 1893 Mach experimented on the rapid flow of air over projectiles and wave propagation. New optical measuring techniques allowed him to obtain remarkable photographs of gas jets and projectiles in flight with accompanying shock waves. His name is associated with the ratio of the speed of a body, or of the flow of a fluid, to the speed of sound in the same medium (Mach number). From the early 1860s Mach stressed the interdependence of physical concepts with the physiological and psychological processes of sensory perception: 'psychophysics'. Within his positivist philosophy, all knowledge was a conceptual organization of sensory experience. Scientific statements must be empirically verifiable and should merely summarize our experience economically; theories dependent on metaphysical constraints or referring to entities not reducible to sensory experience (eg absolute space, absolute time, atoms) were to be rigorously rejected. Mach undertook a searching historical critique of Newtonian mechanics within these terms. The 'Mach principle', so named by **Einstein** in 1918, asserted that the inertia of a body depended on its relationship with the matter of the entire universe: it was meaningless to assign inertia to an isolated body. Einstein admitted the crucial role played by Mach's work in the formation of relativity theory which, ironically, Mach opposed.

MACINTOSH, Charles
(1766–1843)

Scottish industrial chemist, inventor of waterproof fabric and one of the inventors of bleaching powder. He was born in Glasgow and studied at the university there and in Edinburgh, where he was a pupil of **Joseph Black**. By 1786 he had started a chemical works manufacturing sal ammoniac, Prussian blue dye and other chemicals. In 1799 he worked with **Charles Tennant** on the development of bleaching powder. Following up the work of **Berthollet**, they treated slaked lime with chlorine, thus producing a solid bleaching agent which was easier to handle than chlorine in solution. Bleaching powder was used industrially to bleach cloth and paper until the 1920s. Macintosh also helped James Beaumont Neilson perfect the hot blast process of smelting iron. Hot air, rather than cold air, was introduced into the blast furnace, making the smelting much more efficient. In 1823, while investigating uses for the waste products from a gas works, Macintosh noticed that coal-tar naphtha dissolved india-rubber. He bonded two pieces of woollen cloth together with this solution and thus produced the first waterproof cloth. He patented the process in 1823 and, joining forces with Thomas Hancock, began making waterproof garments at a factory in Manchester. The rubber was difficult to sew and tended to crumble in cold weather and become sticky in hot weather. These difficulties were largely overcome with the invention of vulcanization in the late 1830s, after which the production of waterproof clothing and outdoor equipment expanded rapidly. Macintosh died near Glasgow. His name, albeit misspelt, lives on in the mackintosh.

MACKENZIE, Sir James
(1853–1925)

Scottish physician and cardiologist, born in Scone. After an apprenticeship with a chemist, he studied medicine in Edinburgh before settling in Burnley, Lancashire, as a general practitioner. There he developed a 'polygraph' for recording the pulse and its relationship to cardiovascular disease. He described several irregularities of the heartbeat, acquired through his writing an international reputation, and moved to London (1907) to become a cardiological consultant. His work was particularly important in distinguishing atrial fibrillation and in treating this common condition with digitalis. His *Diseases of the Heart* (1908) summarized his vast experience, although he never properly appreciated the possibilities of the electrocardiograph, then being exploited by Sir **Thomas Lewis**. In 1918 he left London to establish the Institute of Clinical Research at St Andrews, designed to encourage general practitioners to make use of their daily experience with common diseases in order to establish better guidelines for their treatment. He was elected FRS in 1915 and knighted in the same year.

MACLAURIN, Colin
(1698–1746)

Scottish mathematician, born in Kilmodan, Argyll. He graduated from the University of Glasgow (1713), and became professor at Aberdeen (1717). In 1725 he was appointed to the Chair of Mathematics at Edinburgh on **Isaac Newton**'s recommendation. He published *Geometria organica* in 1720, and his best-known work *Treatise on fluxions* (1742) gave a systematic account of Newton's approach to the calculus, taking a geometric point of view rather than the analytical one used on the Continent. This is often thought to have contributed to the neglect of analysis in 18th-century Britain. In 1745 Maclaurin organized the defences of Edinburgh against the Jacobite army, but his efforts impaired his health and he died early in the following year.

MACLEOD, John James Rickard
(1876–1935)

Scottish physiologist, born in Cluny, the son of a minister. He graduated in medicine at Aberdeen in 1898, after which he studied biochemistry at Leipzig and later took a diploma in public health at Cambridge. He became Professor of Physiology at Western Reserve University, Cleveland (1903) and at Toronto (1918). Between 1902 and 1922 he published many papers on the control of respiration, and from 1907 a series of papers on aspects of carbohydrate metabolism. In 1913 he published a book on diabetes, and soon after his move to Toronto, *Physiology and Biochemistry in Modern Medicine* (1918), a textbook which went through seven editions before he died. However, Macleod's fame rests upon his involvement with the discovery of insulin. In 1921 he accepted **Banting** into his department for research on the pancreas with **Best**. While Macleod was on holiday, Banting and Best first succeeded in producing a pancreatic extract which could lower sugar levels in the blood of dogs whose pancreases had been removed. On his return, Macleod asked the chemist **Collip** to join the team to work on purifying the extract. They produced a sufficiently pure preparation of insulin and carried out successful clinical trials in early 1922. Insulin therapy soon became the main treatment for diabetes. In 1923 Macleod and Banting were awarded the Nobel Prize for Physiology or Medicine. Banting divided his share with Best, and Macleod divided his with Collip. He returned to Scotland in 1928 as Professor of Physiology at Aberdeen.

MACLURE, William
(1763–1840)

Scots-born American geologist, born in Ayr, the 'father of American geology'. Educated privately, he soon made a substantial fortune as a merchant and entrepreneur. He travelled widely in Europe, Russia and the USA, and lived at times in London, Spain and Mexico. He settled in the USA around 1800. On his travels he always studied the geology and met leading geologists, and he employed a personal cartographer–naturalist (Charles Lesueur, 1778–1846) and organized expeditions with others. His work as an observer and writer in geology made him influential, and he helped to found the Academy of Natural Sciences in Philadelphia, serving as its president from 1817 to 1840. His *Observations on the Geology of the United States* (1817) gives the first full account of the subject, and he went on to study the West Indies and Mexico. Maclure's later writing supported the ideas on evolution offered by **Lamarck**. He believed that primitive rocks had diverse origins, and opposed **Abraham Werner**'s theory of their exclusively sedimentary origin.

MACNAMARA, Dame (Annie) Jean
(1899–1968)

Australian physician, born in Beechworth, Victoria. Educated at Melbourne University, she worked in local hospitals where she developed a special interest in 'infantile paralysis'. During the poliomyelitis epidemic of 1925, she tested the use of immune serum, and convinced of its efficacy, she visited England, the USA and Canada with the aid of a Rockefeller scholarship. With **Burnet**, she found that there was more than one strain of the polio virus, a discovery which led to the development of the **Salk** vaccine. She also supported the experimental treatment developed by Elizabeth Kenny, and introduced the first artificial respirator (iron lung) into Australia. She was created a DBE in 1935, and later became involved in the controversial introduction of the disease myxomatosis as a means of controlling the rabbit population of Australia. In the early 1950s it was estimated that as a result of her efforts the wool industry had saved over £30 million.

MACQUER, Pierre Joseph
(1718–84)

French chemist and physician, born in Paris where he qualified in medicine in 1742. Like many of his distinguished contemporaries he was introduced to chemistry by the lectures of G R Rouelle and began his researches in the early 1740s. Elected to the Academy of Sciences in 1745, he studied dyes, arsenic, platinum and the properties of milk. He began lecturing at the Jardin du Roi in 1770 and was appointed professor in 1777. He helped Jean Hellot, scientific adviser to the famous porcelain factory at Sèvres, to study the effects of firing on hundreds of clays from different parts of France, and succeeded him as adviser in 1766. Macquer is best known as the author of two widely read textbooks and the first dictionary of chemistry; *Élémens de chymie théorique* (1749) and *Élémens de chymie pratique* (1751) provided much-needed, straight-forward, accounts of the state of knowledge at the time and were particularly popular in their English translations. Macquer's greatest work, *Dictionnaire de chymie* (1766), was translated into German, English (by **Keir**), Danish and Italian. Macquer was sympathetic to the revolutionary ideas of **Lavoisier** but died, in Paris, before he could incorporate them into a new edition of the *Dictionnaire*.

MAGENDIE, François
(1783–1855)

French physiologist, born in Bordeaux. Magendie graduated in medicine in Paris in 1808, subsequently practising and teaching medicine in Paris, and becoming physician to the Hôtel-Dieu. Elected a member of the French Academy of Sciences in 1821, he became its president in 1837. In 1831 he was appointed Professor of Anatomy at the Collège de France. He was a pioneer of scientific pharmacology. On the basis of vivisection and a certain amount of self-experimentation, he conducted trials on plant poisons, using animals to track precise physiological effects. He demonstrated the stomach's passive role in vomiting and analysed emetics. Through such researches, he introduced into medicine the range of plant-derived compounds now known as alkaloids. Many possess outstanding pharmacological properties. Magendie demonstrated many of the medicinal uses of strychnine (derived from the Indian vomit nut), morphine and codeine (derived from opium), and quinine (derived from cinchona bark). Magendie's studies were remarkably comprehensive. He investigated the role of proteins in human diet, he was interested in olfaction and he inquired into the white blood cells. He worked extensively on the nerves of the skull; a canal leading from the fourth ventricle is now known as the 'foramen of Magendie'. His demonstration of the separate paths of the spinal nerves extended the findings of **Flourens** and **Charles Bell**. His numerous works included the *Elements of Physiology* (1816–17). Magendie's style of investigation avoided speculation and his results were derived solely from experimental data. He is often regarded as the founder of experimental physiology.

MAGNUS, Heinrich Gustav
(1802–70)

German physicist, born in Berlin, discoverer of the 'Magnus effect'. He entered the University of Berlin in 1822, was awarded a doctorate five years later for a dissertation on tellurium, and then moved to Sweden to work under **Berzelius** who became his lifelong friend. He returned to Berlin in 1828 and continued his researches in the field of chemistry for a while, then seemed to change direction towards physics, and in 1845 he became Professor of Technology and Physics at the University of Berlin. He worked on a very wide range of topics including thermoelectricity, the boiling of liquids, electrolysis (Magnus's rule relates to the electrolytic deposition of metals from solutions of their salts), optics, mechanics, magnetism, fluid mechanics and aerodynamics. It was in this last field that he studied the flow of air over rotating cylinders and in 1853 discovered and evaluated the Magnus effect—the sideways force experienced by a spinning ball, which is responsible for the swerving of golf or tennis balls when hit with a slice. The phenomenon was exploited many years later in the Flettner rotor ship, which used large vertical rotating cylinders as its motive power, and successfully crossed the Atlantic in 1926.

MAGOUN, Horace Winchell
(1907–91)

American neuroscientist, born in Philadelphia. He studied at Rhode Island State College and received his PhD from Northwestern University Medical School in 1934, remaining there as Assistant Professor (1934–7) and Professor of Micro-Anatomy (1937–50). In 1950 he was appointed to the School of Medicine at the University of California at Los Angeles (UCLA) as Professor of Anatomy (1950–62), Graduate Dean (1962–72), and Emeritus Dean and Professor of Psychiatry (1972–91). His early research focused on the structure and function of the hypothalamus, demonstrating its important role in feeding, drinking, body temperature and sleep mechanisms, and he was a pioneer in the development of the field of neuroendocrinology. Later work elucidated the physiological influence of the brainstem reticular formation in the control of sleep, wakefulness, awareness and other components of higher nervous activity, and his 1963 monograph *The Waking Brain* summarizes much of his work on brain-endocrine interactions. He collaborated on or supported many neurological and psychopharmacological projects and was one of the leaders in the creation of neuroscience, the multi-disciplinary approach to the study of the nervous system. His approach is exemplified by the establishment of the Brain Research Institute at UCLA, of which he was a prime instigator, and by his abiding interest in the history of the neurosciences.

MAIDEN, Joseph Henry
(1859–1925)

British-born Australian botanist, born in London and educated at the University of London. Ill-health caused him to emigrate in 1880 to Australia. In 1881 he was appointed as the first Curator of Sydney Technological Museum. From 1896 to 1924 he was Director of the Sydney Botanic Gardens and Government Botanist of New South Wales, and established the National Herbarium and the Botanical Museum (the latter being the first of its kind in Australia). A prolific writer, he published some 300 scientific papers and several books. These included *The Useful Native Plants of Australia* (1889), *Wattles and Wattle Barks* (1890), *Australian Economic Botany* (1892), *A Manual of the Grasses of New South Wales* (1898), *The Forest Flora of New South Wales* (8 vols, 1902–24), *A Census of New South Wales Plants*, and his greatest work, *A Critical Revision of the Genus Eucalyptus* (1903–33). His work was often elaborate, with beautifully executed coloured plates; this lavishness of scale sometimes led to disputes with the printers, causing some of his works to be delayed or even abruptly terminated before completion. Outside official life, somewhat appropriately, he was very active in the Wattle Day movement.

MAIMAN, Theodore Harold

(1927–)

American physicist, constructor of the first working laser, born in Los Angeles, the son of an electrical engineer. After military service in the US navy, he studied engineering physics at Colorado and Stanford universities, then joined the Hughes Research Laboratories in Miami in 1955. The maser (producing coherent microwave radiation) had been devised and constructed in 1953 by **Townes** (and independently by **Basov** and **Prokhorov** in the USSR in 1955). Maiman made some design improvements to the solid-state maser, and turned to the possibility of an optical maser, or laser (Light Amplification by Simulated Emission of Radiation). He constructed the first working laser in the Hughes laboratories in 1960. Lasers have found use in a variety of applications, including spectroscopy, surgical work, such as repair of retinal detachment in the eye, and in compact disc players. Maiman founded the Korad Corporation (1962) to build high-powered lasers and was a co-founder of the Laser Video Corporation (1972) to develop large-screen video displays.

MALPIGHI, Marcello

(1628–94)

Italian anatomist and microscopist, born near Bologna, where he studied philosophy and medicine. He later became professor at Pisa, Messina and Bologna (1666), and from 1691 served as Chief Physician to Pope Innocent XII. Malpighi was an early pioneer of histology, plant and animal, conducting a remarkable series of microscopic studies of the structure of the liver, lungs, skin, spleen, glands and brain, many of which were published in the *Philosophical Transactions* of the Royal Society. Studying the lung, Malpighi showed it was linked to the venous system on one side and to the arterial system on the other, thereby corroborating the insights of **William Harvey** (1578–1657). He also further extended Harvey's work by his discovery of the capillaries. He wrote a treatise on the silkworm, giving the first full account of an insect, the silkworm moth, and investigated muscular cells. In the 1670s he became involved in plant anatomy, discovering stomata in leaves and delineating the formation of the plant embryo.

MALTHUS, Thomas Robert

(1766–1834)

English economist and clergyman, born at The Rookery, near Dorking. He was Ninth Wrangler at Cambridge in 1788, was elected Fellow of his college (Jesus) in 1793, and in 1797 became curate at Albury, Surrey. In 1798 he published anonymously his *Essay on the Principle of Population*, and having travelled to several countries to collect more data, the next edition appeared in 1803, bearing his name. Malthus argued that population increases faster than food supplies, and that population growth should be kept in check. **Charles Darwin** read Malthus in 1838, after his return from his travels on HMS *Beagle*, and was greatly influenced by him, seeing in the struggle for existence a mechanism for producing new species—natural selection. The work was also a strong influence on **Wallace**, who developed a theory of evolution around the same time. In 1805 Malthus became Britain's first Professor of Political Economy, at the newly established East India College in Haileybury. His other works included *An Inquiry into the Nature and Progress of Rent* (1815) and *Principles of Political Economy* (1820).

MALUS, Étienne Louis

(1775–1812)

French physicist, born in Paris. A military engineer in Napoleon's army (1796–1801), he carried out research in optics and discovered the polarization of light by reflection in 1808. This phenomenon provided a very convincing demonstration of the transverse nature of light. Also in 1808 he discovered a fundamental theorem in geometrical optics, now known as the theorem of Malus and Dupin; this theorem is closely related to **Huygens**'s construction, and generalizes it to determine the position of a wavefront after light originating from a point source has been reflected or refracted.

MANSON, Sir Patrick

(1844–1922)

Scottish physician, born in Old Meldrum, Aberdeen, known as 'Mosquito Manson' from his pioneer work with Sir **Ronald Ross** in malaria research. He studied medicine in Aberdeen and then practised in the East in China (from 1871) and Hong Kong (from 1883), where he helped start and was the first Dean of a school of medicine that became the University of Hong Kong. In China, he studied a chronic disease called elephantiasis and showed that it is caused by a parasite spread through mosquito bites. This was the first disease to be shown to be transmitted by an insect vector. In 1890 Manson set up practice in London, where he became the leading consultant on tropical diseases and was appointed medical adviser to the Colonial Office; in 1899 he helped to found the London School of Tropical Medicine. He was the first to argue that the mosquito is host to the malaria parasite (1877), and encouraged Ross in his own researches, acting as his chief London agent. Manson's *Tropical Diseases* (1898) helped define the new specialism, emphasizing the importance of entomology, helminthology and parasitology in understanding the diseases peculiar to tropical climates.

MANTELL, Gideon Algernon
(1790–1852)

English palaeontologist, born in Lewes, Sussex. He studied medicine in London, and returned to Lewes as a practising surgeon. This was a busy and successful career, but he also took time to study the local geology and collected many fossils which were put on show to the public. In 1822 he found a remarkable group of fossils in the freshwater clays and sands of a quarry in the Wealden rocks of the Tilgate Forest, including bones of tortoises, crocodiles and birds, shells, pieces of carbonized wood, large leaves and other remains of plants. His discoveries were examined shortly afterwards by **Lyell**, who was able to recognize the true significance of freshwater deposits many thousands of feet below the marine chalk. He wrote *The Fossils of the South Downs* in 1822. Mantell moved to Brighton in 1833 and was able to complete his *Geology of the south-east of England* by 1837. His collection was sold to the British Museum in 1838, and he followed it to London in 1844. He discovered several dinosaur types, including the first to be fully described; noting the similarity between the fossil teeth and those of the living iguana, he named it 'Iguanodon' (1825). In 1831 he introduced the notion of the 'age of reptiles', one of the earliest pictorial representations of which was produced by the celebrated artist John Martin (1789–1819) for Mantell's *The Wonders of Geology* (1838).

MANTON, Sidnie Milana
(1902–79)

English zoologist, born in London. She was educated at Girton College, Cambridge, and from 1935 to 1942 was Director of Studies in Natural Science there. In 1943 she moved to King's College in London as a lecturer (1943–9) and reader in zoology (1949–60). She retired as Research Fellow of Queen Mary College, London, in 1967. Manton's work greatly improved knowledge of invertebrates. In studies of arthropods she investigated feeding and locomotive mechanisms, and related the different evolutionary forms, introducing the new phylum Uniramia to encompass the Onychophora, Tardigrada, Hexapoda and Myriapoda. She was elected FRS in 1948, and awarded the Linnaean Gold Medal in 1963.

MARCONI, Guglielmo, Marchese
(1874–1937)

Italian physicist and inventor, born in Bologna of wealthy parents, an Italian father and Irish mother. Educated for a short time at the Technical Institute of Livorno, but mainly by private tutors, he became fascinated by the discovery of electromagnetic waves by **Heinrich Hertz** and started experimenting with a device to convert them into electricity. His first successful experiments in wireless telegraphy were made at Bologna in 1895, and in 1898 he transmitted signals across the English Channel. In 1899 he erected a wireless station at La Spezia, but because of the continuing indifference of the Italian government to his work, he decided to establish the Marconi Telegraph Co in London. In 1901 he succeeded in sending signals in Morse code across the Atlantic. He patented in 1902 the magnetic detector, and in 1905 the horizontal directional aerial. He shared the 1909 Nobel Prize for Physics with **Ferdinand Braun**. He later developed short-wave radio equipment, and established a worldwide radio telegraph network for the British government. From 1921 he lived on his yacht, the *Elettra*, and in the 1930s he was a strong supporter of the Italian Fascist leader Mussolini.

MARCUS, Rudolph Arthur
(1923–)

Canadian–American physical chemist, born in Montreal, where he studied chemistry at McGill University (BS 1943, PhD 1946). After postdoctoral work with the National Research Council of Canada and at the University of North Carolina, he joined the chemistry faculty of the Polytechnic Institute of Brooklyn, New York, advancing from assistant professor to professor between 1951 and 1964. From 1964 to 1978 he worked at the University of Illinois, Urbana, and since 1978 he has been Professor of Chemistry at Caltech. He is celebrated for his work on the theory of electron transfer reactions involving molecules and/or ions in solution and the approach he pioneered from the late 1950s has become known as the Marcus theory. Such processes constitute oxidation–reduction reactions and are of importance not only in inorganic chemistry, but also in organic chemistry and biochemistry. Marcus's theory considers in detail the role of the solvent molecules and leads to quantitative expressions for rate coefficients which are of both explanatory and predictive value. The theory has been extended to proton transfer reactions, another important class of processes, particularly in organic chemistry. Marcus has also made an important extension to the theoretical treatment of unimolecular gas reactions in terms of molecular vibrations. He received the Nobel Prize for Chemistry in 1992. In 1987 he was elected a Foreign Member of the Royal Society and in 1991 he became an Honorary Fellow of the Royal Society of Chemistry. His honours from the American Chemical Society include the Debye Award (1988) and the Richards Medal (1990).

MARÉCHAL, André
(1916–)

French optical physicist, born in La Garenne. Educated at the University of Paris and the Institute of Optics, he became assistant professor at the university and institute in 1943, lecturer in 1950 and full professor in 1955; he retired as Professor Emeritus in 1985. From 1968 to 1984 he was also director of the institute. He served as a member of the French National Committee for Scientific Research (1961–88), and was President of the International Commission of Optics (1962–5). Maréchal's distinguished contributions to optics have included work on the diffraction theory of imaging systems, on geometrical optics, and on imagery in partially coherent light. He was awarded the Thomas Young Medal of the UK Institute of Physics in 1965 and the C K Mees Medal of the Optical Society of America in 1977.

MARGGRAF, Andreas Sigismund
(1709–82)

German chemist and agricultural scientist, remembered for his discovery of sugar in beet. He was born in Berlin and studied there and at several other German universities. He was elected to the Royal Prussian Academy in 1738 and became director of its laboratory in 1753. In 1747 he extracted the juice from beet and using a microscope—perhaps the first time that a microscope was used for a chemical identification—showed that the crystals which formed in the juice were identical to cane sugar. This discovery laid the basis for the sugar beet industry but it was **Achard**, his successor at the Academy, who developed it commercially, and the first factory was not opened until 1802. Marggraf made many other important advances in chemistry: for example, he used flame tests to distinguish between sodium and potassium, and showed that magnesium salts burn with a green flame. He also isolated zinc from calamine and carried out valuable investigations of the alkaline earths. He died in Berlin.

MARIOTTE, Edmé
(c.1620–84)

French experimental physicist and plant physiologist, probably born in Chazeuil, Burgundy. Mariotte was prior of Saint-Martin-de-Beaumont-sur-Vingeanne and moved from Dijon to Paris in the early 1670s. Following his election to the Paris Academy of Sciences (1666), Mariotte's skilled experimental work and prolific literary output were tied to the society's activities. He attracted attention as a physiologist, comparing plant sap to the blood circulating in animals. In 1668 he announced his controversial discovery of the eye's blind spot. He studied pendulums and falling bodies (1667–8), and published an exposition of the laws of elastic and inelastic collisions (1673). In the comprehensive review *De la nature de l'air* (1679), Mariotte restated the law bearing his name in France (elsewhere attributed to **Boyle**) and used it to estimate the height of the atmosphere. In the same work he discussed the connection between variations in barometric pressure, the winds and the weather. His hydrodynamical *Traité du mouvement des eaux* (1686) was widely read. Mariotte also discussed scientific methodology, hydrology, optics (including the rainbow), astronomy and the strength of materials.

MARKOV, Andrei Andreyevich
(1856–1922)

Russian mathematician, born in Ryazan. He studied at St Petersburg, where he was professor from 1893 to 1905, before going into self-imposed exile in the town of Zaraisk. A student of **Chebyshev**, he worked on number theory, continued fractions and the law of large numbers in probability theory, but his name is best known for the concept of Markov chain, a series of events in which the probability of a given event occurring depends only on the immediately previous event; this has since found many applications in physics and biology.

MARKOVNIKOV, Vladimir Vasilevich
(1837–1904)

Russian chemist, born in Knyaginino. After studying chemistry at Kazan University under **Butlerov**, he travelled to Germany (1865–7) to study with **Erlenmeyer** and **Kolbe**. He returned to Kazan but in 1871 moved to Odessa and then to Moscow (1873). In Moscow his work became more practical, and he carried out a comprehensive study of the composition of Caucasian petroleum. He discovered a number of compounds existing as isomers and others containing double bonds (olefins). He is best known for his rule that in the addition of a hydrogen halide to an asymmetric double bond, the hydrogen adds to the carbon atom which has the greater number of attached hydrogens. This rule is broken only when free radicals are involved, as was shown by **Kharasch**. Markovnikov retired in 1898 for political reasons but retained a private laboratory at the University of Moscow.

MARSH, James
(1789–1846)

English chemist, born in London. He worked at the Royal Arsenal in Woolwich and was subsequently assistant to **Faraday** at the Royal Military Academy, where his salary was only 30 shillings a week. His most significant work was on poisons; he is best known for devising a sensitive test for arsenic (the Marsh test) which was published in 1836 and quickly translated into French and German. Any arsenic in a suspected substance is converted to the arsine; it then passes through a heated glass tube where it decomposes, leaving a brown arsenic deposit. He also wrote on electromagnetism and invented a percussion tube for ship's cannon made from quills. He died in poverty in London.

MARSH, Othniel Charles
(1831–99)

American palaeontologist, born in Lockport, New York and wealthy by inheritance. He studied at Yale, at New Haven, and in Germany, and became first Professor of Palaeontology at Yale in 1866, without a salary or classes to teach. From 1870 to 1873 he led a series of expeditions through the western territories making spectacular discoveries of vertebrate fossils; this led him into bitter clashes with **Cope** who organized rival dinosaur collecting expeditions. They frequently disagreed about the description and significance of the newly discovered vertebrate faunas, and this ultimately led to the sensational airing of the bitter dispute on the front pages of the *New York Herald* in January 1890, with the result that federal financial support for palaeontology was withdrawn. Marsh discovered (mainly in the Rocky Mountains) over a thousand species of extinct American vertebrates, including dinosaurs and the mammals uintatheres and brontotheres. By 1874 he was able to establish an evolutionary lineage for horses using the fossil remains which he had assembled. He also contributed to the documentation of evolutionary changes with his discovery of Cretaceous birds with teeth. His major contribution to stratigraphical palaeontology was an early discussion of the Miocene–Pliocene boundary. Marsh was the first vertebrate palaeontologist of the US Geological Survey (1882–92). He published over 300 papers, including *Odontornites: a Monograph on the Extinct Toothed Birds of North America* (1880) and *Dinosaurs of North America* (1896).

MARTIN, Archer John Porter
(1910–)

English biochemist, born in London. He trained at Cambridge and then at the Lister Institute where he showed with colleagues that in pigs, unlike rats, nicotinic acid prevented the vitamin deficiency disease pellagra (1937–8), which confirmed the findings of **Elvehjem**. His team also demonstrated the importance of other B vitamins (aneurin and riboflavin). Martin then moved to the Wool Industry Research Association in Leeds (1938–46) where, with **Synge**, he produced classic papers on partition chromatography (1941), illustrated by the use of columns of silica gel for the separation of acetylated amino acids, to separate amino acid derivatives from protein hydrolysates in the analysis of protein structure. Their technique revolutionized analytical biochemistry and enabled the rapid separation of small amounts of complex mixtures of biomolecules not possible by ordinary chemical methods. In addition, Martin introduced electrodialysis for deionizing protein hydrolysates (1947). For this work he was awarded the 1952 Nobel Prize for Chemistry jointly with Synge. Martin subsequently joined the staff of the Medical Research Council (1948–52), and became Director of the Abbotsbury Laboratories (1959–70) and a consultant to the Wellcome Research Laboratories (1970–3). From 1953 he worked on analysis by gas–liquid chromatography and fine engineering, including the development of micromanipulators. He was elected FRS in 1950.

MARTIN, Pierre Émile
(1824–1915)

French metallurgist, born in Bourges, the son of the owner of an iron and steel works. In Sireuil in 1864, with the help of his brother Pierre, he devised an improved bulk steel-making method. Using a furnace specially constructed by Sir **William Siemens**, he melted a bath of pig-iron and fed into this puddled steel or iron, with or without admixtures of steel scrap. With Siemens's heat regeneration method, a sufficiently high temperature was obtained to treat large quantities of metal and to keep it molten throughout the process, enabling it to be cast into ingots when the refining was complete. The open-hearth (or Siemens–Martin) process derived its name from the fact that molten metal lay in a comparatively shallow pool on the furnace hearth. The products of the process won a gold medal at the Paris Exhibition of 1867, and by World War I the open-hearth furnace had overtaken the **Bessemer** converter as the major source of the world's steel. Martin himself was crippled financially by unsuccessful litigation and spent his later years in poverty while others profited from his process, until in 1907 an international benefit fund restored his finances to a level of modest comfort.

MARTYN, David Forbes
(1906–)

Scottish physicist, born in Cambuslang. He studied at Imperial College, London, and remained there for a number of years after obtaining his PhD in 1928. He moved to Australia in 1930 to take up a post at the Radio Research Board, and continued his research at the radiophysics laboratory of the Commonwealth Scientific and Industrial Research Organisation from 1939. Martyn carried out extensive investigations into the interaction of radio waves with the upper atmosphere. He helped explain the 'Luxembourg' effect in which two radio signals of different wavelengths reflected from the ionosphere interfere with each other, resulting in a mixing of the signals. He also investigated the angle of polarization of radio waves reflected from the ionosphere and the way the reflections varied with the time of day. His studies revealed that solar emissions effected the reflective properties of the ionosphere, and provided one of the earliest analyses of the temperature and constituents of the upper atmosphere through radio-wave probing. He was appointed chairman of the Australian National Committee on Space Research (1958) and sat on the UN science and technology committee for peaceful uses of outer space (1962).

MASKELYNE, Nevil
(1732–1811)

English astronomer, born in London. After being educated at Westminster School he went on to study divinity at Trinity College, Cambridge. He joined the Royal Greenwich Observatory in 1755. In 1763 he produced the *British Mariner's Guide* and went to Barbados to test chronometers. In 1765 he was appointed Astronomer Royal. He improved methods and instruments of observation, invented the prismatic micrometer, and made important observations. He went to St Helena to observe the transit of Venus, the aim being to make better estimates of the Earth–Sun distance. In 1767 he founded the *Nautical Almanac*. He also measured the Earth's density from the deflection of the plumb-line at Schiehallion in Perthshire (1774), obtaining a value between 4.56 and 4.87 times that of water. An ordained minister, he was rector from 1775 of Shrawardine, Salop, and from 1782 of North Runcton, Norfolk.

MASON, Sir (Basil) John
(1923–)

English physicist and meteorologist, born in Docking, Norfolk. He graduated from Nottingham University in 1948, and moved to Imperial College, London, where he set up his own section on cloud physics under Sir David Brunt. He worked on the study of rain-making processes, electrification in thunderstorms, ice nucleation and other aspects of cloud physics. His book, *The Physics of Clouds* (1957), is a classic. He was awarded the DSc in 1956 and became the first and only Professor of Cloud Physics (1961–5) at Imperial College. Elected FRS in 1965, he became Director-General of the Meteorological Office (1965–83), where his inspired leadership, scientific ability and administrative skills had full rein: the first operational numerical weather predictions were made, the office carried out research on the effect of Concorde on stratospheric gases and much research was carried out using meteorological radar. From 1983 to 1989 he organized, through the Royal Society and the Swedish and Norwegian academies, a thorough interdisciplinary study of acid rain which finally established the facts. He was very active in international circles, being Chairman of the Global Atlantic Tropical Experiment 1978 and Chairman of the Joint Organising Committee of the World Climate Research Programme. Knighted in 1979, he received many awards including the Royal Medal of the Royal Society (1991).

MASURSKY, Harold
(1922–)

American geologist, born in Fort Wayne, Indiana. After graduating from Yale he joined the US Geological Survey; he later transferred to the Astrogeological Section and began working for NASA. Masursky's main work has been the surveying of lunar and planetary surfaces with special reference to the choice of landing sites for space missions. In 1964 he was a member of the Ranger 9 site selection programme, this spacecraft being the first to take close-up images of the lunar surface. Masursky progressed to the Lunar Orbiter and Apollo programmes. He also participated in the 1971 Mariner Orbiter and 1975 Viking Lander explorations of Mars. These missions led to the discovery of huge volcanoes on Mars and channels on the planetary surface that had clearly been formed by running water. In 1978 Masursky joined the Venus Orbiter Imaging Radar Science Group and was one of the chief scientists associated with the Voyager missions to the outer planets. He suggested that the differences in the reflectivities of the two hemispheres of Saturn's moon Iapetus were produced by internal processes.

MATTHEWS, Drummond Hoyle
(1931–)

English geologist and geophysicist. He was educated at King's College, Cambridge, and started his professional career as a geologist with the Falkland Islands Dependencies Survey (1955–7) before returning to Cambridge, where he became assistant director of research in the department of geophysics in 1966. He became reader in geology in 1971 and subsequently published many important papers in the realm of marine geophysics. He was elected FRS in 1974. With **Vine**, Matthews predicted in 1963 that there should be strips of normally and reversely magnetized oceanic crust on either side of the mid-ocean ridges if sea-floor spreading occurs as postulated by **Harry Hess**, and since the Earth's magnetic field changes polarity periodically. By 1966 sufficient magnetic survey data had accumulated to yield striking proof of their theory. This caused a paradigm shift in geological thinking and led to the widespread acceptance of continental drift and the development of plate tectonic theory.

MATTHIAS, Bernard Teo
(1918–)

German-born American physicist, born in Frankfurt. He studied at Rome University and the Federal Institute of Technology, Zürich, before moving to the USA in 1947, where he became a naturalized citizen in 1951. After a period with Bell Telephone Laboratories he was appointed Professor of Physics at the University of California, San Diego, in 1961. He worked on ferroelectricity, reporting measurements on numerous ferroelectric materials during the late 1940s and early 1950s, and discovering during the same period some new ferroelectrics. By the early 1950s he was also working on superconductivity; in his search for new superconducting materials, he discovered that alloys of metals with five or seven valence electrons were the most effective. Higher temperature superconducting metal alloys discovered as a result of this work were used until superseded by ceramic materials in the late 1980s.

MATUYAMA, Motonori
(1884–1958)

Japanese geophysicist, born in Oita Prefecture, the son of a Zen abbot. He was educated at Hiroshima Normal College where he studied physics and mathematics, and graduated in physics from Kyoto University (1911), where he subsequently became an instructor and then assistant professor (1916). In geophysical work with **Chamberlin** at Chicago (1919), he studied the physics of ice movement and returned to Japan as Professor of Theoretical Geology at the Imperial University (1921). From 1926 he worked on the remanent magnetism of basalts from Japan and Manchuria, and made the first successful link of magnetic reversals with the geological timescale. Matuyama was responsible for the extension of a national gravity survey into Korea and Manchuria (1927–32), and in 1934 commenced a marine gravity survey of the Japan trench. He published numerous papers on the physics of the lithosphere and interior of the Earth, seismology and magnetism, and conducted research into physical methods of locating underground resources.

MAUCHLY, John William
(1907–80)

American physicist and inventor, born in Cincinnati, the son of a physicist at the Carnegie Institution in Chevy Chase, Maryland. He graduated in physics at Johns Hopkins University and after a few years in teaching at Ursinus College, a small private school near Philadelphia, he joined **Eckert** in 1943 at the University of Pennsylvania, where they developed the ENIAC (Electronic Numerical Integrator and Computer). This giant military calculator led to the pair's major contribution to computing: the design of a stored-program machine, the EDVAC (Electronic Discrete Variable Computer), which played a large part in launching the computer revolution in the second half of the 20th century. They founded in 1948 the Eckert–Mauchly Computer Corporation; however, this was not a commercial success and had to be sold in 1950. Following EDVAC, they built UNIVAC, the Universal Automatic Computer first used in 1951 by the US Census Bureau. Although Mauchly and his collaborator relied to a certain extent on the work of others—notably **Atanasoff** (who later was awarded priority by the courts for inventing the computer) and **von Neumann**—it was their conviction that computers had a commercial market that launched the modern data-processing industry in the USA.

MAUNDER, Edward Walter

(1851–1928)

British astronomer, born in London and educated at a school attached to University College London. In 1873 he was recruited through the Civil Service Commission examination as assistant for photography and spectroscopy at the Royal Observatory, Greenwich, a post which he held for 40 years. One of his duties was daily photography of the Sun and the recording of sunspot numbers. From the immense store of data thus built up he established the pattern of latitude drift in sunspots in the course of the sunspot cycle, demonstrated in his well-known 'butterfly diagram'. He also worked on correlations between solar and geomagnetic activity, and studied historical records of the low-sunspot period in the 16th century, the 'Maunder minimum'. He took a part in several eclipse expeditions to photograph the solar corona. His wife, formerly Annie Russell (1868–1946), a mathematics graduate of Girton College, Cambridge, collaborated in his researches. Maunder was the founder of the British Astronomical Association (1890) for amateur astronomers which still flourishes.

MAUPERTUIS, Pierre Louis Moreau de

(1698–1759)

French mathematician, born in St Malo. In 1736–7 he headed a group of French academicians sent to Lapland to measure a degree of the meridian. Frederick the Great, made him president of the Berlin Academy in 1746. In mechanics he formulated the principle of least action in mechanics, which states that a mechanical system evolves in such a way that its action is as small as possible, although he assigned to it an unjustified theological significance; the principle was first stated precisely by **Leonhard Euler**. Maupertuis also formed a theory of heredity which was a century ahead of its time, but his temper provoked general dislike and the special enmity of Voltaire, who satirized him in *Micromégas*, driving him to Basle, where he died.

MAWSON, Sir Douglas

(1882–1958)

English-born Australian explorer and geologist, born in Bradford, Yorkshire. He was educated at Sydney University, where he graduated in engineering (1901) and geology (1904). In 1905 he was appointed lecturer in mineralogy and petrology at the University of Adelaide, where he later became professor (1920–52). In 1907 he joined the scientific staff of Ernest Shackleton's Antarctic expedition during which he mapped the position of the South Magnetic Pole. From 1911 to 1914 he was leader of the Australian Antarctic expedition which charted 2 000 miles of coast; he was knighted on his return. The tribulations of this expedition are recorded in his *Home of the Blizzard* (1915). He also led the joint British–Australian–New Zealand expedition to the Antarctic from 1929 to 1931.

MAXWELL, James Clerk

(1831–79)

Scottish physicist, born in Edinburgh, the son of a lawyer. One of the greatest theoretical physicists the world has known, he was nicknamed 'Dafty' at school (the Edinburgh Academy) because of his gangling appearance. At the age of 15 he devised a method for drawing certain oval curves, which was published by the Royal Society of Edinburgh. He studied mathematics, physics and moral philosophy at Edinburgh University, where he published another paper, on rolling curves, and later graduated from Cambridge University as Second Wrangler. He was appointed Professor of Natural Philosophy at Marischal College, Aberdeen (1856), and King's College, London (1860), but resigned in 1865 to pursue his researches at home in Scotland. In 1871 he was appointed the first Cavendish Professor of Experimental Physics at Cambridge, where he organized the Cavendish laboratory. During his brilliant career he published papers on the kinetic theory of gases, linking the properties of single molecules and bulk matter, and established theoretically the nature of Saturn's rings (1857), later to be confirmed by **Keeler**; he also investigated colour perception and demonstrated colour photography with a picture of tartan ribbon (1861). However, his most important work was on the theory of electromagnetic radiation, with the publication of his great *Treatise on Electricity and Magnetism* in 1873, which treated mathematically **Faraday**'s theory of electrical and magnetic forces considered as action in a medium rather than action at a distance. He showed that oscillating electric charges could generate propagating electromagnetic waves, with a theoretical wavespeed almost exactly the same as the experimentally determined speed of light; this was the first conclusive evidence that light consisted of electromagnetic waves. 'Maxwell's equations', describing mathematically the wave propagation, were later developed by **Heaviside** and independently by **Heinrich Hertz**. Maxwell envisaged electromagnetic radiation carried by the 'ether', the hypothetical medium which pervaded all space, allowing light to travel from the distant stars; but his equations survived unchanged when the revelations of the **Michelson–Morley** experiment finally laid to rest the notion of the ether and **Einstein**'s relativity revolution revised most of classical physics. Maxwell also predicted the possible existence of radiation beyond ultraviolet and infrared, and suggested that electromagnetic waves could be generated in a laboratory—as Hertz was to demonstrate in 1887.

MAY, Robert McCredie
(1936–)

Australian physicist and ecologist, born in Sydney. He was educated at Sydney University where he studied theoretical physics. He taught mathematics at Harvard University (1959–61) and theoretical physics at Sydney, where he held a personal chair (1961–73). He became Professor of Biology at Princeton University from 1973 to 1988 and subsequently Royal Society Research Professor at Oxford. He has worked on the dynamics of animal populations and has applied non-linear models to explain fluctuations in animal numbers. His realization that deterministic chaos underlies not only some biological phenomena but has widespread relevance in explaining, for example, aspects of economics and problems in the social sciences, has opened up fruitful areas of research. He has investigated the factors which influence the abundance and diversity of species, in particular the relationships between parasites and their hosts, the ways in which they have co-evolved, and how infectious diseases regulate natural populations of animals and plants. He wrote *Stability and Complexity in Model Ecosystems* (1973) and *Infectious Diseases of Humans: Transmission and Control* (1991, with R M Anderson).

MAYALL, Nicholas Ulrich
(1906–)

American astrophysicist, born in Moline, Illinois, and educated at the University of California, where he graduated in 1928 and received his PhD in 1934. He served on the staff of Mount Wilson Observatory (1929–31) and Lick Observatory (1933–42), and subsequently worked at MIT Radiation Laboratory (1942–3) and Caltech (1943–5), before returning to Lick Observatory in 1945. After 15 years at Lick, he was appointed Director of Kitt Peak Observatory, Arizona, where he remained until his retirement in 1971. Mayall's principal researches were in the field of optical observations of galaxies. At Mount Wilson in the early 1930s, he collaborated with **Hubble** in making surveys over the sky of faint galaxies, to study their distribution in space. His work at Lick Observatory included optical radial velocity measurements of the Andromeda galaxy, demonstrating its rotation (1950), and (with W W Morgan) the development of a system of classifying galaxies from their composite spectra to indicate their stellar content (1957). In 1961 Mayall was made chairman of a committee to plan and oversee the building of a 4 metre telescope on Kitt Peak. This famous and successful instrument, named the Mayall telescope, was officially inaugurated in 1973 on the occasion of the 5th centenary of the birth of **Copernicus**.

MAYER, (Julius) Robert von
(1814–78)

German physician and physicist, born in Heilbronn. The youngest son of an apothecary, he enrolled at the University of Tübingen in 1832 and received his medical doctorate in 1838. After completing his studies in Paris he travelled to Java in 1840 as a ship's surgeon. Considerations concerning blood and animal heat convinced him that heat and motion were different manifestations of a single indestructible force. His views were published in 1842. A memoir of 1845 explained how Mayer had come to estimate the mechanical equivalent of heat by examining the difference in the two principal specific heats of a gas, assuming no internal work was done during gaseous expansion (Mayer's hypothesis), and enunciated a general principle of conservation for biological, magnetic, electrical and chemical processes (now interpreted as the conservation of energy). In 1848 he proposed that the Sun's heat is continually replenished through the impact of meteors. Mayer initially gained little recognition and became severely depressed, attempting suicide in 1850 and subsequently spending years in asylums. A dispute over priority in connection with the claims of **Joule** saw Mayer championed in England by **Tyndall**.

MAYNARD SMITH, John
(1920–)

English geneticist and evolutionary biologist, born in London. He was educated at Eton and Cambridge, where he graduated as an aeronautical engineer (1941) and at University College London, when he received a degree in zoology in 1951. He taught at University College London (1951–65) before being appointed Professor of Biology at the University of Sussex (1965–85). His early work was in collaboration with **J B S Haldane**, but he went on to develop a new phase of the mathematical understanding of evolutionary processes (*Mathematical Ideas in Biology*, 1968; *Models in Ecology*, 1974; *Evolution of*

Sex, 1978), in particular the application of game theory to behavioural ecology (*Evolution and the Theory of Games*, 1983) where his development of the inclusive fitness concept of **William Donald Hamilton** has been immensely fruitful. His popular book, the *Theory of Evolution* (1958), was widely influential. He was elected FRS in 1977.

MAYOW, John
(1640–79)

English chemist and physiologist, born in Bray, Cornwall. He studied medicine at Oxford, and thereafter practised as a physician, probably dividing his time between London and Bath. He noted the similarities between combustion and respiration, in particular that both use up only a small proportion of the available air. He also noted that the remainder of the air does not support life, extinguishes a lighted candle, and is insoluble in water. Mayow suggested that the function of respiration is to convey life-giving nitrous particles from the air to the blood. He also suggested that respiration is the source of animal heat, and pointed out that the foetus breathes through the placenta. Mayow's originality and the extent of his influence, if any, on **Lavoisier**, are the subject of debate. He died in London.

MAYR, Ernst Walter
(1904–)

German-born American zoologist, born in Kempten. He studied at Berlin and emigrated to the USA in 1932, serving as Professor of Zoology at Harvard from 1953 to 1975. His early work was on the ornithology of the Pacific, leading three scientific expeditions to New Guinea and the Solomon Islands (1928–30), but in his later career he became a leading animal systematist. His main contribution has been as one of the most important architects of the neo-Darwinian synthesis of evolution, which produced a reconciling of the views of (principally) palaeontologists and geneticists. He was the author of one of the key works of the synthesis, *Systematics and the Origin of Species* (1942; extensively revised as *Animal Species and Evolution*, 1963), and also wrote an influential history of biology, *The Growth of Biology Thought* (1982), complemented by *Towards a New Philosophy of Biology* (1988). Mayr was also first to propose the 'founder effect', describing the consequences of the genetic bottleneck which a population experiences when it is 'founded' by a few individuals, for example when an oceanic island is colonized. The founding group is immediately different from the ancestral group as it cannot carry as much inherited variation or have the same distribution of characteristics; this effect is the most rapid way of producing genetic change.

McCLINTOCK, Barbara
(1902–92)

American geneticist, born in Hartford, Connecticut. She received a PhD in botany in 1927 from Cornell, where she worked from 1927 to 1935. Later she held posts at the University of Missouri (1936–41) and Cold Spring Harbor (1941–92). In 1927, with Harriet Creighton, she showed that changes in the chromosomes of maize resulted in physical changes in the colour of the corn kernels; this ultimate proof of the chromosome theory of heredity was published in 1931. In the 1940s she showed how genes in maize are activated and deactivated by 'controlling elements'—genes that control other genes, and which can be copied from chromosome to chromosome. She presented her work in 1951 at a Cold Spring Harbor symposium, but its significance was lost on the attendees who mainly worked with bacteria. It was not until the 1970s, after the work of **Jacob** and **Monod**, that her work began to be appreciated. At a 1976 symposium, McClintock's research was acknowledged with the introduction of the term 'transposon' to describe her 'controlling elements'. Finally in 1983, she was awarded the Nobel Prize for Physiology or Medicine. She continued to work on maize genetics at Cold Spring Harbor until her death in 1992.

McCOLLUM, Elmer Verner
(1879–1967)

American biochemist, born in Fort Scott, Kansas. He studied at Yale, Cincinnati and Manitoba, and became Professor of Biochemistry at Johns Hopkins University, Baltimore (1917–44). In 1913 he published the first description of an accessory food factor (vitamin), thereby providing concrete evidence for the hypotheses of **Frederick Gowland Hopkins** and **Funk**. He explained the necessity for adding milk to a synthetic diet as described by Hopkins in order to sustain normal animal growth, and found that it contained more than one vitamin, distinguishing between vitamins A (fat-soluble) and B (water-soluble). Following the report by **Edward Mellanby** that cod liver oil prevented rickets, McCollum discovered the 'rickets-preventative factor' (1922), vitamin D, by testing more than 1 000 different diets on rats. McCollum's A, B and D were later shown to consist of separable constituents (eg A_1 and A_2), while vitamin B_1 protected against beri-beri and another B vitamin (nicotinamide) protected against pellagra. A prolific writer, he published *Organic*

Chemistry for Medical Students (1916), the popular *The Newer Knowledge of Nutrition* (1918), *The American Home Diet* (1918), *A History of Nutrition* (1957) and his autobiography *From Kansas Farm Boy to Scientist* (1964). He received numerous awards and was elected a Foreign Member of the Royal Society in 1961.

McCREA, Sir William Hunter
(1904–)

Irish theoretical astrophysicist and mathematician, born in Dublin. After being educated at Chesterfield Grammar School in Derbyshire and Trinity College, Cambridge, McCrea lectured in mathematics at Edinburgh University and Imperial College, London, later becoming Professor of Mathematics at Queen's University in Belfast (1936). After World War II he moved to Royal Holloway College, London, and then to the University of Sussex in 1966. In 1934, with **Edward Milne**, McCrea was the founder of modern Newtonian cosmology, applying classical theories of physics with considerable success to the primordial gas cloud that condensed to form the galaxies. He also extended **Mach**'s viewpoint, suggesting that the **Heisenberg**'s uncertainty principle applies to light as it travels, and that our knowledge of the universe therefore deteriorates as we extrapolate both to great distances and back in time. McCrea was the first to emphasize the effect of turbulence in condensing gas clouds and applied this theory to the formation of planets, stars and globular clusters, and specifically to the angular momentum of these condensing systems. He made important contributions to **Dirac**'s 'large number hypothesis' in which the number 10^{39} figures prominently, and to discussions on relativity and the low-probability transitions of electrons between energy states in atoms. In 1975 McCrea proposed that comets may have been formed in the high-density interstellar clouds that are found in galactic spiral arms, and that these comets were then picked up periodically by the solar system. He also suggested that the Earth, Moon and Mars were formed from a single body that differentiated and then split up. He was elected FRS in 1952, and knighted in 1985.

McKENZIE, Dan Peter
(1942–)

British geophysicist, born in Cheltenham. He was educated at King's College, Cambridge, where he received a BA in physics (1963) and a PhD in theoretical geophysics (1966) under **Bullard**. Since 1984 he has been Professor of Earth Sciences at Cambridge. After attending a NASA conference in 1966, McKenzie became convinced of the importance of the new hypothesis of **Harry Hess**, **Vine** and **Matthews** on sea-floor spreading, and in the ensuing three weeks wrote a paper on the relationships of oceanic heat flow and gravity (1967). Influenced by the Bullard fit of the Atlantic continents and his suggested use of Euler rotation poles, he developed a theory of plate tectonics (1967, with Robert L Parker). This manifested in further work on the changing plate geometries at triple junctions, and causes and consequences of plate motions (1969). Plate tectonic theory at last provided a plausible mechanism by which continents could drift; this finally made **Wegener**'s theory acceptable to the scientific community and is the key to understanding crustal tectonics. In research on some of the major problems in geophysics, McKenzie elucidated the formation of sedimentary basins (1978), worked on problems of convection within the mantle (1966–83), the tectonics of continental mountain belts (1978–81) and on aspects of melting within the mantle and its manifestations upon the chemistry and processes in the generation of continents (from 1984). He was elected FRS in 1976.

McLAREN, Anne Laura
(1927–)

British geneticist. She was educated at the University of Oxford, and after postdoctoral research at University College London (1952–5), and the Royal Veterinary College, London (1955–9), she joined the staff of the Agricultural Research Council Unit of Animal Genetics in Edinburgh in 1959. She has been Director of the Medical Research Council's Mammalian Development Unit since 1974. Elected FRS in 1975, she received the Scientific Medal of the Zoological Society of London in 1967. McLaren has published prolifically in the fields of reproductive biology, embryology, genetics and immunology, and is best known for her discovery and isolation of the embryonal carcinoma cell line. This cell type was isolated from the testes of mouse embryos and has the capability of differentiating into many different cell types in culture. It therefore provides the opportunity to study the environmental and genetic requirements which cause embryonic cells to differentiate along particular pathways of development. Because embryonic cells also have similarity to the malignant state, this cell type is also of great value in studying the nature of carcinogenesis.

McLENNAN, John Cunningham
(1867–1935)

Canadian physicist, born in Ingersoll, Ontario. Educated at the University of Toronto where he later became professor (1907–31), he did much research on electricity and the superconductivity of metals. In 1923 he succeeded in liquefying helium. He was an early investigator of atomic and molecular energy levels, making use of spectrographic techniques over a broad range of wavelengths from the ultraviolet to the infrared, and often at low temperatures in the region of 20 kelvin. During the 1920s he published widely on the spectra of many materials including silicon, tin, gold, lead and hydrogen. He also made some of the earliest studies of the effect of temperature on the photoelectric effect.

McMILLAN, Edwin Mattison
(1907–91)

American atomic scientist, born in Redondo Beach, California. He was educated at Caltech and Princeton University, where he took his doctorate in 1932. He joined the staff of the University of California at Berkeley, moving to the Lawrence Radiation Laboratory when it was founded within the university in 1934. In 1940, following up the work of **Fermi** who had split the uranium atom by bombarding it with low-velocity neutrons, McMillan and **Abelson** synthesized an element heavier than uranium (the heaviest of the naturally occurring elements, with an atomic number of 92) by bombarding uranium with neutrons in the Berkeley cyclotron. They called this new silvery metal 'neptunium' after the most distant of the planets then known. The synthesis of neptunium (atomic number 93), the first 'transuranic' element, marked the beginning of a new epoch in science and in world affairs. The following year **Seaborg**, also working at Berkeley, synthesized plutonium, leading on to the development of the atomic bomb. McMillan spent the rest of World War II working on radar and sonar, and on the atomic bomb at Los Alamos. In 1944 he established the principle of phase stability in accelerated particles, and as a result was able to modify the Berkeley cyclotron into a 'synchrocyclotron' which could accelerate particles travelling at unknown speeds as well as particles whose speed could be calculated. This improvement presaged the development of the present-day nuclear accelerators used in the search for the fundamental particles. McMillan was appointed to the chair at Berkeley in 1946 and to the directorship of the Lawrence Radiation Laboratory in 1958, retiring in 1973. From 1968 to 1971 he was Chairman of the US National Academy of Sciences. He was awarded the Nobel Prize for Chemistry, jointly with Seaborg, in 1951.

MEDAWAR, Sir Peter Brian
(1915–87)

British zoologist and pioneering immunologist, born in Rio de Janeiro of an English mother and a Lebanese father. He was educated at Marlborough and in zoology at Magdalen College, Oxford. During World War II he investigated methods of skin grafting for burn victims. From this work arose the realization that homograft rejection occurred by the same immunological mechanism as the response to foreign bodies. He was appointed Professor of Zoology at Birmingham University (1947–51), Jodrell Professor of Comparative Anatomy at University College London (1951–62), and Director of the National Institute for Medical Research at Mill Hill from 1962. His research was concerned with the problems of tissue rejection following transplant operations. In 1960 he shared the Nobel Prize for Physiology or Medicine with **Burnet**, for researches into immunological tolerance. They showed that prenatal injection of tissues from one individual to another resulted in the subsequent suppression of rejection in the recipient to the donor's tissues. This mimicked the situation existing between identical twins. He considered it important to explain science to the lay public and gave the brilliant Reith Lectures on 'The Future of Man' in 1959. His writings included *The Uniqueness of the Individual* (1957), *The Art of the Soluble* (1967), *Pluto's Republic* (1982) and his autobiography *Memoirs of a Thinking Radish* (1986). He was elected FRS in 1949 and knighted in 1965.

MEER, Simon van der
See **VAN DER MEER, Simon**

MÈGE-MOURIÉS, Hippolyte
(1817–80)

French chemist and inventor, born in Draguignan. He patented margarine in its original form in 1869 after several years of research into the purification and preservation of foodstuffs and the food value of animal fats. The French government had offered a prize for a satisfactory and economic substitute for butter, which had become difficult to supply in adequate quantities to the rapidly increasing urban populations and armed forces. Mège-Mouriés's margarine was manufactured from tallow, and although he was awarded the prize, it was not until a process for emulsifying it with skimmed milk and water was patented (by F Boudet) in 1872 that it was sufficiently palatable to be a commercial success. Later, vegetable fats such as cottonseed, soyabean and coconut replaced animal fats as the raw material in all margarines. He died in Neuilly-sur-Seine.

MEITNER, Lise
(1878–1968)

Austrian physicist, born in Vienna. Educated at the University of Vienna, she became professor at Berlin (1926–38) and member of the Kaiser Wilhelm Institute for Chemistry (1907–38), where together with **Hahn** she set up a laboratory for studying nuclear physics. In 1917 she shared with Hahn the discovery of the radioactive element protactinium. In 1938 she fled to Sweden to escape persecution by the Nazis. Shortly afterwards, Hahn observed radioactive barium in the products of uranium bombarded by neutrons. He wrote to Meitner telling her of the discovery, and with her nephew **Otto Frisch**, she proposed that the production of barium was the result of the uranium nucleus being split in two by nuclear fission. Frisch was able to verify the hypothesis within a few days. Meitner worked in Sweden until retiring to England in 1960. Recently nuclear physicists named the element of atomic number 108 after her.

MELLANBY, Sir Edward
(1884–1955)

English pharmacologist, born in West Hartlepool. He was educated at Emmanuel College, Cambridge, where he began biochemical research with **Frederick Gowland Hopkins**. In 1907 he moved to London to complete his clinical studies at St Thomas's Hospital Medical School, where he became a demonstrator in physiology. In 1913 he was appointed Professor of Physiology at King's (later Queen Elizabeth's) College for women, and seven years later became Professor of Pharmacology at Sheffield University and Honorary Physician. He was elected FRS in 1925. Alone or in collaboration, he worked on several important biochemical aspects of nutrition. His research began at a time when the concept of vitamins was still new and not universally accepted. Mellanby's work provided a substantial body of evidence for their significant role in nutrition. From laborious animal experiments and clinical observations he identified vitamin D deficiency as a cause of rickets, work that had almost immediate practical consequences. Mellanby's advocacy of cod-liver oil as a sound source of the vitamin, and the rapid application of his findings soon eliminated the then prevalent disease. His further experiments included studies that revealed that lack of vitamin A during embryonic development could result in serious nerve and bone malformation, and that a bleaching process used for wheat flour was responsible for a form of 'hysteria' found in dogs. He was appointed Secretary of the Medical Research Council (1933–49), and the dedicated assistance of his scientist wife **May Mellanby** allowed him to continue with laboratory research during and after this administrative period of his life.

MELLANBY, Kenneth
(1908–)

Scottish entomologist and environmentalist, born in Barrhead, Renfrewshire. Educated at King's College, Cambridge, he carried out research in medical entomology at the London School of Hygiene and Tropical Medicine. During World War II he worked on the transmission of scabies and scrub typhus, diseases spread in humans by ectoparasitic mites. From 1947 to 1953 he was the first Principal of University College, Ibadan, Nigeria; he returned to the UK to head the Department of Entomology at Rothamstead (1955–61) and to become the first Director of the Nature Conservancy's Monks Wood Experimental Station (1961–74). There he worked on the behaviour of the mole (*The Mole*, 1971) but mainly on problems concerned with the effects of agriculture and industry on the environment, particularly the role of pesticides such as DDT. His publications on this topic include *The Biology of Pollution* (1972), *Farming and Wildlife* (1981) and *Waste and Pollution* (1991). He was appointed CBE in 1954.

MELLANBY, Lady May, née Tweedy
(1882–1978)

British nutritional scientist who spent her early years in Imperial Russia, where her father worked in the oil industry. In 1902 she entered Girton College, Cambridge, and was permitted to attend several lectures not normally open to women. She was awarded the equivalent of a second class honours degree in 1906 (women not being allowed to graduate from Cambridge at this time) and became a research fellow, then lecturer in physiology at Bedford College for women in London. In 1914 she married **Edward Mellanby** whom she had met at Cambridge, and collaborated with him on several nutritional researches. She also developed an important independent, but complementary, research line on dental development, arising from a chance observation in 1917 that her husband's dogs, suffering from experimentally induced rickets, had structural abnormalities in their teeth. From animal experiments she showed that vitamins A and D were essential for the proper postnatal development of teeth, and she was the first woman to present a paper to the British Orthodontics Society in 1919. She travelled widely, examining the teeth of adults and children, and carrying out controlled experiments in children's homes and hospitals, and authored important special reports for the Medical Research Council on the relationship between diet, and dental structure and disease.

MELLONI, Macedonio
(1798–1854)

Italian physicist, born in Parma. As Professor of Physics at Parma (1824–31), he had to flee to France on account of political activities. Returning to Naples in 1839, he directed the Vesuvius Observatory until 1848. He is specially noted for his work on radiant heat, developing with **Nobili** the thermopile (1831–9), to investigate the properties of heat and light. He became convinced that heat and light were different modes of the same process, and introduced the term 'diathermancy' to denote the capacity of transmitting infrared radiation.

MELVILL, Thomas
(1726–53)

Scottish scientist noted for his research on optics, probably born in Glasgow. He studied theology at Glasgow University where he was a friend of Alexander Wilson, later the first Professor of Astronomy. He used a prism to examine the colour of flames, detecting the yellow line of sodium when salt, sal ammoniac and other substances were introduced into burning alcohol. This early attempt at spectroscopy made no impact on the scientific community and no other serious investigations were carried out until **Wollaston** discovered the dark lines in the Sun's spectrum in 1802. Melvill also attempted to explain why different colours of light bend by different amounts when passing from one medium to another, suggesting that they travel at different speeds. His early death may explain his lack of influence; he died in Geneva at the age of 27.

MENAECHMUS
(4th century BC)

Greek mathematician. One of the tutors of Alexander the Great, he may have been the first to investigate conics as sections of a cone.

MENDEL, Gregor Johann
(1822–84)

Austrian botanist, born near Udrau. Entering an Augustinian cloister in Brünn in 1843, he was ordained as a priest in 1847. After studying science at Vienna (1851–3), he returned and later became Abbot in 1868. In the experimental garden of the monastery, Mendel bred peas, and grew almost 30 000 plants between 1856 and 1863. He artificially fertilized plants with specific characteristics, for example crossing species that produced tall plants with those that produced short plants, and counting the numbers of tall and short plants which appeared in the subsequent generations. All the plants of the first generation produced were tall; the next generation consisted of some tall and some short, in proportions of 3:1. Mendel suggested that each plant received one character from each of its parents, tallness being 'dominant', and shortness being 'recessive' or hidden, appearing only in later generations. His experiments led to the formulation of 'Mendel's law of segregation' and his 'law of independent assortment'. These results were published in 1865; although Mendel wrote regularly to **Nägeli**, his ideas were not taken up by others, and he gradually discontinued his work on peas. Later analysis of his results revealed that although he had arrived at the correct conclusions, Mendel's data did not show enough statistical variation to be plausible experimental values, and must have been adjusted in some way to agree with the expected results. In 1877 he was instrumental in introducing regular weather reports for the local farmers. His concepts have become the basis of modern genetics.

MENDELEYEV, Dmitri Ivanovich
(1834–1907)

Russian chemist, famous for drawing up the periodic table which explains the relationship between the elements. He was born in Tobolsk, Siberia, the 14th child of liberal middle class parents. He studied at St Petersburg and Heidelberg, Germany, where he collaborated briefly with **Bunsen** and investigated the behaviour of gases, formulating the idea of critical temperature—the highest temperature at which a gas can be liquefied by pressure alone. Mendeleyev was appointed professor at St Petersburg Technical Institute in 1863 and at the University of St Petersburg in 1866. Russian society then being in a state of upheaval after the abolition of serfdom, he began to study chemical problems relating to agriculture and petroleum, and never lost interest in these practical concerns. In 1869, in the course of writing a textbook, he tabulated the elements in ascending order of their atomic weight and found that chemically similar elements tended to fall into the same columns. Several attempts had already been made to group the elements by their chemical properties and to relate chemical behaviour to atomic weight (eg by **Döbereiner** and **Newlands**). Mendeleyev's great achievement was to realize that certain elements still had to be discovered and to leave gaps in the table where he predicted they would fall. In the second version of his table, published in 1871, the 63 known elements exhibited a striking pattern (or 'periodicity' as Mendeleyev termed it), with families of elements all having atomic weights which varied by integer multiples of eight times the atomic weight of hydrogen. At first the periodic table was largely rejected by the scientific world, but as each new element that was subsequently discovered fitted into it perfectly, scepticism turned to enthusiasm. It also soon became obvious that the table made it possible to predict the properties of the still undiscovered elements. Thus it added enormously to scientific understanding and provided a framework for further research. However, the underlying reason for the periodicity of the elements remained unexplained until the structure of the atom—in particular the arrangement of the electrons around the nucleus—came to be understood. Mendeleyev continued to refine the table for the next 20 years, meanwhile continuing his work on gases, studying solutions, and taking up aeronautical research (he made a solo ascent in a balloon in 1887). In 1890 he was forced to resign his position at the university because the authorities feared the effect of his liberal views on students, but he was subsequently put in charge of a project to produce smokeless fuel and made Director of the Central Board of Weights and Measures. He died in St Petersburg. The transuranic element mendelevium (atomic number 101) is named in his honour.

MERCATOR, Gerardus
(the Latinized form of **Gerhard Kremer**)
(1512–94)

Flemish geographer and map-maker, born in Rupelmonde. He graduated at Louvain in philosophy and theology. He studied mathematics, astronomy and engraving, and produced a terrestrial globe (1536) and a map of the Holy Land (1537). In 1544 he was imprisoned for heresy, but released for lack of evidence. In 1552 he settled at Duisburg in Germany, becoming cosmographer to the Duke of Cleves, and produced maps of many parts of Europe, including Britain. To aid navigators, in 1569 he introduced the map projection that bears his name, in which the path of a ship steering on a constant bearing is represented by a straight line on the map; it has been used for nautical charts ever since. In 1585 he published the first part of an *Atlas* of Europe, said to be the first time that this word was used to describe a book of maps; it was completed by his son in 1595. On the cover was a drawing of Atlas holding a globe on his shoulders, hence the word 'atlas' became applied to any book of maps.

MERCATOR, (German **Kaufmann), Nicolaus**
(c.1620–87)

German mathematician and astronomer. As an engineer he planned the fountains at Versailles, and as a mathematician he was one of the pioneers of the use of infinite series in mathematics. He discovered the series for $\log(1+x)$, and from 1660 lived in England.

MERCER, John
(1791–1866)

English dye chemist, born in Blackburn, Lancashire. Self-educated, he became a partner in a dye works at the early age of 16 and was subsequently employed in the colour shop of Fort Brothers print works at Oakenshaw. In 1813 he discovered how to make a good orange dye for calico, using antimony, and he went on to make yellow dyes from lead chromate, and bronze dyes from manganese compounds. He also improved indigo dyes and the Turkey-red process. He became a partner in Fort Brothers and when the print works was sold profitably, he devoted himself to research. He is chiefly known for inventing the process, named 'mercerization', which gives cotton a lustre resembling silk; it was patented in 1850. The cloth is treated with caustic soda, becoming semi-transparent, stronger and quicker to take up dyes. Mercer also investigated the ferrocyanides and studied catalysis. He was elected FRS in 1852, and died near Oakenshaw.

MERRIAM, Clinton Hart
(1885–1942)

American naturalist, zoologist and early conservationist, born in New York City. A physician by training, he became interested in natural history and travelled with the Hayden Geological Surveys (1872–6), collecting biological specimens from all parts of the USA. In 1888 he helped to found the National Geographical Society (which subsequently became the Fish and Wildlife Service) and was its head between 1885 and 1910. He is best known for devising a scheme of distinct life zones based upon temperature differences, which is described in *Life Zones and Crop Zones of the United States* (1898). He latterly joined the staff of the Smithsonian Institution (1919–37) and collected ethnographic data on the Pacific Coast Indian tribes.

MERRIFIELD, (Robert) Bruce
(1921–)

American chemist, born in Fort Worth, Texas. He obtained both his BA (1943) and PhD (1949) from the University of California at Los Angeles. He joined the Rockefeller

Institute for Medical Research in 1949, and was appointed assistant professor at Rockefeller University in New York City in 1957 and full professor in 1966. His research has been centred upon the laboratory synthesis of proteins. Before Merrifield developed his method of synthesis, the process of linking amino acids was painfully slow, even for the preparation of quite a small protein. He simplified the process by devising a procedure whereby the amino acids were linked in the correct order on the surface of a microscopic bead of polystyrene. This process has now been automated and computer controlled, allowing the ready synthesis of small quantities of quite large proteins. Merrifield was awarded the Nobel Prize for Chemistry for this work in 1984.

MERSENNE, Marin

(1588–1648)

French mathematician and natural philosopher, born in Oize. He became a Minim Friar in 1611, and lived in Paris. Devoting himself to science, he corresponded with all the leading scientists of his day including **Descartes**, **Fermat** and **Pascal**, acting as a clearing house for scientific information. He experimented with the pendulum and found the law relating its length and period of oscillation, studied the acoustics of vibrating strings and organ pipes, and measured the speed of sound. He also wrote on music, mathematics, optics and philosophy.

MESELSON, Matthew Stanley

(1930–)

American molecular biologist, born in Denver, Colorado. Educated at the University of Chicago and in chemistry at Caltech, he held a number of posts there and at the University of California at Berkeley, and later became Associate Professor of Biology (1960–4) and Professor (1964–76) at Harvard, where since 1976 he has been Thomas Dudley Cabot Professor of Natural Science. In 1958, with **Franklin Stahl**, he demonstrated that in bacteria, DNA replicates by a semi-conservative mechanism in which the two strands of one parent DNA molecule each form new daughter DNA without degradation; this gave additional support to the DNA model proposed by **James Watson** and **Crick**. Since this now classic experiment, Meselson has investigated DNA repair and the mechanism of DNA recombination, a process by which the positions of genes along chromosomes are rearranged. This work has included studies of DNA swivelase, which untwists supertwisted DNA strands (1972), the role of methylation in DNA repair (1976), and the role of repeat DNA sequences stimulated by heat shock (1978). He carried out research into the formation of DNA loops in the regulation of the heat shock promoter site, and showed that its location on one particular side of the DNA (helix twist effect) is crucially important (1988).

MESSIER, Charles

(1730–1817)

French astronomer, born in Badonville, Lorraine. He began his astronomical life in 1751 as an assistant to **Delisle** in Paris. He observed the return of Halley's comet in 1759 and from that time onwards was an avid searcher of comets, discovering independently a total of 13 of them. He mapped the faint unmoving nebulous objects in the sky which he could discard in comet-searching, and drew up a catalogue of 103 entries (1781) by which his name is perpetuated in astronomy. Messier's objects, known by the prefix M and their catalogue number, comprise nebulae, galaxies and starclusters. The 'ferret of comets', as he was nicknamed by Louis XV, was elected to the Royal Society (1764) and the Paris Academy of Sciences (1770).

METCHNIKOFF, Elie, originally **Ilya Ilyich**

(1845–1916)

Russian embryologist and immunologist, born in Ivanovka, the Ukraine. After graduating from the University of Kharkov in 1864, he studied invertebrate and fish embryology at several European centres. He received a doctorate from the University of St Petersburg in 1867, and after spells of teaching and research at St Petersburg and Odessa, he took a research post at Messina, Italy. It was here that he began his immunological studies. He observed how mobile cells in starfish larvae surround, engulf and destroy foreign bodies. He called these cells phagocytes, and hypothesized that the role of phagocytes in vertebrate blood is to fight invasion by bacteria. In 1886 Metchnikoff returned to Odessa to become director of the new Bacteriological Institute. He studied the action of phagocytes during infections in dogs, rabbits and monkeys. Harrassed by journalists and physicians for lacking medical qualifications, Metchnikoff left Russia in 1887. **Pasteur** offered him the directorship of a laboratory at the Pasteur Institute in Paris. As the role of phagocytes became accepted, Metchnikoff turned to other problems. After investigating aging and death, he advocated eating large quantitites of yoghurt to promote good health. Metchnikoff was awarded the 1908 Nobel Prize for Physiology or Medicine jointly with **Ehrlich**.

MEYER, Julius Lothar von
(1830–95)

German chemist, born in Varel, Oldenburg. He qualified in medicine at Zürich and subsequently studied and taught at several German universities. In 1876 he was appointed the first Professor of Chemistry at Tübingen. Meyer studied the physiology of respiration, and by 1857 had recognized that oxygen combines chemically with haemoglobin in the blood. Independently of **Mendeleyev**, he examined the relationship between the chemical reactivities of the elements and their atomic weights. In 1864 he arranged 28 elements in six families; in 1870 he arranged 53 elements in nine groups. The same year he showed that atomic volume is a function of atomic weight. Meyer was also a considerable organic chemist and made the revolutionary suggestion that the carbon atoms in benzene might be in the form of a ring; he missed only the fact that the ring contains double bonds. He died in Tübingen.

MEYER, Viktor
(1848–97)

German chemist, born in Berlin. He studied under **Bunsen** in Heidelberg and Berlin, and was professor successively at Zürich, Göttingen and finally at Heidelberg. He discovered a method for determining vapour densities by allowing the vapour of a weighed substance to displace a volume of air, measured by a burette. The apparatus he designed became standard laboratory equipment. He developed a method of synthesizing aromatic acids, and investigated and described the nitroparaffins and their derivatives. Meyer also discovered several new types of organic nitrogen compounds. He discovered oximes, studied their isomerism, and introduced the term 'stereochemistry' for the study of molecular shapes. He suffered from ill-health and depression during the last decade of his life, and died in Heidelberg having taken poison.

MEYERHOF, Otto Fritz
(1884–1951)

American biochemist, born in Hanover, Germany, and trained in medicine at the University of Heidelberg. Initially intending a career in psychiatry, in which he gained a doctoral degree, he was appointed to a clinical position at the Heidelberg Clinic in 1910. A colleague there, **Warburg**, was examining chemical reactions in living cells, especially with reference to cancer, and inspired Meyerhof to change direction and study biochemical mechanisms. After a period working at the Stazione Zoologica in Naples, he moved to the department of physiology at the University of Kiel (lecturer, 1921–8; assistant professor, 1918–24) and was then appointed director of the physiology department at the Kaiser Wilhelm Institute for Biology in Berlin (1924–9); he then moved to a similar position at the Kaiser Wilhelm Institute for Medical Research in Heidelberg (1929–38). His biochemical research was primarily on the bioenergetics of muscle contraction, and in understanding the metabolic pathways that provided the energy-rich compounds that muscles used in such contractions. This was complementary to the biophysical work done by **Hill** and the two men shared the 1922 Nobel Prize for Physiology or Medicine. Meyerhof continued his work on muscle metabolism, most notably working out many of the details of the biochemical pathway by which glucose is utilized by cells, a pathway now known as the Embden–Meyerhof cycle. In 1938, like many other Jewish scientists, he left Nazi Germany, first continuing his work in France, then briefly in Spain, and in 1940 he reached the USA, where a position was created for him at the University of Pennsylvania.

MICHAELIS, Leonor
(1875–1949)

German-born American biochemist, born in Berlin. He became professor at Berlin University (1908–22) and the Nagoya Medical School in Japan (1922–6), before settling in the USA at the Johns Hopkins (1926–9) and Rockefeller (1929–40) institutes. He studied the enzymology and physical properties of pepsin, a digestive enzyme of the stomach, and other proteins such as albumin and gelatin, but is best remembered for his mathematical analysis, with M L Menten (1913), of the rate-controlling steps of enzyme reactions (kinetics). They correctly assumed that the substrate combines with the enzyme to form an intermediate complex, which then breaks down to give products and the unmodified enzyme. The Michaelis–Menten equation, which relates enzyme concentration to the substrate concentration giving half maximum velocity, was further refined by G E Briggs and **J B S Haldane** (1925) according to whether or not the intermediate complex is formed rapidly or slowly relative to its rate of breakdown. Michaelis also contributed over 80 papers on the measurement of pH and electrolytic dissociation. The use of indicators and the hydrogen electrode is described in his *Die Wasserstoffionenkonzentration* (1914).

MICHAELSON, Sidney
(1925–91)

English computer scientist, born in London and educated at Imperial College, London.

From 1949 to 1963 he lectured there, pioneering the design of digital computers incorporating the principle of microprogramming. In 1963 he moved to Edinburgh University, where he founded the Department and held the Chair of Computer Science (1966–91). There he initiated work in the UK on system software to provide shared interactive multiple access to computers and led a diverse range of research activities, from computational theory to VLSI (very large scale integration) design. He was a leader in the field of stylometry, the science of computer-based analysis of literary texts for authorship and chronology, and fought to promote the professional recognition of computer scientists.

MICHEL, Hartmut
(1948–)

German biochemist, born in Ludwigsburg, West Germany. Following his early career at the universities of Tübingen, Würzburg, where he became research group leader (1977–9), Munich and the Max Planck Institute of Biochemistry, he was appointed Director of the Max Planck Institute of Biophysics in Frankfurt in 1987. In 1981 he devised a method of producing a large, well-ordered crystal of the membrane-bound photosynthetic reaction centre of the purple bacterium *Rhodopseudomonas viridis*. Michel, **Huber** and **Deisenhofer** then collaborated to determine its structure by X-ray crystallography. By 1985 they were able to report the complete structure, which confirmed and elaborated predictions about how the energy transfer process in photosynthesis operates. The exact locations of the individual amino acids have still to be determined, but the overall structure appears as a cylinder of seven alpha helices with a central water channel. For this discovery of the structure of a membrane protein, Michel shared the 1988 Nobel Prize for Chemistry with Huber and Deisenhofer. He published *Crystallization of Membrane Proteins* in 1990.

MICHELL, John
(1724–93)

English geologist and astronomer, born in Nottinghamshire. A Fellow of Queen's College, Cambridge, and Professor of Geology (1762–4), he became Rector of Thornhill, Yorkshire (1767), where **William Herschel** was a regular guest. He published an important work on artificial magnets (1750), but is best known as the founder of seismology. After the disastrous Lisbon earthquake of 1755 in which around 70 000 people died, Michell proposed that earthquakes set up wave motions in the Earth. The increased frequency of earthquakes in the proximity of volcanoes led him to suggest that the motions could start as the result of gas pressure as water boils from internal heat. He believed that earthquakes might originate under the ocean floor, and demonstrated methods of locating epicentres. He invented a torsion balance, a device to measure the strength of small forces, and with it intended to measure the value of the gravitational constant. However, he died before he had the opportunity—it was **Cavendish** who finally carried this out in the famous 'Cavendish experiment' and derived from it the mean density of the Earth. Michell also made important contributions to astronomy, demonstrating that many double stars must exist as binary systems, and devising a method to calculate stellar distances.

MICHELSON, Albert Abraham
(1852–1931)

German-born American physicist, born in Strelno (now Strzelno, Poland). His family emigrated to the USA when he was aged four. He graduated from the US Naval Academy in 1873, and after teaching at the Academy, the Case School of Applied Science in Cleveland and Clarke University, he was appointed Professor of Physics at Chicago in 1892. His lifelong passion was precision measurement in experimental physics using the latest technology. He made a number of determinations of the speed of light, the most precise in 1924–6, when over a 22 mile course between mountains in southern California, he determined by optical means the velocity to be 299 796±4 kilometres per second. He is chiefly remembered for the Michelson–**Morley** experiment to determine ether drift, the negative result of which set **Einstein** on the road to the theory of relativity. The interferometer which he invented for this experiment was developed subsequently for spectroscopic studies which revealed the hyperfine structure of spectral lines, and provided measurements of the breadths of spectral lines. The latter investigations showed the appropriateness of **Babinet**'s choice of the red cadmium spectral line as a wavelength standard, and in 1894 Michelson measured the metre and showed that it contained 1 553 163.5 wavelengths of this radiation. He also developed a stellar interferometer for measuring the sizes and separations of celestial bodies which cannot be resolved even with the largest telescopes. His first measurements with this instrument, published in 1898, were of the separation of the double star Capella; later he measured the angular sizes of some of the satellites of Jupiter and the star Alpha Orionis. In 1898 he invented the echelon grating, an ultra-high-resolution device for the study and measurement of hyperfine spectra. Michelson became the first American scientist to win a Nobel Prize when he was awarded the Nobel Prize for Physics in 1907. A member (1888) and President (1923–7) of the US National Academy of Sciences, his many honours included foreign membership of the Royal Society (1902), and the award of the society's Copley Medal (1907). Towards the end of his life he successfully investigated the origin of the iridescent colours exhibited by some birds and insects. Advanced forms of his spectral interferometer and stellar interferometer have recently enabled important advances in physics and astronomy.

MICHIE, Donald
(1923–)

British specialist in artificial intelligence, born in Rangoon, Burma. Educated at Rugby School and Balliol College, Oxford, he served in World War II with the Foreign Office at Bletchley Park (1942–5), where work on the Colossus code-breaking project acquainted him with computer pioneers such as **Turing**, **Newman** and T H Flowers. With Turing he discussed the mechanization of thought processes and chess-playing machines. After a biological career in experimental genetics, he developed the study of machine intelligence at Edinburgh University as Director of Experimental Programming (1963–6) and Professor of Machine Intelligence (1967–84). He is Editor-in-Chief of the *Machine Intelligence* series, of which the first 12 volumes span the period 1967–90. Since 1986 he has been chief scientist at the Turing Institute which he founded in Glasgow in 1984. In publications such as *The Creative Computer* (1984) and *On Machine Intelligence* (1974), he has argued that computer systems are able to generate new knowledge. His research contributions have primarily been in the field of machine learning.

MIDGLEY, Thomas Jr
(1889–1944)

American engineer and inventor, born in Beaver Falls, Pennsylvania. He graduated in mechanical engineering at Cornell University in 1911. During World War I, on the staff of the Dayton (Ohio) Engineering Laboratories (1916–23), he worked on the problem of 'knocking' in petrol engines, and by 1921 found tetra-ethyl lead to be effective as an additive to petrol, used with 1,2-dibromoethane to reduce lead oxide deposits in the engine. He also devised the octane number method of rating petrol quality. Since 1980 there has been rising concern that the lead emitted in vehicle exhausts constitutes a health hazard. As president of the Ethyl Corporation from 1923, he also introduced Freon 12 as a non-toxic non-inflammable agent for domestic refrigerators. Again there is now concern that chlorofluorocarbons (CFCs), such as Freon, cause destruction of the ozone layer in the upper atmosphere with damaging climatic and other effects as a result of the increased passage of ultraviolet radiation. He died tragically by accidental strangulation through the failure of a harness he used to help him rise in the morning, needed because he was a polio victim.

MILANKOVITCH, Milutin
(1879–1958)

Yugoslav geophysicist, born in Dalj. He was educated at the Institute of Technology in Vienna where he received a PhD in 1904. He then moved to the University of Belgrade, where he remained for the rest of his career. During 1914–8 he was held prisoner of war, but was allowed to continue his studies in the library of the Hungarian Academy of Sciences in Budapest. Studying the possible astronomical cycles which produce climatic variations, he attempted to reconstruct the palaeoclimates of the Earth. He realized that the key to past climates was the amount of solar radiation received by the Earth, and that this varies according to latitude and depends upon the Earth's orbit, a 21 000 year precession which determines which hemisphere receives greater radiation, and the tilt of the Earth's rotational axis which changes over a 40 000 year period from 21.8 to 24.4 degrees. Using these parameters he reconstructed the classic theoretical radiation curves for the past 650 000 years for comparison with observed climatic cycles. These are known as the Milankovitch cycles and are still in use.

MILLER, Dayton Clarence
(1866–1941)

American physicist, born in Strongsville, Ohio. He studied at Baldwin–Wallace College and obtained his PhD from Princeton (1890). He moved to the Case School of Applied Sciences, Cleveland, and after a brief period in the mathematics department transferred to physics, becoming professor, a post he held until his death. He was an accomplished flautist and had a keen interest in the physics of music. His invention of the phonodeik (1908) provided a mechanical means of recording sound waves photographically, and he later became an expert in architectural acoustics. He was called upon by the National Research Council during World War I to use the phonodeik to assist attempts to improve methods of locating enemy guns. A considerable part of his research efforts were consumed repeating the crucial experiment of **Michelson** and **Morley** (1887), proposed by **Maxwell**, to detect the stationary ether. Working with Morley (1902–4) and then on his own on Mt Wilson, California, he recorded results that indicated the presence of an ether (thereby refuting **Einstein**'s theory of relativity). The announcement of his results (1925) received considerable attention and he was awarded the annual prize by the American Association for the Advancement of Science. In the 1950s, some 10 years after his death, an appraisal of this work indicated that the most likely origin of his erroneous results lay with diurnal and seasonal variation in the temperature of his equipment.

MILLER, Hugh
(1802–56)

Scottish geologist and writer, born in Cromarty. He was apprenticed to a stonemason at 16, and working with stone for the next 17 years he developed an interest in fossils and devoted his winter months to reading, writing and natural history. He became a bank accountant for a time (1834–9), and later became involved in the controversy over church appointments that led to the Disruption of the Church of Scotland (1843). At the same time he wrote a series of geological articles in the Scottish 'Evangelist' newspaper *The Witness*, later collected as *The Old Red Sandstone* (1841). He made important discoveries of fossil fish from the Devonian rocks of Scotland. Also a pioneer of popular science books, he combated Darwinian evolutionary theory with *Footprints of the Creator* (1850), *The Testimony of the Rocks* (1857) and *Sketchbook of Popular Geology* (published posthumously in 1859). Worn out by illness and overwork, he shot himself in 1856.

MILLER, Jacques Francis Albert Pierre
(1931–)

French–Australian immunologist, born in Nice. Miller graduated in medicine from the University of Sydney in 1955, before working in London at the Chester Beatty Research Institute (1958–65). He studied the aetiology and pathogenesis of mouse leukaemia, and in 1960 obtained his PhD in experimental pathology. The function of the thymus gland, an organ found beneath the breast bone, was unknown until the early 1960s. Removing the gland from animals in experiments produced no apparent effect, and when it was removed from patients with cancer of the thymus, no significant changes were observed. However, Miller tried removing the thymus from newborn mice. He found that they failed to develop properly and died within a few months. He showed that these animals would accept skin grafts from unrelated mice and even rats, and concluded that the thymus gland is an important organ in the control of the immunity system. Later work showed that T-lymphocyte cells are produced in the foetal thymus. Miller returned to Australia in 1966 to become head of the experimental pathology (later thymus biology) unit of the Walter and Eliza Hall Institute, Melbourne.

MILLER, Stanley Lloyd
(1930–)

American chemist, born in Oakland, California. He studied at California University and taught there from 1960. His best-known work was carried out in Chicago in 1953, and concerned the possible origins of life on Earth. Inspired by the theories of **Oparin** and **J B S Haldane**, with **Urey**, he passed electric discharges (simulating thunderstorms) through mixtures containing reducing gases (hydrogen, methane, ammonia and water) which Haldane had suggested were likely to have formed the early planetary atmosphere. After some days, analysis showed the presence of some typical organic substances, including aldehydes, carboxylic acids, amino acids and urea. In later work, Miller used other mixtures containing carbon dioxide and carbon monoxide, and the developmental patterns of different compounds with time were analysed; hydrogen cyanide and cyanogen appeared first as the ammonia concentration declined, while amino acids were formed more slowly. Formation of the Oparin–Haldane 'primeval soup' is now accepted as the most plausible theory for the generation of complex organic molecules on Earth, although the probable subsequent path from these chemicals to a living system is still hotly debated.

MILLIKAN, Robert Andrews
(1868–1953)

American physicist, born in Illinois. He studied at Oberlin College and Columbia University where he received his PhD in 1895. After working at Berlin and Göttingen, he became **Michelson**'s assistant at the University of Chicago, where he was appointed professor in 1910. In 1921 he moved to Caltech where he established the experimental physics laboratory. At Chicago he refined the oil drop technique for measuring the electron's charge that had been developed by **J J Thomson**. He observed the motion of electrically charged drops of oil as they floated between two parallel plates with a voltage applied across them. By measuring the speed of fall of the droplets for different voltages, he was able to show that the charge on each was always a multiple of the same basic unit—the charge on the electron—which he measured very precisely. In studies of the photoelectric effect he confirmed **Einstein**'s theoretical equations and gave an accurate value for **Planck**'s constant. For all these achievements he was awarded the 1923 Nobel Prize for Physics. He also investigated cosmic rays, a term that he coined in 1925.

MILNE, Edward Arthur
(1896–1950)

English astrophysicist, born in Hull, and educated at Hymer's College, Hull, and Trinity College, Cambridge. Assistant Director of the Cambridge Solar Physics Observatory (1920–4), and Professor of Mathematics at Manchester (1924–8) and Oxford (from 1928), he made notable contributions to the study of cosmic dynamics. In 1923, with **Ralph Fowler**, he studied the ionization, temperature and energy flux of stellar surface material and was the first to estimate the electron pressure in a stellar atmosphere. He noticed that a decrease in luminosity might cause the collapse of the star, which could lead to the production of a nova. In 1932 he began to develop his theory of 'kinematic relativity'. One outcome of this was the prediction that the universe is 10 000 million years old.

MILNE, John
(1859–1913)

English seismologist, born in Liverpool. He was educated at King's College and the Royal School of Mines, London. He began his career as a mining engineer in the UK and Germany, then spent two years in Newfoundland and travelled in Egypt, Arabia and Siberia. Always a keen traveller, he joined an expedition in 1874 to locate Mount Sinai and on being appointed Professor of Geology in Tokyo (1875–94), travelled there by camel across Mongolia. In Japan he took up an interest in earthquakes, becoming a supreme authority, for which he was awarded the Order of the Rising Sun. He pioneered modern seismology and introduced precise physical measurements. In 1892, with colleagues, he developed a seismometer to record horizontal components of ground motion which became used on a worldwide basis. From this time seismology advanced rapidly and began to be applied to the study of the internal structure of the Earth. Milne devised methods of locating distant earthquakes and early traveltime curves for seismic wave arrivals, initiated experiments using explosives, and compiled *A Catalogue of Destructive (Japanese) Earthquakes AD 7–AD 1899* (1912). On his retirement to the Isle of Wight (1895) he ran a private seismological observatory, regularly issuing a bulletin summarizing data from a worldwide network of seismological stations which he set up with his own instruments, a forerunner of the International Seismological Summary. He was a prolific writer, publishing *Earthquakes and other Earth Movements* (1886) and *Seismology* (1898).

MILNOR, John Willard
(1931–)

American mathematician, educated at Princeton University where he has taught for most of his life, with spells at Berkeley and Stony Brook, New York. He has worked in many areas of mathematics, but is chiefly a topologist. One of his earliest discoveries was that even familiar objects can be described in unfamiliar ways. For example, many higher dimensional spheres may be defined in alternative ways, so that functions on them are differentiable in one description but not in another. For this highly unintuitive result and others, he was awarded the Fields Medal (the mathematical equivalent of the Nobel Prize) in 1962. He has also worked on the topology of self-intersections of higher dimensional manifolds, which arise naturally in many topics in mathematics. A lucid writer on advanced topics, many of his books have provided introductions to current research, and he has stimulated a number of young mathematicians, including William Thurston who was awarded the Fields Medal in 1978 for his work on the geometrical structure of three-dimensional manifolds, a topic which Milnor has actively promoted.

MILSTEIN, Cesar
(1927–)

Argentinian-born British molecular biologist and immunologist, born in Bahía Blanca. He graduated from Buenos Aires University with a BSc in chemistry in 1945, and worked on the active sites of enzymes in **Sanger**'s laboratory in Cambridge (1958–61), where he obtained a PhD in 1960. In 1961 he returned to Argentina to become Head of the Division of Molecular Biology at the National Institute of Microbiology. When many members of the institute were dismissed following the military coup, Milstein resigned in protest and returned to Cambridge, where he has been on the staff of the Medical Research Council at the Laboratory of Molecular Biology since 1963. He has conducted important research into antibodies, proteins that are produced by the immune system cells in response to foreign molecules called antigens. By fusing cells from tumours of the immune system (myelomas) with cells producing one antibody, to form a 'hybridoma', it became possible to maintain production of the antibody. This technique of 'monoclonal antibodies', developed in 1975 with **Köhler**, has become widespread in the commercial development of new drugs and diagnostic tests. In 1984, Milstein was awarded the Nobel Prize for Physiology or Medicine with Köhler and **Jerne**.

MINKOWSKI, Hermann
(1864–1909)

Russian-born German mathematician, born near Kovno. He was educated at the University of Königsberg, where he received his PhD in 1885. He was later appointed professor at Königsberg (1895), Zürich (1896), where he taught **Einstein**, and at Göttingen (1902), where he was a close friend and colleague of **Hilbert**. Minkowski won a prize for work in the theory of numbers from the Parisian Academy of Sciences when aged only 18, and went on to discover a new branch of number theory, the geometry of numbers. In his most important work he gave a precise mathematical description of space–time as it appears in Einstein's relativity theory; the four-dimensional 'Minkowski space' was described in *Space and Time* (1907). He died of unexpected complications arising from appendicitis.

MINKOWSKI, Rudolph Leo
(1895–1976)

American astrophysicist, born in Strassburg, Germany (now Strasbourg, France). After receiving his doctorate at the University of Breslau, he became Professor of Physics at Hamburg in 1922. He emigrated to the USA in 1935 and started work at the Mount Wilson and Palomar Observatories. In the 1940s he established an empirical classification scheme for supernovae that was based on their spectral properties near maximum light; two types were clearly distinguishable. In 1951, with **Baade**, Minkowski made the first optical identification of a discrete radio source, Cygnus A. They also noticed the unusual elemental abundance in the 350 year old supernova remnant Cassiopeia A (1954). Minkowski was a specialist when it came to the optical identification of high-redshift radio galaxies, and in 1960 he measured a relative redshift (shift in wavelength divided by original wavelength) of 0.46 for the galaxy 3C 295, using the Palomar 200 inch telescope. Minkowski instituted a search for planetary nebulae which increased the number known by a factor of two, and he also supervised the National Geographic Society Palomar Observatory Sky Survey.

MINNAERT, Marcel Gilles Jozef
(1893–1970)

Dutch astronomer, born in Brugge, Belgium, and educated at the universities of Ghent and Leiden where he studied biology and physics (1914–16). In 1918 he started his lifelong association with the University of Utrecht; in 1924 he was put in charge of solar physics and began a programme of the quantitative interpretation of lines in the solar spectrum which culminated in the *Photometric Atlas of the Solar Spectrum* (1940, with Gerardus Franciscus Wilhelmus Mulders and Jacob Houtgast), completed at the beginning of World War II. During this war Minnaert with strong anti-Fascist views was imprisoned for two years by the German occupying authorities. He cooperated with **Moore Sitterly** in *The Second Revision of Rowland's Table of Solar Spectrum Wavelengths* (1966). Minnaert's intellectual interests included a devotion to the universal language esperanto: the bilingual text of his atlas was written in esperanto and English.

MINOT, George Richards
(1885–1950)

American physician, born into a prominent Boston family and educated at Harvard College and Medical School with which, except for three postdoctoral years at Johns Hopkins, he was associated all his working life. Using special staining techniques on blood smears, he began to study anaemia, and was able to demonstrate that some anaemias are caused by the failure of the bone marrow to make enough red blood cells, whereas others are induced when blood cells are destroyed too quickly. From 1925, working with **Murphy**, he examined clinically **George Whipple**'s observation that dogs made anaemic through repeated bleedings improved significantly when fed liver. Minot and Murphy gave large amounts of raw liver to patients diagnosed as suffering from pernicious anaemia, at that time a fatal disease. Their patients rapidly improved, although the 'intrinsic factor' contained in liver was not identified for another 20 years. The 'intrinsic factor' was necessary to the absorption and utilization of folic acid, used by the body in the production of red blood cells. Minot, a diabetic, was one of the earliest patients to benefit from insulin therapy. With Murphy and Whipple, he shared the 1934 Nobel Prize for Physiology or Medicine.

MISES, Richard von
See **VON MISES, Richard**

MITCHELL, Maria
(1818–89)

American astronomer, born in Nantucket into a serious-minded Quaker family. Her father's activities included regulating chronometers for whaling ships in which his daughter took part from an early age. In 1836 the US Coast Survey equipped an observatory at their home as a local station with her father in charge, where Maria, a librarian in the local Athenaeum, had an opportunity to practise astronomy. Her discovery of a comet in 1847 brought her to the public notice and earned for her (1848) the King of Denmark's Gold Medal for first discoverers of telescopic comets, and election to the American Academy of Arts and Sciences, its first woman member. Her first professional commission was the computing of tables of the planet Venus for the American Ephemerides and Nautical Almanac, a duty she performed for 20 years (1849–68). In 1865 she was appointed Professor of Astronomy at the newly founded Vassar College for women at Ploughkeepsie where she was an inspiring teacher and a doughty campaigner in the women's rights and anti-slavery movements. In failing health she retired to her native Nantucket in 1988.

MITCHELL, Sir Peter Chalmers
(1864–1945)

Scottish zoologist and journalist, born in Dunfermline. He began his career as a lecturer at the universities of Oxford and London, where he taught comparative anatomy. In 1903 he was elected Secretary of the Zoological Society and inaugurated a period of prosperity at the London Zoo; he was responsible for the Mappin terraces, Whipsnade, the Aquarium and other improvements. He was a friend and admirer of **T H Huxley** and wrote his biography, *Thomas Henry Huxley: a Sketch of his Life and Work* (1900). His other books included *The Nature of Man* (1904) and *Materialism and Vitalism in Biology* (1930). He was elected FRS in 1906, and knighted in 1929.

MITCHELL, Peter Dennis
(1920–92)

English biochemist, born in Mitcham, Surrey. He graduated from Cambridge and taught there (1943–55) and at Edinburgh University (1955–63) before founding his own research institute, the Glynn Research Laboratories at Bodmin in Cornwall (1964), to extend his studies of the way in which energy is generated inside cells at the molecular level. Scientists knew that the energy comes from the adenosine triphosphate (ATP) molecule, and that somehow ATP is produced from adenosine diphosphate (ADP) by a process known as oxidative phosphorylation, possibly regulated by undiscovered enzymes. In the 1960s Mitchell rejected the idea that oxidative phosphorylation occurs via an active chemical intermediate, and proposed an entirely novel theory in which electron transport from reducing coenzymes to form water causes the formation of a proton gradient across the inner mitochondrial membrane. It was suggested that this 'chemi-osmotic gradient' directly drives the synthesis of ATP from ADP and inorganic phosphate in special membrane structures where protons re-enter the mitochondrion. The mechanism is unknown but possibly involves structural changes in energy-dependent proteins operating in an essentially water-free environment. Although at first greeted with scepticism, his views became widely accepted and are now supported by a considerable body of evidence. Mitchell's theory was formally acknowledged by his award of the unshared Nobel Prize for Chemistry in 1978.

MITSCHERLICH, Eilhard
(1794–1863)

German physical and organic chemist, born in Neuende, near Jeve. He studied Persian at Heidelberg and Paris, medicine at Göttingen, and geology, mineralogy, chemistry and physics in Berlin and in Stockholm with **Berzelius**. He became Professor of Chemistry at the Friedrich Wilhelm Institute in Berlin, and was elected to the Berlin Academy of Sciences in 1852. He contributed to many branches of chemistry, but is best known for his work in crystallography. He discovered that substances with similar chemical compositions may have the same shape of crystal, showing that one element can sometimes substitute for another in the crystal lattice. He demonstrated this phenomenon, which was named 'isomorphism' ('the same shape'), with crystals of potassium arsenate and potassium phosphate and with some of the sulphates. He further noticed that sulphur forms either rhombic or monoclinic crystals and this led him to the discovery of dimorphism, the capacity of some elements to occur in two distinct forms. While visiting Paris in 1823–4 he and **Fresnel** discovered that the optical axes of a biaxial crystal change with a change in temperature. Mitscherlich also investigated the decomposition products of benzaldehyde and benzoin, synthesized artificial minerals by fusing silica with various metallic oxides and wrote a number of successful textbooks. He died in Berlin.

MIVART, St George Jackson

(1827–1900)

English biologist, born in London. He was converted to Roman Catholicism and although he studied for the bar, devoted himself to zoological research. He became Professor of Zoology and Biology at the Roman Catholic University College in Kensington (1874–84) and in 1890 accepted a Chair of Philosophy of Natural History at Louvain. He investigated the anatomy of carnivores and published *The Cat: An Introduction to the Study of Backboned Animals* (1881), a detailed and accurate account. Although he accepted that evolution had occurred, he rejected natural selection as its mechanism and proposed in its place the idea of an innate plasticity which he called individuation. He published several books in support of his views such as *The Genesis of Species* (1871), *Nature and Thought* (1883) and *The Origin of Human Reason* (1889). These works alienated him from the leading evolutionists of the day such as **Charles Darwin** and **T H Huxley**. At Louvain he published articles which were placed on the Vatican index of forbidden readings and he was excommunicated in 1900.

MÖBIUS, August Ferdinand

(1790–1868)

German mathematician, born in Schulpforta. As professor at Leipzig, he worked on analytical geometry, statics, topology and theoretical astronomy. He introduced barycentric coordinates and the 'Möbius net', thus extending Cartesian coordinate methods to projective geometry, and showed how they could be used to express a duality between points and lines (discovered earlier by Joseph Diaz Gergonne). He gave a straight-forward algebraic account of statics, following the work of the French mathematician Louis Poinsot, and in this way used vectorial quantities before vectors as such entered mathematics. He also discovered a novel type of duality in three-dimensional space in this connection. In topology he investigated which surfaces can exist, and became one of the discoverers of the 'Möbius strip' (a one-sided surface formed by giving a rectangular strip a half-twist and then joining the ends together). This shows that a sense of orientation is not necessarily part of the intrinsic geometry of a surface. He also examined in detail the possible types of three-dimensional spaces which can be created by similar gluing constructions.

MOHL, Hugo von

(1805–72)

German botanist, born in Stuttgart. Although he had a classical education, from childhood it was clear that his vocation would combine botany and optics. After studying medicine at Tübingen he was Professor of Physiology at Bern (1832–5) and Professor of Botany at Tübingen (1835–72). Early research included studies of climbing plants, and pioneering work on stomatal movement. He constructed a microscope and published a manual on microscopy, *Mikrographie, oder Anleitung zur Kenntniss und zum Gebrauche des Mikroskops* (1846). His most lasting researches, published as two fundamental papers and in *Die vegetabilischen Zellen* (1846), lay in the field of plant cell structure and physiology, where his meticulous observations were the first attempts at cytochemistry. He differentiated the cell membrane, nucleus, cellular fluid, utricle, and a substance he called protoplasm. This term had earlier been used by the Czech physiologist **Purkinje** to denote the embryonic material of eggs; Mohl was the first person (1846) to use the term protoplasm in plant cell biology. For Mohl, protoplasm was a preliminary substance in cell generation, a quite different sense to the modern usage which dates from **Schultze** (1861). He was also the first to clearly explain osmosis, and discovered that the secondary walls of plant cells are fibrous.

MOHOROVIČIĆ, Andrija

(1857–1936)

Yugoslavian seismologist and meteorologist, born in Volosko, Croatia. He was educated in physics and mathematics at the University of Prague, where one of his teachers was **Mach**. He was later appointed to the Royal Nautical School at Bakar, where he taught meteorology and oceanography, and in 1887 founded the meteorological station at Bakar. Appointed as professor and director at the Zagreb Technical School (1891–1921), he campaigned for the independence of Zagreb observatory from government control, making it a recognized centre for meteorology and geodynamics. Investigating the Croatian earthquake of 1909 in the Kulpa valley, he observed from some distance two distinct seismic wave arrivals. He deduced that the slower of the two arrivals followed the direct route from the earthquake focus to the observation point, while the faster wave is refracted from a discontinuity. From this he concluded that the Earth's crust must overlay a denser mantle and he calculated the depth to this transition. The sharp discontinuity was subsequently found to exist worldwide, and became known as the Mohorovičić discontinuity or 'Moho'. Its depth varies from 30 to 60 kilometres on continents, and from 8 to 10 kilometres beneath oceans.

MOHS, Friedrich
(1773–1839)

German mineralogist, born in Gernrode and educated at the University of Halle. He became professor at Graz (1812), at Freiburg as successor to **Abraham Werner** (1818), and at Vienna (1826). He developed a mineralogical classification system based on a variety of mineral characters, rather than adopting the traditional purely chemical system. The Mohs scale of hardness which he introduced is still in use. Around 1820, he arrived at the concept of the six crystal systems; this mineral classification system was based on the different orientation of crystallographic axes, but the names which he applied were not widely used. His publications included *The Natural History System of Mineralogy* (1821) and *Treatise on Mineralogy* (3 vols, 1825).

MOISSAN, (Ferdinand Frédéric) Henri
(1852–1907)

French chemist, born in Paris. He studied chemistry and pharmacy in Paris and qualified as a pharmacist in 1879. He taught at the School of Pharmacy in Paris, becoming Professor of Toxicology and Inorganic Chemistry in 1886 and moving to the Chair of Inorganic Chemistry at the University of Paris in 1900. Moissan was noted for his teaching and experimental work. He was the first to isolate fluorine (1886). In 1892 he invented the electric arc furnace, which in its simplest form consisted of two blocks of lime with a space in the middle for a crucible and grooves for carbon electrodes. With the high temperatures that could be reached for the first time, he reduced the oxides of uranium and tungsten; prepared carbides, borides and hydrides; and synthesized rubies. His claims to have synthesized diamonds, however, were met with scepticism. Moissan is regarded as the founder of high-temperature chemistry, and both his furnace and his discoveries were soon shown to have many industrial applications. For his work on fluorine he was awarded the Nobel Prize for Chemistry in 1906. He died in Paris.

MOIVRE, Abraham De
See **DE MOIVRE, Abraham**

MOND, Ludwig
(1839–1909)

German-born British chemist, born in Cassel. He studied chemistry at Marburg and then under **Bunsen** at Heidelberg. Keen to apply his chemical knowledge to industry, he left without completing his doctorate and worked in various chemical works, including the Leblanc soda works at Ringenkuhl near Cassel. He settled in Britain in 1862, and while working for John Hutchinson at Widnes developed a process to retrieve sulphur from the waste products of the Leblanc process. In 1873 he joined John Brunner in setting up a factory at Winnington, Cheshire, to manufacture soda by the new ammonia process invented by **Solvay**. Brunner–Mond & Co eventually grew to be the largest soda plant in the world. Mond also developed a new fuel, producer-gas, by burning coal in a mixture of air and steam to produce ammonia, carbon monoxide and hydrogen. Ammonia was then recovered from the mixture, leaving the other two gases. Mond is perhaps best known for the process he invented for purifying nickel. He noticed that nickel valves in a manufacturing plant were corroded by carbon monoxide and discovered that a compound new to science, nickel carbonyl, had been formed, and that it would yield pure nickel on further heating. He put this reaction to use on an industrial scale (the Mond process). Mond died in London and was succeeded in business by his son Robert (later Sir Robert) Mond (1867–1938).

MONGE, Gaspard
(1746–1818)

French mathematician and physicist, born in Beaune. The founder of descriptive geometry, he became Professor of Mathematics at Mézières in 1768, and in 1780 Professor of Hydraulics at the Lycée in Paris. In 1783, independently of **Watt**, **Cavendish** and **Lavoisier**, he discovered that water resulted from an electrical explosion of oxygen and hydrogen. During the Revolution he was minister for the navy, but soon took charge of the national manufacture of arms and gunpowder. He helped to found (1794) the École Polytechnique, and became Professor of Mathematics there. The following year there appeared his *Leçons de géométrie descriptive*, in which he stated his principles regarding the general application of geometry to the arts of construction (descriptive geometry). He was sent by the Directory to Italy, from where he followed Napoleon to Egypt. In 1805 he was made a senator and Count of Pelusium, but lost both dignities on the restoration of the Bourbons.

MONOD, Jacques Lucien
(1910–76)

French biochemist, born in Paris. After he graduated from the University of Paris

(1931), a Rockefeller Fellowship enabled him to work in **Thomas Hunt Morgan**'s laboratory at Columbia University. He returned to France and received his PhD from the Sorbonne for his thesis on bacterial growth. He joined the French Resistance during World War II, and after the war began work at the Pasteur Institute in Paris, becoming Head of the Cellular Biochemistry Department in 1954 and Director in 1971. From 1967, he was also Professor of Molecular Biology at the Collège de France. Monod worked closely with **Jacob** on the genetic control mechanisms of the bacterium *Escherichia coli*. Together they developed the theory of the operon system, whereby a regulator gene binds to a specific section of the DNA strand (the operator), and deactivates the structural genes which code for the protein. They suggested the term 'messenger RNA' (mRNA), which is transcribed from the DNA molecule and is carried to the cytoplasm for protein production. In 1965 Monod and Jacob shared the Nobel Prize for Physiology or Medicine with **Lwoff**. In 1968 in Paris, Monod supported the students in their battles with the university establishment. He published *Chance and Necessity* in 1970, a biologically based philosophy of life.

MONRO, Alexander
(1697–1767)

'Monro Primus', Scottish anatomist, born in London. Himself a surgeon's son, Monro founded a three-generation dynasty of anatomy professors that dominated anatomy teaching at Edinburgh for 126 years. He studied in London (under William Cheselden), in Paris, and in Leiden (under **Boerhaave**). From 1719 he lectured at Edinburgh on anatomy and surgery, serving as professor from 1725 to 1759. He played a key part in founding the Edinburgh Royal Infirmary, and in promoting medicine within the university. His energetic and well-organized teaching, deploying a wide range of effective preparations but no great classroom use of dissection, expedited the rise of Edinburgh as a popular centre for medical training. He wrote *Osteology* (1726), *Essay on Comparative Anatomy* (1744), *Observations Anatomical and Physiological* (1758) and *Account of the Success of Inoculation of Smallpox in Scotland* (1765)—works that reveal a skilled communicator of medical knowledge rather than a profound researcher. It was partly through the orientation of his interests that the Edinburgh ideal of the practitioner integrated the physician and the surgeon.

MONRO, Alexander
(1733–1817)

'Monro Secundus', Scottish anatomist, son of **Alexander Monro** 'Monro Primus'. Educated at Edinburgh University, Monro was groomed by his father to succeed him in the Chair in Anatomy, which he formally did at the tender age of 21 in 1754, though his father continued to teach. Monro pursued his studies in London, under **Hunter** and in Paris. In 1757 he published his *De Venis Lymphaticis Valvulosis*, which demonstrated that the lymphatics were absorbents and quite separate from the circulatory system. This publication sparked a vitriolic priority dispute with William Hunter, which rumbled on for many years and turned into a further acrimonious dispute (1767) with William Hunter's colleague, William Hewson, regarding priority in developing the operation of paracentesis of the thorax in traumatic pneumothorax. An energetic researcher, Monro made public some of his most important findings in *Observations on the Structure and Functions of the Nervous System* (1783). In later life he studied the discharges of electrical fish, the general physiology of fishes (1785), and investigated the eye and ear (1797). He was no less energetic a teacher than his father; he is said to have taught 13 404 students. He also built up a large anatomical and pathological collection which he bequeathed to the university. In 1798 he persuaded the Edinburgh town council to appoint his elder son (also Alexander) as joint professor. In 1808, Secundus retired. 'Tertius' possessed less energy and no originality. He conducted no significant research and his long occupancy (until 1851) contributed to the decline of the Edinburgh medical school.

MONTAGNIER, Luc
(1932–)

French molecular biologist. He was educated at the Collège de Chatellerault, the University of Poitiers and the University of Paris, where he was subsequently appointed to a number of research posts. He became Laboratory Head of the Radium Institute (1965–71) in Paris, and since 1972 has worked at the Pasteur Institute, as Head of the Viral Oncology Unit, professor (since 1985) and Head of the Department of AIDS and Retroviruses (since 1990). Since 1974 he has also been Director of Research at the National Centre for Scientific Research. Montagnier has published widely in molecular biology and virology, and is now credited with the discovery of the HIV virus. He and his team first isolated the HIV virus from a Frenchman with AIDS in 1983. However, an American team led by Robert Gallo claimed to have independently discovered the virus and Montagnier's virus was discredited; the journal *Nature* declined to publish his findings. Around 10 years after the discovery, it was shown that Gallo's original findings were erroneous and that the virus he eventually used to develop and patent an HIV blood test actually came from Montagnier's laboratory.

MONTGOLFIER, Joseph Michel
(1740–1810) and
Jacques Étienne
(1745–99)

French aeronautical inventors, sons of a paper manufacturer of Annonay near Lyons. Joseph developed an early interest in science, while his younger brother became a successful architect before joining the family firm. After some preliminary model experiments, in 1782 they constructed a balloon whose bag was lifted by lighting a cauldron of paper beneath it, thus heating and rarifying the air it contained. The world's first manned balloon flight of 7½ miles in less than half an hour, at a height of 3000 feet, carrying Pilatre de Rozier and the Marquis d'Arlandes, took place in November 1783. Their achievement created great public interest, and many other inventors attempted to follow their example, not always with equal success. Further experiments were frustrated by the outbreak of the French Revolution, Étienne being proscribed, and his brother returning to his paper factory. Joseph later became interested in other applications of science, inventing a type of parachute, a calorimeter and the widely used hydraulic ram, a device for raising small quantities of water to a considerable height. He was subsequently elected to the French Academy of Sciences and created a Chevalier of the Legion of Honour by Napoleon.

MONTUCLA, Jean Étienne
(1725–99)

French mathematician, born in Lyons. In 1758 he wrote the first history of mathematics worthy of the name, *Histoire des Mathématiques*. He then became Inspector of the Royal Buildings and Royal Censor. Impoverished by the Revolution, he began to write a much enlarged second edition of the *Histoire*. Two volumes appeared before his death, and two more were published posthumously in 1802.

MOORE, Stanford
(1913–82)

American biochemist, born in Chicago. He studied chemistry at Vanderbilt University and Wisconsin, and spent his career at the Rockefeller Institute (1939–82). He is best known for inventing, with **Stein**, a column chromatographic method for the sequential elution, identification and quantification of amino acids in mixtures derived from the hydrolysis of proteins or from physiological tissues (1950). By 1958 they had also developed an ingenious automated analyser to carry out all the steps of the analysis of the base sequence of RNA on a small sample. Moore and Stein also studied a novel protease from streptococcus discovered by the English microbiologist Stuart D Elliott. They showed that this enzyme possesses a similar specificity and organization of the catalytic site, including a so-called 'essential thiol group', to the plant protease papain. Other aspects of the molecular structure, however, turned out to be quite different. This was the first example of convergent evolution—two enzymes of similar function arising by different evolutionary paths. Moore and Stein also complemented the structural studies of **Anfinsen** (1954–6) by determining the amino acid composition of bovine pancreatic ribonuclease. All three shared the Nobel Prize for Chemistry in 1972.

MOORE SITTERLY, Charlotte
(1898–1990)

American astronomer and spectroscopist, born in Enciltoun, Pennsylvania, and educated at Swarthmore College where she graduated in 1920. She was assistant to **Henry Russell** (1920–5) at Princeton before moving to Mount Wilson Observatory to work on the revision and extension of the table of wavelengths in the solar spectrum, originally the work of **Rowland** (1895–7). In the revised table, which she published jointly with Charles Edward St John and others (1928), over 58000 wavelengths were tabulated in which 57 chemical elements were reported as present in the Sun. A second revision, with 10000 more lines (1966) was to come out under her direction. She returned to Princeton (1931–45) where she produced her famous *Multiplet Tables of Astrophysical Interest* (1945), and in 1945 joined the Bureau of Standards, Washington (1945–68), where she published her *Ultraviolet Multiplet Tables* (1946) and *Atomic Energy Levels* (1949–58). After her formal retirement, she was attached to the Naval Research Laboratory, Washington, and remained active in spectroscopy until past her ninetieth birthday.

MORGAGNI, Giovanni Battista
(1682–1771)

Italian physician, born in Forli. Graduating in Bologna, Morgagni taught anatomy there and later in Padua. Active throughout his life in anatomical research, his great work *De Sedibus et Causis Morborum per Anatomen*

Indagatis (1761) was not published until he was 80. It was grounded on over 600 postmortems and was written in the form of 70 letters to an anonymous medical confrère. Case by case, Morgagni described the clinical aspects of illness during the patient's lifetime, before proceeding to detail the postmortem findings. His object was to relate the illness to the lesions established at autopsy. Morgagni did not use a miscroscope and he regarded each organ of the body as a complex of minute mechanisms. His book may be seen as a crucial stimulus to the rise of morbid anatomy, especially when physicians also made use of the techniques of percussion, developed by **Auenbrugger**, and ausculation, pioneered by **Laënnec**. Morgagni himself made significant discoveries. He was the first to delineate syphilitic tumours of the brain and tuberculosis of the kidney. He grasped that where only one side of the body is stricken with paralysis, the lesion lies on the opposite side of the brain. His explorations of the female genitals, of the glands of the trachea, and of the male urethra also broke new ground. He is judged the father of the science of pathological anatomy.

MORGAN, Augustus de

See **DE MORGAN, Augustus**

MORGAN, Thomas Hunt

(1866–1945)

American geneticist and biologist, born in Lexington, Kentucky. He graduated in zoology from Kentucky State College in 1886, and received his PhD from Johns Hopkins University in 1890. He became Professor of Experimental Zoology at Columbia University (1904–28) and then at Caltech (1928–45). Morgan started out with many objections to **Mendel**'s theory of heredity, and in 1908, he began work on *Drosophila*, the fruit fly, to test out Mendel's ideas. From his breeding programme, he found that certain traits are linked (for example, only male flies have the 'white eye' characteristic), but that the traits are not always inherited together. From his work, he suggested that certain traits are carried on the X chromosome, that traits can cross-over to other chromosomes and that the rate of crossing-over could be used as a measure of distance along the chromosome. His laboratory became known as the 'fly room', and attracted many researchers. Morgan's co-workers included **Hermann Müller**, **Sturtevant** and C B Bridges, with whom he wrote *The Mechanism of Mendelian Heredity* (1915), which established the chromosome theory of inheritance in confirmation of Mendel's work. Morgan was awarded the 1933 Nobel Prize for Physiology or Medicine. His many other books included *Evolution and Adaptation* (1911), *The Theory of the Gene* (1926) and *Embryology and Genetics* (1933).

MORISON, Robert

(1620–83)

Scottish botanist, born in Aberdeen and educated locally, graduating from Aberdeen University in 1638. His subjects included Hebrew, as his parents wished him to enter the ministry. Times during the reign of Charles I were turbulent, however, and Morison became a Royalist. After recovery from a head wound received in battle, he escaped to France and took his MD at Angers (1648). From around 1649/50 to 1660 he managed the gardens of the Duke of Orléans and travelled throughout France collecting plants. At Blois, Charles II invited Morison to accompany him to England and appointed him as senior physician, King's Botanist and superintendent of the royal gardens. In 1669 he became Professor of Botany at Oxford, shortly after publication of his *Praeludia Botanica* which contained the basis of his classification system. In 1672 he published *Plantarum Umbelliferarum Distributio Nova*. A pioneering work, this was the earliest botanical monograph. The rest of Morison's life was spent compiling *Plantarum Historiae Universalis Oxoniensis* (1680–99). Five of the projected 15 parts on herbaceous plants were published in his lifetime; the remaining 10 parts were posthumously edited by Jacob Bobart (1641–1719). Morison died suddenly when a coach pole struck him in a London street.

MORLEY, Edward Williams

(1838–1923)

American chemist and physicist, born in Newark, New Jersey. He was educated at Williams College in Williamstown, Massachusetts, and from 1860 to 1864 at Andover Theological Seminary. His theological studies were in preparation for becoming a Congregational minister, like his father. However, he also continued scientific studies and when he became pastor of the Congregational Church at Twinsburg, Ohio (1868), he was invited to teach at the nearby Western Reserve College in Hudson. In 1882 this college transferred to Cleveland and became Adalbert College of Western Reserve University. Morley became Professor of Chemistry and Natural History, an appointment he held until his retirement in 1906. He was awarded the Davy Medal of the Royal Society in 1907. Morley had a passionate concern for precise measurement. In Hudson he analysed the oxygen content of the atmosphere with a precision of 0.0025 per cent, and endeavoured to correlate the results for samples taken at different times and places with meteorological records. In Cleveland, he measured the atomic weight of oxygen relative to hydrogen as 15.879, with an uncertainty of only one part in 10000. His later research interests involved collaborative studies with physicists, notably with **Michelson** on the velocity of light and the 'ether drift' problem.

MORRIS, Desmond John
(1928–)

English ethologist, writer and television personality, born in Wiltshire. He was educated at Birmingham and the University of Oxford, where he held a research post from 1954 to 1956. He was head of Granada TV and Film Unit at the Zoological Society of London (1956–9) and subsequently curator of mammals at the Zoological Society (1959–67). He then became Director of the Institute of Contemporary Arts, London (1967–8) and later a Research Fellow at Wolfson College, Oxford (1973–81). During the period when ethology was establishing itself, he carried out important research on the ethology of several animals such as the ten-spined stickleback and tropical finches. An interest in primate behaviour led to the work for which he is widely known. In 1967 he published the semi-popular book *The Naked Ape*. In this he described the behaviour of humans using the approach and techniques of ethology. He used the same formula in several books such as *Manwatching: A Field Guide to Human Behaviour* (1977), which dealt with non-verbal signals, and *The Soccer Tribe* (1981), a study of crowd behaviour. His interest in art led him to examine the ability of chimpanzees to paint and he showed that they appcar to have an innate compositional ability.

MOŚCICKI, Igancy
(1867–1946)

Polish chemist and politician, born in Mierzanów near Płock. He studied at the Polytechnical School in Riga, Latvia, and fled to Britain after being involved in an attempt to assassinate the Governor-General of Warsaw. In 1897 he was appointed Professor of Electrochemistry at Fribourg, Switzerland, and in 1912 became Professor of Electrochemistry at the Polytechnical School at Lwów, Poland. After the end of World War I Mościcki founded research institutions and factories to manufacture fertilizers synthetically from nitric acid, derived from atmospheric nitrogen which had been fixed by electric arcs. In 1926, when the Polish national hero and former revolutionary Jósef Pilsudski overthrew the government, Mościcki supported the move and became President of the Republic, a post he retained until 1939 when he fled from the Nazis and the Russians who both invaded the country. He spent most of the rest of his life in Switzerland, and died in Versoix.

MOSELEY, Harry (Henry Gwyn Jeffreys)
(1887–1915)

English physicist, born in Weymouth and educated at Oxford University where he graduated in 1910. He then joined **Rutherford** in Manchester before returning to the University of Oxford in 1913. Using a crystal diffraction technique, he measured the frequencies of characteristic X-ray lines in the spectra of over 30 different metals, and showed that they varied regularly from element to element in order of their position in the periodic table. He suggested that these regular changes were related to the nuclear charge and would allow the atomic numbers of elements to be calculated. Discontinuities in the spectral series made it clear that a number of elements were missing from the periodic table and allowed prediction of their properties; these elements were sought and soon discovered. Moseley's work was an important step in advancing knowledge of the nature of the atom, firmly establishing that the properties of the elements are determined by atomic number rather than atomic weight. He was killed in action at Gallipoli.

MOSELEY, Henry Nottidge
(1844–91)

English naturalist, born in Wandsworth, Surrey, the son of a mathematician. He entered Exeter College, Oxford (1867), where he first read mathematics, but found it uncongenial and changed to natural sciences. Following a scholarship in Vienna in 1869 he became a medical student at University College London. In 1871 he went to Leipzig to study, and in the same year he was invited to join a government expedition to Ceylon (now Sri Lanka) to observe the solar eclipse. He made spectroscopic measurements and brought back a collection of plants on which he wrote subsequent memoirs. The following year he was appointed naturalist to the *Challenger* expedition (1872–6) led by **Wyville Thomson**, during which he collected marine fauna and land plants. On his return (1876) he accepted a fellowship of Exeter College, enabling him to work on his *Challenger* collections, writing *Notes by a Naturalist on the Challenger* (8 vols, 1879). In 1881 he became Linacre Professor of Human and Comparative Anatomy at Oxford. He was a founder of the Marine Biological Association, discovered a system of tracheal vessels in the arthropod Peripatus, identified the relationships between groups of corals and established the group hydro-corallin.

MÖSSBAUER, Rudolf Ludwig

(1929–)

German physicist, born in Munich. He was educated at the Technical University in Munich and carried out postgraduate research at the Max Planck Institute for Medical Research in Heidelberg before receiving his PhD from the Technical University in 1958. He has held professorships there and at Caltech (since 1961). Before completing his PhD, he observed what is known as the 'Mössbauer effect'. When a gamma ray interacts with nuclei in a material, it usually causes only one nucleus to recoil. This results in a broadening of the resonance in the energy spectrum. Mössbauer noticed that sometimes the resonance in the energy spectrum was very narrow. This was due to the whole of the lattice recoiling from the gamma radiation. The effect has been used to test **Einstein**'s theory of general relativity, to study the properties of nuclei and as an analytical tool in chemistry and biology. Mössbauer shared the 1961 Nobel Prize for Physics with **Hofstadter** for research into atomic structure.

MOTT, Sir Nevill Francis

(1905–)

English physicist, born in Leeds. He studied mathematics at Cambridge and became a lecturer and Fellow there working with **Rutherford**. At the age of 28 he became Professor of Theoretical Physics at Bristol (1933–54), where his group developed the theory of dislocations, defects and the strengths of crystals. It was here that he studied the electronic behaviour of 'Mott transitions'—transitions between metals and insulators. He returned to Cambridge in 1954 to become Cavendish Professor of Physics, decisively shaping the Cavendish Laboratory's research activities. In 1965 he officially retired and returned to full-time research to work on the new area of non-crystalline semiconductors, contributing considerably to the understanding of the fundamental physical and electrical properties of these materials. For his work on the electronic properties of disordered materials he won the 1977 Nobel Prize for Physics jointly with **Philip Anderson** and **van Vleck**. Mott has been one of the major theoretical physicists of this century, opening new and complex areas of solid-state physics and materials science. He was knighted in 1962.

MOTTELSON, Benjamin Roy

(1926–)

Danish physicist, born in Chicago, Illinois, and educated at Purdue University and Harvard. From Harvard he moved to the Institute of Theoretical Physics in Copenhagen (now the Niels Bohr Institute) on a travelling fellowship, where he worked with **Aage Bohr** on developing the collective model of nuclei. Using the idea proposed by **Rainwater** that the deformation of nuclei could be explained by the behaviour of some nucleons that distorted the central symmetric potential, Bohr and Mottelson combined the shell and the liquid drop nuclear models to produce a single more acceptable theory of the nucleus which could describe deformed nuclei. Bohr, Mottelson and Rainwater shared the 1975 Nobel Prize for Physics. From 1953 to 1957 Mottelson held a research position in CERN (the European centre for nuclear research in Geneva) before returning to Copenhagen where he became professor at Nordita (the Nordic Institute for Theoretical Atomic Physics). He took Danish nationality in 1973.

MOUNTCASTLE, Vernon Benjamin

(1918–)

American neurophysiologist, born in Shelbyville, Kentucky. He received his undergraduate education at Roanoke College and his medical training at the Johns Hopkins University School of Medicine. After service with the US Naval Amphibious Force, he joined the faculty at Johns Hopkins (Research Fellow in Physiology, 1946–9; Assistant Professor, 1949–59; Professor of Physiology, 1959–80), where he has been University Professor of Neuroscience since 1980. His research has been concerned with central nervous mechanisms in sensation and perception, and in the late 1950s he first showed that cells in the cerebral cortex of the anaesthetized cat responded specifically to particular types of skin stimulation. He continued to develop this work with reference to the somatic system; it proved to be a major stimulus to the work on the visual system undertaken by **Hubel** and **Wiesel** and led to a better understanding of the neurophysiological basis of higher functions such as learning and memory, much of which is included in his 1978 book (with Gerald Maurice Edelman), *The Mindful Brain*.

MUELLER, Erwin Wilhelm

(1911–77)

German-born American physicist, born in Berlin. He graduated in engineering from the Technical University in Berlin, and worked for industrial laboratories there and at the

Fritz Haber Institute until 1952, when he emigrated to the USA and joined the staff of Pennsylvania State University. He became a naturalized American in 1962. In 1936 he invented the field-emission microscope, consisting essentially of a very fine needle point in a high vacuum from which electrons are emitted; when they strike a suitably positioned fluorescent screen they produce an image of up to one million times magnification, allowing the atomic structure of certain materials to be directly observed. He followed this in 1951 with the field-ion microscope, in which helium ions are emitted in conditions of very low temperature and pressure, resulting in the first photographs affording a direct view of individual atoms in some high melting point metals and alloys and some large organic heat-stable molecules.

MUELLER, Baron Sir Ferdinand Jakob Heinrich von See **MÜLLER, Baron Sir Ferdinand Jakob Heinrich von**

MUETTERTIES, Earl
(1927–84)

American inorganic chemist, born in Elgin, Illinois. He was educated at Northwestern University, Illinois, where he received his BS degree in 1949 and was introduced to chemical research by **Basolo**. He completed his PhD in 1952 before developing a wide range of interests in inorganic chemistry while at the Central Research Department of Du Pont. There his studies included fluorine chemistry, boron hydride chemistry, coordination chemistry and organometallic chemistry. In 1978 he became Professor of Chemistry at University of California, Berkeley. His current research interests include transition metal complexes and the coordination chemistry of metal surfaces. Muetterties won the American Chemical Society Award for Inorganic Chemistry in 1965 and 1979, and was Royal Society of Chemistry Centenary Lecturer in 1981–2.

MÜLLER, (Karl) Alex(ander)
(1927–)

Swiss physicist, born in Basle. He was educated at the Swiss Federal Institute of Technology, Zürich, where he received his PhD in 1958. After five years at the Battelle Institute in Geneva (1958–63) he joined the IBM Zürich Research Laboratory working in the area of solid-state physics. For the discovery of new low-temperature superconductors, Müller shared the 1987 Nobel Prize for Physics with **Bednorz**.

MÜLLER, Baron Sir Ferdinand Jakob Heinrich von
(1825–96)

German-born Australian explorer and botanist, born in Rostock, Schleswig-Holstein. After emigrating to Australia in 1847, he was appointed government botanist for the state of Victoria in 1853, and in the next few years travelled extensively, building up a valuable collection of native flora. He was Director of Melbourne Botanic Gardens from 1857 to 1873. He attempted to make the garden a living botanical textbook, with no concessions to public amenity. This was so unpopular that he was removed from his post by public demand. A separate library and herbarium was set up for him, where he worked until his death. He also explored Western Australia and Tasmania, promoted expeditions into New Guinea and was a member of the first Australian Antarctic Exploration committee. Müller sponsored an expedition in search of the lost explorer Ludwig Leichhardt, and he organized the 1875 trip into the central desert by Ernest Giles. He was responsible for the introduction of the blue gum tree (*Eucalyptus* species) into Europe. He also published a large number of scientific works on the plants of Australia, adding 2 000 species names to its flora, although many of these names have been reduced to synonyms.

MÜLLER, Franz Joseph, Baron von Reichenstein
(1740–1825)

Austrian chemist and mineralogist, born in Nagyszeben, Transylvania (now in Romania). He studied law in Vienna and metallurgy at Schemnitz, Hungary. He worked in the state-owned salt works in Transylvania, becoming the director of state mines in the Tirol from 1775 to 1778 and director of all mining in Transylvania from 1778 to 1802. During 1802–18 he headed the committee which controlled minting and mining in the Austro-Hungarian Empire, and he was created a baron on his retirement. In 1784 he identified an impurity in a gold-bearing ore as a new element, and described its semi-metallic properties. These findings were later confirmed by **Klaproth**, who named the element tellurium. Müller died in Vienna.

MÜLLER, Hermann Joseph
(1890–1967)

American geneticist, born in New York City. He studied at Columbia University, where he received his BSc (1910) and PhD (1916), which he carried out under the supervision of **Thomas Hunt Morgan**. He spent the 1920s at the University of Texas at Austin, where he was appointed Professor of Zoology. In 1933 he moved to Leningrad to work at the Institute of Genetics there. Forced to leave the USSR following his criticisms of **Lysenko** in 1937, he joined a medical unit in Spain during the Spanish Civil War. After a period at the Institute of Animal Genetics of the University of Edinburgh, he finally returned to the USA in 1940. From 1945 to 1967 he was Professor of Zoology at the University of Indiana. With Morgan, Müller helped to establish the chromosome theory of heredity. However, his major work was on the use of X-rays to cause genetic mutations for which he was awarded the Nobel Prize for Physiology or Medicine in 1946. Concerned about the possible dangers of radiation-induced mutations, he campaigned for safety measures in hospitals and against the nuclear bomb tests. He was interested in improving human society and proposed the use of sperm banks for his programmes of 'germinal choice', which he discussed in his book *Out of the Night* (1935).

MÜLLER, Johannes Peter
(1801–58)

German physiologist, born in Coblenz. Son of a successful shoemaker, Müller entered the newly founded Bonn University in 1819 where he excelled. In 1826 he was appointed to the Chair of Physiology at Bonn, and in 1833 moved to Berlin University. His first work covered problems in animal locomotion, but he won fame for his precocious researches in embryology, in 1820 engaging in the Bonn prize question: does the foetus breathe in the womb? His experimentation on a sheep revealed changes in the blood colour entering and leaving the foetus; these indicated that it did indeed respire. He also showed early interest in the eye and vision, investigating the eye's capacity to respond not just to external but also to internal stimuli (the play of imagination as well as organic malfunctions). His later work was wide-ranging; he studied electrophysiology, the glandular system (particularly the relations between the kidneys and the genitals), the human embryo and the nervous system. He was convinced of the need for pathology to make full use of microscopy. He worked on zoological classifiction, dealing especially with marine creatures. In 1840 he proposed the law of specific nerve energies, that is, the claim that each sensory system will respond in the same way to a stimulus whether this is mechanical, chemical, thermal or electrical. However stimulated, the eye always responds with a sensation of light, the ear with a sound sensation. In Müller's view, man did not immediately perceive the external world, but only indirectly, through the effects on his sensory systems; hence introspection was integral to human biology. Müller took a profound interest in the more philosophical problems of life. He upheld the view that vitality was animated by some kind of life force that defied reduction to the purely physical; he was convinced there was a soul separable from the body. Such idealist beliefs were rejected by the next generation of more materialist German physiologists like **Ludwig** and the physicist **Helmholtz**. Müller's *Handbuch der Physiologie des Menschen* (1833–40) was extremely influential, and he himself was probably the most significant life scientist and medical theorist in Germany in the first half of the 19th century.

MÜLLER, Otto Frederick
(1730–84)

Danish biologist, born in Copenhagen. Working in the early years of microscopy, he was the first to describe diatoms and bring to notice the animal kingdom of *Infusoria*. He extended the Linnaean system of classification to describe microscopic plants and animals, and was also the inventor of the naturalist's dredge.

MÜLLER, Paul Hermann
(1899–1965)

Swiss chemist, inventor of the insecticide DDT. He was born in Olten and educated at the University of Basle. From 1925 onwards he worked at the experimental laboratory of Johann Rudolf Geigy, where he later became deputy head of pest control. He is known for his work on insecticides, particularly for discovering and developing DDT (dichlorodiphenyltrichloroethane) which was first marketed in 1942. DDT is extremely toxic to a wide variety of insects, attacking their nervous system and effectively eradicating the carriers of diseases like malaria, yellow fever and typhus, and plant pests such as the Colorado beetle. It was used in tropical areas during World War II and after the war in many parts of the world. In the 1960s it became clear that many species quickly became resistant to it, and that it is such a stable compound that its cumulative effects in the food chain are very destructive; its use was therefore discontinued. For the discovery of DDT, Müller was awarded the 1948 Nobel Prize for Physiology or Medicine. He died in Basle.

MULLIKEN, Robert Sanderson
(1896–1986)

American chemical physicist, born in Newburyport, Massachusetts. He was educated at MIT and at the University of Chicago, where he took a PhD in chemistry in 1921. After holding a fellowship at Chicago and later at Harvard between 1921 and 1925, he became assistant professor of physics at Washington Square College of New York University (1926–8) and then joined the faculty of the University of Chicago, rising to full professor by 1937. From 1961 he was Distinguished Professor of Physics and Chemistry. During World War II Mulliken worked on the development of the atomic bomb. His earliest work was on the isotope effect in the band spectra of diatomic molecules, but he is best known for his share in the creation of molecular orbital theory in the 1930s. This approach to molecular structure and chemical bonding was applied by Mulliken particularly to the systematization of electronic energy states. The theory was adopted and further developed by **Lennard-Jones**, **Coulson** and others. Mulliken also made important contributions to the development of the concept of hyperconjugation and a scale of electronegativity of the elements, and studied donor–acceptor interactions and charge transfer spectra. He was awarded the Nobel Prize for Chemistry in 1966, and became an Honorary Fellow of the Chemical Society in 1956.

MULLIS, Kary Banks
(1944–)

American biochemist, born in Lenoir, North Carolina. He was educated at Georgia Institute of Technology and the University of California. In the early 1980s, while working for Cetus Corporation in California, he discovered a technique known as the 'polymerase chain reaction' (PCR), which allows tiny quantities of DNA to be copied millions of times to make analysis practical. It is now used in a multitude of applications, including tests for the HIV virus and the bacteria which cause tuberculosis, forensic science and evolutionary studies of the genetic material in fossils. For this work Mullis was awarded the 1993 Nobel Prize for Chemistry (jointly with **Michael Smith**). Since 1988 he has worked as an independent consultant for various laboratories.

MUNK, Walter Heinrich
(1917–)

Austrian–American physical oceanographer and geophysicist, born in Vienna. He travelled to the USA in 1932 and was educated at Caltech (BS 1939, MS 1940) and the University of California, where he received a PhD in oceanography in 1947. At Scripps Institution of Oceanography he was appointed assistant professor of geophysics (1947–9), associate professor (1954–9), and professor from 1954. He was also associate director of their Institute of Geophysics and Planetary Physics (1959–82). During World War II his predictions for days of non-high sea swell and surf were used in the allied landings in North Africa and in the Pacific, and probably saved many lives. Using more accurate clocks than **Chandler**, he determined that a July day is two milliseconds shorter than a January day (1961). They attributed variations in day length to seasonal shifts in terrestrial air masses, ocean tides, the distribution of glaciers and changes within the Earth's core. With colleagues, Munk developed a new method for predicting tides, was an initiator of the 'Mohole' deep drill project, and developed ocean acoustic tomography (three-dimensional modelling of the ocean temperature field). The latter method is currently being applied in tests for global warming.

MURCHISON, Sir Roderick Impey
(1792–1871)

Scottish geologist, born in Tarradale. He became a soldier at 15 and left the army in 1816 after the Napoleonic Wars. He had no early interest in science, but following encouragement from Sir **Humphry Davy**, he devoted himself to geology from around 1824. He undertook early work on the geology of north and western Scotland, and during investigations of the geology of Devon and Cornwall he proposed the Devonian system (with **Sedgwick**). Following research in Wales and the Welsh Borders he established the Silurian system (1835) and after travels to Russia (1840–5), he proposed the Permian system. Struck with the resemblance between the Ural Mountains and Australian mountain chains, he foreshadowed the discovery of gold in Australia (1844). He was appointed the second Director of HM Geological Survey and Director of the Royal School of Mines (1855), and was a founder of the British Association for the Advancement of Science. Murchison Falls (Uganda) and Murchison river (Western Australia) are named after him. His works included *The Silurian System* (1839), *The Geology of Russia in Europe and the Urals* (1845–6) and *Siluria* (1854). He was knighted in 1845.

MURPHY, William Parry
(1892–1987)

American physician, born in Stoughton, Wisconsin, and educated at the University of Oregon and Harvard Medical School. Although postgraduate training encouraged his interest in clinical research, Murphy entered private practice, working part-time at Harvard with **Minot** (from 1925) in their investigation of the effect of raw liver in the diets of patients diagnosed as suffering from pernicious anaemia. Murphy was responsible for the clinical management of the patients, whereas Minot concentrated on the experimental design and the analysis of the blood smears. Following the announcement of their dramatic results, Murphy was awarded the 1934 Nobel Prize for Physiology or Medicine jointly with Minot and **George Whipple**. Murphy's research career never flourished. He worked on the standardization of liver extract preparations and other aspects of pernicious anaemia, but he continued in private practice and fell out with Minot over what he interpreted as the latter's lack of support for him.

MURRAY, Sir John
(1841–1914)

Scottish–Canadian marine biologist, born in Cobourg, Ontario, of parents who had left Scotland in 1834. Educated in Canada and at Edinburgh University, he is considered one of the founders of oceanography. During visits to the Arctic islands of Jan Mayen and Spitzbergen on a whaler, he made collections of marine organisms. He was then appointed one of the naturalists to the *Challenger* expedition (1872–6) which, under the leadership of **Wyville Thomson**, explored all of the oceans of the world collecting animals from all depths as well as physical data. He was mainly interested in deep-sea deposits and coral reef formation, writing *On the Structure and Origin of Coral Reefs and Islands* (1880). He assisted in the editing of the *Challenger Reports*, of which there were 50 volumes, and took over as editor-in-chief from Wyville Thomson when the latter's health failed. He was in charge of the important biological collection from the expedition maintained at Edinburgh. Among his publications are a narrative of the expedition and *The Depths of the Ocean* (1912, with J Hjort). Murray also surveyed the depths of the Scottish freshwater lakes. He was knighted in 1898.

MURRAY, Joseph Edward
(1919–)

American surgeon, born in Milford, Massachusetts. Murray attended Harvard Medical School and graduated MD in 1943. He spent 1944–7 in military service at the Valley Forge General Hospital in Pennsylvania. After discharge from the army he joined the staff of the Peter Bent Brigham Hospital in Boston, where he became Chief Plastic Surgeon (1951–86). From the 1940s a team at the Brigham was attempting to overcome the reactions of the immune system to transplanted kidneys. In 1954 Murray and his colleagues first successfully transplanted a kidney between identical twins, which greatly stimulated the clinical and laboratory studies. Animals were subjected to X-rays and drugs in attempts to suppress the immunological reactions. A successful transplant was performed between non-identical twin brothers using the X-ray treatment. In 1961 the first transplant using an unrelated kidney and the immunosuppressive drug azathioprine took place. By the following year this technique was shown to be successful. Soon kidney transplants became common, and systems were established for finding donors. Murray was awarded the 1990 Nobel Prize for Physiology or Medicine with **Donnall Thomas**.

MUSHET, Robert Forester
(1811–91)

English metallurgist, born in Coleford, Gloucestershire. He assisted with and continued his father's researches into the manufacture of iron and steel. In 1856 **Bessemer** patented his new steel-making process, but he soon encountered a major problem with the overoxidation or burning of the metal inside the converter, which removed most of the vital carbon. Working in the Forest of Dean, Mushet correctly identified Bessemer's problem and then solved it by adding his patented 'triple compound' (a type of high-carbon ferromanganese) and then later spiegeleisen. Only through his discovery, also patented in 1856, of the beneficial effects of adding ferromanganese to the blown steel, did the Bessemer process become a commercial success. Ironically, Mushet's patent was allowed to lapse and his vital discovery became common property from which he derived little or no benefit. Bessemer repudiated the validity of Mushet's patent, and only later bestowed upon him a small pension. Mushet was much more successful with his invention of 'R Mushet's Special' steel in 1868, a self-hardening tungsten alloy steel which was later extensively produced in Sheffield, where it spawned a whole family of tool steels.

MUSPRATT, James
(1793–1886)

British chemist and industrialist, born in Dublin of British parents. He was apprenticed to a druggist and then fought in the Peninsular War, writing a record of his dramatic experiences. Back in Dublin he began a small chemical manufacturing business, but when the British salt tax was lifted he moved to Liverpool and began manufacturing soda-ash by the **Leblanc** process. The Liverpool district offered every advantage as both salt and coal were available locally, and there were ships and canals to take the soda-ash to the expanding industries of 19th-century Britain. Muspratt prospered and in 1828 he and Josias Gamble set up a new works at St Helens. In the mid-1800s Muspratt parted with Gamble and founded large new factories employing several thousands of people at Widnes and Flint. He died at Seaforth Hall near Liverpool.

MUSSCHENBROEK, Pieter van
(1692–1761)

Dutch physicist, born in Leiden and educated at the university there. Professor of Physics successively at Duisburg, Utrecht and Leiden (1740–61), he excelled as a teacher and experimenter, and his fame grew with the publication of his lectures in volumes which became larger with each new edition. He invented a pyrometer and in 1746 discovered the principle of the Leyden jar, an early form of electrical capacitor.

MUYBRIDGE, Eadweard,
originally **Edward James Muggeridge**
(1830–1904)

English-born American photographer and inventor, born in Kingston-on-Thames. He emigrated to California in 1852 and became a professional photographer, and eventually chief photographer to the US government. He invented a shutter which allowed an exposure of 1/500 s, and using a battery of between 12 and 24 cameras, was able to show that a trotting horse had all of its feet off the ground at times. In 1880 he devised the zoopraxiscope, a precursor of cinematography, in which the photographs were printed on a rotating glass disc. He used this to demonstrate the trotting horse and after exhibiting it in his Zoopraxographical Hall in Chicago (1893) went on tour in the USA and Europe. In 1884–5 he carried out an extensive survey of the movements of animals and humans for the University of Pennsylvania, publishing the results as *Animal Locomotion* (1887).

N

NÄGELI, Karl Wilhelm von

(1817–91)

Swiss botanist, born in Kilchberg, near Zürich. Professor at Munich (from 1858), he was one of the early writers on evolution. He investigated the growth of cells and originated the micellar theory relating to the structure of starch grains, cell walls and other cell organelles. According to this theory, the molecules making up these substances lie together in a regular arrangement when dry. In contact with water, the molecules become surrounded by an aqueous envelope. He originated the concept of cell organelles, describing the membrane surrounding chloroplasts, and their increase by division. He distinguished between two types of cell formation—vegetative and reproductive—and observed the cell nucleus dividing into two parts before the cell itself divided. His description of 'transitory cytoblasts' was almost certainly the first observation of chromosomes. He emphasized that the cell contents secrete the cell walls at their surfaces, so that the wall is a secondary structure. He distinguished between meristem and permanent tissue, and between primary and secondary meristems. His most significant advance was probably the recognition of phloem as a fundamental tissue, containing sieve-tubes, and forming vascular bundles.

NAKAYA, Ukichiro

(1900–62)

Japanese glaciologist, born in Ishikawa. He graduated from Tokyo University in 1925 and entered the Institute of Physical and Chemical Research. In 1928 he was sent to Europe to study, and he returned to Japan as associate professor at Hokkaido University, where he was appointed full professor in 1932. He remained at Hokkaido, founding the Institute of Low Temperature Science and the department of geophysics there. He investigated snow crystals, the conditions required to form them and introduced the 'Nakaya diagram', showing the classification of crystals. During fieldwork in Greenland every summer between 1957 and 1960 he compared ice and snow crystals on the ice cap. He published *Snow Crystals* in 1957.

NAMIAS, Jerome

(1910–)

American meteorologist, born in Bridgeport, Connecticut. Educated at MIT, Namias showed an early enthusiasm for meteorology which helped him to become an assistant to **Rossby** at MIT. He carried out some original work on sea breezes for Rossby and became interested in subsidence. He realized that the Rossby long waves in the upper atmosphere were more persistent than those associated with baroclinic systems and therefore perhaps more predictable, allowing weather forecasting for up to five days ahead. In 1936 Rossby offered Namias a teaching post at MIT, and after pursuing research on isentropic analysis, he was made chief of the extended forecast division of the Weather Bureau. He held this position until 1971 and during this time the initial five-day forecasting was extended to monthly forecasts of temperature and rainfall for much of the northern hemisphere, and eventually to seasonal forecasts. Namias personally carried out much of the necessary research and was aware that the atmospheric Rossby waves were always likely to be modified by the variations in the underlying surface such as anomalous ocean temperatures, snow cover and deserts. He wrote almost 190 scientific papers, many relating anomalous sea temperature patterns to atmospheric phenomena. He realized that ocean–atmosphere interactions were a two-way process and did much to inspire similar work in Britain. In 1971 he moved to the Scripps Institute of Oceanography of the University of California and continued his research, particularly into methods of seasonal forecasting.

NANSEN, Fridtjof

(1861–1930)

Norwegian explorer, biologist and oceanographer, born in Store-Frøen. He studied at Christiania (now Oslo) University and later at Naples. In 1882 he made a voyage into the Arctic regions in the sealer *Viking*, and on his return was made keeper of the natural history department of the museum at Bergen. In the summer of 1888 he made an adventurous journey across Greenland from east to west. His great achievement was the partial accomplishment of his scheme for reaching the North Pole by letting his ship get frozen into the ice north of Siberia and drift with a current setting towards Greenland. He started in the *Fram*, built for the purpose, in August 1893, reached the New Siberian islands in September, made fast to an ice floe, and drifted north to 84 degrees 4 minutes on 3 March 1895. There he left the *Fram* and pushed across the ice, reaching the highest latitude till then attained, 86 degrees 14 minutes north, on 7 April and over-wintering in Franz Josef Land. Professor of Zoology (from 1897) and Professor of Oceanography (from 1908) at Christiania, he wrote on a wide range of scientific topics, explained the nature of wind-driven sea currents, and made improvements to the design of oceanographic instruments. He furthered the cause of Norwegian independence from Sweden, and was the first Norwegian ambassador in London (1906–8). In 1922 he was awarded the Nobel Peace Prize for Russian relief work, and he did much for the League of Nations.

NAPIER, John
(1550–1617)

Scottish mathematician, the inventor of logarithms, born at Merchiston Castle, Edinburgh. He matriculated at St Andrews University in 1563, travelled on the Continent, and settled down to a life of literary and scientific study. A strict Presbyterian, he published religious works and believed in astrology and divination, and for defence against Philip II of Spain, he devised warlike machines (including primitive tanks). He described his famous invention of logarithms in *Mirifici Logarithmorum Canonis Descriptio* ('Description of the Marvellous Canon of Logarithms', 1614); formulated to simplify computation, his system used the natural logarithm base *e*, but was modified soon after by **Briggs** to use the base 10. Napier also devised a calculating machine, using a set of rods, called 'Napier's bones', which he described in his *Rabdologiae* (1617).

NATHANS, Daniel
(1928–)

American microbiologist, born in Wilmington, Delaware, of Russian-Jewish immigrants. He graduated from the University of Delaware in 1950 before pursuing postgraduate studies at Washington University School of Medicine, where he obtained a medical degree in 1954. From 1955 to 1957 he worked at the National Cancer Institute of the National Institutes of Health. As a guest at the Rockefeller University, New York (1959–62), he researched protein biosynthesis. From 1962 he was Professor at Johns Hopkins University and worked on SV40 virus. Nathans pioneered the use of restriction enzymes (which had recently been isolated by **Hamilton Smith**) to fragment DNA molecules, enabling him to make the first genetic map (of the circular SV40 DNA) and to identify the location of specific genes on the DNA. For this work he shared the 1978 Nobel Prize for Physiology or Medicine with Smith and **Werner Arber**.

NATTA, Giulio
(1903–79)

Italian chemist, born in Imperia, near Genoa. He first studied mathematics at the University of Genoa but switched to chemical engineering as a student at Milan Polytechnic Institute, where he later became assistant lecturer in chemistry. In 1933 he was appointed Professor of Chemistry at the University of Pavia, and two years later he became Director of the Institute of Physical Chemistry in Rome. He was appointed Professor of Chemistry in Turin in 1937 and from 1938 held the post of Professor and Director of the Milan Institute of Industrial Chemistry until his retirement in 1973. His early work concerned heterogeneous catalysts used in a number of industrial processes, and in 1938 he initiated a programme for the production of artificial rubber. His most important work used the organometallic catalysts developed by **Ziegler** for the polymerization of propene to give polypropylene containing uniformly oriented methyl groups. Such polymers have a high melting point and great strength, and have become very important in industrial and commercial fields. In 1963 he shared the Nobel Prize for Chemistry with Ziegler.

NAVIER, Claude Louis Marie Henri
(1785–1836)

French civil engineer, born in Dijon. His father died when he was 14 and he came under the influence of his mother's uncle, the eminent civil engineer Emiland-Marie Gauthey. Educated at the École Polytechnique and the École des Ponts et Chaussées, for much of his life he taught at one or the other of these schools, being principally occupied in developing the theoretical basis of structural mechanics and the strength of materials, as well as the work done by machines. When Gauthey died leaving the unfinished manuscript of a treatise on bridges he completed the work, publishing it with editorial notes in three volumes between 1809 and 1816. He was responsible for the construction of a number of elegant bridges over the river Seine, but one of his most ambitious designs, a suspension bridge in Paris, encountered both engineering and political problems, to such an extent that it was dismantled just before completion. He published a number of important treatises on various aspects of structural mechanics, emphasizing the importance of being able to predict the limits of elastic behaviour in structural materials; his formulae represent some of the greatest advances in structural analysis ever made.

NEEDHAM, Joseph

(1900–)

English biochemist and historian of Chinese science, born in London. He was educated at Cambridge, and remained there as a university demonstrator in biochemistry (1928–33), reader (1933–66), and Master of Gonville and Caius College (1966–76). Inspired by the earlier work of H Spelman, **Julian Huxley** and **de Beer**, he experimented extensively on the nature of the organizers of the so-called 'morphogenic field' in amphibian development. The presence of a hormone (possibly a sterol) was inferred from the observation that dead tissue from the head end of an embryo would organize a second head if transplanted to the head end of a live embryo, but not if transplanted to the tail end. A first grade organizer induced the neural axis, and a second grade organizer produced the eye-cup and lens. Although no firm findings emerged, Needham's work provided the foundation for further studies. His early publications such as *Man a Machine* (1927), *The Sceptical Biologist* (1929) and *Chemical Embryology* (1931) were scientifically oriented, but his historical preoccupations emerged in *A History of Embryology* (1934) and *History is on Our Side* (1945). During World War II he headed the British Scientific Mission in China and became Scientific Counsellor at the British Embassy there. From 1946 to 1948 he was Director of the Department of Natural Sciences, UNESCO. He also published *Chinese Science* (1946) and *Science and Civilisation in China* (12 vols, 1954–84), an exposition of the foremost significance in the history of science and Chinese historical achievement. In addition he published on the history of acupuncture, Korean astronomy and clocks, among a vast body of other work.

NÉEL, Louis Eugène Félix

(1904–)

French physicist, born in Lyons. A graduate of the École Normale Supérieure, he was later Professor of Physics at Strasbourg University (1937–40). In 1940 he moved to Grenoble and became the driving force in making it one of the most important scientific centres in France, becoming Director of the Centre for Nuclear Studies there in 1956. His research has been concerned with magnetism in solids. At a time when three states of magnetism had been identified and explained (dia-, para- and ferromagnetism), he postulated a fourth, antiferromagnetism (1936). He argued for a crystal model in which two lattices with their magnetic fields acting in opposite directions are interlaced. Their opposing magnetic fields would cancel, leaving the crystal with little observable magnetic field. His predictions were verified by experiment in 1938, and fully confirmed by neutron diffraction techniques in 1949. He later explained the strong magnetism found in ferrite materials such as magnetite (1948), demonstrating that if the magnetic field of one of the two lattices (mentioned above) were stronger than the other there would be an observable magnetic field. His work on ferrimagnetic materials saw great application in the coating of magnetic tape, the permanent magnets of motors and the magnetic storage media used by computers. He was awarded the 1970 Nobel Prize for Physics jointly with **Alfvén**. He has also studied the past history of the Earth's magnetic field.

NEHER, Erwin

(1944–)

German biophysicist, born in Landsberg. He was educated in physics at the Technical University of Munich and the University of Wisconsin, gaining his PhD from Munich in 1970. He joined the Max Planck Institute for Psychiatry in Munich (1970–2) and then moved as a research associate to the Max Planck Institute for Biophysical Chemistry in Göttingen in 1972, becoming Director of the Membrane Biophysics Department in 1983. In 1976, after a sabbatical year at Yale University he succeeded, with **Sakmann**, in recording the electric currents through single ion channels in biological membranes, the existence of which were suggested by the work of Sir **Alan Hodgkin** and Sir **Andrew Huxley**. They developed the 'patch-clamp' technique, touching a cell membrane with the tip of a glass pipette filled with a saline solution, and by applying suction through the pipette, creating a seal which isolated a small section of the membrane. This permitted precise biophysical measurements to be made over a discrete area, a method that has revolutionized cell physiology. In 1991 Neher and Sakmann shared the Nobel Prize for Physiology or Medicine for this work.

NERNST, Walther Hermann

(1864–1941)

German physical chemist, born in Briesen, West Prussia. He studied physics at the universities of Zürich, Berlin, Graz and Würzburg (under **Kohlrausch**), and in 1887 he became assistant to **Ostwald** at Leipzig. In

1891 he moved to Göttingen, first as associate professor and from 1894 as Professor of Physical Chemistry. He succeeded Hans Landolt in the Chair of Physical Chemistry at the University of Berlin in 1905. During World War I he engaged in military activities, including gas warfare. In 1922–4 he was President of the Physikalisch Technische Reichsanstalt, but he then returned to the Chair of Physics at the university. He retired in 1933, being out of favour with the Nazi regime. Nernst is regarded as one of the co-founders of physical chemistry, along with Ostwald, **van't Hoff** and **Arrhenius**. His earliest researches were in electrochemistry, and his development of the theory of electrode potential and the concept of solubility product were particularly important. He devised experimental methods for measuring dielectric constant, pH, and other physicochemical quantities. The electrochemical work led to a special interest in thermodynamics and in 1906 he enunciated his heat theorem, which has come to be regarded as a statement of the third law of thermodynamics. This enables equilibrium constants for chemical reactions to be calculated from heat data. He was later much concerned with the quantum theory, and in particular with photochemistry. Nernst devised several inventions, including a metal oxide 'glower' for light bulbs, which was a great improvement on the carbon filament and was widely used until supplanted by the tungsten filament. Nernst received the Nobel Prize for Chemistry in 1920. He was made an Honorary Fellow of the Chemical Society as early as 1911 and became a Foreign Member of the Royal Society in 1932.

NEUGEBAUER, Gerald

(1932–)

American astronomer, born in Göttingen, Germany. Educated at Cornell University and Caltech where he received his PhD in 1960, he later joined the US army and was stationed at the Jet Propulsion Laboratory (1960–2). After returning to work at Caltech and at Mount Wilson and Palomar observatories, he became Professor of Physics at Caltech (1970) and Director of Palomar Observatory (1981). With colleagues, he was responsible for carrying out the first extensive infrared map of the heavens in the 1960s, revealing all strong astronomical infrared emission of wavelength 2.2 micrometres in the sky above Mount Wilson Observatory. The 'Two Micron Survey' produced dramatic results: around 20 000 new infrared sources were discovered, and most did not seem to coincide with previously observed visible sources. New and curious objects were revealed, such as objects thought to be stars in the process of formation, immersed in dense dust clouds. Working with E E Becklin, Neugebauer came across a strange infrared source radiating intensely in a giant cloud of molecular gas in the Orion nebula. Known as the Becklin–Neugebauer object, it is now thought to be a young massive star which is blowing gases outwards at high speed and has supported fusion reactions for perhaps only 10 000–20 000 years. As a member of NASA's Astronomical Missions Board, Neugebauer has played an important role in interplanetary missions such as the Mariner and Viking programmes, and in the design of new infrared telescopes.

NEUMANN, John (Johann) von

See **VON NEUMANN, John (Johann)**

NEWCOMB, Simon

(1833–1909)

Canadian-born American astronomer, born in Wallace, Nova Scotia, the son of a country school teacher. He had little formal education but, after a few years as an apprentice herbalist, he discovered his talent for mathematics while browsing in the libraries of Washington DC. Through his own efforts he equipped himself sufficiently well that he was accepted as a computer at the American Nautical Almanac Office at Washington in 1857, achieving also a degree from the scientific school of Harvard in 1858. He became Professor of Mathematics in the US navy (1861–97) with charge of the naval observatory at Washington. From 1877 he edited the American *Nautical Almanac* and was additionally (1884–93, 1898–1900) Professor of Mathematics and Astronomy at Johns Hopkins University. Newcomb's major work, begun in 1879 and continued throughout his life, was the recalculation of the constants required for the preparation of ephemerides and the drawing up of immense tables of the motions of the planets. He was responsible for the worldwide adoption of a standard system of constants by almanac makers, which served until the middle of the 20th century.

NEWLANDS, John Alexander Reina

(1837–98)

English chemist, born in London. He spent a year at the Royal College of Chemistry and fought with Garibaldi in 1860. From 1868 to 1888 he was chief chemist to a sugar refinery at Victoria Docks, London, and later set up as an independent analyst and consultant. By 1863 Newlands had begun to build on earlier observations by **Döbereiner** and others that there was a relationship between the chemical properties of elements and their atomic weight. In 1865 he drew up a table of 62 elements arranged in eight groups in ascending order of atomic weight in order to illustrate what he described as the 'law of octaves'. Each element in the table was numbered, with hydrogen (the lightest element) as 1, osmium (the heaviest element then known) as 51, and elements which apparently had the same atomic weight sharing the same number. Newlands's hypothesis was given a rough reception, because under his scheme, not all chemically similar elements fell into appropriate groups. It was to be **Mendeleyev** who made the critical leap forward when he realized that spaces should be left for undiscovered elements, thus allowing the known elements to fall into groups which demonstrated a true periodicity. After Mendeleyev published his periodic table, Newlands claimed priority and was eventually awarded the Davy Medal of the Royal Society in 1887. He died in London.

NEWMAN, originally NEUMANN, Maxwell Herman Alexander
(1897–1984)

English mathematician, born in Chelsea, London, the son of a German immigrant. Newman was educated at the City of London School and St John's College, Cambridge, where he read mathematics and became a lecturer in 1927. He was a pioneer of combinatory (later called geometric) topology. During World War II he worked on the code-breaking machine, Colossus, at Bletchley Park, with amongst others **Turing**. This aroused his interest in general-purpose computers. As Professor of Mathematics at Manchester University (1945–64), he was influential in persuading the authorities that a computer should be built there. The Manchester University Mark I computer, engineered by Sir **Frederic Calland Williams** and **Kilburn**, owed something to Newman's influence. He was elected FRS in 1939.

NEWTON, Alfred
(1829–1907)

English zoologist, born in Geneva. In 1866 he was appointed the first Professor of Zoology and Comparative Anatomy at Cambridge. He made visits to Lapland, Spitzbergen, the West Indies and North America on ornithological expeditions, and was instrumental in having the first Acts of Parliament passed for the protection of birds. His ornithological writings included *A Dictionary of Birds* (1893–6) and he was editor of *Ibis* (1865–70) and *Zoological Record* (1870–2).

NEWTON, Sir Isaac
(1642–1727)

English scientist and mathematician, born in Woolsthorpe, Lincolnshire. Educated at Grantham Grammar School and Trinity College, Cambridge, in 1665 he committed to writing his first discovery on fluxions (an early form of differential calculus); and in 1665 or 1666 the fall of an apple in his garden suggested the train of thought that led to the law of gravitation. He turned to study the nature of light and the construction of telescopes. By a variety of experiments upon sunlight refracted through a prism, he concluded that rays of light which differ in colour differ also in refrangibility—a discovery which suggested that the indistinctness of the image formed by the object-glass of telescopes was due to the different coloured rays of light being brought to a focus at different distances. He concluded (correctly for an object-glass consisting of a single lens) that it was impossible to produce a distinct image, and was led to the construction of reflecting telescopes; and the form devised by him is that which reached such perfection in the hands of **William Herschel** and **Rosse**. Newton became a Fellow of Trinity College, Cambridge (1667), and Lucasian Professor of Mathematics in 1669. By 1684 he had demonstrated the whole gravitation theory, which he expounded first in *De Motu Corporum* (1684). Newton showed that the force of gravity between two bodies, such as the Sun and the Earth, is directly proportional to the product of the masses of the bodies and inversely proportional to the square of the distance between them. He described this more completely in *Philosophiae Naturalis Principia Mathematica* (1687); this great work was edited and financed by **Halley**. In the *Principia* he stated his three laws of motion: (1) that a body in a state of rest or uniform motion will remain in that state until a force acts on it; (2) that an applied force is directly proportional to the acceleration it induces, the constant of proportionality being the body's mass ($F = ma$); and (3) that for every 'action' force which one body exerts on another, there is an equal and opposite 'reaction' force exerted by the second body on the first. He also wrote *Opticks* (1703). The part he took in defending the rights of the university against the illegal encroachments of James II procured him a seat in the Convention parliament (1689–90). In 1696 he was appointed Warden of the Mint, and was Master of the Mint from 1699. He again sat in parliament in 1701 for his university. He solved two celebrated problems proposed in June 1696 by **Jean Bernoulli**, as a challenge to the mathematicians of Europe; and performed a similar feat in 1716, by solving a problem proposed by **Leibniz**. He superintended the publication of **Flamsteed**'s *Greenwich Observations*, which he required for the working out of his lunar theory — not without much argument between himself and Flamsteed. He was also involved in several priority disputes with **Hooke** in the latter's capacity as Secretary of the Royal Society. In the controversy between Newton and Leibniz as to priority of discovery of the differential calculus and the method of fluxions, Newton acted secretly through his friends. The verdict of science is that the methods were invented independently, and that although Newton was the first inventor, a greater debt is owing to Leibniz for the superior facility and completeness of his method. In 1705 he was knighted by Queen Anne; he lies buried in Westminster Abbey. Newton also devoted much time to the study of alchemy and theology, and among a mass of more or less worthless material on these subjects he left some substantial discourses on transmutation, a remarkable manuscript on the prophecies of Daniel and on the Apocalypse, a history of creation, and a large number of miscellaneous tracts.

NICHOLS, Ernest Fox
(1869–1924)

American physicist, born in Leavenworth, Kansas, inventor of the Nichols radiometer. He graduated from the Kansas State College of Agriculture (1888), did postgraduate work at Cornell University (1888–92), and became an associate professor at Colgate University (1892–8). In 1920, after periods at a number of American institutions he became director of research in pure science at the laboratories of the National Electric Light Association, Cleveland, a post he held until his death. During a visit to Berlin (1894–6), with the help of **Ernst Pringsheim**, he built a cleverly designed radiometer that was considerably more sensitive than any other available. He used the Nichols radiometer to make measurements in the infrared range of wavelengths in investigations of reflection and refraction, and succeeded in bridging the gap between infrared radiation produced by thermal means and radio waves produced by electrical means. He confirmed the prediction made by **Maxwell**'s laws that light exerts a pressure, and made measurements on the relative temperatures of stars. He died of heart failure whilst giving a paper at a meeting of the US National Academy of Sciences.

NICHOLSON, (Edward) Max
(1904–)

English bureaucrat, ornithologist, and conservation pioneer. He was educated at the University of Oxford, and was one of the founders of the Oxford Ornithological Society (1921); this led to the establishment of the Oxford Bird Census and hence to the British Trust for Ornithology (1932) and the Edward Grey Institute for Field Ornithology (1938). Nicholson was Director-General of the Nature Conservancy (1952–65); in this role he stimulated and established conservation work in the UK and throughout the world. He was the author of *Birds in England* (1926), *How Birds Live* (1927), *Birds and Men* (1951), *The Environmental Revolution* (1970) and *The New Environmental Age* (1987).

NICHOLSON, William
(1753–1815)

English physicist and inventor, born in London. Educated in Yorkshire, at the age of 16 he entered the service of the East India Company. Returning to England in 1776 he became an agent for Josiah Wedgwood, but soon settled in London where he opened a school of mathematics and engaged in a variety of scientific pursuits. He wrote a number of popular textbooks and translated others by such authors as **Fourcroy** and Jean Antoine Chaptal; from 1797 to 1815 he published the *Journal of Natural Philosophy, Chemistry, and the Arts* . . . ; he also engaged in the work of a patent agent, and as engineer to several water supply undertakings in London and south-east England. Only a few months after the first primitive electric battery had been constructed in 1799 by **Volta**, he built the first voltaic pile in England, and soon afterwards noticed that when the ends of the leads from the battery were immersed in water, bubbles of gas were produced. The results of the experiments he performed in collaboration with Anthony Carlisle were reported in his journal in July 1800 and excited a great deal of interest at the time. Among his many inventions were the hydrometer named after him, and a method for printing on linen and other materials, patented in 1790 but never put into practice.

NICOL, William
(1768–1851)

Scottish geologist and physicist, born in Edinburgh, where he lectured in natural philosophy at the university. In 1828 he invented the Nicol prism, which utilizes the doubly refracting property of Iceland Spar, and which proved invaluable in the investigation of polarized light. It was also of fundamental importance in studies of minerals under the microscope. He also devised a new method of preparing thin sections of rocks for the microscope, by cementing the specimen to the glass slide and then grinding it until it was possible to view it by transmitted light, thus revealing the mineral's properties and internal structure. The technique was initially developed to examine the minute details of fossil and recent wood, and Nicol himself prepared a large number of thin sections to this end. Many of these sections were described by Henry Witham in his *Observations on Fossil Vegetables* (1831). Nicol's reluctance to publish delayed the widespread use of thin sections for some 40 years until **Sorby** and others introduced them into petrology.

NICOLLE, Charles Jules Henri
(1866–1936)

French physician and microbiologist, born in Rouen into a medical family, and educated in Rouen and Paris. His aptitude for research was stimulated at the Pasteur Institute by **Émile Roux** and **Metchnikoff**. From 1892 to 1902 he tried unsuccessfully to establish a research laboratory in Rouen. He then became Director of the Pasteur Institute in Tunis, which he and his colleagues turned into a leading research centre, working on the mode of spread, prevention and treatment of a number of diseases, including leishmaniasis, toxoplasmosis, Malta fever and typhus. His discovery that typhus is spread by lice (1909) had important implications during World War I and led to his award, in 1928, of the Nobel Prize for Physiology or Medicine. From 1932 he lectured each year at the Collège de France in Paris, but maintained his base in Tunis. He was a man of wide erudition who wrote novels, short stories and philosophical works.

NIEL, Cornelis Bernardus Kees Van
See **VAN NIEL, Cornelis Bernardus Kees**

NIEPCE, Joseph Nicéphore
(1765–1833)

French chemist and pioneer of photography, born in Chalon-sur-Saône. He served under Napoleon and in 1795 became administrator of Nice. With enough inherited wealth to support himself, he was able to devote himself to research from 1801 onwards. He experimented with the new technique of lithography, using a camera obscura to project an image onto a wall, then tracing round the image in the time-honoured fashion. Being a poor draughtsman, he decided to look for ways of fixing the image automatically. In 1822, using silver chloride paper and a camera, he achieved a temporary image of the view outside his workroom window, but could not fix it. In 1826 he succeeded in making a permanent image using a pewter plate coated with bitumen of Judea, an asphalt which hardens on exposure to light. This historic negative, which Niepce termed a 'heliograph', is now in the Gernsheim Collection at the University of Texas at Austin. From 1829 Niepce collaborated with **Daguerre** in the search for materials which would reduce the exposure time but he died, in Saint-Loup-de-Varennes, before any progress was made. Although Niepce is known principally for his photographic work, he was active in other fields; he invented a method to extract sugar from pumpkin and beetroot, and with his brother Charles, built the Pyreolophore motor.

NIEUWLAND, Julius Arthur
(1878–1936)

Belgian–American chemist whose researches led to the synthesis of neoprene, the first commercially successful synthetic rubber. He was born in Hansbeke, Belgium, and after his parents emigrated to the USA he was educated at the University of Notre Dame, South Bend, Indiana. He then studied for the priesthood at the Catholic University of America, Washington DC. In 1904 he gained his doctorate with a study of acetylene which was to be fundamental to his subsequent research. He spent his teaching career at Notre Dame, as Professor of Botany from 1904 to 1918 and Professor of Organic Chemistry from 1918 to 1936. While studying the reaction between acetylene and arsenic trichloride in the course of his doctoral work, Nieuwland made a highly toxic gas and discontinued his research because of its deadly nature. It was subsequently developed into a poison gas (lewisite) and used in World War I. In 1920 Nieuwland discovered that, in the presence of a catalyst, acetylene could be polymerized—that is, its molecules could be condensed into much larger molecules to form quite different compounds. The product of the polymerization was a mixture of substances including divinyl-acetylene. In 1925 Nieuwland began to develop this reaction in collaboration with the chemists of Du Pont de Nemours, synthesizing neoprene (at first known as Duprene) in 1929. Synthetic rubber was put on the market in 1932. Nieuwland died in Washington DC.

NIRENBERG, Marshall Warren
(1927–)

American biochemist, born in New York City. He was educated at the universities of Florida and Michigan, and worked from 1957 at the National Institutes of Health in Bethesda, Maryland. Following the demonstration of the model of DNA by **James Watson** and **Crick** in 1953, it had been proposed that there are different combinations of three nucleotide bases (triplets or 'codons') in nucleic acid chains in DNA and RNA, each coded for a different amino acid in the biological synthesis of proteins, and that this was the fundamental process in the chemical transfer of inherited characteristics. However, the precise nature of the code remained unknown; there were 64 possible combinations of bases, and only 20 amino acids to be coded. Nirenberg attacked the problem of the 'code dictionary' by synthesizing a nucleic acid with a known base sequence, and then finding which amino acid it converted to protein. With his success, **Khorana** and others soon completed the task of deciphering the full code, which Nirenberg showed was 'universal' by examining *Escherichia coli*, a toad and the guinea pig as representative living organisms (1967). Nirenberg, Khorana and **Holley** shared the Nobel Prize for Physiology or Medicine in 1968 for this work.

NOBEL, Alfred

(1833–96)

Swedish chemist and industrialist, the inventor of dynamite and the founder of the Nobel prizes. Descended from several generations of scientists and inventors, Nobel was born in Bernhard, but spent much of his childhood in St Petersburg, Russia, where his father had an explosives factory and was experimenting with an underwater mine. Alfred was educated by tutors and became fluent in several languages. He studied chemistry in Paris, worked in the USA with Swedish-born John Ericsson, and returned to Sweden in 1859. In 1863 he and his father began to investigate nitroglycerine, an explosive oil that had been prepared by **Sobrero** in 1847. In 1863 Nobel invented the Nobel patent detonator, a metal cap containing an explosive compound, mercury fulminate. When the mercury was detonated, the shock detonated the main charge. The principle of detonating explosives by an initial smaller shock, rather than by heat, transformed blasting techniques. In 1865 Nobel opened the first factory for the manufacture of nitroglycerine but it shortly blew up, killing five people including his younger brother. Forbidden by the government to reopen it, he was reduced at one point to continuing his experiments on a barge. He discovered by chance that kieselguhr, a porous silaceous earth used as a packing material, would absorb large quantities of nitroglycerine and make it safer to handle. The result, which Nobel named dynamite, was first patented in 1867. Nobel later found that nitroglycerine formed a colloidal solution with nitrocellulose in the form of guncotton and that this was more powerful than dynamite, less subject to shock, and resistant to water and corrosion. It became known as gelignite. Nobel's other important achievement, around 1879, was to produce a practically smokeless powder (Ballistite) by adding 10 per cent of camphor to dynamite. He patented many other inventions—he made synthetic gutta-percha and mild steel for armour-plating—and he also had large holdings in oilfields in Russia. He amassed a huge fortune and spent much of his life travelling from country to country monitoring his interests. Towards the end of his life he investigated problems in chemistry and biology at his laboratory in San Remo. He hoped that his explosives would decrease, rather than increase, the chances of war throughout the world and in many respects remained a pacifist. He never married and left his total wealth to fund the rich prizes which bear his name. Since 1901 these have been awarded annually for physics, chemistry, physiology or medicine, literature and contributions to peace (the prize for economics was not established until 1968 and was funded by the Bank of Sweden). Nobel died in San Remo. The transuranic element synthesized in 1958 with an atomic number of 102 was named 'nobelium' in his honour.

NOBILI, Leopoldo

(1784–1835)

Italian physicist, born in Trassilico. After serving for some time in the Italian army as an artillery captain, he was appointed Professor of Physics at Florence, and engaged in research into electricity. He developed the theory that an electric current is a flow of heat or caloric, and took the view that the generation of current by a voltaic pile was not the result of the release of static electricity by electrochemical decomposition, but that the release itself was due to the presence of an electric current; this conclusion was an essential step towards the ultimate acceptance of **Maxwell**'s electrodynamic theories. He invented the thermoelectric couple, in which an electric potential is created by the conjunction of two metals at different temperatures; using a number of thermocouples in series he created the thermopile, which could be used in the measurement of radiant heat. To meausre very small electric currents more accurately, he devised the astatic galvanometer, in which the instrument's moving magnet is shielded from the effects of the Earth's magnetic field.

NODDACK, Ida Eva Tacke
(1896–) and
Walter Karl Friedrich
(1893–1960)

German chemists. Ida Tacke was born in Lackhausen and educated at the University of Berlin-Charlottenburg. Walter Noddack was born in Berlin and educated at Berlin University. They worked together at the Physikalisch Technische Reichsanstalt and continued their collaboration after their marriage in 1926. In 1935 they moved to the Institute of Physical Chemistry at the University of Freiberg; Walter taught at the University of Strasbourg, France, during World War II and they both ended their careers at the Institute of Geochemical Research, Bamberg, Germany. In 1925 they discovered rhenium (atomic number 75) by X-ray spectroscopy. The same year they announced the discovery of element 43 which they called masurium. Its existence was debated until 1937 when Carlo Perrier and **Segrè** demonstrated its presence in a sample of molybdenum which had been bombarded with deuterons, and named it technetium. It is now known that technetium exists in minute quantities in the Earth's crust as a decay product of uranium, and in 1952 it was shown to exist in some stars. In photochemistry the Noddacks investigated the physical properties of sensitizing colouring substances, photochemical problems in the human eye and other subjects.

NOETHER, (Amalie) Emmy
(1882–1935)

German mathematician, born in Erlangen. The daughter of the mathematician Max Noether, she studied at Erlangen and Göttingen. Though invited to Göttingen in 1915 by **Hilbert**, as a woman she could not hold a full academic post at that time, but worked there in a semi-honorary capacity until, expelled by the Nazis as a Jew, she emigrated to the USA in 1933 to Bryn Mawr College and Princeton. She was one of the leading figures in the development of abstract algebra, working in ring theory and the theory of ideals; the theory of Noetherian rings has been an important subject of later research. She developed it to provide a neutral setting for problems in algebraic geometry and number theory, with a view to enabling their essential features to stand out from the technicalities.

NOGUCHI, Hideyo
(1876–1928)

Japanese-born American bacteriologist, born in Inawashiro. He graduated from Tokyo Medical College and worked in the USA from 1900. At the Rockefeller Institute in New York, he successfully cultured the spirochaete bacterium *Treponema pallidum* which causes syphilis. This enabled him to devise a diagnostic skin test for the disease using an emulsion of his culture. As a result of this, he was awarded the Order of the Rising Sun in his home country in 1915. He then went on to show that Oroya fever is caused by the bacterium *Bertonella bacilliformis*, which is transmitted to humans by sandflies. In 1927 he went to West Africa to obtain confirmatory evidence that yellow fever is a viral disease. He succeeded in proving this, but just before his departure to return to New York, he contracted the disease, from which he died shortly afterwards.

NOLLET, Jean Antoine
(1700–70)

French abbé and physicist, born in Pimprez near Noyan. While following an ecclesiastical course of study at Paris he became interested in science, and devoted the rest of his life to research, writing and lecturing, and making scientific instruments. From 1730 he collaborated for a time in electrical researches with **Dufay**, **Réaumur** and others, taking a leading part in the popularization of experimental science in France. In 1748 he became the first Professor of Physics at the Collège de Navarre, in Paris, and in the same year he discovered and gave a clear explanation of the phenomenon of osmosis. He invented an early form of electroscope, and improved the Leyden jar (an early form of capacitor) invented by **Musschenbroek**.

NOROYI, Ryoji
(1937–)

Japanese chemist. Educated at Kyoto University, where he received his BSc (1961) and PhD (1967), he carried out postdoctoral work at Harvard. In 1966 he discovered transition-metal-catalysed asymmetric reactions in homogeneous phases. This discovery has led to his continued interest in the synthesis of chiral molecules using organometallic reagents. One of his major findings is the widely applicable homogeneous asymmetric hydrogenation process, in which different optical isomers may be specifically synthesized. Such syntheses are of particular value in the pharmaceutical industry. His discoveries in this area are important in that the chirality (the optical isomerism) of the catalyst is passed on to the reaction product—the chemical multiplication of chirality. Thus stereoselective organic synthesis may be performed; this is of immense importance in making physiologically active compounds such as pharmaceuticals. Noroyi's newly created tools have opened up efficient routes to such diverse compounds as terpenes, alkaloids, prostaglandins, nucleosides and nucleotides. He won the Chemical Society of Japan award in 1984.

NORRISH, Ronald George Wreyford

(1897–1978)

English physical chemist, born in Cambridge. He won a scholarship to Emmanuel College at Cambridge in 1915, but did not commence his studies until 1919. In the meantime he saw active service in the Royal Field Artillery and spent six months as a prisoner of war. After taking a first class in both parts of the Natural Sciences Tripos, he carried out research under **Rideal**. Following various junior appointments at Cambridge, he became H O Jones Lecturer in Physical Chemistry in 1928, and in 1937 he was promoted to Professor of Physical Chemistry and head of department, a post he held until his retirement in 1965. He was one of the founders of modern photochemistry, and also made advances in the area of chain reactions. His most important innovation (1945), in association with **George Porter**, was flash photolysis. In this technique a very brief flash of intense light causes photochemical change and immediately afterwards the unstable chemical intermediates produced can be studied by means of their absorption spectra. For this work Norrish was awarded the Nobel Prize for Chemistry jointly with Porter and **Eigen** in 1967. He was elected a Fellow of the Royal Society in 1936 and received its Davy Medal in 1958. He received the Faraday Medal of the Chemical Society in 1965 and its Longstaff Medal in 1969, and served as President of the Faraday Society from 1953 to 1955.

NORTHROP, John Howard

(1891–1987)

American biochemist, born in Yonkers, New York. Educated at Columbia University, he became Professor of Bacteriology at the University of California at Berkeley (1949–62). In 1930, using an alcohol/water mixture, he crystallized pepsin, the protein-digesting enzyme of the stomach, showed it to be a protein and estimated its molecular weight. This was followed by the purification of other macromolecules, the gut protease, chymotrypsin, and its precursor, chymotrypsinogen (1935), a trypsin inhibitor from the pancreas (1937), and diphtheria toxin in crystalline form. He isolated the first bacterial virus, and was the first to equate the biological function of an enzyme with its chemical properties. He published *Crystalline Enzymes* (1939) which describes his important discovery of purifying proteins by 'salting out'; he also discovered the fermentation process used in the manufacture of acetone. For their studies of methods of producing purified enzymes and virus products, Northrop, **Wendell Stanley** and **Sumner** shared the 1946 Nobel Prize for Chemistry. From 1947 Northrop studied the autolysis (self-digestion) of pepsin and trypsin, and the effects of plant and animal proteases on living organisms.

NOSSAL, Sir Gustav Joseph Victor

(1931–)

Australian immunologist, born in Bad Ischl, Austria. He arrived in Australia in 1939 and was educated at the universities of Sydney and Melbourne. He was appointed Research Fellow at the Walter and Eliza Hall Institute of Medical Research (1957–9), and worked as Assistant Professor of Genetics at Stanford University (1959–61) before returning to the Hall Institute as Deputy Director (immunology) in 1961. He proceeded to the directorship in 1965, when he also became Professor of Medical Biology at Melbourne University. His research work has been on antibody response in immunity, which has provided strong experimental evidence in support of the clonal selection theory of **Burnet**, and his discovery of the 'one cell–one antibody' rule is crucial to modern work in immunology. He has written several popular books on immunology, and also on the progress of medical science, including *Antibodies and Immunity* (1971), *Medical Science and Human Goals* (1975) and *Reshaping Life: Key Issues in Genetic Engineering* (1984). He was knighted in 1977, and elected FRS in 1982.

NÖTH, Heinrich

(1928–)

German chemist, born in Munich and educated at the University of Munich (PhD 1954), returning there as professor in 1969 after spells at ICI in Billingham and the University of Marburg. He is a main group chemist who has made discoveries in the field of boron–nitrogen chemistry and tetrahydroborate complexes. He has also synthesized many novel ring and cage compounds of aluminium, phosphorus and arsenic. Nöth was partly responsible for developing nuclear magnetic resonance (NMR) spectroscopy as a probe for looking at such inorganic species. He was awarded the Alfred Stock Medal of the German Chemical Society in 1976.

NOYCE, Robert Norton

(1927–90)

American physicist and electronics engineer, born in Burlington, Iowa. As a physics major at Grinell College, Iowa, he learned of the invention of the transistor from Grant Gale, the college's physics professor and a friend of **Bardeen**. He went on to MIT, where he received his PhD in 1953, and two years later he joined **Shockley**'s semiconductor laboratory. In 1957, Noyce, a Swiss-born physicist named Jean Hoerni and six others left Shockley and founded Fairchild Semiconductor in Silicon Valley. Here, using Hoerni's planar process, Noyce developed and perfected the planar integrated circuit, which led directly to the invention of a commercially feasible integrated circuit. With **Kilby** who worked on the microchip independently, Noyce is regarded as the co-inventor of the integrated circuit. In 1961 Fairchild introduced its first chips and Noyce's company prospered: between 1957 and 1967 revenues rose from a few thousand dollars to $130 million, and the number of employees grew from the original eight to 12000. Noyce also co-founded Intel, the chip manufacturer.

NOZOE, Tetsuo
(1902–)

Japanese chemist, born in Sendai City. He was educated at Tohoku Imperial University (1923–6) and then went to Formosa (now Taiwan) to work in a government research institute (1928–9), but soon joined the staff of the newly established Taihoku Imperial University. He remained there until 1942, when World War II disrupted the university. When hostilities ceased the university was re-established as the National Taiwan University. In 1948 he returned to Tohoku Imperial University as professor and has remained there ever since, continuing research and travel long after normal retirement. He has studied the chemistry of a number of groups of natural products, including sapogenins, wool fat and essential oils. However, he is probably best known for his work on seven-membered ring tropylium compounds, as well as other non-benzenoid aromatics. He was President of the Chemical Society of Japan (1975–6) and was awarded the Order of the Sacred Treasure in 1972.

NURSE, Paul Maxime
(1949–)

British microbiologist, born in Norwich. After graduating from the University of Birmingham, his postgraduate work at the University of East Anglia was followed by a brief period at the University of Edinburgh. In 1979 he moved to the University of Sussex, where he held both Science and Engineering Research Council advanced fellowships and Medical Research Council senior fellowships. In 1984 he became head of the Cell Cycle Control Laboratory at the Imperial Cancer Research Fund, London, and he moved to Oxford in 1987 to take up the Iveagh Chair of Microbiology. In 1991 he became Napier Research Professor at Oxford. Nurse's major work on yeast genetics, in particular the regulation of the cell division cycle for control of mitosis, has been rewarded by many guest lectureships and awards. He was elected FRS in 1989.

NUTTALL, Thomas
(1786–1859)

English-born American naturalist, born in Settle, Yorkshire. A printer by trade, in 1808 he emigrated to Philadelphia, Pennsylvania, where he took up botany, accompanied several scientific expeditions between 1811 and 1834, and discovered many new American plants. He wrote *Genera of North American Plants* (1818), and became Curator of the Botanical Garden at Harvard (1822–32). His *Introduction to Systematics and Physiological Botany* was published in 1827. While at Harvard he also turned his attention to ornithology, and published *A Manual of the Ornithology of the United States and Canada* (1832). His two-volume work *North American Silva* was published in 1842. In the same year he returned to England to fulfil the conditions of an inheritance, and remained there until his death.

NYE, John Frederick
(1923–)

English physicist and glaciologist, born in Hove, Sussex. He was educated at King's College, Cambridge, where he received an MA and a PhD (1948). His career began as a demonstrator in the Department of Mineralogy and Petrology, Cambridge (1949–51); he then became a staff member of Bell Telephone Laboratories in the USA (1952–3). In 1953 he was appointed lecturer in physics at the University of Bristol, later becoming reader (1965), professor (1969) and Emeritus Professor (1988). At the Cambridge Cavendish Laboratory, he showed how the photoelastic effect could be used to study arrays of dislocations in crystals, and in 1953 he introduced the theory of continuous distributions of dislocations. He suggested that glacier motion could be treated as non-linear viscous flow and in this way explained many of the characteristics of mountain glaciers and ice sheets, and their response to climatic change. This work provided the foundation for modern glacier mechanics and a quantitative understanding of **Thorarinsson**'s catastrophic glacier floods or 'jökulhlaups'. Nye also showed how fields of waves typically contain dislocations similar to crystal dislocations, and applied mathematical theory to explain the details of these wave refractions.

NYHOLM, Sir Ronald Sydney

(1917–71)

Australian chemist, born in Broken Hill, New South Wales. He carried out much of his research at University College London, where he was professor from 1954 until his untimely death in a road accident in 1971. He was famous for modifying, with **Gillespie**, the valence shell electron pair repulsion (vsepr) theory of **Sidgwick** used for predicting the structures of simple molecules. Elected FRS in 1958, Nyholm won the Corday–Morgan Medal of the Chemical Society in 1950 and the Gold Medal of the Italian Chemical Society in 1968. He was knighted in 1967, and is remembered as an inspiring teacher and promoter of the subject of chemistry.

O

OATLEY, Sir Charles
(1904–)

English electronic engineer and inventor, born in Frome, Somerset. He graduated in physics from St John's College, Cambridge (1925), and shortly afterwards joined the staff of King's College, London. During World War II he was a member of the Radar Research and Development Establishment, and in 1945 returned to Cambridge where in 1960 he became Professor of Electrical Engineering. Efforts by **Zworykin** and others in the early 1940s to construct a practical electron microscope had met with only limited success, but Oatley realized in 1948 that newly developed circuits and components might overcome at least some of the problems. One of his research students, D McMullan, produced a prototype instrument which embodied many of the features of modern electron microscopes. Further development at Cambridge resulted in a scanning electron microscope being manufactured commercially in 1960, capable of producing three-dimensional images at magnifications of 100 000 or more. Oatley was elected FRS in 1969 and knighted in 1974.

OCHOA, Severo
(1905–)

American geneticist, born in Luarca, Spain. He obtained his MD in Madrid (1929) and worked at several research centres, including Heidelberg (with **Meyerhof** on muscle physiology) and Oxford (with **Peters** on thiamine), before emigrating to the USA, where he accepted a post at the Washington University School of Medicine in St Louis. He later settled at the New York University School of Medicine, where he became full professor in 1946. Ochoa isolated two of the enzymes which catalyse part of the **Hans Krebs** cycle, and this led him to study the energetics of carbon dioxide fixation in photosynthesis from 1948. He went on to isolate the enzyme polynucleotide phosphorylase (1955), later used for the first synthesis of artificial RNA, and established its properties and wide distribution in plants and animals. In 1961 he adopted **Nirenberg**'s approach to solving the amino acid genetic code and determined a number of base triplets (codons) based on uridine. He also studied the direction of protein synthesis along the DNA (1965), and the initiation factors associated with binding N-formylmethionine, the first amino acid in a bacterial peptide sequence (1967). For his contributions to the elucidation of the genetic code he was awarded the 1959 Nobel Prize for Physiology or Medicine, jointly with **Arthur Kornberg**.

ODLING, William
(1829–1921)

English chemist, born in London. He qualified in medicine at London University in 1851 and held various positions before becoming Professor of Chemistry at Oxford in 1872, retiring in 1912. He classified the silicates and developed a system of chemical notation which clarified the theory of valence. However, he discounted the advances made by **Berzelius**, and failed to recognize the significance of the distinction between atomic and molecular weight made by **Cannizzaro**, so much of his theoretical work has not endured. He died in Oxford.

ODUM, Eugene Pleasants
(1913–)

American ecologist, born in Newport, New Hampshire, and educated at the universities of North Carolina and Illinois. He was Callaway Professor Emeritus of Ecology and the Alumni Foundation Distinguished Professor Emeritus of Zoology at the University of Georgia at Athens. His research interests have included ecological energetics, estuarine and wetland ecology, and physiological and population ecology. He has stressed the view that ecosystem theory provides a common denominator for man and nature and that neither can be considered in isolation. These views are expounded in his *Ecology and Our Endangered Life-support Systems* (1989). He is also the author of three widely used textbooks, *Fundamentals of Ecology* (1953), *Ecology* (1975) and *Basic Ecology* (1982). His honours include the Institut de la Vie Prize (1975) awarded by the French government, and the Crafoord Prize of the Royal Swedish Academy of Sciences (1987).

OERSTED, Hans Christian
(1777–1851)

Danish physicist, born in Rudkøbing, Langeland, the elder son of an apothecary; due to family circumstances, he lived with his younger brother with neighbours, a German wigmaker and his wife. Aged 11, Oersted began to help out at his father's pharmacy. Although he had little formal education, he learned German, French, Latin and some chemistry, and he had little difficulty passing the entrance examination at Copenhagen University. He was particularly interested in the philosophy of **Kant**, which shaped his scientific attitudes, and the idea that nature's forces had a common origin. This resulted in his epochal discovery in 1820, when professor at the University of Copenhagen, of the magnetic effect produced by an electric current. This paved the way for the electromagnetic discoveries of **Ampère** and **Faraday**, and the development of the galvanometer, in which Oersted also played a part. He made an extremely accurate measurement of the compressibility of water, and succeeded in isolating aluminium for the first time in 1825.

OHM, Georg Simon

(1787–1854)

German physicist, born in Erlangen, Bavaria, the son of a locksmith. He withdrew from the University of Erlangen because his overindulgence in dancing, billiards and ice-skating had incurred his father's displeasure. After a spell in Switzerland teaching mathematics, he completed his studies at Erlangen. He later became professor at Nuremberg (1833–49) and Munich (1849–54). His 'Ohm's law', relating voltage, current and resistance in an electrical circuit, was published in 1827, and was followed by a long struggle before its importance was recognized. **Wheatstone** was an early adherent. His work on the recognition of sinusoidal sound waves by the human ear as pure tones (1843) received similar treatment until it was rediscovered by **Helmholtz**. He was awarded the Royal Society's Copley Medal (1841) and was elected a Foreign Member of the society in 1842. The SI unit of electrical resistance is named after him.

OKAZAKI, Reiji

(1930–75)

Japanese biochemist, born in Hiroshima, and aged 14 when the world's first atomic bomb was dropped on his home town. Having graduated in science at Nagoya University, he remained there as a lecturer and was appointed professor in 1967. Working with bacteria and bacteriophage, Okazaki was the first to identify, by buoyant density measurements, the DNA–RNA fragments named after him (1967). These units of DNA replication, of length about 1000–2000 nucleotides in prokaryotes and 100–200 nucleotides in eukaryotes, resolved the dilemma of how DNA was simultaneously synthesized in opposite directions, but always corresponding to the opposing polarity (the 5′ to 3′ direction) of the two DNA strands. Continuous synthesis occurred on one strand while 'Okazaki fragments', subsequently joined together, built up on the other. With **Arthur Kornberg** he was the first to recognize the 'primer' function of the short RNA sequence to which the DNA is attached, the RNA being subsequently excised and replaced by DNA by the 'Kornberg enzyme'. He also studied the RNA-free, so-called pseudo-Okazaki fragments produced by certain bacterial mutants or derived by degradation of normal DNA. Okazaki was awarded the Asahi Prize in 1970. He developed leukaemia, and his health deteriorated until he died of heart failure.

OKEN, Lorenz, originally Ockenfuss

(1779–1851)

German naturalist and nature philosopher, born in Bohlsbach. He became Professor of Medicine at Jena in 1807, and in 1816 founded the biology journal *Isis*. His forthright expression of his strong scientific and political beliefs led to government interference and his resignation. He was subsequently appointed to professorships at Munich (1828) and Zürich (1832). His theory that the skull is a modified vertebra was later discredited, but was useful in giving scientists an early taste of evolutionary ideas.

OLBERS, Heinrich Wilhelm Matthäus

(1758–1840)

German astronomer, born in Ardbergen near Bremen, the son of a Lutheran minister. He studied medicine at Göttingen and Vienna, and in 1781 set up in medical practice in Bremen. Though a conscientious physician he was mainly interested in astronomy. He devised a method of calculating the orbits of comets (1779), and from the small observatory at his house at Bremen swept the sky whenever it was clear. He discovered the minor planets Pallas (1802)—only the second to be observed after **Piazzi**'s discovery of Ceres in 1801—and Vesta (1807). His greatest discovery was of the comet of 1815 named after him; he found its period to be 70 years, similar to that of comet Halley. The name of Olbers is also associated with the paradox that the sky is dark at night; he pointed out that it would be expected to be bright if the universe of stars were infinite and static. Olbers' paradox is resolved by modern cosmological theories of the expanding universe.

OLDHAM, Richard Dixon
(1858–1936)

Irish geologist and seismologist, born in Dublin, discoverer of the Earth's core. Educated at Rugby and the Royal School of Mines, he was a member of the Geological Survey of India (1878–1903) and for some of this time, Director of the Indian Museum in Calcutta. In 1903 he resigned his post as superintendent of the survey, partly because of ill-health, and returned to England. His important report on the Assam earthquake of June 1897 distinguished for the first time between primary and secondary seismic waves, and was able to characterize many other phenomena of earthquake activity. He proved the generality of his notions about the different types of seismic waves with reference to six other earthquakes in *On the Propagation of Earthquake Motion to Great Distances* (1900), and laid the foundations of what is now one of the principal branches of geophysics. In 1906 he established from seismographical records the existence of the Earth's core. He was the author of *Bibliography of Indian Geology* (1888), *Catalogue of Indian Earthquakes* (1883) and many other works on Indian geology.

OLIPHANT, Sir Mark (Marcus Laurence Elwin)
(1901–)

Australian nuclear physicist, born in Adelaide. He studied there and at Trinity College and the University of Cambridge, where he received his PhD in 1929. He then worked at the Cavendish Laboratory in Cambridge where in 1934, with **Rutherford** and **Harteck**, he discovered the tritium isotope of hydrogen by bombarding deuterium with deuterons. In 1937 he became professor at Birmingham University, where he designed and built a 60 inch cyclotron particle accelerator, completed after World War II. He worked on the Manhattan project at Los Alamos (1943–5) to develop the nuclear bomb, but at the end of hostilities strongly argued against the American monopoly of atomic secrets. In 1946 he became Australian representative of the UN Atomic Energy Commission. He was later appointed research professor at Canberra University (1950–63) and designed a proton synchrotron accelerator for the Australian government. From 1971 to 1976 he served as Governor of South Australia. He was elected FRS in 1937, and knighted in 1959.

OLSEN, Kenneth Harry
(1926–)

American computer engineer and entrepreneur, born in Bridgeport, Connecticut, into an evangelical Scandinavian Protestant family. He studied electrical engineering at MIT, where he obtained a master's degree in 1952 and joined **Forrester**'s pioneering computer group in 1950. Soon afterwards Olsen became an on-site engineer at IBM, but in 1956 he left to establish his own computer company, the Digital Equipment Corporation (DEC) in Maynard, Massachusetts. DEC defined and then exploited a new niche in the growing computer industry—the market for minicomputers, or 'interactive' machines, that were less expensive and easier to use than mainframes. Aided by brilliant engineers, such as **Gordon Bell** from MIT, Olsen launched the PDP-8—the first successful minicomputer—in the early 1960s. By 1986 DEC was the second largest US computer company behind IBM, and Olsen was described by *Fortune* as 'America's most succesful entrepreneur'. In 1992, however, after DEC had suffered heavy losses and stagnating sales in the minicomputer market, Olsen was forced to resign as the chief executive.

O'NIONS, Robert Keith
(1944–)

English geochemist, educated at the universities of Nottingham and Alberta. After postdoctoral work in Norway he became a lecturer in geochemistry at the University of Oxford (1972–5), professor at Columbia University, New York (1975–9), and Royal Society Research Professor at Cambridge from 1979. He has undertaken important work in Earth and planetary geochemistry, with particular reference to chemical and isotope ratio studies of ocean-floor and ocean-island basalts, and their origin in the mantle.

ONNES, Heike Kamerlingh
See **KAMERLINGH ONNES, Heike**

ONSAGER, Lars
(1903–76)

Norwegian–American chemical physicist, born in Christiania (now Oslo). He was trained at the Technical University of Norway as a chemical engineer, but pursued further studies in mathematics in preparation for working on difficult problems in theoretical physics and chemistry. He worked in Zürich with **Debye** from 1926 to 1928 and then went to the USA, where he spent the rest of his life. After periods at Johns Hopkins University and Brown University, Rhode Island, he settled at Yale, where he advanced from assistant professor to associate professor between 1934 and 1945, when he became Gibbs Professor of Theoretical Chemistry. He held this position until 1972, when he moved to the University of Miami as Distinguished University Professor. Onsager's work with Debye was on strong electrolytes, for which he developed an extension of the Debye–**Hückel** theory. However, he is best known for his pioneering work on the thermodynamics of irreversible processes, which he put on a sound basis. The fundamental equations in this field are called the 'reciprocal relations' and are commonly known by his name. For this work he was awarded the Nobel Prize for Chemistry in 1968. The theory of irreversible thermodynamics was developed further by **Prigogine**.

OORT, Jan Hendrik

(1900–92)

Dutch astronomer, born in Franeker. He studied under **Kapteyn** at the University of Groningen, and worked mainly at the Leiden Observatory in Holland (1924–70), becoming director there in 1945. He proved (1927) by observation that our galaxy is rotating, and calculated the distance of the Sun from the centre of the galaxy, initially locating it 300 000 light years away. He found that the Sun has a period of revolution around the galactic centre of just over 200 million years. This enabled him to make the first calculation of the mass of galactic material interior to the Sun's orbit, this being some 10^{11} solar masses. In 1932 he made the first measurement that indicated that there is dark matter in the galaxy, concluding that the visible stars near the Sun could account for only around half the mass implied by the velocity of these stars perpendicular to the galactic plane. Beginning in 1944, Oort traced the structure of the galactic disc by using radio telescopes to detect the 21 centimetre wavelength radiation that is emitted by atomic hydrogen. In 1946 he realized that the filimentary nebulae called the Cygnus Loop is a supernova remnant. In 1950 he extended **Öpik**'s suggestion concerning the huge circular reservoir of comets surrounding the solar system. These have maximum distances from the Sun of some 100 000 astronomical units and are thus susceptible to being perturbed by passing stars. This 'Oort cloud' was the suggested source of long-period comets. In 1956, with Theodore Walraven, he discovered the polarization of the radiation from the Crab nebula indicating that it was produced by synchrotron radiation from electrons moving at high speeds along magnetic field lines.

OPARIN, Alexandr Ivanovich

(1894–1980)

Russian biochemist, born in Uglich, near Moscow. He was educated at Moscow State University, and became head of plant biochemistry at Moscow University in 1929, and an associate organizer then director of the Bakh Institute of Biochemistry of the USSR Academy of Sciences. His first thoughts on the origin of life in *Proiskhozhdenie Zhizny* (1924) became known from **J B S Haldane**'s publication on this subject in 1929, but neither attracted significant attention. In 1952 **Urey** recalled the Oparin–Haldane theory that life slowly emerged from a primeval soup of biomolecules and with **Stanley Miller** revived interest by generating simple biomolecules by spark discharge through a mixture of reducing gases. Oparin suggested that the life-forming process could occur via molecular aggregation to form coacervates, polymer-rich colloidal droplets, that could absorb other biomolecules and spontaneously divide. His simulated living systems, containing one or two enzymes, could mimic fermentation, electron transport and photosynthesis. His other major publications included *The Origin of Life on Earth* (1936) and *The Chemical Origin of Life* (1964). Oparin also worked on practical biochemical problems associated with the production of tea, sugar, bread, tobacco and wine for the Soviet economy.

ÖPIK, Ernst Julius

(1893–1985)

Estonian astronomer, born just north of Rakvere. After studying at Tartu State University he worked at observatories in Tashkent, Moskow and Turkistan, later returning to Tartu where he lectured until 1945. After the upset caused by World War II he moved to the Armagh Observatory in Northern Ireland, where he became the director, a post he held concurrently with a chair at the University of Maryland. Öpik's research concentrated on comets and meteoroids in the solar system. His theory governing the ablation of meteoroids when they burn up and disintigrate on entering the Earth's atmosphere has been applied with considerable success to re-entering space capsules and meteorite-dropping asteroids. Öpik also studied the orbits and orbital perturbations of comets, and in 1932 was the first to predict that there is a huge (60 000 astronomical units radius) cometary cloud surrounding the solar system, a cloud that is occasionally perturbed by passing stars. This idea was reworked by **Oort** in 1950. Öpik designed, in 1934, an ingenious rocking mirror device to improve the accuracy of meteor visual velocity measurements. He erroneously interpreted data from a meteor observing programme in Arizona to indicate that 60 per cent of the sporadic meteoroids were hyperbolic and thus interstellar. Öpik pioneered the measurement of meteoroid size distribution. He also suggested that Apollo asteroids were dormant cometary nuclei. Among his many awards was the Gold Medal of the Royal Astronomical Society (1975).

OPPENHEIMER, (Julius) Robert

(1904–67)

American nuclear physicist, born in New York City. He studied at Harvard where he graduated in 1925, the University of Cambridge and under **Born** at Göttingen, where he received his doctorate in 1927. He returned to the USA and established schools of theoretical physics at Berkeley and Caltech. His work included studies of electron–positron pairs, cosmic-ray theory and deuteron reactions. During World War II, he led a team which pioneered theoretical studies on building the atomic bomb, and he was later selected as leader of the atomic bomb project. He set up the Los Alamos laboratory and brought together a formidable group of scientists. After the war he became Director of the Institute for Advanced Studies at Princeton and continued to play an important role in US atomic energy policy from 1947. He used his political influence to promote peaceful uses of atomic energy and was bitterly opposed to the development of the hydrogen bomb. In 1953 when the US government turned against him and declared him a security risk, he was forced to retire from political activities; later, however, he received the Enrico Fermi Award of the Atomic Energy Commission (1963). He delivered the BBC Reith Lectures in 1953.

OSBORN, Henry Fairfield

(1857–1935)

American palaeontologist and zoologist, born in Fairfield, Connecticut. He studied at Princeton, and became Professor of Zoology at Columbia University and concurrently Curator of Vertebrate Palaeontology at the American Museum of Natural History (1891–1910). Retaining a research professorship at Columbia, he was President of the American Museum of Natural History from 1908 to 1933. Although known as an autocratic leader, he revolutionized museum display with innovative instructional techniques and the acquisition of spectacular specimens, especially dinosaurs. He popularized palaeontology, mounting skeletons in realistic poses with imaginative backdrops. His many publications include *The Age of Mammals* (1910), *Man of the Old Stone Age* (1915) and *The Origin and Evolution of Life* (1917). His major scientific contribution was a vast monograph on *Proboscidea*, published posthumously in two volumes (1935–42).

OSLER, Sir William

(1849–1919)

Canadian–American–British physician, born in Bond Head, Ontario, and educated at Toronto and McGill universities. After graduating in medicine, he toured Britain and Germany for scientific training and in 1874 became Professor of Medicine at McGill. Chairs at the University of Pennsylvania (1884–9) and Johns Hopkins (1889–1904) followed, and he was subsequently appointed to the Regius Chair of Medicine at Oxford. His years at Johns Hopkins were his best; during this time he produced many clinical papers, monographs on cerebral palsy and chorea, and his textbook *The Principles and Practice of Medicine* (1892). This codified the scientific clinical practice of his time and was frequently revised and translated. An advocate of full-time clinical training and research, Osler made a number of bedside and pathological observations of permanent value. He made an early study of the platelets, described hereditary haemorrhagic telangiectasis (Osler–Rendu–Weber disease), polythaemia vera (Vaguez–Osler disease) and infection of the heart valves (endocarditis). He was instrumental in founding the Association of Physicians of Great Britain and Ireland, a society devoted to encouraging clinical research. He was also an elegant stylist, an ardent bibliophile and bibliographer, and advocate of humane values in a world of science. He was made a baronet in 1911. By the time of his death, he was revered throughout the English-speaking world as a kind of patron saint of patient-oriented scientific medicine; the reverence continues unabated through Osler societies, lectures and prizes in many places, including Japan. His last years were clouded by the death of his only son during World War I.

OSTROVSKY, Yuri I

(1926–92)

Russian holographic scientist, born in Baku, Azerbaijan. He was educated at Leningrad State University and became assistant professor at the Leningrad Institute of Mines in 1961. In 1964 he moved to the A I Ioffe Physico-Technical Institute of the USSR Academy of Sciences where he remained until his death. His wife, Professor Galya V Ostrovsky, is also a well-known holographic scientist. Ostrovsky's work lay in the fields of the measurement and interpretation of the intensities of lines in atomic spectra, and the use of holographic techniques for measuring mechanical vibrations and the wear of surfaces. He also made optical studies of hydrodynamic processes such as cavitation and shock waves. He wrote many books on spectroscopy and holography, translated numerous English-language optical books into Russian and was a well-known figure at international conferences on holography and related topics.

OSTWALD, (Friedrich) Wilhelm
(1853–1932)

German physical chemist, born in Riga, Latvia. He studied chemistry at the University of Dorpat (now Tartu), taking the *Candidat* examinations in 1875. After holding various posts as an assistant at Dorpat, he was appointed Professor of Chemistry at the Riga Polytechnic in 1881. In 1887 he moved to Leipzig as Professor of Physical Chemistry, taking early retirement in 1906. He spent the rest of his life on his country estate in Saxony, devoting himself to various scientific, literary and other intellectual activities. With **van't Hoff** and **Arrhenius**, Ostwald is regarded as one of the founders of physical chemistry. At Dorpat he worked on the measurement of chemical affinity by observing changes in the physical properties of solutions as a result of chemical reactions. During his Riga period he used rates of reaction to study chemical affinity and he measured the 'affinity coefficients' of many acids, particularly organic acids, through studies of their catalytic behaviour. His results were greatly illuminated by the electrolytic dissociation theory of Arrhenius, which Ostwald did much to promote. In Leipzig he built up a great school of physical chemistry, which attracted students from all over the world. His studies of electrolytic conductivity (resulting in Ostwald's dilution law) and of the electromotive force of cells were carried out in Leipzig. He founded the journals *Zeitschrift für physikalische Chemie* in 1887 and *Annalen der Naturphilosophie* in 1901. His various books were very influential, notably his *Lehrbuch der allgemeinen Chemie* (2 vols, 1883–7). In his long retirement he worked on the theory of colour perception, an interest which arose from his skill as a landscape painter. He was also an able musician and was interested in various systems of philosophy. For his work on catalysis, Ostwald was awarded the Nobel Prize for Chemistry in 1909. He became an Honorary Fellow of the Chemical Society in 1898 and received its Faraday Medal in 1904.

OTTO, Nikolaus August
(1832–91)

German engineer, born in Holzhausen, Nassau. The son of a farmer, he left school at 16 to work in a merchant's office, but soon moved to Cologne where he became interested in the gas engines of Étienne Lenoir. By 1861 he had built a small experimental gas engine, and three years later he joined forces with Eugen Langen (1833–95), an industrialist who had studied at Karlsruhe Polytechnic, to form a company for the manufacture of such engines. Dissatisfied with the limitations of Lenoir's low-compression low-speed gas engines, Otto devised the four-stroke cycle which bears his name and obtained a patent for it in 1877. His engines were so successful that other manufacturers sought ways in which to evade the restrictions of the patent, and eventually it was discovered that the principle of the four-stroke cycle had been outlined by Beau de Rochas in a patent dated 1862, though he had not built even a prototype engine at the time. Finally in 1886 Otto's patent was cancelled, but by that time over 3000 of his so-called 'silent Otto' engines had been sold, and the Otto cycle remains today the operating principle of the great majority of the world's internal combustion engines.

OUGHTRED, William
(1575–1660)

English mathematician, born at Eton College. Educated there and at Cambridge, he was ordained as a minister and became Rector of Albury. He wrote extensively on mathematics, notably *Clavis Mathematicae* ('The Key to Mathematics', 1631), a textbook on arithmetic and algebra in which he introduced many new symbols including multiplication and proportion signs. He also invented the earliest type of slide rule, and wrote on trigonometry in *Trigonometria* (1657).

OWEN, Sir Richard
(1804–92)

English zoologist and palaeontologist, born in Lancaster. He studied medicine at Edinburgh and at St Bartholomew's Hospital in London, and became curator at the Royal College of Surgeons. In 1856 he was appointed superintendent of the natural history department of the British Museum, and was instrumental in the establishment of the separate British Museum (Natural History) (now the Natural History Museum), becoming its first director in 1881. He was the most prestigious zoologist of Victorian England, and published 400 scientific papers as well as a number of important books, including *British Fossil Mammals and Birds* (1846), *A History of British Fossil Reptiles* (1849–84), and an influential essay on *Parthenogenesis*. He named and reconstructed numerous celebrated fossils, including the giant moa bird *Dinornis*, the dinosaur *Iguanodon*, and the earliest bird, *Archaeopteryx*. He coined the term 'dinosaur' ('terrible lizard'). Owen studied in detail the homologies between apparently dissimilar structures in organisms, and drew the crucial distinction between homologous and analogous organs. However, he remained implacably opposed to evolution; for him, homologies were variants on a divine plan or 'archetype', not evidence of common descent. He was a virulent and outspoken opponent of **Charles Darwin** and **T H Huxley**; Darwin stated that 'his power of hatred was unsurpassed', and put it down to jealousy at the success of the *Origin of Species*. Owen accepted a knighthood in 1884, having previously declined the honour in 1842.

OXBURGH, Sir (Ernest) Ronald

(1934–)

English geologist. Educated at the University of Oxford and Princeton University, he became a lecturer at Oxford (1960–78) and Professor of Mineralogy and Petrology at Cambridge (1978–91). He was also chief scientific advisor to the Ministry of Defence (1988–93). Oxburgh has undertaken important petrological and geochemical research, particularly with his studies of the origin and distribution of radiogenic helium in the Earth's crust. He was elected FRS in 1978, and knighted in 1992.

P

PALADE, George Emil

(1912–)

Romanian cell biologist, born in Iassy. Trained as a doctor in Bucharest, he became Professor of Anatomy there until he emigrated to the USA in 1946. He worked at the Rockefeller Institute, New York (1946–72), and from 1972 headed cell biology at Yale Medical School. Since 1990 he has been Professor of Cellular and Molecular Biology at the University of California, San Diego. In the 1950s Palade developed a method of separating components of the cell known as 'cell fractionation'. Using the newly introduced technique of electron microscopy, he described the components of the cell; the mitochondria, the endoplasmic reticulum, the Golgi apparatus and the ribosomes. He showed that protein synthesis occurs on strands of RNA in the ribosomes which are attached to the membranous endoplasmic reticulum. The proteins are then carried through the cell in sacs, called vacuoles, before being released into the extra-cellular fluid. For his work in cell biology he shared the 1974 Nobel Prize for Physiology or Medicine with **Albert Claude** and **de Duve**.

PALLAS, Peter Simon

(1741–1811)

German naturalist, born in Berlin. The son of a professor of surgery, he studied medicine at Halle, Göttingen and Leiden, but his interests always tended towards natural history. He moved to England in 1761 when he reclassified the worms from the original scheme of **Linnaeus**. He also at this time classified the Zoophytorum, the corals and sponges which, until then, had been considered to be plants. His comparative anatomical methods laid the foundation for modern taxonomy. He published *Miscellania Zoologica* (1766) and *Spicilegia Zoologica* (1967 onwards) and was elected FRS in 1763. Pallas was invited to St Petersburg by the Empress Catherine the Great as professor at the Academy of Sciences; he spent six years (1768–74) exploring the Urals, the Kirghiz Steppes, the Altai Range, part of Siberia and the steppes of the Volga. On his return he wrote a series of works on the geography, ethnography, geology, flora and fauna of the regions he had visited (*Reise durch verschiedene Provinzen des Russischen Reichs*, 3 vols, 1771–6). Although he collected a large number of specimens, his interest was not merely to catalogue them; his observations made over an extreme range of habitats led him to seek causal relationships between the animals and their environment, and he became the first zoogeographer. His major contribution was probably to geology, as he was the first to propose a modern view of the ways in which mountain ranges were formed and developed. Although his health was impaired by the rigors of his first expedition, he made another in 1793–4 when he discovered the remains of mammoths and rhinoceros in the Siberian ice. He subsequently remained in Russia and wrote *Zoographica Russo-Asiatica* (3 vols, 1811–31).

PALMEN, Erik Herbert

(1898–1985)

Finnish meteorologist, born in Vaasa. He was educated at Helsinki University where he obtained a master's degree in astronomy in 1921 and his PhD in meteorology in 1927. In 1922 he joined the Marine Research Institute in Helsinki, where he studied the effect of wind stress on water level changes and on the stratification of water in the oceans. He was director of the institute from 1939 to 1947. Palmen was interested in atmospheric dynamics and the structure of depressions, and was one of the first to discover the jet stream and to realize that it is virtually a global phenomenon. He wrote a classic paper on the general circulation (1947) and his investigation of jet streams led to original studies on how energy is converted from potential to kinetic energy in the atmosphere. He worked with **Rossby** at the University of Chicago (1946–8) and later moved to the University of California at Los Angeles (1953–4). His studies of the subtropical jet stream led to an interest in tropical cyclones, and he was able to show that these can only develop if the ocean temperature is above about 26/27 degrees Celsius, and if the thermal structure of the atmosphere meets certain criteria. Later he worked out a satisfactory energy budget for the whole atmosphere.

PALMITER, Richard De Forest

(1942–)

American molecular geneticist, born in Poughkeepsie, New York. He was educated at Duke University where he received a BA in 1964 and at Stanford University where he received his PhD in 1968. After postdoctoral research at Stanford, at Searle Roche Laboratories in the UK and at Harvard University, he became assistant professor (1974–8) and associate professor (1978–81) at the University of Washington, where he has been Professor of Biochemistry since 1981. Since 1976 he has also held the post of investigator at the Howard Hughes Medical Institute, Seattle. He received the George Thorn Award of the Howard Hughes Medical Institute in 1982. Along with **Brinster**, Palmiter produced the first transgenic mice. The procedure involved the microinjection of the human growth hormone gene into a mouse embryo and reintroduction of the embryo into the mother's uterus. Mice produced in this way were found to be significantly larger than their normal counterparts indicating that the human growth hormone gene had been active. This technique has now been used for a variety of mammalian genes and is a vital tool in the investigation of regulatory mechanisms controlling gene expression.

PANDER, Christian Heinrich
(1794–1865)

Russian-born German anatomist and crucial figure in modern embryology. Born in Riga, Pander studied at Dorpat, and subsequently at Berlin, Göttingen and Würzburg, where he was befriended by **Baer**. He took his MD in 1817. He undertook valuable research on chick development in the egg, in particular demonstrating the embryonic layers named after him, and coining the term blastoderm for the trilaminar structure (from the Greek *blastos* meaning 'germ', and *derma*, 'skin'). Pander never followed up his early findings, leaving the field of embryological research to others, including Baer. He spent much time travelling, in 1820 acting as a naturalist on a Russian mission to Bokhara. In 1826 he was elected a member of the St Petersburg Academy of Sciences, and a year later he retired to his estates around Riga.

PANETH, Friedrich Adolf
(1887–1958)

Austrian chemist, born in Vienna. He studied at Munich, Glasgow and Vienna and carried out research at Prague, Berlin and Königsberg. In 1933 he moved to Britain as the Nazis rose to power. He worked at Imperial College, London, and Durham University, where he was appointed Professor of Chemistry in 1939. In 1953 he returned to Germany and the directorship of the Max Planck Institute. Paneth worked with radio isotopes, collaborating with **Hevesy** on radioactive tracers. He also showed that free radicals exist briefly when organic compounds are decomposed by heat. From 1917 to 1929 he investigated ways of measuring the minute amounts of helium generated by the disintegration of radium, methods of microanalysis which made him famous. This led him to devise methods of dating rocks and meteorites by their helium content and thence to speculations about their formation within the solar system. From 1935 he investigated the trace compounds in the atmosphere and concluded that the atmosphere is uniform for the first 40 miles, but that concentrations of the various components vary above this level owing to the effects of gravity. He died in Mainz, Germany. After his death a trust was established at the Max Planck Institute to further research on meteorites and care for his collection.

PAPANICOLAOU, George Nicholas
(1883–1962)

Greek-born physiologist and microscopic anatomist, born in Kimi, the son of a physician. He received his MD from Athens University (1904) and a PhD from Munich University (1910). He moved to the USA in 1913, becoming assistant in pathology at the New York Hospital and, in 1914, assistant in anatomy at Cornell Medical College. All his research was conducted at these two institutions until 1961, when he was appointed Director of the Miami Cancer Institute, although he died three months later. He became Professor of Clinical Anatomy at Cornell in 1924 and was Emeritus Professor from 1949. Papanicolou's research on reproductive physiology led him to the discovery that the cells lining the wall of the guinea pig vagina change with the oestrus cycle. Similar changes take place in women, but more importantly, Papanicolaou noticed that he could identify cancer cells from scrapings from the cervixes of women with cervical cancer. He subsequently pioneered the techniques, now familiar as the 'pap smear', of microscopical examination of exfoliated cells for the early detection of cervical and other forms of cancer.

PAPIN, Denis
(1647–c.1712)

French scientist, born in Blois. Papin followed his father's profession and studied medicine at the University of Angers. By 1673 he was in Paris assisting **Huygens** in experiments with the new air-pump. He went to London in 1675 where he collaborated in similar investigations with **Boyle** (1676–9). There he created the 'steam digester' (a pressure cooker) which was demonstrated before the Royal Society in 1679. In 1680 he returned to Paris en route to Venice (1681) where he lived for three years, demonstrating experiments for the scientific society of Ambrose Sarotti. He came to London again in 1684 as Royal Society curator of experiments, meanwhile continuing his investigations in hydraulics and pneumatics. Unable to return to France because of his Protestant faith, Papin settled in Marburg as Professor of Mathematics from 1687. In 1690 he published an account of a rudimentary steam engine. Papin described a modification of Thomas Savery's steam-pump in 1707. He returned to England once more but with many former patrons gone he died in obscurity.

PAPPUS OF ALEXANDRIA

(4th century)

Greek mathematician. He wrote a mathematical *Collection* covering a wide range of geometrical problems, some of which inspired **Descartes** and contributed to the development of projective geometry in modern times. It is of great importance for the historical understanding of Greek mathematics. In it he described the economical work of bees, and discussed the isoperimetric problem (the claim that of all curves of a given length the circle encloses the greatest area). He generalized **Pythagoras**'s theorem to triangles that are not right-angled, wrote on the trisection of the angle and devices to square the circle, and offered commentaries on **Euclid**'s *Elements* and **Ptolemy**'s *Almagest*.

PARACELSUS, a name coined for himself by Theophrastus Bombastus von Hohenheim

(1493–1541)

German alchemist, physician and self-styled seer. The name referred to the celebrated Roman physician **Celsus** and meant 'beyond' or 'better than' Celsus. Paracelsus was born in Einsieden, Switzerland, but moved when young to Villach, a mining area in southern Austria. His father was a physician who taught chemistry at the borough school and here the boys were trained to assay local ores, giving Paracelsus an early grounding in metallurgy and chemistry. When he was 14 he began to wander from one university to another in search of inspired teachers, and he is said to have graduated in Vienna and taken his doctorate in Ferrara. Finding universities fossilized in their attitudes and insulated from real life, he then spent many years exploring Europe, including England and Scotland. He served as an army surgeon in the Netherlands and Italy, and was captured by the Mongols during a visit to Russia; finally he travelled through Egypt, Arabia and the Holy Land to Constantinople. In all these lands he studied contemporary medical practice and the medical lore of the common people, always looking for ways to encourage 'the latent forces of Nature' in all healing processes. In 1526 he was appointed town physician in Basle and lecturer in chemistry at the university. Already famous for his erudition and revolutionary views on medicine and religion, he raged against medical malpractices and the fashion for patent medicines. He also criticized the Catholic Church and the new Lutheran doctrines. He taught in German, not Latin, another innovatory move which displeased the authorities who preferred learning to be kept from the populace. In 1528, having antagonized all the vested interests in the town, he had to flee for his life, and spent most of the rest of his life as an itinerant preacher and physician, conducting chemical research wherever he could find a laboratory and writing mystical tracts and works on medicine, chemistry and alchemy. By the end of the 1730s he was to some extent re-established and numbered the rich and influential among his patients, although he continued to treat the poor without charge. He died in Salzburg where his grave became a place of pilgrimage for the sick. Despite his erratic and violent nature, echoed in his writings, Paracelsus had enormous influence, particularly through the emphasis he laid on observation and experiment and the need to assist—rather than hinder—natural processes. He believed that wounds should be allowed to heal naturally without interference except for cleaning. He stated that diseases had external causes and that every disease had its own characteristics, thus reversing the traditional view that disease was generated within the patient and followed an unpredictable course. He made careful studies of tuberculosis and silicosis (both miners' diseases); recognized that there was a connection between goitre and the minerals in drinking water; researched the role of acids in digestion; was the first to recognize congenital syphilis; studied hysteria, and advocated that lunatics should be treated kindly. Paracelsus was also an able chemist who discovered many techniques which became standard laboratory practice, such as concentrating alcohol by freezing it out of its solution. He also prepared drugs with due regard to their purity and advocated carefully measured doses, both important steps forward in medicine.

PARDEE, Arthur Beck

(1921–)

American biochemist, born in Chicago. He trained and worked at the University of California, Berkeley (1942–61), spending some time at Caltech (1943), the University of Wisconsin (1947–9) and the Pasteur Institute, Paris (1957–8), before becoming professor at Princeton (1961–7). Since 1975 he has been Professor of Pharmacology at Harvard. His early work was on biological oxidation processes with Van Potter (of homogenizer fame), notably on the inhibition of succinate dehydrogenase by malonate, and with **Pauling** on tumour metabolism and quantification of the strength of antibody reactions. Pardee became interested in how cells control their own synthetic processes, and in 1958 worked with **Monod** on the lac operon of *Escherichia coli*. He also discovered feedback control of amino acid synthesis (threonine synthetase). In order to study DNA repair and the regulation of DNA synthesis, he developed a gentle method of synchronizing the cell cycle (cyclical growth and division) of initially bacteria and then eucaryotic cells in culture. In this connection he exploited a range of specific inhibitors (bromouracil, chloramphenicol) in relation to such processes as the regulation of bacterial division by protein X (recA gene control) and the requirement for branched chain fatty acids (1971), the regulation of ribonucleotide reductase (1978), the formation of **Okazaki** fragments (1977) and histone synthesis (1978). From around 1980 Pardee investigated the effects of serum peptides, including insulin-like growth factor, and the ubiquitin system (1984).

PARKES, Sir Alan Sterling
(1900–90)

English physiologist, born in Purley and educated at the universities of Cambridge, Manchester and London. He became a lecturer in physiology at University College London (1929–31) and after joining the staff of the National Institute for Medical Research in London (1932–61), he was appointed Mary Marshall Professor of the Physiology of Reproduction at Cambridge (1961–7). His research was concerned with the physiology of reproduction and endocrinology, and with the behaviour of animal cells at low temperatures. His scientific books included *The Internal Secretions of the Ovary* (1929) and *Patterns of Sexuality and Reproduction* (1976), and he edited the third edition of *Marshall's Physiology of Reproduction* (1952). He also wrote autobiographical books *Off-beat Biologist* (1985) and *Biologist at Large* (1988). He was elected FRS in 1933, and knighted in 1968.

PARKES, Alexander
(1813–90)

English chemist, the inventor of celluloid, born in Birmingham. While working for a brass founder in Birmingham he devised a way of electroplating fragile natural objects, such as flowers, and a spider's web which he had electroplated was presented to Prince Albert. Parkes was the first person to show that phosphorus added to metal alloys increased their strength. In 1841 he discovered that rubber could be vulcanized without heat if treated with a solution of carbon bisulphide. This process was used to waterproof fabric and the patent was sold to **Macintosh**. While superintending the construction of a copper-smelting works in south Wales in the early 1850s, Parkes discovered a way to extract silver from lead. In this method zinc and lead are melted together; the zinc then reacts with the silver and any other metals in the lead, and the newly formed compounds—being lighter than silver—float to the surface and can be skimmed off. The Parkes process was soon used worldwide. Around 1855 Parkes was searching for a substitute for ivory and produced a synthetic material from a mixture of chloroform and castor oil. Although it was flexible and durable, it didn't achieve commercial success in Britain. The process was refined in the USA and the product renamed 'celluloid'. Parkes died in West Dulwich, now part of London.

PARSONS, Sir Charles Algernon
(1854–1931)

Irish engineer, born in London, 4th son of the 3rd Earl of **Rosse**. Educated at Dublin and Cambridge, where he was Eleventh Wrangler in 1877, he became an engineering apprentice, then junior partner in a firm concerned in the development of electric lighting. He realized the need for a high-speed engine capable of driving dynamos efficiently, and as reciprocating engines were inherently unsuitable he was led to the concept of the steam turbine, for which he took out the first patents in 1884. He completed his first engine the same year; it produced 10 horse-power (7.5 kilowatts) at 18000 revolutions per minute. Five years later he set up his own company near Newcastle upon Tyne, building ever larger turbo-generators for the fast-growing electricity supply industry. He also built the first turbine-driven steamship, the *Turbinia* (1897), which caused a sensation at Queen Victoria's Diamond Jubilee Naval Review with its top speed of 35 knots, much faster than any other ship at that time. He was elected FRS in 1898, knighted in 1911, and was the first engineer to be admitted to the Order of Merit (1927).

PASCAL, Blaise
(1623–62)

French mathematician, physicist and theologian, born in Clermont-Ferrand, the son of the local President of the Court of Exchequer. His mother having died, the family moved to Paris (1630), where the father, a considerable mathematician, personally undertook his children's education. Blaise was not allowed to begin a subject until his father thought he could easily master it. Consequently it was discovered that the 11 year old boy had worked out for himself in secret the first 23 propositions of **Euclid**, calling straight lines 'bars' and circles 'rounds'. Inspired by the work of **Desargues**, at 16 he published an essay on conics which **Descartes** refused to believe was the handiwork of a youth. It contains his famous theorem on a hexagram inscribed in a conic. Father and son collaborated in experiments to confirm **Torricelli**'s theory, unpalatable to the schoolmen, that nature does not, after all, abhor a vacuum. These experiments consisted in carrying up the Puy de Dôme two glass tubes containing mercury, inverted in a bath of mercury, and noting the fall of the mercury columns with increased altitude. This led on to the invention of the barometer, the hydraulic press and syringe. In 1647 he patented a calculating machine, later simplified by **Leibniz**, which Blaise had built to assist his father in his accounts. His correspondence with **Fermat** in 1654 laid the foundations of probability theory. Following his experience of a religious revelation in 1654, he joined his sister at the Jansenist retreat at Port-Royal (1655), and concentrated on theological works. Around that time he gave up mathematics almost completely, although in one last important contribution he solved the long-standing problem of the area of the cycloid; his publication of this work heralded the invention of the integral calculus.

PASTEUR, Louis
(1822–95)

French chemist, the father of modern bacteriology, born in Dôle. He studied at Besançon and at the École Normale Supérieure, and held academic posts at Strasbourg, Lille and Paris, where in 1867 he became Professor of Chemistry at the Sorbonne. His work was at first chemical, notably on tartrate crystals; he noticed that the crystals appeared in two mirror-image forms, one which would rotate plane polarized light to the left, and the other which would rotate light to the right. He discovered a living ferment, a micro-organism comparable in its powers to the yeast plant, which would, in a solution of paratartrate of ammonia, select for food the 'right-handed' tartrates alone, leaving the 'left-handed'. He went on to show that other fermentations—lactic, butyric, acetic—are essentially due to organisms, not spontaneous generation. He greatly extended **Schwann**'s researches on putrefaction, and gave valuable rules for making vinegar and preventing wine disease, introducing in this work the technique of 'pasteurization', a mild and short heat treatment to destroy pathogenic bacteria. After 1865 he tackled silkworm disease, his research leading to the revival of the silk industry in southern France following a devastating parasitic disease which was killing the silkworms; he also investigated injurious growths in beer, splenic fever, and fowl cholera. Pasteur's 'germ theory of disease' represented one of the greatest discoveries of the 19th century—he began to realize that disease was communicable through the spread of micro-organisms. He showed that it was possible to attenuate the virulence of injurious micro-organisms by exposure to air, by variety of culture, or by transmission through various animals. He thus demonstrated that sheep and cows 'vaccinated' with the attenuated bacilli of anthrax were protected from the evil results of subsequent inoculation with virulent virus; by the culture of antitoxic reagents, the prophylactic treatment of diphtheria, tubercular disease, cholera, yellow fever and plague was also found effective. He introduced a similar treatment for hydrophobia (rabies) in 1885. The Pasteur Institute, of which he became first director, was founded in 1888 for his research.

PAUL, Wolfgang
(1913–)

German physicist, born in Lorenzkirch. He studied in Munich and at the Technical University in Berlin, where he received his doctorate in 1939. He later joined the staff of the University of Göttingen (1944), becoming professor there in 1950. He simultaneously held a teaching post at Bonn University from 1952. He developed the 'Paul trap' to constrain electrons and ions within a small space for study; this allowed important advances in the accuracy with which atomic properties could be measured, and has been used in important tests of modern atomic theory. For this work he shared one-half of the 1989 Nobel Prize for Physics with **Dehmelt**, who had developed a similar technique (the other half of the prize was awarded to **Norman Ramsey**).

PAULI, Wolfgang
(1900–58)

Austrian–Swiss theoretical physicist, born in Vienna. He studied under **Sommerfeld** at Munich University, publishing a highly regarded article on relativity at the age of 20.

He received his doctorate in 1921, then worked at Göttingen University (1921–2) and with **Niels Bohr** at his institute in Copenhagen (1922–3) before becoming professor at Hamburg University (1923–8). In 1928 he moved to Zürich, became a Swiss citizen and was appointed professor at the Federal Institute of Technology. Previously it had been thought that the state of an electron in the atom could be described by three quantum numbers; Pauli resolved the anomalous **Zeeman** effect by demonstrating that a fourth 'spin' quantum number was required, with possible values −½ or +½. He then went on to formulate the 'Pauli exclusion principle' (1924), which states that no two electrons in an atom can exist in exactly the same state, with the same quantum numbers. This gave a clear quantum description of the distribution of electrons within different atomic energy states, and it was for this work that he was awarded the 1945 Nobel Prize for Physics. In 1931 in studies of beta decay he accounted for an apparent violation of energy conservation by suggesting that energy was carried away by a low-mass neutral particle; the 'neutrino' was later discovered. His studies in the early 1950s of the conservation of quantum properties in interactions paved the way for **Lee** and **Yang**'s discovery of parity non-conservation in weak interactions in 1956.

PAULING, Linus Carl

(1901–)

American chemist, born in Portland, Oregon. He was educated at Oregon State College and Caltech, receiving his PhD in 1925. After postdoctoral work in Munich, Zürich and Copenhagen, he was on the chemistry faculty at Caltech from 1927 to 1963, as full professor from 1931. Pauling's early work on crystal structures (1928) led to their rationalization in terms of ionic radii and greatly illuminated mineral chemistry. He then turned to the quantum-mechanical treatment of the chemical bond and made many important contributions, including the concept of the 'hybridization of orbitals', central to understanding the shapes of molecules. This period of his work generated two influential books; *Introduction to Quantum Mechanics* (1935, with E Bright Wilson) and *The Nature of the Chemical Bond* (1939). His interest in complex molecular structures led him into work in biology and medicine; he studied the structures of proteins and antibodies, and investigated the nature of serological reactions and the chemical basis of hereditary disease. During the past 20 years he has advocated the use of vitamin C in combating a wide range of diseases and infections, and his views have generated controversy. He has also been a controversial figure for his work in the peace movement and his criticism of nuclear deterrence policy. Pauling was awarded the Nobel Prize for Chemistry in 1954 and the Nobel Peace Prize in 1962. He became a Foreign Member of the Royal Society in 1948 and was awarded its Davy Medal in 1947. He was elected an Honorary Fellow of the Chemical Society in 1943.

PAVLOV, Ivan Petrovich

(1849–1936)

Russian physiologist, born near Ryazan, the son of a priest. Pavlov studied natural sciences, graduating from St Petersburg in 1875, and medicine, receiving his doctorate 1879. From 1886 he worked at the Military Medical Academy in St Petersburg. He became Professor of Pharmacology (1890), Professor of Physiology (1895), and Director of the Institute of Experimental Medicine (1913). Pavlov's work was concerned with three main areas of physiology: the circulatory system (1874–88), the digestive system (1879–97) and higher nervous activity including the brain (1902–36). He developed a series of surgical preparations for the study of digestion in dogs, and his investigations, which included studies of the nervous control of salivation and the role of enzymes in digestion, provided insights of great value for the clinical pathology of the gut. For his work on digestion, Pavlov was awarded the Nobel Prize for Physiology or Medicine in 1904. Pavlov is most famous for his work which showed that if a bell is sounded whenever food is presented to a dog, it will eventually begin to salivate when the bell is sounded without food being presented. This he termed a 'conditioned' or acquired reflex, established by the involvement of the cortex of the brain in modifying innate reflexes. He regarded this phenomenon as of fundamental importance, and it was the starting point for subsequent studies of experimental psychoses, human psychic disorders, and his theories of animal and human behaviour.

PAYNE-GAPOSCHKIN, Cecilia Helena, née **Payne**

(1900–79)

British-born American astronomer, born in Wendover, Buckinghamshire, the daughter of a barrister. She attended St Paul's Girls' School in London and entered Newnham College, Cambridge, as a scholar in 1919 to study natural sciences. Having listened to a lecture by **Eddington**, she determined to become an astronomer, and on graduating from Cambridge in 1922 was able to fulfil her wish by going to the USA to work under

Shapley at Harvard College Observatory. The topic of her doctoral thesis (1925), *Stellar Atmospheres*, continued to be the special field of research which led to her pioneering work on the determination of the relative abundances of chemical elements in stars of various types and in the universe at large. She remained at Harvard for the rest of her life becoming the first woman professor there (1956). With her husband and colleague Sergei I Gaposchkin, she masterminded an immense programme of identifying and measuring variable stars on photographic plates, involving more than a million observations, which resulted in a catalogue of variable stars published in 1938. A similar project on variable stars in the Magellanic Clouds was later carried out by the Gaposchkins in 1971.

PEANO, Giuseppe

(1858–1932)

Italian mathematician, born in Cuneo. He was educated at the University of Turin, where he later taught and became Extraordinary Professor of Infinitesimal Calculus (1890) and full professor in 1895. He did important work on differential equations and discovered continuous curves passing through every point of a square. This deepened his distrust of intuitive mathematics, and he moved to mathematical logic. He advocated writing mathematics in an entirely formal language, and the symbolism he invented became the basis of that used by **Bertrand Russell** and **Whitehead** in their *Principia Mathematica*. He also promoted Interlingua, a universal language based on uninflected Latin.

PEARSON, Karl

(1857–1936)

English mathematician and scientist, born in London. He turned from the law to mathematics, becoming Professor of Applied Mathematics (1884) and Galton Professor of Eugenics (1911) at University College London. He published *The Grammar of Science* (1892) and works on eugenics, mathematics, and biometrics. He was a founder of modern statistical theory, and his work established statistics as a subject in its own right. His developments in this field included the introduction of the chi-square test of statistical significance, and the concept of standard deviation. Pearson was also motivated by the study of evolution and heredity. In his *Life of Galton* (1914–30) the head of the Eugenics Laboratory applied the methods of his science to the study of its founder. He founded and edited the journal *Biometrika* (1901–36) and wrote on the history of statistics.

PECQUET, Jean

(1622–74)

French anatomist, born in Dieppe. Pecquet studied in Paris, proceeding to Montpellier where he received his MD in 1652. He was to serve as physician to many Paris notables. He built upon, but corrected, **Aselli**'s researches on the lacteals. As Pecquet described in his *Experimenta nova anatomica* (1651), while performing a vivisection experiment on a digesting frog, he was the first to see clearly the thoracic duct, through which the lacteal vessels discharge chyle into the veins. His discovery was warmly welcomed by **Caspar Bartholin**, who clearly distinguished the chylous vessels from the lymphatics; but debate regarding the lymphatic and lacteal system remained intense throughout the rest of the century.

PEDERSEN, Charles

(1904–90)

American chemist, born in Pusan, Korea. Although his mother was Japanese and his father a Norwegian mining engineer, he studied chemical engineering in the USA (University of Dayton, Ohio) and took a master's degree in organic chemistry at MIT. Throughout his life he worked for Du Pont de Nemours as a research chemist, and the work for which he is best known was not published until 1967, almost at the end of his professional life. By accident he prepared a cyclic polyether, a molecule shaped rather like a crown and to which the name 'crown ether' was given. He found that many compounds of this type bind alkali metal ions (sodium, potassium) very strongly, making alkali metal salts soluble in organic solvents. His discovery initiated the study of guest–host chemistry and gave insight into the means by which metal ions are transported across membranes in living organisms. He retired in 1969 and shared the Nobel Prize for Chemistry with **Cram** and **Lehn** in 1987.

PEEBLES, Phillip James Edwin

(1935–)

Canadian–American cosmologist, born in Winnipeg and educated at the University of Manitoba, where he graduated in 1958. He went on to Princeton University as a graduate student, obtaining his PhD in 1962, and made his academic home there; he has been Professor of Physics at Princeton since 1965 and Einstein Professor of Science since 1985. Peebles's first research, begun in 1964 in collaboration with **Dicke** and others, was an attempt to detect by radio astronomy a cosmic background radiation predicted to fill the universe as a remnant of its early 'primeval fireball' phase. Their observation of such radiation (1966) followed shortly after its original discovery by **Penzias** and **Robert Wilson**. Peebles was the first to make reliable calculations of the abundances of helium and deuterium in the fireball, and to study the physics of the early universe (1968). This topic, which has continued to occupy him, includes the analysis of the observed distribution of galaxies in order to determine to what extent they are clustered in space, and the study of the mechanisms of galaxy formation from small irregularities in the expanding universe. His recent studies (1988) consider the implications of observed large-scale velocities of galaxies over and above the general expansion of the universe. Peebles is one of the leading cosmologists of his generation; his books *Physical Cosmology* (1971) and *The Large Scale Structure of the Universe* (1979) are classics in the field. He was awarded the Eddington Medal of the Royal Astronomical Society in 1981, and elected FRS in 1982.

PEIERLS, Sir Rudolf Ernst
(1907–)

German–British theoretical physicist, born in Berlin. Educated in Berlin, he studied under **Sommerfeld** in Munich and **Heisenberg** in Leipzig and became **Pauli**'s assistant in Zürich. Research at Rome, Cambridge and Manchester universities followed, and in 1937 he was appointed professor at Birmingham University. In 1963 he moved to Oxford University, where he was Wykeham Professor of Physics until his retirement in 1974. He studied the theory of solids and analysed how electrons move in them; he also developed the theory of diamagnetism in metals, and in nuclear physics he worked on the problem of how protons and neutrons interact. During World War II, Peierls and **Otto Frisch** studied uranium fission and the neutron emission that accompanies it with a release of energy. In a report in 1940 they showed that a chain reaction could be generated in quite a small mass of enriched uranium, giving an atomic bomb. The British government took this up and Peierls led a group developing ways of separating uranium isotopes and calculating the efficiency of the chain reaction. The work was moved to the USA as part of the combined Manhattan Project (1943). He was awarded the Royal Society's Royal (1959) and Copley (1986) medals, and knighted in 1968.

PEIRCE, Charles Sanders
(1839–1914)

American philosopher, logician and mathematician, born in Cambridge, Massachusetts, the son of the mathematician Benjamin Peirce. He graduated from Harvard in 1859 and began his career as a scientist, working for the US Coast and Geodetic Survey from 1861. In 1879 he became a lecturer in logic at Johns Hopkins University, but he left in 1894 to devote the rest of his life in seclusion to the private study of logic and philosophy. In his scientific work, he developed the theory of gravity measurement using pendulums, and conducted gravity experiments in Europe and North America. He also made an early determination of the metre in terms of a wavelength of light. In philosophy, he was a pioneer in the development of modern, formal logic and the logic of relations, but is best known as the founder of pragmatism, which he later named 'pragmaticism' to distinguish it from the work of the philosopher William James. His theory of meaning helped establish the new field of semiotics, which has become central in linguistics as well as philosophy. His enormous output of papers was collected and published posthumously.

PELL, John
(1610–85)

English mathematician and clergyman, born in Southwick, Sussex. A brilliant student at Cambridge, he was appointed Professor of Mathematics at Amsterdam in 1643 and lecturer at the New College, Breda, in 1646. Employed by Oliver Cromwell, first as a mathematician and later in 1654 as his agent, he went to Switzerland in an attempt to persuade Swiss Protestants to join a Continental Protestant league led by England. In 1661 he became Rector at Fobbing in Essex and in 1663 Vicar of Laindon. In mathematics, he is remembered chiefly for the equation $x^2 = Ay^2 + 1$ for integers x, y and A, mistakenly named after him by **Leonhard Euler**, and for introducing the division sign ÷ into England.

PELLETIER, Pierre Joseph
(1788–1842)

French chemist noted for his work on alkaloids, born in Paris. He qualified as a pharmacist in 1810 and became professor and later assistant director at the School of Pharmacy in Paris, at the same time running a pharmacy and a chemical manufacturing business. His first research was on gum resins and other natural products such as amber and toad venom. From 1817 to 1821 he collaborated with **Caventou**. They investigated the green pigment of leaves, naming it chlorophyll, and won international fame for their studies of alkaloids, a class of complex organic compounds which have powerful physiological effects on human beings and other animals, and are found in some flowering plants. Pelletier and Caventou isolated many alkaloids that are important to medicine and industry, including strychnine, quinine and caffeine. This research marked the beginning of alkaloid chemistry, led to more careful preparation of natural drugs and opened up the possibility of producing them synthetically. When his collaboration with Caventou came to an end, Pelletier continued research on his own, particularly on the alkaloids of opium. In 1823, with **Dumas**, he proved that alkaloids contain nitrogen in a ring structure, and in 1838, with Philippe Walter, he discovered toluene. He died in Paris.

PELTIER, Jean Charles Athanase
(1785–1845)

French physicist, born in Ham. The son of an impoverished shoemaker, Peltier was apprenticed to a clockmaker at the age of 15. By 1806 he had established his own shop in Paris. An inheritance in 1815 made it possible for him to retire. Largely self-taught, he built upon his already wide reading with investigations into phrenology, anatomy, microscopy, meteorology and electricity, many of which were published. A protracted series of experiments led him, in 1834, to observe that at the junction of two dissimilar metals (antimony and bismuth), an electric current produced a rise or fall in temperature, depending on the direction of the current flow (Peltier effect). This phenomenon was given new significance in the subsequent work of **Joule** and **Kelvin** on thermodynamics and thermoelectricity.

PENCK, (Friedrich Carl) Albrecht
(1858–1945)

German geographer, glaciologist and geomorphologist, born and educated in Leipzig. He became Professor of Physical Geography at Vienna (1885–1906) and at Berlin (1906–26), where he became President and Director of the Oceanographic Institution and Museum. He examined the sequence of past Ice Ages, providing a basis for later work on the European Pleistocene. In 1894 he produced his classic *Morphology of the Earth's Surface*; he identified various topographic forms and is believed to have introduced the term geomorphology. Penck also studied glacial geology in the Alps and processes of mountain building in relation to gravity and sedimentation.

PENFIELD, Wilder Graves
(1891–1976)

American-born Canadian neurosurgeon, born in Spokane, Washington. After undergraduate studies at Princeton, he went to Oxford University as a Rhodes Scholar in 1914, and studied physiology under **Sherrington**, but the outbreak of World War I interrupted his studies. Wounded in the war, he returned to the USA, where he finished his medical education at Johns Hopkins University. Further scientific study in Oxford and Spain prepared him for his experimental neurosurgical work, which he developed in conjunction with surgical practice in New York at the Presbyterian Hospital and the College of Physicians and Surgeons at Columbia University. In 1928 he moved to a neurosurgical appointment at McGill University, where he was instrumental in founding the world-famous Montreal Neurological Institute where he became the first director (1934–60). An outstanding practical neurosurgeon, his experimental work on animals and on the exposed brains of conscious human beings helped in understanding the higher functions of the brain, and the causes of symptoms of brain disease such as epilepsy, and the mechanisms involved in speech. He became a Canadian citizen in 1934. Following his retirement in 1960, he began a successful second career as a novelist and biographer.

PENNEY, William George, Baron
(1909–91)

British physicist, born in Sheerness. Educated at London, Wisconsin and Cambridge, he became Professor of Mathematics at Imperial College, London, and worked at Los Alamos on the atom bomb project (1944–5). He later became Director of the Atomic Weapons Research Establishment at Aldermaston (1953–9), and Chairman of the UK Atomic Energy Authority (1964–7). He was the key figure in the UK's success in producing its own atomic (1952) and hydrogen bombs (1957). Knighted in 1952, he was created a life peer in 1967, and became Rector of Imperial College (1967–73).

PENROSE, Lionel Sharples
(1898–1972)

English geneticist, born in London. He studied at St John's College, Cambridge, receiving his BA in 1921. Following postgraduate research in psychology at Cambridge, he worked in psychiatry in Vienna, developing a strong interest in mental illness. Returning to the UK in 1925, he studied medicine and wrote his MD thesis on schizophrenia. During the 1930s he carried out a major survey into the causes of mental illness, the Colchester Survey. As a Quaker and pacifist, he spent World War II in Canada, as Director of Psychiatric Research for Ontario. In 1945 he became Galton

Professor of Eugenics at London. Penrose objected strongly to the attempts of 'eugenics' to improve the characteristics of human populations through the application of genetics, and changed his title to Professor of Human Genetics, having changed the name of the journal of the Galton Laboratory from *Annals of Eugenics* to *Annals of Human Genetics* in 1954. Under his direction (1945–65), the Galton Laboratory became an international centre for human genetics, conducting varied research projects, for example on Down's syndrome, the mapping of genes on chromosomes, and palm and finger prints, using statistical and cytogenetic techniques. He wrote *The Biology of Mental Defect* (1949) and *An Outline of Human Genetics* (1960).

PENROSE, Roger
(1931–)

British mathematical astronomer, born in Colchester, son of **Lionel Penrose**. His university education was at University College London, and he then went on to obtain a doctorate at Cambridge. Since 1966 he has been Rouse Ball Professor of Mathematics at Oxford. Penrose is known for his work on black holes; he showed (jointly with **Hawking** and R Geroch) that once collapse of a very massive star at the end of its life has started, the formation of a black hole singularity—a point of zero volume and infinite density—is inevitable. Penrose also put forward the hypothesis of 'cosmic censorship' which intimates that all singularities are hidden behind the event horizon of the black hole, within a region from which no matter or light can escape. At Oxford Penrose has been working on 'twistor theory' in which the four dimensions of space–time are quantized by imaginary numbers as opposed to real numbers.

PENZIAS, Arno Allan
(1933–)

American astrophysicist, born in Munich, Germany. A refugee with his family from Nazi Germany, he was educated at Columbia University, New York, and joined the Bell Telephone Laboratories in 1961, finally becoming Executive Director of Research and Communication Science there. In 1963 Penzias and his colleague **Robert Wilson** were assigned the task of tracing the radio noise that was interfering with Earth–satellite–Earth communications. Using a 20 foot horn reflector they discovered a 7.3 centimetre wavelength extraterrestrial signal that seemed to show no directional variation and did not vary with the positions of the Sun and stars. They also measured the corresponding 'temperature' (the temperature of a black body which would characteristically emit radiation of this wavelength) of this background radiation, first obtaining a value of 3.5 kelvin but subsequently revising this to 3.1 kelvin. At first it was assumed that the horn reflector was responsible, but in fact Penzias and Wilson had discovered the residual relic of the intense heat that was associated with the birth of the universe following the hot Big Bang. This was the cosmic microwave background radiation predicted to exist by **Gamow** and **Alpher** in 1948. In 1970, with Wilson and K B Jefferts, Penzias discovered the radio spectral line of carbon monoxide at 2.6 millimetres; this has since been used as a tracer of galactic gas clouds. Penzias and Wilson were awarded the Nobel Prize for Physics in 1978, along with **Kapitza**.

PEREY, Marguerite Catherine
(1909–75)

French physicist, born in Villemomble. She was educated in Paris and from 1929 worked at the Radium Institute under **Marie Curie**. She later moved to the University of Strasbourg, becoming Professor of Nuclear Chemistry in 1949 and Director of the Centre for Nuclear Research in 1958. During studies of the radioactive decay of actinium-227, Perey discovered the element francium (originally known as actinium K) in 1939. She became the first female member of the French Academy of Sciences (1962). Her death was thought to be due to prolonged exposure to radiation.

PERKIN, William Henry Jr
(1860–1929)

English chemist, born in Sudbury, the eldest son of **William Henry Perkin Sr**. After study at the Royal College of Science in London, he went to Würzburg (1880–2) to work with **Wislicenus** and to Munich (1882–6) to work with **Baeyer**. He returned to Britain to take up a position at Heriot-Watt College in Edinburgh, and in 1892 became professor at Owens College in Manchester. His final position was as Waynflete Professor of Chemistry at Oxford (1912–29). As a young man in Germany he worked on small ring compounds, and his results were used by Baeyer in his ring strain theory. Subsequent work was entirely concerned with the elucidation of the structures of natural products by degradation studies. He had particular success with a number of alkaloids including berberine, harmine, cryptopine, strychnine and brucine. Although a fine chemist and a skilled practical worker, he did not have the innovative talent of his father or indeed, of his brothers-in-law **Kipping** and **Arthur Lapworth**. He was elected FRS in 1890.

PERKIN, Sir William Henry Sr

(1838–1907)

English chemist, born in Shadwell, London. He became fascinated by chemistry while at school and in 1853 enrolled at the Royal College of Chemistry to study under **Hofmann**. At the age of 17 he became Hofmann's personal assistant, but his first major discovery was made in a private laboratory he fitted out in his home. Here, in 1856, Perkin attempted to synthesize quinine, much in demand for the treatment of malaria, by oxidizing allyltoluidine with potassium dichromate. He was not successful, but after conducting the same experiment with toluidine, he was able to extract a brilliant purple dye, subsequently named mauveine. With the encouragement of the dyeing firm J Pullar and Son, Perth, Perkin at the age of 18 built a factory near Sudbury to manufacture mauveine. A craze for things purple swept Victorian society; Perkin's venture was a great commercial success and initiated the modern synthetic dyestuffs industry, which introduced a new range of colours into human life. He also devised synthetic procedures for the production of the natural dye alizarin. In 1874 Perkin sold his dye works in order to devote himself exclusively to his chemical researches, which reached an important point in 1881 when he observed the magnetic rotatory power of a number of the organic compounds he had made. He continued many investigations into the relationship between molecular constitution and the physical properties of organic compounds. He held no academic post and conducted his research in his private laboratory. Such is his standing among organic chemists that the section of the Royal Society of Chemistry concerned with organic chemistry is called the Perkin Division. He received many honours both in Britain and in the USA, including election to the Royal Society in 1866 and a knighthood in 1906. His three sons all became distinguished chemists.

PERRIN, Jean Baptiste

(1870–1942)

French physicist, born in Lille. He was educated at the École Normale Supérieure in Paris and from 1898 to 1940 was on the physical chemistry staff of the University of Paris, as full professor from 1910. In 1940 he escaped to the USA following the invasion of France; he died in New York in 1942. Perrin's earliest work was on cathode rays and helped to establish their nature as negatively charged particles. However, he is most remembered for his studies of the Brownian movement. Through elegant experimental work he demonstrated that the suspended particles which show Brownian motion in colloidal solutions essentially obey the gas laws and he used such systems to determine a fairly accurate value for the Avogadro number. His book *Les Atomes* (1913), which described this work, became a classic. He was awarded the Nobel Prize for Physics in 1926. Elected a Foreign Member of the Royal Society in 1918, he served as President of the French Academy of Sciences in 1938.

PERSOON, Christiaan Hendrik

(1761–1836)

South African botanist, born at the Cape of Good Hope. He studied theology at Halle, medicine at Leiden, and medicine and natural science at Göttingen. In 1799 he received a PhD at Erlangen, and he moved to Paris in 1802. He corresponded widely and influenced many other botanists. His *Synopsis Fungorum* (1801) is considered to be the basis of modern mycology. Persoon's work is the starting point for the nomenclature of the Gasteromycetes, Uredinales and Ustilaginales. He sought to describe briefly all phanerogams then known in *Synopsis plantarum* (1805–7), and he also produced the incomplete *Mycologia Europaea*. In 1818 he expressed the incorrect view that some fungi grew from spores, while others were formed by spontaneous generation. He gave his botanical collections to the Rijksherbarium in Leiden in 1828, in exchange for a pension. Leiden publishes the mycological journal *Persoonia* named after him.

PERT, Candace, née Beebe

(1946–)

American pharmacologist born in Manhattan, New York. Educated at Bryn Mawr College and Johns Hopkins University Hospital Medical School, she received her PhD in pharmacology in 1974. She stayed at Johns Hopkins as Research Fellow (1974–8) and Research Pharmacologist (1978–82) until her appointment as chief of the section of brain chemistry of the National Institute of Mental Health. Her doctoral research, under the supervision of **Snyder**, was stimulated by the realization that the highly specific effects of synthetic opiates at very small doses indicated that they must bind to highly selective target receptor sites. She began a search for these sites, using radioactively labelled compounds. In 1973 she first reported the presence of such receptors in specialized areas of the mammalian brain. From this arose the suggestion that there might be natural opiate-like substances in the brain that used these sites, as later discovered by **Kosterlitz** and **Hughes**. Her research continues on the chemical characteristics of brain tissue and the relationships of chemicals to neural functioning.

PERUTZ, Max Ferdinand
(1914–)

Austrian-born British biochemist, born in Vienna. After graduating at Vienna, he emigrated to the UK (1936) where, apart from a brief alliance with chymotrypsin, he has worked single-mindedly on the structure of haemoglobin at Cambridge ever since. He became Director of the Medical Research Council (MRC) Unit for Molecular Biology (1947–62) and since 1962 has been Director of the MRC Laboratory for Molecular Biology. His determination of haemoglobin structure reflects the development of X-ray crystallography. By 1939, Perutz had determined the first crystal parameters indicating four haem groups in one plane, and observed effects shown later (1953) by **Kendrew** to reflect structural changes in the molecule. The presence of the alpha helix, discovered by **Pauling** in 1951, was predicted in the same year to occur in haemoglobin by Perutz, **Crick** and Sir **Lawrence Bragg**. By the time Kendrew had published the structure of myoglobin at 2 angstroms resolution, Perutz had determined the haemoglobin structure to 5.5 angstroms. Amino acid sequence studies by others, facilitating comparison between myoglobin and haemoglobin, allowed Perutz to predict the detailed distribution of amino acids in haemoglobin (1964); confirmation at 2.8 angstroms was completed in 1968. Perutz and Kendrew were awarded the 1962 Nobel Prize for Chemistry. Since then, Perutz has established the nature of the subunit interactions in relation to oxygen binding and its control, the effects of genetic variants, the evolutionary development and numerous other aspects of haemoglobin. His publications include *Mechanisms of Cooperativity and Allosteric Regulation in Proteins* (1990). He was elected FRS in 1954.

PETERS, Sir Rudolf Albert
(1889–1982)

English biochemist, born in London. He studied at King's College, London, and at Cambridge, where he assisted **Barcroft** on haemoglobin, **Hill** on muscular contraction, and **Frederick Gowland Hopkins**. In 1917 he moved to Porton Military Research Establishment where he worked with Barcroft on problems of chemical warfare, subsequently receiving his MD at St Bartholomew's Hospital, London. In 1918 he became a lecturer at Oxford, where he was later appointed Whitley Professor of Biochemistry (1923–54), and began research on thiamine and the vitamin B complex. Using the brain from a vitamin-deficient pigeon for assay, he was able to resolve the palliative effects of thiamine confused by others. He independently isolated thiamine from yeast and showed that, as cocarboxylase (thiamine pyrophosphate), it mediates the energy-producing oxidation of pyruvate formed in the glycolysis pathway—this was the first demonstration of how vitamins participated in intermediary metabolism (1937). With **Ochoa**, he partially resolved the cofactors necessary for this (pyruvate dehydrogenase) reaction. His observation that it is inhibited by arsenicals led his laboratory to discover in 1946 that 2,3-dimercaptopropanol (British anti-lewisite) is an effective treatment for victims of the blistering war gas, lewisite, incidentally triggering a worldwide investigation into the reactive sulphydryl groups of enzymes. Peters discovered fluoracetate and fluoroleic acid, poisonous components of certain plants, which, when ingested, form fluorcitrate in the **Hans Krebs** cycle ('lethal synthesis') and inhibit aconitase, the second enzyme of the cycle. He also found that fluoride combined organically in bone (1969). Peters was elected a Fellow of the Royal Society (1935), received its Royal Medal (1949) and was knighted in 1952. He delivered the first Frederick Gowland Hopkins memorial lecture of the British Biochemical Society in 1954.

PETIT, Alexis Thérèse
(1791–1820)

French physicist, born in Vesoul. Petit's record at the École Centrale, Besançon, and the École Polytechnique (1807–9) was outstanding. Graduating with the highest distinction, he was rapidly employed to teach physics at the Lycée Bonaparte (1810). By 1811 he had completed a doctorate on capillary action and at only 23 he became Professor of Physics at the École Polytechnique (1815). Working with his brother-in-law **Arago** on the refraction of light in gases, he became a supporter of the new wave theory of light. From about 1815 Petit worked with his friend **Dulong** towards a competition relating to the laws of cooling. In 1818 the prize was awarded to them for their exemplary experimental determination of the specific heats of solids, the expansion of mercury, and their considered advocacy of the air thermometer. They announced their now famous law in April 1819: for solid elements the product of the specific heat and the atomic weight is the same. Petit died soon afterwards from consumption.

PETTENKOFER, Max von
(1818–1901)

German chemist, born near Neuberg. He was educated at the University of Munich, worked briefly in the court pharmacy, abandoned an unpromising career as an actor in order to marry, and then studied medical chemistry under **Liebig** at Giessen. In 1745 he became an assistant at the Royal Mint and found a way to separate gold, silver and platinum by slow cooling. In 1745 he moved to the Chair of Medical Chemistry at Munich. He noticed in 1750 that there was apparently a connection between the atomic weight of an element and its chemical behaviour. He lacked the funds to make the more accurate assessments of atomic weight that were necessary to verify this, but his observation preceded **Mendeleyev**'s historic formulation of the periodic table by nearly 20 years. Pettenkofer's greatest achievements, however, lay in the field of hygiene and public health. He emphasized the role of chemistry in nutrition, sanitation and forensic medicine. He established the vital role of protein in diet and showed that diabetics used extra protein and fat. After being asked to report on possible health hazards from a new system of central heating for Maximillian II, he began to study the effects of ventilation systems, clothing, building materials and soils on health. He was responsible for purifying the Munich water supply and the introduction of its sewage system. He also advocated heating art galleries, pointing out that the varnish on oil paintings became opaque with damp. Pettenkofer's work on hygiene and epidemiology brought him international renown (shadowed only by his refusal to acknowledge **Koch**'s discovery that cholera was carried in drinking water) and in 1879 he was established as director of the world's first Institute of Hygiene. After a period of illness and depression he committed suicide in Munich.

PEUERBACH, Georg von
See **PURBACH, Georg von**

PFEFFER, Wilhelm Friedrich Philipp
(1845–1920)

German botanist, born near Cassel. Trained as a pharmacist, he became a specialist in plant physiology, was appointed professor successively at Bonn, Basle, Tübingen and Leipzig, and was noted particularly for his researches on osmotic pressure. He experimented with aniline dyes on plant cells, and was able to establish the nature and structure of vacuoles. He concluded that living protoplasts could modify their permeability, allowing non-osmosing substances to pass. He hinted at the role of plasmodesmata in solute transport between cells. He also suggested that streaming movements within cells might be a factor in solute transport, and alluded to the importance of transpiration in plant growth and development. Pfeffer's work on chloroplasts showed the effects of light intensity and other physical and chemical factors on the process of photosynthesis. He also studied nitrogen metabolism in leguminous plants. His *Handbuch der Pflanzenphysiologie* (1881) was a standard work on plant physiology for many years.

PFEIFFER, Richard Friedrich Johannes
(1858–1945)

German bacteriologist, born near Posen (now Poznań, Poland). He studied under **Koch**, and became professor at Berlin (1894), Königsberg (1899) and Breslau (1901). It was during the influenza epidemic of 1889–92 that he discovered the bacillus *Haemophilus influenzae* in patients' throats (1892) and suggested that it was the cause of the disease. However, it was later shown that influenza is caused by a viral infection and that the bacillus is responsible for many of the complications. Pfeiffer's most significant discovery was his observation, for the first time, of a complex immune reaction (1894). He injected live cholera vibrios into guinea pigs which had already been immunized, then extracted some of the germs. Examining the extract under a microscope, he observed the germs becoming motionless then swelling and finally disintegrating, in a process which he named 'bacteriolysis'. He showed that the same process occurred *in vitro*, and that the reaction would cease when heated to over 60 degrees Celsius. This work led **Bordet** to study the immune system, resulting in the discovery of complement. Pfeiffer published books on hygiene and microbiology. He was presumed dead in 1945.

PFLÜGER, Eduard Friedrich Wilhelm
(1829–1910)

German physiologist, born in Hanau. He studied medicine at Marburg and then under **Johannes Müller** in Berlin, receiving his MD in 1855. In 1859 he was appointed Professor of Physiology at Bonn, where he remained for the rest of his career. He became a fine histologist. Like many of his contemporaries, Pflüger placed great stress on precise physiological quantification and improved technology. He contributed to the construction of the mercurial blood-pump. Much of his research focused on the role of physical and chemical forces in the body (eg in regard to gas exchange in respiration). Pflüger devoted much time to the metabolism of the nutritive substances—protein, carbohydrates and fat. Sugars fascinated him; unlike most researchers, he believed pancreatic diabetes was a nervous disturbance. He was a fierce polemicist and became involved in violent controversies.

PIAZZI, Giuseppe
(1746–1826)

Italian astronomer, born in Ponte, northern Italy. He became a Theatine monk (1764), was appointed Professor of Mathematics at Palermo in Sicily in 1780, and founded on behalf of the Bourbon government two observatories, at Naples and at Palermo. After 22 years of observing at Palermo with a vertical circle made by the English optician Jesse Ramsden, he published a monumental catalogue of 7 646 stars (1813). In the course of his observations he discovered on the night of 1 January 1801, the very first minor planet (or asteroid) which he named Ceres after the tutelary deity of Sicily.

PICARD, (Charles) Émile
(1856–1941)

French mathematician, born in Paris. Professor at the Sorbonne (1886–97), he was elected a member of the French Academy of Sciences in 1889. Picard was specially noted for his work in complex analysis, and integral and differential equations. He established that a complex function takes every value, but at most two in the neighbourhood of any essential singularity it may possess. With Georges Simart he wrote the definitive work of its generation on the theory of complex surfaces and integrals over them, generalizing **Riemann**'s work. Their work was complemented by the Italian geometers Guido Castelnuovo and Federigo Enriques from a more geometrical approach, and together they produced the first rich theory of complex surfaces. Picard also introduced the method of 'successive approximations', a powerful technique for determining whether solutions to differential equations exist.

PICCARD, Auguste Antoine
(1884–1962)

Swiss physicist, born in Basle. He became professor at Brussels in 1922 and held posts at Lausanne, Chicago and Minnesota universities. With his brother Jean Felix, he ascended 16–17 kilometres by balloon (1931–2) into the stratosphere. In 1948 he explored the ocean depths off west Africa in a bathyscaphe constructed from his own design. His son Jacques, together with an American naval officer, Donald Walsh, established a world record by diving more than seven miles in the US bathyscaphe *Trieste* into the Marianas Trench of the Pacific Ocean in January 1960.

PICKERING, Edward Charles
(1846–1919)

American astronomer, born in Boston. Educated at Harvard, he became Professor of Physics at MIT. In 1876 he was appointed Professor of Astronomy and director of the observatory at Harvard, a post he held for 42 years. He is remembered for his work on stellar photometry and in 1884 he published the first catalogue of 4 260 visual stellar magnitudes, the *Harvard Photometry*. A revised version containing 45 000 stars was published in 1908. In 1889 he discovered the first spectroscopic binary star; these stars are so close together that they are best distinguished by the **Doppler** movement of their spectral lines. Around 1890 Pickering introduced the colour index, this measuring the difference between a star's photographic (blue) and visual (yellow) magnitudes. The colour index is related to the surface temperature of a star. At this time he also discovered the first variable star in a globular cluster. Pickering oversaw the production of the *Henry Draper Catalogue*, which classified 225 000 stars according to their spectra, and was responsible for the *Photographic Map of the Entire Sky* (1903). This consisted of 55 plates taken at Harvard and its sister observatory at Arequipa in Peru. His brother William Henry Pickering (1858–1938), also an astronomer, discovered Phoebe, the 9th satellite of Saturn (1919); he was in charge of an observation station at Arequipa, and from 1900 was director of a station at Mandeville, Jamaica.

PICKERING, Sir George White
(1904–80)

English medical scientist, born in Whalton, Northumberland. He graduated in natural sciences from Pembroke College, Cambridge (1926), and trained in medicine at St Thomas's Hospital, qualifying in 1928. He was appointed to the staff of the Medical Research Council (1931) and spent eight years working with Sir **Thomas Lewis** at University College Hospital in London. In 1932 he presented results of experiments on headaches and went on to work on hypertension. He studied the causes of increased peripheral resistance, hereditary factors, and aspects of atheromatous disease. He also conducted important experimental work on the mechanism of pain in peptic ulcers. Pickering became lecturer in cardiovascular pathology in 1936, Professor of Medicine at St Mary's Hospital Medical School (1939–56) and Regius Professor of Medicine at Oxford (1956–68). He was a key figure in medical education in Britain from the 1950s, and wrote widely on historical and cultural issues. He was knighted in 1957.

PICKERING, William Hayward
(1910–)

New Zealand–American engineer and physicist, born in Wellington. He emigrated to the USA in 1929 and received his PhD in physics at Caltech in 1936. He undertook some pioneering high-altitude cosmic-ray research with **Millikan**, then in 1944 joined the Jet Propulsion Laboratory at Caltech and developed the first telemetry system to be used in US rockets. From 1954 until he retired in 1976 he was Director of the Jet Propulsion Laboratory, and initiated its space exploration programme that resulted in the launching of the first US satellite, Explorer I, in 1958. He supervised the Ranger lunar-impact flights in 1964–5, and the Mariner flights to Venus in 1962 and Mars in 1964–5. The latter flights provided the first close-up photographs ever taken of the surface of another planet, but were surpassed in 1976 when the two Viking spacecraft succeeded in landing probes on the surface of Mars which continued to transmit data back to Earth for more than two years. He was elected to the US National Academy of Sciences in 1962, and was awarded the National Medal of Science in 1976.

PICTET, Marc-Auguste
(1752–1825)

Swiss physicist, born in Geneva. Pictet qualified in law (1774), but his true interests lay in the natural sciences. He was appointed to the Chair of Philosophy at the Geneva Academy in 1786 on the resignation of his mentor, **Horace Bénédict de Saussure**. Pictet was prominent in the public life of Geneva, respected by Napoleon, and active in the scientific circles of post-revolutionary France. His studies in geology, meteorology and astronomy were characteristically wide-ranging. Best known throughout Europe was the *Essai sur le Feu* (1790) recording researches on heat and hygrometry. There he described how radiant heat could be reflected, like light. Pictet's journal, the *Bibliotèque Britannique*, sustained intellectual contact between Britain and the Continent during the Napoleonic Wars.

PICTET, Raoul Pierre
(1846–1929)

Swiss physicist, born in Geneva. Between 1868 and 1870 he studied physics and chemistry in Geneva and in Paris, returning to Geneva to devote himself to the study of low temperatures in connection with developing refrigeration techniques. He became Professor of Industrial Physics at the University of Geneva in 1879. In 1886 he moved to Berlin to establish an industrial research laboratory and to market his refrigeration inventions. Later he returned to Paris. Working in Geneva in 1877, he liquefied oxygen in bulk by 'cascade' cooling and compression. **Cailletet** in Paris liquefied oxygen independently at around the same time, but in much smaller quantities. Pictet later made an erroneous claim to have liquefied hydrogen.

PIERCE, John Robinson
(1910–)

American electrical engineer, born in Des Moines, Iowa. He graduated successively BS, MS and PhD at Caltech and worked in the Bell Telephone Laboratories from 1936 until 1971, when he returned to Caltech as Professor of Engineering. With wide scientific interests, most of them related in some degree to human or electronic communication, he made important discoveries in the fields of microwaves, radar and pulse-code modulation. In collaboration with **Shockley** he was one of the first to devise an effective electrostatically focused electron multiplier, and later with Rudolf Kompfner he developed the travelling-wave tube which is now an essential part of microwave technology. In the 1950s he was one of the first to see the possibilities of satellite communication, taking a leading part in the development work that resulted in the launch of Echo in 1960 and Telstar in 1962. The plastic surface of Echo was coated with an aluminium film which merely reflected microwave radio signals back to Earth, but Telstar incorporated its own solar power module for signal amplification, allowing live transatlantic television transmissions to be made for the first time. As early as 1943 he had become fascinated by the possibilities of digital communication by pulse-code modulation, and published a paper on the subject in 1948. It was not until the development of the transistor in the 1960s, however, that the necessary low-power decoders and repeaters could be made. Pierce has published many books and technical papers as well as works of science fiction. In 1963 he was awarded the National Medal of Science.

PIMENTEL, George Claude
(1922–89)

American physical chemist, born in Rolinda, California. He was educated at the University of California, Los Angeles (PhD in chemistry, 1949), and advanced from instructor to associate professor of chemistry between 1949 and 1959. From 1959 he was Professor of

Chemistry at the University of California, Berkeley. Pimentel's best-known work was his pioneering of chemical lasers and his development of the method of trapping transient species in solid noble gases for spectroscopic study. He also devised rapid-scan infrared spectrometers, including those carried on NASA's Mariner 6 and 7 probes of Mars in 1969. In the 1960s Pimentel was the editor of a school textbook which was very influential, not only in the USA but (through translation) in many other countries. Among his many honours were the Priestley Medal of the American Chemical Society (1989), of which he was President in 1986, and honorary fellowship of the Royal Society of Chemistry (1987).

PINCUS, Gregory Goodwin
(1903–67)

American physiologist who developed the oral contraceptive pill, born in Woodbine, New Jersey. He graduated in science from Cornell University in 1924, undertook postgraduate study at Harvard (1924–7) and then visited Europe, working at Cambridge and Berlin universities (1927–30). He was a member of the biology faculty at Harvard (1930–8) and of the experimental zoology department at Clark University (1938–45). In 1944 he established, with Hudson Hoagland, the Worcester Foundation for Experimental Research, which became internationally renowned for work on steroid hormones and mammalian reproduction. In 1951, influenced by the birth control campaigner Margaret Sanger, he began work on developing a contraceptive pill. With John Rock and Min Chueh Chang he studied the antifertility effect of those steroid hormones which inhibit ovulation in mammals; in this way refertilization is prevented during pregnancy. Synthetic hormones became available in the 1950s and Pincus organized field trials of their antifertility effects in Haiti and Puerto Rico in 1954. The results were successful, and oral contraceptives ('the pill') have since been widely used, despite concern over some side effects. Their success is a pharmaceutical rarity; synthetic chemical agents do not usually show nearly 100 per cent effectiveness in a specific physiological action, or have such remarkable social effects.

PIPPARD, Sir (Alfred) Brian
(1920–)

English physicist, born in London. He was educated at the University of Cambridge, and after five years as a scientific officer at the Radar Research and Development Establishment (1941–5), he returned to Cambridge where he became Plummer Professor of Physics (1960–71) and Cavendish Professor of Physics (1971–82). He carried out important work on the electromagnetic response of superconductors at very high frequencies, and noticed in 1952 that the predictions made by **Fritz** and **Heinz London**'s electrodynamic equations did not correspond to his own experimental observations. This led him to introduce into one of the London equations a new fundamental parameter called the 'coherence length', highlighting the need to consider macroscopic superconductor volumes (on scales of up to 10^{-6} metres) when framing theoretical descriptions of their electromagnetic properties. The coherence length of a superconducting material became one of the most important descriptive parameters of this group of materials. Pippard is author of many texts including *Elements of Classical Thermodynamics* (1957), *Forces and Particles* (1972) and *Magnetoresistance* (1989). He was elected FRS in 1956 and knighted in 1975.

PLANCK, Max Karl Ernst Ludwig
(1858–1947)

German physicist, born in Kiel. Planck's father was Professor of Civil Law at the University of Kiel. When the family moved to Munich in 1867, Planck entered the Maximilian Gymnasium. Although musically gifted and fascinated, even at an early age, by philology, he decided to study science at the University of Munich from 1874. In Berlin (1877–8) he attended lectures given by **Kirchhoff** and **Helmholtz**, and read **Clausius**'s papers. Returning to Munich he completed his doctorate on the second law of thermodynamics (1879) and then lectured in the university until he became Professor of Theoretical Physics at Kiel (1885). Late in 1888 he succeeded Kirchhoff as Professor of Theoretical Physics at Berlin, where he remained until his retirement (1926). From 1880 Planck had studied the foundations of thermodynamics and also chemical equilibrium. Later, stimulated by experiments in progress at the Physikalisch-Technische Reichanstalt, he combined thermodynamics with electrodynamics in studies of black-body radiation. Where classical theories failed to match experimental results, Planck's formulation (1900) modelled them with great accuracy. He had relied on **Boltzmann**'s statistical interpretation of the second law of thermodynamics; and he had also assumed the 'discontinuous' emission of small discrete packets of energy: 'quanta'. This revolutionary work earned Planck the Nobel Prize for Physics in 1918, by which time conceptions of subatomic processes had changed completely; on such microscopic scales quantum theory took over. **Einstein** soon developed the quantization idea to explain the photoelectric effect (1905), and **Niels Bohr** applied quantum theory to problems of atomic structure (1913). Planck continued to hope for a synthesis of deterministic classical physics and quantum methods, and opposed the extreme indeterminism of **Heisenberg**. In 1930 Planck was elected President of the Kaiser Wilhelm Institute, but he resigned in 1937 in protest against the actions of the Nazi regime. He remained in Germany, however, and was eventually reappointed President of the renamed Max Planck Institute. His life was overshadowed with personal tragedy: his son Erwin was executed for involvement in the 1944 plot against Hitler's life.

PLANTÉ, Gaston
(1834–89)

French physicist, born in Orthy. In 1854 he was appointed a lecture assistant in physics at the Conservatoire des Arts et Métiers, and in 1860 Professor of Physics at the Association Polytechnique pour le Développement de l'Instruction Populaire, both in Paris. He followed up **Ritter**'s discovery of the secondary cell and constructed the first practical lead–acid storage battery or accumulator (1859). His battery was an important early demonstration of the interconvertibility of the various forms of energy; mechanical, electrical and chemical. He first charged his battery with a hand-driven magneto-electric machine, an early form of dynamo (converting mechanical into electrical and then into chemical energy). The charged battery could then be used in turning the magneto-electric machine which also behaved as an electric motor, reversing the process (converting chemical into electrical energy and finally back into mechanical energy).

PLASKETT, John Stanley
(1865–1941)

Canadian astronomer, born in Woodstock, Ontario, and educated at the University of Toronto where he graduated in mathematics, obtained a PhD (1899) and became assistant in the physics department. He joined the staff of the Dominion Observatory in Ottawa in 1903 in advance of its formal opening (1905), and was responsible for designing and making use of spectroscopes to measure radial velocities of stars. In 1917 he moved to Victoria, British Columbia, to take charge of the new Dominion astrophysical observatory which was equipped with a 72 inch reflector designed by him for stellar spectroscopy. He remained director of this observatory until his retirement in 1935. Among his many discoveries (1922) was a pair of the most massive stars then known, named after him. Plaskett is regarded as the founder of modern astronomy in Canada. His son and collaborator, Harry Hemley Plaskett (1893–1980), himself an outstanding solar spectroscopist, became Savilian Professor of Astronomy at Oxford.

PLATEAU, Joseph Antoine Ferdinand
(1801–83)

Belgian physicist, born in Brussels, Professor of Physics at Ghent from 1835. In his study of optics he damaged his own eyesight by looking into the Sun for 20 seconds in order to find out the effect on the eye. By 1840 he was blind, but he continued his scientific work with the help of others. The university appointed him to a research professorship in 1844, which freed him from all teaching duties. His most important researches were concerned with the surface properties of fluids, in the course of which he originated the technique of producing soap films in two- or three-dimensional wire frames, discovered rules governing the geometrical forms of such films, and realized that these are surfaces of minimum area. This stimulated substantial mathematical studies of minimum-area surfaces, but it was not until the 1930s that general mathematical solutions were obtained. Plateau discovered the tiny second drop, named after him, which always follows the main drop of a liquid falling from a surface. In 1873 he published the two-volume work *Statique expérimentale et théorique des liquides summis aux seules forces moléculaires*.

PLATO
(c.428–c.348 BC)

Greek philosopher, indisputably one of the most important philosophers of all time. He was the pupil (or at least the associate) of Socrates and the teacher of **Aristotle**, and this trio were the great figures in ancient philosophy. Plato was probably born in Athens, of a distinguished aristocratic family, but little is known of his early life. After the execution of Socrates, he took temporary refuge at Megara, and then travelled widely in Greece, Egypt, the Greek cities in southern Italy and Sicily. He returned to Athens in c.387 BC to found the Academy, which became a famous centre for philosophical, mathematical and scientific research, and over which he presided for the rest of his life. His corpus of writings consists of some 30 philosophical dialogues and a series of *Letters*. Taken as a whole, his philosophy has had a pervasive and incalculable effect on almost every period and tradition, and was an important stimulus in the development of science. His influence was rivalled only by that of his greatest pupil Aristotle, which was its principal competitor for much of the Hellenistic period, the Middle Ages and the Renaissance.

PLAYFAIR, John
(1748–1819)

Scottish mathematician, physicist and geologist, born in Benvie, near Dundee. He studied at St Andrews University and in 1785 became joint Professor of Mathematics at Edinburgh, where he produced a successful edition of **Euclid**'s *Elements*, but in 1805 he exchanged his appointment for the Chair of Natural Philosophy. He was a strenuous supporter of the **Hutton**'s uniformitarian theory in geology, and travelled widely to make geological observations. His *Illustrations of the Huttonian Theory* (1802) was a landmark in British geological writing. He published his contributions to mathematics in *Elements of Geometry* (1795).

PLAYFAIR, Lyon, 1st Baron Playfair
(1819–98)

Scottish chemist and statesman, born in Chunar, India. He studied medicine at Glasgow and Edinburgh, and under **Liebig** at Giessen. He worked as an industrial chemist in Manchester in association with **Mercer**, and later developed Mercer's ideas on catalysis. He was appointed Professor of Chemistry at the School of Mines in London, and ended his academic career as Professor of Chemistry at Edinburgh. In 1868 he was elected to Parliament, and from 1880 to 1883 he served as Deputy-Speaker in the House of Commons. Playfair described the nitroprussides, worked on vapour densities, and translated the works of **Liebig**, but his principal importance was as an administrator. He was a prominent member of many government committees, for example on public health, the Irish potato famine, and the reform of the civil service. For a period he was inspector of the schools of science in London. He was a member of the committee which organized the Great Exhibition of 1851, and in 1853 he was appointed secretary to the newly formed Department of Science and Art, subsequently helping to establish the Royal College of Science and the South Kensington Museum (which later became the Victoria and Albert Museum and the Science Museum). One of the first scientists to hold important public positions, he worked throughout his life to promote scientific and technical education, and to encourage industry to make use of scientific advances. He was elected FRS in 1848, knighted in 1883 and created a baron in 1892. He died in London.

PLINY, Gaius Plinius Secundus, known as 'the Elder'
(AD 23–79)

Roman writer on natural history. He came from a wealthy north Italian family owning estates at Novum Comum (Como), where he was born. He was educated in Rome, and when about 23 entered the army and served in Germany. He became colonel of a cavalry regiment and a comrade of the future emperor Titus, and wrote a treatise on the throwing of missiles from horseback, and compiled a history of the Germanic wars. He also made a series of scientific tours in the region between the Ems, Elbe and Weser, and the sources of the Danube. Returning to Rome in AD 52, he studied for the bar, but withdrew to Como and devoted himself to reading and authorship. Apparently for the guidance of his nephew, he wrote his *Studiosus*, a treatise defining the culture necessary for the orator, and the grammatical work, *Dubius Sermo*. By Nero he was appointed procurator in Spain, and following his brother-in-law's death (AD 71) became guardian of his sister's son, Pliny the Younger, whom he adopted. Vespasian, whose son Titus he had known in Germany, was now emperor, and became a close friend; but court favour did not wean him from study, and he brought down to his own time the history of Rome by Aufidius Bassus. A model student, amid metropolitan distraction he worked assiduously, and by lifelong application filled the 160 volumes of manuscript which, after using them for his universal encyclopedia in 37 volumes, *Historia Naturalis* (AD 77), he bequeathed to his nephew. In AD 79 he was in command of the Roman fleet stationed off Misenum when the great eruption of Vesuvius was at its height. Eager to witness the phenomenon as closely as possible, he landed at Stabiae (Castellamare), but had not gone far before he succumbed to the stifling vapours rolling down the hill. His *Historia Naturalis* alone of his many writings survives. Under that title the ancients classified everything of natural or non-artificial origin. Pliny adds digressions on human inventions and institutions, devoting two books to a history of fine art, and dedicates the whole to Titus. His observations, made at second-hand, show no discrimination between the true and the false, between the probable and the marvellous, and his style is inartistic, sometimes obscure. But he supplies information on an immense variety of subjects about which, but for him, we should have remained in the dark.

PLÜCKER, Julius
(1801–68)

German mathematician and physicist, born in Eberfeld into an old merchant family. He studied at the universities of Bonn, Heidelberg, Berlin and Paris, and was eventually appointed Professor of Mathematics at Bonn (1836). He was Professor of Physics there from 1847. Plücker started his academic career principally as a mathematician, making contributions to analytical geometry, notably to algebraic curves and line geometry (the analytical geometry of space). From 1846 until 1864 he moved into experimental physics and then returned to geometry. He corresponded with **Faraday**, investigated diamagnetism, originated the idea of spectrum analysis, and in 1859 anticipated **Hittorf**'s discovery of cathode rays, produced by electrical discharges in gases at low pressures. He was elected FRS in 1855 and awarded the Royal Society's Copley Medal in 1868.

POGGENDORFF, Johann Christian
(1796–1877)

German physicist and chemist, born in Hamburg into a wealthy family. His father lost nearly his entire fortune during the French occupation of Hamburg (1806–11) so the young Poggendorff had to work as a pharmacy assistant. He managed to study at the University of Berlin, and became Extraordinary Professor of Chemistry there from 1834. He made discoveries in connection with electricity and galvanism, and his inventions included a 'multiplying' galvanometer (1821), a magnetometer (1827), the compensating circuit for determining electromotive force, a thermopile and the mercurial air-pump. While collaborating with **Liebig** he coined the word 'aldehyde', and the modern chemical notation. He was highly influential as the editor of the journal *Annalen der Physik und Chemie* (1824–74), bringing out 160 volumes during his editorship. He also produced a useful two-volume collection of brief biographies of some 8000 physicists of all countries and periods until 1858, since extended to some 18 volumes.

POGSON, Norman Robert
(1829–91)

English astronomer, born in Nottingham. Though educated for a career in commerce, he followed his scientific bent by learning practical astronomy as an assistant at the private observatory of George Bishop in Regent's Park, London, after which he became assistant under Manuel Johnson at the Radcliffe Observatory, Oxford (1851–8). His particular interest was in the observation of minor planets, of which he discovered several, and variable stars. In connection with his studies of the light curves of these stars he proposed a logarithmic base for defining stellar magnitudes. In this system the ratios of brightnesses of stars are expressed in terms of differences in their magnitudes, an interval of five magnitudes being defined as equivalent to a factor of 100 in brightness. Pogson's scale in due course became and remains the universally adopted standard. Following a short time (1858–60) at John Lee's observatory at Hartwell House, Buckinghamshire, Pogson was appointed Government Astronomer at the Madras Observatory in India where he worked for the rest of his life, mainly on variable stars.

POINCARÉ, Jules Henri
(1854–1912)

French mathematician, born in Nancy. He studied at the École Polytechnique, and as an engineer at the École des Mines became Professor of Mathematics in Paris in 1881. Following the work of Immanuel Fuchs and **Klein**, he created the theory of automorphic functions, which blended new ideas in the theories of groups, non-Euclidean geometry and complex functions, and showed the importance of topological considerations in differential equations. Many of the basic ideas in modern topology—such as triangulation, homology, the Euler–Poincaré formula and the fundamental group—are due to him. In a paper on the three-body problem (1889) he opened up new directions in celestial mechanics, and began the study of dynamical systems in the modern sense. He tried to keep the French tradition of applied mathematics alive through his influential lecture courses on such topics as thermodynamics, optics, magnetism and electricity in the sense of **Maxwell** and **Heinrich Hertz**. In 1905 he came close to anticipating **Einstein**'s theory of special relativity, showing that the **Lorentz** transformations form a group. In his last years he published several articles (later collected as books) on the philosophy of science and scientific method, including *Science et méthode* (1909). He opposed the move towards the axiomatic foundations of mathematics and logic, and advocated a form of intuitionism.

POISSON, Siméon Denis
(1781–1840)

French mathematical physicist, born in Pithiviers. He was educated at the École Polytechnique under **Laplace** and **Lagrange**, and became the first Professor of Mechanics at the Sorbonne, achieving a leading position in the French scientific establishment.

He published extensively on mathematical physics, and although his work was criticized for lack of originality by many of his contemporaries, his contributions to potential theory and the transformation of equations in mechanics by means of Poisson brackets have proved of lasting worth. He is also remembered for his discovery of the 'Poisson distribution', a special case of the binomial distribution in statistics.

POLANYI, John Charles
(1929–)

Canadian physical chemist, born in Berlin, son of **Michael Polanyi**. He grew up in Manchester and studied chemistry at Manchester University, receiving his PhD in 1952. After postdoctoral work at the National Research Council in Ottawa and at Princeton, he joined the chemistry staff at the University of Toronto in 1956, and was made full professor in 1962. In 1974 he was given the title of University Professor. Polanyi has worked extensively on the infrared light emitted during chemical reactions. Analysis of such radiation gives information about the distribution of energy within molecular species and sheds light on the events which occur during reactions. For instance, Polanyi has elucidated details of the various steps of the chain reaction of hydrogen with chlorine. The technique complements the molecular beam method developed by **Herschbach** and **Yuan Tseh Lee**, who shared the 1986 Nobel Prize for Chemistry with Polanyi for these advances. He has also studied the effects on rates of reaction of selectively activating different modes of molecular motion of the reactants, and has written articles on science policy and on control of armaments. He was made a Companion of the Order of Canada in 1977. He was elected a Fellow of the Royal Society in 1971, receiving its Royal Medal in 1989, and became an Honorary Fellow of the Royal Society of Chemistry in 1991.

POLANYI, Michael
(1891–1976)

Hungarian–British physical chemist, social scientist and philosopher, born in Budapest. He qualified in medicine at the University of Budapest in 1913, but he was already seriously interested in chemistry, and had written on thermodynamics and topics in medical biochemistry. He studied physical chemistry with Georg Bredig at the Technische Hochschule, Karlsruhe, on several occasions. During World War I he was a medical officer in the Austrian army. From 1920 to 1923 he worked at the Kaiser Wilhelm Institute for Fibre Chemistry in Berlin, and then moved to the Institute of Physical Chemistry under **Haber**. In 1933, finding the political climate of Germany increasingly unpleasant, Polanyi accepted the Chair of Physical Chemistry at Manchester University. He built up an excellent school of physical chemistry, but his interests were already moving to wider cultural and philosophical matters. In 1948 he gave up the Chair of Physical Chemistry and was given a personal Chair in Social Studies. In 1958 he moved to Oxford as a Senior Research Fellow of Merton College. He was elected FRS in 1944. In Berlin Polanyi worked on X-ray diffraction by fibres and then began his studies of chemical kinetics which continued for some 25 years. He did important experimental work, particularly on reactions of free atoms with small molecules in the gas phase, in the expectation that such reactions would prove amenable to theoretical treatment. He was himself much involved in the development of transition state theory, and this was central to his chemical interests in Manchester. In this theory it is supposed that all reactions proceed via a highly energetic molecular entity, the activated complex, and this idea is fundamental to the understanding of rates and mechanisms of chemical change. His social and philosophical interests are best indicated by the titles of some of his books: *The Contempt of Freedom* (1940), *Full Employment and Free Trade* (1945), *Science, Faith and Society* (1946), *Personal Knowledge* (1958) and *Knowing and Being* (1969). His writings often met with suspicion and criticism in philosophical circles.

PONCELET, Jean Victor
(1788–1867)

French engineer and geometrician, born in Metz. A military engineer during Napoleon's Russian campaign, he was taken prisoner by the Russians on the retreat from Moscow. During this time he thought over the geometry he had learned from **Monge**, and his book on this subject *Traité des propriétés projectives des figures* (1822) made his name and revived interest in the development of projective geometry. He became Professor of Mechanics at Metz (1825–35) and Paris (1838–48). Poncelet sought to found geometry on basic principles as general as those of algebra, so that for example, one argument concerning a line and a conic would suffice for the cases where the line cuts, touches, or does not meet the conic. Disillusioned by hostile contemporary criticism directed at his approach, and despite the many fertile methods his book contained, he turned to the theory of machines and applications of mathematics to technology.

PONS, Jean-Louis

(1761–1831)

French astronomer, born in Peyre, near Dauphine, who at the age of 28 became the porter and door-keeper at the Marseilles Observatory. After tuition by the director he became the *astronome adjoint* in 1813. In 1819 he was appointed as director of the Lucca Observatory in northern Italy, and in 1825 he became director of the Florence Observatory. His main interest was comets, and his tenacity and keen eyesight enabled him to discover 37, this amounting to three-quarters of all the comets discovered between 1801 and 1827.

PONTECORVO, Guido

(1907–)

Italian–British geneticist, born in Pisa, where he studied agricultural science. Thereafter he supervised cattle breeding in Tuscany. He moved to the Institute of Animal Genetics in Edinburgh in 1938, and from 1941 worked at the University of Glasgow, where he became Professor of Genetics in 1956. From 1968 to 1975 he was a member of the research staff at the Imperial Cancer Research Fund, London. In 1950 he described the parasexual cycle in fungi, which allows genetic analysis of asexual fungi. Soon afterwards he proposed that the gene is the unit of function in genetics, an idea developed by **Benzer** and others in 1955.

PONTRYAGIN, Lev Semyonovich

(1908–88)

Russian mathematician, born in Moscow. The loss of his sight in an accident at the age of 14 did not prevent him from graduating from Moscow University, where he became professor in 1935. One of the leading Russian topologists, he worked on topological groups and their character theory, on duality in algebraic topology, and on differential equations with applications to optimal control. His book *Topological Groups* (translated 1939) is still a standard work. At the end of his life he was an influential, and anti-Semitic, member of the Russian mathematical establishment.

POPE, Sir William Jackson

(1870–1939)

English chemist, born in London. After school in London he studied at Finsbury Technical College and the Central Technical College at South Kensington, where he worked with **Henry Armstrong**. In 1897 he became head of the chemistry department of the Goldsmiths' Institute and in 1901 he moved to the Municipal School of Technology in Manchester. His final appointment in 1908, at the early age of 38, was to the Chair of Organic Chemistry at Cambridge. He worked on a number of topics, including mustard gas, camphor, organometallic compounds and photographic sensitizers, but he is best known for his work on optical activity. In 1899 he reported the synthesis and resolution of enantiomeric nitrogen compounds, and at Cambridge repeated this with sulphur and selenium. For his scientific work he was awarded the freedom and livery at the Goldsmiths' Company in 1919, and served as prime warden (1928–9). He was knighted in 1919.

POPOV, Aleksandr Stepanovich

(1859–1905)

Russian physicist, born in Bogoslavsky, the son of an Orthodox priest. After graduating from the seminary in Perm (now Molotov), he studied physics at St Petersburg University, and while still a student he worked at the Elektrotekhnik artel (1881), which ran the first Russian small generating plants and arc light installations. He was appointed as an instructor at the Russian Navy's Torpedo School in Kronstadt (1881), and later professor at the St Petersburg Institute of Electrical Engineering (1905). Independently of **Marconi**, he is acclaimed in Russia as the inventor of wireless telegraphy (1895). He was the first to use a suspended wire as an aerial.

PORTA, Giovanni Battista della

(1535–1615)

Italian natural philosopher, born in Naples. Of a noble family, he was probably self-educated, though few details of his life are known apart from his numerous writings. He was keenly interested in every aspect of Renaissance natural philosophy, including the occult sciences. He wrote on such varied subjects as physiognomy, fortification, palmistry and natural magic, metallurgical technology, crystallography and the classification of plants, besides several comedies. Most of his original work was in optics and the applications of steam. He was one of the first to make a serious study of the camera obscura, and found that the sharpness of the image could be increased by placing a lens at the pin-hole, thus turning it into a remote ancestor of the box camera. He discovered that the condensation of steam in a closed vessel leaves an empty space (the concept of a vacuum was then unknown) into which more liquid may be drawn, and he designed a rudimentary steam-pump to supply water to a fountain—in this case, the ancestor of Thomas Savery's atmospheric steam-pump. Porta founded a number of scientific academies, including one in Naples for the study of the secrets of nature. This may have been the cause of his interrogation by the Inquisition, as a result of which his works were banned from 1592 to 1598. He was admitted in 1610 as a member of the Accademia dei Lincei in Rome.

PORTER, Arthur

(1910–)

British computer scientist. He was educated at the University of Manchester, where he became an assistant lecturer in 1936. Porter became involved at the suggestion of **Hartree** in building analogue differential analysers (including one constructed from 'meccano') based on the design of **Bush** at MIT. In 1937–9 Porter was awarded a Commonwealth Fund Fellowship, which enabled him to study under Bush. After World War II he worked for Ferranti's Canadian subsidiary in Toronto, where he was responsible for important pioneering designs — notably a naval target tracking system and the FP6000, the world's first multi-tasking machine. Porter later became Professor of Industrial Engineering at the University of Toronto (1961–76).

PORTER, George, Baron Porter of Luddenham

(1920–)

English physical chemist, born in Stainforth, Yorkshire. After taking a BSc in chemistry at Leeds University, he became a radar officer in the Royal Naval Volunteer Reserve from 1941 to 1945. He then entered Emmanuel College, Cambridge, for his PhD. From 1949 to 1954 he held junior posts at Cambridge and after a brief stay at the British Rayon Research Association, he became Professor of Physical Chemistry at Sheffield University in 1955. In 1963 he transferred to the Firth Chair of Chemistry. He left Sheffield in 1966 to become Resident Professor and Director of the Royal Institution, where he remained until 1985. His researches have mainly been concerned with extremely rapid reactions in the gas phase, especially photochemical reactions, and with photochemistry generally. In the late 1940s, **Norrish** and Porter developed the technique of flash photolysis, which became important in the study of very rapid gas reactions; with **Eigen**, they were awarded the 1967 Nobel Prize for Chemistry for this work. In flash photolysis a very brief flash of intense light causes photochemical change and immediately afterwards the unstable chemical intermediates produced can be studied by means of their absorption spectra. When the technique was first introduced, the species had to survive for a few milliseconds for study to be possible, but through improvements over the past 40 years the timescale involved has moved from milliseconds to femtoseconds. Porter has been prominent in the refinements of techniques which have made this possible and in exploring the ever-widening range of fast processes which may be studied. In later years he became prominent as a spokesman for science in the UK. While at the Royal Institution he was very successful as a popularizer of science, particularly in the Christmas Lectures shown on television. He was knighted in 1972, admitted to the Order of Merit in 1989, and made a life peer in 1990. He was elected FRS in 1960, received Royal Society's Davy (1971) and Rumford (1978) medals, and served as its president from 1985 to 1990. Porter was President of the Chemical Society from 1970 to 1972, and received the society's Faraday Medal in 1980.

PORTER, Rodney Robert

(1917–85)

English biochemist, born in Newton le Willows, Lancashire. He studied there and with **Sanger** in Cambridge (1946–9), where they developed the 2,4-dinitrofluorobenzene method for determining the N-terminus of a protein, and then worked at the National Institute for Medical Research (1949–60) and St Mary's Hospital Medical School in London (1960–7) before becoming Professor of Biochemistry at Oxford in 1967. His discovery that the plant protease papain cleaves the Y-shaped IgG (immunoglobulin) into the stem and two separate arms (1959), plus the finding that these fragments were solubilized by salt solutions without activity loss, opened the way for a detailed study of immunoglobulin structure, including the location of the antigen binding sites on the ends of the arms and the function of the stem region. He was the first to propose the bilaterally symmetrical four-chain structure which is the basis of all immunoglobulins. **Edelman** carried out complementary structural studies on immunoglobulins in the USA, and for this work they were jointly awarded the Nobel Prize for Physiology or Medicine in 1972. Porter followed this work with studies of the proteases of the related complement system, particularly the components called C1-, C2- and C3-convertase (C4b,2a). He was elected a Fellow of the Royal Society in 1964, and was awarded its Royal (1973) and Copley (1983) medals. In 1985 he was run over and killed crossing the road.

POSIDONIUS

(c.135–c.51 BC)

Greek philosopher, scientist and polymath, born in Apamea in Syria, and nicknamed 'the

Athlete'. He studied at Athens, spent many years on travel and scientific research in Europe and Africa, then settled in Rhodes and became an active citizen there. In 86 BC he was sent as an envoy to Rome, where he settled. He wrote on an enormous range of subjects, including geometry, geography, astronomy, meteorology, history and philosophy (though only fragments of all these survive), and made important contributions to the development of **Zeno**'s doctrines.

POWELL, Cecil Frank
(1903–69)

English physicist, born in Tonbridge, Kent. He was educated at the University of Cambridge, where he received his PhD in 1927. Throughout his career he worked at the University of Bristol, as Wills Professor of Physics (1948–63) and subsequently Director of the Wills Physics Laboratory. A former pupil of **Rutherford** and **Charles Wilson**, he worked on detection techniques for nuclear particles. Powell used specially developed photographic emulsions to study nuclear interactions and improved the techniques used to analyse the particle tracks. Capable of operating without human supervision, the new emulsions were employed in cosmic-ray detectors carried by balloon to high altitudes. A set of emulsions exposed at a height of 3000 metres at the French Pic du Midi observatory in the Pyrennes du Midi showed three connected particle tracks and analysis revealed the existence of the charged pion, a particle which had been predicted by **Yukawa** in 1935. In 1950 Powell was awarded the Nobel Prize for Physics for his development of nuclear emulsions and his part in the discovery of the charged pion. In addition to his scientific work he was one of the leaders of the movement to increase the social responsibility of scientists.

POWELL, John Wesley
(1834–1902)

American geologist, born in Mount Morris, New York, of English parents. Almost wholly self-educated, he served in the American Civil War rising to the rank of major, but lost his right arm during the battle of Pittsburgh Landing. He led daring boat expeditions down the Colorado river through the Grand Canyon (1869), demonstrating that the canyon resulted from the river erosion of rock strata which were being progressively uplifted; the river preserved its level but the mountains were being raised up. From 1874 to 1880 he directed the US Geological and Geographical Survey of Territories, organizing the 'Powell Surveys' to explore and map various parts of the USA. He subsequently became the second director of the combined US Geological Survey (1880–94), and after retirement he continued to work on behalf of the Bureau of Ethnology, which he co-founded in 1884. Powell undertook extensive work on water supplies in arid regions, and also wrote on crustal movements and human evolution.

POYNTING, John Henry
(1852–1914)

English physicist, born in Monton, Lancashire, the younger son of a Unitarian minister. Educated at Manchester and Cambridge, he became Professor of Physics at Birmingham (1880). He investigated electrical phenomena, the radiation pressure on dust in the solar system (important in estimating the absolute temperatures of the Sun and planets, and of space), the phase transition between solid and liquid states (1881), osmotic pressure (1896), and determined the constant of gravitation by a torsion experiment (1891). In 1884 he introduced the Poynting vector giving a simple expression for the rate of flow of electromagnetic energy. He wrote *On the Mean Density of the Earth* (1893), for which he was awarded the Adams Prize at Cambridge, and *The Earth* (1913). With **J J Thomson** he also wrote *Textbook of Physics* (2 vols, 1899, 1914). Poynting invented several instruments, including a double-image micrometer and a saccharometer. Among his earliest studies were two statistical analyses of drunkenness in England (1877, 1878). He was elected FRS in 1888, and awarded the Royal Society's Royal Medal in 1905; in the same year he was elected President of the Physical Society.

PRAIN, Sir David
(1857–1944)

Scottish botanist, born in Fettercairn, Aberdeenshire, educated at Fettercairn Parish School and Aberdeen Grammar School. His early ambition was to enter banking, but his former Fettercairn headmaster intervened and sent him to Aberdeen University. Although his studies were broken by three spells as a schoolmaster, he finally graduated in medicine from Edinburgh University, and in 1884 entered the Indian Medical Service. Unexpectedly, due to the sickness of Lewis Brace, then Curator of the Calcutta Herbarium, Prain was sent to Calcutta Royal Botanic Garden in 1885. When Brace returned, Prain resumed military duties. When Brace again fell ill, Prain became Curator of the Calcutta Herbarium (1887–98) and eventually Superintendent of the Royal Botanic Garden and Cinchona Department (1898–1906). His many publications include

Noviciae Indicae (1905, based on papers published between 1889 and 1904), *The Species of* Pedicularis *of the Indian Empire* (1890), *Bengal Plants* (1903), papers on Labiatae, Leguminosae and Papaveraceae, and reports on Indian hemp, wheats of Bengal, mustards, yams and indigo. From 1906 to 1922 he was Director of the Royal Botanic Gardens, Kew, where he had to turn his attentions to African botany. Knighted in 1912, he served on the committees of innumerable bodies, including the John Innes Horticultural Institute and the Imperial College of Tropical Agriculture, Trinidad.

PRANCE, Ghillean Tolmie

(1937–)

English botanist, born in Brandeston and educated at Malvern College and Keble College, Oxford. From 1963 to 1988 he held various posts at New York Botanical Garden, including B A Krukoff Curator of Amazonian Botany (1968–75), Vice-President (1977–81), Director of Research (1975–81) and Senior Vice-President (1981–8); he was also Director of Graduate Studies at the Instituto Nacional de Pesquisas da Amazônia, Manaus (1973–5). In 1988 he became Director of the Royal Botanic Gardens, Kew. He has made 14 expeditions to Amazonia and has published many papers and several books on tropical botany, especially concerning Amazonia, including *Arvores de Manaus* (1975), *Algunas Flores da Amazônia* (1976), *Extinction is Forever* (1977), *Biological Diversification in the Tropics* (1981), *Key Environments: Amazonia* (1985) and *Manual de Botânica Econômica do Maranhão* (1988) as well as two popular books, *Leaves* (1986) and *Flowers for All Seasons* (1989).

PRANDTL, Ludwig

(1875–1953)

German pioneer of the science of aerodynamics, born in Freising, Bavaria. He studied mechanical engineering in Munich and received his PhD in 1900. The following year he was appointed Professor of Mechanics at the University of Hanover, where he set out to put the teaching of fluid mechanics on a sound theoretical basis. Although apparently destined for a career in theoretical mechanics, while working for the engineering company MAN his interest was redirected to a problem in aerodynamics. In this field he made outstanding contributions to boundary layer theory, airship profiles, supersonic airflow, wing theory and meteorology. His discovery in 1904 of the special properties of the boundary layer, governing the behaviour of fluids in close proximity to a surface, led to, a much better understanding of the phenomena of skin friction and drag on aircraft wings and control surfaces, and resulted in much more attention being paid to the streamlining of aircraft. He showed how the results of tests on model aircraft could be applied to the design of full-scale machines, and took a leading part in the construction of the first German wind tunnel in 1909. He was Professor of Applied Mechanics at the University of Göttingen (1904–53), and Director of the Kaiser Wilhelm Institute (now the Max Planck Institute) for fluid mechanics from 1925.

PRATT, John Henry

(1809–71)

English clergyman, mathematician and geophysicist, born in Ghazipur, India, the son of the secretary of the Church Mission Society. He received an MA at Christ's and Sidney Sussex colleges, Cambridge (1836). His scientific aptitude, coupled with missionary zeal enabled him to follow a dual career. He obtained a chaplaincy with the East India Company (1838), became Chaplain to the Bishop of Calcutta (1844) and was appointed Archdeacon (1850–71). In his geophysical work, Pratt postulated the isostasy principle to account for gravity anomalies resulting from nearby mountains such as those observed by Sir George Everest in his survey of India, suggesting that high mountain ranges have lower density than the underlying crust. From the shape of the Earth he assumed it was essentially fluid, thus the surface topography was due to a number of independent blocks of different densities; the whole crust, having a common depth, would be buoyant relative to a common compensation depth. An opposing theory was put forward by **Airy** (1854), who assumed a common density for the topography, hence for this to achieve isostatic equilibrium at the common compensation depth, the base of the crust would mirror the topography—mountains would have deeper roots. Pratt calculated that the Earth was an oblate spheroid, with a shortening of the polar axes of 26.9 miles (43.3 kilometres); it is now known to be 21 kilometres.

PREGL, Fritz

(1869–1930)

Austrian chemist, born in Laibach (now Ljubljana, Slovenia). He studied medicine at Graz University and spent most of his working life there, becoming Professor of Medical Chemistry in 1913. Finding that traditional methods of analysis were useless when applied to the minute quantities of biochemical materials that he wished to investigate, he devised new techniques for microanalysis, including a balance which could weigh within an accuracy of 0.001 milligrams. His innovations were fundamental to the development of biochemistry, and brought him the Nobel Prize for Chemistry in 1923. He died in Graz.

PRELOG, Vladimir

(1906–)

Swiss chemist, born in Sarajevo. After the assassination of Archduke Ferdinand in Sarajevo, Prelog's family moved to Zagreb, where he attended the local gymnasium. From 1924 until 1929 he studied chemistry at the Institute of Technology in Prague. Unable to fund an academic position, he worked for the chemical company G J Dřiza (1929–35). He then joined the staff of the University of Zagreb where he was successful in synthesizing adamantane, a fascinating molecule related to diamond and previously found only in Czech petroleum. When the Nazis invaded Yugoslavia he went to Zürich as a guest of Leopold Ružička who was able to find him a position at the Federal Institute of Technology. He was appointed associate professor in 1947 and full professor five years later. Prelog spent his professional life studying the shapes of molecules. This study included the phenomenon of chirality (the characteristic of certain molecules that they can exist in left- or right-handed forms, ie optical isomers), particularly important among naturally occurring molecules. He used X-ray crystallography to determine molecular shapes and linked reactivity with stereochemistry. With R S Cahn and **Christopher Ingold** he devised a system of naming stereoisomers based on a set of sequence rules. As a result, he became interested in other problems of stereochemical description, such as group theory, graph theory and chemical topology. He also studied the synthesis and chemistry of a number of natural products, including the enzyme fatty acid synthetase. In 1975 he shared the Nobel Prize for Chemistry with **Cornforth** for his contribution to our understanding of the stereochemistry of enzyme catalysis.

PRESTWICH, Sir Joseph

(1812–96)

English geologist, hydrologist and prehistorian, born in Pensbury, Clapham. He was educated in Paris, Reading and at University College London. Until the age of 60, he was a wine merchant in the family business, his business interest aiding rather than restricting his geological studies with frequent trips to France and Belgium as well as around Britain. He undertook early studies of the stratigraphy of Coalbrookedale, Shropshire, publishing *The Geology of Coalbrookedale* (1836), and on the correlation of the English Eocene with that of France. His principal work was on the stratigraphical position of flint implements and human remains in England and France, helping to confirm the antiquity of early man. His work on *The Water-bearing Strata of the Country around London* (1851) was a standard authority. In 1874 Prestwich became Professor of Geology at Oxford; it was there that he produced his two-volume work on *Geology, Chemical and Physical, Stratigraphical and Palaeontological* (1886–7). He was knighted in 1896, shortly before his death.

PRÉVOST, Pierre

(1751–1839)

Swiss philosopher and physicist, born in Geneva. He occupied chairs of philosophy and physics at Berlin and Geneva. In 1792 he published *Sur l'équilibre du feu*, which marked a significant step forward in understanding the nature of heat. It was widely believed at the time that heat consisted of a fluid called *caloric*, and it was logical to many philosophers that cold should consist of another fluid called *frigoric*. Prévost maintained that only one fluid was involved, and he introduced the concept of dynamic equilibrium in which all bodies are both radiating and absorbing heat, known as the Prévost theory of exchanges. His other writings and translations covered many subjects.

PRIESTLEY, Joseph

(1733–1804)

English clergyman and chemist, renowned for his work on gases, particularly oxygen. He was born in Fieldhead, Leeds, to a dissenting family at a time when membership of the Church of England was a prerequisite for entrance to English universities and most positions of any importance. Intending to become a preacher, he was educated at the dissenting academy at Daventry. There he studied books on philosophy and science as well as theology, but it was to be his religious and political views, rather than his scientific achievements, which shaped his career. After preaching at Downham Market, Norfolk, and Nantwich, Cheshire, where his leanings towards Arianism (denying the divinity of Christ) troubled his congregation, he became a very successful teacher at Warrington Academy replacing the traditional classical syllabus with history, science and English literature. He also began the studies of electricity which earned him a fellowship of the Royal Society in 1766 and the friendship of **Benjamin Franklin** whom he met in London. The following year Priestley was appointed minister at Mill Hill Chapel, Leeds, and soon afterwards began to study gases and to write pamphlets criticizing the government's treatment of the American colonies. Continuing his search towards religious truth, he now leaned towards the Unitarians, a sect founded in 1774 who disregarded the doctrine of the Trinity and placed great faith in reason. From 1772 to 1780 Priestley was librarian and travelling companion to Lord Shelbourne, based at Calne, Wiltshire, and subsequently junior minister at the New Meeting House, Birmingham. He was soon part of the Lunar Society, an informal but influential body of local savants, and included members such as Matthew Boulton, **Watt** and **Erasmus Darwin** among his friends. His revolutionary sympathies continued to attract attention, particularly once the French Revolution began, and in 1791 a mob of townsfolk sacked his house and fired his chapel. Priestley and his family fled to London and three years later emigrated to the USA where he settled in Northumberland, Pennsylvania, where he died 10 years later. Priestley's work on gases made him the most respected pneumatic chemist in Europe. When he began his experiments, only three gases were known: air (not then recognized as a mixture of gases), carbon dioxide and hydrogen. He added another 10, including hydrogen chloride, sulphur dioxide, ammonia and nitrous oxide (laughing gas). His discoveries were made by heating common laboratory substances such as 'oil of vitriol' (sulphuric acid) or 'spirits of salt' (a solution of hydrochloric acid) in a retort, with or without another reagent. He then collected any gas that was evolved over a pneumatic bath containing mercury or water. His most momentous discovery took place in 1774 when he heated mercury in air to form a 'crust'. On further heating this yielded a colourless, odourless gas which made a lighted candle burn more brightly and did not kill his laboratory mice. Priestley recognized that this gas was new to science but he did not confirm that it was an element, nor did he identify it with the new gas prepared by **Scheele**. Because he accepted the traditional view of combustion — that inflammable substances contain invisible fiery essence called phlogiston which they lose on burning — he called this gas 'dephlogisticated air'. When he was in Paris with Lord Shelbourne in autumn of 1774, he repeated this experiment for **Lavoisier** who perceived its significance and in a long series of experiments showed its central role in combustion and named it 'oxygène'. Priestley never accepted Lavoisier's discovery that burning is a chemical reaction in which a substance combines with oxygen and continued to defend the phlogiston theory until his death. Among Priestley's many religious writings are *Letters to a Philosophical Unbeliever* (1774), *Disquisition relating to Matter and Spirit* (1777) and *History of Early Opinions Concerning Jesus Christ* (1786).

PRIGOGINE, Ilya, Vicomte
(1917–)

Belgian theoretical chemist, born in Moscow. After moving to Belgium at the age of 12, he was educated at the Free University of Brussels, where he held a Chair of Chemistry from 1951 to 1987 and is now Emeritus Professor. Since 1967 he has been Director of the Ilya Prigogine Center for Statistical Mechanics, Thermodynamics and Complex Systems of the University of Texas, and since 1987 also Associate Director of Studies at the École des Hautes Études en Sciences Sociales in France. Following the pioneering work of **Onsager**, Prigogine continued the development of the thermodynamics of irreversible processes, and discovered how to treat systems far from equilibrium. His methods are applicable to a wide range of chemical and biological systems. For this work he was awarded the Nobel Prize for Chemistry in 1977. He is the author of several important books, including *Non-equilibrium Statistical Mechanics* (1962), *Thermodynamic Theory of Structure, Stability and Fluctuations* (1971, with P Glansdorff), and *Order out of Chaos — Man's New Dialogue with Nature* (1984). He holds the Grand Cross of the Order of Leopold II, is a Commander of the French Legion of Honour, and holds the Rumford Medal of the Royal Society.

PRINGLE, Sir John
(1707–82)

Scottish physician, born in Roxburgh, the youngest son of a baronet. He studied at St Andrews and Edinburgh universities before spending two years in Leiden, where he received his MD in 1730. Returning to Edinburgh, he practised medicine and held a Chair of Pneumatical and Ethical Philosophy (1734–45). His family connections helped him obtain a commission as physician to the commander of the British military forces on the Continent, and his performance shortly advanced him to the post of Physician-General to British forces in the Low Countries. A subsequent career in London was spectacularly successful, bringing a succession of royal appointments, his own baronetcy and election to the presidency of the Royal Society of London (1772). Throughout his life, Pringle was an active physician despite the many other demands on his time. His many medical observations, still not published today, are kept at the Royal College of Physicians in Edinburgh. On the other hand, his *Observations on Diseases of the Army* (1752) remains a classic of military hygiene, went through many editions and translations, and established many principles for preventing typhus, dysentery and other common diseases of soldiers and others who live in crowded conditions.

PRINGSHEIM, Ernst
(1859–1917)

German physicist, born in Breslau (now Wrocław, Poland). He studied at Heidelberg (1877–8) and Breslau (1878–9) universities, and under **Helmholtz** at Berlin. He received his doctorate in 1882, and in 1905 he was appointed Professor of Theoretical Physics at Breslau. Working with **Lummer**, he studied the radiation emitted by hot bodies, and highlighted inconsistencies which arose from the black-body radiation formulae of **Wien** and **Planck**. This led Planck to return to the problem and to formulate his quantum theory.

PRINGSHEIM, Nathaniel
(1823–94)

German botanist, born in Wziesko, Silesia (now in Poland). Noted for his research on the fertilization of cryptogamic plants, he was professor at Jena for a short time, but for the most part worked privately in Berlin, having inherited wealth from his industrialist father. He worked on the Chlorophyceae (green algae), but also contributed extensively to the study of other cryptogams, including mosses. He was the first scientist to observe and demonstrate sexual reproduction in algae. Observing the freshwater alga *Oedogonium* and the aquatic fungus *Saprolegnia* he reported that the male gamete penetrated into the substance of the ovum, and concluded that fertilization consisted of the material union of two reproductive substances. He introduced a uniform terminology (antheridia, oogonoia, sporangia) for the reproductive structures of the algae, which aided the elucidation of their life cycles. The alternation of generations in mosses was shown by Pringsheim to be homologous to that in pteridophytes.

PROKHOROV, Alexander Mikhailovich
(1916–)

Russian physicist, born in Atherton, Australia, of Russian émigré parents. After the Russian Revolution his family returned to the USSR (1917). He graduated from Leningrad (now St Petersburg) University in 1939, and after serving with the Red Army during World War II, he took a junior post at the Lebedev Physical Institute, rising to become Deputy Director in 1968. In 1952, with his colleague **Basov**, he postulated the possibility of constructing a 'molecular generator' for the amplification of electromagnetic radiation. He later achieved this using a beam of molecular ammonia. He went on to describe a new way in which atomic systems could be employed to produce amplification of microwaves. This led to the development of the maser and eventually the laser (terms that stand for Microwave/Light Amplification by Stimulated Emission of Radiation). For this work he won the 1964 Nobel Prize for Physics jointly with Basov and **Townes**.

PROUST, Joseph Louis
(1754–1826)

French analytical chemist, born in Angers. After studying pharmacy and chemistry in Paris he spent most of his working life in Spain. In the early 1780s he conducted aerostatic experiments with Pilatre de Rozier and **Charles**, and was one of the first people to make an ascent in a balloon, which he did in 1784 with the King and Queen of France among the onlookers. He was appointed Professor of Chemistry at the Royal Artillery College at Segovia and Director of the Royal Laboratories at Madrid (1789–1808), after which he returned to France. Proust made two significant advances in analytical chemistry: he developed the use of hydrogen sulphide as a reagent (important because the differing solubilities of sulphides makes it possible to separate them) and he gave the results of his analyses in terms of percentage weights. By means of the percentages he realized that the proportions of the constituents in any chemical compound are always the same regardless of what method is used to prepare it. He announced this discovery, known as the 'law of definite proportions', in 1794. Not all his contemporaries accepted his findings, his principal adversary in a renowned controversy being **Berthollet**. Although Proust was correct in his observations, the reason why reagents behave in this way did not become clear until **Dalton** formulated his atomic theory in 1803. Proust died in Angers.

PROUT, William
(1785–1850)

English physician, born in Horton, Gloucestershire. He graduated in medicine in Edinburgh in 1811, and settled in London. Taking up physiological chemistry, he furnished his own laboratory and gave a course of chemical lectures (1813). From numerous analyses he deduced the famous 'Prout's hypothesis', that the atomic weights of all the elements are multiples of the atomic weight of hydrogen (1815). He was the first to analyse the constituents of urine and he isolated urea (1818), recognizing that it is secreted into the blood and excreted only by the kidneys. He discovered that reptilian (boa constrictor) urine is composed wholly of highly insoluble uric acid, and observed its relationship to urea and significance for the formation of urinary stone (calculus) in man. Prout also originated several of the revolutionary ideas attributed to **Liebig**, for example, that the various excretions (such as urea, uric acid, carbonic acid) are derived from the waste or destruction of tissues which once formed a constituent part of the organism. He discovered hydrochloric acid in healthy stomach juice (1823), alloxan (used to cause experimental diabetes in animals), and was the first to divide foodstuffs into carbohydrate, fats and proteins (1827). He published the successful *Inquiry into the nature and treatment of Gravel Calculus and other diseases of the Urinary Organs* (1821), *Chemistry, Meteorology and the Function of Digestion, considered with reference to Natural Theology* (1834) and numerous scientific papers. He was elected FRS in 1819.

PTASHNE, Mark Steven
(1940–)

American molecular biologist, born in Chicago. He was educated at Reed College where he obtained his BA in 1961, and at Harvard University where he received his PhD in 1968. He remained at Harvard as a lecturer in biochemistry (1968–71), and since 1971 has been professor there. The concept that genes can be specifically activated or de-activated, which has been essential to our understanding of the normal and abnormal functioning of cells, originated from genetic studies of *Escherichia coli*. These studies implied the existence of a repressor protein that binds to a transcription unit called the 'lac operon' to 'turn off' the b-galactosidase gene when lactose is absent from the cell. The purification of the lactose repressor allowed the mechanisms to be explored by the use of mutant repressor molecules. It was shown by Ptashne and others that the lac repressor inhibits transcription of the lac operon by binding to a specific nucleotide sequence, 21 base pairs long, called the operator. Since the operator overlaps the promoter, binding of the repressor to the operator prevents RNA polymerase binding to the promoter and thus transcription of the adjacent gene sequences. When lactose is present in the cell, a breakdown product dislodges repressor from the operator and allows RNA polymerase attachment to the promoter. Thus the b-galactosidase gene is de-repressed only in the presence of lactose. Similar mechanisms are found in higher organisms, and the lactose system has been of crucial importance as a model.

PTOLEMY (Claudius Ptolemaeus)
(c.90–168 AD)

Egyptian astronomer and geographer, who flourished in Alexandria. His 'great compendium of astronomy' seems to have been denominated by the Greeks *megistē*, 'the greatest', from which the Arab name *Almagest* by which it is generally known was derived. With his *Tetrabiblos Syntaxis* is combined another work called *Karpos* or *Centiloquium*, because it contains a hundred aphorisms—both treat astrological subjects, so have been held by some to be of doubtful genuineness. Then there is a treatise on the fixed stars or a species of almanac, the *Geographia*, and other works dealing with map-making, the musical scale and chronology. Ptolemy, as astronomer and geographer, held supreme sway over the minds of scientific men down to the 16th–17th century; but he seems to have been not so much an independent investigator as a corrector and improver of the work of his predecessors. In astronomy he depended almost entirely on **Hipparchos**. However, as his works form the only remaining authority on ancient astronomy, the system they expound is called the *Ptolemaic System*; the system of **Plato** and **Aristotle**, this was an attempt to reduce to scientific form the common notions of the motions of the heavenly bodies. The Ptolemaic astronomy, handed on by Byzantines and Arabs, assumed that the Earth is the centre of the universe, and that the heavenly bodies revolve round it. Beyond and in the ether surrounding the Earth's atmosphere were eight concentric spherical shells, to seven of which one heavenly body was attached, the fixed stars occupying the eighth. The apparent irregularity of their motions was explained by a complicated theory of epicycles. As a geographer, Ptolemy is the corrector of a predecessor, Marinus of Tyre. His *Geography* contains a catalogue of places, with latitude and longitude; general descriptions; and details regarding his mode of noting the position of places—by latitude and longitude, with the calculation of the size of the Earth. He constructed a map of the world and other maps.

PURBACH, or PEUERBACH, Georg von
(1423–61)

Austrian astronomer and mathematician, considered to be the first great modern astronomer, and teacher of **Regiomontanus**. He became court astrologer to Frederick III and Professor of Mathematics and Astronomy at Vienna. In astronomy he was a proponent of **Ptolemy**'s system of the solid spheres, and his extensive observational work resulted in the publication of a table of lunar eclipses in 1459. In mathematics he is thought to have been the first to introduce sines into trigonometry, and compiled a sine table.

PURCELL, Edward Mills

(1912–)

American physicist, born in Taylorville, Illinois. He studied electrical engineering at Purdue University, Illinois, and received his PhD from Harvard, where he became an associate instructor in 1938. In 1949 he was made a full professor and he became Gerhard Gade Professor of Physics in 1960. During World War II he worked as a group leader at MIT's radiation laboratory. His research has covered nuclear magnetism, radio astronomy, radar, astrophysics and biophysics. Independently of **Felix Bloch**, he developed the technique of nuclear magnetic resonance, and using microwave and radio techniques he was able to tune into resonances when nuclei were placed in a magnetic field. He was awarded the 1952 Nobel Prize for Physics with Bloch for his work. In astronomy he detected the 21 centimetre spectral line due to microwave emissions from neutral hydrogen in interstellar space which had been predicted by the Dutch astronomer **van de Hulst**.

PURKINJE, or PURKYNE, Jan Evangelista

(1787–1869)

Czech physiologist. Born in Libochowitz and educated by monks, Purkinje trained for the priesthood. He then studied philosophy, and finally graduated in medicine, with a dissertation on vision that gained him Johann Wolfgang von Goethe's friendship. He rose to become professor at Breslau and later in Prague. In 1825 he began to use a compound microscope, making important new observations. After 1830, much of Purkinje's work centred on cell observations, deploying compound microscopes to delineate nerve fibres. In 1837 he outlined the key features of the cell theory, to be more fully propounded in 1839 by **Schwann**. Around the same time he described nerve cells with their dendrites and nuclei and the flask-like cells ('Purkinje cells') in the cerebellar cortex. In 1838 he observed cell division, and in the following year promoted the word 'protoplasm' in the modern sense. He made improvements in histology, including early utilization of a mechanical microtome in place of a razor to procure thin tissue slices. Purkinje also experimented on the physiological basis of subjective feelings. He took up the study of vertigo and was interested in the peculiarities of the eyes, performing fascinating self-experimentation on the visual effects of pressure applied to the eyeball. The effect of being able to see in one's own eye the shadows of the retinal blood vessels is now known as 'Purkinje's figure'.

PYTHAGORAS

(6th century BC)

Greek philosopher, sage and mathematician, born in Samos. Around 530 BC he settled in Crotona, a Greek colony in southern Italy, where he established a religious community of some kind. He may later have moved to Megapontum, after persecution. He wrote nothing, and his whole life is shrouded in myth and legend. Pythagoreanism was first a way of life rather than a philosophy, emphasizing moral asceticism and purification, and associated with doctrines of the transmigration of souls, the kinship of all living things and various ritual rules of abstinence (most famously, 'do not eat beans'). He is also associated with mathematical discoveries involving the chief musical intervals, the relations of numbers, the proof of the theorem on triangles which bears his name, and with more fundamental beliefs about the understanding and representation of the world of nature through numbers. The equilateral triangle of ten dots, the tetracys of the decad, itself became an object of religious veneration, referred to in the Pythagoran oath 'Nay, by him that gave us the *tetracys* which contains the fount and root of ever-flowing nature'. The Pythagoreans were the first to recognize that the square root of 2 is an irrational number, ie cannot be expressed as a fraction of two whole numbers, and they are said to have murdered one of the members who disclosed this knowledge to others. It is impossible to disentangle Pythagoras's own views from the later accretions of mysticism and neoplatonism, but he had a profound influence on **Plato** and later philosophers, astronomers and mathematicians.

Q

QUASTEL, Juda Hirsch

(1899–1987)

British biochemist, born in Sheffield. He read chemistry at Imperial College, London, and in 1921 moved to Cambridge to work for his PhD on bacterial metabolism with **Frederick Gowland Hopkins**. In 1930 he became staff biochemist at Cardiff City Mental Hospital, where he began pioneer studies on biochemical aspects of mental disease. He proceeded on three main, complementary fronts: examining the normal and abnormal metabolism of brain tissue; coordinating his laboratory research with clinical investigations, such as the effects of barbiturates on brain metabolism, and developing improved liver function tests for schizophrenia; and investigating the neurochemistry of neuroactive chemicals, especially the effects of amphetamine, and the synthesis of acetylcholine, a chemical then receiving considerable interest because of the contemporary work of **Dale** and **Loewi**. During World War II he worked at Rothamsted Experimental Station (1940–7) on soil biochemistry in a programme to improve soil fertility. Utilizing techniques from his studies on the brain, he examined the structure and metabolic profiles of different types of soil, developed artificial chemical conditioners to improve quality, and produced a particularly powerful selective herbicide. Despite his interest in this work and his willingness to continue it after the war, in 1947 he accepted an invitation to Montreal to return completely to brain chemistry. As Professor of Biochemistry at McGill University and as Deputy Director of a biochemical research unit of Montreal General Hospital, he worked on a wide range of neurochemical problems, as well as on intestinal absorption. On retirement from McGill in 1964 he moved to the Kinsman Laboratories at Vancouver, working most notably on glutamic acid metabolism in the brain.

QUÉTELET, (Lambert) Adolphe (Jacques)

(1796–1874)

Belgian statistician and astronomer, born in Ghent. Educated at the Lycée de Ghent, he became Professor of Mathematics at the Brussels Athenaeum (1819) and Professor of Astronomy at the Military School (1836). In his greatest book, *Sur l'homme* (1835), as in *L'Anthropométrie* (1871), he showed the use that may be made of the theory of probabilities as applied to the 'average man'; he advocated the use of statistics to formulate social laws, and his views on this aroused considerable controversy. His grasp of the mathematical theory of statistics is thought to have been slight, and his methods unsophisticated.

R

RABI, Isidor Isaac

(1898–1988)

Austrian–American physicist, born in Rymanow. He was a graduate of Cornell and Columbia universities, the latter granting him his doctorate in 1927. He was appointed professor at Columbia in 1937 and remained there until his retirement in 1967. In work on atomic and molecular beams, Rabi developed the resonance method for accurately determining the magnetic moments of fundamental particles. For this work he was awarded the 1944 Nobel Prize for Physics. He also assisted with the development of radar and the nuclear bomb, and his work in defining the properties of atoms and nuclei led to the inventions of the laser and atomic clock. He was one of the founders of the Brookhaven National Laboratory, and whilst a member of UNESCO, he originated the movement that established the European nuclear physics research centre (CERN) in Geneva.

RAINWATER, (Leo) James

(1917–)

American physicist, born in Council, Idaho. He was educated at Caltech and Columbia University, and during World War II he contributed to the Manhattan Project to develop the atomic bomb. He became Professor of Physics at Columbia University in 1952 and was Director of the Nevis Cyclotron Laboratory there (1951–3, 1956–61). Rainwater's work led to the unification of two theoretical models of the atomic nucleus. Measurements such as the electric quadropole moment of the nucleus had indicated that some nuclei were deformed, a phenomenon which remained unexplained by either of the two existing nucleus models (the liquid drop model and the shell model), which both assumed that nuclei are symmetric. Rainwater proposed that distortions of the nucleus could be due to some of the nucleons distorting the central symmetric potential. His colleague **Aage Bohr** later developed this idea into the collective model of nuclei with **Mottelson**. Rainwater also worked with **Fitch** on the studies of muonic X-rays which gave the first indications of the nuclear radius. He also developed an improved theory of how high-energy particles scatter as they pass through material to explain aspects of the behaviour of cosmic-ray muons. He shared the 1975 Nobel Prize for Physics with Bohr and Mottelson for the development of the collective model of the nucleus.

RAMAN, Sir Chandrasekhara Venkata

(1888–1970)

Indian physicist, born in Trichinopoly. He was educated at Madras University, then spent 10 years working in the Indian Finance Department before becoming Professor of Physics at Calcutta (1917–33). In 1930 he was awarded the Nobel Prize for Physics for important discoveries in connection with the scattering of light by transparent materials. He showed that the interaction of vibrating molecules with photons passing through altered the spectrum of the scattered light, either increasing or decreasing it by a fixed amount. This 'Raman effect' enabled the probing of molecular energy levels and became an important spectroscopic technique used throughout the world. His other research interests included the vibration of musical instruments and the physiology of vision. He was the first Indian Director of the Indian Institute of Science (1933–48), where he instigated work on light scattering by colloids, Brillouin scattering and the diffraction of light by ultrasonic waves. He was a founder of the Indian Academy of Sciences (1934) and in 1947 founded the Raman Institute in Mysore where he remained as Director until his death. His outstanding contributions to Indian science made him a national hero, and he was knighted by the British government in 1929.

RAMANUJAN, Srinivasa

(1887–1920)

Indian mathematician, born in Eroda, Madras, one of the most remarkable self-taught prodigies in the history of mathematics. The child of poor parents, he taught himself mathematics from an elementary English textbook. Although he attended college, he did not graduate. While working as a clerk, he was persuaded to send over 100 remarkable theorems that he had discovered to **Godfrey Hardy** at Cambridge, including results on elliptic integrals, partitions and analytic number theory. Hardy was so impressed that he arranged for him to come to Cambridge in 1914. There Ramanujan published many papers, some jointly with Hardy. The most remarkable was an exact formula for the number of ways an integer can be written as a sum of positive integers. Having no formal training in mathematics, he arrived at his results by an almost miraculous intuition, often having no idea of how they could be proved or even what the form of an orthodox proof might be. He was elected FRS and a Fellow of Trinity in 1918, but soon returned to India suffering from poor health; he died shortly after.

RAMÓN Y CAJAL, Santiago

(1852–1934)

Spanish physician and histologist, born in Petilla de Aragon in the Spanish Pyrenees. Apparently considered indolent as a child he was apprenticed to both a barber and a cobbler, but was unsuccessful at both occupations, and in some desperation his physician father tried to interest him in medicine. He responded positively, and graduated from Zaragoza University in 1873. He joined the Army Medical Service and served in Cuba where he contracted malaria and was soon discharged through ill-health. He returned to Zaragoza for further anatomical training and in 1883 began his academic career, being appointed Professor of Anatomy at Valencia until 1886, then Professor of Histology at Barcelona (1886–92) and finally Professor of Histology and Pathological Anatomy at Madrid (1892–1922). His major work was on the microstructure of the nervous system and he utilized the specialized histological staining techniques of the Italian **Golgi**. The two men disagreed about their interpretations of neural structure, Ramón y Cajal maintaining that nerve cells were discrete, and that there was no physical continuity between one cell and another, although they shared the 1906 Nobel Prize for Physiology or Medicine. Ramón y Cajal provided detailed histological descriptions of many regions of the brain, including the spinal cord, the cerebellum and the retina and demonstrated several distinct patterns in different parts of the cerebral cortex. He also proposed that the dendrites of a nerve cell receive information which is then transmitted through the axon; and he instigated significant work on the processes of degeneration and regeneration of nerves. He wrote many articles and books in Spanish, some of which have been translated into English, including an autobiographical account *Recollections of my Life* (1937).

RAMSAY, Sir William

(1852–1916)

Scottish chemist who discovered the rare gases of the atmosphere. He was born in Glasgow and studied classics at Glasgow University and chemistry at Tübingen. He then became an assistant at Anderson's College, Glasgow, Professor of Chemistry at University College, Bristol (1880–7), and Principal from 1881 to 1887. He subsequently became professor at University College London (1887–1913). At Glasgow he studied the alkaloids, and at Bristol he worked on the vapour pressure of liquids. In 1892 he set out to discover why nitrogen derived from the air always had a higher atomic weight than nitrogen prepared in the laboratory, a problem which had been highlighted by Lord **Rayleigh** (1842–1919). Ramsay believed that atmospheric nitrogen must contain a small percentage of a heavier gas and proved this to be the case in 1894. Taking a sample of air, he removed the oxygen by sparking and the nitrogen by combining it with heated magnesium, and found a residual inert gas which occupied 1 per cent of the original volume. This was later named 'argon'. In 1895 Ramsay isolated a light inert gas resembling argon by boiling a mineral called cleivite. Spectroscopic analysis showed that this gas was helium, which **Lockyer** and Edward Frankland had discovered in the spectrum of the Sun nearly 30 years earlier. Subsequently helium was shown to exist in the atmosphere, and in 1903 Ramsay and **Soddy** demonstrated that it was formed by the decay of thorium and uranium minerals, a discovery which was to prove critical to the understanding of nuclear reactions. Argon and helium fitted neatly into **Mendeleyev**'s periodic table, but spaces further down the same group suggested that heavier inert gases remained to be discovered in the atmosphere. Working with **Travers**, Ramsay liquefied air, collected the last fraction to evaporate, removed the oxygen and nitrogen, and examined the small residue spectroscopically. In 1898 they found the green and yellow lines of krypton, the crimson of neon and the blue lines of xenon. Further research confirmed the inert nature of these gases and their atomic weights. In 1907, to exclude the possibility that other undiscovered elements were present, Ramsay investigated the residue of 100 tons of liquid air but found nothing. The following year, returning to his earlier studies of radioactivity, he obtained radon — discovered by **Dorn** in 1900 — in sufficient quantities to show that it belonged to the same family as helium and the other inert gases. Ramsay was elected FRS in 1888, knighted in 1904 and awarded the Nobel Prize for Chemistry in the same year. He died in Hazlemere, Buckinghamshire.

RAMSEY, Frank Plumpton

(1903–30)

English philosopher and mathematician, born in Cambridge. He read mathematics at

Trinity College, Cambridge, and went on to be elected a Fellow of King's College when he was only 21. In his tragically short life (he died after an operation) he made outstanding contributions to philosophy, logic, mathematics and economics, to an extent which was only properly recognized many years after his death. To him is due the ingenious idea that large subsets of a set with some structure must also carry at least some of that structure. This has important application in number theory, and especially in combinatorics and graph theory. Typically, results in Ramsey theory assert that complete disorder is impossible. He was much stimulated by his Cambridge contemporaries **Bertrand Russell**, whose programme of reducing mathematics to logic he ingeniously defended and developed, and Ludwig Wittgenstein, whose *Tractatus* he was among the first both to appreciate and to criticize, rejecting the idea of ineffable metaphysical truths beyond the limits of language with the famous remark, 'What we can't say we can't say, and we can't whistle it either'. The best of his work is collected in *Philosophical Papers* (1990).

RAMSEY, Norman
(1915–)

American physicist, born in Washington DC. He was educated at Columbia University and at Cambridge, and after various teaching posts in the USA became Associate Professor of Physics at Harvard in 1947. Since 1966 he has been Higgins Professor of Physics there. Ramsey was awarded the 1989 Nobel Prize for Physics jointly with **Paul** and **Dehmelt** for his development of the 'separated field' method. In this an electromagnetic field applied to a beam of atoms or molecules induces transitions between specific energy states. This allowed the energy of atomic transitions to be measured with great accuracy and led to the development of the caesium clock, the atomic clock which now provides international time standards. Ramsey has also contributed to the development of the hydrogen maser.

RANKINE, William John Macquorn
(1820–72)

Scottish natural philosopher and engineer, born in Edinburgh. His father retired from the army to become a railway engineer. Macquorn studied at Edinburgh University (1836–8), where he achieved distinction in the natural philosophy class of **James Forbes**. In the summer of 1838 he began a civil engineering apprenticeship in Ireland. During the 1840s railway mania he worked in Scotland. In 1849 he began to publish extensively on elasticity and the new mechanical theory of heat, basing his writings on an idiosyncratic 'hypothesis of molecular vortices'. With **Kelvin** and **Clausius** he shaped the new thermodynamics, particularly in its practical dimension (he patented an elaborate air engine with his friend James Robert Napier of the famous Clyde shipbuilding family). Rankine introduced the terms 'actual' (kinetic) and 'potential' energy. Later he proposed an abstract 'science of energetics' which aimed to unify physics. From 1855, as Regius Professor of Civil Engineering and Mechanics at Glasgow, Rankine espoused a fertile harmony between the theoretical and practical worlds which he straddled. His textbooks (on prime movers, machinery, shipbuilding and applied mechanics), calling on first-hand experience of Glasgow industry, were enormously popular and effective in the establishment of an 'engineering science'. Rankine was a keen musician and a prolific writer. His dryly humorous *Songs and Fables* were published posthumously in 1874.

RAOULT, François Marie
(1830–1901)

French physical chemist, born in Fournes, near Lille. He studied in Paris, but financial problems forced him to abandon the course, and from 1853 to 1867 he taught science in various schools, completing his doctorate in 1863. In 1867 he began to teach chemistry at the University of Grenoble, and in 1870 he was promoted to the Chair of Chemistry, which he occupied until his death. His doctoral research concerned heat changes in chemical cells and their relation to electromotive force. Raoult is remembered most, however, for his work on the freezing points and vapour pressures of solutions (1878–92). His findings provided the basis for methods of determining molecular weights. The generalization that the vapour pressure of solvent above a solution is proportional to the mole fraction of solvent in the solution is known as Raoult's law. Raoult also found that electrolyte solutions showed anomalies, which proved later to be important in connection with the ionic dissociation theory of **Arrhenius**. He became an Honorary Fellow of the Chemical Society in 1898.

RAPHAEL, Ralph Alexander
(1921–)

English chemist, born in Dublin. Having studied chemistry at Imperial College, London, as an undergraduate he worked with Sir Ian Heilbron and Sir Ewart Jones for his PhD. After a brief period with May and Baker he returned in 1946 to Imperial College as an ICI Fellow. This was followed by a lectureship in Glasgow, where he developed the synthesis of a number of natural products from acetylenic precursors. Before promotion to the chair in Glasgow (1957), he had a short stay at Queen's University, Belfast. In 1973 he was invited to the Chair of Organic Chemistry at Cambridge. His prowess as a synthetic organic chemist is surpassed only by his skill as an after-dinner speaker. He was elected FRS in 1962 and awarded a CBE for services to chemistry in 1982.

RATHKE, Martin Heinrich

(1793–1860)

German biologist, born in Danzig. Pathke studied medicine at Göttingen and Berlin, where he received his MD in 1818, returning to his native Danzig to practise medicine. He became Professor of Physiology at Dorpat in 1829 and at Königsberg in 1835. Deeply interested in embryology and animal development, in 1829 he discovered gill-slits and gill-arches in embryo birds and mammals. 'Rathke's pocket' is the name given to the small pit on the dorsal side of the oral cavity of developing vertebrates. Building on ideas first developed by his friend **Pander**, he was also interested in the embryonic development of the sexual organs. A fine researcher, his studies of phylogeny never achieved the same prominence as those of **Baer**.

RAY, John

(1627–1705)

English naturalist, born in Black Notley, near Braintree, Essex. He was educated at Catharine Hall and Trinity College, Cambridge, and became a Fellow of Trinity in 1649. He was appointed as a lecturer in Greek (1651), mathematics (1653) and humanity (1655). In 1658 he made his first botanical tour, through the Midlands and north Wales, and in 1660 he published *Catalogus Plantarum circa Cantabrigiam Nascentium*, the first catalogue of the plants of a particular district to be published in England. In 1662 he resigned his fellowship at Cambridge, rather than take the oath of Act of Uniformity after the Restoration. Accompanied and subsidized by a wealthy fellow naturalist Francis Willoughby, he toured extensively in Europe (1663–6), with the aim of jointly preparing a systematic description of the organic world, Ray dealing with plants, Willoughby covering animals. In 1682 Ray published *Methodus Plantarum Nova*, in which he first divided flowering plants into monocotyledons and dicotyledons and demonstrated the true nature of buds. Earlier, in his *Catalogus Plantarum Angliae* (1670), he was the first to separate cryptogams from higher plants. His *Synopsis Methodica Stirpium Britannicorum* (1690) was the first systematic English Flora. After **Morison**'s death (1683), Ray began his major work, *Historia Generalis Plantarum* (3 vols, 1686–1704), which described about 6900 plants. From around 1690 he worked mainly on insects; his zoological work, in which he developed the most natural pre-Linnaean classification of the animal kingdom, has been considered of even greater importance than his botanical achievements. He was elected FRS in 1667.

RAYLEIGH, John William Strutt, 3rd Baron

(1842–1919)

English physicist, born near Maldon, Essex. In 1865 he graduated from Trinity College, Cambridge, as Senior Wrangler and Smith's Prizeman, and was elected a Fellow (1866). He inherited his father's title of third baron in 1873 and continued to work in his laboratory in the family mansion, Terling Place, Essex. He later succeeded **Maxwell** as Professor of Experimental Physics at Cambridge (1879–84). He was Professor of Natural Philosophy at the Royal Institution (1888–1905), President of the Royal Society (1905–8) and became Chancellor of Cambridge University in 1908. In 1871 he married Evelyn Balfour, sister of the Scottish statesman Lord Balfour. Rayleigh's work included valuable studies and research on vibratory motion, in both optics and acoustics. With **Ramsay** he was the discoverer of argon (1894), and for this and his work on gas densities, he was awarded the Nobel Prize for Physics in 1904. He also studied radiation, his research leading to Rayleigh–**Jeans** formula which accurately predicts the long-wavelength radiation emitted by hot bodies. The problems produced by the failure of this theory for short-wavelength radiation were solved by **Planck**'s suggestion that energy is emitted in discrete quanta—this idea, which Rayleigh never found satisfactory, was to dramatically alter the course of physics. His books included *The Theory of Sound* (1877–8) and *Scientific Papers* (1899–1900).

RAYLEIGH, Robert John Strutt, 4th Baron

(1875–1947)

English physicist, born at Terling Place, Essex, son of Lord **Rayleigh** (1842–1919). He was Professor of Physics at the Imperial College of Science from 1908 to 1919. Notable for his work on rock radioactivity, he was elected FRS in 1905 and received the Royal Society's Rumford Medal. His writings include two excellent biographies, one of his father, the other of **J J Thomson**.

RÉAUMUR, René Antoine Ferchault de
(1683–1757)

French natural philosopher, born in La Rochelle. The earlier years of his life are obscure, but in 1703 he moved to Paris and five years later became a member of the Academy of Sciences. He was put in charge of a government-sponsored project to assemble information on all the arts, industries and professions in France, and in gathering the data required for the monumental *Description des arts et métiers* he acquired a very wide knowledge of contemporary science and technology. He developed improved methods for producing iron and steel, being among the first to recognize the importance of carbon as a constituent of steel; the cupola furnace for melting grey iron was developed by him in 1720. He became one of the greatest naturalists of his age, publishing the six-volume *Mémoires pour servir à l'histoire des insectes* (1734–42), the first serious and comprehensive entomological work. In 1740 he produced an opaque form of porcelain which is still known as Réaumur porcelain. His thermometer of 1731 used a mixture of alcohol and water instead of mercury, with 80 degrees between the freezing and boiling points of water—the Réaumur temperature scale. Through his research into digestion in birds and animals he established that it was a chemical rather than a mechanical (grinding) process, and by 1752 he was able to isolate samples of gastric juices and demonstrate the process in the laboratory.

REBER, Grote
(1911–)

American radio engineer, born in Wheaton, Illinois. He was already an enthusiastic radio 'ham' when he began his studies at the Illinois Institute of Technology. He moved his radio telescope from Illinois to Virginia University in 1947, moving again to Hawaii in 1951 and Tasmania in 1954. Hearing of **Jansky**'s discovery of weak radio noise originating outside the solar system, he built the first radio telescope in his own back yard (1937), and for several years after its completion he was the only radio astronomer in the world. The telescope was a steerable parabolic reflector, 9.6 metres in diameter, and turned out to be the prototype of several generations of larger instruments. It had a resolution of 12 degrees, and with it Reber started to map the radio sky at a wavelength of 1.87 metres, publishing his first paper in 1940. He found that the radio map of the sky was quite different to that produced by conventional telescopes. He confirmed Jansky's discovery that the centre of our galaxy, in the direction of Sagittarius, emitted radio waves, and also found two other sources in Cassiopeia and Cygnus. In 1944 he was the first to detect radio emission from the Andromeda galaxy and from the Sun. In 1948 he tried to elicit financial support for a 220 foot radio telescope but without success.

RECHINGER, Karl Heinz
(1906–)

Austrian botanist and traveller, born and educated in Vienna. He first collected plants at the age of four, and his travel instincts were inspired by childhood reading of Sven Hedin's *Von Pol zu Pol* and a journey with his father (the botanist Karl Rechinger, 1867–1952) to the Parndorf plain. From 1927 to 1942 he made several collecting trips to the Aegean Islands; his studies on Aegean phytogeography and botany were published as *Flora Aegaea* (1943) and *Phytogeographia Aegaea* (1951). From 1938 to 1971 he was responsible for the botany department at the Natural History Museum, Vienna, and he latterly became the museum's director (1962–71). Between 1937 and 1977 he made 10 expeditions to Iran, Iraq and Afghanistan. His greatest work has been *Flora Iranica* (published from 1963 with over 160 parts to date), the first Flora to cover Iran, mountainous Iraq and Afghanistan since the time of **Boissier**; large numbers of new species and numerous new genera have been discovered. His other works on south-west Asian botany include *Zur Flora von Armenien und Kurdistan* (1939), *Reliquiae Samuelssonianae* (1959), *Flora of Lowland Iraq* (1964) and *Symbolae Afghanicae* (1954–64).

RECORDE, Robert
(c.1510–58)

English mathematician, born in Tenby. He studied at Oxford, and in 1545 took his MD at Cambridge. He practised medicine in London, and was in charge of mines in Ireland, but died in prison after losing a lawsuit brought against him by the Duke of Pembroke. He wrote the first English textbooks on elementary arithmetic and algebra, which became the standard works in Elizabethan England, including *The Ground of Artes* (1543) and *The Whetstone of Witte* (1557). The books are presented as dialogues, and introduced the equals sign to mathematics.

REDI, Francesco
(1626–97)

Italian physician and poet, born in Arezzo. He studied at Florence and Pisa and became physician to the dukes of Tuscany. At that time it was thought that many organisms appeared through a process of spontaneous generation. Redi, who read the speculation in a book by **William Harvey** (1578–1657) that this was not so, carried out a series of experiments to investigate the phenomenon. In 1668 he took two sets of flasks containing meat and covered one leaving the other set open. He also compared uncovered flasks with flasks covered with gauze. In both cases maggots appeared only in the uncovered flasks. This was onc of the first examples of an experiment in the modern sense where a control is used. The belief in spontaneous generation was so strong that, in spite of these results, Redi still believed it occurred under some circumstances. Indeed it was not until the mid-19th century that the work of **Schwann** and **Pasteur** finally resolved the controversy.

REED, Walter
(1851–1902)

American army surgeon, born in Belroi, Virginia. After training in medicine at the University of Virginia and at Bellevue Hospital Medical School, he entered the Army Medical Corps in 1875. He was later appointed Professor of Bacteriology at the Army Medical College, Washington (1893). Much of his work was on epidemic diseases. He showed that contrary to popular opinion, the outbreak of malaria in Washington was not due to bad water (1896), and that hog cholera is caused by *Bacillus icteroides* (1899). He was appointed head of a commission to study the cause and transmission of yellow fever in Cuba (1899), at that time a US protectorate — hazardous investigations during 1900, in which volunteers attempted to contract the disease under a variety of controlled conditions, proved that transmission of yellow fever was by the mosquito *Aëdes aegypti*. One of the commissioners involved in the experiments died of the disease. Reed demonstrated that the organism carried by the insect was of the type described by Martinus Willem Beijerinck in 1898 — this was the first time that a virus was implicated in human disease. The research led to the eventual eradication of yellow fever from Cuba through the destruction of mosquito breeding grounds.

REES, Charles Wayne
(1927–)

English chemist, born in Egypt. He attended Farnham School and then spent three years as a laboratory technician at the Royal Aircraft Establishment, Farnborough. After graduating from University College, Southampton (1950), he studied there for a PhD and followed this with three years as a Postdoctoral Fellow at the Australian National University (then situated in London). After appointments at Birkbeck College and King's College, London, his first chair was at Leicester University (from 1965). In 1969 he moved to Liverpool University, and he took up the position of Hofmann Professor at Imperial College in 1978. His researches have ranged widely over mechanistic and synthetic organic chemistry, particularly the chemistry of heterocyclic rings, with an emphasis on aromatic and anti-aromatic systems. He is best known for his work on the generation of reactive intermediates (highly reactive, transient chemical species with an unusually large number of non-bonding electrons, such as carbenes and nitrenes). More recently his work has involved the synthesis of cyclic systems with a high proportion of nitrogen and sulphur atoms. His honours have included election to the Royal Society (1974); he became President of the Royal Society of Chemistry (1992–4).

REES, Sir Martin John
(1942–)

British astrophysicist, born in York. He was educated at Trinity College, Cambridge, and became a staff member of the Institute of Theoretical Astronomy there. After holding fellowships in the USA and a short period as professor at Sussex University (1972–3), he returned to Cambridge as Plumian Professor of Astronomy and Experimental Philosophy (1973–91) and Director of the Institute of Astronomy (1977–82, 1987–91). Since 1992 he has been Royal Society Research Professor. Rees has made important contributions to the study of stellar systems, galaxies and the nature of the invisible 'dark matter' known to exist in the universe, but his best-known work is in the study of active galactic nuclei. He demonstrated that the relatively rapid brightness variations observed in quasars and active galaxies could be best understood if the nuclei contained gas which is outflowing at almost the speed of light and predicted that when projected onto the plane of the sky, such motions would be observed as *apparently* faster than the speed of light, or 'superluminal' motion. Observational evidence for this appeared in the 1970s. Rees also showed that the strong radio-emitting regions which lie far from the visible boundaries of some active galaxies could be produced by highly collimated beams of particles moving outwards from the nuclei at almost the speed of light; this has formed part of the current theory of the mechanisms of these mysterious objects. He was elected FRS in 1979 and became President of the Royal Astronomical Society in 1992, the year in which he was knighted.

REGENER, Erich Rudolph Alexander
(1881–1955)

German physicist, born in Schleussenau (now Bydgoszcz, Poland), and educated at Berlin University. He became Professor of Physics at the agricultural college in Berlin and from 1920 at the Technische Hochschule in Stuttgart, but was dismissed in 1937 because his wife was Jewish. In 1937 the Kaiser-Wilhelm-Gesellschaft built a research station for stratospheric physics for him at Lake Constance. In 1944 this was transferred to Weissenau and later became the Institute for Stratospheric Physics. Regener served as its director until his death and was also reinstated at Stuttgart in 1946. His early work was in nuclear physics, where he determined the electric charge to within two parts per thousand of the present standard value. However, his main contributions were to cosmic-ray research, where he developed special detectors to be fitted to high altitude balloons. Using these he discovered that the intensity of cosmic rays increases markedly at a height of 20 kilometres, and in 1933 he was able to link an increase in cosmic-ray activity to an eruption on the solar surface. This demonstrated for the first time that events occurring in stars are a source of cosmic rays.

REGIOMONTANUS, originally **Johannes Müller**
(1436–76)

German mathematician and astronomer, who took his name from his Franconian birthplace, Königsberg (*Mons Regius*). He studied at Vienna, and in 1461 accompanied Cardinal Bessarion to Italy to learn Greek. In 1471 he settled in Nuremberg, where he was supported by the patrician Bernhard Walther. The two laboured at the *Alphonsine Tables*, and published *Ephemerides 1475–1506* (1473), a work used extensively by Christopher Columbus. Regiomontanus established the study of algebra and trigonometry in Germany, and wrote on waterworks, burning-glasses, weights and measures, and the quadrature of the circle. He was summoned to Rome in 1474 by Pope Sixtus IV to help to reform the calendar, and died there.

REGNAULT, Henri Victor
(1810–78)

French chemist and physicist, born in Aix-la-Chapelle (now Aachen, Germany). In his early life he had to contend with poverty, but from 1829 he studied at the École Polytechnique in Paris and later at the École des Mines. He worked with **Liebig** at Giessen and received an appointment at Lyons. In 1840 he succeeded **Gay-Lussac** as Professor of Chemistry at the École Polytechnique and in 1841 he became Professor of Physics in the Collège de France. In 1854 he was appointed as director of the porcelain factory at Sèvres. Regnault's early researches were in organic chemistry. He carried out extensive studies of aliphatic chloro-compounds, and discovered vinyl chloride and other materials which have become industrially important. In 1847 he published a monumental treatise on chemistry. In the 1840s, however, his interests largely moved over to physics. He improved the techniques for determining the specific heats of solids and liquids, devised a method for determining the specific heats of gases, and applied these procedures extensively. He also made precise measurements of the behaviour of gases with respect to pressure, volume and temperature, and showed how real gases deviate from the ideal gas laws. His laboratory at Sèvres was wrecked in the Franco-Prussian War of 1870 and the results of much of his later work were lost. Regnault was admitted to the Legion of Honour in 1850. He received the Rumford (1848) and Copley (1869) medals of the Royal Society, and was made an Honorary Fellow of the Chemical Society in 1849.

REICH, Ferdinand
(1799–1882)

German physicist, born in Bernburg. He was educated at the University of Göttingen and later became Professor at the Freiburg School of Mines. Using spectroscopy and working in collaboration with his assistant **Hieronymous Richter**, Reich discovered the metal indium in zinc-blende in 1863.

REICHENBACH, Karl, Baron von
(1788–1869)

German natural philosopher and industrialist, born in Stuttgart. He was educated at the University of Tübingen, and had interests in metallurgical factories and steel works. He designed a new oven for making charcoal with which it was possible to collect the volatile products. From these he isolated creosote and paraffin in the 1830s. Around 1844, after studying animal magnetism, he believed that he had discovered a new force which he called 'Od', intermediate between electricity, magnetism, heat and light, and recognizable only by the nerves of sensitive persons, usually women. He expended much energy on publications and demonstrations, trying to convert the scientific world to his views, neglecting his businesses and living as a recluse. He also wrote on the geology of Moravia and on meteorites, of which he had a large collection (now in the University of Tübingen). He died in Leipzig.

REICHSTEIN, Tadeus
(1897–)

Swiss chemist, born in Wloclawek, Poland. He spent his early childhood in Kiev before moving to Zürich (1905), where he trained at the State Technical College (Eidgenössische Technische Hochschule). After an industrial interlude on coffee chemistry, he returned to the institute as an assistant to **Ružička**, and was appointed associate professor there in 1937. He then moved to the University of Basle, as head of the pharmacology department (from 1938), the organic division (from 1946) and the university's new Organic Institute (from 1960). His early academic work on carbohydrate chemistry led to the first synthesis of vitamin C (1933) independently of Sir **Norman Haworth**, as well as the sugars l-psicose (1935), l-threose (1936) and d-xyloketose (1937). From 1934 he also began synthesizing new steroids (deoxycorticosterone, 1938), and isolating and identifying the life-maintaining natural steroids of ox adrenal gland. Starting with over 1 000 kilograms of tissue, he eventually isolated 29 steroids in microscopic quantities including deoxycorticosterone (1936), corticosterone, 17-hydroxycorticosterone (cortisol), 11-dehydrocorticosterone, 11-deoxy-17-hydroxycorticosterone and 17-hydroxy-11-dehydrocorticosterone (cortisone). For his outstanding work on the chemistry of the adrenal hormones Reichstein received, with **Edward Kendall** and **Hench**, the 1950 Nobel Prize for Physiology or Medicine. He was elected a Foreign Member of the Royal Society in 1952.

REINES, Frederick
(1918–)

American physicist, born in Paterson, New Jersey. He was educated at the Steven's Institute of Technology and New York University, where he received a PhD in theoretical physics (1944). Between 1944 and 1959 he was a group leader at Los Alamos working on the physics and effects of nuclear explosions. He became head of physics at the Case Institute of Technology, Cleveland, and later Professor of Physics and Dean of Physical Sciences at the University of California at Irvine (from 1966). Together with **Cowan** he proved the existence of nature's most elusive particle—the neutrino. Later, Reines became interested in neutrinos arising from astronomical sources and used large underground detectors for these studies.

REMAK, Robert
(1815–65)

German physician and pioneer in electrotherapy, born in Poznań. Remak studied pathology and embryology as a student of **Johannes Müller** in Berlin. He remained there to develop a general practice and to pursue a university career, although he was prevented from obtaining a senior teaching post because he was a Jew. In his early twenties he did significant work on the microscopy of the nerves. In 1838 he located the lyelin sheath of the main nerves, and also showed that the axis-cylinder (axon) arises in the spinal cord and runs continuously. In this and in further work, he discerned that nerves are not merely—as they had been viewed for centuries—structureless hollow tubes; rather they possessed a flattened solid structure. A pioneer embryologist, Remak was one of the first fully to depict cell division, and to hold that all animal cells came from pre-existing cells. He discovered the 'fibres of Remak' (1830), and the nerve cells in the heart known as Remak's ganglia (1844).

RENSCH, Bernhard
(1900–)

German zoologist, born in Thale and educated at the University of Halle where he studied zoology, botany, chemistry and philosophy. He worked at the zoological museums at Berlin and Münster before becoming Professor of Zoology at the University of Münster, Westphalia, in 1947. He did research on a wide range of problems in the general area of genetics and evolution. An early neo-Darwinian, he was a strong supporter of the idea that random mutations form the basis of evolutionary change. His work in Germany paralleled that done by **Mayr** and **Dobzhansky** in the USA. He investigated kladogenesis, the process whereby the major groups of animals diverged during evolution followed by the appearance of many new species with the ultimate survival of only a few of these. He argued that trans-specific evolution, ie that above the species level, was subject to the same selective factors as at the specific level, a view not generally accepted at that time, particularly among palaeontologists. He also developed the idea that selection acts at all stages of the ontogenetic cycle and therefore evolution must also affect development. While mainly known for his studies in evolution, his early work was with avian and molluscan taxonomy and he also carried out research on animal behaviour and sensory physiology. His many research papers and books were mainly published in German and his major work *Neuere Probleme der Abstammungslehre* (1947) did not appear in translation until 1959 as *Evolution above the Species Level.* His concern for philosophical problems relating to biology is expressed in publications such as *Biophilosophy* (1971) and *Homo sapiens. From Man to Demigod* (1972).

RENWICK, James
(1790–1863)

British–American natural philosopher and engineer, born in Liverpool of Scottish–American parents. Taken to the USA as a child, he graduated from Columbia College (later Columbia University), New York, in 1810. In 1820 he was appointed to the Chair of Natural Philosophy and Experimental Chemistry at Columbia College, which he held until 1853. He was a recognized authority on every branch of engineering, and among the major projects on which he was consulted was the Morris Canal, for which he devised a system of inclined planes to overcome differences in level without the necessity for the time-consuming passage of flights of locks. The most notable of his scientific works were *Outlines of Natural Philosophy* (1822–3) and *Treatise on the Steam-Engine* (1830). His three sons became well-known engineers; his second son James was also a famous architect.

REYNOLDS, Osborne
(1842–1912)

English engineer, born in Belfast of a Suffolk family. After spending two years with a local mechanical engineer to gain some practical experience, he went to the University of Cambridge where he graduated Seventh Wrangler in 1867 and was elected to a fellowship at Queen's College. The following year he was appointed the first Professor of Engineering at Manchester, where he remained for 37 years. He possessed a combination of outstanding abilities in intuitive mechanical design and mathematical analysis, and investigated a wide range of engineering and physical problems. Some of his best work was in the field of hydrodynamics; in 1875, for example, he took out a patent which proposed the use of fixed guide vanes in centrifugal pumps, a feature found in many such pumps today. He also designed the first multi-stage centrifugal pump. Other areas in which he carried out significant research work included cavitation in hydraulic machines, dynamic stresses, thermodynamics and heat transfer, and lubrication. The 'Reynolds number', a dimensionless ratio characterizing the dynamic state of a fluid, takes its name from him. He was elected a Fellow of the Royal Society in 1877 and was awarded its Royal Medal in 1888.

RHETICUS, or RHETIKUS, or RHÄTICUS, originally VON LAUCHEN, Georg Joachim
(1514–76)

Austrian-born German astronomer and mathematician, born in Feldkirch. He became Professor of Mathematics at Wittenberg (1537). He is noted for his trigonometrical tables, some of which went to 15 decimal places. For a time he worked with **Copernicus**, whose *De Revolutionibus Orbium Coelestium* he was instrumental in publishing. His own *Narratio Prima de Libris Revolutionum Copernici* (1540) was the first account of the Copernican theory.

RICHARDS, Dickinson Woodruff
(1895–1973)

American physician, born in Orange, New Jersey. Educated at Yale, he specialized in cardiology, which he taught at Columbia University (1928–61), becoming Professor of Medicine there from 1947. With **Cournand**, Richards developed **Forssman**'s technique of cardiac catheterization into an important procedure for studying blood pressure, oxygen tension and a variety of other physiological variables in health and disease. Their work led to better understanding and treatment of shock, and provided the basis for much of modern cardiology, including the non-invasive surgical treatment of a number of conditions. The three men shared the Nobel Prize for Physiology or Medicine in 1956. Richards was a man with wide cultural attainments and an eloquent advocate for improved standards of medical care for the disadvantaged and elderly.

RICHARDS, Sir Rex Edward
(1922–)

English chemist and administrator, born in Colyton, Devon. He studied chemistry at St John's College, Oxford (DPhil 1947), and from 1947 to 1964 he was a Fellow and Tutor of Lincoln College. In 1964 he became Dr Lee's Professor of Physical Chemistry. He vacated this chair on becoming Warden of Merton College in 1969, a position he held until 1984. Since retiring from Oxford he has been Director of the Leverhulme Trust. His earliest research work was in infrared spectroscopy, but from the late 1940s he carried out pioneering studies in nuclear magnetic resonance spectroscopy (NMR) and contributed greatly to its development into a useful chemical technique. In the late 1960s Richards became interested in the applications of NMR to biological problems, and his work in this field contributed to the development of the medical diagnostic technique of 'magnetic resonance imaging'. Richards's research career ended when he became Vice-Chancellor of the University of Oxford (1977–81). He was elected a Fellow of the Royal Society in 1959 and received its Davy (1976) and Royal (1986) medals. Knighted in 1977, he served as President of the Royal Society of Chemistry in 1992.

RICHARDS, Theodore William
(1868–1928)

American chemist, born in Germantown, Pennsylvania. He was educated at Haverford College and at Harvard, where he was awarded a PhD in 1888. After a year of visiting laboratories in Europe, he became an assistant in analytical chemistry at Harvard, thereafter climbing the academic ladder to full professor by 1901. He held the Erving Chair of Chemistry from 1912 until his death. In his PhD research under Josiah Cooke, Richards redetermined very accurately the mass ratio for the combination of hydrogen and oxygen. This introduction to the methods of atomic weight determination was of lasting influence, and such work was dominant throughout his career. Over the course of 40 years the atomic weights of 25 elements were determined by Richards and his students. His investigation of the variation of the atomic weight of lead with source proved the existence of isotopes (1914). Richards also carried out much work on various topics in physical chemistry, notably thermochemistry (eg heats of reaction, combustion, dilution), electrochemistry (investigating the electromotive force of chemical cells) and physical properties of elements and compounds, particularly measurements of compressibility. He was awarded the Nobel Prize for Chemistry in 1914 for his work on atomic weights. Richards also received the Davy Medal of the Royal Society in 1910 and was made a Foreign Member in 1919. He became an Honorary Fellow of the Chemical Society in 1908 and was awarded its Faraday Medal in 1911.

RICHARDSON, Lewis Fry
(1881–1953)

English philosopher and meteorologist, born in Newcastle upon Tyne. He graduated from Cambridge University in 1903 and became Superintendent of the Eskdalemuir Observatory (1913), but left there to join the Friends Ambulance Unit during World War I. He rejoined the Meteorological Office in 1919, and moved to a teaching post at Westminster Training College (1920–9), later becoming Principal of Paisley Technical College (1929–40). In the first attempt to calculate future weather, Richardson observed the state of the atmosphere at points 200 kilometres apart, then using fundamental equations which he had formulated, calculated the values at the same points six hours ahead. He worked an example based on actual observations on 20 May 1910; the amount of arithmetical calculation involved was enormous, taking around six years. The resulting forecast was a failure, but his ideas formed the platform on which numerical weather prediction was built once large computers became available. His book *Weather Prediction by Numerical Process* (1922) describes his work. Richardson also carried out original research on turbulence, devising the Richardson number which determines whether atmospheric turbulence will increase or decrease. A pacifist, he also studied and wrote on the causes of war.

RICHARDSON, Sir Owen Williams
(1879–1959)

English physicist, born in Dewsbury, Yorkshire. He was educated at Cambridge, where at the Cavendish Laboratory he began his famous work on 'thermionics', a term he coined to describe the phenomenon of the emission of electricity from hot bodies; for this work he was awarded the Nobel Prize for Physics in 1928. He also formulated 'Richardson's law', an empirical formula which relates the rate of electron emission to the absolute temperature of the emitting metal, and became important in the development of the thermionic valve (or electron tube). Richardson was Professor of Physics at Princeton University (1906–13), and at King's College, London (1914–24). In 1924 he relinquished all his teaching duties, and was given the dual appointment of Director for Research in Physics at King's College and Yarrow Research Professor of the Royal Society. He was one of the outstanding pure scientists whose work was of great practical importance in the rapidly growing electronic industry. During World War II he worked on radar (in particular on magnetrons and klystrons), sonar and electronic test instruments. He was knighted in 1939.

RICHET, Charles Robert
(1850–1935)

French physiologist, born in Paris, the son of a surgeon. Richet obtained a medical degree at the University of Paris in 1877, and was appointed to the Chair of Physiology a decade later. During his career he conducted research in many fields including hypnosis, pain, digestion, muscle contraction and animal heat. In pioneering work on serum therapy, he conceived the idea that the blood of animals which are resistant to a harmful bacterium might contain a substance which could be used to confer immunity on a non-resistant animal. Having demonstrated that the principle was sound using a staphylococcus infection in dogs and fowl, Richet and his colleague Jules Héricourt attempted to prepare a serum for the treatment of tuberculosis, but found they could only delay the course of the disease. Some years later, when conducting experiments which attempted to define a toxic dose of an extract of sea-anemone tentacles, Richet and Paul Portier found that dogs that had survived experiments in which they had been injected with the poison were subsequently much more sensitive to it. This was a rather surprising phenomenon, the opposite of what might have been expected from his previous work, and he called this 'anaphylaxis'. He suggested that the mechanism, however, might be similar to that of immunity, resulting from the production of substances in the blood. He confirmed that this is the case by making normal animals anaphylactic by injecting them with serum from anaphylactic animals. It soon became apparent that anaphylactic reactions were not so uncommon, and would have to be taken account of by physicians in serum therapy. Richet was awarded the 1913 Nobel Prize for Physiology or Medicine for this work.

RICHMOND, Sir Mark Henry

(1931–)

British microbiologist, born in Sydney, Australia. He was an undergraduate at Clare College, Cambridge, and after receiving his PhD he became a member of the Medical Research Council scientific staff, moving to a readership in molecular biology at the University of Edinburgh in 1965. His major work has been in the area of antibiotic resistance. Richmond showed the importance of plasmid-mediated and transposon-mediated resistance to antibiotics. In addition, he demonstrated that resistant organisms are very likely to occur in areas of high antibiotic usage, eg in hospitals. In 1968 he took up the Chair of Bacteriology at the University of Bristol, leaving in 1981 to become Vice-Chancellor and Professor of Molecular Microbiology at the University of Manchester. Since 1981 his main activities have concerned university education and the development of science policy. In 1990 he was appointed Chairman of the Science and Engineering Research Council. Active on many national bodies, Richmond was elected FRS in 1980 and knighted in 1986.

RICHTER, Burton

(1931–)

American particle physicist, born in New York City. He studied at MIT where he received a BS in 1952 and his PhD in 1956, and then joined the high-energy physics laboratory at Stanford University, where he became professor in 1967. He was largely responsible for the Stanford Positron–Electron Accelerating Ring (SPEAR), an accelerator designed to collide positrons and electrons at high energies, and to study the resulting elementary particles. In 1974 a team led by him discovered the J/ψ hadron, a new heavy elementary particle whose unusual properties supported **Glashow**'s hypothesis of charm quarks. Many related particles were subsequently discovered, and stimulated a new look at the theoretical basis of particle physics. He shared the 1976 Nobel Prize for Physics with **Ting**, who had discovered the J/ψ almost simultaneously. Richter became a strong proponent of the modern trend in particle physics towards building ever-larger particle accelerator rings.

RICHTER, Charles Francis

(1900–85)

American seismologist, born near Hamilton, Ohio. Educated at Stanford University (AB 1920), he began a PhD at Caltech on atomic theory, but before completing it he was offered a position at the Carnegie Institute of Washington. There he finished his PhD (1928), became fascinated with seismology and met **Gutenberg**, with whom in 1932 he was co-creator of the famous Richter scale, the original instrumental scale for determining the energy released by an earthquake. The magnitude scale ranged from 1 to 9, and a magnitude increase of one unit corresponded to the release of about 30 times the seismic energy. Richter's name has been popularly ascribed to all the later magnitude scales devised and that in use today. In 1937 he returned to Caltech where he spent the rest of his career, as research assistant and Professor of Seismology from 1952. He played a key role in establishing the southern California seismic array and published *Seismicity of the Earth* (1954, with Gutenberg) and *Elementary Seismology* (1958).

RICHTER, Hieronymous Theodor

(1824–98)

German chemist, born in Dresden. He became assistant to **Reich** at the Freiberg School of Mines and subsequently its director in 1875. Using spectroscopy and working in collaboration with Reich, Richter discovered the metal indium in zinc-blende in 1863. He died in Freiberg.

RICHTER, Jeremias Benjamin

(1762–1807)

German chemist who established stoichiometry as a branch of chemistry. Born in

Hirschberg, he spent seven years in the engineering corps of the Prussian army, studied mathematics and philosophy at the University of Königsberg, and after some years as an independent chemical consultant was appointed chemist to the Royal Porcelain Works in Berlin. Richter believed that all chemical reactions are guided by mathematical laws. He analysed a vast number of compounds to determine the proportions of their reagents by weight and described this new mathematical approach as 'stoichiometry'. Although his results demonstrated that reagents do indeed combine in fixed proportions, he never formulated this observation into a law and it was **Proust**, working independently in France and Spain, who first proposed the 'law of definite proportions' in 1794. Richter, in the course of his research on proportions, noticed that when double decomposition takes place between two neutral salts, the compounds formed in the reaction are also neutral. This not only supported his theory of fixed ratios but also helped later chemists in their study of valence. Richter's work was little noticed in his lifetime but its importance became apparent after the atomic theory of **Dalton** provided the explanatory framework for his hypothesis in 1803. He died in Berlin.

RIDEAL, Sir Eric Keightley

(1890–1974)

English physical chemist, born in Sydenham, Kent. The son of Samuel Rideal, public analyst (of the Rideal–Walker test for disinfectants), he took the Natural Sciences Tripos at Trinity Hall, Cambridge. Thereafter he studied in Aachen and in Bonn, earning a PhD at the latter university in 1912 for a thesis on the electrochemistry of uranium. He entered his father's consulting practice, but World War I supervened. During war service he was wounded and discharged in 1916; after this he worked with a team at University College London, on the **Haber** process of ammonia synthesis. In 1920 he was appointed H O Jones Lecturer in Physical Chemistry at Cambridge. From 1930 to 1946 Rideal was Professor of Colloid Science at Cambridge; he then spent three years as Fullerian Professor at the Royal Institution. From 1950 until 1955 he was Professor of Physical Chemistry at King's College, London. Rideal's work during World War I aroused his interest in catalysis, and in his spare time he wrote *Catalysis in Theory and Practice* (1919, with **Hugh Taylor**). In Cambridge he built up a large research group working not only on heterogeneous catalysis, but also on electrochemistry, colloid and surface chemistry, and other topics. In World War II the Colloid Science Laboratory under his direction was much involved in work on explosives, fuels and polymers. Rideal was appointed MBE in 1918 and knighted in 1951. He was elected a Fellow of the Royal Society in 1930 and awarded its Davy Medal in 1951. He served as President of the Chemical Society in 1950–2 and of the Faraday Society from 1938 to 1945.

RIEMANN, (Georg Friedrich) Bernhard

(1826–66)

German mathematician, born in Breselenz. He studied in Göttingen under **Gauss** and in Berlin under **Dirichlet**, whom he succeeded as Professor of Mathematics at Göttingen in 1859. He was forced to retire by illness in 1862 and died of tuberculosis in Italy. His first publication (1851) was on the foundations of the theory of functions of a complex variable, including the result now known as the Riemann mapping theorem. This asserts that any simply connected region (roughly speaking a single region with no holes) can be mapped by a complex function onto a disc so that the boundaries correspond. In this general it is false, but under restriction it is true, and the theory has been a profound contribution to the subject. In a later paper on Abelian functions (1857), he introduced the idea of 'Riemann surface' to deal with 'multi-valued' algebraic functions; this was to become a key concept in the development of analysis. His famous lecture in 1854, 'On the hypotheses that underlie geometry', given in the presence of the aged Gauss, first presented his notion of a 'manifold', an *n*-dimensional curved space, greatly extending the non-Euclidean geometry of **Bolyai** and **Lobachevski**. These ideas were essential in the formulation of **Einstein**'s theory of general relativity, and have led to the modern theory of differentiable manifolds which now plays a vital role in theoretical physics. Riemann's name is also associated with the zeta function, which is central to the study of the distribution of prime numbers; the 'Riemann hypothesis' is a famous unsolved problem concerning this function.

RIESZ, Frigyes (Frédéric)

(1880–1956)

Hungarian mathematician, born in Györ. He studied at Zürich, Budapest and Göttingen, then returned to Hungary as a lecturer at the University of Szeged. In 1945 he was appointed Professor of Mathematics at the University of Budapest. He worked in functional analysis, integral equations and subharmonic functions, and developed a new approach to the **Lebesgue** integral. His work was instrumental in allowing the matrix and wave mechanics methods developed simultaneously during the emergence of quantum theory to be identified as equivalent. Riesz's textbook on functional analysis, *Leçons d'analyse fonctionnelle* ('Lessons of Functional Analysis', 1952) written with Bela Szökefalvi-Nagy (1887–1953), is a classic.

RIGHI, Augusto

(1850–1920)

Italian physicist, born in Bologna, famous for his work on electromagnetic waves. Educated in his home town, he taught physics at Bologna Technical College (1873–80), leaving to take up the newly established Chair of Physics at the University of Palermo. He was Professor of Physics at Padua (1885–9) and later returned to a professorship at Bologna, where he taught until his death. He invented an induction electrometer (1872) capable of detecting and amplifying small electrostatic charges; the principal of the design was very similar to that of **Van de Graaff**'s accelerator (1933). He formulated mathematical descriptions of vibrational motion, a problem addressed by **Lissajous** only a few months earlier. Righi also discovered magnetic hysteresis (1880), and whilst working at Palermo he discovered that bismuth exhibits a Hall effect thousands of times stronger than that of gold. He introduced the term 'photoelectric effect' to describe the increase in the conductivity of air gaps between electrodes when the air is exposed to ultraviolet light. Embarking on a series of experiments that he hoped would demonstrate that Hertzian waves obeyed the laws of classical optics, such as those of interference and diffraction, he generated Hertzian waves with wavelengths of just 26 millimetres. In doing so he was the first person to generate 'microwaves', thereby opening a whole new area of the electromagnetic spectrum to research and subsequent application. His experiments with microwaves did indeed confirm that Hertzian waves behaved very similarly to visible light and showed comparable interference, diffraction, refraction and absorption effects. His *L'ottica delle oscillazioni elettriche* (1897) which summarized these results is considered a classic of experimental electromagnetism. By the turn of the 20th century he had turned to work on X-rays and the **Zeeman** effect. In 1903 he wrote (with B Dessau) the first paper on wireless telegraphy, *La telegrafia senza fila*. He also worked on the conduction of gasses under various conditions of pressure and ionization, and on improvements to the **Michelson–Morley** experiment from 1918.

RITTER, Johann Wilhelm

(1776–1810)

German physicist, born in Samitz, Silesia (now in Poland), the son of a Protestant pastor. He was trained as an apothecary before studying medicine at the University of Jena. He taught at Jena and at Gotha before becoming a full member of the Bavarian Academy of Sciences in Munich (1804). While working in Jena he discovered the ultraviolet rays in the spectrum by means of its darkening effect on silver chloride (1801), but his chief contributions were in electrochemistry and electrophysiology. He demonstrated in 1800 that galvanic electricity was a manifestation of electricity, like static electricity, by means of the electrolysis of water and collected the constituent gases of hydrogen and oxygen (this experiment was first performed with static electricity); made the first dry cell (1802); and a secondary cell, or accumulator (1803); and was the first to propose an electrochemical series. He saw electrical action in terms of chemical behaviour. Although his research was empirical, his conceptual framework was coloured by the German Romantic movement and *Naturphilosophie*, and this adversely affected the acceptance of his work abroad, but he did strongly influence **Humphry Davy**.

ROBBINS, Frederick Chapman

(1916–)

American physiologist and paediatrician, born in Auburn, Alabama, the son of a plant physiologist. Robbins trained at Missouri University and Harvard University Medical School, where he graduated in 1940. He served as intern at the Children's Hospital Medical Centre, Boston (1941–2), before spending four years with the Army Medical Corps. Returning to Boston he completed his paediatric training and then joined **Enders** and **Weller** at the Infectious Diseases Research Laboratory of the Children's Hospital. Robbins, Enders and Weller had all worked on improving techniques for cultivating viruses, and they decided to apply their improvements to a new attempt to cultivate the poliomyelitis virus. Their success, which relied upon the use of antibiotics, led to quicker and cheaper means of diagnosis, and was also an important step in the development of a polio vaccine. For this work the three scientists were awarded the 1954 Nobel Prize for Physiology or Medicine. Robbins was Professor of Paediatrics at Case Western Reserve University, Cleveland, from 1952 until 1980.

ROBERTS, John D

(1918–)

American chemist, born in Los Angeles. He obtained both his AB and PhD from the University of California. After working as a postdoctoral fellow and instructor at Harvard he joined the staff of MIT and became associate professor in 1950. In 1953 he moved to Caltech as professor, and he later became Division Chairman, Provost and Institute Professor Emeritus (1988). His major contributions have been to physical organic chemistry, particularly reaction mechanisms, and to the application of nuclear magnetic resonance spectroscopy to the structure determination of important organic molecules. This form of spectroscopy is probably the single most important development in chemistry since World War II. He is the author, with M C Caserio, of *Basic Principles of Organic Chemistry* (1967). He was awarded the Priestley Medal of the American Chemical Society (1987), the Robert A Welch Award in Chemistry (1990) and the National Medal of Science (1991).

ROBERTS, Richard

(1943–)

British molecular biologist, born in Derby. Educated at the University of Sheffield, he moved to the USA in 1969 and worked at Cold Spring Harbor Laboratory in New York from 1972. Since 1992 he has been research director at New England Biolabs in Beverly, Massachusetts. In 1977, he announced his intriguing discovery that genes contain sections of DNA now known as 'introns' which carry no genetic information. He shared the 1993 Nobel Prize for Physiology or Medicine jointly with **Sharp**, who had independently come to the same conclusions around the same time.

ROBERTS-AUSTEN, Sir William Chandler

(1843–1902)

English metallurgist, born in Kennington, London. Privately educated, he entered the Royal School of Mines aged 18 and was awarded an associateship in 1865. In 1880 he was appointed professor at the Royal School of Mines, two years later becoming chemist and assayer at the Mint. On all scientific and technical operations of the coinage be became a world authority. He was also a pioneer in the developing field of alloys. In 1901 he produced the first nearly correct iron–carbon equilibrium diagram, based on the phase rule and the accepted thermodynamic principles of chemistry. The iron–carbide solid solution ('austenite') formed in cooling steel is named after him. He also demonstrated the possibility of diffusion occurring between a sheet of gold and a block of lead, and experimented with early versions of the pyrometer. Elected FRS in 1875, he was knighted in 1899.

ROBERTSON, Howard Percy

(1903–61)

American mathematician and cosmologist, born in Hoquiam, Washington. He was educated at the University of Washington, where he received his fist degree in 1922, and at Caltech, where he graduated PhD in 1925. Following a spell as Research Fellow at Göttingen, Munich and Princeton, he became Assistant Professor of Mathematics at Caltech (1927–9). He joined the staff of Princeton University in 1928, becoming Professor of Mathematical Physics in 1938, and in 1947 returned to Caltech as Professor of Mathematical Physics, a post which he held until his death. Robertson's researches were in the fields of differential geometry, relativity and theoretical cosmology. The Poynting–Robertson effect (1937) was the result of his rigorous application of relativistic theory to the effect first noted by **Poynting** (1903) of the Sun's radiation pressure on micrometeorites. In cosmology his name is also given to the expression developed by him for the four-dimensional relativistic line-element applicable to a homogeneous isotropic expanding universe (1933).

ROBIN, Gordon de Quetteville

(1921–)

Australian geophysicist and glaciologist, born and educated in Melbourne where he received a BSc in physics (1940) and an MSc (1942). He joined the Royal Australian Naval Volunteer Reserve and served on submarines (1942–5), and later moved to the UK where he became a research student and lecturer at Birmingham University. He moved to the Australian National University as Senior Fellow (1957–8), returning to the UK in 1958 to become Director of the Scott Polar Research Institute, Cambridge (1958–82). He was Secretary (1958–70) and President (1970–4) of the International Scientific Committee on Antarctic Research. During meteorological and glaciological work in the Antarctic, Robin obtained the first seismic traverse from the coast to the Antarctic Plateau and developed a theory of flow and temperature distribution in ice sheets. His research group revolutionized the study of ice sheets by using radio echo sounding techniques and studied the penetration of ocean waves into fields of pack ice. He made the first proposal for studying polar ice sheets by satellite altimetry (1963), and in publications on ice sheets and glaciers investigated their interaction with processes which produce global warming.

ROBINS, Benjamin

(1707–51)

English mathematician and father of the art of gunnery, born into a Quaker family in Bath. He became a teacher of mathematics in London, published several treatises, commenced his experiments on the resisting force of the air to projectiles, studied fortification, and invented the ballistic pendulum to measure the energy of objects in flight. In 1735 he refuted George Berkeley's objections to **Isaac Newton**'s calculus, in a treatise entitled *Newton's Methods of Fluxions*. His *New Principles of Gunnery* appeared in 1742, and inspired **Leonhard Euler** and **Lambert** to investigate the subject. Engineer to the East India Company (1749), he died in Madras. His works were collected in 1761.

ROBINSON, Sir Robert

(1886–1975)

English chemist, born in Chesterfield. The son of a manufacturer of surgical dressings, he was sent to Manchester University in 1902 to study chemistry as it was thought it would be useful to the family firm. He came under the influence of **William Henry Perkin Jr**, and on graduation entered Perkin's research laboratory to study the dyewood colouring matter brazilin. An interest in the chemistry of coloured natural products remained with him for the rest of his life. He also started work on alkaloids while with Perkin. During his time in Manchester, Robinson established friendships with **Arthur Lapworth**, **Weizmann** and **Norman Haworth**. In 1912 he was invited to a chair of organic chemistry in Sydney, where he continued to study topics started in Manchester and also undertook work on material from eucalyptus oils. After only two years he returned to Britain as professor at the University of Liverpool, and entered upon a new and very significant phase of his researches on alkaloids, including his famous synthesis of tropinone. Rather surprisingly he left Liverpool in 1919 to become Director of Research for the British Dyestuffs Corporation based in Huddersfield. However, the move was not a success and in 1920 he resigned to become professor at the University of St Andrews, although he maintained close contact with the dyestuffs industry for many years. He continued research on natural products with significant results and also developed many of his ideas on mechanistic organic chemistry. At this time he proposed a description of aromaticity in terms of a stable sextet of electrons. This was a happy time for Robinson but when the chair at Manchester fell vacant in 1922 he moved there. By this time his fame was great enough to attract many overseas students. After a brief spell at University College London (1928–30) he became Waynflete Professor of Chemistry at Oxford, where he remained until 1955. It was the culmination of his scientific career, but he became increasingly irascible and entered into a long altercation with **Christopher Ingold** on mechanistic organic chemistry. During World War II he played an important role in the development of penicillin. Even after retirement he maintained an interest in chemistry. He was a close friend of Robert Maxwell and founded a number of learned journals. He served on many government committees, was knighted in 1939, awarded the Order of Merit in 1949, and became President of the Royal Society in 1945. He received the Nobel Prize for Chemistry in 1947 for his work on the chemistry of natural products.

ROEMER, Olaus

(1644–1710)

Danish astronomer, born in Aarhuus. He worked at the Paris Observatory (1672–81) where he discovered that the intervals between successive eclipses of a satellite in Jupiter's shadow were less when Jupiter and the Earth were approaching than when they were receding (1675). He concluded that this was due to the finite velocity of light—a new discovery—and from the observed intervals and the known rates of motion of Jupiter and the Earth obtained the first estimate of the velocity of light. Roemer also invented the transit instrument, a telescope movable only in the meridian, which greatly increased the accuracy attainable in the determination of both time and right ascension. In 1681 he returned as royal mathematician and University Professor of Astronomy to Copenhagen, where he remained for the rest of his life.

ROHRER, Heinrich

(1933–)

Swiss physicist, born in Buchs, co-recipient of the 1986 Nobel Prize for Physics with **Binnig** for their development of the tunnelling electron microscope. He was educated at the Swiss Institute of Technology in Zürich, completing his first degree in Physics in 1955 and his PhD in 1960. He moved to the USA to take up a research post at Rutgers University in New Jersey (1961–3). Returning to Zürich in 1963, he joined the IBM Research Laboratory, where he ultimately became manager of the physics department. His work with Binnig on the scanning tunnelling microscope began in 1978; within two years they had constructed their first instrument, and by the end of the third had completed the instrument that won them the Nobel Prize (jointly with **Ruska**). The operation of the microscope is based on a quantum-mechanical effect, namely the tunnelling of electrons across a narrow gap between two surfaces. The microscope uses a needle with a tip of just one atom positioned close above the surface of a sample. With a potential difference applied between the needle and the sample a tunnelling current flows, and the needle can be scanned over the surface of the sample to build up a picture of the surface topography. The microscope is able to resolve features horizontally only 6 angstroms apart, and in the vertical direction can detect distances of 0.1 angstrom, one-thirtieth the size of an average atom. The scanning tunnelling microscope is now found increasingly frequently in laboratories around the world and sees particular application in the development of small solid-state electronic devices. In 1984 Rohrer was co-recipient with Binnig of two international awards, the Hewlett–Packard Europhysics Prize and the King Faisal International Prize.

ROMÉ DE L'ISLE, Jean-Baptiste Louis
(1736–90)

French crystallographer, born in Gray. In 1756 he entered the Royal Corps of Artillery and Engineering and saw active service in the French Indies, where he was captured by the English in 1761 and transported to China. He returned to France in 1764, published his first work on freshwater polyps (1766) and then devoted himself to the study of mineralogy and chemistry. In his *Essai de cristallographie* (1772), he identified 110 crystal forms. In 1779 he became involved in a controversy with **Buffon** concerning the theory of a central terrestrial fire and the eventual cooling of the Earth. His major work, *Cristallographie* (1783), described more than 450 crystal forms. In the course of making terracotta crystal models with his assistant, he devised a contact goniometer and together they discovered the law of the constancy of interfacial angles, whereby the angle between similar faces in a particular mineral is always the same. 1784 saw the publication of his book on the external characteristics of minerals, stating that form, density and hardness were sufficient criteria to permit the identification of any mineral species. He also published a work comparing the weights and measures of antiquity with modern counterparts (1784). He had a strained relationship with **Haüy**, whose ideas became more widely accepted.

ROMER, Alfred Sherwood
(1894–1973)

American palaeontologist, born in White Plains, New York. He studied at Columbia University and became Professor of Vertebrate Palaeontology at Chicago (1923–34), and then Professor of Zoology and Director of the Museum of Comparative Zoology at Harvard (1934–65). He was interested in the evolution of early vertebrates, in particular the transition from water to land, and wrote papers on the importance of the development of the amniotic egg and other preadaptations for terrestrial life. This formed the basis of his *The Vertebrate Story* (1959). One of the standard texts in its area was *Vertebrate Palaeontology* (1933), and *The Vertebrate Body* (1949) was a comprehensive text on comparative anatomy widely used in universities throughout the English speaking world.

RONALDS, Sir Francis
(1788–1873)

English inventor, born in London, the son of a merchant. Interested in the application of electricity, he invented an electric clock whose motive power was the recently introduced high-intensity dry battery, and in the same year (1816, published in 1823) an electrostatic telegraph which he constructed in his garden at Hammersmith. His offer of the telegraph to the Admiralty was refused; the device was rather impractical and practical versions did not appear until the later development of electromagnetic devices. He also invented a system of automatic photographic registration for meteorological instruments (1845). In 1843 Ronalds was made Honorary Director and Superintendent of the Kew Meteorological Observatory, set up by the British Association for the Advancement of Science. He retired in 1852 and lived for many years in Italy. He was elected FRS in 1844, and was knighted in 1870 for his pioneering work on the electric telegraph.

RÖNTGEN, Wilhelm Konrad von
(1845–1923)

German physicist, born in Lennep, Prussia. He studied mechanical engineering at Zürich. After teaching at Strassburg, he was appointed Professor of Physics successively at Giessen (1879), Würzburg (1888), where he succeeded **Kohlrausch**, and Munich (1899–1919). At Würzburg in 1895, while investigating the properties of cathode rays with a 'Crookes's tube', he discovered the electromagnetic rays which he called X-rays (known also as Röntgen rays), because of their unknown properties, and for his work on them he was awarded in 1896, jointly with

Lenard, the Rumford Medal, and in 1901 the first Nobel Prize for Physics. He also did important work on the heat conductivity of crystals, the specific heat of gases, and the magnetic effects produced in dielectrics, predicted by **Heaviside** as a consequence of **Maxwell**'s theory.

ROOZEBOOM, Hendrik Willem Bakhuis
(1854–1907)

Dutch physical chemist, born in Alkmaar. The opportunity of a formal chemical education came to him through employment as an assistant in the chemistry institute of the University of Leiden, where he was awarded a doctorate in 1884. He became lecturer in physical chemistry at Leiden in 1890 and in 1896 he succeeded **van't Hoff** as professor at Amsterdam. He is best known for his many applications of the phase rule to heterogeneous equilibria. The phase rule had been deduced from thermodynamics by **Gibbs** in the 1870s, but chemists were slow to appreciate its importance. The work of Roozeboom and his students did much to remedy this. Their study of the iron–carbon system, fundamental to an understanding of the behaviour of steel, was a remarkable achievement. In 1901 Roozeboom began to publish a multi-volume treatise in German on *Heterogeneous Equilibria from the Standpoint of the Phase Rule*. After his death the work was completed by some of his students.

ROSCOE, Sir Henry Enfield
(1833–1915)

English chemist, born in London, grandson of the English historian William Roscoe. He was educated at University College London, and the University of Heidelberg where, with **Bunsen**, he carried out research on quantitative photochemistry. In 1857 he was appointed Professor of Chemistry at Owens College, Manchester. His students and staff were encouraged to follow their own interests, with the result that Manchester led many new areas of research, and became famous at home and abroad. It had the first Chair of Organic Chemistry in Britain and also taught crystallography, thermal chemistry and gas analysis. Roscoe encouraged links with industry and was energetic in promoting lectures on science for the general public. He was Liberal MP for South Manchester from 1885 to 1895 and Vice-Chancellor of London University from 1896 to 1902. Roscoe's own research was considerable. With Bunsen he showed that when chlorine and hydrogen combine in the presence of light, the extent of the photochemical action varies inversely with the distance from the light source and is directly proportional to its intensity. In 1865 he isolated vanadium from copper ores in the Cheshire mines. Previously vanadium, which had been discovered by **Sefström**, had only been found in very small quantities and its properties were imperfectly known. Roscoe showed that it belongs to the same family as phosphorus and arsenic. He was also the author of influential textbooks. He was elected FRS in 1863, knighted in 1884, President of the British Association for the Advancement of Science in 1887, and a privy councillor from 1909. He died in Leatherhead, Surrey.

ROSE, William Cumming
(1887–1984)

American biochemist, born in Greenville, South Carolina. He studied at Yale and Freiburg, Germany, before returning to the USA where he worked at the University of Texas from 1913 to 1922. He was subsequently appointed professor at Illinois University, where he remained until his retirement in 1955. Following the demonstration by **Frederick Gowland Hopkins** that some amino acids appeared essential for growth, Rose began accumulating purified amino acids for experiments to determine the dietary importance of each of the 20 amino acids that occur in proteins (1924). By 1938 he could reliably state their nutritional significance in the diet, and he identified 10 indispensable amino acids. This led to the currently used division into essential and non-essential amino acids (1948) and his first list of human minimum and recommended daily requirements for each (1949). As a member of numerous nutritional committees he amended these values in 1957. During this work he isolated a new amino acid, threonine (1934), showed that methionine (1937) and valine (1939) are essential in the diet, and demonstrated that valine can form glucose. Rose also revealed that creatinine is formed from creatine, and studied the detoxification of benzoic acid by hippurate formation.

RÖSKY, Herbert
(1935–)

German chemist, born in a small village in East Prussia, who moved to West Germany after World War II and studied at the University of Göttingen. He received his doctorate in 1963 for work on metal fluorides. He then spent a year at the Du Pont Experimental Station in Wilmington, Delaware, where he collaborated with **Muetterties** on phosphorus anion chemistry. Returning to Germany he became Professor of Chemistry at Frankfurt University, and he is now Professor at Göttingen. He is a synthetic inorganic chemist whose principal interest remains the preparation of fluorides of both main group and transition metals. He won the Wöhler Prize in 1960, the Leibniz Award in 1987 and the Alfred Stock Memorial Award in 1990.

ROSS, Sir James Clark
(1800–62)

Scottish polar explorer and naval officer, born in London, the son of a rich merchant and nephew of Arctic explorer Sir John Ross. He first went to sea with his uncle at the age of 12, conducting surveys of the White Sea and the Arctic, and accompanied Sir William Parry on four Arctic expeditions (1819–27). From 1829 to 1833 he was joint leader, with his uncle, of a private Arctic expedition during which he located the magnetic north pole (1831). After conducting a magnetic survey of the British Isles, he led an expedition to the Antarctic (1839–43) on the *Erebus* and the *Terror*, during which he discovered Victoria Land and the volcano Mt Erebus. He was knighted on his return in 1843, and wrote an account of his travels in *Voyage of Discovery* (1847). He made a last expedition in 1848–9, searching for Sir John Franklin's ill-fated expedition in Baffin Bay. Ross Island, the Ross Sea and Ross's gull are named after him.

ROSS, Sir Ronald
(1857–1932)

British physician, born in Almara, Nepal, the son of an army officer. He studied medicine at St Bartholomew's Hospital in London and entered the Indian Medical Service in 1881. He learned bacteriological and microscopical techniques on various furloughs in England, during one of which he met **Manson** (1894). Manson told Ross of his belief that malaria is transmitted through mosquito bites and Ross returned to India determined to investigate this possibility. He discovered the malaria parasite in the stomachs of mosquitoes which had bitten patients suffering from the disease, and by 1898 had worked out the life cycle of the malaria parasite for birds. He returned to England in 1899 to lecture at the newly founded Liverpool School of Tropical Medicine and to lead several expeditions in Africa concerned with mosquito eradication. Elected FRS in 1901, he was knighted in 1911. He moved to London in 1912 where, from 1926, he directed the Ross Institute. He was a gifted if eccentric mathematician who also wrote poetry and romances. His award of the 1902 Nobel Prize for Physiology or Medicine was contested by Giovanni Grassi (1854–1925), an Italian parasitologist who had independently and almost simultaneously worked out the life cycle of the human malaria parasite.

ROSSBY, Carl-Gustaf Arvid
(1898–1957)

Swedish-born American meteorologist, born in Stockholm. He graduated from Stockholm University where he obtained the equivalent of a PhD in 1925. He joined the Bergen School under **Vilhelm Bjerknes** (1919). In 1926 he went to the USA and became professor at MIT (1931–9) and Chicago University (1941–50), before returning to Stockholm University (1950–7). In 1927 he established in California the first weather service for airways in the USA which became the model for others, and during his years at MIT he was largely responsible for the general acceptance of the Norwegian methods of synoptic analysis in the USA. He went further, however, by studying the large-scale wave-like motions in the upper atmosphere. There are usually three to five such 'Rossby waves' in each hemisphere. He demonstrated that the number and spacing of these waves around the hemisphere depended on the strength of the westerly winds. In general, they are slow moving and of considerable value in weather forecasting. Rossby played a valuable role in establishing thorough training courses for meteorologists both at MIT and Chicago. During World War II he was largely responsible for the varied courses needed for different theatres of war. His theoretical work greatly assisted the programming for numerical weather prediction once computers of sufficient speed became available.

ROSSE, William Parsons, 3rd Earl of
(1800–67)

Irish astronomer and landowner, born in York, England, and educated at Trinity College, Dublin, and Oxford University where he graduated in mathematics (1822). With the help only of the workers on his feudal estate of Birr Castle at Birr (then Parsonstown) he constructed on the model of **William Herschel**'s instruments a gigantic metal-mirror telescope 6 feet in diameter and 52 feet in focal length, the 'leviathan of Parsonstown'. With this, the largest telescope ever built until the 100 inch reflector in California (1917), he discovered the spiral structure of the nebula Messier 51 near the tail of the Great Bear (1845), the first ever observation of a spiral galaxy. Drawings of this and of many other nebulae including spirals were published in Rosse's catalogue of nebulae (1850). The giant mirror is preserved in the Science Museum, London.

ROSSI, Bruno
(1905–)

Italian–American physicist, born in Venice. He was educated at the universities of Padua

(1923–5) and Bologna (1925–7) and subsequently took up an assistantship in the physics department at the University of Florence (1928–32). In 1932 he was appointed Professor of Physics at the University of Padua, where he carried out early research on the properties of cosmic rays. He showed that these energetic particles reaching the Earth from space have the capacity to traverse great thicknesses of matter. He also demonstrated that the incident primary radiation from space may collide with atoms such as oxygen and nitrogen in the atmosphere causing nuclear reactions and generating cascades of secondary particles, now called 'showers'. This primary radiation he found to be positively charged (in fact cosmic rays consist mainly of protons), and he was able to show that not all of the particles making up primary cosmic radiation could produce particle showers. In 1939 he moved first to the University of Manchester, and then to the USA, where he spent time continuing his work on cosmic rays at the University of Chicago before becoming Professor of Physics at Cornell University in 1940. Five years later he moved to MIT where he spent the rest of his working life. His work has included investigations of meson decay, and among his many publications were *Cosmic Rays* (1964, with S Olbert) and *Introduction to the Physics of Space* (1970).

RÖTHLISBERGER, Hans
(1923–)

Swiss glaciologist, born in Langnau, near Berne. He entered a teacher training college then changed to study petrology at the Swiss Federal Institute of Technology (ETH) in Zürich, where he received a diploma in 1947. In his doctoral thesis on seismic velocity in sedimentary rocks (1954), he described a new method for grain size determinations. With an interlude as contract scientist to the US army Snow, Ice and Permafrost Research Establishment (1957–61) he held various appointments at ETH, finally becoming head of the glaciology section of the Laboratory of Hydraulics, Hydrology and Glaciology (1980–7). During expeditions to Baffin Island in 1950 and 1953, he undertook seismic soundings across glaciers and with colleagues made the first ascent of Mount Asgard. His main contributions lie in glacial hazard assessment, and in englacial and subglacial drainage of meltwater. Röthlisberger made the important identification of the equivalence of the rate of melting caused by frictional heat of the running water and the rate of conduit closure caused by ice overburden pressure. A meltwater conduit incised in ice at the glacier bed is referred to as a Röthlisberger channel. He published *Seismic Exploration in Cold Regions* (1972), and was President of the International Glaciological Society (1984–7).

ROTHSCHILD, Walter, Lord
(1868–1937)

English eccentric collector, taxonomist and patron, born near Peterborough. He lived most of his life at Tring, Buckinghamshire. Educated at home and briefly at Bonn and Cambridge universities, he became the sponsor of many collecting expeditions, and donated his private collection at Tring to the National History Museum (to pay off the demands of a blackmailer). Rothschild's collections, and the publications based on them, were significant in educating biologists from all over the world about the problems of species distribution, varieties, transitional forms, population variation and classification. He was active in the moves to establish a state of Israel. His life and accomplishments are described by his niece, Miriam Rothschild, in *Dear Lord Rothschild* (1983).

ROUS, (Francis) Peyton
(1879–1970)

American pathologist, born in Baltimore. Educated at Johns Hopkins University and Medical School, he became assistant (1909–10), associate (1910–12), associate member (1912–20) and member (1920–45) of the Rockefeller Institute for Medical Research in New York City. From 1909 he began studying a sarcoma in chickens, which he demonstrated was caused by a virus. He also showed that the sarcoma could be successfully transplanted only in closely related chickens, thus encouraging others to use genetically similar strains of laboratory animals. He returned to this line of work in the 1930s, when a rabbit tumour was also shown to be caused by a virus; Rous found that coal tar and the virus could potentiate each other in making the tumour malignant. The discovery of many other oncogenic viruses from the 1950s made his early work more widely appreciated, and he shared with **Charles Huggins** the 1966 Nobel Prize for Physiology or Medicine. For more than 50 years, Rous was an editor of the *Journal of Experimental Medicine*, published at the Rockefeller Institute.

ROUX, (Pierre Paul) Émile
(1853–1933)

French bacteriologist, born in Confolens, Charente. He studied at the universities of Clermont-Ferrand and Paris, where he became assistant to **Pasteur**, and in 1904 succeeded him as Director of the Pasteur Institute. Roux worked with Pasteur on most of the latter's major medical discoveries. He tested the anthrax vaccine, and did much of the early work on the rabies vaccine. With **Yersin** he showed that the symptoms of diphtheria are caused by a lethal toxin produced by the diphtheria bacillus, rather than by the bacteria themselves. **Behring** and **Kitasato** had shown in 1890 that infected guinea pigs produce an antitoxin in the blood. Using the same principle, Roux obtained large quantities of blood serum containing the antitoxin from horses, and tested it on patients in 1894. In four months the mortality rate had fallen from 51 to 24 per cent. He also made important contributions to research into syphillis.

ROUX, Wilhelm

(1850–1924)

German anatomist and physiologist, born in Jena. Roux studied medicine at Jena University, before spending 10 years (1879–89) at the anatomical institute in Breslau. In 1895 he was appointed head of anatomy at Halle, where he remained until 1921. Roux was prolific both as a researcher and as a propagandist for a particular, highly mechanistic, vision of science. He accomplished extensive practical and theoretical work on experimental embryology within the paradigm of *Entwicklungsmechanik*, or developmental mechanics. A follower of **Haeckel** and **Charles Darwin** and a convinced evolutionist, Roux sought to understand evolutionary processes at the cellular and molecular levels, focusing in particular upon experimental embryology, and applying the findings of physics and chemistry. On the basis of extensive and often violent embryological experimentation, Roux endorsed **Weismann**'s idea that the material of the germ plasm is passed on intact from parent to offspring; he thereby pointed to a physical basis for heredity.

ROWLAND, Henry Augustus

(1848–1901)

American physicist, born in Honesdale, Pennsylvania. It was expected that he would follow family tradition and join the Protestant ministry, but he decided on a career in engineering or science. He was educated at the Rensselaer Polytechnic Institute to which he returned as an instructor in physics in 1872. From 1875 to 1901 he was Professor of Physics at the new Johns Hopkins University and he established there a laboratory on the European model. He was a meticulous experimenter and had a flair for instrument design. His laboratory determination of the ohm was close to the present-day absolute value. He invented the concave diffraction grating used in spectroscopy, discovered the magnetic effect of electric convection, and improved on **Joule**'s work on the mechanical equivalent of heat. Shortly before his death he invented a multiplex telegraph, shown at the Paris Exhibition of 1900. Following his wishes, his ashes were interred in the wall of the basement laboratory which housed the engine with which he ruled his gratings.

ROWLINSON, John Shipley

(1926–)

English physical chemist, born in Handforth, Cheshire. He studied chemistry at Trinity College, Oxford (DPhil 1950). After postdoctoral work at the University of Wisconsin (1950–1), he held various positions at Manchester University until 1961, when he became Professor of Chemical Technology at Imperial College. In 1974 he moved to Oxford as Dr Lee's Professor of Physical Chemistry. Since 1990 he has also been A D White Professor-at-Large at Cornell University. He has made notable contributions to the applications of thermodynamics and statistical mechanics to the understanding of the physical properties of matter in the solid, liquid and gaseous states and to knowledge of intermolecular forces. Rowlinson has written several influential books including *Liquids and Liquid Mixtures* (1959), *The Perfect Gas* (1963) and *The Molecular Theory of Capillarity* (1982, jointly with Benjamin Widom). He was elected FRS in 1970, and received the Royal Society's Leverhulme Medal in 1993. Rowlinson was President of the Faraday Division of the Royal Society of Chemistry from 1979 to 1981, and was awarded its Faraday Medal in 1983.

ROXBURGH, William

(1751–1815)

Scottish botanist, born in Craigie, Ayrshire. He was educated at Edinburgh University, where he studied under John Hope. By Hope's influence he became a surgeon's mate with the East India Company and ultimately assistant surgeon in Madras (1780–93). Stationed at Samulcotta, he cultivated cinnamon, annatto, nutmeg, coffee, peppers and indigo, studied sugar-growing and silkworm-rearing and made large collections of plants, and was appointed the East India Company's 'Botanist in the Carnatic'. Many of his early collections were lost in a flood (1787). He prepared a series of illustrations of local plants, using an Indian artist, published as *Plants of the Coast of Coromandel* (3 vols, 1795–1819). He became Superintendent of Calcutta Botanic Garden (1793–1813), but while there was forced twice to return to Britain because of illness, before his health completely broke down in 1813, when he finally left India. He left behind the manuscript of his *Flora Indica*, which was eventually edited by **Wallich** and partially published (2 vols, 1820–4) by William Carey, with a later complete edition (3 vols, 1832). Roxburgh's book remains valuable for its notes on Indian economic botany and its accurate descriptions.

RUBBIA, Carlo
(1934–)

Italian-born American physicist, born in Gorizia and educated at Pisa, Rome and Columbia universities. From 1960 he worked at CERN (the European nuclear research centre) in Geneva. In 1971 he accepted a Chair in Physics at Harvard University, simultaneously continuing his research at CERN. Rubbia was head of the team that discovered the W and Z bosons which mediate the weak nuclear force. Although indirect evidence for the existence of these particles had been available for some time, this was the first direct proof and put the unified theory of electromagnetic and weak forces (the 'electroweak' theory) on a firm experimental footing. Rubbia pointed out that because the proton and anti-proton have exactly the same mass and opposite charges within experimental limits, they will describe exactly the same orbit but in opposite directions when placed in a magnetic field. This allowed the existing Super Proton Synchrotron (SPS) accelerator to be upgraded to become a collider of protons and anti-protons, achieving enough energy to produce W and Z bosons for the first time. For this work he shared the 1984 Nobel Prize for Physics with **van der Meer**. As Director-General of CERN (1989–93), he was the driving force behind the LEP (an electron–positron collider) and LHC (Large Hadron Collider) projects.

RUBIN, Gerald Mayer
(1950–)

American molecular biologist, born in Boston. He was educated at MIT where he obtained a BS in 1971 and at the University of Cambridge where he received his PhD in 1974. Subsequently he became Helen Hay Whitney Foundation Fellow at Stanford University School of Medicine in California (1974–6) and Assistant Professor of Biological Chemistry at the Sidney Farber Cancer Institute, Harvard Medical School (1977–80). He was then appointed as a staff member at the Carnegie Institute of Washington, Baltimore, from 1980 until 1983, when he moved his current position as John D McArthur Professor of Biological Chemistry at the University of California at Berkeley. He has simultaneously been Investigator at the Howard Hughes Medical Institute since 1987. Rubin was responsible for the first production of transgenic fruit flies. This involves the introduction of specific cloned genes into the germ line of the fruit fly. When used in conjunction with the rapid production of genetic mutations in fruit flies, this technique provides a powerful tool for the study of the regulation of specific gene control. For example, artificially introduced genes can be tracked throughout the development of the fruit fly to determine in which tissue and at what developmental stage specific genes are active. In addition the introduced genes can be mutated in specific ways so that the function of precise DNA sequences on the correct expression of their accompanying gene can be determined.

RUBNER, Max
(1854–1932)

A notoriously taciturn German physician who specialized in physiology and hygiene. Rubner attended the University of Munich, later holding chairs at Marburg and Munich and rising to become **Koch**'s successor in the Chair of Hygiene at Berlin. He worked on many aspects of metabolism, definitively establishing the applicability of the principle of the conservation of energy to living organisms. Impressed by energy physics, Rubner attempted to assess the life-careers of mammals with regard to the energy consumption of their protoplasm and its transformation into growth. Against the background of food shortages during World War I, he increasingly directed his researches into problems of nutrition.

RUDBECK, Olof
(1630–1702)

Swedish physiologist and intellectual, born in Västerås. A bishop's son, Rudbeck studied medicine at the University of Uppsala, and early embarked on research into animal economy. He quickly accepted the views of **William Harvey** (1578–1657) and familiarized himself with recent work on the lymphatic system. Independently of **Pecquet**, he discovered the thoracic duct, through which the lacteal vessels discharge chyle into the veins. There followed a violent priority dispute involving Pecquet, who had published first, and **Thomas Bartholin**, who had separately been working in Copenhagen on the lymphatic system. Further studies at Leiden were followed by appointment to a medical chair in Uppsala. Rudbeck devoted massive energies to the improvement of the buildings and intellectual standing of the university. In later years Rudbeck pursued botanical studies, developing a tradition of which **Linnaeus** was the later beneficiary.

RUE, Warren de la
See **DE LA RUE, Warren**

RUMFORD, Benjamin Thompson, Count
(1753–1814)

English–American administrator and scientist, born in Woburn, Massachusetts. He was an assistant in a store and a school teacher, until in 1772 he married a wealthy widow. He became a major in the 2nd New Hampshire regiment, but left his wife and baby daughter and fled to England in 1776, possibly due to political suspicion. He gave valuable information to the government as to the state of America during the Revolution, and received an appointment in the Colonial Office. In England he began the experiments with gun-powder which continued throughout the rest of his life. He was elected FRS in 1779, and by 1782 he was back in America, with a lieutenant-colonel's commission. After the peace he was knighted, and in 1784 entered the service of Bavaria. In this new sphere he reformed the army, drained the marshes round Mannheim, established a military academy, planned a poor-law system, introduced the cultivation of the potato, disseminated knowledge of nutrition and domestic economy, improved the breeds of horses and cattle, and laid out the English Garden in Munich. For these services he was made head of the Bavarian war department and count of the Holy Roman Empire. In the former capacity he was responsible for the arsenal at Munich, where he observed the immense amount of heat generated by the boring of cannon. He was able to deduce experimentally that there was a relationship between the work done and the heat generated, and this led to his championing of the vibratory (rather than the caloric) theory of heat. He reported his findings to the Royal Society in 1798 in one of the classic papers of experimental science, *An Experimental Inquiry concerning the Source of Heat Excited by Friction*. He also invented the Rumford shadow photometer, designed the so-called Rumford oil lamp, and introduced the concept of the standard candle which became the international standard of luminosity until the middle of the present century. During a visit to England (1795–6) he endowed the two Rumford medals of the Royal Society, and also two of the American Academy, for researches in light and heat. In 1799 he left the Bavarian service, returned to London, and founded the Royal Institution; in 1802 he moved to Paris, married **Lavoisier**'s widow (1804) and lived in her villa at Auteuil, where he died.

RUNGE, Friedlieb Ferdinand
(1795–1867)

German dye chemist and pioneer of chromatography, born near Hamburg. He studied medicine in Berlin, Göttingen and Jena, and then spent three years visiting factories and laboratories throughout Europe. He was appointed to the Chair of Technical Chemistry at Breslau in 1828; from 1831 to 1852 he managed a chemical factory at Oranienburg, and thereafter worked independently as a chemical consultant. At Oranienburg he distilled coal tar and investigated the products which boiled off at different temperatures. In this way he isolated phenol (carbolic acid), aniline (cyanol) and other compounds. In 1834 he patented the process for obtaining the dye 'aniline black' from cyanol. He conducted experiments with a process that can now be identified as paper chromatography, an important technique for separating and identifying the components of a mixture of soluble substances based on the fact that every compound has a different capillarity. He placed a solution of the mixture on adsorbent paper and demonstrated that the components separated out into coloured rings of different radii. This technique was further developed by **Schönbein** and **Tswett**. Runge died in Oranienburg.

RUSKA, Ernst August Friedrich
(1906–88)

German electrical engineer, born in Heidelberg. He was educated at the Technical University of Munich (1925–7) and moved to the Technical University of Berlin in 1931. In the same year he developed the world's first electron microscope. Just four years earlier, **Davisson** and **Germer** of Bell Telephone Laboratories in the USA had demonstrated experimentally that electron beams behave with both particle- and wave-like properties—one of the basic principles of quantum mechanics. The wavelengths of high-energy electrons can be less than 1 angstrom, which is about a third of the size of an average atom and about a thousand times smaller than the wavelength of visible light. The resolving power of a microscope is limited by the wavelength of the illuminating beam, and so by using electron beams, Ruska's electron microscope was able to resolve details on a scale of a few angstroms. This allowed a magnification of one million times, substantially better than the maximum magnification of around 2000 times possible with a light microscope. To focus the electron beams, Ruska used electromagnetic lenses made of current carrying coils. Throughout World War II he continued his development work on the electron microscope in a converted bakery, supported by the German engineering firm Siemens. Most of his career was dedicated to further improving the capabilities of the electron microscope, his areas of work including many aspects of electromagnetic and permanent magnetic lens design and improvements to the transmission and reflection electron microscopes. It wasn't until 1986 that he received a half share of the Nobel Prize for Physics along with the inventors of the tunnelling electron microscope, **Binnig** and **Rohrer**.

RUSSELL, Bertrand Arthur William, 3rd Earl Russell

(1872–1970)

Welsh philosopher, mathematician, prolific author and controversial public figure throughout his long and extraordinarily active life. He was born in Trelleck, Gwent; his parents died when he was very young and he was brought up by his grandmother, the widow of Lord John Russell, the Liberal prime minister and 1st Earl. He was educated privately, and in mathematics and philosophy at Trinity College, Cambridge. He graduated in 1894, was briefly British Embassy attaché in Paris, and became a Fellow of Trinity in 1895. His most original contributions to mathematical logic and philosophy are generally agreed to belong to the period before World War I, as expounded for example in *The Principles of Mathematics* (1903), which argues that the whole of mathematics could be derived from logic, and the monumental *Principia Mathematica* (1910–13, with **Whitehead**), which worked out this programme in a fully developed formal system and stands as a landmark in the history of logic and mathematics. Russell's famous 'theory of types' and his 'theory of descriptions' belong to this same period. Politics became his dominant concern during World War I and his active pacifism caused the loss of his Trinity fellowship in 1916 and his imprisonment in 1918, during the course of which he wrote his *Introduction to Mathematical Philosophy* (1919). Later he wrote on politics and education, and founded a progressive school. The rise of Fascism led him to renounce his pacifism in 1939; his fellowship at Trinity was restored in 1944, and he returned to England after the war to be honoured with the Order of Merit, and to give the first BBC Reith Lectures in 1949. He was awarded the Nobel Prize for Literature in 1950. After 1949 he became increasingly preoccupied with the cause of nuclear disarmament, taking a leading role in the Campaign for Nuclear Disarmament and later the Committee of 100, and engaging in a remarkable correspondence with various world leaders. In 1961 he was again imprisoned for his part in a sit-down demonstration in Whitehall. The last major publications were his three volumes of *Autobiography* (1967–9).

RUSSELL, Sir Frederick Stratten

(1897–1984)

English marine biologist, born in Doncaster, and educated at Gonville and Caius College, Cambridge. He served with distinction in the Royal Naval Air Service during World War I and then joined the scientific staff of the Plymouth Laboratory of the Marine Biological Association, of which he was director between 1946 and 1965. Russell's early research included some pioneering work on the phenomenon of diurnal vertical migration by zooplankters, but it is for his work on medusae (*The Medusae of the British Isles*, 2 vols, 1953, 1970), larval fish (*Eggs and planktonic stages of British marine fish*, 1976) and the long-term dynamics of zooplankton communities that he is best known. Hydoid and scyphozoan cnidarians typically have two markedly different phases in their life-cycle — a free-swimming medusoid (jellyfish) phase and a benthic polypoid phase. By painstaking laboratory rearing, Russell and co-workers succeeded in solving major taxonomic problems to link the medusa to the polyp phase for many species. He initiated long-term sampling and analysis of macrozooplankton off the Eddystone Lighthouse (1924–34, 1946–88), and noted that certain large and readily identifiable plankters could be used as 'indicators' of the source of the water mass with which they were associated. His Plymouth studies led to the elucidation of the natural variations in local plankton communities, and the environmental changes these variations produce; the dynamics of this phenomenon are now known as the 'Russell cycle'. The value of Russell's long-term work lies in its potential use in the assessment of large-scale changes in water mass circulation, and possibly climate change, in addition to predictions for commercially important fish populations. He was elected FRS in 1938 and knighted in 1965.

RUSSELL, Henry Norris

(1877–1957)

American astronomer, born in Pyster Bay, New York. After studies at Princeton and

Cambridge, UK, he was appointed Professor of Astronomy in Princeton in 1905 and six years later director of the university observatory there. A theoretician of the highest calibre, he developed with **Shapley** methods for the calculation of the orbits and dimensions of eclipsing binary stars and for the determination of the distances of double stars. He worked on the theory of stellar atmospheres from the analysis of stellar spectra. His most famous achievement was the formulation of the **Hertzsprung**–Russell diagram (1913) correlating the spectral types of stars with their luminosity which became of fundamental importance for the theory of stellar evolution.

RUSSELL, John Scott

(1808–82)

Scottish engineer, born in what was then the village of Parkhead near Glasgow. He graduated MA from the University of Glasgow at the age of 17 and moved to Edinburgh where he taught mathematics and natural philosophy. In 1832 on the death of **Leslie** he provisionally accepted an invitation of the University of Edinburgh to occupy the Chair of Natural Philosophy, and it was only after some hesitation that he decided not to stand as a candidate for the professorship. Instead he went ahead with his scheme to build and operate steam coaches on the roads of Scotland, and in 1834 a service between Glasgow and Paisley was inaugurated. After only a few months of successful operation, however, an accident involving several fatalities put an end to the project. He had been engaged at the same time in testing ships' hulls of many different shapes and sizes on the Union Canal near Edinburgh, and from the results he developed his wave-line principle of naval architecture, although it was subsequently shown to have been unduly influenced by the narrowness of the waterway. He was, however, one of the first to advocate and adopt a scientific approach to the design of ships, and he took a leading part in the building of **Brunel**'s *Great Eastern*, launched in 1858 from his yard on the Thames. Russell was one of the founders of the Institution of Naval Architects (1860) and served as Vice-President for many years.

RUTHERFORD, Ernest, 1st Baron Rutherford of Nelson

(1871–1937)

New Zealand-born British physicist, one of the greatest pioneers of subatomic physics, born in Spring Grove (later Brightwater) near Nelson, the fourth of 12 children of a wheelwright and flaxmiller. Winning scholarships to Nelson College and Canterbury College, Christchurch, his first research projects were on magnetization of iron by high-frequency discharges (1894) and magnetic viscosity (1896). In 1895 he was admitted to the Cavendish Laboratory and Trinity College, Cambridge, on a scholarship. There he made the first successful wireless transmissions over two miles. Under the brilliant direction of **J J Thomson**, Rutherford discovered the three types of uranium radiations: alpha, beta and gamma rays. In 1898 he became Professor of Physics at McGill University, Montreal, where with **Soddy**, he formulated the theory of atomic disintegration to account for the tremendous heat energy radiated by uranium. In 1907 he became professor at Manchester. While there he asked his students **Geiger** and Ernest Marsden to investigate the scattering of alpha particles from gold foil; they observed that a tiny fraction of those striking the gold foil were deflected back from it. Rutherford compared this astonishing result to firing a pistol at a piece of paper, and finding that the bullet bounced back. He derived a formula to describe the scattering of alpha particles from the gold foil based on a model of the atom where electrons orbit a compact nucleus, and its predictions were verified by Geiger and Marsden. However, the theory ignored the problem that classically the electrons should radiate away all their energy and fall into the nucleus (this was solved by **Niels Bohr**, then at Manchester, using **Planck**'s idea of energy quanta; thus the quantum theory of atoms was born). The new model revealed the importance of the atomic number—the charge on the nucleus. During World War I, Rutherford did research on submarine detection for the admiralty. In 1919, in a series of experiments, he discovered that alpha-ray bombardments induced atomic transformation in atmospheric nitrogen, liberating hydrogen nuclei. The same year he succeeded J J Thomson to the Cavendish professorship at Cambridge. In 1920 he predicted the existence of the neutron, later discovered by his colleague **Chadwick**. Like Niels Bohr, Rutherford's contribution to science was not only his own research but also his influence on the work of his contemporaries. He was awarded the Nobel Prize for Chemistry in 1908. He published nearly 150 original papers, and his books included *Radioactivity* (1904), *Radioactive Transformations* (1906) and *Radioactive Substances* (1930).

RUŽIČKA, Leopold

(1887–1976)

Swiss chemist, born in Vukovar, Croatia. He trained at the Technische Hochschule

Karlsruhe under **Staudinger** on the chemistry of the pyrethrins and other insecticides in the chrysanthemum (1912), and continued this work at the Polytechnic in Zürich (1916), where he was appointed to an unsalaried professorship in 1923. From 1921 he collaborated with a perfume factory on the synthesis of aromatic terpenes (such as civetone, muscone and jasmone). After a short stay in Utrecht (1926–9), he returned as Professor of Organic and Inorganic Chemistry to Zürich, where he remained for the rest of his career. Ružička's work on perfumes introduced him to the multi-membered ring structures (terpenes) which he studied in detail. In the 1930s he discovered their structural relationship to the steroids, which led him to synthesize lanosterol and enunciate in 1953 the biogenic isoprene rule by which these five carbon compounds combine to form steroids. In 1935 he was able to announce the synthesis of the still undiscovered male hormones, testosterone and methyltestosterone. With **Butenandt**, he was awarded the 1939 Nobel Prize for Chemistry; he used the substantial royalties from his discoveries to found an art gallery in Zürich for paintings by Dutch and Flemish 17th century masters.

RYDBERG, Johannes Robert
(1854–1919)

Swedish physicist, born in Halmstad. He was educated at the University of Lund where he remained throughout his career, becoming professor in 1901. He worked on the classification of optical spectra. From his studies he developed an empirical formula relating the frequencies of spectral lines, incorporating the constant known by his name, that was later derived by **Niels Bohr** using his quantum theory of the atom.

RYLE, Sir Martin
(1918–84)

English physicist and radio astronomer, born in Brighton, son of John Ryle, a distinguished physician and Professor of Medicine. He was educated at Bradfield College, Berkshire, and Christ Church, Oxford, graduating in 1939, at the outbreak of World War II. For the duration of the war he was involved in important research in the field of radar. At the end of the war he joined the Cavendish laboratory in Cambridge where he investigated the emission of radio waves from the Sun and improved on the low resolving power of radio telescopes by the introduction of the radio analogue to **Michelson**'s optical interferometer: using two instead of one radio antenna he was able to pinpoint radio sources such as Sunspots with considerable accuracy (1946). He then turned to studies of radio waves from the universe and identified localized radio sources with distant galaxies many millions of light years away. He found that the numbers of radio sources increased as their intensities decreased (1955), indicating that there were more galaxies per unit volume the further one looked into space and back in time, a result which pointed to an evolving universe starting with a Big Bang. He mapped radio sources by his ingenious method of 'aperture synthesis' (developed from 1960 onwards) in which a number of similar small radio telescopes are moved successively to different separations and their output, processed by computer, made to simulate the performance of a huge single telescope. Ryle was one of the most outstanding scientists of his generation. He was awarded a knighthood in 1966, was appointed to the first Chair of radio astronomy in Cambridge in 1969, and in 1972 became the first Astronomer Royal to come from the field of radio astronomy. In 1974 he received the Nobel prize for physics with his colleague **Hewish**.

S

SABATIER, Paul
(1854–1941)
French chemist, born in Carcassonne. After secondary education at the École Normale Supérieure he taught briefly at a lycée in Nîmes and then moved to Paris to work with **Berthelot**. He received his doctorate in 1880 and moved to Bordeaux for a year before taking an established post at Toulouse. Although he was offered posts elsewhere, he chose to remain in Toulouse for the rest of his life. In 1913 he became one of the first scientists to be elected to one of six chairs newly created by the Academy of Sciences for provincial members. He made a number of interesting discoveries in inorganic chemistry, but is best known for his discovery of catalysed hydrogenation of unsaturated organic compounds, such as the conversion of ethene to ethane over reduced nickel. He showed that this technique was applicable to many classes of organic compounds and as a synthetic procedure it has been widely used. For example, catalytic hydrogenation is used to produce margarine from vegetable oils. He proposed a theory (which has stood the test of time) to explain heterogeneous catalysis, involving the formation of unstable intermediates on the surface of the catalysts. He took little interest in the commercial application of his studies but received the Nobel Prize for Chemistry in 1912.

SABIN, Albert Bruce
(1906–93)
Polish–American microbiologist, born in Bialystok, Russia (now in Poland). He was educated at New York University, where he received his MD in 1931, and in 1946 he was appointed Research Professor of Pediatrics at the University of Cincinnati. After working on developing vaccines against dengue fever and Japanese B encephalitis, he became interested in the polio vaccine and attempted to develop a live attenuated vaccine (as opposed to **Salk**'s killed vaccine). He succeeded in persuading the Russians to help with the testing of his live virus, and in 1959 he was able to produce the results of 4.5 million vaccinations. His vaccine was found to be completely safe, possessing a number of advantages over that of Salk: it gave a stronger, longer-lasting immunity and could be administered orally. Consequently there was a widespread international adoption of the Sabin vaccine in the early 1960s. Some years later, he reported a major advance in cancer research, claiming to have evidence in support of the viral origin of human cancer. Later, however, he rejected his own experimental results.

SABINE, Sir Edward
(1788–1883)
Irish soldier, physicist, astronomer and explorer, born in Dublin. Educated at Marlow and the Royal Military Academy at Woolwich, he was commissioned in the Artillery and served in Gibraltar and Canada. He accompanied his lifelong friend Sir **James Clark Ross** as astronomer on John Ross's expedition to find the North-West Passage in 1818 and on William Parry's Arctic expedition of 1819–20. He conducted valuable pendulum experiments to determine the shape of the Earth at Spitzbergen and in tropical Africa (1821–3), and devoted the rest of his life to work on terrestrial magnetism. By means of magnetic observatories established in British colonies, he made the important discovery that there is a correlation between variations in the Earth's magnetism and solar activity, which followed a 10–11 year cycle. He retired from the army in 1877 as a Major-General. Sabine's gull is named after him. He was elected FRS (1818), and served for many years as the Royal Society's president (1861–71). He was knighted in 1869. An adroit politician, he was involved with the reforms of the Royal Society in the 1840s, and was a leading figure in the British Association for the Advancement of Science. His brother, Joseph Sabine (1770–1837), Inspector-General of Taxes, was a noted botanist.

SABINE, Wallace Clement Ware
(1868–1919)
American physicist, born in Richwood, Ohio. After graduating from Ohio State University in 1886, he entered Harvard, receiving the AM in 1888. Apart from service during World War I (he was effectively the chief scientist and authority on instruments for the US air force), Sabine was attached to Harvard throughout his life, from 1908 as Dean of the Graduate School of Applied Science which he had initiated. When Harvard opened its Fogg Art Museum (1895), the

auditorium was found to be 'monumental in its acoustic badness'. Set to remedy this situation, Sabine introduced a quantitative element into the previously empirical study of 'architectural acoustics'. He analysed the problem in terms of the size, shape and materials of a room affecting the reverberation time: his prescription of 22 hair-felt blankets made the theatre functional. By 1898 he had devised the Sabine formula (linking reverberation time, total absorptivity and room volume) and with its aid advised on the new Boston Symphony Hall (1898–1900). The unit of sound absorbing power (the sabin) was named after him.

SACHS, Julius von
(1832–97)

German botanist, born in Breslau (now Wrocław, Poland), one of nine children of a poor engraver. He was befriended by **Purkinje** and enabled to study at the University of Prague, then became botany lecturer at an agricultural college near Bonn; from 1868 he held the post of Professor of Botany at Würzburg. There he carried out important experiments, especially on the influence of light and heat upon plants, and the organic activities of vegetable growth. Sachs is regarded as the founder of modern plant physiology. He took up water culture to establish the mineral requirements of plants. He observed the conversion of sugar into starch in chloroplasts, and suggested that enzymes are involved in the conversion of oil into starch, starch into sugar, and proteins into soluble nitrogen compounds. He contributed the volume *Experimental-Physiologie* to **Hofmeister**'s *Handbuch der physiologischen Botanik* in 1865. His *Lehrbuch der Botanik* (1868) and its English translation *Textbook of Botany* (1875) exerted widespread influence.

SACROBOSCO, Johannes de, or John of Holywood, or Halifax
(fl mid-13th century)

English mathematician, probably born in Halifax. He is said to have studied at Oxford, and taught mathematics at Paris, where he died in 1244 or 1256. He was one of the first to use the astronomical writings of the Arabians, but his elementary knowledge of the Ptolemaic system is evidenced by the fact that he copied mistakes made previously by Alfargani and Albattani. His treatise, *De Sphaera Mundi*, based on **Ptolemy**'s *Almagest* and Arab writings, became the basic (ie elementary) astronomy text of the Middle Ages, and more than 60 editions were printed between 1472 and 1547.

SAGAN, Carl Edward
(1934–)

American astronomer, born in New York City. After studying at Chicago and Berkeley, he worked at Harvard then moved to Cornell, becoming Professor of Astronomy and Space Science in 1970. Interested in most aspects of the solar system, Sagan has done work on the physics and chemistry of planetary atmospheres and surfaces. He has also investigated the origin of life on Earth and the possibility of extraterrestrial life. He was an active member of the imaging team associated with the Voyager mission to the outer planets, and since 1983 he has given considerable thought to the concept of the nuclear winter, a deep winter triggered by some catastrophic event in the Earth's atmosphere. In the 1960s he worked on the theoretical calculation of the Venus greenhouse effect. Sagan and James Pollack were the first to advocate that temporal changes on Mars were non-biological and were in fact due to wind-blown dust distributed by seasonally changing circulation patterns. Through books and a television programme, *Cosmos*, Sagan has done much to interest the general public in this aspect of science. His *Cosmic Connection* (1973) dealt with advances in planetary science; *The Dragons of Eden* (1977) and *Broca's Brain* (1979) helped to popularize recent advances in evolutionary theory and neurophysiology. Sagan is President of the Planetary Society and is a strong proponent of SETI — the search for extraterrestrial intelligence.

SAHA, Meghnad
(1894–1956)

Indian astrophysicist, born in Dacca (now in Bangladesh), the son of a small shopkeeper. He was educated at Presidency College, Calcutta, and subsequently visited Europe on a travelling scholarship. He taught at Allahabad University, and in 1938 was appointed Professor of Physics at Calcutta. He worked on the thermal ionization that occurs in the extremely hot atmospheres of stars, and in 1920 demonstrated that elements in stars are ionized in proportion to their temperature ('Saha's equation'). This thermal ionization theory created a point of departure for the physical interpretation and classification of the spectra of stars. Saha later moved to nuclear physics, and played a prominent role in the creation in India of an institute for its study, which was named after him.

SAINTE-CLAIRE DEVILLE, Henri Étienne
(1818–81)

French chemist, born on St Thomas, West Indies (now part of the Virgin Islands).

During medical studies in Paris, he became attracted to chemistry. From 1845 to 1851 he was Professor of Chemistry at the University of Besançon. In 1851 he returned to Paris as successor to **Balard** at the École Normale Supérieure, and from 1853 he also gave lectures in chemistry at the Sorbonne. He was made an Honorary Fellow of the Chemical Society in 1860. His early work was on turpentine and other natural products, but later his interests moved to inorganic chemistry. In 1849 he isolated nitrogen pentoxide. He devised a process for the large-scale production of aluminium, which involved the reduction of the chloride by sodium (1855). He also developed related processes for the large-scale production of boron, silicon and titanium. Sainte-Claire Deville is also remembered for his extensive measurements of the vapour densities of substances at high temperatures, which led to the recognition of reversible thermal dissociation, of great significance for the theory of chemical equilibrium. His latter days were clouded by ill-health and he took his own life.

SAINT-HILAIRE, Étienne Geoffroy

See **GEOFFROY SAINT-HILAIRE, Étienne**

SAINT-HILAIRE, Isidore Geoffroy

See **GEOFFROY SAINT-HILAIRE, Isidore**

SAINT VINCENT, Gregorius de

(1584–1667)

Flemish mathematician and astronomer, born in Brugge. He was received into the Jesuit Order in Rome in 1607. His major work, the *Opus geometricum* of 1647, contains a method of finding areas under curves and a connection between the hyperbola and the recently invented logarithms.

SAKHAROV, Andrei Dimitrievich

(1921–89)

Soviet physicist and dissident, born in Moscow, the son of a scientist. He graduated in physics from Moscow State University in 1942 and was awarded his doctorate for work on cosmic rays. He worked under **Tamm** at the Lebedev Institute in Leningrad. There he principally worked on nuclear fusion and proposed use of a 'magnetic bottle' to contain the plasma; this has been developed into the Tokamak design for a fusion reactor. He was mainly responsible for the development of the Soviet hydrogen bomb and in 1953 became the youngest-ever entrant to the Soviet Academy of Sciences. He also studied cosmology and proposed that the reason for the dominance of matter over antimatter in the universe is connected to CP violation, a complicated and unexpected phenomenon observed in certain particle interactions. During the early 1960s he became increasingly estranged from the Soviet authorities because of his campaigning for a nuclear test-ban treaty, peaceful international co-existence and improved civil rights within the USSR. In 1975 he was awarded the Nobel Peace Prize, but in 1980, during a Cold War crackdown against dissidents, he was sent into internal exile in the 'closed city' of Gorky. There he undertook a series of hunger strikes in an effort to secure permission for his wife, Yelena Bonner, to receive medical treatment overseas. Under the personal orders of Mikhail Gorbachev, he was eventually released in December 1986. He continued to campaign for improved civil rights and in 1989 was elected to the Congress of the USSR People's Deputies.

SAKMANN, Bert

(1942–)

German electrophysiologist, born in Stuttgart and educated in medicine at Munich. Currently Director of the Max Planck Institute for Medical Research, Heidelberg, his work with **Neher** revolutionized cell physiology with the invention of the 'patch-clamp' recording technique. This made it possible to record the electrical activity of very small areas of membrane, by eliminating the membrane's electrical noise, which improved the sensitivity of previously available methods by a factor of a million. The patch-clamp method of studying minute changes in electrical current caused by the movement of ions has been widely used to study electrical activity in whole cells. Sakmann and colleagues have studied the relationship between the protein structure of ion channels and their function, and the patch-clamp technique has facilitated their studies of neural transmission in the central nervous system, similar to those achieved quite easily in the more accessible peripheral nervous system. Sakmann and Neher's method has promoted a new approach to research in many fields, including studies of nerve impluse propagation along axons; the process of egg fertilization; the regulation of the heartbeat; and investigations into the cellular mechanisms of disease. In 1991 they shared the Nobel Prize for Physiology or Medicine.

SALAM, Abdus

(1926–)

Pakistani theoretical physicist, born in Jhang. Educated at Punjab University and

Cambridge, he became Professor of Mathematics at the Government College of Lahore and at Punjab University (1951–4). He lectured at Cambridge (1954–6) and in 1957 became Professor of Theoretical Physics at Imperial College of Science and Technology, London. His concern for his subject in developing countries led to his setting up the International Centre of Theoretical Physics in Trieste in 1964. In 1979 he was awarded the Nobel Prize for Physics, with **Steven Weinberg** and **Glashow**. Independently each had produced a single unifying theory of both the weak and electromagnetic interactions between elementary particles. The predictions of the 'electroweak' theory were confirmed experimentally in the 1970s and 1980s.

SALISBURY, Edward James
(1886–1978)

English botanist and ecologist, born in Harpenden, Hertfordshire, and educated at University College London. He was a senior lecturer in botany at East London (now Queen Mary) College (1914–18) before joining the staff of University College (1918–43; Professor of Botany, 1929–43). There, he and Felix Eugen Fritsch (1879–1954) jointly prepared many botanical textbooks, including *Plant Form and Function* (1938). He was particularly interested in the quantity of seeds produced by plants, laboriously counting and weighing the seeds of more than 240 species. The results were published in *The Reproductive Capacity of Plants* (1942). From 1943 to 1956 he was Director of the Royal Botanic Gardens, Kew, and initiated a postwar development phase which included the construction of the Australian House (1952). He continued to publish on plant ecology, including two of his best-known books, *Downs and Dunes* (1952) and *Weeds and Aliens* (1961). A hallmark of all his research was his special gift of synthesizing data from many different fields into a unified whole. His ability to popularize his subject shone in his textbooks and in his synthesis of biology, ecology and horticulture, *The Living Garden* (1935). He founded the British Ecological Society in 1917.

SALISBURY, originally MARKHAM, Richard Anthony
(1761–1829)

English botanist and horticulturist, born in Leeds, the son of a cloth manufacturer. Educated at Edinburgh University, he assumed his surname to fulfill the conditions of a bequest. From 1805 to 1816 he was Secretary of the Horticultural Society of London. His illustrated work *Icones Stirpium Rariorum Descriptionibus* was published in 1791. He wrote extensively, producing several works on English botany and classification, including *Prodromus Stirpium* (1796), *Generic Characters in the English Botany* (1806), and *Paradisus Londinesis* (1805–8), but earned opprobrium for unethical professional behaviour and the unwarranted changing of botanical names. His *Genera Plantarum* was edited and published posthumously (1866).

SALK, Jonas Edward
(1914–)

American virologist, born in New York City, the son of a garment worker. He was educated in medicine at the New York University College of Medicine where he obtained his MD in 1939, and subsequently taught there and at several other schools of medicine or public health. In 1963 he became Director of the Salk Institute in San Diego, California, previously known as the Institute for Biological Studies. Some of his early research had been on the influenza virus, but by the time he moved to California, he had attracted worldwide attention for his work on the 'Salk vaccine' against poliomyelitis. In 1949 it became known that there were three types of polio virus, and it was not until 1954 that Salk developed his killed virus vaccine. To be accepted, this had to overcome oppostion which had continued from 1935 when killed and attenuated vaccines given to over 10 000 children had proved ineffective and unsafe. However, the evaluation of the 1954 trial showed that Salk's vaccination was 80–90 per cent effective, and by the end of 1955, over 7 million doses had been administered. Later the vaccine was overtaken by **Sabin** vaccine, which used a live attenuated strain and could be given orally instead of by injection, which Salk's vaccine required.

SAMUELSSON, Bengt Ingemar
(1934–)

Swedish biochemist, born in Halmstad. He entered the medical school of the University of Lund where he worked in **Bergström**'s laboratory. In 1958 he moved with Bergström to the Karolinska Institute in Stockholm. There Samuelsson graduated in medicine (1961) and was appointed Assistant Professor of Medical Chemistry. After a short period of research at Harvard University, he returned to the Karolinska Institute. Bergström had been working on prostaglandins—substances which act, like hormones, as chemical messengers, but which have a more localized action. Samuelsson studied the biosynthesis of prostaglandins showing that they are produced from an unsaturated fatty acid, arachidonic acid, found in certain foodstuffs.

He elucidated the details of the relevant biochemical pathways. Samuelsson and Bergström showed that one group of prostaglandins, known as the E series, relaxes blood vessel walls and lowers blood pressure, while another, the F series, has the opposite effect. The E series may be used in treating some circulatory diseases, while the F series has been used to induce abortion. Samuelsson became Professor of Medical Chemistry at the Royal Veterinary College in Stockholm (1967–72) and then Professor of Chemistry at the Karolinska. He shared the 1982 Nobel Prize for Physiology or Medicine with Bergström and **Vane**.

SANCTORIUS (Santorio Santorio)
(1561–1636)

Italian physician, born in Justinopolis, Venetian Republic (now Koper, former Yugoslavia). From a noble Venetian family, he studied philosophy and medicine at Padua and practised medicine in various locations, before settling in Venice in 1599. He became Professor of Theoretical Medicine at Padua in 1611; by then he had acquired a reputation as an independent thinker, determined to introduce measurement and quantification into physiology and medicine. Sanctorius is best known for his investigations into metabolism. Over a period of around 30 years, he carried out an elaborate series of measurements of his own weight, food intake and excretia, in order to quantify weight loss through insensible perspiration. He invented instruments to measure humidity and temperature; a syringe for extracting bladder stones; a trocar for removing fluid from the abdomen; and a pendulum for measuring the pulse rate. A friend of **Galileo** and other leading figures in the Scientific Revolution, he himself most clearly represents the quantifying spirit in the medicine of his time.

SANDAGE, Allan Rex
(1926–)

American astronomer, born in Iowa City. He studied at Illinois University and Caltech before joining the Hale Observatories, initially as an assistant to **Hubble**. In 1960 he made the first optical identification of a quasar. With Thomas Matthews, a junior colleague, he found a faint optical object at the same location as the compact radio source 3C 48 and found it to have a very unusual spectrum. This was soon shown by **Maarten Schmidt** to be the result of a huge **Doppler** redshift, implying that the object, now known as a quasar, is receding from the Earth at enormous speed. This is now interpreted to suggest that it is extremely remote and as luminous as hundreds of galaxies. Sandage went on to identify many more quasars via this peculiarity of their spectra, and showed that most quasars are not radio emitters.

SANDERS, Howard Lawrence
(1921–)

American marine biologist, born in Newark, New Jersey. He was educated at the universities of British Columbia and Rhode Island, and received his PhD at Yale in 1955. He worked throughout his career at Woods Hole Oceanographic Institution in Massachusetts, where he was appointed as a research associate (1955) and senior scientist (1965). He remained there as Scientist Emeritus following his retirement in 1986. From 1969 to 1980 he simultaneously held posts as Adjunct Professor and research affiliate at the State University of New York, and as Associate Professor at Harvard. His early research concerned the benthic fauna of shallow water invertebrates; Sanders discovered what proved to be the first species of a hitherto undescribed and primitive class of crustacea, the Cephalocarida. Working from the research vessels *Atlantis* and *Chain*, he then focused his attention on obtaining quantitative samples of deep-sea benthos from the north-west Atlantic continental shelf down to abyssal depths. From the earliest samples it became clear that deep-sea communities were characterized by large numbers of very small species which had been overlooked in previous surveys; the patterns of species richness and diversity were extraordinarily high, equalled only by tropical shallow-water ecosystems. These quantitative data, and the development of sampling devices to facilitate the project, revolutionized deep-sea biology. During the mid-1960s, Sanders and colleagues began to formulate the 'stability–time' hypothesis to account for the exceptional levels of species diversity of deep-sea benthic communities. The hypothesis has become a pivotal (if not controversial) concept in marine ecology and was a catalyst in the development of other more general concepts of the structuring of ecological communities. During the 1970s he also became a leading expert on oil spills and their environmental consequences.

SANGER, Frederick
(1918–)

English biochemist, born in Rendcombe, Gloucestershire. He was educated at the University of Cambridge, where he received his BA in 1939 and a PhD in 1943. As a Quaker, he was exempted from military service during World War II. Since 1951 he has been on the staff of the Medical Research Council in Cambridge. In the 1940s Sanger devised methods to deduce the sequence of amino acids in the chains of the protein hormone insulin. By the 1950s he had deduced the sequence of the 51 amino acids in its two-chain molecule, and found the small differences in this sequence in insulin from pigs, sheep, horses and whales. For this he was awarded the Nobel Prize for Chemistry in 1958. He then turned to the structure of nucleic acids, working on RNA and DNA. Using a highly ingenious combination of radioactive labelling, gel electrophoresis and selective enzymes, his group was able, by 1977, to deduce the full sequence of bases in the DNA of the virus Phi X 174, with over 5400 bases, and mitochondrial DNA, with 17000 bases. Such methods led to the full base sequence of the Epstein–Barr virus by 1984. For his nucleic acid work Sanger shared (with **Walter Gilbert** and **Berg**) the 1980 Nobel Prize for Chemistry, and became the first scientist to win two Nobel prizes in this field.

SARPI, Pietro, or Fra Paolo
(1552–1623)

Italian historian, scientist, theologian and patriot, born in Venice. Sarpi entered the Servite Order in his native city in 1565, rising to become Vicar-General in 1599. He studied a broad spectrum of subjects, including mathematics, astronomy, oriental languages, medicine and physiology, and he has been given the credit for various anatomical discoveries relating to the venous valves and the circulation of blood.

SARS, Michael
(1805–69)

Norwegian marine biologist, born in Bergen, son of a sea captain. He studied theology and between 1828 and 1854 was vicar at several Norwegian seaside communities. It was during this period that he carried out much of his research into marine biology. In 1854 he was made Extraordinary Professor of Zoology at Christiania (now Oslo) University. He was one of the founders of marine zoology and made numerous expeditions to collect specimens. He made several trips to the northern seas and to the Mediterranean, to the Adriatic (1851) and to Naples and Messina (1852–3). His particular interests were in the migration of marine organisms and the life cycles of marine invertebrates. Because of the dissimilarity of larval and adult forms, the relationship between the two had scarcely been realized before the work of Sars. He demonstrated for the first time the phenomenon of alternation of generations in the coelenterates, ie the existence of sessile and free-living forms succeeding one another in the same species. He was first to describe the veliger larva of marine molluscs (1837) and the bipinnaria larva of starfish (1844). Using a deep-sea dredge he collected many specimens from depths of up to 450 fathoms with his son Georg, and revealed a hitherto unsuspected fauna. The most spectacular discovery was made from 300 fathoms near Lofoten; the first specimen of a living crinoid, a form previously believed to have become extinct during the Mesozoic era. With his son Georg, who himself became a well-known marine biologist, he published *On Some Remarkable Forms of Animal Life from the Great Deeps off the Norwegian Coast* (1868).

SAUSSURE, Horace Bénédict de
(1740–99)

Swiss physicist and geologist, born in Conches, near Geneva. At the age of 22, following a distinguished college career, he became Professor of Physics and Philosophy at Geneva (1762–88), although his first love was botany. He travelled in Germany, Italy and England; he crossed the Alps by several routes and ascended Mont Blanc (1787). *Voyages dans les Alpes*, describing his observations on mineralogy, geology, botany and meteorology, was published in several volumes between 1779 and 1796. Saussure explained the Alpine topography by erosion and called attention to the evidence of great horizontal disturbance in the strata, but wrongly asserted that a diluvial current rather than glacial action had distributed the boulders around the Alps. The mineral saussurite is named after him.

SAUSSURE, Nicolas Théodore de
(1767–1845)

Swiss botanist, son of **Horace Bénédict de Saussure**. He accompanied his father on many of the latter's expeditions, including the 1787 ascent of Mont Blanc where he made all the meteorological observations. In 1789 he climbed Monte Rosa and corroborated **Mariotte**'s observations on the weight of air which resulted in Mariotte's independent formulation of **Boyle**'s law. He began accumulating observations on plant mineral nutrition, but at the outbreak of the French Revolution he left Geneva for England. Before he left, he had been promised a chair of plant physiology at Geneva Academy, but on his return in 1802, he was instead named Honorary Professor of Mineralogy and Geology. He held this title until 1835 but never taught there. Instead he continued his plant physiological studies, performing a series of fundamental experiments on carbonic acid in plant tissues, the phosphorus content of seeds, the conversion of starch into sugars and the biochemical processes during flower and fruit maturation, as well as on germination chemistry, fermentation and plant nutrition. His most important publication was *Recherches Chimiques sur la Végétation* (1804), which founded the science of phytochemistry. Saussure also did pioneering work in the modern fields of ecology and soil science.

SAVART, Félix

(1791–1841)

French physician and physicist, born in Mézières in the Ardennes, the son of a military engineer. Even during his medical studies at the military hospital at Metz and at the University of Strasbourg, he showed an early interest in the physics of sound, in particular of that of the violin, and he presented several papers on the subject. He succeeded **Fresnel** as a member of the Paris Academy (1827), and was appointed Professor of Experimental Physics at the Collège de France in Paris in 1828. He invented 'Savart's wheel' for measuring tonal vibrations, and the 'Savart quartz plate' for studying the polarization of light. With **Biot** he discovered the law (named after them) defining the intensity of magnetic field produced at a given point near a long straight current-carrying conductor. His brother, Nicolas, an officer in the engineering corps, also studied vibrations.

SAYERS, James

(1912–)

English physicist, born in Ballymena, County Antrim, Northern Ireland. He was a member of the British team associated with the Manhattan project at Los Alamos in World War II, to develop the nuclear bomb. He became Professor of Electron Physics at Birmingham (1946) and in 1949 received a government award for his work on the cavity magnetron valve, which was of great importance in the development of radar.

SCHAEFER, Vincent Joseph

(1906–)

American physicist, born in Schenectady, New York. He worked for a time in the machine shop of the General Electric Company in his home town, then was drawn to an outdoor life and studied at the Davey Institute of Tree Surgery, from which he graduated in 1928. He practised that profession for only a short time before economic necessity drove him back to General Electric where, in the research laboratories, he became assistant to **Langmuir**. During World War II they worked together on the problem of icing on aeroplane wings, and in the course of their experiments discovered that dry ice (solid CO_2) introduced into a cold box containing water vapour resulted in a miniature snowstorm. As a result he was able in 1946 to demonstrate for the first time the possibility of artificially inducing rainfall by seeding clouds with dry ice from an aeroplane. His colleague **Vonnegut** subsequently showed that the same effect could be produced much more conveniently by using silver iodide crystals. Schaefer became Director of Research at the Munitalp Foundation in 1954, and Professor of Physics at the State University of New York, Albany, in 1959.

SCHALLY, Andrew Victor

(1926–)

Polish-born American biochemist, born in Wilno (now Vilnius, Lithuania). He fled from Poland at the time of the German invasion in 1939 and studied at the National Institute for Medical Research in London and McGill University in Montreal. He later worked at the Baylor Medical School (1957–62) and Tulane University (from 1962). Following a suggestion by Geoffrey Harris in Cambridge that special 'releasing factors' from the brain hypothalamus stimulate the release of hormones from the pituitary gland, Schally, and independently **Guillemin**, discovered an assay system for this *in vitro* (1955). Thus began their well-documented, highly competitive race to isolate corticotrophin-releasing hormone (CRH), one of several such factors. Schally used a million pig pituitaries and reported the isolation and structure of other factors, thyrotropin-releasing hormone (TRH) in 1969, luteinizing-hormone-releasing hormone (LH-RH) in 1971, and somatostatin, which inhibits the release of growth hormone (1976). CRH was eventually isolated by W Vale in 1981. Schally's success depended heavily on the participation of **Folkers** and a talented Japanese chemist, Matsuo. He shared the 1977 Nobel Prize for Physiology or Medicine with Guillemin and **Yalow**. Schally also studied the distribution and function of adrenocortical trophic hormone (ACTH), somatostatin and other factors.

SCHAUDINN, Fritz Richard

(1871–1906)

German zoologist and microbiologist, born in Röseningken, East Prussia. He studied philology at the University of Berlin, but later turned to zoology, and after research work in Berlin he became Director of the Department of Protozoological Research at the Institute for Tropical Diseases in Hamburg (1904). Schaudinn demonstrated the amoebic nature of tropical dysentery and distinguished the organism which causes the disease, *Entamoeba histolytica*, from its beneficial relative *Escherichia coli* which inhabits the human intestinal lining. With the dermatologist Erich Hoffmann, he discovered the spirochaete which causes syphilis (1905), *Spirochaeta pallida*, now known as *Treponema pallidum*. A treatment for the disease, first used on patients in 1911, was subsequently developed by **Ehrlich**. In a wide range of investigations, Schaudinn demonstrated that human hookworm infection is contracted through the skin of the feet, researched malaria and made important contributions to zoology.

SCHAWLOW, Arthur Leonard

(1921–)

American physicist, and co-inventor of the laser, born in Mount Vernon, New York. He studied at Toronto and Columbia (1949–51), where he worked with **Townes** and married his sister. He moved to Bell Telephone Laboratories (1951–61) and subsequently became Professor of Physics at Stanford University in 1961. Townes and Schawlow collaborated to extend the maser principle (Microwave Amplification by Stimulated Emission of Radiation) to light, thereby establishing the feasibility of the laser (Light Amplification by Stimulated Emission of Radiation). Although they played a central role in laying down the theoretical framework of the laser and had started to construct one, it was **Maiman** who constructed the first working ruby laser in 1960 at Hughes Research Laboratories in California. From the early 1970s Schawlow worked on the development of laser spectroscopy. Emitting light of a very narrow range of frequencies and in an intense beam, lasers were an attractive light source for investigating the energy levels of electrons in atoms and molecules. The full potential of laser spectroscopy was not, however, realized until a new type of laser using organic dyes had been developed. The wavelengths of the light these dye lasers produced could be tuned over a significant range. Working with German-born physicist Theodor Hänsch, Schawlow made precise measurements of the energy levels the electron can occupy in the hydrogen atom, allowing the value of the Rydberg constant to be determined with unprecedented accuracy. For his work on laser spectroscopy Schawlow shared the 1981 Nobel Prize for Physics with **Bloembergen** and **Kai Siegbahn**.

SCHEELE, Carl Wilhelm

(1742–86)

Swedish chemist who discovered chlorine, preceded **Priestley** in the preparation of oxygen, and identified many important chemical compounds. He was born in Stralsund (now in Germany) and was apprenticed to an apothecary, later working at Malmö, Stockholm, Göteborg and Uppsala, where he arrived in 1770. There he was granted the use of a laboratory and met **Gahn**. In 1775, the year that he was elected to the Stockholm Royal Academy of Sciences, he moved to Köping, where he became the town pharmacist. In the 1760s Scheele began to investigate air and fire, and soon came to doubt the received view, first propounded by **Georg Stahl**, that substances contain a vital essence (referred to as phlogiston) which they lose when they burn. By 1772, according to his laboratory notes, he had realized that air was a mixture of two compounds, one of which encouraged burning and another which prevented it. He prepared a gas identical to the inflammable component of air from a number of compounds, including mercuric oxide and saltpetre (potassium nitrate). His discovery was made known to the scientific world by **Bergman** some months before Priestley announced a similar discovery in August 1774. In a letter written in September the same year, Scheele himself passed on information about his experiments to **Lavoisier**, who subsequently discovered the true nature of combustion and named the new flammable gas ‘oxygine’. Even without oxygen, Scheele’s discoveries would have placed him in the first rank of chemists. In 1770 he encouraged Gahn to look for phosphorus (which had been recently discovered by **Brand**) in animal bones and subsequently developed a method of extracting it, a very useful discovery as previously the only known source had been urine. In 1771 Scheele’s studies of fluorine led to the discovery of hydrofluoric acid. In 1773 he discovered chlorinc in thc course of experiments with pyrolusite (a manganese ore) which he treated with hydrochloric acid. He noticed that the new gas dissolved gold and also acted as a bleach. In 1775 his extensive experiments on arsenic resulted in the discovery of copper arsenide, a green pigment which became known as ‘Scheele’s green’. In his only book, *Abhandlung von der Luft und dem Feuer* (1777, translated as *Chemical Observations and Experiments on Heat, Light and Fire*, 1780), he gave the first accurate description of hydrogen sulphide, which he had prepared by heating sulphur in hydrogen. In 1781 Scheele distinguished between two very similar minerals, plumbago (graphite) and molybdena, discovering the metal molybdenum in the process. His investigations of plant and animal material were fundamental to the development of organic chemistry. He also devised new methods for analysing fragile organic material, avoiding strong heat and often precipitating the new substance from solution. In this way he identified many important organic acids including citric, oxalic, lactic and tartaric acids. He isolated prussic (hydrocyanic) acid in 1783. He died in Köping.

SCHEUTZ, Pehr Georg
(1785–1873)

Swedish lawyer, publisher, and builder of a calculating machine. Born in Jönköping, he was educated in law at Lund. In the 1830s he read about **Babbage**'s 'difference engine' and then designed his own machine, later completed by his son, Edvard Scheutz.

SCHIAPARELLI, Giovanni Virginio
(1835–1910)

Italian astronomer, born in Savigliano, Piedmont. He graduated at Turin in 1854 and later worked under **Wilhelm Struve** at Pulkova before becoming head of the Brera observatory, Milan, from 1860. Schiaparelli worked on the relationship between meteors and comets, and was the first to realize that the daily variation of the meteor count rate was only to be expected if the causative dust particles were moving on orbits similar to those of the short-period comets rather than the planets. He also provided the first identification of a specific meteoroid stream with a specific comet, the pair being the Perseids and comet Swift–Tuttle (1862 III). In 1877, having re-equipped the observatory, Schiaparelli began observations of Mars in order to produce a detailed map. Not only did he detect linear markings on the surface that he termed *canali* (ie channels), but he also noticed that they changed as a function of the Martian season, sometimes splitting into two and sometimes disappearing all together. From his 1882–9 observations of Mercury he concluded that this planet orbited the Sun in such a way that one of its hemispheres always pointed sunwards; he found the same effect occurring for Venus.

SCHICKARD, Wilhelm
(1592–1635)

German polymath, born in Herrenberg. Educated at the University of Tübingen, he was a pioneer in the construction of calculating machines. His 'calculating clock' was designed and built around 1623, but lay forgotten until the discovery of Schickard's papers allowed its reconstruction in 1960.

SCHIMPER, Andreas Franz Wilhelm
(1856–1901)

German botanist, son of **Wilhelm Philipp Schimper**, born in Strassburg (now Strasbourg, France). He was educated there under **de Bary** and later accepted a post at the botanic garden in Lyons, but soon moved to work with **Sachs** at the University of Würzburg. He visited Florida and the West Indies, where his interest in plant geography was first awakened. He also studied epiphytic and carnivorous plants. In 1882 he went to work with **Strasburger** in Bonn. He continued to travel widely, visiting Brazil, Ceylon and Java, then joining the Valdivia expedition to study the planktonic vegetation of the Canaries, Kerguelen, Seychelles, the Congo, Sumatra and Cape Province. Professor at Basle (1898–1901), he was noted as a plant geographer, and divided the continents into floral regions. He also proved, in 1880, that starch is a source of stored energy for plants. He published *Pflanzengeographie auf physiologischer Grundlage* ('Plant Geography on Physiological Principles') in 1898.

SCHIMPER, Karl (or Carl) Friedrich
(1803–67)

German naturalist and poet, born in Mannheim, cousin of **Wilhelm Philipp Schimper**. He studied theology at Heidelberg, and collected plants in the south of France, where he studied medicine. While teaching at Munich he explored the Alps under the service of the King of Bavaria. He is remembered as a pioneer of modern plant morphology. He was notable for his work on phyllotaxis, and in geology for his theory of prehistoric alternating hot and cold periods. Despite his talents, he failed to secure any academic post. Many of his scientific ideas were published as poems; several hundred are known. He experimented with dioecious mosses, and discussed the arrangement of leaves on plant stems according to geometrical principles. He started many projects, but completed almost none. He contributed to F W L Succow's *Flora Manhemiensis*, and to F C L Spenner's *Flora Friburgensis*. The modern concepts of ice ages and climatic cycles were probably originated by Schimper; he also speculated that mountain building was due to the contraction of the Earth.

SCHIMPER, Wilhelm Philipp
(1808–80)

German botanist, born in Dosenhain, cousin of **Karl Friedrich Schimper** and father of **Andreas Schimper**. He studied philosophy, philology, theology and mathematics at the University of Strasbourg, where he became Director of the Natural History Museum in 1835. He was an authority on mosses and co-authored *Bryologia Europaea* (6 vols, 1836–55). He studied the Triassic flora of the Vosges region, and contributed the volume

Paläophytologie to **Zittel**'s *Handbuch der Paläontologie* in 1879. He was a superbly competent observer, making exceptionally detailed descriptions, but he was not an original thinker.

SCHLEIDEN, Matthias Jakob
(1804–81)

German botanist, born in Hamburg, where his father was municipal physician. After studying law at Heidelberg University (1824–7), he established a legal practice in Hamburg, but abandoned the profession soon afterwards, disillusioned and depressed. He began studying natural science at Göttingen in 1833 and later transferred to Berlin. Schleiden's uncle, Johann Horkel (1769–1846), was a botanist who greatly encouraged him to study botany, which he did enthusiastically. At Berlin he met **Humboldt** and **Robert Brown**, and worked in the physiologist **Johannes Müller**'s laboratory, producing many publications and gaining a doctorate from Jena (1839). He lectured at Jena for some years before becoming its titular Professor of Botany (1850–62). After a spell in Dresden he was briefly Professor of Anthropology at Dorpat. Schleiden, with **Schwann**, was responsible for an important stage in the development of cell theory in 1838. Using Brown's discovery of the cell nucleus as a basis, Schleiden explained the role of the nucleus in the formation of new cells. The theory is best explained in his botanical textbook, *Grundzüge der wissenschaftlichen Botanik* (1842), which marks the start of plant cytology.

SCHMIDT, Bernhard Voldemar
(1879–1935)

Estonian optical instrument-maker, born on the island of Naissaar near Tallin. Having worked, on leaving school, as a telegraph operator and photographer, he studied at the Chalmers Institute in Göteborg, Sweden, and joined the engineering school at Mittweida in Germany (1901) where he established a reputation as an optician and where in 1926 he installed a small observatory. In the same year he joined the staff of the Hamburg Observatory in Bergedorf, where he developed his idea of a coma-free mirror and completed his first 0.5 metre Schmidt telescope in 1932. His optical system overcame the aberrations produced by spherical mirrors by introducing a specially shaped correcting plate at their centre of curvature. His invention was of great importance to optical astronomy, as it provided extremely fine image definition over a field of several degrees. The best-known Schmidt telescope is that on Mount Palomar (1949), with an aperture of 1.2 metres and focal length 3 metres, used for the photographic survey of the northern sky. The UK Schmidt telescope of the same dimensions sited in New South Wales, Australia (1973), extended the sky survey to the southern hemisphere.

SCHMIDT, Johannes
(1877–1933)

Danish fisheries biologist, born in Jägerspris. Schmidt's work focused on problems relating to the early life history of marine fish. In 1904 he captured a *Leptocephalus* (a small leaf-shaped fish which later undergoes metamorphosis to become an elver) off the coasts of the Faroes whilst surveying for cod eggs. That observation started him on a 20 year investigation throughout the North Atlantic on the research vessel *Dana*, which culminated in his discovery of the true breeding area of the *Anguilla anguilla* eel. Previously, Italian scientists had erroneously suggested that the spawning grounds of the European eel lay in the vicinity of Messina. Schmidt revealed that breeding actually occurs in the Atlantic Ocean in the vicinity of the Sargasso Sea, at approximately 400 metres depth and at a temperature of 17 degrees Celsius. It is now known that spawning occurs there in February each year; larvae rise to the surface where they are transported by the Gulf Stream towards the coasts of Europe and North Africa. After drifting for approximately 22 months the *Leptocephali* arrive over the European continental shelf in November, and there they metamorphose to elvers and start to seek fresh water. The elvers enter European rivers in the early spring of the following year when they are about 28 months old. No mature eels ever re-enter fresh water and therefore we know that, although spawning adults have never been found, they must die after reproduction at sea.

SCHMIDT, Maarten
(1929–)

Dutch-born American astronomer, born in Groningen. He was educated there and at Leiden, and moved to the USA in 1959 to join the staff of the Hale Observatories. He became Professor of Astronomy in 1964 and Director of the Hale Observatories in 1978. Schmidt made important contributions in the study of galactic structure and dynamics by analysing the distribution of matter in our galaxy, but is best known for his astounding results in the study of quasars. He studied the spectrum of an optically identified quasar and discovered that the peculiarities of its spectrum were caused by a massive **Doppler** redshift; it appeared to be receding from the

Earth at nearly 16 per cent of the speed of light. Such high velocities are now interpreted as implying that quasars are very distant objects, which must therefore be as luminous as hundreds of galaxies to be visible on Earth. He also found that the number of quasars increases with distance from Earth, providing evidence for the 'Big Bang' theory for the origin of the universe.

SCHOENHEIMER, Rudolf

(1898–1941)

German-born American biochemist, born in Berlin. He studied there and taught in Germany for 10 years before moving to the USA in 1933. There, working at Columbia University with **Urey**, he used two new isotopes discovered by Urey (deuterium, to replace stable hydrogen atoms in a molecule, and heavy nitrogen) to trace biochemical pathways (1935). His work showed that many materials of the human body (eg depot fats, proteins and bone) previously regarded as static, are steadily degraded and replaced (1935–7). New dietary lipid was found to be partly stored and partly utilized along with lipid released from the body's fat depots. He distinguished the pathways of unsaturated and saturated fatty acids from the rate of incorporation of deuterium from labelled body water, and showed that the former cannot be intermediates in the production of the latter (1937). Schoenheimer also established that the sterol coprostanol is produced from cholesterol by gut bacteria, and that vitamin D in cows milk is not derived from the potential dietary precursor in plants, ergosterol (1929). He published *The Dynamic State of Body Constituents* (1942). His isotopic tracer techniques, although requiring careful interpretation, became widely used for elucidating biochemical pathways (eg by **Konrad Bloch**). Schoenheimer committed suicide during World War II.

SCHÖNBEIN, Christian Friedrich

(1799–1868)

German chemist, born in Metzingen, Württemberg. He trained as a pharmacist, worked in two chemical factories and taught at a technical institute at Keilhau. He also taught and travelled in England and France, and was professor at Basle from 1835 until his death. He conducted research in many areas, but is mainly remembered for his discovery of ozone in 1839. He noticed that the oxygen obtained by the electrolysis of water had a peculiar smell and discovered that this was due to the presence of a second gas (ozone) produced at the anode. He found that this gas resembled chlorine and bromine in its chemical properties, but believed it to be a compound until Jean Charles Galissard de Marignac and Auguste Arthur de la Rive showed in 1845 that the same gas can be produced by passing electric sparks through pure, dry oxygen. Schönbein then realized that ozone is an allotropic form of oxygen (the same element but in a different physical form). In 1845 he treated cotton-wool with a mixture of sulphuric acid and nitric acid, washing out the excess acid. The result was nitrocellulose (guncotton), a highly inflammable fluffy white substance which was soon widely used as an explosive. It is still used as a propellant, sometimes on its own and sometimes in conjunction with nitroglycerine (invented by **Sobrero** and developed by **Nobel**). By treating cotton-wool with less nitric acid, Scönbein made collodion, which found uses in photography and medicine. He was also one of the pioneers of chromatography, which depends on the fact that the solution of any substance has a characteristic capillarity. When the edge of a paper is dipped into a solution containing a mixture of substances, the substances separate out, each rising by different amounts up the paper. This technique, which he developed from the work of **Runge**, was further refined by **Tswett**. Schönbein died in Sauersberg, near Baden-Baden.

SCHRIEFFER, John Robert

(1931–)

American physicist, born in Oak Park, Illinois. He graduated in electrical engineering and physics at MIT, and studied superconductivity for his PhD under **Bardeen** at Illinois University. Collaboration with Bardeen and **Cooper** led to the BCS (Bardeen–Cooper–Schrieffer) theory of superconductivity. The first steps towards this successful theory were made by Cooper, who showed that although the like charges of two electrons cause them to repel each other, when an electron interacts with the positive charges of a metal lattice it can deform the lattice with the net effect that pairs of electrons experience an overall attraction. These stable couples of electrons became known as 'Cooper pairs'. Using a statistical approach, Schrieffer succeeded in generalizing the theory from a description of the properties of a single Cooper electron pair to that of a solid containing many pairs. In an intensive month of work following Schrieffer's breakthrough, Bardeen, Cooper and Schrieffer were able to show that their theory accounted for all the experimentally observed superconducting phenomena. For the development of the BCS theory all three shared the 1972 Nobel Prize for Physics. Following postdoctoral work in Europe and short-term posts in the USA, Schrieffer held professorships at the University of Pennsylvania, Cornell and the University of California. Since 1992 he has been University Professor at Florida State University. He is the author of *Theory of Superconductivity* (1964), and has also worked on dilute alloy theory, surface physics and ferromagnetism.

SCHROCK, Richard

(1942–)

American organometallic chemist, born in Indiana. He received his BA from University of California at Riverside in 1967 and his PhD from Harvard in 1971. Following a year at Cambridge University, he moved to the Central Research and Development Department of Du Pont. In 1980 he was appointed to a professorship at MIT, where he became Frederick G Keyes Professor of Chemistry in 1989. His research centres upon the chemistry of high-oxidation-state early-transition-metal complexes. In particular he has studied those complexes which contain alkylidene, alkylidyne or dinitrogen ligands. His work is of much significance to industrially important catalytic processes, including the controlled polymerization of olefins and acetylenes. Schrock was the first recipient of the American Chemical Society Award for Organometallic Chemistry in 1985, and is a member of the US National Academy of Sciences.

SCHRÖDINGER, Erwin

(1887–1961)

Austrian physicist, born in Vienna where he was educated by a private tutor before entering Vienna University, receiving his doctorate in 1910. He remained at Vienna as a researcher until 1914, when he joined the Austrian army as an artillery officer. From 1920 he held chairs at Stuttgart (1920), Jena (1920–1), Breslau (1921) and Zürich (1921–7) universities. In 1926 he published a series of papers which founded the science of quantum wave mechanics, and shortly after succeeded **Planck** as Professor of Physics at the University of Berlin. In 1933 during Hitler's rise to power, he accepted an invitation to Oxford (1933–6) but decided to return to Austria as professor at Ganz University (1936–8). When Austria was invaded, he fled to Dublin where he worked at the Institute for Advanced Studies which was created for him (1938–56). He retired in 1956 and returned to Austria as Professor Emeritus at Vienna University. Schrödinger's work was inspired by **de Broglie**'s proposal that particles have a dual nature and in some circumstances will behave like waves: Schrödinger introduced his celebrated wave equation which describes the behaviour of such systems. When applied to the hydrogen atom it correctly predicts the observed energy levels of the electrons, without using the unacceptable assumptions of **Niels Bohr**'s model of the atom. Schrödinger showed that the wave equation was mathematically equivalent to the matrix approaches developed almost simultaneously by **Heisenberg**, **Born** and **Pascual Jordan**. **Dirac** soon developed a more complete theory of quantum mechanics from their foundations, and for this work Schrödinger and Dirac shared the 1933 Nobel Prize for Physics. Born concluded that the wavefunction solutions to Schrödinger's equation describe probability waves, so that for a particle in a given state at a particular time, it is only possible to predict the most likely state of the particle in the future. Schrödinger disliked this statistical approach to quantum-mechanical problems and preferred the more deterministic approach of classical physics. He wrote *What is Life* (1946) and *Science and Man* (1958).

SCHULTZE, Max Johann Sigismund

(1825–74)

German zoologist, born in Frieburg. He studied at Greifswald and Berlin, and taught at Bonn, where he became Director of the Anatomical Institute in 1872. He studied the anatomy of a variety of animals and in particular single-cell ones. This led him to define the cell (1861), with a structure of a nucleus surrounded by protoplasm, as the basic building block of all living organisms. A histologist, he used osmic acid as a stain for nervous tissues and demonstrated the nerve endings in the basilar membrane of the ear. He also showed that the retina of birds possesses two different sensory nerve endings, the rods and cones, to which he ascribed separate functions. He proposed the duplicity theory of vision in 1866, a forerunner of modern theories.

SCHUR, Issai

(1875–1941)

Russian–German mathematician, born in Mogilev. He taught in Berlin from 1916 to 1935, when as a Jew he was forced to retire, in 1939 escaping to Israel. A pupil of **Frobenius**, he worked on representation theory and group characters, and established a simple but central result that still bears his name.

SCHUSTER, Sir Arthur

(1851–1934)

British physicist, born in Frankfurt of Jewish parents. He studied at Heidelberg and Cambridge and became Professor of Applied Mathematics (1881) and Professor of Physics

(1888) at Owens College, Manchester, resigning his chair in 1907 to make way for **Rutherford**. He carried out important pioneering work in spectroscopy which showed that an electric current is conducted through gases by ions, and that once 'ionized', the current could be maintained by a small potential. In terrestrial magnetism, he showed that there were two kinds of daily variations; atmospheric variations caused by electric currents in the upper atmosphere, and internal variations due to induction currents in the Earth. The Schuster–Smith magnetometer is the standard instrument for measuring the Earth's magnetic field. He led the eclipse expedition to Siam in 1875. He was elected FRS in 1879, and was the founder (and the first Secretary) of the International Research Council. He was knighted in 1920.

SCHWABE, Heinrich Samuel
(1789–1875)

German apothecary and amateur astronomer, born in Dessau. For 33 years he made systematic observations of the Sun in a search for a planet revolving within the orbit of Mercury, and discovered in this way an approximately 10 year periodicity in the frequency of sunspots. The average period was later fixed at 11 years. Referring to his serendipitous discovery Schwabe used to compare himself with King Saul in the Bible who 'went out to find his father's asses and found a throne'.

SCHWANN, Theodor
(1810–82)

German physiologist, famous for the cell theory. Born in Neuss, Schwann was educated in Cologne and studied medicine, graduating in Berlin in 1834. He remained in Berlin for four years as assistant to **Johannes Müller**. In his Berlin years, he studied digestion, and isolated from the stomach lining the enzyme pepsin. He later showed the role of yeast cells in producing fermentation; developing the experiments of **Spallanzani**, he thereby cast doubt upon the idea of spontaneous generation, confirming that no micro-organisms were produced and no putrefaction ensued in a sterile broth. Some years later, **Pasteur**'s work finally destroyed the theory of spontaneous generation. Schwann also discovered the 'Schwann cells' composing the lyelin sheath around peripheral nerve axons, and he showed an egg to be a single cell which, once fertilized, evolves into a complex organism. Schwann's most renowned work, however, was on cell theory. Following the botanist **Schleiden**, who had argued that all plant structures are cells, Schwann was persuaded that animal tissues are also based on cells and he became a leading advocate of cell theory. In a major book of 1839 he contended that the entire plant or animal was comprised of cells, that cells have in some measure a life of their own, but that the life of the cells is also subordinated to that of the whole organism. The cell theory became pivotal to 19th-century biomedicine. **Virchow**, who in 1855 said 'all cells arise from pre-existing cells', was to demonstrate how the study of affected cells was central to pathology and physiology. Like Schleiden, Schwann believed cells reproduced by budding from a nucleus; the concept of the formation of cells by division came only later, with **Remak** and Virchow. Wounded by attacks from the chemists **Liebig** and **Wöhler** over his fermentation ideas, Schwann despaired of a career in Germany, in 1838 emigrating to Belgium, where he became professor at Louvain, and in 1848 at Liège. Increasingly solitary and depressed, he did little more science.

SCHWARTZ, Melvin
(1932–)

American physicist, born in New York City. He was educated at the University of Columbia where he received his PhD in 1958 and later became Professor of Physics (1963–6, 1991–). From 1966 he held professorships at Stanford University, but in the early 1980s he left academic research to work in the computer industry until 1991, when he became Associate Director of High Energy and Nuclear Physics at Brookhaven National Laboratory. He was awarded the 1988 Nobel Prize for Physics jointly with **Lederman** and **Steinberger** for carrying out the experiment which demonstrated the existence of the muon neutrino (1962). His other interests have included the study of CP violation.

SCHWARZ, Harvey Fisher
(1905–88)

American electrical engineer, born in Edwardsville, Illinois. He was the co-inventor (with William J O'Brien) of the Decca radio-navigation system for ships and aircraft. He studied electrical engineering at Washington University, St Louis. Working for the General Electric Company in Schenectady, New York, he helped to develop 'Radiola 44', the first domestic radio receiver to use the newly invented screen-grid valve. As chief engineer of Brunswick Radio Corporation, he was sent to Britain in 1932 to design radios and radiograms for manufacture in the UK, and made his home there for the rest of his life although remaining a US citizen. During World War II, working for Decca, he and O'Brien developed a prototype radio-navigation system that was put into operation for the first time during the D-Day landings in the seaborne invasion of Normandy in 1944.

SCHWARZSCHILD, Karl
(1873–1916)

German astronomer, born in Frankfurt; he was the first to predict the existence of 'black holes'. He became interested in astronomy as a schoolboy and had published two papers on binary orbits by the time he was 16. Educated at the universities of Strassburg and Munich, he was appointed Director of the Göttingen Observatory (1901) and the Astrophysical Observatory in Potsdam (1909). He volunteered for military service in 1914 at the beginning of World War I and returned home in 1916 after contracting a rare skin disease, from which he died. His lasting contributions are theoretical and were largely made during the last year of his life. In 1916, while serving on the Russian front, he wrote two papers on Einstein's general theory of relativity, giving the first solution to the complex partial differential equations of the theory. He also introduced the idea that when a star contracts under gravity, there will come a point at which the gravitational field is so intense that nothing, not even light, can escape. The radius to which a star of given mass must contract to reach this stage is known as the Schwarzschild radius. Stars that have contracted below this limit are now known as black holes.

SCHWEIGGER, Johann Salomo Christoph
(1779–1857)

German physicist, born in Erlangen, the son of a prominent theologian and Protestant minister. He was appointed Professor of Chemistry and Physics at the Physikotechnisches Institut in Nuremberg (1811–16), at Erlangen (1818), and at Halle (1819), where he remained until his death. He became best known as the founder of the *Journal für Chemie und Physik* (1811–28), and as one of the inventors of the simple galvanometer (1820), which he called his 'doubling apparatus' as the needle was deflected by a multiturn coil of insulated wire (**Oersted**'s original magnetic needle was deflected by a single wire). **Poggendorff** invented a similar instrument which he called the 'condenser', analogous to **Volta**'s electrostatic condenser for magnifying electrostatic phenomena. Improvements followed rapidly, including by **Ampère**, **Avogadro**, Oersted and **Nobili**, who designed the common astatic form of this instrument (1825), which was made popular in England by **Wheatstone**.

SCHWINGER, Julian
(1918–)

American physicist, born in New York City. He studied at Columbia University, where he received his PhD in 1939. He worked at the University of California at Berkeley under **Oppenheimer** and during World War II at the MIT Radiation Laboratory and the Metallurgical Laboratory of the University of Chicago. He moved to Harvard in 1945 where he became professor in 1947, and since 1972 he has held chairs at the University of California at Los Angeles. Independently of **Feynman** and **Tomonaga**, he developed a theory of quantum electrodynamics describing the interaction of light and matter. Problems had arisen when **Dirac**'s early theoretical work in this field produced infinite values in predictive calculations. Schwinger showed that by reformulating the calculations so that the infinities corresponded to measurable quantities, such as the electron mass and the electron charge, then by replacing these quantities with their known values, finite answers could be found. This mathematical process for avoiding the problem of infinities is known as renormalization and is an important tool in theoretical physics. Schwinger was awarded the 1965 Nobel Prize for Physics jointly with Feynman and Tomonaga for the development of quantum electrodynamics. He went on to study synchrotron radiation.

SCORESBY, William
(1789–1857)

English Arctic explorer and scientist, born near Whitby. As a boy he went with his father, a whaling captain, to the Greenland seas. He studied chemistry and natural philosophy at Edinburgh University, and made several voyages to the whaling grounds which he published in *An Account of the Arctic Regions* (1820), the first scientific accounts of the Arctic seas and lands. In 1822 he surveyed 400 miles of the east coast of Greenland. After a period of study at Cambridge, and after being ordained (1825), he held various charges at Exeter and Bradford but continued his scientific investigations, travelling to Australia in 1856 to study terrestrial magnetism. He became involved in the controversy about the fate of the explorer Sir John Franklin and his companions who had disappeared while searching for the North-West Passage (1845), which induced Scoresby to write *The Franklin Expedition* (1851). He also worked on improving the marine compass, in particular its magnetic needle, and the reliability of such compasses on iron ships. He was elected FRS in 1824.

SCOTT, Dukinfield Henry
(1854–1934)

English botanist, son of Sir George Gilbert Scott, born in London. He studied at Oxford and under **Sachs** at Würzburg, and became Assistant Professor of Botany at the Royal College of Science and later Keeper of Jodrell Laboratory, Kew (1892), devoting himself to plant anatomy and later to palaeobotany. He collaborated with **William Crawford Williamson** in a number of brilliant studies of fossil plants, particularly their fruiting bodies, and established in 1904 the class Pteridospermeae. With Williamson, he demonstrated the close evolutionary relationships between ferns and cycads, important in phylogenetic studies. He published several important textbooks, including *Introduction to Structural Botany* (1894–6), *Studies in Fossil Botany* (1900), *Evolution of Plants* (1911) and *Extinct Plants and Problems of Evolution* (1924).

SEABORG, Glen Theodore
(1912–)

American atomic scientist who synthesized plutonium, born in Ishpeming, Michigan. He studied at the University of California at Los Angeles and took his doctorate at Berkeley where he became an instructor in chemistry in 1939, professor in 1945, and chancellor from 1958 to 1961. Following the work of **Soddy**, his earliest work was on isotopes, discovering many previously unknown isotopes among the common elements. In 1939 he began to follow up the work of **Fermi** in Italy who had attempted and failed to synthesize elements heavier than uranium (the heaviest naturally occurring element, atomic number 92) by bombarding it with neutrons. Instead he had succeeded in splitting uranium into smaller atoms, with the release of a great amount of energy. In 1940 **McMillan** and **Abelson**, working with the Berkeley cyclotron, sythesized the first 'transuranic' element, neptunium (atomic number 93). The following year Seaborg synthesized the next transuranic element in the series (atomic number 94) by bombarding uranium with deuterons in the Berkeley cyclotron. Named plutonium, it has a fissile isotope plutonium-239 whose destructive power Seaborg instantly recognized. Around the same time, Seaborg and his team also discovered that a rare isotope of uranium, uranium-235, is fissile. The USA therefore found itself with two possible sources of power for a nuclear weapon. Seaborg was transferred by the government to the Manhattan Project, a group of scientists gathered at the Metallurgical Laboratory of the Manhattan District of the Corps of Engineers at the University of Chicago to manufacture an atomic bomb. He was part of Fermi's team which achieved the first chain reaction in uranium-235 on 2 December 1942. He was also in charge of developing a technique to separate plutonium after it had been synthesized in uranium-238. It was his laboratory which, in 1945, produced enough plutonium for the first atomic bomb. Seaborg and his team continued research on further transuranic elements and in 1944 synthesized americium (atomic number 95) and curium (96). In 1950, by bombarding these with alpha rays, they produced berkelium (97) and californium (98). They later produced einsteinium (99), fermium (100), mendelevium (101) and unnilhexium (106). The heaviest of these elements, which have a very short half-life (the half-life of unnilhexium, for example, is less than 1 second), could only be prepared in microscopic quantities. In 1951 Seaborg shared the Nobel Prize for Chemistry with McMillan. He was Chairman of the US Atomic Energy Commission from 1961 to 1971.

SECCHI, Angelo
(1818–78)

Italian astronomer, born in Reggio Emilia. He joined the Society of Jesus in 1833 and in 1841 was made Professor of Physics and astronomy at the Jesuit College in Loreto. In 1948 he became Professor of Astronomy at the Roman College of the society but, with all the Jesuits, he was expelled from Italy in the same year and spent a brief exile in England and the USA before returning to become director of the observatory of the Roman College in 1849. He established a new observatory on the roof of the church of Saint Ignatius with a 10 inch refractor which he used particularly for observations of the Sun. At the total solar eclipse of 1860 observed in Spain he succeeded in photographing the prominences and the corona. Using an objective prism he observed several thousand stellar spectra and divided the stars into three types — white, yellow and red — which corresponded roughly to their temperatures. He thus initiated the field of spectral classification which ensured his place as a pioneer of astronomical spectroscopy. His beautifully illustrated book on the Sun, *Le Soleil* (1875), became widely known. Apart from his astronomical researches, Secchi did much work on the phenomena of terrestrial magnetism and meteorology.

SEDGWICK, Adam
(1785–1873)

English geologist, born in Dent, Cumbria. Having graduated in mathematics from

Trinity College, Cambridge (1808), he became Woodwardian Professor of Geology there in 1818. In 1831 he began his geological mapping in Wales and introduced the Cambrian system in 1835. He had carried out studies in the Lake District as early as 1822, but it was not until the Cambrian and Silurian systems had been established in Wales and the Welsh Borders that he was fully able to understand its geology. Sedgwick became embroiled in controversy with **Murchison**; the dispute was finally resolved with the introduction of the Ordovician system by **Charles Lapworth**. His best work was on *British Palaeozoic Fossils* (1854). With Murchison he studied the Lake District, the Alps and south-west England, where they identified the Devonian system.

SEEBECK, Thomas Johann
(1770–1831)

Estonian-born German physicist, born in Tallin. A member of a wealthy merchant family, he went to Germany to study medicine, qualifying in 1802, but spent his time thereafter in research in physics. Like **Oersted**, he was inspired by the German Romantic movement, in particular by Johann von Goethe's anti-Newtonian theory of colours, so that his first research (1806) was on the heating and chemical effects of the colours of the solar spectrum. He investigated optical polarization in stressed glass (1812), but this work had been largely anticipated by **Brewster** and **Biot**. His most significant discovery was that of thermoelectricity, which he called 'thermomagnetism' (1822) as he did not believe that an actual electric current was being generated when heat was applied to a junction of two metals, such as bismuth and copper. Apart from the impact it had on theory, the thermoelectric effect is now much used in thermocouples for temperature measurement.

SEEBOHM, Henry
(1832–95)

English industrialist and ornithologist, born in Bradford, Yorkshire, to a Quaker family of German origin. Early in life he decided to devote himself to business, and in partnership with Georg Dieckstahl he set an iron and steel foundry in Sheffield, which prospered and provided the financial backing for his ornithological interests. He travelled widely to several parts of the world including Greece, Asia Minor, Russia and Japan in pursuit of his hobby, and spent a season in South Africa observing European birds in their winter quarters. He was particularly famed during his lifetime for his trips to Imperial Russia and Siberia. He published several accounts of his travels, *Siberia in Europe* (1880) and *Siberia in Asia* (1884) demonstrating his interest in people as much as in birds. The two components of his life converged as he often carried gifts of Sheffield steel to exchange for bird skins or eggs, and he was renowned for returning to his factory bearing foreign orders garnered on his expeditions. He wrote *A History of British Birds* (1883–5) which became a classic, and acquired a particular interest in the osteology of birds, from which he began to develop a classification scheme. His *Coloured Figures of Eggs of British Birds* (1896) and *The Birds of Siberia* (1901) were published posthumously. He left the bulk of his collections to the Natural History Museum.

SEFSTRÖM, Nils Gabriel
(1765–1829)

Swedish physician and chemist, born in Ilsbo, North Helsingland. He qualified as a physician at Uppsala in 1813. He became a lecturer in chemistry at the Royal Military Academy at Carlberg and was later appointed as professor at the Artillery School at Marieberg, while also teaching at the new School of Mines at Falun. Researching both at Falun and in **Berzelius**'s laboratory in Stockholm, he found a hard white metal in Swedish iron and in the spoil from an ironworks. This metal had previously been described in 1801 by the Spanish mineralogist **del Río** who had not recognized it as an element. Sefström called it vanadium. Only a few grams of it were found and it was Berzelius, not Sefström, who succeeded in examining its compounds and who realized that it resembled chromium. It was many years before its atomic weight and other properties were accurately described by **Roscoe**. It is now important industrially—vanadium is alloyed to steel when very high strengths are required. Sefström died in Stockholm.

SEGRÈ, Emilio
(1905–89)

American physicist, born in Rome. He was educated at Rome University, studying engineering then physics, and obtained his doctorate in 1928. He remained at the University of Rome working with **Fermi** until 1936, when he was appointed director of the physics laboratory in Palermo, but in 1938 he was dismissed from this post under Mussolini's regime. He moved to the University of California at Berkeley where he remained from 1938 to 1972, apart from a period working on the Manhattan atom bomb project during World War II. In 1937 Segrè discovered the first entirely man-made element, technetium. Three years later he was involved in the discoveries of astatine and plutonium (1940). He was also instrumental in devising chemical methods for dividing nuclear isomers. In 1955, using the new bevatron particle accelerator at the University of California, the research team led by Segrè discovered the anti-proton, the anti-particle of the proton (with identical mass but negative electric charge) which had been predicted by **Dirac**. For this work he shared the 1959 Nobel Prize for Physics with **Chamberlain**.

SELYE, Hans Hugo Bruno
(1907–82)

Canadian physician, born in Vienna, the son of a surgeon. He studied medicine in Prague, Paris and Rome, and was assistant in experimental pathology at the German University (1929–31) before emigrating to the USA. After one year as a Research Fellow at Johns Hopkins University, where he worked on detoxification by the liver, he moved to McGill University in Montreal (1933–45) and then became Director of the Institute for Experimental Medicine and Surgery at the University of Montreal in 1945. From 1932 he produced a long series of papers with **Collip** on the hormonal interactions involving the adrenal gland, hypothalamus and pituitary gland, and their effects on osteoblast multiplication, ion regulation and bone formation (introducing the term mineralocorticoid), the blood sugar level, gonadal stimulation by 'pregnancy urine', and lactation and menstruation in relation to development. He also observed the anti-hormone effect resulting from prolonged treatment with pituitary extract. He is best known for his 'general adaptation syndrome', defined as the 'physiological mechanism which raises resistance to damage as such' (1949), which links stress and anxiety, and their biochemical and physiological consequences, to human disorders such as hypertension, nephrosclerosis and rheumatic diseases.

SEMENOV, Nikolai Nikolaievich
(1896–1986)

Russian physical chemist, born in Saratov. In 1917 he graduated from the university at Petrograd (now St Petersburg). From 1920 he worked at the Physico-Technical Institute there, becoming professor in 1928. In 1931 he became a full member of the USSR Academy of Sciences. Also in 1931, the Physico-Technical Institute became the Institute of Chemical Physics of the Academy, and Semenov was appointed its first director. He held this appointment until his death. In 1943 the institute moved to Moscow, and from 1944 Semenov was also professor at Moscow State University. Semenov's earliest researches were on electrical phenomena in gases and solids, and on other topics in molecular physics. However, he is best known for his contributions to chemical kinetics, particularly in connection with chain reactions. He investigated explosion limits and many other features of combustion, flames and detonation. Much of his work was parallel to that of **Hinshelwood**, and they shared the 1956 Nobel Prize for Chemistry. Semenov was the author of several influential books, notably *Chemical Kinetics and Chain Reactions* (1934) and *Some Problems of Chemical Kinetics and Reactivity* (1954). He was awarded the Order of Lenin five times, and became a Foreign Member of the Royal Society in 1958.

SEMMELWEIS, Ignaz Philipp
(1818–65)

Hungarian obstetrician, born in Buda (now Budapest). He studied at the University of Pest and in Vienna. From 1845 he worked in the first obstetrical clinic of the Vienna general hospital. He observed that the first clinic had a much higher rate of puerperal fever than the second obstetrical clinic, run by midwives. His investigations convinced him that this was caused by medical staff and students going directly from the postmortem to the delivery rooms, spreading the putrefactive cause of the disorder. He instituted a rigorous programme of washing hands and instruments in chlorinated lime solution between autopsy work and examining patients, and the mortality rate in the first clinic was reduced to about the same level as the second. He found some support, but also much opposition to his ideas, and he left Vienna in 1850 for Pest. He published his systematic treatise on *The Aetiology, Concept and Prophylaxis of Childhood Fevers* in 1861. His last years were clouded by frustration that his ideas were not more widely accepted, and from mental instability. He died in a mental asylum, ironically from a cut in his finger which turned septic and produced a disease much like puerperal fever. His ideas were peculiar to himself, but in the later bacteriological age he came to be seen as a pioneer of antiseptic obstetrics.

SENEBIER, Jean
(1742–1809)

Swiss botanist, plant physiologist and pastor, born in Geneva. He had an early interest in natural history but his parents wished him to become a minister; in 1765, after presenting a thesis on polygamy, he was ordained pastor of the Protestant Church, Geneva. During a year in Paris he met many scientists. Charles Bonnet enabled him to begin experiments in plant physiology, and to publish a paper on the art of observing (1772) in response to a question posed in 1768 by the Netherlands Society of Sciences, Haarlem. He was pastor of Chancy near Geneva (1769–73) and librarian for the Republic of Geneva from 1773; in 1777 he began translating the works of **Spallanzani**. Senebier was the first to demonstrate the basic principle of photosynthesis; his most important papers were *Action de la Lumière sur la Végétation* (1779) and *Expériences sur l'Action de la Lumière Solaire dans la Végétation* (1788). He was also the first to precisely define experimental method, which he did in *Art d'Observer* (1775, an elaboration of his 1772 paper) and *Essai sur l'Art d'Observer et de Faire des Expériences* (1802). He taught many prominent life scientists, including **Augustin Pyrame de Candolle**, Jean-Antoine Colladon (1758–1830), a precursor of **Mendel**, and **Nicolas Théodore de Saussure**.

SERRE, Jean-Pierre
(1926–)

French mathematician, born in Bages. He studied at the École Normale Supérieure before working at the National Centre for Scientific Research (CNRS) and the University of Nancy, and became professor at the Collège de France in 1956. For his early work in homotopy theory, especially on the computation of the homotopy groups of spheres, and the ideas he introduced into algebraic topology, he was awarded the Fields Medal (the mathematical equivalent of the Nobel Prize) in 1954. Later he turned to algebraic geometry, writing the definitive paper on the theory of sheaves, class field theory, group theory and number theory. He is also the author of numerous elegant books on various branches of mathematics, and was for a time a member of **Bourbaki**.

SERVETUS, Michael
(1511–53)

Spanish theologian and physician. Born in Tudela, he worked largely in France and Switzerland. In *De Trinitatis Erroribus* (1531) and *Christianismi Restitutio* (1553) he denied the Trinity and the divinity of Jesus; he escaped the Inquisition but was burnt by Calvin in Geneva for anti-trinitarian heresy. He lectured on geography and astronomy, and practised medicine at Charlien and Vienna (1538–53). He appears to have prefigured **William Harvey** (1578–1657) in discovering the pulmonary (lesser) circulation of the blood, although his views were little known to contemporaries and carried no biomedical influence.

SEWARD, Sir Albert Charles
(1863–1941)

English palaeobotanist, born in Lancaster. He studied at Cambridge and Manchester, and was a lecturer (1890–1906) and later Professor of Botany at Cambridge (1906–36). He is best known for his works on English palaeobotany, *Wealden Flora* (1894–5), *Jurassic Flora* (1900–3), the four-volume *Fossil Plants* (1898–1919), and a panoramic survey, *Plant Life Through the Ages* (1931). He was knighted in 1936.

SEYFERT, Carl Keenan
(1911–60)

American astronomer, born in Cleveland, Ohio, and educated at Harvard. After working at the McDonald Observatory and the Mount Wilson Observatory, he became Associate Professor of Astronomy and Physics and Director of Barnard Observatory, Vanderbilt University (1946–51); from 1951 he was Professor of Astronomy and Director of the Arthur J Dyer Observatory. Seyfert became famous for his work on a special group of galaxies (named after him) which have very bright bluish star-like nuclei, barely perceivable arms and spectra containing broad high-excitation emission lines; he started to study these systematically in 1943. They are now thought to be the low-luminosity cousins of quasars.

SHANNON, Claude Elwood
(1916–)

American applied mathematician and pioneer of communication theory, born in Gaylord, Michigan. He was educated at Michigan University and at MIT, where he received a PhD in mathematics. A student of **Bush**, he worked on the differential analyser and picked up on Bush's suggestion that he study the logical organization of its relay circuits as a thesis subject. Shannon's master's thesis was one of the most influential ever written, and in 1938 he published a seminal paper (*A Symbolic Analysis of Relay and Switching Circuits*) on the application of symbolic logic to relay circuits. His work not only helped translate circuit design from an art into a science, but its central tenet—that information can be treated like any other quantity and can be manipulated by a machine—had a profound impact on the development of computing. After graduating from MIT, he worked at the Bell Telephone Laboratories (1941–72) (where **Stibitz** had built a binary adder) in the esoteric discipline of information theory. He wrote *The Mathematical Theory of Communications* (1949, with Warren Weaver).

SHAPLEY, Harlow
(1885–1972)
American astronomer, born in Nashville, Missouri. He worked for two years as a newspaper reporter before entering the University of Missouri where he quickly changed from journalism to astronomy. Following three years as a graduate student at Princeton (1911–14) he was appointed to the staff of the Mount Wilson Observatory in California (1914–21). His studies of Cepheid variable stars allowed him to use these as indicators of distance (1914). In this way he established the distances of globular star clusters and discovered that the centre of the globular cluster system is far removed from the Sun (1918). This result placed the Sun near the edge of the stellar system and not at its centre as had been the accepted view. While Shapley's new model of our galaxy was soon adopted, astronomers were not in agreement as to the structure of the universe as a whole; a memorable 'Great Debate' on the question took place between Shapley and **Heber Doust Curtis** in 1921. Shortly afterwards Shapley was appointed Director of Harvard College Observatory (1921) where he remained until his retirement in 1952. The prolific research output at Harvard and the number of distinguished astronomers trained there during that period are mainly due to Shapley's efforts. His own researches include the discovery of the first two dwarf galaxies, companions to our own galaxy, and investigations on the Magellanic Clouds.

SHARP, Phillip Allen
(1944–)
American molecular biologist, born in Kentucky. He was educated at Union College in Barbourville, Kentucky, where he received his BA in 1966, and at the University of Illinois where he obtained his PhD in 1969. He began his career as a Postdoctoral Fellow at Caltech (1969–71), then became a Senior Research Investigator at Cold Spring Harbor Laboratories, New York (1972–4), and Associate Professor at MIT (1974–9). He was appointed Professor of Biology there in 1979, and has also served as Director of the MIT Center for Cancer Research (1985–91) and head of the biology department since 1991. He was also a co-founder of Biogen, where he has been a director since 1978 and chairman of the scientific board since 1987. Sharp invented the technique known as 'S1 nuclease mapping', now used extensively to detect the size of unknown RNA molecules and to delineate the end of the RNA species. This led to his discovery in 1977 that genes are split into several sections, separated by stretches of DNA known as 'introns' which appear to carry no genetic information. The origins of this apparently redundant DNA remain a puzzle, but the discovery has prompted much research on how this phenomenon may be implicated in genetic diseases and speculation that it may provide a mechanism for rapid evolution. Sharp shared the 1993 Nobel Prize for Physiology or Medicine jointly with **Richard Roberts**, who had discovered split genes around the same time.

SHARPEY-SCHÄFER, Sir Edward
(1850–1935)
English physiologist, born in Hornsey. He was educated at University College London. He served as professor there (1883–99) and later at Edinburgh (1899–1933). He gained many distinctions, becoming FRS in 1878 and president of the British Association in 1912, and receiving a knighthood in 1913. An able microscopist, his *Essentials of Histology* (1885) became an essential textbook. One prime area of his research interests focused on neurophysiology and in particular the theory of brain localization. He experimented on nerve section and regeneration. Another field of interest lay in the emergent discipline of endocrinology. Sharpey-Schäfer made extensive studies of the effects of suprarenal and of pituitary extracts. Investigating the role of the pancreas in carbohydrate metabolism, he surmised that the islet tissue must act as an organ of internal secretion of an as yet hypothetical fluid that he named 'insuline'. He championed the materialist doctrine of life and was a passionate defender of vivisection.

SHARPLESS, (Karl) Barry
(1941–)
American chemist, born in Philadelphia. After graduating from Dartmouth College (1963) he studied for a PhD at Stanford University. He then undertook postdoctoral work (1968–70) at Stanford and Harvard before joining the staff of MIT. He was made full professor in 1975. During 1977–80 he held a post at Stanford, but he subsequently returned to MIT, becoming Arthur C Cope Professor in 1987. He later moved to the Scripps Research Institute (1990). In the 1970s Sharpless worked on the metal-catalysed addition of oxygen to double bonds (epoxidation) which led to his most significant work on the asymmetric epoxidation of alkenes. More recently he has developed synthetic strategies for dihydroxylation reactions, of great value in the asymmetric

synthesis of natural products. He was elected to the US National Academy of Sciences in 1985.

SHAW, Sir William Napier

(1854–1945)

English meteorologist, born in Birmingham. He studied mathematics and natural science under **Maxwell** at Emmanuel College, Cambridge, and with **Helmholtz** in Berlin. He was appointed as demonstrator (1879), lecturer (1887) and assistant director (1898) at the Cavendish Laboratory. A member of the Meteorological Council from 1879, he became its secretary in 1900. He was appointed as Director of the Meteorological Office (1905–20), first Professor of Meteorology at Imperial College, London (1920–4) and President of the International Meteorological Committee (1906–21). He completely reorganized the British Weather Service. In a classic paper (with Lempfert) *The Life History of Surface Currents* he developed the analysis of air currents and showed that discontinuities occurred in the atmosphere. He came very close to defining 'fronts', and devised the aerological diagram known as the tephigram which is still in use. His four-volume work *The Manual of Meteorology* (1926–31) stressed the mathematical basis of the subject and gave a complete account of its historical development. He was awarded the Symons Gold Medal in 1910, the Royal Medal of the Royal Society in 1923, and was knighted in 1915.

SHEN KUA

(1031–95)

Chinese administrator, engineer and scientist, born in Hangchow. He made significant contributions to such diverse fields as astronomy, cartography, medicine, hydraulics and fortification. His first appointment was in 1054, and in the following few years he accomplished some notable work in land reclamation. As director of the astronomical bureau from 1072, he improved methods of computation and the design of several observational devices; in 1075 he constructed a series of relief maps of China's northern frontier area, and designed fortifications as defences against nomadic invaders. He surveyed and improved the Grand Canal over a distance of some 150 miles, using stone-filled gabions, wooden piles and long bundles of reeds to strengthen the banks and close gaps. In 1082 he was forced by intrigue to resign from his government posts, and occupied his last years in the writing of *Brush Talks from Dream Brook*, a remarkable compilation of about 600 observations which has become one of the most important sources of information on early Chinese science and technology.

SHEPARD, Francis Parker

(1897–85)

American geological oceanographer, born in Brookline, Massachusetts. He was educated at Harvard (BA 1919) and at the University of Chicago, where he received a PhD in geology in 1922. He joined the staff of the University of Illinois and remained there until 1946, being promoted to professor in 1939. Around 1930 his interests turned towards the sea, apparently following summer cruises aboard his father's yacht off New England. There his sediment sampling failed to reveal theoretically predicted coarse-to-fine grain gradation, and this challenged him to study sea-bed processes. This led to a part-time appointment at Scripps Institution of Oceanography until 1948. The award of a Penrose fund of $10000 to Shepard and colleagues in 1936 marked the beginning of Pacific marine geology and the turning point in his career. During World War II he was employed in San Diego, preparing sediment charts for the navy. In 1948 he was appointed Professor of Submarine Geology at Scripps Institution of Oceanography. Shepard put boundless energy into the study of the environmental conditions under which the ancient marine strata had been laid, making new comparisons with presently forming examples distinguishing sedimentation processes in different environments such as deltas, bays and continental shelves. He published *Submarine Geology* in 1948.

SHEPPARD, Philip Macdonald

(1921–76)

English ecological geneticist. He was educated at Marlborough College and Worcester College, Oxford. After teaching in **Ford**'s department at Oxford (1951–6), he moved to Liverpool University, where he became Professor of Genetics in 1963. Sheppard extended and deepened to Ford's pioneering studies, but also carried out significant work on mimicry and human genetics, and on polymorphism in the land snail *Cepaea*. He showed that the banding patterns on snails, long assumed to have no effect on survival, were highly influential under certain circumstances—this discovery highlighted the dangers of assuming that traits are selectively neutral without direct evidence. He went on to demonstrate the important consequences of linked complexes of genes ('super-genes') in snails, mimetic butterflies, and human characteristics such as the Rhesus blood group system. Sheppard also played a key role in introducing genetical concepts into medicine. He was elected FRS in 1965.

SHERMAN, Henry Clapp

(1875–1955)

American biochemist, born in Ash Grove, Virginia. He trained at Maryland Agricultural College (1893–5) and moved to Columbia University where he became successively Professor of Analytical Chemistry (1905–7), Professor of Organic Analysis (1907–11), Professor of Food Chemistry (1911–24) and Mitchill Professor of Chemistry (from 1924). He devoted his research career to determining the quantitative nutritional requirements for metal ions and vitamins. In 1920 he reported the need for calcium in man, and he went on to establish the daily requirement (1931) and explore calcium's interaction with dietary phosphorus and vitamin D. He found that for rats, double the adequate amount of dietary calcium produced better development, longer reproductive capacity and increased life-span, without arterial calcification, over many generations (1941). From 1932 he studied the requirements for B vitamins in relation to the deficiency disease polyneuritis, and observed that riboflavin in excess caused tumours (1943). His major study on vitamin A identified a suitable weekly dose (1934), its storage in the body (1940) and the requirements for well-being in the rat detailed over more than 60 generations. Sherman also observed iron-deficiency anaemia, and began investigating cobalt in 1946. For most of this work he developed suitable assays. Sherman's prolific publications included *Chemistry of Food and Nutrition* (1911).

SHERRINGTON, Sir Charles Scott

(1857–1952)

English physiologist, born in London. After studying at Caius College, Cambridge, and St Thomas's Hospital in London, he became a lecturer in physiology at St Thomas's and Professor-Superintendent of the Brown Animal Sanatory Institute. In 1895 he became Professor of Physiology at Liverpool, and he was later appointed Waynflete Professor of Physiology at Oxford (1913–35). His career focused on the structure and function of the nervous system, exemplified by his analysis of the reflex arc, summarized in *The Integrative Action of the Nervous System* (1906), a book which constituted a significant landmark in modern neurophysiology. He described the reciprocal innervation of antagonistic muscles, by which the activity of one set of excited muscles is integrated with another set of inhibited muscles; he coined the word 'synapse' to describe the junction between nerve cells; he studied sense organs extensively; and he mapped the motor areas of the cerebral cortex of mammals. In addition to his neurophysiological research, Sherrington was the first person to use diphtheria anti-toxin successfully in Britain; he worked during World War I in a munitions factory and chaired the Industrial Fatigue Board; and produced an influential textbook on experimental physiology, *Mammalian Physiology* (1919). He was President of the Royal Society (1920–5), wrote poetry and medical history, and summarized much of his philosophical approach in *Man on his Nature* (1941). He shared the 1932 Nobel Prize for Physiology or Medicine with **Adrian**.

SHKLOVSKY, Josef Samuilovich

(1916–85)

Soviet astronomer, born in Glukhov, the Ukraine. He worked for a time as a foreman on railway construction in Kazakhstan before becoming a student first at the Far-Eastern State University and later at the University of Moscow, where he graduated in physics in 1938. He proceeded to postgraduate research in theoretical astrophysics at the State Astronomical Institute in Moscow, and was awarded his doctorate in 1949. He was responsible for setting up in 1953 the radio astronomy division of the Astronomical Institute which led to the country's first radio telescope; he was associated with this institution for the rest of his life. He also became involved in the design of equipment for the Soviet space programme, which began with the world's first artificial satellite (Sputnik 1) in 1957. Though successful and innovative in these technical spheres, Shklovsky's outstanding contributions were in high-energy astrophysics. He was among the first to suggest the possibility of synchrotron radiation from astronomical sources, and in 1953 solved the problem of the continuous spectrum of the Crab nebula in terms of synchrotron radiation from energetic electrons in a magnetic field, the result of a supernova explosion in 1054 AD. Other researches concerned radio line emission from ionized hydrogen in space, and the study of faint radio sources. In 1954 Shklovsky was appointed professor at Moscow State University, where he was an inspired teacher and a lucid expositor of astronomy. He collaborated with **Sagan** on the popular book *Intelligent Life in the Universe* (1966); among his other books was *Stars, their Birth, Life and Death* (1975), translated into English in 1978. He was awarded the Lenin Prize in 1959 for his contributions to space research.

SHOCKLEY, William Bradford
(1910–89)

American physicist, born in London, the son of two American mining engineers. Brought up in California, he was educated at Caltech and MIT before starting work at the Bell Telephone Laboratories in 1936. During World War II he directed anti-submarine warfare research and he became Consultant to the Secretary for War in 1945. Returning to Bell Telephones he collaborated with **Bardeen** and **Brattain** in trying to produce semiconductor devices to replace thermionic valves. Using a germanium rectifier with metal contacts including a needle touching the crystal, they invented the point-contact transistor (1947). A month later Shockley developed the junction transistor (for *transfer* of current across a *resistor*). These devices led to the miniaturization of circuits in radio, TV and computer equipment. By the mid-1950s Shockley had founded his own company, the Shockley Semiconductor Laboratory, near Palo Alto, California, to exploit his knowledge of solid-state physics. A poor manager, he was unable to commercialize his brilliant research and many of his most gifted workers, such as **Noyce**, left. Shockley, Bardeen and Brattain shared the Nobel Prize for Physics in 1956. From 1963 to 1974 Shockley was Professor of Engineering at Stanford. He later dabbled in eugenics, his ideas on inherited intelligence amongst the races proving highly controversial.

SHRIVER, Duward Felix
(1934–)

American chemist, born in Glendale, California, and educated at the University of California, Berkeley (BS 1958), and the University of Michigan (PhD 1961). He is now Professor of Chemistry at Northwestern University, Illinois, where he is a member of the Materials Research Center and the Ipatieff Catalysis Center. Famous for his work in organometallic chemistry, his current research covers three areas: organometallic cluster compounds; organometallic chemistry of heterogeneous catalysis; and charge transport in solids. He was the first to synthesize and characterize a C- and O-bonded metal carbonyl as opposed to the widely known metal carbonyls which are bonded solely through the C atom. He has followed this theme by studying the role of C and O bonding in the insertion of alkyl groups into metal carbonyls and in the formation of complexes containing carbide (C) or CCO ligands. This research has stimulated interest in the relationship between organometallic chemistry and heterogeneous catalytic processes which are of great industrial importance. Shriver derived a new understanding of the mode of O_2 to the biological molecule hemerythrin. His book *Manipulation of Air-Sensitive Materials* (1969) is the standard reference work in its field.

SIBBALD, Sir Robert
(1641–1722)

Scottish physician and naturalist, born in Edinburgh. He studied at Edinburgh and Leiden, where he graduated MD in 1661. With Sir Andrew Balfour, and influenced by **Morison**, he founded a physic garden in Edinburgh in 1670, to grow plants for use as *materia medica* for the physicians of the town. This garden formed the nucleus of today's Royal Botanic Garden in Edinburgh. With Balfour and others he founded the Royal College of Physicians of Edinburgh, and helped to obtain its Royal Charter. He became the first Professor of Medicine at Edinburgh in 1685. As physician to King Charles II, he was commissioned to undertake a survey of the natural history and archaeology of Scotland. This work, the *Scotia Illustrata*, was published in 1684. The genus *Sibbaldia* is named in his honour, and appropriately *Sibbaldia procumbens*, which Sibbald described in his book, forms the Royal Botanic Garden's logo. He was knighted in 1682.

SIBTHORP, John
(1758–96)

English botanist, born in Oxford where he was educated before studying medicine at Edinburgh. He then went to Montpellier (1783), where he was elected a member of the Academy of Sciences. Shortly afterwards, he succeeded his father as Sherardian Professor of Botany at Oxford, and in 1784 unsuccessfully bid against his friend **James Edward Smith** for the collections of **Linnaeus**. He left for the Continent to plan a botanical expedition to Greece, aiming to study the plants named by Dioscorides. In Vienna, he examined the Dioscorides Codex and secured the services of the botanical artist **Ferdinand Bauer**. Sibthorp and Bauer reached Crete in 1786. After several months there, they visited other Greek islands, Athens, western Turkey, Cyprus and Athos, returning to England in late 1787. Sibthorp (with Smith) then helped found the Linnaean Society (1788) and published his *Flora Oxoniensis* (1794). On his second expedition to Greece and Turkey (commencing 1794), he developed tuberculosis, from which he died. Apart from *Flora Oxoniensis*, Sibthorp's only work was his share in *Flora Graeca* and *Prodromus Florae Graecae*, both posthumously published by Smith.

SIDGWICK, Nevil Vincent

(1873–1951)

English chemist, born in Oxford. At Christ Church, Oxford, he was a student of **Augustus Vernon Harcourt** and gained a first class in chemistry in 1895. He followed this with a first in classics in 1897. After acting as demonstrator in the Christ Church laboratory for a year, he went to work in **Ostwald**'s laboratory in Leipzig and then with Hans von Pechmann at Tübingen. He returned to be a Fellow and Tutor of Lincoln College, a position he held until retirement in 1947. In the university he became reader (from 1924) and professor (from 1935). Up to 1920 Sidgwick carried out various physico-chemical solution studies involving kinetics, ionic equilibria or phase equilibria. He was greatly stimulated, however, by the nascent electronic theory of valency, and his researches began to bear on this. The most important outcome was his book *The Electronic Theory of Valency* (1927). His later books *The Covalent Link in Chemistry* (1933, from his 1931 Baker Lectures at Cornell University) and *The Chemical Elements and Their Compounds* (2 vols, 1950) were also highly influential. His earlier book *The Organic Chemistry of Nitrogen* (1910) went through later editions under editors. Sidgwick was appointed CBE in 1935, was elected a Fellow of the Royal Society in 1922 and received its Royal Medal in 1937. He was President of the Faraday Society from 1932 to 1934 and of the Chemical Society from 1935 to 1937.

SIEGBAHN, Kai

(1918–)

Swedish physicist, born in Lund, the son of **Karl Siegbahn**. Educated at Stockholm, he was Professor of Physics at the Royal Institute of Technology there until 1954, and thereafter professor at Uppsala University. In the early 1950s he started making high-precision measurements on the energies of electrons emitted from solids exposed to X-rays. Unlike previous observations which had revealed the electron energies to be distributed in bands, Siegbahn found sharp peaks in the bands at energies that were characteristic of the materials he was exposing to the X-rays. These peaks corresponded to the energies required to displace one of the inner bound electrons. Furthermore he found that the peak energies were dependent on the chemical environment of the atoms, ie on the type of chemical bond linking neighbouring atoms or molecules. This technique, which became known as ESCA (electron spectroscopy for chemical analysis), offered a delicate but powerful experimental method for studying the energies of electrons around bonded atoms. The method has been extended by Siegbahn's team for use with liquids and gases as well as solids, and they have also worked on the related technique of ultraviolet photoelectron spectroscopy. He shared the 1981 Nobel Prize for Physics with **Bloembergen** and **Schawlow** for his work in developing high-resolution electron spectroscopy.

SIEGBAHN, Karl Manne Georg

(1886–1978)

Swedish physicist, born in Örebro. He was educated at the University of Lund, and became professor there (1920) and at Uppsala (1923). From 1937 he was Professor of the Royal Academy of Sciences and Director of the Nobel Institute for Physics at Stockholm. Improving on the techniques of **Barkla** and **Harry Moseley**, he succeeded in producing X-rays of various wavelengths and penetrating power, which were labelled K, L, M, N, O, P, and Q in order of increasing wavelength. This discovery reinforced **Niels Bohr**'s shell model of the atom, and the atomic shells were lettered in the same way. For his development of X-ray spectroscopy, Siegbahn was awarded the Nobel Prize for Physics in 1924. In the same year he showed that X-rays could be refracted, like light, by means of a prism.

SIEMENS, (Ernst) Werner von

(1816–92)

German electrical engineer, brother of Sir **William Siemens**, born in Lenthe, Hanover. In 1834 he entered the Prussian Artillery, and in 1844 he took charge of the artillery workshops at Berlin. He developed the telegraphic system in Prussia, discovered the insulating property of gutta-percha, and devoted himself to the construction of telegraphic and electrical apparatus. In 1847 together with Johann Georg Halske, scientific instrument maker at the University of Berlin, he established Siemens & Halske, initially for manufacturing telegraphy equipment, but which evolved into one of the great electrical engineering firms. Besides devising numerous forms of galvanometer and other electrical instruments, he was one of the discoverers of the self-acting dynamo (1867). He determined the electrical resistance of different substances; the SI unit of electrical conductance is named after him. In 1887 he endowed the Physikalisch Technische Reichsanstalt in Berlin, of which **Helmholtz** was the first director. He was ennobled in 1888. One of his sons, Wilhelm (1855–1919), was a pioneer of the incandescent lamp.

SIEMENS, Sir (Charles) William, originally **(Karl) Wilhelm**
(1823–83)

German-born British electrical engineer, brother of **Werner von Siemens**, born in Lenthe, Hanover. In 1843 he visited England to introduce a process for electro-gilding invented by Ernst Werner and himself, and in 1844 he patented his differential governor. He was naturalized in 1859. As manager in England of the firm of Siemens Brothers, he was actively engaged in the construction of telegraphs, designed the steamship *Faraday* for cable-laying, promoted electric lighting, and constructed the Portrush Electric Tramway in Ireland (1883). In 1861 he designed an open-hearth regenerative steel furnace which became the most widely used in the world. Other inventions included a water-meter, pyrometer and bathometer. Siemens was elected FRS in 1863, and was President of the Institution of Mechanical Engineers (1872), the Iron and Steel Institute (1877), and the British Association for the Advancement of Science (1882). He was also the first President of the Society of Telegraph Engineers (1872) and Chairman of the Royal Society of Arts (1882). He was knighted in 1883. He was assisted in England by another brother, Friedrich (1826–1904), who invented a regenerative smelting oven extensively used in glassmaking (1856).

SIERPIŃSKI, Wacław
(1882–1969)

Polish mathematician, born in Warsaw. He studied there, and was professor from 1919 to 1960. The leader of the Polish school of set theorists and topologists, he was a prolific author, publishing more than 700 research papers on set theory, topology, number theory and logic, and several books. In 1919 he founded the still-important journal *Fundamenta Mathematicae* to publish work in these areas.

SIMON, Sir Francis Eugene
(1893–1956)

German physicist, born in Berlin. A year after going to Munich to read physics (1912), Simon was called up for military service. Recommencing study in Berlin, he completed his doctorate with **Nernst** on specific heats at low temperatures. By 1927 he was professor at the university there and in 1931, carrying with him a high reputation, he was appointed Director of the Physical Chemistry Laboratory at Breslau. With the rise of Nazism he left Germany for Oxford and the Clarendon Laboratory at the invitation of Frederick Lindemann. He became reader in thermodynamics (1935) and professor (1945), and only a month before he died, he succeeded Lindemann as Professor of Experimental Philosophy and Director of the Clarendon. Simon had great success in verifying experimentally the third law of thermodynamics, and under his guidance, Oxford became one of the world's leading low-temperature physics centres. From the 1940s Simon enthusiastically and at times controversially advocated the reform of technical education and national energy policy. Involvement in the atomic energy and weapons project (1940–6) earned him a CBE (1946). He was elected FRS in 1941, and knighted in 1954.

SIMON, Sir John
(1816–1904)

English surgeon and public health reformer, born in London, son of a prosperous businessman. He trained for surgery through an apprenticeship and subsequently lectured in surgery and pathology at King's College Hospital and St Thomas's Hospital, London. His *General Pathology* (1850) established his scientific reputation, but he had already also become Medical Officer of Health (1848) to the City of London. His success in monitoring epidemic diseases in the city led to his appointment as Chief Medical Officer to the General Board of Health (1855). From then until his retirement in 1876, Simon combined his epidemiological and scientific skills with political sensitivity, to effect sweeping changes in public health practice in Britain. He demonstrated the advantages of compulsory smallpox vaccination, and worked behind the scenes to influence a succession of Parliamentary bills culminating in the Public Health Act of 1875, which was the most comprehensive legislative health package anywhere in the world. Simon's collected writings were edited by a colleague after Simon's retirement, and his own account of *English Sanitary Institutions* (1890) is still a valuable historical source. He was knighted in 1887.

SIMPSON, George Gaylord
(1902–84)

American palaeontologist, born in Chicago. Educated at the universities of Colorado and Yale, he joined the staff of the American Museum of Natural History in New York City in 1927, and from 1959 to 1970 taught at Harvard. His early work was concerned with the Mesozoic mammals which provide important evidence for the understanding of mammalian evolution. In the 1930s he carried out palaeontological research on the Tertiary mammals from Patagonia, a unique and hitherto little-studied fauna. He is considered one of the leading 20th-century palaeontologists and he proposed a classification of mammals which is now standard. Although mainly concerned with taxonomy, after World War II he devoted himself to demonstrating that the neo-Darwinian ideas of geneticists such as **Mayr** and **Dobzhansky** could be reconciled with the palaeontological evidence. He was particularly concerned with the circumstances which gave rise to the evolution of new species. His influential books *Tempo and Mode in Evolution* (1944) and *The Major Features of Evolution* (1953) were concerned with the fusion of palaeontology and evolutionary genetics. Some of his ideas are presented in popular form in *The Meaning of Evolution* (1949).

SIMPSON, Sir James Young

(1811–70)

Scottish obstetrician and pioneer of anaesthesia, born in Bathgate, West Lothian, the son of a baker. With the financial support of the rest of his large family he went to Edinburgh University at the age of 14, studying arts, then medicine, and becoming Professor of Midwifery in 1840. He originated the use of ether as anaesthetic in childbirth (January 1847), and experimenting on himself and his assistants in the search for a better anaesthetic, discovered the required properties in chloroform (November 1847). He championed its use against medical and religious opposition until its employment by Queen Victoria at the birth of Prince Leopold (1853) signalled general acceptance. He founded gynaecology by his sound tests, championed hospital reform, and in 1847 became Physician to the Queen in Scotland. He was made a baronet in 1866.

SIMSON, Robert

(1687–1768)

Scottish mathematician, Professor of Mathematics at Glasgow from 1711. His life's work was dedicated to the editing and restoration of the work of the ancient Greek geometers; his edition of Euclid's *Elements* (1758) was the basis of nearly all editions for over a century.

SITTER, Willem de

See **DE SITTER, Willem**

SLIPHER, Vesto Melvin

(1875–1969)

American astronomer, born in Mulberry, Indiana. He studied at Indiana University before working for over 50 years at the Lowell Observatory, Arizona, becoming its director in 1926. He obtained the first successful photographs of Mars, and demonstrated that methane is present in the atmosphere of Neptune. The research which led to the discovery of the planet Pluto was carried out under his direction. Primarily a spectroscopist, his spectral studies revealed the presence of gaseous interstellar material, and he suggested that the nebula in the Pleaides cluster of stars is illuminated by starlight reflected off dust grains (1912). By measuring the **Doppler** shift in light reflected from the edges of planetary discs, he determined the periods of rotation of Uranus, Jupiter, Saturn, Venus and Mars in 1912. In his most important work, he extended this method to the Andromeda nebula, not yet perceived as an extragalactic object, and discovered that it is approaching the Earth at around 300 kilometres per second (1912). Similar studies of other nebulae showed this to be an exception, as most were found to be receding from the Earth at very high speed. His results directed **Hubble** to the concept of the expanding universe, in which galaxies are moving apart at relative speeds proportional to their separation.

SMAGORINSKY, Joseph

(1924–)

American meteorologist, born in New York City. He received a master's degree from New York University (1948) and joined the Weather Bureau to work on objective analysis of weather charts and on calculating vertical motion in the atmosphere from horizontal wind and pressure fields. **Charney** invited him to Princeton University in 1950. It was there that the idea of using a very large computer for weather forecasting was developed. After obtaining his PhD (1953), Smagorinsky returned to the Weather Bureau to set up a research group on numerical weather prediction. He headed the General Circulation Research Laboratory, which eventually became the Geophysical Fluid Dynamics Laboratory (GFDL), for 27 years. During this time he was the driving force behind much of the enormous progress made in numerical weather prediction. From the initial, relatively simple, barotropic and primitive equation models, the science progressed so that more complex physical processes such as convection, variations in radiation due to different surfaces and ocean–atmosphere interactions could be included in the programmes. This enabled the original relatively short-period forecast models to be modified for use as climate prediction models. Smagorinsky's work made GFDL a world leader in this field.

SMEATON, John
(1724–92)

English civil engineer, born in Austhorpe near Leeds. The son of an attorney, his father intended that he should enter the legal profession, and at the age of 18 he was sent to London to study law. Even as a boy, however, he was much more interested in science, model-making and mechanical philosophy, and he soon abandoned law and apprenticed himself to a scientific instrument maker. By 1750 he had established his own business in London as an instrument maker, rapidly gaining a reputation for the improvements he devised for a wide variety of instruments, mainly for navigation. Elected FRS in 1753, he won the Royal Society's Copley Medal for his researches into the mechanics of waterwheels and windmills, and firmly established himself as England's first and foremost civil engineer with his novel design for the third Eddystone lighthouse. Two earlier wooden structures had been destroyed by fire and storm; Smeaton designed a masonry tower, completed after three years' work in 1759, which remained in use until 1877, and was re-erected on Plymouth Hoe as a memorial. Through systematic study and experiment, and by designing a boring machine that could turn out more accurate cylinders, he improved the performance of the atmospheric steam engine of Thomas Newcomen. His other chief engineering works included harbours at Ramsgate (1774), Port Patrick and Eyemouth, the Forth and Clyde Canal (1768–88), and bridges at Coldstream and Perth which are still carrying traffic today.

SMITH, Sir Grafton Elliot
(1871–1937)

Australian-born anatomist and anthropologist, born in Grafton, New South Wales. He graduated in medicine from the University of Sydney in 1892. In 1896 he went to England to conduct research at Cambridge, and in 1900 he was made Professor of Anatomy at the new medical school in Cairo. He was later appointed to chairs of anatomy at Manchester (1909) and University College London (1919). Smith became an authority on brain anatomy and human evolution, building his reputation on studies of cranial morphology and the Egyptian practice of mummification. His books, including *Migrations of Early Culture* (1915), *The Evolution of the Dragon* (1919) and *The Diffusion of Culture* (1933), explain similarities in culture all over the world by diffusion from pharaonic Egypt. He was knighted in 1934.

SMITH, Hamilton Othanel
(1931–)

American molecular biologist, born in New York City. He graduated from Johns Hopkins Medical School, and after teaching at the University of Michigan, returned to Johns Hopkins in 1967. He was later appointed Professor of Microbiology (1973) and Professor of Molecular Biology and Genetics (1981) there. During the late 1950s, when he was drafted into the US navy, he began to read about genetics. Initially he began to work on phage (the bacterial virus), but later used the bacterium *Haemophilus influenzae*. He found that bacteria produce enzymes, called endonucleases, which can split the DNA strand of invading phage particles so that they are inactivated. Because the enzymes curtail the activity of the phage, they are also called 'restriction enzymes'. In the 1970s Smith went on to isolate endonucleases which would split a DNA strand at a specific site. More than 100 such enzymes have now been isolated. Because of their site-specificity, the use of these enzymes allows the nucleotide sequence of DNA to be established. Smith shared the 1978 Nobel Prize for Physiology or Medicine with **Werner Arber** and **Nathans**.

SMITH, Henry John Stephen
(1826–83)

Irish mathematician, born in Dublin. He was educated at Rugby School and Balliol College, Oxford, of which he was elected a Fellow. In 1860 he became Oxford's Savilian Professor of Geometry. He was the greatest British authority of his day on the theory of numbers, and wrote a fine six-part report on the subject for the British Association for the Advancement of Science which was influential in communicating new developments to British mathematicians. He was posthumously awarded the prize of the Paris Academy of Sciences for his work on the representation of integers as sums of squares (sharing the prize with the 18 year old **Hermann Minkowski**). He also wrote on elliptic functions and geometry.

SMITH, Sir James Edward
(1759–1828)

English botanist, born in Norwich. Although he inherited a love of botany from his mother, he did not begin studying botanical science until the day **Linnaeus** died, when Smith was 18. He studied medicine at Edinburgh University and London. In 1783, aged only 24, he bought the entire natural history collection of Linnaeus from his widow, bidding against his friend **Sibthorp**, and brought it to London. On his return from a continental tour

(1786–7), he became a founder member and first President of the Linnaean Society of London (1788–1828). After his marriage (1796) he lived in Norwich, travelling to London annually until 1825 to lecture at the Royal Institution. He wrote 3348 botanical articles, his most important works being *English Botany* (36 vols, 1790–1814), *Flora Britannica* (3 vols, 1800–4), *Exotic Botany* (2 vols, 1804–5), *Flora Graeca* (7 vols, 1806–28), *Prodromus Florae Graecae* (2 vols, 1806–13) and *English Flora* (4 vols, 1824–8). He was deacon at the Octagon Chapel, Norwich, and wrote a tract, *A Defence of the Church and Universities of England ...* (1819), in response to Cambridge University's objection to Thomas Mertyn's invitation to him to lecture there, on the grounds of his unitarian beliefs.

SMITH, John Maynard
See **MAYNARD SMITH, John**

SMITH, (Ernest) Lester
(1904–92)

English biochemist, born in Teddington, Middlesex. He was educated at Chelsea Polytechnic, and in 1928 he joined the pharmaceutical firm of Glaxo where he spent his entire career, becoming Senior Research Biochemist. Initially he worked on the dietary requirements for vitamin A and its storage properties as affected by oxidation and light. During World War I he was responsible for the first commercial production of penicillin in England, using bacterial fermentation, and discovered that the biosynthesis of penicillin F (the common form), G or X could be selectively promoted by adjustment of the growth medium. He independently isolated the anti-pernicious-anaemia factor (vitamin B_{12}) in crystalline form in the same year as **Folkers** in the USA (1948), and showed that it contained cobalt. He separated and characterized the various forms of the vitamin, and was the first to prepare the doubly radiolabelled vitamin, containing cobalt-60 and phosphorus-35, for metabolic tracer studies (1952). In 1958 he isolated an antibiotic complex with low host toxicity (called E129) which was effective against streptococci that had developed resistance to other known antibiotics (such as penicillin and streptomycin). He published *Intelligence Comes First* (1975), *Our Last Adventure* (1982) and *Inner Adventures* (1988), and was elected FRS in 1957.

SMITH, Michael
(1932–)

British–Canadian biochemist, born in Blackpool. Educated at the University of Manchester, in 1956 he moved to the University of British Columbia, where he is now professor and Director of the Biotechnology Laboratory. In 1978, he published his discovery of 'site-specific mutagenesis', a technique which allows scientists to alter the genetic code through mutations induced at specific locations — all previous methods of mutation, using radiation and chemicals, produced random mutations. This new method has allowed the production of a whole new range of proteins with diverse functions. Smith was awarded the 1993 Nobel Prize for Chemistry jointly with **Mullis**.

SMITH, Theobald
(1859–1934)

American microbiologist and immunologist, born in Albany, New York, the greatest American bacteriologist of his generation. He received his medical degree from the Albany Medical College, and was subsequently associated with several US institutions, including Harvard University (as professor, 1896–1915) and the Rockefeller Institute for Medical Research (1915–29). He studied both animal and human diseases, and first implicated an insect vector in the spread of disease when he showed that Texas cattle fever is spread by ticks. He distinguished the forms of bacillus causing human and bovine tuberculosis, and laid the scientific foundations for a cholera vaccine. He also improved the production of smallpox vaccine and diphtheria and tetanus antitoxins, and established precise techniques for the bacteriological examination of water, milk and sewage.

SMITH, William
(1769–1839)

English civil engineer and geologist, born in Churchill, Oxfordshire, the son of a blacksmith. In 1787 he became an assistant to a surveyor, and he was later appointed engineer to the Somerset Coal Canal (1794–9). His survey work during canal construction introduced him to a variety of rock sequences of different ages, and in 1799 he produced a coloured geological map of the country around Bath. From 1799 he was a consultant engineer and surveyor, travelling great distances in the course of his work and later settling in London (1804). Smith used fossils to aid his identification of strata and to fix their position in the succession. He produced the first geological map of England, *A Delineation of the Strata of England and Wales, with part of Scotland* (1815), and 21 coloured geological maps of the English counties (1819–24) assisted by his nephew

John Phillips (1800–74). Smith is often regarded as the father of English geology and stratigraphy.

SMITHSON, James Louis Macie

(1765–1829)

English chemist, the benefactor of the Smithsonian Institution in Washington DC. He was born in Paris, the illegitimate son of Sir Hugh Smithson Percy, 1st Duke of Northumberland, and Elizabeth Macie. At first known as Macie, he changed his name in 1801 after the death of his mother. He was educated at Oxford, and showing an early aptitude for mineralogy, he was elected FRS in 1787. He analysed zinc ores and other minerals; zinc carbonate was later named 'smithsonite' after him. He also proposed an innovative balance to weigh very small quantities. However, he is chiefly remembered for the manner in which he disposed of his wealth, rather than for his contribution as a chemist. He left a will stipulating that his estate of £105 000 should pass to his nephew, with the proviso that if the nephew died without heirs the money should be used to found 'at Washington, under the name of the Smithsonian Institution, an Establishment for the increase and diffusion of knowledge among men'. His nephew died in 1835 leaving no children and the Smithsonian Institution was established by Act of Congress in 1846. It soon grew to be a world famous museum and institute for scientific research.

SMYTH, Charles Piazzi

(1819–1900)

British astronomer, born in Naples, son of William Henry Smyth, a British naval officer, and godson of **Giuseppe Piazzi** after whom he was named. On retiring from the navy, William Smyth set up a private observatory at Bedford where he produced his highly popular *Bedford Catalogue* (1844) of double stars which earned him the Gold Medal of the Royal Astronomical Society (1845). Charles Piazzi Smyth was educated at Bedford School and received a sound training in practical astronomy at home. At the age of only 16, he was chosen for the post of chief assistant at the Royal Observatory, Cape of Good Hope, South Africa (1827–37). He succeeded **Henderson** as Astronomer Royal for Scotland and Professor of Astronomy at the University of Edinburgh, where he remained until his retirement in 1888. His most important contribution to astronomy was his expedition to the island of Tenerife (1856) to test the advantages of high-altitude sites for observational astronomy, the first ever scientifically conducted investigation into this question. His results, published by the Admiralty and recorded in his *Tenerife, an Astronomer's Experiment* (1858) proved the superiority of a mountain site for every type of observation. His ambition to erect a station on the Peak of Tenerife went unfulfilled; he did, however, make several excursions to sunny locations abroad to make spectroscopic observations of the Sun. He was a successful early photographer who began with calotypes of South African scenes in 1843, and was the inventor of a flat-fielding lens for cameras (1874). In 1864 he carried out a metrological survey of the great pyramid of Giza, but his bizarre interpretation of the pyramid as a divinely inspired monument—a belief which attracted many followers—gravely tarnished his scientific reputation.

SNELL, George Davis

(1903–)

American geneticist, born in Bradford, Massachusetts, the son of an inventor. He graduated from Dartmouth College in 1926 and conducted graduate studies at Harvard University. After teaching zoology for two years, Snell was awarded a fellowship at the University of Texas where he demonstrated, for the first time, that X-rays can induce mutations in mammals. In 1933 he became Assistant Professor at Washington University, but in 1935 he joined the Jackson Laboratory in Bar Harbor, Maine, where he remained until 1973. In the late 1930s, Snell began to work on the genes responsible for rejection of tissue transplants in mice—which he called histocompatibility genes—later named the major histocompatibility complex (MHC). Research in this area took a dramatic turn when **Dausset** discovered that histocompatibility in humans is also controlled by an MHC. Later **Benacerraf** showed that the MHC controls the production of proteins on cell surfaces which allow certain white blood cells to distinguish them from foreign or abnormal cells. Snell, Dausset and Benacerraf shared the 1980 Nobel Prize for Physiology or Medicine.

SNELL, Willebrod van Roijen, Latin Snellius

(1580–1626)

Dutch mathematician, born in Leiden. He succeeded his father as Professor of Mathematics at Leiden (1613) and discovered the law of refraction known as Snell's law, which relates the angles of incidence and refraction of a ray of light passing between two media of different refractive index. He also extensively developed the use of triangulation in surveying.

SNOW, John

(1813–58)

English anaesthetist and epidemiologist, born in York. He was a young general practitioner when cholera first struck Britain in 1831–2, and his experience then convinced him that the disease was spread through contaminated water. After 1836 he practised in London, and during the cholera outbreaks of 1848 and 1854 there, he carried out some brilliant epidemiological investigations, tracing one local outbreak to a well in Broad (now Broadwick) Street, Soho, into which raw sewage seeped. His additional work implicated the Thames, into which many of London's sewers drained and from which much of London's domestic water was obtained. He showed that houses which were supplied by companies obtaining their water from downstream the Thames had much higher incidences of cholera than those supplied by water drawn from upstream, before the sewage had contaminated it. Snow was also a pioneer anaesthetist. He did fundamental experimental work on ether and chloroform, devised apparatus to administer anaesthetics, and gave chloroform to Queen Victoria in 1853, during the birth of Prince Leopold. During the last 10 years of his life, he was much in demand by surgeons as an anaesthetist.

SNYDER, Solomon Halbert

(1938–)

American psychiatrist and pharmacologist, born in Washington. He received his MD at Georgetown University (1962), worked at the Intern Kaiser Foundation Hospital, San Francisco (1962–3), and then became a research associate at the National Institute of Medical Health in Bethesda, Maryland (1963–5), and resident psychiatrist at Johns Hopkins Hospital in Baltimore (1965–8). He then held a number of professorships at Johns Hopkins Medical School, where since 1980 he has been Distinguished Service Professor in Neuroscience, Psychiatry and Pharmacology. From the mid-1960s Snyder investigated the biochemistry of nervous tissue, determining the stereospecific requirements for catecholamines of synaptosomes (vesicles that contain neurotransmitter substances) from different areas of the brain (1968), and extending earlier observations that mammalian ornithine decarboxylase has an extremely rapid turnover time of 10–30 minutes (1969). This enzyme, which is involved in polyamine synthesis and possibly the regulation of RNA synthesis, links with his major interest in the effects of opiates and psychotropic drugs on the brain and the naturally occurring brain hormones, enkephalins and endorphins. In 1973 he successfully demonstrated the presence of opiate receptors in nervous tissue, opening the way for the study of endogenous algesics. He exploited his receptor binding assay for the study of the functional distribution of neurologically important compounds, such as *met*- and *leu*-enkephalin which have binding potencies similar to morphine (1976–7). More recently he has studied transport across membranes involving cyclic adenosine monophosphate (cAMP) and its link with olfaction (1985–6), inositol triphosphate in calcium ion transport (1988) and the calcium-regulated enzyme nitric oxide synthase which synthesizes the 'vascular endothelium derived relaxing factor', nitric oxide (1990).

SOBRERO, Ascanio

(1812–88)

Italian chemist who invented nitroglycerine, born in Casale, Montferrato. He studied medicine at Turin, and organic chemistry at Paris and with **Liebig** at Giessen. From 1844 he taught at Turin, being appointed to the chair in 1847. In 1844 or 1845 he added glycerol to a mixture of concentrated nitric and sulphuric acids and created nitroglycerine, a colourless, oily, sweet-tasting liquid which is a powerful explosive. Sobrero described its explosive properties and also investigated its effects as a drug. Nitroglycerine was manufactured on a huge scale after **Nobel** invented dynamite. Today it is used as a fuel in rockets and missiles; it is also employed in medicine as it dilates blood vessels and thus eases cardiac pain. Sobrero was also the first person to prepare lead tetrachloride (1850). He died in Turin.

SODDY, Frederick

(1877–1965)

English radio chemist, renowned for his work on radioactivity, born in Eastbourne. He studied at the University College of Wales, Aberystwyth, and at Oxford. In 1900, after two years research at Oxford, he was appointed demonstrator in chemistry at McGill University, Montreal, where he and **Rutherford** studied radioactivity. They realized that the reason radioactive elements display properties at odds with their position in the periodic table is because they have decayed, in part, into other elements. As part of the decay process they emit three types of radiation (now known as alpha, beta and gamma radiation). In 1903, in London working with **Ramsay**, Soddy verified his prediction that radium produces helium when it decays. Lecturing at Glasgow from 1904 to 1914, he showed that uranium decays into radium, work later developed by **Boltwood**. He also showed that a radioactive element may have more than one atomic weight although its chemical properties are identical. Because the chemical properties are unchanged, Soddy argued that all forms of the same element belong in the same place in the periodic table and therefore he named them 'isotopes' ('the same place'). He later demonstrated that isotopes exist among elements which are not radioactive. His discovery of isotopes was of fundamental importance to all physics and chemistry, although the reason for the existence of isotopes did not become clear until **Chadwick** discovered the neutron in 1932. Soddy's other great discovery was that when an atom of a radioactive element emits an alpha particle, it is transformed into an isotope of an element two places down in the periodic table, but when it emits a beta particle it moves one place higher. In 1914 he was appointed to the chair at Aberdeen, where he was largely employed on chemical research connected with World War I. Moving to Oxford in 1919, where he was Dr Lee's Professor of Chemistry until 1936, Soddy reorganized the laboratory facilities and the teaching syllabus. In 1921 he was awarded the Nobel Prize for Chemistry. After his retirement he wrote on ethics, politics and economics, urging fellow scientists only to conduct research in areas which had peaceful applications. He died in Brighton.

SOKAL, Robert Reuven

(1926–)

Austrian-born American entomologist and biometrician, born in Vienna. He studied at St John's in Shanghai and at the University of Chicago, where during 1948–51 he worked in the department of zoology. In 1953 he moved to the University of Kansas, becoming Assistant Professor of Entomology in 1953; more recently he has worked at the State University of New York, Stony Brook. His early researches concerned the genetics of aphids and flies. He has latterly done much valuable work in the fields of biometry and biostatistics. With the English bacterial taxonomist and microbiologist Peter H A Sneath, he has carried out much pioneering work in the development of the principles and methods of numerical taxonomy. This has included the publication of two jointly authored standard works, *Principles of Numerical Taxonomy* (1963) and *Numerical Taxonomy: The Principles and Practice of Numerical Classification* (1973). Sokal's other main works, both jointly written with F James Rohlf, are *Biometry* (1969) and *Introduction to Biostatistics* (1969).

SOLVAY, Ernest

(1838–1922)

Belgian industrial chemist, born in Rebecq-Rognon. He worked in his father's salt-purifying business and then helped to manage his uncle's gas works. He noticed that ammonia, carbon dioxide and salt in solution react to form sodium bicarbonate which could easily be converted into the soda ash required by glass, soap and porcelain manufacturers. At the time soda was manufactured by the **Leblanc** process which created much pollution from sulphur and other by-products. The ammonia–soda reaction had been recorded before, but nobody had managed to adapt it to commercial use. Solvay overcame the difficulty of mixing the reagents properly by treating the ammoniacal brine with carbon dioxide as it was sprayed down a tower. The first sizeable plant was established by Solvay and his brother Alfred at Couillet, near Charleroi, in 1865. The Solvays established plants in many countries and granted licenses for the process in others. It was introduced to Britain by **Mond**. Solvay used the wealth generated by his process to found institutes of physics, chemistry and sociology. He died in Brussels.

SOMERVILLE, Mary Greig, née Fairfax

(1780–1872)

Scottish mathematician and astronomer, born in Jedburgh. The daughter of a naval officer, she was inspired by the works of **Euclid**, and studied algebra and classics, despite intense disapproval from her family. From 1816 she lived in London, where she moved in intellectual and scientific circles, and corresponded with foreign scientists. In 1826 she presented a paper on *The Magnetic Properties of the Violet Rays in the Solar Spectrum* to the Royal Society. 1831 saw the publication of *The mechanism of the heavens*, her account for the general reader of **Laplace**'s *Mécanique Céleste*. This had great success and she wrote several further expository works on physics, physical geography and microscopic science. She supported the emancipation and education of women, and Somerville College (1879) at Oxford is named after her.

SOMMERFELD, Arnold

(1868–1951)

German physicist, born in Königsberg (now Kaliningrad, Russia). He was educated at the University of Königsberg and appointed as professor at Clausthal (1897), Aachen (1900) and Munich (1906). With **Klein** he developed the theory of the gyroscope. He also researched into wave spreading in wireless telegraphy. Sommerfeld generalized the quantization rules developed by **Niels Bohr** so that Bohr's quantum model of the atom could be applied to multi-electron atoms. He also evolved a theory of the electron in the metallic state.

SONDHEIMER, Franz
(1926–81)

English chemist, born in Stuttgart. Of Jewish stock, he moved with his family to England in 1937. His chemical education began at Imperial College, London, in 1944 and he graduated with a PhD in 1948. After collaboration with **Raphael**, he moved to Harvard to work with **Robert Woodward** on steroid synthesis. In 1952 he moved to Syntex SA in Mexico City in succession to **Djerassi**. After four very successful years there he became head of the department of organic chemistry at the Weizmann Institute in Israel, but kept close contact with Syntex. In 1963 he moved to a Royal Society Chair at Cambridge, but he later transferred this to University College London. Although steroid synthesis, particularly those associated with oral contraceptives, had been his area of greatest success, he increasingly worked on large-ring annulene compounds. He was, above all else, a perfectionist and this became an increasing burden to him and led to many periods of deep depression. He took his own life in 1981. He received many honours and was elected FRS in 1967. He collected a fine library of early scientific books.

SORBY, Henry Clifton
(1826–1908)

English geologist and metallurgist, born in Woodbourne, Sheffield. He inherited a modest fortune which allowed him to devote his time to science. Sorby was the first to study rocks in thin section under the microscope, demonstrating that this could reveal their minute structure and composition, as well as much about their mode of origin. He also adapted the technique for the study of metals by treating polished surfaces with etching materials. Using the microscope he observed the alignment of micas which causes cleavage in slates and identified fossils in chalk. In 1858 he published *On the Microscopical Structure of Crystals*, which marks a prominent landmark in petrology. He also wrote on biology, architecture and Egyptian hieroglyphics.

SØRENSEN, Søren Peter Lauritz
(1868–1939)

Danish chemist, born in Havrebjerg. He was educated in Copenhagen, where he later became Director of the Carlsberg Research Laboratory. In 1901 he described how formaldehyde lowers the dissociation constant (pK) of an amino group, so that it could be titrated by standard electrometric procedures. This method, the 'formol titration', is still used today. In 1909, in *Enzyme Studies II* which described the effects of hydrogen ion concentration on enzyme activity, he introduced the term pH to represent the negative logarithm of the hydrogen ion concentration. He also used for the first time the German word 'puffer', which became 'buffer' in English, to describe the ability of certain solutions to resist change in pH. In 1912, he independently established the generalized relationships among acids and bases which later became known as the Henderson–Hasselbalch equation; $pH = pK + \log([base]/[acid])$, where [base] and [acid] denote concentrations. In the following decade, along with others, he prepared extensive lists of dyestuffs that change colour at predetermined hydrogen ion concentrations. In 1923, to unify the mass of data that resulted, he devised the pH scale, arguably the most important single contribution ever made to the life sciences. He made other contributions to pH measurement, the osmotic pressure of protein solutions, and on denaturation.

SOSIGENES
(fl c.40 BC)

Alexandrian astronomer and advisor to Julius Caesar in his reform of the calendar. At that time the error in the calendar was such that three additional months had to be interpolated to correct it. In Caesar's reform the year was to become independent of the Moon, and to consist of 365 days, an extra day being added in February every fourth year (the leap year) to make the average length of the year 365 days. The new Julian calendar began in 45 BC and remained in force until 1582 when Pope Gregory introduced the improved Gregorian system in which centuries are leap years only if divisible by 400.

SOUTHWOOD, Sir (Thomas) Richard (Edmund)
(1931–)

English zoologist and ecologist. He was educated at the University of London (1949–55), and after teaching there he became Director of Imperial College Field Station at Silwood Park (1967–9) and Professor of Zoology and Applied Entomology (1969–79). He then moved to the University of Oxford, where he became Linacre Professor of Zoology in 1979 and Vice-Chancellor in 1989. His early work was concerned with the ecology and systematics of the

Hemiptera. He wrote for the 'Wayside and Woodland' series *Land and Water Bugs of the British Isles* (1959, with D Leston), a volume which married scientific rigour with a popular approach. He has carried out research on the relationships between insects and plants, particularly with regard to the factors which determine the number of species of insect which are associated with a species of plant. He is also interested in the way in which features of habitat influence the evolution of life-history strategies in insects. He was Chairman of the Royal Commission on Environmental Pollution (1981–6) at the time when the recommendation to eliminate lead from petrol was made, Chairman of the Anglo-Scandinavian Committee for the Surface Water Acidification Programme, which is concerned with research into acid rain, and became Chairman of the National Radiological Protection Board in 1985. Elected FRS in 1977, he served as Vice-President of the Royal Society (1982–4), and was knighted in 1984.

SPALLANZANI, Lazaro
(1729–99)

Italian biologist and naturalist. Born in Scandiano in Modena, Spallanzani studied law at Bologna and became a priest. But his first love was natural philosophy. He rose to become Professor of Mathematics and Physics at Reggio (1757), moving on to Modena in 1763 and to the Chair of Natural History at Pavia in 1769. He was an enthusiastic traveller who enlarged the Natural History Museum at Pavia. He is remembered for his skills in experimental physiology, where he pursued wide-ranging and fruitful experimentation. Deeply interested in reproduction, he set about rebutting the long-established theory of spontaneous generation. In the 17th century, **Redi** had shown that insects developed on putrefying flesh only from deposited eggs. Building on Redi's work, Spallanzni showed in 1765 that broth, boiled thoroughly and hermetically sealed, remained sterile. It was not, however, until **Pasteur**'s time a hundred years later that the idea of spontaneous generation was finally abandoned. On the basis of animal experimentation, Spallanzani disputed the traditional opinion that digestion was a kind of cooking by stomach heat, counter-arguing that gastric juice constituted the key digestive agent. He was the first to observe blood passing from arteries to veins in a warm-blooded animal, the chick. He successfully artificially inseminated amphibians, silkworms and a spaniel bitch. A staunch preformationist in embryology, he was convinced the ovum already contained all the organs later materializing in the embryo. He also experimented on the sensory systems of bats, concluding that, even if blinded, bats could still catch insects and fly well enough to avoid even small objects. He was also fascinated by electricity, especially electric fish.

SPEDDING, Frank Harold
(1902–84)

American nuclear scientist, born in Hamilton, Ontario. He studied chemical engineering at the University of Michigan and took his doctorate at the University of California at Berkeley. He spent his working life at Iowa State University at Ames. In 1942 he was co-opted to the Manhattan Project at the University of Chicago—with **Fermi**, **Seaborg** and others—to develop the first atomic bomb. In the laboratory at Ames he found a way to separate the fissile isotope of uranium, uranium-235, from its other isotopes in sufficient quantities for the core of the first atomic pile. He and his co-workers produced 6 tons of uranium in the form of large pellets ('Spedding's eggs') which were dropped into matching holes in the graphite core. The pile went critical on 2 December 1942, in the first man-made nuclear chain reaction. After World War II, Spedding looked for cheaper ways of separating the lanthanides, the group of 'rare earth metals' which have atomic weights from 57 to 71. These elements have many industrial uses but are chemically very similar and therefore difficult to isolate. To overcome the problem, Spedding developed the ion-exchange chromatograph: a mixture of substances is passed through very fine particles of resin in a column and the different substances separate out into different bands. He used the same technique to separate the actinides, heavy metals of atomic weights of 89–103 with properties similar to the lanthanides. He died in Ames.

SPEMANN, Hans
(1869–1941)

German zoologist, born in Stuttgart. Educated in Stuttgart and Heidelberg, he became professor at Rostock (1908–14), Director of the Kaiser Wilhelm Institute of Biology in Berlin (1914–19) and professor at Freiburg (1919–35). He was an experimental embryologist and discovered the 'organizer function' of certain tissues during development. He studied the early stages of newt development and showed that if presumptive skin tissue (ectoderm) is transplanted into presumptive neural tissue, for example, it develops into the latter and not into skin. In other words, the fate of embryonic cells is not programmed at an early stage but rather by the tissues that they are in contact with. He wrote *Embryonic Development and Induction* (1938) and won the Nobel Prize for Physiology or Medicine in 1935.

SPENCE, Peter
(1806–83)

Scottish industrial chemist, born in Brechin. He worked in a grocery in Perth and then in a gasworks in Dundee, meanwhile reading all the scientific books he could find. One of the first people to realize that valuable chemicals could be retrieved from the waste products of gas manufacture, he established a small plant in London, which proved unsuccessful. He then set up at Burgh, in Cumberland, where he concentrated on the production of copperas (ferrous sulphate) and particularly on alum (potassium aluminium sulphate) which had many industrial uses—for example to harden tallow, to remove the grease from printer's blocks, to clear cloudy water, and in the dyeing industry as a mordant. In 1845 Spence took out a patent for a process using shale and sulphuric acid as raw materials, which made the manufacture of alum much cheaper, although it brought serious health hazards in the form of noxious fumes. A few years later Spence moved the plant to Pendleton, near Manchester, where it grew to be the largest alum manufactory in the world. He also began to manufacture other chemicals and to smelt copper at plants in Birmingham and Goole.

SPENCER, Sir (Walter) Baldwin
(1860–1929)

English-born Australian anthropologist and biologist, born in Stretford. In 1884 he graduated in natural sciences from Exeter College, Oxford, and in 1886 he was elected to a fellowship at Lincoln College. Within a year he was made foundation Professor of Biology at Melbourne University, an appointment which he held for 32 years. In 1894 Spencer joined W A Horn's expedition to Central Australia, where he met the telegraphist Francis James Gillen. Gillen had become intimate with the local tribespeople, and Gillen and Spencer collaborated in anthropological studies resulting in a number of invaluable published works, including *Native Tribes of Central Australia* (1889) and *Northern Tribes of Central Australia* (1904). Gillen died in 1912, and in the same year Spencer was appointed Chief Protector of the Aborigines. He was knighted in 1916. He died while on a trip to the world's most southerly settlement, on Tierra del Fuego.

SPENCER, Herbert
(1820–1903)

English philosopher, born in Derby. He had a varied career as a railway engineer, teacher, journalist and subeditor at *The Economist* (1848–53) before devoting himself entirely to writing and study. His particular interest was in evolutionary theory which he expounded in *Principles of Psychology* in 1855, four years before **Charles Darwin**'s *The Origin of Species*, which Spencer regarded as welcome scientific evidence for his own *a priori* speculations and a special application of them. He also applied his evolutionary theories to ethics and sociology and became an advocate of 'social Darwinism', the view that societies naturally evolve in competition for resources and that the 'survival of the fittest' is therefore morally justified. He announced in 1860 a *System of Synthetic Philosophy*, a series of volumes which were to comprehend metaphysics, ethics, biology, psychology, and sociology, and nine of these appeared between 1862 and 1893. He viewed philosophy itself as the science of the sciences, distinguished by its generality and unifying function.

SPENCER JONES, Sir Harold
See **JONES, Sir Harold Spencer**

SPERRY, Roger Wolcott
(1913–)

American neuroscientist, born in Hartford, Connecticut. Educated at Oberlin College, he studied zoology at Chicago University (PhD 1941), then worked as a Research Fellow at Harvard and at the Yerkes Laboratory of Primate Biology (1941–6). He taught at Chicago University (1946–52), and was Hixon Professor of Psychobiology at Caltech from 1954 to 1984. He first made his name in the field of developmental neurobiology, his experiments helping to establish the means by which nerve cells come to be wired up in particular ways in the central nervous system. In the 1950s and 1960s he pioneered the behavioural investigation of split-brain animals, and from 1961 he was able to study human split-brain patients at the White Memorial Medical Centre in Los Angeles. From detailed observations he and his collaborators established that each hemisphere possessed specific higher functions, the left side controlling verbal activity and processes such as writing, reasoning etc; whereas the right side is more responsive to music, face and voice recognition etc. Sperry argued that each hemisphere also contained its own consciousness, and his experiments led him into studies of philosophy and consideration of the mind/brain problem. His work has stimulated new approaches and work in neurology, psychiatry and psychology, as well as the basic neurosciences. He shared the Nobel Prize for Physiology or Medicine in 1981 with **Hubel** and **Wiesel**.

SPITZER, Lyman Jr
(1914–)

American astrophysicist, born in Toledo, Ohio. He was educated at Yale and Princeton, where he was Professor of Astronomy from 1947 to 1979. His interest in energy generation in stars led to his early attempt to achieve controlled thermonuclear fusion, for which he devised a method of 'containing' ionized gas, or plasma, in a magnetic field; the principle continues to form part of experimentation in this area. In 1951, with **Baade**, he suggested that the class of galaxies known as S0 were formed when spiral galaxies collided. With Martin Schwarzschild, he postulated in 1956 the existence of giant molecular clouds in interstellar space long before they were observed. They suggested that these clouds were responsible for inducing random velocities in galactic disc stars, these velocities increasing with the age of the stars. The stability of these clouds when they were far from the galactic plane also indicated that hot gas exists out there. In 1958 Spitzer studied the tidal shock that occurs between cluster stars and the galactic disc; he summarized many of his ideas in *Dynamical Evolution of Globular Clusters* (1987).

SPOTTISWOODE, William
(1825–83)

English mathematician, physicist and printer, born in London. His father was Member of Parliament for Colchester and also a printer. He was educated at Harrow and in mathematics at Balliol College, Oxford, and in 1846 he succeeded his father as head of the printing house of Eyre and Spottiswoode. Spottiswoode did original work on the polarization of light and electrical discharge in rarefied gases. The latter experiments were made with a colossal induction coil (an early form of transformer) made for him by the London instrument-maker, A Apps. With its secondary coil of no less than 280 miles of wire, it was capable of producing discharges with a length of 42 inches. He wrote a series of original memoirs on the contact of curves and surfaces, and the first elementary mathematical treatise on determinants (1851). He was elected FRS in 1871, and served as the Royal Society's President from 1878 until his death. He was buried in Westminster Abbey.

SPRENGEL, Christian Konrad
(1750–1816)

German amateur botanist, born in Brandenburg. He became rector of Spandau, but neglected his duties to make discoveries about the part played in the pollination of plants by nectaries and insects. He was eventually removed from his clerical post for failing to provide even such basic services as sermons on Sundays. He suggested that flowers had adapted specifically to allow for insect pollination. He observed that dichogamy, the floral structure which favours cross-pollination by preventing self-pollination, is extremely frequent. His work influenced entomologists of the time, but became largely neglected until re-examined by **Charles Darwin**. His nephew, Kurt Sprengel (1766–1833), wrote histories of medicine (1803) and botany (1818).

SPRENGEL, Hermann Johann Philipp
(1834–1906)

British chemist and physicist, inventor of the high vacuum pump which bears his name. He was born in Schillerslage, near Hanover, and educated at the universities of Göttingen and Heidelberg. In 1859 he moved to England, later becoming a British citizen. He carried out research at Oxford and in the laboratories of several institutions in London. He mechanized the pump devised by Heinrich Geissler in 1858, making the action of the pump much swifter and more efficient. He described his invention in *On the Vacuum* (1865). Sprengel's pump had far reaching effects; for example, it made possible **Crookes**'s investigations of radiation in a high vacuum (leading to the discovery of the electron by **J J Thomson**), **Ramsay**'s and **Travers**'s work on the rare gases, and **Swan**'s electric light bulbs. Sprengel also developed the pyknometer, a U-shaped vessel for determining the density and expansion of liquids, and researched and wrote extensively on high explosives. He was elected FRS in 1878, and died in London.

SPRUCE, Richard
(1817–93)

English botanist, born in Ganthorpe, near Malton, Yorkshire, where his father was village schoolmaster. He himself became a schoolmaster, first at Haxby, then at St Peter's Collegiate School, York. As a recreation he studied botany, collecting bryophytes and other plants from the North Yorkshire Moors; his first paper (1841) was on the bryophytes of Eskdale. In 1844, when the school at York closed, he decided to make botany his career. From 1845 to 1846 he collected in the Pyrenees, discovering many unrecorded bryophytes. In 1849 Sir **William Jackson Hooker** sent him to South America where, six months later, at Santarem, he met **Wallace**. Spruce spent the following 15 years exploring the Amazon, Orinoco, the Andes and Ecuador, until he returned to England in poor health in 1864. He brought back a collection of more than 7000 plant specimens, maps of three previously unexplored rivers, and vocabularies of 21 Amazonian tribes. His most famous publication was *Notes of a Botanist in the Amazon*, edited by Wallace (2 vols, 1908); he also wrote *The Musci and Hepaticeae of the Pyrenees* (1850), *Palmae Amazonicae* (1869) and *Hepaticae Amazonicae et Andinae* (1884).

SPURR, Josiah Edward
(1870–1950)

American geologist, born in Gloucester, Massachusetts. He was mining engineer to the Sultan of Turkey (1901), geologist in the US Geological Survey (1902) and eventually, Professor of Geology at Rollins College (1930–2). As a result of his work, the age of the Tertiary was estimated at 45 to 60 million years ago. His exploration in Alaska in 1896 and 1898 was commemorated by the naming of Mt Spurr. He also did considerable research on lunar topography and geology, and among other works, he wrote *Geology Applied to Mining* (1904) and *Geology Applied to Selenology* (1944–9).

STAHL, Franklin William
(1929–)

American molecular biologist, born in Boston. He was educated at Harvard and the University of Rochester, and held posts at Caltech (1955–8), the University of Missouri (1958–9) and the University of Oregon, where he became professor in 1970. In 1958 he established with **Meselson** that DNA replicates by a semi-conservative mechanism.

STAHL, Georg Ernst
(1660–1734)

German chemist who developed the phlogiston theory of combustion. He was born in Ansbach, studied medicine at Jena, and in 1687 was appointed court physician to the Duke of Sachsen-Weimar. From 1694 he taught at the newly founded University of Halle and in 1716 became personal physician to Frederick-William I, King of Prussia. Stahl believed that although physiological processes could largely be explained in terms of chemistry, each organism was directed by a life force or 'anima'. This doctrine, which became known as animism, brought him into conflict with his contemporaries who subscribed to the mechanistic explanations prevalent at the time. Stahl was most influential, however, in his attempts to explain combustion. He suggested that when a substance burns it loses a vital essence, which he termed 'phlogiston'. He argued that with metals the process is reversible: metals when heated lose their phlogiston and form a calx, but on further heating the phlogiston rejoins the calx to form the metal again. With organic substances the process is irreversible. Since substances do not burn without air, he argued that air was necessary to absorb the phlogiston. This theory, though erroneous, fitted the experimental facts as they were known at the time and became entrenched in chemical theory. It was only overthrown at the end of the 18th century by the work of **Lavoisier** and others. Stahl died in Berlin.

STANIER, Roger Yate
(1916–)

Canadian microbiologist, born in Victoria, British Columbia. He was educated at Shawnigan Lake School, Canada (where he was profoundly unhappy), and at the University of British Columbia, where he graduated in 1936. He then moved to the USA, obtaining his MA at the University of California at Los Angeles (1940), and his PhD at the University of Stanford (1942), where he worked with **Van Niel**. He was awarded a Guggenheim fellowship to work with Marjorie Stephenson at Cambridge in 1945, and became professor at Berkeley (1947–71), then at the Pasteur Institute. Stanier researched bacterial tryptophan metabolism, and discovered the mandelate pathway and the mechanisms of action of streptomycin. He postulated that the amino acid tryptophan is metabolized to catechol via the intermediates kynurenine and anthranilic acid. He then showed that catechol is degraded to acetyl coenzyme A via the bacterial beta-ketoadipate pathway. He also made major contributions to bacterial chromatophores (coloured pigments)—notably bacteriochlorophyll—and carotenoid pigments. Elected FRS in 1978, he has received many honours, including honorary doctorates and the Carlsberg Medal.

STANLEY, Wendell Meredith
(1904–71)

American biochemist, born in Ridgeville, Indiana. He was educated at Earlham College and Illinois University, where he received his PhD in 1929. He was a Research Fellow in Munich (1930–1) and in 1931 joined the Rockefeller Institute for Medical Research, Princeton, before holding a series of professorships at the University of California from 1940. He was appointed Director of the Virus Laboratory at Berkeley in 1948. After working briefly on zymosterol (yeast sterol) and potassium transport into model cells (1931–2), he isolated the tobacco mosaic virus (1935) using the salt fractionation techniques of **Northrop**, and showed it to contain protein and nucleic acid (1936) — this was described by **Astbury** as 'the most thrilling discovery of the century'. Stanley went on to characterize the physical (shape, birefringence, molecular weight) and chemical (amino acid composition, reactive groups and ribonucleic acid composition) properties of the virus, and with **Fraenkel-Conrat** determined the protein amino acid sequence (1960). Stanley also isolated other plant viruses, compared virus variants using immunological techniques and independently noted that viruses can cause cancer (1949). He shared the 1946 Nobel Prize for Chemistry with Northrop and **Sumner**.

STANLEY, William

(1858–1916)

American electrical engineer, born in Brooklyn, New York City. He patented a carbonized filament incandescent lamp and a self-regulating dynamo which were acquired by Westinghouse's Union Switch and Signal Co (from 1886 Westinghouse Electric). After working for Hiram Percy Maxim (1840–1916), he joined Westinghouse to work on a long-range alternating current system for which one of the principal components to be developed was the transformer. He founded the Stanley Electric Manufacturing Co for the commercial exploitation of alternating current systems.

STARK, Johannes

(1874–1957)

German physicist, born in Schickenhof, the son of a farmer. Educated at Munich, he held teaching posts at Munich, Göttingen, Hanover and Greifswald, before being appointed to chairs at Aachen (1909), Greifswald (1917) and Würzburg (1920). Because of his support for Hitler and his stand against modern theoretical physics which he viewed as a 'Jewish science', he was appointed as President of the Physikalisch Technische Reichsanstalt at Charlottenburg in 1933, and in the following year, of the Notgemeinschaft der Deutschen Wissenschaft (later renamed Deutsche Forschungsgemeinschaft). However, he did not hold these posts for long because of his quarrelsome nature and internal political struggles, and he retired in 1936. In 1947 he was sentenced by a German denazification court to four years in a labour camp. In his early years, Stark made important contributions in physics. He discovered the 'Stark effect' concerning the splitting of spectrum lines by subjecting the light source to a strong electrostatic field, and also the **Doppler** effect in 'canal rays' (rays of positively charged particles emitted from the anode in discharge tubes). He argued that these phenomena reinforced **Einstein**'s theory of special relativity and **Planck**'s quantum theory. He was awarded the Nobel Prize for Physics in 1919. In the 1920s he turned vehemently against both relativity and quantum physics.

STARLING, Ernest Henry

(1866–1927)

English physiologist, born in London. He qualified in medicine in 1889 from Guy's Hospital, where he was then appointed lecturer in physiology, and the following year began a lifelong professional and personal association with **Bayliss**, later his brother-in-law. He moved to University College (Jodrell Professor of Physiology, 1899–1923; Foulerton Research Professor of Physiology, 1923–7) and with Bayliss began a series of experiments on the nervous control of the viscera. In the course of this they discovered the pancreatic secretion *secretin* (1902), and for this and similar chemical messengers they coined the word 'hormone'. His studies of cardiovascular physiology did much to elucidate the physiology of the circulation and the mechanisms of cardiac activity, still known today as 'Starling's law of the heart', and his work on capillary function gave rise to 'Starling's equilibrium' which equated intravascular pressure and osmotic forces across the capillary membrane. During World War I he was director of research at the Royal Army Medical Corps College, and was associated with research on poison gases, and chaired the Royal Society's Food Committee. He wrote many influential texts, including *Principles of Human Physiology* (1912).

STAS, Jean Servais

(1813–91)

Belgian chemist, born in Louvain. He qualified as a doctor at Louvain and in 1837 moved to Paris. Here he investigated carbonic acid and other organic substances with **Dumas**. In 1840 he was appointed to the Chair of Chemistry at the Military School in Brussels, retiring due to ill-health in 1868. Stas's principal work was on atomic weights. Following **Dalton**'s atomic theory, **Prout** had suggested that atomic weights are integer multiples of the atomic weight of hydrogen, and that any deviations were the result of experimental errors. For more than 20 years,

Stas laboured to make accurate determinations of the atomic weight of carbon, nitrogen, chlorine, sulphur, potassium, sodium, lead and silver by chemical methods, setting up chemical reactions and weighing the reagents and their products, and proved conclusively that Prout's proposal was incorrect. The reason for this discrepancy in weights was not understood until **Soddy** discovered isotopes and **Chadwick** discovered the neutron. Stas died in Brussels.

STAUDINGER, Hermann
(1881–1965)

German chemist, born in Worms. After graduating from the local gymnasium, he studied chemistry at the University of Halle but soon transferred to the Technical University at Darmstadt. He obtained his doctorate at Halle in 1903 for a study of malonic esters. He then became assistant to Johannes Thiele in Strassburg, where he discovered keten. On being appointed assistant professor in the Technical University at Karlsruhe he began research on the structure of rubber, and in 1910 found a new and simpler way to synthesize isoprene, the basic unit of rubber. In 1912 he succeeded **Willstätter** at the Federal Institute of Technology in Zürich, where he worked on the synthesis of natural products. After World War I he returned to his work on the structure of rubber and propounded the view that rubber consists of macromolecules rather than aggregates, a suggestion that caused considerable controversy. This controversy continued until 1926, when X-ray crystallographic evidence largely confirmed Staudinger's view. During the 1930s he started the study of complex biological macromolecules and in the 1940s he turned to molecular biology. He was awarded the Nobel Prize for Chemistry in 1953 for his discoveries in the field of macromolecular chemistry. A research institute was established for Staudinger at the University of Freiburg in the 1940s, and he remained there until his retirement in 1956.

STEARN, William Thomas
(1911–)

English botanist, bibliographer and horticulturist, born in Cambridge. His early career was as an apprentice antiquarian bookseller at Cambridge (1929–32) and as librarian of the Royal Horticultural Society (1933–41, 1946–52); the latter post was broken by military service during World War II in Britain, India and Myanma (Burma). From 1952 to 1976 he was a botanist at the British Museum (Natural History), London; now retired, he is still an extremely active consultant. He is the world's foremost authority on **Linnaeus**, with an encyclopedic knowledge of his life and works. His immense output (some 440 works to date) includes the books *Botanical Latin* (1966), *Gardener's Dictionary of Plant Names* (1972), *Peonies of Greece* (1984, with **Peter Davis**) and *Flower Artists of Kew* (1990). His botanical researches have ranged widely, including studies of *Symphytum*, *Vinca*, *Epimedium* and numerous monocotyledonous genera such as *Ornithogalum* and the difficult onion genus *Allium*. Stearn has also published many papers on the history and bibliography of botany.

STEBBINS, George Ledyard
(1906–)

American botanist, born in Lawrence, New York. He studied biology at Harvard University and spent his career at the University of California, at Berkeley (before 1937–50) and Davis (1950–73), where he established the department of genetics. He was the first to apply modern ideas of evolution to botany, as expounded in his *Variation and Evolution in Plants* (1950). From the 1940s he used artificially induced polyploidy (the condition of having more than twice the basic number of chromosomes) to create fertile hybrids; this technique is of value both in taxonomy and in plant breeding. His other books include *Processes of Organic Evolution* (1966), *The Basis of Progressive Evolution* (1969) and *Flowering Plants: Evolution Above the Species Level* (1974). In the latter he proposed a new classification system of flowering plants, closely modelled on that of **Cronquist** but with some modifications.

STEBBINS, Joel
(1878–1966)

American astronomer, born in Omaha, Nebraska. He received his first degree at the University of Nebraska and his doctorate at Lick Observatory. He was appointed as an astronomer on the staff of the University of Illinois and became Professor of Astronomy in 1913. In 1922 he moved to the University of Wisconsin and became Director of the Wasburn Observatory. Soon after the invention of the photocell, in 1911, Stebbins applied electronic photometers to astronomical sources and showed that they were much more accurate for brightness measurement than methods that relied on the eye or photography. In the 1930s he used photoelectric techniques to measure the way in which the dust and gas in the galaxy affected the transmission of starlight, causing a reddening effect. The magnitude of this interstellar reddening helped define the structure and size of the galaxy. With Albert

Whitford he introduced a six-colour photometry system, extending from the ultraviolet to the infrared, which he applied with considerable success to the study of galaxies.

STEENSTRUP, Johannes Iapetus Smith

(1813–97)

Danish zoologist, born in Vang, Norway, and educated in Aalborg and at the University of Copenhagen. From 1846 to 1873 he was Professor of Zoology at Copenhagen and Director of the Zoology Museum there. He is best known for his pioneering studies on cephalopods, but also worked on many marine invertebrates and achieved fame for his demonstration of an alternation of sexual and asexual generations in certain animals. He was a founder of scientific archaeology through his work on the animal and plant remains in Danish peat bogs, recognizing significant climatic changes in prehistoric times and explaining the origin of coastal shell mounds as Stone Age middens. He also wrote on the distribution and extinction of the great auk.

STEFAN, Josef

(1835–93)

Austrian physicist, born near Klagenfurt. He was a brilliant experimenter who became Professor of Physics at Vienna in 1863 after being a school teacher for seven years. In 1866 he was appointed Director of the Institute for Experimental Physics founded in Vienna by **Doppler** in 1850. In 1879 he proposed Stefan's law (or the Stefan–Boltzmann law), that the amount of energy radiated per second from a black body is proportional to the fourth power of the absolute temperature. He used this law to make the first satisfactory estimate of the Sun's surface temperature. He also designed a diathermometer to measure heat conduction, worked on the kinetic theory of heat and on the relationship between surface tension and evaporation (1886).

STEIN, William Howard

(1911–80)

American biochemist, born in New York City. He studied at Harvard and Columbia University, and joined the staff of the Rockefeller Institute, where he became Professor of Biochemistry in 1954. Throughout his career he collaborated closely with **Moore**, particularly in the development of a column chromatographic method for the sequential elution, identification and quantification of amino acid mixtures derived from the hydrolysis of proteins or from physiological tissues. By determining the amino acid composition of bovine pancreatic ribonuclease (1954–6), they complemented the structural studies of **Anfinsen** on this enzyme. They also automated all the steps for the analysis of the base sequence of RNA on a small sample (1958), and studied a novel protease from streptococcus (discovered by the English microbiologist Stuart D Elliott) showing that although this enzyme possesses a similar specificity and catalytic site organization (including a so-called 'essential thiol group') to the plant protease papain, other aspects of the molecular structure are quite different. This was the first example of convergent evolution — two enzymes of similar function arising by different evolutionary paths. Stein, Moore and Anfinsen shared the Nobel Prize for Chemistry in 1972.

STEINBERGER, Jack

(1921–)

German-born American physicist, born in Bad Kissingen. He moved to the USA in 1935, and was educated at the University of Chicago, where he received his PhD in 1948. He held professorships at Columbia University from 1950 to 1972, and from 1968 to 1986 was a staff member at CERN, the European centre for nuclear research in Geneva, where he was a director from 1969 to 1972. With colleagues, he proved the existence of a neutral pion by observing the coincidence of the two gamma rays from its decay at the Berkeley synchroton, and measured the spin and parity of the charged pion. Studies of muons had shown that they behaved like heavy electrons. However, Steinberger showed that, unlike the electron, the muon decay involves not one but two neutrinos. To explain this it was proposed that the muon is not simply a heavy electron but has its own identity, the 'electron charge' being carried by the electron neutrino and the 'muon charge' being carried by the muon neutrino. In 1960 **Schwartz** published a paper proposing an experiment that could establish the existence of the two distinct neutrinos; it was performed at Brookhaven by **Lederman**, Schwartz, Steinberger and their collaborators, and in 1962 they announced that they had observed 20 muon events proving the existence of the two distinct neutrino types. Steinberger went on to study CP violation and carry out further neutrino experiments. He was spokesman for the ALEPH collaboration, one of the four large groups of physicists that built an experiment to run at the Large Electron–Positron (LEP) collider at CERN. Since 1989 this has been precisely testing the electroweak model experimentally. Steinberger was awarded the 1988

Nobel Prize for Physics jointly with Lederman and Schwartz.

STEINER, Jakob
(1796–1863)

German–Swiss geometer, born in Utzendorf. He became professor at Berlin in 1834, and pioneered 'synthetic' geometry, particularly the properties of geometrical constructions, ranges and curves. He possessed tremendous powers of visualization, but his papers were renowned for their obscurity and he frequently withheld proofs, a practice which earned him the knickname of the 'celebrated sphinx'. He also published many ingenious proofs (not all rigorous) of the obvious but elusive result that the circle encloses the greatest area of all curves of a given length.

STEINMETZ, Charles Proteus, originally Karl August Rudolf
(1865–1923)

German-born American electrical engineer, born in Breslau (now Wrocław, Poland). Educated there and at the Technical High School, Berlin, he was forced to leave Germany in 1888 due to his socialist activities and he emigrated to the USA in 1889. He began work in a small electrical factory which was taken over by the General Electric Company in 1893, and he worked for that company for the rest of his life. A hunchback from birth, he was rapidly recognized as a brilliant theoretician, yet was sufficiently practical to be granted over 200 patents for his inventions. Perhaps his finest, and certainly one of his earliest achievements was to work out in complete detail, using complex numbers, the mathematical theory of alternating currents. This enabled the design of AC machines to be made more efficient, and consolidated the victory of AC over DC gained by **Tesla** in fierce competition with **Edison**. In addition to his work for General Electric, he was Professor of Electrical Engineering and later Professor of Electrophysics at Union College, Schenectady, from 1902. Among his many other discoveries were the phenomenon of magnetic hysteresis and a method of providing protection against lightning for high-power transmission lines.

STENO, Nicolaus also known as Niels Stensen
(1638–86)

Danish physician, naturalist and theologian who made major advances in anatomy, geology, crystallography, palaeontology and mineralogy. Born a Protestant in Copenhagen, he converted to Catholicism and settled in Florence. He was appointed personal physician to the Grand Duke of Tuscany in 1666 and Royal Anatomist at Copenhagen in 1672. He became a priest in 1675, and gave up science on being appointed Vicar-Apostolic to North Germany and Scandinavia. He is buried in the crypt of the Medici in Florence. As a physician he discovered Steno's duct of the parotid gland, and investigated the function of the ovaries. A passionate anatomist, he showed that a pineal gland resembling the human is found in other creatures; he used this to challenge **Descartes**'s claim that the gland was the seat of the uniquely human soul. Steno's examination of quartz crystals disclosed that, despite differences in the shapes, the angle formed by corresponding faces is invariable for a particular mineral. This constancy (Steno's law) follows from internal molecular ordering. Steno is perhaps best remembered for his contributions to geology and palaeontology. Having found fossil teeth far inland closely resembling those of a shark he had dissected, he championed the organic nature of fossils against those who believed they were 'sports of nature'. On the basis of his palaeontological views, he also contended that sedimentary strata were laid down in former seas. He sketched what are perhaps the earliest geological sections.

STEPTOE, Patrick Christopher
(1913–88)

English gynaecologist and reproduction biologist, born in Witney. He was educated in London at King's College and St George's Hospital Medical School. Following military service, he specialized in obstetrics and gynaecology, becoming senior obstetrician and gynaecologist at the Oldham Hospitals in 1951. In 1980 he became Medical Director of the Bourn Hall Clinic in Cambridgeshire. He had long been interested in laparoscopy (a technique of viewing the abdominal cavity through a small incision in the umbilicus) and in problems of fertility. He met **Edwards** in 1968, and together they worked on the problem of *in vitro* fertilization of human embryos, which 10 years later resulted in the birth of a baby after *in vitro* fertilization and implantation in her mother's uterus. The ethical issues are still controversial, but the technique has become more common.

STERN, Otto
(1888–1969)

American physicist, born in Sohrau, Germany, and educated at Breslau University where he obtained his doctorate in 1912. He then held posts at Zürich, Frankfurt and Rostock before becoming Professor of Physical Chemistry at the University of Hamburg (1923–33). With the rise of the Nazis he moved to the USA where he became Research Professor of Physics at the Carnegie Institute of Technology in Pittsburgh (1933–45). In collaboration with Walther Gerlach, Stern carried his best-known experiment in 1920–1. By projecting a beam of silver atoms through a non-uniform magnetic field, they showed that two distinct beams could be produced. This provided fundamental proof of the quantum theory prediction that an atom should possess a magnetic moment which can only be oriented in two fixed directions relative to an external magnetic field. For this work he was awarded the Nobel Prize for Physics in 1943. Stern also determined the magnetic moment of the proton.

STEVENS, Nettie Maria
(1861–1912)

American biologist, born in Cavendish, Vermont. She began her career as a librarian, but later entered Stanford University to study physiology. She received a PhD from Bryn Mawr College, Pennsylvania (1903), and was subsequently appointed to research posts there. The college eventually created a research professorship for her, but she died before she could take up the position. Stevens was one of the first to show that sex is determined by a particular chromosome; fertilization of an egg by a sperm carrying the X chromosome will result in female offspring and that Y-carrying sperm will produce a male embryo (a discovery made independently by **Edmund Wilson**). She extended this work to studies of sex determination in various plants and insects, demonstrating unusually large numbers of chromosomes in certain insects and the paired nature of chromosomes in mosquitoes and flies.

STEVIN, Simon
(1548–1620)

Flemish mathematician and engineer, born in Brugge. He held offices under Prince Maurice of Orange, wrote on fortification and book-keeping and invented a system of sluices and a carriage propelled by sails. He was responsible for introducing the use of decimals which were soon generally adopted, and which he had advocated in a somewhat cumbersome notation in his book *De Thiende* (1585). His maxim 'A wonder is not a wonder' was a rallying cry for those who advocated rational experimentation in the new sciences. He wrote influentially on statics and the law of the inclined plane is due to him.

STEWART, Balfour
(1828–87)

Scottish physicist, born in Edinburgh, the son of a tea merchant. He studied at St Andrews and Edinburgh, and became assistant to **Edward Forbes** at Edinburgh and later Director of Kew Observatory (from 1859), and Professor of Physics at Owens College, Manchester (from 1870). He carried out original work on radiant heat (or infrared radiation) and thermal radiation (1858), but unfortunately, this was overshadowed by similar investigations by **Kirchhoff**. Stewart was elected FRS in 1862 and awarded the Royal Society's Rumford Medal. He was one of the founders of spectrum analysis and wrote papers on terrestrial magnetism (explaining both the daily and seasonal variations) and on sunspots.

STEWART, Sir Frederick Henry
(1916–)

Scottish geologist, born in Aberdeen and educated at Aberdeen University. He was a mineralogist with ICI (1941–3) before becoming lecturer in geology at the University of Durham (1943–56). From 1956 to 1982 he was Regius Professor of Geology at the University of Edinburgh. Stewart's principal research interests have been in petrology and mineralogy, particularly evaporite deposits as well as fossil fish. He wrote *Marine Evaporites* (1963) and was co-editor of *The British Caledonides* (1963). He was elected FRS in 1964, and knighted in 1974.

STEWART, Ralph Randles
(1890–)

American botanist, born in West Hebron, New York. He studied botany at Columbia University and during 1911–14 taught biology at Gordon College, Rawalpindi. He returned briefly to the USA (1914–17) and obtained a doctorate from Columbia University for a thesis on the flora of western Tibet. From 1917 to 1960 he again taught at Gordon College, and made many plant collecting trips all over Pakistan, Kashmir and western Tibet; his collections number some 60000 specimens. Stewart became the foremost authority on the botany of Pakistan and the western Himalayas, publishing over 40 papers on the subject. His botany teaching inspired many students, including Syed Irtifaq Ali (b 1930) and Eugene Nasir (b 1908) who began the *Flora of Pakistan* (published from 1970 in over 170 parts) using Stewart's collections as a basis. During 1960–81 he worked as a research associate at the University of Michigan Herbarium, Ann Arbor, and published two important works on the flora of Pakistan: *An Annotated*

Catalogue of the Vascular Plants of West Pakistan and Kashmir (1972) which enumerated 5783 taxa, and *History and Exploration of Plants in Pakistan and Adjoining Areas* (1982). He continues to be active and attended celebrations in Karachi to mark his hundredth birthday.

STIBITZ, George Robert
(1904–)

American mathematician and computer scientist, the son of a theology professor at a small college in Dayton, Ohio. Stibitz attended Denison University, a small liberal arts college in Granville, Ohio, and Cornell University, where he was awarded a doctorate in mathematical physics. By 1937 he was working at Bell Telephone Laboratories, where he utilized telephone relays to build a binary adder (**Shannon**, who later joined Bell, had also noticed the correspondence between relays, binary mathematics and symbolic logic). In 1939 Bell Laboratories supported Stibitz in the building of a more sophisticated 'complex number calculator', the Model I. Though reliable and easy to use, it was not programmable or general-purpose, and did not have a memory. Stibitz later designed program-controlled calculators for the military during World War II, but these were soon to be outmoded by the development of the electronic digital computer.

STIRLING, James,
known as **'the Venetian'**
(1692–1770)

Scottish mathematician, born in Garden, Stirlingshire. He studied at Glasgow and Oxford (1711–16), but left without graduating. His first book, on **Isaac Newton**'s classification of cubic curves, was published in Oxford in 1717. He visited Venice at about this time, returned to Scotland in 1724, and then went on to London, where he taught mathematics. From 1735 he was Superintendent of the lead mines at Leadhills, Lanarkshire, and corresponded with **Maclaurin**. His principal mathematical work was *Methodus differentialis* (1730), in which he made important advances in the theory of infinite series and finite differences, and gave an approximate formula for the factorial function, still in use and named after him.

STOCK, Alfred
(1876–1946)

German chemist, famous for his work on the boron hydrides (boranes). In 1912 he discovered diborane (B_2H_6), and over the next 20 years he pioneered the synthesis and isolation of many 'higher' boranes. In 1926 he found that the reaction of diborane with ammonia yielded the compound $B_3N_3H_6$ (borazine), which is known as 'inorganic benzene' due to some of its chemical and physical properties. Stock is also famous for his development of high-vacuum apparatus for handling chemical samples—indispensable for his work on the highly air-sensitive boranes.

STOKES, Sir George Gabriel
(1819–1903)

Irish mathematician and physicist, born in Skreen, Sligo. He graduated in 1841 from Pembroke College, Cambridge, and in 1849 became Lucasian Professor of Mathematics. From 1887 to 1892 he was Conservative MP for Cambridge University. He first used spectroscopy as a means of determining the chemical compositions of the Sun and stars, published a valuable paper on diffraction (1849), identified X-rays as electromagnetic waves produced by sudden obstruction of cathode rays, and formulated Stokes' law expressing the force opposing a small sphere in its passage through a viscous fluid. He is also remembered for his derivation of Stokes' theorem, a useful result which identifies the equivalence of two particular integral operations in vector calculus. He was made a baronet in 1889.

STOMMEL, Hank (Henry Melson)
(1920–92)

American physical oceanographer, born in Wilmington, Delaware, and educated at Yale where he graduated in physics in 1942. As a conscientious objector during World War II he instructed navy students in geometry and celestial navigation. He subsequently moved to Woods Hole Oceanographic Institution to work in physical oceanography (1944–60), and became Professor of Oceanography at Harvard (1960–2) and at MIT (1963–78). He returned to Woods Hole in 1978. Stommel's research covered the broad field of physical oceanography in both theory and experimental work. He investigated the intensification of oceanic currents due to the **Coriolis** force at the western end of a gyre (circuit), and mapped the thermocline (the base of the warmer surface waters), developing the concept of its action as a boundary layer. He found vertical oscillation in the thermocline as it changed depth seasonally, and was one of the first to investigate ocean circulation at great depths. His books included studies of lost islands and the Gulf Stream.

STONE, Francis Gordon Albert
(1925–)

English chemist, born in Exeter. He received his BSc and PhD from Christ's College,

University of Cambridge, and in 1963 he became the first occupant of the Chair of Inorganic Chemistry at Bristol University. He retired in 1990, when he became Robert A Welch Distinguished Professor of Chemistry at Baylor University, Texas. Stone is an organometallic chemist whose research has centred upon synthetic and mechanistic studies of transition metal complexes; he has published more than 650 articles. One principal interest has been in the field of polynuclear metal compounds, ie the metal 'cluster' compounds where metal atoms occupy the vertices of a polyhedron. He developed the chemistry of metal cluster compounds which contain bridging carbene or carbyne groups and has thus been able to study the reactivity of such groups at di- and tri-metal centres. Such studies have important implications for homogeneous catalytic processes. He was editor of the major work *Comprehensive Organometallic Chemistry* which was published in 1984. Elected FRS in 1976, his achievements have been recognized by many learned societies. He was awarded the Davy Medal of the Royal Society in 1989, the Organometallic Chemistry (1972), Transition Metal Chemistry (1979) and Longstaff (1990) medals of the Royal Society of Chemistry, and the American Chemical Society Award for Inorganic Chemistry in 1985. He has held several Royal Society of Chemistry endowed lectureships.

STONEY, George Johnstone
(1826–1911)

Irish physicist, born in Oakley Park. Educated at Trinity College, Dublin, he became Professor of Natural Philosophy at Queen's College, Galway (1852), and was elected FRS in 1861. He calculated an approximate value for the charge on the electron (1874), a term he himself introduced, and made contributions to the theory of gases and spectroscopy.

STORK, Gilbert
(1921–)

American chemist, born in Brussels. After having obtained his baccalauréat in France, he moved to the USA and graduated from the University of Florida in 1942. He obtained his PhD from the University of Wisconsin (1945), was appointed instructor at Harvard and later became assistant professor. He moved to Columbia University as associate professor in 1953 and was appointed to the Eugene Higgins Chair in 1967. Throughout his career he has been concerned with developing synthetic routes to many different types of compounds. He has received many honours including the National Medal of Science (1983).

STRACHEY, Christopher
(1916–75)

English computer programmer and theorist, born in Hampstead, London. Strachey read mathematics and natural sciences at King's College, Cambridge, and then worked on radar at the research laboratories of Standard Telephones & Cables Ltd (STC). After World War II, he became a school teacher at St Edmond's School, Canterbury, and then at Harrow School. While at Harrow he did programming work on **Turing**'s ACE at the National Physical Laboratory, and on the Manchester University Mark I built by **Kilburn** and Sir **Frederic Calland Williams**. In 1952 he was recruited by Lord Halsbury as adviser to the National Research Development Corporation (NRDC), which placed him at the heart of the developing British computer industry. He ran a private consultancy (1959–65) and was appointed to a personal chair at Oxford in 1971. Recognized as one of the foremost computer architects and logicians of his day, Strachey's influence was apparent in the design of several British computers, such as the Ferranti Pegasus, and in many innovative concepts. He made significant contributions in the areas of time-sharing, whereby users at several consoles could enjoy the facilities of one computer, and in denotational semantics, which sought to understand the meaning of computer languages in a mathematical way.

STRASBURGER, Eduard Adolf
(1844–1912)

German botanist, born in Warsaw. He studied botany in Paris, Bonn and Jena and spent his career at Jena (1869–80) and Bonn (1880–1912). He studied the alternation of generations in plants, the embryo sac found in gymnosperms and angiosperms, and double fertilization in angiosperms. In his book *Cell Formation and Cell Division* (1875) and its later editions he laid down the basic principles of cytology, the study of cells, for which he made Bonn the world's leading centre. His work did much to show that mitosis (normal somatic cell division) in plants is a process essentially similar to that described for animal cells by **Beneden** and others. He observed nuclear fusion in the ovules of gymnosperms and angiosperms, and remarked on the formation of the polar nucleus in their egg-cells. He introduced the terms haploid and diploid to describe respectively the halving and doubling of chromosome numbers in plant generations. Strasburger's *Textbook of Botany for Universities*, written with other botanists (including **Andreas Schimper**) under his guidance, is a classic, much used and widely translated in over 30 editions from 1894 onwards. In its updated form, it is still currently in print.

STRATO, or STRATON, of Lampsacus

(d c.269 BC)

Greek philosopher, the successor to **Theophrastus** as the third head of the Peripatetic School (from about 287 to 269 BC) which **Aristotle** founded. His writings are lost, but he seems to have worked mainly to revise Aristotle's physical doctrines. He had an original theory about the void, its distribution explaining differences in the weights of objects. He also denied any role to teleological, and hence theological, explanations in nature, which led naturally to the position David Hume dubbed 'Stratonician atheism'—the universe is ultimate, self-sustaining and needs no further external or divine explanation to account for it.

STROMEYER, Friedrich

(1776–1835)

German chemist, born in Göttingen, and educated there and in Paris. He taught at Göttingen from 1802, becoming Professor of Chemistry in 1810. He was one of the first teachers to insist that his students had opportunities for practical work and **Leopold Gmelin** and **Bunsen** were among his pupils. He was also the inspector of apothecaries for Hanover. Stomeyer was a noted mineralogist, and in 1817 he discovered cadmium, a silvery white metal element akin to tin, in a sample of zinc carbonate. A rare metal, it is important today because cadmium rods are used to absorb excess neutrons in nuclear reactors. Stromeyer died in Göttingen.

STRÖMGREN, Bengt

(1908–87)

Danish astronomer, born in Götelorge, the son of Elis Strömgren, Director of the Copenhagen Observatory and a distinguished astronomer in the field of classical mathematical astronomy. Bengt studied astronomy under his father's tuition at the University of Copenhagen, and atomic physics and quantum theory under **Niels Bohr** in the nearby Institute of Theoretical Physics. The close cooperation between father and son resulted in their joint publication of an outstanding textbook on astronomy in 1933. In 1936 Strömgren joined the second **Otto Struve** at Yerkes Observatory in the USA in work on the physics of stellar atmospheres and the properties of interstellar gas. His most outstanding work in this field concerns the physics of ionized gas clouds known as HII regions surrounding hot stars. For the period of World War II he returned to Denmark where in 1940 he succeeded his father as Director of the Copenhagen Observatory, but after the war he returned to the USA as Director of Yerkes Observatory and professor at the Princeton Institute for Advanced Studies. In 1967 he returned finally to Denmark where he was granted the palatial residence at Carlsberg formerly occupied by Bohr. In these later years his work was concerned with problems of stellar composition and its correlation with ages of stars. Strömgren was awarded many honours from universities and academies the world over, including the Gold Medal of the Royal Astronomical Society (1962), and he acted both as General Secretary and as President of the International Astronomical Union.

STRUVE, Otto

(1897–1963)

Russian-born American astronomer, born in Kharkov, the son of Ludwig Struve, Professor of Astronomy at the University of Kharkov, and grandson of **Otto Wilhelm Struve**. He was educated at the University of Kharkov, where he graduated in 1919 after his studies were interrupted (1916–18) by service in World War I. In 1919 he joined the White Army in opposition to the revolution and after its defeat escaped into exile, suffering considerable privations before making contact with American astronomers and being offered a post at the Yerkes Observatory in 1921. His family, including his father, perished in the revolution. In 1932 he was appointed Director of Yerkes Observatory and in 1939 he founded the McDonald Observatory of the University of Texas, with charge of both observatories until 1947. He was head of the department of astronomy at the University of Chicago (1947–50) and Director of the Leuschner Observatory at the University of California from 1950 until his death. He was also the first Director of the National Radio Astronomy Observatory (1959–62), and held an appointment at the Princeton Institute for Advanced Studies. Struve was principally a stellar spectroscopist, who performed an immense volume of observational work on stars of various types, on the interstellar medium and on gaseous nebulae. He attracted as staff members or as visiting collaborators many of the world's leading astronomical spectroscopists. He was awarded the Gold Medal of the Royal Astronomical Society in 1944, the fourth member of the family in successive generations to receive this honour, and was President of the International Astronomical Union from 1952 to 1955.

STRUVE, Otto Wilhelm

(1819–1905)

Russian astronomer, born in Dorpat (now Tartu), son of **Wilhelm Struve**, Director of Dorpat Observatory. He entered the University of Dorpat (1835) where he studied under his father, and became assistant at the Pulkova Observatory on his father's appointment as its director in 1839. He remained on the staff of the Pulkova Observatory for the rest of his working life, first as assistant (1839–45), then as assistant director (1845–62) and as director in succession to his father (1862–89). Continuing his father's researches on double stars with Pulkova's 15 inch refractor, he discovered 500 new pairs. His own most important studies were his determination of the constant of precession and of the solar motion through space (1841) for which he was awarded the Gold Medal of the Royal Astronomical Society in 1850. He took part in international projects such as the transits of Venus (1874) for which he organized 31 expeditions within and beyond the Russian empire. The Struve dynasty of astronomers included Otto's sons, (Karl) Hermann (1854–1920), who also received the Gold Medal of the Royal Astronomical Society (1903) and became Director of the Königsberg (1895–1904) and the Berlin-Babelsberg (1904–20) observatories, and (Gustav Wilhelm) Ludwig (1858–1920), Director of Kharkov Observatory (1897–1919) and father of the second **Otto Struve**.

STRUVE, (Friedrich Georg) Wilhelm

(1793–1864)

German-born Russian astronomer, born in Altona near Hamburg, and educated until the age of 14 at the gymnasium where his father was rector. From 1809 to 1813 he studied at the University of Dorpat (now Tartu) in Estonia (then part of the Russian empire), and he was later appointed Professor of Mathematics and Astronomy there (1816–39), becoming also Director of Dorpat Observatory in 1818. At Dorpat he carried out a major programme of double star observations, published in a fundamental catalogue of 3112 double stars entitled *Micrometria Mensurae* (1837). In 1837 he also measured the parallax of the star Vega, one of the three astronomers (the others being **Bessel** and **Henderson**) who in that year first succeeded in making such an observation. In 1835 Struve was summoned by the Russian emperor Nicholas I to superintend the building and equipping of a new observatory at Pulkova near St Petersburg, completed in 1839 with Struve in charge. The lavishly endowed and magnificently equipped observatory became the astronomical capital of the world. Struve's astronomical researches at Pulkova included work on the structure of the Milky Way (1847). He also supervised a huge geodetic survey, completed in 1860, extending from the Baltic to the Caucusus along an arc of meridian through Dorpat. In 1862 he handed over the directorship of Pulkova to his son and assistant, **Otto Wilhelm Struve**, the first of several astronomers among his descendents. He was awarded the Gold Medal of the Royal Astronomical Society in 1827 for his early work on double stars.

STURGEON, William

(1783–1850)

English scientist, born in Whittington, North Lancashire. Initially he followed in his father's trade as shoemaker, but then enlisted in the Royal Artillery and was stationed at Woolwich, where he studied science in his free time. He became a bootmaker in Woolwich (1820), and was appointed a lecturer at the East India Company Royal Military College of Addiscombe (1824) and of the short-lived Adelaide Gallery of Practical Sciences (1832), before becoming Superintendent of the Royal Victoria Gallery of Practical Sciences in Manchester (1840). He ended his career as an itinerant lecturer, and for his services to science was awarded an annuity by the government. He constructed the first practical electromagnet (1825), the first moving-coil galvanometer (1836) and various electromagnetic machines. His *Annals of Electricity* (1836) was the first journal of its kind in Britain.

STURM, Jacques Charles François

(1803–55)

French mathematician, born in Geneva. He discovered the theorem named after him concerning the location of the roots of a polynomial equation. With his friend **Liouville**, he also did important work on linear differential equations. In 1826 he measured the velocity of sound in water by means of a bell submerged in Lake Geneva.

STURTEVANT, Alfred Henry

(1891–1970)

American geneticist, born in Jacksonville, Illinois. He developed an enthusiasm for heredity through devising pedigrees for his father's farm horses, and later became a student of genetics under **Thomas Hunt Morgan** at Columbia University. He received his BA in 1912 and a PhD in 1914. From 1928 he spent his career at Caltech, as Professor of Genetics (1928–47) and Professor of Biology (1947–62). As an undergraduate, Morgan had suggested to him that genes which are far apart on the same chromosome are more likely to be separated by the mechanism of recombination or 'crossing-over'; crossing-over occurs when there is a break in one chromosome which then attaches, or recombines with another chromosome. Using this idea, Sturtevant drew up the first chromosome map of the fruit-fly *Drosophila* in 1911. Later, as part of Morgan's 'fly room' group, he provided the mathematical background for genetic mapping experiments on *Drosophila*. Together with Morgan, **Hermann Müller** and C B Bridges, he established the basis for the chromosomal theory of heredity in *The Mechanism of Mendelian Inheritance* (1915). He also wrote *A History of Genetics* (1965).

SUESS, Eduard

(1831–1914)

Austrian geologist, born in London, the son of a German wool merchant of Jewish extraction. His parents moved to Prague (1843) and then to Vienna (1845) where, after a spell as an assistant in the geological department of the Royal Natural History Museum (1851–7), he rose to great eminence at the university (1857–1901), becoming assistant professor and Professor of Geology. The greater part of his life was devoted to the study of the evolution of the features of the Earth's surface, particularly the problem of mountain building, which presented itself to his mind during his many excursions into the eastern Alps. He also focused attention on the volcanic islands and associated deep-sea trenches in the Pacific. His theory that there had once been a great supercontinent made up of the present southern continents was a forerunner of modern theories of continental drift. His four-volume book *Das Antlitz der Erde* (1885–1909; translated as *The Face of the Earth*, 1904–10) was his most important contribution, ranking alongside **Lyell**'s *Principles of Geology* and **Charles Darwin**'s *Origin of Species* in significance. A man of varied interests and enthusiasms, he was a Radical politician, an economist, an educationalist and a geographer, and sat in the Austrian Lower House.

SUGDEN, Samuel

(1892–1950)

English physical chemist, born in Leeds. After he studied chemistry at the Royal College of Science, London, his career was interrupted by World War I, in which he served briefly in the Royal Army Medical Corps and then as a chemist at Woolwich Arsenal (1916–19). In 1919 he joined Birkbeck College, London, as a lecturer and he later became Professor of Physical Chemistry there (1932–7). He was then appointed Professor of Chemistry at University College London, and remained in that post until his death. During World War II he held various scientific posts in the war effort. Sugden is mainly remembered for devising (in 1924) a function of the molecular volume and surface tension of a liquid which he called the 'parachor'. This was considered to be the molecular volume measured at a standard internal pressure. The parachor appeared to have both additive and constitutive properties in terms of molecular structure, and it was hoped that it would be an important tool in determining the structures of organic molecules. For some years the parachor seemed to be useful, as detailed in Sugden's book *The Parachor and Valency* (1930), but ultimately it did not live up to expectations. Sugden's later work involved studies of paramagnetism in inorganic and organic chemistry, measurements of dipole moments of organic molecules, and pioneering applications of radioactive isotopes in the investigation of reaction mechanisms. He was elected FRS in 1934.

SUMNER, James Batcheller

(1887–1955)

American biochemist, born in Canton, Massachusetts. Educated at Harvard, he became assistant professor (from 1914) and then Professor of Biochemistry (1929–55) at Cornell University, and Director of the Laboratory of Enzyme Chemistry (1947–55). In 1926 he was first to crystallize (the then ultimate criterion of purity) an enzyme (urease), and demonstrated its protein nature. He then determined its kinetic and chemical properties, showing a dependence on reactive sulfhydryl groups on the protein. He raised and purified antibodies to urease (1933–4), and purified plant antibody-like globulins (agglutinins) from Jack bean meal (canavulin and concanavulins A and B) in 1938, thereby establishing a firm basis for the serological investigation of proteins. He also purified enzymes important for carrying out oxidative processes in the body (peroxidase from fig sap, catalase and horseradish peroxidase). These enzymes contain the non-protein components haem and iron, and Sumner investigated the function of these components by the effects various modifications had on the catalytic activity of the enzymes. He introduced the name 'monoamine oxidase' (an important pharmacological enzyme), used potato phosphorylase to prepare glucose 1-phosphate (1944), and isolated bean lipoxidase and rhodanese (1945) among endless other achievements. He shared the 1946 Nobel Prize for Chemistry with **Northrop** and **Wendell Stanley**.

SUN SSU-MO (Sun Simo in *pinyin* romanization)

(c.581–c.682 AD)

Chinese alchemist, born in Huayuan. According to some sources he showed great brilliance at school and as an adult went to live as a Taoist (Daoist) recluse on Mount T'ai-po (Taibo). In AD 659 he was summoned to the court by the Emperor and remained there for some years as an informal adviser. There are many legends associated with the life of Sun Ssu-mo, including his great age of 800 years at the time of his death. He was the author of a substantial text *Tan Ching Yao Chueh* (or *Danjing Yao Jue*) on alchemical procedures for the preparation of elixirs of immortality. He also wrote extensively on religious and philosophical topics.

SUOMI, Verner

(1915–)

American meteorologist and space scientist, born in Evaleth, Minnesota, of Scandinavian parents. After graduating from a teachers' training college at Winona (1939), he went to Chicago University to work on instrument development under **Rossby**. He designed an automatic dewpoint recorder which was carried into the stratosphere by balloon and also worked on a sonic anemometer. In 1948 he went to Wisconsin University and received a PhD (1953) for work on boundary layer processes. A net radiometer which he invented was included in the payload for the first American satellite. His greatest achievement was to design the 'spin-scan' camera for geostationary meteorological satellites: this instrument scans about a quarter of the Earth's surface continuously from one position in space and enables meteorological features such as cloud systems and tropical storms to be monitored, and upper winds to be estimated. It was later modified to operate at infrared wavelengths. Suomi also designed a small radio altimeter for use on constant level balloons, extensively used in the Global Atmospheric Research Program (1974). In the early 1970s he played a large part in the formation of the Man–Computer Interactive Data System to deal with various types of meteorological data. He also worked on radiometers to produce vertical temperature profiles from space.

SUTCLIFFE, Reginald Cockroft

(1904–91)

British meteorologist, born in Wrexham, Wales. He studied mathematics at Leeds University and received a DSc at Bangor, North Wales, in 1927. He joined the Meteorological Office (1928) and worked in Malta in collaboration with **Bergeron** (1928–32). He later became Director of Research at the Meteorological Office (1953), and in 1965 he was appointed as first Professor of Meteorology at Reading University. He was President of the World Meteorological Organisation Commission for Aerology (1957–61) and of the Royal Meteorological Society (1955–6). His greatest contribution to meteorology was in the theory of development. Although handicapped by lack of upper air observations, he tackled the development problem in a systematic three-dimensional way. His most famous paper was *A Contribution to the Theory of Development* (1947) in which he used pressure (instead of height) as the vertical coordinate in his equations. This new departure was subsequently widely adopted and the paper led to practical advances in weather forecasting. Sutcliffe was responsible for the introduction of barotropic and baroclinic numerical weather prediction models into the Meteorological Office forecasting routine, and at Reading University he set up the first undergraduate course with meteorology as a principal subject. He was the recipient of the International Meteorological Organisation Prize (1963) and Symons Gold Medal winner (1955).

SUTHERLAND, Earl Wilbur Jr

(1915–74)

American biochemist, born in Burlingame. He studied medicine at the Washington University School of Medicine in St Louis and received his medical degree in 1942. He served an internship before being called up for service as a surgeon in the army. After World War II, Sutherland became an instructor and Associate Professor in the biochemistry department of Washington University. In 1953 he became Director of the Pharmacology Department of the Western Reserve University in Cleveland. Sutherland's research concerned the conversion of glycogen (the energy store in liver and muscle) into glucose, and the stimulation of this process by the hormones glucagon and epinephrine. He showed that a molecule known as cyclic-AMP promotes the activation of phosphorylase, the enzyme responsible for the glycogen–glucose transformation, and proposed that glucagon and epinephrine act by inducing the cell to produce c-AMP. Sutherland had discovered a new principle—the 'second messenger' theory of hormonal action. In 1963 he became Professor of Physiology at Vanderbilt University. He showed that c-AMP acts as second messenger for many mammalian hormones, and was awarded the 1971 Nobel Prize for Physiology or Medicine.

SUTHERLAND, Sir Gordon Brims Black McIvor

(1907–80)

Scottish physicist, born in Watten, Caithness. He was educated at the University of St Andrews and at Cambridge, where he completed his PhD in 1933. He remained at Cambridge as a Fellow and Lecturer of Pembroke College (1935–49), Assistant Director of Research in Colloid Science (1944–7) and Reader in Spectroscopy (1947–9). He then moved to the USA as Professor of Physics at the University of Michigan (1949–56). Returning to the UK in 1956, he took up the directorship of the National Physical Laboratory at Teddington. In 1964 he became Master of Emmanuel College at Cambridge, where he remained until his retirement. His areas of research included infrared and Raman spectroscopy, his interests lying in the structure of molecules. During the 1930s he published widely on the absorption spectra of numerous molecules, including C_2H_2 and N_2O_4, and was able to explain his observations in terms of the vibrational modes of the molecular bonds. He also investigated various triatomic molecules, including O_3, F_2O and NO_2, again successfully identifying their vibrational modes. He was elected FRS in 1949 and knighted in 1960.

SVEDBERG, Theodor

(1884–1971)

Swedish physical chemist, born in Fleräng, near Valbo. He entered the University of Uppsala in 1904 to study chemistry and was associated with that university for the next 45 years. From 1912 to 1949 he was Professor of Physical Chemistry. Although beyond retiring age, he was Director of the Gustaf Werner Institute of Nuclear Chemistry from 1949 to 1967. Svedberg's early work was on colloid chemistry: he devised an improved method of making metal sols and made extensive studies of them using the ultramicroscope. He also investigated radioactivity. In the 1920s, however, his interest in colloids led him to develop the ultracentrifuge as a means of following optically the sedimentation of particles too small to be seen in the ultramicroscope. The earliest ultracentrifuge exerted a force 5 000 times that of gravity, but between 1924 and 1939 the machine was gradually improved to exert a force over 100 000 times that of gravity. The measurements of the molecular weights of proteins which Svedberg made were particularly important, and for his work on the ultracentrifuge he received the 1926 Nobel Prize for Chemistry. During World War II he developed a synthetic rubber, Sweden's supplies of natural rubber being cut-off by the blockade. Svedberg's work at the Werner Institute involved the applications of a cyclotron in medicine, in radiation physics and in radiochemistry. He was made an Honorary Fellow of the Chemical Society in 1923 and elected a Foreign Member of the Royal Society in 1944.

SVERDRUP, Harald Ulrik

(1888–1957)

Norwegian oceanographer and geophysicist, born in Oslo and related to a number of prominent people, including Johan, prime minister of Norway in the 1870s, and Grieg the composer. Complying with family wishes he entered the University of Norway to study languages, but soon left to begin military service, becoming reserve officer in 1908. On his return to university he took up science and in 1911 became assistant to **Vilhelm Bjerknes**, following him to Leipzig to work on atmospheric circulation (1913–8). World War I forced him to return to Norway where he was engaged by Roald Amundsen as chief scientist on the three-year *Maud* expedition to the North Pole. There he made atmospheric, oceanographic, magnetic and ethnographic observations. In 1922 the *Maud* docked in Seattle for repairs and he took the opportunity to work at the Carnegie Institute, interpreting the magnetic observations, and Arctic tidal dynamics. In 1926 he succeeded Bjerknes as Professor of Geophysics at Bergen. Later he took a research position at the Christian Michelsens Institute in Bergen (1931), and took part in a submarine expedition to the North Pole. He moved to California in 1935 to become Director of Scripps Institution of Oceanography, where his precise determinations of tides and wave heights were valuable in the Pacific during World War II. He returned to Oslo in 1948 as Director of the Norwegian Polar Institute. In 1949 he became Professor of Geophysics at the University of Oslo, and later Dean, then Vice-Chancellor. A unit of volume transport and the Sverdrup Islands in Arctic Canada are named after him.

SWALLOW, John Crossley

(1923–)

English physical oceanographer and geophysicist, born near Huddersfield, Yorkshire. He was educated in physics at St John's College, Cambridge (BA 1945), though his studies were interrupted by military service in the East Indies during World War II. He was captivated by **Bullard**'s lectures and later joined the Department of Geodesy and Geophysics at Cambridge, where he was awarded a PhD in 1954. He spent four years aboard HMS *Challenger* conducting seismic refraction experiments. He was then recruited by **Deacon** to work at the National Institute of Oceanography, where he spent the rest of his career (1954–83), concentrating on physical oceanography. Swallow developed a method for measuring deep currents in the ocean using neutrally buoyant floats, which sink to a predetermined depth, drift freely and can be tracked acoustically from a surface ship. This led to cooperative work with **Stommel**, revealing the deep western boundary current in the North Atlantic (1957) and the presence of strong mesoscale eddies at all depths in mid-ocean (1960). Always a practical scientist, he took part in many oceanographic cruises in the North Atlantic and Mediterranean, observing the winter-time formation of deep water and other phenomena, and the Somali current and equatorial circulation in the Indian Ocean. He was elected FRS in 1968.

SWAMMERDAM, Jan

(1637–80)

Dutch naturalist, born in Amsterdam, the son of an apothecary. He studied medicine at Leiden but never practised. Using a simple microscope, he made many observations of a great range of biological material, his drawings being published after his death in the great *Biblia Naturae* (1737–8). He described in great detail the life cycles of a dozen insect types, the most famous being that of the mayfly and the most complex, the honey bee. He classified the insects on the basis of the type of metamorphosis they undergo during their life cycles, the method which is still in general use. Among his many observations was the presence of the butterfly's wing within the pupa, and he correctly surmised that on hatching the wings are expanded by blood pressure. He was first to describe valves in the lymph vessels and the ovarian follicles in mammals. He also carried out ingenious experiments which demonstrated that contracting muscles change their shape and not their volume, in contradiction with the generally accepted view at that time. He published *Historia Insectorum Generalis* in 1669. All of his considerable biological achievements were achieved between 1663 and 1673; thereafter his father withdrew financial support and he succumbed to religious fanaticism under the influence of Antoinette Bourignon.

SWAN, Sir Joseph Wilson

(1828–1914)

British chemist, inventor and industrialist, notable for his achievements in photography, synthetic textiles and electric lighting. He was born near Sunderland and after leaving school at 13 was apprenticed to a druggist. In 1846 he joined John Mawson who had a pharmaceutical business in Newcastle. In 1856 he took out a patent for improving the wet-plate collodion photographic process and the following year he invented high-speed bromide paper, having observed that heat greatly increased the sensitivity of gelatin and silver bromide emulsion. The patent was bought by George Eastman, founder of Kodak, and helped to make photography cheaper and thus widely popular. By 1848 Swan was experimenting with carbonized paper filaments for electric lamps but it was not until **Hermann Sprengel** developed his mechanized air-pump in 1865 that it became possible to achieve the necessary vacuum in the bulb. Swan gave his first successful demonstration, with a thin carbon rod, in 1879. In 1880 he established a small factory in South Benwell, west of Newcastle. Within three years he was manufacturing 10 000 lamp bulbs a week and a number of famous institutions — for example the British Museum — were illuminated by Swan bulbs. In 1883, he amalgamated his business with **Edison** who had been granted a British patent in 1879, to form the Edison and Swan Electric Light Company. Searching for a better filament for his bulbs, Swan dissolved cellulose in acetic acid and extruded it through narrow jets into a coagulating fluid. Soon afterwards **Chardonnet** adapted this process to make rayon, and it was further developed by **Cross** and **Bevan** who, in conjunction with Courtaulds, laid the foundations of the synthetic textile industry. Swan also made significant improvements to lead-plate batteries by designing cellular lead plates which held the lead oxide more securely. He was elected FRS in 1874, knighted in 1904, and received many other honours. He died in Overhill, Warlingham, Surrey.

SWEDENBORG, Emanuel

(1688–1772)

Swedish mystic, theologian and scientist, born in Stockholm. Born to a family soon to be ennobled, Swedenborg studied at Uppsala, later travelling widely, studying engineering, and returning home in 1716 to work for the Royal Board of Mines. He wrote widely on mathematics and technical matters (longitude, docks, decimal coinage, navigation). His *Opera Philosophica et Mineralia* (1734) expressed his metaphysical interests, and huge works on physiology and anatomy followed. He also developed a religio-geological theory of creation. In 1743–4 he underwent a religious crisis (recorded in his *Journal of Dreams*) and in consequence he resigned his scientific post, in order to be free to expound his mystical views.

SWINBURNE, Sir James, 9th Baronet
(1858–1958)

British chemist and industrialist, notable as an electrical engineer and a pioneer of the plastics industry. He was born in Inverness and educated at Clifton College, Bristol. He began work in a locomotive works in Manchester, and in the early 1880s **Swan** sent him to France and the USA to establish factories to manufacture electric lamps. He then managed **Crompton**'s dynamo works and subsequently worked independently as a consultant and inventor. He was a leading authority on the design of dynamos, and developed instruments to measure alternating currents and very low pressures in highly evacuated vessels. He also wrote on thermodynamics from the point of view of engineering design. Interested in the possibilities of making thread from viscose at a very early stage, he worked with **Cross** and **Bevan** and had a share in the company set up with Courtaulds. He formed a syndicate to develop the synthetic material made by the reaction between phenol and formaldehyde, attempting to patent it in 1907, but found he had been anticipated by **Baekeland**, a Belgian chemist working in the USA. Instead Swinburne established a lacquer company in Birmingham. In 1926 this was bought by Bakelite Ltd and Swinburne became president of the British division of Bakelite, a post he retained until 1948. He died in Bournemouth.

SWINGS, Pol
(1906–83)

Belgian astronomer, born in Ransart near Charleroi. He studied mathematics and physics at the University of Liège (1923–7), finishing with a doctorate in a topic in classical astronomy. He joined the staff of Liège University in 1927 becoming professor in 1936, a post which he retained until his retirement in 1975, apart from the years of World War II spent in the USA. His primary interest was in the identification of spectra of atoms and molecules in astronomical bodies, especially comets which before the space era were amenable to analysis only by spectroscopic means. His *Atlas of Cometary Spectra* (jointly with Leo Haser) was published in 1956. A committed internationalist, he initiated the annual Liège astronomical colloquia (1949), and served as President of the International Astronomical Union (1964–7).

SWINTON, Alan Archibald Campbell
(1863–1930)

Scottish electrical engineer and inventor, born in Edinburgh. Educated there and in France, he was interested in all things mechanical and electrical from an early age, linking two houses some distance apart by telephone at the age of 15, only two years after its invention by **Alexander Graham Bell**. In 1882 he began an engineering apprenticeship in the Newcastle works of William George Armstrong, for whom he devised a new method of insulating electric cables on board ship by sheathing them in lead. In 1887 he moved to London and began to practise as a consulting engineer, specializing in the installation of electric lighting. At the same time he continued to act as consultant to Armstrongs, and later became a director of Crompton and Company, and the Parsons Marine Steam Turbine Company. Having been interested in photography since childhood, he was one of the first to explore the medical applications of radiography; he published the first X-ray photograph taken in Britain in *Nature* (23 January 1896), and was soon in demand as a consultant radiographer. In a letter to *Nature* (18 June 1908) he outlined the principles of an electronic system of television, which he called 'distant electric vision', by means of cathode rays — essentially the system in use today. A member of the Institutions of Civil, Electrical and Mechanical Engineers, and of several other scientific bodies, he was elected FRS in 1915.

SYDENHAM, Thomas
(1624–89)

English physician, 'the English **Hippocrates**', born in Wynford Eagle, Dorset. He served in the Parliamentarian army during the Civil War, and in 1647 went to Oxford. There he studied medicine at Wadham College, and was elected Fellow of All Souls College. In 1651 he was severely wounded at Worcester. From 1655 he practised in London. A great friend of such empiricists as **Boyle** and **Locke**, he stressed the importance of observation rather than theory in clinical medicine. Contemptuous of sterile book-learning, he urged doctors to become close observers at the bedside, where they would learn to distinguish specific diseases, and through trial and error, to find specific remedies. He was much impressed with the capacity of Jesuit's bark (the active principle of which is quinine) to cure intermittent fever (malaria), and believed that other such specific treatments might be found. He wrote *Observationes Medicae* (1667) and a treatise on gout (1683), a disease to which he himself was a martyr, distinguished the symptoms of venereal disease (1675), recognized hysteria as a distinct disorder and gave his name to the mild convulsions of children, 'Sydenham's chorea' (St Vitus's dance), and to the medicinal use of liquid opium, 'Sydenham's laudanum'. He remained in London except when the plague was at its peak (1665). He was a keen student of epidemic diseases, which he believed were caused by atmospheric properties (he called it the 'epidemic constitution') which determined which kind of acute disease would be prevalent each year. Acute diseases, he observed, account for about two-thirds of the afflictions of human beings. By the time of his death, his reputation was growing and his works, with their vivid, homespun qualities, and their astute description of diseases, were often reprinted and translated throughout the 18th century.

SYLVESTER, James Joseph
(1814–97)

English mathematician, born in London. He studied at St John's College, Cambridge, became Second Wrangler in 1837 but, as a Jew, was disqualified from graduating. He became professor at University College London (1837), and the University of Virginia (1841–5). Returning to London, he worked as an actuary, and was called to the bar in 1850. He later returned to academic life as Professor of Mathematics at Woolwich (1855–70) and at Johns Hopkins University, Baltimore (1877–83), where he established the first international journal of mathematics in America. Finally he became Savilian Professor at Oxford (1883–94). With **Arthur Cayley** he was one of the founders of the algebraic theory of invariants which became a powerful tool in resolving physical problems. He also made important contributions to number theory. His mathematical style was flamboyant, and he wrote in haste, continually coining new technical terms, most of which have not survived.

SYLVESTER-BRADLEY, Peter
(1913–78)

English micropalaeontologist, born in Pinhoe, Devon, and educated at Reading University. After graduating he commenced research on the major interest of his life, Jurassic ostracodes. He was initially appointed to the staff at Scale Hayne Agricultural College, where he founded a new geology department, and joined the Royal Navy during World War II. He later became a lecturer at Sheffield University. In 1959 he became F W Bennett Professor of Geology at the University of Leicester. Sylvester-Bradley pursued major studies of Mesozoic ostracodes and oysters, and contributed to theories of evolution and the origin of life. He was also an outstanding leader in research, teaching and administration.

SYMONS, George James
(1838–1900)

English climatologist, born in Pimlico. Symons showed an early fascination with the weather, making observations whilst at school and becoming a reporter for the Registrar-General in 1857. He was particularly interested in rainfall and established a network of voluntary observers in 1860. This became the British Rainfall Organisation and under Symons's enthusiasm the number of observers increased rapidly to reach over 3 500 by 1899. In 1863 he founded a circular which later became *Symons' Monthly Meteorological Magazine*. He was interested in instruments, and invented the brontometer for recording the sequence of phenomena in thunderstorms, as well as organizing a comparison of thermometer screens which resulted in the Stevenson screen being accepted as the world standard. He compiled a catalogue of over 60 000 meteorological books and his comprehensive collection was bequeathed to the Royal Meteorological Society, for which he had served as Secretary and President. He was a member of many scientific committees including the Royal Society Krakatoa Committee, and was elected FRS in 1878. On his death the Royal Meteorological Society opened a memorial fund and the biennial Gold Medal provided from the proceeds is the society's highest award.

SYNGE, Richard Laurence Millington
(1914–)

English biochemist, born in Chester. He trained at Cambridge, where he studied the partition of amino acids in solvent mixtures, and joined the Wool Industry Research Association in Leeds (1941–3). Here he collaborated with **Archer Martin** and exploited his Cambridge research in the development of partition chromatography and the counter-current liquid–liquid separation of mixtures (1941), which revolutionized analytical chemistry. They also showed that mild protein hydrolysis is required to determine the amide content of asparagine and glutamine (1941), and developed methods for the analysis of aldehydes and hydroxyacids. Synge and Martin shared the Nobel Prize for Chemistry in 1952 for their work. In 1944 Synge demonstrated the use of powdered cellulose or potato starch packed in columns for separating amino acids, and in 1948 he moved to the Rowett Research Institute, Aberdeen, where he showed that the arbitrary division between proteins above and below 10 000 molecular weight corresponded to the division between dialysability and non-dialysability through cellophane (1949). Around this time he partially determined the structure of the peptide antibiotic gramicidin. He later identified S-methyl L-cysteine S-oxide in cabbage as the compound responsible for the sulphurous smell of boiled cabbage water (1956), and at the time of possible therapeutic interest. From 1967 until his retirement in 1976, he worked at the Food Research Institute in Norwich. He was elected FRS in 1950.

SZENT-GYÖRGYI, Albert von Nagyrapolt

(1893–1986)

Hungarian-born American biochemist, born in Budapest. He lectured at Groningen, where he discovered hexuronic acid (vitamin C) in the adrenal cortex, and at Cambridge, working on this substance with W H Haworth. He became professor at Szeged (1931–45), where he crystallized vitamin C from paprika, and in consequence, vitamin B_2 (riboflavin). He was later appointed professor at Budapest (1945–7), Director of the Institute of Muscle Research at Woods Hole, Massachusetts (1947–75), and Scientific Director of the National Foundation for Cancer Research, Massachusetts (1975). Szent-Györgyi also discovered the reducing system oxaloacetate to malate involved in the **Hans Krebs** cycle (1935), and made important contributions towards understanding muscular contraction; the enzyme ATPase of myosin (1941) and inhibition by actomyosin; myosin cleavage by trypsin into heavy and light meromyosins (1952); glycerinated fibres, allowing study of a physiologically active biochemical system (1948); and muscle relaxation by adenosine triphosphate (ATP) in the absence of calcium (1953). He was awarded the Nobel Prize for Physiology or Medicine in 1937.

SZILARD, Leo

(1898–1964)

Hungarian-born American physicist, born in Budapest. He studied electrical engineering there, and physics in Berlin, working with **von Laue**. In 1933 he fled from Nazi Germany to England, and in 1938 emigrated to the USA, where he began work on nuclear physics at Columbia. In 1934 he had taken out a patent on nuclear fission as an energy source, and on hearing of **Hahn** and **Meitner**'s fission of uranium (1938), he immediately approached **Einstein** in order to write together to President Roosevelt warning him of the possibility of atomic bombs. Together with **Fermi**, Szilard organized work on the first fission reactor, which operated in Chicago in 1942. He then went to Los Alamos to work on the Manhattan Project, leading to the nuclear fission bomb. After World War II he researched into molecular biology in experimental work on bacterial mutations and theoretical work on aging and memory.

T

TAIT, Peter Guthrie

(1831–1901)

Scottish mathematician and golf enthusiast, born in Dalkeith. He was educated at the universities of Edinburgh and Cambridge, where he graduated as Senior Wrangler in 1852. He became Professor of Mathematics at Belfast (from 1854) and Professor of Natural Philosophy at Edinburgh (1860–1901). He wrote on quaternions, thermodynamics and the kinetic theory of gases, and collaborated with Lord **Kelvin** on a *Treatise on natural philosophy* (1867), the standard work on the natural sciences in English for a generation. His study of vortices and smoke rings led to early work on the topology of knots. He studied the dynamics of the flight of a golf-ball and discovered the importance of 'underspin'.

TAKAMINE, Jokichi

(1834–1922)

Japanese-born American chemist, born in Takaoka. He studied chemical engineering in Tokyo and Glasgow, and in 1887 opened his own factory, the first to make superphosphate fertilizer in Japan. In 1890, having married an American, he moved to the USA and set up an industrial biochemical laboratory there. Takamine worked for some time in the laboratory of **John Abel**. In 1898 he published *Testing diastatic substances*, in which he described the iodine test for following the activity of saliva or other fluids in hydrolysing starch to maltose and glucose. He later published *The blood-pressure-raising principle of the suprarenal glands and its mode of preparation* (1901), the first description of adrenaline isolated in crystalline form. After 1905, when **Starling** first used the word hormone to describe the animal body's 'chemical messengers', it was realized that adrenaline, an intravenous injection of which produces an enormous rise in blood pressure, was the first hormone to be isolated in pure form from a natural source.

TAKHTAJAN, Armen Leonovich

(1910–)

Armenian botanist, born in Šuša, Nagorno Karabakh. From 1932 to 1943 he held various posts at Tbilisi, Georgia, and Yerevan Museum and Yerevan University, Armenia, where he was awarded a doctorate (1943) for a dissertation on the evolution and phylogeny of flowering plants. In 1943 he joined the palaeobotanical section of the Institute of Botany of the Academy of Sciences of the Armenian SSR, Yerevan, but later (1954–84) he worked at the Botanical Institute of the Academy of Sciences of the USSR in Leningrad (now St Petersburg), where he was Director of the Department of Floristics, Systematics and Evolution of Higher Plants (1977–84). He has published several versions of a new system of flowering plant classification, the most definitive being *Sistema Magnoliophytov* (1987). This system is complementary to that of **Cronquist**, with whom Takhtajan corresponded for many years; there are many similarities between the two systems although Takhtajan has a narrower concept of the family and recognizes more of them. Other major works include *Flowering Plants: Origin and Dispersal* (1966) and *Floristic Regions of the World* (1978), a unique synthesis of current thinking on world plant geography.

TALBOT, William Henry Fox

(1800–77)

English physicist and pioneer of photography, born in Melbury House, Dorset, and educated at Harrow and Trinity College, Cambridge. In 1839 he announced his invention of photography ('photogenic drawing'), a system of making photographic prints on silver chloride paper, in the same year as the invention of the daguerreotype by **Daguerre**. In 1841 he patented the calotype, the first process for photographic negatives from which prints could be made; he was awarded the Rumford Medal of the Royal Society in 1842. He also discovered a method of making instantaneous photographs, using electric spark illumination (the first use of 'flash photography'), in 1851. His *Pencil of Nature* (1844) was the first photographically illustrated book to be published. He also published works on astronomy and mathematics, and helped to decipher the cuneiform inscriptions at Nineveh. A 16th-century converted barn at the gates of Lacock Abbey in Wiltshire, where Talbot lived from 1833 onwards, is now a museum of his work and equipment run by the National Trust.

TAMM, Igor Yevgenyevich

(1895–1971)

Soviet physicist, born in Vladivostock, the son of an engineer. He was educated at the universities of Edinburgh and Moscow, and taught at Moscow State University (1924–34) before moving to the Physics Institute of the Academy. Together with **Frank** he developed a theory to describe the 'Cherenkov effect' discovered by **Cherenkov**. They demonstrated that this emission of radiation is due to a particle moving through a medium faster than the speed of light in the medium, and related the type of radiation observed to the particle mass and velocity. Tamm shared the 1958 Nobel Prize for Physics with Cherenkov and Frank for this work. The effect has been utilized in particle detectors, allowing the masses and hence identities of particles to be determined.

TANSLEY, Sir Arthur George
(1871–1955)

English botanist, born in London. Educated at Cambridge, he later lectured at University College London (1893–1906), and then at Cambridge from 1906 to 1923. Sherardian Professor at Oxford (1927–37), he founded the precursor (1904) of the Ecological Society (1914), and was founder-editor of the journal *New Phytologist* (1902). A pioneer British plant ecologist, he published *Practical Plant Ecology* (1923) and *The British Isles and their Vegetation* (1939), and contributed to anatomical and morphological botany as well as physiology. He was President of the British Ecological Society in 1913, Chairman of the Nature Conservancy from 1949 to 1953, and was knighted in 1950.

TARSKI, Alfred
(1902–83)

Polish-born American logician and mathematician, born in Warsaw. Educated in Warsaw, he became professor there (1925–39), then moved to the USA at the University of California at Berkeley (1942–68). He made contributions to many branches of pure mathematics and mathematical logic, including the **Banach**–Tarski paradox, which seemingly allows any set to be broken up and reassembled into a set of twice the size. The paradox hinges upon the use of sets which cannot mathematically be said to have a size at all, the non-measurable sets that arose in studies of the theory of integration. He is best remembered, however, for his definition of 'truth' in formal logical languages, as presented in his monograph *Der Wahrheitsbegriff in den Formalisierten Sprachen* (1933, 'The Concept of Truth in Formalized Languages').

TARTAGLIA, originally FONTANA, Niccolò
(c.1500–57)

Italian mathematician, born in Brescia. From the age of about 12, when he was injured by a French soldier during the invasion of his home town and left with a speech impediment, he was given the name Tartaglia, meaning 'stutterer'. He became a teacher of mathematics in several Italian universities, and settled in Florence in 1524. Tartaglia was one of the first to derive a general solution for cubic equations; he disclosed this result to **Cardano** who claimed priority in the discovery (now known as Cardano's formula), and a long controversy followed; the credit for the very first solution of a type of cubic should probably go to Scipione da Ferro, an Italian mathematician of the previous generation. Tartaglia also published an early work on the theory of projectiles, and translated **Euclid**'s *Elements*.

TATUM, Edward Lawrie
(1909–75)

American biochemist, born in Boulder, Colorado. He studied at the University of Wisconsin, receiving a BA in chemistry in 1931 and a PhD in biochemistry in 1934, and taught at Stanford (1937–45, 1948–57), Yale (1945–8) and Rockefeller University, New York (1957–75). Working with **Beadle** on the bread mould *Neurospora*, he demonstrated the role of genes in biochemical processes. They irradiated *Neurospora* spores with X-rays, and then grew them on a variety of nutritional media. When mutant spores did not grow on minimal medium but did grow when an additional nutrient was supplied, Tatum and Beadle suggested that the spore had one or more blocks in the metabolic pathway for that particular nutrient. They formulated the 'one gene, one enzyme' hypothesis, that a single gene codes for the synthesis of one protein. At Yale, Tatum collaborated with **Lederberg** to show that bacteria reproduce by the sexual process of conjunction. All three shared the 1958 Nobel Prize for Physiology or Medicine. In his later years, at the Rockefeller University, Tatum concentrated on the training and education of students.

TAUBE, Henry
(1915–)

Canadian-born American inorganic chemist, born in Neudorf, Saskatchewan. He studied at Saskatchewan University and received his doctorate at the University of California at Berkeley. In 1942 he became an American citizen and subsequently taught at Cornell and Chicago. He was appointed Professor of Chemistry at Stanford in 1962. Using radioisotopes as tracers, Taube devised new methods for studying the transfer of electrons during inorganic chemical reactions in solution. He also proved the hypothesis that metal ions in solution form chemical bonds with water, and showed that when one metal ion replaces another in a solution of a metallic salt, the acid radical forms a temporary bridge by which electrons are transferred from the replacement ion to the original metallic ion. In 1969 Taube and Carol Creutz synthesized a mixed valence cation, a new type of positively charged ion consisting of two atoms of ruthenium each bonded to five molecules of ammonia and separated by a pyrazine ring. They and their colleagues used it to investigate oxidation–reduction reactions in living tissue. Taube was awarded the Nobel Prize for Chemistry in 1983.

TAUSSIG, Helen Brooke
(1898–1986)

American paediatrician, born in Cambridge, Massachusetts. She received her MD from Johns Hopkins University in 1927 and later became the first woman to become a full professor there. Her work on the pathophysiology of congenital heart disease was done partly in association with the cardiac surgeon **Blalock**, and between them they pioneered the 'blue baby' operations which heralded the beginnings of modern cardiac surgery. The babies were blue because of a variety of congenital anomalies which meant that much blood was passing directly from the right chamber of the heart to the left without being oxygenated in the lungs. Taussig was actively involved in the diagnosis and after-care of the young patients on whom Blalock operated, and their joint efforts helped create a new specialty of paediatric cardiac surgery.

TAYLOR, Brook
(1685–1731)

English mathematician, born in Edmonton. He studied at St John's College, Cambridge, and in 1715 published his *Methodus incrementorum*, containing the theorem on power series expansions which bears his name. He also wrote on the mechanics of the vibrating string, and with real insight on the mathematics of the theory of perspective; although one of his books later inspired William Hogarth to his famous engraving of an impossible view, Taylor's books were generally found to be obscure by later mathematicians.

TAYLOR, Sir Geoffrey Ingram
(1886–1975)

English physicist and applied mathematician, born in London. His father was an artist and his mother was related to **Boole** and Sir George Everest. Taylor graduated from Trinity College, Cambridge, in 1908. Except during the world wars, when he provided assistance to the government at Farnborough and elsewhere, he was based in Cambridge throughout his career. A temporary readership in dynamic meteorology in 1911 was followed by six months on a scientific expedition in the North Atlantic. From 1923 to 1952 he was Royal Society Yarrow Research Professor in Physics at the Cavendish Laboratory. His many original investigations on the mechanics of fluids and solids were applied to meteorology, oceanography, aerodynamics and Jupiter's Great Red Spot. A famous series of papers laid out his statistical theory of turbulence in 1935–8. He proposed in 1934 the idea of dislocation in crystals, a form of atomic misarrangement which enables the crystal to deform at a stress less than that of a perfect crystal. Taylor had a passion for botany, small boats and foreign travel. He was knighted in 1944 and awarded the Order of Merit in 1969.

TAYLOR, Sir Hugh Stott
(1890–1974)

English physical chemist, born in St Helens, Lancashire. After graduating in chemistry from Liverpool University, he worked with **Arrhenius** on acid–base catalysis in Stockholm (1912–13) and with **Bodenstein** on the hydrogen–chlorine reaction in Hanover (1913–14). In 1914 he moved to the USA and thereafter Princeton University was his permanent home, although he remained a British subject. Between 1914 and 1927 he advanced from instructor to full professor, and from 1927 to 1958 he held the David B Jones Chair of Chemistry. Like **Rideal** he did wartime research in London on the **Haber** process (1917–19), and in their spare time they wrote *Catalysis in Theory and Practice* (1919). His pre-war and wartime background set the course of Taylor's researches thereafter. He made many studies of gas reactions involving chains of atoms and free radicals. Particularly in the 1920s, he worked extensively on the kinetics of reactions on surfaces and showed the existence of 'active centres', which are the sites on the surface of a heterogeneous catalyst at which chemical reaction actually occurs. He also identified 'activated adsorption' of a gas on a solid surface. After the isolation of deuterium by **Urey**, Taylor pioneered the use of this 'heavy hydrogen' to elucidate reaction mechanisms.

Taylor became a Commander of the Order of Leopold II of Belgium in 1938 and was knighted in 1953. Elected FRS in 1932, he received the Longstaff Medal of the Chemical Society in 1942 and was made an Honorary Fellow in 1949. He was President of the Faraday Society in 1952–4.

TAYLOR, Joseph Hooton Jr
(1941–)

American astronomer and physicist, born in Philadelphia. Educated at Haverford College and Harvard University, he held various posts at the University of Massachusetts before becoming Professor of Physics at Princeton in 1980. During a systematic search for pulsars, the rapidly rotating dense stars which appear on Earth to emit regular pulses of radio waves, he discovered with graduate student **Hulse** one interesting candidate whose pulse frequency changed periodically (1974). The characteristics of these changes revealed that this was the first discovery of an exotic 'binary pulsar', a pulsar in orbit of another dense neutron star. For this work he shared with Hulse the 1993 Nobel Prize for Physics. Taylor's subsequent observations have given strong evidence in favour of the general relativity prediction that this system of very massive compact objects will create 'gravitational waves', leaking energy from the system and causing the objects to continually move closer together.

TAYLOR, Richard Edward
(1929–)

Canadian physicist, born in Medicine Hat, Alberta, and educated at the University of Alberta in Edmonton and Stanford University. He held posts at the Linear Accelerator Laboratory at Orsay, the Lawrence Berkeley Laboratory and the Stanford Linear Accelerator Center (SLAC), California, where he became professor in 1970 and of which he was associate director from 1982 to 1986. In the 1960s, with **Jerome Friedman** and **Henry Kendall**, Taylor led a group of physicists at SLAC who investigated the structure of the nucleons (protons and neutrons) by scattering high-energy electrons from nuclear targets. These experiments provided data that established the constituents of nucleons, now known as quarks, as real dynamic entities by determining experimentally some of their properties. The three won the 1989 W K H Panofsky prize and the 1990 Nobel Prize for Physics for this work.

TAZIEFF, Haroun
(1914–)

Polish-born French vulcanologist and mountaineer, born in Warsaw. He studied in Russia, France and Belgium, firstly agricultural engineering and then geology. After various short-term posts in the Belgian Congo, he became assistant professor of mining geology in Brussels in 1950. In 1967 he was made head of research at the National Centre for Scientific Research (CNRS), Paris, and subsequently director (1971–81). He became the first French secretary of state for the prevention of natural disasters (1974–86) and was also Mayor of Mirmande (1977–89). Tazieff has investigated many of the world's volcanoes, both active and inactive, and from 1958 to 1974 made 26 expeditions to Nyiragongo, Zaire. He has written around 20 books on volcanoes and world tectonics including *Forecasting Volcanic Events* (1983) and *Sur L'Etna* (1984).

TEILHARD DE CHARDIN, Pierre
(1881–1955)

French Jesuit palaeontologist, theologian and philosopher, born at the castle of Sarcenat, the son of an Auvergne landowner. He was educated at a Jesuit school, lectured in pure science at the Jesuit College in Cairo, and was ordained as a priest in 1911. He was a stretcher bearer during World War I, and subsequently became Professor of Geology (1818) at the Institut Catholique in Paris. Between 1923 and 1946 he accompanied a number of palaeontological expeditions in China, where he directed the 1929 excavations at the Choukoutien Peking Man site. He later worked in central Asia, Ethiopia, Java and Somalia. Increasingly, his anthropological researches did not conform to Jesuit orthodoxy and he was forbidden by his religious superiors to teach and publish. Nevertheless, his work in Cenozoic geology and palaeontology became known and he was awarded academic distinctions, including the Legion of Honour (1946). From 1951 he lived in the USA and worked at the Wenner-Gren Foundation for Anthropological Research in New York. Posthumously published, his philosophical speculations, based on his scientific work, trace the evolution of animate matter to two basic principles: non-finality and complexification. By the concept of 'involution' he explains why *Homo sapiens* seems to be the only species which, in spreading over the globe, has resisted intense division into further species. This leads on to transcendental speculations, which allow him original, if theologically unorthodox, proofs for the existence of God. This work, *The Phenomenon of Man* (1955), is complimentary to *Le Milieu divin* (1957).

TEISSERENC DE BORT, Leon Philippe

(1855–1913)

French physicist and meteorologist, born in Paris. He joined the Central Meteorological Bureau in Paris (1878) where he was in charge from 1880 to 1892. He demonstrated that weather depended greatly on the barometric pressure at certain centres of action, notably the Azores high and the Iceland low. In 1894 he helped to produce an international cloud atlas, and he founded the observatory at Paris (1889) primarily for the study of the upper air using kites and hydrogen-filled balloons carrying instruments for measuring pressure temperature and humidity. He also obtained samples of air from up to 14 kilometres for analysis. During 1902–3, these kites were flown day and night for nine months both in Paris and Holland. He discovered that the temperature of the atmosphere does not continue to fall with height but at a certain height, becomes constant (the tropopause). He thus identified and named the stratosphere. He published a time cross-section of the isotherms above Paris to a height of 10 kilometres from 27 January to 1 March 1901, and later carried out upper air observations in kite and balloon experiments at a number of locations. From these observations he was able to show that the height of the tropopause varies with latitude, being much higher in the tropics. He was awarded the Symons Gold Medal of the Royal Meteorological Society (1908).

TELLER, Edward

(1908–)

Hungarian–American physicist, born in Budapest. He graduated in chemical engineering at Karlsruhe University, and studied theoretical physics at Munich, Göttingen and under **Niels Bohr** at his institute in Copenhagen. He left Germany in 1933, lectured in London and Washington (1935) and contributed profoundly to the modern explanation of solar energy, anticipating the theory behind thermonuclear explosions. In 1940, with **Szilard** and **Wigner**, he met with a US government committee to discuss the feasibility of a nuclear fission bomb. Later, in collaboration with Szilard and **Fermi**, he was involved in the construction of the first nuclear fission pile in Chicago before moving to the Manhattan Project at Los Alamos to develop the fission bomb. He then joined **Oppenheimer**'s theoretical study group at the University of California at Berkeley, and later became director of the newly established nuclear laboratories at Livermore (1958–60). From 1963 he was Professor of Physics at California University. He repudiated any moral implications of his work, stating that, but for Oppenheimer's moral qualms, the US might have had hydrogen bombs in 1947. After Russia's first atomic test (1949) he was one of the architects of President Harry S Truman's crash programme to build and test (1952) the world's first hydrogen bomb. He wrote *Our Nuclear Future* (1958).

TEMIN, Howard Martin

(1934–)

American virologist, born in Philadelphia. As a high school student, he spent summers at the Jackson Laboratory at Bar Harbor, Maine. He attended Swarthmore College, and studied with **Dulbecco** at Caltech, where he completed his PhD on the Rous sarcoma virus (which causes cancer in chickens) in 1959. Since 1969 he has held various professorships at the University of Wisconsin. Temin formulated the 'provirus' hypothesis, that the genetic material of an invading virus is copied into the host cell DNA. In 1970 he isolated the enzyme 'reverse transcriptase' (independently of **Baltimore**), which transcribes RNA into a double-stranded DNA (the provirus), enabling the new DNA to be inserted into the host cell. Viruses which contain this enzyme are 'retroviruses'; RNA forms their genetic material, and they reverse the usual process of DNA being transcribed to RNA. Reverse transcriptase is used to make copies of specific genes, clones, and is widely used for genetic engineering. In 1975, Temin shared the Nobel Prize for Physiology or Medicine with Dulbecco and Baltimore.

TEMMINCK, Coenraad Jacob

(1778–1858)

Dutch ornithologist, born in Amsterdam, the son of the treasurer of the Dutch East India Company. At the age of 17 he became an auctioneer with the company and used this contact to collect exotic birds and animals. When the company was dissolved in 1800, he turned his attentions full-time to the study of natural history and became an accomplished taxidermist. His *Catalogue Systématique du Cabinet d'Ornitholgie et de la Collection de Quadrumanes* (1807) comprised a description of over a thousand specimens from his own collection. He later published *Histoire Naturelle Générale des Pigeons et des Gallinacées* (3 vols, 1813–15), which established him as one of the leading European ornithologists. In 1820 he became the first Director of the Dutch National Museum of Natural History at Leiden which housed his own collection and grew in size and importance under his direction. His *Manuel d'Ornithologie* (1815–40) remained for many years the standard text on European birds.

TENNANT, Charles

(1768–1838)

Scottish chemist and industrialist, born in Ochiltree, Ayrshire. He attended the parish school and was then apprenticed to a silk weaver. He studied bleaching and then set up his own bleachfields at Darnley, near Paisley. At that time, traditional methods of bleaching were being replaced by chlorine, a method introduced in France by **Berthollet**. The chlorine was used in solution and was difficult to handle. In 1799 Tennant took out a patent for a dry bleaching powder made from chlorine and solid slaked lime, an innovation that was probably the invention of **Macintosh**, a fellow chemist and industrialist who was for a short time one of his partners. The powder could be conveniently transported to the expanding textile industry and the chlorine was easily regenerated when required by treating the powder with acid. Demand was such that the St Rollox works, which Tennant established in 1800, grew to be the largest chemical works in the world. By 1835 it covered more than 100 acres and produced sulphuric acid, alkali and soap as well as bleaching powder. Tennant was one of the first men to make a fortune out of the heavy chemical industry. He died in Glasgow.

TENNANT, Smithson

(1761–1815)

English chemist, born in Selby, Yorkshire. He was educated at Edinburgh University, where he was a pupil of **Joseph Black**, and at Cambridge where he became an early supporter of **Lavoisier**. He then travelled in Denmark and Sweden, meeting **Scheele** among other scientists. Around 1796, having bought land near Cheddar, Somerset, he embarked on agricultural research. He analysed lime from many parts of Britain and showed that some limes contain magnesium compounds and that these are injurious to plant life. In 1797 he demonstrated that diamond is a form of carbon by burning the diamond and showing that it produced the same amount of 'fixed air' (carbon dioxide) as an equal weight of charcoal. In 1800 he entered into partnership with a former fellow student, **Wollaston**, to produce platinum vessels, wire and electrodes for chemical research and industry, the merit of platinum being that it is resistant to heat and simple acids. In the course of their research on platina, the naturally occurring form of the metal, Tennant and Wollaston each discovered two new elements. In 1804 Tennant investigated the black residue left when platina is dissolved in aqua regia (a mixture of concentrated nitric and hydrochloric acids), separating and describing two new elements which he named iridium and osmium. Tennant was elected FRS in 1785. In 1815 he was appointed to the Chair of Chemistry at Cambridge; he was killed the following year in a riding accident near Boulogne, France.

TESLA, Nikola

(1856–1943)

Yugoslav-born American physicist and electrical engineer, born in Smiljan, Croatia. He studied at Graz, Prague and Paris, emigrating to the USA in 1884. He left the Edison Works at Menlo Park after quarrelling with **Edison**, worked for a short period with **Westinghouse**, but then concentrated on his own inventions. He was a prolific and highly innovative inventor. Among his many projects, he improved dynamos, and electric motors, invented the high-frequency Tesla coil and an air-core transformer. Tesla was firmly in favour of an alternating current electricity supply, as opposed to direct current initially favoured by Edison. By 1888 he had obtained patents on a whole polyphase AC system which he sold to Westinghouse. He again demonstrated the feasibility of AC, by lighting the 1893 Chicago World Columbian Exposition, and AC transmission was also chosen for the Niagara Falls project (1893–5). He produced artificial lightning of a prodigiousness never since equalled, predicted wireless communication two years before **Marconi**, and experimented with a very low-frequency wireless communication system using the Earth as the conducting medium. Near the end of his life he became an eccentric recluse. His papers have been deposited in the Nikola Tesla Museum in Belgrade.

THALES

(c. 624–545 BC)

Greek natural philosopher, astronomer and geometer, traditionally the founder of Greek and therefore European philosophy. He came from Miletus on mainland Ionia (Asia Minor), as did his intellectual successors **Anaximander** and **Anaximenes**. Thales is believed to have proposed the first natural cosmology, identifying water as the original substance and (literally) the basis of the universe. He is supposed to have visited Egypt, where he developed his astronomical techniques, and to have predicted the solar eclipse of 585 BC. Included in the traditional canon of the *Seven Wise Men*, he attracted various apocryphal anecdotes, for example as the original absent-minded professor who would fall into a well while watching the stars.

THEAETETUS
(c.414–c.369 BC)

Greek mathematician. He was an associate of **Plato** at the Academy, whose work was later used by **Euclid** in Books X and XIII of the *Elements*. Plato named after him the dialogue *Theaetetus*, which was devoted to the nature of knowledge. He is credited with being the first to prove that $\sqrt{n}$ is irrational whenever n is not a perfect square.

THEILER, Max
(1899–1972)

South African-born American bacteriologist, born in Pretoria. He enrolled in the pre-medical course at the University of Cape Town in 1916, and later studied at St Thomas's Hospital Medical School and the School of Tropical Medicine and Hygiene of the University of London. After receiving his medical degree in 1922, he went to Harvard Medical School to work on amoebic dysentery. He settled in the USA and turned his attention to yellow fever, working at Harvard until 1930 and then at the Rockefeller Institute, New York (1930–64). He was Professor of Epidemiology and Microbiology at Yale Medical School from 1964 to 1967. In 1919 **Noguchi** reported that he had isolated a bacterium responsible for yellow fever. Theiler showed in 1926 that in fact yellow fever was caused by a filterable virus. While at the Rockefeller, he perfected the mouse protection test. In this test a mixture of yellow fever virus and human serum was injected into a mouse, and the survival of the mouse indicated that the serum had neutralized the virus and that the serum donor was therefore immune. This enabled an accurate survey of the worldwide distribution of yellow fever. Theiler contracted the disease in 1929, but survived and subsequently became immune. He is also remembered for his discovery of an infection in mice identical to polio—encephalomyelitis, or Theiler's disease. He was awarded the 1951 Nobel Prize for Physiology or Medicine for his work in connection with yellow fever, for which he discovered the vaccine 17D in 1939. This formed the basis of vaccines now used to control the disease.

THÉNARD, Louis Jacques
(1777–1857)

French organic chemist and statesman, born in La Louptière (now Louptière-Thénard), the son of a peasant farmer. At a young age he left home for Paris in search of an education and attended lectures by **Fourcroy** and by **Vauquelin**, who gave him a home in return for his services as a bottle-washer. In 1798 he was appointed demonstrator at the École Polytechnique; he later succeeded Vauquelin in the chair at the Collège de France in 1804, became Dean of the Faculty of Sciences of Paris in 1821, and was Chancellor of the University of France from 1845 to 1952. He was a prominent member of many public bodies, particularly those concerned with the application of science to industry, and received many honours culminating in a peerage in 1832. He also served two terms in the Chamber of Deputies. Thénard made many important discoveries in organic chemistry. He prepared a wide range of esters (neutral products formed by the reaction between an acid and an alcohol), discovering that the reaction was analogous to the reaction between an acid and a base. He investigated cobalt and its compounds, and from alumina and copper arsenate prepared a stable brilliant blue pigment (Thénard's blue) which was used in porcelain manufacture to replace the expensive pigments made from lapis lazuli. Between 1808 and 1811 he collaborated with **Gay-Lussac** to study potassium, an element whose high reactivity made it very difficult to isolate. It had first been isolated by **Humphry Davy**; Thénard and Gay-Lussac prepared it in much larger quantities by fusing potash with iron filings in a gun barrel. While investigating its properties they discovered boron (1808). They proved that sodium hydroxide and potassium hydroxide contain oxygen and hydrogen. They also studied the photochemistry of chlorine, and the composition of organic compounds using potassium chlorate as an oxidizing agent. In 1818 Thénard announced the discovery of hydrogen peroxide, perhaps his greatest achievement. He passed oxygen over barium oxide, made the product (barium peroxide) into a paste, then precipitated the barium out with sulphuric acid. His observation that finely divided metals acted on hydrogen peroxide to produce heat and hydrogen without themselves being affected, together with knowledge of **Döbereiner**'s work on platinum, led him to the study of surface catalysis (although the term 'catalysis', coined by **Berzelius**, was not introduced until several years later). He was also the author of an influential textbook, *Traité Élémentaire de Chimie* (4 vols, 1813–16), which went through six editions and was much translated. He died in Paris.

THEOPHRASTUS
(c.372–c.287 BC)

Greek philosopher, born in Eresus on Lesbos, the son of a fuller. He studied at Athens under **Aristotle**, from whom he had inherited a library, and succeeded him as head of the Peripatetic School (Lyceum) from 322 BC. Theophrastus shared Aristotle's encyclopedic conception of philosophy. Most of his prolific output is lost, but there are still extant important treatises on plants (representative of his interest in natural science), reconstructed fragments of his history of earlier philosophers, and the more literary volume of *Characters*, containing 30 deft sketches of different moral types, which has been widely translated and imitated. His *Historia Plantarum* and *Plantarum Causae* mentioned around 450 species, not described in much detail, but he did allude to differences in floral structure between plants, and to seed germination. He established the fundamental difference of organization between plants and animals. In other observations, he derived insights into the essentials of plant morphology and classification, laying the foundations for the work of later botanists. **Linnaeus** called Theophrastus the father of botany, because so many aspects of modern botany can be traced back to his work, including morphology, anatomy, systematics, physiology, ecology, pharmacognosy, applied botany and plant pathology.

THEORELL, (Axel) Hugo Teodor
(1903–82)

Swedish biochemist, born in Linköping. After training in medicine at the Karolinska Institute, Stockholm, he received his MD for a study of the effect of plasma lipids on the sedimentation of red blood cells (1930), and became a lecturer (1930–2) and assistant professor at Uppsala (1932–6), and Director of the Nobel Institute of Biochemistry at Stockholm (1937–70). He crystallized myoglobin (oxygen storage protein of muscle) and determined its molecular weight (1932). During a period in Berlin, in 1934 he purified the 'yellow ferment' (an electron transfer enzyme), obtained as crude extract there by **Warburg**, and separated the yellow coenzyme (flavine mononucleotide) from the protein. The resulting inactivation was reversed when the two components were reunited. Theorell helped elucidate the nature of the yellow pigment, later identified as riboflavin (vitamin B_2), and established the 1:1 linkage with protein using his newly invented electrophoresis apparatus (1935). On his return to Uppsala he purified diphtheria antitoxin (1937), and crystallized and characterized cytochrome C, another electron carrier, establishing the sulphur linkage between haem and protein (1938–41). He subsequently purified and studied peroxidases and dehydrogenases, and introduced fluorescence spectrometry. Theorell was awarded the 1955 Nobel Prize for Physiology or Medicine, and was elected a Foreign Member of the Royal Society in 1959.

THOM, René Frédéric
(1923–)

French mathematician, born in Montbéliard. He studied at the École Normale Supérieure, and worked at Grenoble and Strasbourg, where he became professor. Since 1964 he has been at the Institut des Hautes Études Scientifiques. In 1958 he was awarded the Fields Medal (the mathematical equivalent of the Nobel Prize). His work has been in algebraic topology, where he was one of the creators of a novel and powerful theory known as cobordism theory, and on the singularity theory of differentiable manifolds, but he is best known for his book *Stabilité structurelle et morphoqenèse* (1972) which introduced 'catastrophe theory'. This has been applied to widely differing situations such as the development of the embryo, social interactions between human beings or animals, and physical phenomena such as breaking waves, and has attracted much publicity as well as some controversy.

THOMAS, (Edward) Donnall
(1920–)

American physician and haematologist, born in Mart, Texas, the son of a general practitioner. In 1937 Thomas entered the University of Texas at Austin, where he studied chemistry and chemical engineering. After receiving his MA in 1943, he entered Harvard Medical School, graduating MD in 1946. After an internship, training in haematology, serving in the army, and posts at MIT and the Brigham Hospital, Boston, Thomas joined the staff of the Mary Imogene Bassett Hospital in Cooperstown. There he began work on bone marrow transplantation in dogs and in humans. This proved a difficult procedure due to problems such as graft rejection and 'graft-versus-host' disease. However, occasional successes occurred with dogs when bone marrow was transplanted between members of the same litter. In 1963 Thomas became professor at the Washington University School of Medicine, Seattle, and here he demonstrated that using new tissue-typing techniques and immunosuppressive drugs, some patients with leukaemia, aplastic anaemia and certain genetic diseases can be cured by bone marrow transplants. He joined the Fred Hutchinson Cancer Research Center in 1975. Thomas shared the 1990 Nobel Prize for Physiology or Medicine with **Joseph Edward Murray**.

THOMAS, Sir John Meurig
(1932–)

Welsh physical chemist, born in Llanelli. He studied chemistry at University College, Swansea, and Queen Mary College, London.

From 1958 to 1969 he advanced from assistant lecturer to reader in chemistry at the University College of North Wales, Bangor, and then moved to the University College of Wales, Aberystwyth, as professor and head of department. From 1978 to 1986 he was Professor of Physical Chemistry at Cambridge. He then became Director of the Royal Institution and of its Davy–Faraday Laboratory. He vacated these posts in 1991, but is still Fullerian Professor at the Royal Institution. Since 1991 he has been Deputy Pro-Chancellor of the University of Wales. Thomas's main field of research has been surface chemistry, particularly heterogeneous catalysis. In recent years he has pioneered in the development of 'uniform' heterogeneous catalysts—solid catalysts in which the active sites are distributed uniformly throughout their bulk. He is the author of *Heterogeneous Catalysis; Theory and Applications* (1991, with W J Thomas) and *Michael Faraday and the Royal Institution* (1991). He was knighted in 1991. Thomas was elected FRS in 1977 and gave the Bakerian Lecture in 1990. He received the Hugo Müller Medal of the Royal Society of Chemistry in 1983 and its Faraday Medal in 1989. He gave the Baker Lectures at Cornell University in 1982–3.

THOMAS, Sidney Gilchrist
(1850–85)

English metallurgist, born in Canonbury in North London. Educated at Dulwich College, he intended to study medicine, but after the death of his father in 1867 he became a police-court clerk. However, he attended evening classes in chemistry at the Birkbeck Institution and studied metallurgy at the Royal School of Mines. From 1870 he decided to tackle a problem that had dogged the **Bessemer** steelmakers: how to remove phosphorus from iron. In May 1878 at a meeting of the Iron and Steel Institute, Thomas was able to announce that, with the help of his cousin **Gilchrist** and Edward Martin, he had solved the problem of 'dephosphorization' by using dolomite for the furnace lining, together with an addition of lime to produce a basic slag that allowed the removal of both phosphorus and sulphur. This method was described as the 'basic Bessemer process' in Britain, but was always known as the 'Thomas process' on the Continent. Within a few years, the same principles were applied to the Siemens open-hearth furnace. Ironically, the main effect of Thomas's invention was to open up the whole range of the world's iron ores, with the loss of the UK's leadership in steelmaking. Ascetic, a pacifist and philanthropist, Thomas died prematurely from a lung complaint (probably TB) and was buried in Paris.

THOMPSON, Benjamin
See **RUMFORD, Benjamin Thompson, Count**

THOMPSON, Sir D'Arcy Wentworth
(1860–1948)

Scottish zoologist and classical scholar, born in Edinburgh and educated at Trinity College, Cambridge. He was Professor of Biology at Dundee (1884–1917) and at St Andrews (from 1917). The ideas for which he is remembered are contained in his *On Growth and Form* (1917), a book read both for its biological content and literary style. His knowledge of mathematics and physics led him to interpret the forms of organs and biological structures on the basis of the physical forces acting upon them during development. He was also able to demonstrate mathematically that the superficial dissimilarity of related animals could be accounted for by differential growth rates, and these ideas resulted in him adopting an anti-Darwinian stance. His *Glossary of Greek Birds* (1895) and *Glossary of Greek Fishes* (1945) derive from his classical interests. He was knighted in 1937.

THOMPSON, Silvanus Phillips
(1851–1916)

English physicist, born in York into a Quaker family. He sat for an external London degree (1869), and was appointed the first Professor of Physics at the newly established University of Bristol (1878) and Professor of Physics and Principal of the City and Guilds Technical College, Finsbury (1885). He was a highly proficient textbook writer on electricity, light and magnetism, wrote a witty, effective little book called *Calculus made Easy* (1910), and was a distinguished historian of science, writing biographies of **William Gilbert**, **Faraday**, **Kelvin**, and Philipp Reis whom he regarded as the true inventor of the telephone. He almost beat **Heinrich Hertz** to the discovery of radio waves, and he made a number of technical contributions. He may well have been appointed Principal of the University of London (1901) if he had not been so outspoken about British conduct in the Boer War.

THOMSON, Elihu
(1853–1937)

English-born American inventor, born in Manchester. He emigrated to the USA and was educated in Philadelphia, where he was a chemistry teacher until he decided on a career as an inventor. He became one of the great pioneers of the electrical manufacturing industry in the USA. He cooperated in 700 patented electrical inventions, which included the three-phase AC generator and arc lighting. With Edwin James Houston, he founded the Thomson–Houston Electric Company (1883), which merged with **Edison**'s firm in 1892 to form the General Electric Company. He declined the Presidency of MIT in 1919, but agreed to be Acting President from 1921 to 1923.

THOMSON, Sir George Paget
(1892–1975)

English physicist, son of **J J Thomson**. He was born and educated in Cambridge, where he became a Fellow of Trinity College. He served in the Royal Flying Corps during World War I, was Professor of Physics at Aberdeen (1922–30) and Imperial College, London (1930–52), and became Master of Corpus Christi at Cambridge (1952–62). In 1927 Thomson and Alexander Reid were the first to notice that a beam of electrons passed through a thin metal foil in a vacuum produced circular interference fringes. This was firm evidence for **de Broglie**'s theory that moving particles have wave-like properties. Thomson went on to analyse the diffraction characteristics of electrons, thereby opening the door to their use in the analysis of surfaces. In 1937 he shared the Nobel Prize for Physics with **Davisson** for the discovery, separately and by different methods, of electron diffraction by crystals. During World War II, Thomson advised the government on the possible relevance of nuclear fission to the making of a superbomb. After the war he supported the introduction of a world atomic authority for the peaceful exploitation of nuclear power. He was scientific adviser to the UN Security Council (1946–7) and for his contributions to electrical science he was awarded the Faraday Medal by the Institution of Electrical Engineers (1960). His works included *The Atom* (1937) and *Theory and Practice of Electron Diffraction* (1939). He was elected FRS in 1930, and knighted in 1943.

THOMSON, James
(1822–92)

Irish–Scottish physicist and engineer, elder brother of Lord **Kelvin**, born in Belfast. His father was a professor of mathematics there, and moved in 1832 to the same position at the University of Glasgow. James Thomson was educated at home until 1832 when he began attending classes at the University of Glasgow, where he became a matriculated student two years later at the age of 12. He graduated MA with honours in mathematics and natural philosophy in 1839, then spent several short periods as an engineering apprentice but was troubled by ill-health until in 1851 he settled as a civil engineer in Belfast. In 1857 he was appointed Professor of Civil Engineering at Queen's College, Belfast, and in 1873 moved to the same chair in succession to **Rankine** at the University of Glasgow. He carried out important researches in fluid dynamics, inventing or improving several types of water-wheels, pumps and turbines. Over a long period he studied the effect of pressure on the freezing point of water, and its influence on the plastic behaviour of ice and the movement of glaciers. He published many papers on that and a wide variety of other subjects including elastic fatigue, ocean under-currents and trade winds. He was elected FRS in 1877.

THOMSON, Sir J(oseph) J(ohn)
(1856–1940)

English physicist, discoverer of the electron, born in Cheetham Hill near Manchester, the son of a Scottish bookseller. He entered Owen's College, Manchester, at the age of 14 with the intention of becoming a railway engineer, but a scholarship took him to Trinity College, Cambridge, where he graduated Second Wrangler in 1880. In 1884 he was elected FRS and succeeded Lord **Rayleigh** (1842–1919) as Cavendish Professor of Experimental Physics, and in 1919 was himself succeeded by his brilliant student, **Rutherford**. Thomson's early theoretical work was concerned with the extension of **Maxwell**'s electromagnetic theories. This led to the study of gaseous conductors of electricity and in particular the nature of cathode rays. By studying the deflections of cathode rays in a highly evacuated discharge tube when electric and magnetic fields were applied, he showed that the rays consist of rapidly moving particles and calculated their charge-to-mass ratio. He went on to measure the particles' specific charge in an experiment similar to that carried out by **Millikan**, and deduced that these 'corpuscles' (electrons) must be more than 1 000 times smaller in mass than the lightest known atomic particle, the hydrogen ion. This discovery was inaugurated by his lecture to the Royal Institution in 1897, and published in the *Philosophical Magazine*. Before the outbreak of World War I, Thomson had successfully studied the nature of positive rays (1911), and this work was crowned by the discovery of isotopes, which Thomson showed could be separated from each other by deflection of positive rays in electric and magnetic fields, a technique now known as mass spectrometry. 'J J' made the Cavendish Laboratory the greatest research institution in the world. Although simplicity of apparatus was carried to 'string and sealing wax' extremes, seven of his research assistants subsequently won the Nobel Prize; Thomson himself was awarded the Nobel Prize for Physics in 1906. He was knighted in 1908, and became the first scientist to be appointed master of Trinity College (1918–40). In 1936 he published *Recollections and Reflections*, and he was buried near Newton in the nave of Westminster Abbey.

THOMSON, William

See **KELVIN, William Thomson, 1st Baron of Largs**

THOMSON, Sir (Charles) Wyville

(1830–82)

Scottish oceanographer, born in Bonsyde, Linlithgow. At the age of 16 he became a medical student at Edinburgh University. Poor health caused him to give up medicine and he left the university in 1850 having taken no examinations. This in no way precluded him from being appointed as lecturer in botany at the University of Aberdeen (1850–1), Professor of Natural History in Queen's College, Cork (1853), Professor of Geology and Professor of Zoology and Botany, both in Belfast (1853–68), and finally Professor of Natural History at the University of Edinburgh from 1870. Specimens collected on deep water expeditions aboard survey vessels and eventually aboard HMS *Challenger* led him to dispute the azoic theory of **James Forbes** and **Alexander Agassiz**, arguing that marine life exists at all depths of the ocean. His book *Depths of the Sea* (1877) was the first general textbook on oceanography. Other topical controversial questions he addressed were mechanisms of evolution, deep-sea temperatures, the Gulf Stream and the continuity of chalk out to sea. He was director of the civilian staff aboard the famous circumnavigating HMS *Challenger* expedition (1872–6), and on its return was appointed Director of the Challenger Expedition Commission and given the onerous task of overseeing the analysis and reporting of the results of the expedition, making Edinburgh the world centre for oceanography at that time. *Voyage of the 'Challenger'—the Atlantic* (2 vols) was published in 1877; poor health prevented him completing more. The Wyville Thomson Ridge was named after him since he had predicted its existence from water temperature measurements. He was knighted in 1876.

'T HOOFT, Gerard

(1947–)

Dutch physicist. He was educated at Utrecht University, where he was later appointed professor. 'T Hooft's work has been concerned with gauge theories of particle physics which attempt to describe the various types of interaction between fundamental particles. The unified theory of the electromagnetic and weak interactions, as proposed by **Steven Weinberg** and **Salam**, originally predicted a force of infinite strength. Whilst a research student, 't Hooft found a way of making the force both finite and calculable; this led to the universal acceptance of the electroweak theory. He has also showed that recent developments in this field predict the existence of a heavy magnetic monopole (so far undiscovered), which **Dirac** also predicted by different reasoning, and has contributed to the theories of quantum gravity.

THORARINSSON, Sigurdur

(1912–83)

Icelandic geologist and glaciologist. His early scientific training was in Stockholm in the 1930s with field studies in Swedish Lapland (1933) and Iceland (1934), and in 1936–8 on the Vatnjökull glacier expedition. He obtained a degree (1939) for studies on the movement and drainage of Hoffellsjökull and his doctorate (1944) for his classical work on tephrochronology, the relative dating of volcanic ash layers. In 1945 he settled in Iceland where he worked as a geologist for the National Research Council and then joined the Museum of Natural History (1947–69). From 1969 he was Professor of Geography and Geology at the University of Iceland. Thorarinsson was the first to use tephrochronology to study the eruption history of volcanoes, the first to describe and analyse catastrophic glacier outburst floods (jökulhlaups), and carried out early work on glacier shrinkage and eustatic changes in sea level. He made the first determination of glacier mass balance and its relation to climatic change, and was also a pioneer in nature conservation in Iceland.

THORPE, Sir (Thomas) Edward

(1845–1925)

English chemist, physicist and historian of science, born near Manchester. He was educated at Owens College, Manchester, where he spent some time as **Roscoe**'s assistant, and at the University of Heidelberg under **Bunsen** and also at Bonn. He was appointed to the Chair of Chemistry first at Anderson's College, Glasgow (1870), then at the Yorkshire College of Science at Leeds (1874) and the Royal College of Science,

London (1885). In 1894 he became Government Chemist—the first such appointment—and he returned to the Royal College of Science from 1909 to 1912. Thorpe discovered several new compounds of chromium, sulphur and phosphorus, including phosphorus pentafluoride which demonstrated that phosphorus could have a valence of five. He also carried out determinations of atomic weights, for example of silicon, gold, titanium and radium, that were more accurate than any others of the time. He travelled to the West Indies and other places to view four eclipses of the Sun, and in collaboration with Sir Arthur Rücker made a magnetic survey of the British Isles. As Director of the Government Laboratory at Clements Inn, London, he was responsible for removing arsenic from some beers, legislating against lead in pottery glazes, eliminating paraffins from tobacco leaf, and substituting red phosphorus for yellow phosphorus in matches (yellow phosphorus having been found to be responsible for the cancer of the jaw which affected many workers in the match industry). Thorpe is also remembered for his work in the history of science, particularly his biography of **Priestley**. He was elected FRS in 1876, made a Companion of the Bath in 1900, and knighted in 1909. He died in Salcombe, Devon.

THORPE, William Homan

(1902–86)

English zoologist, born in Hastings. He studied agriculture at Cambridge, where he developed an interest in agricultural entomology, and investigated the biological control of insect parasites in California (1927–9). He was research entomologist at Farnham Royal Parasite Laboratory of the Imperial Bureau of Entomology (1929–32) and became a lecturer in Entomology at Cambridge University in 1932. He was instrumental in setting up the Ornithological Field Station at Madingley near Cambridge which became the Sub-department of Animal Behaviour in 1960. In 1966 Cambridge's first Chair of Ethology was created for him. Thorne had a strong interest in conservation, and served on many committees concerned with conservation matters. He was one of the founders of ethology and published the influential *Learning and Instinct in Animals* in 1956. His research was mainly concerned with bird song, and working initially with the chaffinch, he demonstrated (with Peter Marler) that song results from the integration of innate and learned components of sound patterning. In 1961 he wrote *Bird-Song: The Biology of Vocal Communication and Expression in Birds*. He was elected FRS in 1951.

THORSON, Gunnar Axel

(1906–71)

Danish marine ecologist, born in Copenhagen. He was educated at Copenhagen University, and subsequently became a leading member of a three-year expedition to east Greenland, during which he undertook research for his PhD on the reproductive ecology of Arctic marine invertebrates. He was later appointed to the staff of the Copenhagen University Zoological Museum (1934) and founded a private laboratory on the island of Ven in the Øresund, from which stemmed the research that led to the publication of his classic and highly influential monograph *Reproduction and larval development of Danish marine bottom invertebrates* (1946). He was elected to the Royal Danish Academy of Sciences in 1955. During World War II the laboratory was moved to Helsingør, and in 1957 Thorson was promoted to full professor; he was later formally appointed Director of the Helsingør Marine Laboratory (a converted torpedo station). Thorson's 1946 monograph is still a cornerstone in marine invertebrate larval ecology, and his geographic categorization of developmental modes of benthic marine invertebrates at all latitudes and depths is now formalized in the literature as 'Thorson's rule'. A second important contribution was his geographic concept of 'parallel marine bottom communities', a hypothesis that similar biotopes at differing latitudes supported ecologically comparable assemblages characterized by the same genera (albeit different species). Despite the conceptual utility of such broad-scale hypotheses, Thorson himself acknowledged that this latter theory was not tenable on its extension to include tropical communities.

THUNBERG, Carl Peter

(1743–1828)

Swedish botanical explorer, born in Jönköping. As a medical student at Uppsala he was taught botany by **Linnaeus** and collected plants for him; after graduation in 1770 he travelled as a ship's surgeon to South Africa, Java and Japan in the company of Francis Masson, collecting and describing 3000 plants, 1000 of which were new to science. In Japan, he was not allowed to collect plants personally, but he organized local collectors. From 1778 he taught botany at Uppsala, and in 1784 he succeeded Linnaeus's son as professor. His Japanese discoveries were published as *Flora Japonica* (1784), and those from South Africa as *Prodromus Plantarum Capensium* (1794–1800) and *Flora Capensis* (1807–23, with Joseph August Schultes). He also wrote and published monographs on

Protea, *Ixia*, *Oxalis* and *Gladiolus*. His description of his voyage was published between 1788 and 1793; translated into English, it was published as *Travels in Europe, Africa and Asia* (1793–5).

TIEGHEM, Philippe Van
See **VAN TIEGHEM, Philippe**

TILDEN, Sir William Augustus
(1842–1926)

English organic chemist, born in London. He was apprenticed to a pharmacist and studied for a year at the Royal College of Chemistry before becoming a demonstrator at the Pharmaceutical Society (1863–72). For the following eight years he taught chemistry at Clifton College, Bristol, moving to the Chair of Chemistry at Mason College and then succeeding **Edward Thorpe** at the Royal College of Science in 1894. Tilden showed that there is only one compound of nitric oxide and chlorine, nitrosyl chloride, and that it is a valuable reagent for investigating the terpenes, a little-understood class of hydrocarbons which occur in the essential oils of many plants. He also discovered that if the hydrocarbon isoprene was prepared from terpenes, it separated out on standing into fragments of a yellowish substance floating in a syrupy fluid. These fragments have properties identical to natural rubber. It was already known that isoprene prepared by other methods interacts with strong acids to give a tough, elastic substance resembling rubber, and Tilden did not pursue his discovery of its spontaneous conversion. In physical chemistry he worked on the solubility of salts at temperatures above 100 degrees Celsius and on the specific heats of metals. He discovered that specific heats alter with temperature, decreasing as temperatures fall (he lowered temperatures to −180 degrees Celsius) and increasing as they rise, with the extent of the shift varying inversely as the atomic weight of the element. This discovery has proved very important in many branches of industry. Tilden was elected FRS in 1880, and knighted in 1909. He died in London.

TIMOSHENKO, Stephen (Stepan Prokofyevich)
(1878–1972)

Soviet–American civil engineer, born in the Ukraine. He was educated at the two technical institutes in St Petersburg, with an intervening period working as a railway engineer, then spent some time studying under **Prandtl** at the University of Göttingen. In 1906 he began lecturing at Kiev University. Dismissed for his pro-Jewish views in 1911, he nevertheless remained in Russia, often under conditions of great difficulty and hardship. Eventually in 1920 he fled to Yugoslavia and then to the USA in 1922. There he found that, in his view, the teaching of mechanics to engineering students was far from satisfactory, and he devoted himself to the improvement of both teaching and research in all aspects of the subject. His own textbooks on the strength of materials and the theory of elasticity, as well as his technical papers, were (and remain) an important contribution towards the achievement of that end. In 1936 he joined the staff of Stanford University in California where he taught engineering mechanics and strength of materials until he was 76 years of age. In 1946 he received the James Watt Gold Medal of the (British) Institution of Mechanical Engineers, and in 1959 he was elected a member of the Soviet Academy of Sciences.

TINBERGEN, Nikolaas
(1907–88)

Dutch ethologist, born in the Hague, brother of the Nobel prize-winning economist Jan Tinbergen. He studied zoology at Leiden, and after World War II taught at Oxford (1949–74). With **Konrad Lorenz** he is considered to be the co-founder of ethology, the study of animal behaviour in relation to the environment to which it is adapted. His best-known studies were on the three-spined stickleback and the herring gull, animals which perform many stereotyped or instinctive behaviour patterns. Much of his work was centred around aspects of social behaviour. He was able to elucidate the evolutionary derivation of many of these behaviours through comparative studies. His concentration on instinctive behaviour brought him into conflict with comparative psychologists who disagreed with the separation between acquired and innate behaviours. His books included his classic *The Study of Instinct* (1951), *The Herring Gull's World* (1953), *Social Behaviour in Animals* (1953) and *The Animal in its World* (2 vols, 1972–3). In 1973 he published a controversial book *Autistic Children* (with his wife Lies), in which he proposed a behavioural causation for autism. He shared the 1973 Nobel Prize for Physiology or Medicine with Lorenz and **Karl von Frisch**.

TING, Samuel Chao Chung
(1936–)

American physicist, born in Ann Arbor, Michigan, where his father was a student at that time. He was raised in China and educated there and in Taiwan, and at Michigan University (1956–62). He worked in elementary particle physics at the European nuclear research centre (CERN) in Geneva and at Columbia University. Later he led a research group at DESY, the German synchrotron project in Hamburg, and from 1967 worked at MIT. In 1974 he was head of a team at the Brookhaven National Laboratory which conducted an experiment in which protons were directed onto a beryllium target, and a product particle, with a lifetime 10 000 times longer than could be predicted by previous discoveries, was observed and named the J particle. At the same time, and independently, **Burton Richter** made the same discovery and named the particle ψ. It is now named the J/ψ particle and for its discovery Ting and Richter shared the 1976 Nobel Prize for Physics. In the 1980s and 1990s, Ting led the L3 collaboration at CERN working at the LEP electron–positron collider where the standard model of particle interactions was precisely tested.

TISELIUS, Arne Wilhelm Kaurin
(1902–71)

Swedish chemist, born in Stockholm. He trained and worked at the University of Uppsala, where he became assistant professor in 1930 and was later appointed Professor of Biochemistry (1938–68). He developed an accurate method for determining diffusion constants of proteins, important for analysing ultracentrifuge sedimentation data (1934). He introduced protein analysis by moving boundary electrophoresis (1930–7), and thus identified serum proteins as albumin and α, β and γ (gamma) globulins; he also showed that antibodies are γ globulins (1937). Tiselius's electrophoretic analysis became the best criterion of protein purity and he found multiple components in crystalline pepsin. He isolated bushy stunt and cucumber mosaic viruses (1938–9), and invented preparative electrophoresis (1943), electrokinetic filtration (1947) and other analytical techniques. From 1944 he developed methods for the chromatographic separation and identification of amino acids, sugars and other molecules using activated charcoal, cellulose, silica, ion exchange and other media. He worked with **Sanger** on the chemistry of insulin (1947) and with **Synge** on chromatographic analysis (1950). Tiselius became Vice-President (1947–60) and President (1960–4) of the Nobel Foundation and was awarded the Nobel Prize for Chemistry in 1948. He was elected a Foreign Member of the Royal Society in 1957.

TITTERTON, Sir Ernest William
(1916–90)

English atomic physicist, born in Tamworth, Staffordshire. Educated at Birmingham University, he was a research officer for the Admiralty during World War II, before becoming in 1943 a member of the British mission to the USA to participate in the Manhattan Project to develop the atomic bomb. He was a senior member of the timing team at the first atomic test in 1945, and advisor on instrumentation at the Bikini Atoll tests in 1946, before returning to Los Alamos, New Mexico, as head of the electronics division until 1947. Titterton then worked at the Atomic Energy Research Establishment at Harwell until 1950 when he became Professor of Nuclear Physics at the Australian National University, Canberra. He was involved in the British nuclear tests at Maralinga, South Australia, until 1957, and subsequently held various research and advisory appointments in the field of nuclear energy. He was knighted in 1970.

TIZARD, Sir Henry Thomas
(1885–1959)

English chemist and administrator, born in Gillingham, Kent. He studied chemistry at Magdalen College, Oxford, and worked with **Nernst** in Berlin (1908–9). Returning to Oxford, he was soon appointed Fellow and Tutor of Oriel College, but within a few years his career was interrupted by World War I. After brief service in the army, he transferred to the Royal Flying Corps and spent much of the war as an experimental pilot. In 1918–19 he was Assistant Controller of Experiments and Research for the RAF. He returned to Oxford and became reader in thermodynamics, but soon left to be Assistant Secretary of the Department of Scientific and Industrial Research, of which he became Permanent Secretary in 1927. From 1929 to 1942 he was Rector of Imperial College and from 1942 to 1946 he was President of Magdalen College, Oxford. He then became Chairman of the Defence Policy Research Committee and of the Advisory Council on Scientific Policy, from which he retired in 1952. His personal scientific work included some electrochemistry before 1914 and important work on aircraft fuels carried out with David Pye around 1920, which led ultimately to the system of octane rating, which expresses the anti-knocking characteristics of a fuel. In the 1930s and 1940s he was increasingly involved as an adviser to the British government in the scientific aspects of air defence, particularly in connection with radar. He was Chairman of the Aeronautical Research Committee from 1933 to 1943 and led a scientific mission to the USA in 1940. His influence probably made the difference between victory and defeat in the Battle of Britain in 1940. He received many military honours, and was elected FRS in 1926.

TOBIAS, Phillip Vallentine
(1925–)

South African anatomist and physical anthropologist, born in Durban. He studied at Witwatersrand University, receiving a BSc in 1946, and an MB and BCh in 1950. He has remained at Witwatersrand nearly all his life, becoming a lecturer in anatomy in 1951 and Professor of Anatomy and Human Biology in 1959. Since 1979 he has been Director of the Palaeoanthropological Research Unit. Tobias has worked on cytogenetics and human genetics, the human biology of the living peoples of Africa, and palaeoanthropology. He has studied and described hominid fossils in many parts of Africa including those of the Olduvai Gorge, Tanzania. With **Louis Leakey** and J R Napier he described and named the species *Homo habilis*. His work has led to about 800 publications, including three major works: *Australopithecus (Zinjanthropus) boisei* (1967), *The Brain in Hominid Evolution* (1971) and a two-volume work on *Homo habilis* (1991).

TODD, Alexander Robertus, Baron of Trumpington
(1907–)

Scottish chemist, born in Glasgow. He studied at the University of Glasgow and obtained his first doctorate at Frankfurt on the chemistry of bile acids in 1931. He was awarded an 1851 Exhibition Senior Studentship and worked with **Robinson** at Oxford for a second doctorate on the chemistry of natural pigments (1933). He spent two years in the medical chemistry department in Edinburgh, working on the chemistry of vitamin B, and then moved to the Lister Institute of Preventive Medicine in London. In 1938 he was offered a position at what was then a small college in Pasadena, later to be widely known as Caltech, but fortunately for British chemistry, the Chair of Organic Chemistry at Manchester became vacant at the same time and he held that post from 1938 to 1944. His final post was as professor at Cambridge where he remained until his retirement in 1971. All his researches concerned the chemistry of natural products, including vitamins B_1, E and B_{12}; the constituents of cannabis species; insect colouring matters; factors influencing obligate parasitism; and various mould products. However, the work for which he was awarded the Nobel Prize for Chemistry in 1957 concerned the structure and synthesis of nucleotides. The four chemical bases (adenine, guanine, uracil and cytosine) of DNA had been known for some time. Todd and his co-workers established the manner in which sugar molecules and phosphate groups are attached to these bases to form nucleotides, the building blocks of DNA. This work was a necessary preliminary to **Crick** and **James Watson**'s proposal of the double helix as the structure of DNA, and to an understanding of the biological processes of growth and inheritance. Todd received many academic honours. He was elected to the Royal Society in 1942 and was later its president (1975–80). He was knighted in 1954, and was awarded the Order of Merit and made a Life Peer in 1977. He was also the first Chancellor of Strathclyde University. A man of strong personality, he was known affectionately at Cambridge as Todd Almighty, later Lord Todd Almighty. As a trustee of various charities and as a member of many government committees he has played a substantial part in promoting scientific activity both in Britain and abroad.

TOLANSKY, Samuel
(1907–73)

English physicist, born in Newcastle upon Tyne. He was educated at the university in his home town and at Imperial College, London, and from 1934 held teaching posts at the University of Manchester. In 1947 he moved to become Professor of Physics and then head of department at Royal Holloway College of the University of London. His research interests included nuclear physics, in particular the areas of hyperfine structure and line spectra, using the latter to investigate the nuclear spin of tin, bromine and iodine. He also carried out work in multiple-beam interferometry and studied the surface structure and properties of diamonds. His publications included *History and Use of Diamonds* (1962), *High Resolution Spectroscopy* (1948) and *Surface Microtopography* (1960). He was elected FRS in 1952.

TOMBAUGH, Clyde William
(1906–)

American astronomer, born in Streator, Illinois. Too poor to attend college, he built his own 9 inch telescope, and in 1929 became an assistant at the Lowell Observatory. In 1933 he won a scholarship to the University of Kansas and received an MA in 1936. **Lowell** had predicted the existence of an outermost planet, which he named Planet X, from his estimates of the perturbation of the orbits of Uranus and Neptune. Tombaugh joined the search team that was run by **Slipher**, Director of Lowell Observatory from 1926. Tombaugh devised the blink comparator which enabled him to detect if anything had moved in the sky between the taking of two celestial photographs, a few days apart. On 18 February 1930, thanks to his meticulous technique, he discovered Pluto in the constellation of Gemini. It was too faint to be the expected Planet X, and Tombaugh spent another eight years looking, but without success. In 1946 he became astronomer at the Aberdeen Ballistics Laboratories in New Mexico, and he was later appointed astronomer (1955–9), associate professor (1961–5) and professor (from 1965) at New Mexico University.

TOMONAGA, Sin-Itiro
(1906–79)

Japanese physicist, born in Kyoto. He was educated at Kyoto Imperial University where he was a classmate of **Yukawa**. After graduating (1929) they both remained at Kyoto as unpaid assistants. Tomonaga then joined Yoshio Nishina at Riken, the Institute for Physical and Chemical Research in Tokyo (1932). Nishina had studied in Copenhagen for seven years and passed on the spirit of the Danish approach to quantum mechanics to Tomonaga. During the next five years Tomonaga published many papers on positron creation and annihilation, and one on high-energy neutrino–neutron scattering in which he emphasized the increased probability of a neutrino reaction with increasing energy. In 1937 he moved to Leipzig to work with **Heisenberg**, and his work here on a model of the nucleus earned him a DSc from Tokyo University (1939). He returned to Riken (1939) and during World War II did minor research on radar. During this time he produced his most important work, a relativistic quantum description of the interaction between a photon and an electron. For the resulting theory of 'quantum electrodynamics' he shared the 1965 Nobel Prize for Physics with **Feynman** and **Schwinger**, who had derived the same results independently. He increasingly became involved in scientific administration, becoming President of the Science Council of Japan (1951) and President of Tokyo University (1956).

TONEGAWA, Susumu
(1939–)

Japanese molecular biologist, born in Nagoya. Tonegawa studied chemistry at the University of Kyoto, graduating in 1963. He joined the department of biology of the University of San Diego where he received his PhD in 1968. In 1971 he accepted an appointment at the Institute for Immunology in Basle, Switzerland. Tonegawa applied the restriction enzyme and recombinant DNA techniques to the problem of the origins of antibody diversity. There were two schools of thought—one was that the genes controlling the production of all the different antibodies that an animal can make are inherited. The other was that in the formation of the antibody-manufacturing cells, the B-lymphocytes, the genes somehow undergo changes allowing them to produce a new wide range of antibodies. Tonegawa's research confirmed the second view, and provided details of the mechanism by which the genes are changed. In 1981 he returned to the USA as Professor of Biology at MIT, where he has applied the techniques of molecular biology to another aspect of the immune system—the action of the T-lymphocytes. Since 1988 he has also been Howard Hughes Medical Institute Investigator. Tonegawa was awarded the 1987 Nobel Prize for Physiology or Medicine.

TORREY, John
(1796–1873)

American botanist, born in New York City. He qualified in medicine and taught physical sciences at the West Point Military Academy, joining the US army as assistant surgeon in 1824. He was Professor of Chemistry at West Point, the US Military Academy, and at Cornell, before becoming Chief Assayer at the US Assay Office in New York from 1854 to 1873. He founded the New York Lyceum of Natural History, and became Emeritus Professor of Botany and Chemistry at Columbia College in 1856. However, throughout his life his main interest was botany, and he prepared several floras for North America and also collected over 50000 plant species; his collection formed the basis for the herbarium of the New York Botanical Gardens. The genus *Torreya* in the yew family is named after him, as well as the Torrey Botanical Club. His publications include *A Flora of the Northern and Middle Sections of the United States* (1824), *A Flora of North America* (1838–43, with **Asa Gray**) and *Flora of the State of New York* (1843).

TORRICELLI, Evangelista
(1608–47)

Italian physicist and mathematician, probably born in Faenza. In 1627 he went to Rome, where he devoted himself to mathematical studies, and later made the acquaintance of **Galileo**'s *Dialoghi delle nuove scienze* (1638). He was inspired by that work to develop some of its propositions in his own treatise *De Motu* (1641), which led Galileo to invite him to become his amanuensis; on Galileo's death in 1642 he was appointed mathematician to the grand-duke and professor to the Florentine Academy. He discovered that it is because of atmospheric pressure that water will not rise above 33 feet in a suction pump. To him are owed some of the fundamental principles of hydromechanics, and in a letter written in 1644 he gave the first description of a mercury barometer or 'torricellian tube'. His skill in grinding near-perfect lenses enabled him to build remarkably effective telescopes, and he even made a simple microscope by using a small glass sphere as a lens. He published a large number of mathematical papers on such topics as conic sections, the cycloid and logarithmic curves, and he determined the point still known as Torricelli's point on the plane of a triangle for which the sum of the distances from the vertices is a minimum.

TOURNEFORT, Joseph Pitton de
(1656–1708)

French botanist, born in Aix, and educated by Jesuit priests, as his family had destined him for the church. From his father's death (1677) until 1683, Tournefort studied at Montpellier. He became Professor of Botany at the Jardin du Roi, Paris (1683–1708), where he enlarged the living collections by undertaking a series of European expeditions (1685–9), from Holland to the Iberian Peninsula. By 1689 he had become one of Europe's foremost botanists. In 1694 he published a three-volume textbook, *Éléments de Botanique*, translated into Latin as *Institutiones Rei Herbariae* (1700). From 1700 to 1702 he travelled in the Levant with the artist Claude Aubriet; his account of the journey, *Relations d'un Voyage du Levant* ... was published posthumously (1717). Tournefort's fundamental contribution to botany was the creation of the modern concept of the genus, defining it by diagnosis and carefully distinguishing between the describing and the naming of a genus. His generic concept was followed by **Linnaeus**, **Adanson** and **Antoine Laurent de Jussieu**. Most of the 725 plant genera he recognized are still accepted. Tournefort made outstanding contributions towards the establishment of objectivity in taxonomy. Although his own classification was highly artificial, his methods ultimately led to the later development of natural classification. He was also interested in mineralogy and shells, and was a noted physician who played a key role in the emancipation of botany from medicine.

TOWNES, Charles Hard
(1915–)

American physicist, born in Greenville, South Carolina. He was educated in his home town at Furman University, then at Duke University and Caltech, where he completed his PhD in 1939. During World War II he worked at the Bell Telephone Laboratories, designing radar bombing systems and navigational devices. He also made the first studies of the microwave spectra of gases. In 1948 he joined Columbia University where he used short-wavelength radar to probe the electrical and magnetic interaction between the rotating motion of molecules and the 'spinning' nuclei within. In need of an intense source of microwaves to extend these investigations, Townes turned to the problem of designing such a source. This led him in 1951 to use highly energized oscillating ammonia molecules to produce microwaves. Gaseous ammonia molecules were first excited by pumping energy into them either electrically or thermally. By passing through the gas a weak beam of microwaves with the same frequency as the natural molecule oscillating frequency, Townes triggered the ammonia molecules to emit their own microwave radiation. Intense, coherent radiation was produced by the molecules with only a narrow range of frequencies, in the first operational maser (Microwave Amplification by Stimulated Emission of Radiation), the forerunner of the laser. For his work on the maser and fundamental work in the field of quantum electronics that underpinned the maser–laser principles, Townes was joint winner of the Nobel Prize for Physics with **Basov** and **Prokhorov** in 1964. He was appointed to professorships at MIT (1961–7) and the University of California at Berkeley (1967–86), and has been active in developing microwave and infrared astronomy techniques. In 1968, with his associates at Berkeley, he discovered the first polyatomic molecules in interstellar space (ammonia and water).

TOWNSEND, Sir John Sealy Edward
(1868–1957)

Irish physicist, born in Galway. The son of a civil engineering professor at Queen's College in Galway, Townsend graduated from Trinity College, Dublin, in 1890. After teaching mathematics in Ireland he went to Trinity College, Cambridge (1895), as one of **J J Thomson**'s first research students at the Cavendish Laboratory. He became Wykeham Professor of Physics at Oxford in 1900. By 1897 Townsend had succeeded in making a direct determination of the elementary electrical charge, and his main area of research continued to be the kinetics of ions and electrons in gases. After 1908 he concentrated on the study of the properties of electron clouds. His investigations into the electron's mean free path were later seen to have implications for an understanding of its wave-like nature within quantum theory. Townsend was knighted in 1941.

TRAUBE, Ludwig
(1818–76)

German pathologist, brother of **Moritz Traube**. Born in Ratibor, he was educated at Breslau and then at Berlin. Overcoming much petty anti-Semitism, he became professor at the Berlin Friedrich-Wilhelm Institute (1853) and at Berlin University (1872). Much influenced by **Purkinje** and **Johannes Müller**, he developed the study of experimental pathology in Germany, using animal experimentation. His most important work focused upon the pathology of fever, and the effects of various drugs and other stimuli and inhibitors upon muscular and nervous activity. He explored the effects of digitalis and other drugs in the management of heart disease and described the rhythmic variations in the tone of the vasoconstrictor centre (now known as the Traube–Hering waves). Traube won an unmatched reputation for his researches in experimental pathology, while remaining an unregenerate therapeutic nihilist.

TRAUBE, Moritz
(1826–94)

German wine merchant and chemist, born in Ratibor, Silesia (now part of Poland). He studied chemistry in Berlin and Giessen, and was encouraged by **Liebig** and by his elder brother **Ludwig Traube** to pursue the study of fermentation. In 1849 he took over the family wine business in Ratibor, transferred it to Breslau in 1866, and ran it until 1886. Both in Ratibor and in Breslau he carried out research in his private laboratory, mainly on fermentation. He showed that an 'unorganized ferment' (later called an enzyme) produced by yeast was responsible for fermentation. Traube studied and classified various enzymes. He also showed that protein was not the source of muscle energy, contrary to the views of **Liebig**. In 1867 he discovered that a membrane of cupric ferrocyanide is permeable to water but not to certain solutes. Around 10 years later such a semi-permeable membrane was used by **Pfeffer** in his studies of osmotic pressure, which were interpreted by **van't Hoff**.

TRAVERS, Morris William
(1872–1961)

English chemist, renowned for his work on the rare gases. He was born in London and educated at the universities of London and Nancy, France. He was a demonstrator (1894–8) and later assistant professor (1898–1903) at University College London, before moving to the chair at University College, Bristol. During 1906–11 he did much to establish the Indian Institute of Science at Bangalore, of which he became director, and during World War I he was put in charge of Duroglass Ltd at Walthamstow. He was later President of the Society of Glass Technology. He worked as a consultant chemical engineer, returning to Bristol from 1927 to 1939. During World War II he was a consultant on explosives to the Ministry of Supply. At University College Travers helped **Ramsay** to determine the properties of argon and helium. They found helium in meteorites while heating the meteorites in search of new gases. In May 1898 Travers evaporated some liquid air and found that a spectroscopic analysis of the least volatile fraction showed lines never observed before. This was krypton. Liquefying air again in June, Ramsay and Travers collected the most volatile fraction and discovered neon. In July they discovered xenon, the least volatile of all. As a spin-off from his work on the rare gases, Travers developed an apparatus for liquefying hydrogen (though after **James Dewar**) and helped to set up experimental liquid air plants in several European countries. In 1920 he began work on high-temperature furnaces and fuel technology, and in 1927 he established a research group at Bristol to work on organic gases at high temperatures. He also wrote a biography of Ramsay (1956) and arranged 24 volumes of his papers. Travers was elected FRS in 1904. He died in Stroud, Gloucestershire.

TREVIRANUS, Gottfried Reinhold
(1776–1837)

German biologist and anatomist. Born in Bremen, the brother of Ludolf Christian Treviranus, he studied medicine and mathematics at the University of Göttingen, before being appointed Professor of Mathematics at Bremen. His *magnum opus*, *Biologie* (1802–22), summarized all that was known at the time about the phenomena of life and proved extremely influential in introducing the new concept of 'biology' to the German public. Treviramus carried out important histological and anatomical research on vertebrates, investigating the reproductive organs of worms, mulluscs and arachnids. He was famous for his anatomical study of the louse.

TRUMPLER, Robert Julius
(1886–1956)

Swiss-born American astronomer, born in Zürich and educated at the university there and at the University of Göttingen, Germany (1911). He worked with the Swiss geodetic survey before moving to the USA, where he served on the staff of the Allegheny Observatory (1915–19) and Lick Observatory (1919–38). From 1938 until his retirement in 1951 he was professor at the University of California at Berkeley. At Lick Observatory he studied the dimensions and brightnesses of open star clusters in the Milky Way and explained the disproportionate faintness of the more distant ones as the effect of absorption of light in interstellar space (1930). This important discovery led to a reassessment of the distance scale of our galaxy. He also demonstrated that the light of distant clusters is reddened as well as dimmed, an effect caused by small grains of dust in the spiral arms of the galaxy. Trumpler was elected to the US National Academy of Sciences in 1932.

TSIOLKOVSKY, Konstantin Eduardovich

(1857–1935)

Russian astrophysicist and pioneer of rocket propulsion, born in the village of Izheskaye in the Spassk district. Almost totally deaf from the age of nine, and largely self-educated in science, in 1881 he worked out the kinetic theory of gases, unaware that **Maxwell** had already done so more than a decade earlier. By 1895 his published papers had begun to mention the possibility of space flight, and three years later he pointed out that this would require the development of liquid fuel rocket engines. In 1903 he published his seminal work, *Exploration of Cosmic Space by means of Reaction Devices*, which established his reputation as the father of space flight theory. Unfortunately he lacked the resources to carry out any experimental work, and the first liquid fuel rocket was launched by **Goddard** in the USA in 1926. Tsiolkovsky continued to publish scientific papers, and also gave his ideas on space travel wider circulation by writing a number of works of science fiction. Towards the end of his life the Soviet government became interested in space flight and his work was belatedly given the recognition it deserved. Around 22 years after his death it was intended to launch the first Soviet satellite on the hundredth anniversary of his birth on 17 September 1857; in fact Sputnik I was 29 days late, but honoured his memory none the less.

TSWETT, or TSVETT, Mikhail Semenovich

(1872–1919)

Russian organic chemist, pioneer of chromatography. He was born in Asti, Italy, and studied at Geneva and Kazan before being appointed as assistant at Warsaw University in 1903. From 1908 to 1917 he taught botany and microbiology at Warsaw Technical University which moved to Nizhni Novgorod during World War I. In 1917 Tswett was appointed Professor of Botany and Director of the Botanical Gardens at Yuryev (Tartu) University, which was transferred to Voronezh in 1918. As a student he investigated chlorophyll and by 1900 had established that it contains at least two green pigments. However, traditional methods of organic analysis proved too destructive for delicate organic materials, and he began to look for a method of separating substances physically in an unchanged state. He built on earlier work by **Runge** and **Schönbein** which made use of the fact that different pigments have different capacities for being adsorbed by paper. Tswett ground leaves in a mixture of ether and alcohol, shook the mixture with distilled water to remove the alcohol, and allowed the solution to filter through a tube filled with ground calcium carbonate, which after much trial and error he had found to be the best adsorbent for his purpose. The pigments separated into different layers in the tube, and when the powder was extracted and cut up, they could be washed out separately for study. By this means Tswett added chlorophyll c to chlorophyll a and b, and discovered a family of yellow pigments which he called 'carotenoids'. His method, which he named 'chromatography', did not attract much interest until the 1930s. Since then it has developed into a number of highly specialized and widely used techniques which are employed when complex mixtures have to be separated or substances purified. Tswett died in Voronezh.

TULASNE, Louis René

(1815–85) and

Charles

(1816–84)

French mycologists, born in Azay-le-Rideau. Louis and his brother Charles carried out important researches on the structure and development of fungi, and wrote *Selecta Fungorum Carpologia* (1861–5), which is notable for its many fine illustrations. Their work was the first exact study of the smut and rust fungi (Uredinales and Ustilaginales). They followed this with a long series of papers on different fungi, especially underground species. They also studied the development of the ergot fungus on rye (1853), spore formation and germination in *Puccinia*, *Ustilago* and others, and the sexual organs of *Peronospora*.

TURING, Alan Mathison

(1912–54)

English mathematician, born in London. Educated at Sherborne, he read mathematics at King's College, Cambridge and also studied at the Institute for Advanced Study in Princeton. In 1936 Turing made an outstanding contribution to the development of computer science in his paper *On Computable Numbers* (1936), in which he outlined a theoretical 'universal' machine (or 'Turing' machine) and gave a precise mathematical characterization of the concept of computability. In World War II he was a member of the Bletchley Park code-breaking team, working on 'Colossus' (a forerunner of the modern computer), before joining the National Physical Laboratory in 1945. Here Turing was able to put his theoretical ideas on computing into practice, with his brilliant design for the Automatic Computing Engine (ACE). Frustrated with the slow progress in constructing the ACE, in 1948 he accepted a post at Manchester University, where work on the Manchester Mark I computer was in full swing under **Kilburn** and Sir **Frederic Calland Williams**. Turing made contributions to the programming of the machine, researched some complicated theories in plant morphogenesis and explored the problem of machine 'intelligence'. Prosecuted for an alleged homosexual offence in 1952, he committed suicide by swallowing cyanide two years later. His stature as a major pioneer of computer science has since grown steadily.

TURNER, James Johnson

(1935–)

English inorganic chemist, born in Darwen, Lancashire. He was educated at King's College, Cambridge, where he received his BA (1957) and PhD (1960). He then worked for a year with **Longuet-Higgins** and for two years at Berkeley, California, with **Pimentel**. He was a lecturer in chemistry at Cambridge University until 1972, when he moved to Newcastle University as professor. Since 1979 he has been Professor of Inorganic Chemistry at Nottingham University. Turner's main discoveries have been concerned with the characterization of intermediates in organometallic photochemistry (using matrix isolation to trap such species within a solid host at very low temperatures, low-temperature solutions, and very fast infrared spectroscopy to study such species in solution at room temperature on a very rapid timescale). In this way he has been responsible for the elucidation of many reaction pathways in inorganic photochemistry. Recently he turned his attention to the use of very fast spectroscopy to monitor the excited states of coordination compounds, and to the study of extremely fast organometallic exchange processes by observing infrared bandwidths. He was elected FRS in 1992.

TURNER, William

(c.1510–68)

English clergyman, physician and naturalist, known as the 'father of British botany', born in Morpeth, Northumberland. A Fellow of Pembroke Hall, Cambridge, he became a Protestant, and to escape religious persecution in England travelled extensively abroad, studying medicine and botany in Italy. He formed a close friendship with **Gesner** in Zürich, and became the author of the first original English works on plants, including *Libellus de re Herbaria Novus* (1538), the first book in which localities for native British plants were recorded. In his *Names of Herbes* (1548) he stated that he had intended to produce a Latin herbal, but had been persuaded to wait until he had observed the plants growing in England. His major work is *A new Herball*, published in three instalments (1551–62). This book demonstrated Turner's independence of thought and observation, but used woodcuts prepared for the octavo edition of Fuschs's *De Historia Stirpium* of 1545. He was Dean of Wells (1550–3), but left England during the reign of Mary I; he was restored to Wells in 1560. He named many plants, including goatsbeard and hawkweed. The basis he laid for 'a system of nature' was developed by **Ray** in the following century.

TWORT, Frederick William

(1877–1950)

English bacteriologist, born in Camberley, Surrey. He studied medicine in London, and became Professor of Bacteriology there in 1919. He was the last Superintendent of the Brown Institution of the University of London. Twort studied Jöhne's disease and methods of the culture of acid-fast organisms, and extracted an 'essential substance' (later shown to be of the vitamin K group) from dead tubercle bacilli. In the early part of this century he discovered a 'transmissible lytic agent', named some years later by **d'Hérelle** as 'le bactériophage', a virus for attacking certain bacteria. In many ways the discovery of the invasion of bacteria by viruses formed the beginnings of molecular biology.

TYNDALL, John

(1820–93)

Irish physicist, born in Leighlin-Bridge, County Carlow. Largely self-educated, he was employed on the ordnance survey and as a railway engineer, before studying physics in England and at Marburg in Germany under **Bunsen**. He became professor at the Royal Institution in 1854. In 1856 he and **T H Huxley** visited the Alps and collaborated in *The Glaciers of the Alps* (1860), when he made the first ascent of the Weisshorn. In 1859 he began his researches into the action of radiant heat on gases and vapours. He later investigated the acoustic properties of the atmosphere, and the behaviour of light beams passing through various substances, in the course of which he discovered in 1869 the 'Tyndall effect', the scattering of light by colloidal particles in solution, thus making the light beam visible when viewed from the side. His suggestion that the blue colour of the sky is due to the greater scattering of the shorter wavelength blue light by the colloidal particles of dust and water vapour in the atmosphere was confirmed by the theoretical studies of Lord **Rayleigh** (1842–1919). A prolific writer on scientific subjects, his presidential address to the British Association in 1874 in Belfast was denounced as materialistic. He died from accidental poisoning with chloral.

TYRRELL, George Walter

(1883–1961)

English igneous petrologist, born in Watford. He studied geology at the Royal College of Science, London, and joined the staff of the University of Glasgow (1906) as an assistant to **John Gregory** at the start of an active career as an outstanding teacher and researcher. His move to Scotland was rewarded with fertile ground for petrological studies which he pursued with vigour. He was a pioneer of igneous petrogenesis, and undertook important studies of rocks from Scotland, particularly Arran, as well as Rockall, Iceland, Jan Mayen, Antarctica, South America and Africa. He was the author of the widely distributed textbook *Principles of Petrology* (1926), as well as *The Geology of Arran* (1928), *Volcanoes* (1931), *The Earth and its Mysteries* (1953) and other petrological works.

TYRRELL, Joseph Burr

(1858–1957)

Canadian economic and structural geologist, stratigrapher, explorer and mining engineer, born in Weston, Ontario. He was educated at Upper Canada College, Toronto, and in pursuit of a legal career, he graduated in arts from the University of Toronto (1880); on medical advice, however, he changed course to pursue an outdoor career. He undertook pioneering geological work with the Canadian Geological Survey in Manitoba, Saskatchewan and Alberta, and was involved in important studies of the glaciation of north central Canada and the Yukon goldfields. He discovered coal and fossil dinosaurs in Alberta. In 1893 he undertook a 3 200 mile trek across northern Canada covering some 600 miles on foot, and in 1898 he reported for the government on the geological situation in Yukon, where the Klondike goldrush was at its height. He worked in industry from 1906, developing the Kirkland Lake gold deposits. The Tyrrell Museum in Alberta was named in his honour.

TYSON, Edward

(1651–1708)

English physician, born in Bristol. Educated at Magdalen Hall, Oxford, he graduated BA in 1670 and MA in 1673. Influenced by the naturalist Robert Plot, he undertook botanical and zoological studies while in Oxford, where he also commenced medical studies, gaining his MB in 1677. Tyson set up in practice in London, being appointed physician to Bridewell and Bethlehem hospitals, while continuing to pursue anatomical experiments and dissections. He published papers in the *Philosophical Transactions* of the Royal Society of London on pathological subjects. In 1690 he published a monograph on the anatomy of the porpoise which expatiated on his wider natural history interests. Tyson saw comparative anatomy as the discipline that would reveal the underlying structural unity of nature, whose plan was given by the 'great chain of being'. The porpoise was, he believed, transitional in morphological terms between fish and land creatures. Tyson published similar anatomical work on the lumpfish, rattlesnake and shark. In 1697, with William Cowper, he dissected an opposum, drawing attention to its peculiar reproductive system. He is remembered for his 1699 work on what he called the 'orang-outang' (in reality a chimpanzee brought back from Malaya). He viewed this primate as intermediate between man and the apes on the great chain of being. Tyson's work helped spark a fierce and continuing debate about man's relations to other primates.

U

UHLENBECK, George Eugene
(1900–88)

Dutch–American physicist, born in Batavia (now Jakarta), Indonesia. He studied at Leiden and was awarded his PhD in 1927. From 1927 until 1960 he worked at the University of Michigan, where he was appointed Professor of Theoretical Physics in 1939. To explain the results of the **Stern**–Gerlach experiment which showed that when a beam of particles pass through a magnetic field some are deflected in one direction and some in the opposite direction, and the observation of close doublets of spectral lines, Uhlenbeck and his fellow student **Goudsmit** proposed that electrons in atoms can have intrinsic spin angular momentum as well as orbital angular momentum. Initially this was not accepted as physicists found it difficult to ascribe rotation to electrons, but later **Dirac**'s theory of relativistic quantum mechanics showed that spin is an intrinsic property of electrons.

ULAM, Stanislaw Marcin
(1909–)

American mathematician and research scientist, born in Lwów, Poland (now Lvov, the Ukraine). Educated at the Lwów Polytechnic Institute where he received an MA in 1932 and a DSc in 1933, a common interest in set theory led to contact with **von Neumann**. In 1936 von Neumann invited Ulam to the USA, where he secured a post for him at the Institute for Advanced Study, and later involved him in the atomic bomb project at Los Alamos in 1944 (Ulam had been naturalized in the previous year). This required massive calculations, and Ulam and von Neumann utilized existing calculating machines and applied probabilistic (the so-called 'Monte Carlo') methods. After World War II, Ulam continued his interest in using machines to solve mathematic and scientific problems, and held several professorships. He maintained a lifelong friendship and correspondence with von Neumann, and took great interest in his later work on artificial intelligence.

ULUGH-BEG
(1394–1449)

Ruler of Turkestan from 1447. A grandson of Tamerlane who drew learned men to Samarkand, he founded an observatory there in 1420, and between 1420 and 1437 prepared new planetary tables and a new star catalogue, the latter being the first since that of **Ptolemy**. Positions were given with precision; this was the first time that latitude and longitude were measured to minutes of arc and not just degrees. His instruments must have been excellent and included a quadrant of 60 feet in radius. He also wrote poetry and history. After a brief reign, he was defeated and slain by a rebellious son.

UNSÖLD, Albrecht Otto Johannes
(1905–)

German astrophysicist, born in Bolheim, Württemberg, and educated at the realgymnasium in Heidenheim and at the universities of Tübingen and Munich. At Munich he obtained a doctorate in theoretical physics under **Sommerfeld** (1928) and continued on the academic staff until his appointment as lecturer at the University of Hamburg (1930–2), where his colleagues included **Baade** and **Rudolph Minkowski**. In 1932 he was appointed Professor of Theoretical Physics and director of the observatory at the University of Kiel where he established a flourishing school of theoretical astrophysics and has remained, apart from short visits abroad, until now; he is currently Emeritus Professor. Unsöld's principal field of research has been the physics of stellar atmospheres: he was the discoverer of the hydrogen convection zone (1931) which explains how heat energy is transported upwards to the Sun's photosphere, and his fundamental *Physik der Sternatmosphären* ('Physics of Stellar Atmospheres', 1938) remains a classic on the subject. He is also the author of the successful college textbook *The New Cosmos*, first published in German and English in 1966. Unsöld was awarded the gold medal of the Royal Astronomical Society in 1957.

UNVERDORBEN, Otto
(1806–73)

German chemist, born in Dahme, near Potsdam. He studied for a year at the Pharmaceutical Institute at Erfurt and later entered the family manufacturing business at Dahme. All of his scientific work took place before he went into business. He experimented with the destructive distillation of organic substances, including resins, shellacs and animal oils. He discovered guaiacol by the destructive distillation of wood, and also aniline (which he termed 'crystalline') by the dry distillation of indigo. Aniline, which was later found to be a primary aromatic amine, became important to the German dye industry. It was extracted from coal tar and gave Germany a virtual monopoly of synthetic dyes until World War I. Today most of the aniline produced is used in the manufacture of urethane polymers, rubber and chemicals for agriculture. Unverdorben also investigated the fluorides, and mistakenly believed he had synthesized chromium hexafluoride. He died in Dahme.

URBAIN, Georges
(1872–1938)

French chemist who discovered many of the rare earths. He was born in Paris and educated at the Éole de Physique et de la Chimie in Paris, where **Pierre Curie** was on the staff, and at the University of Paris. He was Professor of Analytical Chemistry at the Sorbonne from 1906 to 1928, when he became Director of the Institute de Chimie de Paris. Between 1895 and 1912 he performed more than 200 000 fractional crystallizations in which he separated the elements samarium, europium, gadolinium, terbium, dysprosium and holmium. In 1907 he discovered lutetium in ytterbium, previously thought to have been in a pure form, and in 1922 isolated hafnium at the same time as **Hevesy** and **Coster** working in Copenhagen. Urbain lent samples of the rare earths to **Marie Curie** who was investigating the radioactive properties of all the known elements. In parallel with his work on the rare earths, Urbain wrote on isomorphism and phosphorescence. He died in Paris.

UREY, Harold Clayton
(1893–1981)

American chemist, born in Walkerton, Indiana. He taught in rural schools (1911–14) and then studied at Montana State University (BS, 1917) and the University of California, Berkeley (PhD in chemistry, 1923). In 1923–4 he worked with **Niels Bohr** in Copenhagen. From 1924 to 1929 he was an associate in chemistry at Johns Hopkins University and he was on the chemistry faculty of Columbia University, New York, from 1929 to 1945 (as full professor from 1934). He worked at the chemistry department and Institute for Nuclear Studies of the University of Chicago from 1945 to 1958, and thereafter continued to be scientifically active in retirement for many years at the University of California, La Jolla. Urey's earliest researches were on atomic and molecular spectra and structure, but he is chiefly remembered for his discovery in 1932 of heavy hydrogen (deuterium), jointly with Ferdinand Brickwedde and George Murphy. Subsequently he made many studies of the separation of isotopes and isotopic exchange reactions. During World War II he was prominent in the attempts to separate uranium-235 for the atomic bomb as part of the Manhattan Project. He later advocated an international ban on nuclear weapons. After 1945 his research interests moved to geochemistry and cosmochemistry, and he wrote *The Planets* (1952) and *Some Cosmochemical Problems* (1961). Urey was awarded the Nobel Prize for Chemistry in 1934. He received the Davy Medal of the Royal Society in 1940 and became a Foreign Member in 1947. He was made an Honorary Fellow of the Chemical Society in 1945, and also received the Priestley Medal of the American Chemical Society.

USSING, Hans Henrikson
(1911–)

Danish biophysicist who graduated in physiology from the University of Copenhagen in 1934. He joined an expedition to study plankton, but came across a technical problem which he discussed with his former professor **Krogh**. Krogh used the opportunity to divert him instead into a study of permeability problems using 'heavy water', water containing the deuterium isotope of hydrogen. After examining permeability parameters in frog skin, Ussing started to analyse protein structure and synthesis, using deuterium as a tracer, but was disrupted by the German invasion of Denmark. During World War II he became a lecturer in biochemistry, and after liberation returned to the zoophysiology department to study active and passive transport mechanisms across biological membranes. He developed a comprehensive theoretical framework for working with radioactive tracers, techniques he pioneered as isotopes were then coming into widespread use in biology. His experiments particularly utilized a frog skin preparation mounted in a specially constructed apparatus, known as an Ussing chamber, so that each surface of the skin was bathed in a separate fluid. The chemical manipulation and analysis of the medium on each side provided information about transport processes across the skin. In 1951 Ussing proposed a cyclical carrier mechanism of permeability, linking active sodium and potassium transport across the membrane, and during the following decade active transport of sodium ions was demonstrated across many membrane preparations. The fundamental properties of biological tissues that Ussing discovered have been important in several areas of medicine, such as absorption, diffusion and secretion in tissues like the gut and kidneys. Apart from a year as a Rockefeller Fellow at the University of California (1948) he remained in Copenhagen, as Research Professor and head of the isotope division of the department of zoophysiology (1951–60) and head of the Institute of Biological Chemistry (1960–80), where he continues as Emeritus Research Professor.

V

VAN ALLEN, James Alfred
(1914–)

American physicist, born in Mount Pleasant, Iowa. He graduated from Iowa Wesleyan College in 1935, then spent time at numerous institutions before becoming head of the physics department at the State University of Iowa. During World War II he developed the radio proximity fuse, a device fitted to explosive projectiles that made use of radio waves to detect the proximity of targets, detonating the explosives once the target came within a certain distance. In anti-aircraft fire this meant that a direct hit was no longer necessary to detonate a shell. This work gained Van Allen expertise in the miniaturization of electronics, which was of great use to him in experiments carried out after the war into the properties of the Earth's upper atmosphere. Using V2 rockets acquired from Germany at the end of the war, Van Allen measured the cosmic-ray intensity at altitudes of around a hundred miles, using radio telemetry to convey the experimental data to Earth. In the 1950s he began work on 'rockoons', rockets launched from balloons. He was involved in the launching of the USA's first satellite, Explorer I (1958), and had ensured that his cosmic-ray detector was amongst the payload. This detector and those in the subsequent Explorer satellites launched in the same year revealed the startling result that above a certain altitude there was much more high-energy radiation than anyone had expected. The satellite observations showed that the Earth's magnetic field traps Earth-bound high-speed charged particles in two doughnut-shaped zones which have become known as the Van Allen belts. Van Allen, who has produced a large number of scientific papers and received numerous scientific awards, has been a member of several US governmental committees concerned with space exploration.

VAN ANDEL, Tjeerd Hendrik
(1923–)

Dutch-born American marine geologist and geological archaeologist, born in Rotterdam. He was raised in Indonesia and educated at the University of Groningen. His career began with an academic appointment at the State Agricultural University at Wageningen (1948–50) followed by a spell in industry as a sedimentologist with Shell Oil Company (1950–6). From 1957 to 1964 he was associated with the Scripps Institute of Oceanography, University of California, before becoming Professor of Geology at the School of Oceanography of Oregon State University (1968–76). He was appointed Wayne Loel Professor of Geology at Stanford University in 1976. Van Andel has devoted his professional life to the investigation of the undersea world and has been involved in tectonic ocean mapping, deep-sea drilling, mineral resource assessment and palaeo-oceanography. He has published extensively on recent sediments of the continents and oceans, the origin and nature of the continental shelf, the geology and geophysics of the mid-ocean ridges, palaeoclimatology, mineral resource assessment and geo-archaeology. He was a member of the first scientific expedition to view and map the Mid-Atlantic Ridge from a deep-sea submersible (1974).

VAN DE GRAAFF, Robert Jemison
(1901–67)

American physicist, born in Tuscaloosa, Alabama. An engineering graduate of the University of Alabama (1923), he continued his studies at the Sorbonne (1924) where **Marie Curie**'s lectures inspired him to study physics. During his PhD research at Oxford, he conceived the design of an improved type of electrostatic generator, in which electric charge could be built up on a hollow metal sphere. At Princeton in 1929 he constructed the first working model of this generator (later to be known as the Van de Graaff generator) in which the charge was carried to the sphere by means of an insulated fabric belt; in this way, potentials of over a million volts could be achieved. At MIT, Van de Graaff adapted his generator for use as a particle accelerator using the high voltages to precisely control the acceleration of charged nuclear particles and electrons to high velocities. This Van de Graaff accelerator became a major research tool of atomic and nuclear physicists. The generator was also employed to produce high-energy X-rays, useful in the treatment of cancer. During World War II he was the Director of the MIT High Voltage Radiographic Project which developed X-ray sources for the examination of the interior structure of heavy ordnance. He was a co-founder of the High Voltage Engineering

Corporation which manufactured particle accelerators, and resigned from MIT in 1960 to devote his time to the corporation.

VAN DE HULST, Hendrik Christoffell
(1918–)

Dutch astronomer, born in Utrecht, where he was educated and obtained his PhD in 1946. After several years at universities in the USA, he became Professor of Astronomy at Leiden and Director of Leiden Observatory in 1970. In 1944 he suggested that interstellar hydrogen might be detectable at radio wavelengths, due to the 21.1 centimetre (1420.4 megahertz) radiation emitted when the orbiting electron of a hydrogen atom flips between its two possible spin states. Due to World War II it was not until 1951 that such emissions were first detected, by **Purcell** and Harold Ewen. The technique has since proved invaluable in detecting neutral hydrogen in both our own and other galaxies. Van de Hulst was also an expert on the scattering and absorption of dust in the interstellar medium, and was the first to suggest that the diameter of a typical interstellar dust grain is around the same as the wavelength of visible light. With C Allen, he showed that the solar F (Fraunhofer) corona is due mainly to forward scattering by dust (1946), this dust being the innermost component of the zodiacal cloud.

VAN DER MEER, Simon
(1925–)

Dutch physicist and engineer, born in the Hague and educated at the Technical University, Delft. He worked at the Philips research laboratories in Eindhoven (1952–5) before becoming senior engineer for CERN, the European nuclear research centre in Geneva. He developed a method known as 'stochastic cooling' to produce a higher intensity beam of anti-protons in accelerators than had been produced before. This technology made possible the experiments which led to the discovery of the field particles W and Z, which transfer the weak nuclear interaction. Van der Meer shared the 1984 Nobel Prize for Physics with **Rubbia** for their separate contributions to this discovery.

VAN DER WAALS, Johannes Diderik
(1837–1923)

Dutch physicist, born in Leiden. The son of a carpenter, at 25 he entered the University of Leiden, graduating in 1865. After teaching physics at Deventer and the Hague, he studied again at Leiden: his doctoral dissertation *On the Continuity of the Liquid and Gaseous States* (1873) was quickly seen to be of major significance in the study of fluids and academic recognition followed. Van der Waals convincingly accounted for many phenomena concerning vapours and liquids observed by **Thomas Andrews** and others, notably the 'critical temperature'. Simple kinetic theories of gases assumed non-interacting molecules with no volume; by postulating the existence of intermolecular forces and a finite molecular volume, van der Waals derived a new equation of state (the van der Waals equation) which agreed much more closely with experimental data, particularly under extreme conditions. This work led to a Nobel Prize for Physics (1910), and was a guide for the future liquefaction of permanent gases such as hydrogen and helium. The weak attractions between molecules (van der Waals forces) were named in his honour. As Professor of Physics at Amsterdam University (1877–1907) he gained distinction as a teacher and as an advocate of the chemical thermodynamics of **Gibbs**.

VANE, Sir John Robert
(1927–)

English pharmacologist, born in Tardebigg, Worcestershire. He studied at Birmingham and Oxford before taking up pharmacology appointments at Yale (1953–5) and the Institute of Basic Medical Sciences, Royal College of Surgeons, London (1955–73), where he was appointed professor in 1966. He subsequently moved to the Wellcome Research Laboratories, Kent (1973–85). Working on adrenergic receptors of the nervous system and the role of the lung in drug uptake and metabolism, he devised a bioassay for the detection (1967) and (in conjunction with a blood platelet aggregometer) characterization of labile and bioactive arachidonic acid (essential fatty acid) metabolites. He investigated the interconversion of prostaglandins (associated with the swelling, reddening and pain of tissue damage) by isomerases (1970), and reported the inhibition of prostaglandin synthesis by asprin (1971). In 1976, studying thromboxane biosynthesis (prostaglandin metabolites causing platelet aggregation and vasoconstriction), he discovered prostacyclin (PGI_2), the short-lived antagonist of platelet aggregation (with a half-life of approximately 20 seconds), which is synthesized from prostaglandin endoperoxidase in the blood vessel wall. The ratio of PGI_2 to PG edoperoxidase appears crucial for the control of thrombus formation in coronary disease and (with renin) for the control of blood pressure in the kidney; in the latter connection Vane also worked on angiotensin and bradykinin. For this work he shared the 1982 Nobel Prize for Physiology or Medicine with **Bergström** and

Samuelsson. Vane was elected FRS in 1974, served as the Royal Society's Vice-President (1985–7) and was awarded its Royal Medal (1989). He was knighted in 1984.

VAN NIEL, Cornelis Bernardus Kees
(1898–)

Dutch microbiologist, born in Haarlem. He studied chemistry at the Technological University in Delft, and later worked with **Kluyver** on iron and sulphur bacteria, and propionic acid bacteria. In 1929 he accepted a position at the Hopkins Marine Station of Stanford University. He made major contributions to the study of photosynthesis in bacteria, particularly in the *Thiorhodaceae* and *Athiorhodaceae*. Van Niel showed that the green and purple sulphur bacteria do not use water as the exclusive hydrogen donor (as in plants), but use hydrogen sulphide and other reduced sulphur compounds instead; this explains both their dependence on these reduced compounds and their inability to produce oxygen. He was able to delineate the light and dark reactions in *Athiorhodaceae*, where oxidation proceeds by way of reactions which do not require the presence of light.

VAN'T HOFF, Jacobus Henricus
(1852–1911)

Dutch physical chemist, born in Rotterdam. He was educated at the polytechnic in Delft, the University of Leiden, Bonn (under **Kekulé**), Paris (under **Wurtz**) and Utrecht, where he took his doctorate in 1874. After a brief period teaching physics and chemistry at the veterinary school in Utrecht, he became Professor of Chemistry, Mineralogy and Geology at the University of Amsterdam (1878). In 1896 he moved to Berlin as a member of the Prussian Academy of Sciences and as professor at the university there. Van't Hoff is rightly regarded as one of the founders of physical chemistry, but his first work was in organic chemistry. He postulated that the four bonds of carbon are directed towards the corners of a tetrahedron (suggested independently by **Le Bel** around the same time). This idea, which provided a basis for explaining the optical activity of certain organic compounds, was published in 1874, but was not recognized as of fundamental importance until later. From 1877 he began to devote himself to physical chemistry and in his *Etudes de Dynamique Chimique* (1884), he developed the principles of chemical kinetics and applied thermodynamics to chemical equilibria. The equation for the effect of temperature on equilibria is commonly called the van't Hoff isochore. His important work on osmotic pressure was published in 1886, and this work was further developed in the next decade in connection with the theory of electrolytic dissociation of **Arrhenius**. With **Ostwald** he founded the journal *Zeitschrift für physikalische Chemie* in 1887. He later studied the phase relationships of the Stassfurt salt deposits. He was awarded the first Nobel Prize for Chemistry in 1901. Van't Hoff was elected a Foreign Member of the Royal Society in 1897, having been awarded its Davy Medal in 1893. He became an Honorary Fellow of the Chemical Society in 1888.

VAN TIEGHEM, Philippe
(1839–1914)

French botanist and biologist, born in Bailleul. Under **Pasteur** he produced a dissertation on ammoniacal fermentation. His second dissertation was on the plant family Araceae. He became professor at the École Normale Supérieure in 1864, where he began work on pure cultures of isolated bacterial strains. Well known for his studies of myxomycetes and bacteria, he produced a new classification of plants based on their gross anatomy. He established the relationship of the blue algae to the bacteria, and showed that coal originated by a fermentation process. He also elucidated the principles of plant symmetry, and produced two important textbooks in which his ideas on plant anatomy and plant physiology were expounded. The *Traité de Botanique* was published in 1884, and the *Éléments de Botanique* in two volumes between 1886 and 1888.

VAN VLECK, John Hasbrouck
(1899–1980)

American physicist, born in Middletown, Connecticut, the son of a mathematics professor. He largely founded the modern theory of magnetism, by applying **Dirac**'s theory of quantum mechanics to the magnetic properties of atoms. Unusually for this field, he was trained in the USA; after study at Wisconsin and Harvard, he took up posts at Minnesota, Wisconsin and finally Harvard (1934–69). In the late 1920s and early 1930s his research was in dielectric and magnetic susceptibilities, culminating in his classic text, *The Theory of Electric and Magnetic Susceptibilities* (1932). He also elucidated chemical bonding in crystals and studied the crystal fields and ligand fields, electric fields experienced by the electrons of an ion or atom due to the presence of neighbouring ions or atoms. These fields influence the energy levels permitted in the system and therefore have far reaching effects on the optical, magnetic and electrical properties of the material.

During World War II van Vleck contributed to the exploitation of radar, showing that atmospheric water and oxygen molecules would cause troublesome absorption at certain wavelengths. In 1977 his pioneering research was recognized with the joint award of the Nobel Prize for Physics, with **Philip Anderson** and **Mott**.

VARMUS, Harold Elliot
(1939–)

American molecular biologist, born in Oceanside, New York. He was educated at Amherst College, Virginia, where he received a BA in 1961, Harvard University where he received his MA in 1962, and Columbia University, New York, where he graduated MD in 1966. He began his career as a surgeon in the US Public Health Service (1968–70) then moved to the University of California Medical Centre, San Francisco, as a lecturer (1970–2), assistant professor (1972–4) and associate professor (1974–9). Since 1979 he has been Professor of Microbiology and Immunology there, simultaneously holding the posts of Professor of Biochemistry and Biophysics (from 1982) and American Cancer Society Professor of Molecular Virology (from 1984). In 1989 he was awarded the Nobel Prize for Physiology or Medicine (jointly with **Bishop**) for his contribution to the discovery of oncogenes. Oncogenes are normal cellular genes which direct various aspects of cellular growth and differentiation. If their production is altered in some way, for example by mutation or viral activation, the faulty protein gives rise to cancer in the cell. The discovery of these genes has been of vital importance in understanding the mechanisms of cancer.

VASKA, Lauri
(1928–)

American chemist, born in Rakvere, Estonia. He attended the universities of Hamburg and Göttingen in Germany before moving to the USA where he received his PhD at the University of Texas. In 1964 he was appointed as professor at Clarkson College of Technology, Potsdam, New York. Vaska is famous for his work on oxygen complexes of transition metals, including his synthesis of the so-called Vaska's complex, $(PPh_3)_2Cl(CO)(O_2)$. This complex takes up O_2 in a reversible manner and is thus used as a model for the uptake and transport of O_2 by haemoglobin in the blood.

VAUQUELIN, Nicolas-Louis
(1763–1829)

French chemist, born in Saint-André-d'Hébertot, Normandy. He was an assistant to pharmacists in Rouen and in Paris where he met **Fourcroy**, becoming his assistant around 1784. This was the beginning of a lifelong friendship and scientific collaboration. Among the posts Vauquelin held during the turbulent years of the French Revolution were those of Inspector of Mines, Professor of Assaying at the School of Mines, and official assayer of the precious metals of Paris. From 1804 to 1809 he held the Chair of Applied Chemistry at the Museum of Natural History and from 1811 to 1822 he was Professor of Chemistry at the Faculty of Medicine of the University of Paris. He is chiefly remembered for the analyses of organic substances that he carried out with Fourcroy and for the discovery of chromium. He obtained a rare Siberian mineral (crocite), which he boiled with potassium carbonate, then reduced the resulting yellow salt with carbon. He also found a new compound, beryllia (beryllium aluminium silicate), but the metal—later named 'beryllium'—was not isolated until 1828 when **Wöhler** and Antoine Bussy both prepared it. Vauquelin and Fourcroy established a chemical factory in Paris, Vauquelin taking the biggest share of its management. He died in Saint-André-d'Hébertot.

VAVILOV, Nikolai Ivanovich
(1887–1943)

Russian botanist and plant geneticist, born in Moscow, brother of the physicist Sergei Vavilov (1891–1951). He trained in Moscow and at the John Innes Agricultural Institute, Merton, Surrey. During World War I, he was asked to go to Iran, where he discovered many crop plant varieties and took seed samples back to Russia—the initial nucleus of his internationally renowned World Collection. He undertook further journeys to the Pamir Mountains and to Afghanistan, and then to the Mediterranean, Abyssinia, the Far East and the Americas. He assembled the world's largest collection of seeds, numbering some 200 000 specimens, over 40 000 of them being varieties of wheat. In 1923 he established a network of 115 experimental stations across the USSR to sow the collection over the widest possible range. He published extensively on the centres of origin of crop plants, formulating the principle of diversity which postulates that, geographically, the centre of greatest diversity represents the origin of a cultivated plant. His commanding international reputation was challenged by the politico-scientific 'theories' of **Lysenko**, who denounced him at a genetics conference (1937) and gradually usurped his position. Arrested in 1940, he died of starvation in a Siberian labour camp.

VENING MEINESZ, Felix Andries
(1887–1966)

Dutch geophysicist, born in the Hague. In 1910 he graduated in civil engineering from Delft Technical University. His first appointment was with the Netherlands State Committee to undertake a gravity survey of the Netherlands where, in order to overcome the problems of making measurements on vibrating peaty subsoil, he designed his classical pendulum apparatus with two pendulums swinging in the same plane; for the theory of this work he was awarded a doctorate in 1915. This led him to investigate means of measuring gravity at sea. Working on a naval submarine to avoid wave turbulence during a cruise to Indonesia in 1923, he achieved marine gravity determination to an accuracy of 1 mgal, comparable to land measurements. In other long submarine voyages which followed, he aimed to determine the Earth's shape, but also discovered that Airy isostasy prevails over the oceans, and that a belt of negative isostatic anomalies exists parallel to trenches, where he calculated an elastic crustal thickness of 35 kilometres. His speculative thoughts on mantle convection, downbuckling and crustal shear failure showed great insights into Earth processes, which with his book *The Earth's Crust and Mantle* (1964) were preludes to modern theories of plate tectonics, although he did not believe that continental drift had occurred during recent geological times. He was Professor of Geophysics at the Rijksuniversiteit Utrecht (1937–66) where the geophysical laboratory is named after him.

VENN, John
(1834–1923)

English logician, born in Drypool, Hull. A Fellow of Caius College, Cambridge (1857), he developed **Boole**'s symbolic logic, and in his *Logic of chance* (1866) the frequency theory of probability. He is best known for 'Venn diagrams', pictorially representing the relations between sets, though similar diagrams had already been used by **Leibniz** and **Leonhard Euler**.

VENTURI, Giovanni Battista
(1746–1822)

Italian physicist, born in Bibiano, near Reggio. He was ordained priest at the age of 23 and in 1773 became Professor of Geometry and Philosophy at the University of Modena. Later he was appointed professor at Pavia, although from 1796 he lived mainly in Paris. In addition to work on sound and colours, he published studies of the geological material contained in the notebooks of Leonardo da Vinci (1797). Venturi is remembered for his work on hydraulics (published 1797), particularly for the effect named after him (the decrease in the pressure of a fluid in a pipe where the diameter has been reduced by a gradual taper) which he first investigated in 1791. The Venturi flow-meter, based on this phenomenon, was invented by the American engineer Clemens Herschel (1842–1930).

VERNADSKY, Vladimir Ivanovich
(1863–1945)

Russian mineralogist, geochemist and biogeochemist, born in St Petersburg. He studied at the University of St Petersburg, where **Mendeleyev**'s brilliant lectures awakened a strong desire for knowledge and its application. From 1886 he was curator of the university's mineral collection, and in 1888 he travelled abroad to work in Munich and with **Le Chatelier** in Paris. He returned to Russia in 1890 to take up a research post at Moscow University, and became Professor of Mineralogy there in 1898. In 1914 he was appointed director of the geological and mineralogical museum of the Academy of Sciences in St Petersburg; he also spent a period in the Ukraine (1917–21), where he founded the Ukranian Academy of Sciences. Vernadsky conducted important research on rock-forming silicates and aluminosilicates and their structures, and on the relationship between crystal form and physiochemical structure. Later in life he concentrated on the study of the chemical composition of plants and animals. He introduced the term 'biosphere' in 1926, and as a result of his biogeochemical studies concluded that all of the main gases in the Earth's atmosphere were generated by living organisms. He was one of the first to recognize the potential of nuclear power as an important source of energy and argued for responsibility on the part of scientists.

VERNIER, Pierre
(1584–1638)

French scientific instrument-maker, born in Ornans near Besançon. He spent most of his life serving the king of Spain in the Low Countries and in 1631 invented the famous auxiliary scale named after him to facilitate an accurate reading of a subdivision of an ordinary scale.

VERNON HARCOURT, Augustus George
(1834–1919)

English chemist, born in London. While at Balliol College, Oxford, he was first a student of Benjamin Collins Brodie and then his assistant. From 1859 to 1902 he was Dr Lee's

Reader in Chemistry at Christ Church, Oxford, where he converted a former anatomy museum into a chemical laboratory. He had many students who were later distinguished, including **Dixon**, **David Chapman** and **Sidgwick**. Vernon Harcourt was elected FRS in 1868 and was President of the Chemical Society from 1895 to 1897. In various capacities he also served the British Association for the Advancement of Science (he was a nephew of **William Vernon Harcourt**, who was essentially the founder of the association). His lifelong research interest was the study of rates of chemical change, pursued in collaboration with the mathematician William Esson. Around 1864–6 they discovered the law of mass action, at about the same time as the Norwegians **Guldberg** and **Waage**. Vernon Harcourt also did much work related to the manufacture of coal-gas, and his pentane lamp provided the British standard of luminosity for many years. He also devised an inhaler for the controlled administration of chloroform as an anaesthetic.

VERNON HARCOURT, William Venables

(1789–1871)

English chemist and clergyman, one of the founders of the British Association for the Advancement of Science, born in Sudbury, Derbyshire. He was educated at home and then spent five years in the Royal Navy before going to Christ Church, Oxford, in 1807. At Bishopsthorpe, which was his first parish, he set up his own chemical laboratory, taking advice from **Wollaston** and **Humphry Davy**. Later he moved to other ecclesiastical posts. He worked on the effect of heat on inorganic compounds, and on methods of making achromatic lenses by combining glasses with different dispersions; in 1824 he was elected FRS. His principal importance, however, lay in the encouragement he gave to others. He became the first President of the Yorkshire Philosophical Society and played a major part in the establishment of the British Association for the Advancement of Science. At the time, the Royal Society was becoming more restrictive and professional in its membership: the idea behind the British Association was that it should be open to everyone interested in science. Today any member of the general public may still join simply by contacting the association's headquarters in London and paying the small annual subscription. The association held its first meeting in 1831 and Vernon Harcourt was elected its president in 1839. In 1861 he succeeded to the family estates at Nuneham, Oxfordshire, where he died 10 years later.

VERRIER, Urbain Jean Joseph Le

See **LE VERRIER, Urbain Jean Joseph**

VESALIUS, Andreas

(1514–64)

Belgian anatomist, one of the first dissectors of human cadavers. Born in Brussels, a pharmacist's son, Vesalius studied in Paris, Louvain and Padua, where he took his degree in 1537. He was appointed Professor of Surgery at Padua University. An ardent champion of dissection, his lectures were somewhat novel, in that he performed dissections himself, instead of following the normal custom of leaving this to an assistant while the professor read from a textbook. He made use of drawings in his lectures. In 1538 he published his six anatomical tables, still largely Galenist, and in 1541 he edited **Galen**'s works. Comprehensive anatomizing enabled him to point out many errors in the traditional medical teachings derived from Galen. For instance, Vesalius insisted he could find no passage for blood through the ventricles of the heart, as Galen had assumed. His greatest work, the *De Humani Corporis Fabrica* ('On the Structure of the Human Body', 1543), was enriched by magnificent illustrations. With its excellent descriptions and drawings of bones and the nervous system, the book set a completely new level of clarity and accuracy in anatomy. Many structures are described and drawn in it for the first time, eg the thalamus. Perhaps upset by criticism, Vesalius left Padua to became Court Physician to the emperor Charles V and his son Philip II of Spain. He died on the way back from a pilgrimage to Jerusalem.

VIETA, Franciscus, or VIÈTE, François

(1540–1603)

French mathematician, born in Fontenay-le-Comte. He became a privy councillor to Henri IV of France and decoded an important Spanish cypher. His *In Artem Analyticam Isagoge* (1591) is probably the earliest work on symbolic algebra, and he devised methods for solving algebraic equations up to the fourth degree. He also wrote on trigonometry and geometry, and obtained the value of π as an infinite product. He believed that his algebra was essentially the method used by the Greeks, but not transmitted in their published works. His influence on **Descartes** is a matter of dispute, and may have been slight as Descartes himself claimed, because Vieta's work was mostly published posthumously and was not widely available; however, Vieta undoubtedly was a major influence on **Fermat**, whose earliest work was written in Vieta's notation.

VIGNEAUD, Vincent du
(1901–78)

American biochemist, born in Chicago. After training in the USA he spent some time in Berlin and Edinburgh before becoming head of the George Washington School of Medicine (1932–8) and then professor and departmental head at Cornell University Medical College (1938–67). A prolific scientist, Vigneaud published on a diversity of topics, particularly related to the identification of amino acids and the resolution and interconversion of their isomeric forms, both chemically and enzymically by racemase enzymes which he discovered. From the early 1930s he studied the chemistry and dietary requirements of the sulphur amino acids, eventually discovering the metabolic pathway from methionine to homocysteine, which then reacts with serine to form cysteine (1942); oxidation of the latter was found to give taurine. Similarly he resolved the interrelated roles of methionine, choline, lecithin and other compounds as donors of methyl groups in intermediary metabolism, and in a long series of experiments, he elucidated the nutritional importance of thiamine and its biosynthesis involving cysteine and pimelic acid. Vigneaud synthesized thiamine (1942), penicillin and the two neurohypophysial peptide hormones, oxytocin and vasopressin (1953–4). For this last achievement he was awarded the 1955 Nobel Prize for Chemistry.

VILLEMIN, Jean-Antoine
(1827–92)

French physician and experimentalist, born in Prey, Vosges. He studied medicine in Strasbourg and Paris, where he received his MD in 1853. A modest man, he continued in medical practice in Paris, but in addition operated a private laboratory where he worked assiduously in his spare time. Among his most fundamental observations was the discovery, in the 1860s, that material taken from the lung of a person with tuberculosis would, when inoculated into an animal, produce tuberculosis in the animal. This work pointed towards a specific infective agent, which **Koch** discovered in 1882.

VINE, Frederick John
(1939–88)

English geophysicist, educated at St John's College, Cambridge. He became associate professor at the department of geological and geophysical sciences at Princeton University (1967–70), then reader (1970–4) and Professor of Environmental Science (1974–88) at the University of East Anglia. He undertook important work with **Matthews** in the interpretation of marine magnetic anomalies and their use in confirmation of the sea-floor spreading hypothesis, as well as palaeomagnetism, plate tectonics and energy resources.

VIRCHOW, Rudolf
(1821–1902)

German pathologist and politician, and founder of cellular pathology. Born in Schivelbein, Pomerania, Virchow graduated in medicine in Berlin, and then secured a junior post in Berlin's great hospital, the Charité. He rose to became Professor of Pathological Anatomy at Würzburg (1849–56). He quickly proved himself a skilful pathologist. In 1845 he recognized leukaemia, and proceeded to study animal parasites, inflammation, thrombosis and embolism. Deeply politically committed, his involvement on the liberal side in the revolutions of 1848 led to his dismissal from his Würzburg post, though he was subsequently reinstated. In 1856 he returned to Berlin as Professor of Pathological Anatomy. In the 1850s Virchow enthusiastically adopted **Schwann** and **Schleiden**'s cell theory, applying it to pathological investigations. He argued that disease originated in cells, or at least was the response of cells to abnormal circumstances. His suggestions led to much fertile work, aided by refinements in microscopes, new dyes for selective staining and the development by **His** of the microtome for making thin sections. With Virchow modern pathology begins. However, he did not see eye-to-eye with his French contemporary **Pasteur** on the germ theory of disease. Virchow saw disease as a matter of the continuous cellular change, rather than as a result of attack by invasive agents. His *Cellularpathologie* (1858) established that tumours and all other morbid structures contained cells derived from previous cells. Virchow was also sceptical about **Charles Darwin**'s evolutionary theory, treating it as a hypothesis only. He remained lastingly politically active. Sitting as a liberal member of the Reichstag (1880–93), he opposed the Chancellor so forcefully that Bismarck challenged him to a duel in 1865 (it was not fought). In practical politics, his efforts in public health in Berlin led to improved water and sewage purification. An enthusiastic archaeologist, he worked on the 1879 dig to discover the site of Troy.

VIRTANEN, Artturi Ilmari
(1895–1973)

Finnish biochemist, born in Helsinki. He was educated at the University of Helsinki and after holding a number of industrial posts associated with the butter and cheese industry, he became Professor of Biochemistry, first at the Finland Institute of Technology (1931–9) and then at the University of Helsinki (1939–48). He studied the bacterial metabolism of sugars to form succinate and lactate, and observed that proteases are released into the growth medium (1931). This related to his earlier finding that root nodules of leguminous plants release nitrogous substances (1927). He discovered aspartase (1932), that legume bacteria can convert aspartate to beta-alanine (1937) and the transamination of aspartate with pyruvate to give alanine (1940). He implicated the formation of hydroxylamine at an early stage in nitrogen fixation by legume root nodules, and established its conversion to aspartate via an oxime intermediate (1938). For these discoveries he was awarded the Nobel Prize for Chemistry (1945). Virtanen also worked on the nutritional requirements of plants, and the plant biosynthesis of carotene and vitamin A. He also isolated and characterized haemoglobin and other pigments from legume nodules, and showed that silage can be preserved by dilute hydrochloric acid. He published *Cattle Fodder and Human Nutrition* in 1938.

VLECK, John Hasbrouck van

See **VAN VLECK, John Hasbrouck**

VOGEL, Hermann Carl

(1841–1907)

German astronomer, born in Leipzig and educated at the university in his native city. He was appointed assistant at the Leipzig Observatory, where Johann Karl Friedrich Zöllner stimulated his interest in astrophysics and in 1870 recommended him to a position at a newly founded private observatory at Bothkamp near Kiel. Using a 28 centimetre refractor, Vogel made some pioneering studies of the spectra of the major planets which established his reputation as an astrophysicist. In 1874 he was called in by the Prussian government as adviser in the planning of the new astrophysical observatory in Potsdam, and in 1882 was appointed its director. Introducing photographic methods into stellar spectroscopy, Vogel was the first to achieve sufficient accuracy to measure radial velocities of stars. In 1889 he discovered the binary nature of Algol from such measurements, thereby opening up the important field of spectroscopic double stars. He had somewhat less satisfaction with the 'great Potsdam double refractor' with an aperture 80 centimetres and a focal length of 12 metres installed in 1899 whose performance did not match that of the large mirror telescopes being built in California. Nevertheless, the Potsdam astrophysical observatory became under his guidance one of the world's leading centres of astrophysics. Vogel continued in office until his death, though in his later years he became a recluse whose chief interest was to play the magnificent organ installed in his official residence at the observatory.

VOGEL, Hermann Wilhelm

(1834–98)

German chemist, known for his contribution to colour photography and colour printing. He was born in Doberlug, Lower Lusatia, and was Professor of Photochemistry, Spectrum Analysis and Photography at the Royal Technical College, Berlin, from 1879 until his death. In 1874 he discovered that certain organic dyes can make silver bromide dry plates sensitive to light in the wavelengths that they themselves absorb. He then developed the orthochromatic photographic plate which was sensitive to red and yellow light. He also took out a patent for three-colour printing which specified that each colour should be printed at a slightly different position, thus avoiding the blurred result when the dots from the different colours of screen were printed on top of each other. He died in Berlin.

VOGT, Peter

(1932–)

German-born American microbiologist. He was educated at the University of Tübingen and moved to the USA as an assistant professor of pathology (1962–6) at the University of Colorado and associate professor until 1967, when he became associate professor (1967–9) and then Professor of Microbiology (1969–71) at the University of Washington. He was Hastings Distinguished Professor of Microbiology at the University of Southern California from 1978 to 1980, and has been Chairman of the Microbiology Department there since 1980. A major field of interest for Vogt has been oncogene transduction by retroviruses. Oncogenes are normal cellular genes which can be activated in a variety of ways to become carcinogenic. One way this can occur is via infection of a cell by a retrovirus. Retroviruses are RNA viruses which take over a cell's machinery to produce a DNA copy which then becomes incorporated into cellular DNA. When this happens the retrovirus occasionally picks up a cellular oncogene or a corrupted copy of this gene that may drastically affect subsequently infected host cells. Vogt showed that there are basically two ways by which oncogenes can be activated by retroviral transduction. The gene sequence may be altered so that it codes for protein with abnormal function or it can be brought under control of powerful viral enhancers or promoters to overproduce a normal gene product. In either case the involvement of retroviruses is clearly of vital importance in certain carcinogenic processes.

VOLTA, Alessandro Giuseppe Anastasio, Count

(1745–1827)

Italian physicist, inventor of the electric battery, born in Como. He was appointed Professor of Physics at Como (1775) and at Pavia (1778). In 1795 he became Rector of Pavia University, but he was dismissed in 1799 for political reasons; later reinstated by the French, he retired in 1815. He was summoned to show his discoveries to Napoleon, and received medals and titles at home and abroad. His main contributions were in electrostatics, current electricity and gas chemistry. He invented the electrophorus (1775), the precursor of the induction machine of which the Wimshurst of the early 1880s became the best-known example, the condenser (1778), the candle flame collector of atmospheric electricity (1787), the calibrated straw electrometer (1780s), and the electrochemical battery, or 'voltaic pile' (1800), which was the first source of continuous or current electricity. It was inspired by a controversy he had with **Galvani** concerning the nature of animal electricity. He also invented an electric spark eudiometer (1776), and his famous 'inflammable air' (hydrogen) electric pistol (1777). **Lavoisier** followed Volta's suggestion that the mixtures of air and hydrogen should be sparked over mercury (and not water), and identified the resultant to be water (1782). His name is given to the SI unit of electrical potential difference, the volt.

VOLTERRA, Vito

(1860–1940)

Italian mathematician, born in Ancona. He was professor at Pisa, Turin and Rome. In 1931 he was dismissed from his chair at Rome for refusing to sign an oath of allegiance to the Fascist government, and he spent most of the rest of his life abroad. He worked on integral equations, where he introduced the idea of studying spaces of functions that proved exceptionally fertile. He also worked on mathematical physics and the mathematics of population change in biology, where he put forward the Lotka–Volterra equations, a pair of differential equations that describe a simple predator–prey population model. Through his breadth of interest and energy, he became a leading representative of Italian mathematics abroad.

VON BRAUN, Wernher

(1912–77)

German-born American rocket pioneer, born in Wirsitz. He studied engineering at Berlin and Zürich and founded in 1930 a society for space travel which carried out experiments at a rocket-launching site near Berlin. Since rockets were outside the terms of the Versailles Treaty, the German army authorities became interested and by 1936, with Hitler's backing, von Braun was director of a rocket research station at Peenemünde, where he perfected the infamous V-2 rockets first launched against Britain in September 1944. A total of 4300 were fired, more than a thousand of which landed on London. At the end of World War II he surrendered, with his entire development team, to the Americans. He became a naturalized American in 1955 and a director of the US army's Ballistic Missile Agency at Huntsville, Alabama, and was chiefly responsible for the manufacture and successful launching of the first American artificial earth satellite, Explorer I, in 1958. From 1960 to 1970 he was Director of the Marshal Space Flight Center, where he developed the Saturn rocket for the Apollo 8 Moon landing (1969). His books include *Conquest of the Moon* (1953) and *Space Frontier* (1967).

VON KLITZING, Klaus

(1943–)

German physicist, born in Schroda. He was educated at the Technical University in Munich and at Würzburg University, where he received his doctorate in 1972. He was then appointed professor at Munich in 1980, and in 1985 became Director of the Max Planck Institute, Stuttgart. In 1977 he presented a paper on two-dimensional electronic behaviour in which the quantum Hall effect, which occurs in semiconductor devices at low temperatures, was clearly implied, but few realized its significance and von Klitzing only appreciated what had occurred in 1980. This caused a major revision of the theory of electric conduction in strong magnetic fields and also provided a highly accurate laboratory standard of electrical resistance. For this work he was awarded the 1985 Nobel Prize for Physics.

VON LAUE, Max Theodor Felix

(1879–1960)

German physicist, born near Koblenz. In 1905 he was offered an assistantship by **Planck**; they became lifelong friends, and an especially appreciated honour must have been the award of the Max Planck Medal from the German Physical Society (1932).

After university posts in Zürich (1912), Frankfurt (1914) and Berlin (1919) he was appointed advisor to the Physikalisch Technische Reichsanstalt, and Deputy Director of the Kaiser Wilhelm Institute for Physics. He lost his influence during the Nazi era because of his opposition to the regime, but was reinstated after World War II, and aged 71 was appointed Director of the former Kaiser Wilhelm Institute for Chemistry and Electrochemistry in Berlin-Dalhlem (1951). He applied the concept of entropy to optics, and demonstrated that **Fizeau**'s formula for the velocity of light in flowing water followed from **Einstein**'s theory of special relativity (of which he was an early adherent). In 1912 he discovered that X-rays are diffracted by the three-dimensional array of atoms in crystals; for this work he was awarded the 1914 Nobel Prize for Physics. He died as the result of a car accident.

VON MISES, Richard
(1883–1953)

American mathematician and philosopher, born in Lemberg, Austria-Hungary (now Lvov, the Ukraine). He was professor at Dresden (1919), Berlin (1920–33), Istanbul, and from 1939 at Harvard, where he became Gordon McKay Professor of Aerodynamics and Applied Mathematics in 1944. An authority in aerodynamics and hydrodynamics, he set out in *Wahrscheinlichkeit, Statistik und Wahrheit* ('Probability, Statistics and Truth' 1928) a frequency theory of probability which has had wide influence, even though not generally accepted.

VONNEGUT, Bernard
(1914–)

American physicist, born in Indianapolis. Educated at MIT, he worked under **Schaefer** at the General Electric Company (1945–52) then at the A D Little Company until 1967, when he became Professor of Atmospheric Science at New York State University. In 1947 he improved a method for artificially inducing rainfall, by using silver iodide as a cloud-seeding agent.

VON NEUMANN, John (Johann)
(1903–57)

Hungarian-born American mathematician, born in Budapest. Educated in Berlin and Budapest, he taught at Berlin (1927–9), Hamburg (1929–30) and Princeton (1930–3) before becoming a member of the newly founded Institute for Advanced Study at Princeton (1933). In 1943 he became a consultant to the Manhattan Project at Los Alamos for the construction of the first atomic bomb, and in 1954 he joined the US Atomic Energy Commission. His best-known mathematical work was on the theory of linear operators, but he also gave a new axiomatization of set theory, later used by **Gödel**. In addition he formulated a precise mathematical description of the recently developed quantum theory (1932), and worked on **Lie** groups. His work during World War II led him to study the art of numerical computation and to design some of the earliest computers, and his theoretical description of a programmable computer governed computer architecture until quite recently. In *The Theory of Games and Economic Behavior* (1944), written with Oskar Morgenstern, he created a theory applicable both to games of chance and to games of pure skill, such as chess. These ideas have since become important in mathematical economics and operational research. He went on to invent the idea of self-replicating machines (arguably later shown to be exemplified through the action of DNA in cells) and cellular automata.

VORONOFF, Serge
(1866–1951)

Russian physiologist, born in Voronezh. Educated in Paris, he became Director of Experimental Surgery at the Collège de France. Later working in Switzerland, he built on **Metchnikoff**'s work on longevism and developed a theory connecting gland secretions with senility. Pioneering endocrinological surgery, Voronoff specialized in grafting animal glands (especially monkey glands) into the aging human body, with a view to restoring potency and ensuring long-life. These experiments won him considerable notoriety. He also grafted the thyroid from monkeys into mentally backward and defective children in an attempt to restore them to normalcy.

VRIES, Hugo Marie de
See **DE VRIES, Hugo Marie**

W

WAAGE, Peter
(1833–1900)

Norwegian chemist, born near Flekkefjord. He entered the University of Christiania (now Oslo) in 1854 to study medicine, but soon changed to science. After working with **Bunsen** at Heidelberg in 1859–60, he returned to Christiania as lecturer in chemistry and in 1866 succeeded Adolf Strecker as Professor of Chemistry. Waage is chiefly remembered for his work with his brother-in-law **Guldberg**, which established the law of mass action governing the influence of reactant concentrations on rates of reaction (1864). He later turned to practical problems relating to nutrition and public health, and he also engaged in social and religious work.

WAALS, Johannes Diderik van der
See **VAN DER WAALS, Johannes Diderik**

WADATI, Kiyoo
(1902–)

Japanese seismologist. After graduating from Tokyo Imperial University in 1925 and receiving a DSc in 1932, he entered the Central Meteorological Observatory (now the Japan Meteorological Agency). His most important contributions have been to advances in the detection of deep earthquakes, particularly those lying on an inclined plane dipping deep within the Earth beneath Japan. Similar deep, inclined seismic zones were being located by **Benioff** around the same time. These are now known as Wadati–Benioff zones and show the motion of downgoing oceanic crust as it subducts into the mantle at an island arc. Also interested in Antarctic research, Wadati carried out fieldwork at the Showa base between 1973 and 1974.

WADDINGTON, Conrad Hal
(1905–75)

English embryologist and geneticist, born in Evesham. He graduated from the University of Cambridge in 1926 with a degree in geology, and after a brief interlude as a palaeontologist, turned to embryology. From 1947 to 1970 he was Professor of Animal Genetics at Edinburgh University. He studied the effects of chemical messengers in inducing embryonic cells to form particular tissues during development, and was especially concerned with the ways in which both genes and environmental influences control the development of embryos in a stepwise manner, the process of epigenesis. He held the unfashionable view that environmentally induced effects could be incorporated in a heritable manner. Waddington introduced the term ‘canalization’ for the process of developmental stability where a particular phenotype is expressed in spite of the presence of genes with a different potential. His *Organisers and Genes* (1940) is concerned with the relationship between Mendelian genetics and experimental embryology; he also wrote a standard textbook of embryology, *Principles of Embryology* (1956). He was interested in the popularization of science and in forging links between science and the arts, particularly the visual arts. His more popular books included *The Ethical Animal* (1960) and *Biology for the Modern World* (1962).

WAERDEN, Bartel Leendert van der
(1903–)

Dutch mathematician, born in Amsterdam, where he obtained his doctorate in 1926. He was professor at Groningen (1928–31), Leipzig (1931–45), Johns Hopkins (1947–8), Amsterdam (1948–51) and Zürich (1951–62). Waerden worked in algebra, algebraic geometry and mathematical physics, and published books on the history of science and mathematics in the ancient world such as *Science awakening* (1954). His classic textbook *Moderne Algebra* (1931) was influential in publicizing the new algebra developed by **Hilbert**, Ernst Steinitz, **Artin**, **Noether** and others, and his book on the application of group theory to quantum mechanics (1932) showed its relevance to physics. He also wrote a series of 20 papers on algebraic geometry devoted to showing the power and rigour of the new algebraic methods in resolving old questions about curves and surfaces that had been bedevilled by over-generous intuition.

WAGER, Lawrence Rickard
(1904–65)

English geologist, petrologist and explorer, born in Batley, Yorkshire. He was educated at Pembroke College, Cambridge, where he became interested in petrology under the influence of **Harker**, and was appointed as a lecturer at the University of Reading (1929–43). He was an excellent rock-climber and mountaineer, and in 1933 climbed Everest to 28000 feet without oxygen, some 20 years before John Hunt's successful summit attempt. He took part in a series of major scientific expeditions including several to East Greenland (1930–53), where he undertook his most significant geological work, mapping a major area and carrying out his classical study of the petrology and geochemistry of the Skaergaard layered igneous complex. He served with distinction in the photographic reconnaissance section of the Royal Air Force during World War II, and subsequently resumed his geological career as Professor of Geology at Durham (1944–50) and then at Oxford from 1950. He was elected FRS in 1946. Wager's important research on various aspects of igneous and metamorphic petrology included notable studies of crystal nucleation and the origin of rhythmic layering in the Rhum Tertiary igneous complex of Scotland, and the nature of the Marscoite hybrid rock suite on Skye. In 1955 he was instrumental in the establishment of the Oxford radiometric age determination laboratory.

WAGNER-JAUREGG, or WAGNER VON JAUREGG, Julius

(1857–1940)

Austrian neurologist and psychiatrist, born in Wels and educated in Vienna. He became professor at Graz (1889) and Vienna (1893). Although his chairs were in psychiatry, he remained more interested in general medical aspects of psychiatric disorders, such as the relationship between cretinism and goitre. He won the 1927 Nobel Prize for Physiology or Medicine for his discovery in 1917 of a treatment for general paralysis (a late stage of syphilis) by infection with malaria. This was based on an older observation that patients with a variety of serious mental disorders occasionally improved after they had suffered from a bout of febrile illness. He devised a series of experiments on patients suffering from dementia and other forms of psychiatric disease, noting that the best results were obtained in those with general paralysis, several of whom showed sufficient amelioration that they could be discharged from the asylum. This 'fever therapy' was hardly ideal and was abandoned when antibiotics and other better treatments became available.

WAKSMAN, Selman Abraham

(1888–1973)

American biochemist, born in Priluka, the Ukraine. After receiving private tuition in Russia he moved to the USA in 1910 and became a US citizen in 1915, graduating in the same year at Rutgers University, where he spent most of his research life, becoming Professor of Microbiology in 1930. From 1915 he worked on the microbial breakdown of organic substances in the soil and the nature of humus. From this work emerged a new classification of microbes (1922) and methods for their scientifc cultivation (1932). He began a metabolic characterization of actinimycete fungi, and observed that they produce antibacterial substances (1937). From 1939 he searched for antibiotics of medical importance and discovered the rather toxic anticancer drug actinomycin (1941), the first anti-tuberculosis drug streptomycin (1944), another anti-streptococcal drug neomycin (1949), the anti-trichomonad streptocin and several other anti-bacterial agents. For these worldly benefits he was awarded the Nobel Prize for Physiology or Medicine in 1952. He also worked extensively on marine bacteria and the enzyme alginase, and wrote *Enzymes* (1926), *Principles of Soil Microbiology* (1938), the autobiographical *My Life with the Microbes* (1954), *The Actinomycetes* (3 vols, 1959–62), *The Conquest of Tuberculosis* (1964) and other works.

WALD, George

(1906–)

American biochemist, born in New York City. He studied zoology at New York University and at Columbia University, and worked under **Warburg** in Berlin. Subsequently he worked at Harvard (1932–77), where he became Professor of Biology in 1948. His early work was divided between the possible role of carotenoids in photosynthesis (soon abandoned) and the way in which visual purple, the retinal pigment of the eye, responds to stimulation by light. He established in 1935 that visual purple is converted by light to a yellow compound which slowly changes to a colourless compound (vitamin A). His subsequent work was directed mainly towards elucidating the details of this process, and in 1956 he made the key discovery that only one of six geometric isomers of vitamin A (11-*cis*-retinal) combines with the protein opsin to form visual purple (rhodopsin). He found that light transformed 11-*cis*-retinal to all-*trans*-retinal (retinene$_1$) which caused the opsin to change shape and release the all-*trans*-retinal, which was reduced to vitamin A_1 then slowly oxidized back to rhodopsin. He discovered a similar system using vitamin A_2 in fish, the visual mechanism proving common to the eyes of all known animals. Wald also established the nutritional relationship between vitamin A, night blindness and vitamin-deficient retinopathy. For these discoveries he shared the 1967 Nobel Prize for Physiology or Medicine with **Granit** and **Hartline**.

WALDEN, Paul

(1863–1957)

Latvian chemist, born in Rosenbeck. He studied at Riga Polytechnical School where he collaborated with **Ostwald** on ionic conductivity in solution. He became professor at the school in 1894, and two years later discovered the Walden inversion for which he is best known. From a study of changes in optical activity he found that substitution on an enantiomer brings about inversion of configuration, ie the R form is converted into the S form. A complete explanation of this came only with the mechanistic studies of **Christopher Ingold**. He also produced many electrochemical studies. He became director of St Petersburg Academy of Sciences chemical laboratory, and in 1919 moved to Germany and became professor at Rostock. He continued lecturing at Tübingen until the age of 90.

WALDEYER-HARTZ, Wilhelm

(1839–1921)

German histologist and anatomist, born in Hehlen and educated in Göttingen, Greifswald and Berlin. Inspired to a career in medical science by **Henle**'s lectures, Waldeyer-Hartz became professor at several universities, including Breslau, Strassburg and Berlin. He established his reputation with his histological studies of cancers, which he classified according to their embryological cells of origin. Thus, carcinomas come from epithelial cells, whereas sarcomas originate from the connective (mesodermal) tissues. He was asked to give a diagnosis for Emperor Frederick III's tumour of the vocal cords. His other work included studies of the histology of the spinal cord, comparative neuroanatomy of the ape's brains and the embryological development of the tonsils, and a synposis of surgical anatomy. He coined the words 'neuron' and 'chromosome'.

WALKER, Sir James

(1863–1935)

Scottish physical chemist, born in Dundee. He began his career in the flax and jute industry, but forsook this to study science at Edinburgh University, from which he graduated in 1885. From 1887 to 1889 he worked with **Baeyer** at Munich and then with **Ostwald** at Leipzig. He returned to Edinburgh as an assistant, but went on to University College London, to work with **Ramsay** in 1892. From 1894 to 1908 he was Professor of Chemistry at University College, Dundee, and he was then professor at Edinburgh until his retirement in 1928. Walker's researches were mainly on the physical chemistry of aqueous solutions, and his papers covered such topics as ionization constants, kinetics and osmotic pressure. In 1895 he carried out a pioneering study of the kinetics and mechanism of the conversion of ammonium cyanate into urea, a topic to which he subsequently reverted many times. His *Introduction to Physical Chemistry* (1899) was influential in shaping chemical education. During World War I he organized the manufacture of TNT in Edinburgh. Walker was elected a Fellow of the Royal Society in 1900 and received its Davy Medal in 1923. In 1921–3 he was President of the Chemical Society. He was knighted in 1921.

WALLACE, Alfred Russel

(1823–1913)

Welsh naturalist, born near Usk, Gwent. He worked as a surveyor and as a teacher in Leicester, but his passion was natural history, and he was inspired by **Charles Darwin**'s *Origin of Species* and Robert Chambers's evolutionary treatise *Vestiges of Creation*. Together with **Bates**, he travelled and collected in the Amazon basin (1848–52). During the return voyage, the ship caught fire and sank, and his vast collection of living and preserved specimens was lost. Undaunted, he planned a new expedition, this time to the Malay Archipelago (1854–62). He was an indefatigable collector, describing thousands of new tropical species, and was the first European to observe orang-utans in the wild. He also made pioneering contributions to the ethnology and linguistics of the native peoples in the regions he visited. His *Geographical Distribution of Animals* (1876) became the founding text of zoogeography, and he is remembered for 'Wallace's line', the division in the Malay Archipelago between the Asian and Australian floras and faunas. While in Malaysia, he conceived the idea of the 'survival of the fittest' as the key to evolution, applying **Malthus**'s ideas on checks to population growth to the natural variation he had observed as a naturalist. In 1858 he wrote up his discovery in an article entitled *On the tendency of varieties to depart indefinitely from the original type*, and sent it to Darwin, who was sent into turmoil at the duplication of his own unpublished ideas. This precipitated the joint Darwin–Wallace paper at the Linnaean Society in 1858, and the hurried publication of Darwin's *Origin of Species* the following year. On his return to England, Wallace generously allowed Darwin the credit, and even entitled his own book on evolution *Darwinism* (1889). He wrote several other highly influential books, including *Contributions to the Theory of Natural Selection* (1870) and *Island Life* (1882), as well as inspired accounts of his voyages, *Travels on the Amazon* (1869) and *The Malay Archipelago* (1872). He was an outspoken advocate of socialism, pacifism, women's rights and other causes, and encapsulated his views in *Social Environment and Moral Progress* (1913). He was also active in psychic and spiritualist circles, ultimately causing many scientific colleagues to distance themselves from him. Though he remained a staunch believer in natural selection, he did not believe that this process had created the higher faculties of the human brain, which he regarded as miraculously endowed.

WALLACH, Otto
(1847–1931)

German chemist, born in Königsberg, Prussia (now Kaliningrad, Russia). After study at the Potsdam Gymnasium, he entered the University of Göttingen in 1867. He worked there for his doctorate with Hans Hübner on the positional isomerism of toluene compounds. In 1870 he moved to Bonn as assistant to **Kekulé** and remained there for 19 years. From 1889 until his retirement in 1915 he worked as Director of the Chemical Institute in Göttingen. Throughout his life he studied the composition of essential oils obtained from plants, a topic which Kekulé suggested to him. From these oils he isolated many compounds belonging to a class he called terpenes, showing that they consisted of five carbon atom fragments known as isoprene units. The compounds differ in the way the units are arranged. One of his greatest achievements was the elucidation of the structure of a-terpineol (1895). Even in his own lifetime the value of his work was greatly appreciated, and it initiated much work which continues to this day. He was awarded the Nobel Prize for Chemistry in 1910. As well as a distinguished chemist he was a serious art collector.

WALLER, Augustus Volney
(1816–70)

English physiologist, born near Faversham. Waller studied medicine in Paris, receiving his MD in 1840. He set up in general practice in Kensington. Elected FRS in 1851, he devoted himself to full-time research for five years, working in Bonn, and later under **Flourens** at the Jardin des Plantes in Paris, before being appointed Professor of Physiology at Queen's College, Birmingham. Waller was a fine microscopist, remembered for his patient anatomical investigations of the nervous system. He paid special attention to the processes of nerve degeneration and regeneration, discovering the Wallerian degeneration of nerve fibres. He also conducted innovative investigations of the autonomic nervous system, particularly analysing the dilation of the iris under light stimuli. His experiments on the vasoconstrictor properties of nerves from the ciliospinal region extended the researches of **Bernard** and **Brown-Séquard**.

WALLICH, Nathaniel
(1786–1854)

Danish botanist, born in Copenhagen, where he studied under Martin Vahl and graduated in medicine. In 1807 he became surgeon to the Danish colony at Serampore, India. This became British in 1813 and Wallich joined the British medical staff. In 1815 he was appointed Superintendent of the Calcutta Botanical Garden, and so began a very active, distinguished botanical career. In 1820, with William Carey (1761–1834), he began to publish **Roxburgh**'s *Flora Indica*, with much additional matter written by himself. He collected many plants on expeditions to Nepal (1820) and western India (1825), and was one of the earliest botanists to collect in Myanma (Burma) (1826–7). Returning home on the grounds of ill-health in 1828, he brought back some 8 000 specimens; 9 148 species are represented in his *A Numerical List of Dried Specimens of Plants in the East India Company's Museum* (1828). Between 1830 and 1832 he published his most important work, *Plantae Asiaticae Rariores* (3 vols). He went back to India (1832–47) and explored Assam, paying particular attention to wild tea plants, before returning to England.

WALLIS, Sir Barnes Neville
(1887–1979)

English aeronautical engineer and inventor, born in Derbyshire. After winning a scholarship to Christ's Hospital, London, and training as a marine engineer at Cowes, he joined the Vickers Company in 1911 and two years later was transferred to the design office of Vickers Aviation. After a short period of military service during World War I, he rejoined the company as chief designer in their airship department at Barrow-in-Furness, for whom he later designed the airship R100 which made its maiden flight in 1929 and successfully crossed the Atlantic. From 1923 he also acted as chief designer of structures at Vickers Aviation, Weybridge, where he designed the Wellesley and Wellington bombers with their revolutionary geodetic fuselage structure, the bombs which destroyed the German warship *Tirpitz* and V-rocket sites, and the 'bouncing bombs' which destroyed the Möhne and Eder dams in Germany during World War II (1943). From 1945 to 1971 he was chief of aeronautical research and development for the British Aircraft Corporation at Weybridge. In the early 1950s he was responsible for the design of the first variable-geometry (swing-wing) aircraft, the experimental Swallow; the same design principle was later incorporated in the US Air Force's General Dynamics F-111 multi-purpose fighter which first flew in December 1964, and in the Panavia Tornado which is in service with many of the world's air forces. He was elected FRS in 1945, and knighted in 1968.

WALLIS, John
(1616–1703)

English mathematician, born in Ashford, Kent. He graduated at Cambridge, and took holy orders, but in 1649 became Savilian Professor of Geometry at Oxford. During the Civil War he sided with the parliament, was secretary in 1644 to the Westminster Assembly, but favoured the Restoration. Besides the *Arithmetica Infinitorum* (1656), in which he offered a remarkable method for finding areas under curves in terms of infinite sums (soon replaced by the more rigorous calculus), he wrote on the binomial theorem and gave an infinite product for π. He also wrote on proportion, mechanics, the quadrature of the circle (in opposition to Thomas Hobbes), grammar, logic, theology, and the teaching of the deaf and dumb. In addition Wallis was an expert on deciphering, and edited the work of some of the Greek mathematicians. He wrote a tendentious and xenophobic history of algebra, and is also remembered as one of the founders of the Royal Society.

WALTON, Ernest Thomas Sinton
(1903–)

Irish physicist, born in Dungarvan, Waterford. He studied at Trinity College, Dublin, graduating in 1926. In 1927 he went to the Cavendish Laboratory, Cambridge, where he studied under **Rutherford** and was awarded his PhD in 1934. He later became Professor of Natural and Experimental Philosophy at Trinity College, Dublin (1947–74). With **Cockcroft**, he produced the first artificial disintegration of a nucleus by bombarding a lithium nucleus with protons accelerated across a potential of 710 kilovolts (1932). This was the first successful use of a particle accelerator, and by studying the energies of the two alpha particles produced they were able to verify **Einstein**'s theory of mass–energy equivalence. The use of particle accelerators was crucial for the understanding of the substructure of nuclei and later the nucleons themselves. Cockcroft and Walton were awarded the 1951 Nobel Prize for Physics in recognition of this work. In 1952 he was appointed chairman of the School of Cosmic Physics and the Dublin Institute for Advanced Studies.

WANG, An
(1920–90)

Chinese-born American physicist and computer company executive, born in Shanghai. He graduated in science from Jiao Tong University in Shanghai (1940), and in 1945 emigrated to the USA, where he studied applied physics at Harvard. Known as 'the Doctor', Wang had both technological genius and entrepreneurial ability. He played a major role in inventing magnetic core memories for computers, and his patents provided enough income for him to break into the commercial world. In 1951 he founded Wang Laboratories in Boston, Massachusetts, which expanded rapidly through the success of another invention attributed to Wang—the electronic calculator. Wang became a leading manufacturer of minicomputers, competing alongside the likes of **Olsen**'s DEC. His company was also renowned for its word-processing software. However, in the 1980s personal computers and workstations ate into Wang's business; in 1983 he handed control of the business to others, including his son Frederick, who was appointed Wang's president in 1987. It was a disastrous move: after Frederick was ousted in 1989, An Wang returned briefly after cancer surgery to attempt to turn the company round. However, in 1992 (two years after his death) the company which had enjoyed revenues of $3 billion and employed 31 000 in its heyday filed for bankruptcy.

WARBURG, Otto Heinrich
(1883–1970)

German biochemist, born in Freiburg, Baden. Educated at Berlin and Heidelburg, he won the *Pour le Mérite* (the German equivalent of the Victoria Cross) during World War I. He worked at the Kaiser Wilhelm (later Max Planck) Institute in Berlin from 1913, becoming director there in 1953. He was the first to discover the important role of iron, in association with oxidase enzymes, in nearly all cells. He called the iron-containing enzyme that catalysed direct electron transfer to oxygen *Atmungsferment* and showed that its 'reduction' was inhibited by cyanide, and that it was probably a haem protein. In 1926 he demonstrated that oxygen uptake by yeast is inhibited by carbon monoxide and exploited **J B S Haldane**'s finding that carboxyhaemoglobin is dissociated by light to determine from its absorption characteristics that this is also a haem protein, subsequently equated by **Keilin** with cytochrome oxidase. He also discovered that the oxidation of glucose 6-phosphate by a yeast preparation ('old yellow enzyme') requires the cofactor riboflavin (1932), and that the 'new yellow enzyme', discovered by **Hans Krebs**, contains a different electron carrier (flavinadeninedinucleotide). With E Negelin he showed that the green alga *Chlorella* produces one molecule of oxygen from the absorption of four quanta of red light (1923)—an efficiency of 65 per cent, disputed by others but confirmed by Warburg in 1950. The gas manometer developed by Warburg in 1926, from **Barcroft** and Haldane's blood-gas manometer for measuring metabolic reactions by the amount of oxygen or carbon dioxide taken up or released, was crucial to the discoveries of Krebs and others, and continued in use until replaced by the oxygen electrode around 1970. Warburg also engaged in cancer research. He was awarded the 1931 Nobel Prize for Physiology or Medicine, but as a Jew was prevented from accepting it by Hitler.

WARD, Nathaniel Bagshaw
(1791–1868)

English physician and botanist, born in London. He was involved with the administration of the Chelsea Physic Garden, where he worked on methods of plant cultivation. He published *On the Growth of Plants in closely glazed Cases* (1842), and invented the 'Wardian case' which enabled live plants to be transported successfully on long voyages, and also their cultivation in Victorian drawing rooms. Among the many plants which owe their establishment to the Wardian case, the most significant is the tea plant, which **Fortune** successfully brought from China to India. Ward published *Aspects of Nature* in 1864.

WARMING, (Johannes) Eugenius Bülow
(1841–1924)

Danish botanist, born in the North Frisian island of Manö. He studied botany in Munich under **Nägeli** and others. Professor at Stockholm (1882–5) and Copenhagen (1885–1911), he wrote important works on systematic botany (1879) and plant ecology (*Plantesamfund*, 1895), being regarded as a founder of the latter, although the term had been coined earlier by **Haeckel**. His work demonstrated that groups of species could form a well-defined unity, such as a meadow ecosystem. He produced two excellent textbooks, *Haandbog i den Systematiske Botanik* (1879) and *Den Almindelige Botanik* (1880). Between 1863 and 1866 he travelled in Brazil, making the most detailed and thorough study of a tropical area then produced.

WASSERMAN, August Paul von
(1866–1925)

German bacteriologist, born in Bamberg. He studied medicine at Erlangen, Vienna, Munich and Strassburg, where he graduated in 1888, and worked on bacteriology and chemotherapy at the Robert Koch Institute in Berlin from 1890. In 1906 he discovered and gave his name to a blood-serum test for syphilis. An infected person will produce syphilis antibodies in the blood; in the Wasserman test these will react with known antigens to form a chemical complex. The test is still widely used in diagnosis.

WATERSTON, John James
(1811–83)

Scottish natural philosopher and engineer, born in Edinburgh. He attended the University of Edinburgh, and following an engineering apprenticeship he practised as a surveyor in London (1832), later moving to a position in the hydrographer's department of the Admiralty. As naval instructor to the cadets of the East India Company in Bombay he taught navigation and gunnery from 1839, only returning to Scotland permanently in 1857. He wrote on astronomy, solar radiation, chemistry, the physiology of the central nervous system, sound and a novel kinetic theory of gases and liquids. Waterston's famous speculative memoir on gases linked heat with molecular motion and included a calculation of the ratio of specific heats at constant temperature and constant volume. Submitted to the Royal Society in 1845, it was dismissed by the referees as 'nothing but nonsense'. In 1892 it was reproduced in complete form by Lord **Rayleigh** (1842–1919). Many of the key ideas had by then been published by **Clausius** and **Maxwell**.

WATSON, David Meredith Seares
(1886–1973)

English zoologist, educated in chemistry and geology at the University of Manchester. After early research in palaeobotany, his interests drifted to palaeontology, to which he devoted his research, despite having no formal zoological training. Before World War I he travelled widely in South Africa and Australia collecting fossil vertebrates. From the reptilian remains in this material he traced the evolution of the mammalian skeleton from that of primitive reptiles. He later became a leading authority on early amphibians. From 1921 he was Jodrell Professor of Zoology and Comparative Anatomy at University College London, where he remained until his retirement in 1951. He was elected FRS in 1922.

WATSON, James Dewey
(1928–)

American biologist, born in Chicago. He received a BSc in zoology from the University of Chicago, and as a postgraduate at Indiana University he studied under **Hermann Müller** and **Luria**. He spent the period 1951–3 at the Cavendish Laboratory in Cambridge, UK, and from 1955 taught at Harvard, where he became Professor of Biology in 1961. Since 1968 he has been Director of the Cold Spring Harbor Laboratory in New York, and he has also served as Director of the National Center for Human Genome Research (1989–92). While in Cambridge in 1951, Watson worked with **Crick** on the structure of DNA, the biological molecule contained in cells which carries the genetic information. They published their model of a two-stranded helical molecule in 1953, showing that each strand consists of a series of the nucleotide bases (adenine, thymidine, guanine and cytosine) wound around a common centre. The strands were shown to be linked together with hydrogen bonds, adenine on one strand pairing with thymidine on the other, with similar pairing between guanine and cytosine. For this work, Watson was awarded the 1962 Nobel Prize for Physiology or Medicine jointly with Crick and **Wilkins**. Since the mid-1980s, he has been an active supporter of the Human Genome Initiative, which aims to locate all genes in the human body and determine their DNA sequences. He wrote a personal account of the discovery of the DNA structure in *The Double Helix* (1968), and the textbooks *The Molecular Biology of the Gene* (1965) and *Recombinant DNA* (1984).

WATSON, Sir William
(1715–87)

English scientist, born in London, the son of a cornchandler. He became an apothecary and was one of the earliest experimenters on electricity. He discovered (among others) the importance of insulating conductors to increase charge, improved the Leyden jar (the first form of capacitor or condenser), was an early supporter of **Benjamin Franklin**'s electrical theories (1748), and was the first to investigate the passage of electricity through a rarefied gas. In botany, he did much to introduce the Linnaean system to Britain. He was elected FRS (1841), and awarded the Royal Society's Copley Medal (1745). In 1757 he was awarded an MD from Halle University, became a licentiate of the Royal College of Physicians, and was appointed physician to the Foundling Hospital (1762). He was chiefly interested in epidemic children's diseases, about which he wrote a pamphlet comparing methods of inoculation against smallpox. He was knighted in 1786.

WATSON-WATT, Sir Robert Alexander
(1892–1973)

Scottish physicist, born in Brechin. He was educated at Dundee and the University of St Andrews, where he remained to teach from 1912 to 1921. Pursuing meteorological research he used the reflection of short-wave radio waves to locate thunderstorms as early as 1919. By 1935 Watson-Watt had perfected a system that was able to locate aeroplanes in this way; it was called 'RAdio Detection And Ranging', abbreviated to radar. By the autumn of 1938 radar stations were set up in Britain and by the time the Battle of Britain took place in 1940, it was possible to detect oncoming German planes at all times of day and night and in all weather. Watson-Watt became scientific adviser to the air ministry in 1940, and the following year visited the USA where he helped complete the job of establishing US radar systems. He was elected FRS in 1941, and knighted for his role in the development and introduction of radar in 1942. In 1958 he published *Three Steps to Victory*.

WATT, James
(1736–1819)

Scottish engineer and inventor, born in Greenock. He went to Glasgow in 1754 to learn the trade of a mathematical instrument maker, and there, after a year in London, he set up in business. The university made him its mathematical instrument maker from 1757 to 1763. He was employed on surveys for the Forth and Clyde canal (1767), the Caledonian and other canals, and was engaged in the improvement of harbours and in the deepening of the Forth, Clyde and other rivers. As early as 1759 his interest had been aroused in the possibilities of developing the use of steam as a motive force. In 1763–4 a model of Thomas Newcomen's engine was sent to his workshop for repair. He easily put it into order, and seeing the defects in the working of the machine, hit upon the expedient of the separate condenser. This was probably the greatest single improvement ever made to the reciprocating steam engine, enabling its efficiency to be increased to about three times that of the old atmospheric engines. After an abortive enterprise funded by John Roebuck, he entered into a partnership with Matthew Boulton of Soho, near Birmingham (1774), when (under a patent of 1769) the manufacture of the new engine was commenced at the Soho Engineering Works. Watt's soon superseded Newcomen's machine as a pumping engine, and between 1781 and 1785 he obtained patents for the sun-and-planet motion, the expansion principle, the double-acting engine, the parallel motion, a smokeless furnace, and the governor. He described a steam locomotive in one of his patents (1784), but discouraged William Murdock from further experiments with steam locomotion. In 1785 he was elected FRS. The SI unit of power is named after him, and horse-power, the original unit of power, was first experimentally determined and used by him in 1783. He retired in 1800, and died at Heathfield Hall, his seat near Birmingham. His son, James (1769–1848), a marine engineer, fitted the engine to the first English steamer to leave port (1817), the *Caledonia*.

WEATHERALL, Sir David John
(1933–)

English molecular geneticist, educated at the University of Liverpool and Oxford University, where he received his MA in 1974. After holding various medical posts in the UK and the Far East, he became a researcher at the Johns Hopkins Medical School, Baltimore (1960–2, 1963–5). Subsequently he moved to a lectureship at the University of Liverpool, where he was later appointed Professor of Haematology (1971–4). He later became Nuffield Professor of Clinical Medicine at the University of Oxford (1974–92), where he has been Regius Professor of Medicine since 1992. Weatherall has worked for many years on the genetics and clinical aspects of the thalassaemias, a group of inherited anaemias in which regulation of the globin genes is perturbed in some way. The clinical outcome is greatly influenced by the detailed knowledge of the causative lesion, by early detection of the problem and by the ability to predict the outcome of a pregnancy. He was elected FRS in 1977, knighted in 1987, and received the Royal Medal of the Royal Society in 1989.

WEBER, Ernst Heinrich
(1795–1878)

German physiologist, born in Wittenberg, brother of **Wilhelm Weber**. He studied at the University of Wittenberg and later at Leipzig. In 1818 he was appointed to the Chair of Human Anatomy; the Chair of Physiology was added in 1840. Weber undertook extensive comparative embryological and palaeontological studies, especially on the middle ear of mammals. He also demonstrated that the digestive juices are the specific products of glands, thereby opining up major new fields of physiological and chemical research. A key focus of his interest lay in the study of sensory functions, especially skin sensitivity. Novel investigations probed what was later to be called the sensory 'threshold'. He devised a method of determining and quantifying the sensitivity of the skin, enunciated in 1834, and gave his name to the Weber–**Fechner** law of the increase of stimuli.

WEBER, Wilhelm Eduard
(1804–91)

German scientist, born in Wittenburg. His father was Professor of Theology at the University of Wittenberg. He shared with his two brothers an interest in science. With his elder brother **Ernst Heinrich Weber** he published a treatise on wave motion, and with his younger brother, Eduard (1806–71), who became Professor of Anatomy at Leipzig, he studied the mechanism of walking (1836). The brothers were brought up in a scientific environment; one of the family's acquaintances in Wittenberg was the acoustician **Chladni**, and at Halle University Weber studied under **Schweigger**. He was appointed as a lecturer in physics at Halle (1828), and from 1831 to 1837 he was Professor of Physics at Göttingen. He left because of political troubles, was appointed to a chair at Leipzig in 1843, but returned to his old position in Göttingen in 1849, and also became the director of the astronomical observatory. His initial research was on the acoustical theory of reed organ pipes, but **Gauss** inspired Weber to join him in the study of geomagnetism. Their laboratories in Halle were among the first establishments to be connected by electric telegraph (1833), and they established a network of magnetic observatories to collate the measurements. In 1845 Weber developed an electrodynamometer for the absolute measurement of an electric current, in 1849 the mirror galvanometer (the technique of using a light beam as a 'weightless pointer' had already been used by **Poggendorff** (1826), and was to be further developed by **Kelvin**), and he proposed a system of electrical units analogous to those proposed by Gauss for magnetism. His most important contribution was to determine with **Kohlrausch** the ratio of the electrodynamic and electrostatic units of charge, later used by **Maxwell** to support his electromagnetic theory. His name is given to the SI unit of magnetic flux.

WEDDERBURN, Joseph Henry Maclagan
(1882–1948)

Scottish-born American mathematician, born in Forfar. He graduated in mathematics at Edinburgh in 1903, visited Leipzig, Berlin and Chicago, and returned to Edinburgh as a lecturer (1905–9). In 1909 he moved to Princeton, New Jersey, but he returned to fight in the British army during World War I. After the war he settled at Princeton University until his retirement in 1945. His work on algebra included two fundamental theorems known by his name, one on the classification of semi-simple algebras, and the other on finite division rings.

WEEKS, Willy (Wilford Frank)
(1929–)

American glaciologist and geophysicist, born in Champaign, Illinois. He graduated from the University of Illinois with a BA in geology (1951), and received a PhD in geochemistry (1956) from the University of Chicago. Having already received a commission in the US Air Force, he was called to active service in 1955 at the Cambridge Research Center, Boston. While there an opportunity arose for him to study sea ice along the Labrador coast; this marked the start of a long-term interest in this topic. Upon receiving his discharge in 1957, he accepted an assistant professorship of Earth sciences at Washington University, St Louis (1957–62), and he later transferred to the Cold Regions Research and Engineering Laboratory in Hanover, New Hampshire (1962–86). Since 1986 he has been professor at the Geophysical Institute of the University of Alaska, Fairbanks, and chief scientist of the Alaska Synthetic Aperture Radar Facility. He has taken part in numerous field studies near both poles and made extensive studies of a wide variety of aspects of ice and snow.

WEGENER, Alfred Lothar
(1880–1930)

German meteorologist and geophysicist, born in Berlin and educated at the universities of Heidelberg, Innsbruck and Berlin. He first worked as an astronomer, then joined his brother Kurt at the Prussian Aeronautical Observatory in Tegel where they undertook a 52½ hour balloon flight to test the accuracy of a clinometer. In 1906 he joined a Danish expedition to north-east Greenland and learned the technique of polar travel while making meteorological observations. On his return he became a lecturer in astronomy and meteorology at the University of Marburg. His second expedition to Greenland (1912) was almost wrecked by calving of the ice. During army service in World War I, Wegener was wounded and was never fit for active service again. After the war he joined the German Marine Observatory in Hamburg and was also Professor of Meteorology in Graz, Austria (1924). *Die Entstehung der Kontinente und Ozeane* ('The Origin of Continents and Oceans') for which he is famous was first published in 1915, based upon his observations that the continents may once have been joined into one supercontinent (Pangaea), which later broke up, the fragments drifting apart to form the continents as they are today. Wegner provided historical, geological, geomorphological, climatic and palaeontological evidence, but at that time no logical mechanism was known by which continents could drift and the hypothesis remained controversial until the 1960s, when the structure of oceans became understood. He died in Greenland during his fourth expedition there.

WEIDENREICH, Franz
(1873–1948)

German anatomist and anthropologist, born in Edenkoben. He studied medicine at Munich, Kiel, Berlin and Strassburg. He taught anatomy at Strassburg (1903–18) and Heidelberg (1919–24), and was Professor of Anthropology at Frankfurt from 1928 to 1933. In 1934 he left Nazi Germany and worked for seven years in China (1935–41) at the Peking Union Medical College, collaborating with **Teilhard de Chardin** on fossil remains of Peking Man. From 1941 to 1948 he worked at the American Museum of Natural History in New York City. His early work was concerned with blood, bone, teeth and connective tissue. Later studies of hominid fossil remains led him to espouse an orthogenetic view of human evolution, which he summarized in *Apes, Giants and Man* (1946).

WEIERSTRASS, Karl Theodor Wilhelm
(1815–97)

German mathematician, born in Ostenfelde. Educated at the universities of Bonn and Münster, he became professor at Berlin in 1856. He published relatively little but became famous for his lectures, in which he gave a systematic account of analysis with previously unknown rigour, basing complex function theory on power series in contrast to the approach of **Cauchy** and **Riemann**. He made important advances in the theory of elliptic and Abelian functions, constructed the first accepted example of a continuous but nowhere-differentiable function, and showed that every continuous function could be uniformly approximated by polynomials. His name became a byword among mathematicians for rigour, and many of his most profound ideas grew out of his attempts to present a completely systematic, self-contained account of contemporary mathematics. As with many such attempts, later students found Weierstrass's paths reasonably easy to follow but difficult to extend, and the ideas of Cauchy and Riemann, once excluded from Berlin, had to be reintroduced.

WEIL, André

(1906–)

French mathematician, brother of the philosopher and religious writer Simone Weil, born in Paris. He studied at the University of Paris, and spent two years in India and some time in Strasbourg (1933–40), the USA (1941–2, 1947–58) and Brazil (1945–7), before settling at Princeton in 1958. One of the most brilliant mathematicians of the century, he has worked in number theory, algebraic geometry and topological group theory. He was one of the founders of the **Bourbaki** group, and has written on the history of mathematics. A study of **Gauss**'s work helped lead him to conjectures concerning arithmetic on algebraic varieties (in simple terms the solution of polynomial equations in integers) that attracted the attention of the best of Bourbaki and others. The last and most profound of these, a generalized **Riemann** hypothesis, was solved by Pierre Deligne in work that earned him the Fields Medal (the mathematical equivalent of the Nobel Prize), while the original Riemann hypothesis still remains unresolved. Weil did much to extend the theory of algebraic geometry to varieties of any dimension, and to define them over fields of arbitrary characteristics.

WEINBERG, Robert Allan

(1942–)

American biochemist, born in Pittsburgh, Pennsylvania. He received his PhD from MIT and then held a series of fellowships in the USA and Europe. From 1973 to 1982 he was first assistant professor and then associate professor in the Department of Biology and Center for Cancer Research at MIT, where he is currently Professor of Biochemistry. Since 1984 he has also been a member of the Whitehead Institute for Biomedical Research in Cambridge, Massachusetts. Weinberg studies the causes of cancer, both as a result of acquisition of cancer-susceptible genes (oncogenes) and the loss of tumour suppressor genes. He discovered the tumour suppressor gene Rb1, whose loss is associated with a rare childhood cancer in which tumours develop in the retina. The Rb1 gene has now been cloned and shown to encode a protein normally expressed in the retina and possessing similar features to those found in many DNA binding proteins; it also binds to the proteins encoded by oncogenes of some DNA viruses. Although the retina cancer is rare, there is increasing evidence that loss or inactivation of tumour suppressor genes also plays a part in many common cancers occurring in adult life.

WEINBERG, Steven

(1933–)

American physicist, born in New York City. He was educated at Cornell before spending a year at the Niels Bohr Institute in Copenhagen, then obtained his doctorate at Princeton University. He then held appointments at Columbia University, the University of California at Berkeley, MIT and Harvard University, before becoming Professor of Physics at the University of Texas in 1986. In 1967 he produced a theory that unified the electromagnetic and weak forces, an achievement analogous to that of **Maxwell** and **Einstein** in classical electromagnetism. This incorporated the prediction of a new weak interaction due to 'neutral currents', whereby a chargeless particle, now known as the Z, is exchanged giving rise to a force between particles. This was duly observed in 1973, giving strong support to the theory, also independently developed by **Salam** and now known as the Weinberg–Salam or electroweak theory. **Glashow** extended the theory to include a new concept known as 'charm', and all three shared the 1979 Nobel Prize for Physics. The combined theory is known as 'the standard model' and has recently been precisely tested by experiments at the LEP electron–positron collider at the European nuclear research centre, CERN, in Geneva. Weinberg is also noted for his ability to explain complex physical concepts in non-technical terms, a gift demonstrated in his prize-winning popular book, *The First Three Minutes* (1977), which explains how the physics of the first three minutes of the universe have shaped what we observe today.

WEISMANN, August Friedrich Leopold

(1834–1914)

German biologist, born in Frankfurt. He studied medicine at Göttingen from 1852 to 1856. In 1867 he became Professor of Zoology at the medical school of the University of Freiburg and subsequently at a new Institute of Zoology there. He investigated the development of the two-winged flies, the Diptera, describing the neuro-humoral organ which bears his name, the Weismann ring. He was a strong supporter of **Charles Darwin**'s theory of evolution through natural selection. In an attempt to disprove the idea of acquired characters proposed by **Lamarck**, he amputated the tails from mice during five successive generations and found, not surprisingly, that there was no reduction in the propensity to grow tails. His early work on the development of the Hydrozoa led him to develop his germ-plasm theory. He appreciated that the information required for the development and final form of an organism must be contained within the germ cells, the egg and sperm, and be transmitted unchanged from generation to generation. He realized that this would account for the phenomenon of sex. He also noted that some form of reduction division, which we now know to occur during meiosis, must occur if the genetic material were not to double on each generation. His theories were developed in a series of essays, translated as *Essays upon Heredity and Kindred Biological Problems* (1889–92). His *Vorträge über Descendenztheorie* (1902) was an important contribution to evolutionary theory.

WEISS, Robin (Robert Anthony)

(1940–)

English molecular biologist. He was educated at University College London, and became a lecturer in embryology there from 1963 to 1970. After holding short-term research posts in the USA, he joined the staff of the Imperial Cancer Research Fund Laboratories in London (1972–80). From 1980 to 1989 he was Director of the Institute of Cancer Research, and since 1990 he has been head of the institute's Chester Beatty Laboratories. Weiss has made important contributions to studies of the role of retroviruses in the causation of cancer and of the HIV virus, particularly the mechanisms by which the virus enters the mammalian cell.

WEISSMAN, N Charles

(1931–)

Swiss molecular biologist, born in Budapest. He was educated at the Kantonales Gymnasium in Zürich and at Zürich University, where he became assistant to **Karrer** (1960–1). In 1963 he moved to the USA, where he became Assistant Professor of Biochemistry (1964–7) at the New York University School of Medicine. In 1967 he returned to Switzerland as Professor Extraordinarius in Molecular Biology at the University of Zürich (1967–70), where he is currently Director of the Institute of Molecular Biology. He was also President of the Roche Research Foundation (1971–7) and has been a member of the scientific board of Biogen since 1978. Weissman worked extensively on the mechanisms by which genes are transcribed, in particular identifying the DNA sequences which are recognized by RNA polymerase and other transcription factors to give correct high-level expression of genes. More recently, he has concentrated on the structure and function of the insulin genes, and was responsible for cloning them; this allowed the therapeutic use of highly purified gene copies in the treatment of diabetes.

WEIZMANN, Chaim

(1874–1952)

German chemist, born in Motol, Russia. His scientific education began at the gymnasium in Pinsk and he later moved to Darmstadt (1893–4) and Berlin (1895–8). He obtained his doctorate from the University at Fribourg, Switzerland, and then moved to the University of Geneva, where he produced a number of commercially profitable patents on dyestuffs. By this time he was already an important figure in the Zionist movement. In 1904 he moved to Manchester to work with **William Henry Perkin Jr**, partly because he felt that Britain would do more to establish a Jewish national homeland in Palestine. He continued his work on dyestuffs and commenced a series of studies of fermentation. In 1912 he found a bacterium *Clostridium acetobutylium* which would convert carbohydrate into acetone. This process proved of great importance in World War I as acetone is used in large quantities to plasticize the propellant cordite. Partly out of gratitude for the development of the acetone process, the government agreed to the Balfour Declaration promising British help in establishing a Jewish homeland. This conflicted with promises given by T E Lawrence to the Arabs. The Daniel Sieff (later Weizmann) Institute of Science in Rehovot was founded in 1934 and Weizmann continued his researches there on industrial chemistry. In 1948 be became the first President of Israel and his scientific work ceased.

WEIZSÄCKER, Baron Carl Friedrich von

(1912–)

German physicist, born in Kiel and educated at the universities of Berlin, Göttingen and Leipzig, where he was **Heisenberg**'s assistant, before moving on to Berlin, where he became assistant to **Meitner**. He was appointed associate professor at the University of Strasbourg before becoming Professor of Philosophy at Hamburg. Independently of **Bethe**, he proposed that the source of energy in stars is chain nuclear fusion reactions (1938) and described the 'carbon cycle' sequence of reactions involved. In a development of **Laplace**'s nebular hypothesis, he also suggested a possible scenario for the formation of the planets.

WELCH, William Henry
(1850–1934)

American pathologist and bacteriologist, born in Norfolk, Connecticut. Descended from a line of physicians, Welch studied medicine at the College of Physicians and Surgeons in New York, and received his MD in 1875. After a period of research in Germany with **Ludwig** and **Cohnheim**, he became Professor of Pathology and Anatomy at Belleue Hospital Medical College in New York in 1879. Later he was appointed Professor of Pathology at Johns Hopkins University (1884), and following his retirement he became founding-director of the university's School of Hygiene and Public Health. Welch did much to establish a thriving world-renowned medical research centre at Johns Hopkins, introducing many new techniques from Europe. Around 1891 he demonstrated the pathological effects induced by diphtheria toxin; his greatest personal discovery came in 1892 with his identification of the causative agent of gas gangrene, later named *Clostridium welchii* in his honour.

WELLER, Thomas Huckle
(1915–)

American virologist, born in Ann Arbor, Michigan, the son of a pathology professor at Michigan University. Weller studied medical zoology at Michigan and graduated in 1936. He entered Harvard University Medical School where he conducted research under **Enders**, who was working on methods for the cultivation of animals cells. After graduating in 1940, Weller was appointed to the staff of the Children's Hospital in Boston, and in 1942 joined the US Army Medical Corps. During World War II he conducted research in the tropics on the blood trematode *Schistosoma*, a subject he continued to investigate in his years at Harvard. After the war Enders invited Weller and **Robbins** to join him at the newly created Infectious Diseases Research Laboratory at the Boston Children's Hospital. Weller and his colleagues developed new techniques for cultivating the poliomyelitis virus which made it possible for other workers to develop the polio vaccine. For this achievement Weller, Enders and Robbins shared the 1954 Nobel Prize for Physiology or Medicine. Weller isolated the aetiological virus of chickenpox and of shingles, demonstrating that the same virus caused both diseases. He also isolated the causative agent of German measles, and discovered a new viral aetiology of congenital damage, a virus he named 'cytomegalovirus'. In 1954 he was named Strong Professor and head of the Department of Tropical Public Health at Harvard, a post which he held until his retirement in 1981.

WERNER, Abraham Gottlob
(1749–1817)

German geologist, born in Wehrau, Silesia (now in Poland). A teacher at Freiburg in Saxony from 1775, he was one of the first to frame a classification of rocks and gave his name to the Wernerian (or Neptunian) theory of deposition. The controversy between the Neptunists and Plutonists became one of the great geological debates of the late 18th century. In essence Werner advocated that crystalline igneous rocks were formed by direct precipitation from seawater, as part of his overall system of strata from the crystalline 'primitive rocks' succeeded by the 'transition rocks', resting on highly inclined strata, the well-stratified and flat-lying 'floetz rocks' and finally the poorly stratified alluvial series. The Plutonists, led by **Hutton**, were able to demonstrate the intrusive nature of such rocks. Werner was one of the great geological teachers of his time and many scholars, including Johann von Goethe and **Buch**, travelled to Freiburg to study under him.

WERNER, Alfred
(1866–1919)

Swiss inorganic chemist, the founder of coordination chemistry. He was born in Mulhouse, France, and studied at the Polytechnical School in Zürich, returning to France (1891–2) to work with **Berthelot** at the Collège de France. He was appointed assistant professor at Zürich in 1893 and full professor from 1895 to 1915. Werner was the first person to demonstrate that isomerism applies to inorganic as well as to organic chemistry (isomerism being the phenomenon in which compounds with the same chemical formula differ in the arrangement of their atoms within the molecules). In Werner's day, theories of valence in inorganic chemistry were unsatisfactory, being dominated by theories of ionic dissociation and the idea that valence was constant. Werner suggested that metals have a primary electrovalence in the centre of the molecule, surrounded by a fixed number of secondary valences which bind neutral molecules such as ammonia, water or organic amines. The secondary, or coordinate, bonds formed one of several possible geometric shapes—for example tetrahedra or octahedra—leading to the possibility of isomerism in inorganic compounds. Searching for evidence for his theory between 1893 and 1896, Werner and Arturo Miolati electrolysed salts and measured their conductivity. Over the next 20 years Werner worked towards further proof of his theory and was able, finally, to demonstrate different spatial arrangements in cobalt salts in 1914. His views, at first regarded with hostility, gradually won acceptance and were confirmed later by X-ray diffraction. They revolutionized inorganic chemistry and opened up many new areas of research. Werner was awarded the Nobel Prize for Chemistry in 1913. He died in Zürich.

WERNICKE, Carl
(1848–1905)

German neurologist and psychiatrist, born in Tarnowitz, Upper Silesia (now in Poland). He qualified in medicine at the University of Breslau, and trained under the distinguished neuropathologist Theodor Meynert, who also influenced Sigmund Freud. Wernicke moved to Karl Westphal's psychiatric and neurological clinic in 1876, two years after publishing his major work, *Der Aphasische Symtomencomplex* ('The Aphasic Syndrome'). The form of aphasia, loss of speech, which he described was marked by a severe defect in the understanding of speech, and it became known as sensory aphasia. This was in contast to the motor aphasia proposed by **Broca**, which involved loss or defect in the expression of speech. Sensory aphasia includes a wide range of symptoms, including disorders in word usage and word choice, which in severe cases degenerates to incomprehensible gibberish. From postmortem studies of brains of his patients he showed that his type of aphasia was typically localized in the left temporal lobe, now known as Wernicke's area, although he also noted bilateral lesions. He tried to correlate his anatomical and functional findings to produce a theory of language and its disorders, incorporating Broca's work in an attempt to describe language generation and dysfunction within the brain. Using his results and those of Broca, **Hitzig**, Gustav Theodor Fritsch and others, he suggested that fundamental neural properties were discretely localized, and that more complex functions arose out of interactions and connections between different areas. He established a clinic in Berlin specializing in diseases of the nervous system, where he worked until 1885 when he returned to Breslau (Associate Professor of Neurology and Psychiatry 1885–90; Professor 1890–1904). In 1904 he moved to Halle as professor, but he died as a result of an accident the following year. As well as papers on aphasia, Wernicke wrote on a wide variety of psychiatric and neurological issues, and described a form of encephalopathy resulting from dietary thiamine deficiency (common in alcoholics), which bears his name.

WESTINGHOUSE, George
(1846–1914)

American engineer, born in Central Bridge, New York, the son of a manufacturer of farm machinery. At the age of 15 he ran away from school to fight for the North in the American Civil War, and later served for a short time in the US navy, but decided in 1865 to return to his father's workshop. In October of the same year he took out the first of his more than 400 patents, for a railway steam locomotive. He invented many other devices connected with railways, but the most important was the air-brake system he patented in 1869, which became known as the Westinghouse air brake. This allowed the brakes on all the coaches or wagons throughout the length of a train to be applied simultaneously under the engine driver's control, and greatly increased the speed at which trains could safely travel. In the same year he founded the Westinghouse Air Brake Company, and subsequently devised a number of improvements which made the air brake even more effective. His works was situated in Pittsburgh, and he took an active part in the distribution and utilization of the natural gas deposits then beginning to be exploited in the area. He was a pioneer in the use of alternating current for distributing electric power, and founded the Westinghouse Electrical Company in 1886, attracting **Tesla** to work with him after leaving the employment of **Edison**. In 1895 he successfully harnessed the power of the Niagara Falls to generate sufficient electricity for the town of Buffalo, 22 miles away.

WEYL, Hermann
(1885–1955)

German mathematician, born in Elmshorn. He studied at Göttingen under **Hilbert**, and became professor at Zürich (1913), and Göttingen (1930). Refusing to stay in Nazi Germany, he moved to Princeton in 1933. Among important contributions to the theory of **Riemann** surfaces, he gave the first rigorous account of the surfaces while extending the original insights of Riemann. Inspired by a brief period with **Einstein** in Zürich, he wrote on the mathematical foundations of relativity and quantum mechanics, and subsequently on the representation theory of **Lie** groups. Modern attempts to create a gauge theory of particle interactions may be traced back to Weyl's concepts of measurement, and he wrote profoundly on how mathematics and quantum mechanics interrelate. He also wrote on the philosophy of mathematics, on the spectral theory of integral operators (which have played an essential role in quantum theory), and on algebraic number theory. His book *Symmetry* (1952) is an elegant and largely non-technical account for the general reader of the relation between group theory and symmetry in pattern and design.

WHEATSTONE, Sir Charles

(1802–75)

English physicist, born in Gloucester. He first became known as a result of his work in acoustics. He invented the concertina (1829), and published important papers on the figures of **Chladni** (the patterns produced by a fine powder spread over the surface of a vibrating plate), the resonance of columns of air, the transmission of musical sound through rigid linear conductors, and on the experimental proof of **Daniel Bernoulli**'s theory of the vibration of air in musical instruments. In 1834 he was appointed Professor of Experimental Physics at King's College, London, a position he held for the rest of his life, in spite of his unwillingness to lecture. In 1837 he and Sir William Cooke took out a patent for an electric telegraph, and in conjunction with the new London and Birmingham Railway Company installed a demonstration telegraph line about a mile long. With the needs of the rapidly expanding railways providing the impetus and the finance, by 1852 more than 4000 miles of telegraph line were in operation throughout Britain. Wheatstone built the first printing telegraph in 1841, and in 1845 devised a single-needle instrument. In 1838, in a paper to the Royal Society of which he had become a Fellow in 1836, he explained the principle of the stereoscope (later improved by **Brewster**). He invented a sound magnifier for which he introduced the term 'microphone'. Wheatstone's bridge, a device for the comparison of electrical resistances, was brought to notice (though not invented) by him. He was knighted in 1868.

WHEELER, David John

(1927–)

English computer scientist and programmer. He was educated at Trinity College, Cambridge, and as a Research Fellow at Cambridge between 1951 and 1957, he joined **Wilkes**'s EDSAC team in establishing the world's first computing service. Wheeler's expertise on the programming side, alongside the work of Stanley Gill, was of paramount importance, especially since a keynote of the Cambridge approach was user-convenience. The pioneering work of Wilkes, Wheeler and Gill was later published in an influential book, *The Preparation of Programs for an Electronic Digital Computer* (1951). The economy and elegance of the EDSAC programming, largely the work of Wheeler, was much in advance of any US or British group and the book was very influential. Wheeler worked in the USA at the University of Illinois between 1951 and 1953, before returning to Cambridge University, where he was appointed Professor of Computer Science in 1978. He was elected FRS in 1981.

WHEELER, John Archibald

(1911–)

American theoretical physicist, born in Jacksonville, Florida. He was educated at Johns Hopkins University, where he received his PhD in 1933, and spent the following two years in Copenhagen. He was appointed professor at Princeton (1947) and later at the University of Texas (1976). Wheeler worked with **Niels Bohr** on the theory of nuclear fission, and on the hydrogen bomb project. He also contributed to the search for a unified field theory, and studied with **Feynman** the concept of action at a distance.

WHIPPLE, Fred Lawrence

(1906–)

American astronomer, born in Red Oak, Iowa. He studied at California University, and became Professor of Astronomy at Harvard in 1945. An expert on the solar system (his *Earth, Moon and Planets* published in 1941 is a standard work), he is known especially for his work on comets. In 1950 he put forward his icy conglomerate model of the cometary nucleus in which the fount of cometary activity is a single 'dirty snowball', a few kilometres across. The spin of this nucleus, coupled with the delay in the transmission of heat through the dusty crust, leads to a jet effect which can change the orbit of the comet. He was also the first to define the term micrometeorite, realizing that below a certain size, a dust particle incident on Earth's atmosphere would be such an efficient heat radiator that it would be retarded by atmospheric friction without becoming molten. He used the rate of decay of meteors as an indicator of the temperature profile of the atmosphere. Whipple (with Fletcher Watson) was responsible for Harvard's two-station meteor programme. This showed that the vast majority of meteor orbits were similar to those of comets and asteroids and that hyperbolic (ie interstellar) meteoroids were not an obvious component of the meteoroid flux. Whipple was the prime-mover behind the production and use of the Baker Super-Schmidt meteor cameras. He was also a pioneer in the use of these cameras to observe the decay of satellite orbits, and from this obtained measurements of atmospheric density. His meteoroid bumper shield was used to dissipate the energy of impacting dust particles on the Giotto mission to Halley's comet, which confirmed his 'dirty snowball' model. In 1967 he wrote a keynote paper on the origin and evolution of dust in the solar system.

WHIPPLE, George Hoyt
(1878–1976)

American pathologist, born in Ashland, New Hampshire, the son of a general practitioner. Whipple graduated with a BA from Yale University in 1900, and received a medical degree from Johns Hopkins University in 1905. He joined the staff at Johns Hopkins as an assistant in pathology and began experiments on liver damage in dogs which developed into studies of the relationships between the liver, bile and haemoglobin. In 1914 he was appointed Director of the Hooper Foundation for Medical Research at the University of California. With his colleagues, he developed methods of making dogs anaemic. They showed that feeding liver to such anaemic animals was followed by a pronounced increase in haemoglobin regeneration. In 1921 Whipple became Professor of Pathology and Dean of Medicine at University of Rochester in New York. Until 1925 he was pre-occupied with administrative duties, but his co-worker Frieda Robbins supervised further research on haemoglobin during this period. In 1926 a liver extract was produced in cooperation with Eli Lilly. Whipple laid the groundwork for **Minot** and **Murphy**'s successful treatment of pernicious anaemia with liver in 1926—until then this had been a fatal disease. He shared the 1934 Nobel Prize for Physiology or Medicine with Minot and Murphy, and remained as Dean of Medicine at Rochester until 1952.

WHITE, Gilbert
(1720–93)

English naturalist and clergyman, born in Selbourne, Hampshire. He studied at Oriel College, Oxford (1739–43), became a Fellow of the college in 1744, took holy orders in 1747 and became junior proctor in 1752. In 1756 he became curate of Selbourne and in 1761 curate of the neighbouring parish of Faringdon. The college sinecure of Moreton Pinkney in Northamptonshire made him financially secure. His fame is based upon his *Natural History and Antiquities of Selborne* (1789); this was a natural history of a parish comprising the journal for a whole year and resulted from a series of letters written by White. In its original form, dealing only with the natural history of Selborne, the journal was completed in 1769 but the inclusion of additional letters on antiquarian and parish subjects delayed its publication for almost 20 years. The journal is still read, not only because of its charming style and literary merit, but because of its acute observation on the habits and lives of a wide range of birds, mammals and insects.

WHITEHEAD, Alfred North
(1861–1947)

English mathematician and Idealist philosopher, born in London. He was educated at Sherborne and Trinity College, Cambridge, where he was Senior Lecturer in Mathematics until 1911. He became Professor of Applied Mathematics at Imperial College, London (1914–24), and Professor of Philosophy at Harvard (1924–37). Extending the Booleian symbolic logic in a highly original *Treatise on Universal Algebra* (1898), he contributed a remarkable memoir to the Royal Society, *Mathematical Concepts of the Material World* (1905). Profoundly influenced by **Peano**, he collaborated with his former pupil at Trinity, **Bertrand Russell**, in the *Principia Mathematica* (1910–13), the greatest single contribution to logic since **Aristotle**. In his Edinburgh Gifford Lectures, 'Process and Reality' (1929), he attempted a metaphysics comprising psychological as well as physical experience, with events as the ultimate components of reality. Other more popular works included *Adventures of Ideas* (1933) and *Modes of Thought* (1938). He was awarded the first James Scott Prize of the Royal Society of Edinburgh (1922).

WHITNEY, Josiah Dwight
(1819–96)

American geologist, born in Northampton, Massachusetts. He graduated at Yale, and in 1840 joined the New Hampshire Survey. He worked in Michigan from 1847 to 1849, and in the Lake Superior region with **James Hall** (1811–98). Following his studies of mining problems in Illinois, he published *Mineral Wealth of the United States* (1854). He was appointed professor at Iowa University in 1855, State Geologist of California in 1860, and professor at Harvard in 1865. Whitney produced important work on the *Auriferous Gravels of the Sierra Nevada* (1879–80), in which he recognized that the gold deposits were not marine deposits as had been supposed, but were the products of erosion and deposition of pre-existing gold bearing mineral veins. He also wrote on the *Climate*

Changes of Later Geological Time (1880, 1882). Mount Whitney in southern California is named in his honour.

WHITTAKER, Sir Edmund Taylor
(1873–1956)

British mathematical physicist, born in Birkdale, Lancashire. He was educated at the University of Cambridge, where he held a teaching post from 1896, and in 1906 was appointed Professor of Astronomy at Dublin University and Astronomer Royal for Ireland. He later became Professor of Mathematics at Edinburgh (1912–46). In his earlier studies of differential equations, Whittaker formulated a general solution to **Laplace**'s equation in three dimensions. He gave great stimulus to the mathematical development of relativity theory, and introduced an important general integral representation of harmonic functions. His publications covered a diverse range of topics, including quantum mechanics, electromagnetism, the history of science and philosophy. Elected FRS in 1905, he was knighted in 1945.

WHITTINGTON, Harry Blackmore
(1916–)

English palaeontologist. After studying at the University of Birmingham, his active research career commenced with a fellowship at Yale Peabody Museum (1938), where he worked on North American and European trilobites. He became a lecturer at Judson College, Rangoon (1940), until forced to flee from the advancing Japanese army to Szechuan. He then became lecturer and later professor at Ginling Women's College (1943). From 1945 he held posts at the University of Birmingham (1945–50) and Harvard (1950–66) before returning to Britain as Woodwardian Professor of Geology at Cambridge (1966–83). He wrote *The Burgess Shale* (1985) and *Trilobites* (1992).

WHITTLE, Sir Frank
(1907–)

English aeronautical engineer and inventor, born in Coventry. Son of a designer and craftsman, his interests in engineering and invention were fostered during the times he spent as a boy in his father's workshop. He developed an early interest in aeronautics, and conceived the idea of trying to develop a replacement for the conventional internal combustion aero engine. He joined the RAF as an apprentice (1923), and studied at the RAF College, Cranwell, and at Cambridge University (1934–7). By the time he was 21 he could see that the most promising new form of propulsion for aircraft would probably consist of a high-speed jet of hot gases. He began research into jet propulsion before 1930, while still a student, and after a long fight against official inertia his engine was first flown successfully in a Gloster aircraft, code-named E.28/39, in May 1941. The actual aircraft has been preserved in the Science Museum, London, although it was in Germany in 1939 that the world's first flights of both turbo-jet and rocket-powered aircraft took place. Whittle was elected FRS in 1947, and knighted in 1948 on his retirement from the RAF. He has since acted as consultant and technical adviser to a number of British firms, and in 1977 was elected a member of the faculty of the US Naval Academy at Annapolis, Maryland.

WIELAND, Heinrich Otto
(1877–1957)

German chemist, born in Pforzheim. He studied chemistry at the universities of Munich, Berlin and Stuttgart and received a PhD from the University of Munich (1901), where he subsequently became lecturer, and in 1909, associate professor. Four years later he was appointed professor at Munich Technical University. During World War I, while on leave of absence, he worked on chemical weapons at the Kaiser Wilhelm Institute. After the war he returned to Munich until 1921, when he moved to the University of Freiburg for three years. He returned to the University of Munich in 1924 as chairman, a position he held until his retirement in 1950. He made many contributions to the development of organic chemistry. His initial studies involved nitrogen compounds and he also made extensive investigations of oxidation reactions. However, he is most famous for his studies of the bile acids (substances stored in the bladder which aid the digestion of lipids). It was for this work that he received the Nobel Prize for Chemistry in 1927. He subsequently studied the chemistry of a number of naturally occurring substances, including butterfly pigments.

WIEN, Wilhelm Carl Werner Otto Fritz Franz
(1864–1928)

German physicist, born in Gaffken in East Prussia, the son of a farmer. He was a slow starter academically and was taken out of school to learn agriculture, but he resumed his studies and went to the universities of Göttingen, Berlin (where he studied under **Helmholtz**) and Heidelberg. He became Helmholtz's assistant at the Physikalisch Technische Reichanstalt in Charlottenberg (1890), was appointed to professorships at Aachen (1896), Giessen (1899), Würzburg (1899) and finally Munich (1920). He was the successor of **Röntgen** at both Würzburg and Munich. His chief contribution was on black-body radiation. Advancing the work of **Boltzmann** (1884), he showed that the wave-length at which maximum energy is radiated is inversely proportional to the absolute temperature of the body (1893). His attempt to formulate an equation that would fit the observed distribution of all possible frequencies (both short-wavelength/high-frequency and long-wavelength/low-frequency radiation) was unsuccessful (1896), but cleared the way for **Planck** to resolve this with the quantum theory (1900). In 1911 Weber was awarded the Nobel Prize for Physics for his work on the radiation of energy from black bodies. His researches also covered hydrodynamics, and X-rays and cathode rays. He showed that cathode rays consisted of negatively charged particles moving at a very high velocity (1897–8), and that 'canal rays' were positively charged particles which were deflected by electric and magnetic fields (1905). He edited the *Annalen der Physik* (1906–28).

WIENER, Norbert
(1894–1964)

American mathematician, born in Columbia, Missouri. A youthful prodigy, he studied zoology at Harvard and philosophy at Cornell; in Europe he studied with **Bertrand Russell** at Cambridge and at Göttingen. He was later appointed Professor of Mathematics at MIT (1932–60). At MIT he worked on stochastic processes and harmonic analysis, inventing the concepts later called the Wiener integral and Wiener measure for application to physical problems such as Brownian motion. During World War II he studied mathematical communication theory applied to predictors and guided missiles. His study of the significance of feedback in the handling of information by electronic devices led him to compare this with analogous mental processes in animals in *Cybernetics, or control and communication in the animal and the machine* (1948) and other works. His frankly egocentric autobiography *I am a mathematician — the later life of a prodigy* was published in 1956. For all his gifts, he seems to have had deep doubts about his own mathematical ability.

WIESEL, Torsten Nils
(1924–)

Swedish neurophysiologist, born in Uppsala. He studied medicine at the Karolinska Institute in Stockholm, and then went to the USA for postdoctoral work with **Kuffler**, initially at Johns Hopkins Medical School (Fellow in Ophthalmology, 1955–8; Assistant Professor of Ophthalmic Physiology, 1958–9), and then at Harvard Medical School (Assistant Professor of Neurophysiology, 1959–60; Associate Professor, 1960–7; Professor of Physiology, 1967–8; Professor of Neurobiology, 1968–74; Robert Winthrop Professor of Neurobiology, 1974–83). Working with Kuffler he met **Hubel**, and together they studied the way in which the brain processes visual information, following on from the work of **Hartline**, **Granit** and Kuffler himself. They demonstrated that there is a hierarchical processing pathway, of increasingly sophisticated analysis of visual information by nerve cells from the retina to the cerebral cortex. Their detailed results were not only scientifically important, they also had almost immediate clinical relevance for the treatment of children with visual problems. In 1981 Wiesel and Hubel shared the Nobel Prize for Physiology or Medicine with **Sperry**.

WIGGLESWORTH, Sir Vincent Brian
(1899–)

English biologist, born in Kirkham, Lancashire. He was educated at the University of Cambridge and St Thomas's Hospital, and became Reader in Entomology at London (1936–44) and Cambridge (1945–52), where he was subsequently appointed Quick Professor of Biology (1952–66). From 1943 to 1967 he was also Director of the Agricultural Research Council Unit of Insect Physiology. In a series of remarkable studies of insect metamorphosis, Wigglesworth demonstrated the production and secretory cell sources of hormones which selectively activate different genetic components of insects during various stages of their life cycles. He also succeeded in artificially inducing such changes experimentally by manipulating the associated hormone levels. This work and his many other wide-ranging studies of insects led to a much greater understanding of their physiology and interactions with the environment. Wigglesworth was elected FRS in 1939, and knighted in 1964.

WIGNER, Eugene Paul
(1902–)

Hungarian-born American theoretical physicist, born in Budapest. He was educated at Berlin Technische Hochschule, where he was awarded a degree in chemical engineering (1924) and a doctorate in engineering (1925). He moved to the USA in 1930, and apart from two years at the University of Wisconsin (1936–8) he worked at Princeton University throughout his academic career, becoming Thomas D Jones Professor of Theoretical Physics in 1938. Wigner made a number of important contributions to nuclear physics and quantum theory. In 1927 he introduced the idea that the quantum property known as parity is conserved in nuclear interactions. He believed this to be true for all nuclear reactions, but **Lee** and **Yang** later demonstrated parity non-conservation in the weak interaction. In the 1930s, Wigner demonstrated that the strong nuclear force which binds protons and neutrons in nuclei has very short range, and is independent of any electric charge. He is especially known for the Breit–Wigner formula which describes cross-sections (probabilities) for resonant nuclear reactions, and the Wigner theorem concerning the conservation of the angular momentum of electron spin. His name is also given to the most important class of mirror nuclides (Wigner nuclides). Wigner's calculations were used by **Fermi** in building the first nuclear reactor in Chicago (1942) and he was instrumental in convincing the US government of the need for a nuclear bomb. He received the Fermi award in 1958, the Atom for Peace award in 1959, and the Nobel Prize for Physics in 1963 for his work in furthering quantum mechanics and nuclear physics.

WILCKE, Johan Carl
(1732–96)

Swedish physicist, born in Wismar, Germany. Wilcke moved to Sweden with his parents in 1739. He entered Uppsala University to study theology in 1750 but concentrated instead on mathematics and physics. Whilst in Berlin he investigated electrical phenomena with his friend **Aepinus**, preparing a doctoral dissertation which he defended at Rostock in 1757. From 1759 he lectured on experimental physics at the Royal Swedish Academy in Stockholm. A painstaking evaluation of existing data culminated in his comprehensive map of the Earth's magnetic inclination (1768). Wilcke is also known for scientific instrument design, and above all, for experimental work into the nature of heat. In 1772 he measured the heat required to melt snow at its freezing point (the latent heat of fusion); and in 1781 he drew up a list of specific heats for different substances, giving precise details of the experimental methods he had used to determine them. Wilcke began these experiments on specific heats independently of **Joseph Black**.

WILKES, Maurice Vincent
(1913–)

English computer scientist, born in Dudley. He was educated at the University of Cambridge, where he was a Mathematical Tripos Wrangler at St John's College. He conducted research in physics at the Cavendish Laboratory, and after the outbreak of World War II worked in radar (as did many of the computer pioneers, such as **Kilburn** and Sir **Frederic Calland Williams**). Wilkes directed the Mathematical (later Computer) Laboratory at Cambridge (1946–80), where he was best known for his pioneering work with the EDSAC (Electronic Delay Storage Automatic Calculator). This stored-program computer was unabashedly based on the US designs of **Eckert** and **Mauchly** and the theoretical work of **von Neumann**, which Wilkes had become familiar with on a US visit in 1946. Wilkes's intention—besides building the computer—was to provide a useful and reliable computing service. Helped by a team which included **David Wheeler**, this service, the first in the world, was available by early 1950—ahead of the Americans. Besides important software advances, Wilkes's work also included fundamental work on processor controls (microprogramming). The EDSAC influenced the design of the Lyons LEO through the laboratory's close links with the catering firm of J Lyons. Developments at the laboratory continued into the late 1950s, when a new computer, EDSAC II, was designed and built. After 1980 Wilkes became a computer engineer for **Olsen**'s Digital Equipment Corporation until 1986. He was elected FRS in 1956, and published his *Memoirs* in 1985.

WILKINS, Maurice Hugh Frederick
(1916–)

British physicist, born in Pongaroa, New Zealand. Educated at St John's College, Cambridge, he worked on uranium isotope separation at the University of California in 1944. Opposed to the use of the atomic bomb, he turned away from nuclear physics after World War II. After reading **Schrödinger**'s book *What is Life* he developed an interest in biology and the application of physical techniques to biological research. In 1946 he joined the Medical Research Council's Biophysics Research Unit at King's College, London, where he applied the techniques of X-ray crystallography developed by **William** and **Lawrence Bragg** to biological molecules. **Crick** and **James Watson** deduced their double helix model of DNA from Wilkins and **Rosalind Franklin**'s X-ray data of DNA fibres. Crick, Watson and Wilkins were awarded the 1962 Nobel Prize for Physiology or Medicine for their work on DNA. Wilkins went on to become director of the Medical Research Council's Biophysics Research Unit at King's College, London (1970–2).

WILKINSON, Sir Geoffrey

(1921–)

English chemist, renowned for his work on organometallic compounds. He was born in Springside near Manchester and studied at Imperial College, London. During World War II he worked on the Canadian branch of the atomic bomb project with the National Research Council of Canada. He then moved to the Lawrence Radiation Laboratory of the University of California at Berkeley and was appointed assistant professor at Harvard in 1951. From 1955 to 1988 he was Professor of Inorganic Chemistry at Imperial College. Wilkinson's early research was on the chemistry of the transition elements. While at Harvard he studied ferrocene, a new type of compound synthesized by Thomas Kealy and Peter Ludwig Pauson, which consists of an iron atom attached to two five-sided rings of carbon and hydrogen but whose exact spatial structure and bonding had not then been determined. Using nuclear magnetic resonance spectroscopy, Wilkinson and **Robert Woodward** showed that the iron atom is sandwiched between the two rings and bonded to each of the five carbon atoms in both rings, a type of structure entirely new to chemistry which explained the great stability of the molecule. Wilkinson and his colleagues went on to synthesize other organometallic sandwich molecules, often using the transition elements. Since then many thousands of similar compounds have been synthesized in other laboratories. They have led to new lines of research in organic, inorganic and theoretical chemistry, and to the development of new catalysts used in the production of plastics and low-lead fuels. They are also employed in pharmaceuticals, for example L-dopa, used to treat Parkinson's disease. Wilkinson was elected FRS in 1965 and was knighted in 1976. In 1973 he shared the Nobel Prize for Chemistry with **Ernst Fischer** who had worked independently in Germany on organometallic sandwich compounds.

WILKINSON, James Hardy

(1919–86)

English mathematician and computer scientist. He was educated at Sir Joseph Williamson's Mathematical School, Rochester, and at Trinity College, Cambridge. During World War II he worked at the Mathematical Laboratory, Cambridge, and at the Armament Research Department, Fort Halstead. In 1946 he joined the Mathematical Division of the National Physical Laboratory (NPL) and became involved with **Turing**'s construction of the ACE. His chief effort was directed to producing programs for the several versions of the ACE designed by Turing and resulted in some of the earliest floating-point programs. He was also involved in the logical and electronic design of the Test Assembly, a preliminary version of the pilot ACE initiated by **Huskey**. Before he left the NPL in 1980, Wilkinson also published work on rounding errors in algebraic processes. In 1977 he became Professor of Computer Science at Stanford University.

WILLIAMS, Sir Alwyn

(1921–)

Welsh geologist and palaeontologist, born in south Wales and educated at the University of Wales in Aberystwyth. He was awarded a Harkness fellowship to work at the US National Museum of Washington DC (1948–50), and subsequently became a lecturer (1950–4), Professor of Geology (1954–74) and Pro-Vice-Chancellor (1967–74) at Queen's University, Belfast. In 1974 he was appointed Lapworth Professor of Geology at the University of Birmingham, and from 1976 until his retirement in 1988 he was Principal and Vice-Chancellor of the University of Glasgow. Williams's principal research interests are in Scottish Lower Palaeozoic stratigraphy and palaeontology, and he has made important contributions to the studies of brachiopods. He was elected FRS in 1967 and knighted in 1983.

WILLIAMS, Sir Frederic Calland

(1911–77)

English electrical engineer, born in Romily, Cheshire. After attending Stockport Grammar School, he studied engineering at Manchester University, where he received his DSc in 1939, and Oxford University. During World War II Williams worked for the Telecommunications Research Establishment at Malvern, where he was recognized as a world authority on radar. While there he had begun experimental work on cathode-ray tube storage, which he put to brilliant effect when he accepted the Chair in Electrotechnics at Manchester University in 1946. With the help of **Kilburn**, he utilized cathode-ray tubes to build the world's first electronic random access memory for a computer. The aptly named Williams tube became the basis for a prototype machine, built by Williams and Kilburn, which ran the world's first stored-program on 21 June 1948. This event, recorded Williams, was 'the breakthrough and sparks flew in all directions'. This prototype was developed by Williams and Kilburn into the famous Manchester University Mark I, a commercial version of which Ferranti produced to Williams's specification. Williams tubes came to be widely used in early commercial computers during the 1950s in Britain and the USA. Soon, however, Williams—once described as a typical example of the British string-and-sealing-wax inventor—lost interest in computers and handed the work to Kilburn. He transferred his attention to other electrical engineering projects, such as induction machinery. He was elected FRS in 1950, and knighted in 1976.

WILLIAMS, Robert Joseph Paton

(1926–)

English inorganic chemist, born in Wallasey, Merseyside. He attended Merton College, Oxford, and published his first paper (on the Irving–Williams series describing the stabilities of certain transition-metal complexes) before obtaining his BA degree in 1948. His DPhil was completed in 1950. He then went for one year to Uppsala, Sweden, where he contributed to the development of the use of gradient elution analysis in chromatography. His lifelong interest in bioinorganic chemistry—the function of inorganic compounds in biology, already begun in his DPhil thesis—became his major research topic on his return to Oxford in 1951. He published his first paper on the subject in 1953 when the field was only starting to become a recognized area of research. In 1956 he became a lecturer in chemistry at Oxford University and carried out work on the stability, redox equilibria and spectra of metal complexes, and electron transfer in mixed oxidation state compounds. He also investigated the physical, chemical and biochemical properties of vitamin B_{12}. During a year at Harvard Medical School he worked on copper- and zinc-containing enzymes. He has proposed the way in which protons can drive biological energy capture as polymerized phosphate, and has developed nuclear magnetic resonance (NMR) spectroscopy as a means to study proteins in solution. He was elected FRS in 1972, and from 1974 to 1991 served as Napier Royal Society Research Professor.

WILLIAMS, Robley Cook

(1908–)

American biophysicist, born in Santa Rosa, California. He studied physics at Cornell University and became Assistant Professor of Astronomy (1935–45) and Professor of Physics (1945–50) at the University of Michigan. His research interests then were in the spectroscopic analysis of the surface temperatures of stars. In 1944, his colleague **Wyckoff** was dissatisfied with the electron-microscopic techniques available to examine the size and shape of viruses, and discussed their limitations with Williams. As an astronomer, familiar with the practice of measuring lunar mountains from calculations of the shadow cast in known conditions of illumination, Williams suggested enhancing the structure of the virus so as to make it cast a shadow. Treated with a heated tungsten filament coated with gold, viruses could be made to cast measurable shadows, which in turn led to further information about their size and structure. This technique of metal-shadowing became widely used for the electron-microscopic examination of biological materials. These interests in electron microscopy diverted Williams towards biological problems and he moved to the University of California at Berkeley as Professor of Biophysics (1950–9), Professor of Virology (1959–64) and Professor of Molecular Biology (1964–76). He has made major contributions to knowledge of virus structure, including a reconstitution, in the mid-1950s, of the tobacco mosaic virus, from its constituent proteins and nucleic acids.

WILLIAMSON, Alexander William

(1824–1904)

English chemist, born in Wandsworth. He studied chemistry at Heidelberg and Giessen (1844–6), and then moved to Paris where he had a private laboratory. He was appointed Professor of Analytical Chemistry (1849), and later Professor of General Chemistry, at University College London. He was an inspiring teacher, but in 1855 his enthusiasm began to decline and his interest turned from chemistry to sources of hot air: steam engines and university committees. He resigned in 1887. He is most famous for his work on the synthesis of ethers. He supported the radical theory of organic chemistry and saw ether as of the water type. He was elected FRS in 1855.

WILLIAMSON, William Crawford

(1816–95)

English botanist, surgeon, zoologist and palaeontologist, born in Scarborough, the son of the Curator of Scarborough Museum. He trained in medicine and became Curator of the Museum of the Manchester Natural History Society (1835–8). He was later appointed the first Professor of Natural History and Geology (later of botany) at Owens College, Manchester (1851–92). Williamson was the first to investigate thoroughly the plant remains (coal balls) in coal. At the time, however, the full significance of his work in fossil botany was not appreciated, and after 41 years of teaching at Owens College (later Manchester University) he was refused a pension. He is regarded as the founder of modern palaeobotany. He made several contributions to *Fossil Flora of Great Britain* (1831–7).

WILLIS, John Christopher

(1868–1958)

English botanist, born in Birkenhead. He was educated at University College, Liverpool, and Cambridge University. In 1892 he became a lecturer there and began investigating the origin of gynodioecism in the Labiatae; this research led into more general studies on floral biology. In 1894 he became senior assistant to **Bower** at Glasgow University and began compiling his *Dictionary of the Flowering Plants and Ferns* (1896). This was originally an assemblage of taxonomic facts based on his own summaries made as a student, but in its later editions, by **Airy Shaw**, it became primarily a list of all known vascular plant generic names. Willis succeeded Henry Trimen (1843–96) as director of the botanic gardens at Peradeniya, Ceylon (1896–1900). While there, he monographed the family Podostemaceae; this work profoundly altered his views on natural selection, which he discredited as an origin for the family's endemic species. This led to the development of his 'age and area' hypothesis in which he claimed that, other things being equal, the area occupied by a taxon is directly proportional to its age. His theories were published as *Age and Area* (1922), *The Course of Evolution* (1940) and *The Birth and Spread of Plants* (1949).

WILLIS, Thomas

(1621–73)

English physician, one of the founders of the Royal Society (1662). Born in Great Bedwyn, Wiltshire, Willis studied classics and then medicine at Oxford, graduating BA at Christ Church, Oxford, in 1636 and MB in 1646. He briefly served in the Royalist army in the Civil War. He was one of the small group of natural philosophers including **Boyle** who met in Oxford in 1648–9 and who were to become founder members of the Royal Society of London. He practised medicine in Oxford, becoming Sedleian Professor of Natural Philosophy there (1660–75); however, his fame and wealth derived from a fashionable medical practice in the metropolis. His main work was on the anatomy of the brain and on diseases of the nervous system. His *Cerebri anatome* (1664) was the principal study of brain anatomy of its time. In this he offered new delineations of the cranial nerves and described cerebral circulation, discovering the ring of vessels now called the 'circle of Willis'. He overcame some of the difficulties of studying soft brain tissues by injecting the vessels with wax. Willis produced faithful descriptions of such fevers as typhus, typhoid and puerperal fever. He also pioneered the clinical and pathological analysis of diabetes, demonstrating that the excessive urine manufactured by the diabetic possesses a sweet taste. He was also the first to recognize that spasm of the bronchial muscles was the essential characteristic of asthma.

WILLSTÄTTER, Richard

(1872–1942)

German chemist, born in Karlsruhe. After graduation from the Realgymnasium in Nuremberg (1890), he moved to the University of Munich and produced a doctoral thesis on the chemistry of cocaine. From Munich he was appointed to a chair at the Eidgenössische Technische Hochschule in Zürich. He also worked (1912–16) at the Kaiser Wilhelm Institute in Dahlem, Berlin, before returning to become professor in Munich. As a protest against anti-Semitism he resigned this post in 1924 and later moved to Locarno, Switzerland, where he remained until his death. Although a man of wide scientific interests, most of his chemical studies concerned the structure and synthesis of natural products. Following his doctoral work on cocaine, a series of studies during 1894–8 led him to revise many formulae of the tropine alkaloids, a topic to which he occasionally returned. A by-product of this work was the synthesis of the cyclic but completely non-aromatic compound 1,3,5,7-cyclo-octatetraene. His most significant work concerned the structure of the chlorophylls. He showed that they were magnesium complexes of porphyrins, and laid the basis for the complete elucidation of their structures by **Hans Fischer**. He also studied the anthocyanin pigments of flowers. His final researches concerned enzymes, which he erroneously thought were small molecular weight compounds absorbed onto colloids. He was awarded the Nobel Prize for Chemistry in 1915 for his work on the chemistry of natural products.

WILSON, Charles Thomson Rees

(1869–1959)

Scottish pioneer of atomic and nuclear physics, born in Glencorse near Edinburgh. Educated at Manchester and at Cambridge, where later he became Professor of Natural Philosophy (1925–34), he was noted for his study of atmospheric electricity, one by-product of which was the successful protection from lightning of Britain's wartime barrage balloons. His greatest achievement was to devise the cloud chamber. While working in Scotland, he observed that shadows cast by the Sun on the mountain mist were surrounded by a coloured halo. In trying to reproduce this in the laboratory he discovered that water droplets would form around ions in air that was saturated with water vapour. He used this effect to develop the cloud chamber, in which water droplets form around the track of ionization left by radiation passing through a chamber of saturated water vapour. The movement and interaction of ionizing radiations can thus be followed and photographed. This principle was also used by **Glaser** to develop the bubble chamber. Wilson shared the 1927 Nobel Prize for Physics with **Compton**, and in 1937 received the Copley Medal of the Royal Society.

WILSON, Edmund Beecher
(1856–1939)

American zoologist and embryologist, born in Geneva, Illinois, and considered to be one of the founders of modern genetics. He studied at Yale and Johns Hopkins universities, and after several teaching posts became Da Costa Professor of Zoology at Columbia University in New York. His early research was concerned with cell lineage and the formation of tissues from precursor cells. He demonstrated that in molluscs, isolated cleavage cells eventually developed into the tissues they would have become in the intact embryo. He emphasized the significance of cells as the building blocks of life and argued against the vitalistic ideas prevalent at the end of the 19th century. His major contribution was to show the importance of the chromosomes, particularly the sex chromosomes, in heredity and cell structure. His *The Cell in Development and Inheritance* (1896, 1925) was instrumental in the synthesis of cytology and Mendelian genetics.

WILSON, Edward Osborne
(1929–)

American biologist, born in Birmingham, Alabama. He studied there and at Harvard, where since 1956 he has been Baird Professor of Science and Curator of Entomology at the Museum of Comparative Zoology. He has been a major figure in the development of sociobiology, the investigation into the biological basis of social behaviour. His early researches into the social behaviour, communication and evolution of ants resulted in the publication of *The Insect Societies* (1971), an overview of insect social behaviour in which Wilson outlines his belief that the same evolutionary forces have shaped the behaviours of insects and other animals including human beings. His book *Sociobiology: the New Synthesis* (1975) was acclaimed for its detailed compilation and analysis of social behaviour in a wide range of animals. However, the last chapter, 'From Sociobiology to Sociology', proved controversial. In it he emphasized the importance of the genetic component in controlling a range of human behaviours, including aggression, homosexuality, altruism and differences between the sexes. His ideas probably redressed the balance that previously existed in favour of environmental determinants of human behaviour and they stimulated considerable research efforts. His books include *On Human Nature* (1978), for which he received the Pullitzer Prize, *The Ants* (1990, with B Holldobler) and *The Diversity of Life* (1992).

WILSON, Ernest Henry
(1876–1930)

English plant collector, botanist and prolific writer, born in Chipping Campden, Gloucestershire. He was apprenticed to nurseries at Solihull when very young, and was recommended to the Curator of Birmingham Botanic Garden at 16. Thereafter he studied at Birmingham Technical School and at Kew. The Director of Kew, Sir William Thiselton-Dyer, recommended him to Veitch's Nurseries as a plant collector. He left for central China in 1899; the expedition was so successful that after his return in 1902 a second soon followed (1903–5). In 1906 he joined the Arnold Arboretum and made further expeditions to China (1906–11), Japan (1914, 1917–19) and Australia, New Zealand, India and Africa (1919–21). He discovered over 3 000 new species and introduced more than 1 000 to cultivation. His last post was as Keeper of the Arnold Arboretum (1917–30). His published works included *A Naturalist in Western China* (1913), *Cherries (*Prunus*) of Japan* (1916), *Conifers and Taxads of Japan* (1916), *Lilies of Eastern Asia* (1925), *The Vegetation of Korea* (1918–20) and *Plant Hunting* (1927).

WILSON, Kenneth Geddes
(1936–)

American theoretical physicist, born in Waltham, Massachusetts, son of Harvard Professor of Chemistry E Bright Wilson. He was educated at Harvard University and Caltech, where he received his PhD in 1961. He worked at Harvard and at the European nuclear research centre, CERN, in Geneva (1962–3). In 1963 he moved to Cornell University where he became professor in 1971. He is now professor at Columbus, Ohio. Wilson applied ingenious mathematical methods to the understanding of the magnetic properties of atoms, and later used similar methods in the study of phase changes between liquids and gases, and in alloys. Analysis of how a system changes from one phase to another, such as the onset of ferromagnetism in a magnet cooling below its Curie point, had until then been virtually impossible due to the length-scale of the interactions varying over several orders of magnitude. Leo Kadanoff had suggested that the effective spin of a block of atoms should be found and then a renormalization, or scaling, transformation made to calculate the effective spin of a larger block composed of known smaller blocks. Wilson developed this method towards a general theory, allowing the properties of a system of large numbers of interacting atoms to be predicted from observations of individual atoms, and his technique is used in many differing fields of study. For this work he was awarded the Nobel Prize for Physics in 1982. More recently, he has applied his technique to the strong nuclear force that binds quarks in the nucleus.

WILSON, Robert Woodrow

(1936–)

American physicist, born in Houston, Texas. He was educated at Rice University and Caltech. He then joined Bell Laboratories in New Jersey and became head of the radiophysics research department in 1976. There he collaborated with **Penzias** in using a large radio telescope designed for communication with satellites; they detected in 1964 a radio noise background coming from all directions with an energy distribution corresponding to that of a black body at a temperature of 3.5 kelvin. **Dicke** and **Peebles** suggested that this radiation is the residual radiation from the Big Bang at the universe's creation, which has cooled to 3.5 kelvin by the expansion of the universe. Such a cosmic background radiation had been predicted to exist by **Gamow**, **Alpher**, **Bethe** and Robert Herman in 1948. Wilson and Penzias (jointly with **Kapitza**) shared the 1978 Nobel Prize for Physics for their work, which can reasonably be claimed to be one of the most important contributions to cosmology in this century. In 1970 he continued his collaboration with Penzias and they discovered (with K B Jefferts) the 2.6 millimetre wavelength radiation from interstellar carbon monoxide.

WILSON, (John) Tuzo

(1908–93)

Canadian geophysicist, born in Ottawa. He was educated at the University of Toronto, where by changing courses to physics and geology he became the first Canadian geophysics graduate (1930). On a scholarship to Cambridge (1931–2), he studied under **Harold Jeffreys** then went on to Princeton, where he received a PhD in geology in 1936. His thesis fieldwork included making the first solo ascent of Mount Hague (1935), and his first appointment was with the Geological Survey of Canada. World War II was spent in the Canadian army (1939–48), where he reached the rank of Colonel. In 1946 he was appointed professor at the University of Toronto. When palaeomagnetic and heat-flow evidence supported **Harry Hess**'s 1960 theory of sea-floor spreading, Wilson was not only converted, but became a central figure in its promotion. Expert at the overview, his ideas about permanent hotspots in the Earth's mantle (1963), his elucidation of oceanic transform faults (1965) and his hypothesis on mountain building (1966) were major steps towards plate tectonic theory. He became principal of the new Erindale campus at Toronto (1967–74) then Director-General of the Ontario Science Centre. He co-authored *Physics and Geology* (1959), one of the first geophysical textbooks. The Wilson Range in Antarctica is named after him, and the cyclical opening and closing of oceans is referred to as the Wilson cycle. He was awarded an OBE in 1946 and elected FRS in 1968. Mount Tuzo in British Columbia is named after his mother Henrietta Tuzo, who was the first to climb it (1906).

WINDAUS, Adolf

(1876–1959)

German chemist, born in Berlin. He commenced medical studies at the University of Berlin in 1895, but both there and subsequently at the University of Freiburg, he became increasingly interested in chemistry. He abandoned medicine and wrote a doctoral thesis on the cardiac poisons of digitalis in 1899. Following a year of military service he returned to Freiburg where he became lecturer, and later professor. In 1913 he was appointed Professor of Medical Chemistry at the University of Innsbruck, but two years later he moved to the University of Göttingen, and remained there for the rest of his professional life. His most important research was on the structure of cholesterol, a substance widely distributed in the body and associated with a number of cardiovascular disorders. At this point his work overlapped with that on bile acids by **Wieland**. In the 1920s he turned to a study of vitamin D, which is structurally related to cholesterol. For his work on cholesterol and vitamins he was awarded the Nobel Prize for Chemistry in 1928. During the 1930s he continued to study the structure of natural products, including vitamin B, and colchicine (a drug used in cancer chemotherapy). In 1938 he ceased his research and he retired in 1944.

WINKLER, Clemens Alexander
(1838–1904)

German chemist, born in Freiberg. His father managed a plant for extracting cobalt, and after studying at the School of Mines in Freiberg, Winkler held various industrial posts. He was appointed Professor of Analytical and Technical Chemistry at the School of Mines in 1873, retiring in 1902. Winkler looked for a way to make sulphuric acid from the sulphurous gases produced by smelting cobalt ore. Finding methods of gas analysis inadequate, he designed a gas burette, later known by his name. He prepared sulphuric acid for the dye industry, making use of the well-known properties of platinum as a catalyst, but instead of using the metal in a finely divided state, he increased its surface area (and hence its efficiency) further by presenting it in the form of platinized asbestos. In 1885, while analysing argyrodite (a newly discovered silver sulphide from the Freiberg mines), he discovered that silver and sulphur accounted for only 93 per cent of the compound and suspected the existence of a new element. The following year he isolated germanium, a silvery grey metalloid, which had all the properties which **Mendeleyev** had predicted for the missing element at number 32 in the periodic table. The discovery demonstrated the predictive powers of the periodic table and helped to make it more generally accepted. Winkler was a respected teacher and wrote many useful books on analytical chemistry. He died in Dresden.

WINSTEIN, Saul
(1912–69)

American chemist, born in Montreal, Canada. He moved to the USA in 1923, graduated BA and MA from the University of California at Los Angeles (UCLA) and obtained his PhD from Caltech in 1938. After two postdoctoral years and a year as instructor at Illinois Institute of Technology, he returned to UCLA where he was professor from 1947 until his sudden death in 1969 at the height of his career. Following in the footsteps of Sir **Christopher Ingold** he made substantial contributions to our understanding of organic solvolytic reactions including neigbouring group participation, solvent ionizing power, nucleophilicity, salt effects and ion-pair phenomena. He pioneered the concept of homoaromaticity but is most famous for his work on non-classical carbonium ions, over which he sustained a long controversy with **Herbert Brown**. Although the controversy was never fully resolved, Winstein's concept of delocalized positive charge is clearly of considerable importance. He received many honours, including the US National Medal of Science (1970).

WISLICENUS, Johannes Adolf
(1835–1902)

German chemist, born in Klein-Eichstedt, near Querfurt. While he was a student at the University of Halle, an attempt was made to arrest his father for publishing a liberal book of Biblical criticism, and the family fled (1853) to the USA. He worked in a laboratory at Harvard but returned to Germany to complete his studies in 1856. Subsequently he became professor at the Zürich Oberen Industrieschule (1861), at the University of Zürich (1864), and at the Eidgenössische Technische Hochschule in Zürich (1870). In 1872 he moved to Würzburg and in 1885 to Leipzig. Some of his early work was on human physiology, and between 1863 and 1873 he studied lactic acid and related compounds. He enthusiastically applied the ideas of **van't Hoff** and **Le Bel** on the tetrahedral bonding to carbon to explain geometrical isomerism. He also developed a number of important synthetic procedures.

WITHERING, William
(1741–99)

English physician, born in Wellington in Shropshire and educated at the University of Edinburgh. He practised medicine at the County Infirmary in Stafford, and later in Birmingham, where he became Chief Physician in the General Hospital. In 1776 he published a British Flora, *Botanical Arrangement of all the Vegetables Naturally growing in Great Britain*, which was arranged according to **Linnaeus**'s system, and which also contained an introduction to botany. He became acquainted with **Priestley**, and began studying chemistry and mineralogy as well as botany. He wrote *An Account of the Foxglove* (1785), introducing digitalis, extracted from that plant, as a drug for cardiac disease. He was the first to see the connection between dropsy and heart disease. Following his involvement in riots after the French Revolution, he fled from his job and home in Birmingham, and concentrated on bringing out a third volume of the *Botanical Arrangement* in 1792, dealing mainly with fungi and other cryptogamic plants.

WITTIG, Georg
(1897–1987)

German chemist, born in Berlin. He entered the University of Tübingen in 1916 but soon left to serve in World War I. In 1920 he recommenced the study of chemistry for a degree at the University of Marburg where he later joined the staff. He was appointed associate professor at the Technical University of Brunswick in 1932 and five years later moved to the University of Freiburg. In 1944 he was made professor at the University of Tübingen, and in 1956 he moved to Heidelberg. His early work on the solution chemistry of radicals and carbanions established him as a chemist of great skill, but he is most famous for a serendipitous discovery that some ylides (organometallic compounds containing both positive and negative charges) react smoothly with aldehydes and ketones with the creation of an olefinic double bond. This procedure has been of enormous value in the laboratory synthesis of numerous important compounds, including vitamin A, vitamin D, steroids, and prostaglandin precursors. For this work he shared the 1979 Nobel Prize for Chemistry with **Herbert Brown**. He continued publishing until the age of 90.

WÖHLER, Friedrich
(1800–82)

German chemist, born in Eschersheim. He was educated at the universities of Marburg and Heidelberg, and qualified as a doctor of medicine in 1823, specializing in gynaecology, but he never practised. He taught chemistry at industrial schools in Berlin (1825–31) and Kassel (1831–6), and became Professor of Chemistry at Göttingen in 1836, remaining there until his death. However, one of the formative experiences of Wöhler's life was the year he spent with **Berzelius** in Stockholm, and this led to a lifelong friendship. His friendship with **Liebig** was equally important to him, and from their common interest in cyanates came Wöhler's most famous discovery. In 1828 he attempted to prepare ammonium cyanate from silver cyanate and ammonium chloride, but instead obtained urea. This was an example of the production of a natural product from non-organic materials, a result which did not accord with the theory of vitalism, which claimed that natural products could be made only in living things. Wöhler was excited by his discovery but subsequent claims that the theory of vitalism was overturned by this one experiment are exaggerations. Wöhler's urea was not the first natural product to be made in the laboratory and it was soon followed by many more, leading to a gradual erosion of the theory of vitalism. Equally important as the synthesis of urea was Wöhler's preparation of aluminium in 1827. The Danish scientist **Oersted** claimed to have extracted the metal from alumina in 1825, but it is doubtful whether the metal he obtained was pure aluminium. Wöhler used a different procedure and the product (still extant) is essentially pure metal. For this work he was honoured by Napoleon III but there are still arguments, mainly nationalistic, about priority. Wöhler collaborated with Liebig on a number of occasions and together they made a substantial contribution to the foundation of modern organic chemistry. From 1840 onwards he undertook numerous administrative and government duties, and his research output diminished. He was an inspiring teacher and maintained a keen interest in chemistry well into his old age. He was responsible for the translation of the influential annual reports prepared by Berzelius, occasionally moderating the latter's strident language.

WOLF, Emil
(1922–)

American optical physicist, born in Prague. He moved to the UK in 1940, and was educated at the University of Bristol. His first research was into the characteristics of high-performance optical systems, partly using methods due to **Zernike**; subsequently he worked on optical coherence in Edinburgh, Manchester and Rochester (USA). He was appointed to a full professorship at the University of Rochester in 1961, and became Professor of Physics there and Professor of Optics at the nearby Institute of Optics in 1978. His work on coherence theory paved the way for the development of quantum optics by **Glauber** in 1963. A reformulation of coherence theory by Wolf in 1982 led to the remarkable prediction that the spectrum of a radiation field may change as it propagates—even through the vacuum of interstellar space. This prediction of 'spectral non-invariance' contradicted unstated presuppositions which were strongly held to in every branch of physics, but it was almost immediately confirmed experimentally in optics, and shortly thereafter in acoustics. Wolf was co-author with **Born** of *Principles of Optics* (1959), and editor of the series *Advances in Optics* (30 vols, 1961–92). In 1977 he was Frederick Ives medallist of the Optical Society of America, and he served as the society's president from 1978 to 1981.

WOLF, Maximilian Franz Joseph Cornelius
(1863–1932)

German astronomer, born in Heidelberg, where while still a schoolboy be began his well-known observations of asteroids. After completing his mathematical studies at the University of Heidelberg (1888) followed by two years at Stockholm University, he established a private observatory in which he continued his searches for asteroids using photographic methods. On being appointed Professor of Astronomy at the University of Heidelberg (1893) he established a well-equipped observatory on the Königstuhl which became famous for the high quality of its photographs of star clusters, bright and dark nebulae, and star clouds of the Milky Way. The 'Wolf diagram', a useful and widely used method of discovering interstellar dust by star counting, owes its name to him.

WOLLASTON, William Hyde
(1766–1828)

English chemist, born in East Dereham, Norfolk, into a family of scientists and physicians. He graduated in medicine at Cambridge in 1793 and was awarded a fellowship at Caius College. He practised in London until 1800, when he entered into a partnership with **Smithson Tennant** to produce platinum. Platinum is resistant to heat and simple acids, and therefore had uses in some chemical apparatus and in boilers used during the manufacture of sulphuric acid. At that time, nobody—except for **Chabaneau** on a very small scale—had succeeded in producing it in a malleable form. By 1805 Wollaston, who was always the more active partner, had evolved a successful technique: the platinum sponge obtained by heating the ore slowly was gently powdered, sieved, washed, allowed to settle in water, pressed, dried, heated and hammered. He built up a lucrative business making platinum boilers, wire and other apparatus, keeping the process secret until just before his death when he described it in a lecture to the Royal Society. He also conducted experiments in many areas including physiology, pathology, botany and pharmacology. In the course of studying crystallography he made improvements to the goniometer, an instrument designed by **Haüy** for measuring the angles between crystal faces. He wrote extensively on atomic theory, sometimes opposing and sometimes modifying the new and epoch-making theories of **Dalton**. Wollaston added to Dalton's advances by being one of the first scientists to realize that the arrangement of atoms in a molecule must be three-dimensional, and he also came close to formulating the 'law of definite proportions' which is credited to **Proust**. He was elected FRS in 1793. He died in London, leaving some of his considerable wealth to the Royal Society and the Geological Society of London to promote scientific research. The mineral wollastonite, one of the three forms of calcium silicate, was named in his honour.

WOOD, Robert Williams
(1868–1955)

American physicist, born in Concord, Massachusetts. Educated at Harvard, in Chicago and in Berlin, he was Professor of Experimental Physics at Johns Hopkins University from 1901 to 1938. He carried out research on optics, atomic and molecular radiation and sound waves. One of his most notable researches was his study of resonance radiation, which led to the spectroscopic technique of optical pumping developed by **Kastler**. Wood also pioneered the production of phase gratings and zone-plate lenses, anticipating the much later science of 'diffractive optics'. He wrote *Physical Optics* (1905), some fiction, and illustrated nonsense verse in *How to Tell the Birds from the Flowers* (1907).

WOODS, Donald Devereux
(1912–64)

British microbiologist. He was educated in Ipswich and at Trinity Hall, Cambridge, where he received his PhD in 1937. After a period as reader in microbiology at the University of Oxford, he became Iveagh Professor at Oxford in 1955. His work resulted in significant advances in chemotherapy, particularly the introduction of the concept of competitive inhibition, and the elucidation of the mode of action of the sulphanilamide drugs. He was elected FRS in 1952.

WOODWARD, Sir Arthur Smith
(1864–1944)

English geologist, born in Macclesfield and educated at the University of Manchester. He was appointed as assistant (1881–92), assistant keeper (1892–1901) and keeper of geology (1901–24) at the Natural History section of the British Museum. Woodward did notable work on fossil fish developing from his initial brief of cataloguing the museum's fossil fish collection. From 1885 to 1923 he made some 30 expeditions abroad, in his vacation time and largely at his own expense, visiting museums, meeting palaeontologists and making collections throughout Europe, the USA, the Middle East and South America. One of Britain's most prolific ever geologists, the four volumes of his *Catalogue of Fossil Fishes in the British Museum (Natural History)* (1889, 1891, 1895, 1901) were numbered amongst some 650 publications. In spite of this, Woodward is chiefly remembered for his part in the controversy over the Piltdown Man. The amateur geologist Charles Dawson gave him the cranial fragments found at Piltdown from 1908 to 1912 for identification. Together with parts of a jawbone unearthed later, these were accepted by many anthropologists as the 'missing link' in **Charles Darwin**'s theory of evolution, and as such one of the greatest discoveries of the age. Following scientific tests, the skull was denounced as a fake in 1953; Woodward's firm conviction that the remains were human had been the main reason for the success of the hoax. In 1898 he published *Outlines of Vertebrate Palaeontology*. He was knighted in 1924.

WOODWARD, Robert Burns
(1917–79)

American chemist, born in Boston. Having been interested in chemistry since the age of eight when he was given a chemistry set, his formal chemical education at MIT resulted in a PhD by the age of 20. He then moved to Harvard (1937) and by 1944 had become associate professor, working on the chemistry of the antimalarial drug quinine and the new wonder drug penicillin. For the next 20 years he executed the syntheses of a dazzling array of biological compounds, including strychnine, cholesterol, lanosterol, lysergic acid, reserpine, chlorophyll, colchicine and cephalosporin C. His feel for the art and architecture of constructing complex molecules was astounding. At the same time he became famous for his lectures, delivered without a single note or a visual aid apart from blackboard and coloured chalk, sometimes lasting more than three hours with only a short pause at half-time while the speaker drank a refreshing martini. The audience rarely took notes but photographed the blackboard at the end. He was awarded the Nobel Prize for Chemistry in 1965 for the totality of his work in the art of synthetic chemistry, structure determination and theoretical analysis. But his finest work was still to come. Woodward and the Swiss chemist **Eschenmoser** set out to synthesize vitamin B_{12} in a collaborative venture. The work took over 10 years and was completed in 1976. In the course of this synthesis Woodward conceived the idea that molecular orbitals could affect the products obtained in cyclization reactions and he invited a young Harvard theoretician **Hoffmann** to collaborate. This led eventually to the Woodward–Hoffmann rules for the conservation of orbital symmetry. Unfortunately Woodward had died from a heart attack before the award of a Nobel Prize for Chemistry for this work. He received almost every honour and award possible for an organic chemist and the pharmaceutical company Ciba–Geigy established the Woodward Research Institute for him in Basle. Many think of Woodward as the greatest synthetic organic chemist of all time. It is reported that at the 1978 meeting of the Association of Harvard Chemists he was carried into the lecture hall in a fashion no less regal than that of a 13th-century caliph.

WOOLLEY, Sir Richard van der Riet
(1906–86)

British astronomer, born in Weymouth. In his early years he lived in South Africa and studied at the University of Cape Town before returning to England and entering Cambridge University (1925–8). On graduating in mathematics as Wrangler he became a research student under **Eddington** and was awarded the Isaac Newton studentship (1931). He was appointed chief assistant at the Royal Observatory, Greenwich (1933–7), returning to Cambridge as John Couch Adams Astronomer (1937–9). In 1939 just before the outbreak of World War II he was appointed Government Astronomer and Director of the Commonwealth Observatory, later known as the Mount Stromlo Observatory, in Canberra, Australia. During the years of World War II his energies were diverted to optical design of military instruments, but at the same time he was able to establish the observatory as an important institution for observations of the southern skies. His researches at Greenwich and Canberra were mainly concerned with solar and stellar atmospheres; he published two important books on that subject, *Eclipses of the Sun and Moon* (1937, with Sir Frank Dyson) and *The Outer Layers of a Star* (1953, with Douglas Walter Noble Stibbs). In 1956 Woolley, as 11th Astronomer Royal, succeeded **Jones** with charge of the Royal Greenwich Observatory at Herstmonceux in Sussex, where he supervised the completion and erection there of the 98 inch (2.5 metre) Isaac Newton telescope (now re-sited on the Canary island of La Palma). Woolley was involved in two further major telescopic projects; the Anglo-Australian 3.8 metre telescope and the United Kingdom 1.2 metre Schmidt telescope, both sited at the Sidings Springs Observatory in New South Wales, Australia. On retiring from the Royal Greenwich Observatory in 1971, Woolley, who had received a knighthood on becoming Astronomer Royal, accepted the appointment of director of the newly established South African Astronomical Observatory (1972–6).

WRIGHT, Sir Almroth Edward
(1861–1947)

English bacteriologist, born in Yorkshire. Educated in Dublin, Leipzig, Strassburg, Marburg and Sydney, he was subsequently appointed to an army medical school, where he developed a vaccine against typhoid fever.

He later became Professor of Experimental Pathology at St Mary's Hospital, London (1902). Wright was known specially for his work on the parasitic diseases, and for his research on the protective power of blood against bacteria. He was an important influence on the work of his student Sir **Alexander Fleming**, and took considerable pride in the publicity it afforded for St Mary's Hospital Medical School. He was knighted in 1906.

WRIGHT, Orville
(1871–1948) and
Wilbur
(1867–1912)

American pioneers of aviation. Orville was born in Dayton, Ohio, and his elder brother Wilbur was born near Millville, Indiana. They were the sons of a bishop of the United Brethren Church, who encouraged them to be independent and use their talents to the full, and they both developed a keen interest in mechanical devices at an early age. To earn a living they founded in 1892 the Wright Cycle Company, but all the time their real interests lay in emulating and surpassing the aerial exploits of the German gliding pioneer Otto Lilienthal, killed when one of his gliders crashed in 1896. Convinced that it was vital to put as much effort into learning how to control aircraft as would be needed to build them, they embarked in 1899 on a careful programme of model glider flights, followed in 1900 by short piloted flights in a glider of 5 metres wingspan. They then built a small wind tunnel in which they tested more than 200 models, enabling them to compile the first accurate tables of lift and drag for different wing configurations. Their third piloted glider was thoroughly tested at Kitty Hawk, North Carolina, in September 1902, and it was so successful that they decided to attempt powered flight the following summer. After searching in vain for a suitable engine and propeller they set about making their own, and on 17 December 1903 Wilbur and Orville each made two flights, the last and longest of 59 seconds duration covering 255 metres. For the next three years the Wrights were the only men in the world to achieve a succession of controlled, powered and sustained flights; in October 1905 Wilbur flew non-stop for a distance of almost 40 kilometres. Encouraged by this, they abandoned their cycle business, and having patented their flying machine and its controls, formed an aircraft production company (1909). In 1915, Wilbur having died of typhoid fever three years earlier, Orville sold his interests in the business in order to devote himself to aeronautical research. The world acknowledged their success with numerous honours and awards.

WRIGHT, Sewall
(1889–1988)

American geneticist, born in Melrose, Massachusetts. He was educated at Lombard College, the University of Illinois and Harvard. He began his career with the US Department of Agriculture, and following studies of breeding methods to improve livestock quality, he developed a mathematical description of evolution. He showed that within small isolated populations, certain genetic features may be lost randomly if the few individuals possessing the genes happen not to pass the genes on to the next generation. This 'Sewall Wright effect' allows evolution to occur without the involvement of natural selection.

WRIGHT, Thomas
(1711–86)

English natural philosopher, born in Byer's Green near Durham. He was trained as a scientific instrument-maker, and also gained a reputation as a teacher of mathematics. He is known for his cosmological speculations in his *Original Theory of the Universe* (1750) in which he put forward a model of the Milky Way as an indefinitely long slab filled with stars, with the Sun inside it. The model explained the high density of stars in the great circle of the Milky Way and the low density in directions at right angles to it. In later models he described the universe in terms of spheres of stars, and in terms of rings like those of Saturn which he predicted were composed of many small satellites. His work which was of a metaphysical nature was known to **Kant** who referred to it in his writings. Though Wright preceded **William Herschel** in point of time, the latter's disc theory of the stellar system, based on observational data, was quite independently developed.

WRÓBLEWSKI, Zygmunt Florenty von
(1845–88)

Polish physicist, born in Grodno (now in Byelorussia). The son of a lawyer, he entered Kiev University in 1862 but as an activist in the Polish insurrection of 1863 he was exiled to Siberia for six years. After being amnestied he settled in Berlin, studying in the laboratory of **Helmholtz**. He obtained a PhD in Munich (1874) and then, whilst assistant at the University of Strassburg (1875–6), studied gaseous diffusion. Financial assistance from the Krakow Academy of Sciences enabled Wróblewski to study in Paris under **Sainte-Claire Deville** and to visit London, Oxford and Cambridge. He became Professor of Physics at Krakow in 1882, although he spent the first year in Paris with Deville working on aqueous solutions of carbon dioxide under high pressure. This was the starting point for his famous series of investigations on the liquefaction of gases. With the chemistry professor at Krakow, he liquefied air on a large scale for the first time (1883). A year later he obtained a mist of liquid hydrogen. Wróblewski died in 1888 from severe burns after a kerosene lamp set light to his clothes.

WU, Chien-Shiung

(1912–)

Chinese-born American physicist, born in Shanghai. She studied at the National Centre University in China, and from 1936 in the USA, at the University of California at Berkeley. From 1946 she was on the staff of Columbia University, New York, where she was appointed professor in 1957. In 1956 Wu and her colleagues carried out an experiment to test **Tsung-Dao Lee** and **Yang**'s hypothesis that parity is not conserved in weak decays; this would result in the decay not being mirror symmetric, ie having a preferred direction. The experiment was performed by cooling cobalt-60 nuclei in a magnetic field so that their spins were aligned. They observed that the electrons resulting from the beta decay of the nuclei were emitted preferentially in one direction, proving that the principle of parity conservation was indeed violated in weak interactions. This was later explained by the V–A theory of weak interactions proposed by **Feynman** and **Gell-Mann**.

WURTZ, Charles Adolph

(1817–84)

French chemist, born in Strasbourg. He studied medicine in Strasbourg, but worked with **Liebig** in Giessen on the chemistry of hypophosphorous acid. In 1844 he went to Paris and became assistant to **Dumas**, who he later succeeded as professor in the École de Médecine, and in 1857 he became professor at the Sorbonne. He was one of the founders of what later became the French Chemistry Society. Apart from his work on phosphorous acids, he is best known for his synthetic route to the larger alkanes and for the discovery of glycol (1,2-ethanediol). He wrote a number of successful textbooks.

WYCKOFF, Ralph Walter Graystone

(1897–)

American biophysicist, born in Geneva, New York. He studied at Cornell University, gaining a PhD in chemistry. He worked as a physical chemist in the geophysics laboratory of the Carnegie Institute (1919–27) before moving to the subdivision of biophysics at the Rockefeller Institute (1927–38). After a spell in commercial laboratories he moved to the University of Michigan (1943–5) and then to the National Institutes of Health until 1959. His postwar career included a period on attachment to the American Embassy in London (1952–4) and in 1959 he was appointed Professor of Physics and Microbiology at the University of Arizona. Whilst at the Rockefeller, Wyckoff developed new ultracentrifugation techniques to purify viruses, such as that causing equine encephalomyelitis, which facilitated the production of pure virus preparations from which effective vaccines against the disease could be developed. In 1944, whilst working in epidemiology, he collaborated with his Michigan colleague, the astronomer **Robley Williams**, in developing the metal shadowing method for providing three-dimensional imaging of viruses in the electron microscope, which has since been widely used. His achievements include several studies of molecular structure, such as an examination of crystals, other macromolecules and several viruses. He investigated the effects of radiation on cell structure and continued to refine investigative techniques such as centrifugation methods and electron microscopy procedures. He also developed improved routines for the purification of viruses.

WYNNE-EDWARDS, Vero Copner

(1906–)

English zoologist, born in Leeds. He was educated at the University of Oxford and the Marine Biology Laboratory in Plymouth (1927–9), and taught at Bristol (1929–30) and at McGill University, Montreal (1930–46). He travelled widely in Canada carrying out research in avian biology on Baffin Island, in the Yukon Territory and at the Mackenzie river. He returned to Britain in 1946 and became Regius Professor of Natural History at Aberdeen. In 1962 he published *Animal Dispersion in Relation to Social Behaviour*, in which he argued that animal dispersal had evolved in order to regulate population density. He suggested that individuals of a species altruistically reduce their birth rates so as to benefit the species as a whole. To support his theory he produced numerous examples from throughout the animal kingdom. The mechanism he proposed to explain this was group selection; groups which exhibited such altruistic behaviour would be at an advantage in competition with populations whose members behaved selfishly and bred without regard to population pressure. Altruism he believed was mediated via territorial behaviour and dominance hierarchies. These ideas aroused considerable controversy and adverse criticism but were nevertheless instrumental in stimulating advances in sociobiology and behavioural ecology. It is generally accepted that while group selection may occur under exceptional circumstances, it has not been a major factor in evolution and most of Wynne-Edward's examples have subsequently been explained adequately on the basis of selection acting at the level of the individual. He wrote *Evolution through Group Selection* in 1986. He served on the Red Deer Commission as Vice-Chairman (1959–68) and the Royal Commission on Environmental Pollution (1970–4), and was elected FRS in 1970.

X

XENOCRATES
(c.395–314 BC)

Greek philosopher and scientist, born in Chalcedon on the Bosphorus. He was a pupil of **Plato** and in 339 BC became head of the Academy which Plato had founded. He is recorded as travelling with **Aristotle** after Plato's death in 348 BC to do research under the patronage of Hermeias, tyrant of Atarneus in north-west Asia Minor, and as joining some Athenian embassies on foreign diplomatic missions. He wrote prolifically on natural science, astronomy and philosophy, but only fragments of this output survive. He generally systematized and continued the Platonic tradition but seems to have had a particular devotion to threefold categories, perhaps reflecting a Pythagorean influence: philosophy is subdivided into logic, ethics and physics; reality is divided into the objects of sensation, belief and knowledge; he distinguished gods, men and demons; he also probably originated the classical distinction between mind, body and soul.

XENOPHANES
(6th century BC)

Greek philosopher, poet and religious thinker, born in Colophon in Ionia (Asia Minor) where he probably lived until the Persian conquest of the region in 546 BC. He seems then to have lived a wandering life round the Mediterranean, perhaps settling in Sicily for a while and visiting Elea in Southern Italy. He wrote poetry, fragments of which survive, and seems to have been an independent and original thinker, though later traditions tried to claim him as a member either of the Ionian or the Eleatic school. He attacked the anthropomorphism of popular religion and Homeric mythology (pointing out that each race credits the gods with their own physical characteristics, and that animals would do the same), posited by way of reaction a single deity who somehow energizes the world ('without toil he shakes all things by the thought of his mind'), and made some bold speculations about the successive inundations of the Earth based on the observation of fossils. His rather bizarre astronomical theories suggested that a new Sun rises each day and that there is a different Moon for each region or zone of the flat Earth, all the heavenly bodies having been created from clouds which were set on fire.

Y

YALOW, Rosalyn, née Sussman

(1921–)

American biophysicist, born in New York City. Yalow was the first woman to graduate in physics from Hunter College, New York (1941). She obtained a PhD from the College of Engineering of the University of Illinois in 1945. She taught physics at Hunter College until 1950, and in 1947 became consultant to the Radioisotope Unit at the Bronx Veterans Administration (VA) Hospital. From 1950, Yalow collaborated with Solomon Berson, and during the course of research on diabetes they developed 'radioimmunoassay' (RIA). This is an ultrasensitive method of measuring concentrations of substances in the body which relies upon 'labelling' molecules with radioactive isotopes. Yalow and Berson found that in adult diabetics, the rate of clearance of injected labelled insulin from the blood is surprisingly low, and suggested that antibodies inactivating the insulin are formed. In 1977, for her work on RIA, Yalow shared the Nobel Prize for Physiology or Medicine with **Guillemin** and **Schally**. Besides diabetes, Yalow has used RIA in work on dwarfism, leukaemia, peptic ulcers and neurotransmitters in the brain. Since 1969 she has been Chief of the Radioimmunoassay Reference Laboratory and since 1973 Director of the Solomon A Berson Research Laboratory of the VA Medical Centre.

YANG, Chen Ning

(1922–)

Chinese-born American physicist, born in Hofei, the son of a professor of mathematics. He was educated in Kuming, China, before gaining a scholarship to Chicago in 1945 to work under **Teller**. He became professor at the Institute for Advanced Studies, Princeton (1955–65), and from 1965 was Director of the Institute for Theoretical Physics at New York State University, Stony Brook. In 1956 with **Tsung-Dao Lee**, who had been his fellow student at Chicago, he made a thorough analysis of all the known data in particle physics and concluded that the quantum property known as parity was unlikely to be conserved in weak interactions, and they suggested a simple experiment which would prove it. In the same year (1956), a similar experiment by a group of physicists from Columbia University and the American National Bureau of Standards, headed by **Wu**, confirmed that the 'law' of parity is indeed violated in the case of weak interactions. For this prediction, Lee and Yang were awarded the 1957 Nobel Prize for Physics, and the Einstein Commemorative Award from Yeshiva University in the same year. With Robert Mills, Yang also developed a non-Abelian gauge theory which proved to be an important development in the theories of fundamental interactions for elementary particles and fields.

YANOFSKY, Charles

(1925–)

American geneticist, born in New York City. He studied at New York's City College and at Yale, and since 1961 has been Professor of Biology at Stanford, working on gene mutations, particularly on the trytophan operon of *Escherichia coli*. He has shown that the sequence of bases in the genetic material DNA acts by determining the order of the amino acids which make up proteins, including the enzymes which control biochemical processes. In 1967 he showed that the amino acid sequence of the protein for synthesizing tryptophan is colinear with the genetic map of the gene. He went on to describe how the production of the amino acid tryptophan is controlled by a process called attenuation (1977). Attenuation results in the synthesis of a smaller modified tryptophan molecule, and occurs when the levels of tryptophan are high.

YERSIN, Alexandre Émile John

(1863–1943)

Swiss-born French bacteriologist, born in Rougemont and educated at Lausanne, Marburg and Paris. He carried out research at the Pasteur Institute in Paris, working along with **Émile Roux** on diphtheria antitoxin. In Hong Kong in 1894 he discovered the plague bacillus at the same time as **Kitasato**. He developed a serum against it, and founded two Pasteur institutes in China. He also introduced the rubber tree into Indo-China.

YODER, Hatten Schuyler Jr

(1921–)

American experimental petrologist, born in Cleveland, Ohio. He was educated at the University of Chicago and MIT. Thereafter he spent his career as Petrologist (1948–71) and Director (1971–86) at the Geophysical Laboratory of the Carnegie Institution of Washington, where he has been Director Emeritus since 1986. Yoder undertook important experimental petrological studies of phase equilibria in mineral systems, properties of minerals at high pressures and temperatures, hydrothermal mineral synthesis and experimental heat transfer in silicates.

YONGE, Charles Maurice
(1899–1986)

English zoologist and marine biologist, born in Wakefield. Educated at Oxford, where he started reading history, and Edinburgh, where he graduated in zoology, he carried out research in marine laboratories at Millport, Plymouth and Naples. At Plymouth he studied the feeding behaviour and nutritional physiology of the oyster, a long-standing interest culminating in publication of the book *Oysters* in 1960. He was leader of the Great Barrier Reef Expedition in 1928 during which 20 scientists spent 13 months on the reef studying all aspects of the ecology and physiology of the marine organisms. Yonge researched coral physiology, in particular the role of the symbiotic zooanthellae. He edited the six volumes on the work of the expedition published by the British Museum (Natural History). He was Professor of Zoology at Bristol (1933–44) then Glasgow (1944–70), and on retiral from the latter he continued his research at Edinburgh. His major interest was in the bivalve molluscs. Among the aspects of their biology he investigated were the role of the crystalline style in digestion, filter feeding, structure and function of the mantle cavity and ability to bore into rock. He wrote several popular books including *The Sea Shore* (1949) and *Collins Pocket Guide to the Sea Shore* (1958, with J H Barrett). He was elected FRS in 1946. His public services included membership of the Fisheries Advisory Committee (1936–56) and the Colonial Fisheries Advisory Committee (1949–60, Chairman from 1955).

YOUNG, James
(1822–83)

Scottish industrialist who founded the shale-oil industry in Scotland. He was born in Glasgow, worked as a joiner with his father, and studied chemistry part-time at Anderson's College where he met **Lyon Playfair** and David Livingstone, both to be lifelong friends. He became assistant to his teacher **Graham** in 1832 and moved with him to University College London in 1837. He then turned to industry, working with **Muspratt** at Newton-le-Willows and **Charles Tennant** in Manchester. In 1848 he set up a small oil refinery at Alfreton, Derbyshire, to exploit an oil seepage in a disused coal mine. Two years later he joined with Edward William Binney and Edward Meldrum to manufacture naphtha and lubricating oils from oil-shale. Their plant at Bathgate, West Lothian, began producing paraffin oil and solid paraffin in 1856, following Young's discovery that low-temperature distillation of shale yields the maximum amount of these products. The business prospered and Young sold it in 1866 for £400 000, having previously bought out his partners. He was appointed President of Anderson's College in 1868, where he founded a Chair of Technical Chemistry. He also helped to finance Livingstone's second and third expeditions, and after Livingstone's death paid for his servants to come to Britain. He was elected to the Royal Society of Edinburgh in 1873 and died in Wemyss Bay, Ayrshire.

YOUNG, John Zachary
(1907–)

English zoologist, born in Bristol and educated at Magdalen College Oxford, where he was strongly influenced by his tutor **Julian Huxley**. He graduated in 1928 and went to the Stazione Zoologica, Naples, as Oxford's biological scholar. There, whilst working on the innervation of the gastro-intestinal tract of fish, he became interested in the nervous system of cephalopods, the class of animals that includes octopuses, squids and their relatives. He returned to Oxford as university demonstrator in zoology (1933–45) and was then appointed to the Chair of Anatomy at University College London (1945–74), the first non-medically qualified Professor of Anatomy in Britain. Young's research has centred on the anatomy and physiology of the nervous system. In 1933 whilst visiting the Woods Hole Marine Laboratory in Massachusetts, he reported that the squid had an unusually large diameter nerve fibre, about one hundred times greater than mammalian nerves. This giant nerve fibre subsequently became the basic research material of later neurophysiologists such as **Alan Hodgkin** and **Andrew Huxley**. During World War II he studied regeneration in damaged nerves, and subsequently returned to investigations of the central nervous system, and mechanisms of learning and memory, especially in the octopus. He has proposed neural models to account for memory processes, summarized in *The Memory System of the Brain* (1966) and *Programs of the Brain* (1978). In addition, he has written several influential textbooks including *The Life of Vertebrates* (1950), *The*

Life of Mammals (1957) and *Introduction to the Study of Man* (1971). He was elected FRS in 1945.

YOUNG, Thomas
(1773–1829)

English physicist, physician and Egyptologist, born in Milverton, Somerset, the son of a mercer and banker. A precocious student, he was interested in classical languages, higher mathematics, natural history and natural philosophy (physics), and being of independent means, he could follow his varied interests as he wished. He studied medicine at London, Edinburgh, Göttingen and Cambridge, and became a physician in London in 1800, but devoted himself to scientific research. In 1801 he was appointed Professor of Natural Philosophy at the Royal Institution, but three years later was forced to resign as his lectures were found to be too technical for a popular audience, and he resumed his medical practice. He was appointed physician to St George's Hospital (1811), and held this position until his death. He held several public offices related to science, spent several periods as Consultant to the Admiralty, and was Secretary to the Royal Commission on Weights and Measures (1816–21), Secretary of the Board of Longitude (1818–28) and Superintendent of the *Nautical Almanac* (1818–29). He supplemented his income as an anonymous author of a wide variety of scientific articles. He was elected FRS (1794), and was involved in the Royal Society's affairs as Foreign Secretary and Member of Council. He became best known in the 19th century for his undulatory (wave) theory of light as expounded in his *Outlines of Experiments and Enquiries respecting Sound and Light* (1800) and in *A Course of Lectures on Natural Philosophy and the Mechanical Arts* (1807). He combined the wave theory of **Huygens** and **Isaac Newton**'s theory of colours to explain the interference phenomenon of colours produced by ruled gratings, thin plates, and the colours of the supernumerary bows of the rainbow, but his theory was unacceptable during his lifetime as it was regarded 'anti-Newtonian'. He also did valuable work in insurance, haemodynamics and Egyptology, and made a fundamental contribution to the deciphering of the inscriptions on the Rosetta Stone.

YUKAWA, Hideki,
originally **Hideki Ogawi**
(1907–81)

Japanese physicist, born in Tokyo. He was educated at Kyoto Imperial University and after graduating in 1929, remained at Kyoto where he was appointed lecturer. He married Sumi Yukawa in 1932 and assumed her family name. In 1933 he became a lecturer at Osaka Imperial University and received his doctorate there in 1938. The following year he returned to Kyoto as Professor of Theoretical Physics (1939–50) and he later became Director of the Kyoto Research Institute for Fundamental Physics (1953–70). In 1935 he published his first paper in which he proposed his theory of nuclear forces, suggesting a strong short-range attractive interaction between nucleons (neutrons and protons) that would overcome the electrical repulsion between protons. This interaction was propagated by the exchange of massive particles between the nucleons, the mass of the exchange particle determining the range of the force. The existence of these intermediate particles was confirmed by **Cecil Powell**'s discovery in 1947 of the π-meson or pion. Yukawa also predicted that a nucleus may absorb one of the innermost electrons in the atom (1936); such 'K-capture' was soon observed. During World War II he played a minor role in military research whilst continuing his scientific work. For his theory of strong nuclear forces, and his work on quantum theory and nuclear physics, he was awarded the Nobel Prize for Physics in 1949, the first Japanese to be so honoured.

Z

ZEEMAN, Pieter
(1865–1943)

Dutch physicist, born in Zonnemaire, Zeeland. He was educated under **Lorentz** at the University of Leiden, where he received his doctorate in 1893. He became a lecturer at Leiden in 1897, and in 1900 was appointed as Professor of Physics at Amsterdam University. In 1896 he studied the effects of a magnetic field on sodium and lithium light sources, and observed that the spectral emission lines were broadened. He demonstrated that this was due to the splitting of spectrum lines into two or three components and the phenomenon became known as the Zeeman effect. This was consistent with Lorentz's classical theory of light produced by vibrating electrons. More generally, atoms display several closely spaced lines; this was later explained with the development of quantum theory. The magnetic field differentiates a single atomic energy level into several components of slightly different energy associated with different quantized orientations of the total electron magnetic moment with respect to the field. Zeeman also investigated the absorption and motion of electricity in fluids, magnetic fields on the solar surface, the **Doppler** effect and the effect of nuclear magnetic moments on spectral lines. In 1902 he shared with Lorentz the Nobel Prize for Physics for the discovery and explanation of the Zeeman effect, and he was awarded the Rumford Medal of the Royal Society in 1922.

ZEL'DOVICH, Yakov Borisovich
(1914–)

Soviet astrophysicist, born in Minsk. He graduated from the University of Leningrad in 1931 and moved to the Soviet Academy of Sciences, becoming a full Academician in 1958. In the late 1930s he concentrated on nuclear physics and specifically (with Y B Khariton) the chain reaction in uranium fission. During the 1940s he investigated the mechanisms responsible for the oxidation of nitrogen during an explosion, and the problems of flame propagation and gas dynamics. In the 1950s he turned to cosmology, and studied the production of the initial hydrogen-to-helium ratio and the degree of isotropy in the early stages of the universe. He also predicted that it would be possible to find black holes associated with X-ray-emitting binary stars. In 1972, with Rashid Sunyaev, he discovered how the total energy carried by the microwave background radiation could be increased as it passed through the medium between the galactic clusters. This effect has important cosmological applications and can be used to independently estimate the Hubble constant, which measures the rate at which the expansion of the universe varies with distance.

ZENO OF ELEA
(c.430–c.490 BC)

Greek philosopher and mathematician. Little is known of his life: he was a native of Elea, a Greek colony in southern Italy, where he lived all or most of his life. He was a disciple of Parmenides of Elea, and in defence of his monistic philosophy against the Pythagoreans he devised his famous paradoxes which purported to show the impossibility of motion and of spatial division, by showing that space and time could be neither continuous nor discrete. The paradoxes are: 'Achilles and the Tortoise', 'The Flying Arrow', 'The Stadium' and 'The Moving Rows'. **Aristotle** attempted a refutation but they were revived as raising serious philosophical issues by **Bertrand Russell**.

ZERMELO, Ernst Friedrich Ferdinand
(1871–1953)

German mathematician, born in Berlin. He studied mathematics, physics and philosophy at Berlin, Halle and Freiburg, and was professor at Göttingen (1905–10) and Zürich (1910–16). From 1926 to 1935 he was an honorary professor at Freiburg im Breisgau. Although he worked in physics and the calculus of variations among other subjects, he is now best remembered for his work in set theory. Following **Georg Cantor**'s pioneering work, Zermelo gave the first axiomatic description of set theory in 1908; though later modified to avoid the paradoxes discovered by **Bertrand Russell** and others, it remains one of the standard methods of axiomatizing the theory. He also first revealed the importance of the axiom of choice, when he proved in 1904 that any set could be well-ordered, a key result in many mathematical applications of set theory.

ZERNIKE, Frits
(1888–1966)

Dutch physicist, born in Amsterdam. He was educated at Amsterdam University and became Professor of Physics at Groningen (1910–58). Zernike developed (from 1935) the phase-contrast technique for the microscopic examination of transparent—frequently biological—objects. For this work he was awarded the Nobel Prize for Physics in 1953. Related to this was his invention of the 'coherent background' technique for revealing the presence of phase variations in interference and diffraction patterns, and for studying very weak fringes in such patterns. He made important pioneering contributions to the understanding of optical coherence, showing how it could be measured from observable features of the light field. He also discovered a modified subset of the **Jacobi** polynomials, which is particularly appropriate for the study of optical system performance and led to a new formulation of the wave theory of lens aberrations; these expressions are now known as Zernike polynomials.

ZIEGLER, Karl
(1898–1973)

German chemist, born in Helsa. The son of a Lutheran minister, he studied chemistry at the University of Marburg where he obtained his PhD in 1920. He was briefly a lecturer in Frankfurt and in 1927 took up the post of professor at the University of Heidelberg. He moved to the Chemical Institute Halle-Saale in 1936 and became Director of the Kaiser Wilhelm Institute for Coal Research at Mülheim-in-Ruhr in 1943. He remained there until his retirement in 1969. His initial research showed that organometallic compounds effect polymerization by the generation of radicals. After World War II he concentrated on studies of organo-aluminium compounds, and found that zirconium acetylacetonate added to triethyluminium catalysed the polymerization of ethene to give polyethylene of very high molecular weight. Equally important was his discovery that trialkylaluminiums and titanium chloride polymerized ethene at ambient temperature and pressure. Collaboration with **Natta** led to the development of a family of catalysts and gave birth to a new era in the plastics industry. The royalties from these industrial processes partly fund an endowment at the Kaiser Wilhelm Institute. Ziegler shared the Nobel Prize for Chemistry with Natta in 1963.

ZINDER, Norton David
(1928–)

American geneticist, born in New York City. He studied at Columbia University and with **Lederberg** at Wisconsin, and became Professor of Genetics at Rockefeller University in 1964. In 1951, studies using mutants of the bacterium *Salmonella* led him to describe the process of bacterial transduction. Transduction refers to the transmission of a bacterial gene from one bacterium to another by means of a viral phage particle which carries the gene. It led to an explanation for the spread of drug resistance in bacteria, and offered a mechanism of inserting specific genes into a host cell bacterium. Recently, Zinder has been Chairman of the Program Advisory Committee for the Human Genome Project in the USA.

ZINSSER, Hans
(1878–1940)

American bacteriologist and immunologist, born in New York City, the son of a prosperous German immigrant. He was educated at Columbia University and its College of Physicians and Surgeons (MD 1903). His predilection for science led him into bacteriology and immunology, which he taught at Columbia and Stanford universities before going to Harvard in 1923. Zinsser worked on many scientific problems, including allergy, the measurement of virus size and the cause of rheumatic fever. Above all, however, he clarified the rickettsial disease typhus, differentiating epidemic and endemic forms (the endemic form is still known as Brill–Zinsser's disease), researches which he brilliantly described in his popular book *Rats, Lice and History* (1935). His *Textbook of Bacteriology* (1910) and *Infection and Resistance* (1914) became classics. A highly cultured man, he wrote poetry and essays, and left an evocative autobiography, *As I Remember Him* (1940), written in the third person while he was dying from leukaemia.

ZITTEL, Karl Alfred von
(1839–1904)

German palaeontologist, born in Bahlingen, Baden, the youngest son of a clergyman. He was educated at Heidelberg and Paris, and commenced geological research in Dalmatia during a spell as a voluntary assistant with the Geological Survey of Austria. In 1862 he became an assistant at the Mineralogical Museum in Vienna where he undertook some teaching. The following year he was appointed professor at Kahlsruhe. From 1880 he held professorships at Munich, and in 1890 he became Keeper of the State Geological Collections. During his career he also served as President of the Bavarian Academy. A distinguished authority on his subjects and their history, he was a pioneer of evolutionary palaeontology and was widely recognized as the leading teacher of palaeontology in the 19th century. His five-volume *Handbuch der Paläontologie* (1876–93) was arguably his greatest service to science, and it remains as one of the most comprehensive and trustworthy palaeontological reference books. His important textbook *Grundzüge der Paläontologie* first appeared in 1895, with an English translation *Textbook of Palaeontology* being produced in 1900. It was later revised by **Arthur Smith Woodward** (1925).

ZOHARY, Michael
(1898–1983)

Israeli botanist and phytogeographer, born in Galicia (now part of Poland). He emigrated to Palestine in 1920 and worked as a road builder. In 1922 he entered Jerusalem Teacher's Seminary and became interested in botany. He later joined the botany department at the Institute of Agriculture, Tel Aviv (1925); with Naomi Feinbrun and Alexander Eig, he formed the first team of botanists to study the flora and vegetation of Palestine. From 1929, Zohary's team was based at the Hebrew University of Jerusalem. He obtained a doctorate from the University of Prague (1936), based on studies of seed dispersal ecology in Palestinian plants in which he introduced the theory of antiteleochory, whereby seed germination of desert plants is ensured by dispersal near the parent plant. He became interested in evolutionary trends in seed dispersal (especially in Cruciferae, Compositae and Leguminosae), studied the taxonomy of many genera, and co-authored a world monograph of *Trifolium* (1984). His fundamental geobotanical researches culminated in the publication of the monumental *Geobotanical Foundations of the Middle East* (2 vols, 1973). Zohary's other main publications were *Flora Palaestina* (4 vols, 1966–86, with Feinbrun), *Plant Life of Palestine* (1962) and *Plants of the Bible* (1982).

ZONDEK, Bernhard
(1891–1967)

Israeli gynaecologist and endocrinologist, born in Wronke, Germany. He trained in medicine at the University of Berlin, where he became a member of staff in the Department of Obstetrics and Gynaecology (lecturer, 1923–6; associate professor, 1926–9; director, 1929–33). He left Nazi Germany in 1933 to become Professor of Obstetrics and Gynaecology at the Hebrew University, Jerusalem (1934–61). His research interests, influenced by his elder brother Hermann Zondek, an endocrinologist, were predominantly in reproductive endocrinology. In collaboration with Selmar Aschheim he developed the first reliable pregnancy test in 1928. They later discovered that the anterior pituitary gland produced hormones called gonadotrophins, which in turn stimulated other endocrine glands, such as the ovary, to release their hormones. This work provided important evidence of control mechanisms in reproduction, which has had widespread significance in the development of medical and social attitudes towards questions of fertility, infertility, contraception and abortion.

ZOTTERMAN, Yvnge
(1898–1982)

Swedish neurophysiologist, born in Stockholm and educated in medicine there at the Karolinska Institute. After clinical studies in the University of Uppsala he did national service in the Swedish Royal Navy, and for many years subsequently spent part of each year with the navy working on physiological problems of deep-sea diving. In 1919 he travelled to Cambridge for a few months to study physiology, and returning the following year he began working with **Adrian**, on recording and analysing the nerve impulses from sensory nerve endings. He returned to Sweden in 1927, continuing his research on the sensory function of the skin, particularly thermal and pain sensation, and also did pioneering work on the physiology of taste. During World War II he went to the USA on a short mission to gain support for Finland, and began work on industrial physiology and on human physiology, studying the Swedish lumberjack. After the war he became Head of the Department of Physiology at the Veterinary College of Stockholm, and after formal retirement he initiated a multi-disciplinary working group in 1973 studying the problems of the elderly. Much of his life and work is summarized in his two-volume autobiography *Touch, Tickle and Pain* (1969, 1971).

ZSIGMONDY, Richard Adolf
(1865–1929)

Austrian chemist, famous for his contribution to colloidal chemistry. He was born in Vienna and attended the university there, moving to Munich to take his doctorate. He worked briefly in Berlin on the chemistry of glass, studying the colloidal inclusions which give glass its colour and opacity. After teaching in Graz for five years he was employed by the Schott Glass Manufacturing Company in Jena (1897–1900), where he invented Jena milk glass. From 1900 to 1907 he worked independently in his own laboratory and as a result of his discoveries was appointed Professor of Inorganic Chemistry at Göttingen

(1900–29). Zsigmondy made colloidal solutions, particularly gold sols, his life's study. He invented the ultramicroscope, in which intense light projected from the side shows, against a dark background, the light scattered by particles. This made it possible to see individual particles (in the same way that dust is visible in a sunbeam). This microscope, developed in conjunction with the Zeiss Company of Jena, led to great advances in colloidal chemistry, then in its infancy, although it has now been replaced by the electron microscope and ultracentrifuge. With his microscope Zsigmondy discovered that changes of colour occur in colloidal solutions when particles coagulate, and that protective agents such as gum arabic and gelatin preserve colour by preventing coagulation. He also showed that the valuable paint known as purple of Cassius, whose composition had long puzzled chemists, contains a mixture of gold and stannic acid particles in colloidal solution. Zsigmondy's later work was on silica and soap gels which he investigated by another new technique, ultrafiltration, in which the substances to be separated are drawn through a membrane by a decrease in pressure. In 1925 he became the first person to be awarded a Nobel Prize for colloidal chemistry. He died in Göttingen.

ZUCKERMAN, Solly, Baron
(1904–93)

British zoologist, born in Cape Town, South Africa. He carried out research with chacma baboons near Cape Town before moving to England, where he taught at Oxford University from 1932. During World War II he investigated the biological effects of bomb blasts. He became Professor of Anatomy at Birmingham (1946–68) and Secretary of the Zoological Society of London in 1955. He was chief scientific adviser to the British government from 1964 to 1971, and wrote official reports on aspects of farming, natural resources, medicine and nuclear policy. The results of his research on chacma baboons and captive hamadryas baboons at London Zoo were published in two influential books, *The Social Life of Monkeys and Apes* (1932) and *Functional Affinities of Man, Monkeys and Apes* (1933). He was the first primatologist to consider that such studies could provide insights into the origins and behaviour of humans. He argued that as female primates were prepared to mate over a prolonged period, sex was the original social bond, a view reinforced by his observations of male hamadryas fighting to the death over females at London Zoo. Subsequent research has shown that the majority of primates other than humans are seasonal breeders and fatal encounters are rare in the wild. Zuckerman published his autobiography, *From Apes to Warlords*, in 1978.

ZUSE, Konrad
(1910–)

German computer pioneer, born in Berlin. He was educated at the Berlin Institute of Technology before joining the Henschel Aircraft Company in 1935. In the following year, tiring of the drudgery involved in solving complex linear equations, he began building a calculating machine in his spare time, a task which occupied him until 1945. With the help of a friend, Helmut Schreyer, he built a number of prototypes, the earliest of which were constructed in the living room of his parents' home. Zuse's most historic machine was the Z3. Consisting of a tape reader, an operator's console and two cabinets packed with 2600 relays, it had a small memory (capable of storing only 64 22-bit numbers), but was fast enough to multiply two rows of digits in only 3–5 seconds. It was the first operational general-purpose program-controlled calculator. Its inventor built up his own firm, Zuse KG, until it was bought out by another firm in the 1960s. Zuse became Honorary Professor at Göttingen University in 1966.

ZWEIG, George
(1937–)

Russian-born American physicist, born in Moscow. He was educated at the University of Michigan and Caltech, where he received his PhD in 1963. He then worked at the European nuclear research centre, CERN, in Geneva (1963–4), before returning to Caltech, where he became professor in 1967. Independently of **Gell-Mann**, he developed the theory of quarks as the fundamental building blocks of hadrons, the particles which experience strong nuclear forces. They suggested that three types exist, giving rise to the three different observed properties associated with new particles that were being discovered at that time, although there are now believed to be six types of quark.

ZWICKY, Fritz
(1898–1974)

American–Swiss astronomer and physicist, born in Varna, Bulgaria. He was educated at the Federal Institute of Technology at Zürich, graduating with a PhD in physics in 1922. In 1925 he went on a fellowship to Caltech, where he remained all his life, becoming successively Professor of Theoretical Physics (1927–42) and Professor of Astrophysics until his retirement in 1968. Zwicky's fruitful and wideranging researches included work on cosmic rays and rocket design as well as many branches of astronomy. He was one of the first to recognize the power of the recently invented Schmidt telescope as a means of exploring the universe on a large scale, and from 1936 onwards used the 18 inch Schmidt telescope on Mount Palomar to produce, among other researches, his catalogue of clusters of galaxies. He was the author of a highly original book on that subject, *Morphological Cosmology* (1957), and the discoverer of compact galaxies (1963), objects of exceptionally high surface brightness which are hence intrinsically very luminous. He was awarded the Gold Medal of the Royal Astronomical Society for his cosmological researches in 1973.

ZWORYKIN, Vladimir Kosma

(1889–1982)

American physicist, born in Mourom, Russia, the son of a river-boat merchant. After graduating in electrical engineering at the Petrograd Institute of Technology (1912), he studied under **Langevin** in Paris, served as a radio officer with the Russian army during World War I, but settled in the USA because of the Russian Revolution (1919). He joined the Westinghouse Electric and Manufacturing Co in Pittsburg (1920), and took a doctorate at the University of Pittsburg (1926). From 1929 he pursued a career with the Radio Corporation of America, rising to Director of Electronic Research (1946) and Vice-President and Technical Consultant (1947–54). He is chiefly remembered for applying the cathode-ray tube to television, a development which he patented in 1928. Paul Gottfried Nipkow patented the first practical television system in Germany in 1884, followed up in England by **Baird** in 1926, but in the final analysis such electromechanical systems were too slow to be successful. What was required was electronic scanning based on the cathode-ray tube invented by **Ferdinand Braun** in 1897. Before Zworykin, this had already been proposed by the Russian physicist Boris Rosing in 1907 and independently by the English physicist Alan Campbell-Swinton in 1908, but the technology was not available to make it a reality. By 1938 Zworykin had developed the first practical television camera which he called the 'iconoscope', but its application was delayed by World War II. He made a number of other important contributions to electronic optics with his scientific team at RCA, including the electron microscope (1939), the sniperscope (1941) and secondary-emission multipliers in scintillation coulters for measuring radioactivity. In his retirement he worked primarily on the medical application of electronics.

Nobel Prizewinners

Physics

1901 RÖNTGEN, Wilhelm Konrad von
1902 LORENTZ, Hendrik Antoon
ZEEMAN, Pieter
1903 BECQUEREL, Antoine Henri
CURIE, Marie
CURIE, Pierre
1904 RAYLEIGH, John William Strutt
1905 LENARD, Philipp Eduard Anton
1906 THOMSON, Sir J(oseph) J(ohn)
1907 MICHELSON, Albert Abraham
1908 LIPPMANN, Gabriel
1909 BRAUN, (Karl) Ferdinand
MARCONI, Guglielmo, Marchese
1910 VAN DER WAALS, Johannes Diderik
1911 WIEN, Wilhelm
1912 DALÉN, Nils Gustav
1913 KAMERLINGH-ONNES, Heike
1914 VON LAUE, Max
1915 BRAGG, Sir (William) Lawrence
BRAGG, Sir William Henry
1917 BARKLA, Charles Glover
1918 PLANCK, Max Karl Ernst
1919 STARK, Johannes
1920 GUILLAUME, Charles Édouard
1921 EINSTEIN, Albert
1922 BOHR, Niels Henrik David
1923 MILLIKAN, Robert Andrews
1924 SIEGBAHN, Karl Manne Georg
1925 FRANCK, James
HERTZ, Gustav Ludwig
1926 PERRIN, Jean Baptiste
1927 COMPTON, Arthur Holly
WILSON, Charles Thomson Rees
1928 RICHARDSON, Sir Owen Willans
1929 DE BROGLIE, Louis-Victor Pierre Raymond
1930 RAMAN, Sir Chandrasekhara Venkata
1932 HEISENBERG, Werner Karl
1933 DIRAC, Paul Adrien Maurice
SCHRÖDINGER, Erwin
1935 CHADWICK, Sir James
1936 ANDERSON, Carl David
HESS, Victor Francis
1937 DAVISSON, Clinton Joseph
THOMSON, Sir George Paget
1938 FERMI, Enrico
1939 LAWRENCE, Ernest Orlando
1943 STERN, Otto
1944 RABI, Isidor Isaac
1945 PAULI, Wolfgang
1946 BRIDGMAN, Percy Williams
1947 APPLETON, Sir Edward Victor
1948 BLACKETT, Patrick Maynard Stuart, Baron
1949 YUKAWA, Hideki
1950 POWELL, Cecil Frank
1951 COCKCROFT, Sir John Douglas
WALTON, Ernest Thomas Sinton
1952 BLOCH, Felix
PURCELL, Edward Mills
1953 ZERNIKE, Frits
1954 BORN, Max
BOTHE, Walther
1955 KUSCH, Polykarp
LAMB, Willis Eugene
1956 BARDEEN, John
BRATTAIN, Walter Houser
SHOCKLEY, William Bradford
1957 LEE, Tsung-Dao
YANG, Chen Ning
1958 CHERENKOV, Pavel Alekseevich
FRANK, Ilya Mikhailovich
TAMM, Igor Yevgenyevich
1959 CHAMBERLAIN, Owen
SEGRÈ, Emilio
1960 GLASER, Donald Arthur
1961 HOFSTADTER, Robert
MÖSSBAUER, Rudolph Ludwig
1962 LANDAU, Lev Davidovich
1963 GOEPPERT-MAYER, Maria
JENSEN, (Johannes) Hans (Daniel)
WIGNER, Eugene Paul
1964 BASOV, Nikolai Gennadiyevich
PROKHOROV, Alexander Mikhailovich
TOWNES, Charles Hard
1965 FEYNMAN, Richard
SCHWINGER, Julian
TOMONAGA, Sin-Itiro
1966 KASTLER, Alfred
1967 BETHE, Hans Albrecht
1968 ALVAREZ, Luis Walter
1969 GELL-MANN, Murray
1970 ALFVÉN, Hannes Olof Gösta
NÉEL, Louis Eugène Félix
1971 GABOR, Dennis
1972 BARDEEN, John
COOPER, Leon Neil
SCHRIEFFER, John Robert
1973 ESAKI, Leo
GIAEVER, Ivar
JOSEPHSON, Brian David
1974 HEWISH, Antony
RYLE, Sir Martin
1975 BOHR, Aage Niels
MOTTELSON, Benjamin Roy
RAINWATER, (Leo) James
1976 RICHTER, Burton
TING, Samuel Chao Chung

1977 ANDERSON, Philip Warren
MOTT, Sir Nevill Francis
VAN VLECK, John Hasbrouck
1978 KAPITZA, Peter (Pyotr Leonidovich)
PENZIAS, Arno Allan
WILSON, Robert Woodrow
1979 GLASHOW, Sheldon Lee
SALAM, Abdus
WEINBERG, Steven
1980 CRONIN, James Watson
FITCH, Val Logsdon
1981 BLOEMBERGEN, Nicolaas
SCHAWLOW, Arthur Leonard
SIEGBAHN, Kai
1982 WILSON, Kenneth Geddes
1983 CHANDRASEKHAR, Subrahmanyan
FOWLER, William Alfred
1984 RUBBIA, Carlo
VAN DER MEER, Simon
1985 VON KLITZING, Klaus
1986 BINNIG, Gerd Karl
ROHRER, Heinrich
RUSKA, Ernst August Friedrich
1987 BEDNORZ, (Johannes) Georg
MÜLLER, (Karl) Alex(ander)
1988 LEDERMAN, Leon
SCHWARTZ, Melvin
STEINBERGER, Jack
1989 DEHMELT, Hans
PAUL, Wolfgang
RAMSEY, Norman
1990 FRIEDMAN, Jerome
KENDALL, Henry
TAYLOR, Richard
1991 DE GENNES, Pierre-Gilles
1992 CHARPAK, Georges
1993 HULSE, Russell
TAYLOR, Joseph Hooton Jr

Chemistry

1901 VAN'T HOFF, Jacobus Henricus
1902 FISCHER, Emil Hermann
1903 ARRHENIUS, Svante August
1904 RAMSAY, Sir William
1905 BAEYER, Johann Friedrich Wilhelm Adolf von
1906 MOISSAN, Henri
1907 BUCHNER, Eduard
1908 RUTHERFORD, Ernest
1909 OSTWALD, (Friedrich) Wilhelm
1910 WALLACH, Otto
1911 CURIE, Marie
1912 GRIGNARD, (François Auguste) Victor
SABATIER, Paul
1913 WERNER, Alfred
1914 RICHARDS, Theodore William
1915 WILLSTÄTTER, Richard
1918 HABER, Fritz
1920 NERNST, Walther Hermann
1921 SODDY, Frederick
1922 ASTON, Francis William
1923 PREGL, Fritz
1925 ZSIGMONDY, Richard Adolf
1926 SVEDBERG, Theodor
1927 WIELAND, Heinrich Otto
1928 WINDAUS, Adolf Otto Reinhold
1929 EULER-CHELPIN, Hans Karl August Simon Von
HARDEN, Sir Arthur
1930 FISCHER, Hans
1931 BERGIUS, Friedrich
BOSCH, Carl
1932 LANGMUIR, Irving
1934 UREY, Harold Clayton
1935 JOLIOT-CURIE, Frédéric
JOLIOT-CURIE, Irène
1936 DEBYE, Peter Joseph Wilhelm
1937 HAWORTH, Sir Walter Norman
KARRER, Paul
1938 KUHN, Richard
1939 BUTENANDT, Adolf Friedrich Johann
RUŽIČKA, Leopold
1943 HEVESY, George Karl von
1944 HAHN, Otto
1945 VIRTANEN, Artturi Ilmari
1946 NORTHROP, John Howard
STANLEY, Wendell Meredith
SUMNER, James Batcheller
1947 ROBINSON, Sir Robert
1948 TISELIUS, Arne Wilhelm Kaurin
1949 GIAUQUE, William Francis
1950 ALDER, Kurt
DIELS, Otto
1951 McMILLAN, Edwin Mattison
SEABORG, Glenn Theodore
1952 MARTIN, Archer John Porter
SYNGE, Richard Laurence Millington
1953 STAUDINGER, Hermann
1954 PAULING, Linus Carl
1955 VIGNEAUD, Vincent du
1956 HINSHELWOOD, Sir Cyril Norman
SEMENOV, Nikolai Nikolaevich
1957 TODD, Alexander Robertus
1958 SANGER, Frederick
1959 HEYROVSY, Jaroslav
1960 LIBBY, Willard Frank
1961 CALVIN, Melvin
1962 KENDREW, Sir John Cowdery
PERUTZ, Max Ferdinand
1963 NATTA, Giulio
ZIEGLER, Karl
1964 HODGKIN, Dorothy Mary
1965 WOODWARD, Robert Burns
1966 MULLIKEN, Robert Sanderson
1967 EIGEN, Manfred
NORRISH, Ronald George Wreyford
PORTER, Sir George
1968 ONSAGER, Lars
1969 BARTON, Sir Derek Harold Richard
HASSEL, Odd
1970 LELOIR, Luis Frederico
1971 HERZBERG, Gerhard
1972 ANFINSEN, Christian Boehmer
MOORE, Stanford
STEIN, William Howard

1973 FISCHER, Ernst Otto
WILKINSON, Sir Geoffrey
1974 FLORY, Paul John
1975 CORNFORTH, Sir John Warcup
PRELOG, Vladimir
1976 LIPSCOMB, William Nunn
1977 PRIGOGINE, Ilya
1978 MITCHELL, Peter Dennis
1979 BROWN, Herbert Charles
WITTIG, Georg
1980 BERG, Paul
GILBERT, Walter
SANGER, Frederick
1981 FUKUI, Kenichi
HOFFMANN, Roald
1982 KLUG, Sir Aaron
1983 TAUBE, Henry
1984 MERRIFIELD, (Robert) Bruce
1985 HAUPTMAN, Herbert Aaron
KARLE, Jerome
1986 HERSCHBACH, Dudley Robert
LEE, Yuan Tseh
POLANYI, John Charles
1987 CRAM, Donald James
LEHN, Jean-Marie
PEDERSEN, Charles
1988 DEISENHOFER, Johann
HUBER, Robert
MICHEL, Hartmut
1989 ALTMAN, Sydney
CECH, Thomas
1990 COREY, Elias James
1991 ERNST, Richard Robert
1992 MARCUS, Rudolph Arthur
1993 MULLIS, Kary Banks
SMITH, Michael

Physiology or Medicine

1901 BEHRING, Emil von
1902 ROSS, Sir Ronald
1903 FINSEN, Niels Ryberg
1904 PAVLOV, Ivan Petrovich
1905 KOCH, Robert
1906 GOLGI, Camillo
RAMÓN Y CAJAL, Santiago
1907 LAVERAN, Charles Louis Alphonse
1908 EHRLICH, Paul
METCHNIKOFF, Elie
1909 KOCHER, Emil Theodor
1910 KOSSEL, Albrecht
1911 GULLSTRAND, Allvar
1912 CARREL, Alexis
1913 RICHET, Charles Robert
1914 BÁRÁNY, Robert
1919 BORDET, Jules Jean Baptiste Vincent
1920 KROGH, (Schack) August (Steenberg)
1922 HILL, Archibald Vivian
MEYERHOF, Otto Fritz
1923 BANTING, Sir Frederick Grant
MACLEOD, John James Rickard
1924 EINTHOVEN, Willem
1926 FIBIGER, Johannes Andreas Grib
1927 WAGNER-JAUREGG, Julius
1928 NICOLLE, Charles Jules Henri
1929 EIJKMAN, Christiaan
HOPKINS, Sir Frederick Gowland
1930 LANDSTEINER, Karl
1931 WARBURG, Otto Heinrich
1932 ADRIAN, Edgar Douglas
SHERRINGTON, Sir Charles Scott
1933 MORGAN, Thomas Hunt
1934 MINOT, George Richards
MURPHY, William Parry
WHIPPLE, George Hoyt
1935 SPEMANN, Hans
1936 DALE, Sir Henry Hallett
LOEWI, Otto
1937 SZENT-GYÖRGYI, Albert von Nagyrapolt
1938 HEYMANS, Corneille Jean François
1939 DOMAGK, Gerhard
1943 DAM, Carl Peter Henrik
DOISY, Edward Adelbert
1944 ERLANGER, Joseph
GASSER, Herbert Spencer
1945 CHAIN, Sir Ernst Boris
FLEMING, Sir Alexander
FLOREY, Sir Howard Walter
1946 MÜLLER, Hermann Joseph
1947 CORI, Carl Ferdinand
CORI, Gerty Theresa Radnitz
HOUSSAY, Bernardo Alberto
1948 MÜLLER, Paul Hermann
1949 EGAS MONIZ, António Caetano de Abreu Freire
HESS, Walter Rudolf
1950 HENCH, Philip Showalter
KENDALL, Edward Calvin
REICHSTEIN, Tadeusz
1951 THEILER, Max
1952 WAKSMAN, Selman Abraham
1953 KREBS, Sir Hans Adolf
LIPMANN, Fritz Albert
1954 ENDERS, John Franklin
ROBBINS, Frederick Chapman
WELLER, Thomas Huckle
1955 THEORELL, (Axel) Hugo Teodor
1956 COURNAND, André Frédéric
FORSSMAN, Werner
RICHARDS, Dickinson Woodruff
1957 BOVET, Daniel
1958 BEADLE, George Wells
LEDERBERG, Joshua
TATUM, Edward Lawrie
1959 KORNBERG, Arthur
OCHOA, Severo
1960 BURNET, Sir (Frank) Macfarlane
MEDAWAR, Sir Peter Brian
1961 BÉKÉSY, Georg von
1962 CRICK, Francis Harry Compton
WATSON, James Dewey
WILKINS, Maurice Hugh Frederick
1963 ECCLES, Sir John Carew
HODGKIN, Sir Alan Lloyd
HUXLEY, Sir Andrew Fielding

NOBEL PRIZEWINNERS

1964 BLOCH, Konrad Emil
LYNEN, Feodor Felix Konrad
1965 JACOB, François
LWOFF, André
MONOD, Jacques
1966 HUGGINS, Charles Brenton
ROUS, (Francis) Peyton
1967 GRANIT, Ragnar Arthur
HARTLINE, Haldan Keffer
WALD, George
1968 HOLLEY, Robert William
KHORANA, Har Gobind
NIRENBERG, Marshall Warren
1969 DELBRÜCK, Max
HERSHEY, Alfred Day
LURIA, Salvador Edward
1970 AXELROD, Julius
EULER, Ulf Svante von
KATZ, Sir Bernard
1971 SUTHERLAND, Earl Wilbur Jr
1972 EDELMAN, Gerald Maurice
PORTER, Rodney Robert
1973 FRISCH, Karl von
LORENZ, Konrad Zacharias
TINBERGEN, Nikolaas
1974 CLAUDE, Albert
DE DUVE, Christian René
PALADE, George Emil
1975 BALTIMORE, David
DULBECCO, Renato
TEMIN, Howard Martin
1976 BLUMBERG, Baruch Samuel
GAJDUSEK, Daniel Carleton
1977 GUILLEMIN, Roger
SCHALLY, Andrew Victor
YALOW, Rosalyn
1978 ARBER, Werner
NATHANS, Daniel
SMITH, Hamilton Othanel
1979 CORMACK, Allan MacLeod
HOUNSFIELD, Sir Godfrey Newbold
1980 BENACERRAF, Baruj
DAUSSET, Jean
SNELL, George Davis
1981 HUBEL, David Hunter
SPERRY, Roger Wolcott
WIESEL, Torsten Nils
1982 BERGSTRÖM, Sune Karl
SAMUELSSON, Bengt Ingemar
VANE, Sir John Robert
1983 McCLINTOCK, Barbara
1984 JERNE, Niels Kai
KÖHLER, Georges Jean Franz
MILSTEIN, Cesar
1985 BROWN, Michael Stuart
GOLDSTEIN, Joseph Leonard
1986 COHEN, Stanley
LEVI-MONTALCINI, Rita
1987 TONEGAWA, Susumu
1988 BLACK, Sir James Whyte
ELION, Gertrude Belle
HITCHINGS, George Herbert
1989 BISHOP, (John) Michael
VARMUS, Harold Elliot
1990 MURRAY, Joseph Edward
THOMAS, (Edward) Donnall
1991 NEHER, Erwin
SAKMANN, Bert
1992 FISCHER, Edmond
KREBS, Edwin Gerhard
1993 ROBERTS, Richard
SHARP, Phillip Allen

Index

INDEX

INDEX

INDEX

INDEX

INDEX